Handbook of Hydrothermal Technology

Handbook of Hydrothermal Technology

Second Edition

K. Byrappa
Masahiro Yoshimura

ELSEVIER

AMSTERDAM • BOSTON • HEIDELBERG • LONDON
NEW YORK • OXFORD • PARIS • SAN DIEGO
SAN FRANCISCO • SINGAPORE • SYDNEY • TOKYO

William Andrew is an imprint of Elsevier

William Andrew is an imprint of Elsevier
The Boulevard, Langford Lane, Kidlington, Oxford OX5 1GB, UK
225 Wyman Street, Waltham, MA 02451, USA
First edition 2006
Second edition 2013

Library of Congress Cataloging-in-Publication Data
A catalog record for this book is available from the Library of Congress

British Library Cataloguing in Publication Data
A catalogue record for this book is available from the British Library

ISBN: 978-0-12-375090-7

For information on all William Andrew publications
visit our web site at books.elsevier.com

Typeset by MPS Limited, Chennai, India
www.adi-mps.com

Printed and bound by CPI Group (UK) Ltd, Croydon, CR0 4YY

Dedication

To our mothers
Puttamma
Katsuko

Contents

Preface

The term *hydrothermal* is purely of geological origin. It was first used by the British geologist, Sir Roderick Murchison (1792 – 1871), to describe the action of water at elevated temperature and pressure in bringing about changes in the Earth's crust, and leading to the formation of various rocks and minerals. Geologists carried out the earliest work on the hydrothermal technique in the nineteenth century in order to understand the genesis of rocks and minerals by simulating the natural conditions existing under the Earth's crust. However, materials scientists popularized the technique, particularly during 1940s. Schafhautl, who obtained quartz crystals upon freshly precipitated silicic acid in a Papin's Digester, carried out the first hydrothermal synthesis in 1845. Subsequently, hydrothermal synthesis of a wide variety of minerals was carried out, especially in Europe.

The largest-known single crystal formed in nature (beryl crystal of >1000 kg) and some of the largest quantities of single crystals created in one experimental run (quartz crystals of >1000 kg) are both of hydrothermal origin.

The first successful commercial application of hydrothermal technology began with mineral extraction or ore beneficiation in the nineteenth century. With the beginning of the synthesis of large single crystals of quartz by Nacken (1946) and zeolites by Barrer (1948), the commercial importance of the hydrothermal technique for the synthesis of inorganic compounds was realized.

The general acceptance of plate tectonics theory during the mid-1970s garnered much interest in geochemical processes at plate boundaries, which led to the discovery of hydrothermal activity in the deep sea directly on the Galapagos Spreading Center in 1977, and a large number of other spectacular submarine hydrothermal systems of global significance to ocean chemistry and geochemistry. In fact, this discovery has led to a new thinking in marine biology, geochemistry, and economic geology, and has spawned an entirely new term, namely, hydrothermal ecosystem, which means water-containing terrestrial, subterranean, and submarine high temperature environments, which are the sites of investigations for many palaeobiologists and biologists looking for primitive forms of life. It is strongly believed that the roots of life on Earth can be found in hydrothermal ecosystems. These ecosystems may also serve as an analogue for the possible origin of life on Mars, where a similar environment might have existed or still exists.

Earth is a blue planet of the universe where water is an essential component. Circulation of water and other components such as entropy (energy) are driven by water vapor and heat (either external or internal). Water has a very important role in the formation of material or transformation of materials in nature, and hydrothermal circulation has always been assisted by bacterial activity.

From the mid-1970s, exploration of the advantages of hydrothermal reactions, other than the hydrometallurgical and crystal growth aspects, began in Japan, particularly with reference to the ceramic powder processing. A team of researchers from the Tokyo Institute of Technology, Japan, did pioneering work in ceramic processing such as powder preparation, reaction sintering, hot isostatic processing, and so on.

From 1990, the hydrothermal technique has offered several new advantages like homogeneous precipitation using metal chelates under hydrothermal conditions, decomposition of hazardous and/or refractory chemical substances, monomerization of high polymers like polyethylene terephthalate, and a host of other environmental and chemical engineering issues dealing with recycling of rubbers and plastics (instead of burning), and so on. The solvation properties of supercritical solvents are being extensively used for detoxifying organic and pharmaceutical wastes and also to replace toxic solvents commonly used for chemical synthesis. Similarly, it is used to remove caffeine and other food-related compounds selectively. In fact, the food and nutrition experts in recent years are using a new term: *hydrothermal cooking*. These unique properties take the hydrothermal technique altogether in a new direction for the twenty-first century and one can forecast a slow emergence of a new branch of science and technology for sustained human development. We have collected a more-or-less complete list of publications in hydrothermal technology and provided the statistical data to show the growing popularity of the technique (see Figures 1.12 – 1.14). The main disadvantage of the hydrothermal system, as believed earlier, is the black-box nature of the apparatus, because one cannot observe directly the crystallization processes. However, in recent years, remarkable progress has been made in this area through the entry of physical chemists, and the modeling of the hydrothermal reactions and the study of kinetics of the hydrothermal processes, which have contributed greatly to the understanding of the hydrothermal technique. One can understand the hydrothermal chemistry of the solutions more-or-less precisely, which provides a solid base for designing hydrothermal synthesis and processing at much lower pressure and temperature conditions. The hydrothermal technique exhibits a great degree of flexibility, which is being rightly exploited by a large scientific community with diverse interests. Hydrothermal processing has become a most powerful tool especially with the emergence of nanotechnology, and also for transforming various inorganic compounds and treating raw materials for technological applications.

Today, the hydrothermal technique has found its place in several branches of science and technology, and this has led to the appearance of several related techniques, with strong roots to the hydrothermal technique, involving materials scientists, earth scientists, materials engineers, metallurgists, physicists, chemists, biologists, and others. The use of additional energy sources along with hydrothermal or solvothermal techniques such as microwave–hydrothermal, mechanochemical–hydrothermal, hydrothermal–electrochemistry, hydrothermal–acoustic waves, and hydrothermal with stirring led to the development of a new processing technology, namely, multienergy processing of materials with highly enhanced reaction kinetics, which takes us to the chemistry at the speed of light and instant hydrothermal systems. The understanding of thermodynamics of the hydrothermal systems, in situ surface modification, new solvent types, organic cappings, and ligands has made the instant hydrothermal vending machine not a distant reality. Thus, the importance of hydrothermal technology from geology to technology has been realized. In view of such a rapid growth of the hydrothermal technique, it was imperative to revise the first edition of the *Handbook of Hydrothermal Technology*, and this second edition will cover all the recent developments in this frontier area of research keeping in mind that this is the technology for the twenty-first century. There are thousands of articles and reviews published on various aspects of science of hydrothermal research but, before 2001, the most comprehensive works on this topic were limited to reviews and edited books, and the first edition of this book had covered information up to the year 2000. But from 2000 onward, the popularity of hydrothermal technology has grown tremendously and it is closely related to the success accomplished in nanotechnology. A revision was therefore essential, and the authors have made a sincere effort to compile all the data available up to mid-2012 in this fast-growing field of hydrothermal technology of materials processing.

The first author, Dr. K. Byrappa, edited a book entitled *Hydrothermal Growth of Crystals* in 1990 for Pergamon Press, Oxford, United Kingdom. During early 1995, the authors conceived the idea of writing this handbook and began collecting old records from various sources. The first edition of the book was published during early 2001, and that first edition dealt with all the aspects of hydrothermal research: covering historical development, natural hydrothermal systems, instrumentation, physical chemistry of hydrothermal growth of crystals, growth of some selected crystals like quartz, berlinite, potassium titanyl phosphate, and calcite, and hydrothermal synthesis of a host of inorganic compounds like zeolites, complex coordinated compounds (silicates, germanates, phosphates, tungstates, molybdates, etc.) and simple oxides, native elements, and the hydrothermal processing of materials with an emphasis on future trends of hydrothermal technology in the twenty-first century. This second edition is more comprehensive with all the latest developments in the field of hydrothermal research.

Many publications on hydrothermal research, especially in the Russian journals, were not easily accessible to us. Our Russian colleagues in the field of hydrothermal research extended great cooperation in this regard. Especially, the help rendered by the late Dr. L. N. Demianets, Dr. V. I. Popolitov, and Dr. O. V. Dimitrova is highly appreciated. The authors are very grateful for the help and encouragement extended by the senior people in this field like Prof. Shigeyuki Sōmiya (Professor Emeritus), Prof. N. Yamasaki (Tohoku University, Japan), the late Dr. Bob Laudise (AT & T Bell Labs.), the late Prof. Rustom Roy (Penn State University), Prof. C. N. R. Rao (President, JNCASR, Bangalore, India), Prof. R. Chidambaram (Former Chairman, Atomic Energy Commission, India, and Principal Scientific Advisor, Government of India), Dr. Krishan Lal (Former Director, National Physical Laboratory, New Delhi, and President, Indian National Science Academy), the late Dr. B. P. Radhakrishna (Former Director, Department of Mines & Geology), the late Prof. J. A. Rabenau (Max-Planck Institute of Physics, Stuttgart), Prof. Ichiro Sunagawa, Prof. J. A. Pask, Prof. Paul

Hagenmuller, Prof. H. L. Barnes, the late Prof. Saito Shinroku, and Prof. Toshiyuki Sata. Both Prof. Sōmiya and Prof. Yamasaki spent several hours with the authors and provided fruitful discussion. Their suggestions and comments have been included in this book. We would like to extend our special thanks to the most active hydrothermal researchers of the present day, like Prof. S. Hirano, Prof. M. Hosaka, Prof. K. Yanagisawa, Prof. T. Moriyoshi (Director, RIST, Takamatsu), Prof. S. Komarneni, Prof. R. E. Riman, Prof. T. Adschiri, Prof. M. Goto, Prof. Y. Funazukuri, Prof. S. Mathur, Prof. Brian Korgel, Dr. Wojciech Suchanek, Prof. Dirk Ehrentraut, Prof. Yoshitake Masuda, Prof. Ed Lester, Prof. Martyn Poliakoff, Prof. Thomas Fehr, Prof. S. H. Yu, Prof. T. Watanabe, Prof. Don Wang, Prof. Hao Wang, Prof. Qi-Sheng Chen, Dr. Don Palmer, Dr. Dave Wesolowski, Dr. S. Taki, Prof. K. Arai, Prof. T. B. Brill, Prof. T. M. Seeward, Prof. Yuri Gogosti, Prof. K. Bowen, Prof. David A. Payne, the late Prof. Fred F. Lange, and Prof. Fathi Habashi, who have helped us by providing their results and data for inclusion in this book. Our colleagues at the Tokyo Institute of Technology, like Prof. Masato Kakihana, Prof. Nabuo Ishizawa, Prof. Eiichi Yasuda, Prof. Akira Sawaoka, Prof. Masatomo Yashima, Prof. Yasuo Tanabe, Prof. Masanori Abe, Prof. Kazunari Domen, and Prof. Kiyoshi Okada, extended great cooperation in our task. Other Japanese friends, like Dr. Yasuro Ikuma, Prof. Tsugio Sato, Prof. Koji Kajiyoshi, Dr. Atsuo Ito, Dr. E. H. Ishida, Prof. Zenbee Nakagawa, Prof. Yusuke Moriyoshi, Prof. Yasumichi Matsumoto, Prof. Tadashi Ohachi (Doshisha University, Kyoto), Dr. Kohei Soga (Tokyo University), Prof. Toshio Yamaguchi, and a host of other friends in the field of hydrothermal research, have helped us extensively in completing this great work. We would like to express our sincere thanks to our friends and colleagues from India (the late Prof. M. N. Vishwanathiah, Prof. J. A. K. Tareen, Prof. T. R. N. Kutty, Prof. S. B. Krupanidhi, Prof. K. T. Jacob, Prof. P. N. Satish, Prof. B. Basavalingu, Prof. H. L. Bhat, Dr. S. Srikantaswamy, Prof. G. S. Gopalakrishna, Dr. K. M. Lokanath Rai, Prof. S. Ananda, Prof. R. Somashekar, Prof. C. Ranganathaiah, Prof. V. Ravishankara Rai, Dr. B.V. Suresh Kumar), Russia (Prof. V. S. Balitsky, Prof. V. A. Kuznetsov, Prof. D. Yu Pushcharovsky, Dr. Oleg Karpov, Dr. G. I. Dorokhova, and Dr. E. Strelkhova), Korea (Prof. Choy Jin-ho, Prof. Y.W. Lee, and Prof. Z. H. Lee), Spain (Prof. Juan Ma Garcia Ruiz, Prof. Rafael Rodriguez Clemente, Prof. Salvador Gali, Dr. Jaime Moralez Gomez,), Poland (Prof. Keshra Sangwal), and China (Prof. Shen-tai Song, Prof. Y.-T. Qian, and Prof. S. Feng), who were a great source of inspiration and help while writing this book. Here, it is extremely difficult to list all the names of the students, post-docs, and research associates from our groups who have helped us greatly. Among them, the help rendered by Ms. Namratha Keerthiraj, Senior Research Fellow, who spent several months working on this project, is highly acknowledged. Also the facilities provided by Prof. J. D. Liao of National Cheng Kung University, Taiwan, during the preparation of this second revised edition were much appreciated here. Also our special thanks to Prof. V. G. Talawar, Vice Chancellor of the University of Mysore, and Prof. Ishwar Goudar, Visiting Professor and Library Advisor, University of Mysore, for their help in the preparation of this book. We are highly obliged to all our family members—Prof. K.T. Sunitha Byrappa, Shayan and

Nayan (Prof. Byrappa's family members), Mrs. Akiko Yoshimura, Sayaka, Ayumi, and Hirono (Prof. Yoshimura's family members) for their great patience and cooperation during the preparation of this book.

Finally, we would like to express our deep gratitude to Mr. Matthew Deans and Mr. David Jackson, of Elsevier, for bringing out this book, and honestly without their constant pursuance this book would not have materialized. It is also an added pleasure to acknowledge the role played by Ms. Pauline Wilkinson, of Elsevier, and the typesetters for composition. If we have omitted any important topic or any other names of the friends, or citation of the important references in this book, it is not intentional. We have tried our best to cover as much as we could of the hydrothermal research based on the most exhaustive up-to-date literature search. Hopefully, this book will be a most valuable textbook and reference for the students and researchers in the field of hydrothermal technology, both at the beginner and advanced levels.

K. Byrappa
M. Yoshimura

1 Hydrothermal Technology—Principles and Applications

1.1 Introduction

The hydrothermal technique has been most popular, garnering interest from scientists and technologists of different disciplines, particularly from 1990s. The term *hydrothermal* is of purely geological origin. It was first used by the British geologist, Sir Roderick Murchison (1792–1871), to describe the action of water at an elevated temperature and pressure in bringing about changes in the earth's crust leading to the formation of various rocks and minerals [1]. The majority of the minerals formed in the postmagmatic and metasomatic stages in the presence of water at elevated pressure and temperature conditions are said to be "of hydrothermal origin." This covers a vast number of mineral species including ore deposits. It is well known that the largest single crystal formed in nature (beryl crystal of >1000 g) and some of the largest quantities of single crystals created by man in one experimental run (quartz crystals of several hundreds of grams) are both of hydrothermal origin.

An understanding of the mineral formation in nature under elevated pressure and temperature conditions in the presence of water led to the development of the hydrothermal technique. In 1839, the German chemist Robert Bunsen contained aqueous solutions in thick-walled glass tubes at temperatures above 200°C and at pressures above 100 bar [2]. The crystals of barium carbonate and strontium carbonate that he grew under these conditions mark the first use of hydrothermal aqueous solvents as media. It was successfully adopted by Schafthaul [3] to obtain quartz crystals upon the transformation of freshly precipitated silicic acid in Papin's digestor. Thus, the hydrothermal technique became a very popular means to simulate the natural conditions existing under the earth's crust and synthesize them in the laboratory. These studies dealing with laboratory simulations have helped the earth scientists to determine complex geological processes of the formation of rocks, minerals, and ore deposits. As the subject became more and more popular among geologists, new branches of geology emerged, such as "Experimental Mineralogy" and "Experimental Petrology."

The first successful commercial application of hydrothermal technology began with mineral extraction or ore beneficiation in the previous century. The use of sodium hydroxide to leach bauxite was invented in 1892 by Karl Josef Bayer

Handbook of Hydrothermal Technology.

(1871–1908) as a process for obtaining pure aluminum hydroxide which can be converted to pure $A1_2O_3$ suitable for processing to metal [4]. Even today, over 90 million tons of bauxite ore is treated annually by this process [5]. Similarly, ilmenite, wolframite, cassiterite, laterites, a host of uranium ores, sulfides of gold, copper, nickel, zinc, arsenic, antimony, and so on, are treated by this process to extract the metal. The principle involved is quite simple, very effective, and inexpensive, as shown below, for example:

$$Al(OH)_3 + OH^- \rightarrow [AlO(OH)_2]^- + H_2O$$
$$AlOOH + OH^- \rightarrow [AlO(OH)^2]$$

The above process is easy to achieve and the leaching can be carried out in a few minutes at about 330°C and 25,000 kPa [6].

Further importance of the hydrothermal technique for the synthesis of inorganic compounds in a commercial way was realized soon after the synthesis of large single crystals of quartz by Nacken [7] and zeolites by Barrer [8] during late 1930s and 1940s, respectively. The sudden demand for the large size quartz crystals during World War II forced many laboratories in Europe and North America to grow large size crystals. Subsequently, the first synthesis of zeolite that did not have a natural counterpart was carried out by Barrer [8] in 1948 and this opened altogether a new field of science, viz., molecular sieve technology [9]. The success in the growth of quartz crystals has provided further stimuli for hydrothermal crystal growth [10].

Today, the hydrothermal technique has found its place in several branches of science and technology, and this has led to the appearance of several related techniques with strong roots attached to the hydrothermal technique. So we have: hydrothermal synthesis, hydrothermal growth, hydrothermal alteration, hydrothermal treatment, hydrothermal metamorphism, hydrothermal dehydration, hydrothermal decomposition, hydrothermal extraction, hydrothermal sintering, hydrothermal reaction sintering, hydrothermal phase equilibria, hydrothermal electrochemical reaction, hydrothermal mechanochemical, microwave hydrothermal, hydrothermal sonochemical, and so on. This involves: materials scientists, earth scientists, materials engineers, metallurgists, physicists, chemists, biologists, and others. Although the technique has attained its present high status, it has passed through several ups and downs owing to the lack of proper knowledge pertaining to the actual principles involved in the process. Hence, the success of the hydrothermal technique can be largely attributed to the rapid advances in the apparatus involved (new apparatus designed and fabricated) in hydrothermal research and also to the entry of a large number of physical chemists who have contributed greatly to the understanding of hydrothermal chemistry [11]. Further, the modeling and intelligent engineering of the hydrothermal processes have also greatly enhanced our knowledge in the field of hydrothermal research [12,13].

In recent years, with the increasing awareness of both environmental safety and the need for optimal energy utilization, there is a case for the development of nonhazardous materials. These materials should not only be compatible with human life but also with other living forms or species. Also, processing methods such as fabrication, manipulation, treatment, reuse, and recycling of waste materials should be environmentally friendly. In this respect, the hydrothermal technique occupies a unique place in modern science and technology. The rapid development in the field of hydrothermal research during 2000s. or so, motivated the present authors to bring out this handbook covering almost all aspects of hydrothermal research. Although there is a growing interest among scientists from various branches of science, at the moment there are no books or monographs available in the field of hydrothermal technology. Most of the major works, even by the pioneers in this field, have been confined to reviews and edited books. Table 1.1 lists important reviews and edited books in the field of hydrothermal research.

1.2 Definition

In spite of the fact that the hydrothermal technique has made tremendous progress, there is no unanimity about its definition. The term hydrothermal usually refers to any heterogeneous reaction in the presence of aqueous solvents or mineralizers under high-pressure–high-temperature (HPHT) conditions to dissolve and recrystallize (recover) materials that are relatively insoluble under ordinary conditions. Morey and Niggli [14] defined hydrothermal synthesis as "... in the hydrothermal method the components are subjected to the action of water, at temperatures generally near though often considerably above the critical temperature of water (~370°C) in closed bombs, and therefore, under the corresponding high pressures developed by such solutions." According to Laudise [15], hydrothermal growth means growth from aqueous solution at ambient or near-ambient conditions. Rabenau [16] defined hydrothermal synthesis as the heterogeneous reactions in aqueous media above 100°C and 1 bar. Lobachev [17] defined it as a group of methods in which crystallization is carried out from superheated aqueous solutions at high pressures. Roy [18] declares that hydrothermal synthesis involves water as a catalyst and occasionally as a component of solid phases in the synthesis at elevated temperature (>100°C) and pressure (greater than a few atmospheres). Byrappa [19] defines hydrothermal synthesis as any heterogeneous reaction in an aqueous media carried out above room temperature and at pressure greater than 1 atm. Yoshimura [20] proposed the following definition: reactions occurring under the conditions of high temperature and high pressure (>100°C, >1 atm) in aqueous solutions in a closed system.

The above definitions hold good for materials synthesis, metal leaching, and treatment of waste materials. However, there is no definite lower limit for the pressure and temperature conditions. The majority of the authors fix the hydrothermal synthesis, for example, at above 100°C temperature and above 1 atm. But, with the vast number of publications under mild hydrothermal conditions in recent years, we propose to define hydrothermal reaction as "any heterogeneous chemical

Table 1.1 List of the Books and Reviews in the Field of Hydrothermal Research

S. No.	Author	Title	Publishers
Books			
1.	Labachev, A.N. (Ed.)	*Hydrodrothermal Synthesis of Crystals*	Nauka, Moscow (1971)
2.	Ulmer, G.C. (Ed.)	*Research Techniques for High Pressure and High Temperature*	Springer-Verlag, New York (1971)
3.	Labachev, A.N. (Ed.)	*Crystallization Processes under Hydrothermal Conditions*	Consultants Bureau, New York (1973)
4.	Ikornikova, N. Yu	*Hydrothermal Synthesis of Crystals in Chloride Systems*	Nauka, Moscow (1975)
5.	Kuzmina, I.P. and Khaidukov, N.M.	*Crystal Growth from High Temperature Aqueous Solutions*	Nauka, Moscow (1977)
6.	Demianets, L.N., Labachev, A.N., and Emelechenko, G.A.	*Germanates of Rare Earth Elements*	Nauka, Moscow (1980)
7.	David T. Rickard Frans E. Wickman (Ed.)	*Chemistry and Geochemistry of Solutions at High Temperature and Pressure*	Pergamon, New York (1981)
8.	Somiya, S. (Ed.)	*Proc. 1st Int. Symp. Hydrothermal Reactions*	Japan (1982)
9.	Litvin, B.N. and Popolitov, V.I.	*Hydrothermal Synthesis Inorganic Compounds*	Nauka, Moscow (1984)
10.	Popolitov, V.I. and Litvin, B.N.	*Growth of Single Crystals under Hydrothermal Conditions*	Nauka, Moscow (1986)
11.	Ulmer, G.C. and Barnes, H.L. (Eds)	*Hydrothermal Experimental Techniques*	John Wiley & Sons, New York (1987)
12.	Nesterov, P.V. (Ed.)	*Progress in Science and Technology, Crystal Chemistry of Germanates of Tetravalent Metals*	VINITI, Moscow (1989)
13.		*Hydrothermal Reactions ISHR—89*	Nauka, Moscow (1989)
14.	Occelli, M.L. and Robson, H.E. (Eds)	*Zeolite Synthesis*	ACS Symp. Series 398 Am. Chem. Soc., Washington, DC (1989)
15.	Somiya, S. (Ed.)	*Hydrothermal Reactions for Engineering*	Elsevier, Applied Science (1989)
16.	Byrappa, K. (Ed.)	*Hydrothermal Growth of Crystal*	Pergamon, Oxford (1991)

(*Continued*)

Table 1.1 (Continued)

S. No.	Author	Title	Publishers
17.	Cuney, M. and Cathelineau, M. (Eds)	*Proc. 4th Int. Symp. Hydrothermal Reactions*	Nancy, France (1993)
18.		*Proc. 1st Int. Conf. Solvothermal Reactions*	Japan (December 6–8, 1996)
19.		*Proc. 2nd Int. Conf. Solvothermal Reactions*	Japan (December 18–20, 1996)
20.	Palmer, D.A. and Wesoloski, D.J.	*Proc. 5th Symp. Hydrothermal Reactions*	Gatlinburg, USA (1997)
21.	G. Demazeau	*Proc. 3rd Int. Conf. Solvothermal Reactions*	Bordeaux, France (1999)
22.	K. Yanagisawa & Qi Feng	*Proc. Joint 6th Int. Symp. Hydrothermal Reactions & 4th Int. Conf. Solvothermal Reactions*	Kochi, Japan (2000)
23.	R.E. Riman	*Proc. 5th Int. Symp. Solvothermal Reactions*	East Brunswick, New Jersey, USA (2002)
24.	S. Feng, J.S. Chen, and Z. Shi	*Proc. 7th Int. Symp. Hydrothermal Reactions (Hydrothermal Reactions and Techniques, World Scientific Publications, 2003)*	Changchun, China (2003)
25.	K. Byrappa and M. Yoshimura	*Proc. 6th Int. Symp. Solvothermal Reactions (Special Edition of Journal of Materials Science, Vol. 41 No. 5 (2006))*	Mysore, India (2004)
26.	K. Byrappa and T. Adschiri	*Proc. Joint 8th Int. Symp. Hydrothermal Reactions & 7th Int. Conf. Solvothermal Reactions (Special Edition of Journal of Materials Science, Vol. 41 No. 5 (2006))*	Sendai, Japan (2006)
27.	Edward Lester	*Proc. 1st Int. Conf. Solvothermal and Hydrothermal Reactions (ISHA-2008)*	Nottingham, UK (2008)
28.	Don Wang	*Proc. 2nd Int. Conf. Solvothermal and Hydrothermal Reactions (ISHA-2008)(Special Edition of the Journal Research in Chemical Intermediates (2011))*	Beijing, China (2010)

(Continued)

Table 1.1 (Continued)

S. No.	Author	Title	Publishers
Reviews			
1.	Morey, G.W.	*Hydrothermal Synthesis*	*J. Am. Ceram. Soc.* (1953)
2.	Roy, R. and Tuttle, O.F.	*Investigation under Hydrothermal Conditions*	*Phy. Chem. Earth* 1:138 (1955)
3.	Ballman, A.A. and Laudies, R.A.	*Solution Growth*	In: *Art and Science of Growing Crystals* (Gilmann, J.J., Ed.) John Wiley & Sons, New York (1963)
4.	Laudise, R.A.	*Hydrothermal Growth of Crystals*	In: *The Growth of Single Crystals*, Prentice-Hall, New York (1970)
5.	Kuznetsov, V.A. and Lobachev, A.N.	*Hydrothermal Method for the Growth of Crystals*	*Sov. Phys. Crystallogr.* 70:775–804 (1973)
6.	Nassau, K.	*Synthetic Emerald: The Confusing History and The Current Technology*	*J. Crystal Growth* 35:211–222 (1976)
7.	Rabenau, A.	*The Role of Hydrothermal Synthesis in Preparative*	*Angew. Chem. Int. Engl. Ed.* 24:1026–1040 *Chemistry* (1985)
8.	Laudise, R.A.	*Hydrothermal Crystal Growth —Some Recent Results*	In: *Advanced Crystal Growth*, Prentice-Hall, New York (1987)
9.	Komareni, S., Fregeau, E., Breval, E., and Roy, R.	*Hydrothermal Preparation of Ultrafine Ferrites & their Sintering*	*J. Am. Ceram. Soc.* (1988)
10.	Somiya, S.	*Hydrothermal Reactions in Inorganic Systems*	In: *Advance Materials Frontiers in Material Science & Engineering* (Somiya, S., Ed.); *Trans. Mat. Res. Soc. Jpn.*, Vol. 19B, Elsevier Science, B.V. (1994)
11.	Yoshimura, M. and Suda, H.	*Hydrothermal Processing of Hydroxyapatite: Past, Present & Future*	In: *Hydroxyapatite and Related Compounds*, Brown, P. W. and Constantz, B. (Eds), pp. 45–72, CRC Press (1994)
12.	Byrappa, K.	*Hydrothermal Growth of Crystals*	In: *Handbook of Crystal Growth, Vol. 2a* (Hurle, D. T. J., Ed.), Elsevier Science B.V. (1994)
13.	Leco catalogue	Leco Corporation	Tem-Press Division, USA (1999)

(*Continued*)

Table 1.1 (Continued)

S. No.	Author	Title	Publishers
14.	Byrappa, K.	*Growth of Quartz Crystals*	In: *Bulk Crystal Growth of Electronic, Optical and Optoelectronic Materials*, (Peter Capper, ed.) John Wiley & Sons, UK. Chapter 13 (2004), pp. 387–404
15.	Byrappa, K	*Hydrothermal Processing of Advanced Materials*	Kirk–Othmer Encyclopedia of Chemical Technology, John Wiley & Sons, UK (2005)
16.	Byrappa, K. and Adschiri, T.	*Hydrothermal Technology for Nanotechnology*	In: *Progress in Crystal Growth and Characterization of Materials*, Vol. 53 (2007), pp. 117–166
17.	Yoshimura, M. and Byrappa, K.	*Hydrothermal Technology: Past, Present and Future*	J. Mater. Sci., Vol. 43 (7) (2008), pp. 2085–2103
18.	Byrappa, K. Ohara, S. and Adschiri, T.	*Nanoparticles Synthesis Using Supercritical Fluid–Towards Biomedical Applications*	Advanced Drug Delivery Reviews, Elsevier, The Netherlands, Vol. 60 (2008) 299–327
19.	Byrappa, K.	*Novel Hydrothermal Solution Routes of Advanced High Melting Nanomaterials Processing*	*J. Ceram. Soc. Japan* Vol. 117(3) (2009), pp. 236–244
20.	Adschiri, T. and Byrappa, K.	*Supercritical Hydrothermal Synthesis of Organic–Inorganic Hybrid Nanoparticles*	In: *Nanohybridization of Organic–Inorganic Materials*, Eds: Atsushi Muramatsu, Publishers: Springer–Verlag, Germany (2009), pp. 217–250
21.	Byrappa, K.	*Hydrothermal Growth of Polyscale Crystals*	In: *Springer Handbook of Crystal Growth*, Eds: G. Dhanaraj, K. Byrappa, M. Dudley and V. Prasad, Springer-Verlag, Germany (2010) pp. 599–653
22.	Shu-Hong Yu	*Hydrothermal/Solvothermal Process: Powerful Routes for Controlled Synthesis of Functional Materials*	ISHA Newsletter Vol. 5, No. 2 (2010) pp. 9–32
23.	Demazeau, G.	*Solvothermal and Hydrothermal Processes: The Main Physico-Chemical Factors Involved and New Trends*	Res. Chem. Intermed., Vol. 37 (2011) pp. 107–123.

reaction in the presence of a solvent (whether aqueous or nonaqueous) above room temperature and at pressure greater than 1 atm in a closed system." However, with an ever-increasing application of hydrothermal technology in nanotechnology, the definition of hydrothermal needs a major revision; since in nanomaterials preparation most often solution-based precursors and gels are employed to obtain high-quality nanomaterials under hydrothermal conditions. Thus, the hydrothermal can be defined as: "any heterogeneous or homogeneous chemical reaction in the presence of a solvent (whether aqueous or nonaqueous) above room temperature and at pressure greater than 1 atm in a closed system." [21] In addition to this nonunanimity, there is also a lot of confusion with regard to the very usage of the term hydrothermal. For example, chemists prefer to use a broader term, viz., solvothermal, meaning any chemical reaction in the presence of a solvent in supercritical or near-supercritical conditions [22]. However, this term has been introduced recently, and in fact, the early work in this direction was carried out by geologists using CO_2 [23]. Several international conferences and symposia have already been organized in different countries [24–26]. Similarly, there are several other terms like *glycothermal*, *alcothermal*, *ammonothermal*, *lyothermal*, *carbonothermal*, and so on, depending upon the type of solvent used in such chemical reactions [27]. Table 1.2 gives different terminologies under hydrothermal and related processes involved.

However, the purpose behind using these different solvents in the chemical reactions is essentially to bring down the pressure–temperature conditions. In this context, Yoshimura [28] has proposed a new term, *soft solution processing*, for processes in which the pressure and the temperature conditions reach near or just above ambient conditions. Though this term has a broader meaning, it covers only a portion of the hydrothermal research and refers mainly to any solution processing at or near ambient conditions. Thus, in the present book, the authors retain a broader term, hydrothermal, throughout the text and use other terms only when such occasion arises.

As mentioned above, under hydrothermal conditions, the reactants which are otherwise difficult to dissolve go into solution as complexes under the action of mineralizers or solvents; hence, one can expect the conditions of chemical transport reactions. Therefore, some workers even define hydrothermal reactions as special cases of chemical transport reactions. Owing to the specific physical properties, particularly the high solvation power, high compressibility, and mass transport of these solvents, one can also expect the occurrence of different types of reactions like:

i. Synthesis of new phases or stabilization of new complexes.
ii. Crystal growth of several inorganic compounds.
iii. Preparation of finely divided materials and microcrystallites with well-defined size and morphology for specific applications.
iv. In situ fabrication of materials with desired size, shape, and also dispersibility in case of nanomaterials.
v. Leaching of ores in metal extraction.
vi. Decomposition, alteration, corrosion, and etching.

Table 1.2 Different Terminologies Under Novel Solution Routes [27]

Conventional Hydrothermal	Solvothermal	Supercritical Hydrothermal	Related Terminologies	Multienergy Hydrothermal
Aqueous solvent Refers to conditions above the atmospheric temperature and pressure. Suitable for high-quality bulk, fine nanocrystals	Nonaqueous solvents Low to high temperature conditions. Suitable for good-quality bulk, fine nanocrystals	Critical to supercritical conditions Both aqueous and nonaqueous solvents Suitable for fine and nanocrystals, rapid	Ammonothermal, glycothermal, lyothermal, alcothermal, carbonothermal, etc. depending upon the specific solvent used Spray pyrolysis Suitable for fine to nanocrystals and thin films	Hydrothermal in combination with extra energy like microwave, electrochemical, sonar, mechanochemical, biomolecular, sol−gel, etc. Extremely efficient for thin films, fine to nanocrystals. Epitaxy, etc. Very fast processing

1.3 Mineralizers

In any hydrothermal system or reaction confined to any one of the processes described in Section 1.1, the role played by the solvent under the action of temperature and pressure is very important. In case of hydrothermal processing, the solvent plays multiple roles and its role has to be understood clearly in order to work out a mechanism for materials processing under hydrothermal conditions. It has been interpreted in various ways by many workers. Yoshimura and Suda [20] have described these processes to understand the action of solvent, for example, water on solid substances under elevated pressure and temperature conditions. This process is represented in Table 1.3 and Figure 1.1. Figure 1.1 shows the actions of hydrothermal fluids on the solid substances from physical action as the P, T medium to chemical action as the reactant. Accordingly, we can expect the solvent to work as a medium providing the suitable environment, also it can act as an absorbent, as a solvent and reactant. Each action has a specific role to play in hydrothermal reactions. Depending upon the action, the application and apparatus are selected for the hydrothermal processing of materials.

Through proper interpretation of the above-listed processes, one can easily develop a required hydrothermal process corresponding to the material synthesis or crystal growth or materials process using a suitable solvent to increase the solubility of the desired compound. Water is the most important solvent and it was popularly used as a hydrothermal mineralizer in all the earlier experiments. However, several compounds do not show high solubility for water even at supercritical

Table 1.3 Action of Hydrothermal Fluid (High-Temperature and High-Pressure Aqueous Solution/Vapor) on Solid-State Materials

S. No	Classified	Action	Application
1.	Transfer medium	Transfer of kinetic energy, heat and pressure forming, etc.	Erosion, machining abrasion, HIP
2.	Adsorbate	Adsorption/desorption at the surface	Dispersion, surface diffusion, catalyst, crystallization, sintering, ion exchange, etc.
3.	Solvent	Dissolution/precipitation	Synthesis, growth, purification, extraction, modification, degradation, etching, corrosion, etc.
4.	Reactant	Reaction	Formation/decomposition (hydrates, hydroxides, oxides), corrosion, etc.

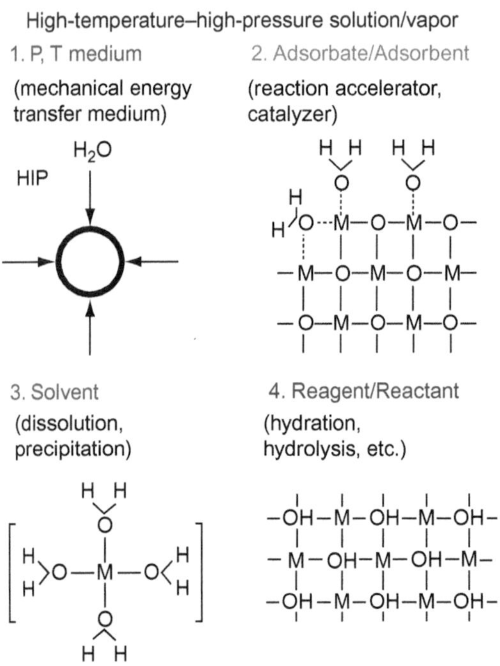

Figure 1.1 Action of hydrothermal fluid on solid substances [21].

temperature, and hence the size of the crystals or minerals obtained in all the early hydrothermal experiments of the nineteenth century did not exceed thousandths or hundredths of a millimeter. Therefore, the search for other suitable mineralizers began in the nineteenth century itself. A variety of aqueous and nonaqueous solutions were tried to suit the preparation of a particular compound [29]. The selection of the mineralizers and their role in hydrothermal systems with suitable examples are discussed in great detail in Chapter 3. The knowledge acquired through the use of several new mineralizers has helped to implement this hydrothermal technique as an effective one in preparative chemistry. Table 1.4 shows the use of hydrothermal processing in various fields of materials synthesis, crystal growth, and materials processing [20]. Before going into the details of this, it is appropriate to discuss the natural hydrothermal systems.

1.4 Surfactants

These are usually organic fatty acids, usually amphiphilic, containing a blend of surface active agent that is both hydrophobic and hydrophilic. The alignment and aggregation of surfactant molecules at the surface acts to alter the surface properties of water at the water–air or water–oil interface. The use of surfactants, chelates, and other organic cappings is becoming very popular in the processing of advanced materials to control the nucleation of a desired phase, such that phase homogeneity, size, shape, and dispersibility could be achieved during the crystallization of the fine particles. The surfaces of the materials could be altered to hydrophobic or hydrophilic depending upon the applications [30]. Hence, surfactants are also called surface modifiers. Today, this approach is playing a key role in nanotechnology to prepare highly dispersed, oriented, and self-assembled particles of even complex and multicomponent materials. Also, in crystal growth, the use of surfactants helps in modifying the crystal size and morphology. Thermodynamics of the surfactant systems are of great importance, theoretically and practically [31]. This is because surfactant systems represent systems between ordered and disordered states of matter. Surfactant solutions may contain an ordered phase (micelles) and a disordered phase (free surfactant molecules and/or ions in the solution). In surface modification, the *pH* of the medium, isoelectric point (*iep*), and dissociation constant (pK_a) of the surfactants are very important. At pH below pK_a, the modifier does not dissociate. Moreover, below iep, the surface of metal oxide nanoparticles is surrounded by positive charges (major) and hydroxylic groups (minor). Under

Table 1.4 Development of Hydrothermal Processing

1. Crystal synthesis and growth:	Oxide, sulfide, fluoride… (1978–)
2. Preparation of fine to nanocrystals with controlled composition, size, and shape:	PZT, ZrO_2, $BaTiO_3$, HAp, metal oxides, ferrite (1978–)
3. Crystallized thin/thick films:	$BaTiO_3$, $SrTiO_3$, $LiNbO_3$ (1989–)
4. Etching and corrosion:	Oxide, nitride, and carbide
5. Polishing and machining:	Oxide, nitride, and carbide
6. Combined with electrical, photo-, radio-, and mechanoprocessing:	Synthesis, modification, and coating
7. Organic and biomaterials } :	Hydrolysis and extraction
8. Nonaqueous solution } :	Polymerization and synthesis
9. Continuous system } :	Decomposition and wet combustion

these conditions, there is no chemical reaction occurring between the modifier and the material surface, but it is only through a strong hydrogen bonding the modifier can attach to the nanoparticle surface. On the contrary, at higher pH than pK_a, dissociation of the modifier takes place and results in a chemical reaction between the dissociated part of modifier and OH_2^+ from the particles' surface. Thus, by dehydration reaction, the modifier attaches to the surface of the particles or crystal. Mass balances, charge balance in the actual system, pH, and the surfactant can be fixed for most of the systems by considering the chemical reactions [32]. The role of surfactants is described in detail in Chapter 10.

1.5 Natural Hydrothermal Systems

The beginning of hydrothermal research is firmly associated with the study of the natural systems by earth scientists who were interested in understanding the genesis of various rocks, minerals, and ore deposits through laboratory simulations of the conditions existing in the earth's crust. Therefore, it is appropriate to discuss briefly the research on natural hydrothermal systems and its contribution to the development of this field to its present status. Starting from the earliest experiment by Schafthaul in 1845 on the synthesis of quartz [3], over 130 mineral species were synthesized by the end of nineteenth century, and the experiments were carried out on various geological phenomena ranging from the origin of ore deposits to the origin of meteorites. Today, it is being popularly used by geologists to solve several existing problems in petrology, geochemistry, mineralogy, ore genesis, and paleontology. The impetus for the experimental investigations during the nineteenth century was provided not only by a desire to explain geological phenomena but also by greatly improved equipment and techniques fostered by the industrial revolution. Such investigations helped in unraveling the hitherto unknown natural geological processes of mineral formation. Also they have helped in finding uses for the artificially synthesized single crystals like ruby, emerald, sapphire, quartz and diamond in the gemstone industry.

In order to understand the formation of several ore deposits including that of noble metals, it is necessary to discern the physicochemical conditions which govern the transport and precipitation mechanisms of these metals in hydrothermal solutions. Several thermodynamic models have been proposed to explain these mechanisms in nature. Relatively, much is known, for example, about the hydrothermal chemistry of gold [33,34]. Similarly, the behavior of common rock-forming minerals in a variety of electrolytic solutions has been studied in detail. Here, the authors present only the salient features of these works to provide the background for hydrothermal technique since the main theme of this handbook is crystal growth and materials processing. Besides, a quantitative model of the transport and deposition mechanisms is still impeded by a dearth of reliable high temperature and high pressure, experimentally based, thermodynamic data. However, there is some remarkable progress being made in this direction, thanks to the advances in thermodynamic modeling, the direct sampling and analyses of the natural geothermal systems, their extinct analogues, epithermal ore deposits, and other geologic environments, primarily by analyses of fluid inclusions [35]. Therefore, an understanding of the mineral–water reaction kinetics is essential to quantifying the behavior of natural and engineered earth systems. Although several studies have been carried out on the behavior of most hydrothermal systems, the predictions based upon rates from deionized water are unlikely to be representative of processes occurring in complex fluids of natural systems which contain numerous dissolved ions. These constituents have both significant rate-inhibiting and enhancing effects on behavior, even when present in very small quantities [36,37].

In nature, the most common minerals in soils and rocks are generally in contact with water at a wide pH range. In extreme cases, the pH can be nearly 2 in the presence of sulfides, which oxidize to give H_2SO_4, and as high as 10 in the presence of alkaline salts like Na_2CO_3. Here, the authors briefly discuss the gold deposition in hydrothermal ore solutions. In this case, the major role is played by the chloride and sulfur-containing ligands [38,39]. The dominant gold complexing ligands are usually sulfide species. The stability constants for gold(I) chloride complexes (e.g., at 250°C) are up to 20 orders of magnitude smaller than those of Au(I) hydrosulfide complexes and, therefore, the latter predominate in nature [40]. Despite this observation, the stability constants for Au(I) hydrosulfide complexes under high-temperature and high-pressure environments are not yet well defined. This is particularly true for the low pH region where no satisfactory data are available.

Benning and Seward [41] have proposed three sets of experimental conditions of pH range for gold deposition in nature:

$$Au_{(s)} + H_2S + \quad\quad + H_2O + H_2(gas) \quad pH \approx 4$$
$$Au_{(s)} + H_2S + NaHS \quad + H_2O + H_2(gas) \quad pH \approx \text{neutral}$$
$$Au_{(s)} + H_2S + H_3PO_4 \quad + H_2O + H_2(gas) \quad pH < 4$$

The solubility of gold increases with increasing temperature, pH, and total dissolved sulfur. At near neutral pH, an inverse correlation between solubility and pressure has been observed, whereas in acid pH solutions above 150°C, an increase in pressure increases the solubility. The equilibrium constants for the uncharged complex AuHS show that this species plays an important role in the transport and deposition of gold in ore-depositing environments which are characterized by low pH fluids.

Recently, some thermodynamic modeling in the chloride systems $Au-NaCl-H_2O$ and $Au-NaCl-CO_2-H_2O$ shows that the gold solubility decreases in the presence of CO_2 due to decreasing dielectric permeability of the CO_2-bearing solution [42]. This model has been experimentally verified at 350°C and 50 MPa in 1 M KCl + 0.1 M HCl solutions in the presence of 3 M CO_2 and without CO_2. It was found that the gold concentration in chloride CO_2-bearing solutions is one order lower in magnitude than in systems without CO_2. Similarly, the silver bearing systems show distinct influence of CO_2 in comparison with gold system. CO_2 has a negative solubility effect on gold and silver crystallization, and CO_2 acts as a nonpolar component of crustal fluids in the crystallization of many ore deposits [43]. Thus, the recent hydrothermal solution speciation (solvation to ion pairing and complexing) study has greatly contributed to the knowledge and better understanding of various geological problems.

1.6 The Behavior of Volatiles and Other Incompatible Components Under Hydrothermal Conditions

The physical and thermodynamic properties of silicate melts depend upon melt structure. The structure of a melt is determined both by its composition and the ambient conditions and, with the notable exception of liquid immiscibility, may vary continuously with changes in these parameters. The structures of crystalline silicates vary only within restricted limits. As a result, variation in mineral–melt equilibria caused by changes in either compositions or external parameters may very well reflect the change in properties of the melt phase to a greater extent than those of the crystalline silicates.

The behavior of phosphorus pentoxide is complex. Phosphorus pentoxide depolymerizes pure SiO_2 melts by entering the network as a fourfold coordinated cation but polymerizes melts in which an additional metal cation, other than silicon, is present. The effect of this polymerization is apparent in the widening of the granite–ferrobasalt two-liquid solvus. In this complex system, P_2O_5 acts to increase phase separation by further enrichment of the high charge density cations Ti, Fe, Mg, and Ca, in the ferrobasaltic liquid. Phosphorus pentoxide (P_2O_5) also produces an increase in the ferrobasalt–granite REE liquid distribution coefficients. These distribution coefficients are close to 4 in P_2O_5-free melts but close to 15 in P_2O_5-bearing melts.

Several attempts to understand the internal evolution of highly fractionated pegmatites focused on the roles of H_2O and other components, especially rare alkalis, B, P, F, are being carried out from time to time. Such studies yield a great amount of data on the role of these volatiles under hydrothermal conditions and also distinguish rare earth pegmatites from other rocks [44,45]. These volatiles exert a significant control on fluid properties, solidus and liquidus temperatures.

1.6.1 *Water*

Water is an important constituent of any hydrothermal system. It exhibits unique properties, especially under supercritical conditions. These properties have been exploited appropriately in the recent years to disintegrate toxic organics and recycle or treat waste materials. In nature also, water plays an important role in the formation of various rocks and minerals and in the creation of life (origin of life). As a component of granitic melts, H_2O depresses solidus and liquidus temperatures [46,47], lowers melt viscosities [48–51], and promotes coarse grain size [52].

1.6.2 Fluorine and Chlorine

Fluorine, as H_2O lowers solidus and liquidus temperatures, enhances cation diffusivities and decreases melt viscosities [53–55]. One important difference between F and H_2O is that the DCF (Determination of Moisture Content) vapor/melt is L1 in metaluminous and peraluminous granitic bulk compositions [56,57]. Fluorine decreases the solubility of the melt [58,59] so that, at equal H_2O content of volatile-undersaturated magma at the same pressure and temperature, H_2O is higher in F-bearing melts than in F-absent melt. Metal–fluoride complexing in aqueous vapor may be important in rare-metal transport and the formation of hydrothermal ores [60,61].

1.6.3 Boron

The pronounced effects of boron in hydrous silicate melt are well known. Boron lowers the solidus and it increases the solubility of H_2O in silicate melts. Lewis acid–base properties suggest that the solubility of H_2O/mole B_2O_3 in melts should increase with increasing boron content because of changes in B–O coordination. Boron also decreases the viscosity of silicate melts, presumably through melt depolymerization caused by the synergistic network-modifying effects of boron and water [62].

In silicate melts, boron forms two common oxyanions: trigonal planar BO_3^{3-} and tetrahedral BO_4^{5-}. The [BO_4^{5-}/BO_3^{3-}] ratio increases with melt alkalinity [63,64] and boron content [65]. Although boron exhibits strong interaction with silicate melt, nonideal mixing between borate and silicate melt components is reflected by stable liquid–liquid or metastable glass–glass immiscibility over a wide range of bulk compositions in a hydrous silicate system [66]. The miscibility gap between metal-rich borate and essentially pure silica liquids increases with the increasing field strength of added metal cations, with a consequent rise in the consolute temperature. Limited experimental evidence also indicates that the solubilities of other high charge-density cations (e.g., Group IV and Group V elements) are significantly higher in borosilicate melts than in simple aluminosilicate melts [67].

As in fluorine-bearing systems, the addition of boron to silicate melts leads to an expansion of the liquidus field of quartz [68]. This behavior may reflect the removal of cations from coordination with SiO_4^{4-} of the melt framework, resulting in higher SiO_2 through increased polymerization of SiO_4^{4-} tetrahedra [69]. Among Group I cations, boron has a tendency to depress the activities of the higher field-strength ions. The increased solubility of H_2O in borosilicate melts corresponds to a lower H_2O that may stem from direct hydrolysis of borate oxyanions.

1.6.4 Phosphorus

Phosphorus exhibits limited solubility in silicate melt [70–73], hence the addition of phosphorus promotes phosphate-silicate liquid immiscibility. The phase equilibria data in the systems, $SiO_2–P_2O_5$, $P_2O_5–M_xO_y$, and $P_2O_5–M_xO_y–SiO_2$, show that phosphorus has an affinity for H, and it increases the solubility of H_2O in silicate melts, possibly by $P{=}O + H_2O \rightarrow (HO)–P–(OH)$ [74]. Although experimental evidence of the interaction of phosphorus with other Group I elements is lacking, the common pegmatite assemblage $LiAlPO_4$ (OH,F) + $NaAlSi_3O_8$ (rather than $NaAlPO_4$(OH,F) + $LiAlSi_2O_6$ + H_2O) suggests that P is more compatible with the smaller, more acidic cations of Group I [53–57].

1.6.5 Behavior of Alkalis

The zonation of rare element pegmatites is manifested largely by heterogeneous distributions of alkali aluminosilicates. Theoretical and experimental investigations of the interactions of Group I elements with aluminosilicate melts provide a basis for understanding the zonation of alkali aluminosilicates in pegmatites.

Both De Jong and Brown [59] and Navrotsky et al. [66] proposed that the smaller alkalis Li and Na should exhibit a greater tendency to destabilize silicate melts than K, Rb, and Cs.

A number of salient pegmatite characteristics can be explained by the effects of high concentrations of boron, fluorine, and phosphorus on phase equilibria in hydrous aluminosilicate melts.

1.6.6 Crystallization Temperatures

Comparatively, high concentrations of B, P, F, and Group I elements serve to depress pegmatite magma liquid to approximately 650°C (within the stability fields of petalite or spodumene), and solids to <500°C. The low liquidus temperatures permit rare-element pegmatite magmas to migrate to metamorphic conditions of the andalusite–cordierite–staurolite facies [75,76]. The physical migration of magma may be facilitated also by the lower melt viscosities of the H_2O-, B-, and F-rich pegmatite system. Because of the low temperature interval of crystallization, however, rare element pegmatite magmas may experience rapid increases in melt viscosity with only slight cooling, as glass transition temperatures are approached (e.g., as in the macusanite analogue). As a result of increased kinetic barriers to crystallization, internal disequilibrium may prevail (an important and poorly defined parameter in pegmatite crystallization is the rate of cooling), but evidence from wall-rock studies indicates that rare element pegmatite magmas are hosted by rocks at a temperature <500°C [77]. Common textural features of rare element pegmatites, such as graphic intergrowths and radial or banded fabrics, can be interpreted as disequilibrium phenomena in supercooled liquids or glasses [78]. Experiments with dry macusanite, however, present alternative possibilities. In these experiments, pegmatitic fabrics, mineral assemblages, and zonation have been generated at or near equilibrium conditions with high concentrations of B, P, and F but low water content.

Numerous experimental investigations of aqueous systems at high temperature and high pressure have been undertaken using conductivity, potentiometry, spectrophotometry, solubility, PVT (Price and Volume Trend technical analysis indicator), and calorimetry, neutron diffraction, EXAFS (Extended X-ray Absorption Fine Structure), and other related methods. These studies yield a vast amount of information on cation-oxygen pairing and their increased or decreased distances with varying temperature. Likewise, the anion-water distances, for example, as in the case of iodide-oxygen (water) bond lengths, indicate that the solvation shell expands slightly with increasing temperatures. Similarly, molecular dynamics simulation studies on the alkali metals hydration in high temperature water up to 380°C have been carried out recently, demonstrating a small expansion of the first hydration shell around the chloride with increasing temperature [79]. More or less complete data are available on the formation of simple, neutral ion pairs for dilute alkali metal halide solutions at HPHT conditions. However, the understanding of the formation of polynuclear species is still not clearly understood. Some workers have studied the ion pairing and cluster formation in a 1 M NaCl solution at 380°C and near critical pressure [80]. These studies indicate the presence of simple monocationic ions and ion pairs together with triple ions such as Na_2Cl^- and $NaCl_2^-$ as well as the $Na_2Cl_2^-$ and more complicated polynuclear species. Similar studies on other solutions are available in the literature. All these studies have greatly contributed to the understanding of the geochemical system wherein the metal-complexing by other ligands is the most important aspect. However, there is a major lack of overall experimental data pertaining to the metal complex equilibria in supercritical aqueous systems as well as in binary solvent systems such as H_2O-CO_2. Such data are of enormous importance to the understanding of the geochemistry of element transport by hydrothermal fluids active in the earth's crust.

During 1994, a new concept, viz., *geothermal reactor*, introduced by Japanese workers, is slowly catching the attention of hydrothermal engineers [81,82]. The principles of geothermal reactors include the direct use of geothermal energy as a heat source or driving force for chemical reactions. It helps to produce hydrothermal synthesis of minerals and a host of inorganic materials, extraction of useful chemical elements contained in crustal materials such as basalt, and use them as raw materials for hydrothermal synthesis. Thus, the concept of geothermal reactor leads to the construction of a high-temperature and high-pressure autoclave underground. This has several advantages over the conventional autoclave technology.

Figure 1.2 shows the schematic sketch of a typical geothermal reactor for mineral synthesis [82]. The major advantages of the geothermal reactor are:

1. Synthesis of ceramic materials by hydrothermal reaction is possible without using fuel or electricity as main energy source.
2. The system does not discharge the used heat.
3. Outer tube with a slit in the bottom must be strong enough to keep the inner space of the tube against the pressure by the wall of formation, but inner double tube does not need the strength against the inner pressure as usual autoclaves.
4. Area of the plant is small because the vertical long reactor is under the ground.
5. It is highly useful to study the alteration of various rocks occurring in the earth's crust in the presence of fluids of various compositions. Such studies have been carried out for granite and basalts [75].

Such geothermal reactors are in use in the Miyagi prefecture of Japan, and the researchers are involved in the studies related to the rock–hydrothermal fluid interactions and also synthesis of some materials (K. Nakatsuka, personal communication).

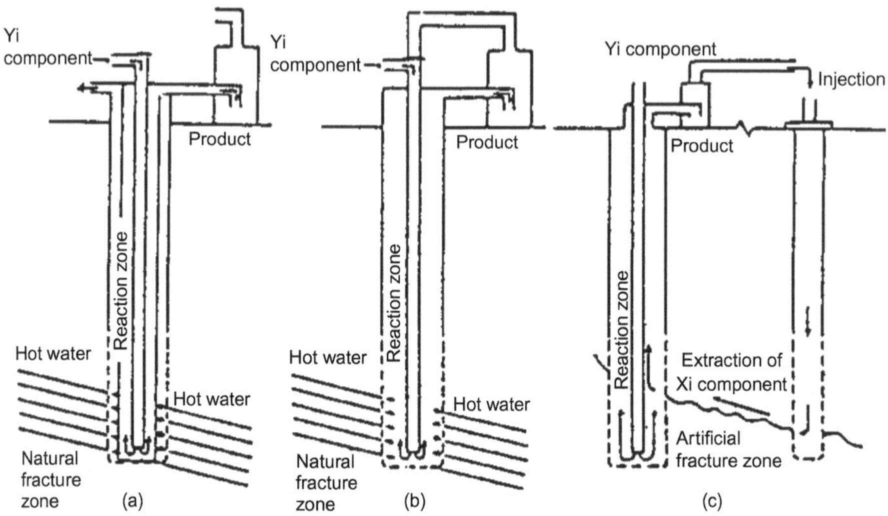

Figure 1.2 Geothermal reactors for mineral synthesis [82].

Similarly, enhanced geothermal systems (EGS) are available in several countries. These EGS are renewable energy and provide better understanding of hot rock–fluid interactions, through more realistic simulations that mimic extreme conditions not currently achievable with other systems [83].

The main disadvantage of the geothermal reactor is that the flow characteristics of high-temperature slurry accompanied with the chemical reaction must be well understood for controlling the reaction. The volumetric capacity of the geothermal reactor is very large compared to that of usual autoclave, and the operation must be continuous. The merit of the geothermal reactor and its cost of operation can be realized only if the target material is to be developed in large quantities.

1.7 Submarine Hydrothermal Systems

The general acceptance of plate tectonics theory some 2½ decades ago has garnered much interest in geochemical processes at plate boundaries which has led to the discovery of hydrothermal activity in the deep sea directly on the Galapagos Spreading Centre in 1977 [84] and a large number of other spectacular submarine hydrothermal systems (like the Red Sea Rift Valley, Juan de Fuca Ridge-North Pacific Ocean; 21°N East Pacific Rise, Kamchatka, Kurile Islands, Atlantis II Deep in the Red Sea, Lake Kivu, Lost City hydrothermal field, and so on) of global significance to ocean chemistry and geochemistry [85]. The global distribution of known hydrothermal vents is shown in Figure 1.3. The studies related to the temperature and chemical anomalies of these hydrothermal vents hint that many more sites exist throughout the world's oceans. In fact, this discovery has led to a new thinking in marine biology and geochemistry, and in economic geology and has spawned an entirely new term, viz.,

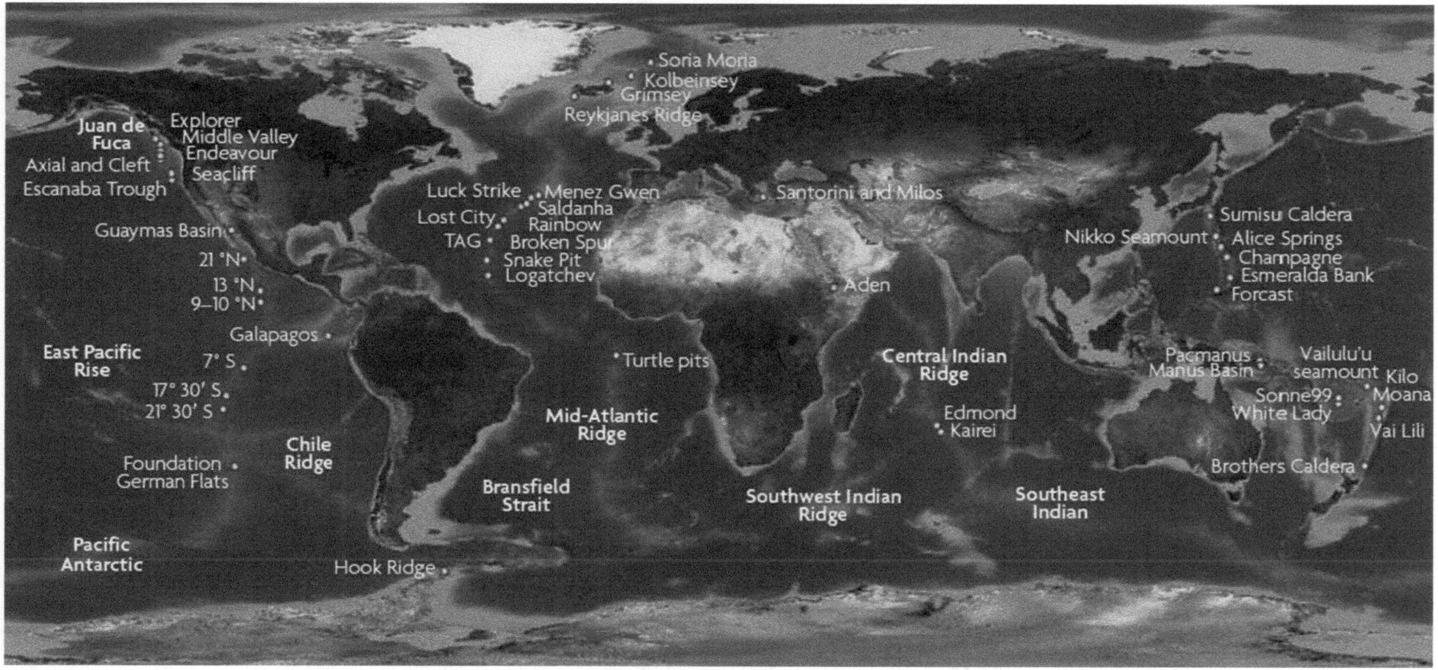

Figure 1.3 Global distribution of known hydrothermal vents.
Source: Courtesy of D. Fornari and T. Shank, Woods Hole Oceanographic Institute, Massachusetts, USA.

hydrothermal ecosystems, which means water-containing terrestrial, subterranean, and submarine high-temperature environments which are the sites of investigations for many paleobiologists and biologists looking for primitive forms of life. It is strongly believed that the roots of life on earth can be found in hydrothermal ecosystems. These ecosystems may also serve as an analogue for the possible origin of life on Mars, where a similar environment might have existed or still exist.

The conditions at the hydrothermal ecosystems mimic, to some extent, the conditions on early earth because of the presence of both heat and water. These conditions were abundant at around 3.5 Ga, when there was much greater vulcanism and a higher ambient temperature on earth. Dick Henley [86] describes the geochemical activity of the hydrothermal ecosystems as a biotic factory. According to Everett Shock [87], life thrives in submarine hydrothermal conditions because they have a (geologically supplied) source of chemical disequilibrium which brings in redox reactions. Further, he states that life originated at warmer temperatures. This was also supported by simple experiments on organic synthesis under hydrothermal conditions. These higher temperatures would mix various elements and supply the energy for the formation of simple compounds. Abundant mineral deposits at hydrothermal ecosystems imply that they provide a fossil record of their biological inheritance. Often the minerals deposited are precious metals such as gold, silver, copper, and zinc. These are common outpourings from hydrothermal vents. Thermal waters usually contain high concentrations of dissolved components which are deposited when the hot spring discharge gases are released and the temperatures fall, leading to the deposition of mostly calcium carbonate, silica, iron oxides, and manganese oxides. Thus, any living organism that once existed in the vents become fossilized at the very site it lived in. All these theories proposed that the life existed some 3.5 billion years ago.

The spectacular nature of the submarine hydrothermal ecosystem with features such as *black smokers*, *white smokers*, and peculiar ecosystems, that are independent of sunlight as a source of reducing power, has focused much interest on hydrothermal processes for the explanation of an array of geochemical processes and phenomena [88]. The submarine hydrothermal systems reveal that most primitive organisms found in modern environments are thermophiles (e.g., *archea*). Many scientists believe in a "redox neutral" in the primitive atmosphere. The most important aspect is the possibility that the iron vapor and reduced carbon liberated from impacting objects like meteorites would leave the ocean, reducing for a long period. In addition, those submarine hydrothermal systems are the only environments where primitive life would have been protected against postulated meteoritic impacts and partial vaporization of the ocean. Supercritical fluids, like H_2O, CO_2, or CH_4, are the main constituents of any hydrothermal system. They serve as excellent solvents of organic compounds and would probably be of great potential for several of the chemical reactions eventually leading to the origin of life. Further, the pressure and the temperature gradients existing in natural hydrothermal systems have a dramatic effect on the properties of the hydrothermal fluids. This is known from the current experimental data on many electrolytic solutions and some nonaqueous solutions [89].

It is strongly believed that not all submarine hydrothermal vents are hot. The discovery of the Lost City hydrothermal field (LCHF) near the Mid-Atlantic ridge in the year 2000 characterized by carbonate chimneys that rise 60 m above the ultramafic sea floor [90] is highly relevant to the understanding of the origins of life. This alkaline vent bears H_2-rich water of about 40–90°C [91,92]. This chimney provides insights into past mantle geochemistry. ^{14}C radioisotopic dating indicates that hydrothermal activity has been ongoing for at least 30,000 years [93] whereas uranium–thorium dating indicates that venting has been active for ~100,000 years [94]. Such studies indicate that the submarine hydrothermal systems operate today in much the same manner as they are believed to have functioned in the pre-Cambrian period. They provide a wide range of closely linked gradients in both physical and chemical conditions obviously having a direct link with magmatic heat. Thus, the biological communities (biotic family) occupying vast and relatively stable soft bottom habitats of the sea are characterized by low population densities, high species diversity, and low biomass, compared to those inhabiting the unstable conditions of submarine hydrothermal ecosystems which exhibit high densities and biomass, low species diversity, rapid growth rates, and high metabolic rates.

Despite such an understanding of the hydrothermal ecosystem, there are many unclear aspects like period of formation and depth. As mentioned earlier, the available evidences indicate that the ancient ocean was a chemically reducing environment, probably warm or hot (>30°C to >100°C) [95], shallow (maximum depth of 1000 − 2000 m), and considerably more active, tectonically and hydrothermally, than it is today [96]. The model proposed by Abbott and Hoffman [97] suggests that, at the time of formation of the ocean (4.2 Ga), it is likely that the whole earth was covered with water and that hydrothermal activity was at least five times greater than at present. The overall rate of sea floor spreading and subduction has declined steadily over the evolution of the earth. Figure 1.4 shows the geological and geochemical features of the hydrothermal circulation of seawater through the oceanic crust at mid-ocean ridges [98]. Rona et al. [99] have found two types of vents occurring along spreading ridge crests of the East Pacific Rise and Juan de Fuca Ridge in the Pacific. The fluids emanating from these vents are enriched with magmatically derived elements and complexes. Depending upon the degree of mixing, we can expect a variety of chemical components in different quantities which serve as the primary energy source for the extensive microbial communities as represented in Figures 1.5 and 1.6. All the biologically important trace elements are also known to be associated with submarine hydrothermal vents. However, the actual reactions involved in the production of various chemical components are not clearly known. It is well known that both H_2 and CO are common gases associated with volcanic activity. The major source of H_2 is believed to be the oxidation of magnetite to hematite, while CO is believed to originate within the magma from C–O–H–S equilibrium [101]. Similarly, the concentration of CH_4 and helium gases has been discussed in the submarine hydrothermal vents. Oxygen and other electron acceptors (NO^{3-}, SO_4^{2-}, and PO_4^{3-}) and a variety of many unknown chemical components have allowed for the abundant growth of diverse microorganisms in the modern oceans.

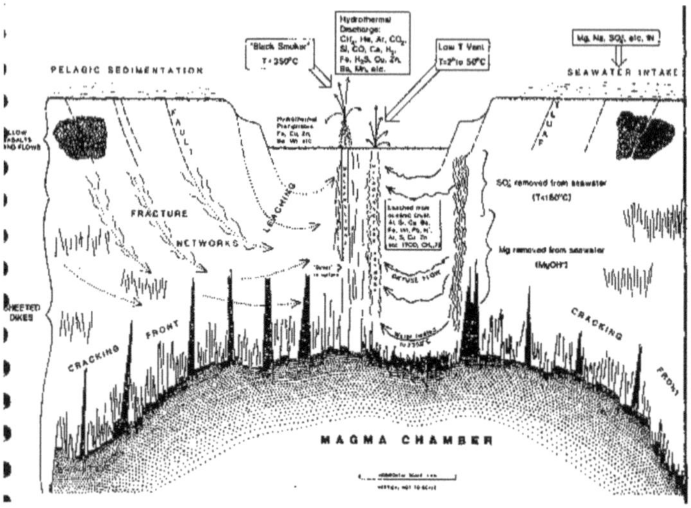

Figure 1.4 Major processes during the hydrothermal circulation of seawater through the oceanic crust at mid-ocean ridges.
Source: Courtesy of H.W. Jannasch [98].

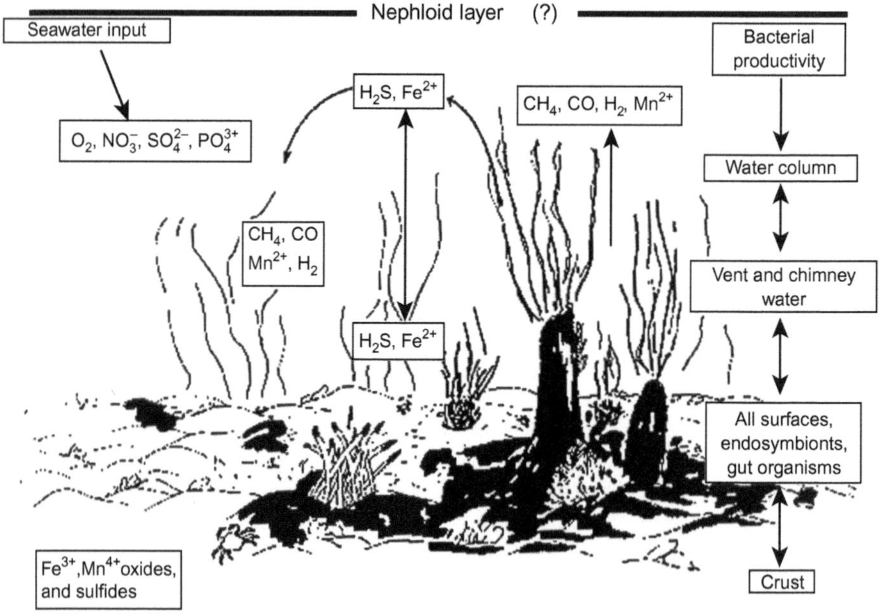

Figure 1.5 Microbial community [88].

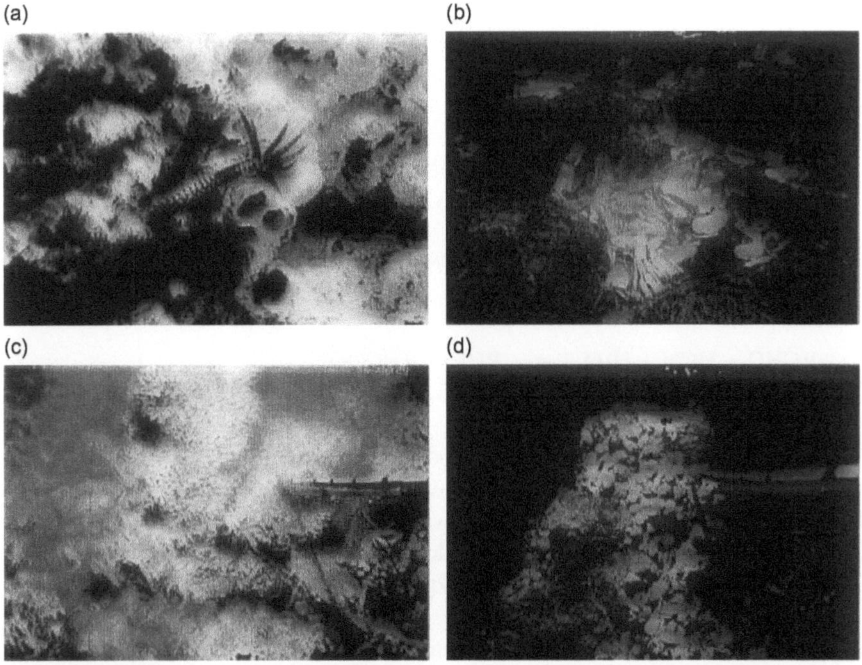

Figure 1.6 Microbial community. (a) The Pompei worm *Alvinella* at 9°N on the East Pacific Rise. The densely packed tubes form the wall of a chimney. One worm has emerged: to the upper right, the tips of the tentacles of several others can be seen at the tube openings. (b) A dead vent field on the Galapagos Rise near Rose Garden. Partially dissolved shells of vesicomyid clams fill the extinguished vent opening. But there are a few survivors—mussels filter-feeding for ambient suspended particles. (c) Barnacles encrust the edge of a vent opening at the Mariana back-arc spreading center. Alvinoconchid snails live in the openings where temperatures were measured at 10 − 25°C. Bresiliid shrimp and bythograeid crabs scavenge off the surfaces of rock and other animals. The cloudiness of the water is caused by suspended bacteria. (d) The tip of a chimney at the Mariana vent emitting 250°C effluent. Animals avoid this hot spot but are able to live only a few centimeters below it, where the temperature is only 25°C. Alvinoconchid snails, bresiliid shrimp, bythograeid crabs, and limpets can be seen.
Source: Photographs are reproduced from Ref. [100].

In order to prove the above theories and discoveries, several workers have proposed the thermodynamics of strecker synthesis in hydrothermal systems and also the hydrothermal synthesis of several amino acids under laboratory conditions. Schulte and Shock [101] and Shock [102] have worked out the strecker synthesis to produce biomolecules (amino and hydroxy acids) from starting compounds (ketones, aldehydes, HCN, and ammonia). They have evaluated their work quantitatively using thermodynamic data and parameters for the revised Helgeson−Kirkham−Flowers (HKF) equation of state. Although there is an overwhelming thermodynamic drive to form biomolecules through streckers synthesis under hydrothermal conditions, the availability and concentration of starting components limit the efficiency and productivity of strecker reactions. Some representative reactions for strecker synthesis are given below [102]:

$$\begin{array}{lcl} \text{RCHO} + \text{NH}_3 + \text{HCN} & = & \text{RCH(NH}_2\text{)CN} + \text{H}_2\text{O} \\ \text{aldehyde/ketone} & & \text{amino nitrile} \end{array}$$

$$\begin{array}{lcl} \text{RCH(NH}_2\text{)CN} + \text{H}_2\text{O} & = & \text{RCH(NH}_2\text{)CO(NH}_2\text{)} \\ \text{amino nitrile} & & \text{amino amide} \end{array}$$

$$\begin{array}{lcl} \text{RCHO} + \text{HCN} + \text{H}_2\text{O} & = & \text{RCH(NH}_2\text{)COOH} \\ \text{aldehyde/ketone} & & \text{amino acid} \end{array}$$

There are several reports on the laboratory hydrothermal synthesis of amino acids to justify the origin of life on the earth. It is appropriate to mention that the organic synthesis on the whole is not new and, in fact, it began in the previous century. However, the organic synthesis under hydrothermal conditions with reference to the origin of life began in a systematic way only during 1980s, after the discovery of hydrothermal activity in the deep sea on a Galapagos spreading rich in thermophile organisms (Figure 1.5a and b) during 1977. The reader can get additional information from the works of Fox and Windsor [103], Marshall [104], Shock [105,106], and Hennet et al. [107]. These workers have synthesized amino acids in a temperature range of 150–275°C from aqueous solutions containing KCN, NH_3, HCHO, CO_2, H_2, O_2, NaCN, NH_4HCO_3. Miller [108], and Shock in Ref. [102], have shown that the autotrophic synthesis of all 20 protein-forming amino acids was energetically favored in hot (100°C), moderately reduced, submarine hydrothermal solutions relative to the synthesis in cold (18°C), oxidized, surface seawater [108]. Although the above studies do not support the challenge raised by several others over the life in submarine hydrothermal ecosystems, these studies definitely have set a new trend in hydrothermal research.

This field of research is growing very fast and a large number of publications are coming on diversified aspects of submarine hydrothermal ecosystems. The reader can refer to some of the most important reviews for further studies [100,108,109] (Figure 1.7).

1.8 Hydrothermal Crystal Growth and Materials Processing

After the successful application of the hydrothermal technique in hydrometallurgy in the previous century, the other important application of this method began with the artificial production of bulk single crystals of quartz and also with the synthesis of zeolites during late 1930s and 1940s. The hydrothermal method of crystal growth has several advantages. It is very important for its technological efficiency in

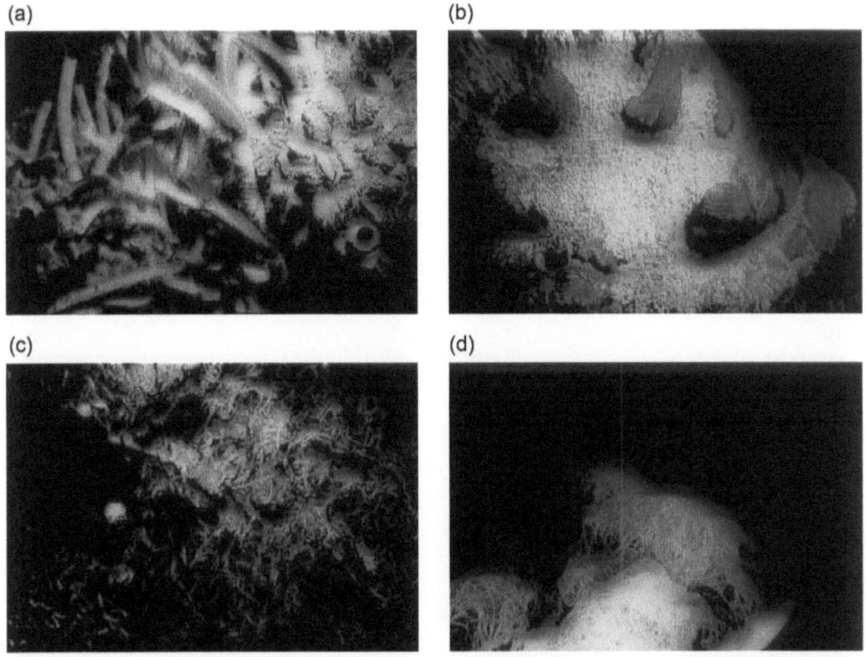

Figure 1.7 Hydrothermal activity in the deep sea. (a) The Rose Garden vent on the Galapagos Rise in 1979. Vestimentiferan tube worms and mussels dominate. Bresiliid shrimp and archacogastropod limpets walk on them. (b) Bresiliid shrimp at a Mid-Atlantic Ridge vent. (c) Serpulid polychaetes and a dandelion in the near field at the Rose Garden. Most of the serpulids are retracted into their tubes, but the tentacles of several are spread for capturing food particles from the water. (d) Spaghetti worms cover pillow lava at the periphery of Rose Garden. (e) The huge thicket of vestimentiferan tube worms at the main vent fissure at Rose Garden in December, 1979. Mussels live among the worms, principally at the base of the tubes. (f) The population of vestimentiferans has been drastically reduced, and mussels have proliferated to form a great pile over the fissure opening. (g) The rift valley on the East Pacific Rise at 9°N in April 1991. A recent lava flow has blanketed the valley. Wide-scale venting has encouraged bacterial growth, but there are not yet any animals grazing on the mats. (h) The same general area in December, 1993, a little over two and a half years later, showing vigorous colonization by vestimentiferans. The bacterial mats have disappeared.
Source: Reproduced from Ref. [100].

developing bigger, purer, and dislocation-free single crystals. The method has been widely accepted since 1960s and practically all inorganic species, starting from native elements to the most complex oxides, silicates, germanates, phosphates, chalcogenides, carbonates, and so on, have been obtained by this method. The technique is being employed on a large scale to prepare piezoelectric, magnetic, optic, ceramic, and a host of other materials both as single crystals and polycrystalline materials. In the recent years, several new advantages of the hydrothermal technique

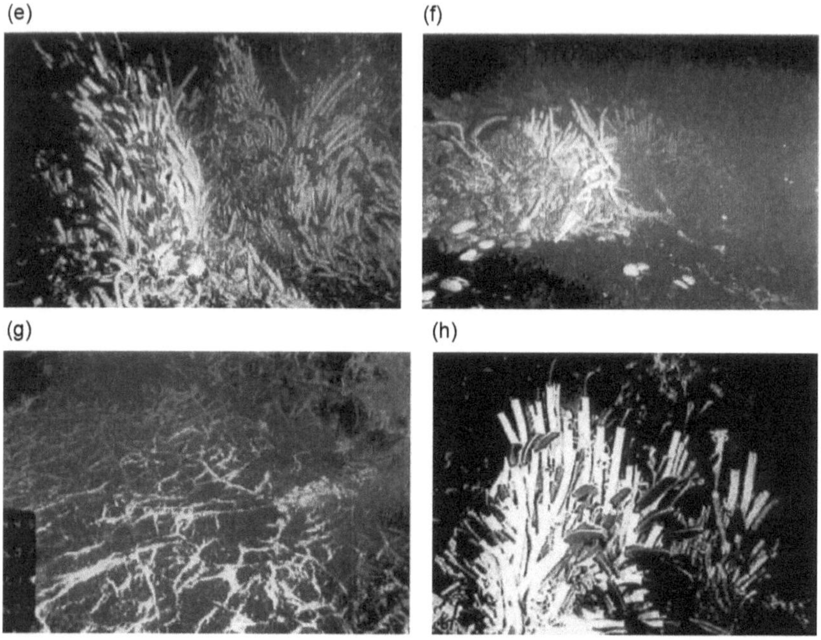

Figure 1.7 (Continued)

have been described by many workers and they are discussed in Chapters 5–10 in great detail. The hydrothermal technique, in contrast to other conventional techniques, offers several advantages:

i. Compounds with elements in oxidation states that are difficult to obtain, especially important for transitional metal compounds, can be obtained in a closed system by the hydrothermal method (e.g., ferromagnetic chromium (IV) oxide).
ii. The hydrothermal method is also useful for the so-called low temperature phases, e.g., α-quartz, α-berlinite, and others.
iii. For the synthesis of metastable compounds, such as subiodides of tellurium, Te_2I, the hydrothermal method is unique.

Commercial production of quartz and zeolites began during 1940s, but it was during 1940s and 1950s that the study of hydrothermal phase equilibria became quite popular owing to the appearance of new designs in autoclaves, particularly the Tuttle cold-cone sealed autoclaves which became popular as test tube–type autoclaves [110]. The main advantage of the phase equilibria studies under hydrothermal conditions is that it is a closed system; one can study the influence of temperature, pressure, and compositional variations individually. These studies helped to understand the phase formations in many inorganic systems, which had direct bearings on the natural systems as well. They also helped in the crystallization of several new phases which do not have natural analogues. Moreover, these studies laid a firm foundation for the growth of some of the most complex inorganic single crystals which did not have analogues in nature. During 1960s, there was a question about the search and growth of hitherto unknown compounds of photosemiconductors, ferromagnets, lasers, piezoelectrics, and ferrielectrics, so the hydrothermal method gained greater attention. This also led to the studies concerning the solubility, solvent–solute interaction, kinetics, and thermodynamic aspects of crystal growth. A large number of groups appeared in Europe, Asia, and North America, and also the number of crystals obtained by hydrothermal method increased exponentially. Further, the studies concerning the phase equilibria paved a way for the emergence of a new technology, viz., *ceramics processing technology* [111].

From the mid-1970s, exploration of the advantages of hydrothermal reactions other than the hydrometallurgical and crystal growth aspects began in Japan, particularly with reference to ceramic powder processing. A team of researchers from the Tokyo Institute of Technology, Japan, did pioneering work in ceramic processing such as powder preparation, reaction sintering, hot isostatic processing, and so on. Powder prepared by the hydrothermal technique is not agglomerated, but is fine-grained, highly pure, with controlled morphology, has a narrow size distribution, and consists of single crystals. The technique has several advantages like high reaction rate of powders, good dispersion in liquid, almost pollution free, does not require very expensive and highly sophisticated equipment, energy-saving process, and many times it produces new phases. These aspects are discussed in the subsequent chapters of this book.

In the last decade, the hydrothermal technique has offered several new advantages like new homogeneous precipitation using metal chelates under hydrothermal conditions, decomposition of hazardous and/or refractory chemical substances, monomerization of high polymers like polyethylene terephthalate, and a host of other environmental engineering and chemical engineering issues dealing with the recycling of rubbers and plastics instead of burning, and so on. The solvation properties of supercritical solvents are being extensively used for detoxifying organic and pharmaceutical wastes and also for replacing toxic solvents commonly used for chemical synthesis. Similarly, they are used to remove caffeine and other food-related compounds selectively. In fact, a new term, *hydrothermal cooking*, is being used by the food and nutrition experts in recent years [112]. These unique properties take the hydrothermal technique altogether in a new direction toward the twenty-first century and one can forecast a slow emergence of a new branch of science and technology for sustained human development. An understanding of the structure, dynamics, and reactivity of water and other aqueous electrolyte solutions with the advancement in instrumentation like neutron diffraction, X-ray, synchrotron scattering and diffraction, Raman and NMR (Nuclear Magnetic Resonance) spectroscopic approaches have greatly contributed to the emergence of this new branch of science and technology employing hydrothermal techniques. As is well known, water is environmentally the safest material and the cheapest of all solvents. It can act as a mineralizer or a catalyst under elevated pressure–temperature conditions. The thermodynamic and transport properties of supercritical water are remarkably different from those of ambient water. The solubility of nonpolar species increases, whereas that of ionic and polar compounds decrease. As a result of the drop of the polarity of water, the molecular mobility increases due to a decrease in the solvent viscosity (η). Drastic changes of ionic hydration are brought about by the decrease in the dielectric constant (ε) and density (ρ). Table 1.5 gives the values of dielectric constant, density, and viscosity of water varying with temperature and pressure in comparison with the ambient values of 78.3, 0.997 g/cm^3, and 0.890 mPa, respectively [113]. The temperature dependence of the saturation vapor pressure is illustrated up to the critical point in Figure 1.8 [113]. Below 150–200°C, the saturation vapor pressure is relatively low. Above this temperature, however, the vapor pressure rises rather sharply. Such anamolous properties of water with pressure–temperature variations are associated with the hydrogen bonding, which has attracted the attention of solution chemists. Thus, recent advances in hydrothermal solution chemistry have paved a new path to materials processing techniques with minimum energy consumption in comparison with larger energy consumption to create melt, vapor, gas, or plasma. Figure 1.9 shows free energy versus temperature diagram in a single component system. Similarly,

Table 1.5 Values of ε_r, ρ (g/cm^3), and η (mPa/s) of Water at High Temperatures and High Pressures

p **(MPa)**	**Parameter**	***t*** **(°C)**					
		200	**250**	**300**	**350**	**400**	**450**
10	ε_r	35.1	27.4	20.4	1.2	1.2	1.1
	ρ	0.871	0.806	0.715	0.045	0.038	0.034
	η	0.136	0.108	0.087	0.022	0.025	0.027
20	ε_r	35.3	28.0	21.2	14.1	1.6	1.4
	ρ	0.878	0.816	0.733	0.600	0.101	0.079
	η	0.139	0.110	0.091	0.070	0.026	0.028
30	ε_r	35.9	28.4	22.0	15.7	5.9	2.1
	ρ	0.885	0.826	0.751	0.646	0.357	0.148
	η	0.141	0.113	0.094	0.076	0.044	0.031
40	ε_r	36.3	28.9	22.6	16.7	10.5	3.8
	ρ	0.891	0.835	0.765	0.672	0.523	0.271
	η	0.114	0.115	0.097	0.080	0.062	0.039
50	ε_r	36.6	29.3	23.1	17.6	12.2	6.6
	ρ	0.897	0.843	0.777	0.693	0.278	0.402
	η	0.146	0.118	0.099	0.083	0.068	0.051
60	ε_r	37.0	29.7	23.6	18.2	13.3	8.5
	ρ	0.903	0.850	0.788	0.711	0.612	0.480
	η	0.148	0.120	0.101	0.086	0.073	0.059
70	ε_r	37.3	30.0	24.0	18.8	14.2	9.9
	ρ	0.909	0.857	0.798	0.726	0.638	0.528
	η	0.150	0.122	0.104	0.089	0.077	0.065

Source: Cited from Ref. [113].

Figure 1.10 shows the pressure–temperature range for material preparation or processing. A portion of the hydrothermal field falls within the biocompatability range, which can be further supported by recent findings in the field of submarine hydrothermal or biohydrothermal research.

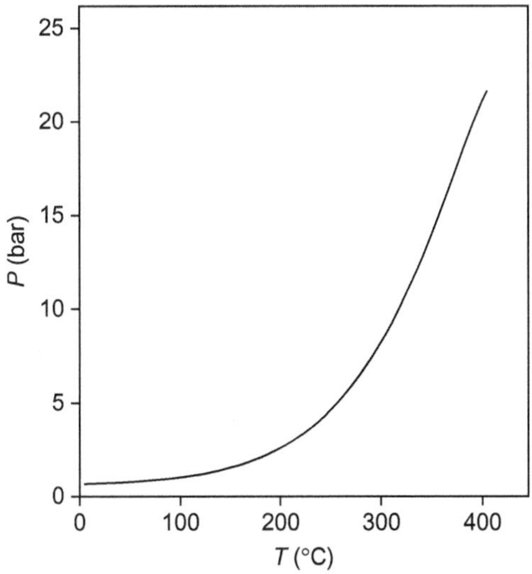

Figure 1.8 Temperature dependence of saturation vapor pressure.

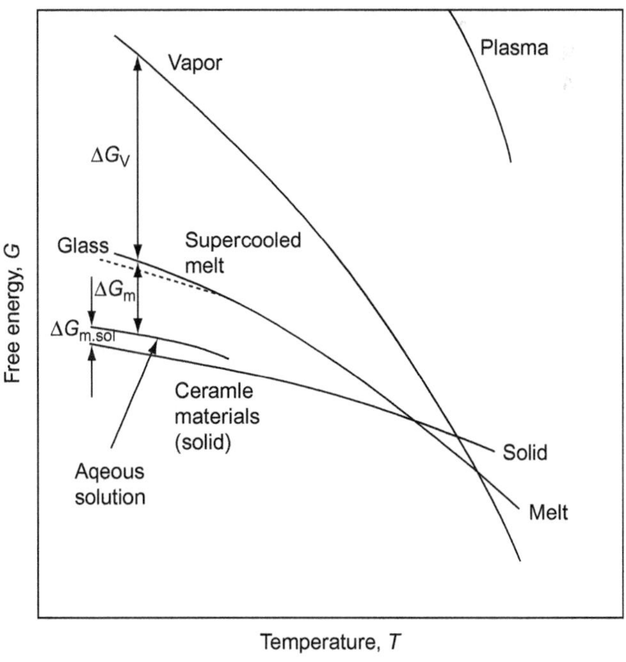

Figure 1.9 Schematic energy diagram (G–T) in a single component system.

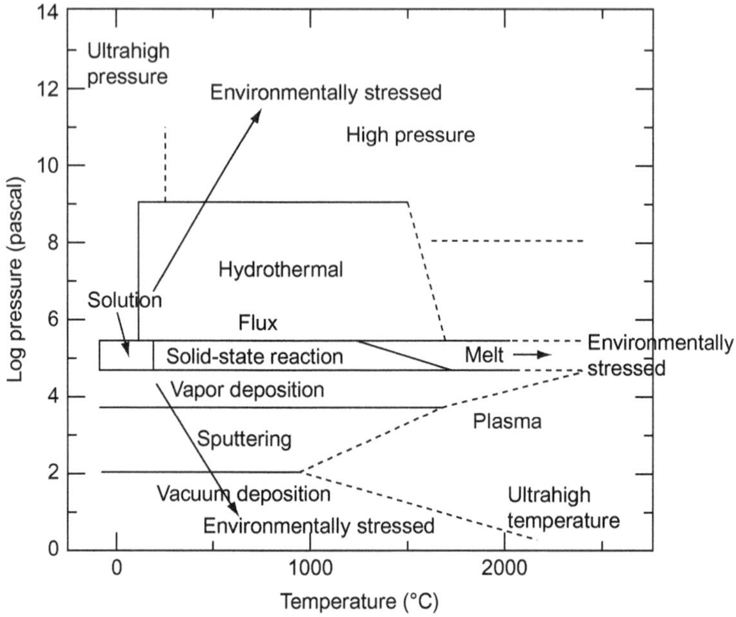

Figure 1.10 Pressure–temperature range for material preparation.

1.9 Statistics of Publications and Research in Hydrothermal Technology

The statistical data on hydrothermal research give a picture of its developments. Throughout the course of its evolution from geoscientists to modern technologists, the hydrothermal technique has captured the attention of scientists and technologists from different branches of science. Today, it is a highly interdisciplinary subject and the technique is popularly used by geologists, biologists, physicists, chemists, ceramists, hydrometallurgists, materials scientists, engineers, and so on. Figure 1.11 shows various branches of science either emerging out from the hydrothermal technique or closely linked up with the hydrothermal technique. One could firmly say that this family tree will keep expanding its branches and the roots in the years to come.

In view of such a rapid growth of the hydrothermal technique, it is becoming imperative to have a highly specialized book on this topic. There are thousands of articles published on various aspects of hydrothermal research. In the initial days of the hydrothermal research, there was a dearth of diversified journals, thus much of the hydrothermal research data from the nineteenth century to the beginning of the second half of this century was published in essentially *Earth Science Journals*, *Chemistry Annals*, and *Bulletins*. Since then, success in the growth of

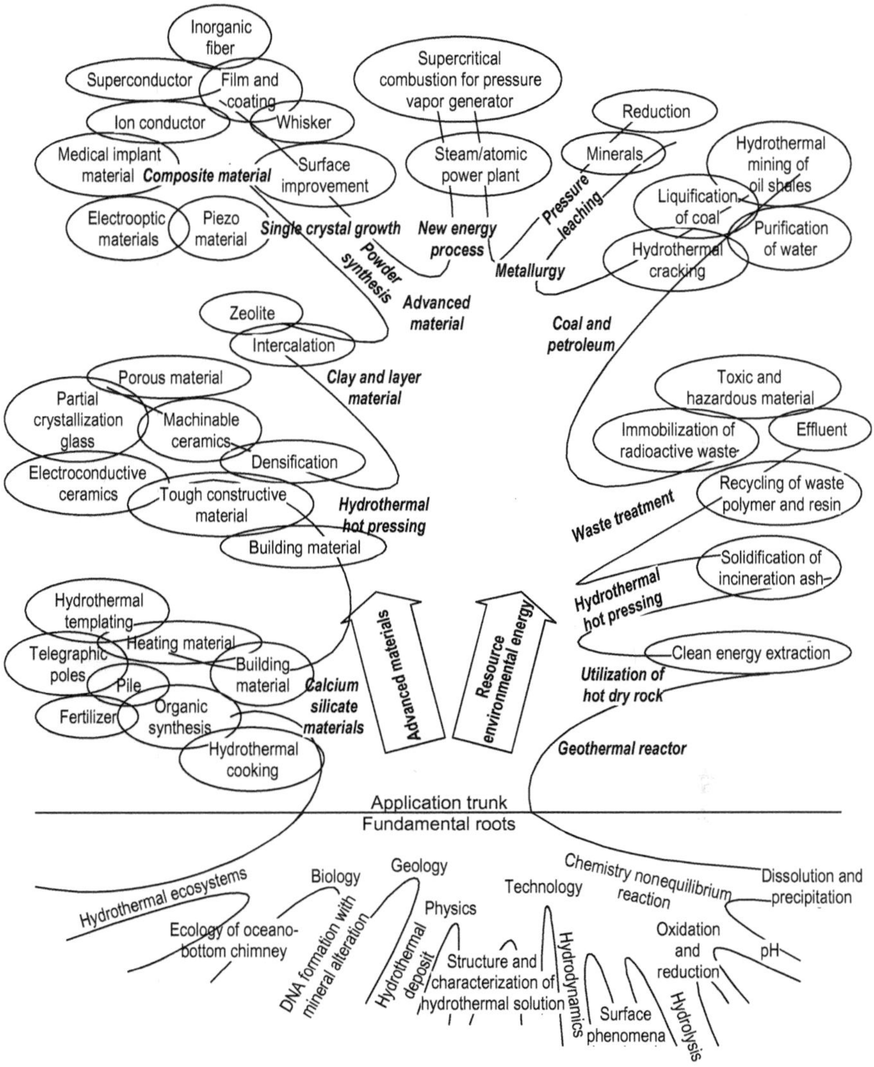

Figure 1.11 Tree showing the interdisciplinary nature of hydrothermal technology. *Source*: Courtesy of N. Yamasaki, modified by K. Byrappa.

large single crystals of quartz and, subsequently, a large number of other minerals and inorganic compounds, as well as the progress achieved in the instrument side in attaining higher pressure and higher temperature conditions, led to the slow diversity of the field. Today, a large quantum of work on hydrothermal research is being published in more than 100 journals, and the number is still growing. Owing to the limitation of space and with an intention to focus on the main theme of the book, we give below only the important journals which publish the work on hydrothermal research:

- Journal of Crystal Growth
- Journal of Materials Research
- Journal of Materials Science
- Journal of Materials Science Letters
- Materials Research Bulletin
- Materials Physics and Chemistry
- Advanced Materials
- Materials Science Forum
- Materials Letters
- Journal of Solid State Chemistry
- European Journal of Solid State and Inorganic Chemistry
- Progress in Crystal Growth and Characterization of Materials
- Crystal Research and Technology
- Crystallography (Russian Journal)
- Izvestia Academy Nauk Inorganic Materials (Russian)
- Solid State Communication
- Doklady Academy Nauk (Russian)
- Journal of American Ceramic Society
- Chemistry of Materials
- Journal of Applied Physics
- Physics and Chemistry of Earth
- Journal of American Chemical Society
- American Mineralogist
- Canadian Mineralogist
- Geochimica Cosmochimica Acta
- Zeolites
- Journal of Ceramic Society of Japan
- Bio-materials
- Japanese Journal of Applied Physics
- Ceramics Transactions
- Materials Science and Engineering
- Materials Science Research
- Materials Technology
- Journal of Microporous Materials
- Materials Research Innovations
- Journal of Nanotechnology
- Nanoscale
- Journal of Materials Chemistry
- Crystal Growth and Design
- Research on Chemical Intermediates
- High Pressure Research
- Industrial Engineering Chemistry Research
- Angewandte Chemie International Edition
- Journal of Supercritical Fluids
- Inorganic Chemistry

Although there are thousands of research publications on hydrothermal research, so far, the most comprehensive works are limited to reviews and edited books. No books have been available until the publication of this work. In connection with the hydrothermal technique for crystal growth and materials processing, the books and reviews listed in Table 1.1 are very useful.

We have made a very sincere effort to collect a more-or-less complete list of publications related to the field of hydrothermal technology, especially with a higher reliability in the last decade. Therefore, it is interesting to discuss briefly the statistics of this field of research. Figure 1.12a shows the number of publications year-wise from 1989 up to 1999, Figure 1.12b and shows the number of publications year-wise from 2000 up to mid-2011, and this number is increasing steadily with the entry of scientists from other branches of science. The preparative chemists and ceramists have dominated this field, particularly with the recent advances in the advanced materials and electronic ceramics. It is interesting to note that the hydrothermal technique of material synthesis, although began in the previous century, has gained its momentum during the postwar period. Prior to that, the majority of the compounds synthesized under hydrothermal conditions were essentially natural analogues, as the main thrust was on the study of the origin of rocks, minerals, and ores through laboratory simulations (Figure 1.13a,b shows the number of papers on hydrothermal research in materials science during 1989-2011).

With the availability of the improved equipment and also with the knowledge on the preparative chemistry routes to synthesize many inorganic compounds, with or without natural analogues under hydrothermal conditions, the popularity of the technique grew fast. Today, the number of compounds without any natural analogue synthesized under hydrothermal conditions is more than the number of compounds with natural analogues synthesized. With the advent of new mineralizers, a wide variety of organic and inorganic compounds hitherto unknown are being prepared by the hydrothermal technique. The technique is being popularly used for the crystallization of materials, crystal growth, materials processing, thin film preparation, and so on. As is evident from the synthesis of silicates, phosphates, and oxides followed by the ceramics are very popular materials being obtained under hydrothermal conditions. The popularity of this field is probably connected with the better understanding of the hydrothermal solution chemistry and the effective use of several new mineralizers which have virtually changed the scenario, particularly with reference to the pressure and the temperature conditions of synthesis of materials which were obtained under very HPHT conditions. For instance, rare earth tungstates, rare earth silicates, and so on are being obtained under mild hydrothermal conditions [114].

Hydrothermal research had its origin in Europe and later spread its activity to North America during the early twentieth century. Hydrothermal research in Asia began in the 1920s in Japan. Today, China and Japan have emerged as leaders in this field of research on a par with the United States. However, hydrothermal research is becoming quite popular in several other countries, particularly in the last four decades. The countries engaged in hydrothermal research are listed below in alphabetical order: Australia, Belgium, Brazil, Bulgaria, Chile, China, Canada,

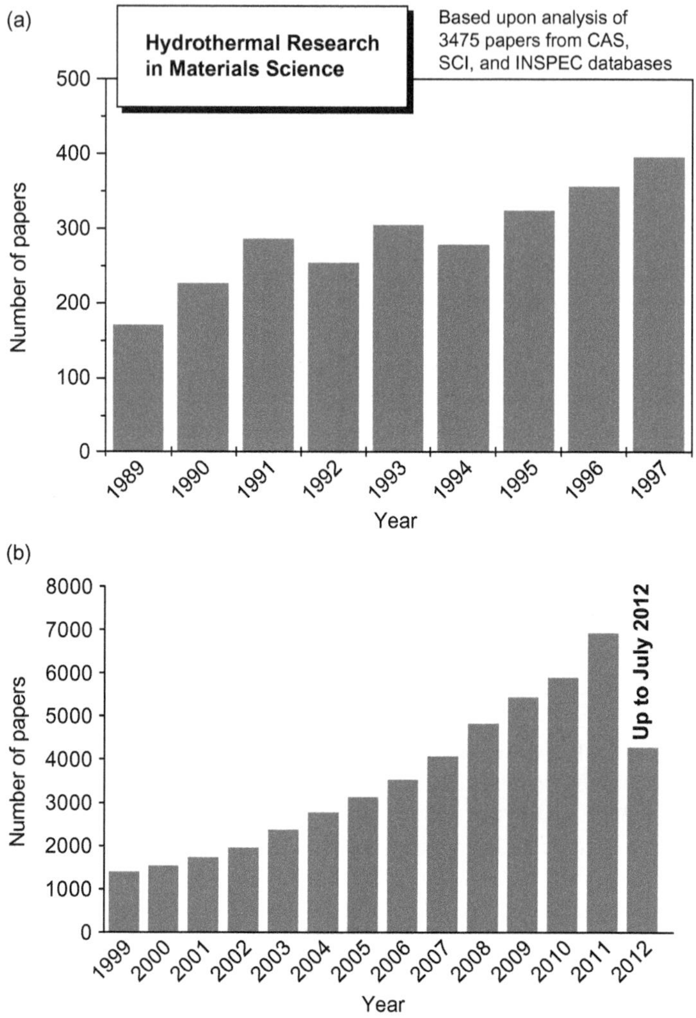

Figure 1.12 (a) Number of publications year-wise during 1989–1999. (b) Number of publications year-wise from 2000 up to mid-2011.

Denmark, France, Germany, Holland, Italy, India, Japan, Korea, Norway, Poland, Russia, Switzerland, Spain, Sweden, Taiwan, United Kingdom, Ukraine, and United States. Among these countries, over 75% of the hydrothermal research is going on in the United States, China, and Japan. Figure 1.14 shows the countries actively engaged in hydrothermal research.

The above statistical data on hydrothermal research clearly illustrate a growing popularity of the technique, covering various branches of science. The main disadvantage of the hydrothermal system, as believed earlier, is the black box nature of

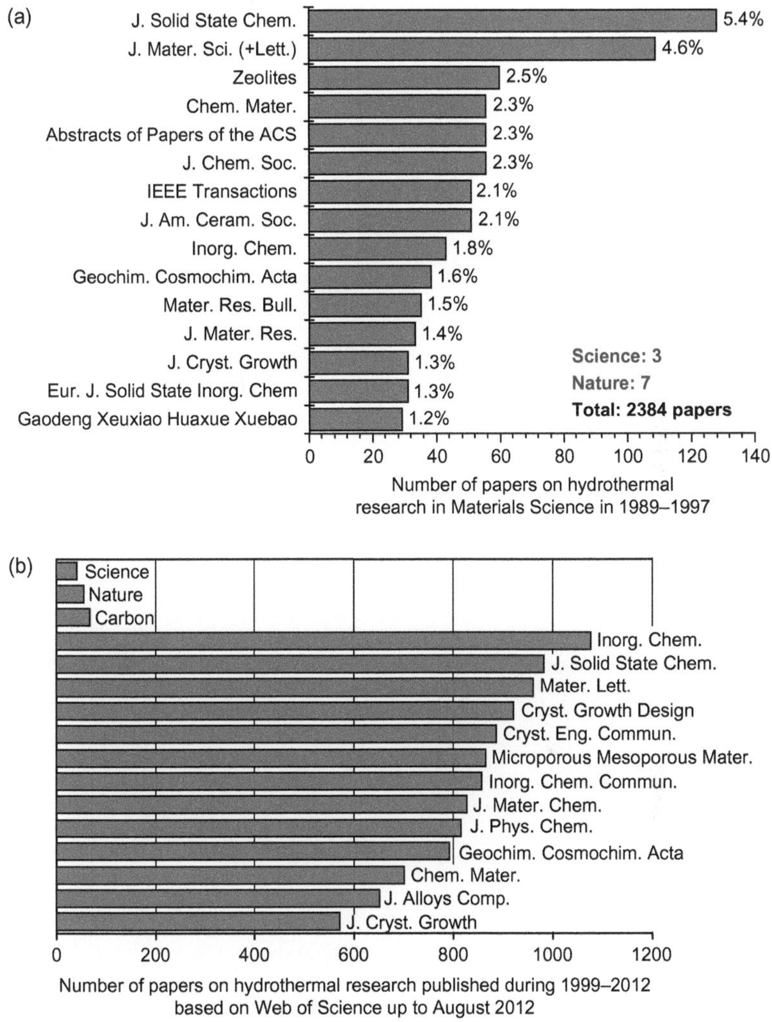

Figure 1.13 (a) Number of papers on Hydrothermal Research in Materials Science during 1989–1997. (b) Number of papers on Hydrothermal Research in Materials Science during 2000–mid-2011.

the hydrothermal autoclave, because one cannot make any direct observation of the crystallization processes. Though several attempts were made to observe the crystallization processes through special ruby windows for these autoclaves, not much success has been achieved because of the high pressure and high temperature of the system. However, in recent years, remarkable progress was made in this area through the entry of physical chemists and the modeling of the hydrothermal reactions, and the study of the kinetics of the hydrothermal processes have contributed

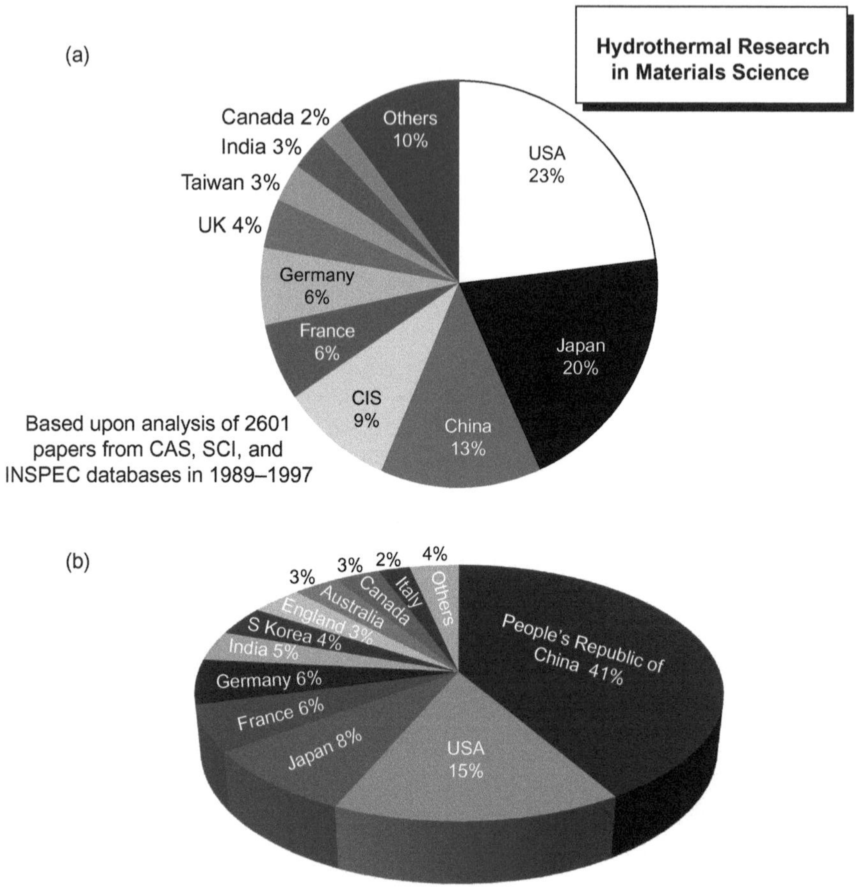

Figure 1.14 (a) Number of papers on Hydrothermal Research in Materials Science country-wise during 1989–1999. (b) Number of papers on Hydrothermal Research in Materials Science country-wise during 2000–mid-2011.

greatly to understanding of the hydrothermal technique. One can now understand the hydrothermal chemistry of solutions more or less precisely, which provides a solid base for hydrothermal synthesis and processing at much lower pressure and lower temperature conditions [12,35]. These developments are slowly removing the concept of the black box for the hydrothermal system, and one can make use of thick-walled silica autoclaves to carry out hydrothermal experiments up to a temperature of 300°C and pressure up to several hundreds of bars, to facilitate direct observation of the hydrothermal processes taking place in a given system. Already some groups in Russia, Japan, and the United States have begun work in this direction and numerous publications have appeared. These developments have given the hydrothermal technique an edge over other techniques for the preparation of pure materials as it is a closed system providing a more controlled diffusion.

Besides, the role of pressure together with temperature helps enhanced molecular arrangements including those of organic molecules as supported by the synthesis of organic amino acids and other life-forming complexes, and also the natural submarine hydrothermal ecosystems. Thus, the hydrothermal technique exhibits a great degree of flexibility or adaptability which is being rightly exploited by a large scientific community with diversified interests.

1.10 Hydrothermal Materials Processing

Hydrothermal materials processing is becoming a popular field of research, particularly after the successful development of ceramic processing technology during 1970s. From 1990 onwards, it has become more popular and an efficient technique for the processing of advanced nanomaterials including carbon polymorphs and hybrid nanomaterials. Different carbon polymorphs have been obtained under subnatural conditions, and these processes involve the use of organic solvents in the system [115]. Also the in situ surface modification under hydrothermal conditions is more attractive. A wide range of nanomaterials including nanocomposites have been prepared using the hydrothermal technique. In this respect, hydrothermal technology has become a boon to nanotechnology for the preparation of high-quality materials with size, shape, and surface properties. Further, there is a growing interest in enhancing hydrothermal reaction kinetics using microwaves, acousto-wave simulations, mechanical mixing, electrochemical reactions, and also addition of biomolecules. These processes have made the hydrothermal technique more unique in the twenty-first century to ceramists and preparative chemists because of the enhanced kinetics. The duration of the experiments is reduced by at least four orders of magnitude, which makes the technique more economic. In fact, this approach is popular as multienergy processing of advanced materials, and it takes materials processing to an altogether new avenue called chemistry at the speed of light and instant hydrothermal systems [27]. The microwave hydrothermal technique is especially handy for the synthesis of lead zirconium titanate (PZT) and other oxide ceramics. Added to this, the crystal size, morphology, and level of agglomeration of the different ceramic oxides can be controlled through careful selection of the ratio of starting materials, pH, time, and temperature. Submicron size powders of TiO_2, ZnO, SnO_2, ZrO_2, Fe_2O_3, $KNbO_3$, $BaTiO_3$, $PbTiO_3$, and their solid solutions have been prepared in this way. This has made the technique a more valuable one in the low-temperature production of fine ceramic powders, clays, and zeolites [18,116,117].

Similarly, ultrasonic energy and sonochemical simulation are used to prepare novel materials and accelerate the chemical reactions. In this way, hydrothermal reactions can be carried out at very rapid rates, and several new phases and high-temperature phases can be obtained in minutes.

Hydrothermal epitaxy began during 1970s in connection with the development of magneto-optic and magnetic bubble domain devices, especially thin and highly crystallized magnetic films of yttrium iron garnets and their solid solutions on gadolinium gallium garnet substrates [118,119]. It did not capture the attention of specialists owing to the higher experimental pressure and temperature conditions involved ($T > 500°C$, $P > 1$ kb). However, from 1990 onwards, the work in this direction has been revived. With the use of electrochemical reactions under hydrothermal conditions, it has become possible to grow highly pure and dislocation-free crystalline thin films of various perovskite-type oxides like $BaTiO_3$, $SrTiO_3$, $CaTiO_3$, and so on, under much lower pressure (<100 bar) and temperature (100–200°C) conditions over a variety of substrates [120–122]. The method offers a high success rate and there are efforts to obtain functionally gradient materials through hydrothermal electrochemistry. With an increase in the demand for the composites, the hydrothermal technique offers an excellent facility for the coating of various compounds with metals and ceramics. For example, the coating of hydroxyapatite (HAp) or bonelike apatite layers on chemically treated Ti metals, organic polymers and other metals has emerged out as a promising field [123–125]. This is discussed in detail in Chapter 10.

The hydrothermal technique has become very useful for the production of whiskers, which are finding extensive applications in modern technology. For example, whiskers of HAp are being produced under mild hydrothermal conditions. Although, the crystal growth of whiskers with controlled morphology is not a new field, it was already on the verge of extinction during the last decade. A few years ago, some reports on the growth of whiskers which are biocompatible under mild hydrothermal conditions appeared, giving a fresh lease on life to this field of research concerning whiskers crystal growth [126,127]. Among the popular whiskers crystal growth, the most popular one is the biocompatible HAp whiskers of nanometer size and this can be prepared at temperatures $<200°C$ and pressure <2 MPa within a few hours. Whiskers prepared by the hydrothermal method have a more controlled and better morphology, uniformity in size and purity in composition compared to the HAp whiskers produced by other methods. The whiskers obtained by other methods usually contained contaminants like carbonates and heterogeneity in shape and size. The hydrothermally prepared whiskers are finding promising applications mostly as reinforcements in composites and thermal insulation. They are highly biocompatible, as they contain nontoxic species, such as Ca, P, and OH^-, and are noncarcinogenic, chemically inert both outside and inside the living bodies. This is discussed in greater detail in Chapter 10.

The hydrothermal hot pressing (HHT) technique is becoming a very simple and most effective processing technique for the preparation of solid bodies from inorganic powders under mild hydrothermal conditions [128–130]. This technique is especially useful for the solidification of thermally decomposable powders such as carbonates, sulfates, and nitrates at high temperatures. The solidification process is thought to be similar to liquid-phase sintering under pressure. In this technique, the starting powder containing water is continuously compressed from the outside of an autoclave under hydrothermal conditions. The powder is then treated at autogeneous pressure, while it is compressed at much higher pressure than the vapor pressure inside the autoclave. The water still remaining in the compacts after the HHT can be easily drained out by drying in air. By this means, porous ceramics can be easily prepared. This technique becomes handy for the solidification of radioactive wastes [131] and sludge ashes [132].

Corrosion is a major problem in industries and power stations. Corrosion afflicts all metals, alloys, and even advanced ceramics. The commonly observed problems of corrosion include exfoliation, stress-corrosion cracking, intergranular corrosion, and dealloying. From the perspective of engineering, the prediction, monitoring, and mitigating of the corrosion problem is highly cumbersome. Once the attack is initiated, it makes localized corrosion a major problem in industries, power plants, and reactors. In this respect, the supercritical solvents oxidation, particularly of water, can be of great help to solve many of the problems associated with corrosion and erosion. In these systems, the toxic and corrosive ions are treated and made to form a solid crystalline coating of the metal surfaces as oxides or as oxychlorides. Several aqueous and nonaqueous solvents have been tried on a variety of metals, alloys, and ceramics, which greatly contribute toward solving the major industrial corrosion problem. The reader can get additional information from the works in Refs [133–140].

Hydrothermal processing has become a most powerful tool, especially from 1990s, to transform various inorganic compounds and also to treat the raw materials for technological applications. The hydrothermally treated raw materials become very highly reactive as a function of solvent used in the preparation of advanced materials. One can find an extensive literature data available on this topic covering a wide range of oxides and hydroxides, carbosils, and so on. Similarly, the hydrothermal transformation, for example, montmorillonite and other variety of clays into the most useful zeolites with larger pore volumes, is a subject of great interest, because of its efficiency and economy [117]. In the last couple of years, the hydrothermal technique has been widely employed for metal intercalated nanocomposites preparation. Several varieties of clay minerals which have two-dimensional layer structures [140] like montmorillonite, beidellite, hectorite, saponite, nontronite, and so on, are attractive materials because of their properties; like cation exchange and swelling–shrinking in the interlayers. They readily produce pillared clays upon hydrothermal treatment. The preparation of pillared clays involves essentially the cation exchange with hydroxy polymeric cations followed by dehydration and dehydroxylation of the polymeric species to prop the layers apart with ceramic oxide pillars in the interlayers [141]. These pillared layers are successfully intercalated with transitional metals. Likewise, the hydrothermal technique is very useful for stabilizing the metastable phases. Also, for example, the water-rich natural phyllosilicates, which have lower thermal stability, can be made more stable chemically and thermally through the replacement of $(OH)^-$ groups by oxygen atoms [142]. By these means, several phyllosiloxides have been prepared by many workers [143,144].

All the above information on hydrothermal technology provides enough support for the growing popularity of this technique for materials synthesis, crystal growth, and materials processing. Table 1.6 lists all the advantages of hydrothermal technique in comparison with other popular techniques frequently employed in the processing of nanomaterials [21].

The scope of this book has been restricted to recent developments under each one of the applications described above with the exception of the natural hydrothermal systems which are not within the purview of the present handbook. Although we have given the historical aspects of the technique in Chapter 2, the emphasis has been focused mainly on the crystal growth and materials processing including nanomaterials through the ages.

This chapter clearly reveals the fact that the hydrothermal technique is drawing the attention of scientists from different branches of natural science. The future of this technique lies in this kind of multidisciplinary approach. More and more new findings and applications in this field are not only contributing to the scientific knowledge of the hydrothermal technique but are also posing new problems. The role of organics both as solvents and as surfactants in hydrothermal systems, for example, has to be studied more seriously, which would definitely twist the geological thinking to a greater extent. It is well known that the organics in hydrothermal systems not only lead to the formation of new phases or new structures, and stability of metastable phases but can also bring down the pressure and the temperature conditions of crystallization. Similarly, they also modify the surface properties and provide control over the size, shape, and dispersibility in case of nanomaterials. In nature, we can expect a very wide range of chemical components including hydrocarbons, which greatly contribute to the crystallization of rocks, minerals, and ore deposits. The role of these organics in the earth's crust has not been understood properly by geoscientists in the context of thermodynamics and kinetics of crystallization. It is expected that such studies will definitely propose much lower temperature and lower pressure conditions of crystallization for various rock bodies and add many more new questions to experimental petrology. Thus, the future of

Table 1.6 Comparison of Various Powder Synthesis Methods [21]

Parameter	Solid-State Reactions	Coprecipitation	Sol–Gel	Polymerizable Complex	Hydrothermal
Cost	Low to moderate	Moderate	High	High	Moderate
State of development	Commercial	Demonstration/ commercial	R&D	R&D	R&D/demonstration
Compositional control	Poor	Good	Excellent	Excellent	Good/Excellent
Morphology control	Poor	Moderate	Moderate	Moderate	Good
Powder reactivity	Poor	Good	Good	Good	Good
Purity (%)	<99.5	>99.5	>99.9	>99.9	>99.5
Calcination step	Yes (multiple)	Yes	Yes	Yes	No
Milling step	Yes (multiple)	Yes	Yes	Yes	No

hydrothermal research (both crystal growth and materials processing) is not toward higher pressure and higher temperature conditions, but definitely toward the lowering of the pressure and the temperature conditions, which will facilitate in situ observation. This will also provide new stimuli for the fabrication of simpler (may be even silica and composites based) and larger autoclaves. As the design and fabrication of the autoclaves becomes simpler and easier, the cost will automatically go down, thereby making them more attractive to specialists. In the last quarter century, as we see from the statistics discussed earlier in this chapter, more new applications for hydrothermal technique are being reported continuously. It is predicted that it will merge with soft solution chemistry and soft solution processing by being more compatible with the biosphere. It will come out as an important tool for the development of advanced materials with limited resources and energy consumption. These materials are less hazardous to living species, also environmentally friendly without firing or sintering at ultrahigh temperatures, or melting or plasma or vaporizing. Because the energy consumed in attaining the melt or vapor or plasma conditions is very high, it proves to be environmentally unfriendly. Hence, the future of hydrothermal technology lies in promoting the lower pressure and lower temperature conditions for crystal growth and materials processing including nanomaterials. We have made an attempt to expose the reader to the above-mentioned aspects of hydrothermal growth and materials processing in the forthcoming chapters.

References

[1] Sir Roderick Murchison (1840s), cited by S. Somiya.

[2] R. Bunsen (1839), cited by R.A. Laudise, The Growth of Single Crystals, Prentice-Hall, Englewood Cliffs, NJ, 1970, pp. 278–281.

[3] K.F.E. Schafthaul, Gelehrte Anzeigen, Bayer. Akad. 20 (557) (1845) 569, 575, 592.

[4] R.W. Goranson, Solubility of water in granite magmas, Am. J. Sci. 22 (1931) 481–502.

[5] F. Habashi, Recent advances in pressure leaching technology, in: N. Yamasaki, K. Yanagisawa (Eds), Proceedings of the First International Conference on Solvothermal Reactions, 13–16, Takamatsu, Japan, December 5–7, 1994.

[6] F. Habashi, A Textbook of Hydrometallurgy, Libraire Universitaire du Quebec, Quebec, Canada, 1993.

[7] R. Nacken, Artificial quartz crystals, etc., US Office of Technical Services Report, PB-18–748 and 28–897, 1946.

[8] R.M. Barrer, Syntheses and reactions of mordenite, J. Chem. Soc. 2158 (1948).

[9] R.M. Barrer, Some features of ion exchange in crystals, in: N. Yamasaki, K. Yanagisawa (Eds), Molecular Sieves, Society of Chemical Industry, London, 1962, pp. 1258–1266.

[10] R. Nacken, Hydrothermal synthese als Grundlage fur Zuchtung von Quarz-Kristallen, Chem. Z. 74 (1950) 745–749.

[11] K. Byrappa, Hydrothermal growth of crystals, in: D.T.J. Hurle (Ed.), Handbook of Crystal Growth, Elsevier Science B.V., England, 1994, pp. 465–562.

[12] M.M. Lencka, A. Anderko, R.E. Riman, Hydrothermal precipitation of lead zirconate titanate solid solutions: thermodynamic modeling and experimental synthesis, J. Am. Ceram. Soc. 78 (1995) 2609–2618.

[13] J.O. Eckert Jr., C.C. Hung-Houston, B.L. Gersten, M.M. Lencka, R.E. Riman, Kinetics and mechanisms of hydrothermal synthesis of barium titanate, J. Am. Ceram. Soc. 79 (1996) 2929–2939.

[14] G.W. Morey, P. Niggli, The hydrothermal formation of silicates, a review, J. Am. Chem. Soc. 35 (1913) 1086–1130.

[15] R.A. Laudise, The Growth of Single Crystals, Prentice-Hall, Englewood Cliffs, NJ, 1970, pp. 278–281

[16] A. Rabenau, The role of hydrothermal synthesis in preparative chemistry, Angew. Chem. Int. Engl. Ed. 24 (1985) 1026–1040.

[17] A.N. Lobachev (Ed.), Crystallization Processes under Hydrothermal Conditions, Consultants Bureau, New York, NY, 1973.

[18] R. Roy, Acceleration the kinetics of low-temperature inorganic syntheses, J. Solid State Chem. 111 (1994) 11–17.

[19] K. Byrappa (Ed.), Hydrothermal Growth of Crystals, Pergamon Press, Oxford, UK, 1992.

[20] M. Yoshimura, H. Suda, Hydrothermal processing of hydroxyapatite: past, present, and future, in: P.W. Brown, B. Constanz (Eds), Hydroxyapatite and Related Materials, CRC Press, Inc., Boca Raton, FL, 1994, pp. 45–72.

[21] M. Yoshimura, K. Byrappa, Hydrothermal processing of materials: past, present and future, J. Mat. Sci. 43 (2008) 2085–2103.

[22] R. Roy, Fifty-year perspective on hydrothermal research, in: Proceedings of the Workshop on Solvothermal and Hydrothermal Reactions, Sun Messe Kagawa, January 22–24, 1996, pp. 1.1–1.20.

[23] G. Tamma, The State of Aggregation, D. VanNostrand Company, Princeton, NJ, 1925.

[24] Proceedings of the First International Conference, Solvothermal Reactions, Takamatsu, Japan, December 5–7, 1994, pp. 1–223.

[25] Proceedings of the Workshop on Solvothermal and Hydrothermal Reactions, Sun Messe Kagawa, Japan, January 22–24, 1996, pp. 1–133.

[26] Proceedings of the Second International Conference on Solvothermal Reactions, Takamatsu, Japan, December 18–20, 1996, pp. 1–242.

[27] K. Byrappa, Novel hydrothermal solution routes of advanced high melting nanomaterials processing, J. Ceram. Soc. Japan 117 (2009) 236–244.

[28] M. Yoshimura, Why, and how about advanced inorganic materials, Eur. J. Solid State Inorg. Chem. 32 (1995) I–IV.

[29] B.N. Litvin, D.A. Tules, Apparatus for hydrothermal synthesis and growth of single crystals, Hydrothermal Synthesis of Crystals, Nauka, Moscow, 1968 (in Russian).

[30] T. Mousavand, Synthesis of Organic–Inorganic Hybrid Nanoparticles by In situ Surface Modification under Supercritical Hydrothermal Conditions, Ph.D. Thesis, Tohoku University, Sendia, Japan, 2007.

[31] S. Ohara, T. Mousavand, T. Sasaki, M. Umetsu, T. Naka, T. Adschiri, Continuous production of fine zinc oxide nanorods by hydrothermal synthesis in supercritical water, J. Mater. Sci. 43 (2008) 2393–2396.

[32] K. Namratha, M.B. Nayan, K. Byrappa, Hydrothermal synthesis and photocatalytic properties of modified and unmodified zinc oxide nanoparticles, Mater. Res. Innov. 15 (2011) 36–42.

[33] T.M. Seward, The stability of complexes of silver in hydrothermal solutions up to 350°C, Geochim. Cosmochim. Acta 40 (1976) 1329–1341.

[34] T.M. Seward, Metal complex formation in aqueous solutions at elevated temperatures and pressures, in: D.T. Rickard, F.E. Wichman (Eds), Chemistry and Geochemistry of Solutions at High Temperatures and Pressures, Pergamon Press, New York, NY, 1979, pp. 113–132. Proceedings of the Noble Symposium, 13/14.
[35] E.U. Franck, Survey of selected non-thermodynamic properties and chemical phenomena of fluids and fluids mixtures, in: D.T. Rickard, F.E. Wichman (Eds), Chemistry and Geochemistry of Solutions at High Temperatures and Pressures, Pergamon Press, New York, NY, 1979, pp. 65–88. Proceedings of the Noble Symposium, 13/14.
[36] D.C. Hurd, Factors affecting the solution rate of biogenic opal in seawater, Earth Planet. Sci. Lett. 15 (1972) 411–417.
[37] P.M. Dove, D.A. Crerar, Kinetics of quartz dissolution in electrolyte solutions using a hydrothermal mixed flow reactor, Geochim. Cosmochim. Acta 54 (1990) 955–959.
[38] R.W. Henley, Solubility of gold in hydrothermal chloride solutions, Chem. Geol. 11 (1973) 73–87.
[39] H. Honma, N. Shikazono, M. Nakata, Hydrothermal synthesis of gold, electrum and argentite, Can. Min. 29 (1991) 217–221.
[40] T.M. Seward, Hydrothermal solution speciation (solvation to ion pairing and complexing), in: D.A. Palmer, D.J. Wesolowski (Eds), Proceedings of the Fifth International Symposium on Hydrothermal Reactions, Gatlinburg, USA, July 20–24, 1997, pp. 7–9.
[41] L.G. Benning, T.M. Seward, Hydrosulphide complexing of Au^{1+} in hydrothermal solutions from 150–400°C and 500–1500 bar, Geochim Cosmochim Acta 60 (1996) 1849–1871.
[42] G.R. Kolonin, G.A. Palyanova, G.P. Shironosova, K.G. Morgunov, The effect of carbon dioxide on internal equilibria in the fluid during the formation of hydrothermal gold deposits, Geochem. Int. 35 (1997) 40–50.
[43] G.A. Palyanova, G.P. Shironosova, L.V. Laptev, G.R. Kolonin, Experimental checking of CO_2 influence on gold and silver solubility in high-temperature complex fluids, in: D.A. Palmer, D.I. Wesolowski (Eds), Proceedings of the Fifth International Symposium on Hydrothermal Reactions, Gatlinburg, USA, July 20–24, 1997, pp. 255–256.
[44] R.H. Jahns, C.W. Burnham, Experimental studies of pegmatites genesis: I.A. Model for the derivation and crystallization of granitic pegmatites, Econ. Geol. 64 (1969) 843–864.
[45] R.H. Jahns, Internal evolution of pegmatite bodies, in: P. Cerny (Ed.), Granitic Pegmatites in Science and Industry, Mineral Association of Canadian Short Course Handbook, 8, 1982, pp. 293–327.
[46] O.F. Turtle, N.L. Bowen, Origin of granite in the light of experimental studies in the system $NaAlSi_3O_4 - KalSi_3O_4 - SiO_4 - H_2O$, Geol. Soc. Am. Mem. 74 (1958).
[47] W.C. Luth, R.H. Jahns, O.F. Tuttle, The granite system at pressures of 4 to 10 kbars, J. Geophys. 69 (1964) 759–773.
[48] C.W. Burnham, Water and magmas: a mixing model, Geochim. Cosmochim. Acta 39 (1975) 1007–1084.
[49] C.W. Burnham, Magmas and hydrothermal fluids, in: Barnes (Ed.), Geochemistry of Hydrothermal Ore Deposits, second ed., John Wiley & Sons, New York, NY, 1979, pp. 71–136.
[50] H.R. Shaw, Obsidian—H_2O viscosities at 1000 and 2000 bars in the temperature range 700 to 900°C, J. Geophys. 68 (1963) 6337–6343.
[51] H.R. Shaw, Comments on viscosity, crystal settling and convection in granitic magmas, Am. J. Sci. 263 (1965) 120–152.

[52] P.M. Fenn, The nucleation and growth of alkali feldspars from hydrous melts, Can. Mineral. 15 (1977) 135–161.

[53] P.J. Wyllie, O.F. Tuttle, Experimental investigation of silicate system containing two volatile components. Part II. The effects of NH_3 and HF, in addition of H_2O, on the melting temperatures of albite and granite, Am. J. Sci. 259 (1961) 128–143.

[54] D.A.C. Manning, The effect of fluorine on liquidus phase relationships in the system Qz–Ab–Or with excess water at 1 Kb, Contrib. Mineral. Petrol. 76 (1981) 206–215.

[55] D.B. Dingwell, The structure and properties of fluorine-rich silicate melts: implications for granite petrogenesis (abstract), in: R.P. Toylor, D.F. Strong (Eds), Granite-Related Mineral Deposits, Canadian Institute of Mining, Extended Abstracts of Conference on Granite-Related Mineral Deposits, Halifax, 1985, pp. 72–81.

[56] L.N. Kogarko, L.D. Krigman, N.S. Sharudilo, Experimental investigations of the effect of alkalinity of silicate melts on the separation of fluorine into the gas phase, Geochem. Int. 8 (1968) 782–790.

[57] J.D. Webster, J.R. Holloway, R.L. Hervig, Phase equilibria and volatile partitioning in a Be-, U- and F-enriched vitrophyre from spor mountain, Utah. Geochim. Cosmochim, Acta 51 (1987) 389–402.

[58] D.A.C. Manning, J.S. Martin, M. Pichavant, C.M.B. Henderson, The effect of F, B and Li on melt structures in the granite system: different mechanisms? in: C.M.B. Henderson (Ed.), Progress in Experimental Petrology, National Environment Research Council Publication Series D, vol. 25, 1984, pp. 36–41.

[59] B.H.W.S. De Jong, G.E. Brown Jr., Polymerization of silicate and aluminate tetrahedra in glasses, melts and aqueous solutions—II. The network modifying effects of Mg^{+2}, K^+, Na^+, Li^+, OH^-, F^-, Cl^-, H_2O, CO_2 and H_3O^+ on silicate polymers, Geochim. Cosmochim. Acta 44 (1980) 1627–1642.

[60] J.C. Bailey, Fluorine in granitic rocks and melts: a review, Chem. Geol. 19 (1977) 1–42.

[61] P. Cerny, R.E. Meintzer, A.J. Anderson, Extreme fractionation in rare-element granite pegmatites: selected examples of data and mechanisms, Can. Mineral. 23 (1985) 381–421.

[62] D. London, Magmatic–hydrothermal transition in the Tanco rare-element pegmatite: evidence from fluid inclusions and phase equilibrium experiments, Am. Mineral. (71) (1986) 376–395.

[63] P.J. Bray, NMR studies of borates, in: L.D. Pye, V.D. Frenchette, N.J. Kriedl (Eds), Borate Glasses: Structure, Properties, Applications, 1, Plenum Press, New York, NY, 1978.

[64] W.J. Dell, P.J. Bray, S.Z. Xiao, IIB NMR studies and structural modeling of Na_2O–B_2O_3–SiO_2 glasses of high soda content, J. Non-Cryst. Solids 58 (1983) 1–16.

[65] R. Oestrike, K. Geisinger, A. Navrotsky, G.L. Turner, R.J. Kirkpatrick, Structure and thermochemistry of glasses along the join $NaAlSi_3O_8$–$NaBSi_3O_8$: the effect of boron, Geol. Soc. Am. Abstr. Prog. 18 (1986) 709.

[66] A. Navrotsky, K.L. Geisinger, P. McMillan, G.V. Gibbs, The tetrahedral framework in glasses and melts—inferences from molecular, thermodynamics and physical properties, Phys. Chem. Min. 11 (1985) 284–298.

[67] J.E. Mendel, The Storage and Disposal of Radioactive Waste as Glass in Canisters. PNL-2764, Pacific Northwest Laboratories, Richland, Washington, DC, 1978.

[68] D. London, Internal differentiation of rare-element pegmatites: effects of boron, phosphorus and fluorine, Geochim. Cosmochim. Acta 51 (1987) 403–420.

[69] P.C. Hess, The role of high field strength cations in silicate melts, in: I. Kushiro, I. Perchuk (Eds), Advances in Physical Geochemistry, Springer-Verlag, New York, NY, 1986.
[70] F.J. Ryerson, P.C. Hess, The role of P_2O_5 in silicate melts, Geochim. Cosmochim. Acta 44 (1980) 611–624.
[71] B.O. Mysen, F. Ryerson, D. Virgo, The structural role of phosphorus in silicate melts, Am. Mineral. 66 (1981) 106–117.
[72] M. Huffman, A. Navrotsky, F.S. Pintchovski, Thermochemistry and structure of low pressure chemically vapor deposited and bulk $SiO_2-P_2O_5$ and SiO_2-GeO_2 glasses, J. Electrochem. Soc. 133 (1986) 431–439.
[73] C.W Burnham, H. Nekvasil, Equilibrium properties of granite pegmatite magmas, Am. Mineral. 71 (1986) 239–263.
[74] P.J. Wyllie, O.F. Tuttle, Experimental investigation of silicate systems containing two volatile components, Part III: the effects of SO_3, P_2O_5, HCl and Li_2O, in addition to H_2O, on the melting temperatures of albite and granite, Am. J. Sci. 262 (1964) 930–939.
[75] P. Cerny, Petrogenesis of granitic pegmatites, in: P. Cerny (Ed.), Granitic Pegmatites in Science and Industry, Mineralogical Association of Canada, Quebec, Canada, 1982, pp. 405–461.
[76] D. London, Experimental phase equilibria in the system $LiAlSiO_4-SiO_2-H_2O$: a petrogenetic grid for lithium-rich pegmatites, Am. Mineral. 69 (1984) 995–1004.
[77] G.B. Morgan, Alteration of Amphibolitic Wallrocks Around the Tanco Rare Element Pegmatite, Manitoba. M.S. Thesis, University of Oklahoma, VI, 1986.
[78] P.M. Fenn, On the origin of graphic granite, Am. Mineral. (71) (1986) 325–330.
[79] A.S. Quist, W.L. Marshall, Electrical conductances of aqueous sodium chloride solutions from 0 to 800°C and to pressures to 4000 bars, J. Phy. Chem. 72 (1968) 684–703.
[80] W.L. Marshall, J.D. Frantz, Electrical conductance measurements of dilute, aqueous electrolytes at temperatures to 800°C and to pressures to 400 bars: techniques and interpretations, in: G.C. Ulmer, H.L. Barnes (Eds), Hydrothermal Experimental Techniques, John Wiley & Sons, New York, NY, 1987, pp. 261–292.
[81] H. Takahashi, K. Nakatuka, Advanced geothermal energy utilization for geochemical energy utilization for geochemical reactor, in: T. Moriyoshi (Ed.), Proceedings of the First International Conference on Solvothermal Reactions, Takamatsu, Japan, December 5–7, 1994, pp. 71–75.
[82] K. Nakatsuka, Geothermal reactor—concept and perspectives, in: T. Moriyoshi (Ed.), Proceedings of the Second International Conference on Solvothermal Reactions, Takamatsu, Japan, December 18–20, 1996, p. 156.
[83] PARC, 3333 Coyote Hill Road, Palo Alto, CA 94304, USA. <www.parc.com>.
[84] J.B. Corliss, Metallogenesis at oceanic spreading centres, J. Geol. Soc. London. 136 (1979) 621–626.
[85] D.J. Fornari, R.W. Embley, Tectonic and volcanic controls on hydrothermal processes at the mid-ocean ridge: an overview based on near-bottom and submersible studies, in: S.E. Humphris, R.A. Zierenberg, L.S. Mullineause, R.E. Thomson (Eds), Seafloor Hydrothermal Systems Physical, Chemical, Biological and Geological Interactions, American Geophysical Union, Washington, DC, 1995, pp. 1–46.
[86] D. Henley, cited by Gillings, A., Evolution of hydrothermal ecosystems on Earth (and Mars?), Bioessays 18 (1996) 515–517.
[86a] D. Fornari, T. Shank, Surveying a subsea lava flow using the Autonomous Benthic Explorer, Woods Hole Oceanographic Institute, Massachusetts, USA, 2003.

[87] E.L. Shock, Chemical environments of submarine hydrothermal systems, Orig. Life Evol. Biosph. 22 (1992) 67–107.

[88] J.A. Baross, S.E. Hoffman, Submarine hydrothermal vents and associated gradient environments as sites for the origin and evolution of life, Orig. Life 15 (1985) 327–345.

[89] B.R.T. Simoneit, Aqueous organic geochemistry at high temperature/high pressure, Orig. Life Evol. Biosph. 22 (1992) 43–65.

[90] W. Martin, J. Baross, D. Kelley, M.J. Russell, Hydrothermal vents and the origin of life, Nat. Rev. Microbiol. 6 (2008) 805–814.

[91] D.S. Kelley, J.A. Baross, J.R. Delaney, Volcanoes, fluids, and life at mid-ocean ridge spreading centers, Annu. Rev. Earth Planet Sci. 30 (2002) 385–491.

[92] G. Proskurowski, M.D. Lilley, D.S. Kelley, E.J. Olson, Low temperature volatile production at the Lost City hydrothermal field, evidence from a hydrogen stable isotope geothermometer, Chem. Geol. 229 (2006) 331–343.

[93] G.L. Früh-Green, D.S. Kelley, S.M. Bernasconi, J.A. Karson, K.A. Ludwig, D.A. Butterfield, et al., 30,000 years of hydrothermal activity at the Lost City vent field, Science 301 (2003) 495–498.

[94] K.A. Ludwig, D.S. Kelley, C. Shen, H. Cheng, R.L. Edwards, U/Th geochronology of carbonate chimneys at the Lost City hydrothermal field, Eos. Trans. AGU, **86**, 2005.

[95] U.R. Costa, W.S. Fyfe, R. Kerrich, M.W. Nesbitt, Archean hydrothermal talc evidence for high ocean temperatures, Chemical Geol. 30 (1980) 341–349.

[96] S.E. Hoffman, J.A. Baross, Workshop on the Early Earth: The Interval from Accretion to the Older Archean, vol. 34, Lunar and Planetary Institute, Houston, TX, 1984.

[97] D.H. Abbott, S.E. Hoffman, Archean plato tectonics revisited 1: heat flow, spreading rate and age of subducting oceanic lithosphere and their effects on the origin and evolution of continents, Tectonics 3 (1984) 429–448.

[98] H.W. Jannasch, Microbial interactions with hydrothermal fluids, in: S.E. Humphris, R.A. Zierenberg, L.S. Mulineaux, R.E. Thomson (Eds), Seafloor Hydrothermal Systems, American Geophysical Union, Washington, DC, 1995, pp. 273–296.

[99] P.A. Rona, K. Bostrom, L. Laubier, K.L. Smith Jr. (Eds), Hydrothermal Processes at Seafloor Centers, NATO Conference Series, Plenum Press, New York, NY, 1983, p. 796.

[100] B.E. Humphris, R.A. Zierenberg, L.S. Mullineaux, R.E. Thomson (Eds), Seafloor Hydrothermal Systems, Physical, Chemical, Biological and Geological Interactions, American Geophysical Union, Washington, DC, 1995.

[101] M.D. Schulte, E.L. Shock, Thermodynamics of strecker synthesis in hydrothermal systems, Orig. Life Evol. Biosph. 25 (1995) 161–173.

[102] E.L. Shock, Organic acids in hydrothermal solutions: standard molar thermodynamic properties of carboxylic acids and estimates of dissociation constants at high temperatures and pressures, Am. J. Sci. 295 (1995) 496–580.

[103] S.W. Fox, C.R. Windsor, Synthesis of amino acids by the heating of formaldehyde and ammonia, Science 170 (1970) 984–985.

[104] W.L. Marshall, Possible geochemical production of biological precursors: amino acids and other amines (also hydrocarbons) from aqueous ammonium carbonate solutions and metal carbides at 200°C–300°C, EOS 68 (1987) 458.

[105] E.L. Shock, Geochemical constraints on the origin of organic compounds in hydrothermal systems, Orig. Life Evol. Biosph. 20 (1990) 331–367.

[106] E.L. Shock, Chemical environments of submarine hydrothermal systems, Orig. Life Evol. Biosph. 22 (1992) 66–107.
[107] J.C. Hennet, N.G. Holm, M.H. Engel, Abiotic synthesis of amino acids under hydrothermal conditions and the origin of life: a perpetual phenomenon? Naturwissenschaften 79 (1992).
[108] S.L. Miller, A production of amino acids under possible primitive earth conditions, Science 117 (1953) 528–529.
[109] N.G. Holm, A.G. Cairna-Smith, R.M. Daniel, J.P. Ferris, R.J.C. Hennet, E.L. Shock, et al., Molecular Self-Assembly and the Origin of Life, Orig. Life Evol. Biosph. 22 (1992) 181–190.
[110] O.F. Tuttle, A new hydrothermal quenching apparatus, Am. J. Sci. 246 (1948) 628–635.
[111] S. Somiya (Ed.), Hydrothermal Preparation of Fine Powders, Advanced Ceramics III, Elsevier Applied Science Publishers, UK, 1990.
[112] P.E.L.E. Tuitemwong, D.V.C. Fung, K. Tuitemwong, Effect of processing temperature on microbiological and chemical quality of soy milk produced by rapid hydration hydrothermal cooking, J. Food Process. Preserv. 17 (1993) 153–175.
[113] IAPWS Release on the Values of Temperature, Pressure and Density of Ordinary and Heavy Water Substances at Their Respective Critical Points, 1992; available from the IAPWS Secretary, Dr. Barry Dooley, Electric Power Research Institute, 3412 Hillview Avenue, Palo Alto, CA 94303, USA.
[114] K. Byrappa, Crystallization process in some phosphate systems under hydrothermal conditions, in: Proceedings of the Second International Conference on Solvothermal Reactions, Takamatsu, Japan, December 18–20, 1996, p. 84.
[115] B. Basavalingu, K. Byrappa, P. Madhusudan, Hydrothermal synthesis of nanosized crystals of diamond under sub-natural conditions, J. Geol. Soc. India 69 (2007) 665–670.
[116] S. Komarneni, R. Roy, Q.H. Li, Microwave-hydrothermal synthesis of ceramic powders, Mater. Res. Bull. 27 (1992) 1393–1405.
[117] S. Komarneni, M.Z. Hussein, C. Liu, E. Breval, P.B. Malla, Microwave-hydrothermal processing of metal clusters supported in and/or on montmorillonite, Eur. J. Solid State Inorg. Chem. 32 (1995) 837–849.
[118] E.D. Kolb, R.A. Laudise, Phase equilibria of $Y_3Al_5O_{12}$, hydrothermal growth of $Gd_3Ga_5O_{12}$ and hydrothermal epitaxy of magnetic garnets, J. Crystal Growth 29 (1975) 29–39.
[119] M.J.G. Van Hout, J.C. Verplanke, J.M. Robertson, Hydrothermal synthesis of single crystal thin films of magnetic garnets and their analysis, Mat. Res. Bull. 10 (1975) 125–132.
[120] M. Yoshimura, S.-E. Yoo, M. Hayashi, N. Ishizawa, Preparation of $BaTiO_3$ thin film by hydrothermal electrochemical method, Jpn. J. Appl. Phys. 28 (1989) L2007–L2009.
[121] R. Basca, P. Ravindranathan, J.P. Dougherty, Electrochemical, hydrothermal and electro-chemical-hydrothermal synthesis of barium titanate thin films on titanium substrates, J. Mater. Res. 7 (1992) 423–428.
[122] K. Kajiyoshi, Y. Hamaji, K. Tomono, T. Kasanami, M. Yoshimura, Microstructure of strontium titanate thin film grown by the hydrothermal–electrochemical method, J. Am. Ceram. Soc. 79 (1996) 613–619.
[123] W. Suchanek, M. Yoshimura, Processing and properties of hydroxyapatite-based biomaterials for use as hard tissue replacement implants, J. Mater. Res. 13 (1998) 1–24.

[124] W.S. Cho, M. Yoshimura, Hydrothermal synthesis of $PbTiO_3$ films, J. Mater. Res. 12 (1997) 833–839.
[125] Y. Fujishiro, A. Fujimoto, T. Sato, A. Okuwaki, Coating of hydroxyapatite on titanium plates using thermal dissociation of calcium–EDTA chelate complex in phosphate solutions under hydrothermal conditions, J. Colloid Interface Sci. 173 (1995) 119–127.
[126] W. Suchanek, H. Suda, M. Yashima, M. Kakihana, M. Yoshimura, Biocompatible whiskers with controlled morphology and stoichiometry, J. Mater. Res. 10 (1995) 512–529.
[127] W. Suchanek, M. Yashima, M. Kakihana, M. Yoshimura, Processing and mechanical properties of hydroxyapatite reinforced with hydroxyapatite whiskers, in: Proceedings of the Second International Symposium on Apatite, Japanese Association of Apatite Science, 1997, pp. 15–20.
[128] K. Yanagisawa, M. Sasaki, M. Nishioka, K. Ioku, N. Yamasaki, Preparation of sintered compacts of anatase by hydrothermal hot-pressing, J. Mater. Sci. Lett. 13 (1994) 765–766.
[129] K. Sato, T. Hashida, H. Takahashi, N. Yamasaki, Development of high strength calcium aluminate-phosphate cement by hydrothermal hot-pressing, J. Mater. Sci. Lett. 16 (1997) 1464–1468.
[130] K. Yanagisawa, K. Ioku, N. Yamasaki, Formation of anatase porous ceramics by hydrothermal hot-pressing of amorphous titania spheres, J. Am. Ceram. Soc. 80 (1997) 1303–1306.
[131] N. Yamasaki, K. Yanagisawa, M. Nishioka, A Kanahara, Hydrothermal hot-pressing method: apparatus and application, J. Mater. Sci. Lett. 5 (1986) 355–356.
[132] M. Nishioka, K. Yanagisawa, N. Yamasaki, Solidification of sludge ash by hydrothermal hot-pressing, Res. J. Water Pollut. Contr. Fed. 62 (1991) 926–927.
[133] E. Heitz, Corrosion of metals in organic solvents, in: M. Fontana, R. Staehle (Eds), Advances in Corrosion Science and Technology, 4, Plenum Press, New York, NY, 1974.
[134] M. Sakakibara, N. Saito, H. Nishihara, K. Aramaki, Corrosion of iron in an hydrous methanol, Corrosion Sci. 34 (1993) 391–409.
[135] G.Y. Gogotsi, M. Yoshimura, Formation of carbon films on carbides under hydrothermal conditions, Nature 367 (1994) 628–630.
[136] R.G. Kelly, Small scale corrosion, Electrochem. Soc. Interface Summer (1997) 18–23.
[137] T. Sato, K. Tamura, A. Okuwaki, Corrosion behavior of silicon carbide ceramics in caustic alkaline solutions of high temperature, Br. Ceram. Trans. J. 91 (1992) 181–182.
[138] F. Habashi (Ed.), Handbook of Extractive Metallurgy, vols 1–4, Wiley-VCH, Germany, 1997.
[139] L. Lay, Corrosion Resistance of Technical Ceramics, Her Majesty's Stationery Office, London, UK, 1984.
[140] S. Yamanaka, Design and synthesis of functional layered nanocomposites, Am. Ceram. Soc. Bull. 70 (1991) 1056–1058.
[141] P.B. Malla, P. Revindranthan, S. Komarneni., R. Roy, Intercalation of copper metal clusters in montmorillonite, Nature 315 (1991) 555–557.
[142] G. Demazeau, Solvothermal reactions: a new opening in materials science either for basic research or industrial applications, in: Proceedings of the First International

Conference on Solvothermal Reactions, Takamatsu, Japan, December 5–7, 1994, pp. 5–9.

[143] G. Demazeau, The solvothermal synthesis: a way to a new chemistry, in: M. Yoshimura (Ed.), Proceedings of the International Workshop on Soft and Solution Processing for Advanced Inorganic Materials, Tokyo, Japan, February 26–27, 1996, pp. 4–12.

[144] G. Demazeau, Hydrothermal and solvothermal processes: the main physico-chemical factors involved and new trends, Res. Chem. Intermed. 37 (2011) 107–123.

2 History of Hydrothermal Technology

2.1 Introduction

In the ancient times, Romans used to celebrate the year-end and year-beginning with great festivities in honor of the God Janus, in whose honor the month January is named. The God Janus is a deity with two heads: one looking to the past and the other looking to the future (Figure 2.1) [1]. Here, we have made a sincere attempt to explore and review the early history of the hydrothermal technology that has an extremely promising future in view of its high efficiency in the processing of nanomaterials.

As mentioned earlier, Sir Roderick Murchison, a British geologist, was the first to use the term *hydrothermal* in the mid-nineteenth century [2]. The first publication on hydrothermal research appeared in 1845 (Figure 2.2). This reports successful synthesis of tiny quartz crystals upon the transformation of freshly precipitated silicic acid in Papin's digestor by Schafthaul [3]. Following this, mineralogists, especially from France, Germany, and Italy, started synthesizing various other minerals. Although the size of the crystals or minerals they obtained did not exceed thousandths or hundredths of a millimeter, their main objective was to create or simulate, in the laboratory, the natural conditions existing in the earth's crust. If, in nature, minerals and mineral assemblages were formed at elevated temperature and pressure conditions in the presence of volatiles—mainly water, it will be quite impossible to simulate the natural conditions or processes responsible for the formation of rocks and minerals except under the *hydrothermal conditions*. Thus, the early interest in hydrothermal research was merely in the synthesis of a particular mineral or in obtaining compounds similar to natural minerals rather than in carrying out any systematic investigation pertaining to phase relations or geochemistry of the earth's interior. Even the conditions under which they worked were, in general, simple and sometimes lacking precision. The minerals obtained by the earlier workers are chiefly those which are stable or at any rate, phanerostable, over a wide range of conditions, for example, quartz and feldspar. As the resultant products obtained were very tiny, accurate chemical analysis or identification of the phase by optical microscopic methods was quite doubtful and many a times led to speculations. Besides, the experiments were carried out in glass tubes in which it was difficult to attain higher pressure–temperature conditions; also water was used in most cases as a solvent. Therefore, the crystals obtained were very small, and the purity of the crystals could not be studied satisfactorily.

The majority of the early hydrothermal experiments carried out during the 1840s to early 1900s mainly dealt with the nanocrystalline products, which were

Handbook of Hydrothermal Technology.

Figure 2.1 The God Janus.

discarded as failures due to the lack of sophisticated electron microscopic techniques available during that time to observe such small-size products. Thus, the whole focus was on the processing of bulk crystals or bulk materials. Many times when bulk crystals or single crystals were not obtained as products of several millimeters in size, the experiments were considered failures and the materials were washed away. Prior to X-ray techniques, chemical techniques were mainly employed in identifying the products. It was only after the application of X-rays for crystal studies that the researchers slowly began to study the powder diffraction patterns of the resultant products, and by the 1920s a systematic understanding of the products began. Before that the experiments were considered as failures. The experiments were concluded by stating that the solubility was not suitable for growing crystals [4]. Until the works of Giorgio Spezia in 1900 [5], hydrothermal technology did not gain much importance in the growth of bulk crystals, as the products in the majority of the cases were very fine grained without any X-ray data. Most of the experiments during the nineteenth century under hydrothermal conditions, with some exceptions, were related to silicate synthesis. Therefore, the contribution from the glass tubes to the resultant product could not be precisely understood. After the introduction of steel autoclaves and suitable metal linings, attempts were initiated to reach higher pressure–temperature conditions to obtain other compounds and purer phases. However, no attention was paid to the chemistry of the solvent, the frequent appearance of metastable phases, solubility relations, kinetics, phase equilibria, and related phenomena. Thus, the frequent appearance of metastable phases also complicated the earlier studies.

Most of the work on hydrothermal research during the nineteenth century was confined only to Europe, especially to the Mineralogy, Petrology, and Geochemistry laboratories in Germany, France, Switzerland, and Italy. Bunsen [6] carried out hydrothermal experiments for the first time using thick-walled glass tubes to contain high-temperature and high-pressure liquids and prepared strontium and barium carbonates. Wohler [7] recrystallized apophyllite by heating apophyllite in water solutions at 180 – 190°C under 10 – 12 atm pressure. In fact, all the early hydrothermal experiments, until 1881, were carried out in simple glass tubes with sealed ends, of course, with an exception of one or two attempts with steel tubes. Even the term

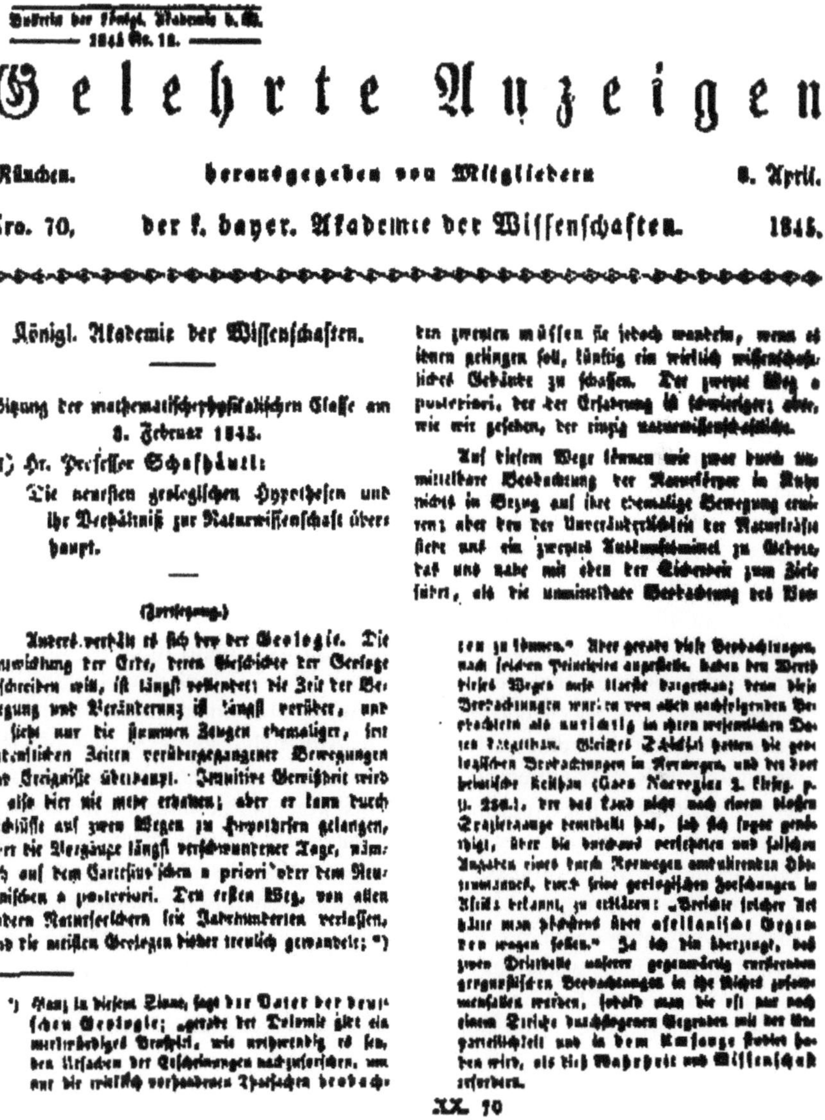

Bulletin der königl. Akademie d. W.
1845 Nr. 18.

Gelehrte Anzeigen

München. herausgegeben von Mitgliedern 8. April.
Nro. 70. der k. bayer. Akademie der Wissenschaften. 1845.

Königl. Akademie der Wissenschaften.

Sitzung der mathematisch-physikalischen Classe am 8. Februar 1845.

1) Hr. Professor Schafhäutl:
Die neuesten geologischen Hypothesen und ihr Verhältniß zur Naturwissenschaft überhaupt.

(Fortsetzung.)

XX. 70

Figure 2.2 The first publication on the hydrothermal research.
Source: Courtesy of Prof. S. Somiya.

hydrothermal bomb was not used by earlier workers and they simply referred to them as autoclaves or tubes. The greatest contribution in the nineteenth century was by H. De Senarmont [8], the founder of hydrothermal synthesis in geoscience. He used glass tubes containing gel, SiO_2 and H_2O, HCl, or CO_2, enclosed in steel tubes, and heated at $200 - 300°C$. The resultant product was a six-sided quartz prism with pyramidal termination. When the solution contained $NaHCO_3$, CO_2, and some realgar, very little quartz was obtained. It was probably Daurree [9], a French mineralogist, who first used a steel tube to synthesize quartz and wollastonite at about 400°C with water as a solvent. The most important aspect of his experiments was the pressure balance he introduced for the first time. The starting substances were taken in a sealed glass tube with water as a mineralizer, and this glass tube was placed inside a steel tube. The space between the glass tube and steel tube was filled with water. He carried out a series of experiments with this design and a varied concentration of the starting materials to obtain several other minerals. Similarly, he conducted experiments on the recrystallization of feldspars, micas, and pyroxenes.

Daurree's other significant contribution was his attempt to use natural hot spring water from Plombieres as a mineralizer and this led to the deposition of a small quantity of quartz over 2 days. In addition, he treated kaolin with Plombieres water under hydrothermal conditions and obtained feldspar and a small amount of quartz. Here, it is interesting to note that, after the discovery of hydrothermal activity in the deep sea on the Galapagos during the 1970s, active research is going on using the natural submarine hydrothermal vent fluid to study the origin of life (organic synthesis). This led to the establishment of the relation between the submarine hydrothermal activity and rich mineralization of a very wide spectrum from native elements to sulfides, oxides, and so on [10–12]. The literature survey on the early hydrothermal research shows that several attempts were made to transform zeolite minerals and kaoline in the nineteenth century [7 – 13], but it was only during 1940 that Barrer [14] succeeded in synthesizing zeolite for the first time, that is, *analcime*, using the hydrothermal technique. Although Deville St. Claire (1857) attempted to transform bauxite into corundum under hydrothermal conditions using NaOH as mineralizer, the experimental results were not definite. Perhaps, he was the first one to use a mineralizer other than water [15]. In fact, many of the earlier workers, including von Chroustshoff [16], found that the glass tubes they were using were frequently attacked under hydrothermal conditions. However, they do not mention anything about precautions taken in this regard, as they were more concerned in obtaining a product corresponding to some natural minerals. During 1873, von Chroustshoff first proposed the gold lining of steel autoclaves to prevent corrosion. By introducing gold lining for the steel autoclaves, he could achieve a temperature of 350°C to synthesize the tridymite phase of quartz and increase experimental duration [17].

With the introduction of steel autoclaves along with noble metal linings, the tendency to reach higher pressure–temperature conditions than the usual 5–10 atm began. However, the majority of the works up to 1880 remained pertaining to quartz, feldspar, and related silicates. Thus, many of the earlier authors even define hydrothermal technique as the "one used in the synthesis of silicates." In 1880, Meunier [17] carried out an experiment using water vapor passed over $AlCl_3$ and Mg contained in an open tube, heated to a low red heat and obtained mainly spinel. In some parts of the tube, he even observed tiny crystals of periclase and corundum.

Followed by this work, Hannay [18] claimed to have synthesized artificial diamond by the hydrothermal technique. Similarly, Moissan [19] also claimed to have synthesized diamond artificially, as large as 0.5 mm, from charcoal. Though the success of these experiments was treated as dubious, they certainly provided a further stimulus for hydrothermal research, particularly the development of high-pressure techniques. It is interesting to note that the results of some of these earlier works were treated as dubious. This is probably in connection with the small size of their resultant products. The first ever large-size crystals obtained by the earliest workers were those of hydrated potassium silicate, which were about $2-3$ mm long, by Friedel and Sarasin [20]. In this case, while growing orthoclase and feldspar, hydrated potassium silicate was obtained as an additional phase containing surplus potassium silicate. Here, it is appropriate to quote one or two earlier experiments for the curiosity of the readers and to show their experimental conditions. Chroustshoff [21] obtained quartz by heating a colloidal solution of SiO_2 to 250°C for 6 months. He started his experiments with 4 tubes, 3 of which soon burst. Some crystals as large as 8 mm long and 3 mm wide quartz crystals with $(10\bar{1}0)$, $(10\bar{1}1)$, and $(01\bar{1}1)$ faces were crystallized. Perhaps, this is the first systematic work on the growth of quartz. Although, the experimental duration was quite long, almost 6 months, he could obtain quartz crystals as large as 8 mm long and 3 mm wide. In another experiment, Chroustshoff [22] used a thick-walled evacuated glass tube of 25 cc volume, containing a mixture of colloidal SiO_2 solution, colloidal $Fe(OH)_3$, colloidal $Fe(OH)_2$, lime water, freshly precipitated $Mg(OH)_2$, and several drops of a NaOH–KOH solution which he heated for 3 months at 550°C. The resultant product contained, among other things, long, thick, dark-colored prismatic crystals of hornblende, 1 mm long, 0.5 mm thick, with (010), (110), and (011) faces. The most interesting point to be noticed here is that the experimenter obtained a mineral in 3 months under hydrothermal conditions having no bearing on its petrogenesis and no relation to any known equilibrium. Morey [23] quotes this as a horrible experience of the experimenter. Here, the reader should notice the duration of each experiment and the size and type of the minerals obtained. This is definitely related to the lack of knowledge in areas of solvent chemistry, kinetics, and solubility of the compound. Despite very low growth rates achieved in the nineteenth century, the earlier workers in hydrothermal research continued to synthesize a very wide range of mineral species. According to Morey and Niggli [24], over 80 mineral species are supposed to have been synthesized during nineteenth century. The list includes quartz, feldspars, mica, leucite, nephelite, epidote, hornblende, pyroxene of minerals from the silicate group, and several nonsiliceous minerals like corundum (Al_2O_3), diaspore ($Al_2O_3 \cdot H_2O$), and brucite ($Mg\ (OH)_2$).

For the sake of convenience, the authors describe the further development of the hydrothermal research in coherence with the autoclave design, development, and pressure–temperature range. It is well known that the study of hydrothermal processes is merely the study of certain aqueous systems at high temperature, usually near to and often above the critical temperature of pure water; therefore, considerable amounts of pressure are developed at these temperatures by water or aqueous solutions. The experimental difficulties of such a study are many but are largely connected with the choice and control of working conditions and the selection of a suitable autoclave to withstand the pressure at high temperatures over a long period of time. It was soon after Chroustshoff [16] proposed the gold lining for the steel autoclaves that many earlier researchers began to work on new designs and new lining materials for autoclaves to obtain the most ideal conditions for the synthesis of several high temperature oxide minerals. Probably, Friedel and Sarasin [20] termed their hydrothermal autoclave as *hydrothermal bomb* because of the high-pressure working conditions in their experiments. Obviously, new metals and alloys available during that time were tested. For example, De Schulten [25] even used a copper bomb to synthesize analcite by heating a mixture of sodium silicate, sodium aluminate, and lime water at 180°C for 18 h. Similarly, Sir Ramsay and Hunter [26] used a cast iron bomb. Bruhns [27] used steel bombs lined with platinum, with a cover held down by bolts and made tight by means of a copper washer, which was protected by the action of the mineralizer on platinum. Other contemporary workers used this design. Perhaps, the simplest design of the autoclave, made of a nickeled gun-barrel or silver-lined steel tube closed by a screw cap and copper washer, was developed by Doetler [28]. Using this type of autoclave, Doetler has done extensive work by obtaining several mineral recrystallized species like chabazite, okenite, heulandite, analcite, and natrolite. However, in most of these works from 1870s to 1890s, using the above-mentioned type of autoclaves, experimental uncertainties were introduced owing to the lack of a perfectly tight closure, which resulted in the leakage of water, thereby causing variations in the pressure conditions (and hence the concentration) of water during the course of experiments. Friedel [29] obtained corundum crystals by heating a solution of NaOH with an excess of Al_2O_3 at a higher temperature at that time, 530 − 535°C.

Toward the end of the nineteenth century, Spezia [5] from the Torino Academy of Science began his classical work on the seeded growth of quartz. His contribution to the field of hydrothermal research is remembered even today. The early works showed that plates of quartz kept at 27°C for several months with water under a pressure 1750 − 1850 atm did not lose its weight and also showed no etch figures. Thus, he concluded that pressure alone has no influence on the solubility of quartz [30]. Subsequently, Spezia [30] studied the action of Na_2SiO_3 in the solubility of quartz with temperature. He found that when alkali was present in the system, SiO_2 separated out as quartz and the rhombohedral faces of the quartz were easily attacked, and on the other hand, on the same faces were the greatest depositions of SiO_2 from Na_2SiO_3 solutions. The more rapid growth of the quartz crystals along the *c*-axis was explained.

During the nineteenth century, much work was done on the geological side of hydrothermal research and the entire work was confined to Europe, particularly Germany, France, Italy, and Switzerland. It was only toward the turn of the nineteenth century that the science of hydrothermal technology moved to North America, which was assisted by the American Industrial Revolution. Further, the establishment of the Geophysical Laboratory at the Carnegie Institute of Washington, USA, in 1907, probably marked the most important milestone in the history of hydrothermal technology. Although a great deal of research was carried out on hydrothermal technology in the nineteenth century, the facilities for large-scale hydrothermal research before the end of the World War II were virtually nonexistent, except the Carnegie Institute of Washington, where Bowen began his hydrothermal research and later moved to the University of Chicago, Bridgman's high-pressure laboratory at Harvard University during 1940s and 1950s, Kennedy's at the University of Chicago, and Tuttle's at Penn State University. This situation dramatically changed with the development of the test tube–type pressure vessel by Tuttle in which high pressures up to 5 kbar could be maintained at temperatures of 750°C for a long period of time. However, before going into the contributions of these great personalities, we shall discuss the early contributions during the turn of the nineteenth century and the beginning of twentieth century. As mentioned earlier, the entire activity on hydrothermal research was concentrated in Europe and there was no activity in North America or Asia, including Japan, China, India, and Taiwan, which are included today in the top 10 countries actively engaged in hydrothermal research. Perhaps, the first North American published work on hydrothermal research was by Barus [31], who essentially worked on the impregnation of glass with water to such an extent that it melted below 200°C. He used steel bombs for these experiments. Subsequently, Allen [32] obtained very fine quartz crystals by heating a solution of $MgCl_2 \cdot 6NH_4Cl$ and Na_2SiO_3 for 3 days at $400 - 450$°C. Also, large crystals of 2 mm long, often barrel shaped with short rhombohedral terminal faces, were obtained. Allen used steel bombs closed by a Cu disk held in place by a screw on steel cap, used either $MgCl_2 \cdot 6NH_4Cl$ solutions or a mixture of $MgCl_2$ solution with $NaHCO_3$, and either amorphous SiO_2 or Na_2SiO_3, heated for 3–6 days at $375 - 475$°C to obtain magnetite, quartz, fosterite, and so on. However, the early American hydrothermal research was carried out using the autoclaves of European designs. Around the turn of the century, it was again the German workers who were active in the field of new designs, especially Tammann [33a] and Boeke [33b] from Tottingen who carried out extensive studies with reference to many carbonate and silicate reactions in the newly designed bombs at higher pressure and higher temperature conditions with CO_2 pressure. Perhaps, this is the beginning of the solvothermal research, although the term *solvothermal* was introduced by French chemists in 1971 [34]. The autoclaves used by Tammann and Boeke were all heavy steel vessels, 10 in. diameter, 18 in. long, with soft metal gaskets which formed the seal when compressed by a set of radially deployed screws and nuts. The heating was internal with a resistance furnace. Today, the use of CO_2 as a solvent is one of the most popular routes in hydrothermal technology, which is also popularly called green technology, owing to the near-ambient conditions of materials processing using CO_2 as a solvent. A great range of pharmaceutical products in the nanosize range are being processed using continuous flow reactors in the presence of CO_2 solvents [35].

With the establishment of the Geophysical Laboratory at the Carnegie Institute of Washington, USA, Bowen [36] established a fine hydrothermal research laboratory. Obviously, all the major contributions to the field of hydrothermal research from the United States of America in the early twentieth century are solely from this laboratory and have received worldwide recognition. Bowen took up a systematic study of the phase equilibria in high temperature silicate systems and, around the 1920s, he framed the most important petrologic principle, viz., *Bowen's Reaction Principle*, as shown in Figure 2.3.

These ideas of Bowen have been, without question, one of the rare landmarks in the history of petrology. The main opposition for this principle was that the extrapolation of experimental phase equilibrium results was based on the data he obtained from systems made from relatively few and chemically pure components to natural systems. Subsequently, Bowen continued his research in this field, worked on several complex systems, and revised his earlier proposed reaction series slightly. While Bowen was working on the petrologic problems, Morey, Niggli, and Fenner were actively studying the silicate systems with water as one of the components. Morey [37] first designed a simple, gasketed, sealed steel autoclave of 25–100 ml volume in the year 1913. It became very popular because of its simplicity and ease to handle. Later, the autoclave that bore his name was widely used for working up to the temperature of 450°C and 2 kbar pressure. The details of the design and construction are discussed in Chapter 3. The first American autoclave designs were those of Adams and Smyth, again from Geophysical Laboratory of the Carnegie

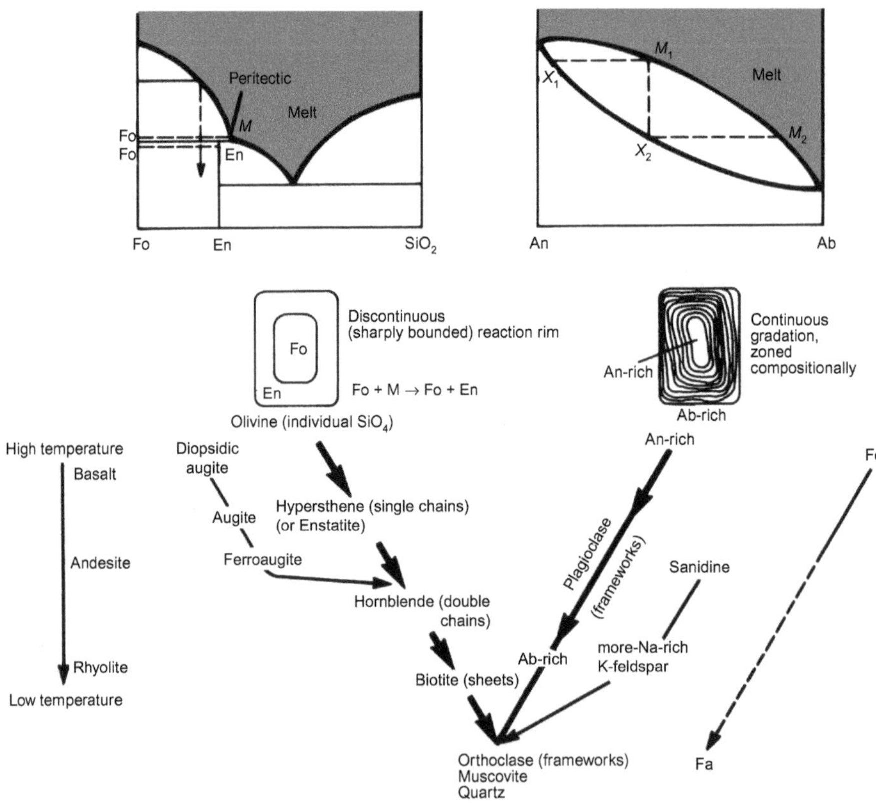

Figure 2.3 Bowen's reaction principle [36].

Institute of Washington in 1923 [38], which later led to the modified versions by Goranson [39] in 1931. Figure 2.4 shows the schematic diagram of the autoclave design of Adams and Smyth. Before the development of these autoclaves, hydrothermal research was confined only to lower pressure and lower temperature limits because the externally heated vessels failed as a result of creep and oxidation of the alloy used. On the other hand, the pressure vessels designed by Adams and Smyth were internally heated and externally cooled by circulating water. These developments in the 1920s are regarded as important milestones in hydrothermal research. They set a new trend in the search for new alloys, and extended pressure

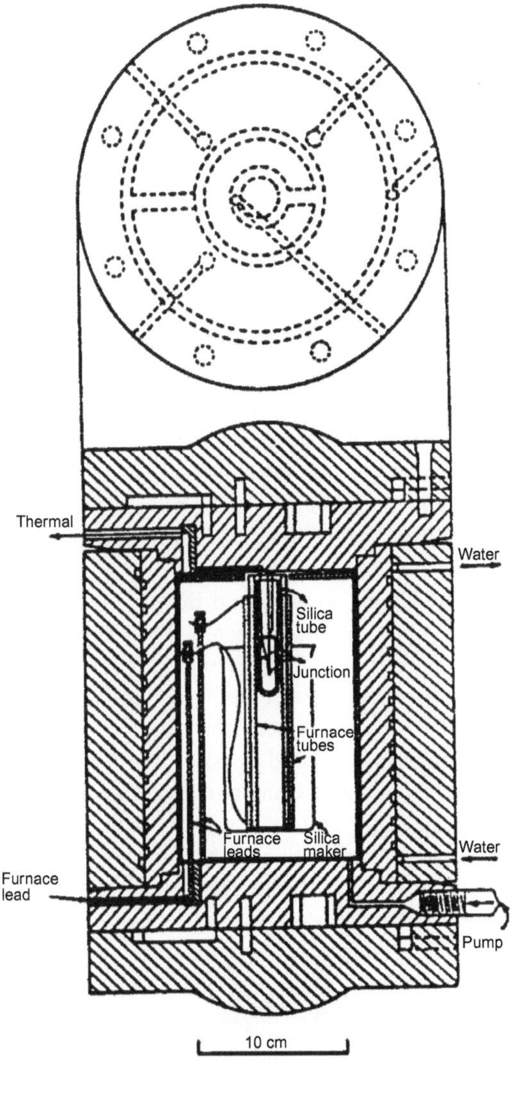

Figure 2.4 Schematic diagram of the autoclave design of Adams and Smyth [38].
Source: Courtesy of the American Chemical Society, Washington, DC.

and temperature limits. Using this set up, Goranson [39] carried out a systematic study pertaining to the solubility of water in a liquid having the composition of the Stone Mountain Granite, and this investigation was considered as the most outstanding advancement in the application of hydrothermal laboratory studies to petrologic problems. The driving forces for the volcanic activity were related to the crystallization of hydrous silicate liquids.

During this time, hydrothermal research began in Eastern Europe and Asia on a small scale, and much later in China and India—maybe after World War II. Here, the reader should observe that the hydrometallurgy was quite an established field of research in many countries in Europe, North America, Eastern Europe, and Asia, particularly after the discovery of bauxite leaching by Karl Josef Bayer (1871 − 1908) [40]. While writing the history of hydrothermal technology, the authors paid more attention to crystal growth and materials science rather than metallurgy or metal extraction. Therefore, historical references are not made for hydrometallurgical processes. As the authors of this monograph are from Japan and India, a brief history of the hydrothermal research in these two countries is given. Today, Japan is the second most active country in the world for hydrothermal research next to the United States of America, and India stands in the fifth position (based on the statistics prepared from Chemical Abstracts, Science Citation Index, and INSPEC data bases from 1989 to 2011) with respect to the hydrothermal research in Materials Science (Figure 1.14a and b). More details on the statistics can be obtained from Chapter 1. According to Prof. Shigeyuki Somiya [41], in Japan the hydrothermal work was first initiated in 1926 by Dr. Tominosuke Katsurai, and the work was related to hydrometallurgy. He worked on the soda treatment of aluminum ore through hydrothermal extraction. Following this, Prof. Shoichiro Nagai reported the synthesis of calcium silicate in 1931 [42]. Perhaps, this was the beginning of hydrothermal synthesis in Japan. Before going into the further developments which took place during the World War period, let us discuss briefly the progress achieved by the earliest workers of twentieth century in the area of hydrothermal phase equilibrium studies.

Pioneers like Bridgman, Cohen, Morey, Niggli, Fenner, and Bowen in the early twentieth century changed the scenario of hydrothermal research. They carried out an impressive amount of basic research along with their European counterparts, especially Tammann. Much of this early work was concerned with the setting up the pressure scale and testing. The results were predicted from thermodynamics. However, the total research effort was small and the study passed into a period of dormancy except for phase equilibria studies in some systems relevant to the natural systems. This was connected with the need for materials with a combination of high strength and corrosion resistance at high temperatures. Thus, the interest was again mostly geological, and the bulk of research is from the geological science perspective. Most of the early geological workers appreciated the necessity for considering very simple chemical systems before there was any hope of understanding the complexities of natural chemical systems. Thus, the early work of the Geophysical Laboratory of the Carnegie Institute of Washington was heavily biased toward pure chemistry [43].

Here, it is appropriate to discuss one or two important systems studied in the early twentieth century. Perhaps, the first systematic and pioneering study of the hydrothermal system began with the publication of the work on the ternary system $H_2O-K_2SiO_3-SiO_2$ by Morey and Fenner [44].

This work gives a more precise account of the pressure and temperature conditions throughout the experiment. Besides, nutrient materials were prepared from highly pure substances and consequently the composition of the charges used in each experiment was precisely known. The paper carried great significance to geological science as it demonstrated clearly the possibility of a complete solubility between anhydrous silicate phases and water, which in turn assisted in framing a mechanism for pegmatite crystallization in the presence of a volatile. Morey and Fenner [44] studied this system from 200°C to >1000°C [42]. The work comprises the determination of the composition and properties of the various stable solid phases which can coexist with solution and vapor within the above temperature range, the composition of the solutions in equilibrium with the solid phases, the change in composition of these solutions with temperature, and the approximate determination of the corresponding three-phase pressures. Morey and Fenner used the famous Morey autoclave in all their work on phase equilibria studies in different systems. The isothermal polybaric saturation curves and the isobaric polythermal saturation curve at 1 atm pressure of H_2O vapor is shown in Figure 2.5.

The most significant aspect of this work is that a new trend can be observed here with reference to the theoretical relations governing the equilibrium in binary and ternary systems containing a volatile component, and a short discussion of the proper application of the term *solubility*. Besides Morey and his group, another group from the same laboratory led by Bowen was very active during that time. The study of the PVT relations of water became popular during the late 1920s. The works of van Nieuwenburg and Blumendahl [45] followed by Keenan and

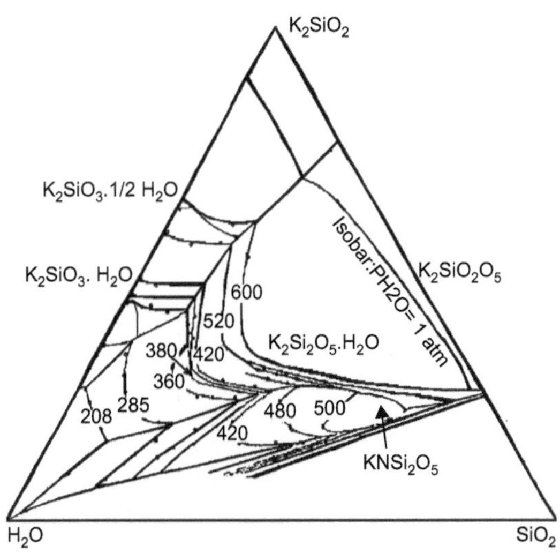

Figure 2.5 The isothermal polybaric saturation curves and the isobaric polythermal saturation curve at 1 atm pressure of H_2O vapor [44].

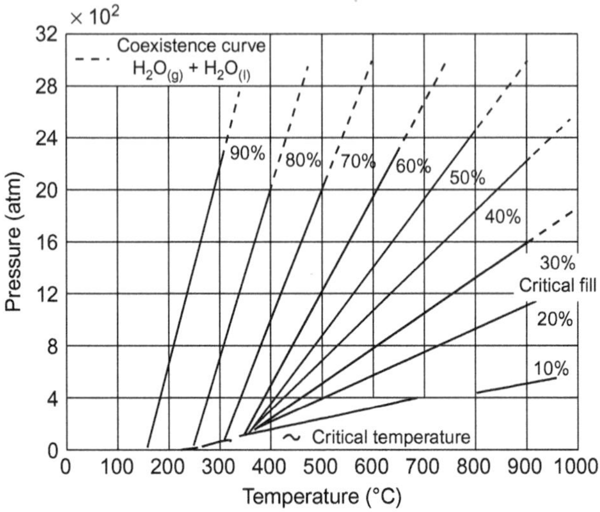

Figure 2.6 PVT diagram of water [47].

Keyes [46] are the most significant ones. Later, Kennedy [54] oriented his famous PVT data on water. Kennedy's diagram is shown in Figure 2.6.

Hitherto, the hydrothermal researchers were quite satisfied with the synthesis of inorganic compounds that resemble or are similar to natural minerals. They did not bother much about the size of the crystals obtained. The growth rate was extremely low because of the lack of accurate knowledge on the proper solvent, and with such slow growth rates, the hydrothermal research was less promising. Only a few people were engaged in this research studying the phase relations in various systems including that of water, and the interest was mainly on Geoscience. However, when Brazil imposed an embargo on the supply of high-purity quartz which was a strategic material for the telecommunications purpose during World War II, hydrothermal researchers began to think seriously about the growth rate of crystals. At that time, Brazil was the only source country in the world for electronic grade natural quartz, and even today, the situation is the same. Thus, the crisis, owing to the shortage of pure quartz, forced many researchers to jump into the field of hydrothermal research with a sole objective of growing large-size single crystals of quartz. Many countries like the United States of America, United Kingdom, Germany, and the erstwhile Soviet Union were in the field and the success achieved was immense. Therefore, the period from the late 1930s to 1940s is referred to as the "golden period" in the hydrothermal research, not only because of the enhanced research activity, but also because of the many new discoveries with reference to autoclave designs and other important technological materials.

The authors feel that it is highly appropriate to discuss briefly the historical contribution made by the German hydrothermal researchers with reference to the increased growth rate and growth of large-size single crystals of quartz. In contrast to the slow rate of growth achieved by the earliest workers like De Senarmont and Spezia, the captured German reports show that Nacken [48], using natural α-quartz as seed crystals and vitreous silica as the nutrient, had grown quartz crystals in an isothermal system and had succeeded in obtaining large single crystals of quartz from a small seed. Prof. Richard Nacken (1884–1971) worked on the synthesis of various minerals from 1916 onwards, but left this field. In 1927 or 1928, he started working only on the hydrothermal growth of quartz crystals. On Nacken's work, Sawyer writes (cited by Bertaut and Pauthenet, 1958) that, "... Nacken made quartz crystals of 1 in. diameter by using hydrothermal method and the conditions are given as ..." followed by some biographical data. Similarly, Nacken's emphasis on the quartz growth has also been documented by Sawyer. Almost at the same time, Nacken made emerald single crystals by the hydrothermal method and also beryl or corundum crystals for watch bearings. He prepared a large number of synthetic emeralds by using a trace of chromium to produce the color and could obtain hexagonal prisms of emerald weighing about 0.2 g in a few days [49]. However, much of the work carried out by Nacken [50] during late 1930s and 1940s remained as intelligence reports since he could not publish much. During 1950, Nacken [51] published his work on quartz that remains a classic even today. Figure 2.7 shows a photograph of the quartz crystals obtained by Nacken [51]. As mentioned earlier, Doetler was the first one to use closed nickeled gun-barrel autoclaves, way back in 1890. It attained its prominence during World War II when several laboratories from many countries began working on the growth of large-size crystals of quartz. This is also partly connected to the availability of alloys containing Mo and

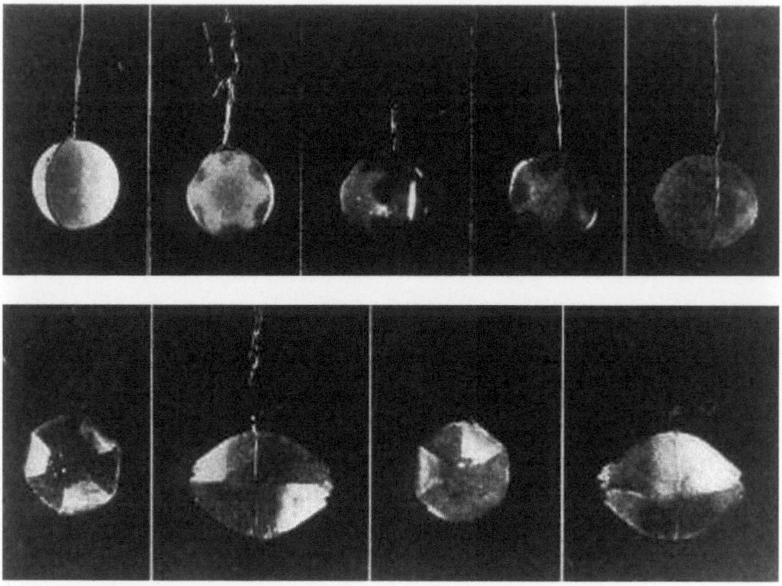

Figure 2.7 Photographs of the quartz crystals obtained by Nacken [51].
Source: Courtesy of the VEB Deutscher Verlag, Germany.

other important metals, which provided better strength to the alloys. Many countries attempted to grow large single crystals of quartz using old cannon barrels with closed ends during early 1940s. In this regard, the contributions of Lobachev and Shtenberg are important. By the mid-1940s, many laboratories, particularly from the United States of America, Germany, Russia, and the United Kingdom, were actively engaged in the growth of large quartz crystals. At this time, Tuttle [52] designed a cone-in-cone seal for a small vessel with no threads to seize. However, the Tuttle apparatus could not achieve high-pressure and high-temperature conditions, and it did not become very popular.

Later, Rustom Roy [53] modified another vessel that had been designed by Tuttle but never exploited. It proved to be much simpler and highly versatile. This modification by Roy is popularly known as the "test-tube bomb," which is made up of Hastelloy; Rene 41; Stellite 25 or Udimet, with $1 - 11/2$ in. o.d. $\times$ 8 in. long, having a $3/16 - 1/4$ in. diameter hole drilled down to ½ in. of the bottom. It was threaded at one end and closed at the other. A more detailed description of the design and construction of this autoclave is given separately in Chapter 3. The greatest advantage of this test-tube bomb was not only the ease of operation but also a large number of such vessels could be served by one pressure source. This integrated system of several vessels, each with its own gauge with one pressure generator, became standard hydrothermal laboratory apparatus. Figure 2.8 shows a

Figure 2.8 Picture of an assembly of 10 autoclaves in the laboratory of K. Byrappa.

picture of such an assembly of 10 autoclaves of test-tube type, with one pressure generator source, operating simultaneously.

Many laboratories throughout the world have this apparatus for studies concerning phase equilibria, materials synthesis, crystal growth, particles preparation, materials processing, diamond synthesis, and so on. The recent modifications of these test-tube bombs have been extended further for higher pressure and higher temperature conditions. The credit goes to the pioneers of this apparatus, Tuttle, Roy, and Licastro, who could set up a Tem-Press Company to manufacture these autoclaves at Penn State University. Thus, during the 1950s, the trend in hydrothermal research was toward the development of new designs, again to attain higher pressure and higher temperature limits with the better alloys available. This also led to the study of phase relations in several systems within a broader pressure–temperature range. In this regard, the contribution of Kennedy [54], Morey and Hesselgesser [55], Tuttle and England [56], and Della Roy and Rustom Roy [57] are very important. The SiO_2–H_2O system was studied in great detail. Followed by this, many other laboratories in Europe, North America, erstwhile Soviet Union, and Japan took up the study of phase equilibria in several simple and complicated systems [57]. These studies laid a firm foundation for the ceramists and physical chemists, after many years, to develop a new technology. At this time, Prof. Yamasaki, Prof. Kumitomi, Prof. Ohara, and Dr. Akitsu were doing an excellent work in hydrothermal research in Japan. Some of them were closely associated with the industries like Toyo Communication Co. Ltd. and Nippon Dempa Co. Ltd., which resulted in the successful growth of large-size quartz single crystals. Though they started in a very small way during the postwar period, Japan is now the largest producer of commercial quartz in the world. Japan alone produces >50% of

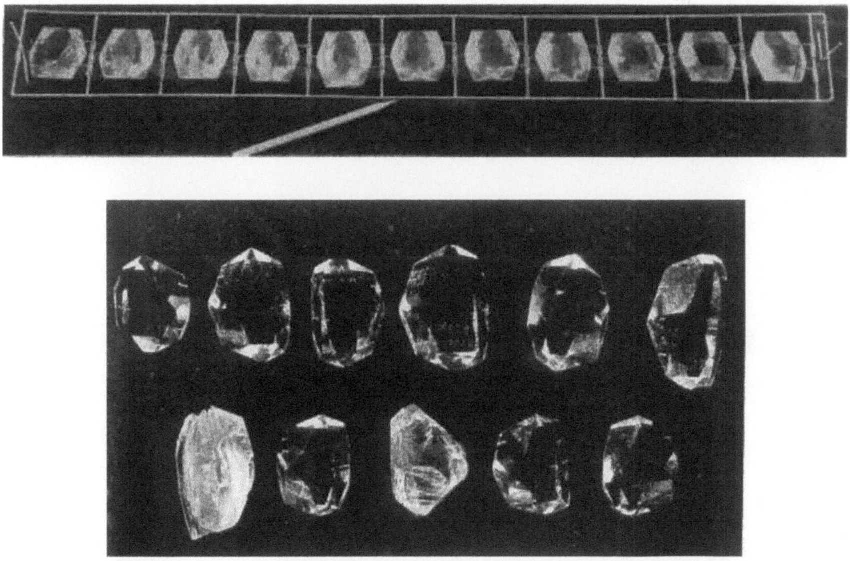

Figure 2.9 Quartz crystals obtained by Walker [59].

the world's annual production; it has the largest autoclaves in the world [58]. Figure 2.9 shows the quartz crystals obtained by Walker during 1950s [59].

Figure 2.10 shows the quartz crystals obtained at AT&T Bell Labs during the 1970s (R.A. Laudise, personal communication), and Figure 2.11 shows the growth of quartz crystals in the world's largest autoclave located at the Toyo Communication Co. Ltd., Japan, during the 1980s [56]. In one experimental run, about 4000–4500 kg of quartz was produced. The early success in producing commercial quartz crystals in Japan during the postwar period paved the way for several others in the field. Some prominent ones are: Prof. T. Noda, Prof. Doimon, Prof. R. Kiyoora, and Dr. Y. Itoh. When the North American laboratories were concentrating mainly on the development of new designs of autoclaves, the study of phase equilibria, and commercial production of quartz, Japan concentrated essentially on the growth of quartz crystals and achieved its goal in becoming the world's largest producer.

The study of ternary and other more complex systems during the 1950s was very important, not only from the point of view of geological science, but also from the inorganic chemistry point of view, as several new compounds which were hitherto unknown or not found in nature could be crystallized. Some of the important systems studied under this category are: $MgO-SiO_2-H_2O$, $Al_2O_3-SiO_2-H_2O$,

Figure 2.10 Quartz crystals obtained at AT&T Bell Labs.
Source: Photograph courtesy of R.A. Laudise.

$BaO-Al_2O_3-H_2O$, $SrO-SiO_2-H_2O$, $K_2O-SiO_2-H_2O$, $Na_2O-SiO_2-H_2O$, $CaO-SiO_2-H_2O$, $CaO-Al_2O_3-H_2O$, $CaO-Al_2O_3-SiO_2-H_2O$, and so on. More accurate phase equilibria data were obtained for the earlier studied systems. At this time, many laboratories in the erstwhile Soviet Union attained a special prominence which led to the rapid growth of this branch of science not only in terms of laboratories but also in terms of research personnel.

At one time, in the erstwhile Soviet Union there were more than 1000 researchers in this field. During the 1960s, an intensive study of the hydrothermal process of synthesis and growth of single crystals which did not have the analogues in nature began. During the 1970s, there was a quest for the search and the growth of hitherto unknown compounds of photosemiconductors, ferromagnets, lasers, piezoelectrics, and ferroelectrics and, in this regard, hydrothermal technology attracted a great attention. Many other nations, including China and India, got into the field of hydrothermal growth of crystals. Several established laboratories in the world began to study systematically various aspects of the hydrothermal growth of crystals such as the physicochemical principles, kinetics, designing new apparatus, and growing new compounds. This marks a significant change in the trend of hydrothermal research. Specific aspects of the hydrothermal method as a modeling tool

Figure 2.11 Growth of quartz crystals in the world's largest autoclave.
Source: Photograph courtesy of Dr. Taki.

to understand the natural processes of mineral formation changed dramatically into an important method characteristic for inorganic chemistry. During the early 1980s, a new sealing for the autoclave was designed, that is, *Grey-Loc* sealing, which facilitates the construction of the very large-size autoclaves with a volume of 5000 l [60]. The world's largest autoclave is located in Japan and works with such a Grey-Loc sealing mechanism. The details of the sealing mechanism are discussed in Chapter 3.

Toward the end of the 1970s, on the whole, the hydrothermal field experienced a declining trend for two reasons: There was no scope for further work on the growth of large-size single crystals of quartz on the one hand, and on the other hand, large-scale attempts to grow bigger crystals of other compounds investigated during the 1960s and 1970s failed. It was unanimously decided that the hydrothermal technique is not suitable for the growth of large crystals other than quartz. The focus at this time was on Czechrolskii, metal-organic chemical vapor deposition (MOCVD), and molecular beam epitaxy (MBE). This is largely connected to the general approach of the hydrothermal researchers to grow large single crystals without looking into the hydrothermal solvent chemistry and kinetics of the crystallization process. Although Franck started working on the study of the behavior of electrolytic solutions with temperature and pressure conditions way back in the 1950s [61,62], it did not attract the hydrothermal researchers' attention immediately. They thought that the conditions under which Franck worked were confined to the lower pressure and lower temperature conditions, whereas the hydrothermal technique actually belonged to the higher pressure and higher temperature conditions at least until recently. In fact, Ballman and Laudise [63], in their famous article published in the book, *Art and Science of Growing Crystals* edited by Gilman, write that "crystal growth is more an art than science." However, the Nobel Symposium organized by the Swedish Academy of Sciences, during September 17 − 21, 1979, on "The Chemistry and Geochemistry of Solutions at High Temperatures and Pressures" is remembered as an eye opener. The presence of pioneers in the field of hydrothermal physical chemistry like Franck, Seward, Helgeson, and Pitzer drew the attention of hydrothermal crystal growers and a new trend was set to look into the hydrothermal solvent chemistry and the physical chemistry of the hydrothermal systems [64 − 67]. Following this, Japan organized the first ever International Hydrothermal Symposium during April 1982 [68], which was attended largely by specialists from different branches of science like physical chemistry, inorganic chemistry, solid-state physics, materials scientists, organic chemists, hydrometallurgists, and hydrothermal engineers. This is the dawn of modern hydrothermal research. Since then, new avenues in the field of hydrothermal research are being explored. The modeling of the hydrothermal systems, study of the hydrothermal crystallization mechanism, thermodynamics, and kinetics of the reactions began. New designs of the reactors to suit the specific requirements of research into areas like crystal growth and materials processing were developed. Unfortunately, this period also marks the fall of the Russian domination in the field of hydrothermal research. Japan emerged as a leader in hydrothermal research on par with the United States. The decade-wise evolution of the hydrothermal research

Table 2.1 Evolution of Hydrothermal Technology with Time

Period	Focus	Equipment	Remarks
1850–1900	Mineral synthesis, imitation of natural conditions	Simple reactors, glass reactors, digestors	Lower growth rate, tiny particles, geological interest
1900–1940	Mineral synthesis, improvement in PT conditions, German domination	Morey autoclaves, flat closures	Lower growth rate, silicates, carbonates, Germany, Russia, France, the United States of America, geological interest
1940–1950	Large-size and large-scale production of quartz, beginning of zeolites, clays, and micas	Test-tube type (cold-cone sealed), welded closure modified Bridgman type	Cold-cone-seal type autoclaves made revolution, PVT diagrams systems
1950–1960	Phase diagrams for natural systems	Morey, Tuttle–Roy, welded closures, modified Bridgman	The dawn of modern hydrothermal research
1960–1970	Synthesis of technological materials, new inorganic compounds without natural analogues	New designs from erstwhile Soviet Union, commercialization of the autoclaves, improved sealing, larger size of the autoclaves	Russian school dominated, Japanese laboratories appeared
1970–1980	A variety of new materials synthesis, ceramic processing in a bigger way, advanced materials	New designs, improved PT conditions, Grey-Loc sealing, large autoclaves	Appearance of many hydrothermal laboratories in several countries
1980–1990	Decline in interest on hydrothermal research. Importance of the technique in materials science, physical chemistry of hydrothermal solutions		Japan organized First International Conference on Hydrothermal Reactions. Beginning of the entry of physical chemists
>1990	Diversification of hydrothermal technique, age of solvothermal, physical chemistry of	Design of new reactors to suit the specific applications: batch reactors, flow reactors, and so on	Entry of organic chemists, environmental scientists, fall in Russian domination,

(Continued)

Table 2.1 (Continued)

Period	Focus	Equipment	Remarks
	hydrothermal solutions		and beginning of the Japanese domination
>2000	Nanomaterials processing for nanotechnology, nanobiotechnology, rapid processing of materials, multienergy processing of materials, modeling of hydrothermal processes	Design of new flow reactors, instant hydrothermal systems, multienergy processing systems, and so on	Entry of biologists, beginning of Chinese domination

with regard to its objective and apparatus used is given with appropriate remarks in Table 2.1.

Today, the hydrothermal technique is being applied widely by various specialists from different branches of science including organic chemistry, biochemistry, biotechnology, food and nutrition, environmental safety, nanotechnology, and nanobiotechnology as shown in Figure 1.11. These developments in hydrothermal research form the main theme of this book.

References

[1] F. Habashi (Ed.), A History of Metallurgy, Metallurgic Extractive, Laval University, Quebec, Canada, 1994.

[2] Sir Roderick Murchison (1840s), cited by S. Somiya (Ed.), Hydrothermal Reactions for Materials Science and Engineering, Elsevier Applied Science, London, 1989.

[3] K.F.E. Schafthaul, Gelehrte Anzeigen Bayer. Akad. 20 (1845) 557,561, 569, 575, 593.

[4] K. Byrappa, T. Adschiri, Hydrothermal technology for nanotechnology, Prog. Cryst. Growth Charact. Mater. 53 (2007) 117–166.

[5] G. Spezia, Sull accresimento del Quarzo, Atti Accad. Sci. Torino 35 (1900) 95–107.

[6] R. Bunsen, Bemerkungen Zu Cinigen Einwürten Gegen Mehrere Ansichten über die Chemisch-geologischon Erscheinungen in Island, Ann. Chim. Phys. 65 (1848) 70.

[7] Wohler Grundriss der Organischen Chemie (1840), cited by R. Bunsen, Ann. Chim. Phys., 65 (80) (1848).

[8] H. De Senarmont, Résumé du cours de physique, Ann. Chim. Phys. 32 (1851) 142–145.

[9] M. Daurree, Sur le Metamorphisme et Recherches Experimentales sur Quelques-uns, Ann. Mines 12 (1857) 289–326.

[10] J.B. Corliss, Metallogenesis at Oceanic Spreading Centres, J. Geol. Soc. London 136 (1979) 621–626.

[11] J.B. Corliss, Hot springs and the origin of life, Nature 347 (1990) 624–625.

[12] N.G. Holm, E.M. Andersson, Abiotic synthesis of organic compounds under the conditions of submarine hydrothermal systems: a perspective, Planet. Space Sci. 43 (1995) 153–159.
[13] J. Lamberg, Ueber silicatumwandlungen, Z. Deut. Geol. Ges. 28 (1876) 519–621.
[14] R.M. Barrer, Syntheses and reactions of mordenite, J. Chem. Soc. 2158 (1948).
[15] St.C. Deville, De la pressance du vanadium clous un mineral alumina, Ann. Chim. Phys. 61 (1857) 309–314.
[16] K. von. Chroustshoff, Studies in Hydrothermal Alteration, Ann. Chemist. 3 (1873) 281–286.
[17] S. Meunier, Production et cristallisation d'un silicate anhydre (enstatite) en presence de la vespeur d'eau á la pression ordinaire, Compt. Rend. 99 (1880) 349–351.
[18] J.B. Hannay, On the artificial formation of the diamond, Proc. R. Soc. London 30 (1880) 178–189.
[19] Moissan, cited by A.D. Edgar, Experimental Petrology, Basic Principles and Techniques, Clarendon Press, Oxford, UK, 1973, p. 5.
[20] C. Friedel, E. Sarasin, Compt. Rend. 92 (1881) 1374–1378.
[21] K. von Chroustshoff, Studies in Hydrothermal Alteration, Bull. Soc. Min. 10 (1887) 31–36.
[22] K. von Chroustshoff, Studies in Hydrothermal Alteration, Compt. Rend. 122 (1890) 677–679.
[23] G.W. Morey, Hydrothermal synthesis, J. Amer. Ceram. Soc. 36 (1953) 279–285.
[24] G.W. Morey, P. Niggli, The hydrothermal formation of silicates, a review, J. Am. Chem. Soc. 35 (1913) 1086–1130.
[25] A. De Schulten, Sur la Reproduction de l'analcime, Die Römischen Grundherrschaften. Eine agrarhistorische Untersuchung, Bull. Soc. Min. 5 (1882) 7–9.
[26] W. Ramsay, Report on the Conditions under which Sedimentary Material is convertible into. Metamorphic Rock Hunter, Rept. Brit. Assn. 1882 (1882) 239–240.
[27] W. Bruhns, Apparatus, a steel bomb, lined with platinum, Neues Jahrb. Min. Geol. II (1889) 62–65.
[28] C. Doetler, Recrystallization of apophyllite, Neues Jahrb. Min. Geol. I (1890) 118–139.
[29] G. Friedel, Hydrothermal synthesis of corundum, Bull. Soc. Min. 14 (1891) 7–10.
[30] G. Spezia, La pressione e' chimicamente inattive nella solubilite e riecostituzione del quarzo, Atti Accad. Sci. Torino. 40 (1904 – 1905) 254–262.
[31] C. Barus, Am. J. Sci. 6 (270) (1898), cited by G. W. Morey, P. Niggli, The Hydrothermal Formation of Silicates—A Review, J. Am. Chem. Soc., 35 (1913) 1086 – 1111.
[32] E.T. Allen, The lime–silica series of minerals by Day and Shepherd, Am. J. Sci. 22 (1906) 297.
[33a] G. Tammann, The States of Aggregation, D. Van Nostrand Company, New York, USA Basel, 1925.
[33b] H.E. Boeke, Die Schmelzerscheinungen und die umkehrbare unwandlung des calcium-carbonats: Neues Jahrb. Mineralogie Geologie, 1 (1912) 91-212.
[34] G. Demazeau, A. Marbeuf, M. Pouchard, P. Hagenmuller, Sur une serie de compose's oxygenes du nickel trivalent derive's de la perovskite, J. Solid State Chem. 3 (1971) 582–589.
[35] K. Byrappa, S. Ohara, T. Adschiri, Nanoparticles synthesis using supercritical fluid—towards biomedical applications, Adv. Drug Deliv. Rev. 60 (2008) 299–327.
[36] N.L. Bowen, The behavior of inclusions in igneous magmas, J. Geol. 30 (1922) 513.

[37] G.W. Morey, Neue kristallisierte silikate von kalium und natrium. Darstellung und allgemeine eigenschaften, Z. Anorg. Chem. 86 (1914) 305–324.
[38] F.H. Smyth, Adams, The system calcium oxide–carbon dioxide, J. Amer. Chem. Soc. 45 (1923) 1172.
[39] R.W. Goranson, Solubility of water in granite magmas, Amer. J. Sci. 22 (1931) 481–502.
[40] K.J. Bayer (1887), cited by F. Habashi, A Text Book of Hydrometallurgy, Libraire Universitaire du Quebec, Canada, 1993, p. 10.
[41] Katsurai (1926), cited by S. Somiya (Ed.), Hydrothermal Reactions for Materials Science and Engineering, Elsevier Applied Science, London, 1989.
[42] S. Nagai (1931), cited by S. Somiya (Ed.), Hydrothermal Reactions for Materials Science and Engineering, Elsevier Applied Science, London, 1989.
[43] A.J. Ellis, W.S. Fyfe, Hydrothermal chemistry, Rev. Pure Appl. Chem. 7 (1957) 261–316.
[44] G.W. Morey, C.N. Fenner, The ternary system $H_2O-K_2SiO_3-SiO_2$, J. Amer. Chem. Soc. 39 (1917) 1173–1229.
[45] C.J. Van Nieuwenburg, H.B. Blumendahl, Isotherms of water from 350 to 480 and for pressures up to 600 kg/cm^2, Rec. Trav. Chim. 51 (1932) 707–714.
[46] J.H. Keenan, F.G. Keyes, Thermodynamic Properties of Steam, John Wiley & Sons, 1936.
[47] G.C. Kennedy, PVT relations in water at elevated temperatures and pressures, Am. J. Sci. 248 (1950) 540–564.
[48] R. Nacken, Artificial Quartz Crystals, etc., US Office of Technical Services Report PB-6948 (1946).
[48a] E. F. Bertaut and F. Forrat, Structure and Ferrigmatnetism of the Llmenite Compound MnNi03, Laboratoire d'Electrostatique et de Physique du Metal, Institut Fourier, Grenoble, France J. Appl. Phys. 29, 247 (1958).
[49] K. Nassau, Synthetic emerald: the confusing history and the current technologies, J. Cryst. Growth 35 (1976) 211–222.
[50] R. Nacken, Artificial Quartz Crystals, etc., US Office of Technical Services Reports PB-18,748 and 28,897 (1946); F. Bertaut, R. Pauthenet: Proc. Instn Elect. Engrs Suppl. B 104 (1957) 261.
[51] R. Nacken, Hydrothermal synthese als grundlage fur zuchtung von quarz-kristallen, Chem. Z. 74 (1950) 745–749.
[52] O.F. Tuttle, A new hydrothermal quenching apparatus, Am. J. Sci. 246 (1948) 628–635.
[53] R. Roy, Fifty-year perspective on hydrothermal research, in: S. Somiya (Ed.), Proceedings of the Workshop on Solvothermal and Hydrothermal Reactions, (N. Yamasaki and K. Yanagisawa, Eds.) Sun Messe Kogawa, January 22 – 24, 1996, pp. 1.1–1.20.
[54] G.C. Kennedy, A portion of the system silica-water, Econ. Geol. 45 (1950) 623–629.
[55] G.W. Morey, J.M. Hesselgesser, The solubility of some minerals in superheated steam at high pressure, Econ. Geol. 46 (1951) 821–835.
[56] O.F. Tuttle, J.L. England, A preliminary report on the system $SiO_2 - H_2O$, Bull. Geol. Soc. Am. 66 (1955) 149–152.
[57] D.M. Roy, R. Roy, Synthesis and stability of minerals in the system $MgO-Al_2O_3-SiO_2-H_2O$, Am. Min. 40 (1955) 147–178.
[58] S. Taki, Improvement of growth process and characterization of quartz crystals, Prog. Cryst. Growth Charact. 23 (1991) 313–339.

[59] A.C. Walker, Hydrothermal synthesis of quartz crystals, Ind. Eng. Chem. 36 (1953) 250−256.
[60] J. Asahara, K. Nagai, S. Harada, Synthetic quartz crystals by large autoclaves, in: S. Somiya (Ed.), Proceedings of the First International Symposium Hydrothermal Reactions, Gakujutsu Bunken Fukyu-Kai, Japan, March 22−26, 1982, pp. 430 − 441.
[61] E.U. Franck, I. Hochverdichteter Wasserdampf, Ele—Ktrolyticshe Leitfahigkeit in KCl−H_2O Losungen bis 750°C, Zeit. Physik. Chem. 8 (1956) 92−106.
[62] E.U. Franck, Experimental Investigation of Fluids at High Pressures and Elevated Temperatures, North American Treaty Organization (NATO) Advanced Study Institute, Series C, C41 (High Pressure Chemistry), 1978.
[63] A.A. Ballman, R.A. Laudise, Solution growth, in: J.J. Gilman (Ed.), Art and Science of Growing Crystals, John Wiley & Sons, New York, NY, 1963, p. 231.
[64] E.U. Franck, Survey of selected non-thermodynamic properties and chemical phenomena of fluids and fluids mixtures, in: D.T. Rickard, F.E. Wickman (Eds), Chemistry and Geochemistry of Solutions at High Temperatures and Pressures, Proceedings of the Noble Symposium, vol. 13/14, Pergamon Press, New York, NY, September 1979, pp. 65 − 88.
[65] T.M. Seward, Metal complex formation in aqueous solutions at elevated temperatures and pressures, in: D.T. Rickard, F.E. Wickman (Eds), Chemistry and Geochemistry of Solutions at High Temperatures and Pressures, Proceedings of the Noble Symposium, vol. 13/14, Pergamon Press, New York, NY, September 1979, pp. 113 − 132.
[66] H.C. Helgeson, Prediction of the thermodynamic properties of electrolytes at high pressures and temperatures, in: D.T. Rickard, F.E. Wickman (Eds), Chemistry and Geochemistry of Solutions at High Temperatures and Pressures, Proceedings of the Noble Symposium, vol. 13/14, Pergamon Press, New York, NY, September 1979, pp. 133 − 178.
[67] K.S. Pitzer, Characteristics of very concentrated aqueous solutions, in: D.T. Rickard, F.E. Wickman (Eds), Chemistry and Geochemistry of Solutions at High Temperatures and Pressures, Proceedings of the Noble Symposium, vol. 13/14, Pergamon Press, New York, NY, September 1979, pp. 249 − 272.
[68] Proceedings of the First International Symposium Hydrothermal Reactions, in: S. Somiya (Ed.), Gakujutsu Bunken Fukyu-Kai, Japan, March 22 − 26, 1982.

3 Apparatus

3.1 Introduction

Advances in any scientific field of research depend largely on the equipment available. Crystal growth or materials processing under hydrothermal conditions requires a pressure vessel capable of containing highly corrosive solvent at high temperature and high pressure. Experimental investigations under hydrothermal conditions require facilities that must operate routinely and reliably under extreme pressure–temperature conditions. Thus, the experimenter has to face a variety of difficulties, and often, peculiar problems pertaining to the design, procedure, and analysis.

Designing a suitable or ideal hydrothermal apparatus, popularly known as an autoclave, is the most difficult task, and perhaps impossible to define, because each project has different objectives and tolerances. In Chapter 2, we described the historical development of hydrothermal technology closely associated with the development of the instrumentation. If we look into the historical development in the design and fabrication of autoclaves, the entire activity was concentrated in Europe alone during the nineteenth century. It was only after the American industrial revolution that activity slowly started spreading into the other parts of the world. Here, we discuss only the autoclaves or hydrothermal reactors used worldwide; they were all developed during the twentieth century. There is a great variety of apparatus existing in hydrothermal technology. Most of the earlier workers had dealt with this aspect in their own way. Hence, we have made a thorough survey of all the available literature and present it here in this section in a simple way to make the reader understand easily the problems of designing, fabrication, and maintenance.

The literature survey clearly indicates two schools of autoclave design: (1) Western school and (2) Russian school. The former is very important as it dominates most of the earliest designs, which later became the standard international designs throughout the world. Most of these designs are relatively simple and quite old (designed prior to the 1970s), whereas, the Russian school had dominated this field from the 1970s and came out with many new designs. The Special Construction Bureau at the Institute of Crystallography, Academy of Sciences, Moscow, Russian Federation, has designed several new autoclaves to suit a specific task of either growing single crystals or studying physicochemical aspects such as solubility and PVT behavior. Besides, several erstwhile Soviet hydrothermal laboratories have produced their own designs to suit their requirements and the type of crystal or material under investigation. Obviously, most of these autoclaves are only local and have not been adopted

Handbook of Hydrothermal Technology.

worldwide. Hence, for the benefit of the reader, we consider only the standard autoclaves used worldwide and some important designs from the Russian school. With the disintegration of the erstwhile Soviet Union, there is not much development in the field of design and fabrication of new autoclaves. The Japanese groups are coming out with new designs of autoclaves for applications in crystal growth and materials processing. With the recent advances in hydrothermal technology for nanotechnology, the design and fabrication of new reactors focus mainly on the rapid production of nanomaterials with a minimum residence time. Although there is no significant development in designing and fabrication of hydrothermal autoclaves for crystal growth and bulk materials processing, there is a substantial progress in the designing of new reactors such as flow reactors, multienergy processing system reactors, hydrothermal cells for in situ time-resolved synthesis for tomographic energy-dispersive diffraction imaging (TEDDI) or neutron diffraction analyses, and instant hydrothermal systems [1].

An ideal hydrothermal autoclave should have the following characteristics:

- **i.** Inertness to acids, bases, and oxidizing agents.
- **ii.** Easy to assemble and dissemble.
- **iii.** A sufficient length to obtain a desired temperature gradient.
- **iv.** Leak-proof with unlimited capabilities to the required temperature and pressure.
- **v.** Rugged enough to bear high-pressure and high-temperature experiments for long duration, so that no machining or treatment is needed after each experimental run.

The most commonly used autoclaves in hydrothermal research are listed in Table 3.1 [2]. The majority of these autoclaves are externally heated pressure vessels and their pressure–temperature range cannot be extended further due to lack of suitable refractory alloys. However, internally heated pressure vessels are now commercially available up to 10 kbar and 1400°C. The list given in

Table 3.1 Autoclaves

Type	Characteristic Data
Pyrex tube 5 mm i.d. 2 mm wall thickness	6 bar at 250°C
Quartz tube 5 mm i.d. 2 mm wall thickness	6 bar at 300°C
Flat-plate seal, Morey type	400 bar at 400°C
Welded Walker–Buehler closure 2600 bar at 350°C	2 kbar at 480°C
Delta ring, unsupported area	2.3 kbar at 400°C
Modified Bridgman, unsupported area	3.7 kbar at 500°C
Full Bridgman, unsupported area	3.7 kbar at 750°C
Cold-cone seal, Tuttle–Roy type (batch reactors)	5 kbar at 750°C
Piston cylinder	40 kbar at 1000°C
Belt apparatus	100 kbar at >1500°C
Opposed anvil	200 kbar at >1500°C
Opposed diamond anvil	Up to 500 kbar at >2000°C
Continuous flow reactors	2 kbar at 600°C

Table 3.1 does not include some of the recent designs and modifications or upgrades as their applications are restricted to specific studies.

3.2 Selection of Autoclave and Autoclave Materials

When selecting a suitable autoclave, the first and foremost parameter is the experimental temperature and pressure conditions and the corrosion resistance in that pressure–temperature range in a given solvent or hydrothermal fluid. If the reaction is taking place directly in the vessel, the corrosion resistance is, of course, a prime factor in the choice of autoclave material. The most successful materials are corrosion-resistant, high-strength alloys, such as 300 series (austenitic) stainless steel, iron, nickel, cobalt-based superalloys, and titanium and its alloys. Several workers have reviewed the properties of some alloys used commonly in the making of hydrothermal autoclaves [3,4]. Tables 3.2–3.4 give certain properties of these alloys. The critical property for a material used in a hydrothermal autoclave is its creep-rupture strength, a measure of the length of time until rupture of a stressed material occurs at a given temperature. Similarly, the ultimate tensile strength and yield strength are to be taken into account. In the 300 series of

Table 3.2 Properties of Certain Alloys [3]

	Resistance to Aqueous OH	Rupture Stress (atm)	
		1000 h, 800°C	1000 h, 600°C
Low-carbon steel	+		170
Tool steel	+		
4140 (similar to EN19)	+		170
Stainless type 340[c]	–	270	1100
19-9-DL (Universal Cyclops Steel Co.)	?	680	2200
Croloy15-15N	+[a]	610	2260
Timken17-22-A	+[a]	–	–
Inconel X[d]	?	1200	4500
Stellite[e]	?	–	–
Udimet500[e]	?	2100	–
Silver	+[b]		
Platinum	+[b]		
Platinum–10% iridium	+[b]		

[a]Denotes the balance of metal in the alloy.
[b]Cr and Ni values in stainless steels are values <2 wt%.
[c]A product of Allegheny Ludlum Steel Corporation, Pittsburgh, Pennsylvania.
[d]A product of Inco Alloys International, Inc., Huntington, West Virginia.
[e]A product of Cabot Corporation, Kokomo, Indiana.
Source: Data from Ref. [4]. Courtesy of the Academic Press, Orlando, FL.

Table 3.3 Compositions of High-Strength Alloys for Autoclaves

	Metal Content (wt%)												
	Cr	**Ni**	**C**	**Mn**	**Mo**	**Si**	**Co**	**V**	**Al**	**Ti**	**Fe**	**Others**	**Note**
Stainless steel[a]													
304	19	9	0.1	2	–	1	–	–	–	–	ba[b]		
310	25	20	0.2	2	–	1	–	–	–	–	ba		
316	17	12	0.1	2	3	1	–	–	–	–	ba		
410	12	–	0.2	1	–	1	–	–	–	–	ba		
Titanium													
RMI-55	Commercially pure titanium												Alpha
Ti6-4	–	–	0.1	–	–	–	–	4	6	ba	0.25		Alpha–beta phase
Ti38644	6	–	0.05	–	4	–	–	8	3	ba	0.03	Zr-4	Beta phase
Ti 17	4	–	–	–	4	–	–	–	5	ba	–	Zr-2, Sn-2	Alpha–beta
Superalloys													
A-286	15	25	0.1	2	1	1	–	0.2	0.4	2	ba		Iron base
Waspalloy[c]	19	ba	0.1	1	4	0.7	13	–	1	3	2	Cu-0.5	Nickel base
Inconel 702[d]	16	ba	0.1	1	–	0.7	–	–	3	0.5	2	Cu-0.5	Nickel base
Hastelloy B[e]	1.0	ba	0.05	1	28	1	2.5	0.4	–	–	5	–	Nickel base
Hastelloy C[e]	16	ba	0.1	1	16	1	2.5	0.4	–	–	5	W-4	Nickel base
HS-21	27	3	0.3	1	5	1	ba	–	–	–	2		Co base (satellite)

[a]Cr and Ni values in stainless steels are values <2 wt%.
[b]Denotes the balance of metal in the alloy.
[c]A product of Allegheny Ludlum Steel Corporation, Pittsburgh, Pennsylvania.
[d]A product of Inco Alloys International, Inc., Huntington, West Virginia.
[e]A product of Cabot Corporation, Kokomo, Indiana.
Source: Data from Ref. [4].

Table 3.4 High-Temperature Strength Properties of Autoclave Alloys[a]

Melt	Tensile Strength (°C)			Yield Strength (°C)			Creep-Rupture Strength (h)		Creep Strength (h)		Thermal Expansion (10^6)
	25	400	600	25	400	600	1000	10,000	1000	10,000	
Stainless steel											
304	85	66	49	38	21	16	49	38	24	18	14
310	92	81	71	40	29	24	39	36	23	14	16
316	85	72	66	38	24	21	66	64	31	19	16
410	89	72	34	32	22	13	34	26	29	14	11
Carbon steel											
0.3% C	64	57	25	36	25	16	19	12	–	10	12
Superalloys											
A-286	143	–	103	94	–	88	88	76	–	81	–
Waspalloy	180	–	170	115	–	100	85[b]	–	–	–	–
Inconel	141	–	105	85	–	75	65[b]	–	–	–	–
Hastelloy C	121	–	98	58	–	43	55[c]	–	–	–	–
HS-21	101	–	73	82	–	39	60[d]	–	–	–	–
Titanium alloys											
RMI-55	65	30	20	55	18	10	20[e]	–	–	–	8
Ti6-4	130	90	60	120	70	55	50	30	30	20	8
Ti38644	170	150	120	160	130	90	140[f]	–	100	–	5
Ti 17[g]	–	–	–	(150–170)			–	–	–	–	–

[a]Strength in 100 pounds per square inch (kpsi), yield strength is for 0.2 elongation, creep strength is for 1% elongation unless otherwise noted.
[b-e]For 1000 h at 650°C.
[f]For 100 h at 450°C.
[g]For 100 h at 350°C and 0.2% elongation.

stainless steels, 316 has the greatest creep-rupture strength. The superalloys are usually much stronger than SS316, even at higher temperatures. Pure titanium metal, though acid resistant, is weak and difficult to handle, but some of the titanium alloys have strengths comparable to stainless steels. Pure titanium, because of its superior corrosion resistance, can be used as an autoclave up to about 300°C or more. Its relatively low creep strength becomes inadequate. For example, after 36 h at 300°C in 0.5 M HCl + 1.5 M NaCl mineralizer, titanium was slightly corroded/tarnished but was distinctly corroded in 6 M HCl at 300°C. It is soluble, even at room temperature, in phosphoric acid and is moderately resistant to acidic sulfate solutions [5]. Extra strength for alloys is required because the actual strengths of alloys under normal operating conditions are less than their strengths under ideal conditions owing to various corrosion processes that occur during hydrothermal reactions as well as to structural changes due to partial annealing of the alloy during conditions of prolonged high-temperature service. This is further complicated by the geometry of the vessel, because the effective stress is more intense in areas of high curvature. In addition, it is noted that the creep-rupture strengths decrease significantly with increasing temperature, thereby lowering the maximum permissible pressure of an experiment at higher temperature.

Alloys of either stainless steel or titanium have a tendency to form an impermeable oxide surface, which prevents continued oxidation. Thus, the corrosion resistance of an alloy depends upon the permeability, reactivity, and solubility of this oxide layer in the corrosive fluid. Obviously, a layer so permeable that solutions can penetrate and contact the unoxidized metal can react with the underlying metal. The corrosion can be crevice corrosion, stress corrosion cracking, and intergranular attack. The crevice corrosion can be prevented through proper agitation of the vessel and polishing of the interior surface.

Intergranular corrosion can be prevented through the use of low-carbon stainless steel, or by alloying with small amounts of metals forming very stable carbides. Stress corrosion can be retarded through proper annealing of the alloy used for making autoclaves, and through the use of molybdenum-containing austenitic steels [4].

The hydrothermal experimenter should pay special attention to the systems containing hydrogen under hydrothermal conditions. Hydrogen at high temperature and/or high pressures can have a disastrous effect on alloys used in the making of autoclaves. It reduces the strength of the autoclaves through any one of the following processes: hydrogen embrittlement, irreversible hydrogen damage, or metal-hydride formation. These problems could be overcome through careful selection of alloys containing small additives such as Ti, Mo, V, heating in an H_2-free atmosphere, and using alloys with low thermodynamic activity.

The autoclave selection is usually done by considering the above-discussed aspects accordingly, for the type of material or compound under investigation, the medium in which the reaction is taking place, and the experimental pressure–temperature conditions. For example, the first author of this handbook experienced a dangerous situation while using hastelloy high-pressure autoclave during the growth of α-Al_2O_3 fine crystals using glycol-based solvent. Generally, hastelloy contains a high amount of nickel, which can act as a catalyst and triggers an exothermic reaction in such an organic system. Although this experiment was carried out at 300°C in a high-pressure hastelloy autoclave with a pressure–temperature range of 4 kbar and 500°C, the autoclave leaked at 300°C. However, researchers have successfully synthesized α-Al_2O_3 at 300°C and lower pressure using much cheaper SS316 autoclaves. Some crystals can be grown readily within the autoclave without any lining, liners, or cans. For example, the growth of quartz can be carried out in low-carbon steel autoclaves. The low-carbon steel is corrosion resistant in systems containing silica and NaOH, because the relatively insoluble NaFe-silicate forms and protectively coats the ground vessel. In contrast, the growth of berlinite crystals requires a Teflon lining or beakers because phosphorus is highly corrosive; it can even corrode platinum if used for a long time. Therefore, the corrosion resistance of any metal under hydrothermal condition is very important. For example, turbine engineers have long known that boiler water with pH > 7 is less corrosive than slightly acidic water, especially for alloys containing Si. Several conventional methods of studying corrosion are known. Usually low-carbon steel is used in the fabrication of simple acid digestion bombs, Morey-autoclaves, modified Bridgman autoclaves, and other related ones. Inconel, Udimet, and Stellite are highly useful because of their high strength at high temperature. These alloys are used in the fabrication of cold-cone-seal bombs employed in phase-equilibria studies at pressures up to 5.4 kbar and temperatures up to 950°C. The recent TZM (Titanium Zirconium Molybdenum) autoclaves are used at pressures up to 10 kbar.

Although the use of glass vessels began in the nineteenth century, for low-temperature and low-pressure experiments thick glass vessels are still used. The quartz glass pressure vessel for hydrothermal studies was first used in 1964 by Speed and Filice [6]. Recently, in 1991, Popolitov [7] used a thick-walled silica vessel with a Teflon capping for experiments up to 200°C and <500 bar pressure. The construction details of various autoclaves will be given separately in the forthcoming section.

3.3 Liners

In the majority of hydrothermal experiments, the mineralizer used is highly corrosive and it can attack the vessel, which is inimical to obtaining high-purity crystals for devices. It requires a suitable lining for the inner wall of the autoclave or separate liners placed in the autoclave. Even in the case of quartz growth, where an acmite-protective layer is found all along the inner wall of the autoclave, the purity of the substance is doubtful. Therefore, for the growth of high-purity quartz, a silver lining for the autoclave is provided. Similarly, other metal linings are provided depending upon the solvent medium. For example, a copper lining is used in many of the borate systems. Copper partly dissolves in dilute bases at 400°C, and copper crystals are deposited in the cooler region of the autoclave. Similarly, nickel lining can be used, but it is also readily attacked by the solvent. Hence, noble metal lining, liners, or capsules are used successfully for alkaline and neutral media. Titanium is much more corrosion resistant, but it is difficult to handle. Litvin and Tules [8] have summarized the applicability of various liners as listed in Table 3.5.

The use of glass tube as a liner was first applied by De Senarmont as early as AD 1851 for the growth of carbonates, sulfates, and sulfides. Figure 3.1 shows the sealed glass tube used by De Senarmont [9]. The tubes were enclosed in water-filled fused gun-barrels and heated up.

Daurree [10] first introduced the pressure balance arrangement between the small glass tube containing the nutrient and the steel tube hosting this glass tube liner. An improved version using pyrex glass and again water as a pressure transmitter was reported by Allen et al. [11]. In fact, most of the earlier workers during the nineteenth century, including von Chroustshoff, found that the glass tube they were using was much attacked under hydrothermal conditions; however, they did not mention anything about precautions taken to overcome this effect, as the

Table 3.5 Materials Used as Reactor Linings

Material	Temperature (°C)	Solutions	Remarks
Titanium	550	Chlorides, hydroxides, sulfates, sulfides	Corrosion in NaOH solution >25% in NH_4Cl solution >10% (at 400°C)
Armco iron	450	Hydroxides	Gradual oxidation producing magnetite
Silver	600	Hydroxides embrittlement	Gradual recrystallization and partial dissolution
Platinum	700	Hydroxides, chlorides, sulfates	Blackening in chlorides in the presence of sulfur ions; partial dissolution in hydroxides
Teflon	300	Chlorides, hydroxides	Poor thermal conduction
Tantalum	500	Chlorides	Begin to corrode in NH_4Cl solution 78%
Pyrex	300	Chlorides	
Copper	450	Hydroxides	Corrosion reduced in the presence of fluoride ions and organic compounds
Graphite	450	Sulfates	Pyrolytic graphite most suitable for linings
Nickel	300	Hydroxides	
Quartz	300	Chlorides	
Gold	700	Hydroxides, sulfates	Partial dissolution in hydroxides

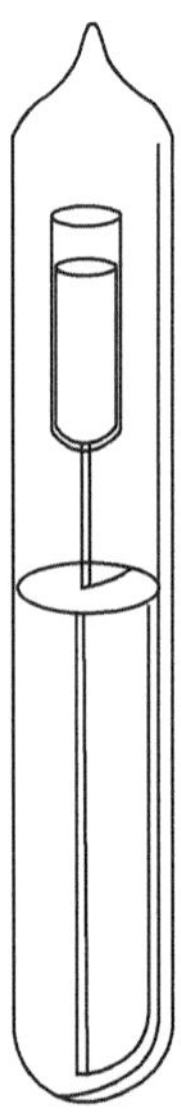

Figure 3.1 Sealed glass tube used by De Senarmont [9].

primary interest was only in synthesizing a product resembling some natural mineral. It was von Chroustshoff who first proposed the gold lining steel autoclaves to prevent corrosion. By doing so, he could attain a temperature of 350°C to synthesize tridymite phase of quartz and increased experimental duration. Subsequently, the use of steel autoclave and noble metal lining became very popular, and the tendency to reach higher pressure–temperature conditions also started. Bruhns [12] used steel bombs with a platinum lining and cover held down by bolts and made tight by means of a copper washer, which was protected from the action of the mineralizer by platinum.

Doetler [13] used a nickeled gun-barrel or a silver-lined steel tube in his experiments. The quartz glass tube liners of 15 mm diameter can easily withstand the outside pressure of 3 kbar at 500°C. However, the internal pressure has to be kept a little below than that of the outside value. Rabenau and Rau [14] have described this technique in detail. For larger autoclaves, like the Morey flat and the modified Bridgman closures, noble metal or other suitable metal lining is provided. The main difficulty in such a design is that the lining should be free of cracks, pinholes, or ridges, and it should be honed near to a mirror finish. For Tuttle cold-cone-seal autoclaves, capsules made of noble metals are used. The capsule tube, charged with a desired amount of nutrient, is clamped tightly at the top and is welded shut. During the experiment the capsule collapses but does not rupture. The vessel is then heated to the desired temperature at constant pressure. When the desired temperature is reached, the capsule expands to its original volume and is not subjected to tension and virtually no capsule failures occur. For some experiments, the capsule needs only to be pressed together, the capsule length being shorter for phase-equilibria studies and longer where a temperature gradient for crystal growth is desired. In large diameter vessels, platinum liners are usually used. These liners have silver caps to avoid platinum seizing.

The liner is filled to a given percent and closed, and the space between the liner and the inner wall of the autoclave is filled to the same percent or slightly higher; this provides a pressure balance and the liner supports virtually no pressure. Pressure imbalance can cause rupture of the liner. For the growth of high-purity crystals in solution in highly corrosive media, Teflon beakers or liners are used. Similarly, for special studies pertaining to reaction kinetics, solubility and materials processing under mild hydrothermal conditions or pressure–temperature conditions below 250 bar and 300°C, Teflon is the most popularly used lining material. Several new autoclave designs with Teflon lining or coating for such studies have been reported in the literature. The Teflon liner or beaker should sit exactly inside the autoclave without leaving any gap, pressure balancing technique can be used alternatively. A slightly higher pressure of nearly 1–2% of fill outside the Teflon liner helps in the sealing of Teflon liners as the cap and liner cannot be welded. In addition, in some cases where the experiment is carried out for longer duration, effective sealing of the Teflon liner can be obtained with a sharp ringlike mechanism on the liner as shown in Figure 3.2. As the temperature rises, the Teflon expands and hermitic sealing can be obtained. The greatest disadvantage with liners or cans is the difficulty associated with the measuring of the actual temperature

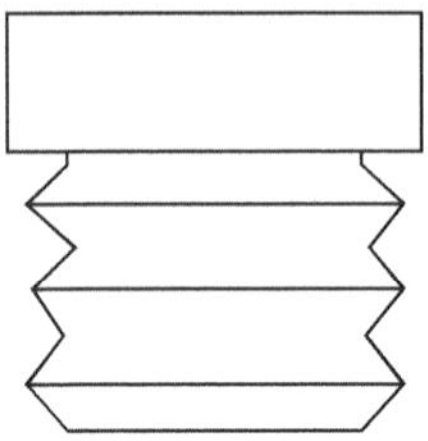

Figure 3.2 Sealing of the Teflon liner.

inside. Certainly, temperature gradients are appreciably lower in the autoclaves with liners than in the autoclaves without liners. This difference becomes more prominent when Teflon liners are used. The greatest disadvantages of Teflon lining is that beyond 300°C, it cannot be used because Teflon dissociates, affecting the pH of unbuffered, near-neutral solutions [15]. This coating tends to tear and generally must be reapplied after one or two experiments. Despite all these difficulties, single crystal growth and materials processing using Teflon are being carried out extensively at various laboratories.

Corrosion resistance can easily be improved at modest cost by coating the interior of the reaction vessels by painting, electroplating, or vapor deposition. As mentioned earlier, chromium and gold linings are often applied to stainless steel vessels to retard corrosion. Such linings are not permanent because of the intrinsic permeability of the thin linings along crystal boundaries or cracks that develop during temperature cycling owing to the differing thermal expansion of the lining and the vessel. When more acidic solutions are used, most of the alloys corrode readily and they must be provided with titanium lining, high-strength titanium alloys, coated alloys, or other alternatives.

Let us consider briefly the internal arrangement of liners in some typical hydrothermal crystal growth experiments. In the growth of quartz crystals, the liner has an assembly as shown in Figure 3.3 [16]. The lower portion of the liner is filled with the required nutrient material. In the upper portion, a seed holder is placed and this is the growth zone (cooler zone). In between the growth zone and the nutrient zone, a baffle having a definite percent of opening is placed at an appropriate position. The positioning of the baffle and the percent of opening together play an important role in the growth kinetics. This is the most popular liner assembly in hydrothermal crystal growth. In contrast to this, in the hydrothermal growth of berlinite, the internal assembly of the liner differs from that of quartz and is shown in Figure 3.4. The liner material is usually Teflon. As berlinite has a negative temperature coefficient of solubility, the nutrient material is placed in the upper portion (cooler zone) and the seed holder is placed at the lower portion (hotter zone). In between the two zones, a baffle with a desired percent of opening is placed in addition to the vertical positioning of these liners. There are several modifications, especially in the horizontally mounted autoclaves. Jumas et al. [17] proposed the liner internal assembly as shown in Figure 3.5 for the composite gradient method of growing berlinite crystals.

Ashby et al. [18] proposed a variable-volume silver liner for high-pressure hydrothermal crystal growth. The schematic diagram of the variable-volume silver liner is

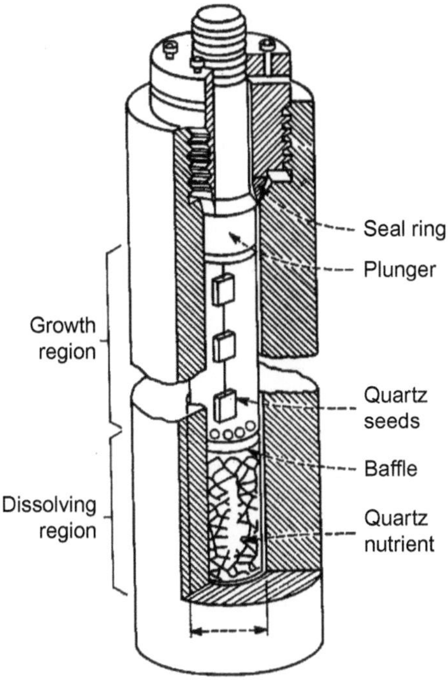

Figure 3.3 Growth of quartz crystals in the liner [16].
Source: Courtesy of R.A. Laudise.

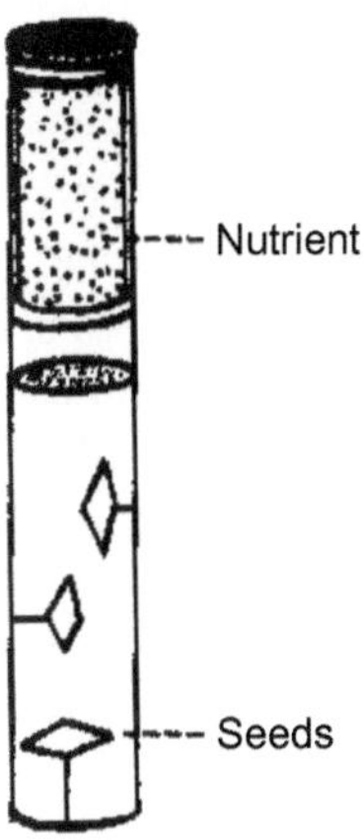

Figure 3.4 Hydrothermal growth of berlinite.

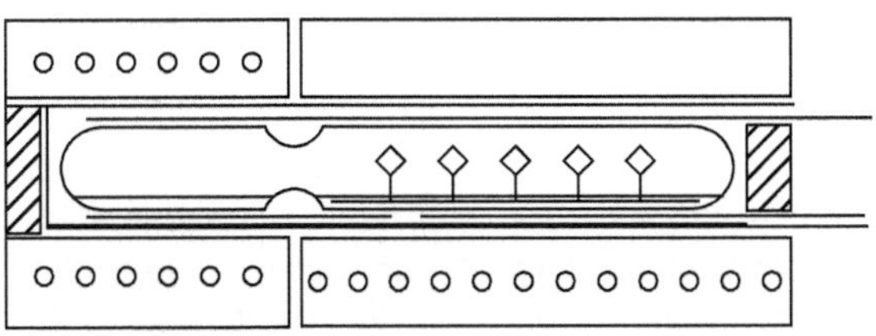

Figure 3.5 Liner internal assembly [17].

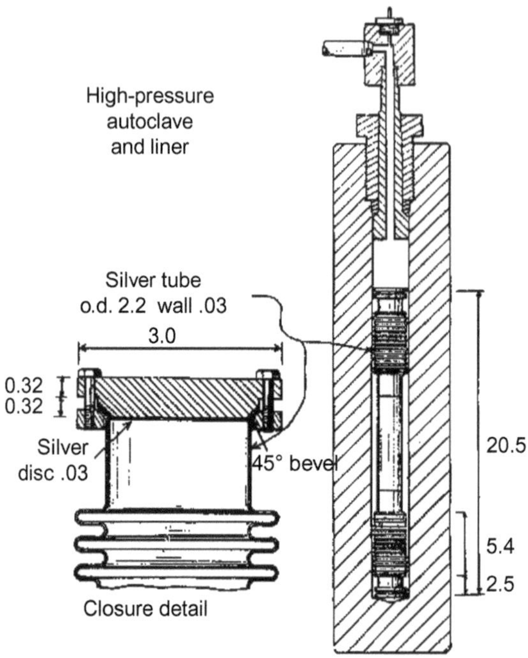

Figure 3.6 The schematic diagram of a high-pressure autoclave with a variable-volume silver liner [18].

shown in Figure 3.6. Here, silver bellows segments are formed from dead-soft cup stock by hydraulic pressure applied internally. As the tube expands outward into a die, it is folded longitudinally. The intrinsic advantage of these liner lies in the fact that the percent fill in the liner may be adjusted without opening the liner. Thus, in practice the liner is charged with nutrient, a seed, and crystal-growing medium. Such silver liners have been used by Ashby et al., successfully, hundreds of times up to 400°C for the hydrothermal growth of crystals. There are several such modifications in the liner assembly to suit specific crystal growth systems. A silver liner ladder assembly consisting of fixed and floating liners has been used in the hydrothermal crystal growth of oxides for optical applications [19]. The authors also propose that a wide variety of metals including platinum and gold (liners) can be formed into bellows. Hence, internal assembly of the liner and the lining material are very important in crystal growth and materials processing.

3.4 Temperature and Pressure Measurements

Several modern electronic controllers are commercially available today to regulate temperature and pressure in hydrothermal experiments. Pressure vessels heated externally can be placed in a suitable isothermal zone or multiple heating zone furnaces controlled by thyristor-based controllers. Highly sophisticated temperature programmers–controllers are commercially available for programming the

Table 3.6 Some Common Thermocouple Wires and Their Temperature Ranges

Wires	Temperature Range (°C)	Maximum Temperature (°C)
Chromel–Alumel (90% Ni–10% Cr) (95% Ni, 5% Al, SiMn)	−200 to 1200	1350
Pt–$Pt_{90}Rh_{10}$ Pt–$Pt_{87}Rh_{13}$ Pt–$Pt_{80}Rh_{20}$	0 to 1450	1700
Copper–Constantan (60–45% Cu, 40–55% Ni)	−200 to 350	600
Chromel–Constantan Iron–Constantan	−200 to 750	1000

Source: From Ref. [20].

experimental temperature, temperature gradient, heating and cooling rates, and so on. A variety of alloys are used for furnace windings to reach the desired temperatures in the hydrothermal range. The most common are nichrome and Kanthal. These are more durable and are less subject to damage if sheathed in magnesia insulation plus Inconel or similar high-temperature alloys. The ratio of maximum voltage over the required amperage equals the necessary resistance of the furnace winding. The necessary resistance divided by the length of the winding provides the unit length with which the furnace should be constructed. Tables of resistance per unit length with measurements for each of the available commercial heater wires are often supplied by the manufacturer. To control temperature gradients in the reaction vessel, the windings should be more closely spaced near the ends of the furnace, especially at the end where more pressure tubing is connected to compensate for the higher heating fluxes from these areas.

The temperatures in hydrothermal experiments are measured using suitable thermocouples that are classified according to the pair of metals as noble metal thermocouples and base metal thermocouples. They are made of chromel and alumel or platinum and rhodium alloys owing to their stability and resistance to oxidation throughout the hydrothermal temperature range. Table 3.6 gives the list of some commonly used thermocouple wires and their temperature ranges [19]. Thermocouples are best inserted into holes drilled into the walls at both ends of the vessel to allow temperature gradients in the vessel to be monitored. Such external temperature measurements should be calibrated by inserting a thermocouple directly into a closed bottom, hollow tube extending into the vessel. The thermocouple calibration is accomplished by measuring the elements at the melting points of reference materials. Table 3.7 shows the melting points of standards used for thermocouple calibration [20].

A computer can be used to record the experimental temperatures uninterruptedly, and it requires an analogue-to-digital converter (ADC) to transform the thermocouple millivolt signal to a digital signal to be processed and displayed by the

Table 3.7 Some Melting Point Standards for Thermocouple Calibration

Material	Temperature (°C)
Tin	231.9
Lead	327.4
Antimony	630.5
Sodium chloride	801.0
Silver[a]	960.8

[a]Calibration strongly impaired by minor amounts of dissolved oxygen.
Source: From Ref. [21].

computer. With a rack-type system of 10–20 vessels, such a system can record run temperatures, in 1 s, of each vessel several times.

The pressure in hydrothermal experiments is measured by different means. In the initial experiments in a given system, standard calibrated bourdon gauges are used and are recalibrated at regular intervals against either a dead weight tester or master gauge. Pressure inside a vessel can be monitored through a pressure transducer to separate the corrosive contents of the vessel from the pressure gauge. It is useful to have a valve between the vessel and the transducer to act as a micropressure generator to check for hysteresis in the transducer and to isolate the vessel during evacuation at the start of the run. A stainless steel or titanium capillary tube is used for the connection with the pressure gauge because of its easy installation and the small volume of trapped fluid that it contains. For routine pressure measurements, Kennedy's PVT diagram and Bain's PVT data are used because of their simple procedure [21,22]. Whether heating or cooling, identical pressure values should be observed, and differences of a few tenths of a bar indicate either hysteresis in the pressure transducer, or that a gas consuming or evolving reaction is taking place. Automatic pressure controllers and recorders are also available commercially. For example, in the Tuttle cold-cone-seal autoclaves unit (Tem-Press, Leco Corp., USA; Figure 3.7a and b), the required pressure is fixed in the gauge which will maintain the pressure at that level for a fairly long duration. However, pressure measurements with a temperature gradient have not been studied in detail, but it is certain that, under hydrothermal conditions, there can be no pressure gradient. Hence, the local temperature and local density anywhere in the system would determine the pressure of the entire system. With a computer, the experimental pressures can be continuously monitored by the EMF (Electromagnetic Field) signal of a pressure transducer (i.e., strain gauge or manganin cell).

In addition to monitoring pressure and temperature, computers offer the potential to control run conditions. As an example, the computer could check the millivoltage signal from the thermocouple of each vessel and compare the signal to that of the desired temperature. The control action could be provided through on/off relays controlling power to the furnaces. Thus, the computers can make the recording and controlling more handy in hydrothermal experiments in a racklike system.

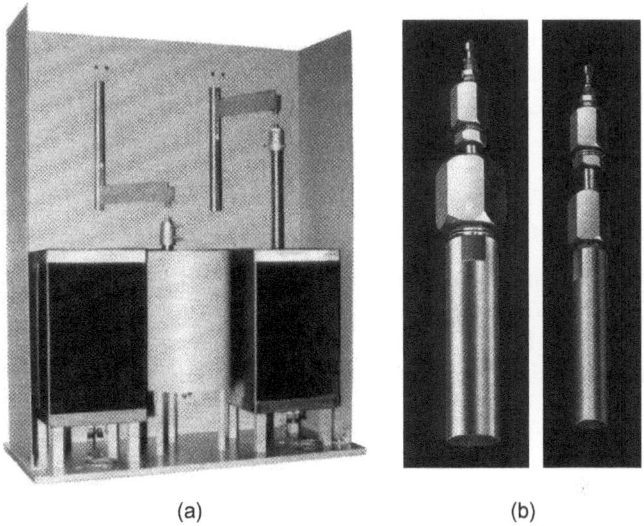

Figure 3.7 Commercially available Tuttle cone-seal autoclave system (a); Tuttle cone- seal autoclave assembly (b).

3.5 Autoclaves and Autoclave Designs

This forms the most important aspect of hydrothermal research. The advances in the hydrothermal research go in parallel with the autoclave design. The earliest autoclaves were very simple in design and generally made up of glass; the availability of suitable metals and alloys was limited. Besides, the main objective was to synthesize some compound that resembled some natural mineral. Thus, the pressure–temperature range was rather low. It was only after the first successful application of stainless steel as the autoclave material [10] that the design and fabrication of autoclaves for higher pressure–temperature limits was actually initiated. Such efforts resulted in the attempts to grow diamond and corundam crystals during the 1870s [23–25]. The actual impetus for the design of new autoclaves came during the 1940s with the development/discovery of several new alloys. The complexity in the designing of the autoclaves also began. Further, during the 1970s, the designing of autoclaves (using a suitable alloy) for specific purposes like the study of kinetics, solubility, and direct or visual observation of the hydrothermal reactions began. With the entry of scientists from different branches of science into hydrothermal research, the diversity and adaptability of hydrothermal research began to make it a more interdisciplinary science than just the study of mineral or phase equilibria and crystal growth. The materials processing under hydrothermal conditions got a new impetus. All these developments resulted in the designing of a large number of new autoclaves, which are together grouped under novel designs. Byrappa [26] has reviewed various autoclave designs in detail. Here, it is extremely difficult to discuss the development of each and every autoclave design, as the number exceeds 100. However, a majority of them are only to suit a specific aspect of hydrothermal research and hence are not versatile. Therefore, for the sake of simplicity, we have discussed only the more universal autoclave designs under the two broad classes: (1) conventional autoclave designs and (2) novel autoclave designs.

3.5.1 Conventional Autoclave Designs

These are the more universal designs commercially available for general hydrothermal research. The designing of the conventional autoclaves began in mid-nineteenth century, and simple glass vessels were the earliest members of this category. Let us discuss some conventional autoclaves one by one.

Glass Vessels

The glass vessel in hydrothermal research has advantages of ease of observation and resistance to acid solutions (except HF acid). The disadvantages are the low pressure–temperature conditions and ease of attack by basic solutions. Hence, its use has been generally neglected in hydrothermal work, mainly because the early interest concerned materials requiring basic media and pressure–temperature conditions beyond the limits of glass. When the solvent and pressure–temperature conditions permit and the visual observation enables, glass vessels have much to their recommendation and should be used more frequently in hydrothermal work. Before the design of Morey-autoclaves in 1913, the hydrothermal vessels consisted of tubes loaded with solid and water and sealed either by welding both the ends of the tube or by means of a flat washer or screw. However, such vessels are very difficult to maintain and dangerous to operate. Croxall et al. (1979) from General Electric Company, UK, have successfully grown berlinite crystals using glass vessels [27].

Sealed quartz glass tubes were first tried successfully as pressure vessels in 1964 by Speed and Filice [6]. Quartz autoclaves are good for low pressure–temperature hydrothermal experiments in silica-saturated systems. These tubes provide an inexpensive means for making simultaneous runs along an isotherm with varying bulk composition, pH, partial gas pressure, etc., within a single heater. Figure 3.8 shows

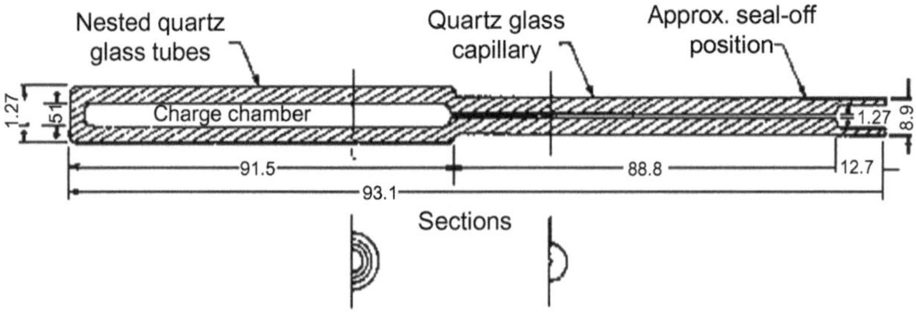

Figure 3.8 The cross sections of quartz glass tube [6].

the cross sections of quartz glass tube. The lower half or charge chamber is made by nesting three close-fitting quartz glass tubes of 1.25 mm wall thickness and appropriate internal diameters and fusing the tubes together in vacuum. Quartz glass is the most suitable composition because of its strength at high temperatures, low thermal coefficient of expansion, and absence of possibly reactive components, whereas Vycor and Pyrex both contain 3% or more B_2O_3 in addition to significant quantities of Al_2O_3 and the alkalies, and Pyrex softens at about 600°C [27]. Speed and Filice [6] give the temperature and internal pressure of quartz glass tube failure in Table 3.8.

The use of quartz glass autoclaves, however, was not popular until recently because of the problems associated with higher pressure–temperature conditions. With the recent discovery of new mineralizers (mixed solvents, high molar acid solutions, organic solvents, etc.), there is a growing tendency for using the glass or thick-walled silica tubes in several hydrothermal experiments. The new mineralizers, and a better understanding of hydrothermal technology, have brought down the crystallization temperature with increased solubility for compounds hitherto obtained only at high pressure–temperature conditions. All these advances prompted Popolitov [7] to develop and test a hydrothermal autoclave for visual examination. The schematic diagram of the visual autoclave is given in Figure 3.9. The apparatus consists of a cylindrical reactor made of transparent quartz glass with the following parameters: compression strength = 6556 kg/cm^2, tensile strength = 2734 kg/cm^2, bending strength = 1134 kg/cm^2, and shock bending strength = 1.08 kg/cm^2. These data show that the quartz glass is a good material for a transparent reactor, provided that no shocks will occur during its operation. This condition requires certain precautions to be followed during charging, discharging, and sealing of the reactor.

The reactor has been fabricated in the following way: A cylindrical block of quartz glass of special purity having a diameter of 85–1000 mm was drilled using a diamond bore with the diameter 50–55 mm. Then, the inner and outer walls were ground to the specified dimensions and polished. The hydrostatic pressure admissible for this reactor was not calculated before hand, but its strength was checked. To do so, the reactor was filled with water, sealed by a special valve, placed into a furnace, and heated. The growing pressure due to the heating was measured with a manometer. The reactor was heated until the pressure inside it achieved 100 atm, but the reactor did bear this pressure. The external heating of steel autoclaves introduces stresses on its walls due to heating. By contrast, quartz

Table 3.8 Temperatures and Internal Pressures of Quartz Glass Tube Failure [6]

Tube	Charge	Temperature (°C)	Pressure (bar)
1	Saturated NaBr solution	543 ± 7	146 ± 4
2	Saturated NaCl solution	511 ± 13	345 ± 13
3	Saturated NaCl solution	493 ± 13	323 ± 12
4	H_2O	474 ± 5	414 ± 11

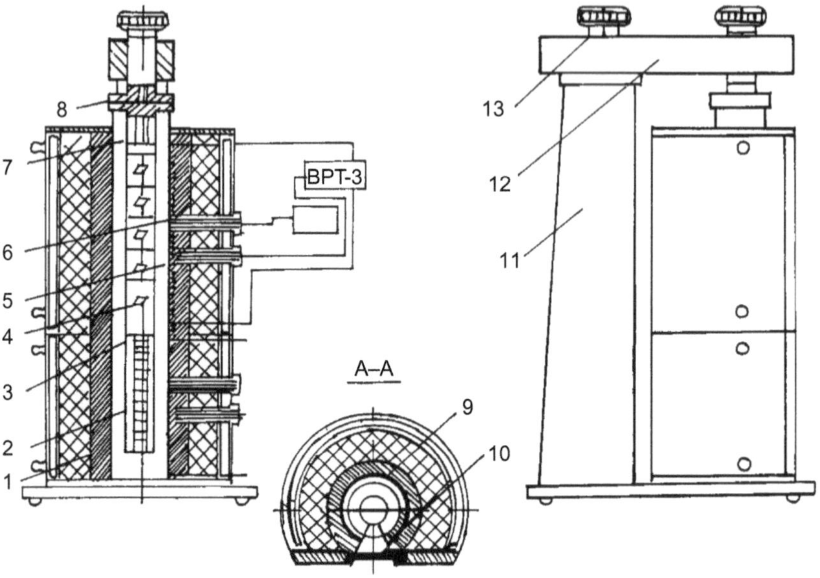

Figure 3.9 The schematic diagram of the visual autoclave [7]. (1) Quartz reactor, (2) nutrient container, (3) baffle, (4) seed holder, (5) seeds, (6) Teflon tube, (7) solid cylinder, (8) sealing disk, (9) heater, (10) thermocouple, (11) autoclave frame, (12) upper mount, and (13) adjustable nuts.

glass has a very small coefficient of thermal expansion and hence the stresses induced by temperature gradients are negligible. To make the quartz reactor tight, a valve, which meets the following requirements, should be used:

i. The valve must be self-sealing, i.e., stress in the seal must grow as the pressure increases, and no strong force that could damage the brittle reactor must be exerted on it when sealed.
ii. The sealing material must be plastic; if so, the initial stress needed for self-sealing is low and the reactor is safe.
iii. The valve must be chemically resistant with respect to the chemical agents or the solvents used.

The Teflon ring is placed in an annular groove whose diameter is greater than that of the ring. Also, the ring's outer diameter will be greater than the inner diameter of the reactor block. Hence, stresses appear in the ring due to its deformation, and this is sufficient for the initial sealing.

The main advantage of quartz autoclaves is their high stability with respect to an acid medium which permits us to use a condensed solution, i.e., the concentration of substance to be crystallized may be enhanced without increasing the temperature. In this case, a considerable gain can be obtained for the processes which run under lower temperature and lower pressure, which turns out to be cheaper and safer. The apparatus can also be used in alkali solutions and hydrofluoric acid media by providing a thin Teflon coating to the inner walls of the quartz glass. This quartz vessel can be used to synthesize and grow single crystals at temperatures up to 300°C and pressures up to 100 atm. The greatest advantage is the facility to observe all the processes taking place inside the autoclave including dissolution of the mixture, movement of the solution, and growth of the crystal. By employing the appropriate thermocouple sheath, the temperature distribution in the autoclave can be measured accurately. It is convenient to study the physicochemical processes and hydrodynamics of the solution both in the homogeneous and binary phase "gas–liquid" reagents.

Steel Autoclaves

The success of the earliest experiments in the hydrothermal growth of minerals using very simple autoclaves during the nineteenth century provided an impetus for Morey at the Geophysical Laboratory of Carnegie Institute, Washington, to design a highly versatile autoclave to work in the medium pressure range [28]. This was further developed by him and his coworkers to operate at 500 bar and 400°C [29]. Followed by this, several other designs appeared, and today there are many designs to suit any particular aspect of hydrothermal crystallization. All the conventional autoclaves can be divided into four types and are popularly known as: (1) Flat-plate closures developed by Morey and coworkers; (2) cold-cone-seal closures designed by Tuttle [30] and popularized by Roy and coworkers; (3) welded closures designed by Walker and Buehler [31]; and (4) unsupported area closures developed by Bridgman and modified by several commercial autoclave vendors [32]. These autoclaves have been discussed by many authors [2,3,33].

Flat-Plate Closure Autoclave

This is popularly known as a *Morey-autoclave* since Morey designed this simple gasketed, sealed steel autoclave of 25–100 ml capacity. It is widely used in hydrothermal research and the cross section of a typical Morey-autoclave is shown in Figure 3.10. The usual dimensions of a Morey-autoclave are 10–20 cm length and 2.5 cm inner diameter. The problems encountered by Morey and coworkers in the initial experiments were alleviated in a later design by Morey 1913, where in the closure is made by a Bridgman unsupported area seal gasket made of copper or silver or Teflon. Higher pressure–temperature conditions may be obtained with an Inconel vessel. At low pressures, sealing is achieved by compression when the main nut or closure nut is tightened. The autoclave generates an autogeneous pressure depending on the degree of filling, the fluid, and the temperature. The autoclave is limited to ~450°C and 2 kbar in routine use. In the later versions, the pressure can be directly measured and adjusted during an experiment by providing an axial hole through the closure nut, but this caused, sometimes, compositional changes as the material was transported to the cooler region. A thermocouple is inserted in the well close to the sample, and the vessel is placed inside a

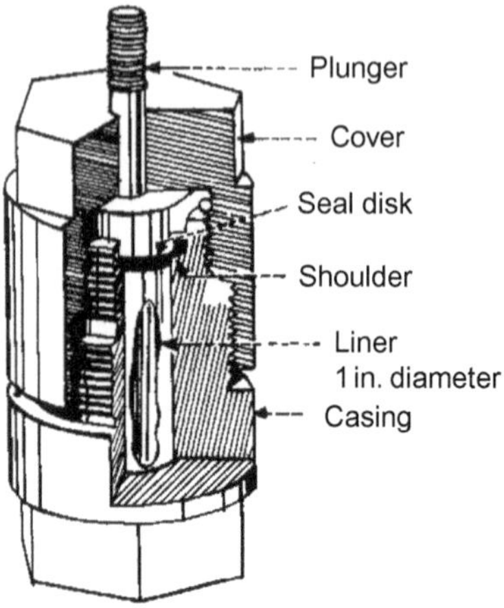

Figure 3.10 Cross section of a typical Morey-autoclave [28].

suitable furnace so that the entire Morey-autoclave and closure lie within the element of the furnace. At the end of the run, the vessel is quenched in a jet of air followed by dipping in water and the closure seal is broken.

Cone Closure

By 1948, Tuttle at the Geophysical Laboratory had designed an ingenious cone-in-cone seal for a small vessel with no threads to seize. Although this "Tuttle apparatus" could go to much higher temperatures, and the pressure could be independently controlled by a drilled lower seat, it had very small volumes, and it did not catch on.

These autoclaves are known as *cold-cone-seal autoclaves*, and also as *Tuttle autoclaves* (Figure 3.11) [34]. A *Tuttle bomb* consists of a stainless steel, low-carbon steel, or more exotic alloy such as Inconel, Udimet, or Stellite (Co alloy), test tube closed by a cone-in-cone seal. As with the Morey vessels, here also the entire vessel assembly is kept inside the furnace. The seal is closed by weights acting through heat-resistant alloys above and below the vessels. The cone at the lower end of the support rod is machined to 59°, the cone or the vessel to 60°; this results in a ring seal. At pressures exceeding 2 kbar and temperatures above 800°C, these cones have to be machined after each experiment. Pressure is transmitted to the sample through an axial hole in the lower support rod, and temperature is measured by placing a thermocouple in a well of the vessel within 3 mm of the sample. A vertical split furnace is mounted on a hinge allowing it to be swung clear of the vessel during loading or quenching. This vessel can be used up to 800°C at 4 kbar.

In 1949, Tuttle described a modification to his earlier design, which is much easier to use. This consists of a longer vessel in which the open end and seal are outside the furnace; hence the term "cold-cone seal" (although in the real sense it is far from being cold). Pressure is transmitted to the sample, which is contained in a sealed capsule, through a hole in the closure. The capsules are normally made up of noble metals. These vessels may be operated closure up, closure down, or horizontally. The ratio of the vessel diameter to the hole diameter (wall diameter) determines the

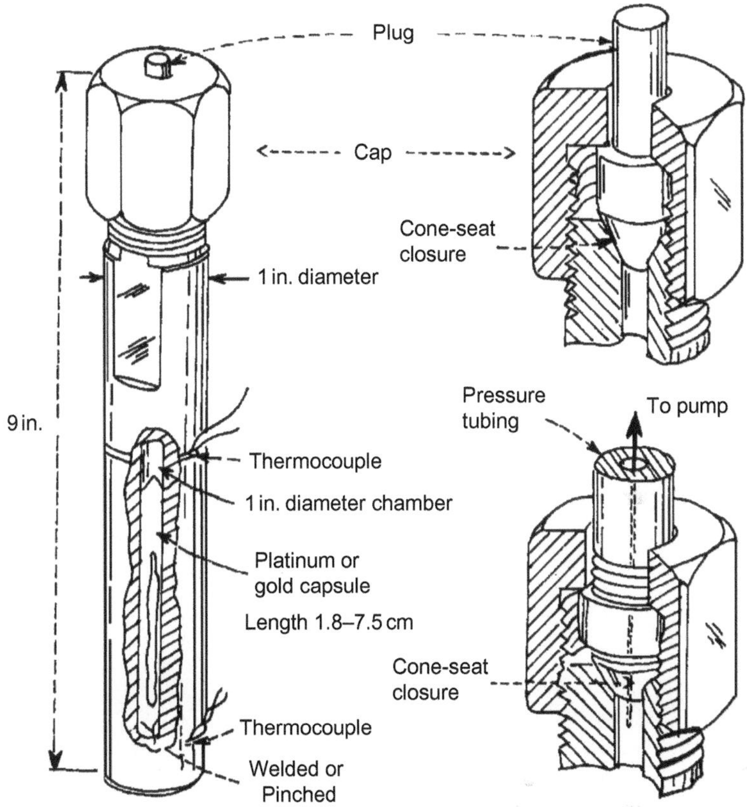

Figure 3.11 Tuttle autoclaves [34].

strength of the vessel. In most standard Tuttle vessels, this ratio is 4%. Using Stellite 25, experiments can be carried out at 900°C and 1 kbar or 750°C and 3 kbar for long-term use (from hours to weeks). Roy and Tuttle have actually made many hydrothermal experimental runs of several months in the late 1950s. Moreover, the reaction could be "quenched" by lowering the furnace quickly and surrounding the bomb with a container of water.

Around 1950, Roy and Osborn [35] had also designed a simple universal pressure intensifier for compressing virtually all gases (H_2, N_2, CO_2, and NH_3) or liquids like H_2O from the compressed gas tank pressures of about 100–200 bar to 5 kbar. And from 1950 onward, "test-tube racks" of hydrothermal vessels including such compressors were used worldwide for hydrothermal research, mainly in the geochemical community. Figure 3.12 shows the advertisement of Leco Company, Tem-Press Research Bellefonte, Pennsylvania, which has been set up by Tuttle, Roy, and Licastro to make these key tools of hydrothermal research available worldwide. This continues to the present day, mainly due to the overriding convenience and simplicity.

Figure 3.12 Advertisement of Leco Company, Tem-Press Research, Bellefonte, Pennsylvania.

Tuttle's original design underwent several improvements in closure design using new alloys which have extended the pressure–temperature range of cold-seal vessels. Luth and Tuttle [36] described a modified Tuttle-type closure which permits vessels to go up to 8 in. length, 1.25 in. o.d., and 0.25 in. i.d. to be used up to 10 kbar pressure and at temperatures as high as 750°C. Rene 41 (Ni-based alloy) is used for this vessel. The most comprehensive reviews of design and operation of cold-seal equipment are given by Huebner in 1971 [37] and Edgar in 1973 [38].

Similarly, Roy modified another vessel which had been designed by Tuttle but never exploited. It proved that Roy and Tuttle's improved version was much simpler and much more convenient than all others. This Roy modification is popularly known as the "test-tube bomb" and it is typically 1" to 0.5" o.d. × 8" in. long superalloy (Hastealloy; Rene 41, Stellite 25, or Udimet) rod with a 3/16" to 1/4" diameter hole drilled down to 1/2 in. of the bottom. It was threaded at one end and the pressure seal was made there with a 59–60° cone seat against which the connector from the pressure pump was forced by a nut and thread.

This vessel made quantum advances on two fronts. First, it raised the temperature limit to ≈800°C with special steel and 950°–1000°C with superalloys. Second, the connection to the pressure system of pumps and gauges meant that pressure and temperature were decoupled. Third, because of this simplicity, several vessels could be served by one pressure source. Hence, an integrated system of several vessels, each with its own gauge with one pressure generator became a standard hydrothermal laboratory apparatus. By sealing the reactants from inside with gold or platinum tubes, one to four of which could be simply inserted into the "test tube," one could have a simple apparatus for maintaining a precise composition, with no contamination at any chosen pressure (up to 5 kbar) and temperature (up to 900°C) condition for periods which could extend from hours to weeks. We actually made many hydrothermal runs for several months in the late 1950s. Moreover, the reaction could be quenched by lowering the furnace quickly and surrounding the bomb with a container of water.

The advent of new materials, particularly molybdenum alloys, has extended the temperature capabilities of cold-seal vessels to about 1150°C at a pressure above 4 kbar (the so-called TZM vessels). The prospects of new refractory alloys extending up to this range are encouraging. These cold-seal vessels are safe, inexpensive, simple, and operationally routine.

All the above-described autoclaves are externally heated. In Morey and modified Bridgman-type autoclaves, the pressure created inside the vessel is calculated through the PVT relation, and in only a few cases an external pressure gauge is used. In the Tuttle's cold-seal vessel, the pressure is built up by an external pump, and the pressure medium is usually distilled water with a little glycol added to inhibit corrosion. For higher fluid pressure, a hydraulic intensifier or air-driven pump is used with argon or nitrogen as the pressure medium. Normally, several vessels run from a single pump, where each vessel is connected to a high-pressure line by a valve and T-junction. The valve (two way) isolates the vessels from the line. Each vessel is connected to a separate Bourdon gauge with a safety glass dial and blowout block. An auto-controller is normally used. The types of pressure tubing, valves, and fittings used depend on the operating pressures. Normally, stainless steel alloy is used up to 13 kbar, and recommended size is 1.4 in. o.d. and 1/16 in. inner diameter. Although there has been rather little evolution in the basic design of cold-seal pressure vessels, marked changes have occurred in controllers and pumps.

Welded Closure

This is a "work horse" vessel in most hydrothermal laboratories because the disposable liner permits a good deal of exploratory work to be done in systems where corrosion might ruin even a more complex vessel. It is a high-pressure vessel designed by Walker and Buehler [31] with the temperature range up to 400°C and pressure of 3 kbar. The vessel consists of an inner liner welded at both ends and an outer tube of heavier construction. Figure 3.13 shows the construction of a welded closure [39].

At the conclusion of an experiment, one end of the welded liner is sawn off to remove the charge. The liner is then driven out of the outer casing and discarded

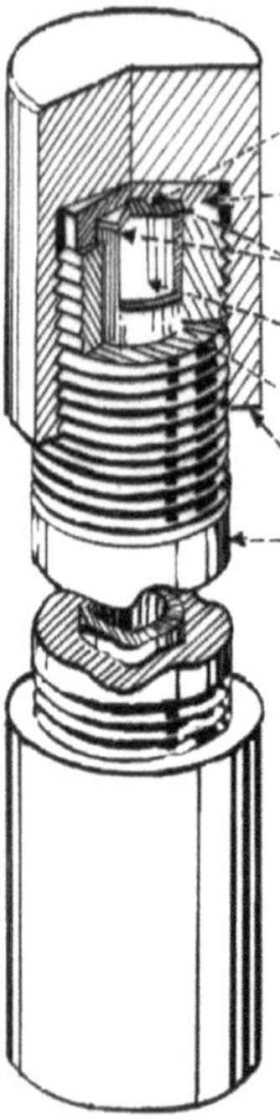

Figure 3.13 Construction of a welded closure [31].

and a new liner is inserted. Thus, for a sequence of experiments, rather extensive machine shop services are needed. However, welded liners are still relatively inexpensive when compared with an entire vessel.

Modified Bridgman Autoclave

A further advancement in pressure–temperature capabilities, size, and design among the steel autoclaves is the modified Bridgman high-pressure autoclave. This is embodied in several commercially available autoclaves. Figure 3.14 shows a cross section. The initial seal that closes the vessel is made by mechanically tightening the plunger against the deformable gasket (Figure 3.15 shows a deformed copper ring) by the setscrews in the head. The unsupported area closure makes its seal. The force of the generated pressure acts upon the piston, moving it upward against the deformable gasket and creating a perfect sealing. Such vessels can be used quite successfully in the temperature of 500°C and pressure of 3.7 kbar. Construction materials such as stainless steel or Inconel are chosen according to the requirements of the particular experimental conditions. For general applications, quartz is commonly grown under hydrothermal conditions directly in steel autoclaves. Autoclave attack is minimal because the common mineralizers react with iron to form sodium iron silicates, which adhere to the autoclave wall and inhibit further corrosion. The iron content in the solution is so low that Fe in the grown quartz crystals is typically only a few parts per million so that crystals of excellent acoustic quality and watch grade quality can be obtained [15,38]. However, almost all other hydrothermal crystals require

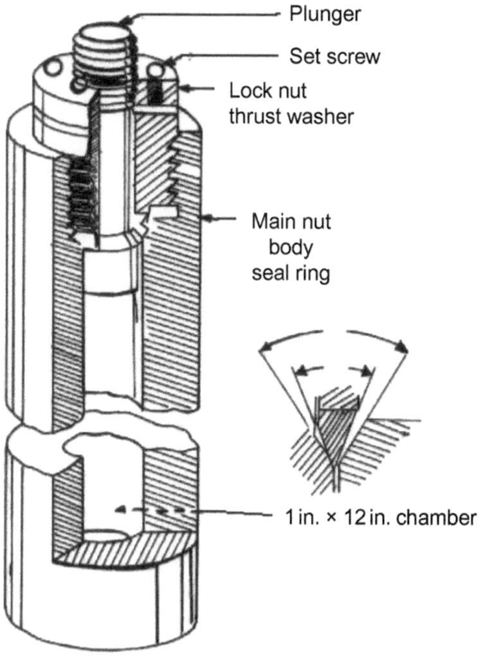

Figure 3.14 Modified Bridgman high-pressure autoclave [16].
Source: Drawing courtesy of R.A. Laudise.

Figure 3.15 Deformed copper ring.

growth in a low-carbon steel or mild steel liner or noble metal or copper liner. The pressure inside the liner is counterbalanced with pressure between the liner outer wall and inner wall of the autoclave. A required amount of the nutrient material and the solvent is introduced into the liner with a definite percent fill. Then the pressure inside the autoclave is calculated through PVT relations. The modified Bridgman vessels are commercially used for crystal growth and they are further modified to grow very large single crystals of industrial importance [40]. Sealing is the most important aspect in these autoclaves, and it consumes much time and effort in the initial runs.

Table 3.9 Historical Development of Large Autoclaves [42]

Installation	1959	1963	1965	1973	1984
Sealing type	Flange	Modified Bridgman		Grey-Loc	
Inside diameter (mm)	120	180	300	400	650
Depth (m)	2	3	5	8	14
Volume (l)	22	76	353	1005	4650
Output (kg)	7	25	120	450	3000

Most of the above closures, which are called *self-energized closures*, have sealing gaskets generally made of plastic materials and, most frequently, made of annealed copper. At temperatures below 300°C Teflon can be used. If the closure is provided with a cooling system, rubber gaskets are used and the sealing of such autoclaves does not require as much mechanical effort as with a copper gasket. In fact, much of the success in hydrothermal growth experiments is due to the equipment. A dramatic advance in upgrading the capacity of the autoclave (e.g., as in the case of quartz growth) is the result of new sealing systems from the original flange-type seals to the modified Bridgman seal which in turn has been superseded by Grey-Loc seals [41]. Large-capacity autoclaves of 1000–5000 l volume are in use in Japan with this type of sealing system. Table 3.9 shows the evolution of sealing with reference to the growth of quartz crystals [42]. In most hydrothermal experiments, corrosive solutions are used and special protective inserts are used nowadays to prevent contamination of the growth zone by corrosion products. A floating-type insert or variable-volume insert has been developed at the Institute of Crystallography, Moscow. Such an insert occupies only a part of the autoclave interior while the remaining space is filled with water. These inserts are usually made of carbon-free iron, copper, gold (for alkaline media), titanium, platinum, various grades of glass, molten quartz (for acid media), Teflon, and so on.

Laudise et al. [43] have proposed a pressure balance under hydrothermal conditions. The pressure–temperature relations in the can and in the space between the can and the autoclave system are critical. The water fill is chosen so that the can is usually under compression when the autoclave is heated to the desired temperature since the can is less likely to fail under compression. The goal is to have pressure balance at the operating temperature (or to have the can under slight compression). An unbalanced pressures result in can cause dimples, bubbles, and often ultimate failure. Dimples can restrict flow, crush crystals, and cause imperfections in crystals if they grow against the dimpled wall. To understand the balance strategy examination of H_2O and mineralizer solution, PVT data is essential. The authors have studied this while growing $K(TiO)PO_4$, a nonlinear optical material, in pressure-balanced gold cans and suggest several strategies for pressure balance in order to avoid dimple formation. Figure 3.16 shows the minimized volume between the autoclave and noble metal lining. Usually, autoclave bottoms

Figure 3.16 The minimized volume between the autoclave and the noble metal lining [43].

are hemispherical and can bottoms are flat. A hemispherical metal "space filler" can be used to fill this volume. Similarly, a readily deformed region in the liner (a bellows, a soft metal section, or a thinner-walled separate portion) is used where deformation will preferentially take place and will be of less importance. If the deformation is large during cooling, rapid cooling is advantageous as long as thermal shock does not crack large grown crystals. Such studies are very important in the self-energized closures.

All the above-discussed conventional autoclaves are externally heated ones. Though these autoclaves enjoy a great advantage of simplicity and convenience, they are limited with reference to temperature and pressure by the high-temperature strengths of alloys used. In contrast to this, another group or class of equipment in which the pressure vessel is cooled and the sample is heated internally had been designed by several workers. We shall discuss one or two such designs. In this category, one of the earliest designs was proposed in 1923 by Smyth and Adams [44] and was modified later in 1931 by Goranson [45] and again in 1950 by Yoder [46]. This type of vessel is capable of attaining an experimental temperature as high as 1600°C and pressure >10 kbar. Thus, much of the later geochemical research has been carried out worldwide using such apparatus. Figure 3.17 shows the schematic diagram of the internally heated high-pressure apparatus. It consists of a heavy-walled, heat-treated vessel of tool steel (usually 41–50 Cr–Mo), which contains the pressure, and an internal furnace and sample support. This vessel is admirably suited for the study of reactions by dynamic methods like differential thermal analyses at a series of pressures. In Yoder's modification, it is possible that a temperature of 2000°C may be attained although no data have been recorded at temperatures above 1500°C. It cannot be used directly with water as the pressure transmitting fluid, but by sealing charges containing water in the platinum tube. Studies of such systems have also been made up to these temperatures [45].

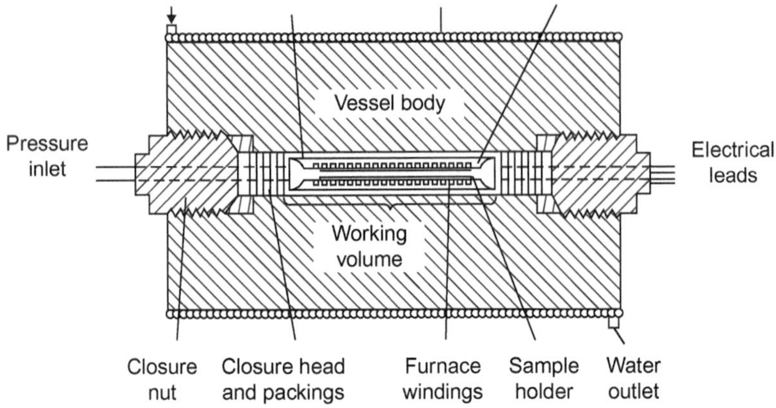

Figure 3.17 Schematic diagram of the internally heated high-pressure apparatus [38].

However, vessels of this type are both expensive and cumbersome for routine and small-scale hydrothermal research.

Similar to the above-discussed high-pressure autoclaves, there are many other designs, but most of them are for dry systems and also for growing crystals from flux in high-pressure medium. For example, the belt apparatus, piston cylinder apparatus, opposed anvil or diamond anvil systems, or high-pressure autoclave system for flux growth [47–51]. However, this equipment is more useful for the geological sciences than for crystal growth or materials processing and hence we do not discuss these designs in this book.

3.5.2 Novel Autoclaves

The novel designs of the autoclaves are for specific studies such as controlled growth, kinetics, solubility, PVT studies, hydrothermal–electrochemistry, conductivity of electrolytes, visual examination, HHP, microwave-plus-hydrothermal, and sonar-plus-hydrothermal. The actual motive behind the designing of such novel equipment is that, in the hydrothermal method, little information is available as to the processes taking place in the autoclave and also it is extremely difficult to observe these processes inside metal reactors or autoclaves. Hence, several earlier workers called the hydrothermal vessels/reactors as “black boxes.” Some of these novel designs have come from the Special Construction Bureau, Institute of Crystallography, Moscow, during the 1970s. It is very difficult to discuss each and every design in this book, as there are too many novel autoclave designs. Therefore, we shall restrict our discussion to only a few popular and unique designs.

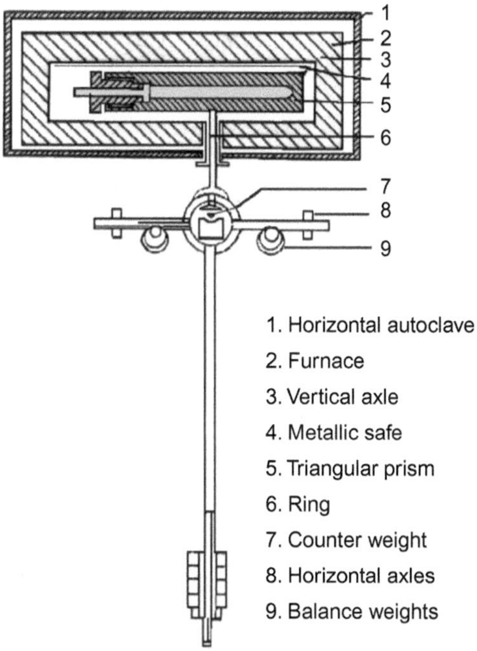

Figure 3.18 Fundamental arrangement of horizontal autoclaves [52].

Horizontal Autoclaves for Controlled Growth of Crystals

This autoclave is highly sensitive to the displacement of mass from the nutrient zone to the growth zone through the arrangement of a highly sensitive balance system. Figure 3.18 shows the fundamental arrangement. The main part of the unit is a horizontal autoclave mounted at the center of gravity on the vertical axle. The autoclave freely swings in the furnace and the deflection of the axle from the vertical indicates a shift of the mass inside the autoclave from one side to the other. Special weighing scales make it possible to estimate the amount of displaced material, and additional weights help to return the autoclave to a horizontal position.

Thus, by carrying out experiments in the balance system, one can obtain continuous information regarding the motion of mass in the autoclave and know when homogenization of the medium sets in. Also, information regarding whether recrystallization of the material is taking place, or at what velocity the recrystallization is proceeding can be obtained. If crystals are being grown on seeds with a prespecified area, it would be possible to calculate the rate of growth of the crystals and to regulate this by varying the temperature difference between the dissolution and growth zones.

Shternberg [52] designed and fabricated this autoclave. He has also standardized this system to study the spontaneous crystallization of sphalerite, recrystallization of sodalite, solubility isochore of potassium niobate crystals, and so on.

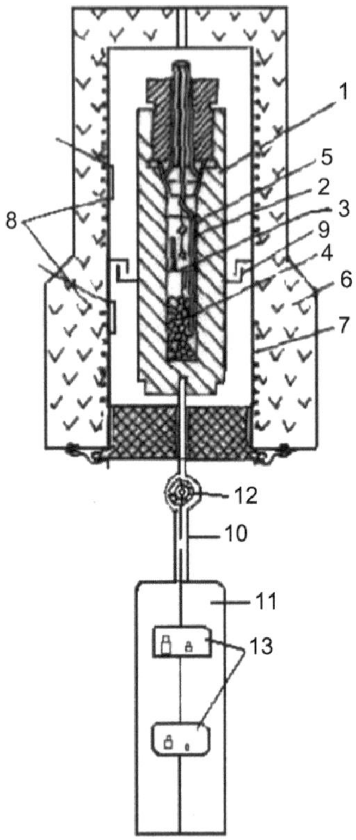

Figure 3.19 Vertical autoclave mounted in the device [52]. (1) Autoclave, (2) thermocouple, (3) baffle, (4) nutrient, (5) seed holder, (6) furnace, (7) electric winding, (8) two sets of thermocouples, (9) bellows-type barrier, (10) axle, (11) counter weight, (12) roller bearing, and (13) two balance shelves.

Pendulum Apparatus

Mass transfer of material under hydrothermal conditions may be determined continuously or periodically from the motion of the center of gravity of a vertical autoclave mounted in the device shown schematically in Figure 3.19 [52]. The oscillations of the autoclave are recorded and counted by means of a photorelay and pulse counter. Crystal growth can be monitored by observing the change in the oscillation period of the system. If the crystals are growing, the center of gravity of the autoclave will move upward, and the oscillation period of the system will increase. The pendulum apparatus designed by Shternberg is highly useful for kinetic studies.

Hydrothermal Reaction Cell for Kinetic Studies

Recently, Clark et al., and Evans et al. [53,54] developed a large volume reaction cell for kinetic studies using energy-dispersive powder diffraction. This apparatus offers a powerful method of measuring reaction rates for processes that involve crystalline materials. In addition, this helps to carry out multiphase quantitative analysis on a series of powder diffraction spectra collected during the course of a chemical reaction to determine the amount of each crystalline component as a function of time [55]. This has several advantages when a number of processes are occurring simultaneously as the powder method can determine the amount of each phase and the rate at which it takes place, in a mixture of phases. The hydrothermal cell consists of a stainless steel container with a Teflon lining. The top of the cell is connected to a control manifold, which contains a pressure bleed valve, a bursting disk assembly, and a pressure transducer which is used to measure the pressure inside the cell. The cell operates up to 40 bar at a maximum temperature of 230°C and is protected by a thin metal foil-bursting disk, which bursts at pressure beyond 65 bar. The sample holder is a cylinder with a volume of 30 cm^3. The cell wall thickness is 3 mm and is reduced to 1 mm at the beam entry and exit points. Figure 3.20 shows the schematic diagram of a hydrothermal reaction cell.

PVT Apparatus

Several designs have been made for PVT measurements in highly corrosive systems. These devices usually consist of standard pressure manometers. Kolb and Laudise [56] have designed a PVT apparatus (Figure 3.21) for highly corrosive fluids (e.g., berlinite systems). Also, this has been used to measure the pressure in the H_2O–SiO_2–NaOH system. The pressure takeoff is lined with platinum and connected to a water capillary of small capacity which goes to the pressure gauge. Several other designs have been proposed by the Special Construction Bureau, Institute of Crystallography, Moscow.

Apparatus for Solubility Determination

The solubility of substances under hydrothermal conditions is fundamental information concerning their crystal growth. Solubility measurements are usually carried out by the weight loss method. This approach is effective only for compounds having positive temperature coefficients of solubility. For compounds with negative solubility coefficients, the possibility of further weight loss of the specimen due to increased solubility during the quench, or even at room temperature, must be considered [57]. In such cases, solubility measurements are carried out in specially designed autoclaves. However, the solubility and phase-equilibria studies are usually carried out in Tuttle cold-cone-sealed autoclaves or small autoclaves without a pressure gauge. Sometimes, other available autoclaves are also used if the quenching can be carried out effectively. The quenching ensures the maintenance of conditions in the autoclave corresponding to equilibrium conditions at test temperature. The Special Construction Bureau, Institute of Crystallography, Russian Academy of Sciences, Moscow, Russia, has designed several solubility measurement apparatuses consisting of single autoclave or multiple autoclaves (by creating a liquid thermostat). Figure 3.22 shows an assembly for studying solubility simultaneously in two autoclaves [58].

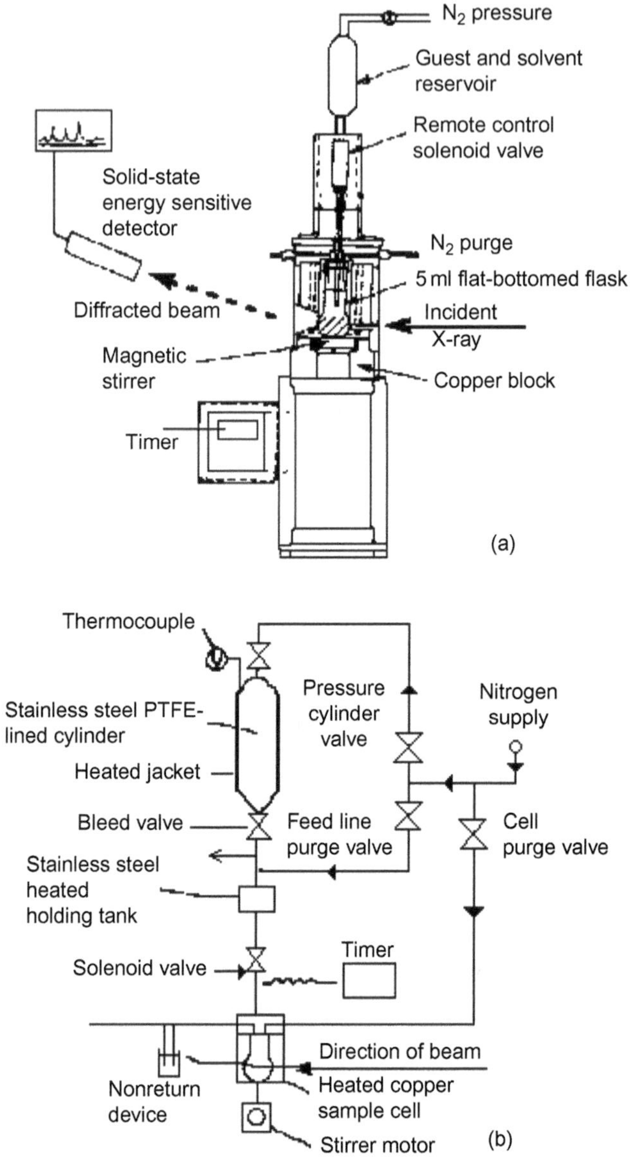

Figure 3.20 The schematic diagrams of : a) hydrothermal reaction cell designed by Clark et al. [53]; b) hydrothermal reaction cell designed by Evans et al. [54].

More recently, Yanagisawa et al. [59] designed an autoclave for solubility measurement. This autoclave has a relatively simple design, but gives accurate solubility data under relatively mild hydrothermal conditions in which the vapor phase coexists with the liquid phase. The method basically consists of separation of the

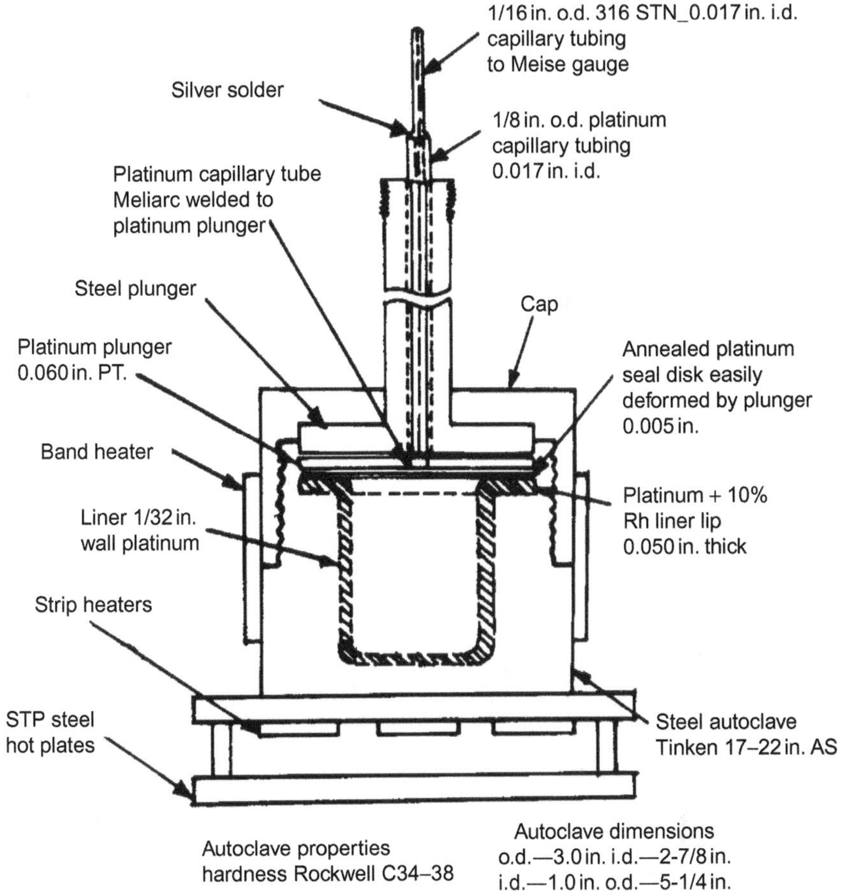

Figure 3.21 PVT apparatus [56].

solid specimen from the solution under hydrothermal conditions in order to avoid regrowth or further weight loss during quenching. Figure 3.23 shows the schematic diagram of the autoclave designed for solubility measurements by Yanagisawa et al. [59]. The autoclave was fabricated using stainless steel with a Teflon liner. The chamber was separated into two parts by platinum mesh. A solution was transferred to the bottom of the chamber, and crystals were placed on the platinum mesh. The position of the platinum mesh and the volume of the solution were selected so that the crystals did not have contact with the solution under hydrothermal conditions for solubility measurements. The autoclave was connected to a rotation shaft in a hot air oven with an electric fan to minimize the temperature difference in the oven. The contents of the autoclave are agitated slowly, and this helps in establishing equilibrium. About 3 min are needed to

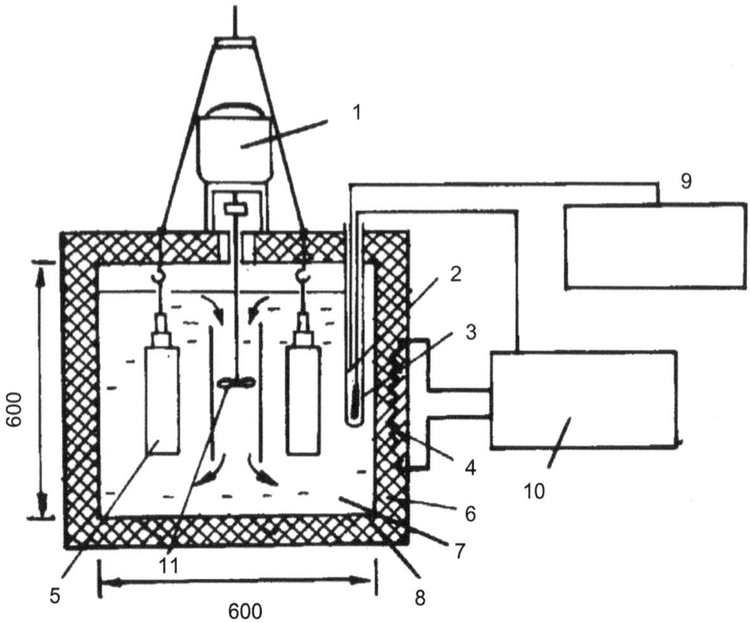

Figure 3.22 An assembly for studying solubility simultaneously in two autoclaves [58]. (1) Motor, (2) thermocouple, (3) sensor, (4) heater, (5) autoclave, (6) body of furnace, (7) salt solution, (8) thermal insulation, (9) potentiometer, (10) electronic regulator, and (11) stirrer.

complete the separation of the crystals from the solution, and then it must be cooled to room temperature by an electric fan or cooling fan. The crystals are separated from the solution under the hydrothermal condition vapor phase during cooling. The solubility calculated from the weight loss of the crystals must be accurate as the crystals do not dissolve in water vapor, because this method totally hinders the regrowth of crystals with a positive temperature coefficient of solubility, and redissolution of the crystals with a negative temperature coefficient. Using this autoclave, the authors have accurately determined the solubility of berlinite in phosphoric acid and calcinite in ammonium chloride solution and their results match well with the earlier published data [60–62]. Figure 3.24a and b shows the solubility data for berlinite.

Rocking Autoclaves

These are more versatile designs for carrying out a variety of hydrothermal experiments including measurements of mineral and gas solubilities, mineral phase stabilities, kinetics of mineral transformations, stable isotope fractionation factors, experiments involving crystal growth, and the formation of synthetic fluid

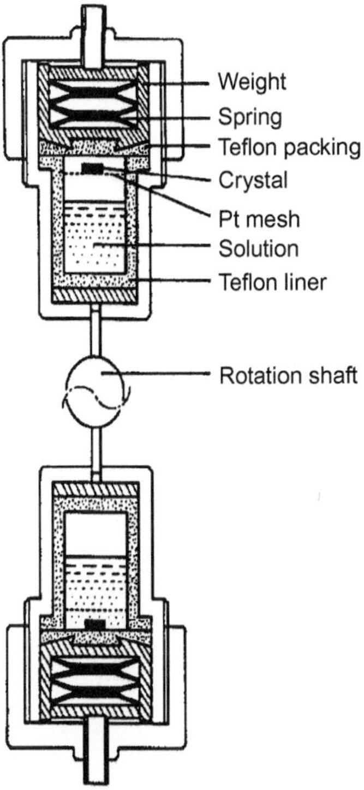

Figure 3.23 Schematic diagram of the autoclave designed for solubility measurements [59].

inclusions. Some of these experiments are not only hazardous but also have errors in applications and equipment failure. Figure 3.25 shows a schematic diagram of a rocking autoclave, which can be used for various types of studies [63]. These autoclaves are made of stainless steel or other useful alloys and usually contain self-energized closures. There can be two types of rocking autoclaves: (1) fixed-volume systems and (2) flexible-cell systems. These systems are placed in a furnace that rotates on an axis of 180°, approximately 10 times each minute. The purpose of rocking the pressure vessel is for temperature stability and homogeneity as well as to mix the contents of the reaction cell, thereby accelerating chemical exchange between solids and fluids and the attainment of equilibrium [15]. These rocking autoclaves provide an extremely reliable and effective means of conducting hydrothermal experiments under supercritical conditions [64]. A reader can get more information on rocking autoclaves in Refs. [65] and [66].

Multichamber Autoclave

Multichamber autoclaves are very useful in carrying out hydrothermal experiments with separately placed charge nutrients and offer some advantages in the synthesis of special materials. The reaction products occur through ion diffusion. Yanagisawa et al. (1997) have designed and tested these multichamber autoclaves

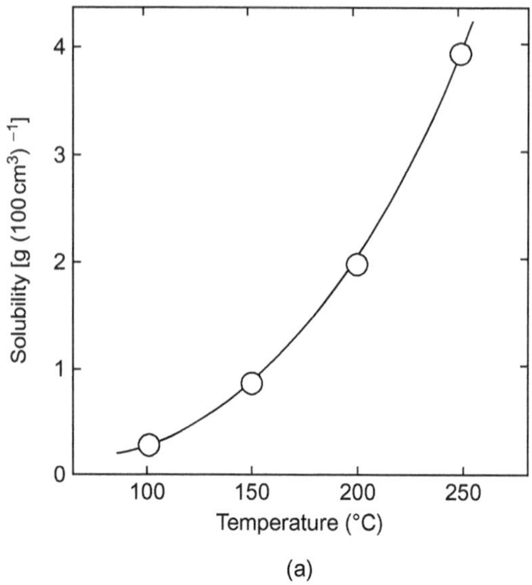

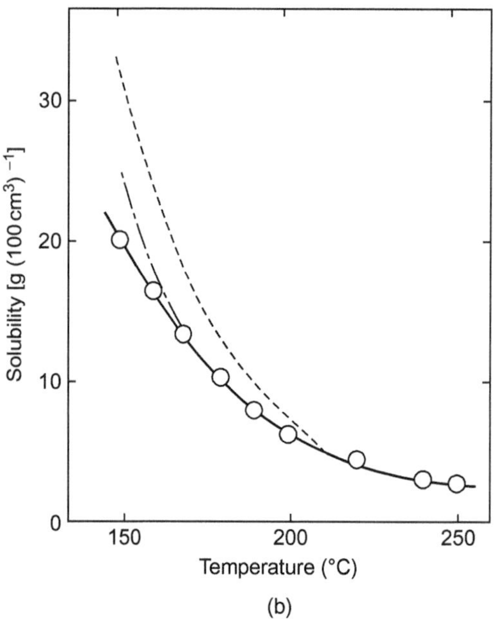

Figure 3.24 The solubility data for berlinite a) data from [60]; b) data from [61].

for the synthesis of xonotlite whiskers by ion diffusion. Figure 3.26 shows the schematic diagram of the multichamber autoclave [67]. It is made up of stainless steel with a total inner volume of 50 cm^3 and has two separate chambers each with volume 12.5 cm^3. The authors placed silica and calcium source at the bottom of each chamber and poured decarbonated distilled water into the autoclave so as not to suspend the raw materials. The silica source was brought into contact with the

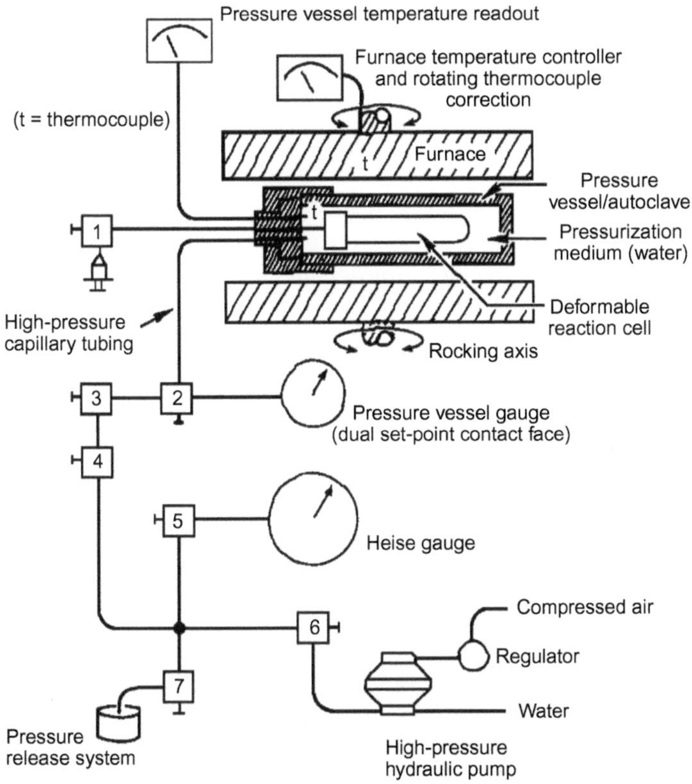

Figure 3.25 Schematic diagram of a rocking autoclave [63].

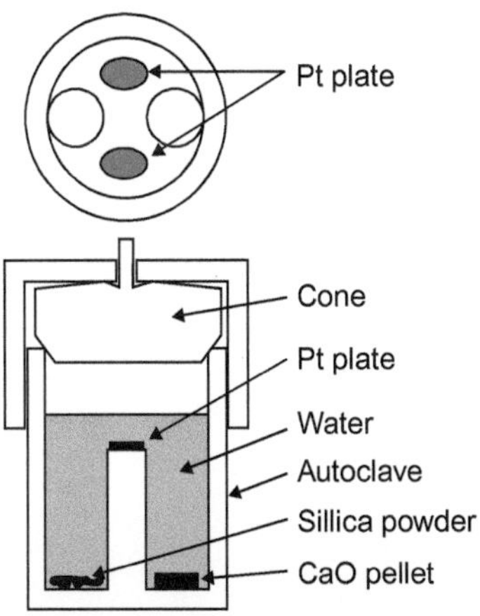

Figure 3.26 Schematic diagram of the multichamber autoclave [67].

calcium source via water; thus, ion species dissolved from the two sources diffuse into each other in the autoclave to produce xonotlite (hydrated calcium silicate). Platinum plates are placed in the autoclaves to collect the reaction products formed by ion diffusion. The design must have originated from the solution growth and gel growth techniques. This autoclave can work up to a temperature of 250°C and a pressure of usually <100 bar. However, this design is very useful for the synthesis of some special compounds through ion diffusion.

Autoclaves for the Sampling of Hydrothermal Fluids

The "black box" nature of the hydrothermal system hinders the ready understanding of the hydrothermal crystallization processes taking place in a given set of conditions in an autoclave. This is one of the reasons for the slow raise in the popularity of hydrothermal research, especially in the previous century. The earliest autoclave designs did not have any provisions for the withdrawal of the sample or fluid from the autoclave during the experimental run. As the experiment progresses, the solvent–solute interaction yields several metastable and stable phases, accompanied by a continuous change in the characteristic properties of the media, e.g., the pH and redox potential. It is possible to understand these changes only if we can withdraw the hydrothermal solution from the autoclave and subject it to several studies, or by introducing a desired gaseous pressure in the autoclave through an external pressure source. Only in the last few decades have such provisions been made in the autoclave designs, and they are commercially available. Figures 3.27–3.29 show the commercially available autoclaves with facilities to draw out the sample or fluid. These are popularly known as *Start Parr autoclaves* (Japan and USA) [68] and *Berghoff autoclaves* [69]. Also some Japanese manufacturers have come out with similar designs [70,71]. Several configurations of valves permit the sampling of

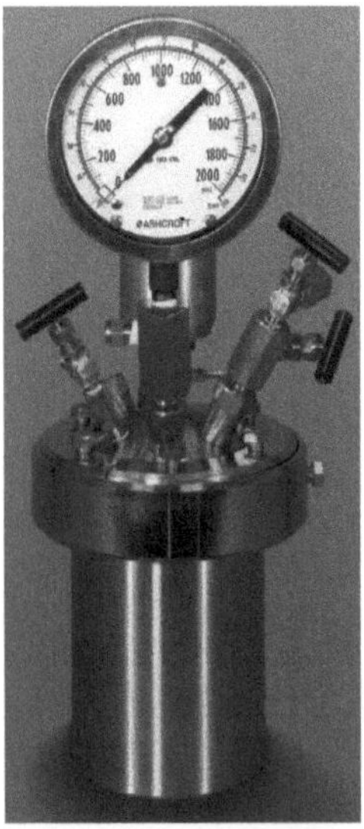

Figure 3.27 Commercially available autoclaves.
Source: Photograph courtesy of Parr Instruments Co., IL, USA [68].

fluids from hydrothermal vessels. The hydrothermal experiments in these autoclaves permit us periodic sampling at constant temperature and pressure, and total fluid chemical analysis. If mineral buffers are used to constrain the activity of pertinent aqueous species, these species should, if at all possible, be monitored directly to determine the system and provide a means to evaluate buffer reactivity and equilibria. Such studies have been carried out with excessively dissolved O_2, and with excessively dissolved H_2 and CH_4, to evaluate reaction progress and equilibria in the system Fe_3O_4–Fe_2O_3–S–C–NaCl–H_2O at 300°C and 500 bar. The dissolved concentrations of several redox-sensitive species were periodically monitored [66].

It is important to measure the total weight of the solution withdrawn from the vessel, including both liquid and vapor phases. Partitioning of water from the liquid to the vapor phase can significantly change the effective concentration of dissolved solutes from their initial room temperature values. For example, a solution of 1 M NaCl at room temperature would increase 10% in effective concentration to 1.1 M, in a vessel at 350°C, half-filled with liquid, because of the water partitioned out of the liquid into the gas phase. After each sample is taken, the salt and gas concentrations must be recalculated for the effect of partitioning into the increased vapor

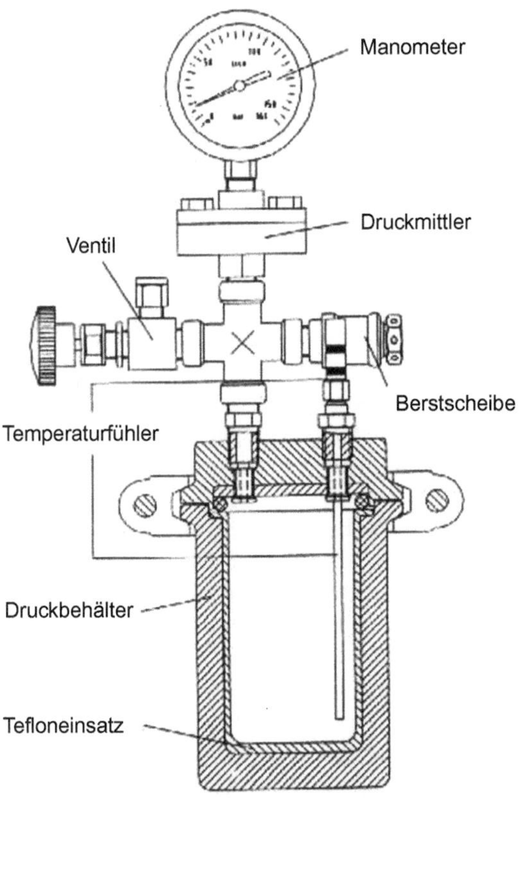

Figure 3.28 Commercially available autoclaves.
Source: Courtesy of M/S Toshin Kogyo Co. Ltd., Japan.

Figure 3.29 Commercially available autoclaves.
Source: Courtesy of M/S Toshin Kogyo Co. Ltd., Japan.

volume, for the temperature of the next sample. As the whole procedure is highly complicated and cumbersome, the sampling studies are done only in special cases, when required. This is extremely useful in the solution speciation calculations to know the regular changes in the chemical equilibria, pH, and so on.

Hydrothermal Autoclaves for Electrical Conductance Measurements of Electrolyte Solutions

Although the electrical conductance of the dilute, aqueous electrolytes, was highly desired in hydrothermal research in the early days, it was not popular among hydrothermal researchers owing to the lack of suitable autoclaves to work with such provisions under hydrothermal conditions. The electrical conductance measurements determine the extent of ionic behavior versus the association of a dissolved aqueous electrolyte on the basis of conductance theory. It was not until early 1950s that Fogo et al. [72,73] developed an apparatus and studied the conductivity of aqueous NaCl solutions at temperatures up to 400°C and pressures to 300 bar. They were the first to obtain the values of ionization constants of NaCl as a function of density in the supercritical region. Followed by this, Prof. E.U. Franck from Gottingen, Germany, carried out the most pioneering research work from 1950s to 2000 [74–76]. The other notable contributions are by T.M. Seward from ETH, Zurich [77,78], W.L. Marshall from Oak Ridge National Laboratory, USA [79,80], Ryzhenko and Smolyakov of Russia [81,82], and Helgeson of USA [83,84], who not only popularized this field but also designed some novel autoclaves for such studies. Figures 3.30 and 3.31 show the schematic diagrams of the interior of a conductance cell, and the conductance cell and electrodes developed by Quist and Marshall which are in use at the Oak Ridge National Laboratory, USA, for studies up to 800°C and pressure to 4 kbar. Though this vessel was initially designed by

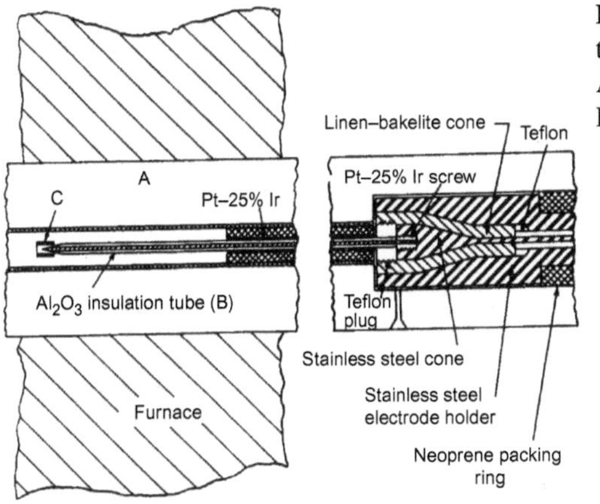

Figure 3.30 The schematic diagram of the interior of a conductance cell [85]. Actual conductance cell is shown in Figure 3.31.

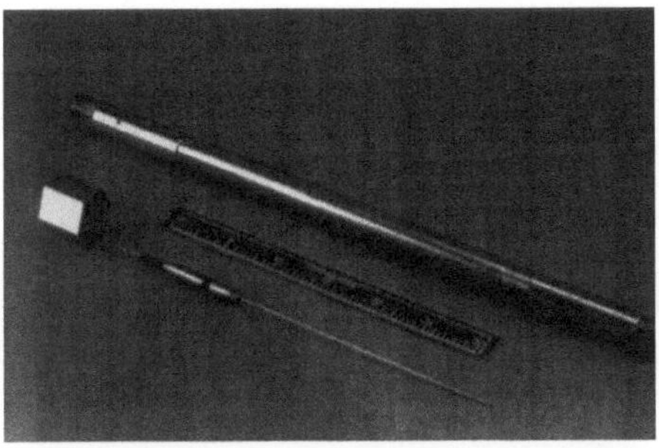

Figure 3.31 A conductance cell [85]. Schematic diagram is shown in Figure 3.30.

Franck in 1956 [74], later it was improved greatly by Quist and Marshall during 1968 [85]. The cell assembly is made of 61 cm long platinum—25% iridium-lined pressure vessel with a platinum–iridium wire running from one end of the vessel through a 30.5 cm long aluminum oxide tube to a platinum–iridium cylinder in the center. The wire passes out of the vessel through a Bakelite (a trademark of Union Carbide Corporation) insulator. The linings of the vessel and the small cylinder are both electrolytically coated with "platinum black" (a mixture of fine-grained platinum and platinum oxide), and both linings act as two electrodes [86].

Hydrothermal–Electrochemical Method

The hydrothermal–electrochemical method is an attractive technique for the synthesis of several complex oxide thin films, including the perovskite type. The method makes positive use of the reaction between species included in the electrolytic solution and the substrate being used as an electrode of the electrolytic cell under hydrothermal conditions. For example, perovskite-type oxide, ABO_3, has to be synthesized on an electrode of B-site metal being oxidized anodically in an alkaline solution containing an A-site element at lower temperatures (typically, 100–200°C) than those needed for conventional methods. Hawkins and Roy [87] introduced electrodes into a hydrothermal bomb and studied the influence of an electric field on the synthesis of kaolinite. They showed a very substantial effect (>100°C) of lowering the crystallization temperature of kaolinite. Yoshimura popularized this concept by extending it to the synthesis of electroceramics, particularly perovskite-type oxides [88–90]. Basca et al. [91] reported its advantages and disadvantages compared with a pure hydrothermal method, and Bendale et al. [92] investigated the thin film formation at lower temperature (29–100°C). Yoshimura's group has worked extensively on this and successfully prepared perovskite-type titanates such as $SrTiO_3$, $CaTiO_3$, and $BaTiO_3$ thin films up to several micrometers in thickness, almost linearly, at fairly lower temperatures (<90°C). They have also studied the growing interface, the transport mechanism of the atoms across the growing film without concentration gradients.

Figure 3.32 shows the schematic diagram of the autoclave designed in Yoshimura's laboratory at the Tokyo Institute of Technology, Japan, for hydrothermal–electrochemical reactions, using the three-electrode technique of Bard and Faulkner [93]. This has a provision to measure the potentials of electrodes in aqueous solutions at high temperatures and high pressures on a thermodynamically meaningful scale, regardless of the electrolysis conditions [93–95]. The preprocessed Ti substrate and the platinum plate were suspended as the working electrode (anode) and the counter electrode (cathode), respectively, by 0.5 mm diameter wires of the same metal keeping an interval of 30 mm between them in the electrolytic cell containing 500 ml of the solution. The Ti electrode is kept at a desired potential until a prescribed quantity of electricity passes through it. The Ti

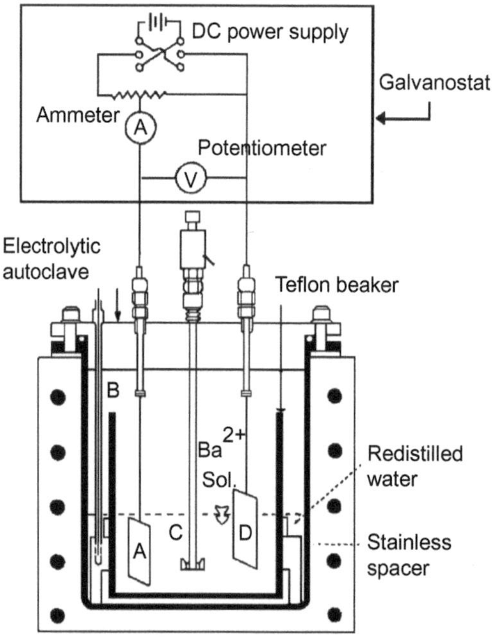

Figure 3.32 Schematic diagram of the autoclave designed by Yoshimura's laboratory [88].

electrode potential, the electrolysis current, and the quantity of electricity passed through the Ti electrode were recorded as functions of the electrolysis time to investigate the effect of oxygen evolution at the Ti electrode on the microstructure of the growing films in a series of experiments. The pressure of the electrolytic cell was controlled at higher values than the saturated vapor pressure. Similarly, the galvanostatic electrolysis can be used to control the rate of oxygen evolution.

Autoclave for Ammonothermal Synthesis

The first investigations on metal ammonia solutions were carried out by Weyl in 1864 [96] and Seely in 1871 [97]. It is well known that alkali and alkaline earth metals (except Be), as well as lanthanides like europium and ytterbium, dissolve in liquid ammonia under normal pressure. These interesting metal ammonia solutions have inspired many preparative chemists to synthesize materials in ammonia solution under hydrothermal conditions. The reaction products obtained in liquid ammonia under normal pressure are mostly microcrystalline or even amorphous. But this problem can be solved by working with supercritical ammonia as both solvent and reactant at pressures up to 6 kbar and temperatures up to 520°C. Thus, specially designed autoclaves are needed. As many of the compounds prepared by this method are sensitive to moisture and oxygen, special glove boxes have been constructed. Today, this ammonothermal synthesis is very popular for the synthesis of gallium nitride [98]. Figure 3.33 shows

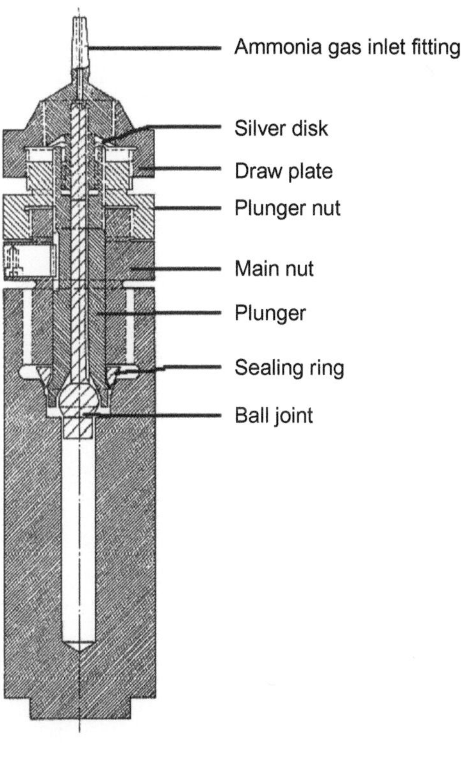

Figure 3.33 Schematic diagram of the autoclave designed by Jacobs and Schmidt [99].

the schematic diagram of the autoclave designed by Jacobs and Schmidt [99] to work in the region of supercritical ammonia. The autoclave is made up of a special alloy vacumelt ATS 340. Above 527°C, this alloy begins to flow. On the other hand, the workability of this alloy is quite difficult. The closing mechanism is a Bridgman-type closing, and the sealing rings are usually made up of copper or brass. Chalcogenides and polychalcogenides can also be synthesized by the reaction of the alkali metals with sulfur, selenium, or tellurium in ammonia under high pressure. Similarly, some hydroxides and hydrosulfides can be synthesized. Besides the synthesis, the ammonothermal technique has potential in the treatment or disintegration of organic compounds.

Microautoclave

Yamasaki and coworkers [100] have developed a microautoclave lined with Inconel-600 with a reaction chamber volume of 18 ml for hydrothermal oxidation studies. The oxygen gas can be charged into the reaction chamber through the

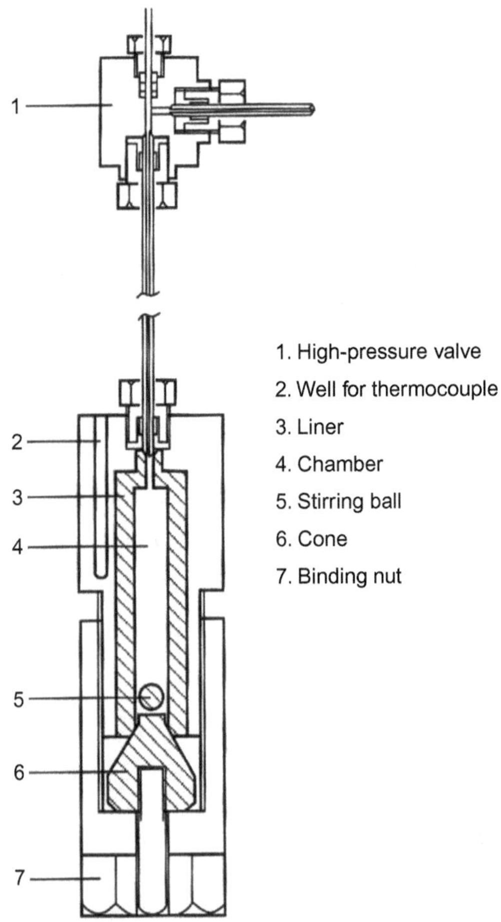

Figure 3.34 Schematic diagram of the microautoclave [100].

high-pressure value as an oxidant. Figure 3.34 shows the schematic diagram of the microautoclave for using high-pressure gas. An Al_2O_3 stirring ball of diameter 9.0 mm was added to the autoclave and oxygen gas (2.0 MPa) was charged through the pressure valve as the oxidant. This setup has been used by the designers to study the oxidation behavior of lead metal, and this autoclave is very effective for the treatment of lead. As the oxidation of sprayed liquid lead is carried out, usually at high temperature (250 − 400°C), by conventional processes, it results in vaporization of the lead leading to an inevitable environmental pollution problem. In this connection, this novel equipment comes as a helping device.

Hydrothermal Autoclaves for Visual Examination

The hydrothermal process cannot be monitored by "looking" into a steel autoclave. Systematic errors appear in the analysis of the crystallization conditions because the post-factum information is difficult to interpret and is mainly based on consecutive approximations which are only sufficient if the crystal growth is reproducible. Nevertheless, in this case, it is difficult to point out a parameter that controls the entire process because the greater the number of nonmeasured parameters in the method, the more difficult it is to optimize the solution. In order to facilitate the direct observation and control of dissolution of solids, their synthesis, recrystallization, and growth on seeds, several hydrothermal autoclaves with quartz and sapphire windows have been designed both in the West and in Russia. However, their potential has not been fully exploited as the designs are highly complex, especially the earliest designs of Turlakov et al. [101] and Kuznetsov and Lobachev [102], which are quite cumbersome to operate. The objective of these earlier authors was to study the physicochemical properties of hydrothermal solutions, the states of crystalline substances in them, the solution chemistry, and so on, by direct observation. Their studies also included investigation of the convection currents in the autoclave. For this purpose, the walls of the autoclaves are drilled with two diametrically opposite holes, which are covered with transparent crystals, or windows made of quartz or sapphire. If the interior of the autoclave is illuminated through one such window, one can observe what is happening in the autoclave through the other. By replacing the eye with a camera, movie camera, or video camera, one can record the processes taking place inside the autoclave; and by using a spectrograph, the state of the solution can be investigated. Figure 3.35 shows the schematic diagram of the autoclave with transparent windows [103]. Several problems arose while using these autoclaves under moderate to higher pressures and temperatures due to the differential expansion of steel and sapphire, which can be partially overcome by machining the cap from a material with a coefficient of thermal expansion similar to that of the windows. For example, niobium has been successfully used to about 750°C in combination with sapphire windows [104]. The experiments showed that when the autoclave is not under higher pressure–temperature conditions, the structure could easily support an internal pressure of up to 1 kbar. However, it did not always remain sealed when subjected to high temperatures and high pressures of and to an active soda

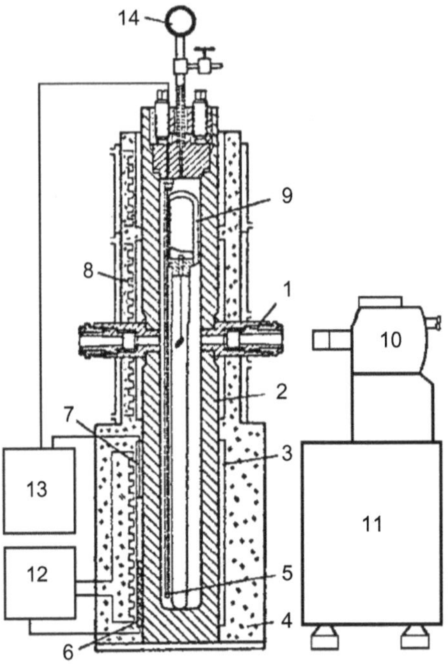

Figure 3.35 Schematic diagram of the autoclave with transparent windows [103]. (1) Window, (2) body of autoclave, (3) copper casing, (4) thermal insulation, (5) measuring thermocouples, (6) resistance thermometer, (7) thermocouples, (8)heater, (9) seal, (10) camera, (11) device for automatic single frame filming, (12) three channels proportional regulator, (13) 12-point EPR-09-Rd apparatus, and (14) pressure gauge.

solution. The cause of leakage was either the cracking of sapphire windows or the solution squeezing out through the sleeve sapphire seal. At temperatures >300°C, the surfaces of sapphire cylinders in contact with the solution were etched and their transparency was impaired.

In recent years, several improvements have been made in the designs, and special provisions have been made to perform optical/spectroscopic measurements. For example, the Laser Raman and other vibrational spectroscopic cells have been fabricated in some of the recent autoclave designs [105]. It is worth mentioning here that the recent advances in the instrumentation, particularly to study the optical/vibrational spectroscopic measurements under hydrothermal conditions, have contributed greatly to understanding the nature or behavior of hydrothermal solutions during the crystallization process and the complexation processes taking place inside the autoclave. Thus, the hydrothermal autoclave is no longer a "black box." Further, such studies have greatly helped in reducing the pressure–temperature conditions of growth for a large variety of inorganic compounds. With the reduction in the pressure–temperature conditions of growth, one can use thick glass-walled or thick silica glass autoclaves, which permit visual examination.

Hydrothermal Hot Pressing

HHP technology is a method by which hard solid bodies from powders can be produced in a short time and at a relatively low temperature (approximately 200–300°C) under saturated vapor pressure [106]. Using the HHP technique, one

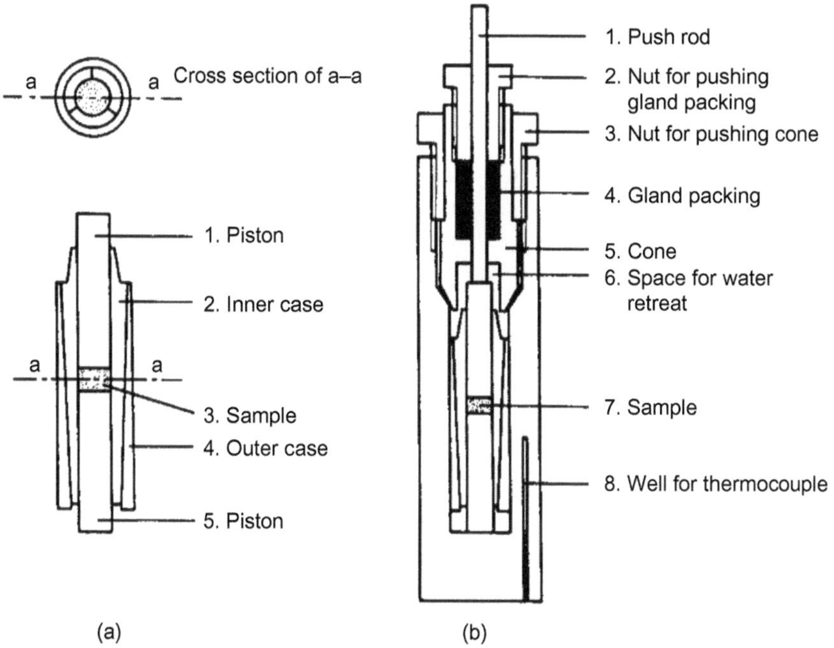

Figure 3.36 Schematic diagram of (a) die and (b) autoclave for HHP [107].

can obtain sintered compacts of pure oxides or inorganic powders which it is not possible to obtain by other ordinary sintering processes due to the transformation. Silicate powders such as borosilicate and silica glass can easily form a solid body by HHP technology with compression at 20 − 30 MPa when the proper amount of water or alkaline solution is added as a mineralizer.

In this technique, a starting powder containing water is continuously compressed from outside an autoclave under hydrothermal conditions. The powder is hydrothermally treated at autogeneous pressure while it is compressed at much higher pressure than the vapor pressure inside the autoclave. Figure 3.36 shows the schematic diagram of an autoclave for HHP. It is a cylinder made of steel with a cylindrical chamber 1 cm in diameter. The term *reaction pressure* means compressive pressure from outside the autoclave. The effective pressure for samples in the autoclave should be smaller than the reaction pressure because of vapor pressure in the autoclave. The piston had a space into which water from the starting sample was released by the HHP treatment. The gland packing made of Teflon between the piston and the push rod prevented leakage.

This technology has been used successfully for solidification of various industrial wastes such as sludge ash [108], radioactive wastes [107], and concrete wastes [109]. One of the advantages of HHP technology is that the hydrothermal reaction occurs in a closed system, which prevents the evaporation of volatile materials [110].

Vertical Autoclaves Used in Hydrometallurgy

High-pressure hydrometallurgy has made considerable progress in the last decade owing to advances in autoclave designs, which are quite large and are on an industrial scale [111]. These autoclaves may be in the form of cylinders, vertically mounted or horizontally laid, spherical, or in the form of a long horizontal tube. The agitation in these autoclaves is accomplished with the injection of high-pressure steam [112]. These *steam agitated autoclaves* are used mainly for leaching bauxite by NaOH within the temperature range 140–150°C and pressure 2500–3500 kPa. The vessels are usually fabricated from welded steel cylinders with spherical ends. Diameters vary from 1.5 to 2 m and heights from 6 to 12 m. Autoclaves of similar designs, but with acid-resistant brick lining, are used for leaching oxidized ores, e.g., laterites by concentrated H_2SO_4 at 250°C and 4000 kPa. These autoclaves have undergone an enormous increase in size for processes treating pyrite and arsenopyrite concentrates, containing gold particles and other related sulfide ores. Some autoclaves, especially with spherical shapes, have provisions for rotation at 8–15 rpm. In continuous operations, high-pressure membrane piston pumps are used for introducing pulps into autoclaves. Figure 3.37 shows schematic diagrams of representative autoclaves [113]. The reader can get more information in the works of Habashi [111,112,114].

Flow Reactors

The supercritical hydrothermal method is an extension of hydrothermal technology. The supercritical hydrothermal technology deals with the reactions at temperatures just near or above the critical temperature. Although supercritical fluid technology SCF (Skp, Cullin, F-box) technology has been in use from late nineteenth century as a tool to understand the natural mineralization, the actual impetus for this technology as a tool to process a wide range of materials began in 1980s. With the invention of Green Chemistry in the early 1990s, there was a surge in the popularity of SCF technology. Green Chemistry is important because supercritical fluids, particularly supercritical CO_2 ($scCO_2$) and, to a more limited extent, scH_2O are perceived as environmentally more acceptable replacements for the petroleum-based solvents, which are currently used in the world's chemical industries [115]. The supercritical hydrothermal method has witnessed a seminal growth in its application to the processing of a variety of materials owing to the availability of a wide range of solvents. The use of carbon dioxide has lots of advantages and replacing organic solvents in a number of chemical processes, including nanoparticle fabrication, food processing (such as decaffeination of coffee beans), chemical manufacturing, extraction, dry cleaning, semiconductor wafer cleaning, polymer processing, recycling, waste treatment, organic decomposition of dangerous and

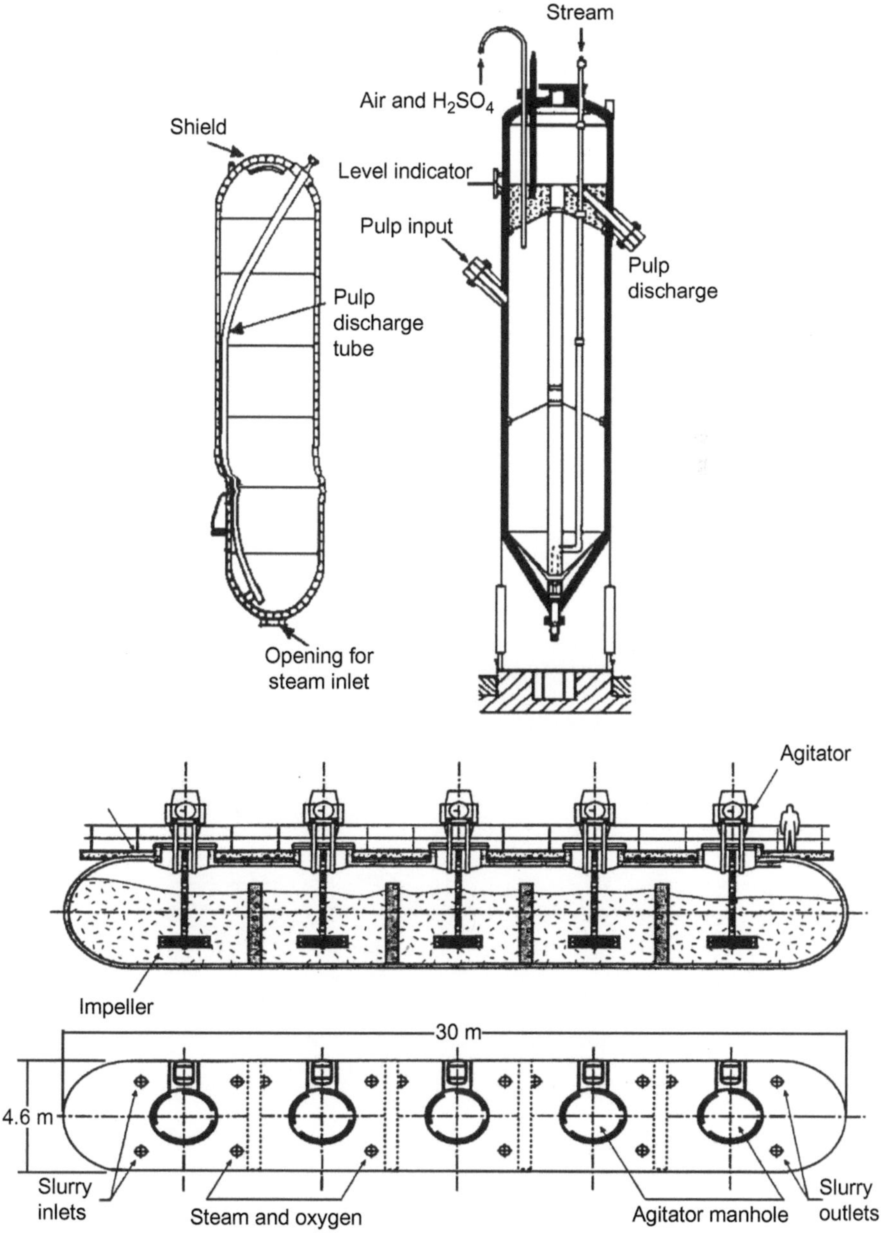

Figure 3.37 Schematic diagrams of autoclaves [113].

toxic wastes, polymeric wastes or hazardous waste, the recovery of inorganic compounds, particles coating, and drug manufacturing [116]. The supercritical hydrothermal method with water as a solvent is highly suitable for the nanoparticle fabrication of a wide range of high-melting inorganic compounds and also for

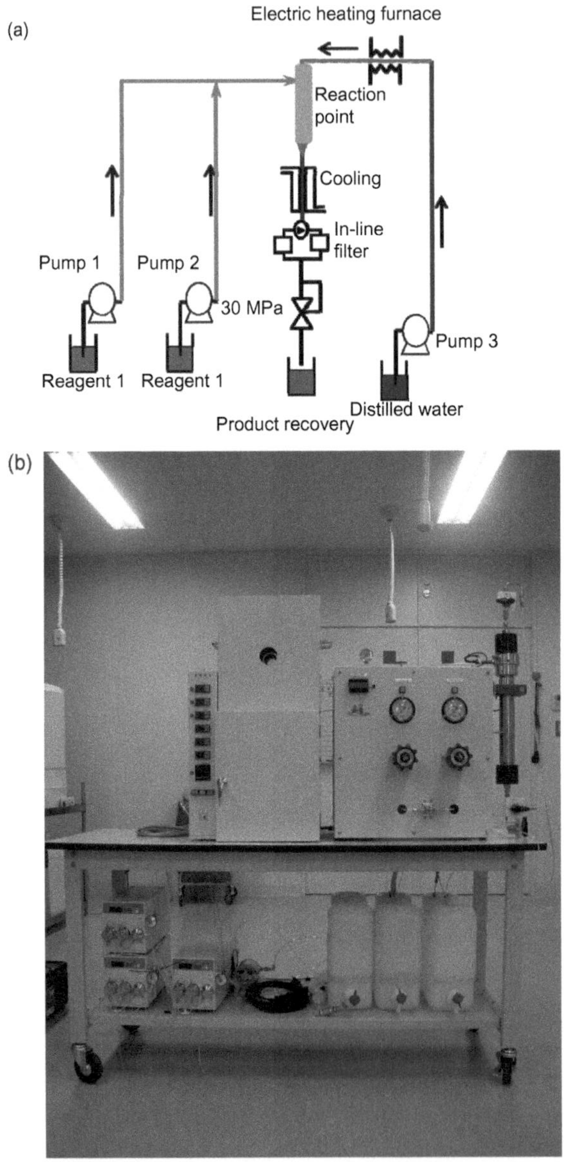

Figure 3.38 (a) Schematic diagram of a typical flow reactor [118]. (b) Commercial flow reactor "MimoCho," ITEC, Japan.

organic–inorganic hybrid nanoparticles. There are several variants in the supercritical fluid technology like static supercritical fluid (SSF) process, rapid expansion of supercritical solutions (RESS), particles from gas-saturated solutions (PGSS), gas antisolvent (GAS) process, precipitation from compressed antisolvent (PCA), aerosol solvent extraction system (ASES), supercritical antisolvent (SAS) process, solution-enhanced dispersion by supercritical fluids (SEDS), supercritical antisolvent process with enhanced mass transfer (SAS-EM), depressurization of an expanded liquid organic solution (DELOS), supercritical assisted atomization (SAA), hydrothermal synthesis under supercritical conditions via flow reactor (HTSSF), hydrothermal synthesis under supercritical conditions via batch reactor (HTSSB), supercritical fluids drying (SCFD), and supercritical fluid extraction emulsions (SFEE) [117]. Accordingly different designs are available for each variant of supercritical fluid processing of materials. Here, only the characteristic and general SCF flow reactors are shown in Figure 3.38a and b. An aqueous metal salt solution is first prepared and fed into the apparatus in one stream. In another stream, distilled water is pressurized and then heated to a temperature above the desired reaction temperature. The pressurized metal salt solution stream and the pure scH_2O stream are then combined at a mixing point, which leads to rapid heating and subsequent reactions in the reactor. After the solution leaves the reactor, it is rapidly quenched. In-line filters are used to remove larger particles. Pressure is controlled with a back-pressure regulator. Fine particles are collected in the effluent. By this rapid heating method, the effect of the heating period on the hydrothermal synthesis is eliminated: thus specific features of supercritical hydrothermal synthesis can be elucidated. There are several advantages in supercritical hydrothermal flow reactors, which provide nanoparticles with desired shape, size, and composition in a shortest possible residence time. Hence, this technique has an edge over the conventional methods in synthesizing a wide range of inorganic nanoparticles and also hybrid organic–inorganic nanoparticles within a very short time or shortest residence time. Figure 3.38b shows the commercial flow reactor prepared jointly by Adschiri's Laboratory and ITEC Company, Japan [118,119].

Autoclaves for Accelerating the Kinetics of Hydrothermal Reactions

There is a growing interest in recent years in enhancing the hydrothermal reaction kinetics through superimposing electric fields or mechanical forces, or mechanochemically including high uniaxial pressure with or without shear mechanical grinding, sonochemically (or acoustic wave stimulation, AWS), with microwave fields, and so on [120]. These techniques are being popularly used to synthesize a great variety of materials and also materials processing. Burns and Bredig [121] had established the remarkable effect that grinding calcite in a mortar converts it into aragonite. Dachille and Roy [122] generalized the finding to many other phases such as PbO_2 and MnF_2 showing that mechanochemical effects produce:

1. High-pressure phases stable above or even about 10 – 15 kbar at room temperature.
2. Enhanced kinetics by nearly two orders of magnitude in similar high-pressure solid-state reactions, when shearing stresses are superimposed on pressures up to 100,000 atm.

It is obvious that the grinding and mechanical mixing and subsequent fracture generate fresh, highly reactive surfaces, and lead to substantial improvement in the reaction kinetics—especially in the conversion of highly stable oxides to make clays. This was made possible by a simple innovation in the apparatus. In effect, the Morey-autoclaves were converted into ball mills or rod mills [123,124]. The autoclave is rotated around its long axis inside the furnace. An increase in the rate of reaction by two or three orders of magnitude can be attained at a given pressure and temperature. It is not clear whether this should be ascribed to strain energy stored in the lattice or merely to breakage of bonds.

Boldyrev et al. [124], through his extensive work with mechanochemical effects, concludes that for many reactions it is possible to increase the kinetics by about two orders of magnitude.

During the early 1980s, a considerable number of publications have appeared in the field of sonochemical breakdown of liquid phases. Suslick [125] gives an excellent review on this topic and shows that metallic phases may be melted and/or corroded in aqueous suspension at "room temperature." The general understanding is that collapsing bubbles during cavitation could generate temperatures of the order of 5000 K and modest local pressures. Roy and his group [126] attempted to synthesize novel materials or combination reactions, which could be accelerated at least by two orders of magnitude by using AWS. These AWS techniques have an interesting effect on size. In another case, ultrasonic energy was used to accelerate the formation of hydroxylapatite [127]. Abu-Samra et al. [128] and Kingston and Jassie [129] reported that all inorganic mineral phases can be dissolved faster in (dilute) acids by several orders of magnitude when a microwave field is superimposed upon mild ($<200°C$, $10-20$ bar) hydrothermal conditions. This technique is used worldwide today for chemical analysis as well as for dissolving samples. Komarneni and Roy [130,131], in a series of recent papers, have been applying hydrothermal-plus-microwave conditions for material synthesis, and their study enhances, by two orders of magnitude, the reaction kinetics of many reactions from the precipitation of metals from polyols to the synthesis of ferrites, phosphors, and so on. There is also some advantage in morphology control and it sometimes leads to the crystallization of new phases like hydrate of Al_2O [132]. Recently, Komarneni [133] has reviewed the importance of this. Figure 3.39 shows a schematic diagram of a typical hydrothermal-plus-microwave setup [132]. As the popularity of this technique is growing with the new results poured into literature, there is already a commercial production of these autoclaves.

There are several other designs of hydrothermal autoclaves to suit a specific purpose. For example, Strubel (1975) has designed a Teflon-lined autoclave with a stirrer [137]. Similarly, Shternberg [52] has designed a double autoclave (exoclave) in order to avoid having a mechanical compression system for producing a desired pressure in the reaction chamber. Also, this permits independent selection of temperature and pressure. The exoclave is designed for use at temperatures up to 700°C and pressures up to 3000 kPa/cm^2.

In recent years, there is an increasing tendency for *in situ* time-resolved hydrothermal synthesis for TEDDI using synchrotron beam and similarly in situ

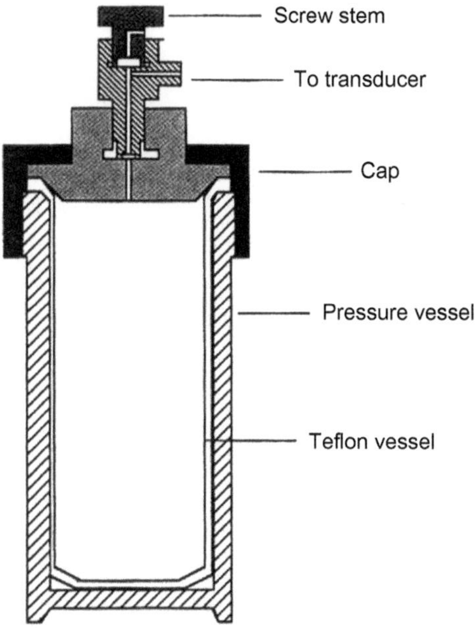

Figure 3.39 Schematic diagram of a typical hydrothermal-plus-microwave [132].

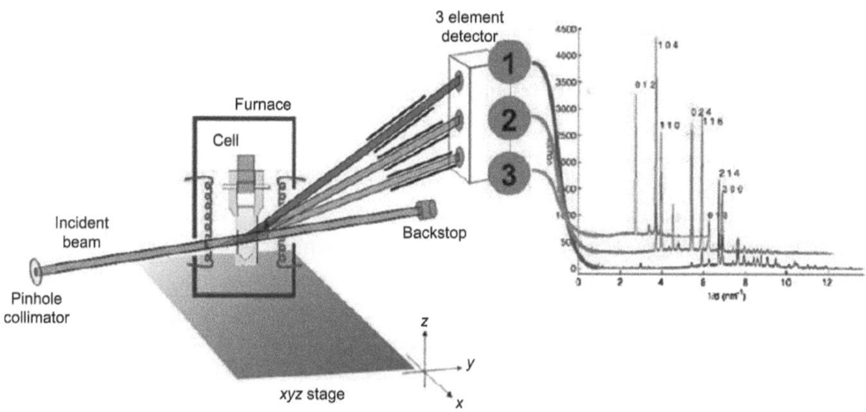

Figure 3.40 Schematic diagram of experimental setup for time-resolved hydrothermal synthesis for TEDDI using synchrotron beam [135].

time-resolved neutron diffraction analyses of hydrothermal synthesis. Accordingly, the hydrothermal reactors or cells are designed to suit such studies, which are very popular among zeolites, clays, and other complex inorganic compounds [135–137]. Figure 3.40 shows such a setup used for zeolite synthesis.

In recent years, high-pressure systems (about 40 kbar, temperature up to 1500°C) have been used in the synthesis of new compounds. Cylindrical platinum

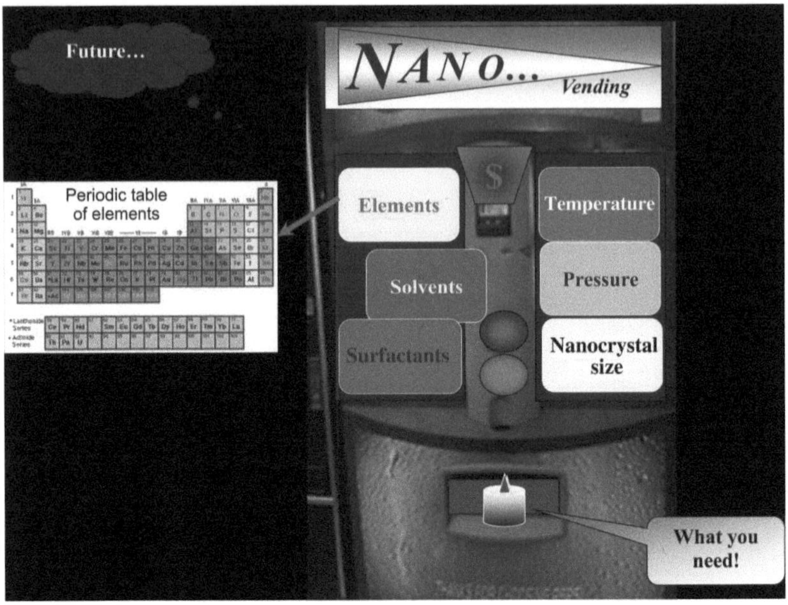

Figure 3.41 Nanovending machine for obtaining desired nanoparticles instantly with a controlled size and shape [1].

cells of 4 mm diameter and 6 mm height serve as the reaction vessels. However, these high-pressure vessels are not very important for the growth of crystals and processing of materials.

The progress in thermochemical computation, autoclave designing, multienergy processing, and selection of an appropriate solvent and surfactant has led to the materials processing at the speed of light or chemistry at the speed of light. Accordingly, the researchers are visualizing the instant hydrothermal systems or a new concept of nanovending machine to obtain instantly the desired nanoparticles with a controlled size and shape [1]. Figure 3.41 shows the materials scientist's imagination for such a nanovending machine [1]. In fact, this would be the future of hydrothermal technology with respect to nanotechnology.

3.6 Safety and Maintenance of Autoclaves

Safety is the prime factor in carrying out experiments under hydrothermal conditions. It is estimated that, for a 100 cm^3 vessel at 20,000 psi, the stored energy is about 20,000 J. Hydrothermal solutions—either acidic or alkaline—at high temperatures are hazardous to human beings if the autoclave explodes. Figure 3.42a and b show Stellite Tuttle cold-cone-sealed autoclaves and stainless steel acid digestion bombs which ruptured at high pressure. Therefore, vessels should have

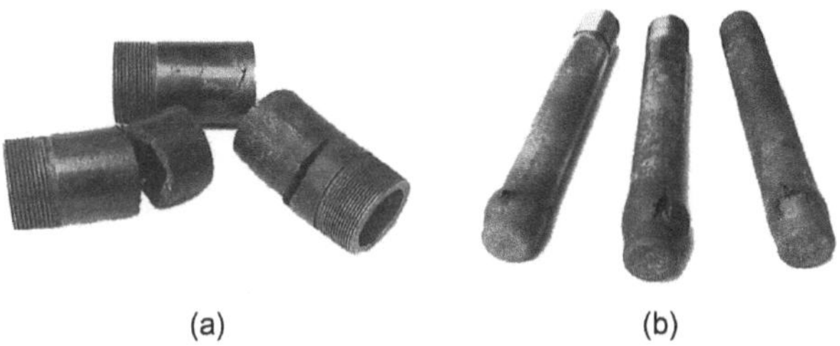

Figure 3.42 (a) Stellite Tuttle cold-cone-sealed autoclaves and (b) stainless steel acid digestion bombs which ruptured at high pressure.

rupture disks calibrated to burst above a given pressure. Such rupture disks are commercially available for various ranges of bursting pressure. Provision should be made for venting the live volatiles out in the event of rupture. A blow-down tank has to be commissioned in the case of a large-size hydrothermal laboratory. Proper shielding of the autoclave is needed to direct the corrosive volatiles away from the personnel. In the case of a large autoclave, the vessels are placed with proper shielding in a pit. For safety purposes, when hydrothermal autoclaves are operated simultaneously at a given temperature, a total static overpressure caused by a simultaneous rupture of all the operating autoclaves should not exceed 0.5 psi, which is determined by the strength of the walls and the dimension of the laboratory. In several laboratories, make-up pressure units are installed for safety.

The hydrothermal laboratory should have a well-equipped workshop for the proper maintenance of the autoclaves. After each run, the autoclave should be properly treated before using it for the next run. The contacts, especially at the sealing sections, should be kept sharp without any lint, by polishing the surfaces at the contact after each run to avoid scratches. The gasket must remain flat in its seat without tilting. The autoclave threads should be maintained carefully and are always painted with pipe thread lubricant to reduce friction and prevent seizing of the autoclaves at high temperatures. Some workers even place a light coating on the seal rings to ensure that they are properly sealed. Graphite grease or graphite in water medium and suspended molybdenum sulfide or copper lubricants are used. For runs above 200°C, graphite lubricant is more effective because other lubricants may decompose.

With a new or recently remachined vessel, it is advisable first to clean out the residual cutting oil with an organic solvent and to heat distilled water in the vessel one or more times, until the odors of sulfur gases or organic species are no longer present. For vessels that have been used with H_2S, many cycles may be required. Once a vessel has contained reduced sulfur, it should not be used for sulfur-free experiments unless it has been vacuum cleaned while hot. Such a degassed vessel previously used for reduced systems can then be heated, at about 350°C open to the atmosphere, to promote an oxide surface layer for corrosion protection. Proper precaution should be taken while sealing the vessel. In the case of cold-seal equipment, explosive vessel failure is of paramount concern. Such failure is likely to occur in the event of rapid quenching of the entire vessel and overheating of the furnace. In spite the simple, rapid quench it provides, immersing the vessel in water is to be avoided. Nowadays, numerous remote quenching designs have been developed

[138]. Several authors have discussed, in detail, major hazards with hydrothermal autoclaves [37,139].

Overheating of the furnace resulting from the malfunction of the control system is a common reason for vessel failure. Hence, furnaces should be properly shielded. Burns from hot vessels represent another major hazard with cold-seal laboratories. If it is necessary to handle a hot vessel, it should be done by grasping the relatively cool closure assembly wearing special insulated gloves.

Note that any vessel to be heated must never be filled completely with an aqueous solution. Fluids expand on heating and are capable of generating extreme pressures when further heated beyond the PT limits of the autoclave. It is critically important to calculate the thermal expansion of the aqueous phase and to allow at least 20% of the total volume of the vessel for uncertainty or temperature overturn. An apprenticeship in an operating hydrothermal laboratory is strongly recommended to learn, to anticipate, and to cope with the dangers of such experiments. For those planning to set up a hydrothermal laboratory, it is advisable to visit other existing laboratories to review safety procedures as well as designs. Therefore, autoclave maintenance and safety make hydrothermal experiments quite expensive.

References

[1] K. Byrappa, Novel hydrothermal solution routes of advanced high melting nanomaterials processing, J. Ceram. Soc. Japan 117 (2009) 236–244.

[2] A. Rabenau, The role of hydrothermal synthesis in preparative chemistry, Angew. Chem. (English Ed.) 24 (1985) 1026–1040.

[3] R.A. Laudise, J.W. Nielsen, in: F. Seitz, D. Turnbull (Eds.), Hydrothermal Crystal Growth, Solid State Physics, vol. XII, Academic Press, New York, NY, 1961.

[4] D. Clauss, Engineer's Guide to High-Temperature Materials, McGraw-Hill, New York, NY, 1969, pp. 1–612.

[5] C.J. Smithells, Metals Reference Handbook, 1, Plenum Press, New York, NY, 1967. p. 378.

[6] R. Speed, A. Filice, Quartz glass pressure vessels for hydrothermal studies, Am. Min. 49 (1964) 1114–1119.

[7] V.I. Popolitov, Hydrothermal growth of crystals under visual examination, in: K. Byrappa (Ed.), Hydrothermal Growth of Crystals, Pergamon Press, Oxford, 1991, pp. 255–297.

[8] B.N. Litvin, D.A. Tules, Apparatus for hydrothermal synthesis and growth of monocrystals, in: A.N. Lobachev (Ed.), Crystallization Processes Under Hydrothermal Conditions, Studies in Soviet Science, Consultant Bureau, New York, NY, 1973, p. 139.

[9] H. De Senarmont, Recherches sur les Propriétés optiques biréfringentes des corps isomorphes, Ann. Chim. Phys. 32 (1851) 129–132.

[10] M. Daurree, Sur le Metamorphisme et Recherches Experimentales Sur Quelques-uns, Ann. Mines 12 (1857) 289–326.

[11] E.T. Allen, J.L. Crenshaw, J. Johnston, Am. J. Sci. Fourth Ser. 33 (1912) 167.

[12] W. Bruhns, Apparatus, a steel bomb, lined with platinum, Neues Jahrb. Min. Geol. II (1889) 62–65.
[13] C. Doetler, Recrystallization of apophyllite, Neues Jahrb. Min. Geol. 1 (1890) 118–139.
[14] A. Rabenau, H. Rau, Crystal growth and chemical synthesis under hydrothermal conditions, Philips Tech. Rev. 30 (1969) 14–21.
[15] H.L. Barnes, Investigations in hydrothermal sulfide system, in: G.C. Ulmer (Ed.), Research Techniques for High Pressure and High Temperatures, Springer-Verlag, NewYork, NY, 1971, pp. 317–355.
[16] R.A. Laudise, What is materials chemistry? in: S.R. Marder, J.E. Sohn, G.D. Stucky (Eds.), Materials for Nonlinear Optics, Chemical Perspectives, American Chemical Society, Washington, DC, 1991, pp. 411–433.
[17] J.C Jumas, A. Goiffon, B. Capelle, A. Zarka, J.C Doukhan, J. Schwartzel, et al., Crystal growth of berlinite, $AlPO_4$: physical characterization and comparison with quartz, J. Cryst. Growth 80 (1987) 133–148.
[18] C.T. Ashby, J.W. Berry, A.J. Deutschman, A variable-volume silver liner for high pressure hydrothermal crystal growth, Am. Min. 55 (1970) 1800–1802.
[19] C.D. McMillen, Hydrothermal Crystal Growth of Oxides for Optical Applications, Clemson University, Clemson, SC, 2007, pp. 31–33.
[20] W.F. Roeser, Temperature: Its Measurement and Control in Science and Industry, Reinhold, New York, NY, 1941. p. 180
[21] R.W. Charles, R. Vidale, Temperature calibration of a new rapid quench vessel, Am. Min. 67 (1982) 175–179.
[22] G.C. Kennedy, Pressure–volume–temperature relations in water at elevated temperatures and pressures, Am. J. Sci. 248 (1950) 540–564.
[23] R.W. Bain, Steam Tables, Her Majesty's Stationary Office, Edinburgh, 1964, pp. 1–146.
[24] J.B. Hannay, On the artificial formation of the diamond, Proc. R. Soc. London 30 (1880) 188–189.
[25] G. Friedel, Hydrothermal synthesis of corundum, Bull. Soc. Min. 14 (1891) 7–10.
[26] K. Byrappa, Hydrothermal growth of crystals, in: D.T.J. Hurle (Ed.), Handbook of Crystal Growth, Elsevier Science B.V., England, 1994, pp. 465–562.
[27] D.F. Croxall, L.R.A. Christie, B.J. Isherwood, A.G. Todd, J. Brich, The growth and assessment of berlinite single crystals, in: F. W. Ainger (Ed.), Proceedings of the Second European Conference on Crystal Growth, Lancaster, UK, 1979. J. Mat. Sci, vol. 6, 21, (1986) pp. 2202-2206..
[28] G.W. Morey, P. Niggli, The hydrothermal formation of silicates, a review, J. Am. Chem. Soc. 35 (1913) 1086–1130.
[29] G.W. Morey, E. Ingerson, The pneumatolytic and hydrothermal alteration and synthesis of silicates, Econ. Geol. 32 (1937) 607–761.
[30] O.F. Tuttle, Two pressure vessels for silicate–water systems, Geol. Soc. Am. Bull. 60 (1949) 1727–1729.
[31] A.C. Walker, E. Buehler, Growth large quartz crystals, Ind. Eng. Chem. 42 (1950) 1369–1375.
[32] R.A. Laudise, Hydrothermal synthesis and crystal growth—history, status and prospects, in: P.M. Dryburgh, B. Cockayne, K.G. Barraclough (Eds.), Advanced Crystal Growth, Prentice-Hall, London, 1987, p. 267.
[33] R.A. Laudise, Hydrothermal growth, in: P. Hartman (Ed.), Crystal Growth: An Introduction, North-Holland Publ. Co., 1973, pp. 162–197.

[34] O.F. Tuttle, A new hydrothermal quenching apparatus, Am. J. Sci. 246 (1948) 628–635.
[35] R. Roy, E.F. Osborn, Some simple aids in hydrothermal investigation of mineral system, Econ. Geol. 47 (1952) 717–721.
[36] W.C. Luth, O.F. Tuttle, Externally heated cold-seal pressure vessels for use to 10,000 bars and 750°C, Am. Min. 48 (1963) 1401–1403.
[37] J.S. Huebner, Buffering techniques for hydrostatic systems at elevated pressures, in: G.C. Ulmer (Ed.), Research Techniques for High Pressure and High Temperature, Springer-Verlag, New York, NY, 1971, pp. 123–177.
[38] A.D. Edgar, Experimental Petrology: Basic Principles and Techniques, Clarendon Press, Oxford, 1973. p. 217
[39] A.C. Walker, Hydrothermal synthesis of quartz crystals, Ind. Eng. Chem. 36 (1953) 250–256.
[40] R.A. Laudise, The Growth of Single Crystals, Prentice-Hall, Englewood Cliffs, NJ, 1970, pp. 278–281
[41] J. Asahara, K. Nagai, S. Harada, Synthetic quartz crystals by large autoclaves, in: S. Somiya (Ed.), Proceedings of the First International Symposium on Hydrothermal Reactions, Gakujutsu Bunken Fukyu-Kai, Japan, March 22 – 26, 1982, pp. 430–441.
[42] S. Taki, Improvement of growth process and characterization of quartz crystals, Progr. Cryst. Growth Char. 23 (1991) 313–339.
[43] R.A. Laudise, P.M. Bridenbaugh, T. Iradi, Pressure balance under hydrothermal conditions, J. Cryst. Growth 140 (1994) 51–56.
[44] F.H. Smyth, L.H. Adams, The system calcium-oxide–carbon dioxide, J. Am. Chem. Soc. 45 (1923) 1172.
[45] R.W. Goranson, Solubility of water in granite magmas, Am. J. Sci. 22 (1931) 481–502.
[46] H.S. Yoder, High–low quartz inversion up to 10,000 kg/cm^3, Trans. Am. Geophys. Union 31 (1950) 827–835.
[47] F.R. Boyd, J.L. England, The quartz–coesite transition, Geophys. Research 65 (1960) 741–749.
[48] B. Vodar, J. Kieffer, Influence of pressure on the elastic modulus, the yield strength, and the deformation of metals and alloys, in: A. Pugh (Ed.), Mechanical Behavior of Materials Under Pressure, Elsevier, Barking, Essex, 1970, p. 1.
[49] H.T. Hall, Some high pressure, high temperature apparatus design considerations: Equipment for use at 100,000 atm. and 300 °C, The effect of high pressure on the crystal structure of LaOs2 and CeOs2, Rev. Sci. Instrum. 29 (1958) 267.
[50] W.A. Bassett, A.H. Shen, M. Bucknum, A new diamond anvil cell for hydrothermal studies to 2.5 Gpa and from −190 to 1200°C, Rev. Sci. Instrum. 64 (1993) 2340–2345.
[51] S. Mroczkowski, Development of high-pressure, high-temperature technique for growing single crystals from the flux, J. Cryst. Growth 24/25 (1974) 683–687.
[52] A.A. Shternberg, Controlling the growth of crystals in autoclaves, in: A.N. Lobachev (Ed.), Crystallization Processes Under Hydrothermal Conditions, Consultant Bureau, New York, NY, 1973, pp. 225–240.
[53] S.M. Clark, A. Nield, T. Rathbone, J. Flaherty, C.C. Tang, J.S.O. Evans, et al., Development of large volume reaction cells for kinetic studies using energy-dispersive powder diffraction, Nucl. Inst. Methods Phys. Res. B97 (1995) 98–101.
[54] J.S.O. Evans, R.J. Francis, D. O'Hare, S.J. Price, S.M. Clark, J. Flaherty, et al., An apparatus for the study of the kinetics and mechanism of hydrothermal reactions by

in situ energy dispersive X-ray diffraction, Rev. Sci. Instrum. 66 (1995) 2442–2445.

[55] S.M. Clark, E. Doorhyee, A quantitative kinetic study of the I ↔ II phase transition of ammonium chloride, J. Phys. Condens. Matter. 4 (1992) 8969–8974.

[56] E.D. Kolb, R.A. Laudise, Pressure–volume–temperature behavior in the system H_2O–H_3PO_4–$AlPO_4$ and its relationship to the hydrothermal growth of $AlPO_4$, J. Cryst. Growth 56 (1982) 83–92.

[57] E.D. Kolb, R.L. Barns, R.A. Laudise, J.C. Grenier, Solubility, crystal growth and perfection of aluminum orthophosphate, J Cryst. Growth 50 (1980) 404–418.

[58] N.Y. Ikornikova, A.N. Lobachev, A.R. Vasenin, V.M. Egrov, A.V. Autoshin, Apparatus for precision research in hydrothermal, in: A.N. Lobachev (Ed.), Crystallization Processes under Hydrothermal Conditions, Consultant Bureau, New York, NY, 1973, p. 241.

[59] K. Yanagisawa, M. Xian-Pin, M. Nishioka, K. Loku, N. Yamasaki, Development of solubility measurement method under hydrothermal conditions, J. Mater. Sci. Lett. 12 (1993) 1800–1802.

[60] J.M. Stanley, Hydrothermal synthesis of large aluminium phosphate, Cryst. Industr. Eng. Chem. 46 (1954) 1684–1689.

[61] M. Poignant, L. Le, Y. Toudic, Etude de la Solubilite du Phosphate d'Alunimium ($AlPO_4$) dans des Solutions Hydrothermales d'Acide Orthophosphorique (H_3PO_4), Mat. Res. Bull. 14 (1979) 603–612.

[62] D.R. Kinloch, R.F. Belt, R.C Puttbach, Hydrothermal growth of calcite in large autoclaves, J. Cryst. Growth 24/25 (1974) 610–613.

[63] N. Yamasaki, M. Fujiki, M. Nishioka, K. Ioku, K. Yanagisawa, N. Kozai, S. Muraoka, Hydrothermal decomposition of TBP and fixation of its decomposed residue by hydrothermal hot pressing technique, in: Proceedings of the Third International Symposium of Advance Nuclear Energy Research—Global Environment and Nuclear Energy, Japan Atomic Energy Research Institute, Tokyo, 1991, pp. 1–5.

[64] W.E. Seyfried Jr., D.R. Janecky, Heavy metal and sulfur transport during subcritical and supercritical hydrothermal alteration of basalt: influence of fluid pressure and basalt composition and crystallinity, Geochim. Cosmochim. Acta 49 (1985) 2545–2560.

[65] W.L. Bourcier, H.L. Barnes, Rocking autoclaves for hydrothermal experiments I. Fixed-volume systems, in: G.C. Ulmer, H.L. Barnes (Eds.), Hydrothermal Experimental Techniques, John Wiley & Sons, New York, NY, 1987, pp. 189–215.

[66] W.E. Seyfried Jr., D.R. Janecky, M.E. Berndt, Rocking autoclaves for hydrothermal experiments II, the flexible reaction-cell system, in: G.C. Ulmer, H.L. Barnes (Eds.), Hydrothermal Experimental Techniques, John Wiley & Sons, New York, NY, 1987, pp. 216–239.

[67] K. Yanagisawa, Q. Feng, N. Yamasaki, Hydrothermal synthesis of xonotlite whiskers by ion diffusion, J Mat. Sci. Lett. 16 (1997) 889–891.

[68] Parr Autoclaves Co., USA, Catalogue, 1999.

[69] Berghoff Autoclaves Ltd., Germany, Catalogue, 1999.

[70] Toshin Kogyo Co., Ltd., Japan, Catalogue, 1999.

[71] Nitto Koatsu Co., Japan, Catalogue, 1999.

[72] J.K. Fogo, C.S. Copeland, S.W. Benson, Pressure counterbalance apparatus for the measurement of the electrical conductivity of aqueous solutions above their critical temperatures, Rev. Sci. Instrum. 22 (1951) 765–769.

[73] J.K. Fogo, S.W. Benson, C.S. Copeland, The electrical conductivity of supercritical solutions of sodium chloride and water, J. Chem. Phys. 22 (1954) 212–216.

[74] E.U. Franck, Hochverdichteter Wasserdampf I. Elektrolytische Leitfahigkeit in KCl−H_2O Losungen bis 750°C, Zeit. Physik. Chem. 8 (1956) 92−106.
[75] E.U. Franck, Experimental Investigations of Fluids at High Pressures and Elevated Temperatures, North American Treaty Organization (NATO) Advanced Study Institute, 1978 Series C, C41 (High Pressure Chemistry).
[76] E.U. Frauck, Survey of selected non-thermodynamic properties and chemical phenomena of fluids and fluids mixtures, in: D.T. Rickard, F.E. Wickman (Eds.), Chemistry and Geochemistry of Solutions at High Temperatures and Pressures, 13/14, Pergamon Press, New York, NY, 1979 Proceedings of Noble Symposium.
[77] T.M. Seward, The stability of chloride complexes of silver in hydrothermal solutions up to 350°C, Geochim. Cosmochim. Acta 40 (1976) 1329−1341.
[78] T.M. Seward, Metal complex formation in aqueous solutions at elevated temperatures and pressures, in: D.T. Rickard, F.E. Wickman (Eds.), Chemistry and Geochemistry of Solutions at High Temperatures and Pressures, 13/14, Pergamon Press, New York, NY, 1979Proceedings of Noble Symposium.
[79] W.L. Marshall, Conductances and equilibria of aqueous electrolytes over extreme ranges of temperature and pressure, Pure Appl. Chem. 18 (1968) 167−186.
[80] W.L. Marshall, Complete equilibrium constants, electrolyte equilibria and reaction rates, J. Phy. Chem. 74 (1970) 346−355.
[81] B.N. Ryzhenko, A conductance cell for use at high temperatures and pressures, Geokhimiya 1963 (1963) 41.
[82] B.S. Smolyakov, Limiting equivalent ionic conductances up to 200°C, International Corrosion Conference Series, Proceedings of the High Temperature High Pressure Electrochemistry in Aqueous Solutions Conference (NACE-4), in: D. de G. Jones and R.W. Staehle, (Eds.), National Association of Corrosion Engineers ,Houston, TX, 1976, pp. 177−181.
[83] H.C. Helgeson, D.H. Kirkham, Theoretical prediction of the thermodynamic behavior of aqueous electrolytes at high pressures and temperature: I: summary of the thermodynamic/electrostatic properties of the solvent, Am. J. Sci. 274 (1974) 1089−1198.
[84] H.C. Helgeson, D.H. Kirkham, Theoretical prediction of the thermodynamic behavior of aqueous electrolytes at high pressures and temperature, III: equation of state for aqueous species at infinite dilution, Am. J. Sci. 276 (1976) 97−240.
[85] A.S. Quist, W.L. Marshall, Electrical conductances of aqueous sodium chloride solutions from 0 to 800°C and to pressures to 4000 bars, J. Phy. Chem. 72 (1968) 684−703.
[86] W.L. Marshall, J.D. Frantz, Electrical conductance measurements of dilute, aqueous electrolytes at temperatures to 800°C and pressures to 4000 bars, techniques and interpretations, in: G.C. Ulmer, H.L. Barnes (Eds.), Hydrothermal Experimental Techniques, John Wiley & Sons, New York, NY, 1987, pp. 261−292.
[87] D.B. Hawkins, R. Roy, Electrolytic synthesis of kaolinite under hydrothermal conditions, J. Am. Ceram. Soc. 45 (1962) 507−508.
[88] M. Yoshimura, S.-E. Yoo, M. Hayashi, N. Ishizawa, Preparation of $BaTiO_3$ thin film by hydrothermal electrochemical method, Jpn. J. Appl. Phys. 28 (1989) L2007−L2009.
[89] W.-S. Cho, M. Yashima, M. Kakihana, A. Kudo, T. Sakata, M. Yoshimura, Room-temperature preparation of highly crystallized luminescent $SrWO_4$ film by an electrochemical method, J. Am. Ceram. Soc. 78 (1995) 3110−3112.
[90] K. Kajiyoshi, Y. Hamaji, K. Tomono, T. Kasanami, M. Yoshimura, Microstructure of strontium titanate thin film grown by the hydrothermal−electrochemical method, J. Am. Ceram. Soc. 79 (1996) 613−619.

[91] R. Basca, P. Ravindranathan, J.P. Dougherty, Electrochemical, hydrothermal, and electrochemical-hydrothermal synthesis of barium titanate thin films on titanium substrates, J. Mater. Res. 7 (1992) 423–428.
[92] P. Bendale, S. Venigalla, J.R. Ambrose, E.D. Verink Jr., J.H. Adair, Preparation of barium titanate films at 55°C by "An Electrochemcial Method," J. Am. Ceram. Soc. 76 (1993) 2619–2627.
[93] A.J. Bard, L.R. Faulkner, Electrochemical Methods—Fundamentals and Applications, John Wiley & Sons, New York, NY, 1980.
[94] K. Kajiyoshi, K. Tomono, Y. Hamaji, T. Kasanami, M. Yoshimura, Growth of strontium titanate thin films of controlled thickness by the hydrothermal–electrochemical method, J. Am. Ceram. Soc. 77 (1994) 2889–2897.
[95] D.D. Macdonald, A.C Scott, P. Wentrcek, External reference electrodes for use in high temperature aqueous systems, J. Electrochem. Soc. 126 (1979) 908–911.
[96] W. Weyl, Uber Metallammonium-Verbindungen, Ann. Physik. 121 (1864) 606–610.
[97] C.A. Seely, On ammonium and the solubility of metals without chemical action, J. Franklin Inst. 61 (1871) 110.
[98] J.J. Callahan, Q.S. Chen, Hydrothermal and ammonothermal growth of ZnO and GaN, in: G. Dhanraja, K. Byrappa, V. Prasad, M. Dudley (Eds.), Springer Handbook of Crystal Growth, Springer, Heidelberg, Germany, 2010, pp. 655–689.
[99] H. Jacobs, D. Schmidt, High-pressure ammonolysis in solid-state chemistry, in: E. Kaldis (Ed.), Current Topics in Materials Science, vol. 8, North Holland Publishing Co., Amsterdam, 1982, pp. 383–427.
[100] H. Wakayama, Y. Fukushima, N. Yamasaki, Wet oxidation of waste oil, in: R. Yamanoto et al. (Eds.), Advanced Materials 1992, Trans. Mat. Res. Soc. Japan, 18A, 1994, pp. 775–778.
[101] V.N. Turlakov, G.M. Safronov, G.V. Kleshchev, A.V. Simonov, Experimental autoclave with transparent windows for studying hydrothermal solutions, Izvestia Akad. Nauk. Inorg. Mater. 6 (1970) 1216–1217.
[102] V.A. Kuznestov, A.N. Lobachev, Hydrothermal method for the growth of crystals, Sov. Phys. Crystallogr. 17 (1973) 775–804.
[103] N.Y. Ikornikova, A.N. Lobachev, A.R. Vasenin, V.M. Egorov, A.V. Antoshin, Apparatus for precision research in hydrothermal experiments, in: A.N. Lobachev (Ed.), Crystallizatioin Processes Under Hydrothermal Conditions, Consultants Bureau, New York, NY, 1973, pp. 241–255.
[104] M. Buback, E.U. Franck, Measurements of the vapor pressures and critical data of ammonium halides, Berichteder Bunsengesellschaft fur Physikalische Chemie 76 (1972) 350–354.
[105] M. Hosaka, S. Taki, Raman spectral studies of SiO_2–NaOH–H_2O system solution under hydrothermal conditions, J. Cryst. Growth 100 (1990) 343–346.
[106] K. Yanagisawa, M. Sasaki, M. Nishioka, K. Ioku, N. Yamasaki, Preparation of sintered compacts of anatase by hydrothermal hot-pressing, J. Mat. Sci. Lett. 13 (1994) 765–766.
[107] N. Yamasaki, K. Yanagisawa, M. Nishioka, S. Kanahara, Solidification of radioactive waste by hydrothermal hot pressing method, J. Mat. Sci. Lett. 5 (1986) 355.
[108] M. Nishioka, K. Yanagisawa, N. Yamasaki, Solidification of sludge ash by hydrothermal hot pressing method, J. Water Pollut. Contr. Fed. 62 (1991) 926.
[109] Y. Nakane, T. Hashida, H. Takahashi, N. Yamasaki, Strengthening of hydrothermal hot pressed concrete wastes by the addition of fresh cement, J. Ceram. Soc. Japan 103 (1995) 513–516.

[110] N. Yamasaki, L. Xiang, Q. Feng, Formation of PbO powder by hydrothermal oxidation of granular Pb metal, J. Mater. Sci. Lett. 15 (1996) 2153–2157.
[111] F. Habashi, A Textbook of Hydrometallurgy, Libraire Universitaire du Quebec, Quebec, Canada, 1993.
[112] K.G. Thomas, Research, Engineering. Design, and Operation of a Pressure Hydrometallurgy Facility for Gold Extraction, CANMET, Ottawa, Canada, 1994.
[113] F. Habashi, Industrial autoclaves for pressure hydrometallurgy, in: N. Yamasaki, K. Yanagisawa (Eds.), Proceedings of the Second International Conference on Solvothermal Reactions, Takamatsu, Japan, December 18–20, 1996, pp. 65–67.
[114] F. Habashi (Ed.), Handbook of Extractive Metallurgy, 4 vols., Wiley-VCH, Germany, 1997.
[115] K. Byrappa, S. Ohara, T. Adschiri, Nanoparticles synthesis using supercritical fluid technology—towards biodmedical applications, Adv. Drug Deliv. Rev. 60 (2008) 299–327.
[116] H. Uchida, M. Amiya, Y. Iwai, Y. Arai, Separation of dimethylnaphthalene isomers by supercritical phase absorption using zeolite, in: N. Yamasaki, K. Yanagisawa (Eds.), Proceedings of the Second International Conference on Solvothermal Reactions, Takamatsu, Japan, December 18–20, 1996, pp. 127–130.
[117] T. Adschiri, K. Arai, Hydrothermal synthesis of metal oxide nanoparticles under supercritical conditions, in: Y.-P. Sun (Ed.), Supercritical Fluid Technology in Materials Science and Engineering, Marcel Dekker, New York, NY, 2002, pp. 311–325.
[118] T. Adschiri, K. Byrappa, Supercritical hydrothermal synthesis of organic–inorganic hybrid nanoparticles, in: A. Muramatsu, T. Miyashita (Eds.), Nanohybridization of Organic–Inorganic Materials, Springer, Heidelberg, Germany, 2009, pp. 247–280.
[119] K. Byrappa, T. Adschiri, Hydrothermal technology for nanotechnology, Prog. Cryst. Growth Charact. Mater. 53 (2007) 117–166.
[120] R. Roy, Accelerating the kinetics of low-temperature inorganic syntheses, J. Solid State Chem. 111 (1994) 11–17.
[121] J.H. Burns, M.A. Bredig, Transformation of calcite to aragonite by grinding, J. Chem. Phys. 25 (1956) 1281.
[122] F. Dachille, R. Roy, High-pressure phase transformations in laboratory mechanical mixers and mortars, Nature 186 (34) (1960) 71.
[123] V.V. Boldyrev, A historic view on the development of mechanochemistry in Siberia, Chem. Sustain. Dev. 10 (2002) 3–10.
[124] V.V. Boldyrev, A. Kabibullin, N.V. Khkosova, E.G. Avaakumov, Hydrothermal reactions under mechanochemical treatment, J. Mater. Synth. Process. 4 (1996) 377–381.
[125] K.S. Suslick, Sonochemistry, Science 247 (1990) 1439–1445.
[126] V. Srikanth, R. Roy, S. Komarneni, Acoustic-wave stimulation of the leaching of layer silicates, Mater. Lett. 15 (1992) 127–129.
[127] Y. Fang, D.K. Agarwal, D.M. Roy, R. Roy, Fabrication of porous hydroxyapatite ceramics by microwave ceramics, J. Mater. Res. 7 (1992) 490–494.
[128] A. Abu-Sharma, J.S. Morris, S.R. Koirtyohann, Wet ashing of some biological samples in microwave oven, Anal. Chem. 47 (1975) 1475–1477.
[129] H.M. Kingston, L.B. Jassie (Eds.), Introduction to Microwave Sample Preparation, American Chemical Society, Washington, DC, 1988.
[130] S. Komarneni, R. Roy, Q.H. Li, Microwave-hydrothermal synthesis of ceramic powders, Mater. Res. Bull. 27 (1992) 1393–1405.
[131] S. Komarneni, R. Pidugu, Q.H. Li, R. Roy, Microwave-hydrothermal processing of metal powders, J. Mater. Res. 10 (1995) 1687–1692.

[132] S. Komarneni, Enhanced reaction kinetics under microwave-hydrothermal conditions, in: N. Yamasaki, K. Yanagisawa (Eds.), Proceedings of the Second International Conference on Solvothermal Reactions, Takamatsu, Japan, December 18–26, 1996, pp. 97–99.
[133] S. Komarneni, Microporous metal intercalated clay nanocomposites, in: K. Ishizaki, L. Sheppard, S. Okada, T. Hamasaki, B. Huybrechts (Eds.), Ceramic Transactions, Porous Materials, 31, American Ceramic Society, Westerville, OH, 1993.
[134] G. Strubel, Dt. Gemmol. Ges. 24 (1975) 138–144.
[135] V. Middelkoop, S.D. Jacques, M.G. O'Brien, A.M. Beale, P. Barnes, Hydrothermal/autoclave synthesis of AlPO-5: a prototype space/time study of crystallization gradients, J. Mat. Sci. 43 (2008) 2222–2228.
[136] A.J. Celestian, J.B. Parise, A. Clearfield, Crystal growth and ion exchange in titanium silicates, in: G. Dhanaraja, K. Byrappa, V. Prasad, M. Dudley (Eds.), Springer Handbook of Crystal Growth, Springer, Heidelberg, Germany, 2010, pp. 1637–1662.
[137] S. Bernstein, K.T. Fehr, The formation of 1.13 nm tobermorite under hydrothermal conditions: 1. The influence of quartz grain size within the system $CaO–SiO_2–D_2O$, Prog. Cryst. Growth Charact. Mater. Elsevier, Oxford (2012).
[138] R. Vidale, Pore solution compositions in a pelitic system at high temperatures, press and salinities, Am. J. Set. 283A (1983) 298–313.
[139] G.K. Jacobs, D.-M. Kerrick, A simple rapid-quench design for cold–seal pressure vessels, Am. Min. 65 (1980) 1053–1056.

4 Physical Chemistry of Hydrothermal Growth of Crystals

4.1 Introduction

This is perhaps the least known aspect of the hydrothermal growth of crystals. No book on crystal growth gives a complete picture of the theoretical aspects of hydrothermal growth. Similarly, no author has produced a definitive discussion of the physical chemistry of hydrothermal growth of crystals, covering aspects of physico-chemical and hydrodynamic principles, solutions, solubility, phase equilibria, thermodynamics, kinetics, modeling or intelligent engineering of the hydrothermal reactions, and so on. The Nobel Symposium organized by the Royal Swedish Academy of Sciences during 1978, followed by the First International Symposium on Hydrothermal Reactions organized by the Tokyo Institute of Technology in 1982, helped in setting a new trend in hydrothermal technology by attracting physical chemists in large numbers [1,2]. Hydrothermal physical chemistry today has enriched our knowledge greatly through a proper understanding of hydrothermal solution chemistry, which holds a key to the intelligent engineering of materials. This is also popularly known as thermochemical modeling or thermochemical computation. There are several groups all over the world working on various aspects of physical chemistry under hydrothermal conditions.

4.1.1 *Physicochemical and Hydrodynamic Principles of the Hydrothermal Growth of Crystals*

This is one of the fundamental aspects of hydrothermal growth of crystals. There are two aspects to be discussed here:

i. The physicochemical studies on various aqueous solutions studied within a wide range of pressure–temperature conditions.
ii. The physical chemistry of the hydrothermal growth of crystals.

Each one of these has been studied extensively. However, the correlation of the two works has not been done with enough accuracy.

Ezersky et al. [3,4] have studied the hydrodynamics under hydrothermal conditions using a shadowgraph technique. This facilitated understanding of the spatio-temporal structure of hydrothermal waves in Marangoni convection. However, a lot more experimental and theoretical developments are needed to clarify and complete the description of the hydrothermal waves. Thus, an emphasis is made on some of the basic physicochemical aspects of the hydrothermal growth of crystals.

During the crystallization process from saturated solutions, the mol number of the solute compounds changes and, therefore, the free energy is represented as a function of not only the temperature and pressure but also the mol numbers:

$$G = f(T, P, n_1, n_2, \ldots, n_i) \tag{4.1}$$

$$dG = (\delta G/\delta T)_{p,n} dT + (\delta G/\delta P)_{T,n} dP + (\delta G/\delta n_1)_{P,T,n} dn_1 + (\delta G/\delta n_1)_{P,T,n} dn_2 \tag{4.2}$$

where

n_i = constant mol number of all components,
n_j = constant number of components except one.

The variation in the free energy can be expressed in terms of the chemical potential:

$$(\delta G/\delta n_i)_{P,T,n} = \mu_i \tag{4.3}$$

When P and T are kept constant, Eq. (4.2) becomes:

$$dG_{P,T} = \mu_i \, dn_1 + \mu_i \, dn_2 + \cdots, \quad \text{or } dG_{P,T} = (\Sigma dn_i)_{P,T} \tag{4.4}$$

The integration of Eq. (4.4) gives:

$$G_{P,T} = \Sigma_i n_i \tag{4.5}$$

Differentiation of Eq. (4.5) gives:

$$\begin{aligned} dG_{P,T} &= n_1 \, d\mu_1 + n_2 \, d\mu_2 + \cdots + n_i \, d\mu_i + \mu_i \, d\mu_1 + \mu_i \, d\mu_2 + \cdots + \mu_i \, d\mu_i \\ &= \Sigma n_i \, d\mu_i + \Sigma \mu_i \, dn_i \end{aligned} \tag{4.6}$$

By equating Eqs (4.4) and (4.6) we get:

$$\Sigma n_i \, d\mu_i = 0 \tag{4.7}$$

Equation (4.7) is considered as the overall equilibrium condition in the system with variable mol numbers under constant P and T.

Let us consider a hydrothermal system containing n_A mol of solid A and partially soluble (n_{1A} mol) in n_B mol solvent. Then the free energy of the system corresponds to Eq. (4.5):

$$G = n_A \mu_A + n_{1A} \mu_{1A} + n_B \mu_B \tag{4.8}$$

If the process of dissolution of the solid components continues, then the free energy dG becomes:

$$G + dG = (n_A - dn)\mu_A + (n_{1A} + dn)\mu_{1A} + n_B\mu_B \tag{4.9}$$

The change in the free energy corresponding to this, under constant P and T, becomes:

$$dG = (\mu_{1A} - \mu_A)dn \tag{4.10}$$

or

$$dG/dn = \mu_{1A} - \mu_A \tag{4.11}$$

If $\mu_{1A} < \mu_A$, then dG/dn is negative and the process occurs on its own, since the free energy decreases. When $\mu_{1A} > \mu_A$, the crystallization of the substance from solution B takes place and the chemical potential decreases, and when $\mu_{1A} = \mu_A$, the process reaches the equilibrium state. Consequently, the solute component changes over from the phase wherein its chemical potential is higher into the phase where its chemical potential is lower. The phase transition stops when the chemical potential of the component becomes equal in both phases. It is observed that the above-considered mechanism is realized only when, in the nonequilibrium process, the changes in dG do not depend upon the chemical potential and concentration of other components of the solution. However, consideration of the above relationship may help in understanding the physicochemical principles of the hydrothermal process.

Most hydrothermal crystal growth experiments are carried out under the conditions of temperature gradient in standard autoclaves. The growth of a single crystal into the seed can be carried out in two ways:

1. Recrystallization of the solid substance, including its dissolution in the liquid phase, convective mass transfer of the dissolved part of the substance to the growth zone or seed.
2. Dissolution of the mixture of the nutrient components with the help of their convective mass transport into the growth zone and interaction of the dissolved components on the seed surface.

The methodology of the growth of single crystals on a seed is the establishment of growth conditions in which the process is represented by the sum of macro- and microprocesses occurring between the interface boundary of the solution and the crystal. The composition and concentration of the solution, temperature and pressure, hydrodynamic conditions, and surface contact of the phases are some of the basic physical and chemical parameters which determine the regime and rate of dissolution of the nutrient, mass transport, and the possibility of the formation of new phases.

The isothermal chemical reaction can be expressed in terms of the van't Hoff equation:

$$\Delta G_{P,T} = RT \ln \pi a' - RT \ln K_{ai} \tag{4.12}$$

where

$\pi a' = \pi' \cdot (a_i', y_i)$,
$K_a = \pi' \ (a_i', y_i)$, is the thermodynamic equilibrium constant,
y_i = activity coefficient,
a_i = initial nonequilibrium activity of the nutrient component in the solution,
a_i = equilibrium activity of the nutrient components in the solution.

From Eq. (4.12) it is observed that ΔG depends upon the relative values of $\pi a'$ and K_a when

$$\pi a' < K_a; \quad \Delta G < 0 \tag{4.13}$$

$$\pi a' = K_a; \quad \Delta G = 0 \tag{4.14}$$

$$\pi a' > K_a; \quad \Delta G > 0 \tag{4.15}$$

For example, in Eq. (4.13) the process is irreversible and the reaction takes place from left to right. In the case of Eq. (4.14), the reaction takes place under equilibrium condition and, for example, in Eq. (4.15) the process becomes reversible. As a result, the changes in ΔG in Eq. (4.12), for example, can be represented as follows:

$$\Delta G = RT \ \ln(a_i/a_i')$$

where a/a_i' is the analogue of the relative supersaturation.

If the activity coefficient $y \approx 1$ or, in the given region, y does not depend upon the concentration then, for example in Eq. (4.12), the activity coefficient can be replaced by the mol fraction. Then the ratio a_i/a_i' is replaced by S_i/S_i', where S_i' = undersaturated solubility.

Equilibrium concentration of the supersaturated solution should be achieved. The ratio S_i/S_i' is called the relative supersaturation or degree of supersaturation.

In recent years, there has been good progress on the experimental side. Much credit goes to the Russian hydrothermal crystal growers. Balitsky and Bublikova [4a] have studied the physicochemical foundation of malachite synthesis. Similarly, Kuznetsov [4b] has reviewed the physical chemistry of hydrothermal crystal growth and the crystallization of some A_2B_6 crystals. Further, Popolitov [4c] has reviewed the physical chemistry of the hydrothermal growth of tellurium dioxide crystals. The reader can get valuable information in the book *Hydrothermal Growth of Crystals* [5]. These studies on the said works, along with the pioneering works by Hegleson, Pitzer, Frank, Seward, Kaufman, etc. set a new trend for thermochemical computation leading to the intelligent engineering of materials through the generation of stability field diagrams for a variety of systems followed by their experimental validation.

4.2 Basic Principles of Phase Formation Under Hydrothermal Conditions

A *phase* is defined as a part of the system that is homogeneous throughout and is physically separable from other phases by distinct boundaries. The number of components in a system is the minimum number of chemical constituents that must be specified to describe the composition of all phases present. Phase relationships are conveniently represented by means of phase diagrams. Phase diagrams and the phase rules apply only to systems at equilibrium. These data are extremely important to any crystal grower for successful growth of crystals. Gibbs deducted a quantitative relationship for phase rule:

$$F = C - P + 2$$

where

F = degrees of freedom,
C = number of components,
P = number of phases.

The above relation holds true if pressure, temperature, and composition are the only variables. If other variables such as magnetic, electric, or gravitational fields are considered, the numerical value 2 on the right-hand side of the equation will increase by the number of new variables considered.

The hydrothermal method is considered as one of the best methods to study phase relationships in numerous systems, and the PTX diagrams thus obtained are published from time to time. Phase equilibria have been described in detail by several workers [6,7]. Hence, we restrict our discussion to some basic systems and nonconventional phase diagrams.

Among the binary systems, the H_2O—CO_2 system is important with regard to the role of H_2O and CO_2 vapor pressures in native mineral formation and in the hydrothermal synthesis of inorganic compounds. Figure 4.1 shows the phase diagram for the H_2O—CO_2 system [6]. It is found that, below the critical temperature of water, the dependence V versus P has a gap at the pressure corresponding to the pressure of water at its critical density. The isochores of real mixtures are linear up to the immiscibility dome that gives a possibility to experimental data in the calculation of pressures of mineral formation from the gas—liquid inclusions.

The nonconventional phase diagrams plotted for equilibrium conditions are popular, especially among Russian workers. These diagrams are popularly known as *composition diagrams* or NC diagrams, where N is the nutrient composition and C is the concentration of the solvent mineralizer. Hundreds of phase diagrams of compositions are available, especially in the Russian literature. An example of such a diagram is given in Figure 4.2 corresponding to the system Na_2O—ZrO_2—SiO_2 [9]. Similarly, the distribution of the silicon—oxygen radical in the rare earth silicates is shown in the NC diagram in Figure 4.3 corresponding to the system Na_2O—RE_2O_3—SiO_2—H_2O under hydrothermal conditions [10]. The solid lines

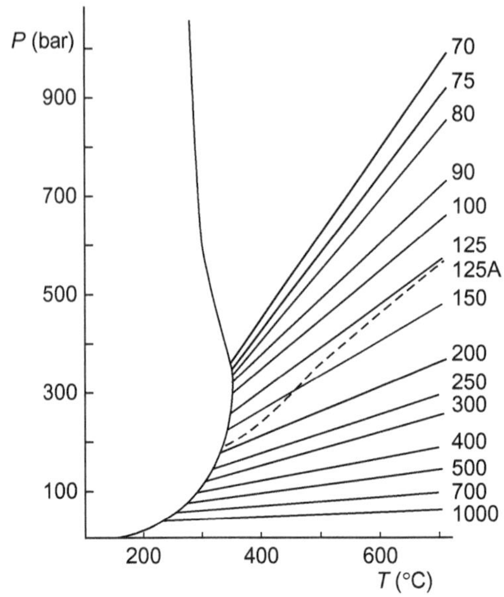

Figure 4.1 Phase diagram for the H_2O-CO_2 system [8].

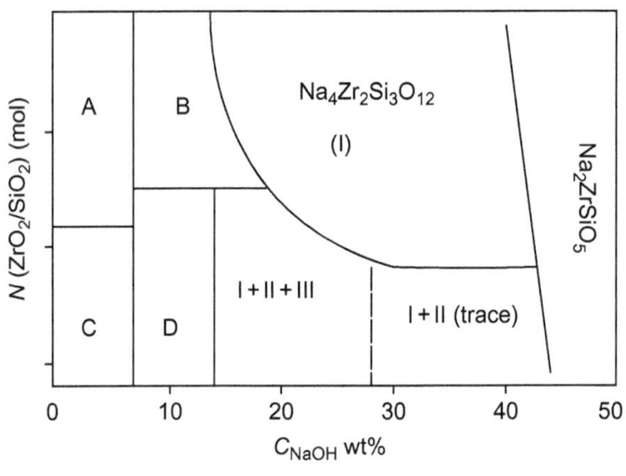

Figure 4.2 NC diagram for the system $Na_2O-ZrO_2-SiO_2$ [9].

indicate the regions of monophase crystallization and dashed lines indicate the beginning of crystallization of the excess component. Similarly, temperature versus concentration diagrams (TC diagrams) are also popular for some of the systems, because the variation in temperature of the system can lead to the formation of various polymorphic modifications and several other phases which are not normally observed in NC diagrams. Figure 4.4 shows the TC diagram for the system $K_2O-La_2O_3-P_2O_5-H_2O$ [11]. These nonconventional diagrams are relatively easy to obtain and are highly useful for the growth of single crystals as they clearly depict the growth conditions. Similarly, just as in the case of conventional phase

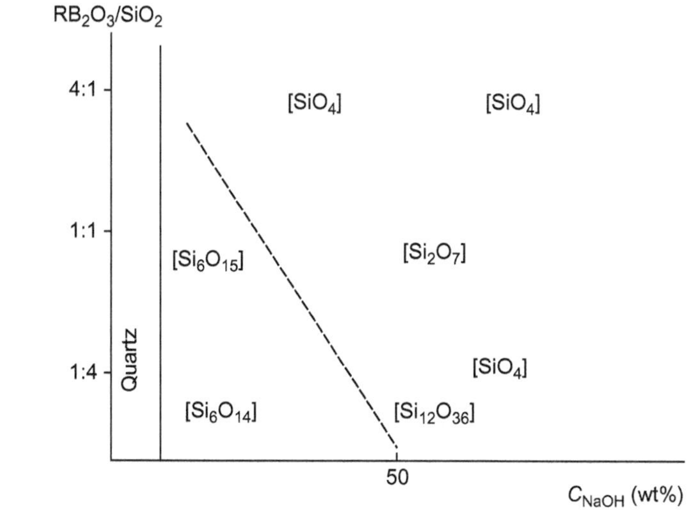

Figure 4.3 NC diagram for the system $Na_2O-RE_2O_3-SiO_2-H_2O$ under hydrothermal conditions.

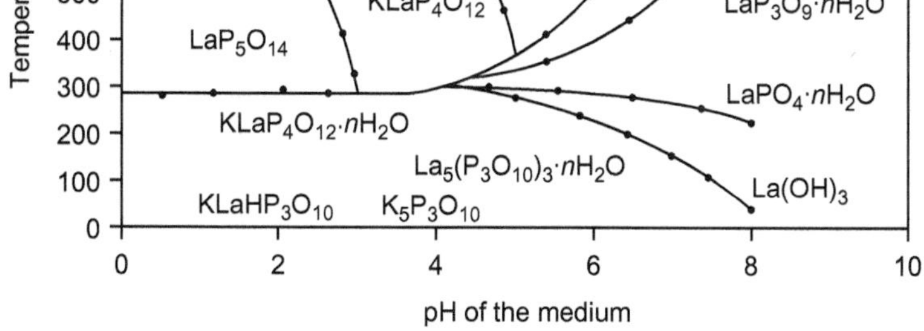

Figure 4.4 TC diagram for the system $K_2O-La_2O_3-P_2O_5-H_2O$ [11].

diagrams—even in the phase diagrams of crystallization, the components can be represented, for example, in a ternary system as a triangle. Figure 4.5 shows the phase diagram of crystallization in the system $LiO_2-Ta_2O_5-H_2O$ at 750°C and 1700 atm [12].

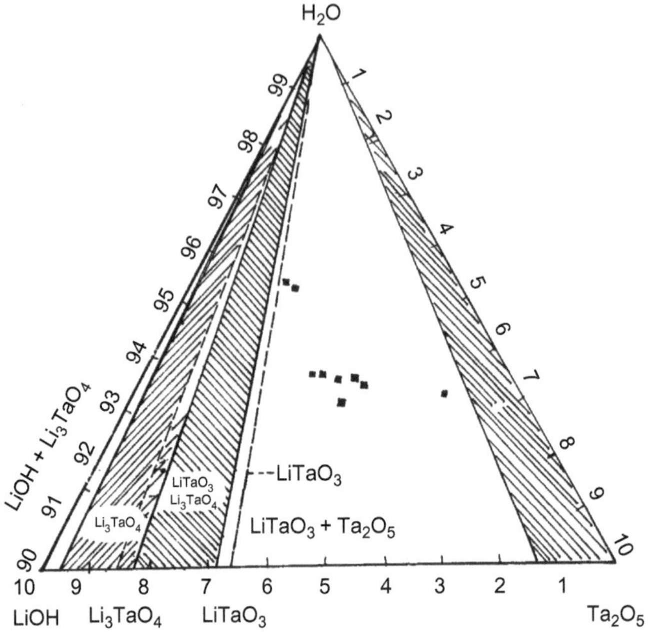

Figure 4.5 The phase diagram of crystallization in the system $LiO_2-Ta_2O_5-H_2O$ at 750°C and 1700 atm.

4.3 Solutions, Solubility, and Kinetics of Crystallization

This is one of the most important aspects of the hydrothermal growth of crystals. Initial failures in the hydrothermal growth of a specific compound are usually the result of lack of proper data on the type of solvents, the solubility, and solvent–solute interaction. Despite the large amount of literature data already available, it is still a highly attractive field in hydrothermal research, geoscience, and so on. For example, geologists are now in a position to tell us much about the conditions of formation of hydrothermal ore deposits (e.g., temperature, pressure, oxidation potential, pH, and fluid salinity), but in order to fully understand the mechanisms of transport and deposition of ore minerals, information about metal complex formation, stability, and stoichiometry in high-temperature and high-pressure aqueous electrolyte solutions, thermodynamic properties of electrolytes at high pressures and high temperatures, phase behavior, transport phenomena, and electrical and optical properties is necessary [13–15]. Unfortunately, most of the studies have been performed only at the saturated vapor pressure of the system and very few combined high-temperature and high-pressure experiments have been designed to yield thermodynamic data.

Recent advances in computer technology, which have seen both speed and storage capacities dramatically increased, offer the possibility that chemical speciation modeling of multicomponent electrolyte solutions can be made much more reliable than it has been. Automation of the whole calculation process involving the assessment and selection of data, the determination of basis sets, the formation of equations, and the methods by which results are evaluated are now within reach. This should be seen as the provision of technical assistance rather than as a substitute for human expertise. Nevertheless, by reducing errors and by ensuring that assumptions are recorded systematically, many of the problems that have been so apparent in the literature may be better controlled in the future [16,17].

Sterner et al. [18] have developed a new computer algorithm, INSIGHT, which facilitates the nonlinear thermodynamic analysis of large heterogeneous data sets for the purpose of developing accurate equations of the state of aqueous solutions and their interactions with other substances. The code is capable of incorporating a diverse array of primary thermophysical data including: osmotic and vapor pressure, freezing point depression, solubility [fixed pH and pCO_2], electromotive force data, speciation information obtained from spectroscopic measurements and molecular dynamic simulations. Also it can in corporate several other parameters such as enthalpies of mixing, dilution and solution heat capacity, volumetric, and compressibility data. These data, together in a global, polythermal polybaric, thermodynamic analysis simultaneously solve chemical equilibria within the aqueous phase via Gibbs energy minimization.

A hydrothermal solution is generally considered as a thermodynamically ideal one, yet in the case of strong and specific interaction between the solute and the solvent or among the components of the soluble substance in them, significant deviations from Raoult's law occur. Consequently, real hydrothermal solutions differ from ideal solutions and their understanding requires knowledge of the influence of the solvent in the process of dissolution and crystallization of various compounds. Obviously, as shown in most of the experiments, the type of solvent and its concentration determine a specific hydrothermal process and its important characteristics such as the solubility of the starting materials, quantity of the phases, their composition, output of the phases, kinetics, and growth mechanism of single crystals. The value of the changes in the Gibbs free energy in reaching an equilibrium condition varies with the transition from one solvent to another and it can be shown as:

$$\Delta G = \Delta H - T\Delta S = RT \ \ln \ K, \quad \text{or } K = \exp(-\Delta H/T)\exp(\Delta S/R) \qquad (4.16)$$

where K = equilibrium constant.

The above expression shows the influence of enthalpy and entropy on the equilibrium constant, so that the enthalpy and entropy of the solubility (at constant P and T) are different for different solvents. Also, the solubility of one and the same solid substance changes with the solvent.

At the moment, there is no theory which can explain and estimate the solubility in real solutions. However, many of the problems connected with solubility can be explained on the basis of overall physicochemical principles or laws.

In some cases, it is better to use the empirical rule which agrees with the fact that solubility becomes high in solvents with higher dielectric constant (ε) and types of chemical bond which are closer to those of the solute substance. Deviation from this group has a place in the case where specific interaction between solid substance and the solvent occurs.

The synthesis and recrystallization of a specific compound and the growth of single crystals on the seed are all carried out using different solvents on the basis of physicochemical considerations. The following conditions are adopted in selecting the most suitable mineralizers:

i. Congruence of the dissolution of the test compounds.
ii. A fairly sharp change in the solubility of the compounds with changing temperature or pressure.
iii. A specific quantitative value of the absolute solubility of the compound being crystallized.
iv. The formation of readily soluble mobile complexes in the solution.
v. A specific redox potential of the medium ensuring the existence of ions of the required valence.

Additionally, the solvent should have the desired viscosity, insignificant toxicity, and very weak corrosion activity with respect to the apparatus.

These factors fulfill the requirement of the hydrothermal mineralizer in addition to determining the values of solubility of the compound under investigation.

The interaction between the solute and the solvent is called *solvation*. Solvent molecules become linked to the dissolved molecules or ions, and the resulting species containing solvent molecules are *solvates*. Depending upon the reactants, the linkage between the solvent and the solute may exhibit different specific (coordination, hydrogen bond formation, etc.) and nonspecific (electrostatic) factors, or solvation processes of different strength, and hence, solvates include very varied formations. Therefore, it is not possible to find a single physical parameter characterizing the solvent which, in itself, could rationalize the solvation process. Simple ions may differ considerably in their electronic structures, charges, and radii, and even larger differences are exhibited by solvent molecules representing different types of chemicals. Hence, solvated ions may also be very different as regards the number of solvent molecules bound to the ion (the *salvation number*), the forces giving rise to the binding, and the resulting physical and chemical properties like spatial requirements (there are various ways of determining the solvation number—viscosity measurements, pH, conductivity measurement, density measurements, and so on). The specific solvation process is the result of various types of complex formation between the solvent and solute components. In the case of homogeneous aqueous solutions of specific solvation, an activated complex theory can be considered [19], according to which the rate constant, ξ, of the reaction type $A + B \equiv X^{\neq} \rightarrow$reaction products ($X^{\neq}$ is the activated complex) is determined by:

$$\xi = (kTK^{\neq}/h)(\nu_A \nu_B/\nu^{\neq}) \tag{4.17}$$

The concentration of the activated complex X may be determined by the following equation:

$$[X^{\neq}] = K^{\neq}(C_A)(C_B)\nu_A \nu_B/\nu^{\neq} \tag{4.18}$$

where h is the Planck constant, $K^{\neq}$ is the equilibrium constant of the activated complex, C_A and C_B are the concentrations of the components A and B, and ν_A and ν_B are the activity coefficients of the nutrients, and $\nu^{\neq}$ is the activity coefficient of the active complex.

In nonspecific solvation when the medium is not homogeneous, i.e., the solvent–solute interaction is not homogeneous and continuous, such an interaction plays a significant role for ionic compounds only [20,21]. The classical electrostatic model links the rate of reaction constant ξ (in a given known dielectric value ε), with the rate constant (in the medium with $\varepsilon = \infty$), Eq. (4.18):

$$\ln \xi = \ln \xi_0 + \frac{e^2}{2\varepsilon kT}\left[\frac{Z_A^2}{r_A} + \frac{Z_B^2}{r_B} - \frac{(Z_A + Z_B)}{r_{\neq}}\right] + {}^3\!/\!_4\varepsilon kT\left(\frac{\mu_B^2}{r_B^3} - \frac{\mu_{\neq}^2}{r_{\neq}^3}\right) \tag{4.19}$$

where r is the radius of the particle, μ is the dipole moment of the particles, Z is the charge number, k is the Boltzmann constant, T is the temperature, and e is the electron charge.

This equation shows that the log of the rate constant depends linearly upon the reverse (negative) values of the dielectric constants.

Let us consider briefly the role of water as a solvent in the hydrothermal growth of crystals. Water is the major component of hydrothermal solutions, always with a varied chemical composition in the laboratory and in nature.

All the solutions used in hydrothermal experiments vary from one another in their properties, ability to dissolve and crystallize, and in the nature of the linking between water and electrolyte. Moreover, the properties of each solution depend upon physicochemical aspects and the structure of the pure water. The formation of associates and complexes in the aqueous solutions of electrolytes is possible because of the presence of structural water, i.e., consisting of water molecules with directional hydrogen bonding. Above the critical point, the density of water varies greatly with a little change in temperature and pressure. Because of the drastic change in density, all the fluid properties change greatly around the critical point, including the dielectric constant that is a controlling factor of reaction rate, equilibrium, and solubility of metal oxides. Therefore, it is necessary to understand the basic principles, which insist upon the understanding of the properties of water, including density, dielectric constant, and ion product, varying greatly around the critical point of water and result in a specific reaction atmosphere [22]. The experimental PVT-behavior of water has been reviewed and summarized by several workers [23–26]. Figure 4.6 shows the PVT relation of water [22]. The reaction rate, equilibrium, phase behavior, solubility of metal oxides, and distribution of soluble chemical species change greatly at the critical point range. Various models have been proposed to describe the variation of reaction rate or equilibrium over a range of supercritical state [27,28]. Due to the variation in the properties of water, phase behavior changes greatly around the critical point. Since supercritical water is a high density steam, light gases like oxygen or hydrogen form a homogeneous phase with supercritical water. Figure 4.7 shows the critical loci for binary systems

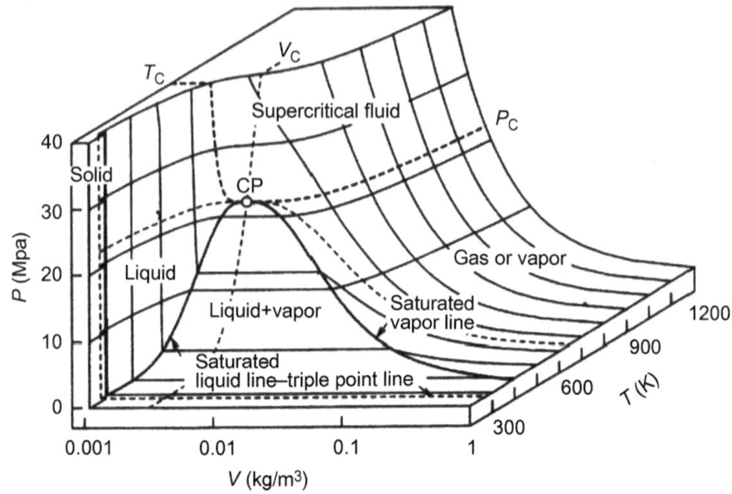

Figure 4.6 PVT surface of water: CP—critical point [22].

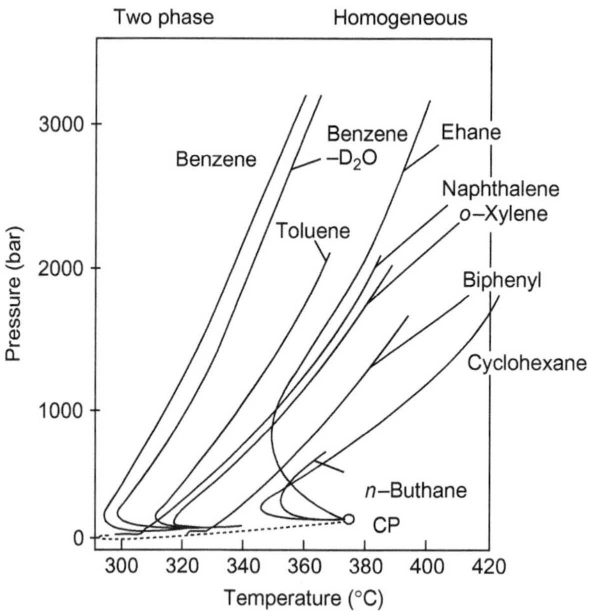

Figure 4.7 Phase behavior of binary systems for water–organic compounds [29].

of organic compound and water [29]. In the right-hand side of these curves, a homogeneous phase is formed at any composition. The organic compounds and water are not miscible at a low-temperature range but form a homogeneous phase at higher temperatures, which is because of the reduced dielectric constant of water.

It should be noted that when the pressure is very high (i.e., the density of fluid is as high as of liquid), even if both water and organic compound are in supercritical state, phase separation occurs, just like liquid–water and liquid–oil phase separation. Several chemical changes arise from changes in ionic dissociation of the solution. Therefore, it is better to understand the characteristics of the pure water under hydrothermal conditions, especially the hydrogen ions show an influence on the solubility of various compounds under hydrothermal conditions.

4.4 Thermodynamic Principles of Solubility

Thermodynamic principles enable one to determine how to design a reaction to yield phase-pure phases. Without knowledge of how to do this, it is impossible to distinguish a process that is being controlled by thermodynamics versus kinetics. A large number of publications have appeared on this aspect based on both the theoretical and experimental data. Several models have been proposed in order to understand the hydrothermal synthesis of a variety of compounds like ABO_4 (A = alkaline-earth elements, B = Ti, Zr, Hf), HAp, sulfides of gold, silver, iron, and copper [30–35]. Thermodynamic studies yield rich information on the behavior of solutions with varying pressure–temperature conditions. Some of the commonly studied aspects are solubility, stability, yield, dissolution–precipitation reactions, and so on under hydrothermal conditions. The thermodynamic principles of solubility are discussed here in brief. It is well known that:

$$\Delta G^0_{\mathrm{sol}} = \Delta H^0_{\mathrm{sol}} - T\Delta S^0_{\mathrm{sol}} \tag{4.20}$$

where $\Delta G^0_{\mathrm{sol}}$, $\Delta H^0_{\mathrm{sol}}$, and $\Delta S^0_{\mathrm{sol}}$ are the changes in the corresponding free energy, enthalpy, and entropy of dissolution.

If $\Delta G^0_{\mathrm{sol}} < 0$, then dissolution takes place. When $\Delta G^0_{\mathrm{sol}} = 0$, the system exists in the equilibrium state; when $\Delta G^0_{\mathrm{sol}} > 0$, crystallization is thermodynamically possible.

With a change in the enthalpy, it is observed that either:

i. In the entire temperature interval, $\Delta H^0_{\mathrm{sol}} < 0$.
ii. In the low-temperature region, $\Delta H^0_{\mathrm{sol}} > 0$, whereas in the high-temperature region, $\Delta H^0_{\mathrm{sol}} < 0$.
iii. In the entire temperature region, $\Delta H^0_{\mathrm{sol}} > 0$.

If entropy is considered in Eq. (4.20) then with an increase in the temperature, just like $\Delta H^0_{\mathrm{sol}}$, the $\Delta S^0_{\mathrm{sol}}$ may also change. Here also we can expect three possibilities:

1. In the entire temperature region, $T\Delta S^0_{\mathrm{sol}} < 0$.
2. In the low-temperature region, $T\Delta S^0_{\mathrm{sol}} > 0$, whereas in the high-temperature region, $T\Delta S^0_{\mathrm{sol}} < 0$.
3. In the entire temperature interval, $T\Delta S^0_{\mathrm{sol}} > 0$.

In the first case, the entropy helps in the dissolution process. In the second case, it can be both helping and reducing, and in the third case, it reduces the dissolution process.

In the majority of cases, the solubility of the solid substances in a liquid phase is limited. Solubility is an important physicochemical and technological parameter whose estimation determines any dissolution process. It acts as a factor which strongly influences the kinetics or rate of dissolution. Solubility depends upon the nature of the substance, its aggregate state, temperature, pressure, and a series of other factors. Quantitative dependence of the solubility of the solid substance on the temperature at a given constant pressure is given by:

$$(\delta \ln S/\delta T)_P = (Q_2/RT^2)(\delta \ln a_2/\ln S_{\mathrm{PT}}) \tag{4.21}$$

where

S = solubility of the solid substance,
a_2 = activity of the solid substance,
Q_2 = heat of dissolution of 1 mol of the substance in the saturated solution,
R = gas constant.

The solubility of the solid substance may be expressed as a set of two processes: (1) Melting of the solid substance and (2) mixing of two liquids.

In the case of ideal solubility of the solid substance, the heat of the second process is equal to zero and $\overline{Q}_2$ is equal to the heat of melting of the solid substance: $\overline{Q}_2$ = % melting and does not depend upon the selection of the solvent, where if $\delta \ln a_2/\delta \ln S = 1$, then Eq. (4.21) can be written as:

$$(\delta \ln S/\delta T)_P = \%\,\mathrm{melting}/RT^2 \tag{4.22}$$

such that the right-hand side is >0, i.e., with an increase in the growth temperature, ideal solubility of the solid substance in an ideal solution increases. The dependence of the solubility of any substance in the solution and the pressure is expressed as:

$$(\delta \ln S/\delta T)_T = (V^0_{\mathrm{solid}} - V_{\mathrm{liquid}}/RT)(\delta \ln a_2/\delta \ln S)_{PT} \tag{4.23}$$

where V_{liquid} = partial molar volume of the soluble substance in the solution, V^0_{solid} = molar volume of that substance in the solid phase. In the case of an ideal solution, $\ln a_2 = \ln S$, $\delta \ln a_2/\delta \ln S = 1$, and Eq. (4.23) becomes:

$$(\delta \ln S/\delta P)/T = (V^0_{\mathrm{solid}} - V_{\mathrm{liquid}})/RT \tag{4.24}$$

Thus, in the case of ideal solutions of the solid substance in the investigated media, the solubility decreases with an increase in the pressure, if the molar volume of the soluble substance increases during its melting. Similarly, if $V^0_{\mathrm{solid}} - \overline{V}_{\mathrm{liquid}}$, then the solubility should increase with an increase in the pressure (in the case when the partial volume of the liquid is equal to the volume of the melting salt).

In recent times, many aspects of hydrothermal chemistry have been experimentally studied, particularly in the past decade. The credit goes to workers like Franck, Seward, Hegleson, and others [13–15,36]. However, most of these studies do not give an insight into the complete physical chemistry of the hydrothermal media as they represent mainly the lower PT conditions. Helgeson has reviewed exhaustively the thermodynamic properties of electrolytes at high pressures and high temperatures.

An understanding of the theory and the experimental results on hydrothermal physical chemistry is essential to the crystal grower. Here, in the experimental work, some of the basic properties of the hydrothermal medium like viscosity, dielectric constant, compressibility, and coefficient of expansion are discussed briefly in the crystal growth context. Since diffusion is inversely proportional to solvent viscosity, we would expect very rapid diffusion in hydrothermal growth.

This leads to the growth of perfect single crystals with well-developed morphology. We can expect higher growth rates, a narrower diffusion zone close to the growing interface and less likelihood of constitutional supercooling and dendritic growth. It is, thus, no wonder that quartz growth rates as high as 2.5 mm/day without faults or dendritic growth have been observed [37]. Figure 4.8 shows the

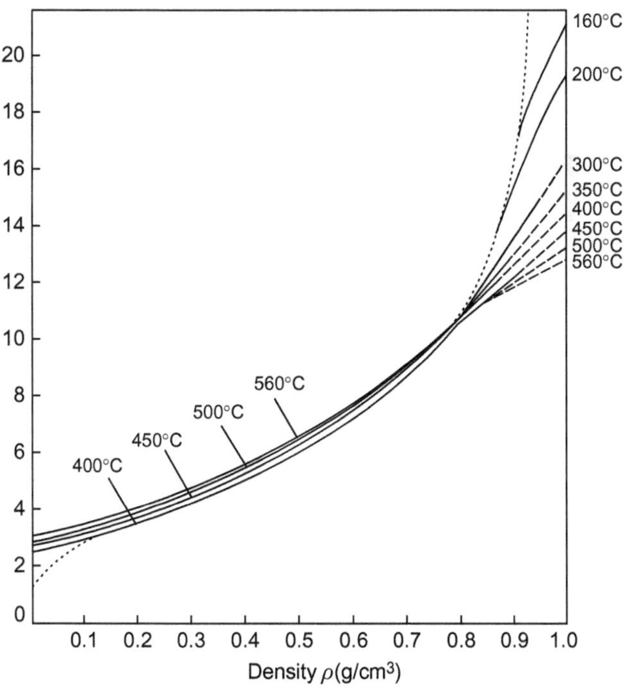

Figure 4.8 Viscosity of water as a function of density and temperature [36].

viscosity of water as a function of density and temperature [36]. It has been demonstrated in many cases that the mineralizer solutions (typically 1 M NaOH, Na_2CO_3, NH_4F, K_2HPO_4, etc.) are close to the properties of water. For 1 M NaOH at room temperature, $\eta_{solution}/\eta_{H_2O} = 1.25$ [38], one can expect that the viscosity of hydrothermal solutions can be as much as two orders of magnitude lower than "ordinary" solutions. The mobility of molecules and ions in the supercritical range is much higher than under normal conditions. Also, electrolytes, which are completely dissociated under normal conditions, tend to associate with rising temperatures [39].

The dielectric constant is one of the important properties of a solution. A knowledge of the temperature dependence of the relative dielectric constant is of great importance for understanding the hydration/dehydration phenomena and reactivities of a variety of solutes in supercritical water. More than 40 years ago, Hasted et al. [40] observed that the dielectric constant of electrolyte solutions (ε) can be regarded as a linear function of molarity up to 1–2 M, depending on the electrolyte. Franck [36] has discussed ionization under hydrothermal conditions and made most careful and complete conductivity studies, and shown that the conductance of hydrothermal solutions remains high in spite of a decrease in ε, because that effect is more than compensated for by an increase in the ion mobility brought about by decreased viscosity under hydrothermal conditions. Figure 4.9 shows the dielectric constant of water [41]. Thermodynamic and transport properties of supercritical water are remarkably different from those of ambient water. Supercritical water is unique as a medium for chemical processes. The solubility of nonpolar species increases, whereas that of ionic and polar compounds decreases as a result of the drop of the solvent polarity, and the molecular mobility increases due to a decrease in the solvent viscosity (η). Drastic changes of ionic hydration are brought about by the decrease in the dielectric constant (ε) and density (ρ). Largely as a consequence

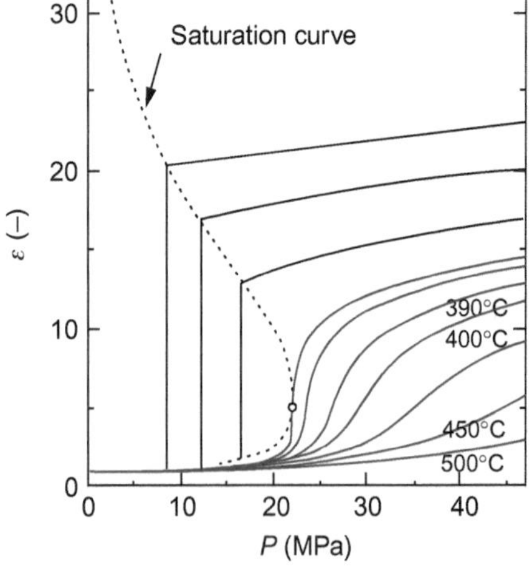

Figure 4.9 Dielectric constant of water [41].

of the dramatic decrease in the dielectric constant of water with increasing temperature at constant pressure and/or decreasing pressure at constant temperature, "completely" dissociated electrolytes at low temperatures and pressures may become highly associated in the supercritical region. Computer simulation studies have been carried out on this topic, which are helpful for the interpretation from the molecular point of view. Viral expansion of Debye–Hückel theory affords a basis for extrapolating thermodynamic observations of "completely" dissociated electrolytes to high pressures and high temperatures, where they may be highly associated. However, the Debye–Hückel equation fails to represent adequately the observed activity coefficients for dilute solutions of 2:2 electrolytes. Landolt-Börnstein [42] have estimated the compressibility of water from the PVT curves. The hotter and less dense the fluid is, the larger is its compressibility. It is an important factor used in the calculation of density change with pressure. The coefficient of thermal expansion is also an important parameter that helps in the growth of flawless and strain free crystals. It can be calculated from PVT data. Water has a much increased coefficient of expansion under hydrothermal conditions. PVT measurements for water and other solvents are rather rare. Water above critical temperature ($t_c = 374.2°C$) and critical pressure ($p_c = 22.1$ MPa) is called *supercritical water.* At the critical point, the phases of liquid and gas are not distinguishable. There are extensive studies on the structure of water under ambient conditions, but the structure of water at high temperatures and high pressures are not well investigated, mainly due to experimental difficulties. In recent years, there has been increased interest in the study of the structure, dynamics, and reactivity of supercritical and subcritical water from the chemical, physical, geological, and engineering points of view. There are several reasons for this trend. Water at high temperatures and high pressures is believed to have played a major role in creating organisms on the earth, the compounds necessary for them, and fossil fuels produced after their death.

Seward [43], Nakahara et al. [44], Yamaguchi et al. [45], Nakahara et al. [46], and Ohtaki et al. [47] have carried out extensive experimental work on the aqueous systems at high temperatures and high pressures using conductivity, potentiometric, spectrophotometric, solubility, PVT, and calorimetric methods. However, insight into the configurational aspects of ion–solvent interactions under hydrothermal conditions has come mainly from neutron diffraction studies and more recently from synchrotron X-ray absorption (EXAFS) spectroscopic measurements [43–47]. These latter studies have confirmed that cation-oxygen (*nearest-neighbor water*) distances decrease with increasing temperature and are accompanied by an associated decrease in the number of first shell solvated water molecules as has been shown for Ag^+, Sr^+, and other cations [43].

Water is an environmentally safe material and cheaper than other solvents, and it can act as a catalyst for transformation of desired materials by tuning temperature and pressure. The larger the reaction barrier height, the larger the temperature effect on the reaction. For example, reactions whose Arrhenius activation energies are 42, 83, and 125 kJ mol^{-1} in water at 25°C are accelerated by factors of 8.2×10^3, 6.6×10^7, and 5.4×10^{11}, respectively, at 400°C [48]. Thus, a new reaction pathway, which is not available in ambient water due to a high barrier, low solubility, or low-molecular diffusion rate, can be made accessible thermally in supercritical water. Studies of the structure dynamics and reactivity of supercritical water lead to a better understanding of how complicated compounds are adaptively evolved from simple ones in hydrothermal reactions. It is important to elucidate how the structure and dynamics of supercritical water are controlled by intermolecular interactions (density) as well as kinetic energies (*T*). This is a new frontier of solution chemistry. The PVT data for water up to 1000°C and 10 kbar is known accurately enough (within 1% error) [49–51]. At very high PT conditions (1000°C and 100 kbar), water is completely dissociated into H_3O^+ and OH^-, behaving like a molten salt, and has a higher density of the order of $1.7 - 1.9\ g/cm^3$. Figure 4.10 shows the temperature–density diagram of water, with pressure as a parameter [51].

When aqueous solutions are heated up to the solvent critical point, the partial mol volume of "hydrophilic" species becomes largely negative and that of "hydrophobic" ones largely positive due to a divergent behavior of the solvent compressibility. It is well known that, as distances of the intermolecular charge separation decreased along the reaction coordinate, these "hydrophilic" species were dehydrated in supercritical water due to the low dielectric constant of the solvent. The desolvation effect was pronounced in the solvent with a low dielectric constant over a long range. One can regard this dehydration effect on the ionic and dipolar species as the catalytic effect of supercritical water because of its ability of lowering the height of the reaction barrier. Similarly, the conductance data for dilute alkali metal halide solutions at elevated temperatures and pressures have been usually interpreted in terms of the formation of simple, neutral ion pairs. It has nevertheless been recognized that higher order polynuclear species (e.g., triplets and quadruplets) may form in electrolyte solutions having a low dielectric constant at

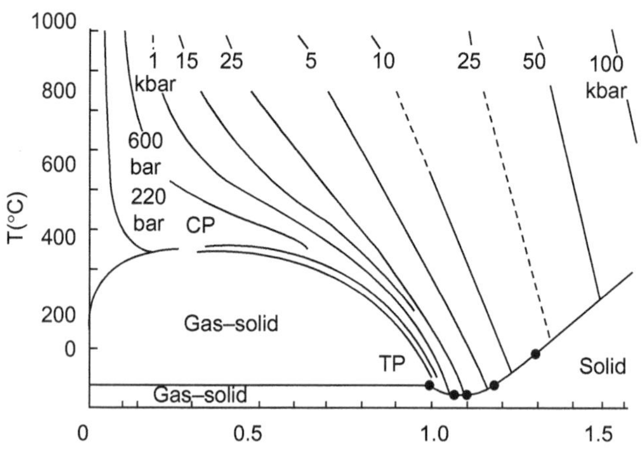

Figure 4.10 Temperature–density diagram of water, with pressure as a parameter [51].

ambient temperatures. By analogy, one might therefore expect the formation of such polynuclear ions or clusters in near-critical region and at supercritical temperatures and pressures in aqueous electrolyte solutions. The molecular dynamics simulations of ion pairing and cluster formation in a 1 M NaCl solution at 380°C and near-critical pressures indicating the presence of simple monatomic ions and ion pairs together with triple ions such as Na_2Cl^+ and $NaCl_2^-$ as well as the Na_2Cl_2 and more complicated polynuclear species. Combined molecular dynamics simulations and EXAFS measurements on 1 M $SrCl_2$ solutions up to 300°C also indicate the presence of Sr^{2+}, Cl^-, $SrCl^+$, $SrCl_2$ as well as cluster species such as $Sr_2Cl_2^{2+}$ and $Sr_2Cl_2^+$ [52]. Insight into ion pair and cluster formation is of importance to the understanding of many phenomena operating in hydrothermal systems in the earth's crust, including mineral equilibria, stable isotope fractionation, and element transport.

Benning and Seward [53] have studied the stepwise metal complex formation of the $AuHS^0$ and $Au(HS)_2^-$ complexes and their thermodynamic data in sulfide aqueous solutions up to 400°C and 1500 bar. However, a major source of uncertainty in the study of metal hydrosulfide complexes at high temperatures and high pressures has been the lack of reliable data for K_1, the first ionization constant for H_2S. There is a major lack of experimental data pertaining to metal complex equilibria in supercritical aqueous systems as well as in binary solvent systems such as $H_2O{-}CO_2$. Walther and Schott [54] applied the dielectric-constant approach to study the speciation and ion pairing at high temperature and high pressure. These authors have proposed many theoretical and empirical relations to account for various aspects of solubility behavior and speciation. The calculation of the dielectric-constant dependence of ion pairing at high temperatures and high pressures allows the prediction of speciation in complex fluids at high temperature and high pressure for reduced activities of H_2O. Using this approach, the behavior of aqueous silica in complex solutions at high temperatures and high pressures has been studied. Such data are of enormous importance to understanding the geochemistry of element transport by hydrothermal fluids migrating in the earth's crust [43].

In the hydrothermal growth of crystals, the PVT diagram of water proposed by Kennedy [55] is very important (Figure 2.6). Usually, in most routine hydrothermal experiments, the pressure prevailing under the working conditions is determined by the degree of filling and the temperature. When concentrated solutions are used, the critical temperature can be several hundred degrees above that of pure water [56]. The critical temperatures are not known for the usually complex solutions at hand; hence, one cannot distinguish between sub- and supercritical systems for reactions below 800°C. Although the temperature in the growth zone and the actual vapor pressure are not known to 100% accuracy, the PVT diagram of Kennedy is used by most hydrothermal crystal growers routinely. The PVT relations in $AlPO_4$ and SiO_2 systems have been reviewed [57–59]. Also, the P–T behavior of the quartz–water system has been well discussed by Brebrick [60].

4.5 Kinetics of Crystallization Under Hydrothermal Conditions

Hydrothermal crystallization is only one of the areas where our fundamental understanding of hydrothermal kinetics was totally lacking until recently. Because, in majority of the cases, we have little knowledge about the intermediate phases forming in solution, our fundamental understanding of hydrothermal crystallization kinetics is at an early stage. In the absence of predictive models, we must empirically define the fundamental role of temperature, pressure, precursor, and time on crystallization kinetics of various compounds. Insight into this would enable us to understand how to control the formation of solution species, solid phases, and the rate of their formation. In this respect, the recent progress in the area of time-resolved hydrothermal synthesis for TEDDI using synchrotron beam and neutron beam has contributed significantly towards the understanding of the intermediate phases and in turn the hydrothermal mechanism of formation of various inorganic phases and their kinetics. There are several groups worldwide working on such studies covering the hydrothermal synthesis of zeolites, clays, and so on. In combination with the thermodynamics of the reaction system, one can precisely understand various parameters related to the hydrothermal synthesis of a desired phase. There are several groups led by professors like A. Clearfied (USA), T. Fehr (Germany), and G. Sankar and P. Barnes (UK) [61–65].

The importance of kinetics of crystallization studies was realized with the commercialization of the synthesis of zeolites during 1950s and 1960s. Following this, a large number of publications appeared from the Soviet laboratories on kinetics studies related to the growth of single crystals under hydrothermal conditions. During this period, the anisotropy of the rate of growth of the faces of the crystals during growth on a seed acquires particular importance, since it makes it possible to orient the seed crystal platelets in such a way that the growth surface coincides with the rapidly growing face [66–70]. At the same time, Laudise and coworkers have studied the crystallization kinetics of quartz and zincite [66–73].

These studies were based on the experience in the growth of crystals in aqueous solutions which shows clearly the relationship between the anisotropy of the rates of process: Temperature, supersaturation, and the presence of "foreign" components, including the solvent under hydrothermal conditions. These parameters can be varied over very wide limits.

The influence of the solvent is specific for each crystal. In the case of ZnO, for example, the presence of NH_4^+ ions in the solution produces a marked increase in the rate of growth of the prismatic faces, which under ordinary conditions does not exceed hundredths of millimeters per day [73]. For calcite under conditions of increased oxygen activity, the faces of the scalenohedra, $(21\bar{3}1)$ and $(10\bar{1}0)$, become predominant [74].

Various general laws characterizing the relationship between the anisotropy of the rates of growth of the faces and the chemical nature of the solvent have been studied by several workers. Some major conclusions have been drawn based on the studies on the growth of various crystals.

Supersaturation in hydrothermal experiments is given by the magnitude of the temperature gradient (ΔT) between the dissolution and growth zones, so that the relationship between the rate of growth and the magnitude of ΔT is usually investigated. If the solubility changes with temperature according to a law which is close to linear with ΔT, the growth rate is also practically linear. For all crystals, which have been studied under hydrothermal conditions, the rates of growth of the faces increase linearly with supersaturation. The high activation energies of the growth of the faces, combined with other factors—the anisotropy of the rates of growth of the faces, the marked dependence of the rate of crystallization on the composition of the solution, etc.—provide strong evidence in support of the suggestion that in hydrothermal crystallization under conditions of excess mass transfer, a primary role is played by surface processes taking place directly at the crystal–solution interface. It should be remembered that the activation energy of diffusion in solutions usually does not exceed 4–5 kcal/mol [75], and the activation energies of dissolution rarely exceed 10 kcal/mol. These values are much lower than the activation energies of growth, indicating that diffusion in the solution and dissolution of the charge do not limit the rate of crystallization with increase in the concentration of the solution. The rate of the crystallization can increase in two ways. For crystals that do not contain components of the solvent, the rate increases sharply at low concentrations and remains practically unchanged at high concentrations. Similarly, in some cases, an increase in the rate of growth of the faces with increase in pressure has been observed. Pressure apparently does not have any significant direct effect on the rate of growth of crystals, but it may have an influence through other parameters: mass transfer and solubility.

Kuznetsov [67] reported the effects of temperature, seed orientation, filling, and temperature gradient, ΔT, on the growth kinetics of corundum crystals. The activation energies for growth on the $(11\bar{2}0)$ and $(10\bar{1}1)$ faces, together with some preliminary evidence on the rate-limiting step in the hydrothermal crystallization of Al_2O_3 are given.. Figures 4.11 and 4.12 show the growth rates as functions of fillings—x: $(10\bar{1}1)$; o: $(11\bar{2}0)$, and relation of log R to $1/T$ for the filling factors given

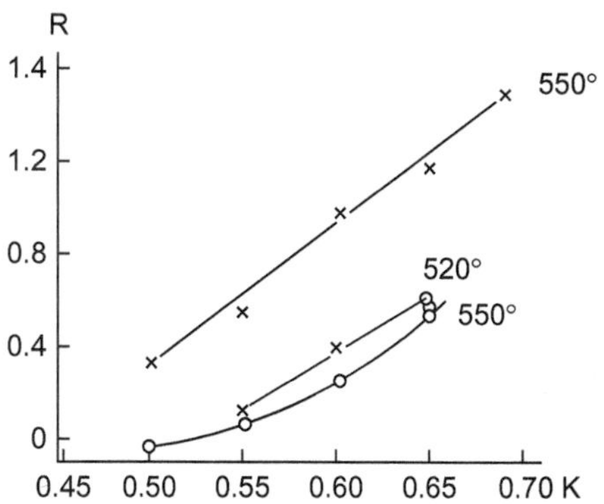

Figure 4.11 Growth rates as a function of filling—x: (1011); o: (1120) [67].

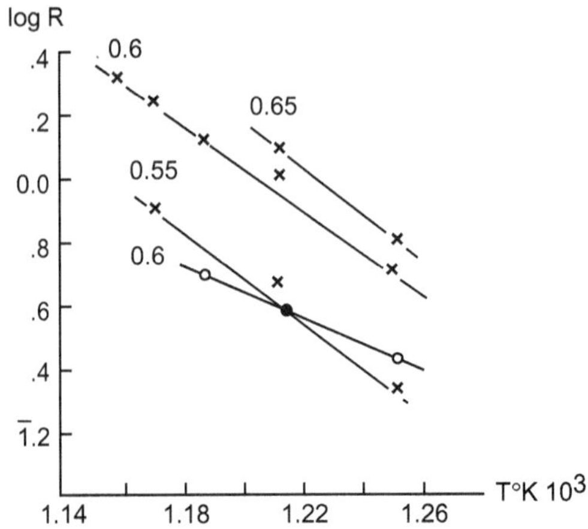

Figure 4.12 Relation of log R to $1/T$ for the filling factors—x: (1011); o: (1120) [67].

on the curves—x: $(10\bar{1}1)$; o: $(11\bar{2}0)$. The activation energies calculated from Figure 4.12 are 32 kcal/mol for $(10\bar{1}1)$ face and 17.5 kcal/mol for $(11\bar{2}0)$ face [67].

Riman and coworkers [30,31,76,77] have investigated the crystallization kinetics of ABO_3 (A = alkaline-earth elements, B = Ti, Zr, Hf) and their solid solutions. The fundamental role of temperature, pressure, precursor, and time on crystallization kinetics of perovskite oxides have been studied in detail. Early work by Battelle Laboratories set the stage for the development of hydrothermal processes that provide excellent morphological control for a variety of ceramic chemistries [78]. The morphology control and size of lead zirconate titanate and lead titanate precipitated from organic mineralizer solutions have been worked out in detail. Kaneko and Imoto [79] have investigated the effects of pressure, temperature, time, and Ba:Ti ratio on the kinetics of a hydrothermal reaction between barium hydroxide and titania gels to produce barium titanate powders. Ovramenko et al. [80] conducted kinetic studies to compute an activation energy (E_a) of 21 kJ/mol. In contrast, Hertl [81] calculated activation energy of 105.5 kJ/mol. However, this discrepancy may be due to the difference in source of titania precursor as well as other reaction conditions (e.g., temperature). Computations performed in the composition temperature–pressure space facilitate the construction of stability diagrams, speciation diagrams, and yield diagrams (Figure 4.13) [82]. In addition, the authors have carried out kinetic analysis based on reaction progress, yielded into the reaction-rate regime for various perovskite-type oxides.

Other materials of commercial interest in the present-day context whose kinetics of crystallization have been studied in detail are zeolites and HAp [83–85]. Various approaches or models have been proposed to understand the reaction mechanism of these compounds. The models based on the dissolution process, growth process, surface diffusion, structure directing chelates, or amines in the synthesis of zeolites, etc., have been used as the parameters in the study of crystallization kinetics. These studies have helped in the preparation of phase-pure zeolites and HAp particles with a perfect control over the morphology, grain size, rate of crystallization, and so on.

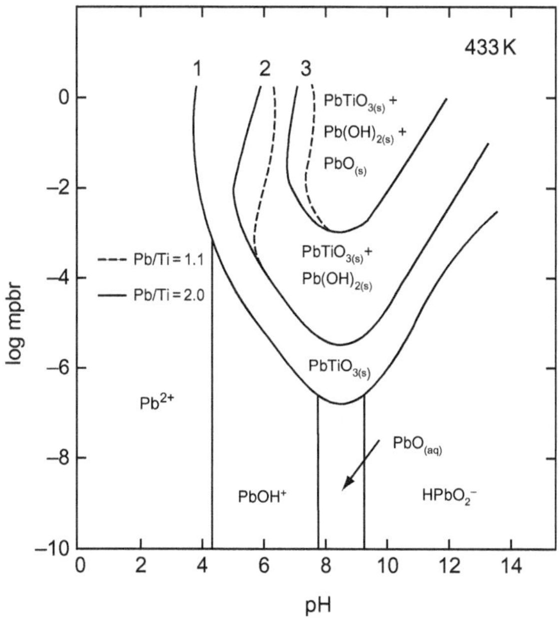

Figure 4.13 Stability diagrams, speciation diagrams, and yield diagrams [82].

In the recent years, microwave hydrothermal is being popularly used to greatly enhance the reaction kinetics. Also, efforts have been made to have better control over the particle size, shape, yield, and purity [86–88]. However, the reaction kinetics of most of the compounds under microwave hydrothermal have not been understood precisely.

4.5.1 Experimental Investigations of Solubility

The selection of a suitable solvent and the theoretical aspects of solubility have been discussed in the previous section. Here, let us consider the experimental aspect, its determination, and role in growing single crystals. There are two principal methods for determining the solubility under hydrothermal conditions. The first one is a sampling technique and has been described in detail by Morey and Hesselgesser [89]. It is especially useful where two fluid phases are present. In this technique, a suitably designed valve is arranged so that a fluid sample small enough so as not to perturb equilibrium can be withdrawn and chilled from an isothermal hydrothermal autoclave at operating conditions. If only one fluid phase is present, a weight loss method has been proved to be somewhat less cumbersome [90]. It is quite difficult to study the solubility of crystals under hydrothermal conditions if the compound does not form any metastable solutions over a wide temperature range. If there is a considerable metastable range of this kind, however, we may use the least troublesome method of studying solubility at high temperatures and high pressures—the method of weight loss. If a weighed single crystal (or crystals) is placed in the capsule under conditions where the phase is stable and where the fluid phase is liquid alone or gas alone, the weight loss will be a measure of solubility. However, special care must be exercised to avoid overshooting, to ensure that the desired temperature is reached during warm up, to avoid losing material before weighing, or weighing material that has precipitated during the quenching. In many cases, the solubility obeys the van't Hoff equation and is linearly dependent on solution density [91]. The ratio of solubility to mineralizer concentration often gives clues to the species present [92]. Several additional difficulties were encountered with the silicates, aluminosilicates, phosphates, and so on. They were first noticed while studying the solubility of quartz. These difficulties principally arise from phase separation of the solution under certain conditions, and the formation of a "heavy" phase enriched with silicon [93]. For example, it was shown by the authors of Ref. 94, in connection with the solubility of albite in H_2O at 200–350°C, that under these conditions there was a partial decomposition of the albite forming analcime. Hence, in this case, the true solubility of albite in water was not really being determined. Therefore, to obtain the true solubility of any compound which forms the heavy phases or other metastable phases, it has to be determined under conditions similar to its conditions of synthesis, so as to determine its optimum growth conditions. Russian workers at the Institute of Crystallography have designed an apparatus to suit the solubility study of various compounds (Figure 3.22).

Let us consider the solubility studies of some selected compounds like zincite and malachite. Solubility of a compound may be positive or negative. If the solubility of a compound increases with increasing temperature, it is positive. If the solubility of the compound decreases with increasing temperature, it is negative. Some compounds show positive solubility up to a certain temperature and then show negative solubility and vice versa. Quartz and berlinite form the best examples. The majority of the earliest hydrothermal experiments, carried out in the nineteenth century, had pure water as the mineralizer. Today we use a great variety of mineralizers, especially aqueous solutions of salts, acids, bases, etc., at very low concentrations. Likewise, there are a great numbers of nonaqueous solvents. Hannay and Hogarth used an alcoholic solution as a mineralizer a century ago [95]. Usually nonaqueous solvents are preferred when the required solubility cannot be reached in an aqueous medium. However, the composition of the effective solvents used in growing crystals is fairly standard. Pure water can be a good solvent at very high-temperature and high-pressure conditions. The use of various acids, salts, and bases considerably reduce the PT conditions. The commonly used nonaqueous solvents are NH_3, HF, HCl, HBr, Br_2, S_2Cl_2, S_2Br_2, $SeBr_2$, $H_2S + (C_2H_5)_3$, NH_4Cl, C_2H_5OH, CS_2, CCl_4, C_6H_6, CH_3NH_2, etc. In the past decade, the growth conditions have been further reduced with the use of some acid solvents. For example, high molar acids like HCl, H_2SO_4, H_3PO_4, HNO_3, HCOOH are freely used in the growth of various compounds and this has brought down the growth temperature to below 300°C. In recent years, mixed solvents have been used in the growth of some compounds like $AlPO_4$. The mixed solvents are found to be highly effective. The reduction in the growth temperature considerably reduces the working pressure, which in turn helps in the use of simple apparatus. Even silica autoclaves can be used, which facilitate direct visibility of the growth medium just as in any other low-temperature solution growth. This has greatly attracted the attention of crystal growers in the last 2 years and a great variety of compounds which hitherto were produced only at high temperature and high pressure have been today obtained at lower PT conditions. Hence, the mineralizer plays an important role in the hydrothermal growth of crystals and it has changed the tendency and perspective of the method. All these aspects can be very well described with reference to quartz, berlinite, sodalite, and α-TeO_2. The reader is advised to refer to the works of Pearson [96,97] regarding the empirical selection of solvents based on the hard and soft acids and bases.

Solubility of Zincite

The solubility of zincite has been studied in great detail during the 1970s by several workers [72,73,89]. Measurements of the solubility of zincite has been performed first in noncomplexing solutions (HTr, NaTr) at various ionic strengths (0.03–1.0 mol NaTr) and temperatures (50 – 290°C) to determine the equilibrium quotient of the following reaction:

$$ZnO_{cr} + 2H^+ \rightarrow Zn^{2+} + H_2O \quad (4.25)$$

Figure 4.14 represents an example of the solubility obtained at 200°C and 0.1 mol by approaching equilibrium from undersaturation (triangles), whereby the pH was changed by titration of acid into the cell during the experiments. The solid

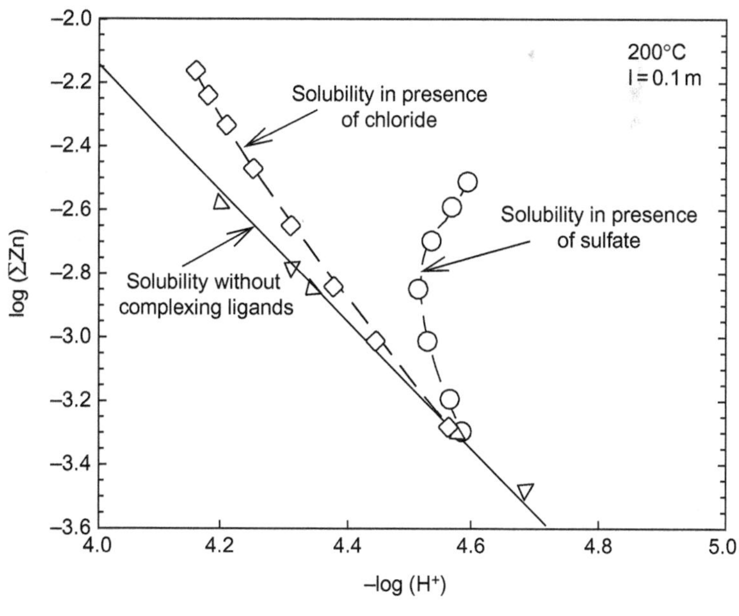

Figure 4.14 Solubility of zincite obtained at 200°C [98].

line represents the best fit of the data with a slope of −2 dictated by Reaction (4.25). The solubility is shown in this figure to be reversed by a back titration (inverse triangles) with a basic titrant causing ZnO to precipitate.

The dissociation quotients of Reaction (4.25) were regressed against a variety of temperature–ionic-strength functions using the ORGLS (software developed at Oak Ridge National Laboratory, USA) general least squares program of Busing and Levy [99]. At infinite dilution, good agreement is obtained with previous experimental and model values (i.e., Khodakovsky and Yelkin [100], Ziemniak [101]) except for the values obtained which deviate from the values at 250 and 250°C.

In this figure, comparison is also made between these solubilities and those obtained in experiments performed by titration of solution containing chloride (diamond) and sulfate (circle) into the cell. It is clearly shown that the effect of addition of either chloride or sulfate is to increase the concentration of zinc. At this temperature, ionic strength, and stability constants of the reactions:

$$Zn^{2+} + SO_4^{2-} \rightarrow ZnSO_4^0 \quad (4.26)$$

$$Z^{2+} + Cl^- \rightarrow ZnCl^+ \quad (4.27)$$

$$Zn^{2+} + 2C1^- \rightarrow ZnCl_2^0 \quad (4.28)$$

were obtained. The dashed curves in Figure 4.14 are generated from these stability constants and are shown to reproduce the experimental points [98]. For Reaction (4.26), there are no published experimental data with which to compare these results. In order to compare with literature data, the constants for Reactions (4.27) and (4.28) are extrapolated to infinite dilution at 200°C. Good agreement with Bourcier et al. [102] and Ruaya and Seward [103] is obtained for the dichloro-zinc complex stability constant. However, the result for the monochloro-zinc complex disagrees markedly.

Solubility of Malachite

The growth of malachite single crystals has posed a challenge in the past decade. Several original methods, which allow nearly all textural varieties found in natural malachite, have been devised. Malachite solubility in pure water and in electrolytic aqueous solutions has not been understood precisely. Symes and Kester [104] studied the peculiarities of malachite dissolution in pure and seawater in order to reveal the conditions necessary for its precipitation. Sulyapnikov and Shtern [105] determined the maximum copper concentration during malachite dissolution in solutions of $NaHCO_3$ and $KHCO_3$ of different concentrations at temperatures up to 200°C and different amounts of CO_2. Their data has shown that, under normal conditions, the copper content in these solutions changes, strongly depending on the time of the solution mixing with the powder of the basic copper carbonate. Balitsky and Bublikova [106] have investigated malachite solubility in 1,2,3 M solutions of ammonium hydroxide at temperatures 20, 25, 50, and 75°C ($P = 1$ bar). Figure 4.12

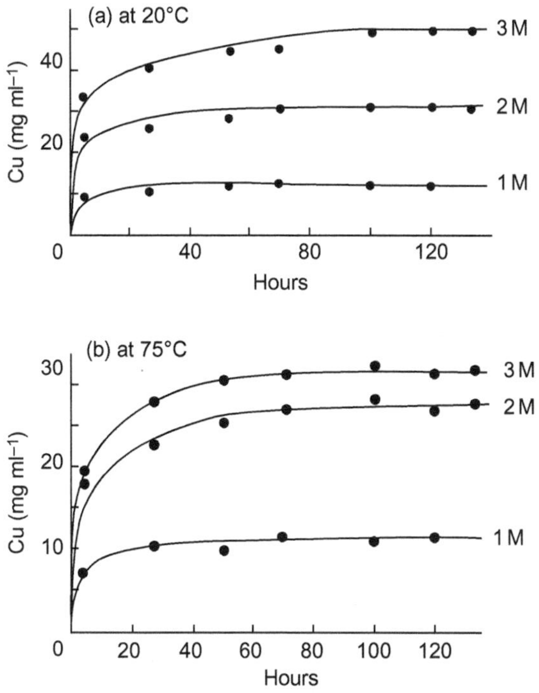

Figure 4.15 Kinetic curves of malachite dissolution in ammonia aqueous solutions at 20°C and 75°C [106].

shows kinetic curves of malachite dissolution in ammonia aqueous solutions at 20°C (a) and at 75°C (b) respectively. As can be seen from Figure 4.15, equilibrium copper contents in the ammonia solutions increase as ammonia concentration increases and they decrease as temperature rises.

These examples give a more or less complete picture with regard to the type of solvent, the PT conditions, and the progress made in the search for new solvents for various inorganic compounds under hydrothermal conditions. All these studies yield crystal growth kinetics, the phase-pure product.

Thus, hydrothermal physical chemistry is a fast growing field that is going to play a key role in the twenty-first century in the preparation of future materials. The physicochemical data will provide an insight into the actual hydrothermal reaction mechanism and in turn will bring down the PT conditions of the preparation of materials.

4.6 Thermodyanmic Calculations for the Intelligent Engineering of Materials

Hydrothermal research has significantly changed in the twenty-first century with researchers giving up the Edisonian trial and errors approach for designing hydrothermal experiments, which is highly time consuming and very expensive due to frequent failures in experimental results. It is well known that the hydrothermal chemical reactions are either controlled by thermodynamic or kinetics. Present generation researchers are using such thermodynamic data for a given precursor system to design the experiments. There are several commercial available softwares having such a thermodynamic database for a wide range of compounds, which can be utilized to calculate the thermodynamic properties for a given system. Thermodynamic processing variables, such as temperature, pH, concentrations of reactants, and additives, determine not only the processing space for a given material, but also influence both reaction and crystallization kinetics. To some extent even the size and morphology could also be predicted using such a thermodynamic model, especially based on the concentration and temperature of the hydrothermal medium.

Thermodynamic modeling has been successfully used for the synthesis of a wide range of metal oxides, HAp, PZT family of materials, amongst others. Riman and coworkers [76,77,107–111] have done a pioneering work in this area for the past two decades. It is even popularly called Lencka–Riman model for hydrothermal materials processing. A few other groups are also working on the thermodynamic modeling of hydrothermal reactions and kinetics [112–115]. For the benefit of readers, only a selected systems have been discussed here briefly.

The optimal approach to control the chemical phases under hydrothermal conditions is to calculate the stability and yield conditions by a thermodynamic model. Commercial softwares are available for both aqueous and nonaqueous systems with different ranges of applicability in terms of species concentration, pressure, and temperature. Usually, any thermodynamic model requires the following information for all species in the solution: the standard Gibbs energies ΔG_f^o and enthalpies ΔH_f^o of formation, entropies S_o at a reference temperature (298.15 K), and the partial molar volumes V^o and heat capacities C_p^o as functions of temperature. Equilibrium equations for all the species in the solution are set up. The model considers all the activity coefficients of aqueous species, for example, the data bank from OLI Systems Inc. [116], and in the absence of thermodynamic values, estimation methods are used to obtain such values [77,117]. The results of the calculation of the model are the equilibrium concentrations of all species as a function of pH, temperature, and initial reagent concentrations. The key to representing the standard-state properties over substantial temperature and pressure ranges is the accurate knowledge of the heat capacity and volume. For this purpose, the Helgeson–Kirkham–Flowers–Tanger (HKFT) equation of state is used [27,118]. This equation accurately represents the standard-state thermodynamic functions for aqueous, ionic, and neutral species as functions of both temperature and pressure. In the case of solid compounds, solubility data in pure water as well as in alkaline or acidic solutions at different temperatures are the best source of data for standard-state properties. These data are available in books of Seidell and Linke [119,120]. Whenever possible, regression of solubility data are performed using the OLI software [116] to obtain accurate and consistent thermochemical data for solid compounds.

The main objective of these thermodynamic modelings is to calculate the optimum synthesis conditions for the formation of phase-pure materials with controlled size and shape to some extent. This information can be obtained from the knowledge of stability and yield diagrams calculated for hydrothermal systems involving ceramic materials. Repetitive application of the model for different input concentrations of the different precursors as well as pH-adjusting agents (mineralizers) makes it possible to construct stability and yield diagrams. Stability diagrams show the ranges of equilibrium conditions for which various aqueous and solid species are stable in the hydrothermal system. Equilibrium conditions include the total molality of a selected aqueous metal precursor, solution pH, and temperature. The total molality of the aqueous metal precursor is the sum of the equilibrium concentrations of all dissolved metal species and does not include the species that precipitate from the solution as solids. While the stability diagrams provide information about the conditions for incipient precipitation of various solid phases, they do not specify the reaction conditions required for an assumed yield of reaction product. To accomplish this goal, the idea of yield diagram was conceived [121]. The reader can get more valuable information in the works [108,109]. The theoretical predictions are validated through selected experiments using the calculated stability and yield diagrams as a guide. The hydrothermal reactions are performed to investigate whether the location of the various stability and yield regions agrees with the experimental findings.

Figure 4.16 shows calculated yield of lead zirconate titanate ($PbZr_{0.6}Ti_{0.4}O_3$) at 433 K with the ratio of lead acetate to a sum of titanium and zirconium dioxide equal to 1.1. Phase-pure can form in a relatively wide region of input precursor and

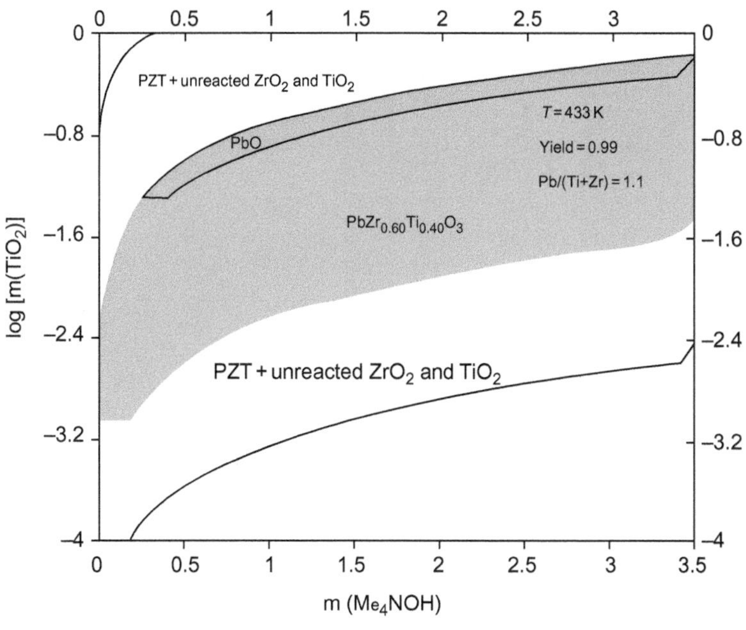

Figure 4.16 Calculated yield diagram of $PbZr_{0.6}Ti_{0.4}O_3$ at 433 K. $PbAc_2$, TiO_2, and ZrO_2 are used as starting materials for the calculations [108].

mineralizer molalities (cf. shaded area in Figure 4.16). $PbZr_{0.6}Ti_{0.4}O_3$ precipitates along with PbO in the small region shown on the top of the phase-purity region. The stability area of PbO is relatively small and limited to a high concentration of precursors. In highly alkaline solutions, lead oxide dissolves. When the ratio of metals in precursor materials is greater than one, the area of the perovskite phase-purity region widens. However, at some reaction conditions it causes contamination of ceramic materials with additional phases.

Figure 4.17 shows the stability field diagram for $CaO-P_2O_5-NH_4NO_3-H_2O$ system at 200°C as a function of precursor concentration and pH. The pH values are calculated from the model for the temperatures shown in Figure 4.17. Initial room temperature pH of the starting slurries have been recalculated to pH's at the experimental temperatures. The diagram shows stability fields for HAp and monetite (Mt). Experimental points represent phase assemblages according to the legends. At the phosphorus concentrations at 200°C, the HAp stability field extends to an equilibrium pH as low as 2.9. The latter pH corresponds to a room temperature pH of about 3.8. Powders run at 200°C and which were prepared from slurries with room temperature pH's lower than about 3.8 yielded only monetite [110].

These thermodynamic calculations help in obtaining advanced ceramic powders with controlled size and shape. Hence, especially for the nanomaterials processing it is highly useful. However, there are several other additional parameters which influence the kinetics of the hydrothermal reactions like the stirring rate, external fields like magnetic field, sonar, and mechanochemical grinding.

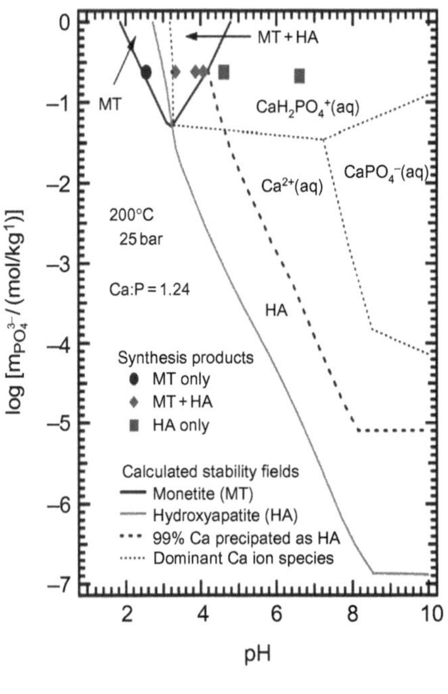

Figure 4.17 Calculated stability field diagram for the HAp system at 200°C and 25 bar with Ca:P ratio at 1.24.

References

[1] F.E. Wickman, D. Rickard (Eds.), Chemistry and geochemistry of solutions at high temperatures and pressures, Proc. Nobel Symp. Phys. Chem. Earth, 13/14, 1981, 562.

[2] S. Somiya, Proceedings of the First International Symposium on Hydrothermal Reactions, Gakujutsu Bunken Fukyu-Kai, Tokyo, 1983, pp. 975.

[3] A.B. Ezersky, A. Garcimartin, J. Burguete, H.L. Mancini, C. Perez-Garcia, Hydrothermal waves in Marangoni convection in a cylindrical container, Phys. Rev. E. 47 (1993) 1126–1131.

[4] A.B. Ezersky, A. Garcimartin, H.L. Mancini, C. Perez-Garcia, Spatiotemporal structure of hydrothermal waves in Marangoni convection, Phys. Rev. E. 48 (1993) 4414–4422.

[4a] V.S. Balitsky and T.M. Bublikova, Handbook of Hydrothermal Technology: A Technology for Crystal Growth and Materials Processing, William Andrew, (2001)

[4b] V.V. Kuznetsov. Intomodesma evenica in Permian Inoceramus-like mollusks of the genus Intomodesma Popov. Paleontological Journal 25 (4): (1991) 159–164.

[4c] V.I. Popolitov, Hydrothermal growth of crystals under visual examination. 21 (1), (1991) 255–297.

[5] K. Byrappa (Ed.), Hydrothermal growth of crystals, Prog. Cryst. Growth Charact. Mater. 21 (1991) 1–365.

[6] L. Kaufman, H. Nesor, Theoretical approaches to the determination of phase diagrams, Ann. Rev. Mater. Sci. 3 (1973) 1–30.

[7] R.F. Brebrick, Phase equilibria, in: D.T.J. Hurle (Ed.), Handbook of Crystal Growth, vol. 1a, North-Holland Publishing Co., Amsterdam, 1993, p. 91.

[8] I.V. Zakirov, Experimental Investigations of PVT Relations in the H_2O-CO_2 System, Ph.D. Thesis, Moscow State University, Moscow, 1987.

[9] G.D. Ilijushin, L.N. Demianets, V.V. Iljukhin, N.V. Belov, Phases and crystallization in the system Na_2O - ZrO_2 - SiO_2 Dokl. Akad. Nauk USSR (in Russian) 217 (1983) 1133.

[10] O.V. Dimitrova, Investigations of the Phase Formations in the System $Na_2O-RE_2O_3-SiO_2-H_2O$ Under Hydrothermal Conditions, Ph.D. Thesis, Moscow State University, Moscow, 1985.

[11] K. Byrappa, J.R. Paramesha, Crystal growth and characterization of rare earth phosphates, Mater. Sci. Forum 315–317 (1999) 514–518.

[12] E.D. Kolb, R.A. Laudise, The phase diagram, $LiOH-Ta_2O_5-H_2O$ and the hydrothermal synthesis of $LiTaO_3$ and $LiNbO_3$, J. Cryst. Growth 33 (1976) 145.

[13] S.D., Hamann, Properties of electrolyte solutions at high pressures and temperatures, in: D.T. Rickard, F.E. Wickman (Eds.), Proceedings of the Nobel Symposium—Organized by the Royal Swedish Academy of Sciences, Pergamon Press, New York, September 17–21, 1979, pp. 90–111.

[14] T.M. Seward, Metal complex formation in aqueous solutions at elevated temperatures and pressures, Proc. Nobel Symp. Phys. Chem. Earth 13/14 (1981) 113–132.

[15] H.C. Helgeson, Prediction of the thermodynamic properties of electrolytes at high pressures and temperatures, Proc. Nobel Symp. Phys. Chem. Earth 13/14 (1981) 113–177.

[16] D. Read, T.W. Broyd, CHEMVAL/MIRAGE report, Verification of Specification Models, W.S. Atkins, CEC Contract FI, IW0077, 1988.

[17] P.M. May, Towards more reliable chemical specification models of multicomponent electrolyte solutions, in: D.A. Palmer, D.J. Wesolowski (Eds.), Proceedings of the Fifth International Symposium on Hydrothermal Reactions, Gatlinburg, TN, USA, July 20–24, 1997, pp. 32–34.

[18] S.M. Sterner, A.R. Felmy, J.R. Rustad, K.S. Pitzer, Thermodynamic analysis of aqueous solutions using INSIGHT, in: D.A. Palmer, D.J. Wesolowski (Eds.), Proceedings of the Fifth International Symposium on Hydrothermal Reactions, Gatlinburg, TN, USA, July 20–24, 1997, pp. 35–38.
[19] P.W. Atkins, Physical Chemistry, ELBS, Oxford University Press, Oxford, 1986. p. 745.
[20] G.A. Krestov, Thermodynamics of solvation, Ellis Horwood, New York (1998) p. 16.
[21] K. Deidler, Kinetics of Organic Reactions M, Mir, Moscow, 1966. p. 237
[22] M. Uematsu, E.U. Franck, Static dielectric constant of water and steam, J. Phys. Chem. Ref. Data 9 (1980) 1291–1297.
[23] G.C. Kennedy, W.T. Holser, Pressure–volume–temperature and phase relations of water and carbon dioxide, in: S.P. Clark Jr. (Ed.), Handbook of Physical Constants, Geological Society of America, Washington, DC, 1966, pp. 371–383.
[24] H. Köster, E.U. Franck, Das spezifische volumen des Wassers be: Hohen Drücken bis 600°C und 100 kbar, Ber. Bunsenges 73 (1969) 716–722.
[25] K. Tödheide, Water at high temperatures and pressures, in: F. Franks (Ed.), Water A Comprehensive Treatise, vol. 1, Plenum Press, New York, NY, 1972.
[26] P.T. Cummings, Molecular simulation of supercritical water and of ionic association in supercritical aqueous solutions, in: D.A. Palmer, D.J. Wesolowski (Eds.), Proceedings of the Fifth International Symposium on Hydrothermal Reactions, Gatlinburg, TN, USA, July 20–24, 1997, pp. 153–156.
[27] H.C. Helgeson, D.H. Kirkham, G.C. Flowers, Theoretical prediction of the thermodynamic behavior of aqueous electrolytes at high pressures and temperatures: IV. Calculation of activity coefficients, osmotic coefficients, and apparent molal and standard and relative partial molal properties to 600°C and 5 kb, Am. J. Sci. 281 (1981) 1249–1516.
[28] G.M. Anderson, S. Castet, J. Schoot, R.E. Mesmer, The density model for estimation of thermodynamic parameters of reactions at high temperatures and pressures, Geochim. Cosmochim. Acta 55 (1991) 1769–1779.
[29] E.U. Franck, Special aspects of fluid solutions at high pressure and sub- and supercritical temperatures, Pure Appl. Chem. 53 (1981) 1401–1416.
[30] R.E. Riman, M.M. Lencka, L.E. McCandlish, B.L. Gersten, A. Andrenko, S.B. Cho, Intelligent engineering of hydrothermal reactions, in: D.A. Palmer, D.J. Wesolowski (Eds.), Proceedings of the Fifth International Symposium on Hydrothermal Reactions, Gatlinburg, TN, USA, July 20–24, 1997, pp. 74–78.
[31] M.M. Lencka, R.E. Riman, Thermodynamics of the hydrothermal synthesis of calcium titanate with reference to other alkaline-earth titanates, Chem. Mater. 7 (1995) 18–25.
[32] M.M. Lencka, R.E. Riman, Thermodynamic modeling of hydrothermal synthesis of ceramic powders, Chem. Mater. 5 (1993) 61–70.
[33] W.E. Brown, L.C. Chow, Thermodynamics of apatite crystal growth and dissolution, J. Cryst. Growth 53 (1981) 31–41.
[34] L.G. Benning, T.M. Seward, Hydrosulphide complexing of Au^{1+} in hydrothermal solutions from 150–400°C and 500–1500 bar, Geochim. Cosmochim. Acta 60 (1996) 1849–1871.
[35] M. Woitsekhowskaya, R. Roberts, J. Cline, J.J. Hemley, K. Weaver, Gold and arsenic in iron sulfides from the Getchell deposit, Nevada: thermodynamic evaluation, in: D.A. Palmer, D.J. Wesolowski (Eds.), Proceedings of the Fifth International Symposium on Hydrothermal Reactions, Gatlinburg, TN, USA, July 20–24, 1997, p. 301.

[36] E.U. Franck, Water and aqueous solutions at high pressures and temperatures, Pure Appl. Chem. 24 (1970) 13–30.
[37] R.A. Laudise, A.A. Ballman, J.C. King, Silica, synthetic quartz crystals, J. Phys. Chem. Solids 26 (1965) 1305.
[38] sixty fourth ed., R.C. Weast (Ed.), Handbook of Physics and Chemistry, D261, CRC Press, Boca Raton, FL, 1983.
[39] H.C. Helgeson, Prediction of the Thermodynamic properties of Electrolytes at High Pressures and Temperatures, Phys. Chem. Earth 13/14 (1981) 133.
[40] J.B. Hasted, D.M. Ritson, C.H. Collie, Dielectric properties of aqueous ionic solutions, Parts I and II, J. Chem. Phys. 16 (1948) 1.
[41] M., Uematsu, W., Harder, E.U., Franck, The dielectric constant of water. Technical paper 38p (WLKg. Gp. 3), International Association for the properties of steam, Mtg Kyoto Mtg, Jpn, 1976.
[42] Landolt-Börnstein, Numerical data and functional relationships, Gr. IV, in: G. Beggerow, K.L. Schafer (Eds.), High-Pressure Properties of Matter, vol. 4, Springer, Berlin, 1980.
[43] T.M. Seward, Hydrothermal solution speciation (Solvation to ion pairing and complexing), in: D.A. Palmer, D.J. Wesolowski (Eds.), Proceedings of the Fifth International Symposium on Hydrothermal Reactions, Gatlinburg, TN, USA, July 20–24, 1997, pp. 7–9.
[44] M. Nakahara, M. Nobuyuki, C. Wakai, NMR study of hydrogen bonds and water structure in super- and subcritical conditions, in: N. Yamasaki, K. Yanagisawa (Eds.) Proceedings of the Second International Conference on Solvothermal Reactions, Takamatsu, Japan, December 18–20, 1996, pp. 9–12.
[45] T. Yamaguchi, H. Fujita, K. Hidaka, H. Wakita, Structure of water and methanol at subcritical and supercritical states by x-ray diffraction, in: N. Yamasaki, K. Yanagisawa (Eds.) Proceedings of the Second International Conference on Solvothermal Reactions, Takamatsu, Japan, December 18–20, 1996, pp. 13–15.
[46] M. Nakahara, T. Yamaguchi, H. Ohtaki, The structure of water and aqueous electrolyte solutions under extreme conditions, Recent Res. Dev. Phys. Chem. 1 (1997) 17–49.
[47] H. Ohtaki, T. Radnai, T. Yamaguchi, Structure of water under subcritical and supercritical conditions studied by solution X-ray diffraction, Chem. Soc. Rev. (1997) 41–51.
[48] M. Nakahara, Very high temperature and high pressure reactions, Encyclopedia of Experimental Chemistry, vol. 11, Jikken Kagaku Koza, Muruzen, Tokyo, 1993 (in Japanese).
[49] S.D. Haman, Physical chemistry of hydrothermal growth of crystals, Phys. Chem. Earth 13/14 (1981) 89.
[50] K. Todheide, Ber. Bunsenges, hydrothermal solutions, Physik. Chem. 86 (1982) 1005.
[51] E.U. Frank, Properties of water, Int. Corros. Conf. Ser. 109 (1973). p.45.
[51a] T. Driesner, The Effect of Pressure on Deuterium-Hydrogen Fractionation in High-Temperature Water. Science 277, (1997) 791–794.
[52] T.M. Seward, C.M.B. Henderson, J.M. Charnock, Driesner, An EXAFS study of solvation and ion pairing in aqueous strontium solutions to 300°C, Geochim. Cosmochim. Acta 63 (1998) 2409–2418.
[53] L.G. Benning, T.M. Seward, Hydrosulphide complexing of Au^{1+} in hydrothermal solutions from 150°–400°C and 500–1500 bar, Geochim. Cosmochim. Acta 60 (1996) 1849–1871.
[54] J.V. Walther, J. Schott, The dielectric constant approach to speciation and ion pairing at high temperature and pressure, Nature 332 (1988) 635–638.

[55] G.C. Kennedy, Pressure–volume–temperature relations in water at elevated temperatures and pressures, Am. J. Sci. 248 (1950) 540–543.
[56] V.I. Popolitov, Hydrothermal synthesis of inorganic compounds, D.Sc. Thesis, Hydrothermal Synthesis of Complex Oxides for Hydrothermal Growth, Institute of Crystallography, Russian Academy of Sciences, Moscow, 1990.
[57] E.D. Kolb, P.L. Key, R.A. Laudise, E.E. Simpson, Pressure-Volume-Temperature Behavior in the System H_2O-NaOH-$SiO_2$and its Relationship to the Hydrothermal Growth of Quartz, Bell. Sys. Tech. J. 61 (1983) 639.
[58] W.L. Marshall, E.V. Jones, J. Jones, Liquid–vapour critical temperatures of aqueous electrolyte solutions, Inorg. Nucl. Chem. 36 (1974) 2313.
[59] R.A. Laudise, W.A. Sunder, R.F. Belt, G. Gashurov, Solubility and P–V–T relations and the growth of potassium titanyl phosphate, J. Cryst. Growth 102 (1990) 427.
[60] R.F. Brebrick, Phase equilibria, in: D.T.J. Hurle (Ed.), Handbook of Crystal Growth, vol. 1a, Elsevier Science, B.V., The Netherlands, 1993.
[61] A.J. Celestian, J.B. Parise, A. Clearfield, Crystal growth and ion exchange in titanium silicates, in: G. Dhanarja, K. Byrappa, V. Prasad, M. Dudley (Eds.), Springer Handbook of Crystal Growth, Springer, Heidelberg, Germany, 2010, pp. 1637–1662.
[62] S. Bernstein, K.T. Fehr, The formation of 1.13 nm tobermorite under hydrothermal conditions: 1. The influence of quartz grain size within the system CaO–SiO_2–D_2O, Prog. Cryst. Growth Charact. Mater. (2012)58 84–91.
[63] V. Middelkoop, S.D. Jacques, M.G. O'Brien, A.M. Beale, P. Barnes, Hydrothermal/autoclave synthesis of AlPO-5: a prototype space/time study of crystallization gradients, J. Mat. Sci. 43 (2008) 2222–2228.
[64] K. Simmance, G. Sankar, R.G. Bell, C. Prestipino, W van Beek, Tracking the formation of cobalt substituted AlPO-5 using simultaneous in situ X-ray diffraction and X-ray absorption spectroscopy techniques, Phys. Chem. Chem. Phys. 12 (2010) 559–562.
[65] S.D.M. Jacques, K. Pile, P. Barnes, An in situ synchrotron X-ray diffraction tomography study of crystallization and preferred crystal orientation in a stirred reactor, Cryst. Growth Des. 5 (2005) 395–397.
[66] V.A. Kuznetsov, A.N. Lobachev, Hydrothermal method for the growth of crystals, Sov. Phys. Crystallgr. 17 (1973) 775–804.
[67] V.A. Kuznetsov, Hydrothermal crystallization kinetics of corundum, Sov. Phys. Crystallogr. 10 (1966) 561–564.
[68] V.A. Kuznetsov, Kinetics of hydrothermal crystallization of corundum, II. Effect of solvents on crystallization, Sov. Phys. Crystallogr. 12 (1968) 608–611.
[69] A.N. Lobachev, L.N. Demianets, I.P. Kuzmina, E.N. Emelianova, Investigation of the solubility and crystallization kinetics of sodium-zinc germanate (Na_2ZnGeO_4) under hydrothermal conditions, J. Cryst. Growth 13/14 (1972) 540–544.
[70] V.I. Popolitov, A.N. Lobachev, Kinetics and structure aspect of the growth of single crystals of bismuthinite under hydrothermal conditions, Sov. Phys. Crystallogr. 81 (1973) 131–132.
[71] R.A. Laudise, A.A. Ballman, Hydrothermal synthesis of zinc oxide and zinc sulfide, J. Phys. Chem. 64 (1960) 688.
[72] A.A. Laudise, E.D. Kolb, A.J. Caporaso, Hydrothermal growth of large sound crystals of zinc oxide, J. Amer. Ceram. Soc. 47 (1964) 9–12.
[73] E.D. Kolb, R.A. Laudise, Hydrothermally grown ZnO crystals of low and intermediate resistivity, J. Amer. Ceram. Soc. 49 (1966) 302–305.
[74] N.Y. Ikornikova, Hydrothermal synthesis of crystals in chloride systems, D.Sc. Thesis, Hydrothermal Synthesis and Characterization of Transition Metal (Mn and V) Oxides

Containing Phosphates, Institute of Crystallography, Academy of Sciences, Moscow, 1970.
[75] O.Y. Samoilov, The Structure of Aqueous Solutions and the Hydration of Ions, Akad. Nauk USSR Publication, Moscow, 1957.
[76] R.E. Riman, M.M. Lencka, L.E. McCandlish, B.L. Gersten, A. Andrenko, S.B. Cho, Intelligent Engineering of Hydrothermal Reactions, in: Proceedings of the Second International Conference on Solvothermal Reactions, Takamatsu, Japan, December 18–20, 1996, pp. 148–151.
[77] M.M. Lencka, A. Andrenko, R.E. Riman, Hydrothermal precipitation of lead zirconate titanate solid solutions: thermodynamic modeling and experimental synthesis, J. Am. Ceram. Soc. 78 (1995) 2609–2618.
[78] E.P. Staumbagh, J.F. Miller, Hydrothermal precipitation of high quality inorganic oxides, in: S. Somiya (Ed.), Proceedings of the First International Symposium on Hydrothermal Reactions, Gakujutsu Bunken Fukyu-Kai, Tokyo, Japan, 1983, pp. 858–871.
[79] S. Kaneko, F. Imoto, Synthesis of fine-grained barium titanate by a hydrothermal reaction, Nippon Kagaku Kaishi 6 (1975) 985–990.
[80] N.A. Ovramenko, L.I. Shvets, F.D. Ovcharenko, B.Y. Kornilovich, Kinetics of hydrothermal synthesis of barium metatitanate, Izv. Akad. Nauk USSR Inorg. Mater. 15 (1979) 1982–1985.
[81] W. Hertl, Kinetics of barium titanate synthesis, J. Amer. Ceram. Soc. 71 (1988) 879–883.
[82] J.O. Eckert Jr., C.C. Hung-Houston, B.L. Gersten, M.M. Lencka, R.E. Riman, Kinetics and mechanisms of hydrothermal synthesis of barium titanate, J. Amer. Ceram. Soc. 79 (1996) 2929–2939.
[83] T. Lindner, H. Lechert, Chelate ligands as mineralizing agents in hydrothermal synthesis of faujasite-type zeolites: a kinetic study, Zeolites 16 (1996) 196–206.
[84] A.Y. Sheikh, A.G. Jones, P. Graham, Population balance modeling of particle formation during the chemical synthesis of zeolite crystals, assessment of hydrothermal precipitation kinetics, Zeolites 16 (1996) 164–172.
[85] K. Onuma, A. Ito, T. Tateishi, T. Kameyama, Growth kinetics of hydroxyapatite crystal revealed by atomic force microscopy, J. Cryst. Growth 154 (1995) 118–125.
[86] S. Komarneni, R. Roy, Q.H. Li, Microwave-hydrothermal synthesis of ceramic powders, Mater. Res. Bull. 27 (1992) 1393–1405.
[87] Y. Fang, D.K. Agarwal, D.M. Roy, R. Roy, Fabrication of porous hydroxyapatite ceramics by microwave processing, J. Mater. Res. 7 (1992) 490–494.
[88] S. Komarneni, Enhanced reaction kinetics under microwave-hydrothermal conditions, in: N. Yamasaki K. Yanagisawa (Eds.) Proceedings of the Second International Conference on Solvothermal Reactions, Takamatsu, Japan, December 18–20, 1996, pp. 97–100.
[89] G.W. Morey, J.M. Hesselgesser, Thermochemistry of gaseous SiO(OH), $SiO(OH)_2$, and SiO_2, Am. J. Sci. 367 (1952).
[90] R.A. Laudies, Kinetics of hydrothermal quartz crystallization, J Am. Chem. Soc. 81 (1959) 562.
[91] R.A. Laudise, Physical Chemistry of Hydrothermal Growth of Crystals, in: F.A. Cotton (Ed.), Progress in Inorganic Chemistry, 3, Wiley-Interscience, New York, NY, 1962.
[92] D.J. Marshall, R.A. Laudise, Crystal growth by hydrothermal technique, in: H.S. Peiser (Ed.), Crystal Growth, Pergamon, New York, NY, 1966, p. 557.

[93] L.V. Bryatov, Some Problems Relating to the Solubility and Growth of Quartz Crystals, Ph.D. Thesis, Institute of Crystallography, Academy of Sciences, Moscow, 1955.

[94] A.P. Frederickson, I.E. Cox, Mechanism of solution of quartz in pure water at elevated temperatures and pressures, Am. Mineral. 39 (1954) 738.

[95] J.B. Hannay, Hogarth, On the solubility of solids in gases, J. Proc. R. Soc. London 30 (1880) 178.

[96] R.G. Pearson, Hard and soft acids and bases, HSAB, Part 1: Fundamental principles, J. Chem. Education 45 (1968) 581–587.

[97] R.G. Pearson, Hard and soft acids and bases, HSAB, Part 2: Fundamental principles, J. Chem. Education 45 (1968) 645–652.

[98] P. Benezeth, D.A. Palmer, D.J. Wesolowskii, Potentiometric studies of aluminum hydroxide and zinc oxide solubility to high temperature, in: D.A. Palmer, D.J. Wesolowski (Eds.), Proceedings of the Fifth International Symposium on Hydrothermal Reactions, Gatlinburg, TN, USA, July 20–24, 1997, pp. 149–152.

[99] W.R. Busing, H.A. Levy, ORNL-TM-271 Report, 1962.

[100] I.L. Khodakovsky, A.Y. Yelkin, Measurement of the solubility of zincite in aqueous NaOH at 100, 150, and 200°C, Geokhimica 10 (1957) 1490–1498.

[101] S.E. Ziemniak, Metal oxide solubility in high temperature aqueous solutions, J. Sol. Chem. 21 (1992) 745–760.

[102] W.L. Bourcier, K.G. Knauss, K.J. Jackson, Aluminum hydrolysis constants to 250°C from boehmite solubility measurements, Geochim. Cosmochim. Acta 57 (1993) 747–762.

[103] J.R. Ruaya, T.M. Seward, The stability of chlorozinc(II) complexes in hydrothermal solutions up to 350°C, Geochim. Cosmochim. Acta 50 (1986) 651–661.

[104] J.L. Symes, D.R. Kester, Thermodynamic stability studies of the basic copper carbonate mineral, malachite, Geochim. Cosmochim. Acta 48 (1984) 2219.

[105] D.S. Shlyapnikov, E.K. Shtern, Dokl. Acad. Nauk. USSR 207 (1972) 966.

[106] V.S. Balitsy, T. Bublikova, Hydrothermal growth of malachite, in: K. Byrappa (Ed.), Hydrothermal Growth of Crystals, Prog. Cryst. Growth Charact. Mater., vol. 21, 1991, p. 139.

[107] B. Gersten, M. Lencka, R.E. Riman, Engineered low temperature hydrothermal synthesis of phase pure lead based perovskites using EDTA complexation, Chem. Mater. 14 (2002) 1950–1961.

[108] M.M. Lencka, R.E. Riman, Thermodynamics of multicomponent perovskite synthesis in hydrothermal solution, in: K. Byrappa, T. Ohachi (Eds.), Crystal Growth Technology, Springer, Germany, 2003, pp. 271–297.

[109] B.L. Gersten, Growth of multicomponent perovskite oxide crystals: synthesis conditions for the hydrothermal growth of ferroelectric powders, in: K. Byrappa, T. Ohachi (Eds.), Crystal Growth Technology, Springer, Germany, 2003, pp. 299–333.

[110] R.E. Riman, W.L. Suchanek, K. Byrappa, C.W. Chen, P. Shuk, C.S. Oakes, Solution synthesis of hydroxyapatite designer particulates, Solid State Ionics 151 (2002) 393–402.

[111] W.L. Suchanek, R.E. Riman, Hydrothermal synthesis of advanced ceramic powders, Adv. Sci. Techn. 45 (2006) 184–193.

[112] R. Piticescu, C. Monty, D. Millers, Hydrothermal synthesis of nanostructured zirconia materials: present state and future prospects, Sensor. Actuator. B 109 (2005) 102–106.

[113] X. Han, D. Zhang, Z. Zhong, F. Yang, N. Wei, K. Zheng, et al., Theoretical design and experimental study of hydrothermal synthesis of $KNbO_3$, J. Phys. Chem. Solids 69 (2008) 193–198.

[114] J. Joo, B.Y. Chow, M. Prakash, E.S. Boyden, J.M. Jacobson, Face-selective electrostatic control of hydrothermal zinc oxide nanowire synthesis, Nat. Mater. 10 (2011) 596–601.

[115] S.J. Nejad, A. Golzary, Experimental design for the optimization of hydrothermal synthesis of samarium oxide (Sm_2O_3) nanoparticles under supercritical water conditions, Int. J. Chem. Engg. Appl. 2 (2011) 243–247.

[116] Environmental and Corrosion Simulation Program (ESP/CSP) version 6.7, OLI Systems, Inc., Morris Plains, NJ, 2010.

[117] B.L. Gersten, Influences of the A and B Site Cation Species in the Kinetics of Hydrothermal Synthesis of ABO_3 Perovskite-Type Materials, Ph.D. Thesis, Rutgers University, NJ, USA, 1999.

[118] J.C. Tanger, H.C. Helgeson, Calculation of the thermodynamic and transport properties of aqueous species at high pressures and temperatures: revised equations of state for the standard partial molal properties of ions and electrolytes, Am. J. Sci. 281 (1981) 1249–1428.

[119] A. Seidell, third ed., Solubilities of Inorganic and Metal Organic Compounds, vol. 1, D. Van Nostrand Company, Inc., New York, NY, 1940.

[120] E.F. Linke, A. Seidell, fourth ed., Solubilities of Inorganic and Metal Organic Compounds, vol. 2, American Chemical Society, Washington, DC, 1965.

[121] M.M. Lencka, R.E. Riman, Hydrothermal synthesis of perovskite materials: thermodynamic modeling and experimental verification, Ferroelectrics 151 (1994) 159–164.

[122] M. Uematsu, Thermophysical properties of supercritical fluids, in: Y. Arai, T. Sako, Y. Takebayashi (Eds.), Supercritical Fluids—Molecular Interaction, Physical Properties, and New Applications, Springer, Germany, 2002, pp. 71–78.

5 Hydrothermal Growth of Some Selected Crystals

The hydrothermal technique has produced a wide variety of minerals and crystals, both in nature and in laboratory in all sizes from nanosize to bulk size. In fact, it is the only technique to synthesize some of the inorganic compounds like quartz and berlinite. However, the method has some limitations as far as the size is concerned. Except quartz, the method has not yielded large single crystals. Although berlinite, gallium berlinite, and lithium tetraborate show many superior piezoelectric properties compared to α-quartz, it is not possible to obtain them as large as α-quartz crystals. Further, it is difficult to discuss here the growth of all the compounds of hydrothermal origin. Hence, the scope of this chapter has been restricted only to the growth of some bulk crystals of technological importance.

5.1 Quartz

Quartz is one of the most abundant minerals and occurs as both essential and accessory constituents of rocks. References to quartz are known from 1505 onward. It may have been derived from the Saxon word *querkluftertz*, or cross-vein-ore, which could easily have become condensed to *querertz* and then to *quartz* [1].

Quartz, SiO_2, exists in both crystalline and amorphous forms in nature. The crystalline form of quartz has over 22 polymorphic modifications. The three principal crystalline forms of SiO_2 (quartz, tridymite, and cristobalite) each possess a well-defined field of stability under equilibrium conditions. The transformations from one to another are, however, somewhat sluggish, so that the higher temperature forms, cristobalite and tridymite, can exist metastably below their inversion temperatures. Each crystalline form of quartz, metastable tridymite and metastable cristobalite, has furthermore low- and high-temperature modifications designated as α- and β-, respectively [2–4]. Among them, α-quartz, which is stable below 573°C at atmospheric pressure, is the most popular and technologically important. The structure of α-quartz is shown in Figure 5.1 [5,6] The α-quartz has trigonal symmetry, belongs to the enantiomorphous crystal class-32, and its space group is $P3_121$ or $P3_221$ according to its right- or left-handedness. Its structure was among the first to be investigated by X-ray techniques as early as 1914 by Bragg [6]. The structure of quartz is made up of SiO_4 tetrahedra, which are linked by sharing each of their corners with another tetrahedron. In the three-dimensional framework thus formed, every Si^{4+} has four oxygens and every oxygen has two

Handbook of Hydrothermal Technology.

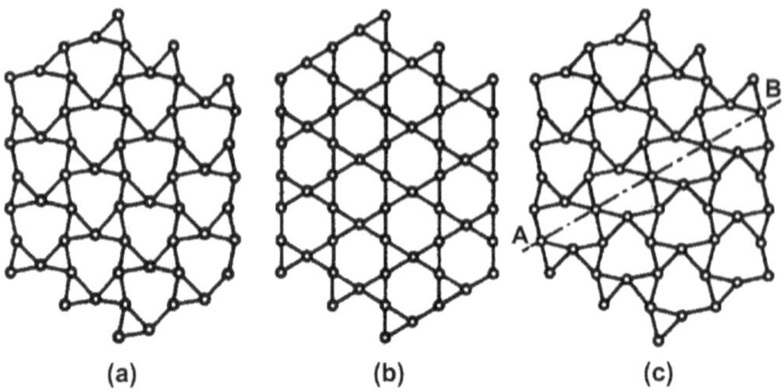

Figure 5.1 Structure of α-quartz [5].

silicons as nearest neighbors. There are several works on the hydrothermal synthesis of other varieties of silica—like coesite, β-quartz, stisshovite, cristobalite, and tridymite [7−10]. However, the scope of this chapter has been restricted to the growth of large size single crystals of quartz by the hydrothermal technique. In this respect, it is only the α-quartz, which could be grown as large bulk single crystals, whereas the other varieties of silica have been obtained so far only as fine crystalline products, or tiny crystallites.

The α-quartz is one of the most extensively studied materials. It is the most popular and technologically very important among all the hydrothermally grown crystals or materials. Quartz, as a piezoelectric crystal, has the ability to convert electric waves into mechanical waves and to reverse the process. Because of this property, quartz is widely used in filters, timing and frequency control applications, optical fibers, dielectric applications, and so on. In the recent years, quartz "tuning forks" have become essential for timing functions in electronic watches and in timing circuits for computers and telecommunications. Quartz crystal is one of the most transparent materials over a wide range of optical frequencies from UV (ultraviolet) to IR (infrared) regions, and it has double refraction (birefringence) and optical rotation power. Because of these properties, the quartz crystal is utilized as an optical device on a large scale. Typical applications are in optical low pass filters (OLPF) for video cameras and waveplates for optical pickup. Table 5.1 gives a wide range of applications for α-quartz.

The principal source of electronic grade natural quartz is Brazil. Today, the electronic industries are largely inclined to use synthetic quartz, because natural quartz crystals are generally irregular in shape, automatic cutting is cumbersome, and the yield is low. Over 3000 tons of quartz are produced annually for a variety of applications. These applications range from optical components (due to its high transparency) to precise time and frequency oscillators based on its piezoelectric properties. The important countries contributing to the world production are the United States, Japan, China, Korea, Taiwan, Poland, Belgium, France, United Kingdom, Germany, Russia, Brazil, and so on. At one time, Japan alone produced more than 50% of the worldwide production, followed by the United States till recently [11]. Now, China has emerged as one of the major contributors to the world

Table 5.1 Applications for α-Quartz

Industrial equipment	Precision oscillators, optical fibers, dielectric materials, radiocommunications, cable communications, electronic applications, measurement equipment, pagers, security systems (alarms)...
Consumer equipment	Electronic hand calculators, watches, clocks, timers, cable TVs, color TVs, video recorders, RF converters, transceivers, radio equipment, microphones, electronic appliances, microphones, microcomputer and computer terminals, TV-game machines, telephones, copy machines...

production of quartz ahead of Japan and the United States. More recently, several other Japanese companies such as Kyocera, Shin-Etsu, and Asahi Glass Co. have started producing quartz for highly specialized applications. Figure 2.11 shows a photograph of synthetic quartz crystals pulled out from the world's largest autoclave located at the Toyo Electric Co., Japan [12]. The inner diameter of the autoclave is 65 cm. To quote an example, the combined production quantity of quartz crystal units, filters, and oscillators in 1980 was 390 million pieces in Japan and it was remarkably increased to 2.13 billion pieces in 1990. About 50% of the quartz produced goes into devices in the automotive industries, 30% of production goes into frequency control devices, and the remaining 20% goes into optical devices. The total cost involved in the production of quartz is:

End users	:	≈ US$1,000,000,000,000
Equipment	:	≈ US$5,000,000,000
Components	:	≈ US$100,000,000

Thus, quartz takes the first place in value and quantity of single-crystal piezoelectric materials produced. Much of the interest related to quartz today is confined to industry, while only scant publications emerge from research laboratories on new applications of quartz and the preparation of nanoscale silica particles. Otherwise, it is mainly research and development (R&D) based industries that carry out research on the production of high-quality quartz and its applications, and these results are not published [13,14].

The hydrothermal growth of quartz has been described and reviewed in several earlier works [15–19]. Hence, we will only consider the conventional method of the growth of α-quartz and its kinetics data in brief. Instead, we discuss the more recent developments in the growth of high-purity and low dislocation quartz.

The −α−β transition in quartz takes place at about 573°C, and the crystal growth of quartz insists upon a growth technique suitable to the low-temperature crystal growth process. This led to the development of hydrothermal process. In fact, much of our knowledge of the hydrothermal technique has resulted from the success in the growth of quartz crystals. The first publication in hydrothermal research is again pertaining to the synthesis of quartz by Schafhaul during 1845 [20]. The best known source of natural quartz, even today, is Brazil, which has large occurrences of high-quality electronic grade quartz. As a matter of fact, Brazil was the only supplier of large crystals of high-quality natural quartz for electronic applications until World War II. Shortages in the United States and Europe caused by German submarine activity prompted efforts to synthesize large crystals of quartz, in the laboratory, using seeds. Spezia (1900–1909) was the first to attempt the seeded growth of quartz crystals and to carry out a systematic study of the solubility. However, the growth rate was poor. It was only during World War II that Nacken could achieve higher growth rates for quartz by the hydrothermal technique. The pilot scale production of quartz began during mid-1950s in the United States and erstwhile USSR. Several countries attempted the artificial growth of large size single crystals of quartz. Especially, after the publication of the works of Nacken in 1950 [21] and the captured German reports [22,23], several laboratories began studying quartz crystal growth almost simultaneously and the hydrothermal technique became a very popular tool for developing many complex inorganic compounds. Industrial scale production occurred essentially during 1960s and 1970s, with a peak development between 1968 and 1976, especially in the United States, Japan, and Europe.

In the growth of α-quartz by the conventional method, the autoclaves used by most of the workers were modified Bridgman-type autoclaves. The design and construction of the modified Bridgman autoclave is described in Chapter 3. The typical laboratory size autoclave is 1″ diameter × 1′ length. The pressure autoclave used in Europe and the United States is 13″ diameter and >10′ length with 350 l internal volume, which gives some 150 kg of quartz per cycle. But an evolution concerning vessel's dimensions has to be noted, especially in Japan, where autoclaves with 1000–5000 l of internal volume are in operation producing 500–2000 kg of quartz per cycle. The evolution of the autoclaves, the sealing involved, and the volume are discussed in Chapter 3. These larger internal volume autoclaves are provided with Grey-Loc type of sealing. Parallel to the increase in output, electronic components manufacturers have been looking, for the past 2½ decades, for a material with steadily increasing performances for professional applications in the telecommunications field, civilian as well as military.

In the growth of α-quartz, available nutrient material such as small particle size α-quartz, silica glass, high-quality silica sand, or silica gel is placed in a liner made up of iron or silver with a suitable baffle and a frame to hold the seed plates. A mineralizer solution with a definite molarity is poured into the liner to make the required percent fill (% fill). The increased solubility in the presence of mineralizer increases the supersaturation without spontaneous nucleation and consequently allows more rapid growth rates on the seeds. Figure 3.3 shows the cross section of the modified Bridgman autoclave used in the growth of quartz crystals. The commercial autoclaves used have 10″ inner diameter and are 10 feet long unlined. These autoclaves can work at conditions up to 30,000 psi and 400°C. Most of these experiments are carried out for 25–90 days to obtain full size crystals, 4 cm in the *Z*-direction and 12.5–15 cm in the *Y*-direction. Temperature gradient is varied according to the nutrient used. 1N NaOH (1 M) or Na_2CO_3 (1 M) are the most commonly used mineralizers. The solubility change with temperature is smaller in NaOH and slightly larger in Na_2CO_3. The temperature of the autoclave at the nutrient zone is usually kept at 355–369°C and, in the growth zone, it is kept at 350°C. The addition of lithium improves the growth rate and small amounts of Li salts are routinely added to the solution [22]. The solubility is also, to some extent, a function of increasing pressure. The pressure is controlled by the percentage fill in the autoclave, and it is usually about 80% for hydroxyl mineralizer (20,000 psi internal pressure). In most of the experiments, the percent opening of the baffle is 20%; even lower percentages are used by several workers. However, the actual percent opening of baffle area and its geometry are not disclosed, especially by commercial growers.

The optimum growth conditions for synthesis of quartz based on the work in Bell Laboratories are [24,25]:

Dissolution temperature	–	425°C
Growth temperature	–	375°C
Pressure	–	15,000–25,000 psi
Mineralizer concentration	–	0.5–1.0 M NaOH
Temperature gradient (ΔT)	–	50°C
% Fill	–	78–85%
Growth rate in (0001)	–	1.0–1.25 mm/day

The quality of the grown crystals is also a function of the seed orientation and its quality. Strained seeds generally produce a strained growth region [26]. The seeds are polished to a very fine finish before use. Most high-quality crystals are grown using seeds with surfaces perpendicular to the *Z*-direction since the *Z*-growth region is the lowest in aluminum concentration. Though the main part of quartz production consists of *Y*-bar crystals, i.e., small crystals ($Z = 20$–25 mm, 64 mm seed) capable of several *Y*-bars per crystal, the pure *Z*-bars are also produced, representing 10–20% of this production. In medium- and high-quality grades, we notice a rise in demand for crystals of very large dimensions and upper-medium quality, especially in the United States, for manufacturing wafers used in surface wave applications [11]. Earlier, most of the seeds used were natural quartz cut into a definite orientation, but in recent years this practice is only used when a high-quality crystal is desired.

The growth of quartz crystals has been understood precisely with reference to the growth temperature, temperature gradient, percent of fill, solubility, percent of baffle open, orientation and nature of seed, and type of nutrient. Also, many kinetic studies have been carried out [15,16]. Figures 5.2 and 5.3 show the solubility of quartz with temperature, growth rate as a function of seed orientation, and growth rate as a function of percent of fill. Figure 5.4 shows hydrothermally grown quartz crystals.

The type of crystal to be grown depends on the application, as different properties are required in each case. For optical use, high uniformity, low strain, and low inclusion counts are needed since all of these can affect the transparency. For surface acoustic wave devices, large pieces are needed which can take a very high-quality surface finish. The quality of the material required for resonators used in time and frequency devices varies with the application. The more precise the need, the more

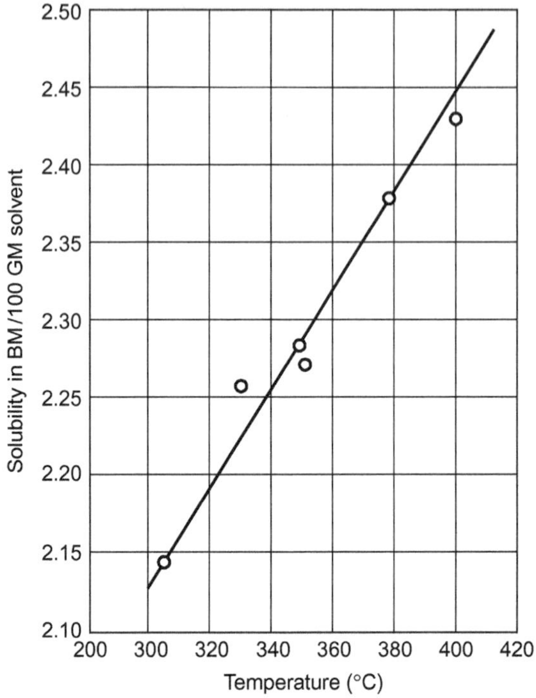

Figure 5.2 Solubility of quartz with temperature [16].

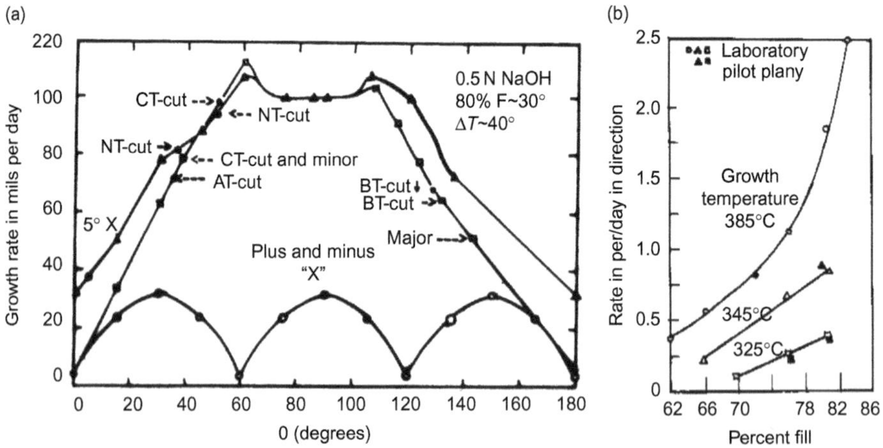

Figure 5.3 Growth rate of quartz as (a) a function of seed orientation and (b) a function of percent fill [16].

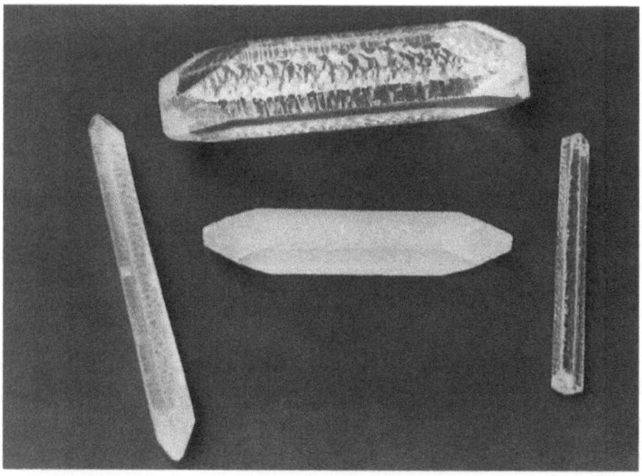

Figure 5.4 Hydrothermally grown quartz crystals [19].

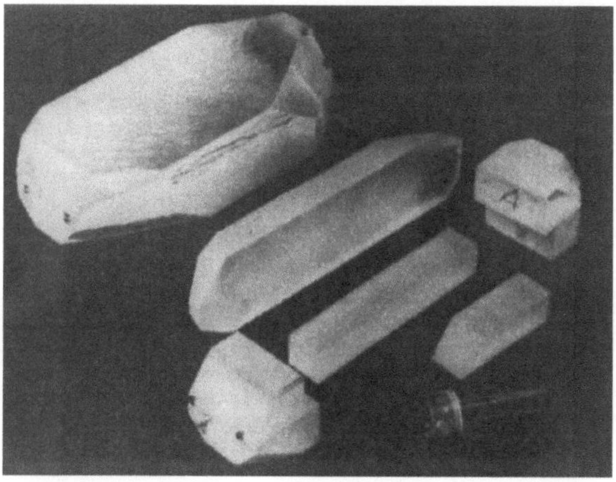

Figure 5.5 Fabrication of resonators from a single crystal [18].

stringent are the requirements. For most applications, a truly high-quality material is not needed. For high precision uses, such as in navigational devices and satellites, a very high-quality material must be used. Most of the recent research on quartz growth is for improved resonator performance; which requires the growth of high-quality and low dislocation quartz, which is discussed in Section 5.2.6. Figure 5.5, shows the fabrication of resonators from a single crystal.

5.2 Growth of High-Quality (and Dislocation-Free) Quartz Crystals

It is well known that the hydrothermal growth of quartz in the industrial sector is at least 45 years old. Though many improvements have been brought about, it is still more an art than a truly reproducible industrial and scientific process. We discussed in Chapter 1 the shortcomings in the hydrothermal growth of crystals, especially with reference to the theoretical knowledge. This applies even to quartz growth under hydrothermal conditions. The growth of high-quality quartz crystals insists on so many parameters like phase relationships, nutrient's solubility, thermodynamics (state equations, kinetics of reaction, impurities repartition), and crystal defects characterization. The other aspect is that the needs for cheaper material in a wide range of applications like clocks and microprocessors result in being content with a medium-quality product, and the low profitability precludes financing long and expensive studies. Though much of the recent publications on quartz is on the possibility of obtaining/growing high-quality quartz crystals for stringent electronic applications, this group is still in minority. Except for quartz, no other material coming from hydrothermal synthesis has sizable industrial use; even though $AlPO_4$, $GaPO_4$, $Li_2B_4O_7$, and microcrystallites show interesting promise, industry does not seem ready to make an important R&D effort concerning these products. Thus, it is difficult to get financial support to acquire a better knowledge of hydrothermal growth, which is highly intricate and necessitates an important effort for a long duration of time. Private industry cannot deal alone with the costs of such a program; even in the United States, research programs are government sponsored, especially in the defense field [11].

Several criteria are used to evaluate the quality of quartz crystals. The most commonly used criterion is the Q value or quality factor, which is a measure of the acoustic loss of the material. It is important for a resonator to have high electrical Q value and superior frequency–temperature characteristics. In a piezoelectric resonator, electrical energy and mechanical energy are interconvertible. In such a case, Q is expressed as:

$$Q = \frac{[X]}{R}$$

where X is the inductive or capacitive reactance at resonance and R is the resistance.

The quality factor, Q, can be considered as the inverse of the fraction of the energy lost per cycle. The highest values of Q are required in order to prevent loss of energy into coherent phonons. The acoustic Q for natural α-quartz crystal varies in the range of $1-3 \times 10^6$, while for synthetic quartz crystals, the value drops down to 2.105–1.106. Thus, in the last two decades, the main objective among quartz crystal growers has been to improve Q, which in turn leads to the production of a low concentration of physicochemical and structural defects. We shall discuss the growth of such quartz crystals of high quality through the study of growth rate, mineralizers, solubility liner material autoclave, seed effect, nutrient effect, and finally the recent advances in the processing of defect-free quartz for stringent electronic applications.

5.2.1 Growth Rate

Growth rate is determined by the ratio of increase in thickness of seed and duration of the run. The growth rate along the main crystallographic axes, R_c, is determined by dividing the thickness of the layer grown on the seed, $(h_1 - h_o)/2$, by the run duration t, and is expressed in mm per day. A much more precise procedure is to determine R_c as a function of weight increment provided the surface area does not change during the run, $R_c = (P_t - P_o)/2S_t\delta$, where $\delta = 2.65\ g/cm^3$ is the density of quartz. The growth rate increases with the run duration, the increment being relatively faster during the early hours and days [27]. The growth rate depends upon various factors like growth temperature, experimental pressure (% fill), impurities concentration, presence of defects, seed orientation, solvent/mineralizer, and % of baffle opening. It has also been known for some time that, qualitatively, acoustic Q is inversely proportional to the growth rate and directly related to chemical impurities which will be discussed separately. The internal friction (inverse of mechanical Q) is dependent on growth rate of synthetic quartz crystals. This was first demonstrated by Brown [28]. Chakraborty (1977) [29] has studied the dependence of mechanical Q on the growth rate of quartz crystals using different mineralizers and concludes that the inverse relationship (exponential) between mechanical Q and growth rate is independent of the nature of the solvent in which the crystals have been grown. Thus, the mechanical Q is dependent not only on growth rate but also upon various other operational variables. For example, for growth on surface normal to (0001) (basal plane growth) or on samples 5° from (0001) (+5° X-cut surface), high Q quartz ($Q > 10^6$) can be grown at rates below 20 mil/day (0.5 mm/day), while even in the presence of Li^+, Q's above 10^6 have ordinarily not been obtained at growth rates much above 60 mil/day (1.5 mm/day). Techniques for obtaining high Q at high rates have obvious economic importance. In particular, rates above 100 mil/day (2.5 mm/day) with $Q\text{'s} > 10^6$ would provide significant savings in both capital and operating expenses in commercial quartz growth [30].

5.2.2 Seed Effect

The seed plays a predominant role in the quality of the resulting crystal. For higher frequency applications, a smaller X-axis dimension is needed. Material with growth along the Z-axis (Z-growth material) is desired for resonators as it has been shown that this material is about an order of magnitude lower in aluminum concentration [31]. The thickness of the seed is usually between 1 and 2 mm. Until recently, it was necessary to use natural seeds for the preparation of low dislocation crystals. This is an extremely tedious process since only a small portion of natural crystal is of sufficient quality for seed use, resulting in a complicated selection process. The seeds must also be of sufficient size for useful crystal growth. Christie et al. (1983) [32] have reported that small seeds can be fastened together to make longer seeds, but this has not always been successful. Most seeds used today are fabricated from synthetic quartz crystals. This is a much simpler process since the seeds can be cut parallel from the original seed. Most crystals grown from synthetic seeds, however, contain a large number of dislocations of the order of several hundred per square centimeters. This results in the formation of etch channels in the resonator, which weaken it mechanically and cause problem when electronic devices are deposited on the surface [33]. The studies of Armington and Larkin (1985) [34] have shown

Table 5.2 Dislocation Density in Different Seeds

Seed	Mineralizer	Etch Channel Density
Z	Hydroxide	253
Z	Carbonate	247
X+	Hydroxide	1
X+(reused)	Hydroxide	14

Figure 5.6 Dislocation density in quartz [34].

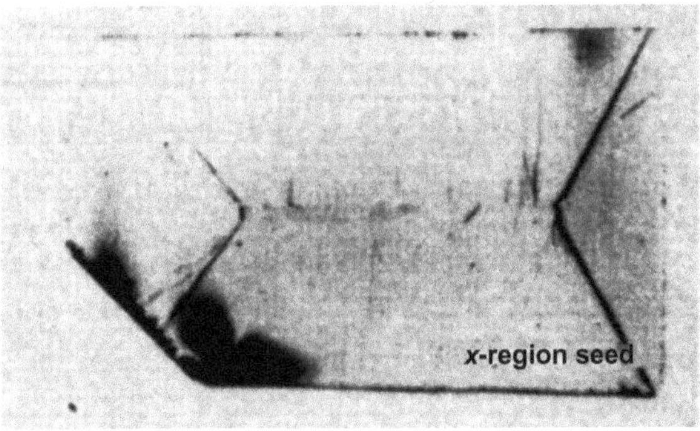

Figure 5.7 Dislocation density in quartz [34].

that, when a seed perpendicular to the *Z*-axis (but from the *X*-growth region) is used, the dislocation density can be reduced to below 10, and sometimes to 0 dislocations per square centimeter (Table 5.2; Figures 5.6 and 5.7). Although the reasons for this are not entirely understood, it may be related to the fact that, unlike dislocations in the *Z*-region where they form at the seed and grow close to the *Z*-axis, most dislocations in the *X*-region grow at an angle of at least 45° and are usually close to 90° from the *Z*-axis. Studies have shown that aluminum in the seed does not migrate into the growth regions. Thus, the purity of the crystals produced using seeds cut from the *X*-region is as good as crystals produced from the usual *Z*-seed.

5.2.3 Nutrient Effect

The quality of the nutrient has a profound effect on the purity of the grown crystal. It appears that most of the impurities incorporated into the crystal come from the nutrient and not from the autoclave walls. The search for new sources of nutrient is not a new field of research. In the nineteenth century, earlier workers tried several varieties of nutrient materials to obtain α-quartz under hydrothermal conditions. However, the results were not satisfactory owing to the lack of knowledge on the solubility data for quartz, which resulted in a very low growth rate. In this respect, some serious efforts were initiated by Kolb et al. during 1976 [35]. They examined new sources of nutrient to replace Brazilian α-quartz for hydrothermal crystallization. It was found that the acoustic *Q* strain depends upon the source of the nutrient and also its geographic region/location. Both vein and appropriately chosen pegmatite quartz can be used as a nutrient. High purity sand can be used as the nutrient provided process conditions are altered so as to compensate for its effective lower surface area. Non-α-quartz nutrients such as silica glass and silica gel produce higher supersaturation initially and have fast growth rates leading to poor quality of crystal growth. However, alterations in process conditions can be made to reduce initial growth rates and prepare reasonable quality crystals. Therefore, in the growth of high-quality quartz crystals, *Z*-growth material and recrystallized glass is used as the nutrient material. If future requirements become more stringent, particularly for radiation damage, it will be necessary to reduce the aluminum content even further. In this case, recrystallized glass will be the leading candidate, as higher purity glass is available. This generally requires two runs, since the glass must first be converted to α-quartz.

There are several laboratories throughout the world working to find a suitable nutrient for high-quality quartz growth. Hosaka and Miyata (1991) [36] have obtained high-quality α-quartz crystals using cristobalite as the nutrient. The authors have used high-purity α-cristobalite powder compacted into grains of 0.5−1.00 mm or into lumps of approximately 1 cm in diameter for the hydrothermal synthesis of α-quartz.

Alpha-cristobalite is a polymorphic form of silica as are quartz and tridymite and has a higher solubility than quartz [37]. Alpha-cristobalite powder [38] has the following advantages: It can be obtained in a high-purity state with low contents of Al and alkaline metallic ions, and can be prepared with relative ease as particles of uniform size. As compared with the conventional growth of synthetic quartz crystals using Brazilian lascas, the use of α-cristobalite powder may be expected to allow:

i. Synthesis of high-purity large quartz crystals.
ii. Synthesis of microquartz crystals having uniform grain size.

Hosaka (1991) [38] had attempted to crystallize microquartz crystals by HHP method using α-cristobalite powder as a source material. Figure 5.8 shows α-quartz crystals grown at high (left) and low (right) fillings using α-cristobalite as the nutrient. The percentage fill or pressure greatly influences the growth rate. However, a considerable amount of work has to be carried out for the industrial production of α-quartz using cristobalite as the nutrient.

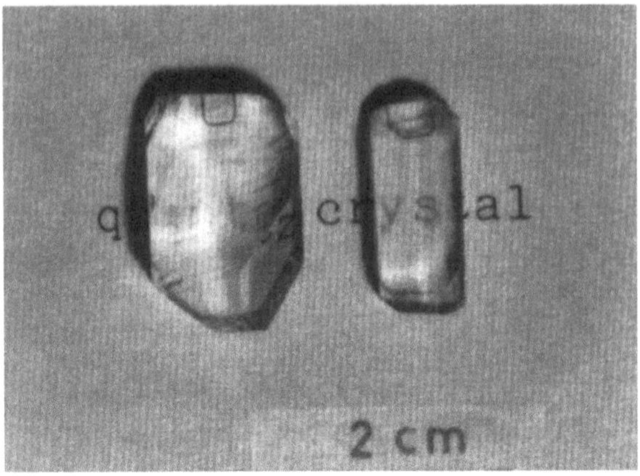

Figure 5.8 Alpha-quartz crystals [38].

5.2.4 Solubility

Solubility is one of the most important aspects in hydrothermal crystal growth. The early literature surveyed clearly shows that the very slow growth rate achieved until the works of Nacken (Germany 1950) [21] and Wooster and Wooster (1946) [40] were published was mainly attributed to the lack of knowledge of the solubility of quartz. The same applies to other compounds also. The first systematic study of the solubility of quartz was carried out by Spezia (1905) [39]. Since then, a lot of progress has been achieved in the understanding of the solubility in general for various inorganic compounds. During 1960s, new methods of investigating the solubility and the new experimental setup for the determination of solubility under hydrothermal conditions were proposed [41,42].

The solubility of quartz in pure water was found to be too low for crystal growth (0.1–0.3 wt%), but the solubility could be markedly increased by the addition of OH^-, Cl^-, F^-, Br^-, I^-, and acid media which act as mineralizers. For example, the reactions:

$$SiO_2 + 2OH^- \rightarrow SiO_3{}^{2-} + H_2O$$

$$2OH^- + 3SiO_2 \rightarrow Si_3O_7{}^{2-} + H_2O$$

show the formation of various complexes or species during the hydrothermal crystallization of quartz. Hosaka and Taki [43] have used Raman spectra to identify and quantify such species.

In pure aqueous solutions (even at 400°C and 25,000 psi), the solubility of quartz is too low to allow growth to take place in any reasonable time. Alkaline additions, such as NaOH, Na_2CO_3, KOH, and K_2CO_3 are all effective as mineralizers in this pressure and temperature range. A small increase in molarity results in only a slight increase in the growth rate, whereas a large increase begins to produce an additional phase along with quartz. The minimum molarity for a good growth rate is about 0.25 M for NaOH. Concentrations of about 4.0 M for NaOH and 2.0 M for KOH form sodium or potassium silicates along with α-quartz. Laudise and Ballman (1961) [44] have measured the solubility of quartz in 0.5 M NaOH as a function of the percentage fill and temperature. The quartz solubility dependence on the percentage fill is shown in Figure 5.9.

An important result of solubility determinations is the delineation of the pressure, temperature, and composition regions where the temperature coefficient (TC) of solubility is negative. These regions are to be avoided in the growth of quartz, since they require a different setup; otherwise, they result in the loss of seed crystals.

In recent years, mixed solvents are being used even in the growth of quartz. High-quality large quartz crystals have been obtained in NaCl and KCl solutions, NaOH and Na_2CO_3 solutions, (10%) [45,46]. Most of the literature data available on the solubility of α-quartz deals with the data in various solutions at elevated temperatures and pressures. It was recently observed that, in the growth of quartz crystals under hydrothermal conditions, the evaluation of pressure generation and its effect on the dissolution in different aqueous solutions are the important steps. Hence the authors have studied the solubility of α-quartz in NaOH (1 M) + Na_2CO_3 (1 M) using high-pressure conditions from 200 to 350 MPa at temperatures of 400°C [46]. The impetus for this work was provided by the fact that high-frequency applications of α-quartz require sheets with small thickness of the same order of size as the defects (such as inclusions, etch pits, and dislocations). Thus, the pressure can make new solvents viable for hydrothermal growth, especially by reducing their concentrations. Figures 5.10 and 5.11 show the solubility limit comparison between NaOH (1 M) and Na_2CO_3 (1 M) versus pressure at

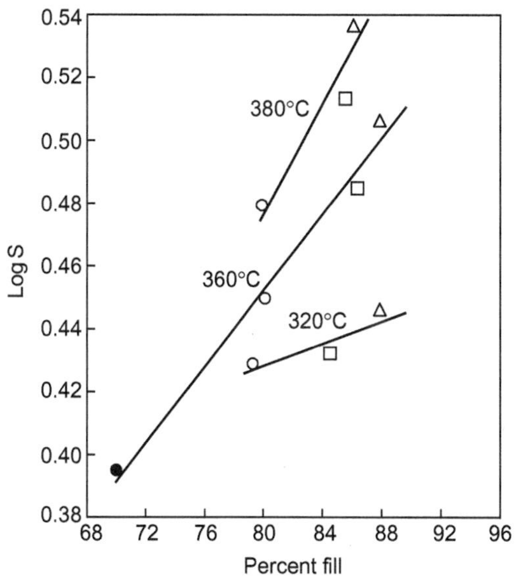

Figure 5.9 Quartz solubility dependence on % fill [44].

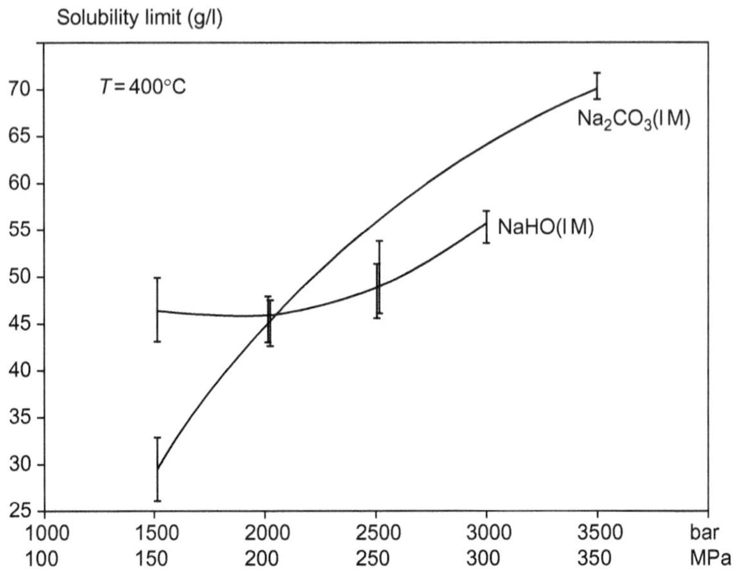

Figure 5.10 Solubility limit comparison between NaOH (1 M) and Na_2CO_3 (1 M) versus pressure at 400°C [46].

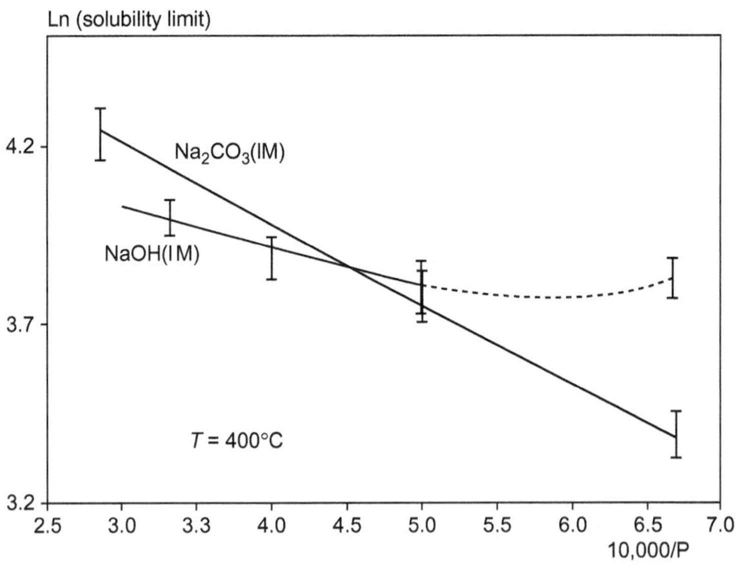

Figure 5.11 Log solubility (S_1) in NaOH (1 M) and in Na_2CO_3 (1 M) versus $1/P$ [46].

400°C and log solubility (S_1) in NaOH (1 M) and in Na_2CO_3 (1 M) versus $1/P$. These experimental results show that Na_2CO_3 appears to be a better solvent than sodium hydroxide in high-pressure domain for a temperature close to 400°C. From Figure 5.11, the enthalpy values have been calculated and the values are as follows:

$$\Delta H_T = 2395 \pm 5 \text{ cal/mole for NaOH (1 M)}; \quad 200 \le P \le 350 \text{ MPa}$$

$$\Delta H_T = 4001 \pm 2 \text{ cal/mole for } Na_2CO_3 \text{ (1 M)}; \quad 150 \le P \le 350 \text{ MPa}$$

5.2.5 Defects Observed in Synthetic α-Quartz Single Crystals

The defects present in the synthetic α-quartz crystals play an important role in determining their quality and in turn their applications. Much of the recent works on quartz is essentially focused on the defect studies. All the defects present in quartz can be classified into two types:

1. The physicochemical defects
2. The structural defects.

Physicochemical defects are mainly induced by the nature of the nutrient or the solvent and also by chemical contamination of the solution by the metal constituting the reaction vessel. Quite often, solid or liquid inclusions like $NaFe^{3+}Si_2O_6$ (acmite) are inserted in the α-quartz. The most important impurity in quartz is H^+, which easily fits interstitially in the large (1 Å) channels that lie parallel to the c-axis to charge compensate for Al^{3+} which goes to an Si^{4+}. It enters the lattice

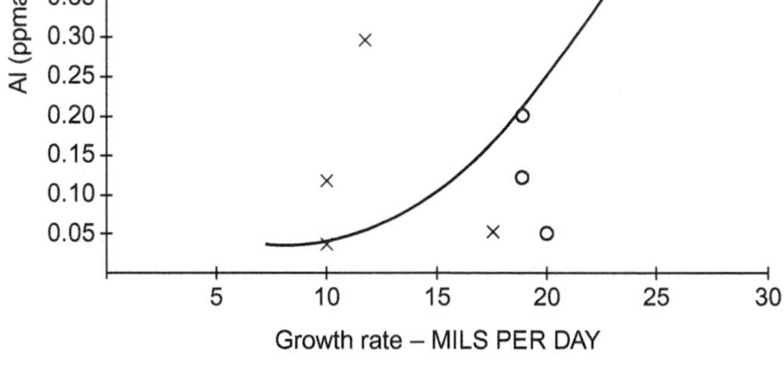

Figure 5.12 Effect of growth rate on Al^{3+} concentration in the Z-region of growth crystals [18].

from the growth solutions as OH^-, the O^{2-} being incorporated in the SiO_2 lattice. In thermodynamical conditions, different cations (in particular Li^+) can be in the interstitial positions, in particular for compensating the $Si^{4+} \rightarrow M^{3+}$ cationic substitution into the α-SiO_2 lattice. Figure 5.12 shows the effect of growth rate on Al^{3+} concentration in the *Z*-region of grown crystals using a hydroxide mineralizer [47]. Aluminum concentration was determined by EPR [18]. Similarly, Laudise [48a,b] has studied the dependence of the effective partition coefficient for OH^- impurity in quartz on growth rate (Figure 5.13). Lithium salt is added to the solvent to check the aluminum concentration in the grown crystal. The entry of Al occurs in the use of (OH) and (CO_3) based mineralizers. Table 5.3 gives the distribution of impurities in different sectors of synthetic quartz, which has been studied by Yoshimura et al. (1979) [49], and Iwasaki and Kurashige (1978) [50].

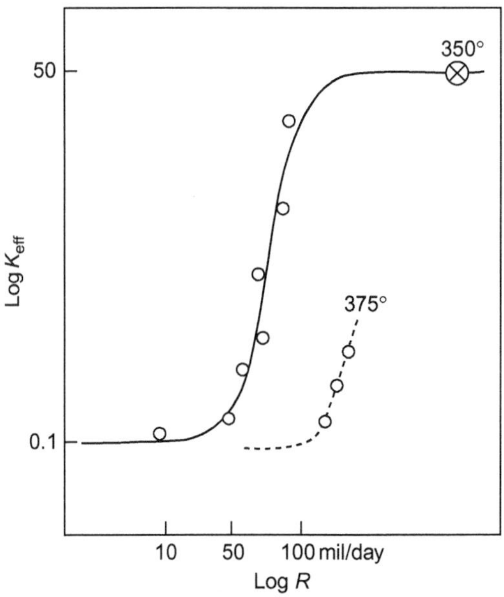

Figure 5.13 Effective partition coefficient for OH^- impurity in quartz on the growth rate [48a,b].

Table 5.3 Impurities in Synthetic Quartz Crystals (ppm)

Sector	Al	Na	Li
Z	5	1	0.5
+*X*	31	9	5
−*X*	122	40	5
S	85	26	16

The structural defects in quartz are basically the dislocations. Their origin can be due, generally, to foreign particles, thermodynamical parameters governing the crystal growth, or hydrodynamic conditions inside the reaction vessel.

The asymmetry in the dissolution behavior of quartz under hydrothermal conditions is an interesting topic in order to enhance growth rate on certain directions [51,52].

5.2.6 Processing of α-Quartz for High-Frequency Devices

The study of the type of defects present on silica surfaces would be of great importance for a better understanding of electrical or piezoelectric properties. The development of high-frequency devices (24 MHz → 100 MHz or more) induces some strong constraints concerning the shaping of the α-SiO_2 material. The low-defect, high-purity synthetic quartz should have the following characteristics [53]:

Parameter	Desirable	So far Achieved
Etch channel density	$<10/cm^2$	<86
Inclusion density	<10/bar	
Impurity concentrations (ppb)		
Al	<200	700
Li	<300	300
Na	<500	1640
K	<40	300
Fe	<100	1800
Q (3500/3800 cm^{-1})	$<2.5 \times 10^6$	$>2.5 \times 10^6$
Strain	None	Variable
Fringe distortion	<0.05 RMS	Variable

The existence of defects, either physicochemical or structural, in synthetic quartz crystals leads to the critical modifications of devices. There are three different parameters on which the quality of quartz depends [54]:

i. Chemical impurities, e.g., $(OH)^-$ distribution, induce a large decrease of the acoustic Q. The substitution $O^{2-} \rightarrow (OH)^-$ being coupled with the cationic one ($Si^{4+} \rightarrow M^{3+}$), the M^{3+} impurities can play an important role concerning the chemical aging of resonators and its behavior versus ionizing radiations.

ii. Crystal defects like dislocations, etching channels, and fractures, which influence the acoustic distribution.

iii. Inclusions (iron, sodium, acmite, aluminum, and so on).

As the application of quartz goes to higher and higher frequency devices, the thickness of the quartz plate drops. For example:

24 MHz range applications: $\sim 70\ \mu m$ thickness

100 MHz range applications: $\sim 6 - 7\ \mu m$ thickness.

The thinning down of the quartz plate depends upon the frequency measurements and can be expressed as:

$$F\ (\text{MHz}) = \frac{K\ (\text{MHz}\ \mu H^{-1})}{e\ (\mu m)}$$

where e is the plate thickness, K is a constant characteristic of the material and of its orientation, and F is the resonance frequency.

Sweeping is one of the most popularly used techniques in recent years to enhance the performance of quartz resonators. Sweeping or solid-state electrolysis or electro-diffusion is generally performed under a vacuum, or air, or hydrogen, or desired atmosphere [55–57]. During sweeping, the crystal is placed in an electric field and heated. Figure 5.14 shows the schematic diagram of the sweeping apparatus. Then, there is a migration of the impurities, and some modifications are induced within the crystal. According to the parameters used by the manufacturer during the sweeping process, and the quality of the as-grown crystals, the results could be very different. Sweeping reduces the formation of etch tunnels. The effect of sweeping is to remove lithium and sodium deposited interstitially during growth. These ions are usually trapped along an angstrom-wide tunnel, which is parallel to the Z-axis in the quartz crystal lattice. In an interstitial position they interact with aluminum impurities substitutional in the lattice to form Al–Li centers, which have shown to be weakly bonded and are the cause of low radiation tolerance in a resonator. In the sweeping process, these are replaced by Al-OH or Al-hole centers, which have a much higher radiation tolerance.

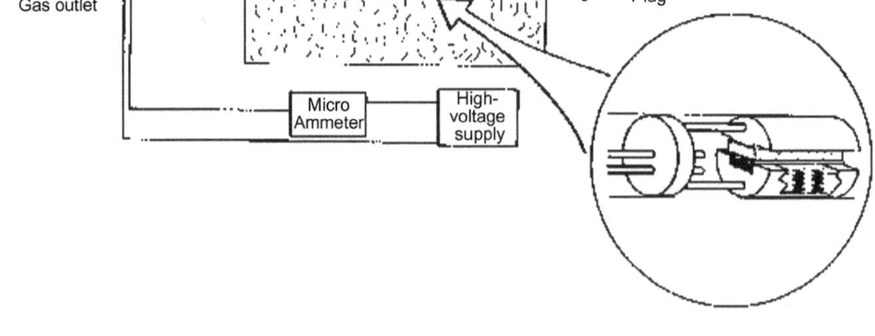

Figure 5.14 Schematic diagram of the sweeping apparatus [57].

Sweeping is influenced by the electric field, strength, electrode type, and atmosphere. It takes less time in air due to the availability of water in the atmosphere [58]. Electrodes, usually platinum, or gold or aquadog are deposited on each of the Z surfaces of a lumbered bar. The crystal is then subjected to a field of 1000–2000 V/cm (thickness) at a temperature of 500–550°C. The current is continuously measured and the sweeping is considered completed when the current has dropped to a constant value. It takes usually 5–7 days to remove alkalies. In some quartz, the sweeping may take 7 days or even more. Sweeping has also been used to dope quartz crystals in order to determine the effect of the dopant on the properties of the crystallization of quartz. A vast amount of literature data shows that sweeping cannot become a routine process even if all the same experimental conditions are ever applied. The quality of the swept crystal can change in such a way that the modifications are sometimes unexpected. So, each industrially swept crystal must be controlled before using. Some of the controlling techniques are the X-ray topography and the IR spectroscopy. These techniques help to characterize the physicochemical defects and the α_{3500} value (Q factor). The reader can refer to the works of Refs [12,59,60].

There are several other techniques employed for the fabrication of piezoelectric high- to ultrahigh-frequency devices based on quartz resonators. The important ones are chemical polishing and ion beam etching (IBE). The reverse thermodynamic relations are employed for the refinement of chemical polishing. Several solvents like HF and NH_4HF_2, NaOH, $KOH \cdot xH_2O$, $NaOH \cdot xH_2O$, etc. can be used [61].

Similarly, the industrial chemical etching process is specially dedicated to large thickness removals without damaging the blank surface texture. This is most useful for frequency applications of quartz, because the mechanical grinding and lapping introduce surface stresses. Fluoride media is most popularly used for this type of chemical polishing [62]. Cambon et al. (1994) [63] have tried industrial chemical etching successfully in the temperature range of 150–180°C using concentrated NaOH solvents. During the chemical etching process, several factors influence the process: kinetics, etching temperature, etching time, plate orientation, SiO_2 concentration, solvent concentration, wafer carrier geometry, and so on. Using this process, about 3200 quartz plates can be processed. The resonators manufactured by this process have demonstrated a high level of performance, even higher than those obtained by mechanical means.

Future avenues in this direction are improvements on theoretical knowledge, improvements in the quality of solvents, nutrients, and seeds, autoclaves' technology, material used for autoclaves, dimensional ratio (diameter/length), crystallization zone/dissolution zone, basic questioning of the validity of presently used growth parameters (P, T), and better understanding of complementary treatments like sweeping and chemical etching.

In recent years, there have been some efforts to grow crystalline quartz films with AT-cut plane by means of catalyst-enhanced vapor-phase epitaxy under atmospheric pressure. It has been proved that quartz films grown by vapor-phase epitaxy show good oscillation characteristics at room temperature [64]. Using electro-diffusion, several types of structures were obtained inside and on the cathodic face of the quartz. The study of structures growth by electro-diffusion in crystal lattices with special characteristics (such as quartz structural channels) seems very interesting. It is possible to obtain different types of structures, metallic and nonmetallic with special properties [65]. Besides, quartz substrates are popularly used for the growth of thin films of various compounds for special applications [66].

Thus, any work on quartz is still a promising field, and it will have some important consequential effects on hydrothermal growth, in general, and facilitate the development of other materials production.

5.3 Berlinite

Berlinite, $AlPO_4$, is named after N.J. Berlin (1812), a pharmacologist of the University of Lund. Its physical properties are given in Table 5.4 [67]. Berlinite replaces quartz in electronic devices because its large mechanical coupling factors are greater than α-quartz, and its resonant frequency is nearly independent of temperature for certain orientations [68–70]. Berlinite is also interesting because its presence in Ca aluminum phosphate bioglass ceramics is believed to lead to higher bioactivity of bone implants [71–73]. Table 5.5 gives a tentative comparison with quartz.

Berlinite occurs in nature associated with augelite, attacolite, and other phosphates as small crystals at the Westana Iron mines near Nasum, Kristiaanstad, Sweden [73]. Always berlinite occurs in association with metamorphic or hydrothermally altered rocks. Exceptionally, for the first time occurrence and formation of berlinite have been reported under sedimentary conditions was found in a highly unusual cave setting of Cioclovina Cave, Romania [74]. However, the synthetic berlinite is more popular and abundant, and it can be obtained as fairly big crystals. The first successful growth of berlinite crystals was by Jahn and Kordes (1953) [75], followed by Stanely (1954) [76]. Stanely carried out the experiments in a sealed borosilicate glass vessel placed in a low-pressure environment. Mason (1950) [77] carried out the piezoelectric measurements for a small crystal of berlinite. However, the importance of berlinite only came into light in 1976 when Chang and Barsch [68] studied the piezoelectric properties of berlinite and reported large mechanical coupling constants and that its resonant frequency was nearly independent of temperature for certain orientations.

The main problem connected with the growth of berlinite is the negative TC of solubility, which insists on some special growth conditions, because of the lower solubility at higher temperatures and higher solubility at lower temperatures. One of the principal problems in a systematic growth of $AlPO_4$ crystals is the growth rate relationship in H_3PO_4 and the lack of good solubility data (because of the negative TC of solubility) at the constant fill conditions, which are essential for successful growth [78].

Although much of progress has been achieved in the growth of berlinite crystals, our knowledge on $AlPO_4$ is still comparable to that of quartz some 30 years ago, particularly with reference to solubility, reproducibility, crystal perfection, phase equilibria data, crystal size, and so on.

Table 5.4 Physical Properties of Berlinite and Quartz (Schwarzenbach, 1966) [69]

Properties	$AlPO_4$	Quartz
Density	2.64 (natural) 2.56 (artificial)	2.655
Hardness	6–7	7
n_e	1.529 ± 0.003	1.544
n_o	1.519 ± 0.003	1.535
Birefringence	+ 0.01	+ 0.009
a	4.94291	4.9138
c	10.94761	5.4052
c/z	5.4738 Å	
c/a	2.21481	1.10
$c/2a$	1.1074	
Space group	$P3_121$ or $P3_221$	$P3_121$ or $P3_221$
z	3	3

Table 5.5 Tentative Comparison with Quartz (Thickness *Y* Rotated Resonators at Same Frequency)

Useful Coupling Coefficient (AT Filter Trapped Resonators)	Enhanced by 1.4–1.5
Shift of oscillators or bandwidth of filters (AT-cut)	Twice
Angular sensitivity (first-order FTC)	Reduced
Thermal stability (higher order FTC)	Better
Q factor propagation losses	Already sufficient (may be comparable)
Thickness of plates (AT)	Reduced by 1.15
Electrode dimensions (AT)	Reduced by 1.32 (TT)
(Same plating 2D TT of TS)	Reduced by 1.23 (TS)
Nonlinear properties	To be determined

"Dry" berlinite has similar C, E, EPS constants, first-order TC and much reduced higher order TC.

Today it is possible to grow more than 500 kg of dislocation-free high-purity quartz in a single experimental run. However, in the case of berlinite, it cannot be more than a few hundred grams. The major problem is the highly corrosive phosphoric acid media in which the growth takes place.

5.3.1 Crystal Chemical Significance of the Growth of $AlPO_4$ Crystals

Berlinite is isostructural with quartz, $Si(SiO)_4$, and the artificial compound $AlAsO_4$. It shows the same thermal inversion as SiO_2 [79], although, at slightly lower temperatures, it crystallizes in hexagonal system—trigonal trapezohedral class-32.

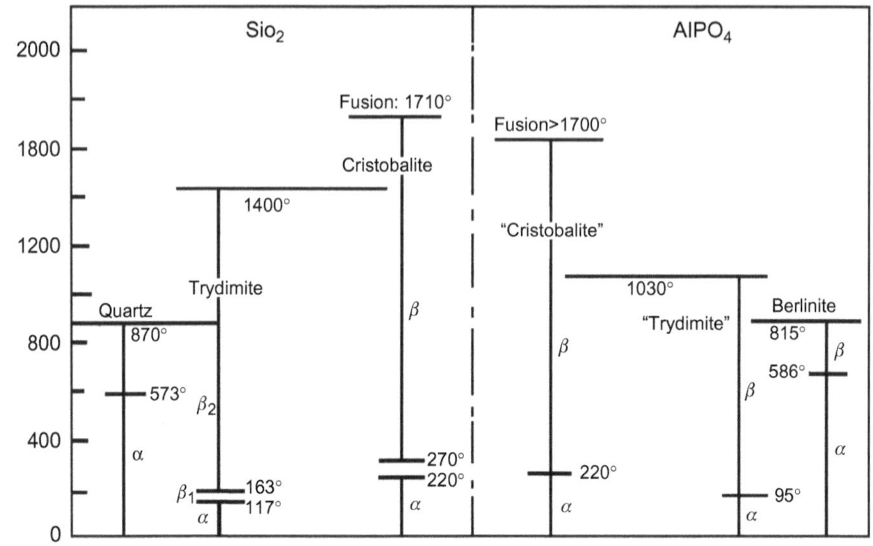

Figure 5.15 Six polymorphic modifications of berlinite analogs to six polymorphic modifications of quartz.

Artificial crystals are very similar to quartz with $(10\bar{1}1)(01\bar{1}1)$ and (1011) faces. The natural crystals are massive and granular. As mentioned above, berlinite is isostructural with quartz but has a doubled *c*-parameter. It consists of alternating AlO_4 and PO_4 tetrahedra linked with vertices. Sosman [80] has reviewed all the forms of silica reported up to 1960. According to him, there are 23 polymorphic modifications of quartz, and 11 of them are the basic ones. Berlinite, although isostructural and isoelectronic with quartz, shows only six polymorphic modifications analogous to six major varieties of silica. Figure 5.15 shows six polymorphic modifications of berlinite analogous to six polymorphic modifications of quartz. The structure of all the six forms of berlinite can be described under three subgroups, i.e., quartz, tridymite, and cristobalite with higher and lower temperature forms.

The growth of any compound does not depend only upon the growth techniques and growth parameters like the starting materials, flux agents, mineralizer, temperature, pressure, rate of heating or rate of cooling, but also upon its crystal chemical parameters. A systematic study of the synthesis of ABO_4 compounds (A = Al, Fe, Cr, Co, Ga, B, Mn; B = P, As and V) and quartz with reference to their crystal chemical significance yields very interesting data which help to understand the growth technique of ABO_4 compounds [81,82].

5.3.2 Solubility of Berlinite

Solubility is an important parameter in the successful growth of any compound in the form of large single crystals. The lack of solubility data is responsible for the

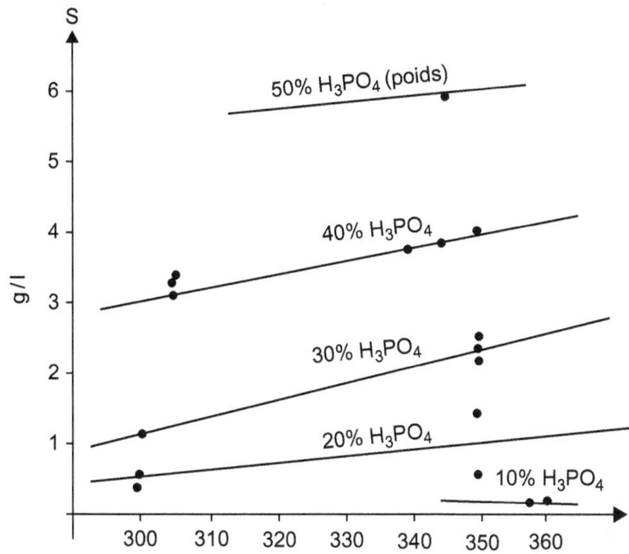

Figure 5.16 Solubility of berlinite in orthophosphoric acid above 300°C [75].

earlier failures in the growth of $AlPO_4$ crystals. It was only in the late 1970s and early 1980s that reports on the solubility of AlPO4 appeared in the literature. The problem associated with the growth of large single crystals of $AlPO_4$ is directly related to the lack of a systematic study of the solubility and partly to the P_2O_5 pressure. Although, several reports have appeared on the solubility measurements with reference to various parameters, still there is no unanimity in the results. The solubility of berlinite was first determined by Jahn and Kordes (1950) [75] in orthophosphoric acid above 300°C and it was found positive (Figure 5.16). Subsequently, Stanley (1954) [76] reported a negative solubility for berlinite in 6.1 M H_3PO_4 (Figure 5.17). The solubility of $ALPO_4$ varies widely with the type of solvent used. Some authors claim that solubility of $AlPO_4$ in HCl is similar to that in H_3PO_4. The most important difference is the higher solubility at comparable mineralizer concentration.

The authors [83] have made an attempt to study the solubility of $AlPO_4$ in some new solvents like HCOOH, NH_4Cl, Na_2CO_3, $NH_4H_2PO_4$, NaF, KF, and LiF. The solubility of $AlPO_4$ (in wt%) as a function of temperature and at a pressure of 2 kpsi in 2 M HCOOH solution is shown in Figure 5.18.

It has been shown that the solubility of $AlPO_4$ in H_2SO_4 is retrograde with respect to temperature and higher than in H_3PO_4 and HCl at comparable temperature, pressure, and acid molarity. However, the viscosity of sulfuric and phosphoric acid solutions is greater than that of HCl. Hence, the addition of HCl to H_2SO_4 would form a more efficient solvent for the growth of $AlPO_4$ [84]. Figure 5.19 shows solubility as a function of temperature in $H_2SO_4 + HCl$ mixtures. The

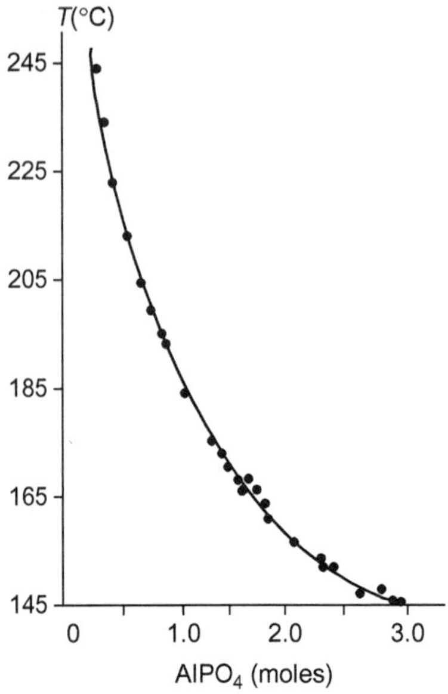

Figure 5.17 Negative solubility for berlinite in 6.1 M H_3PO_4 [76].

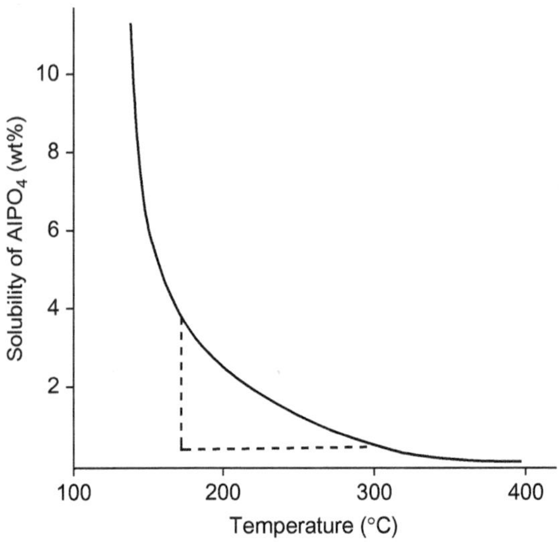

Figure 5.18 Solubility of $AlPO_4$ (in wt%) as a function of temperature [83].

authors have also carried out thermodynamic interpretation of the solubility data. The dissolution of berlinite in the mixtures of acids H_2SO_4 and HCl is a process of complex formation between Al^{3+} and the ligands present in the solution: OH^-, H_2O, Cl^-, HSO_4^-, SO_4^-, $H_2PO_4^-$. At low temperature, the complexes are similar to those of $AlPO_4$ in HCl; i.e., with Al$-$Cl complexes predominating. Even polynuclear complexes might form a broader metastable zone under these circumstances.

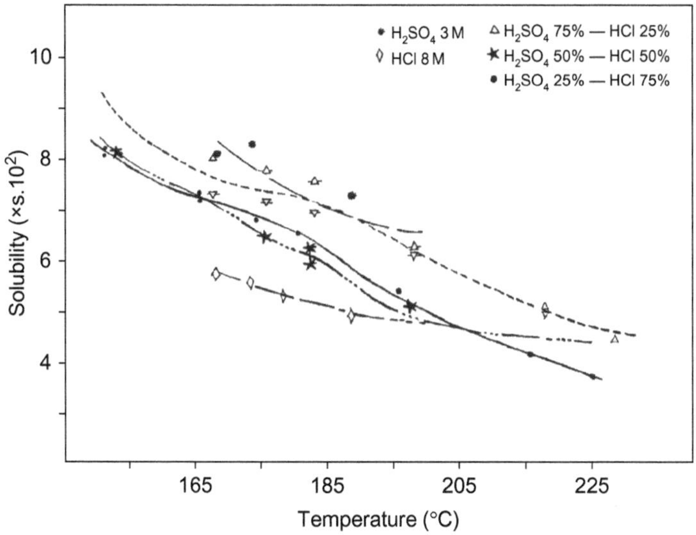

Figure 5.19 Solubility as a function of temperature in $H_2SO_4 + HCl$ mixtures [19].

With the rise in temperature, the association of Cl^- with H^+ increases the yield of HCl and destabilizes the $[Al^{3+}]$ complex. Under these circumstances, sulfate and monohydrogen phosphates replace chloride ligands and force a change in the coordination number of Al from six to four due to the steric effect. Similarly, H^+ also tends to destabilize the complex with sulfate and becomes $H_2PO^-{}_4$ hegemonic ligand as its association constant with H^+ does not vary significantly with a rise in temperature. Finally, the nucleation takes place by polymerization of tetrahedral $[Al^{3+}]$ complexes.

Table 5.6 gives comparative data of the solubility of $AlPO_4$ obtained for various mineralizers under varying experimental condition [83].

5.3.3 Crystal Growth

Berlinite could be best grown only by the hydrothermal technique. The α-berlinite is stable up to 584°C, and below 150°C the hydrates and $AlPO_4 \cdot H_2O$ are stable. Therefore, the growth temperature should be greater than 150°C but less than 584°C. However, the solubility and PVT studies have clearly shown that crystal growth must be carried out at less than 300°C because of the reverse solubility. Several versions of the hydrothermal growth of berlinite single crystals have been tried to suit the solubility.

Nutrient preparation. Berlinite, in small crystal form, can be obtained by hydrothermal reactions of H_3PO_4 with different chemicals containing aluminate

Table 5.6 Comparative Data of Solubility

Solvent	*T* (°C)	*P* (kpsi)	Time (days)	Remarks
3 M HCOOH	300	1	5	–
3.05 M NaOH	170	10	3	Residue not $AlPO_4$
2 M NH_4Cl	300	1	5	High solubility
4.62 M Na_3PO_4	300	10	3	Residue not $AlPO_4$
2 M HCOOH	300	11	5	Good solubility
3 M Na_2HPO_4	300	10	5	1% solubility
2.5 M NH_4Cl	300	1	5	Fairly good solubility
3.05 M CH_2COOH	170	10	3	Negligible solubility
2 M $NaHCO_3$	300	1	5	Fairly good solubility
2 M NaCl	300	1	5	Negligible solubility
3.05 M NaCl	170	10	3	4% solubility
2 M NaF	300	1	5	Good solubility
2 M KF	300	1	5	Good solubility
2 M LiF	300	1	5	Good solubility
3.05 M NH_4HF_2	300	10	5	Residue not $AlPO_4$
2 M $NH_4H_2PO_4$	300	1	5	Negligible solubility
2 M HCl	300	1	5	Good solubility
3.05 M NH_4HF_2	170	10	5	Residue not $AlPO_4$
3.05 M HNO_3	170	10	5	Negligible solubility
3.05 M NaCl	300	10	4	Negligible solubility
1.0 M Na_2CO_3	300	1	5	Negligible solubility

[76,85–87]. Normally, high-purity alumina is used as the nutrient, because of the purity of the grown crystals. The crystallization reaction is:

$$Al_2O_3 + 2H_3PO_4 \rightarrow 2AlPO_4 + 3H_2O$$

These fine crystals of aluminum orthophosphate are in turn used as the nutrient to grow bigger single crystals of berlinite.

The TC of solubility of $AlPO_4$ is negative in the temperature–pressure range below 300°C. Thus, in the usual hydrothermal temperature gradient where the bottom of the autoclave is hotter than the top, $AlPO_4$ would be expected to nucleate preferably in the bottom region, so one would like to place the seed in the lower hotter region and nutrient in the upper cooler region.

The most important modifications suggested from time to time in the growth of berlinite crystals are as follows:

i. Crystal growth by slow heating method
ii. Crystal growth by composite gradient method
iii. Crystal growth by temperature gradient
iv. Growth on seeds.

Chai et al. [86] suggested the following nutrient preparation conditions for large (kg) quantities, where the temperature is cycled:

Solvent H_3PO_4 (molar)	6.1
% Fill	81
$AlPO_4$ charge (gm)	45 presaturation at 25°C
Initial temperature (°C)	150
Final temperature (°C)	275
Differential temperature, dT (°C)	1–2
Heating rate (°C)/day	10
Method of heat rate increase	10
Yield (gm)	
>60 mesh[a]	19.5 gm (70% yield)
<0 mesh	8.1 gm
Total yield	27.6 gm
Run duration (day)	15

[a]60 mesh = 0.25 mm.

Crystal growth by slow heating method. The method was previously described in Refs [83,88–92], and it was mainly used for improving the quality of seeds and to enhance the size of the crystal with conditions similar to those used to obtain nucleations. A major drawback of this method is the impossibility of obtaining crystals of sufficient size in one operation due to the limited quantity of aluminum phosphate available in the solution.

Synthesis of berlinite crystals under lower temperature and pressure conditions using slow heating method was carried out using acid digestion bombs (T = 100–250°C; P = 100–300 atm) and smaller Morey autoclaves (T = 200–250°C; P = 300–1000 atm). The nutrient was taken in Teflon liners.

The experiments at higher temperature (T > 500°C and P > 100 bar) yielded the β-berlinite. With an increase in the pressure, other high-pressure modifications of aluminum orthophosphate belonging to the cristobalite form were obtained.

Crystal growth by composite gradient method. In order to gain more flexibility in adjusting the growth rates for Z seeds and to avoid the preliminary work of crystal growth, nutrients of sufficient size (that did not cause crystallites to drop on crystals) needed to be obtained [93]. A modification of the horizontal gradient method was originally proposed in Ref. [94]. Berlinite crystals have been obtained by this method successfully [94]. The autoclave used in the initial experiment is shown in Figure 5.20. The following growth conditions were chosen by the authors [93,94] after experimental trials:

- Silica glass autoclave: 100 cm, diameter 30 mm
- Ratio hot/cold zone: x/1
- Solution: H_3PO_4, $H_2O/AlPO_4$ (6.5/1.1)
- Growth temperature: 170°C
- Horizontal temperature difference: 5°C < T < 30°C
- Construction diameter: 10 mm
- Powdered nutrient
- Filling: 80%.

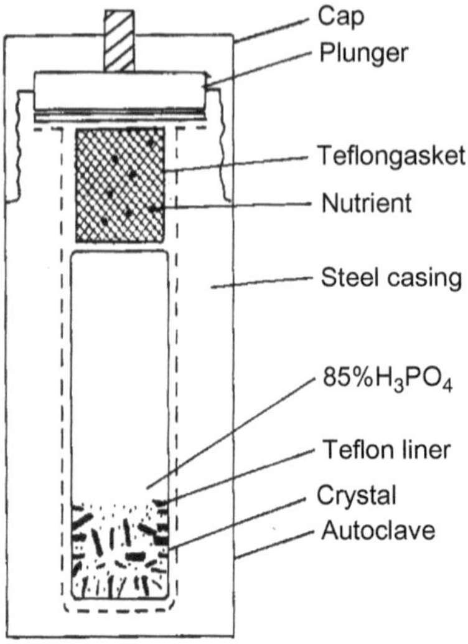

Figure 5.20 Autoclave used in the berlinite synthesis [83].

Growth with reverse temperature gradient. This is a more popular and widely used method in the growth of berlinite crystals. Here, the nutrient is kept in a casket at the upper portion of the autoclave, which is cooler. The nutrient is nothing but the crystalline powder or fine grains of $AlPO_4$ obtained by a slow heating method. The temperature of the bottom of the liner is kept at a slightly higher temperature and this is the zone of crystallization, where berlinite can be crystallized either spontaneously or on seeds [95].

Crystal growth on seeds. Spontaneously nucleated seeds and oriented seeds cut from grown crystals are mounted on a platinum frame which is placed in the bottom (hotter) region of an autoclave, and the nutrient (~60 mesh particle size prepared as described earlier) is placed above in a platinum gasket in the upper region (or in a Teflon gasket). Thus, the seeds are in the warmer supersaturated region at the bottom of the autoclave and the temperature gradient achieved by cooling the top allowed proper convection. The schematic diagram showing the position of seeds, nutrient, and the baffle is represented in Figure 5.21.

In the growth of berlinite on seeds, the orientation of the seed is very important because the growth rate and perfection depend upon the seed orientation. According to Laudise (1985) [85], the growth rates are shown for seeds whose faces are indicated below:

(0001) (basal plane)	0.25–0.50 mm/day
$(10\bar{2}\bar{1})$ (X-cut)	0.23–0.30 mm/day
$(10\bar{1}0)$ (Y-cut)	0.12–0.15 mm/day
$(011\bar{1})$ (minor rhombohedral face)	0.12 mm/day
$(101\bar{1})$ (major rhombohedral face)	0.15 mm/day

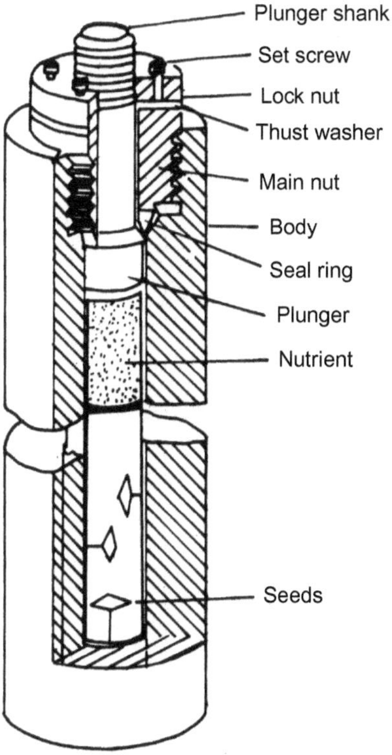

Figure 5.21 Schematic diagram showing the position of seeds, nutrient, and the baffle in the growth of berlinite crystals [85].

The relative growth rates are in general agreement with the morphology of the equilibrium form, as judged from observing spontaneously nucleated crystals, which are bounded by small prism $(10\bar{1}0)$ faces and terminated by minor and major rhombohedral faces. Thus, it would be expected that the rates on prism, major and minor faces, are smaller than for $(11\bar{2}0)$ and (0001) faces, since the latter two do not exist as bounding equilibrium surfaces. In general, the morphology of the spontaneously nucleated crystals depends upon various parameters like pressure, temperature, volume, and composition of the nutrient and mineralizer [96–98]. For example, impurities in the powder come into the corresponding nutrient and cause twinning. The starting material like sodium aluminate or aluminum hydroxide gives needle-like crystals, whereas pure $AlPO_4$ gives rhombohedral and pyramidal crystals [83].

Yaroslavskii and Popolitov (1990) [99] have studied the growth of berlinite on seeds in detail using fine crystalline charge and also aluminophosphate glass charge as nutrients. They found that dissolution rate of the aluminophosphate glass charge is three times greater than that of the crystalline charge. The solubility growth rate and crystal quality have been studied in great detail using different solvents like H_3PO_4, HCl, HNO_3, and H_2SO_4. If the acid concentrations are increased above 2.5 M, the spontaneous crystals form, and the mother liquor is captured by the edges of the growing crystal. The growth rates differ greatly between the fine

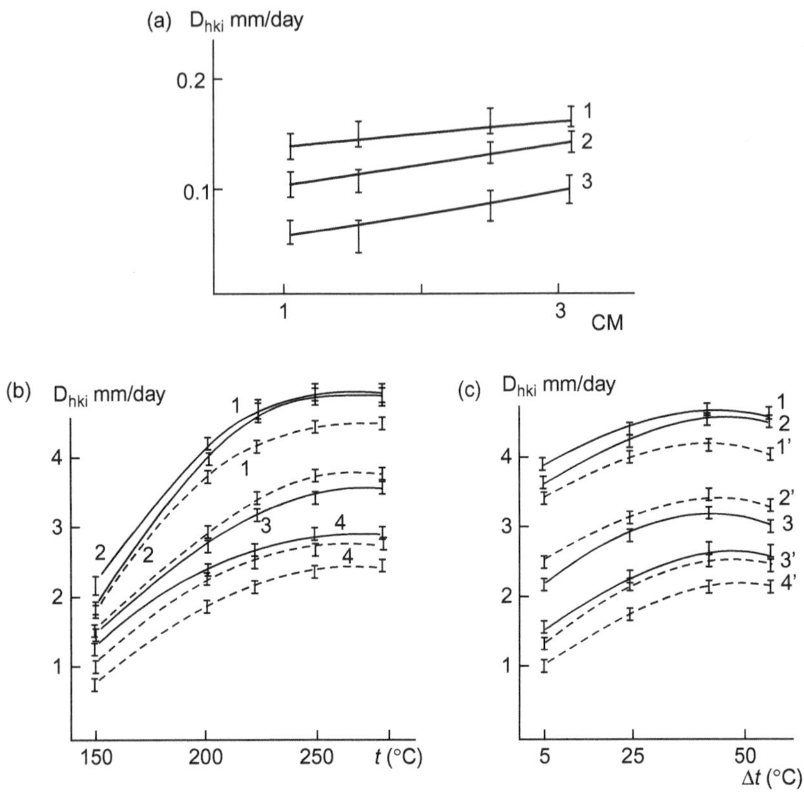

Figure 5.22 Solution concentration, growth temperature, and temperature gradient versus growth rate (a–c) [99].

crystallization charge and aluminophosphate charge. The temperature gradient has been varied from 5°C to 50°C. When the ΔT is high (40–50°C), it introduces macro inclusions and gas–liquid inclusions and are clearly visible upon passing bright light through the crystal. The growth rates observed by the authors are rather high at 0.35–0.50 mm/day in H_3PO_4, 0.35–0.45 mm/day in HCl, 0.25–0.35 mm/day in HNO_3, and 0.2–0.3 mm/day in H_2SO_4 solvent media. The growth rate of α-berlinite on $[10\bar{1}0]$ and [0001] seed plates against the solvent concentration, growth temperature, and temperature gradient are shown in Figure 5.22a–c. The growth rate from aluminophosphate glass is 1.5–2 times greater than that from a fine crystalline charge under the same conditions for mild process parameters.

5.3.4 Morphology

Although $AlPO_4$ crystals obtained spontaneously are small in size, they exhibit distinct morphological features. The crystal size varies from 0.5 to 4 mm and the common forms are hexagonal, rhombohedral, rod shaped, needle like, and occasionally rounded. The morphological forms of $AlPO_4$ mainly depend upon the nutrient composition and the solvent in action. Data on the dependency of the morphological features, and the nutrient composition, as well as the solvent used are given in Table 5.7.

A typical morphology of $AlPO_4$ is shown in the schematic diagram (Figure 5.23) consisting of major and minor rhombohedral faces. Figure 5.24a and b shows the characteristic photographs of the berlinite crystals.

Table 5.7 Morphology of Aluminum Orthophosphate Crystals

Nutrient Composition	***T*** **(°C)**	**Morphology**
$Al_2O_3 + H_3PO_4$ mineralizer HCl (1.5 M)	280	Hexagonal and rhombohedral
$Al_2O_3 + H_3PO_4$ mineralizer NaCl (2 M)	300	Rhombohedral
$Al_2(OH)_3 + H_3PO_4$ mineralizer NaF (2.5 M)	300	Rods, needles, or acicular
$Al_2O_3 + H_3PO_4$ mineralizer NH_4Cl (4 M)	300	Small hexagonal, rhombohedral and equi-dimensional
$Al_2O_3 + H_3PO_4$ mineralizer HCOOH (2 M)	280	Hexagonal and rhombohedral
$Al_2O_3 + H_3PO_4$ mineralizer $NaHCO_3$ (3 M)	300	Hexagonal, rhombohedral, and slightly irregular
$AlCl_3 + H_3PO_4$ mineralizer HC1 (2 M)	270	Hexagonal

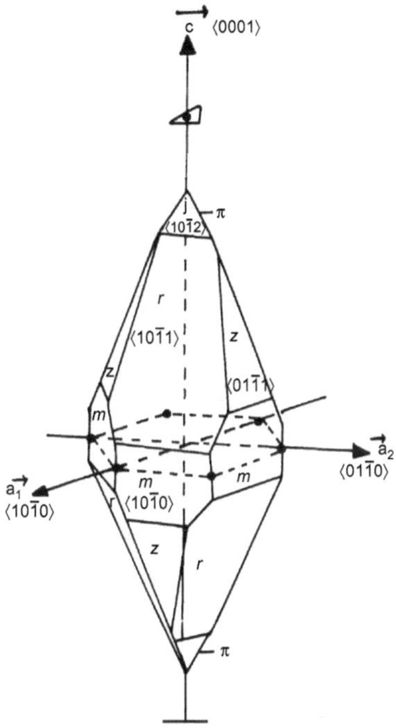

Figure 5.23 Morphology of $AlPO_4$ consisting of major and minor rhombohedral faces [90].

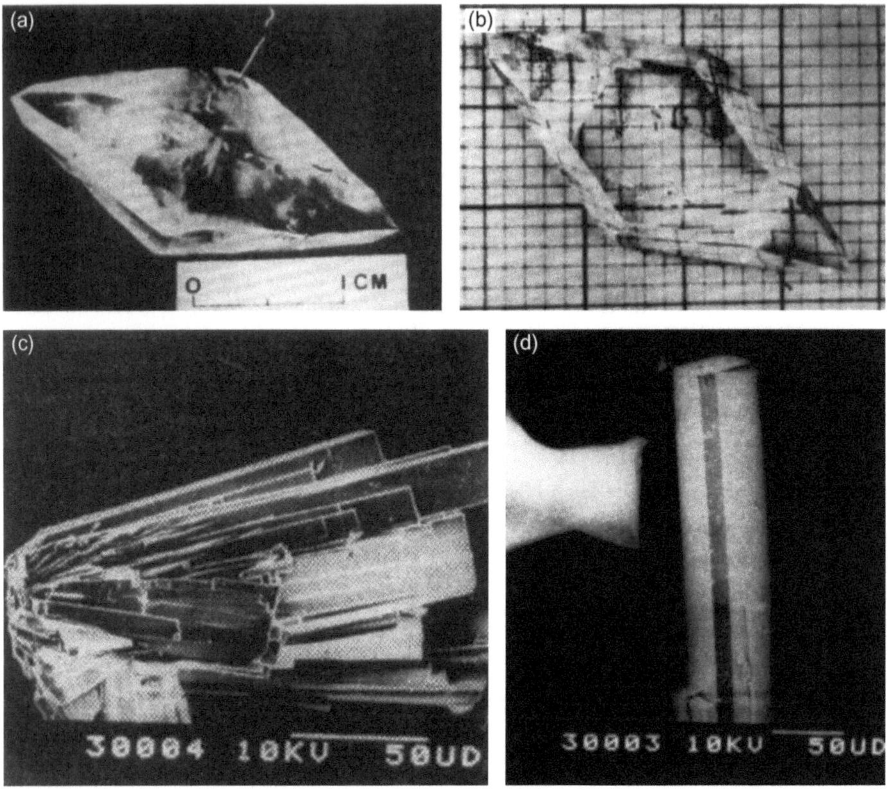

Figure 5.24 Characteristic photographs of the berlinite crystals (a–d) [83,90].

It is observed that these admixtures play an important role in the process of crystallization of $AlPO_4$. In the actual experiments without any admixtures, well-developed rhombohedral and hexagonal crystals have been obtained, but with the addition of various admixtures the crystal changes to fibrous, rod-shaped, acicular, massive, equidimensional and irregular shapes (Figure 5.24c and d). In the case of Zr and Ti as the admixtures, fibrous or acicular crystals of berlinite have been obtained; such results were earlier obtained only in experiments containing $Al(OH)_3$ as the starting component [83,98].

Growth defects in synthetic berlinite samples have been studied by several workers through X-ray topography [96,100]. The berlinite crystals, in general, show internal structural defects along with a high density of defects on X-ray Lang topography. It can be seen that crystals which appear transparent from outside contain many internal defects. Figure 5.25 shows a typical Lang X-ray topograph taken in (2110) face [90]. Line defects were attributed to the disorder in crystal structure and more likely to the impurities existing in the boundary of repetitive overgrowth. However, it is possible to distinguish various types of contrasts that

Figure 5.25 Lang X-ray topograph taken in (2110) face [90].

differ by their temperature behavior. Dislocation-related contrasts vary slightly with temperature, but they appear more strongly, especially after the second thermal cycle. The variation of contrast present in the topography for the growth bands can be interpreted in terms of localized variations of the deformation in the sample with temperature. These variations appear in the same temperature range as the anomalies observed in propagation loss, frequency, and temperature characteristics. These anomalies have been attributed to water impurities in these hydrothermally grown crystals [101,102]. This explanation is qualitatively valid and in good agreement with the changes of the deformation contrast observed during the thermal cycles. The contrast due to the water impurity content is observed in selected areas and is predominantly located at growth bands that correspond to the adsorption of water on some growing faces during the cycle.

A combination of selected growth conditions reducing the water content and using heat treatment to control the water distribution could be of great interest for improving the piezoelectric device application of berlinite. The choice of the seed is fundamental for the production of good crystal and a selection of excellent slices for this use is absolutely necessary when this condition is fulfilled. The X-ray topographs obtained on different crystals have shown that the quality was improved but the remaining defects could be classified into two classes. This is due to the defects present in the seed which have been developed in the grown crystals (dislocations and sometimes twins) and those which were carried during the start of the growth (bundles of dislocations) or during the growth (growth bands).

No fundamental differences were found by X-ray topography between the horizontal gradient, vertical gradient, and slow heating methods, when good seeds and optimized conditions were used [99]. The two principal imperfections readily observable are *crevice flawing* and *cracks*. When growth is on a nonequilibrium face, there is a tendency for higher growth rates and the formation of *hillocks* which tend to protrude across the solute-depleted diffusion layer near the growing face. The tips of such hillocks experience a high supersaturation and tend to grow faster than the regions between hillocks. This phenomenon is the hydrothermal analogue of constitutional supercooling and may be viewed as a kind of dendritic growth [103,104]. Entrapment of solution between hillocks leads to bubbles of solutions in the grown material, which are observed as *viels*. Thus, growth at slower rates and on equilibrium faces produces a greatly reduced tendency for crevice flawing.

Cracks are often observed in hydrothermally grown material. In the case of quartz [27,105], we found that they were associated with strain caused either by strain in the seed propagating into the grown crystal or by strain associated with dislocations arising at inclusions in the new growth. It was further observed in quartz that disorder regions in seeds propagated into crystals and strain-free seeds were much less likely to produce cracked growth.

At an annealing temperature which is high enough, berlinite becomes milky, and in TEM pictures one observes a large density of bubbles with a mean size of about 1000 Å; while in untreated, as-grown berlinite, no such bubbles are visible. Furthermore, the growth of these bubbles is accompanied by a severe increase in the dislocation density of the order of 10^5 cm^{-2} or less, but in the heat-treated crystals this density reaches 10^9 cm^{-2}. An increase of at least six orders of magnitude is produced, similar to wet quartz, which exhibits the same dramatic dislocation multiplication. This situation can be observed in berlinite, where bubbles with a mean diameter of 250 Å can be observed together with a very high density of dislocation [100,105]. Morris and Chai (1998) [106] have studied the imperfect low-angle boundaries and fracture in hydrothermally grown berlinite crystals obtained on multiple-seed plate arrays.

5.3.5 Thermal Behavior

The stability range of any material with a device potential is an important parameter. Byrappa et al. (1986) [107] have reported the differential thermal analysis for containing various admixtures (Figure 5.26) [108]. For pure berlinite, the inversion temperature was reported to be at 586°C and this is similar to that reported by Beck (1949) [79]. However, there are reports fixing this transition temperature at 581°C [108] and at 579°C [109]. Such a variation in the $\alpha-\beta$ inversion temperature can be explained by the effect of the starting materials, the presence of admixtures in the final products, and even the metals from the mineralizers which also act as admixtures.

Studies on the thermal expansion of berlinite have not been reported much in the literature. Byrappa and Prahallad (1989) [110] have reported the thermal expansion of

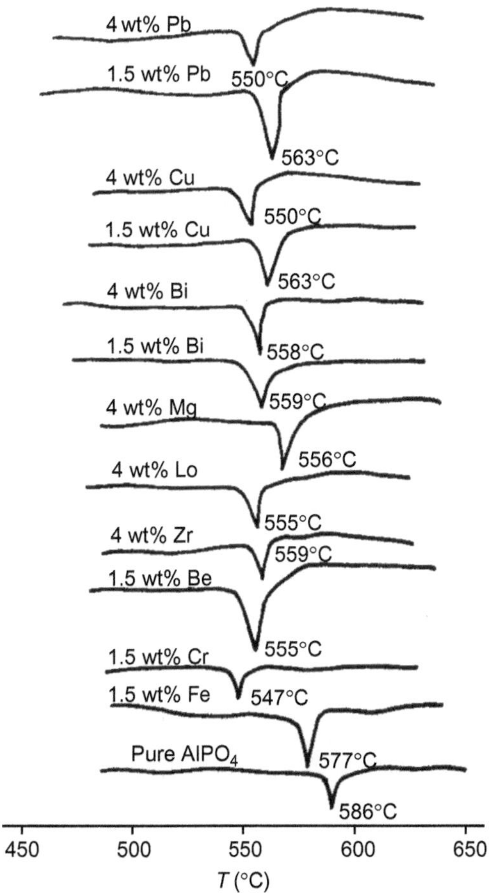

Figure 5.26 Thermal behavior of $AlPO_4$ [107].

Table 5.8 Thermal Expansion of α-Berlinite Containing Admixtures [2,11]

Admixture	Ionic Radii	Atomic Weight	*a*	*c*	*v*
–	–	–	1.63×10^{-5}	1.068×10^{-5}	4.38×10^{-5}
Sr	0.63	51.996	1.90×10^{-5}	1.89×10^{-5}	5.77×10^{-5}
Fe	0.64	55.847	1.98×10^{-5}	1.13×10^{-5}	5.16×10^{-5}
Cu	0.72	63.546	1.76×10^{-5}	1.81×10^{-5}	5.39×10^{-5}
Nd	0.99	144.24	2.02×10^{-5}	2.64×10^{-5}	6.78×10^{-5}
La	1.016	138.905	1.75×10^{-5}	2.64×10^{-5}	6.22×10^{-5}

– concentration of the admixture in the nutrient was 1.5 wt%

berlinite in detail. The thermal expansion values vary depending upon the percentage of concentration of impurities in the crystal.

The values of thermal expansion for α-berlinite containing various admixtures are given in Table 5.8, and the values are, in general, very low for these materials because of the covalent bonding. Alpha-berlinite is relatively stable compared to β-berlinite with reference to the action of the dopants. The difference in α_a (max) − α_a (min) = $0.39 \times 10^{-5}\ C^{-1}$ and α_c (max) − α_c (min) = $1.572 \times 10^{-5} C^{-1}$, which shows that the *c*-axis is more susceptible to thermal expansion than the α-axis. The coefficient of thermal expansion is quite high when rare earths are used as admixtures.

5.3.6 Piezoelectric Properties of Berlinite

The piezoelectric properties of berlinite have been studied in detail by several authors. Chang and Barsch (1976) [68], and Ballato and Iafrate (1976) [102] pointed out the attractive properties of *Y* rotated and doubly rotated cuts of berlinite in terms of its large coupling coefficient (twice that of quartz), with a constraint of zero first-order frequency temperature coefficient (FTC). Several groups appeared throughout the world immediately to study the growth and characterization (with reference to its first-, second-, and third-order FTCs) of berlinite. Figure 5.27 shows the comparison of *Y* quartz and *Y* berlinite resonators. It can be noted that the *Q* factor is greater by an order of magnitude than for resonators. In Figure 5.27, also given the frequency response of the *Y* quartz plate of a similar design whose *Q* factor is not so different and whose coupling coefficient is smaller. The higher order TCs measured with different crystals, even with different plates of the same crystal, show values much larger than for quartz resonators of the corresponding orientation. Table 5.9 gives the comparison between quartz and berlinite. However, it has been demonstrated that the water dissolved in berlinite crystal induces low values

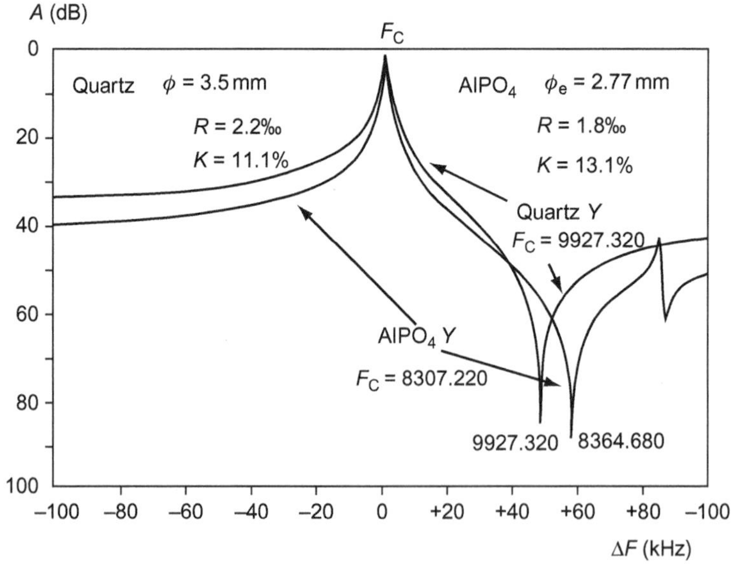

Figure 5.27 Frequency response of the *Y* quartz plate [68].

of two factors and impairs significantly the intrinsically very favorable thermal behavior of this material.

Jumas et al. (1987) gave an excellent comparison of piezoelectric properties of $AlPO_4$ to quartz [96]. According to them, the coupling coefficients measured with AT resonators with filter-type response were about 11%. For quartz, the maximum values were 7.4%, and this permits one to obtain filters with twice the bandwidth. The electric constant of berlinite is lower by a factor of 1.2−1.35 than the corresponding constant of quartz, but the density is very similar. The frequency constants fr are lower than those for the corresponding orientation of quartz by a factor 1.1−1.7; for the AT-cuts the factor is 1.15.

The first-, second-, third-, etc. order FTCs of berlinite are mostly dependent on the TC of the pertinent elastic constants and the dilation coefficient [111]. As the latter are similar to those of quartz [68], most of the better thermal behavior observed with "dry" berlinite, as compared to quartz, results from smaller values of the TC of elastic constants. For some other constants, similar things may also be expected. In Figure 5.27, the results of Poignant are given, indicating a similar behavior of the TC of C33 for quartz and berlinite [90]. It is also expected that for other analogues of quartz with similar $\alpha-\beta$ phase transition and reduced elastic constants, similar or better thermal behavior of resonators could be obtained.

This point is of great importance for frequency generation and filtering applications of resonators, for which it is fundamental to obtain a frequency response free of unwanted (unharmonic) modes. For identical cut, resonant frequency, and plating, it can be shown that smaller dimensions of electrodes are required for berlinite than for quartz to obtain a frequency response without spurious unharmonic modes (trapping of only one mode).

Shannon and coworkers (1993) [112,113] have studied in detail the dielectric properties of berlinite with reference to the presence of $(OH)^-$. It has been well established that water in berlinite is found in both macroscopic and microscopic inclusions. Because crystals are grown from H_3PO_4 or H_3PO_4/HCl solutions, it is likely that the inclusions also contain this acidic solution and consequently, high

Table 5.9 Tentative Comparison with Quartz (Thickness *Y* Rotated Resonators at the Same Frequency)

Useful coupling coefficient (AT filter trapped resonators)	Enhanced by 1.4−1.5
Shift of oscillators or bandwidth of filters (AT-cut)	Twice
Angular sensitivity (first-order FTC)	Reduced
Thermal stability (higher order FTC)	Better
Q factor propagation losses	Already sufficient (may be comparable)
Thickness of plates (AT)	Reduced by 1.15
Electrode dimensions (AT)	Reduced by 1.32 (TT)
(Same plating 2D TT of TS)	Reduced by 1.24 (TS)
Nonlinear properties	To be determined

"Dry" berlinite has similar C, E, EPS constants, first-order TC and much reduced higher order TC.

concentrations of H_2PO_4, $HPO^-{}_4$, and/or Cl^- ions. Quantitative NMR determination of proton content unequivocally determines the H_2O and hydroxyl content, whereas complementary IR absorbency studies of the 3300 cm^{-1} band confirm the molar absorption coefficient of 81 lmol/cm originally determined for liquid H_2O by Thompson (1965) [114]. The dielectric constant variability of berlinite prepared by different investigators is believed to result from (i) a variation in H_2O content in microscopic and macroscopic inclusions and (ii) unidentified crystalline defects.

On the whole the research on the bulk crystal growth of berlinite has declined significantly except for some recent publications on the special physical characteristics like, for example, variations in the structure and molecular dynamics with pressure, as it shows memory glass property under high pressure [115]. Besides, berlinite attracts the attention of researchers as a substrate material for a variety of nitrides [116].

5.4 Gallium Phosphate

Gallium phosphate, like aluminum phosphate, belongs to the family of quartz and quartz-like MXO_4, where M = B, Al, Ga, Fe, Mn, and X = P, As [117]. Among them, $AlPO_4$ and $GaPO_4$ have been studied extensively with reference to their crystal growth, crystal chemistry, and physical properties, especially piezoelectric properties [82,118–123]. Both $AlPO_4$ and $GaPO_4$ exhibit piezoelectric characteristics which are superior to those of conventional α-quartz. As evident from Table 5.10, for the same AT-cut, $GaPO_4$ exhibits a larger coupling coefficient which makes $GaPO_4$ a most promising material in the field of wide-band filters. In addition to this, the α–β phase transition for $GaPO_4$ occurs at 976°C, and hence, it can be used in a very wide range of temperatures for filters and sensor devices [124–127]. In fact, $GaPO_4$ is the most promising one and more extensively being studied today among all other piezoelectric crystals.

Table 5.10 Comparison of Some Piezoelectric Characteristics [117,123]

Parameter	Quartz	Aluminum Phosphate (Berlinite)	Gallium Phosphate
Coupling coefficient[a] k (%)	8.5	11.0	>16.0
Surtension coefficient[a] Q	3.10^6	10^6	$>5.10^4$
α–β Phase transition (°C)	573	584	No

[a]AT-cut

5.4.1 Crystal Growth of Gallium Phosphate

The synthesis of gallium orthophosphate does not differ much from the synthesis of berlinite. It is usually carried out by the hydrothermal technique at temperatures less than 250°C. On the whole, the growth of aluminum and gallium phosphates is carried out using reverse temperature gradient, or slow temperature, increasing essentially in the acid media like H_3PO_4 [128], H_2SO_4, $HCl + H_2SO_4$, $H_3PO_4 + H_2SO_4$ [121,126,127], HCl [126,129], KF, NaOH [129]. The starting material/nutrient for $GaPO_4$ growth is prepared usually by solid-state reactions with Ga_2O_3 and $NH_4H_2PO_4$, twice at 1000°C for 24 h and subsequently hydrothermally treated at 150°C for 30 h in 4 M H_3PO_4 solution. The resultant product is generally a single phase of $GaPO_4$. However, in recent years, the use of organic solvents and surfactants has brought down the pressure and temperature conditions of synthesis. But the size of the crystals is a major issue in solvothermal growth. Like berlinite, the gallium phosphate also shows a retrograde solubility in all the solvents below 400°C, and above this temperature, the solubility becomes positive [118,122,129,130]. Hence, the nutrient is kept at the low-temperature zone (i.e., dissolution region), and the seed crystal is kept at higher temperature zone (i.e., crystallization region). The temperature gradient (5−40°C) is selected according to the quality and the growth rate required. Figure 5.28 shows the schematic representation of the growth of

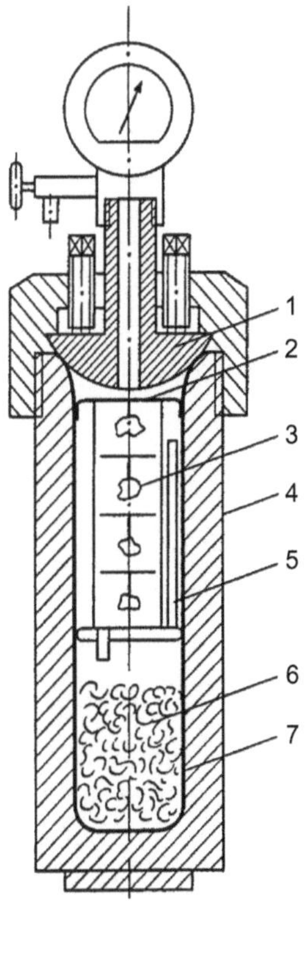

Figure 5.28 Schematic representation of the growth of berlinite crystals [128].

berlinite crystals and the same experimental setup can be used for the growth of gallium orthophosphate crystals. The growth of gallium orthophosphate can be carried out at 180°C for 7–10 days. By this arrangement, it was expected that the dissolved nutrient would be efficiently used to grow single crystals on the seed without the occurrence of spontaneous nucleation. As the lattice parameters of both berlinite and gallium phosphate are very close, some authors have used even berlinite crystals with a definite orientation/cut as seeds and obtained $GaPO_4$ epitaxy on $AlPO_4$ seeds using 6 M H_2SO_4 solvent, at a temperature greater than 200°C [126]. The experimental temperature is critical in the case of gallium phosphate because the crystals obtained at a temperature less than 200°C show the presence of water, which deteriorates the piezoelectric characteristics. Hence, it is advisable to use acid mineralizers with a boiling temperature of 225–240°C. Practically, there is no water in such a solvent, and it consists of orthophosphoric acid (90%) and pyro-phosphoric acid (10%). The solubility is presaturated with $GaPO_4$ to avoid the dissolution of the seeds. To keep the solution from boiling during the crystal growth, inert gas is pumped into an autoclave up to 50 bar [131].

Cochez et al. (1994) [126] have proposed the following scheme for the growth of α-$GaPO_4$ as the best method:

Thin exitaxy of $GaPO_4$ on berlinte seeds by VTG in 6M H_2SO_4, $T_c > 200°C$

↓

Flat crystals

↓

Crystal growth by SHT and composite gradient Methods in 15M H_3PO_4, $150 < T_c < 180°C$

↓

Cut in X seeds

↓

Crystal growth of X $GaPO_4$ seeds by VTG in 6M H_2SO_4, $T_c < 200°C$

↓

Crystals with low (OH) content

The first stage of $GaPO_4$ crystal growth process is the epitaxy of $GaPO_4$ on berlinite seeds. The second stage consists of growing two-dimensional crystals through the slow heating method, or the horizontal gradient method, or the composite method (i.e., the combination of the first two). In the third stage, the *X* seeds are grown which helps in reducing the $(OH)^-$ content in the crystals.

Popolitov and Yaroslavskii (1990) [129] have studied the crystallization processes of α-$GaPO_4$ under hydrothermal conditions. They have studied the rate of crystallization, (i.e., the amount of $GaPO_4$ crystals formed per unit time), action of solvent, its concentration, and the temperature regime. Gallium phosphate glass (prepared by sol–gel method) has been used as the nutrient with Ga:P ratio = 1:1.5. The experiments were carried out in the temperature range 200–300°C, $P = 20–35$ MPa using Teflon liners, with a temperature gradient of 4–5°C. The yield of $GaPO_4$ single crystals as a function of growth temperature and concentration of different solvents used is shown in Figures 5.29 and 5.30. All the curves shown in Figures 5.29 and 5.30 are described by a simple exponential equation $M = KC^n$, where K is the rate of bulk crystallization, C is the solvent concentration, and n is the formal order of bulk crystallization. Table 5.11 gives the values of K, C, and n for aqueous HCl, HNO_3, KF, and NaOH at 200°C, 250°C, and 300°C.

Also, the authors [129] have calculated the activation energy for the formation of $GaPO_4$ single crystals from the Arrhenius equation. Similarly, the solubility of α-$GaPO_4$ has been studied by several workers within a wide range of PT conditions ($T = 200–500$°C, $P = 50–1000$ atm) [120,122,126,132]. Figure 5.31 shows the solubility of $GaPO_4$ in the phospho-sulfuric acid solution 15 M H_3PO_4 + 9 M H_2SO_4 in terms of % in H_2SO_4. There is an increase in the solubility with a progressive enrichment in H_2SO_4 in the solvent. However, it always remains lower than in pure H_2SO_4. The solubility remains retrograde in all the studies up to 400°C.

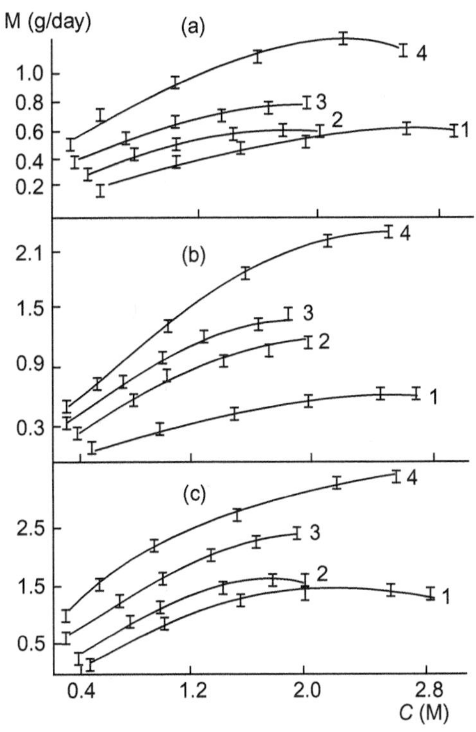

Figure 5.29 Yield of $GaPO_4$ single crystals as a function of growth temperature and concentration of different solvents [129].

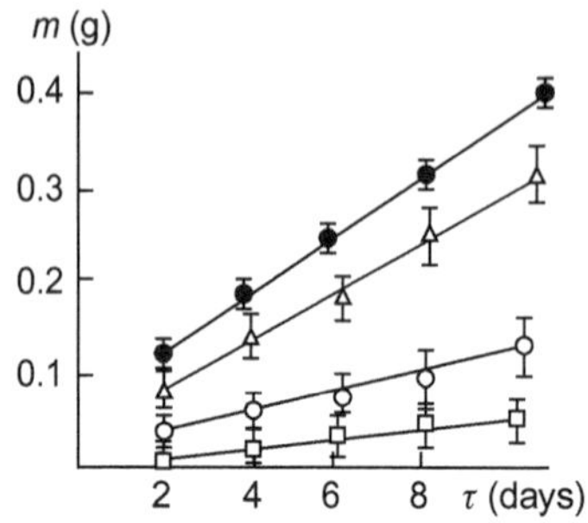

Figure 5.30 Yield of $GaPO_4$ single crystals as a function of time [129].

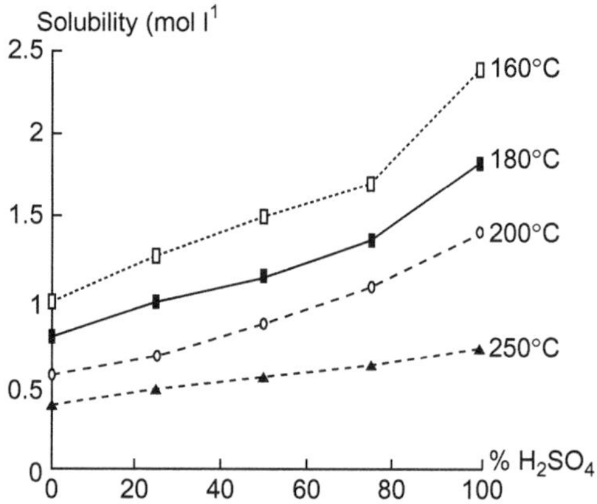

Figure 5.31 Solubility of $GaPO_4$ in the phospho-sulfuric acid solution 15 M H_3PO_4 + 9 M H_2SO_4 [122].
Source: Courtesy of the Gauthier-Villars/ESME, Paris.

Table 5.11 Crystallization Kinetics for $GaPO_4$

Solvent	**HCl**			**HNO_3**			**KF**			**NaOH**		
T (°C)	200	250	300	200	250	300	200	250	300	200	250	300
K	0.22	0.48	0.76	0.11	0.28	0.41	0.09	0.14	0.26	0.04	0.05	0.15
N	0.61	0.58	0.57	0.58	0.50	0.47	0.20	0.22	0.18	0.30	0.28	0.27

5.4.2 Morphology

The morphology of the $GaPO_4$ single crystals is shown in Figure 5.32 [117]. The commonly appearing faces are $[10\bar{1}1]$, $[01\bar{1}1]$, $[10\bar{1}0]$, $[10\bar{1}2]$. Figure 5.33 shows the characteristic photographs of $GaPO_4$ crystals obtained by hydrothermal method. The growth rates along various crystallographic directions decrease in the following sequence:

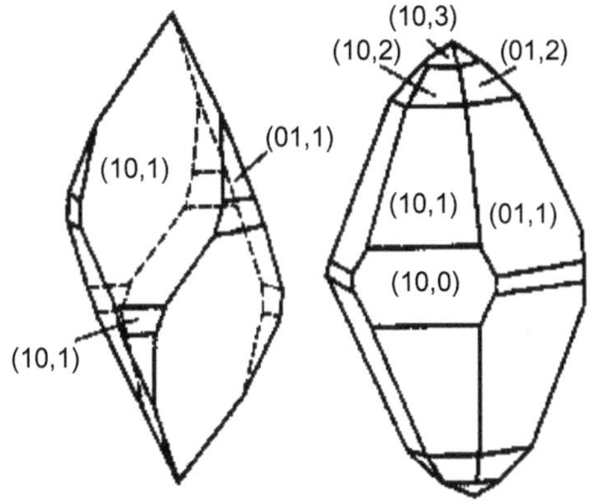

Figure 5.32 Morphology of the $GaPO_4$ single crystals [129].

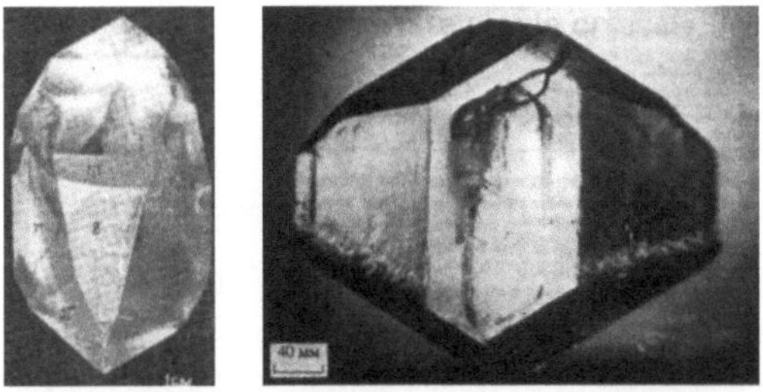

Figure 5.33 Characteristic photographs of $GaPO_4$ crystals obtained by the hydrothermal method [131].

$$\gamma(0001) \gg \gamma(01\bar{1}2) > \gamma(10\bar{1}0) > \gamma(01\bar{1}1) > \gamma(10\bar{1}2) > \gamma(10\bar{1}1)$$

The common twins observed in $GaPO_4$ are according to the Dauphine, Brazil and Laydolt's twinning laws. Twinning in GaPO4 has been discussed in great detail in Ref. [131]. Also X-ray topographic studies on $GaPO_4$ have been carried out by Capelle et al. [127] to study twinning in GaPO4 crystals.

5.4.3 Dielectric Properties of Gallium Phosphate

The influence of the growth rate on the quality of the $GaPO_4$ crystals has been studied through IR spectrometry by Cochez et al. (1993) [122]. As conventionally done for quartz and berlinite, the $(OH)^-$ impurity content was monitored by IR measurements in the wave number range of 4000–2500 cm^{-1}, using the relation:

$$\alpha = 1/d \text{ cm} \log (T_{3800}/T_{3167})$$

where T_{3800} and T_{3167} are respectively the transmittances at 3800 and 3167 cm^{-1}, respectively. The strongest absorption band of Al-OH for berlinite appears at 3291 cm^{-1} and it is shifted to 3167 cm^{-1} for Ga-OH in case of $GaPO_4$ (Figure 5.34). The α values obtained by the authors are 1.6 and 2.2 for two of their

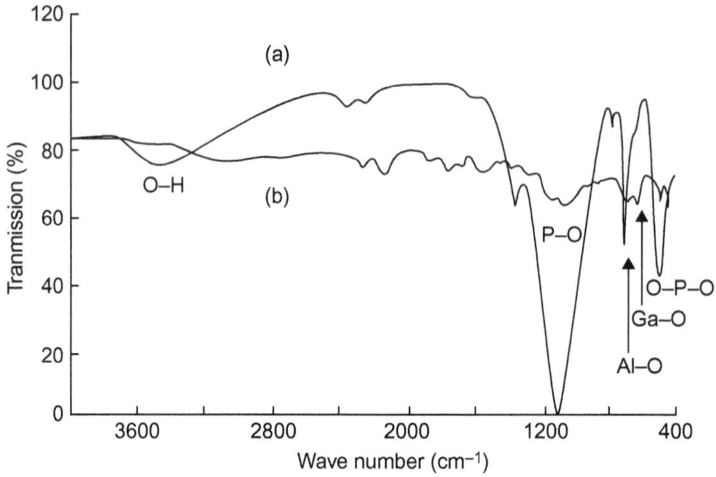

Figure 5.34 FT-IR spectra of (a) $AlPO_4$ (wet) and (b) as-grown single crystals of $GaPO_4$ [122].
Source: Courtesy of the Gauthier-Villars/ESME, Paris.

characteristic $GaPO_4$ samples. It can be observed that the $(OH)^-$ content increases with the temperature gradient.

The dielectric properties have been studied in detail by Hirano et al. (1990) for $GaPO_4$ crystals [119]. According to the authors, the frequency independence of ε' at room temperature in this crystal indicates that mainly electronic ionic polarizations contribute to the dielectric constant of $GaPO_4$. These two polarizations always exist below 10^{13} Hz.

Figure 5.35 shows the extrapolation of possible piezoelectric properties to other quartz-like materials [117,133]. Table 5.12 gives the extrapolation of AT-cut characterizations from well-known quartz and quartz-like materials.

Cochez et al. (1994) [126] have well studied and extrapolated the relations that exist between the crystal structure distortions, expressed in terms of the M–O–X angle (where M = B, Al, Ga, Fe, Mn, and X = P, As) value and different physical properties. For instance, piezoelectric characteristics of the AT-cut and dielectric constant anisotropy of three known materials (SiO_2, $AlPO_4$, and $GaPO_4$) have been linearly related to their M–O–X angle value successfully. Conversely, knowledge of the M–O–X value allows the prediction of physical properties for quartz-like crystals that have not been synthesized. Consequently, using only M–O–X angles, one can describe the structural strains, the α–β transition existence, and some physical properties, such as density, piezoelectric characteristics, and elastic constants.

Based on this reasoning, $GaAsO_4$ and GeO_2 appear to be the most promising materials for piezoelectric applications, unfortunately, the concomitant increase in structural distortions from quartz to $GaAsO_4$ and GeO_2 make their crystal growth very difficult. Thus, from the crystal growth point of view, $GaPO_4$ seems to be the

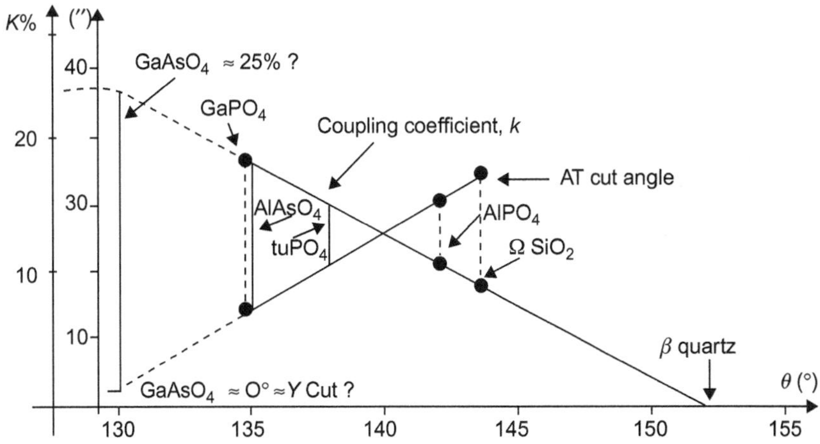

Figure 5.35 Extrapolation of possible piezoelectric properties to other quartz-like materials [117]. *Source*: Courtesy of the Academic Press, Orlando, FL.

Table 5.12 Extrapolation of AT-Cut Characteristic from Well-Known Quartz and Quartz-Like Materials

Material	SiO_2	$AlPO_4$	$FePO_4$	$AlAsO_4$	$GaPO_4$	GeO_2	$GaAsO_4$
AT-cut angle (°)	−35.25	−33.02	≈ −20	≈ −15	≈ −15	≈ 0	≈ 0
Coupling coefficient *k* (%)	8.5	11.0	≈ 15	≈ 18	≈ 18	≈ 25	≈ 25
c/*a*	2.20	2.22	2.23	2.23	2.26	2.27	2.28

best compromise for industrial development because of its high AT-cut coupling coefficient and large temperature range stability with a few drastic packing distortions. If the size of the $GaPO_4$ crystals grown by the hydrothermal technique is increased at least by one-fourth the value of the quartz, then $GaPO_4$ would be the best piezoelectric material of the twentieth century and probably of the early twenty-first century as well.

5.5 Potassium Titanyl Phosphate

Potassium titanyl phosphate, $KTiOPO_4$ (KTP), is one of the best nonlinear optical materials increasingly being used commercially for the second harmonic generation (SHG) and parametric oscillation since its discovery and introduction in 1976 [134]. It all began when there was a search for a suitable crystal to convert efficiently the near-IR light at 1.06 μm, which is useful in laser surgery. However, from the surgeon's point of view, there is a difficulty as the near-IR is not visible clearly. Thus, the solution for this problem was to use a second crystal which efficiently converts the near-IR to the visible—a harmonic generator crystal [135,136]. The requirement was for a material with high conversion efficiency and appropriate index of refraction characteristics to allow phase matching. The AT&T Bell Labs group [135] tried with $LiNbO_3$.

Subsequently, the DuPont group discovered KTP to be phase matchable most efficiently at 1.064 μm to an easily visible green light at 0.53 μm. It has very high laser damage resistance, coupled with low optical losses and it is transparent over a wide range of wavelengths with favorable angular, spectral and temperature bandwidths and has high conversion efficiencies [134,137,138]. Also, it was found that KTP's conversion efficiency was twice that of $LiNbO_3$ and hence it is the most commonly used 0.53 μm harmonic generator. The additional advantage is that the preparation of $LiNbO_3$, which could be grown from the melt, was very expensive. Also, the starting materials for the growth of $LiNbO_3$ were equally expensive. In contrast to this, the starting materials for the growth of KTP are rather cheaper and the growth techniques are relatively inexpensive. The lower cost preparative methods could have an impact on its use, and indeed, more generally on the commercialization of nonlinear optical materials and electro-optic devices and systems on the whole. KTP crystals are mechanically, chemically, and thermally stable and nonhygroscopic. It is remarkable for both high nonlinear optical coefficients and high conversion efficiency. The material has excellent physical properties and can be used in a type II phase-matching configuration. In the recent years, KTP has been developed for quasi-phase-matched guided-wave devices to access the visible spectrum.

KTP belongs to the family of compounds that have the formula unit $MTiOXO_4$, where M = K, Rb, Tl, NH_4 or Cs (partial) and X = P or As. All members belong to the orthorhombic space group, P_{na21}, point group, mm2, with the following lattice cell parameters: $a = 12.814$, $b = 6.404$, $c = 10.616$ Å, $V = 371.115$ Å^3. An important crystallographic property with reference to the optical, chemical, and solid-state behavior of KTP is that there are two formula units in the asymmetric unit so that the proper solid-state formula is $K_2(TiOPO_4)_2$. The structure of KTP is composed of helices of TiO_6 octahedra linked via phosphate bridges. This leads to an open framework structure in which the charge-balancing cations are incorporated within the channels (Figure 5.36) [139]. The nature of the $KTiOPO_4$ host framework suggested the possibility that modification of the internal structural characteristics, and hence the electro-optic properties, might be achieved by gas-phase absorption and description. The K-ion exists in a high-coordination number site and is weakly bonded to both the Ti octahedral and P-tetrahedral channels existing along the Z-axis [001] direction, where K can diffuse, with a diffusion coefficient, several orders of magnitude greater than in the $X-Y$ plane.

In spite of the above-mentioned unique properties of KTP, its application has been limited by a shortage of crystals of sufficient size and quality. In particular, the crystal growth processes have tended to be plagued by spurious nucleation problems and the inclusions of the solvents. The first attempt to prepare KTP crystals was by Ouvard during 1890 [140], who melted TiO_2, $K_4P_2O_7$, and K_3PO_4. Masse

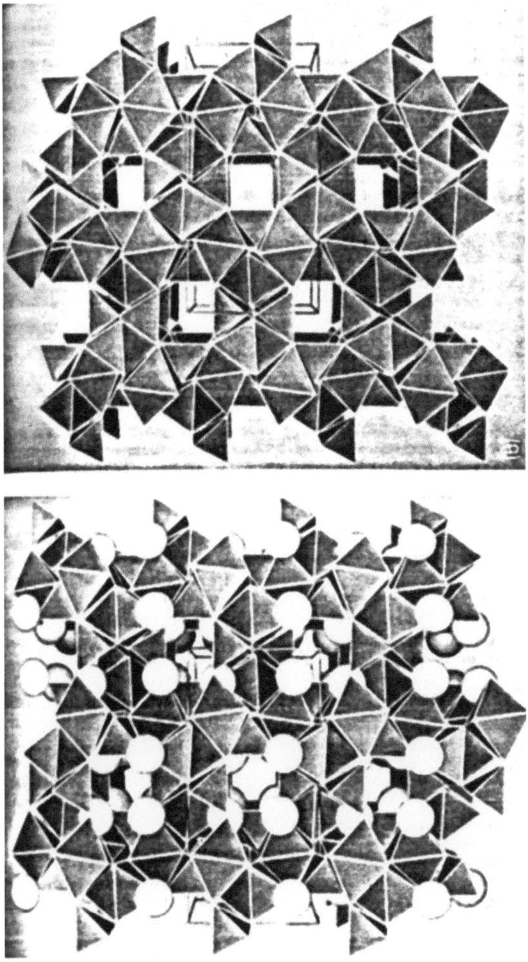

Figure 5.36 Structure of $KTiOPO_4$ [139].

and Grenier (1971) [141] prepared KTP in the polycrystalline form. However, the single crystals of KTP were first obtained by Zumsteg et al. (1976) [134] at the DuPont Experimental Station.

5.5.1 Crystal Growth of KTP

KTP decomposes before melting at 1148°C. Thus, incongruent melting and glass formation make it impossible to grow KTP crystals directly from the melt. Similarly, growth from an aqueous solution at low temperature (< 300°C) is impossible because of the low solubility of KTP. Therefore, large single crystals of KTP can be grown by both high-temperature and -pressure hydrothermal and flux techniques. There are several problems with reference to the quality of the crystals grown by the flux method. When pure phosphate self-fluxes are used, the main

Table 5.13 Difference Between Flux and Hydrothermally Grown KTP Crystals

Flux-Grown KTP	Hydrothermally Grown KTP
– Temperature is constantly lowered and crystals grow through continuous change in solubility	Temperature is constant and crystal growth occurs through continuous transport of nutrient by convection currents
– Growth temperature is high (> 950°C)	Growth temperature is moderate (< 500°C)
– Pressure atmospheric	Pressure is <1.4 bar
– Growth rate is ≥ 1 mm/day	Growth rate is <0.6 mm/day
– Experimental duration: 7–10 days	Experimental duration: several weeks (~30 days)
– Incorporation of fluxes into the lattice and concentration of nonstoichiometric defects occurs	Incorporation of $(OH)^-$ into the lattice, as crystals are grown at lower temperature
– Reduced stoichiometry, thermo-optic effect, poorer quality, susceptible to optical damage (less damage resistant)	Increased chemical stoichiometry, uniform refractive index, high quality, higher resistance to optical damage (most damage resistant)
– Cost of production—less expensive	Cost of production—very expensive
– Limited or slow incorporation in laser devices	More useful for laser applications
– Poor in perfection, purity, and homogeneity	Better, perfect, purity, and homogeneity
– Suffers from blackening at the end part of the negative pole when placed in DC field	Does not suffer from blackening at the end part of the negative pole when placed in DC field
– Higher ionic conductivity	Lower ionic conductivity
Growth morphology is more or less the same	

problem is the high melt viscosity. Its tendency to spuriously nucleate and also to leave behind major structural defects in the crystals has motivated the researchers to look for better fluxes. We have summarized the advantages and disadvantages of both flux and hydrothermal methods and quality of KTP crystals in Table 5.13. Similarly, the main problem with the use of the hydrothermal technique for the growth of KTP crystals is the higher pressure–temperature conditions of growth. In fact, when Zumsteg et al. (1976) [134] and Berlien and Gier (1976) [142] synthesized KTP by the hydrothermal method, the pressure–temperature conditions were quite high ($P = 1.5–3$ kbar, $T = 650–700$°C). The crystal size was not more than a few millimeters since the experiments were carried out in small platinum or gold capsules. Liu et al. (1982, 1984) [143,144] have reported the growth and properties of hydrothermally grown KTP. They prepared nutrient by the reaction of KH_2PO_4 and TiO_2 at 1250°C. The hydrothermal "flux" was 1.5 KH_2PO_4, 1.0 TiO_2 in Ag^- or Au^- lined autoclaves at a pressure of 24 – 28 kpsi (1.66–1.93 kbar) and 520 – 560°C in a gradient of 30 – 70°C. Growth rate on (011) was 0.2 – 1.8 mm/week. The higher pressure–temperature conditions do not yield large size crystals and the cost of the experiments involved is rather high. This led to serious efforts from several groups, particularly with reference to the PVT relations in several related systems like $K_2O-TiO_2-P_2O_5$, which ultimately brought down the pressure–temperature conditions to the tune of $P = 700$ bar, $T = 325 - 400°C$. Similarly the use of the solvothermal technique along with the organic additives and surfactant havs brought down the pressure–temperature conditions for KTP growth. It has renewed interest in KTP crystal growth. On the whole, the credit goes to Laudise and Belt for contributing extensively on the PVT relations of the KTP systems and also for growing large size single crystals of KTP by the hydrothermal technique [24,145–147]. The earlier hydrothermal experiments on KTP were carried out using platinum- or gold-lined autoclaves with Tuttle seals (cold-cone seal closures). The crystalline KTP nutrient in appropriate dimensions and quantity was introduced into the bottom part (dissolution zone), and the seeds (cut in proper directions) were suspended on a frame in the upper part (growth zone) of the liner cavity. A baffle separated them so as to control the solution convection for establishing a suitable temperature gradient. The mineralizers used in the earlier experiments were usually KF and K_2HPO_4 taken with a definite molarity. The percentage fill (70 − 80%), percent open area of the baffle (5–10%) and temperature gradient ($\Delta t = 10 - 80°C$) were selected appropriately. Under such experimental conditions, the crystals are usually $16 \times 15 \times 7$ mm^3.

At present, several variants of hydrothermal growth of KTP single crystals are known [24,145,148–150]. Laudise et al. (1990) [148] have carried out PVT measurements on 2 M K_2HPO_4 and 2 M K_2HPO_4 saturated with KTP (Figure 5.37a and b). Zhang et al. (2006) [149] succeeded in growing KTP crystals with high damage threshold by hydrothermal method.

It was important that growth be conducted in the absence of a gas–liquid interface, otherwise seeds in the gas phase do not grow, and bubbling, boiling, and bumping contribute to poor quality deposition. They made nutrient by the reaction of KH_2PO_4 and TiO_2 in platinum at 1250°C. The hydrothermal flux was 1.5 KH_2PO_4 to 1.0 TiO_2 in gold- or silver-lined autoclaves at $P = 1.66 - 1.93$ kbar, $T = 520 - 560°C$, and $\Delta T = 30 - 70°C$. Growth rates on (011) were 0.2 − 1.8 mm/week. Laudise and coworkers have discovered much lower PT conditions for KTP growth, permitting the use of ordinary, unlined steel autoclaves in a K_2HPO_4 solvent. By using the reaction $KH_2PO_4 + TiO_2 \rightarrow KTiOPO_4 + H_2O$ to form KTP, they carried out a phase stability study, the results of which are shown in Figure 5.38. Figure 5.38 shows only the water-rich corner of the complete ternary. Everywhere along the line A-H_2O, mole ratio $KPO_3/TiO_2 = 1.00$, the same as in KTP. Arrow 1 in Figure 5.38, at approximately 65 mol% $KPO_3/KPO_3 + TiO_2$ ($KPO_3/TiO_2 \cong 1.8$), is at or close to the boundary of the phase fields α + KTP and KTP at 600°. Alpha (α) is not a completely characterized phase, probably less K than in KTP. The boundary is probably nearly the same position at 500°C. Arrow 2 in Figure 5.38, at approximately 50 mol% $KPO_3/KPO_3 + TiO_2$ ($KPO_3/TiO_2 \cong 1.0$), is the boundary between the phase fields KTP and KTP + TiO_2 (anatase) at 600°C. It coincides with the line A-H_2O. Arrow 3, at approximately 58 mol% KPO_3/KPO3 + TiO_2($KPO_3/TiO_2 \cong 1.4$), indicates that the boundary

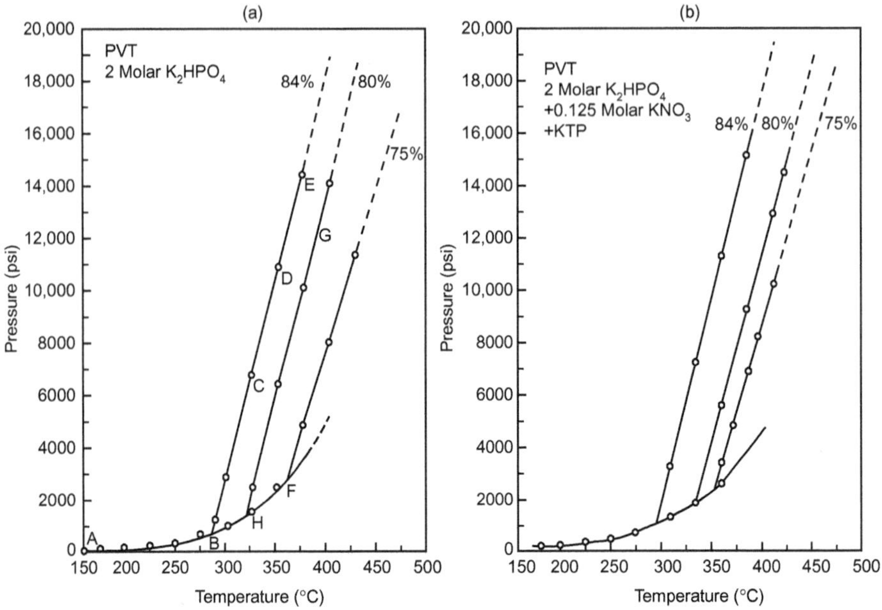

Figure 5.37 PVT relations in KTP system [148]. (a) Pressure for various fills as a function of temperature in 2 M K_2HPO_4. (b) Pressure for various fills as a function of temperature in 2 M K_2HPO_4 + 0.125 M KNO_3 saturated with KTP.

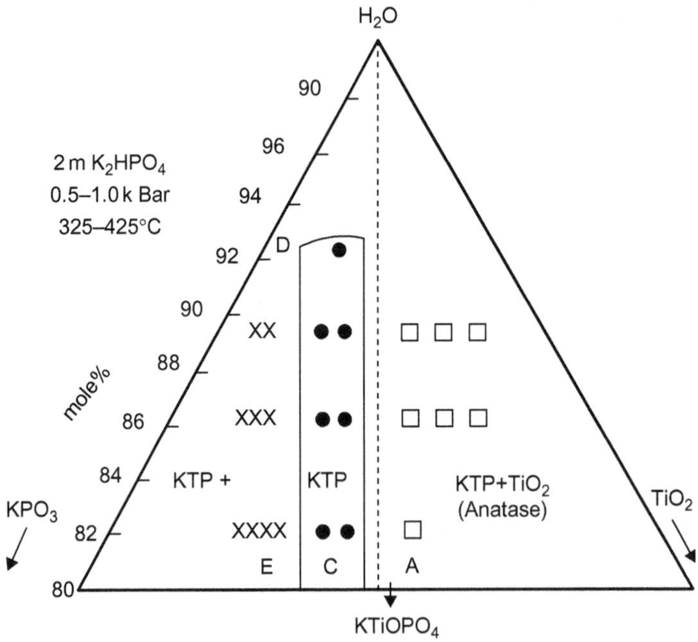

Figure 5.38 Phase diagram for a KTP system [148].

between the phase fields KTP and KTP + TiO_2 (anatase) at 500°C and lies at stoichiometry: richer in KPO_3 than A-H_2O. It moves to the left at lower temperatures. Assuming the validity of the boundaries defined by the arrows 1, 2, and 3, the appropriate phase boundaries have been drawn in Figure 5.38 so as to terminate in the complete phase diagram on the KPO_3–TiO_2 axis at K(TiO)PO. Thus, the lines EC(1)G, F(3)H, and D(2)A terminate at KTP. The line EF schematically represents the solubility of KTP at 500°C. It does not intersect A-D-H_2O, so KTP is incongruently saturating at 500°C, i.e., the mole ratio of KPO_3/TiO_2 in the liquid, where KTP is the only stable solid phase, not equal to the ratio in KTP. Somewhere in the TiO_2-rich part of the diagram, anatase would probably be the sole stable phase and, in the KPO_3-rich region, E would probably be the sole stable phase. The authors did not try to find these boundaries because they are not directly relevant to KTP growth.

Typical conditions used in growth runs are as follows:

Autoclave	– Pt-lined, 1″ ID × 6″ IL (2.5 cm × 15 cm)
Mineralizer	– 2 M K_2HPO_4 + 0.5 M KPO_3
% Fill	– 75%
Crystallization temperature	– 375°C
Nutrient (dissolving) temperature	– 425°C
Temperature differential (ΔT)	– 50°C
Seed orientation (201), (011)	– rates from 2.6 to 5.7 mil/day (~0.07 – 0.14 mm/day), (010) rate generally higher, quality good
(Seed orientation based on a = 12.80 Å, b = 6,400 Å, c = 10.580 Å) [148]	
Spontaneous wall nucleation	– moderate; crystals clear but not too well formed

The growth rates are for the total increase in crystal thickness or time of the run. Increasing ΔT to 75°C and increasing KPO_3 decreased the growth rate, but growing without KPO_3, i.e., with the mineralizer (2 M K_2HPO_4) alone, did not markedly affect the growth rate or quality. Growth in KH_2PO_4 or KPO_3 alone was very slow.

In the growth of KTP, like quartz, the seed orientation is very important. The growth rates are different in different seed orientations and the highest rate for KTP has been obtained on the (011) seed crystal [151].

The crystals grown at lower growth rates are superior in quality. Therefore, proper care has to be taken to achieve the optimum growth rate under hydrothermal conditions by selecting appropriate growth parameters, because a combination of several growth parameters act upon the growth rate and crystal quality of KTP.

5.5.2 Solubility of KTP

The solubility of KTP is an important aspect. The higher experimental temperature–pressure adopted in the earlier works was due to the lack of accurate data on the solubility of KTP and added to that the solubility of KTP is rather low in most of the conventional solvents.

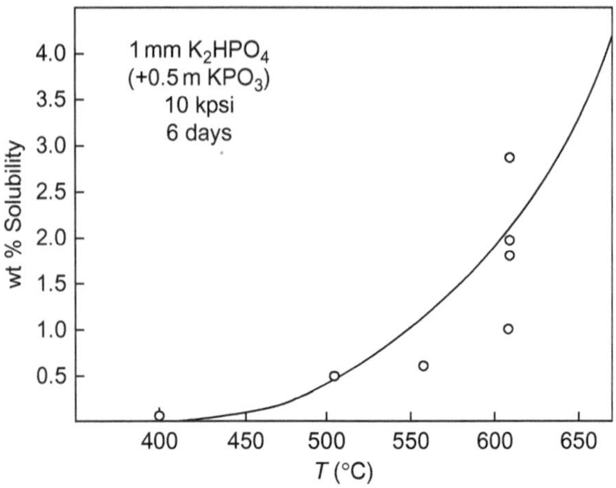

Figure 5.39 Solubility of KTP (in wt%) in 1 M $K_2HPO_4 + 0.5KPO_3$ at 10 kpsi [148].

The solubility of KTP in 1 M $K_2HPO_4 + 0.5KPO_3$ solutions has been studied in detail [148]. Figure 5.39 shows the solubility of KTP (in wt%) in 1 M $K_2HPO_4 + 0.5KPO_3$ at 10 kpsi. The solubility data indicate good solubility for growth at 600°C and a positive TC.

KTP crystals, using a complex mixture of water-soluble phosphate glasses, KTP, solvents, and water, have been obtained by several researchers. The glass composition is K_2O (42 − 70), TiO_2 (1 − 18), P_2O_5 (26 − 55) (mol%). The optimum ratio K/P is equal to 0.8 − 2.5, optimum glass composition is K_2O (55 − 60), P_2O_5 (38 − 42), TiO_2 (1 − 5) (mol%).

Byrappa (1996) [152] has carried out an extensive study on the solubility and growth of KTP under lower temperature and pressure conditions. The main objective of this study was to find a suitable mineralizer solution in order to increase the growth rate, solubility, and crystal quality, but at reduced PT conditions. As is well known from the literature survey, the hydrothermal technique is most suitable for the growth of KTP crystals as it produces high-quality defect-free crystals. However, the main problem with the use of the hydrothermal technique for the growth of KTP crystals is the higher PT conditions of growth. In fact, when the early hydrothermal experiments were carried out, the PT conditions were rather high ($P = 1.5–2$ kbar; $T = 650–700$°C). The higher PT conditions do not yield large size crystals and the cost of the experiments involved is rather high. Subsequently, the inflow of the research data particularly on the PVT relations in several related systems of $K_2O–TiO_2–P_2O_5$ has brought down the PT conditions to the tune of $P = 10$ kpsi and $T = 375 − 425$°C. Recently, the material aspects, structural, physical, and chemical properties, and device feasibility of the KTP family of crystals have been reviewed [153].

Investigators have carried out hydrothermal experiments within a wide range of PT conditions using Morey- and Tuttle-type autoclaves. The nutrient materials were taken in Teflon liners for experiments at lower PT conditions (using the Morey type of autoclaves) and in platinum liners for experiments at higher PT conditions (using Tuttle type of autoclaves). A series of mixed mineralizers like $H_2SO_4 + KOH$, $KF + HCl$, $HF + KF$, $H_2SO_4 + HCl + KF$ in different ratios were tried in order to bring down the PT conditions of the crystallization of KTP crystals. The pressure in the system was calculated from PVT relations. In the case of lower PT experiments, the temperature was 275°C and the pressure was approximately 100 bar. The experiments were carried out for 8 − 10 days and spontaneously nucleated small crystals of KTP were obtained. The resultant product was washed thoroughly in distilled water to remove the excess solvents and the crystals were collected and dried. The size of the crystals was around 1 mm, colorless, transparent, and exhibiting a vitreous luster. The faces were developed in most of experiments. The solubility measurements were carried out on KTP crystals using the above said mixed mineralizers under hydrothermal conditions. Figure 5.40 shows the solubility curves for KTP. Figure 5.41 shows a representative photograph of a KTP crystal obtained by the hydrothermal method. Similar experiments were carried out under elevated PT conditions ($P = 500$ bar and $T = 325$°C). The crystals were better in quality compared to the crystals obtained under lower PT conditions.

Belt et al. (1989) [151] have reported the growth of KTP single crystals in 4 l capacity autoclaves at temperatures of 475°C and 1.5 kbar. Gold liners were used here exclusively as corrosion resistance to the phosphates. Baffle openings were adjusted in the range of 20% with $T = 30 - 40$°C and the total run durations were 40 − 50 days. The growth rates sustained were 0.5 − 1.0 mm/side/week on (011) seed crystals [123].

Belt and coworkers from Airtron Division of Litton Systems have increased the autoclave volume further and developed the low-temperature hydrothermal process in 5 l systems. These large autoclaves have produced the largest KTP single crystals available in the world (Figure 5.42). These crystals are typically $4 \times 5 \times 8$ cm^3 and easily provide finished parallelepipeds of 1 − 1.5 cm^2 apertures and up to 2 cm long when material is extracted from either side of a seed. The schematic diagram of the large KTP crystals grown in 5 l autoclaves is shown in Figure 5.43. The rectangle in the center is seed plate, and the numbers identify prominent planes [134].

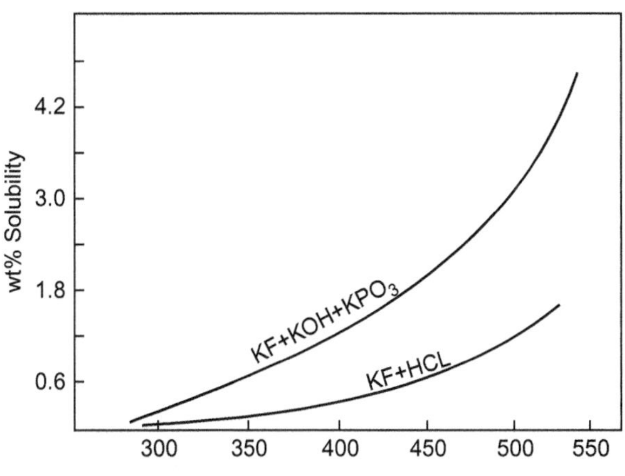

Figure 5.40 Solubility curves for KTP [152].

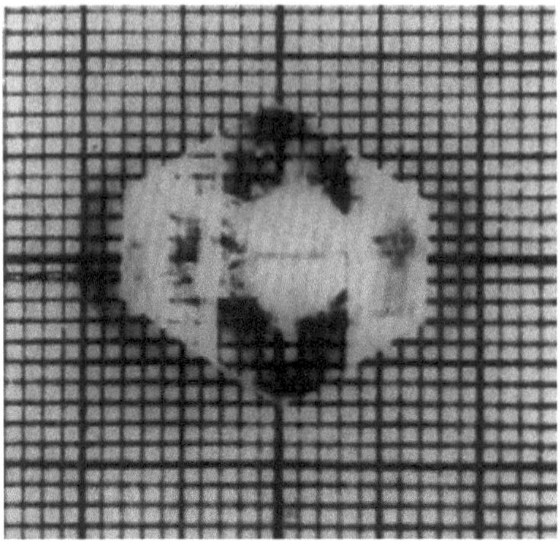

Figure 5.41 Representative photograph of a KTP crystal obtained by the hydrothermal method [148].

Figure 5.42 KTP single crystals grown using large autoclaves (5 l).
Source: Photo courtesy of R. F. Belt.

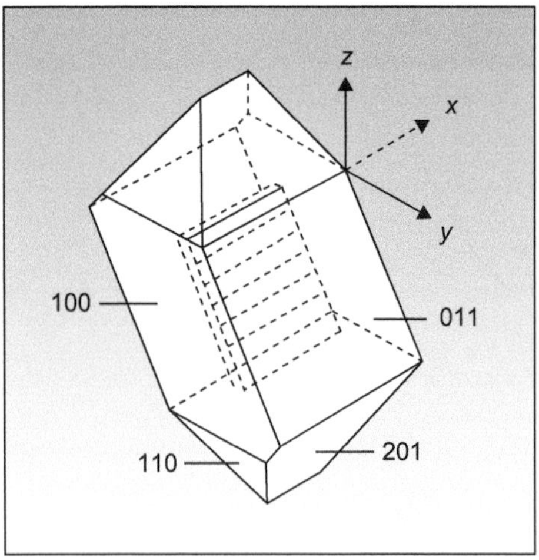

Figure 5.43 Schematic diagram of the large KTP crystals grown in 5 l autoclaves [147].
Source: Courtesy of the Penn Well Publishing Co., the United States.

A successful large-scale hydrothermal growth process affects the size, cost, yield, and quality of KTP crystals. The cost of an autoclave is governed by its size, the type of alloys used, and availability of heat-treated billets from which the autoclaves are fabricated. Thus, lower temperature experiments can use autoclaves made from more-common steel alloys, in turn lowering the cost.

The darkening of KTP crystals caused by the lack of oxygen in the growth process can be overcome by introducing an oxidizing agent, e.g., $1-5$ wt% of H_2O_2, into the mineralizer solution.

5.5.3 Morphology

The morphology of KTP crystals is quite interesting and it does not vary much from flux grown to hydrothermally grown. KTP usually grows as multifaceted crystal with 14 faces typical for the KTP structure. The commonly seen crystal faces are {100}, {011}, {201}, and {110}. These faces possess different growth rates and their relative sequence remains constant, even though the absolute growth rate of each face may vary with growth temperature gradient, solvent concentration, and so on. The average growth rates for various faces of hydrothermal KTP are given in Table 5.14.

The {100} face has the largest area and, thus, the smallest growth rate among the faces. The width and height of the steps in KTP are inversely proportional to the growth rate or to the ΔT in different runs. Since KTP grows as multifaceted crystal, cutting and polishing of the crystal are quite cumbersome. Therefore, the growth of KTP with increased volume at lower cost is very important.

As given in Table 5.13, the properties of KTP crystals vary with the growth method. Hydrothermally grown KTP crystals generally show an absorption band between 3550 and 3600 cm^{-1} due to the presence of $(OH)^-$. Generally, if a crystal used for a nonlinear interaction has even a small amount of absorption, the efficiency of the nonlinear interaction will be decreased. Absorption of the laser radiation occurs through the volume of the crystals illuminated by the laser beam. Thus, heating of the crystal occurs through this volume also. Cooling, on the other hand, occurs by conduction. Consequently, heating the surface of the crystal establishes thermal gradients within the crystal. As the refractive index depends on the temperature through the thermo-optic effect, gradient variations in the refractive index

Table 5.14 Average Growth Rate for Various Faces of Hydrothermal KTP

Growth Rate (10^{-3} mm/day)			
{100}	**{201}**	**{011}**	**{110}**
90	97	123	156
65	72	88	110
40	45	53	71

will also be generated. Variations in the refractive index will result in nonideal phase matching throughout the volume of the nonlinear interaction. As the nonlinear interaction is quite sensitive to the variations in phase matching, the heating of the crystal can cause a significant degradation in the efficiency of the nonlinear interaction [154].

Several authors have measured the ionic conductivity in pure and doped KTP crystals [155,156]. The ionic conduction property of these crystals allows ion-exchanged waveguides to be formed readily in them.

5.6 Potassium Titanyl Arsenate

It has been shown that the nonlinear optic and electro-optic coefficients of potassium titanyl arsenate (KTA) crystals are significantly superior to those of KTP crystals [141]. Similarly, the arsenate isomorphs, like KTA, RTA, and CTA, unlike the phosphate isomorphs, are prone to ferroelectric multidomain formation which renders these crystals useless in practical applications. KTA belongs to the class of 40 compounds which are all structurally characterized by corner-linked octahedrally coordinated titanium chains connected with tetrahedrally coordinated phosphorus or arsenic bridges.

KTA can be prepared by methods analogous to those used for KTP. Durand and El-Brahimi (1986) [157] have reported the structure of KTA crystals. The reputed effective nonlinear coefficient (d_{eff}) of KTA is 1.6 times that of KTP. This favorable property has encouraged several groups to carry out the hydrothermal growth of KTA at temperatures less than 600°C. Belt and Ings (1993) [158] have hydrothermally grown KTA crystals using KH_2AsO_4 and KOH mineralizers. Flux-grown KTA crystals were used as the seed crystals by these workers. These authors have also studied the PVT relationship, solubility, and the growth rates and growth morphology in detail; with the pressure-balancing method, growth rates of $0.2 - 0.4$ mm/side/day can be achieved on {011} seeds. The experiments have been carried out in large size autoclaves of 4–5 l capacities. Figure 5.44 shows the characteristic photographs of KTA single crystals obtained by Belt and Ings under hydrothermal conditions. The ideal morphology of KTA crystals is shown schematically in Figure 5.45. The experimental conditions are given in Table 5.15.

Some authors have studied the ionic conductivity in KTA crystals and usually the ionic conductivity is one order of magnitude lower than KTP crystals. The ionic conductivity is found to be one dimensional, i.e., along the *Z*-direction only.

The structure of the KTP family has the typical formula $\{K^+\}$ $[Ti^{4+}]O$ $(P^{5+})O_4$, where the flower brackets indicate nine- or eightfold coordination, the square brackets sixfold octahedral coordination, and the parentheses fourfold tetrahedral coordination. Different isovalent and aliovalent substitutions in a given structure are quite interesting, and could, perhaps, result in different structural and optical properties. Cheng et al. [159] have found that stability of KTA crystals is higher

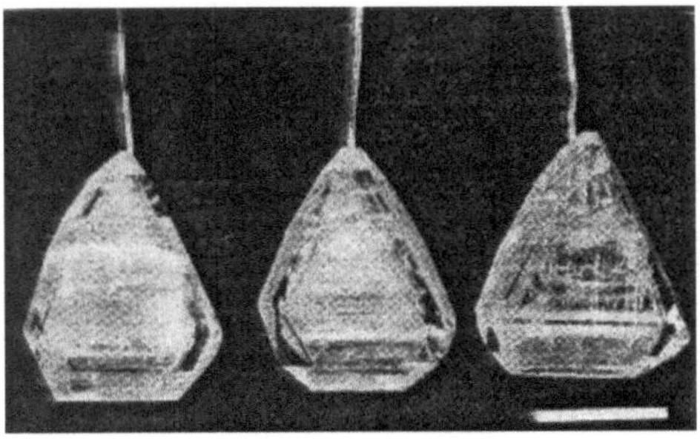

Figure 5.44 Characteristic photographs of KTA single crystals obtained by Belt and Ings under hydrothermal conditions [158].

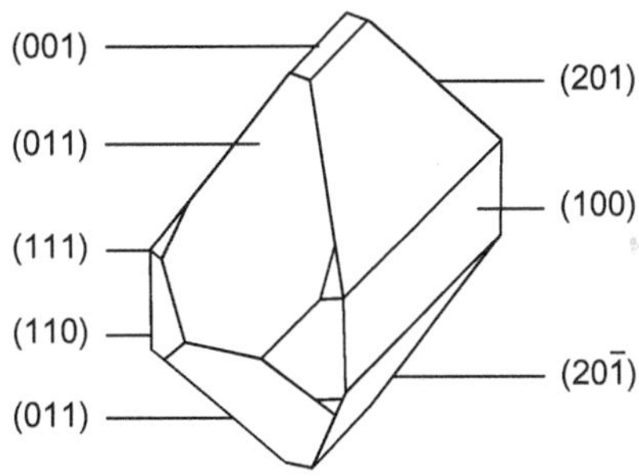

Figure 5.45 Ideal morphology of KTA crystals is shown schematically [158].

compared to other KTP-like compounds, Nb-doped KTP crystals are observed to be stable, and K_{1-x} Ti_{1-x} Nb_xO PO_4 single crystals with $x \approx 0.25$ have been obtained without any charge compensator. Partial charge compensation of Nb^{5+} in K_{1-x} Ti_{1-x} Nb_xO PO_4 crystals by Ga^{3+}, Fe^{3+}, and Mg^{2+} has been reported. However, complete substitution of Ti^{4+} in single crystals of KTP with any aliovalent cation pair has not been achieved by any method.

On the contrary, complete substitution of Ti^{4+} with Nb^{5+}–Mg^{2+} and Nb^{5+}–Zn^{2+} cation pairs was reported to be possible in the case of KTA crystals. This result by Chani et al. (1997) [160] suggested that it might be possible to prepare new aliovalent analogous of KTA containing Nb^{5+}–M^{3+} (M = Al, Cr, Ga, Fe, In) cation pairs in octahedral positions of the KTP structure. KTA crystal was the chosen basic material for this purpose as it was noted to be the most stable compared to KTP [159].

Lattice parameter values, in general, were found to increase with increasing M^{3+} cation radius, as expected. The results of compaction measurements indicate that Ti-free crystals with the structure of KTP have tendency to create potassium vacancies. This result is in good agreement with that reported for nonsubstituted KTP crystals. Structural analogues of KTP corresponding to K[Ta,M]O AsO_4 were prepared by solid-state reaction. However, no KTP phase formation was observed in K[Nb,M]O PO_4 and K[Ta,M]O PO_4 systems.

Table 5.15 Summary of the Experimental Parameters of the Hydrothermal Growth Experiments in Production-Sized Autoclaves

Run No.	Mineralizer	% Fill	Nutrient	Nutrient Temperature (°C)	Pressure (atm)	Seeds	Gradient (°C)	Run time (weeks)	Results	
		Inner	Outer							
1	4 M KH_2ASO_4	80	57	Flux-grown KTA	560	1700	Unoriented flux growth KTA	28	2	Slight growth on KTA seeds
2	4 M KH_2ASO_4	78	58	Flux-grown KTA	590	1820	(201), (011), and (100) flux-grown KTA	50	15	0.1 mm/side/week on (201) and (011) seeds, 0.2 mm/side/week on (100) seeds
3	4 M KH_2ASO_4	77	57	Flux-grown KTA	590	1700	(201) and (011) flux-grown KTA	30	2	0.2 mm/side/week on (201) and (011) seeds
4	4 M KH_2ASO_4	77	57	Flux-grown KTA	590	1700	(201) flux-grown KTA	40	4	0.4 mm/side/week on (201) and (011) seeds, 0.2 mm/side/week on (100) seeds

The growth of KTP family of materials on the whole is fascinating and challenging for both technologists and academicians.

5.7 Calcite

Calcite, $CaCO_3$, is an important carbonate mineral and it occurs commonly in nature as both well-developed crystals and amorphous material. The pure and optically clear calcite is called Iceland spar. Calcite is polymorphous and exists in at least five modifications. The two polymorphs commonly found in nature are calcite and aragonite. In addition, there are two synthetic forms known only at high pressures: calcite II and calcite III. Vaterite (γ-$CaCO_3$) is a metastable hexagonal form which crystallizes at ordinary temperatures and pressures [161].

Calcite single crystals form an important optical material owing to its large birefringence and transparency over a wide range of wavelengths. This property makes it one of the significant materials for polarized devices such as optical isolaters and *Q* switches. It also exhibits antiferromagnetic properties. Although there are many deposits of calcite in the world, the optically clear quality calcite (e.g., Iceland spar) has been depleted in recent years, leading to its shortage in nature. However, the demand for optically clear calcite single crystals is increasing greatly with the development of laser devices such as optical isolators [162,163], and similarly, the chemical impurities and growth defects depend upon the geological settings and often lead to variations of refractive index and internal light scattering centers, and low optical damage resistance. Hence, the artificial means of obtaining optically pure calcite is becoming very popular. Many laboratories throughout the world have attempted to obtain calcite, but most of them have suffered from size or quality problems. In fact, the first publication on the synthesis of calcite appeared in 1883. Lemberg [164] heated basalt for 6 months with distilled water. It took up 2.43% H_2O, which was not removed over H_2SO_4 in 2 weeks; the solution remaining in the tube had a slightly alkaline reaction. The same basalt heated with K_2CO_3 solution for 9 months gave a glass rich in H_2O and $CaCO_3$ crystals. A glass made from palagonit from Vidoe heated for 3 months with distilled H_2O was little attacked, but the product contained 8.61% H_2O. Similarly, melted labradorite, heated for 13 months with K_2CO_3 solution gave a glass rich in K, and the Ca replaced had formed $CaCO_3$. Further, Lemberg obtained $CaCO_3$, while the MgO remained combined with the SiO_2 after heating augite with K_2CO_3 solution for 1 year at 100°C. When anorthite, with a little augite, was heated for 189 h at 180–190° with a Na_2CO_3 solution, the product consisted of calcite crystals, amorphous substances, and a columnar mineral similar to cancrinite. Later, Friedel and Sarasin (1885) [165] obtained small calcite crystals by heating $CaCl_2$, precipitated $CaCO_3$, and 60–70 ml H_2O for 10 h at 500°C; with 20 gm $CaCl_2$ and a few grams $CaCO_3$, the crystals were large enough for goniometric measurements. Crystals obtained were simple rhombohedra. There was no aragonite in the product. However, all these earlier works on the synthesis of calcite were purely of geological/mineralogical interest, and the size of the crystals obtained was in no way nearer to the fabrication of devices. The most significant problem for the growth of calcite crystals is that the growth techniques are restricted due to the dissociation of $CaCO_3$ at high temperatures. Several techniques have been employed to grow calcite single crystals: like low-temperature solutions [166,167], gels [168,169], fluxes or eutectics [170,171], top seeded solution growth [172], traveling solvent zone melting [173], HHP [174–176], and hydrothermal [162,163,177–182]. Calcite crystals grown by most of these methods encounter several problems including the presence of large thermal stress, CO_2 dissociation, and alkaline carbonate flux inclusions which cause serious defects and contamination by metal ions. On the other hand, the hydrothermal method is one of the most promising techniques for growing unstable crystals such as a carbonate compound, with the advantage of suppressing the dissociation of CO_2 and yielding high-quality crystals in a homogeneous ambient at relatively low temperatures [183]. Also, this technique is closer to the conditions in which calcite grows in nature.

Many researchers have tried to grow calcite single crystals at relatively low temperatures since calcite crystals dissociate to form CaO and CO_2 above 900°C under atmospheric pressure. Varieties of solvents have been used for hydrothermal growth of calcite, but none of them have succeeded in growing large single crystals. The solubility of calcite is very interesting. It has been studied in water under various CO_2 pressures [184,185]. The CO_2 pressure would be concerned with the dissolution and precipitation of calcite in nature. Thus, hydrothermal growth under variable CO_2 pressures is also very interesting from the point of view of geochemistry. The solubility is positive for increasing CO_2 pressure and negative for increasing temperature. This behavior of calcite has posed a real problem in the search for a suitable solvent to obtain higher/optimum growth rates. Despite this, a large number of reports concerning calcite growth, hydrothermal reaction of calcite in chloride, and other chloride solutions have appeared. Because natural calcite crystals are formed in both chloride and carbonate hydrothermal solutions, these chloride solutions are almost analogous to natural carbonate thermal springs. Ikornikova [178] has done an extensive work on the aspects of solubility, designing an apparatus to grow calcite crystals with changing CO_2 concentration as the pressure is reduced at constant temperature, crystal growth kinetics, and mechanism.

For hydrothermal crystal growth of calcite, several inorganic solvents like NaCl, LiCl, $CaCl_2$, $NaNO_3$, $Ca(NO_3)_2$, NH_4NO_3, K_2CO_3, and carbonic acid have been employed as mineralizers. The growth conditions vary from 150°C and 15 MPa, to 600°C and 200 MPa. No one has used organic salt solutions to grow calcite crystals under hydrothermal conditions. Yamasaki and coworkers [182] have tested the solubility of calcite in ammonium acetate solution. However, there is no unanimity with respect to solubility owing to the change in the sign of the TC of CO_2 solubility in H_2O, as shown in the isobaric diagram (Figure 5.46). The diagram clearly shows two regions: I and II, lying on the two sides of the 200 bar isobar, which is close to the isobar of the critical pressure of water. The pH of the system falls with increasing P_{CO_2} under standard conditions. This is connected to the fact that, with increasing CO_2 solubility, the concentration of both the hydrocarbonate and the

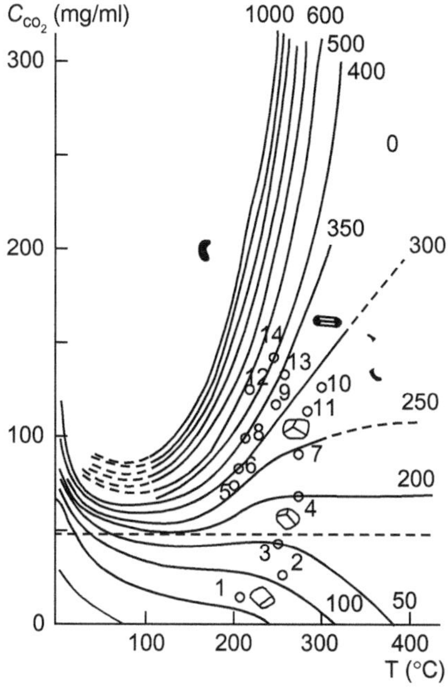

Figure 5.46 Isobaric diagram for CO_2 [178].

hydrogen ions increases. The solubility of calcite in aqueous solutions of carbon dioxide was studied between 10–300°C and 1–100 bar [184,185]. At the same time, for T = constant, the solubility rises exponentially with increasing P_{CO2}. Later experiments at 200–600°C and 200–1400 bar in solutions of 10–100 mg/ml CO_2 showed that absolute solubilities of calcite at high temperatures and pressures were almost an order of magnitude lower than those measured at pressures up to 100 bar [186]. The following groups have carried out extensive studies on the solubility of calcite: (i) Ikornikova of Russia, during 1960s and 1970s—using chloride solutions; (ii) Belt, the United States, during 1970s—using carbonate solutions; (iii) Hirano, Japan, during 1980s and early 1990s—using nitrate solutions; (iv) Kodaira, Japan, during 1990s—using H_2O–CO_2; (v) Yamasaki, Japan, during 1990s to date—using ammonium acetate and other organic solvents. Each one of the last three groups claims the superiority of their solvents over the others. Figures 5.47 and 5.48 show the solubility curves for calcite in different solvents. Yanagisawa et al. (1996, 2001) [182,187] have studied extensively the effect of pH and increase of carbon number in alkyl group of the ammonium monocarboxylates solvents on the growth of calcite and also the effect of dopant metals on the dislocation density of calcite crystals. It was found that calcite crystals could not grow in the CH_3COONH_4 solutions with a low pH. This has been explained by the change in the existing ion species in the solution. A high growth rate was obtained in the solutions with a pH of 7.0–7.5. Beyond a pH of 7.5, the growth rate falls again. The pH of the solution

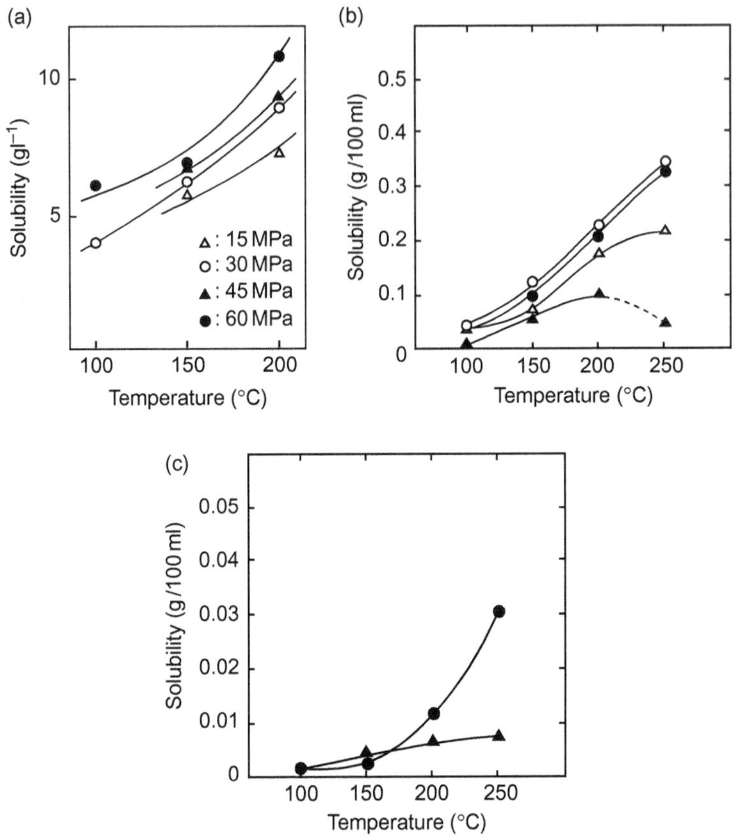

Figure 5.47 Solubility curves for calcite in different solvents: (a) in 1 M NH_4NO_3 solution as a funciton of temperture under varying pressures; (b) in 1 M NH_4NO_3 and (c) in 3 M NH_4NO_3 [183].

decreased with the increase in temperature, and it reached 6.0 at 200°C. This result suggested that H_2CO_3 was the main dissolved species in the solution at 250°C. When calcite dissolves in water, the carbonate ion must change its ionic form to HCO_3^-, which is very essential for the crystal growth of calcite, whereas H_2CO_3 does not contribute to the growth of calcite crystals. The formation of the ionic species HCO_3^- has been noticed by several workers like Kikuta and Hirano [186] in the NH_4NO_3 solutions from FT-IR studies. This suggests that calcite dissolves in the solvent below 200°C as follows:

$$NH_4NO_3 + CaCO_3 + H_2O \rightarrow Ca^{2+} + HCO_3^- + NO_3^- + NH_4^+ + OH$$

$$2NH_4F + CaCO_3 + H_2O \rightarrow CaF_2 + HCO_3^- + 2NH_4^+ + OH^-$$

$$(NH_4)_2SO_4 + CaCO_3 + H_2O \rightarrow CaSO_4 + HCO_3^- + 2NH_4^+ + OH^-$$

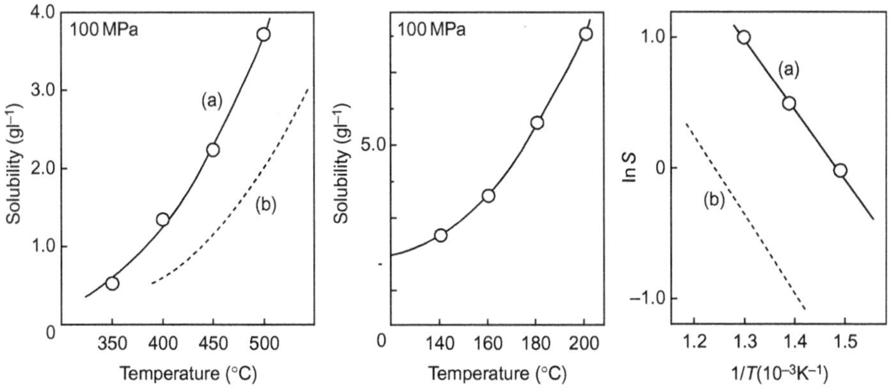

Figure 5.48 Solubility curves for calcite in different solvents: I. a) in 1.5 M Ca(NO3)2 solution; b) in 3M NaCl solution; II. in 3M NH4NO3 solution; III. Log S (solubility) as a function of 1/T: a) in 1.5M Ca(NO3)2; b) 3M NaCl [180].

This work strongly confirms that a small amount of additives, such as phosphate, sulfate, borate, and ethylene glycol, effectively arrested the growth of calcite. Similarly, the use of some organic additives like polypeptides is becoming very popular. Nefyodova et al. (2003) [188] have studied the optical properties of hydrothermal calcite crystals with respect to the dopant metals and dislocation density.

5.7.1 Crystal Growth

Solubility of calcite can be both positive and negative for increasing CO_2 pressure and increasing temperatures, respectively. Therefore, the growth of calcite crystals insists on a special experimental setup. Figure 5.49 shows the arrangement of the apparatus designed by Ikornikova for growing calcite crystals in aqueous solutions of CO_2 with a negative TC of solubility, together with curves illustrating the height distribution of temperature and CO_2 concentration [178]. The crystals obtained by this means, with seed crystals at the bottom and nutrient at the top followed by a temperature gradient between the two zones separated through a baffle, have rhombohedral faces covered with stepped pyramids. Similarly, Ikornikova has designed another apparatus for synthesizing calcite crystals in aqueous solutions of carbonic acid under isothermal conditions. The autoclave used is a rocking autoclave. Figure 5.50 shows the schematic diagram of the apparatus used in the synthesis of calcite. The autoclaves in both the cases are either glass or Teflon lined [178,181]. In this case, usually at pressure under 150 kg/cm^2, rhombohedral calcite crystals without any pinacoid are formed. The pinacoid appears on crystals formed at pressures greater than 200 kg/cm^2.

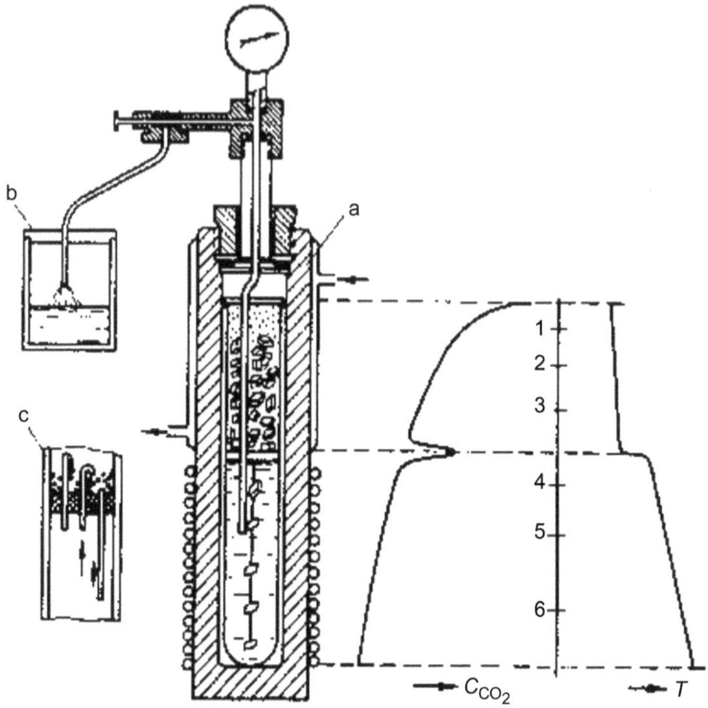

Figure 5.49 Arrangement of the apparatus for the growth of calcite [178].

Kinloch et al. (1974) [162] have obtained calcite single crystals in large autoclaves (23 cm internal diameter × 280 cm length)—commercial production using a 6 M K_2CO_3 mineralizer. As K_2CO_3 attacks autoclaves made of ferrous alloys, the autoclaves were provided with cylindrical silver and platinum cans. The following are the experimental conditions used by these workers:

Nutrient	– 45 kg of optical grade crystals of Mexican origin
Seeds	– Natural calcite crystals cutting rhombohedra with 10–20 cm^2 surfaces and 1–2 cm thick
Solvent	– 6 m K_2CO_3
Nutrient temperature	– 435–445°C
Seed temperature range	– 410–425°C
% Fill	– 86%(1.72 kbar)
Baffle opening	– 8%
Temperature gradient	– 5–10°C
Experimental duration	– 30–50 days
Growth rate obtained	– 50 μm/day/face for {1011} face

Much of the recent works on the growth of calcite single crystals are being carried out under moderate to lower temperatures (200–150°C) using several new solvents like ammonium nitrate and ammonium acetate. The results are far better than crystals obtained under higher temperatures because of the reduced thermal strain. Recent workers use Teflon liners. A considerable amount of progress has been achieved in the single crystal growth of calcite with reference to the solubility, growth rate, reaction kinetics, seed orientation, morphology, surface morphology, optical and laser quality, and so on. Figure 5.51 shows the schematic diagram of the autoclave for the hydrothermal crystal growth of calcite commonly used in recent years. It is well confirmed that the ammonium nitrate and acetate solvents are more effective than chloride solvents from the viewpoint of calcite crystal growth. The calcite crystals dissolve in ammonium nitrate or acetate solutions, accompanied by the evolution of ammonia. This reaction increases the dissolution of calcite in the hydrothermal solution which gives an advantage of lowering the pressure, temperature conditions of the growth, and in turn facilitates the use of Teflon liners. Great improvements are achieved by using a baffle plate and band heater to control the temperature gradient and the transport behavior. Higher temperature gradients give higher growth rates, especially at higher growth temperatures. Usually the growth rate varies from 25 to 213 μm/day. The rhombohedral

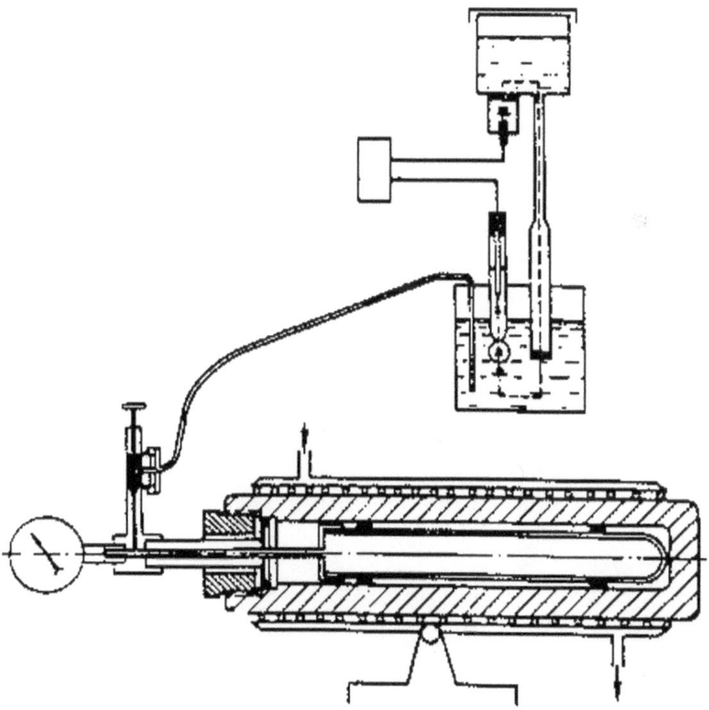

Figure 5.50 Schematic diagram of the apparatus used in the synthesis of calcite [178].

face (1011) is the fastest growing face. Hence, natural calcite crystals cleaved at (1011) plane are used generally as seed crystals. Yanagisawa et al. (1996) [182] have obtained the highest growth rate reported so far for calcite, 213 μm/day using ammonium acetate solvent at 230°C and a temperature gradient of 20°C. However, the optical quality of the calcite crystal was relatively poor. Only under experimental conditions, such as temperature gradient of 15°C and growth temperature of 235°C, was the growth of transparent calcite crystals achieved at 57 μm/day. Figure 5.52 shows the hydrothermally grown calcite crystals by different authors. When as-cleaved crystals are used, the calcite crystals obtained are usually opaque or not transparent, and the surfaces show a greater degree of defects like growth hillocks, a lot of small secondary grown crystals, and strains. The crystals grown on the etched seed crystals are usually transparent and the surfaces are very smooth. Figure 5.53 represents the characteristic photographs of calcite single crystal grown in NH_4NO_3 solution using as-cleaved and etched seeds. Similarly, Figure 5.54 represents the surface morphology of the calcite crystals using as-cleaved

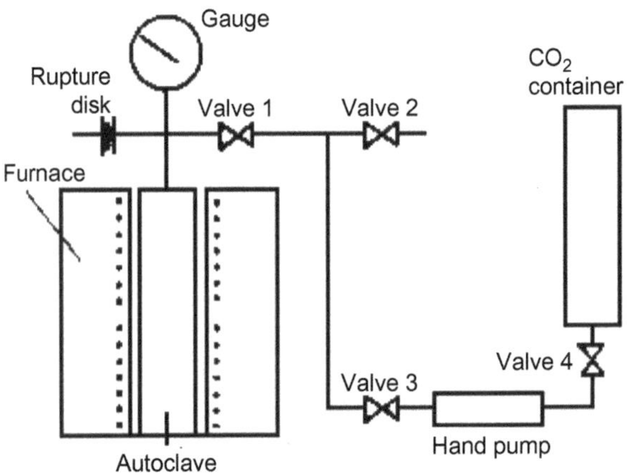

Figure 5.51 Schematic diagram of the autoclave [181].

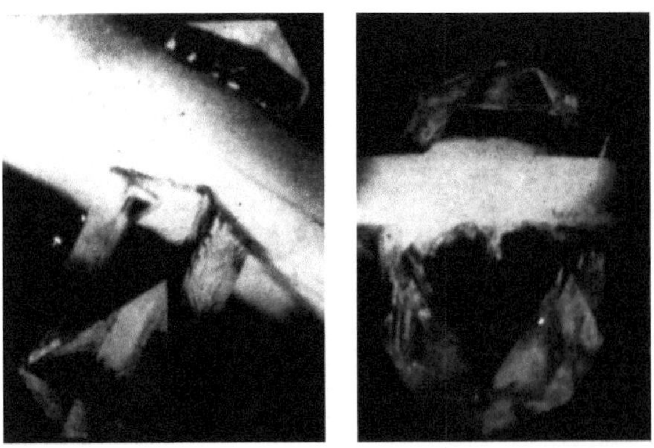

Figure 5.52 Hydrothermally grown calcite crystals [178].

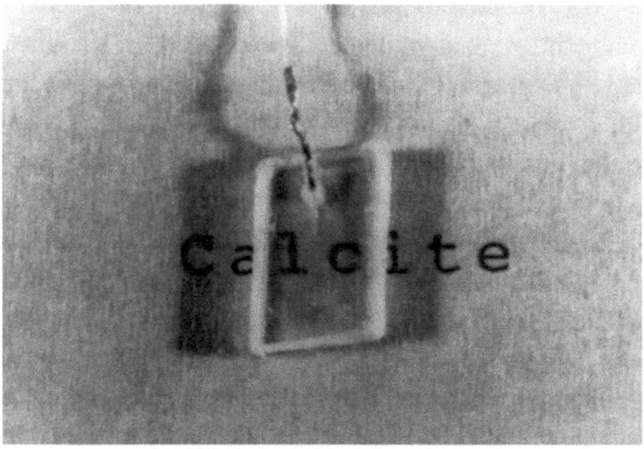

Figure 5.53 Characteristic photograph of calcite single crystal grown in NH_4NO_3 solution [180].

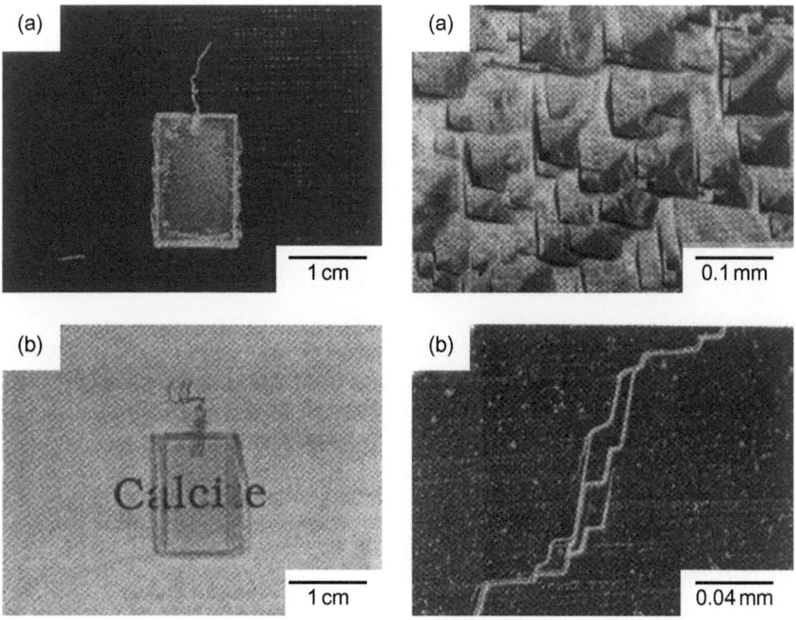

Figure 5.54 Surface morphology of the calcite crystals using as-cleaved and etched seeds [180].

and etched seeds. Lee and Chung (1998) [189] have reattempted the use of the solvent NH_4Cl, first reported by Ikornikova [177], to obtain calcite single crystals to suppress the spontaneous mineralizer. It was observed that the addition of such an organic additive played an important role in suppressing the formation of spontaneous nuclei; usually a higher percentage fill (81–83%) was selected in all the experiments and the pressure–temperature conditions were kept low. Hence, Teflon was used as the liner.

5.7.2 HHP of Calcite

The HHP is being applied in recent years for the solidification of $CaCO_3$. The HHP technique is a method by which hard solid bodies of powders can be produced in a short time and at a relatively low temperature under saturated vapor pressure. This process is very similar to diagenesis [190] in geology; therefore, a diagenesis process, such as for limestone, may be resurrected by using this technique. The results, using the HHP technique, may bring in some new information to the sedimentary process and a new technology of forming fabrication processes. Yamasaki et al. (1993) [175] have obtained $CaCO_3$ by HHP with seawater, and they have designed a "quasi-diagenesis" experiment. Initially, the $CaCO_3$ powder was synthesized in seawater, and this powder was then solidified by the HHP technique. Various ions of seawater, Cl^-, Na^+, K^+, and SO_4^{2-}, were absorbed on the surface of the synthesized $CaCO_3$ sample powder during the precipitation process, and these ions formed soluble salts such as NaCl and $CaSO_4$ under HHP conditions, i.e., the solubility of the synthetic $CaCO_3$ powder surface may have increased rapidly in the presence of $CaSO_4$ and NaCl solution under hot pressing conditions. Consequently, it was revealed that the small grains were dissolved, and then recrystallization occurred, or the grains adhered to the surface of larger grains and formed large grain growth during HHP. The solidification process was carried out by a dissolution–recrystallization mechanism during the hot pressing process.

Yamasaki et al. (1993) [175] have solidified the powders of calcium carbonate with sodium carbonate by mechanical compression under hydrothermal conditions. The effects of hydrothermal conditions such as temperature, pressure, and ratio of Ca/Sr on the reaction process and microstructure of the solidified bodies have been studied.

The growth of calcite single crystals is becoming very popular in recent years with the fast decrease in the availability of natural calcite. Although the growth of calcite has been better understood in the last 5–6 years, still there is no unanimity with reference to the solvent and experimental conditions. There is no doubt that calcite crystal growth is one of the flourishing and challenging topics.

5.7.3 Growth of Related Carbonates

There are several works in the literature dealing with the growth of other carbonates like $MnCO_3$ (rhodochrosite), $FeCO_3$ (siderite), $CaCO_3$ (spherocobaltite), $CdCO_3$ (otavite), and $NiCO_3$. Most of these are the important carbonate minerals, and their growth carries a great significance to the geological literature.

$MnCO_3$ (rhodochrosite). The crystallization of rhodochrosite crystals takes place in LiCl solutions (1–25 wt%) at temperatures 400–450°C, pressure 680–2500 atm, and $\Delta T = 10$–50°C [191]. In the experiments with 20–25% LiCl solutions at $\Delta T = 10$°C, $T = 430$°C, and $P = 960$ atm, Ikornikova has obtained transparent crystals of rhodochrosite with a size of 5–7 mm.

$FeCO_3$ (siderite). The synthesis of siderite is slightly more difficult than $MnCO_3$ having to do with the stability during the variation in the oxygen potential.

Ikornikova (1970) [192] has obtained transparent crystals of $FeCO_3$ in NH_4Cl solutions at $T = 300-350°C$, $P = 600-800$ atm, and $P_{CO2} = 300$ atm.

$CaCO_3$ (spherocobaltite). The growth of spherocobaltite crystals has been carried out on the seeds in the LiCl solution. Autoclaves of 300 cm^3 using titanium liners have been used. Separate experiments have been carried out in 1.2–1.7 l autoclaves with platinum liners. Spontaneously grown crystal was used as a seed crystal. However, detailed studies on the growth of spherocobaltite are not known [191,192].

$CdCO_3$ (otavit). Monocrystals of otavit were first obtained by Ikornikova [192]. Crystals were grown in LiCl (5–20%) and NH_4Cl (1–5%) solutions in the temperature interval 300–550°C, $\Delta T = 15-20°$, $P = 700-900$ atm. The size and quality of the crystals in the growth zone were regulated through the concentration of solution and temperature gradient.

$NiCO_3$. Monocrystals of carbonate of nickel could be obtained at $T > 150-500°C$ and $P > 1500$ atm using $NiCO_3$, $Ni(OH)_2$. During the evolution of CO_2 in the autoclave, the crystallization of $NiCO_3$ takes place at a much lower pressure and in a NH_4Cl (2–10%) solution [178].

Figures 5.55–5.57 show photographs of some selected hydrothermally synthesized carbonates in the hydrothermal laboratory of Byrappa.

The growth of carbonates, in general, is quite interesting academically, technologically, and geologically.

Figure 5.55 $ZnCO_3$ crystal.

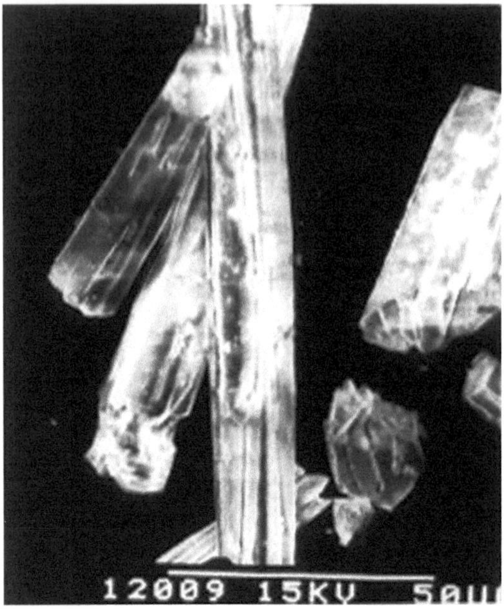

Figure 5.56 $SrCO_3$ crystal.

Figure 5.57 $BaCO_3$ crystal.
Source: Photographs courtesy of J. A. K. Tareen and B. Basavalingu.

5.8 Hydroxyapatite

Apatite is a common mineral in igneous, sedimentary, and metamorphic rocks [193], which has the general chemical formula of $A_{10}(BO_4)_6 \times {}_2$, where A = Ca, Sr, Ba, Fe, Pb, Cd, and contains many rare earth elements: $BO_4 = PO_4^{3-}$, VO_4^{3-}, SiO_4^{4-}, AsO_4^{3-}, CO_3^{2-}; $X = OH^-$, Cl^-, F^-, CO_3^{2-}. Water exists in different forms [194,195]. The presence of the elemental phosphorus, which is a major component of apatite, was first reported in the seventeenth century and the phosphorus in mineral pyromorphite (lead apatite) was discovered in 1779 [196,197]. Since the first synthesis of apatite done in the middle of nineteenth century [198], a lot of studies in the geochemical, crystallochemical, biological, and other fields have been carried out. In fact, most of the geochemical and crystallochemical studies were carried out on natural apatites, whereas biological or other recent studies were mainly done using synthetic apatites. However, many efforts have been made to prepare large single crystals of HAp with only limited success. The interest in the synthesis of HAp is linked with its importance as a major constituent of the inorganic component in bones and teeth where it occurs as tiny crystals. Since it has been recognized to be one of the best biocompatible materials, there have been many reports on the biological aspects [199–201]. HAp crystals have selective adsorption ability of proteins depending on crystal planes and are used as an adsorbent in liquid chromatography [202]. Recent interest in the material has spread to the fields of electronics and surface science (or adsorption chemistry), where sintered or powder HAp is expected to function as, for example, gas sensors, catalysts, and chromatographic absorbers. The fabrication processing of HAp has also progressed with the growing importance of the biological applications of HAp as artificial bone and teeth [203]. Table 5.16 gives the physicochemical properties of apatites [204]. However, a detailed knowledge of chemical and physical properties of HAp, including those relevant to the behavior in a biological environment, depends on the availability of sufficiently large single crystals of known chemical composition. Even today most of the studies on chemical properties, mechanical properties, biological responses, and the prosthetic application of HAp are based on polycrystalline bodies with random crystal arrangement including sintered bodies, powders, granules and coatings.

Daubree (1851) was the first to synthesize apatite by passing phosphorus trichloride vapor over red-hot rime [198]. Since then a number of reports on the preparation of HAp by various methods have appeared in the literature. Yoshimura and Suda (1994) [204] have reviewed the preparation methods for HAp and grouped all the methods into three categories, as given in Table 5.17. First is the conventional *solid-state reaction method* at high temperature, the second is the *wet chemical method* at relatively low temperature and the third, which is a very important method today, is the *hydrothermal method* using high-temperature–high-pressure aqueous solutions.

The solid-state reaction method has generally been used for processing ceramic powders and for studying phase stabilities. The powders prepared by this method, however, usually have irregular forms with a large grain size and often have

Table 5.16 Physicochemical Properties of Synthetic Hydrothermal or Natural FAp [204]

Properties	**Values**	**Comments**
Theoretical formula	$Ca_{10}(PO_4)_6(OH)_2$	Composition varies with sample preparation
Space group	$P6_{3/m}$ (hexagonal) $P2_{1/b}$ (monoclinic)	mono. ↔ hexa. transition may occur at ca. 200°C
Lattice parameters (LP)	$a = 9.41-9.44$ Å $c = 6.84-6.94$ Å	Large discrepancies in LP of "Wet chemically" HAp
Theoretical density	3.16 g/cm^3	Varies with composition
Moh's hardness	3(CO_3-Ap) 5(HAp)	Water and CO_3 in apatite lattice lower hardness
Heat capacity	184.07 cal/K mol at 298.16 K 950°C 180.16 cal/K mol at 298.15 K	"Wet" HAp following calcined at "wet" HAp following calcined at 1100°C
Thermal expansion coefficient	$11-14 \times 10^{-6}K^{-1}$	"wet" HAp expands nonlinearly
Melting point	1614°C	"synthetic" HAp, hot-stage microscope method
Surface energy	S(001) = 95 ± 25 MPa	Natural FAp, slow cleavage method
Stiffness coefficient at room temperature (mbars)	C11 = 1.434, C33 = 1.805 C44 = 0.415, C12 = 0.445 C13 = 0.575, C66 = 0.495	Natural FAp, ultrasonic pulse superposition
Dielectric constant	7.40–10.47	Varies with composition
Refractive indices	$n_w = 1.649-1.651$, $n_c = 1.642-1.644$	relationship between indices and composition is shown in Ref. [1].
Optical frequency	2.71 (‖c) 2.69 (⊥c)	Varies with composition

heterogeneity in the composition due to incompleted an reaction resulting from small diffusion coefficients of ions within solids.

The wet chemical method is a relatively easier method for obtaining HAp. Besides, the reaction in any living organisms may occur in aqueous solutions at low temperatures. Many experiments on the preparation of HAp by this method have been reported which investigate the formation mechanism of any calcium phosphates in vitro, in addition to in vivo. The powders prepared by this method, however, do not seem to be appropriate as the starting materials for ceramics because they are usually less crystallized, heterogeneous in composition, and irregularly formed. Sol–gel, alkoxide, and other chemical methods have also been

Table 5.17 Preparation Techniques for HAp [204]

Techniques	Starting Materials	Synthesis Conditions	Comments
Solid-state reaction	$Ca_3(PO_4)_2 + CaCO_3$ $Ca_2P_2O_7 + CaCO_3$	900–1300°C, usually with water vapor	Ca/P: 1.67, large grain size, irregular forms, inhomogeneous flowing
Wet chemical method	$Ca(NO_3)_2 + (NH_4)_2 HPO_4$ $Ca(OH)_2 + H_3PO_4$	RT: 100°C, pH: 7–12	Ca/P < 1.67 fine irregular crystals with low crystallinity inhomogeneous
Hydrothermal method	Wet chemically prepared HAp, other calcium phosphates, seeding	100–200°C (1–2 MPa), 300–600°C (1–2 kbar)	Ca/P = 1.67 homogeneous, fine single crystals or large crystals
Gel growth method	Gel + Ca^{2+} + PO_4^{3-}	RT: 60°C, pH: 7–10	Large Monetite, Brushite, OCP, but small HAp
Flux growth method	CaF_2, $CaCl_2$ as flux Ca $(OH)_2$ as flux	1325°C (FAp, ClAp) HAp	Large crystals with little lattice strain

used. CVD, plasma spray method and other (electro) chemical methods have been mainly used to form HAp layers on substrates.

The hydrothermal method enables us to prepare well-crystallized, compositionally homogenous, uniform, and easily sinterable powders due to the great effects of the high-temperature–high-pressure aqueous solutions.

The growth of HAp crystals has been carried out by several other methods like the gel-growth method [205]. There are several reports on the melt growth and flux growth of apatites of different compositions. Johnson (1961) [206] has used the melt growth for the preparation of single crystals of apatite from the stoichiometric melt. These crystals grown from the melt at high temperatures, however, are usually severely strained due to the large temperature gradients existing during the growth. Prener (1967) [207] used the flux growth method, which is an excellent method since the fluxes like CaF_2, $CaCl_2$, and $Ca(OH)_2$ mixed with the starting apatite powders make lower liquidus temperature, far below those needed for the melt growth method, resulting in the production of less strained apatite crystals. Oishi and coworkers [208–210] have obtained chlorapatite and fluorapatite (FAp) crystals by the flux method using NaCl, KF, and Ca_2ClPO_4 fluxes. Masuda et al. (1990) [211] have used the sol–gel method to obtain HAp from metal alkoxides. Brendel et al. (1992) [212] have used the polymerized method to achieve HAp coatings. Likewise, many other methods are utilized in the processing of apatites. Thus, for the growth of HAp, the volatility of water at the necessary high temperatures makes hydrothermal conditions unavoidable.

5.8.1 Crystal Structure of Apatite

Most of the studies concerning the crystallochemical, geochemical and phase stability-related aspects of apatite have been carried out on natural or hydrothermally synthesized large single crystals. The crystal structure of FAp was first determined by Naray-Szabo (1930) [213] and by Mehmel (1930) [214] independently. Followed by these, several publications appeared on the structure of various forms of apatite [215–217]. The general agreement on the crystallographic aspects is as follows:

- **i.** Apatite shows ionic bonding character, and it is made up of a close packing of large oxygen ions, resulting in the hexagonal crystal system [218].
- **ii.** The space group of FAp is $P6_{3/m}$ [213–217].
- **iii.** Although the space group of HAp is believed to be $P6_{3/m}$, HAp prepared at high temperature is $P2_{1/b}$ (monoclinic) at room temperature [213], nearly identical with chlorapatite (ClAp).
- **iv.** A phase transition in HAp is suggested to occur at approximately 200°C, probably due to the order–disorder orientation at $(OH)^-$ ions along the *c*-axis [219,220].

5.8.2 Phase Equilibria

The phase equilibria in the system $CaO–P_2O_5–H_2O$ have been extensively studied by the solid-state reactions method under the atmospheric pressure of water vapor by Van Wazer (1958) [221], and in aqueous systems at temperatures lower than 100°C by Brown et al. (1991, 1992) [222,223]. Biggar (1966) [224] has studied the phase equilibria in the system $CaO–P_2O_5–H_2O$ in the temperature range 700–950°C, and *P* of 1 kbar. Feng and Rockett [225] (1979) have studied the system $CaO–P_2O_5–H_2O$ at 200°C. Figure 5.58 shows the $Ca(OH)_2–Ca_3(PO_4)_2–H_2O$ at 1000 bar with 50 wt%

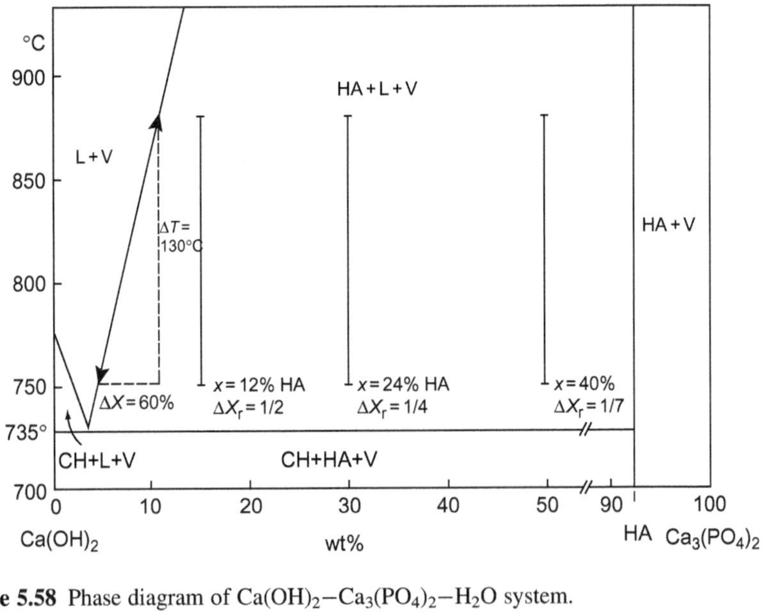

Figure 5.58 Phase diagram of $Ca(OH)_2–Ca_3(PO_4)_2–H_2O$ system.

H_2O. Skinner (1974) [226] found that the field of stability of HAp alone with water in the triangle $CaO-P_2O_5-H_2O$ was confined to a narrow band joining the representative point of stoichiometric HAp to the water representative point of the phase diagram. There are several other phase equilibria studies on the apatite system and the reader can refer to the works [204,227]. Andrade et al. (1997) [228] carried out a detailed thermodynamic analysis of the system $Ca-P-H_2O$ based on the development of the Eh−pH and Pa_C−pH (where $Pa_{Ca} = -\log_{10} a_{Ca}$) diagrams at different temperatures and different activities of phosphorous and calcium in aqueous solution. The HAp precipitated at ambient temperature (30°C) was hydrothermally treated at 150°C. Figures 5.59−5.63 show the Eh−pH diagrams of the $Ca-P-H_2O$ system at 25°C and 300°C, for 1.67 mol activity of the Ca and 1 mol activity of pressure in the aqueous solution. Figures 5.61 and 5.62 present the Pa_{Ca}−pH diagrams (where $Pa_{Ca} = -\log_{10} a_{Ca}$), for the same system with $a_{Ca} = 1.67$ aP and show that the pH value for the point of minimum solubility of the HAp clearly decreases with an increase in the temperature of the system. At each temperature level, HAp predominates in the higher pH range, while $Ca_3(PO_4)_2$, $Ca_2P_2O_7$, and $CaH_6P_2O_9$ have their predominate

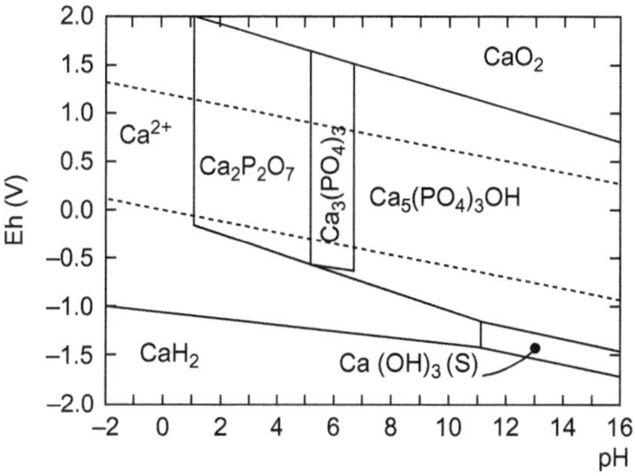

Figure 5.59 Eh−pH diagram of the $Ca-P-H_2O$ system at 25°C for 1.67 mol activity of Ca and $a_{Ca} = 1.67a_P$ [228].

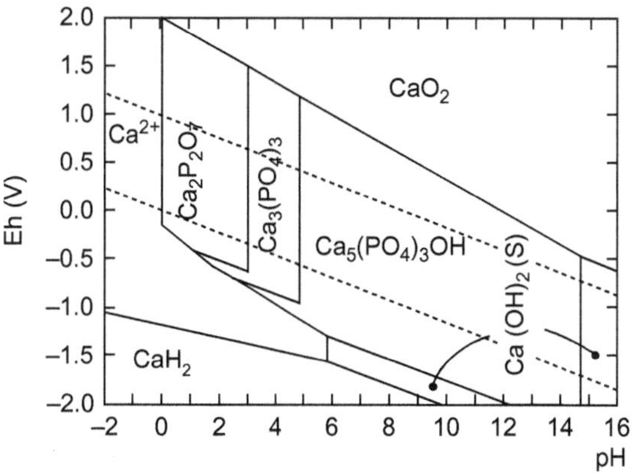

Figure 5.60 Eh−pH diagram of the $Ca-P-H_2O$ system at 300°C for 1.67 mol activity of Ca and $a_{Ca} = 1.67a_P$ [228].

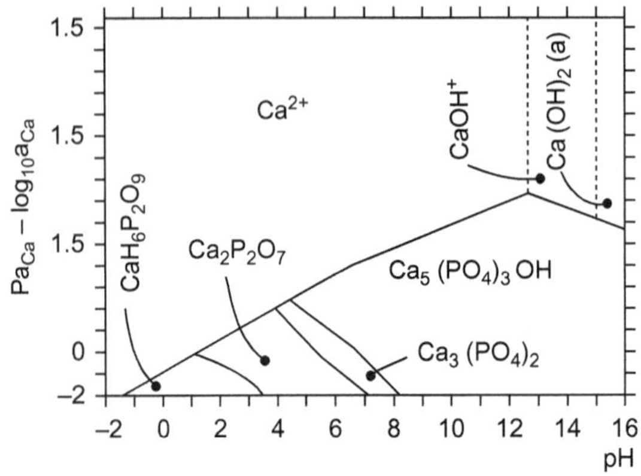

Figure 5.61 Pa_{Ca}–pH diagrams of the Ca–P–H_2O system at 25°C for $a_{Ca} = 1.67a_P$ [228].

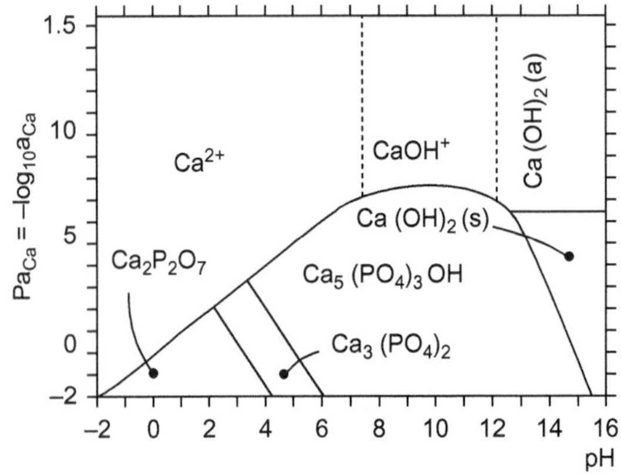

Figure 5.62 Pa_{Ca}–pH diagrams of the Ca–P–H_2O system at 300°C for $a_{Ca} = 1.67a_P$ [228].

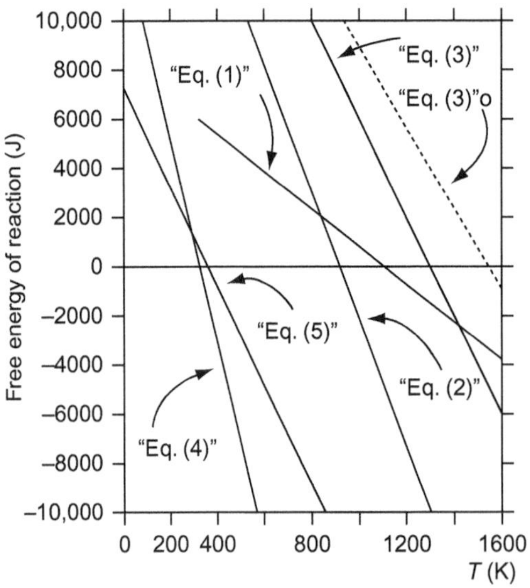

Figure 5.63 Free energy of reaction against temperature for some calcium phosphates according to Eqs (1)–(5) for water vapor fugacity equal to 0.03 atm, except for the dashed line, Eq. (3), which corresponds to water vapor fugacity equal to 1.0 atm [228].

pH range successively in the more acid direction. The stability of the calcium phosphates at higher temperatures can be seen in Figure 5.63. The equation numbers indicated in Figure 5.63 refer to the following reactions:

1. $3CaHPO_4$ (c) = $Ca_3(PO_4)_2$ (c) + H_3PO_4 (c, l)
2. $3CaH_2PO_4$ (c) = $Ca_3(PO_4)_2$ (c) + $4H_3PO_4$ (c, l)
3. $Ca_{10}H_2P_6O_{26}$ (c) = 3 $Ca_3(PO_4)_2$ (c) + CaO (c) + H_2O (g)
4. CaH_5PO_6 (c) = $CaHPO_4$ (c) + $2H_2O$ (g)
5. $CaH_6P_2O_9$ (c) = $Ca(H_2PO_4)_2$ (c) + H_2O (g)

The Eh–pH and Pa_{Ca}–pH diagrams have shown that HAp may be obtained at both atmospheric and hydrothermal conditions.

Section 4.6 discusses the intelligent engineering of crystal growth processes using thermochemical computation, and Figure 4.17 represents the stability field diagram generated for HAp crystal growth under hydrothermal conditions.

5.8.3 Crystal Growth

The crystal growth of HAp is a highly challenging area. The growth of large size crystals and the control of crystal morphology have evaded the crystal growers for decades. There are a large number of publications on the preparation of large HAp single crystals. Today a large variety of organics are employed in the HAp crystal growth for this purpose; to alter the crystal morphology to suit various applications. Table 5.18 gives the representative works on the hydrothermal synthesis of HAp

Table 5.18 Hydrothermal Synthesis of HAp Large Single Crystals [204]

Starting Materials	Synthetic Conditions	Comments
$CaHPO_4$ wet chemically prepared HAp	300°C, 1250 lb/in^2, for 10 days; 500°C, 354 bar, for 10 days	0.3 mm in length, 2.35 mm in length, 0.2 mm in diameter
$CaHPO_4$	350°C, 100 MPa	2 mm
$Ca_3(PO_4)_2 \cdot Ca(OH)_2$ or $Ca_3(PO_4)_2 \cdot Ca(OH)_2$–$H_2O$ system	15,000 psi, held at 850–900°C, then cooled at 20°C/h to ~750°C, quenched	4 mm in length
synthesized HAp as nutrient and HAp seed	300–670°C, 45,000–60,000 psi	$7 \times 7 \times 3$ mm^3, identify the retrograde solubility of HAp
$Ca_3(PO_4)_2 \cdot Ca(OH)_2$–$CaCO_3$–$H_2O$ system, $Ca(OH)_2$ as flux	Oscillating temperature technique, 750–800°C, 14,500–17,500 psi	8×0.5 mm^2
$CaHPO_4 \cdot 2H_2O$	100–300°C, 2 kbar	0.1 mm
Wet chemically, followed by Soxhlet extracted HAp	430 – 500°C, 2 kbar	3.5 mm, check the retrograde solubility of HAp

large single crystals [229–238]. Morey and Ingerson (1937) [229] first attempted the hydrothermal synthesis of HAp crystals. Since then, several reports have appeared in the literature. Roy (1971) [234] grew prismatic single crystals of HAp, up to 4 mm length hydrothermally, from melts in the system $3CaO{-}P_2O_5{-}Ca(OH)_2$ and $3CaO{-}P_2O_5{-}Ca(OH)_2{-}H_2O$. Mengeot et al. (1973) [235] first succeeded preparing HAp single crystals of up to $7 \times 3 \times 3$ mm^3 in size by controlling the temperature gradient. This work also revealed for the first time that HAp exhibited retrograde solubility in hydrothermal fluids between 300°C and 670°C under pressures of 3–4 kbar. With the appearance of this report, most of the researchers placed nutrients at the cooler zone. Here, dissolution occurs and HAp crystals grow at the hotter region where the seed crystals are placed. This arrangement facilitated the growth of larger HAp single crystals. However, the preparation of large single crystals of HAp has encountered several difficulties such as: the tendency to incorporate impurities, the stability of nonstoichiometric composition, and the restricted field of crystallization in the phase diagram $CaO{-}P_2O_5{-}H_2O$, where HAp and H_2O only exist.

In the recent years, Ito and coworkers [239,240] have done excellent work on the single crystal growth of HAp. Single crystals of carbonate-containing HAp have been grown hydrothermally by gradually heating with a temperature gradient applied to the vessel, using $CaHPO_4$, H_2O, and carbon dioxide as dry ice. The hydrothermal reaction for the formation of HAp can be written as follows:

$$10CaHPO_4 + 2H_2O \rightarrow Ca_{10}(PO_4)_6(OH)_2 + 4H_3PO_4$$

Crystal growth experiments were performed, slowly hydrolyzing $CaHPO_4$ by gradually heating a pressure vessel with a temperature gradient applied.

Kikuchi et al. (1994) [241] have obtained Sr-substituted HAp crystals of size $0.2 \times 0.2 \times 5$ mm^3 under hydrothermal conditions. The starting materials such as calcium and strontium hydrogen phosphate ($SrHPO_4$) were gradually heated with a temperature gradient. The crystal growth occurred at low pH.

In most of the works, the effects of CO_3^{2-} on the morphology of the crystals have been reported: Needle-like crystals are predominant if the carbonate content is higher than 0.1%, whereas equiaxial crystals occur if the carbonate content is below 0.1%. Hata et al. (1983) [242] synthesized Cd HAp single crystals, partly substituted with the first series transition elements for Cd ion, to investigate the substitution effects on the crystal growth and structural changes. These crystals form a superstructure of pure Cd HAp. Nagata et al. (1995) [243] have studied the synthesis of HAp crystals in the presence of methanol and the morphological variations in HAp crystals. It was found that the products obtained from the slurries without methanol were rod-like or granular crystals. Addition of methanol to the slurries caused an increase in the ratio of plate-like crystals to rod-like or granular crystals. When the weight of methanol added was equal to the weight of the slurries, only plate-like crystals were obtained.

Although several workers have reported a negative TC of soluble for HAp, there is no systematic study either on the experimental or theoretical side. In this respect,

Vereecke and Lemaitre (1990) [244] reported the calculation of the solubility diagrams and the effects of some parameters such as T ($< 100°C$), pH; partial CO_2 pressure, and ionic strength. According to this report, the solubility of HAp is lowered with temperature indicating the retrograde solubility of HAp below 100°C.

Onuma et al. (1998) [245] and Kanzaki et al. (1998) [246] have carried out the direct growth rate measurement of HAp single crystals by Moire phase shift interferometry and AFM. They carried out direct investigations of the growth kinetics for each crystal face of HAp. It was found that the a-face grew by step flow combined with two-dimensional nucleation, and the rate-determining process of the growth was found to be the incorporation of a growth unit at the step front, from the measurement of step velocities. The growth rate of the a-face was estimated to be of the order of 10^{-4} nm/s from the step advancing rate. This value is about 3–4 orders of magnitude smaller than those for soluble inorganic crystals [245]. In contrast, the growth kinetics of HAp (0001), c-face, remain unclear. These authors have concluded that the volume diffusion process for growth is strong during the initial stages. It is believed that a strong bond site exists in the [0001] direction and there is no strong bond site in the $[10\bar{1}0]$ direction, which makes the rate-determining process between c- and a-faces different. Figure 5.64 shows the carbonate-containing HAp single crystals grown on the polycrystalline HAp disk and Figure 5.65 shows carbonate-containing HAp single crystals.

Like zeolites, HAp is also an important technological material owing to its widespread applications attracting the attention of scientists from interdisciplinary branches of science—including physicists, chemists, ceramists, medical doctors, and engineers. In fact, the growth of nanosize HAp crystals has progressed so much that the nanocrystals of desired size, morphology, and surface chemistry could be obtained at ease. Also, several International Symposia on Apatite have

Figure 5.64 Carbonate-containing HAp single crystals grown on the polycrystalline HAp disk. *Source*: Photo courtesy of A. Ito.

Figure 5.65 Carbonate-containing HAp single crystals.
Source: Photo courtesy of A. Ito.

been held. The importance of this material as a biomaterial, composite material, whiskers, fine particles, and so on are discussed in Chapter 10 and 11.

References

[1] S.I. Tomkeieff, On the origin of the name "quartz", Min. Mag. 26 (1942) 172.
[2] R.B. Sosman, The Properties of Silica, American Chemical Society, Chemical Catalogue Co., Inc., New York, NY, 1927.
[3] E.C.T. Chao, J.J. Fahey, J. Littler, D.J. Milton, Stisshovite, SiO_2, a very high pressure new mineral from meteor crater, Arizona, J. Geophys. Res. 67 (1962) 419–423.
[4] C.S. Ramsdell, The crystallography of "coesite", Amer. Min. 40 (1955) 975.
[5] P. Wei, The structure of α-quartz, Zeit. Krist. 92 (1935) 355.
[6] W.H. Bragg, The X-ray spectra given by crystals of sulphur and quartz, Proc. Roy. Soc. 89 (1914) 575.
[7] M. Hosaka, T. Miyata, I. Sunagawa, Growth and morphology of quartz crystals synthesized above the transition temperature, J. Cryst. Growth 152 (1995) 300–306.
[8] J. Arndt, N. Rombach, Synthesis of coesite from aqueous solutions, J. Cryst. Growth 35 (1976) 28–32.
[9] T.I. Dyuzheva, L.M. Lityagina, N.A. Bendeliani, N.A. Nikolaev, G.I. Dorokhova, Hydrothermal synthesis of coesite (SiO_2), Crystallogr. Rep. 43 (1998) 511–513.
[10] T.F.W. Barth, The cristobalite structure: I high-cristobalite; II low-cristobalite, Am. J. Sci. 23 (1932) 35024:97.
[11] X. Buisson, R. Arnaud, Hydrothermal growth of quartz crystals in industry, present status and evolution, J. de Phesique IV, 4[C_2] (1994) 25–32.

[12] S. Taki, Improvement of growth process and characterization of quartz crystals, in: N. Nizeki (Ed.), The Role of Crystal Growth for Device Development, Prog. Crystal Growth Charact. Mater., 23, 1991, pp. 313–339.
[13] K. Byrappa, Growth of quartz crystals, in: P. Capper (Ed.), Bulk Crystal Growth of Electronic, Optical and Optoelectronic Materials, John Wiley & Sons, Ltd., Chichester, UK., London, UK, 2005, pp. 387–406.
[14] K. Byrappa, Hydrothermal growth of polyscale crystals, in: Dhanraj, Byrappa, Prasad, Dudley (Eds.), Springer Handbook of Crystal Growth, Springer, Heidelberg, Germany, 2010, pp. 599–653.
[15] R.A. Laudise, J.W. Nielsen, in: F. Seitz, D. Turnball (Eds.), Solid State Physics, 12, Academic Press, New York, 1961.
[16] A.A. Ballman, R.A. Laudise, in: J.J. Gilman (Ed.), Art and Science of Growing Crystals, Wiley, New York, NY, 1963, p. 231.
[17] V.A. Kuznetsov, A.N. Lobachev, Hydrothermal method for the growth of crystals, Soviet Phys. Crystallogr. 17 (1973) 775–804 (Translated English edition).
[18] A.F. Armington, Recent advances in the growth of high quality quartz, in: K. Byrappa (Ed.), hydrothermal growth of crystals, Prog. Crystal Growth Charact. Mat., 21, 1990, pp. 97–111.
[19] K. Byrappa, Hydrothermal growth of crystalsvol. 2 in: D.T.J. Hurle (Ed.), Handbook of Crystal Growth, 464, Elsevier Science Publishers, The Netherlands, 1994.
[20] K.F.E. Schafhaul, Gelehrte Anzeigen Bayer Akad. 20 (1845) 557.
[21] R. Nacken, Hydrothermal syntheses als grundlage fur zuchtung von quarz-kristallen, Chem. Z. 74 (1950) 745–749.
[22] R. Nacken, Artificial Quartz Crystals, etc., US Office of Technical Services Report, PB-6948, 1946.
[23] R. Nacken, Artificial Quartz Crystals, etc., US Office of Technical Services Report, PB-18-748 and 28-897, 1946.
[24] R.A. Laudise, Hydrothermal crystal growth—some recent results, in: P.M. Dryburgh, B. Cockayne, K.G. Barraclough (Eds.), Adv. Crystal Growth, Prentice Hall, New York, NY, 1987, pp. 267–288.
[25] R.A. Laudise, R.A. Sullivan, Pilot plant production of synthetic quartz, Chem. Eng. Prog. 55 (1959) 55–59.
[26] R.L. Barns, E.D. Kolb, R.A. Laudise, E.E. Simpson, K.M. Kroupa, Production and perfection of "*Z-face*" quartz, J. Cryst. Growth 34 (1976) 189–197.
[27] G.K. Kirov, Initial growth stages of man-made quartz crystals and growth mechanism of their (0001) face, Cryst. Res. Technol. 27 (1992) 335–342.
[28] C.S. Brown, Internal friction in synthetic quartz, Proc. Phys. Soc. (London) 75 (1960) 459.
[29] D. Chakraborty, Dependence of mechanical Q on growth rate of quartz single crystals, J. Cryst. Growth 41 (1977) 177–180.
[30] N.C. Lias, E.E. Grudenski, E.D. Kolb, R.A. Laundise, The growth of high acoustic Q quartz at high growth rates, J. Cryst. Growth 18 (1973) 1–6.
[31] H.G. Lipson, A.F. Armington, Aluminum a–d hydroxide distribution in cultured quartz grown from $^{+}$x seeds, J. Cryst. Growth 80 (1987) 51–59.
[32] I.R.A. Christie, D.F. Croxall, B.J. Isherwood, UK Patent 2079175B, December 7, 1983.
[33] A.F. Armington, J.J. Larkin, J.J. O'Conner, J.E. Cormier, J.A. Horrigan, Crystal Growth for Hydro Thermal Handbook, in: G.E. McGuire, S.M. Rossingel, R.F. Bunshah (Eds.), Proceedings of the 35th Annual Frequency Symposium, Arthur Ballato (Ed.), Effect of seed treatment on quartz dislocations, El A, Washington, DC, 1981, p. 297.

[34] A.F. Armington, J.J. Larkin, The growth of high purity, low dislocation quartz, J. Cryst. Growth 71 (1985) 799–806.
[35] E.D. Kolb, K. Nassau, R.A. Laudise, E.E. Simpson, K.M. Kronpa, New sources of quartz nutrient for the hydrothermal growth of quartz, J. Cryst. Growth 36 (1976) 93–100.
[36] M. Hosaka, T. Miyata, Hydrothermal growth of α-quartz using high-purity α-cristobalite as feed material, Mat. Res. Bull. 28 (1993) 1201–1208.
[37] V.A. Avakov, B.N. Vinogradov, Solubility of SiO_2 polymorphs, Izv. Vyssh. Ucheb. Zaved. Khim. Teknol. 17 (1972) 879–882.
[38] M. Hosaka, Synthesis of micro α-quartz crystals by hydrothermal hot-press method, J. Cryst. Growth 112 (1991) 291.
[39] J.G. Spezia, La Pressione e chimicament inattiva nella solubilita e ricostituzione del Quarzo, Atti Accad. Sci. Torino Contribuzioni di Geologia Chimica 40 (1905) 254–262.
[40] N. Wooster, W.A. Wooster, Preparation of synthetic quartz, Nature 157 (1946) 297.
[41] A.A. Shternberg, Controlling the growth of crystals in autoclaves, in: A.N. Lobachev (Ed.), Crystallization Processes Under Hydrothermal Conditions, Consultants Bureau, New York, NY, 1973, pp. 225–240.
[42] L.N. Demianets, E.N. Emelyanova, O.K. MelniKoV, Solubility of sodalite in aqueous solutions of NaOH under hydrothermal conditions, in: A.N. Lobachev (Ed.), Crystallization Processes Under Hydrothermal Conditions, Consultants Bureau, New York, NY, 1973, pp. 125–150.
[43] M. Hosaka, S. Taki, Hydrothermal growth of quartz crystals in NaCl solution, J. Cryst. Growth 100 (1990) 343–346.
[44] R.A. Laudise, A.A. Ballman, The solubility of quartz under hydrothermal conditions, J. Phys. Chem. 65 (1961) 1396–1400.
[45] M. Hosaka, S. Taki, Hydrothermal growth of quartz crystals in NaCl solution, J. Cryst. Growth 52 (1981) 837–842.
[46] F. Lafon, G. Demazean, Pressure effects on the solubility and crystal growth of α-quartz, J. de Physique IV, 4[C_2] (1994) 177–182.
[47] J.J. Martin, A.F. Armington, Effect of growth rate on quartz defects, J. Cryst. Growth 62 (1983) 203–206.
[48a] R.A. Laudise, What is materials chemistry? in: S.R. Marder, J.E. Sohn, G.D. Stucky (Eds.), Materials for Nonlinera Optics (Chemical Perspectives), American Chemical Society, Washington, DC, 1991, pp. 410-433.
[48b] R.A. Laudise, A.A. Ballman, J.C. King, Impurity content of synthetic quartz and its effect upon mechanical Q, J. Phys. Chem. Solids 26 (1965) 1305-1308.
[49] J. Yoshimura, T. Miyazaki, T. Wada, K. Kohra, T. Ogawa, S. Taki, J. Cryst. Growth 46 (1979) 691–700.
[50] F. Iwasaki, M. Kurashige, $BaTiO_3$ thin film by hydrothermal electrochemical method, J. Appl. Phys. 17 (1978) 817–820.
[51] P.M. Dove, N. Han, J.De Yoreo, Mechanism of classical crystal growth theory explain quartz and silicate dissolution behavior, Proc. Nat. Acad. Sci. U.S.A. 102 (2005) 15357–15362.
[52] G.T. Ostapenko, B.M. Mitsyuk, Asymmetry of growth and dissolution on basal, minor rhombohedral and prism faces of quartz, J. Cryst. Growth 294 (2006) 330–338.
[53] G.R. Johnson, R.A. Irvine, J.W. Foise, A parametric study of the variables involved in quartz growth, in: IEEE Proceedings of the 44th Annual Symposium on Frequency Control, Baltimore, MD, 1990, pp. 216–221.

[54] G. Demazeau, F. Lafon, Hydrothermal crystal growth of α-quartz: new specificities correlated to applications, J. de Physique IV, 4[C_2] (1994) 13–18.
[55] H.G. Lipson, A. Kahan, J. O' Connor, Aluminum and hydrogen defect centers in vacuum swept quartz, in: IEEE Proceedings of the 37th Annual Symposium on Frequency Control, Bedford Mas., New York, 1983, pp. 169–173.
[56] J.G. Gualtieri, The influence of temperature and electric field on the etch-channel density in swept cultured quartz, in: IEEE Proceedings of the 39th Annual Symposium on Frequency Control, New York, 1985, pp. 247–254.
[57] J. Martin, Electro-diffusion or sweeping of ions in quartz, in: IEEE Proceedings of the 41st Annual Symposium on Frequency Control, Philadelphia, USA, 1987, pp. 167–174.
[58] B. Fraser, Factors influencing the acoustic properties of vitreous silica, in: W.P. Mason (Ed.), Physical Acoustics 5 (54) Academic Press, NY, (1968) Ch. 2.
[59] H. Bahadur, Sweeping and irradiation effects on hydroxyl defects in crystalline natural quartz, IEEE Trans. Ultrason. Ferr. 41 (1994) 820–833.
[60] P. Zecchini, K. Yamni, B. Viard, Modifications induced by sweeping and observed in quartz crystals according to their α_{3500}-value, J. de Physique IV, 4[C_2] (1994) 189–194.
[61] M. Deleuze, O. Cambon, A. Goiffon, A. Ibanez, E. Philippot, Controlled dissolution of quartz material, Part I, controlled dissolution of quartz plates in potassium and sodium hydroxides: chemical thinning down of SC cuts, J. de Physique IV, 4[C_2] (1994) 79–84.
[62] K. Brauer, E. Muller, Correlation between parameters of plastic deformation in doped NaCl Crystals, Cryst. Res. Technol. 19 (1984) K101–109.
[63] O. Cambon, M. Deleuze, J.P. Michel, J.P. Aubry, A. Goiffon, E. Philippot, Controlled dissolution of quartz material, Part II, quartz chemical etching applied to blanks industrial manufacturing, J. de Physique IV, 4[C_2] (1994) 85–91.
[64] N. Takahashi, T. Nakumura, S. Nonaka, Y. Kubo, Y. Sinriki, K. Tamanuki, Growth of crystalline quartz films with AT-cut plane by means of catalyst-enhanced vapor phase epitaxy under atmosphere pressure, J. Phys. Chem. Solids 66 (2005) 1145–1149.
[65] B. Iliescu, I. Enculescu, I. Pera, G. Alexe, S. Polosan, A. Stanculescu, Chemical composition of structures obtained inside quartz crystals by sodium electrodiffusion, Cryst. Res. Technol. 36 (2001) 403–410.
[66] Y. Murakami, S. Chiashi, Y. Miyauchi, M. Hu, M. Ogura, T. Okubo, et al., Growth of vertically aligned single-walled carbon nanotube films on quartz substrates and their optical anisotropy, Chem. Phys. Lett. 385 (2004) 298–303.
[67] D. Schwarzenbach, Verfeinezung der Struktur der Tiefquarz-Modification von $AlPO_4$, Z Krist. 123 (1966) 161–185.
[68] Z.P. Chang, G.R. Barsch, Elastic constants and thermal expansion of berlinite, IEEE Trans. Sonics Ultrason. 23 (1976) 127–130.
[69] R.M. O'Connell, P.H. Corr, Temperature compensated cuts of berlinite and β-eucryptite: for saw devices, in: Proceedings of the 31st Annual Frequency on Control Symposium, Atlantic City, NJ, 1977, pp. 182–186.
[70] E.D. Kolb, A.M. Glass, R.L. Rosenberg, J.C. Grenier, R.A. Laudise, Dielectric and piezoelectric properties of aluminum phosphate, in: Proceedings of the 35th Ultrasonic Symposium, New York, NY, 1981.
[71] J. Vogel, W. Holand, W. Vogel, US Patent 4698318, assigned to Jenaer Glaswerk, October 6, 1987.

[72] W. Vogel, J. Vogel, W. Holand, P. Wange, Development of bioactive silica-free phosphate glass ceramics for use in medicine, wiss. Z. friedrich schiller univ., Jena Math. Nat. Reihe 36 (1987) 841–854.
[73] C.W. Blomstrand, Om westana mineralier, Öfversigt at Kongliga Vetenskaps-Akademiens Handlingar 3 (1868) 197–212.
[74] B.P. Onac, W.B. White, First reported sedimentary occurrence of berlinite ($AlPO_4$) in phosphate-bearing sediments from Cioclovina Cave, Romania, Am. Miner. 88 (2003) 1395–1397.
[75] V.M. John, Kordes, Chem. Earth 70 (1953) 75–89.
[76] J.M. Stanley, Hydrothermal synthesis of large aluminum phosphate crystals, Ind. Eng. Chem. 468 (1954) 1684–1689.
[77] W.P. Mason, Piezoelectric Crystals and Their Application Ultrasonic, Van Nostrand, New York, NY, 1950.
[78] E.D. Kolb, R.A. Laudise, Hydrothermal synthesis of aluminum orthophosphate, J. Cryst. Growth 43 (1978) 313–319.
[79] W.R. Beck, Crystallographic inversions of the aluminum orthophosphate polymorphs and their relation to those of silica, J. Am. Ceram. Soc. 32 Chicago, Ill., (Refractories Division, No. 22), (1949) 147–153.
[80] R.S. Sosman, Phases in Silica, Rutger University Press, New Brunswick, NJ, 1964.
[81] S. Srikantaswamy, Synthesis and Characterization of ABO_4 Crystals, Ph.D. Thesis, University of Mysore, Mysore, India, 1988.
[82] K. Byrappa, S. Srikantaswamy, Recent progress in the growth and characterization of aluminum orthophosphate crystals, Prog. Cryst. Growth Charact. 21 (1990) 199–254.
[83] K. Byrappa, V. Venkatachalapathy, B. Puttaraj, Crystallization of orthophosphate, J. Mater. Sci. 19 (1984) 2855–2862.
[84] K. Byrappa, S. Srikantaswamy, G.S. Gopalakrishna, V. Venkatachalapthy, Infrared spectra of aluminum orthophosphate crystals, J. Mater. Sci. Lett. 5 (1986) 203–205.
[85] R.A. Laudise, in: Kaldis (Ed.), Crystal Growth of Electronic Materials, Elsevier Science Publ., 1985, p. 159.
[86] B.H.T. Chai, M.L. Shand, I. Bucher, M.A. Gillee, Hydrothermal growth of polyscale crystals, in: G. Dhanaraj, K Byrappam V. Prasad, M. Dudley (Eds.), Proceedings of the IEEE Ultrasonic Symposium New Orleans, LA, 1979, p. 557.
[87] B.H.T. Chai, US Patent, 4,382,840 (May 10, 1983).
[88] D.F. Croxall, L.R.A. Christie, B.J. Isherwood, A.G. Todd, J. Brich, The growth and assessment of berlinite single crystals, in: G. McGuire (Ed.), Proceedings of the Second European Conference on Crystal Growth, Lancaster, England, C 118, September 10–15, 1979.
[89] H. Schulz, R. Kluckou, in: H.S. Peiser (Ed.), Crystal Growth, Pergamon Press, New York, NY, 1967.
[90] H. Poignant, Hydrothermal Synthesis of Berlinite, Ph.D. Thesis, University of Rennes, Rennes, France, 1979.
[91] E.D. Kolb, J.C. Grenier, R.A. Laudise, Solubility and growth of $AlPO_4$ in a hydrothermal solvent: HCl, J. Cryst. Growth 51 (1981) 178–185.
[92] K. Bryappa, S. Srikantaswamy, G.S. Gopalakrishna, V. Venkatachalapath, Influence of admixtures on the crystallization and morphology of $AlPO_4$ crystals, J. Mater. Sci. 21 (1986) 2202–2206.
[93] V. Krauss, G. Lehann, Z. Naturforsch 309 (1975) 28.
[94] J. Detaint, E. Phillipport, J.C. Jumas, J. Schwartzel, A. Zarka, B. Capelle, J.C. Doukhan, A transmission electron microscope study of lattice defects and water

precipitation in α-berlinite AIPO4, in: 39th Annual Frequency Control Symposium, IEEE, Philadelphia, PA, 1985.

[95] K. Byrappa, J. Shashidhara Prasad, S. Srikantaswamy, Growth and properties of new polymorphic modification of $AlPO_4$, J. Cryst. Growth 79 (1986) 232.

[96] J.C. Jumas, A. Goiffon, B. Capelle, A. Zaraka, J.C. Doukhan, J. Schwartzel, et al., Crystal growth of berlinite, $AlPO_4$: physical characterization and comparison with quartz, J. Cryst. Growth 80 (1987) 133–148.

[97] H. Poignant, L.Le. Marechal, Y. Toudic, Etude De La Solubilite du Phosphate d'aluminium ($AlPO_4$) dans des Solutions Hydrothermales d'acide orthophosphoriaue H_3PO_4, Mater. Res. Bull 14 (1979) 603–612.

[98] K. Hasegawa, K. Minegishi, S. Somiya, Effects of starting materials and temperature on the hydrothermal synthesis of aluminum orthophosphate, in: S. Somiya (Ed.), Proceedings of the First International Symposium on Hydrothermal Growth, Tokyo, 1982, pp. 509–518.

[99] I.M. Yaroslavskii, V.I. Popolitov, Growth of single crystals of aluminum monophosphate (berlinite), Izv. Akad. Nauk SSSR, Neorg. Materialy 26 (1990) 1055–1059.

[100] A. Zarka, B. Capelle, Observations of growth defects in berlinite by high temperature X-ray topography, Appl. Cryst. 18 (1985) 533–534.

[101] J. Detaint, H. Poignant, Y. Toudic, Experimental thermal behavior of berlinite resonators, in: 34th Symposium on Frequency Control, May, Munchen, Germany 1980, New York, pp. 28–30.

[102] R.F. Steinberg, M.K. Roy, A.K. Ester, B.H.T. Chai, R.C. Morris, Advances in crystal growth and characterizations of gallium orthophosphate, GaPO4, in: 36th Annual Ultrasonic Symposium, IEEE, Dallas, TX, November 1984, pp. 14–16.

[103] R.A. Laudise, The Growth of Single Crystals, Prentice Hall, Englewood Cliffs, NJ, 1970.

[104] R.A. Laudise, in: N.B. Hanvy (Ed.), Treatise on Solid State Chemistry, 5, Plenum, New York, NY, 1970.

[105] J. Detaint, E. Philippot, J.C. Jumas, J. Schwartzel, A. Zarka, B. Capelle, et al., Crystal growth, physical characterization and BAW devices applications of berlinite, Third Annual Frequency Control Symposium, Philadelphia, PA, May 29–31, 1985.

[106] R.C. Morris, B.H.T. Chai, Imperfect low-angle boundaries and fracture in hydrothermally grown berlinite crystals, J. Cryst. Growth 191 (1998) 108–112.

[107] K. Byrappa, S. Srikantaswamy, G.S. Gopalakrishna, V. Venkatachalapathy, Influence of admixtures on the alpha-beta berlinite inversion, J. Mater. Sci. Lett. 5 (1986) 347–348.

[108] M. Troccaz, C. Berger, M. Richard, L. Byraud, Etude de la transformation de phase α–β de la variété "Phosphoquartz" de l'orthophosphate d'aluminium $AlPO_4$, Bull. Soc. Chim. de France 11 (1967) 4256.

[109] R. Lang, W.R. Datars, C. Calvo, Phys. Lett. 30A (1969) 340.

[110] K. Byrappa, U.D. Prahallad, Thermal expansion of berlinite, J. Mat. Sci. Lett. 8 (1989) 1667–1669.

[111] R. Bechman, A.D. Ballato, T.T. Lukaszek, High order temperature coefficients of the elastic stiffness and compliance of α-quartz, Proc. IREE 30 (1982) 1812.

[112] R.D. Shannon, A.J. Vega, B.H.T. Chai, G.R. Rossman, Effect of H_2O on dielectric properties of berlinite: I. Dielectric constant, J. Phys. D Appl. Phys. 26 (1993) 93–100.

[113] J.J. Fontanella, M.C. Wintersgill, R.D. Shannon, B.H.T. Chai, Effect of H_2O on dielectric properties of berlinite: II. Dielectric loss, J. Phys. D Appl. Phys. 26 (1993) 101–105.

[114] W.K. Thompson, Infrared spectroscopic studies of aqueous systems, I, Trans. Faraday Soc. 61 (1964) 1635–1640.

[115] N. Garg, S.M. Sharma, A molecular dynamical investigation of high pressure phase transformations in berlinite (α-$AlPO_4$), J. Phys. Condens. Matter 12 (2000) 375–397.

[116] M. Moret, S. Ruffenach, O. Briot, B. Gil, M. Pauthe, The epitaxial growth of indium nitride using berlinite ($AlPO_4$) and other piezoelectric crystals of the quartz family as substrates, Appl. Phys. Lett. 95 (2009) 041910.

[117] E. Philippot, D. Palmier, M. Pintard, A. Goiffon, A general survey of quartz and quartz-like materials: packing distortions, temperature and pressure effects, J. Solid State Chem. 123 (1996) 1–13.

[118] S. Hirano, K. Miwa, S. Naka, Growth of gallium orthophosphate crystals, J. Cryst. Growth 79 (1986) 215–218.

[119] H. Nakae, K. Kihara, M. Okuno, S. Hirano, The crystal structure of the quartz-type form of $GaPO_4$ and its temperature dependence, Zeit. Krist. 210 (1995) 746–753.

[120] S. Hirano, P. Kim, Physical properties of hydrothermally grown gallium orthophosphate single crystals, J. Mater. Sci. 25 (1990) 4772–4775.

[121] S. Hirano, P. Kim, H. Orihara, Dielectric properties of hydrothermally grown gallium orthophosphate single orgstals, J. Mat. Sci. 25 (1990) 2800–2804.

[122] M. Cochez, A. Ibanez, A. Goiffon, E. Philippot, Crystal growth and infrared characterization of $GaPO_4$ in phospho-sulphuric media, Eur. J. Solid State Inorg. Chem. 30 (1993) 509–519.

[123] K. Jacobs, P. Hofmann, D. Klimm, J. Reichow, M. Schneider, Structural phase transformations in crystalline gallium orthophosphate, J. Solid State Chem. 149 (2000) 180–188.

[124] J.D. Foulon, J.C. Giuntin, E. Philippot, Dielectric properties of an α-quartz type material: $GaPO_4$, Eur. J. Solid State Inorg. Chem. 31 (1994) 245–256.

[125] S. Hirano, P. Kim, H. Orinara, Dielectric properties of hydrothermally grown gallium orthophosphate single crystals, J. Mat. Sci. 25 (1990) 2800–2804.

[126] M. Cochez, J.D. Foulon, A. Ibanez, A. Goiffon, E. Philippot, B. Cappelle, et al., Crystal growth and characterization of a quartz-like material: $GaPO_4$, J. de Physique IV, 4[C_2] (1994) 183–188.

[127] B. Capelle, A. Zarka, J. Schwartzel, J. Detaint, E. Philippot, J.P. Denis, Characterization of piezoelectric materials: old and new crystals, J. de Physcique IV, 4[C_2] (1994) 123–134.

[128] O.V. Zvereva, Yu.-M. Mininzon, Growth elucidation of gallium orthophosphate under hydrothermal conditions, Kristallografia 37 (1992) 1051–1054.

[129] V.I. Popolitov, I.M. Yaroslavskii, Crystallization of gallium monophosphate, Izv. Akad. Nauk SSSR, Neorg. Materialy 26 (1990) 892–894.

[130] P. Yot, O. Cambon, D. Balitsky, A. Goiffon, E. Philippot, B. Capellie, et al., Advances in crystal growth and characterizations of gallium orthophosphate, GaPO4, J. Cryst. Growth 224 (2001) 294–302.

[131] O.V. Zvereva, Yu, M. Mininzon, L.N. Demianets, Hydrothermal growth of OH^- free $AlPO_4$ and $GaPO_4$ crystals the way of twin reducing, J. de Physique IV, 4[C_2] (1994) 19–24.

[132] A.S. Shigarov, O.V. Zvereva, Yu.M. Mininzon, Investigation of the solubility of gallium orthophosphate, $GaPO_4$ under hydrothermal conditions, Kristallografia 39 (1994) 717–719.

[133] POLECER, Guide on the best piezoelectric properties of lead-free single crystals, Prepared for POLECER. Thematic Network Workpackage 9 (30) (2004).

[134] F.C. Zumsteg, J.D. Bierlein, T.E. Gier, K_xRb_{1-x} $TiOPO_4$ a new nonlinear optical material, J. Appl. Phys. 47 (1976) 4980–4985.
[135] A. Ashkin, G.D. Boyd, J.M. Dziedzic, R.E. Smith, A.A. Ballman, J.J. Levinstein, et al., Optically-induced refractive index in homogeneities in $LiTaO_3$ and $LiNbO_3$, Appl. Phys. Lett. 9 (1966) 72–77.
[136] M. Okada, S. Ieiri, Influence of self-induced thermal effects on phase matching in nonlinear optical crystals, IEEE J. Quantum Electron. QE-7 (1971) 560–563.
[137] J.D. Bierlein, US Patent 3949323, Assigned to E.I. Du Pont de Nemours and Company, April 6, 1976.
[138] T.E. Gier, US Patent 4231838, Assigned to E.I. Du Pont de Nemours and Company, November 4, 1980.
[139] I. Tordjman, R. Masse, J.C. Guitel, Structure cristalline du monophosphate $KTiPO_5$, Z. Kristallogr. 139 (1974) 103–115.
[140] L. Ouvard, Compt. Rend. (Paris) 121 (117) (1890), as reported in The Gmelin Handbook, vol. 41, T, 8th edition, maintained by Elsevier MDL, 1951.
[141] R. Masse, J.C. Grenier, Phase transformation of $AlPO_4$:Fe^{3+} by electron paramagnetic resonance, Bull. Soc. France Miner. Cryst. 94 (1971) 437.
[142] J.D. Bierlien, T.E. Gier, US Patent 3949323, April 6, 1976.
[143] Y.S. Liu, L. Drafall, D. Dentz, R. Belt, Nonlinear optical phase-matching properties of $KTiOPO_4$ (KTP), G.E. Tech. Inf. Series Rep. No. 82 CR DO16, General Electric Company, Schenectady, NY, 1982.
[144] Y.S. Liu, D. Dentz, R. Belt, High-average-power intracavity second-harmonic generation using $KTiOPO_4$ in an acousto-optically Q-switched Nd:YAG laser oscillator at 5 KHz, Opt. Lett. 9 (1984) 76.
[145] R.A. Laudise, R.J. Cava, A.J. Caporaso, Phase relations, solubility and growth of potassium titanyl phosphate, KTP, J. Cryst. Growth 74 (1986) 275–281.
[146] R.A., Laudise, US Patent 4654111, March 31, 1987.
[147] R.F. Belt, T. Iradi, Hydrothermal growth produces large KTP nonlinear crystals, Laser Focus World (Nov. 1993) 155–162.
[148] R.A. Laudise, W.A. Sunder, R.F. Belt, G. Gashurov, Solubility and P–V–T relations and the growth of KTP, J. Cryst. Growth 102 (1990) 427–433.
[149] Z.L. Zhang, L.X. Huang, W.N. Zhou, G. Zhang, H.D. Hou, Q.F. Ruan, et al., Growth of KTP crystals with high damage threshold by hydrothermal method, J. Cryst. Growth 292 (2006) 364–367.
[150] S.Q. Jia, H.D. Niu, J.G. Tan, Y.P. Xu, Y. Tao, Hydrothermal growth of KTP crystals in the medium range of temperature and pressure, J. Cryst. Growth 99 (1990) 900–904.
[151] R.F. Belt, Low temperature hydrothermal growth of potassium titanyl phosphate (KTP) in 4 liter autoclaves, Extended Abstract ICCG-9, Sendai, Japan, August 20–25, 1989, p. 428.
[152] K. Byrappa, Board of Research in Nuclear Sciences, Department of Atomic Energy, Technical Report, Bombay, 1996, p. 4.
[153] M.N. Satyanarayan, A. Deepthy, H.L. Bhat, Potassium titanyl phosphate and its isomorphs: growth, properties and applications, Crit. Rev. Solid State 24 (1999) 103-189.
[154] D.J. Gettemy, W.C. Harker, G. Lindholm, N.P. Barnes, Some optical properties of KTP, $LiIO_3$, and $LiNbO_3$, IEEE J. Quantum Electron. 24 (1988) 2231–2237.

[155] J.D. Bierlein, A. Ferretti, L.H. Brixner, W.Y. Hsu, Fabrication and characterization of optical wave guides in $KTiOPO_4$, Appl. Phys. Lett. 50 (1987) 1216–1219.

[156] J.D. Bierlein, H. Vanherzeele, A. Ballman, Linear and nonlinear optical properties of flux growth KTP, Appl. Phys. Lett. 54 (1989) 783–787.

[157] J. Durand, M. Brahimi, Structure et propriétés d' optique nonlinéaire de $KTiOAsO_4$, Rev. Chim. Miner. 23 (1986) 146–148.

[158] R.F. Belt, J.B. Ings, Hydrothermal growth of potassium titanyl arsenate (KTA) in large autoclaves, J. Crystal Growth 128 (1993) 956–962.

[159] L.K. Cheng, L.T. Cheng, J. Galperin, P.A. Morris Hotsenpiller, J.D. Bierlein, Crystal growth and characterization of $KTiOPO_4$ isomorphs from the self-fluxes, J. Cryst. Growth 137 (1994) 107–115.

[160] V.I. Chani, K. Shimamura, S. Endo, T. Fukuda, Substitution of Ti^{4+} with $Nb^{5+}-M^{3+}$ (M = Al, Cr, Ga, Fe, In) in Crystals of $KTiOAsO_4$, J Cryst. Growth 17 (1997) 117–122.

[161] W.A. Deer, R.A. Howie, J. Zussman, An Introduction to the Rock- Forming Minerals, Longman Group Limited, England, 1966.

[162] D.R. Kinloch, R.F. Belt, R.C. Puttbach, Hydrothermal growth of calcite in large autoclaves, J Cryst. Growth 24/25 (1974) 610–613.

[163] S. Hirano, K. Kikuta, Solubility and hydrothermal growth of calcite single crystals in hitrate solutions, J. Cryst. Growth 79 (1986) 223–226.

[164] J. Lemberg, Ueber silicatumwandlungen, Z. Deut. Geol. Ges. 35 (1883) 557–618.

[165] C. Friedel, E. Sarasin, Bull. Soc. Min. 8 (1885) 304–305.

[166] P.M. Gruzensky, in: H.S. Peiser (Ed.), Crystal Growth, Pergamon, Oxford, 1967, p. 365.

[167] J. Kaspar, Growth of Crystals, vol. 2, Consultants Bureau, New York, NY, 1959.

[168] H. Morse, D.H. Donnay, Bull. Soc. France Miner. 54 (1931) 19–25.

[169] H.J. Nickl, H.K. Henisch, Growth of calcite crystals in gels, J. Electrochem. Soc. 116 (1969) 1258–1264.

[170] J.F. Nester, J.B. Schroeder, Growth of synthetic calcite single crystals, Am. Miner. 52 (1967) 276–280.

[171] C. Belin, J.J. Brissot, R.F. Jesse, The growth of calcite single crystals by travelling solvent zone melting, J. Cryst. Growth 13/ 14 (1972) 597–600.

[172] J.F. Balascio, W.B. White, Growth of single crystal calcite by top-seeded solution growth technique, J. Cryst. Growth 23 (1974) 101–105.

[173] C. Belin, On the growth of large single crystals of calcite by travelling solvent zone melting, J. Cryst. Growth 34 (1976) 341–345.

[174] N. Yamasaki, T. Weiping, Hydrothermal hot pressing of calcium carbonate with sea water, J. Mat. Sci. Lett. 12 (1993) 516–519.

[175] N. Yamasaki, T. Weiping, K. Yanagisawa, Solidification of $CaCO_3$ containing $SrCO_3$ by hydrothermal hot pressing, J. Mat. Res. 8 (1993) 1972–1976.

[176] N. Yamasaki, T. Weiping, H. Lei, K. Hosoi, Solidification of aragonite-type $CaCO_3$ powder containing chitosan with acetic acid by hydrothermal hot pressing, J. Mat. Sci. Lett. 14 (1995) 1751–1753.

[177] N.Yu. Ikornikova, Growth characteristics of calcite crystals in aqueous solutions of carbonic acid, in: A.N. Lobachev (Ed.), Crystallization Processes Under Hydrothermal Conditions, Consultants Bureau, New York, NY, 1973, pp. 93–112.

[178] N.Yu. Ikornikova, Hydrothermal Synthesis of Crystals in Chloride Systems (in Russian), Nauka, Moscow, 1975.

[179] M. Higuchi, A. Takeuchi, K. Kodaira, Hydrothermal growth of calcite single crystals from $H_2O-CO_2-CaCO_3$ system, J. Cryst. Growth 92 (1988) 341–343.
[180] S. Hirano, K. Kikuta, Hydrothermal growth of calcite single crystal in $Ca(NO_3)_2$ andNH_4O_3 solutions, J. Cryst. Growth 94 (1989) 351–356.
[181] M. Higuchi, D. Miyauchi, A. Takeuchi, K. Kodaira, Hydrothermal growth of calcite single crystals from $H_2O-CO_2-CaCO_3$ system under controlled CO_2 pressure, J. Cryst. Growth 110 (1992) 39–43.
[182] K. Yanagisawa, Q. Feng, K. loku, N. Yamasaki, Hydrothermal single crystal growth of calcite in ammonium acetate solution, J. Cryst. Growth 163 (1996) 285–294.
[183] S. Hirano, T. Yogo, K. Kikuta, Y. Yoneta, Hydrothermal growth of calcite crystals in nitrate solutions, J. Ceram. Soc. Jpn. 101 (1993) 113–117.
[184] S.D. Malinin, Geochemical Investigations at High Temperatures and Pressures, Nauka, Moscow, 1965.
[185] A.I. Ellis, The solubility of calcite in NaCl solutions at high temperatures, Am. J. Sci. 261 (1963) 259–267.
[186] K. Kikuta, S. Hirano, Hydrothermal growth and dissolution behaviour of calcite single crystal in nitrate solutions, J. Cryst. Growth 99 (1990) 895–899.
[187] K. Yanagisawa, K. Kageyama, Q. Feng, I. Matsushita, Improvement of quality of hydrothermally grown calcite single crystals, J. Cryst. Growth 229 (2001) 440–444.
[188] I.V. Nefyodova, N.I. Leonyuk, I.A. Kamenskikh, Hydrothermal synthesis and optical properties of calcite single crystals, J. Optoelectron. Adv. Mater. 5 (2003) 609–613.
[189] Y.-K. Lee, S.J. Chung, Hydrothermal growth of calcite single crystals in NH_4Cl solution, J. Cryst. Growth 192 (1998) 350–353.
[190] M. Gorai, K. Syuutou, Tikaku Ganseki Koukutsu, Kyuyitsu Publishing House, Japan, 1981.
[191] N.Yu. Ikornikova, Formation and crystal growth of trigonal carbonates, Hydrothermal Synthesis of Crystals, Nauka, Moscow, 1968.
[192] N.Yu. Ikornikova, Crystal growth elucidation of the calcite crystals in aqueous solutions of HCl, Investigations of the Crystallization Processes Under Hydrothermal Conditions, Nauka, Moscow, 1970.
[193] W.A. Deer, R.A. Howie, J. Zussman (Eds.), Rock-Forming Minerals, Vol. 5: (Non-Silicates), Longmans, Green and Co. Ltd., London, UK, 1962.
[194] D.R. Simpson, Problems of the composition and structure of the bone minerals, Clin. Ortho. Relat. Res. 86 (1972) 260–286.
[195] D.W. Holcomb, R.A. Young, Thermal decomposition of human tooth enamel, Calcif. Tissue Int. 31 (1980) 189–201.
[196] W.H. Waggaman, Phosphoric Acid, Phosphates and Phosphatic Fertilizers, second ed., Reinhold Publishing Corp., New York, NY, 1952.
[197] A.D.F. Toy, The chemistry of phosphorus, Inorganic Chemistry, vol. 3, Pergamon Press, Oxford, UK, 1973.
[198] A. Daubree, Experiences sur la production artificielle de l'apatite, de latopaze, et de quelques autres métaux fruoriféres, Compt. Rend. Acad. Sci. Paris 32 (1851) 625.
[199] P. Ducheyne, T. Kokubo, C.A. Van Blitterswijk (Eds.), Bone-Bonding Biomaterials, Rees Healthcare Communications, Harrow, UK, Leiderdorp, The Netherlands, 1992.
[200] H. Aoki, Science and medical applications of hydroxyapatite, Japanese association of apatite science, JAAS, Tokyo, Japan (1991).
[201] L.L. Hench, Bioceramics: from concept to clinic, J. Am. Ceram. Soc. 74 (1991) 1487–1510.

[202] K. Yanagisawa, H. Toya, Qi Feng, N. Yamasaki, *In-situ* formation of HApcrystals under hydrothermal conditions, Phosphorus Res. Bull. 5 (1995) 43–46.
[203] T. Kanazawa (Ed.), Inorganic Phosphate Materials, Elsevier Science Publishers, Amsterdam, 1989.
[204] M. Yoshimura, H. Suda, Hydrothermal processing of hydroxyapatite: part, present and future, in: P.W. Brown, B. Constanz (Eds.), Hydroxyapatite and Related Materials, CRC Press, Inc., Boca Raton, Florida, 1994, pp. 45–72.
[205] J.W. Me Cauley, R. Roy, Controlled nucleation and crystal growth of various calcium carbonate phases by the silica gel technique, Am. Miner. 59 (1974) 947–963.
[206] P.D. Johnson, Some optical properties of powder and crystal holophosphatephosphorous, J. Electrochem. Soc. 108 (1961) 159–162.
[207] J.S. Prener, The growth and crystallographic properties of calcium fluor- and chlorapatite crystals, J. Electrochem. Soc. 114 (1967) 77–83.
[208] S. Oishi, T. Kamiya, High-temperature solution growth of chlorspodiosite and chlorapatite crystals, J. Chem. Soc. Jpn. 10 (1993) 1129–1132.
[209] S. Oishi, T. Kamiya, Flux growth of fluorapatite crystals, J. Chem. Soc. Jpn. 9 (1994) 800–804.
[210] S. Oishi, I. Sugiura, Growth of chlorapatite crystals from a sodium chloride flux, Bull. Chem. Soc. Jpn. 70 (1997) 2483–2487.
[211] Y. Masuda, K. Matsubara, S. Sakka, Synthesis of HAp from metal alkoxides through Sol–gel technique, J. Ceram. Soc. Jpn. 98 (1990) 1255–1266.
[212] T. Brendel, A. Engel, C. Russel, Hydroxyapatite coating by a polymeric route, J. Mater. Sci. Mater. Med. 3 (1992) 175–179.
[213] S. Naray-Szabo, The structure of apatite $(CaF)Ca_4(PO_4)_3$, Zeit. Krist. 75 (1930) 387–398.
[214] M. Mehmel, Uber die Struktur des Apatits, Zeit. Krist. 75 (1930) 323–331.
[215] S.B. Hendricks, M.E. Jefferson, V.M. Mosley, The crystal structures of some natural and synthetic apatite-like substances, Zeit. Krist. 81 (1932) 352.
[216] A.S. Posner, A. Perloff, A.F. Diorio, Refinement of the hydroxyapatite structure, Acta. Cryst. 11 (1958) 308–309.
[217] M.I. Kay, R.A. Young, A.S. Posner, Crystal structure of hydroxyapatite, Nature 204 (1964) 1050–1052.
[218] J.C. Elliott, The problems of the composition and structure of the mineral components of the hard tissues, Clin. Ortho. Relat. Res. 93 (1973) 313–345.
[219] J.C. Elliott, P.E. Mackie, R.A. Young, Monoclinic hydroxyapatite, Science 180 (1973) 1055–1057.
[220] H.B. Van Rees, M. Mengeot, E. Kostiner, Monoclinic-hexagonal transition in hydroxyapatite and deuterohydroxyapatite single crystals, Mat. Res. Bull. 8 (1973) 1307–1310.
[221] J.R. Van Wazer, Phosphorus and its Compounds, Vol. 1, Chemistry, Wiley-Interscience, New York, NY, 1958.
[222] P.W. Brown, Phase relationships in the ternary system $CaO{-}P_2O_5{-}H_2O$ at 25°C, J. Am. Ceram. Soc. 75 (1) (1992) 17–22.
[223] P.W. Brown, N. Hocker, S. Hoyle, Variations in solution chemistry during the low-temperature formation of hydroxyapatite, J. Am. Ceram. Soc. 74 (8) (1991) 1848–1854.
[224] A.M. Biggar, Min. Mag. 35 (1966) 1110–1115.
[225] S.S. Feng, T.J. Rockett, The system $CaO{-}P_2O_5{-}H_2O$ at 200°C, J. Am. Ceram. Soc. 62 (1979) 619–620.

[226] H.C.W. Skinner, Studies in the basic mineralizing system, $CaO-P_2O_5-H_2O$, Calcif. Tiss. Res. 14 (1974) 3–4.
[227] W. Eysel, D.M. Roy, Hydrothermal flux growth of hydroxyapatites by temperature oscillation, J. Cryst. Growth 20 (1973) 245–250.
[228] M.C. Andrade, T.F.T. Ogasawara, Silva, Hydrothermal crystallization of HAp, in: T. Umegaki (Ed.), Proceedings of the Second International Symposium on Apatite, Japan Association of Apatite Science, Tokyo, Japan, 1997, pp. 41–47.
[229] G.W. Morey, E. Ingerson, The pneumatolitic and hydrothermal alteration and synthesis of silicates, Econ. Geol. 32 (1937) 607–760.
[230] E. Hayek, J. Lechtleitner, W. Bohler, Hydrothermal synthesis of hydroxylapatite, Angew. Chem. 67 (1955) 326.
[231] A. Perloff, A.S. Posner, Preparation of pure hydroxyapatite crystals, Science 124 (1956) 583–584.
[232] H. Jullmann, R. Mosebach, Zur synthese, licht-und doppelbrechung des hydroxylapatits, Naturforsch 21B (1966) 493–494.
[233] J.F. Kirn, H. Leidheiser Jr., Progress in efforts to grow large single crystals of hydroxyapatite, J. Cryst. Growth 2 (1968) 111–112.
[234] D.M. Roy, Crystal growth of hydroxyapatite, Mat. Res. Bull. 6 (1971) 1337–1340.
[235] M. Mengeot, M.L. Harvill, O.R. Gilliam, Hydrothermal growth of calcium hydroxyapatite single crystals, J. Cryst. Growth 19 (1973) 199–203.
[236] W. Eysel, D.M. Roy, Hydrothermal flux growth of hydroxyapatites by temperature oscillation, J. Cryst. Growth 20 (1973) 245–250.
[237] H. Aoki, K. Kato, M. Shiba, Synthesis of hydroxyapatite under hydrothermal conditions, in: kinetic analysis of the reaction process from monetite into hydroxyapatite, J. Dent. Apparatus Mater. 13 (27) (1972) 170–176.
[238] J. Arends, J. Schuthof, W.H. van der Linden, P. Bennema, P.J. van en Berg, Preparation of pure hydroxyapatite single crystals by hydrothermal recrystallization, J. Cryst. Growth 46 (1979) 213–220.
[239] A. Ito, S. Nakamura, H. Aoki, M. Akao, K. Teraoka, S. Tsutsumi, et al., Hydrothermal growth of carbonate-containing hydroxyapatite single crystals, J. Cryst. Growth 163 (1996) 311–317.
[240] A. Ito, K. Teraoka, S. Tsutsumi, T. Tateishi, Single crystal hydroxyapatite: preparation, composition and mechanical properties, Bio-ceramics 9 (1996) 189–192.
[241] M. Kikuchi, A. Yamazaki, R. Otsuka, Crystal structure of Sr-Substituted HAp, synthesized by hydrothermal method, J. Solid State Chem. 113 (1994) 373–378.
[242] M. Hata, F. Marumo, Synthesis and superstructure of $(Ca,M)_5$ $(PO_4)_3OH(M = Mn, Fe,Co,Ni,Cu,Hg)$, Miner. J. 11 (1983) 317–330.
[243] F. Nagata, Y. Yokogawa, M. Toriyama, Y. Kawamoto, T. Suzuki, K. Nishizawa, Hydrothermal synthesis of HAp crystals in the presence of methanol, J. Ceram. Soc. Jpn. 103 (1995) 70–73.
[244] G. Vereecke, J. Lemaitre, Calculation of the solubility diagram in the system Ca $(OH)_2-H_3PO_4-KOH-HNO_3-CO_2-H_2O$, J. Cryst. Growth 104 (1990) 820–832.
[245] K. Onuma, N. Kanzaki, A. Ito, T. Tateishi, Growth kinetics of the HAp (0001) face revealed by phase shift in terferometry and atomic force microscopy, J. Phys. Chem. B 102 (1998) 7833–7838.
[246] N. Kanzaki, K. Onuma, A. Ito, K. Teraoka, T. Tateishi, S. Tsutumi, Direct growth rate measurement of HAp single crystal by moire phase shift interferometry, J. Phys. Chem. B 102 (1998) 6471–6476.

6 Hydrothermal Synthesis and Growth of Zeolites

6.1 Introduction

Zeolites form an important group of hydrated aluminosilicate minerals of the alkalies and alkaline earths, with an infinitely 3D anion network and with the atomic ratio O:(Al + Si) = 2. The history of zeolites began with the discovery of stilbite in 1756 by a Swedish mineralogist, Crönstedt [1], who heated this silicate mineral, which fused readily with marked intumescence. The name zeolite signifies *boiling stone*, the meaning derived from the two Greek terms, *zein*, to boil, and *lithos*, stone, an allusion to the intumescence of most zeolites with a borax lead, and refers to the frothy mass which will result when a zeolite is fused in the blowpipe. Volatile zeolitic water forms bubbles within the sinter or melt [2].

Zeolite minerals were first recognized as minor but were later regarded as widespread components in amygdales and fissures, chiefly in basic volcanic rocks. They are also found in veins and other later-stage hydrothermal environments. Big crystals of zeolites can develop in these amygdales and fissures through the mineralizing action of trapped or circulating solutions on the alkaline matrix. It is these crystals which, because of their beauty and diversity of form, usually comprise zeolite collections [3]. Zeolites are also found in some igneous rocks and occur as alteration products of aluminosilicates, such as the felspars or felspathoids. Their occurrence is not just restricted to basalt matrices, although the scale and variety of natural depositions have been appreciated only over the last three decades. They occur as authigenic minerals in sandstones and other sedimentary rocks. Loew [4] reported chabazite in a sedimentary tuff bed near Bowie, Arizona. Similarly, Several kinds of zeolites were identified by Coombs [5].

6.2 Mineralogy of Zeolites

Zeolites belong to the tectosilicates group covering the felspars and felspathoid family of minerals. However, the structure of felspars is a more compact framework, with or without relatively small cavities. Thus, the cations and framework are strongly interdependent so that the cations cannot easily move unless framework bonds are broken, and replacement of Na or K by Ca necessarily involves

Handbook of Hydrothermal Technology.

a change in the Si:Al ratio, whereas the structure of felspathoids have a more open framework so that the ions can be moved, removed, or interchanged through the channels/cavities without disturbing the framework. The cavities are usually filled with water molecules.

Finally, the zeolites' structures contain larger cavities/channels and, therefore, many exhibit, to a greater extent, the properties of ion exchange and molecular adsorption. Furthermore, these cavities or channels in zeolites are usually filled with water molecules. These are relatively loosely bound to the framework and cations and, like the cations, these can also be removed and replaced without disrupting framework bonds. Accordingly, the felspars are anhydrous aluminosilicate minerals with a density of 2.6–2.7, followed by felspathoids (density 2.3–2.5) and zeolites (density 2–2.3). The hardness decreases from felspar to the zeolite end of the minerals owing to the above-stated structural features. Thus, the characteristic feature of zeolite mineralogy is the presence of OH^- molecules in the structural channels, which are loosely bonded to the cation framework. The cations can be removed and replaced by other metals without disrupting framework bonds.

There are over 46 naturally occurring zeolite minerals, belonging to some major structure types. The channel systems in various zeolites are formed by different combinations of linked rings of tetrahedra. The wider the channels at their narrowest parts, the larger the cation that can be introduced into the structure. A grading of the zeolites according to the openness of their structure is given in Table 6.1.

When zeolites are heated, water is given off continuously rather than in separate stages at definite temperatures, and the dehydrated or partially dehydrated mineral can reabsorb water to its original amount when it is again exposed to water vapor. Dehydrated zeolites can also absorb other liquids or vapors in place of water, e.g., ammonia, alcohol, NO_2, and H_2S. The base-exchange properties of zeolites were first reported by Eichorn [6] and have been reviewed by Hey [7] who also investigated in detail the base exchange of various members of the group. The colloidal properties of the zeolites have been reviewed by Marshall [8]. McBain [9]

Table 6.1 Numbers of Tetrahedra in Rings and Approximate Diameters of Channels in Various Felspathoids and Zeolites

Minerals	No. of Tetrahedra in Rings	Minimum Diameter of Widest Channel
Sodalite, Nosean	4 and 6	2.2 Å
Analcite	4 and 6	2.2 Å
Harmotome, Phillipsite	4 and 8	3.2 Å
Levyne	4, 6, and 8	3.2 Å
Erionite	4, 6, and 8	3.6 Å
Chabazite	4, 6, and 8	3.9 Å
Heulandite	5, 6, and 8	–
Gmelinite	4, 6, 8, and 12	6.4 Å
Faujasite	4, 6, and 12	9 Å

introduced a new term, *molecular sieve*, to zeolites to describe porous materials which can act as sieves on a molecular scale, which is not just limited to aluminosilicate zeolites but also to a large number of other zeolites such as aluminophosphate zeolites, vanadophosphosphate zeolites, and titanophosphate zeolites. The zeolites and related materials, like phosphate-based molecule sieves, are microporous solids with a broad range of physicochemical properties, which are listed in Table 6.2 [10].

The commercial success of molecular sieves is largely due to the continual discovery of new materials whose diverse properties have allowed process improvements and the development of new technologies. Thus, the ability to control the materials' properties through synthetic efforts has been and will continue to be of paramount importance. Davis and Lobo [10], Feijen et al. [11], Auerbach et al. [12], Cundy and Cox [13], and Yu [14] have reviewed and discussed all the major issues related to the synthesis and recent applications of zeolites. Accordingly, the zeolites are no longer of interest to only those who are working in ion exchange. Distinction between the microscopic structure and the macroscopic properties of these materials allows one to use molecular sieves to recognize, discriminate, and organize molecules with a precision that can be <1 Å. Also, it has expanded to other exciting areas such as supramolecular catalysis, photochemistry, nanochemistry, and electrochemistry. In this chapter, the major developments in the field of zeolites, covering aspects of their crystallization mechanism and applications, are discussed.

6.3 Crystal Chemistry of Zeolites

There are 46 known zeolite minerals in nature and over 204 synthetic zeolites known in the literature. The number of synthetic zeolites is increasing day by day, particularly after the discovery of aluminophosphate zeolites by Wilson et al. [15]. Despite the remarkable structural diversity in aluminophosphate zeolites, the product compositions are very similar:

$$xR + Al_2O_3 + (1.0 - 0.2)P_2O_5 + yH_2O$$

Table 6.2 Physicochemical Properties of Zeolites and Molecular Sieves [10]

Property	Range
Pore size	$\sim$4–13 Å
Pore shape	Circular, elliptical
Dimensionality of pore system	1D, 2D, 3D
Pore configuration	Channels, cages
Surface properties	Hydrophilic, hydrophobic (high silica)
Void volume	Less than $\sim$50%
Framework oxide composition	Si, Al, P, Ga, Ge, B, Be, Zn:minor Ti, Fe, Co, Cr, V, Mg, Mn:minor

where *R* is an amine or quaternary ammonium ion. This organic species seems to exert a unique structure-directing influence during the crystallization of zeolites. Actually, when the first atlas of zeolite structure types was published in 1970, some 27 zeolite structure types were reported [16]. During 1978, the structure commission of International Zeolite Association (IZA) published about 38 entries in the *Atlas of Zeolite Structure Types* by Meier and Olson [17]. Later, in 1987, the second edition of the *Atlas*, comprising 64 structure types, was published [18]. The third edition of the *Atlas* in 1992, compiled by Meier and Olson [19], covers up to 1991 and carries about 85 entries of zeolite structure types. The *Atlas* has become a frequently quoted work of reference and such a publication has to be frequently updated. The current sixth edition of the *Atlas* published in 2007 has come out with many more new entries of zeolite structure types, and a further new feature of this edition is that characteristic building units are listed for each of the framework types [20]. The list is expanding, and since the publication of the sixth revised edition, several more entries have been added by the Zeolites Structure Commission http://www.iza-structure.org/data-bases/.

There are several zeolites' structure types reported from phosphates, and they apparently constitute two distinctive categories of microporous materials. The microporous zeolite-type materials are listed in Table 6.3.

The primary building block of the zeolite structure is a tetrahedron of four oxygen atoms surrounding a central Si atom (SiO_4). These are connected through their corners of shared oxygen atoms to form a wide range of small secondary building units (SBUs). They are interconnected to form a wide range of polyhedra, which in

Table 6.3 Microporous Zeolite-Type Materials[a]

ABW	**ACO**	**AEI**	**AEL**	**AEN**	**AET**	**AFG**	**AFI**	**AFN**	**AFO**	**AFR**	**AFS**
AFT	**AFX**	**AFY**	**AHT**	**ANA**	**APC**	**APD**	**AST**	**ASV**	**ATN**	**ATO**	**ATS**
ATT	**ATV**	**AWO**	**AWW**	**BCT**	**BEA**	**BEC**	**BIK**	**BOF**	**BOG**	**BPH**	**BRE**
BSV	**CAN**	**CAS**	**CDO**	**CFI**	**CGF**	**CGS**	**CHA**	**-CHI**	**-CLO**	**CON**	**CZP**
DAC	**DDR**	**DFO**	**DFT**	**DOH**	**DON**	**EAB**	**EDI**	**EMT**	**EON**	**EPI**	**ERI**
ESV	**ETR**	**EUO**	**EZT**	**FAR**	**FAU**	**FER**	**FRA**	**GIS**	**GIU**	**GME**	**GON**
GOO	**HEU**	**IFR**	**IHW**	**IMF**	**IRR**	**ISV**	**ITE**	**ITH**	**ITR**	**-ITV**	**ITW**
IWR	**IWS**	**IWV**	**IWW**	**JBW**	**JRY**	**JST**	**KFI**	**LAU**	**LEV**	**LIO**	**-LIT**
LOS	**LOV**	**LTA**	**LTF**	**LTJ**	**LTL**	**LTN**	**MAR**	**MAZ**	**MEI**	**MEL**	**MEP**
MER	**MFI**	**MFS**	**MON**	**MOR**	**MOZ**	**MRE**	**MSE**	**MSO**	**MTF**	**MTN**	**MTT**
MTW	**MVY**	**MWW**	**NAB**	**NAT**	**NES**	**NON**	**NPO**	**NPT**	**NSI**	**OBW**	**OFF**
OSI	**OSO**	**OWE**	**-PAR**	**PAU**	**PHI**	**PON**	**PUN**	**RHO**	**-RON**	**RRO**	**RSN**
RTE	**RTH**	**RUT**	**RWR**	**RWY**	**SAF**	**SAO**	**SAS**	**SAT**	**SAV**	**SBE**	**SBN**
SBS	**SBT**	**SFE**	**SFF**	**SFG**	**SFH**	**SFN**	**SFO**	**SFS**	**SFV**	**SGT**	**SIV**
SOD	**SOF**	**SOS**	**SSF**	**SSY**	**STF**	**STI**	**STO**	**STT**	**STW**	**-SVR**	**SZR**
TER	**THO**	**TOL**	**TON**	**TSC**	**TUN**	**UEI**	**UFI**	**UOS**	**UOZ**	**USI**	**UTL**
UWY	**VET**	**VFI**	**VNI**	**VSV**	**WEI**	**-WEN**	**YUG**	**ZON**			

[a]Database of zeolite structures, by Structure Commission of the International Zeolite Association (IZA-SC) 2012.

turn connect to form the infinitely extended frameworks of the various specific zeolite crystal structures. Different combinations of the same SBU may give numerous distinctive structural polyhedra formed from smaller ring units. The net negative charge on the framework is balanced by the presence of cations, in most cases Ca, Na, or K, which are situated in cavities within it. The zeolite framework is structurally almost independent of the (Na, Ca, K) cations, and, as the latter do not fill all the cavities, replacements of the type Ca$\leftrightarrow$2(Na, K) can also occur. Figures 6.1 and 6.2 show the basic structure of zeolite frameworks [21,22]. The structural complexity of zeolites is because of the various ways in which the tetrahedral groups are linked by the common sharing of oxygen ions to form polynuclear complexes. Considerable variation in the chemical composition results from the substitution of cations, such as those listed in Table 6.4 [22]. They give rise to zeolite and zeolite-like materials of a very wide diversity. Thus, some simple criteria are used to establish the zeolite frameworks. The most important one is the framework density (FD), which means the number of temperature atoms (T-atoms) per 1000 Å^3. Figure 6.3a and b shows the distribution of these values for porous and dense frameworks whose structures are well established [23]. In Figure 6.3a,

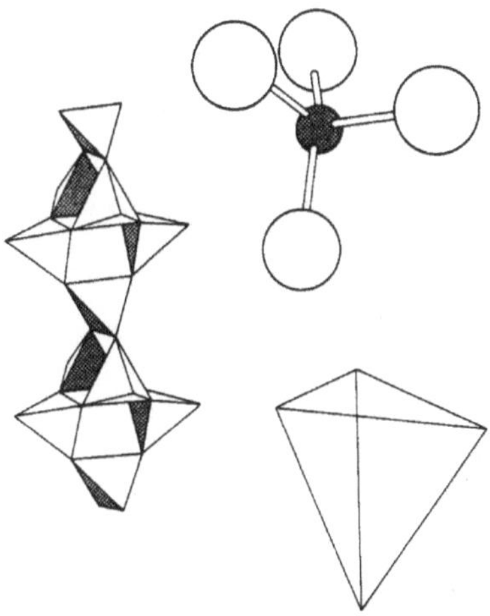

Figure 6.1 Basic structure of zeolite frameworks [21].

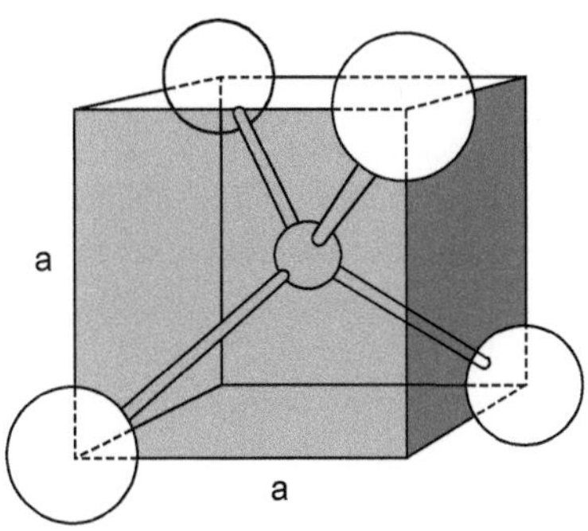

Figure 6.2 Basic structure of zeolite frameworks [22].

Table 6.4 Coordination of Cation with Oxygen in Silicate Structures [22]

Ion	Radius (Å)	Radius Ratio	Coordination No.	Bond Strength
B^{+3}	0.20	0.20	3,4	1 or 3/4
Be^{+2}	0.31	0.25	4	1/2
Li^{+}	0.60	0.34	4	1/4
Si^{+4}	0.41	0.37	4,6	1
P^{+5}	0.34	0.34	4,6	5/4 or 5/6
Al^{+3}	0.50	0.41	4,5,6	3/4 or 1/2
Ge^{+4}	0.44	0.43	4,6	1 or 2/3
Ga^{+3}	0.62	0.46	4,6	3/4 or 1/2
Mg^{+2}	0.65	0.47	6	1/3
Na^{+}	0.95	0.54	6,8	1/6 or 1/8
Ca^{+2}	0.99	0.67	7,8,9	1/4
K^{+}	1.33	0.75	6,7,8,9,10,12	1/9
NH_4^{+}	1.50	0.85	9,12	1/9 or 1/12

a distinct gap between zeolite-type and dense tetrahedral framework structures is seen. The boundary ranges from just under 20 to over 21 T-atoms per 1000 $Å^3$, depending on the type of smallest rings present. Figure 6.3b shows the correlation between the pore size of molecular sieves and the diameter (σ) of various molecules. While, for zeolites, the observed values range from about 11.1, for structures with the largest pore volume, to around 20.5. To date, framework densities of less than 12 have only been encountered for interrupted framework, like in cloverite, and hypothetical networks [24,25]. The FD is obviously related to the pore volume, but does not reflect the size of the pore openings, and to some extent it depends on the composition for structural and chemical reasons. The list of zeolite structure types also comprises interrupted frameworks of zeolite-like materials, provided the framework atoms, other than oxygen, are tetrahedrally coordinated, at least in the calcined form.

The chemistry of zeolites is based on the tetrahedron (TO_4), where, T = Si, Al, or P. However, zeolites are so diverse that secondary structural units based on small grouping of linked tetrahedra are needed in describing and systematizing their topologies. A detailed structural classification of natural and synthetic zeolites has been proposed by several workers like Breck [21], Smith [24], and Barrer [25]. However, Meier [26] was the first to propose a classification based on second building units contained in the structure. The classification based upon SBUs is given below. Table 6.5 gives the building units in zeolite structures [27]. Some of the synthetic zeolites are structurally related to minerals and are included in the above classification; however, for many of the synthetic zeolites structures are not known, therefore, in many cases the framework may be considered as made up of larger polyhedral units, such as the truncated cubo-octahedron (a-cage), truncated octahedron (b-cage), double 8 with 4-ring bridges (g-cage), double 8 ring (d-cage), and cancrinite [19].

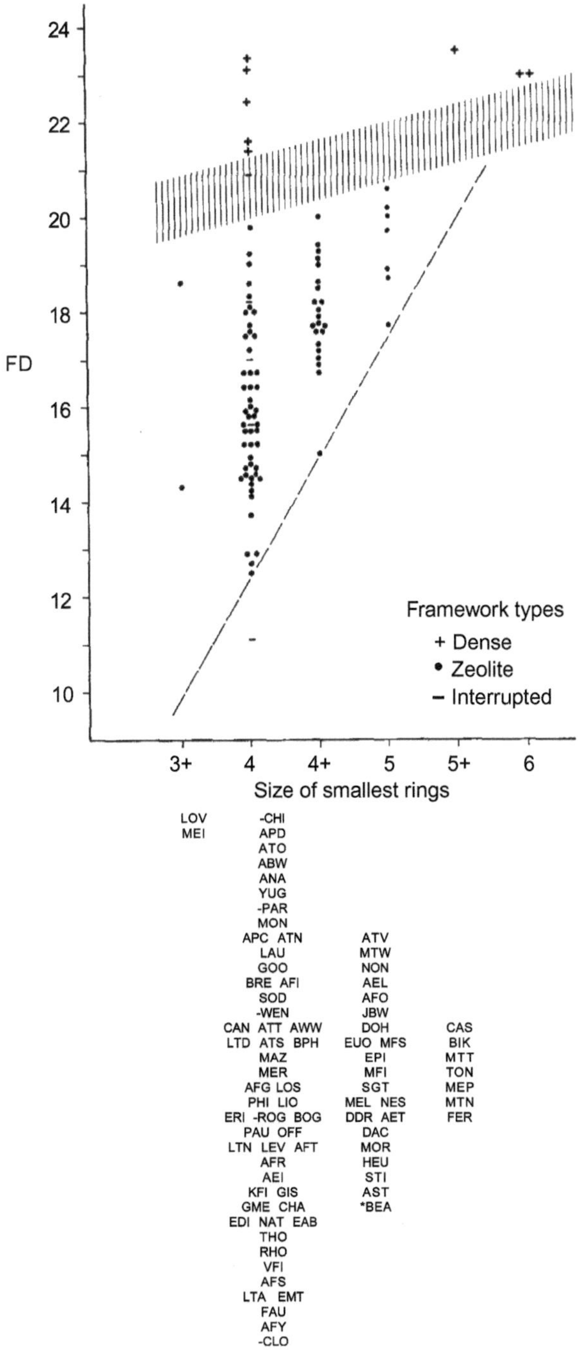

Figure 6.3 (a) Distinct gap between zeolite-type and dense tetrahedral framework structure [23]. (b) Correlation between pore size of molecular sieves and the diameter (σ) of various molecules [10].

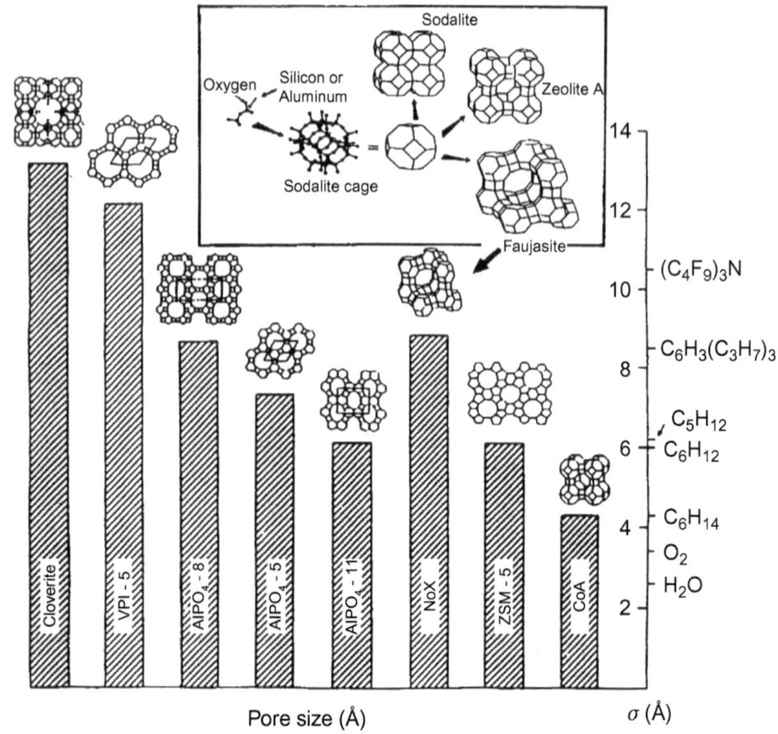

Figure 6.3 Continued

Group	SBUs
1	Single 4 ring (S4R)
2	Single 6 ring (S6R)
3	Single 8 ring (S8R)
4	Double 4 ring (D4R)
5	Double 6 ring (D6R)
6	Double 8 ring (D8R)
7	Complex 4-1, T_5O_{10} unit
8	Complex 5-1, T_8O_{16} unit
9	Complex 4-4-1, $T_{10}O_{20}$ unit

Qinhua and Aizhen [28] have summarized the structure of zeolites as follows: the primary building block $(SiO_4)^{4-}$ tetrahedron (Figure 6.4a) and the tetrahedra are connected through their corners of shared oxygen atoms to form a wide range of small SBUs (Figure 6.4b). They are interconnected to form a wide range of polyhedra (Figure 6.4c), which in turn connect to form the infinitely extended frameworks of the various specific zeolite crystal structures. In Figure 6.4a–d, showing the

Table 6.5 Building Units in Zeolite Structures [27]

Primary Building Units (PBUs)	Tetrahedra (TO_4)
	Tetrahedra of four oxygen ions with a central ion of Si^+, Al^+, P^+, etc.
SBUs	Single rings: S-4, S-5, S-6, S-8, S-10, S-12
	Double rings: D-4, D-6, D-8
Larger symmetrical polyhedra	Truncated octahedron (T.O.) or sodalite unit
	11-hedron or cancrinite unit
	14-hedron or gmelinite unit

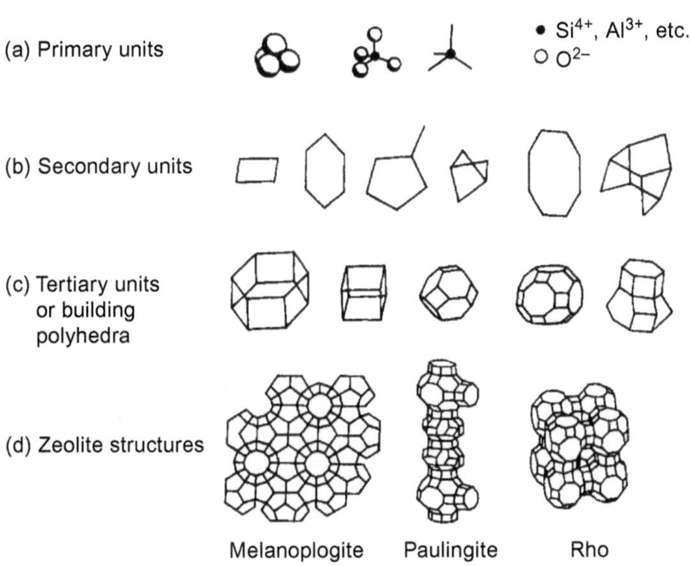

Figure 6.4 Development of zeolite structures [28].

structure of zeolites, the corners of the polyhedra represent Si or Al atoms and the connecting lines represent the shared oxygen atoms. Individual structures may comprise only one basic unit or many of them. A record is held by the mineral paulingite, which contains five such polyhedra.

Different combinations of the same SBU may give numerous distinctive zeolite structures. Figure 6.5 shows an example of three different zeolites that have the same structural polyhedron (cubo-octahedron), but probably form from smaller ring units.

The IUPAC commission on zeolite nomenclature has framed rules and designations consisting of three letters that have been used through the assignment of structure-type codes [29]. This is subjected to review and clearance by the IZA

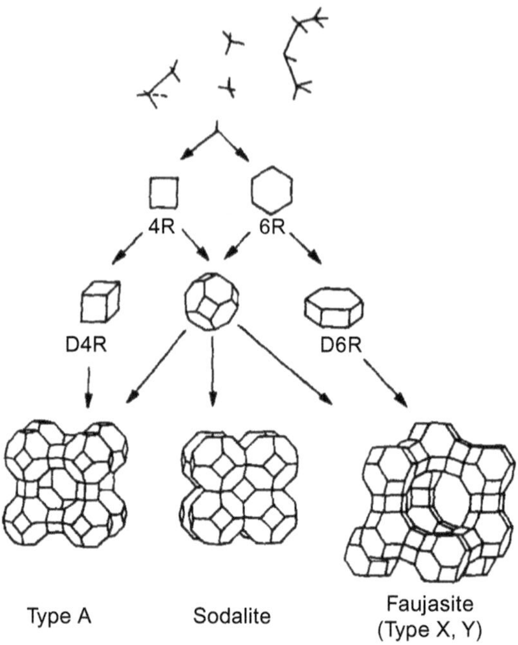

Figure 6.5 Three different zeolites.

Structure Commission according to a decision of the IZA Council (taken at the time of the seventh IZC in Tokyo, 1986). These mnemonic codes should not be confused or equated with actual materials. Structure types do not depend on composition distribution of the various possible T-atoms (e.g., Si, Al, P, Ga, Ge, B, and Be), cell dimensions, or symmetry. The codes are generally derived from the names of the type materials and they do not include numbers and characters other than capital Roman letters. Structural criteria alone do not provide an unambiguous numbering scheme; therefore, to facilitate later additions and simple indexing, the structure types have been arranged in alphabetical order according to the structure-type codes. For interrupted frameworks, the three-letter code is preceded by a hyphen for each structural type, the information given in bold type also includes the full type name, the maximum topological symmetry (i.e., the highest symmetry special group), a listing of the topologically distinct T-atoms, their number unit cell, and the maximum point symmetry of the respective temperature sites (in parentheses). The reader can get more detailed information in Refs [16–20].

6.4 Comparison Between Natural and Synthetic Zeolites

Ever since the discovery of stilbite by a Swedish chemist, Crönstedt (1756), a large number of publications have appeared both on natural and on synthetic zeolites. In the last century, Wöhler (1848) attempted to recrystallize apophyllite by heating it in water solutions at 180–190°C under 10–12 atm pressure. There are thousands of reports on various aspects of both natural and synthetic zeolites.

All the natural zeolites can be included in the following genetic types [30]:

1. Crystals resulting from hydrothermal or hot-spring activity involving reaction between solutions and basaltic lava flows.
2. Deposits formed from volcanic sediments in closed alkaline and saline lake systems.
3. Similar formations from open freshwater lakes or groundwater systems acting on volcanic sediments.
4. Deposits formed from volcanic materials in alkaline solids.
5. Deposits resulting from hydrothermal or low-temperature alteration of marine sediments.
6. Formations which are the result of low-grade burial metamorphism.

The above reactions occur in open systems and may be as vast in scale as in the range of variables such as pressure, temperature, and time. Zeolites' formation, either in nature or in the laboratory, takes place under conditions in which water is present in considerable amounts, often at elevated temperatures and hence, under hydrothermal conditions. Thus, zeolites have been made in the laboratory only by the hydrothermal method because the open aluminosilicate host framework must be stabilized during growth by being filled with guest molecules. In particular, zeolites never form under acid hydrothermal conditions. The sodalites and cancrinites, like zeolites, have open frameworks, but their synthesis can be achieved pyrolytically as well as hydrothermally because nonvolatile salts can act like zeolitic water and as stabilizers by occupying the pore spaces within the aluminosilicate framework. In hydrothermal synthesis using alkaline salt solutions, water and salts may compete as stabilizers so that each may be present within the intracrystalline pores [31]. However, the laboratory synthesis of zeolites differs much from that of the natural process, as the laboratory synthesis involves a closed system. Early attempts to synthesize zeolites centered around imitating geological conditions and it was mainly the recrystallization of zeolites taking place until the first-synthesized analcime. In 1948, the first synthesis of a zeolite that did not have a natural counterpart was carried out by Barrer.

Laboratory synthesis has evolved by duplicating the conditions under which natural zeolites were produced. However, the one condition that we can never duplicate is crystallization of time, which covers 1000 years or more. The laboratory systems operate at high pH (> 12, usually >14 for aluminosilicates, and >7 for aluminophosphates zeolites) and high temperature and, thereby, produce smaller, less-perfect, crystals. So, the majority of the synthetic zeolites are formed under nonequilibrium conditions and are metastable phases produced at supersaturated conditions.

Experimental results indicate that, the less stable the phase, the greater the chance that, initially, it will nucleate and grow fast, but its chance of subsequent survival is less.

Synthetic zeolites represent metastable structures that may, under given conditions, be transformed into other thermodynamically more stable types of zeolites. Also, the synthetic zeolites contain both inorganic and organic cations. Although more than 150 synthetic zeolites have been reported, many important types have a number of natural mineral counterpart. Conversely, synthetic counterparts of many zeolite minerals are not yet known. In recent years, some minerals having zeolite structure have been reported; one such example is cavansite, Ca(VO) (Si_4O_{10}) $4H_2O$, which was reported by Staples et al. [32], from cavities and veinlets in basalts and tuffs in Oregon, USA. Its crystal structure was reported by Evans [33,34]. Subsequently, this mineral was reported from several other localities. Powar and Byrappa [35] have reported cavansite from India and they have carried out detailed X-ray, thermal, and infrared studies. The paragenesis of the mineral cavansite is closer to zeolites. It has a 3D/vanadosilicate network analogous to gismondine. However, based on a detailed infrared spectroscopic study, it was found that in cavansite, water molecules occur both as loosely held and as tightly bound crystal water [35]. Thus, the possibility of the existence of such new minerals having the structures of zeolites cannot be ruled out.

In recent years, the compositional–structural range of molecular sieves have been significantly increased by the discovery of crystalline microporous aluminophosphate-based frameworks, beginning with the aluminophosphates or $AlPO_4$-*n* molecular sieves, followed by silicoaluminophosphates, titanium aluminophosphates, metal aluminophosphates, and metal silicophosphates, designated as SAPO-*n*, TAPO-*n*, MeAPO-*n*, and MeAPSO-*n*, respectively.

In 1961, d'Yvoire [36] reported the synthesis of several aluminophosphate hydrates such as H_1, H_2, H_3, and H_4. These materials were prepared from dilute aqueous solutions of alumina and phosphoric acid with a P_2O_5:Al_2O_3 ratio of 2.7.

Much of the success in the laboratory synthesis has been achieved by duplicating the conditions under which natural zeolites were produced. In nature, water plays a key role in the crystallization of zeolites, whereas in the laboratory the zeolites' structure formation insists on the presence of the organic molecules. Tables 6.6–6.8

Table 6.6 Synthetic Counterparts of Natural Zeolites—Alkaline Metal [37]

Zeolite	Alkali	Investigator	Date
Mordentie	Na	R.M. Barrer	1948
Analcime	Na	R.M. Barrer	1949
Phillipsite	Na	R.M. Barrer	1951
Cancrinite	Na	R.M. Barrer	1952
Natrolite	Na	R.M. Barrer	1952
Faujasite	Na	R.M. Milton	1956
Gmelinite	Na	R.M. Barrer	1959
Chabazite	Na	R.M. Milton	1960
Erionite	K, Na	D.W. Breck	1960
Clinoptilolite	Li	L.L. Ames, Jr.	1963
Ferrierite	Na	E.E. Senderov	1963
Gismondine	Na	A.M. Taylor and R. Roy	1964
Bikitalite	Li	D.J. Drysdale	1971
Edingtonite	K	R.M. Barrer	1974

give the natural zeolites whose structures were successfully duplicated from silica/alumina gels and alkali metal, alkaline earth metal, or alkyl ammonium cations, respectively [37]. One can expect more examples in the future with respect to the discovery of natural counterparts of existing synthetic zeolites. Table 6.9 gives a list of several synthetic zeolites with structures reasonably well defined but with no

Table 6.7 Synthetic Counterparts of Natural Zeolites—Alkaline Earths [37]

Zeolite	Alkali	Investigator	Date
Heulandite	Ca	R. Roy	1960
Scolecite	Ca	R. Roy	1960
Wairakite	Ca	R. Roy	1960
Epistilbite	Ca	R. Roy	1960
Garronite	Ca	R.M. Barrer	1961
Thomsonite	Ca	R.M. Barrer	1961
Yugawaralite	Sr	R.M. Barrer	1964
Clinoptilolite	Sr	R.M. Barrer	1964
Harmotone	Ba	R.M. Barrer	1964

Table 6.8 Synthetic Counterparts of Natural Zeolites—Alkylammonium [37]

Zeolite	Alkali	Investigator	Date
Offretite	TMA, [a]Na, K	H.E. Robson	1966
Levynite	$[CH_3NN]^+$, K	G.T. Kerr	1969
Maxxite	TMA, [a]Na	R.M. Barrer	1970

Table 6.9 Natural Zeolites with No Synthetic Counterpart

Zeolite	Year Discovered
Stilbite	1756
Laumontite	1785
Mesolite	1813
Brewsterite	1822
Herschelite	1825
Kehoeite	1893
Gonnardite	1896
Dachiardite	1905
Stellerite	1909
Viseite	1942
Paulingite	1960
Merlinoite	1977

counterparts in natural zeolites. New synthetic zeolites' phases will continue to grow in numbers faster than the natural zeolites.

6.5 Synthesis of Zeolites

Efforts to synthesize zeolites can be traced as far back as 1848 when Wöhler [38] first recrystallized apophyllite by heating it in water solutions at 180–190°C under 10–12 atm. Followed by this, several more attempts were made during the 1860s when Claireville [39] synthesized zeolites by the hydrothermal method in 1862. Today, over 3 million tons of zeolites are produced annually for use in laundry detergents, cat litter, filters, molecular sieves, selective catalysis, medical, and so on with China alone producing more than 2 million tons followed by countries such as South Korea, Japan, Jordan, Slovakia, Turkey, and United States of America. The hydrothermal reaction merges into a low-temperature reaction and indeed, in zeolite synthesis, the trend has been inter alia for greater experimental convenience, to grow zeolites at or below 100°C provided the reaction rates are adequate. It was only during the 1940s that great attention was devoted to zeolite synthesis as a result of the pioneering work by Barrer [40] and Milton [41]. Even now, the area of zeolite synthesis is still expanding and hence is resulting in the discovery of synthetic zeolites with new topologies and new catalytic sorption and separation properties. Generally, three processes are used to produce zeolites and are as follows:

1. Preparation of molecular sieve zeolites as high-purity crystalline powders.
2. The conversion of clay minerals into zeolites.
3. Processes based on the use of other naturally occurring raw materials.

It is not surprising that much effort is made to unravel the mechanisms responsible for the formation of zeolites from its precursors. Using this knowledge, concepts may be derived for *tailor-made zeolite synthesis*, making the *art* of creating zeolite synthetics a real scientific issue. Hydrothermal synthesis of aluminosilicate zeolites involves a few elementary steps by which a mixture of Si and Al species, metal cations, organic molecules, and water is converted via an alkaline supersaturated solution into a microporous crystalline aluminosilicate. The complex chemical processes involved in this transformation can be denoted as *zeolitization*. Since the 1940s, systematic studies on synthesis have been disclosed. During 1948, Barrer [40] synthesized the first zeolite that did not have a natural counterpart. However, zeolite technology was initiated during the late 1940s on a large scale, essentially by the groups of Barrer and Milton. They developed hydrothermal synthesis using reactive alkali-metal aluminosilicate gels at low temperatures ($\sim$100°C) and pressure (autogeneous). By 1958, under the leadership of Milton, the Linde Division of Union Carbide had successfully synthesized nearly all the commercially important zeolites [41–45]. The next advance in zeolite synthesis was again due to Barrer [40]. In 1961, Barrer and Denny [46] reported zeolites' synthesis using alkylammonium cations. They noted that the addition of alkylammonium ions to sodium aluminosilicate gels increased the framework Si/Al ratio [46]. At about the same time, Kerr [47], at Mobil Oil Company, began using organic molecules in zeolite synthesis. The use of organic materials in zeolite synthesis was quickly expanded after the initial work of Barrer and Denny [46].

In 1961, d'Yvoire [36] reported the synthesis of several aluminophosphate hydrates, followed by the scientists of Union Carbide Laboratories who synthesized a new generation of aluminophosphate zeolites in the late 1970s [37]. Initial success resulting in the discovery of the aluminophosphate zeolites was reported by Wilson et al. [15]. Specific reference to these developments (listed in Table 6.10) may be found in several historical reviews [48,49]. The addition of Si and P framework elements resulted in SAPO zeolites which was also reported by Wilson et al.

Table 6.10 Development in Zeolite Science and Technology [40]

Early History		
1756	A.F. Cronstedt	Described natural zeolites
1840	A. Damour	Observed reversible dehydration
1858	H. Eichhorn	Reported ion-exchange properties of zeolites
1862	H. Claireville, de St.	Early hydrothermal synthesis (levyne)
1876	Lew	Reported sedimentary zeolites in tuff deposits
1896	G. Friedel	Observed organic liquid sorption; proposed open spongy framework structure
1909	F. Grandjean	Reported gas sorption (NH_3, air, H_2)
1925	O. Weigel and F. Stermhoff	Observed molecular sieve effect
1930	W.H. Taylor and L. Pauling	Structure determination of analcime and natrolite
1932–1940s	R.M. Barrer	Systematic studies of synthesis, ion-exchange sorption and dehydration
Industrial Era		
1949–1954	Union Carbide	Synthesis of zeolite A and zeolite X (synthetic faujasite)
1954	Union Carbide	Commercialized zeolite-based gas purification and separation processes; early catalytic applications
1955–1962	Mobil	Commercialized rare earth zeolite X cracking catalyst
1958–1964		Discovery and commercialization of significant sedimentary zeolite deposits in Western United States and Japan
1958	Mobil	Zeolite synthesis with organic cations
1962–1980s	Mobil/Union Carbide	Synthesis of high silica zeolites
1982	Union Carbide	Synthesis of microporous $AlPO_4$ molecular sieves
Recent Developments		
1985–1994	Academic and Industrial Researchers	Synthesis of "large pore" (i.e., VPI-5, cloverite) and mesopore (MCM-41) materials, new silica zeolites and clathrates, advanced materials applications

[50]. The metal aluminophosphate (MeAPO) zeolites with the framework containing aluminum, phosphorus, and metal cation (Me), where Me = Mg^{2+}, Mn^{2+}, Fe^{2+}, Fe^{3+}, Co^{2+}, and Zn^{2+}, were reported by Flanigen et al. [51]. As evident from the literature survey, there are contrasting reports on the synthesis of zeolites. Different chemical routes are tested by researchers, each claiming superiority over the other and yet without any clear understanding of the mechanism of formation of zeolites. Several authors have discussed these issues explicitly in a large number of reviews and books published from time to time [10–14].

Until recently, zeolite synthesis has been mainly an empirical science in which a large number of experiments were used to change synthetic parameters systematically with a hope of obtaining new phases and crystal composition. The zeolite synthesis is now being gradually replaced by methods based on the use of new routes like computational modeling, thermochemical calculations and characterization techniques, which will provide a better understanding of the gel chemistry, nucleation, structure building, crystal growth, crystallization kinetics, and structure-directing phenomenon. Boisen et al. [52] generated a series of framework structures using a molecular-based potential energy function and simulated annealing strategies. Friedrichs et al. [53] systematically enumerated a series of crystalline networks using a mathematical tiling method. Xu et al. [54] systematically enumerated a series of 2D layered aluminophosphate zeolites with $Al_3P_4O_{16}^{3-}$ stoichiometry using a divide-and-conquer algorithm and a genetic algorithm. Draznieks et al. [55] performed novel methods of prediction of inorganic structures developed through automated assembly of the SBUs. Yu et al. [56] attempted toward rational design and synthesis of aluminophosphates with 2D layer and 3D open-framework structures using building block construction strategy to construct new 2D networks and 3D open frameworks. A computer modeling method was also used to systematically enumerate the theoretical network structures. The interesting part of this work is the templating abilities of various amines in the formation of 2D layer, and 3D open-framework aluminophosphates can be evaluated in terms of nonbonding interaction energies of the guest templates and the host inorganic networks or frameworks. Li et al. [57] attempted a computational method for the design of zeolite frameworks with specified pore geometry through the constrained assembly of atoms around a specified pore structure. For the first time, a new concept of forbidden zones corresponding to a porous pattern has been described. Using such a program, for example, various structure topologies have been generated through computer a program, through the enumeration of combinations of different pore sizes, different numbers of unique T-atoms (where T = Si, Al, P), and different site symmetries [57]. All these attempts could be treated as an approach to understand the assembly of zeolite structure framework through theoretical means. Xu and coworkers [58] described the template-designed syntheses of open-framework zinc phosphites with extra-large 24-ring channels using the rational selection of the templating agents or structure-directing agents (SDAs). They utilized organic amines that are predicted by computer simulations as the templates or SDAs, such as cyclohexylamine (COHA), cyclopentylamine (CPA), and *n*-propylamine, three new open-framework zinc phosphites, including $(C_6H_{14}N)_2[Zn_3(HPO_3)_4]$ (ZnHPO-CJ2), $(C_5H_{12}N_2)[Zn_3(HPO_3)_4]$ (ZnHPO-CJ_3), and

$(C_3H_{10}N)_2[Zn_3(HPO_3)_4]$ (ZnHPO-CJ4), have been hydrothermally prepared with desired extra-large 24-ring channels analogous to $(C_4H_{12})_2[Zn_3(HPO_3)_4]$ (ZnHPO-CJ1). Such work would be highly useful for the designed synthesis of extra-large micropore-structured materials. Yu and Xu [59] proposed rational synthesis including the use of a template to direct the structure, the use of heteroatoms as a framework substituent, and the use of computational data mining. This has helped them to develop innovative methods toward the synthesis of target structures with specific channel structures, such as extra-large pores and chiral channels. Recently, Wang et al. [60] have reviewed the needs and trends in the rational synthesis of zeolitic materials. The geometrical characteristics of the zeolite frameworks affecting their catalytic performances have been demonstrated. In spite of such a tremendous development in the rational approaches toward the design and synthesis of zeolite frameworks, targeting novel structures and functions remains a formidable task. The challenge in rational synthesis lies in the unknown mechanism for their formation: The relationship between the synthetic parameters and structural characteristic of the products is not clear. Yu and Xu [59] have built up a ZEOBANK, a database of zeolite structures and a database for their synthesis. This ZEOBANK allows us to use data mining to find new methods for guiding the synthesis of zeolitic materials.

The general route whereby zeolites can be synthesized in a laboratory or in a plant scale is shown in Table 6.11 [61]. The SDAs listed are only examples of a wide range of materials that have been used. Zeolites have only been synthesized by the hydrothermal method. It involves silica, alumina, and/or phosphorus species, metal cations, organic molecules, and water, which are converted via an alkaline supersaturated

Table 6.11 Molecular Sieve Zeolite Preparation Process [61]

Molecular Sieve Process	**Zeolite Preparation Process**	
	Reactants	**Products**
Hydrogel	Reactive oxides	High-purity powders
	Soluble silicates	Gel perform
	Soluble aluminates	Zeolite in gel matrix
	Caustic	
Clay	Raw kaolin	Low- to high-purity powder
Conversion	Metakaolin	Binderless, high purity
	Calcined kaolin	Preform zeolite in clay
	Soluble silicate	Derived matrix
	Caustic sodium	
	Chloride	
Others	Natural SiO_2	Low- to high-purity powder
	Acid-treated clay	Zeolite on ceramic support
	Amorphous mineral	Binderless preforms
	Volcanic glass	
	Caustic	
	Al_2O_3, $3H_2O$	

solution into microporous crystalline zeolites. The complex chemical processes involved in this transformation can be denoted as *zeolitization*. The process of zeolitization is thermally activated and usually takes place at elevated temperatures in order to achieve a high yield of crystals in an acceptable period of time.

In spite of the success of the last 60 years in the synthesis of zeolites, the factors controlling the synthesis of porous crystalline aluminosilicates or aluminophosphates are not well defined. There are contrasting opinions on the mechanism of their formation and the role of organics in the hydrothermal system. Several laboratories around the world have synthesized zeolites with large and small pores by using a combination of inorganic and organic materials to guide the structures as they are formed. Meanwhile, several other researchers especially from Germany and France discovered that, by carefully controlling the processing parameters during the growth, they could form a large-pore zeolite without the need for expensive organic compounds typically used to guide their architecture as they grow. This motivates for the cheaper catalysts [62]. Tsapatsis and colleagues [63] at the University of Minnesota have discovered a breakthrough method for chemical separations by using ultrathin zeolite membranes, which is highly useful for a wide variety of chemically selective filters. In this connection, the IZA Synthesis Commission has given an outline of a thorough report of a molecular sieve synthesis and the following product characterization [61]. The *priority* designation is a consensus opinion of the committee as to the importance that we attach to the item. The simplest experiments will lack details in some of these categories but not in all cases. The author considered whether the particular procedure is unique in any way and might be significant in the overall result. Negative answers to the implied question can be helpful.

Similarly, the IZA Synthesis Commission gives two more outlines dealing with the background material and synthesis report of a hydrothermal zeolite. The background material covers aspects like source materials, gel preparation, crystallization, product recovery, and characterization. These aspects have been discussed in great detail by several authors from time to time.

Synthetic zeolites are generally made by mixing solutions of aluminates and silicates, often with the formation of a gel, and by maintaining the mixture at temperatures of 100°C or more for selected periods [40,61].

The synthesis of zeolites involves several steps, as shown below:

Reactant

↓

Reactant mixtures

↓

Nucleation

↓

Crystal growth

The mechanisms of zeolite formation are very complex due to the plethora of chemical reactions, equilibria, and solubility variations that occur through the heterogeneous synthesis mixture during the crystallization process.

The process of zeolitization is thermally activated and usually takes place at elevated temperatures in order to achieve a high yield of crystals in an acceptable period of time. The variables in the synthesis of zeolites are temperature, alkalinity (pH), and chemical composition of the reactant mixtures. These variables do not necessarily determine the products obtained in hydrothermal reactions because nucleation appears to be kinetically rather than thermodynamically determined and controlled. The kinetic variables include the treatment of reactants prior to crystallization, their chemical, and physical nature [10,64].

With reference to the source materials, one would hope that a mole of SiO_2 would be equivalent to another SiO_2, from whatever source, when compounded into a gel. Experience teaches us that this is not the case; SiO_2 and Al_2O_3 are to some degree polymeric before gelation. Any reagent contains some unsuspected impurities, which can be active on the synthesis. To duplicate the work, one has to repeat with the same source material. The starting material cannot be just *silica*, but *precipitated silica, fumed silica*, or *silica sol*, from a designated supplier. For very high silica phases, the alumina content of the silica source may be very important. Besides, one has to carefully prepare the gel and the step-by-step procedure has to be followed as noted. These steps include the order of mixing, the device used to mix, the aging time and temperature, nucleation, and batch composition. For example, the ingredients are not added sequentially but combined into two (or three) subgroups that are blended to make the final gel. After a careful preparation of the precursor, a suitable reaction vessel has to be selected to provide the desired temperature, pressure, agitation, and the facility to withdraw an occasional sample. Several companies from the United States of America, Japan, and Germany manufacture autoclaves with all such facilities for the synthesis of zeolites. Figure 3.27 shows the Parr autoclave commonly used in zeolite synthesis. For the commercial production of zeolites, large-size autoclaves or batch reactors are usually employed. The rate of heating is very important. If possible, a periodic/continuous monitoring facility is necessary to obtain the values of pH, percent of crystallinity, and the rate of crystallization. After carrying out the experimental runs, the filtration, washing, drying, and yield estimation are to be done carefully. For a better reproducibility of the results, the above aspects are to be kept in mind.

Several techniques are used to characterize the zeolite phases obtained. The most commonly used techniques are X-ray powder diffraction, Rietvield technique of structure refinement, neutron scattering, NMR, IR, thermal, SEM, laser Raman, measurement of sorptive capacity, particle size and pore size distribution, and so on.

Davis and Lobo [10] have proposed two extremes of the mechanisms of zeolite synthesis: (i) the solution-mediated transport mechanism and (ii) the solid-phase transformation mechanism. Figure 6.6 shows the schematic illustrations of the solution-mediated transport and solid-hydrogen transformation crystallization mechanisms. The first one involves the diffusion of aluminate, silicate, and/or aluminosilicate species from the liquid phase to the nucleation site for crystal

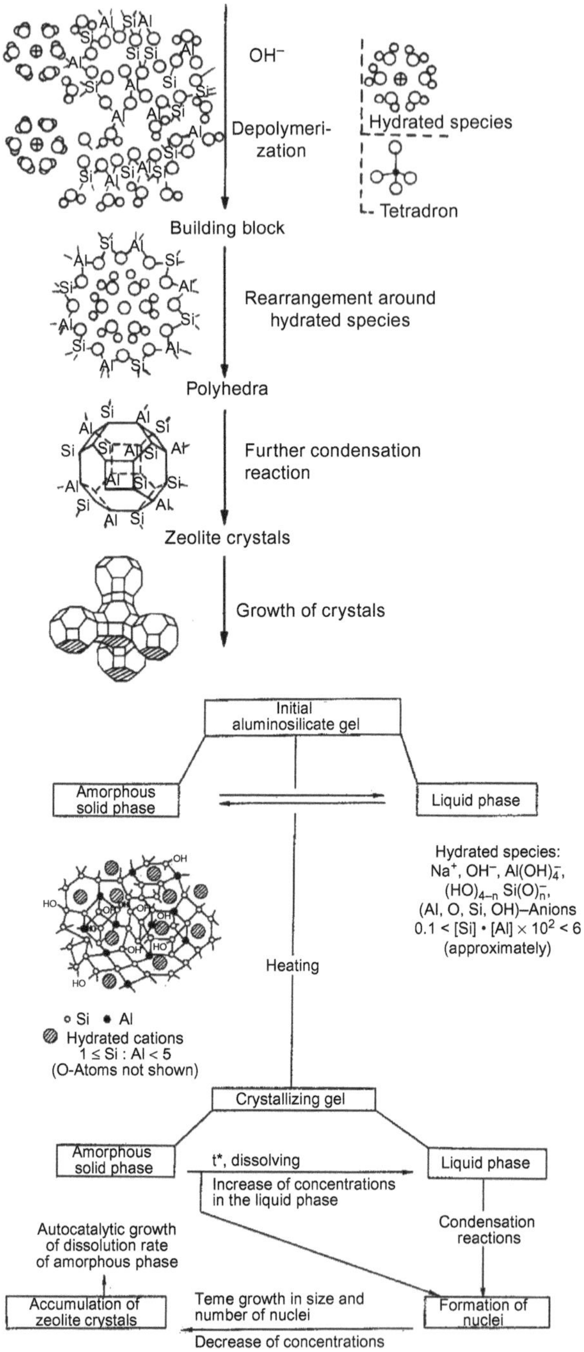

Figure 6.6 (a) Schematic illustrations of the solution-mediated transport. (b) Solid-hydrogen transformation crystallization mechanism [10].

growth. There are many examples of solution-mediated transport in the crystallization of aluminum-rich zeolites. Ueda et al. [65] have prepared zeolites Y, S (GME), and P (GIS), which were synthesized from clear solutions, that is, without the presence of hydrogel (solid phase). The solid-phase transformation mechanism suggests that the solid hydrogel reorganizes in forming the zeolite structure. Xu et al. [66] have well illustrated this process by the synthesis of ZSM-35 (FER) and ZSM-5 from nonaqueous reaction mixtures. There are several such reports on the synthesis of various zeolites using both of the proposed mechanisms of zeolite crystallization. In some cases, zeolites like Z.Y were found to crystallize by a combination of the two mechanisms. As a particular structure can be formed via different crystallization processes, extreme caution is necessary when attempting to generalize conclusions from one zeolite to a class of zeolites or molecular sieves. Here, the authors discuss some of the important aspects/factors influencing zeolite synthesis. For the sake of convenience, the authors deal with aluminosilicate zeolites followed by aluminophosphate zeolites and the family of microporous compounds.

6.5.1 Molar Composition

The chemical composition of a synthesized hydrogel is expressed generally in terms of an oxide formula, SiO_2 (P_2O_5), Al_2O_3, bMxO, cNyO, dR, and eH_2O, in which *M* and *N* stand for (alkali) metal ions and *R* for organic templates. The relative amount of Si, P, Al, M, N, and R is one of the key factors determining the outcome of the crystallization. Next to the nature of the template used (inorganic and organic cations), the ratios SiO_2/Al_2O_3, P_2O_5/Al_2O_3, $MxO(NyO)/SiO_2$, $MxO(NyO)/P_2O_5$, R/SiO_2, R/P_2O_5, and H_2O/SiO_2 or H_2O/P_2O_5 can intervene during zeolitization. Such a variation greatly influences the crystallization processes right at the level of nucleation and crystallization kinetics, the nature of the crystallization material, the lattice Al content and distribution, the crystal size, and morphology [67,68]. For example, the synthesis of high-silica zeolites requires the addition of organic molecules into the reaction mixture, an exception is ZSM-5 which can be synthesized without the use of organic reagents in a very narrow range of Na^+ and Al^{3+} concentrations, and the reaction temperatures are normally higher ($\sim$100–200°C) than those used to crystallize aluminum-rich zeolites [69]. Similarly, aluminum-rich zeolites, e.g., Z.A, X, P, sodalite, chabazite (CHA), and edingtonite (EDI), have pore volumes in the range $\sim$0.4–0.5 cm^3 void/cm^3 of crystal and are formed near 100°C, whereas the alkali-metal cations in the synthesis of zeolites act as the source of hydroxyl ions (alkali-metal hydroxides commonly used in zeolites synthesis) and have a limited structure directing role as well [70]. Recently, it has been observed that the cation–water complexes stabilize small aluminosilicate anions that are responsible for forming unique zeolite structures primarily through electrostatic and steric factors [71].

Figure 6.7 shows a crystallization diagram for the system $Na_2O–Al_2O_3–SiO_2–H_2O$ for a temperature of 363 K and $(Al_2O_3 + SiO_2)/H_2O = 0.0005$. An interesting factor that can be seen from this diagram is that the content of alkali has a much stronger influence on the kind of crystallizing

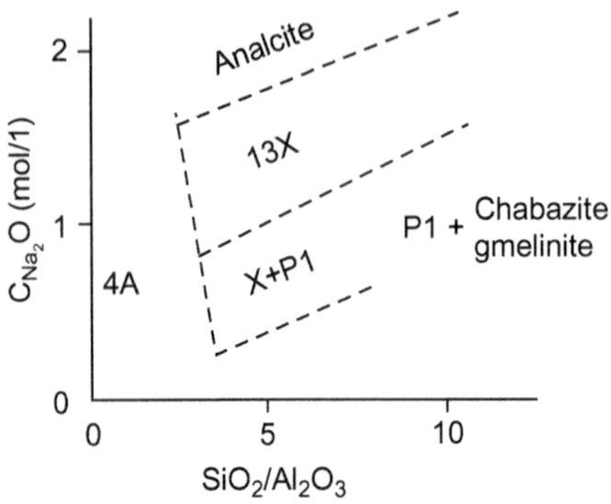

Figure 6.7 Crystallization diagram for the system $Na_2O—Al_2O_3—SiO_2—H_2O$ for a temperature of 363 K and $(Al_2O_3 + SiO_2)/H_2O = 0.0005$ [28].

Table 6.12 Primary Influence of the Compositions of Reaction Mixture [72]

Mole Ration	Primary Influence
SiO_2/Al_2O_3	Framework composition
H_2O/SiO_2	Rate, crystallization mechanism
OH^-/SiO_2	Silicate molecular weight, OH^- concentration
Na^+/SiO_2	Structure, cation distribution
R_4N^+/SiO_2	Framework aluminum content

zeolite than the SiO_2/Al_2O_3 ratio. The primary influences of compositions of reaction mixture are given in Table 6.12 [72].

The earliest synthetic zeolites were essentially Al-rich or had a low Si to Al ratio and their Al concentrations were equivalent to Si; however, Al—O—Al linkages have not been observed in zeolites [73]. Loewenstein [74] first rationalized the absence of Al—O—Al linkages in tectoaluminosilicates on the basis of clusters of negative charges. Thus, if Al atoms do not have Al atoms in their second coordination sphere, then Si/Al1. Low-silica zeolites are defined as having 1Si/Al < 2, while intermediate Si/Al zeolites as having 2 < Si/Al5. Table 6.13 gives a number of molecular sieve structures and representative materials according to their Si/Al ratio. In the late 1960s and early 1970s, high-silica zeolites with Si/Al > 5 were synthesized. As the Si/Al ratio increases, the properties of zeolites are significantly altered. At higher Si/Al ratios, the crystals are hydrophobic. In the early 1980s, silica-free molecular sieves were based on the aluminophosphate compositions [50,75—77]. The role of organics and their concentrations are discussed in detail separately in the forthcoming sections.

6.5.2 *The Aging of Hydrogel*

After the preparation of the precursor or the hydrogel, using desired ratio of the starting materials as explained in the previous section, the hydrogel is usually kept for a certain period of time below the crystallization temperature. This aging

Table 6.13 Compositions of Zeolites and Molecular Sieves

Si/Al ≤ 2	**2 < Si/Al ≤ 5**	**5 < Si/Al**	
Low Silica	***Intermediate Silica***	***High Silica***	***Phosphate***
ABW, Li-A (BW)	BHP, linde Q	BEA, zeolite beta	AEI, $AlPO_4$-18
AFG, afghanite[a]	BOG, boggsite[a]	DDR, decadodecasil 3R	AEL, $AlPO_4$-11
ANA, analcime[a]	BRE, brewsterite[a]	DOH, dodecasil 1H	AET, $AlPO_4$-8
BIK, bikitaite[a]	CAS, Cs-aluminosilicate	EUO, EU-1	AFI, $AlPO_4$-5
CAN, cancrinite[a]	CHA, chabazite[a]	FER, ferrierite	AFO, $AlPO_4$-41
EDI, edingtonite[a]	CHI, chiavennite[b]	LEV, NU-3	AFS, MAPSO-46
FAU, Na-X	DAC, dachiardite[a]	MEL, ZSM-11	AFT, $AlPO_4$-52
GIS, gismondine[a]	EAB, EAB	MEP, melanophlogite	AFY, CoAPO-50
GME, gmelinite[a]	EMT, hexagonal faujasite	MFI, ZSM-5	APC, $AlPO_4$-C
JBW, Na-J	EPI, epistilbite[a]	MFS, ZSM-57	APD, $AlPO_4$-D
LAU, laumontite[a]	ERI, erionitea	MTN, dodecasil 3C	AST, $AlPO_4$-16
LEV, levyne[a]	FAU, faujasite[a], Na-Y	MTT, ZSM-23	ATF, $AlPO_4$-25
LIO, liottite[a]	FER, ferrierite[a]	MTW, ZSM-12	ATN, MAPO-39
LOS, losod	GOO, Goosecreekite[a]	NON, nonasil	ATS, MAPO-36
LTN, Na-Z-21	HEU, heulandite[a]	SGT, sigma-2	ATT, $AlPO_4$-12 TAMU
NAT, natrolite[a]	KFI, ZK-5	SOD, sodalite	ATV, $AlPO_4$-25
PAR, partheite[a]	LOV, lovdarite[b]	TON, theta-1	AWW, $AlPO_4$-22
PHI, phillipsite[a]	LTA, ZK-4	ZSM-48	BPH, beryllophosphate-H
ROG, roggianite[a]	LTL, linde L		CAN, tiptopite[a]
SOD, sodalite	MAZ, mazzite[a]		CHA, SAPO-47
THO, thomsonite[a]	MEI, ZSM-18		CLO, cloverite
WEN, wenkite[a]	MER, merlinoite[a]		ERI, $AlPO_4$-17
	MON montasommaite[a]		FAU, SAPO-37
	MOR, mordenite[a]		GIS, MgAPO-43
	OFF, offretite[a]		RHO, pahasapaite[a]
	PAU, paulingite[a]		SOD, $AlPO_4$-20
	RHO, rho		VFI, VPI-5
	STI, stilbite[a]		
	YUG, yugawaralite[a]		

[a]Natural materials.
[b]Beryllosilicates (natural).

(ripening) period is very crucial in the synthesis of a desired zeolite, and it differs from aluminosilicate zeolites to aluminophosphate zeolites [78]. One of the important steps occurring during the aging period is the (partial) dissolution or depolymerization of the silica sol, which is promoted by the alkaline conditions for zeolite synthesis [40]. Cook and Thompson [79] have simulated the hydrothermal zeolite synthesis with the aging of the amorphous gel at low temperature followed by elevated temperature crystallization by population balance methods.

The amorphous gel is formed at room temperature, homogenized, and left to age for a period of time at room temperature. The subsequent crystallization at elevated temperature after aging proceeds more quickly than in the nonaged case. Several authors have noted that germ nuclei formed within the gel (or the solution) during gel aging grow to observable sizes upon subsequent high-temperature synthesis [40,80]. Zhdanov and Samulevich [80] reported that aging zeolite Na-A gels for up to 3 days resulted in a more rapid conversion upon heating and in smaller final crystal sizes. They concluded that nuclei form during aging and then grow to observable size during heating. Subotic and Bronic [81] reported experimental values of percent crystallinity versus time for zeolite Na-A syntheses at 80°C, for which some of the amorphous gel/water solutions had been aged at 25°C for as many as 17 days prior to crystallization. Upon heating the aged solutions, those aged for longer times were shown to crystallize faster after a shorter induction time. These data have been plotted in Figure 6.8. The evolution of the average zeolite crystal size with time, based on the computer simulations is shown in Figure 6.9. As the aging time increases, the number of nuclei increases, thereby decreasing the final average crystal size. This result is perfectly logical, as with more nuclei present, the reagent mass is distributed among more crystals, resulting in smaller crystals on average. Figure 6.10 shows the number of nuclei formed per day by dividing values of Ns (No. of nuclei formed during aging).

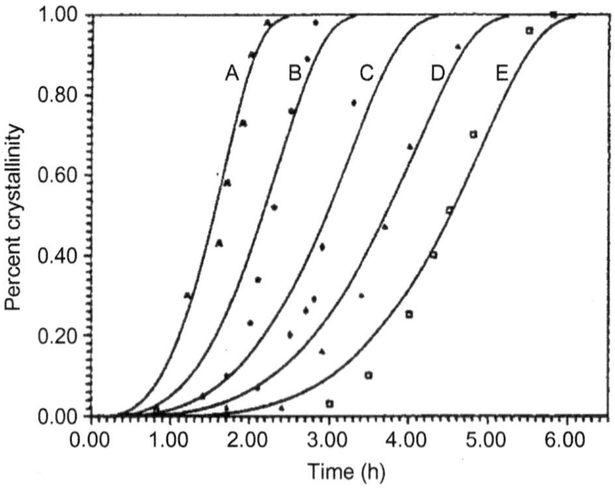

Figure 6.8 Crystallization curve simulated assuming nuclei form during reaction time [67].

Wu and Chao [82] have studied systematically the gel aging through Raman spectroscopy and derivative thermogravimetry DTG (Digital TV Group). Figures 6.11 and 6.12 show the room temperature for Raman spectra of the gel phase obtained during aging at ambient temperature and reaction at 115°C during the synthesis of faujasite zeolites [82]. In the early stages of aging, insoluble silica and unreacted $Al(OH)_4^-$ retained in the gel, A, give bands at ~441 cm^{-1} and ~617 cm^{-1}, respectively. At the end of aging, the $Al(OH)_4^-$ species in the solid part and Al species in the solution part of gel A are found almost to disappear, and this disappearance might be a result of the formation of the aluminosilicate hydrogel, which is considered to be the precursor of zeolite. Upon heating, the insoluble silica dissolves progressively so that the concentration of silicates in solution measured by inductively coupled plasma-atomic emission spectroscopy ICP-AES

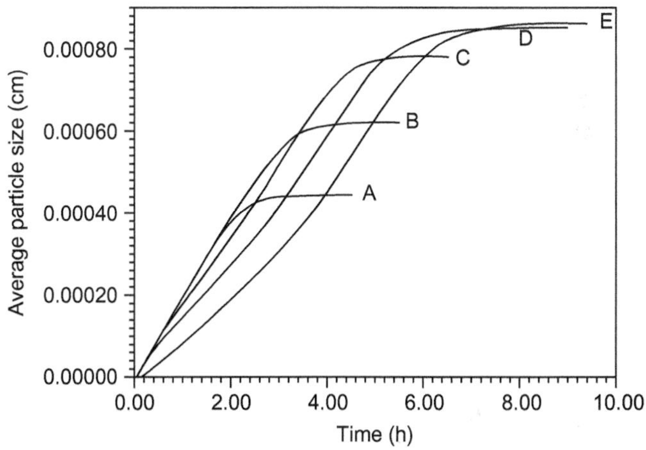

Figure 6.9 Evolution of the average zeolite crystal size with time [81].

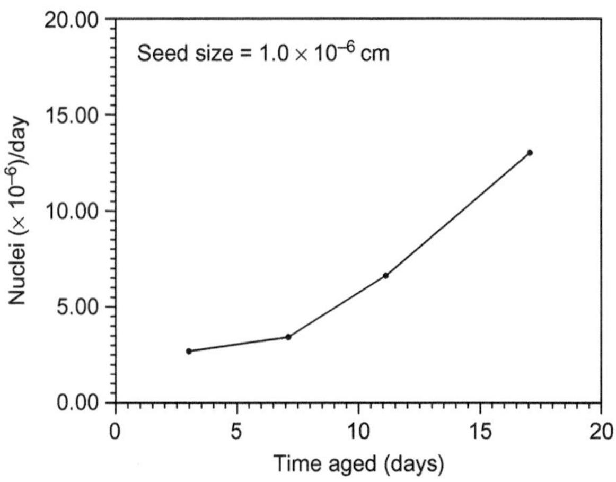

Figure 6.10 Number of nuclei formed per day by dividing values of Ns [81].

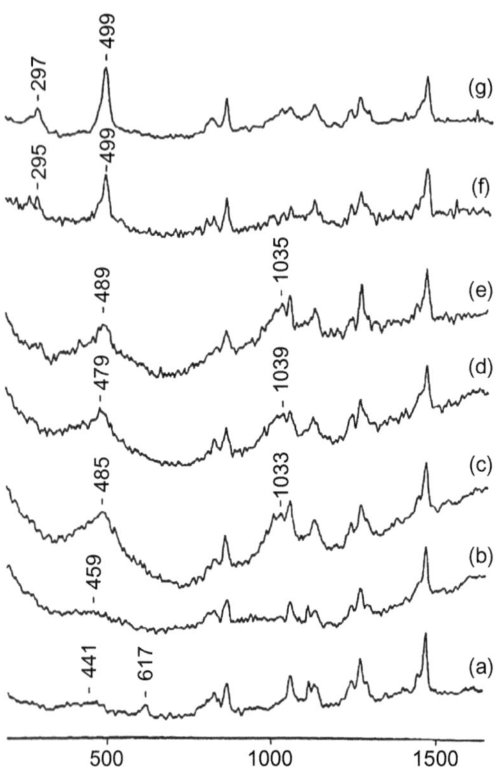

Figure 6.11 Raman spectra of the gel phase obtained during aging at ambient temperature, and reaction at 115°C during synthesis of faujasite zeolites [82].

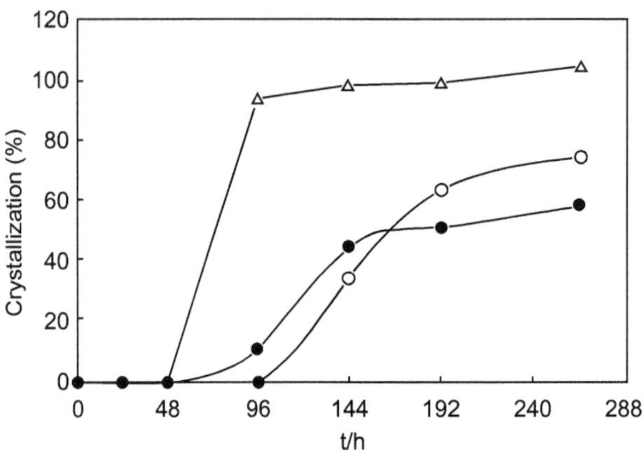

Figure 6.12 Crystallizaton curve of EMT (○: E-4 or gel A, Δ: A-1) and FAU (●: E-8 or gel) for the 18-crown-6 ether preparations [82].

increases significantly. After the sample has been heated for 1–5 days, the cyclic tetrameric aluminosilicate and silicate anions are observed at 479–489 cm^{-1}.

The DTG curves have been recorded for gel after different hours of aging at room temperature by Wu and Chao [82] and the results are shown in Figures 6.13 and 6.14. The thermogravimetry/differential scanning calorimetry TG/DSC (thermal analysis with online mass spectrometry) showed three distinct stages of

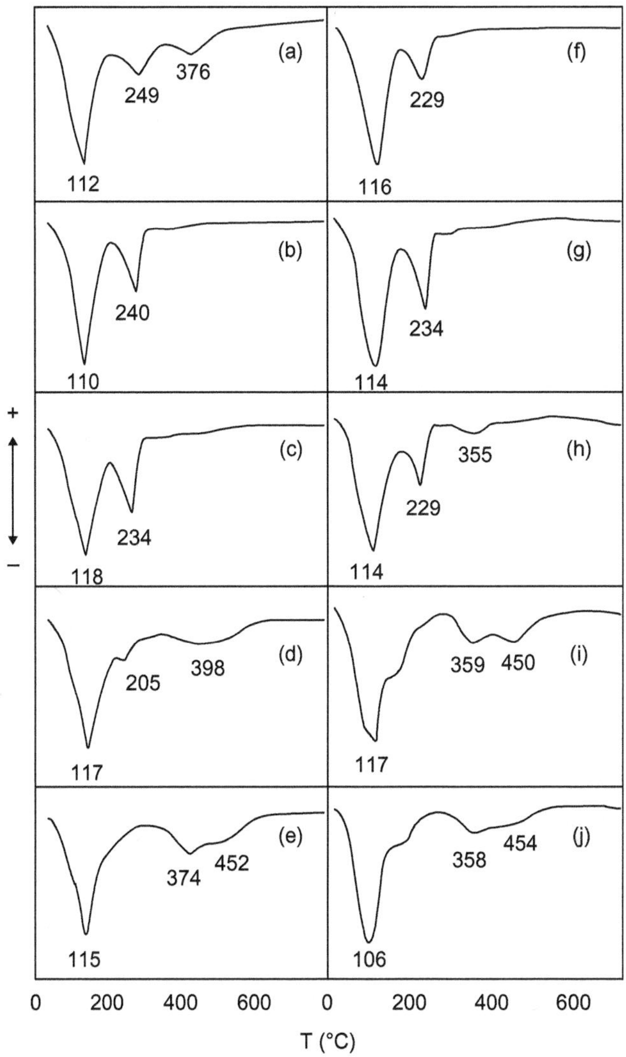

Figure 6.13 DTG curves of different gels (a) 24 h of aging at room temperature and (b) 1 day, (c) 3 days, (d) 5 days and (e) 7 days of heating at 115°C, and gel B after (f) 24 h of aging at room temperature and (9) 1 day, (h) 3 days, (i) 5 days and (j) 7 days of heating at 115°C [82].

weight loss on the faujasites, 25−200°C, 200−300°C, and 300−500°C. The first endotherm is mainly due to the desorption of water, while the second and the third result from the exothermic decomposition of organics in faujasite zeolites.

Similarly, Bell [83] and Gilson [84] have attempted to identify the nature and relative amounts of the silicate and aluminosilicate species present in the mother liquor using ^{29}Si and ^{27}Al NMR spectroscopy. However, ^{29}Si NMR technique has an important limitation, i.e., silicate species containing more than 12 Si atoms have

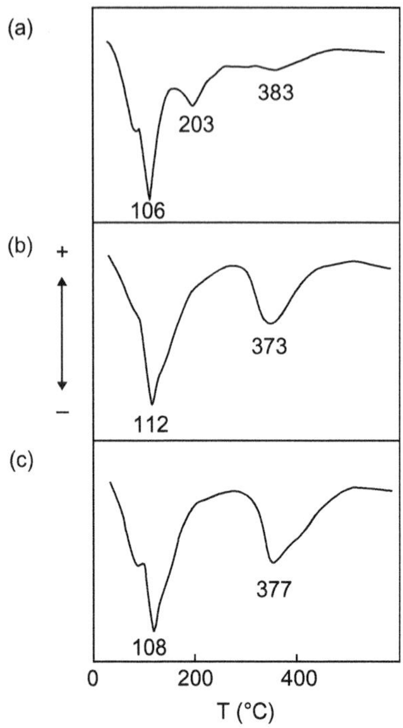

Figure 6.14 DTG curves of different gels (a) 24 h of aging at room temperature, (b) 1 day and (c) 3 days of heating at 115°C [82].

never been unequivocally identified [77]. This information has been used to understand the effect of aging of hydrogel in zeolite synthesis.

These studies indicate that the zeolite nuclei, the hydrogel or precursor, and the liquid solution are in equilibrium during crystallization, and because of the existence of this equilibrium, the specific precursor or hydrogel can play a very important role in influencing the form of nuclei and final productions. Also, the encapsulation of organic templates or inorganic cations in the hydrogel seems to determine the type of zeolite production. Thus, gel aging controls the crystallinity and size of the zeolites synthesized to a large extent.

6.5.3 *Water in Zeolite Synthesis*

As mentioned earlier, zeolites could be synthesized only by the hydrothermal method. The versatility of hydrothermal chemistry owes much to the mineralizing role of water [40,85]. The factors that promote reactivity in aqueous magmas include the following [86]:

1. Stabilization of porous lattices as zeolites by acting as space fillers, referred to above.
2. Through its presence, especially at high pressures, water may be incorporated into hydrous glasses, melts, and solids. Through chemisorption into siliceous materials, Si—O—Si and Al—O—Si bonds hydrolyze and reform. Chemical reactivity is enhanced and magma viscosity is lowered.

3. High pressures of water can modify phase equilibrium temperatures.
4. Water is a good solvent, a property that assists disintegration of solid components of a mixture and facilitates their transport and mixing.

Water is important as a guest molecule in zeolite structures with relatively high Al contents and consequently aqueous media favor their formation while salts have a parallel role in the stabilization of sodalite and cancrinite. Highly siliceous zeolites, such as silicalite and ZSM-5 (Zeolite Socony Mobil-5), which have zeolitic structures, but contain very little Al^{III} and M^{m-}, are less hydrophilic than aluminous zeolites and their structures can be stabilized by certain organic guest molecules, notably amines, alcohols, and amino-alcohols. In general, the zeolitic water can be removed leaving the unchanged hydrous zeolite. In hydrothermal systems, the good solvent powers of water promote mixing, transport of materials, and facilitates nucleation and crystal growth. Water stabilizes zeolite structures by filling the cavities and forming a type of solid solution. The stabilizing effect is such that the porous aluminosilicates will not form in the absence of a *guest* molecule, which may be a salt molecule as well as water. However, the water concentration or the degree of dilution is of minor importance for the synthesis of ZSM-5, which can crystallize out of gels with an extremely wide range of H_2O/SiO_2 ratios (from 7 to 22). During 1970s, Flanigen and Patton [87] introduced a new route for zeolite synthesis that involved the use of F^- as the mineralizing agent. This method was further developed by Guth et al. [88]. Replacement of OH^- by F^- allows the crystallization of zeolites, pure-silica ZSM-5, MFI (Membrane Fouling Influence), FER (iron-sulfur cluster), MTT (3-(4,5-Dimethylthiazole-2-yl)-2,5-diphenyltetrazolium bromide, a yellow tetrazole), MTN (Metriol Trinitrate), and TON (Turnover Number), at neutral or acidic conditions.

6.5.4 Temperature and Time

Temperature and time have a positive influence on the zeolite formation process, which occurs over a considerable range of temperatures. Based on geological evidence, an upper limit of 623 K has been suggested. Analcime has been obtained at 639 K, mordenite up to 703 K, clinoptilolite up to 643 K, and ferrierite at 648 K [40]. A rise in temperature will increase both the nucleation rate and the linear growth rate (expressed as $K = 0.5\ \delta l/\delta t$ with l = crystal size); hence, the crystallinity of the samples normally increases in time. As far as time is concerned, zeolite synthesis is governed by the occurrence of successive phase transformations (Ostwald rule of successive phase transformation). The thermodynamically least favorable phase will crystallize first and will be successively replaced in time by more stable phases [40]. The best example is the crystallization sequence of amorphous→faujasite→Na-P (gismondine type).

The temperature, however, can also influence the type of product that has to be crystallized. A rise in temperature leads to the crystallization of more dense products as the fraction of water in the liquid phase, which has to stabilize the porous products by filling the pores, will drop. Therefore, the existence of an upper limit for the formation of zeolites is to be expected. The use of a nonvolatile pore space occupying (filling) species would, in principle, allow a high-temperature synthesis of open, porous structures [11].

Temperature can obviously affect the rate of nucleation and crystal growth. The rate of nucleation and the linear rate of crystal growth are shown as follows [28]:

$$dN/dt = A[\exp(Et)^{-1}]$$
$$r = kt$$

The coefficients *K*, *A*, and *E* in Table 6.14 and *K* in Table 6.15 all increase with temperature, indicating that linear rates of crystal growth (coefficients *K*) and rates of nucleation (coefficients *A* and *E*) both increase with rising temperatures.

When the temperature is elevated, the induction time of s-shaped crystallization curves is shortened. It is also observed that the temperature strongly affects the induction time. Zhdanov and Samulevich [80] have developed a method to analyze nucleation and crystal growth:

$$\frac{d\ln(1/\theta)}{d(1/T)} = \frac{En}{R}$$

where θ is the induction time, i.e., a point on the crystallization curve where the conversion into the crystallization phase just begins. The apparent activation energy for the nucleation of phillipsite was given as 13.5 and 14.3 kcal/mole with and without stirring respectively, while for the Al-free end member of the ZSM-5 series (silicalite 1), this energy was given as 9.1 kcal/mole [90,91]. The *En* value for Na-Y was obtained as 17 kcal/mole [91].

6.5.5 Alkalinity (pH)

Alkalinity of the media plays a vital role in crystal growth, materials synthesis/preparation, and processing on the whole. It influences the supersaturation, kinetics,

Table 6.14 K, A, and E, for Zeolite Na-A as Function of Alkalinity and Temperature [89]

T (k)	H_2O/Na_2O	*K* (μ m h^{-1})	$A \times 10^6$	*E*
343	20	0.050	20.0	0.102
343	30	0.027	13.2	0.033
343	40	0.017	9.0	0.115
333	20	0.03	1.0	0.062
343	20	0.05	20.0	0.102
353	20	0.06	200.0	0.132

Table 6.15 Linear Rates $K = 0.5\ \Delta l/\Delta t$ for Growth of Na-X [80]

T (k)	343	353	363	373
K (μ m h^{-1})	0.0175	0.0375	0.0625	0.1071

morphology, shape, size, and crystallinity of the particles or materials as the OH^- anions fulfill the crucial role of mineralizing agents. The pH is influenced by the reactants and their concentrations/ratios, followed by temperature and time. Furthermore, with the introduction of organics, the pH changes rapidly in the system; hence, pH is the key parameter in determining the crystallization rate. An increase in OH^- concentration will generally bring about an accelerated crystal growth and a shortened induction period before viable nuclei are formed.

In zeolite synthesis, the pH of the alkaline solution is usually between 8 and 12. The major role of pH is to bring the Si and Al oxides or hydroxides into solution at an adequate rate. It is well known that OH^- is a powerful mineralizer. Water as a mineralizer has been discussed in detail in Chapter 4. Water has much increased the coefficient of expansion under hydrothermal conditions. The PVT measurements for water and other solvents are rather rare. The PVT data for water up to 1000°C and 10 Kbar are known accurately within 1% error [85]. Figure 4.10 shows the temperature–density diagram of water, with pressure as a parameter [85]. At a very high PT condition (1000°C and 100 kbar), water is completely dissociated into H_3O^+ and OH^- behaving like a molten salt, and has a higher density of the order of 1.7–1.9 g/cm^3.

The OH^- ion is a good complexing agent, which can bring amphoteric oxides and hydroxides into solution. An increase in pH will accelerate crystal growth and shorten the induction period (period before formation of viable nuclei) by an enhanced reactant concentration [89,92]. For example, the values shown for Linde zeolite, Na-A, (Table 6.15) indicate that the nucleation rate dN/dt, for both coefficients A and E increase strongly with increasing pH. The solubility of silica increases nearly exponentially with the concentration of alkali and, in the resultant mixture, a range of silicate anions may appear with various degrees of oligomerization. In the case of alumina at high pH minimal oligomerization, there is always the dominant anion, $Al(OH)_4^-$. The high pH causes supersaturation of silicate and aluminate and the formation of a large number of nuclei. The growth of these nuclei proceeds until the aluminum in the gel is exhausted so that the alkaline media enable the ready mixing of reactants and also facilitate nucleation and crystal growth. Lechert [93] has given the importance of pH values in the crystallization of zeolites.

6.5.6 Structure-Directing and Composition-Determining Species (Templating)

Templating has become an important tool for material preparation and processing. Both organic and inorganic agents can be used as templating agents. Generally speaking, this processing route uses channels, layers, and cavity spaces of nanoporous host structures to grow nanoscale objects [94]. In some aspects, templating is related to biomimetics as the controlled nucleation, and growth of inorganic phases in the organic host is an important step of biomineralization [95]. There exists a variety of organic and inorganic host materials with 1D, 2D, and 3D structures as reviewed in Ref. 94. The concept of templating in molecular sieve synthesis is not new.

In zeolite synthesis, the term *templating* is used in a loose sense to refer to the filling of the void spaces of a zeolite structure by the bulky organic/inorganic cations frequently used as structure-directing or structure-stabilizing agents. They contribute to the formation of the zeolite lattice during the zeolitization by influencing the gelation and/or nucleation processes and by lowering the chemical potential of the lattice formed by the inclusion of the templates during zeolite synthesis. Templating contributes to the stability through new interacts (hydrogen bonds and electrostatic and London dispersion interactions) and further controls the formation of a particular topology through its geometry (form and size) [96]. Thus, the action of a template appears to have both electronic and steric components. It is evident that changes in the cation density provoked by geometrical or physical properties of the template will be reflected in the chemical composition (Si/Al ratio) of a given topology. In the early 1960s, the TMA (Thermal-Mechanical Analysis) cation was introduced as the first organic cation to be used in zeolite synthesis [47,97].

Thus far, it has not been possible to predict the actual role of templates and which template is reported to obtain a particular structure and composition. Therefore, in selecting possible templates, one has to bear in mind some general criteria regarding the templating potential in zeolitization, such as solubility in the solution, stability under synthesis conditions, steric compatibility, and possible framework stabilization, and the most important practical issue is to remove the template without destroying the framework. The synthesis of omega zeolite is strongly favored by the presence of TMA^+ ions; thus, templates can be structure-making or structure-breaking agents [98,84]. Also, it is well known that neutral molecules, as well as cations or ion pairs, are able to fulfill structure- and composition-directing functions. Each one of these can be organic or inorganic in nature. Recently, Feng et al. [99] have carried out the amine-directed synthesis for some of the zeolites. It is clear from the literature data that multiple small organic cations cooperatively template the extra-large micropores, in contrast, highly siliceous zeolites template as one bulk large molecule. On the basis of structural analysis of the extra-large micropore compounds, Li et al. [58] proposed the following formation mechanism (Figure 6.15): In the initial reaction, the inorganic and

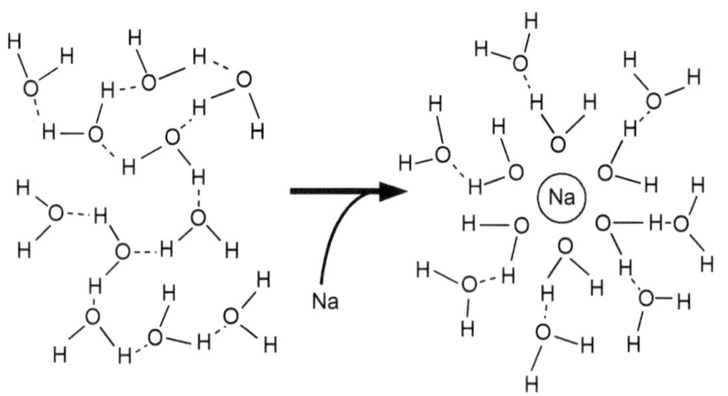

Figure 6.15 Schematic diagram of water–cation interaction [40].

organic species randomly distribute in the gel. Then a self-assembly process occurs among these reaction species driven by the nonbonding interactions (e.g., electrostatic and H-bonding) between the inorganic and organic species. The arrangement of the organic and inorganic species gradually becomes orderly, where several small organic amines assemble together with their hydrophobic ends pointing inward and the hydrophilic ends interacting with the inorganic species. The formation of such ordered inorganic–organic composite species provides the fundamental structural units for nucleation and crystal growth. However, the self-assembly process of the inorganic–organic species needs to be further investigated [58].

The cations influence greatly the type of zeolite that crystallizes from aqueous alkaline magmas. Also, in the laboratory, synthesis of zeolite cations occupy an important place, not only as structure- and composition-directing agents but also as agents influencing the rate of zeolite synthesis [28,40]. The species such as Na^+, Li^+, Cs^+, K^+, Rb^+, Ca^{2+}, Sr^{2+}, tetra-alkylammonium cations (e.g., TMA^+, TEA^+, TPA^+, dihydroxyethyl-dimethylammonium), dialkyl and trialkyl amines, and phosphonium compounds are frequently used. Sometimes, neutral molecules, like crown-ethers, are used to form charged species upon complexation of an alkali metal cation.

Structure-making cations are small cations, like Na^+ and Li^+, which interact strongly with water molecules because of their high charge density. As a result of this, the original hydrogen bonds are broken and the water molecules will be (reorganized) organized around the cations. This water–cation interaction is shown schematically in Figure 6.16.

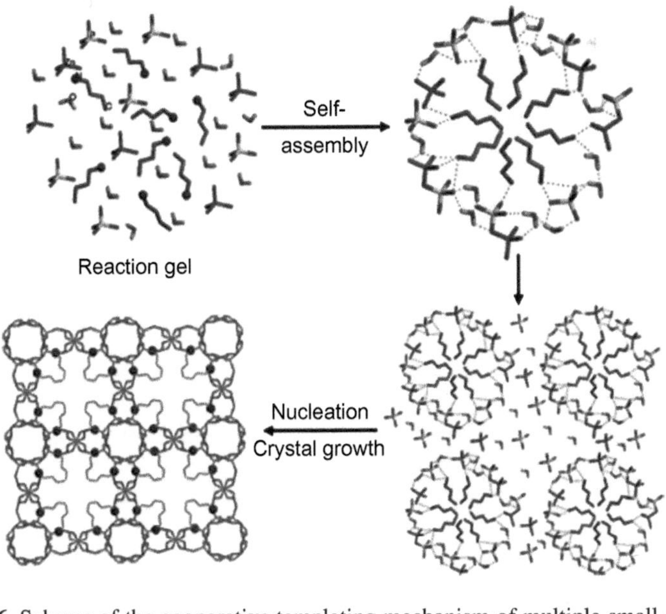

Figure 6.16 Scheme of the cooperative templating mechanism of multiple small amines in the formation of extra-large micropore [58].

The structure-breaking cations are large cations, like NH_4^+, K^+, and Rb^+, which also interact with the water molecules, break the original hydrogen bonds, and, therefore, are known to break the water structure [84]. However, there are some large cations, like tetra alkylammonium, which seem to organize the water structure; the other templating neutral molecules used are amines, ethers, alcohols, di, and triols, and so on [96,100].

The presence of salts such as NaCl, KCl, KBr, CaF_2, $BaCl_2$, and $BaBr_2$ during zeolite synthesis can be an important factor as these ion pairs may be occluded (possibly next to water molecules) into the pore system, thereby stabilizing the pore structure. These nonvolatile species increase the silica, which is soluble with an increase in temperature, and open several new pathways for zeolites synthesis and also for new products besides improving crystallinity and yield.

As mentioned earlier, the use of organics in the zeolite synthesis has opened up a debate whether their presence is required at all or the zeolite synthesis can be carried out even without these organics in the system. Application of templating methods to produce zeolite materials with hierarchical bi- or trimodal pore size is a popular area of research, although several nontemplated approaches also exist. In general, mesopores can be templates in zeolite materials by the use of solid templating, supramolecular templating, or indirect templating [101]. The researchers are contemplating the development of room temperature synthesis of mesoporous zeolites through an appropriate and highly efficient template selection [102]. Recently, Na et al. [103] have come out with a breakthrough in zeolite synthesis, which leads to the direction of zeolite structures into hierarchically nanoporous architectures. The organics also influence the size and morphology of the zeolites in addition to the structure-building activity. The surface characteristics are also altered significantly thus making the surface either hydrophobic or hydrophilic for specific applications.

The reader can go through the excellent reviews given in Refs [101]–[105] on the role of organic substances in the synthesis of zeolites. Table 6.16 indicates that organic bases facilitate the formation of several novel structures, especially those with extremely silica-rich zeolites, such as the pentasils, ZSM-5 and ZSM-11, and also zeolite Nu-1 and Fu-1 [40]. Table 6.17 shows the organic zeolite structure relationships [89]. In recent years, several zeolites, like ZSM-5, mordenite, and other related zeolite structures, could be synthesized in the absence of any organic compound; here the hydrated Na^+ ions are able to function as templates for the formation of SBUs.

In recent years, templating has found several new applications [106]. Among the inorganic materials, zeolites have been widely used as hosts for a variety of materials, including conjugated polymers, selenium, chains, and so on. Zeolites, as well as glass or alumina nanoporous membranes, have been used to fabricate organic and inorganic tubules or fibrils [107]. Carbon nanotubes can serve as removable templates for other inorganic nanorods/nanowires, e.g., GaN, Cu, or V_2O_5 [108]. The mesoporous materials can be fabricated with a variety of shapes and sizes, from nanostructures to almost millimeter-size objects [109].

Table 6.16 Syntheses of Some Zeolites in the Presence of Organic Bases [28]

Cations	Zeolite Type	Cations	Zeolite Type
TMA^+	Sodalite	$Li^+ + Cs^+ + TMA^+$	Cancrinite
	Gismondine		Li-ABW
$Na^+ + TMA^+$	Sodalite		Zeolite A
	Cancrinite		Edingtonite (K-F)
	Gismondine		Analcime
	Zeolite A (isotypes N-A,α,ZK-4)		Offretite (O)
			Zeolite (ZK-5) (KFL)
	Funijusite		Zeolite ZSM-2
	Zeolite TMA-E(EAB)		Gismondine
	Muzzite (zeolites W, ZSM-4)		Sodalite
	Zeolite (Na, TMA)-V	$Na^+ + K^+ +$	Erionite
	Benzyltrimethylammonium (isotypes N and Z-21)		Losod
	Zeolite Nu-1	$Na^+ + N$	Losod
	Zeolite Fu-1	$Na^+ + ONO$	
$Ca^{2+} + TMA^+$	Sodalite	Na^+ + neopentyltrimethyl-ammonium	Losod
	Gismondine		

(*Continued*)

Table 6.16 (Continued)

Cations	Zeolite Type	Cations	Zeolite Type
Ba^{2+} + TMA^+	Erionite		
	Zeolite L	Na^+ + CH_3N^- N^+CH_3	Zeolite ZK-5 (KFL)
	Erionite		Levynite
	Offretite	Na^+ + primary *n*-alkylamines (C_2—C_{12})	Zeolite ZSM-5 (MFL)
		Na^+ + isopropylamine, dipropylamine	Zeolite ZSM-5
	Chabazite	or dibutylamine	
Na^+ + $NH_2(CH_2)_6NH_2$	Zeolite ZSM-5 (MFL)		
NH_4 + tetrapropylammonium	Zeolite ZSM-5		
Na + tetrapropylammonium	Zeolite ZSM-5		
Na^+ + tetrapropylammonium	Zeolite ZSM-8		
Na^+ + tetrapropylammonium	Zeolite ZSM-11 (MFL)		
Na^+ + benzyltriphenylammonium	Zeolite ZSM-11		
Na^+ + tetrapropylammonium	Zeolite ZSM-11		
NH_4^+ + tetrapropylammonium	Zeolite ZSM-11		
	Zeolite ZSM-10		
K^+ + CH_3N	NCH_3		
Na^+ + K^+ + tetrapropylammonium	Zeolite ZSM-8		
with $Cr(OH)_3$, $Fe(OH)_3$, $Al(OH_3)$			

Table 6.17 Organic Zeolite Structure Relationships [89]

Organic	Structure	Structure Type
TEA	β	?
	ZSM-8	ZSM-5
	ZSM-12	?
	ZSM-20	Faujasite
	Mordenite	Mordenite
	ZSM-25	?
Methyltriethylammonium	ZSM-12	?
TPA	ZSM-5	ZSM-5
n-Propylamine	ZSM-5	ZSM-5
TBA	ZSM-11	ZSM-11
Choline	ZSM-38	Ferrierite
	ZSM-34	Erionite–Offretite
	ZSM-43	?
	CZH-b	?
TMA + TBA	ZSM-39	ZSM-39
TMA + *n*-propylamine	ZSM-39	ZSM-39
	ZSM-48	ZSM-12?
Pyrrolidine	ZSM-35	Ferrierite
	ZSM-21	Ferrierite
	ZSM-23	?
1,2-Diaminoethane	ZSM-5	ZSM-5
	ZSM-21	Ferrierite
	ZSM-35	Ferrierite
1,3-Diaminoethane	ZSM-35	Ferrierite ZSM-5
	ZSM-5	
1,4-Diaminoethane	ZSM-35	Ferrierite, ZSM-5
	ZSM-5	
1,5-Diaminoethane	ZSM-35	ZSM-5
1,6-Diaminoethane	ZSM-5	ZSM-5
1,7-Diaminoethane	ZSM-11	ZSM-11
1,8-Diaminoethane		ZSM-11
ZSM-11		
	ZSM-48	
1,9-Diaminoethane	ZSM-11	ZSM-11
1,10-Diaminoethane	ZSM-11	ZSM-11
DDO	ZSM-10	?
	ZK-5	?
MDO	ZK-20	Levynite
MQ	LZ-132	Levynite
	Nu-	Levynite
TQA	ZSM-18	?
BP	LOSOD	LOSOD
Dihexamethylene-triamine	ZSM-30	?
Neopentylamine	Mordenite	Mordenite

6.5.7 Nucleation

This forms an important part of zeolite synthesis and it is one of the most complicated aspects because the chemical precursors here are more complex and varied compared to the chemical precursors in the formation of many ionic solids, like M^+Cl (M^+ = Li, Na, K, Rb, Cs). Nevertheless, the stages of nucleation are more or less the same in all cases, i.e., small aggregations of precursors give unstable germ nuclei; some of these embryos become large enough to be stable nuclei and spontaneous deposits of more material on such nuclei result in larger crystallites.

The analysis of metastable solution structure is intimately related to nucleation theory. There are several approaches for understanding the nucleation: the classical theory and the modern nucleation theory [110,111].

In the first case, if a solution is in a metastable state (supersaturated solution), it will, sooner or later, enter another state, which is stable. Gibbs was the first to propose that the work of the critical nucleus formation can be considered as a measure of the metastable state stability. Gibbs' idea was further developed by Volmer–Doring, Frenkel, Zeldovich, and Lifshitz–Spyozov using thermodynamic concepts. The modern theoretical models of nucleation are based on the field-theoretic approach to nonequilibrium dynamics of the metastable state. These models provide an alternative to the classical theories.

Nucleation, on the whole, occurs from supersaturation and it can be of two types: Homogeneous and heterogeneous nucleation; the former occurs spontaneously, and the latter is induced largely by the presence of impurities or foreign particles present in the solution.

Barrer [40] has summarized the general properties of nucleation that can be applied to zeolite synthesis also.

1. Rates of nucleation increase with the extent of undercooling, i.e., with increasing metastability; however, viscosity also tends to increase, often rapidly as temperature falls, so that the effects of the degree of undercooling and of viscosity oppose one another in influencing nucleation rates that can then pass through a maximum as temperature falls.
2. An incubation period is observed, particularly in condensed phases, during which nucleation cannot be detected. Even in seeded solutions, metastable regions of supersaturation can occur within which nucleation is not detectable. In many phase studies involving solutions, well-defined composition boundaries are found beyond which nucleation occurs freely so that under appropriate conditions the rate of nucleation increases extremely rapidly with the degree of supersaturation.
3. The extent of the incubation time can be changed significantly by very small changes in composition.
4. The onset of nucleation often depends on the previous history of the system.

The formation of nuclei takes place during the entire process of crystallization, but the rate of nucleation increases only during its initial stage. During the formation of viable nuclei, different kinds of germ nuclei (embryos) will form by chemical aggregations of the precursor species mentioned above and disappear again upon depolymerization [40]. As a result of such fluctuations, in time the germ nuclei will grow and eventually form different kinds of nuclei with dimensions having the critical size to become viable, i.e., nuclei on which crystal growth occurs spontaneously.

The nucleation theory with reference to the zeolite synthesis can be approached from different angles. Subotic and Bronic [81] have developed a theory guided by their experimental observation. In this connection, the theory proposed by Zhdanov [64] and Zhdanov and Samulevich [80] infers that zeolite nuclei must be hidden in the amorphous gel and released during gel dissolution. The basis for their theory is the following empirical equation:

$$f(t) = \frac{mz(t)}{mz(tq)} = Kf$$

where f is the percent of zeolite relative to the final equilibrium value, t is the crystallization time, q is the constant power, and K is the constant coefficient.

Warzywoda and Thompson [112] proposed the autocatalytic nucleation that can be incorporated in the population balance formalism, which allows the inclusion of gel and solution material balances. Using the population balance formalism, one can simulate crystal size distribution behavior and extend these simple analyses to other crystallizer types. However, it has been shown that the predicted nucleation rate profile for such a process is that it is unlikely, as observed experimentally. Nuclei released from the gel are nucleated much later in the crystallization process than observed experimentally [112].

The net free energy of the formation of a spherical nucleus can be expressed in terms of degree of supersaturation (S), the density (ρ), and surface energy (σ) of the nucleus and is as follows [112]:

$$\Delta g = \frac{16\pi\sigma^3(MW)^2}{3(RT\rho \ln S)^2}$$

It is obvious from the above equation that from a saturated solution ($S = 1$, $\ln S = 0$), no spontaneous nucleation can occur as the value for Δg equals infinity. For a supersaturated solution ($S > 1$), Δg has a finite negative value and hence spontaneous nucleation is possible [112].

The nucleation rate, J (number of viable nuclei formed per unit of time) is commonly described as an Arrhenius rate equation [113]:

$$J = A \exp\left(\frac{-\Delta g}{RT}\right)$$

The above relation predicts an exponential increase in the nucleation rate for an increasing supersaturation degree once a critical degree of supersaturation has been reached.

Experimentally, the rate of nucleation can be derived from crystal size distribution measurements of the final crystallization product and size increase measurements of the largest crystals during the course of crystallization [11,20]. As the nucleation and crystal growth are assumed to consume the same precursor species, the nucleation rate is expected to go through a maximum and decline again after a certain period of time when the consumption of precursor species by crystal growth will limit their availability for forming new nuclei [113].

Several kinetic models have been proposed for the crystallization of zeolite based on the above considerations. A reader can get more information in Refs [28,40,112,114]. Figure 6.17a shows the growth in the number of nuclei and change in the rate of their formation during crystallization A and Figure 6.17b shows the crystallization of zeolites as a function of time. Figure 6.18a shows the crystal size growth of Na-X zeolite during crystallization of the gel and the

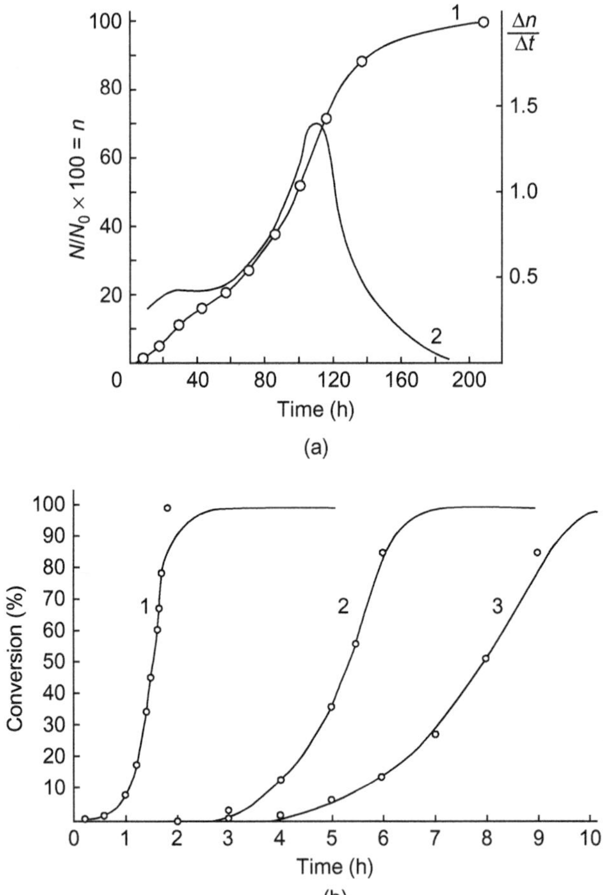

Figure 6.17 (a) Growth of number of nuclei (1) and change in rate of their formation (2) during crystallization of zeolite A. (b) Crystallization of zeolites as a function of time: (1) zeolite A, 373 K, (2) zeolite X, 373 K, and (3) mordenite, 573 K.

histogram of crystal size distribution in the final product Figure 6.18b shows the derivative of curves (1) and (2) of Figure 6.18a nucleation kinetic curve, and (3) the crystal mass growth curve for Na-X zeolite calculated from the data in Figure 6.18a and b.

In recent years, there has been considerable progress made in the field of accelerating the kinetics of low-temperature inorganic synthesis [115]. Several techniques

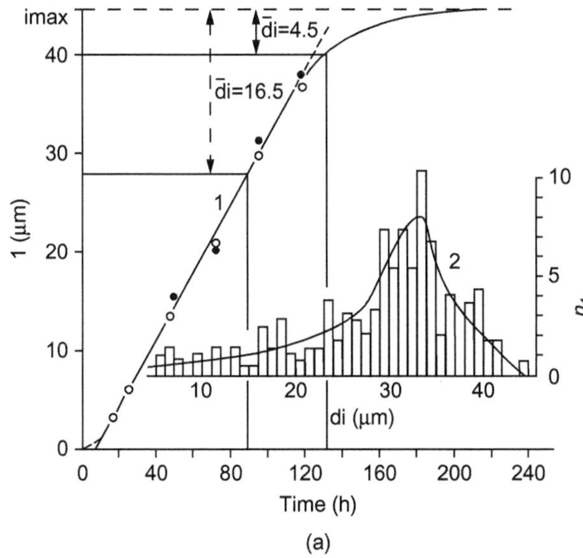

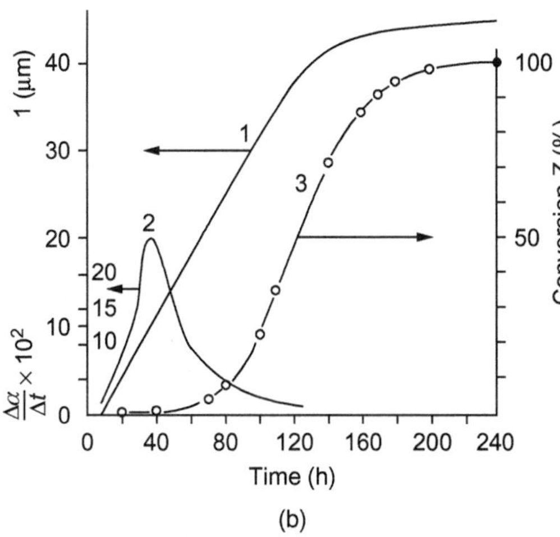

Figure 6.18 (a) Crystal size growth of Na-X zeolite during crystallization of the gel (1) and the histogram of crystal size distribution in the final product (2). (b) (1) Derivative of curves 1 and 2 of (a), (2) nucleation kinetic curve, and (3) the crystal mass growth curve for Na-X zeolite [28].

such as sonochemical (acoustic wave stimulation), ionizing and nonionizing radiation, microwave-hydrothermal, hydrothermal-electrochemical, hydrothermal-mechanochemical, including very high-uniaxial pressure and shear are being used extensively to accelerate the reaction kinetics. In order to have a coherent approach to correlate the kinetics of crystallization with crystal size distribution, Figure 6.19 shows the conceptual diagram of nucleation and crystallization of zeolite A. Carmona et al. [116] have explored the crystallization of Virginia Polytechnic Institute No. 5 (VPI-5), a microporous $AlPO_4$ phase, using conventional boiling and refluxing systems, conventional hydrothermal reactors, microwave-activated refluxing systems, and microwave hydrothermal reactors. They explained the microwave activation techniques in terms of the influence of microwaves in the change of coordination of Al in the gel from octahedral Al (o) to tetrahedral Al (T), to yield the specific Al (o)/Al (T) ratio in the gel precursor necessary to direct the crystallization of VPI-5 and in terms of the rapid heating obtained by using microwaves necessary to reduce the probability of the formation of intermediate metastable phases [116].

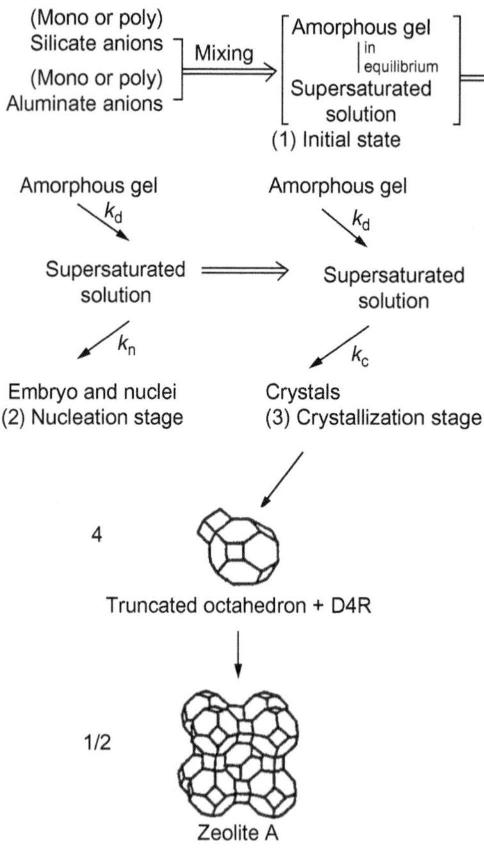

Figure 6.19 Conceptual diagram of nucleation and crystallization of zeolite A [28].

6.6 Crystal Growth

The synthesis of zeolites at temperatures around 100°C usually produces crystals in the size range 0.1–10 μm. In applications as catalysts, sorbents, and ion exchangers, this size range is fully suitable. The small size indicates that once nucleation begins there is a rapid formation of numerous tiny particles all competing for the available chemical nutrients. In order to carry out the single crystal X-ray diffraction studies, zeolites should be 100 μm, or even more, in diameter. Similarly, for other examinations, like molecule and ion-selective membranes or other electrochemical studies, bigger crystals are required. Therefore, crystal growth of zeolites can be both challenging and interesting as well. Natural zeolites occur as large single crystals; however, these are often impure, rare, and require conditions that cannot be reproduced in the laboratory [49]. Large zeolite crystals are difficult to produce in the laboratory by conventional hydrothermal synthesis where crystal formation is controlled to a large extent by the solubility of the reagent gel particles, the rate of generation of nucleation centers, and the solubility of the resulting products [64]. The low solubility of the crystallization products leads to a high degree of supersaturation, causing high nucleation leading to the formation of small crystals. A careful monitoring of the reactants, dilution temperature, agitation, and crystallization time has yielded some large crystals of some zeolite types. Similarly, the use of slow-release complexing agents seeding sub- and supragravity conditions to increase crystal size has met with limited success. There are several works on the crystal growth of zeolites.

Crystal growth succeeds nucleation, i.e., the nuclei will grow by addition or condensation of precursor species toward full-grown crystals. The autocatalytic nature of the first stage of the crystallization reflects the self-accelerating behavior of a crystallization process.

Crystal size can be influenced by experimental factors which alter the numbers of viable nuclei that develop so that if these numbers are reduced a greater share of the chemical nutrients is available for each growing crystal. The use of seed crystals can be similarly employed in the case of zeolites, Na-Y and Na-X. The smaller the number of seeds involved, the larger the average size of the final crystals because a larger amount of chemical nutrient is available for each crystal before these nutrients are exhausted. Ghobarkar [117], Ghobarkar and Schaf [118–120], Kuperman et al. [121], Mel'nikov et al. [122], and Demianets et al. [123] have carried out extensive work on hydrothermal crystal growth of several zeolites and studied their morphology in detail. Figures 6.20–6.23a–j show the characteristic photographs of various zeolite crystals obtained under hydrothermal conditions. Kuperman et al. [121] have used a nonaqueous solvent, a mineralizing agent, reagent quantities of water, an optional organic template, silica or silica and alumina, or alumina and phosphorous, and oxide sources to obtain giant crystals of zeolites and molecular sieves [121]. A close examination of the reactant products, under a polarizing microscope, did not show any twining. Ghobarkar and Schaf [119,120] have carried out hydrothermal experiments and found that at 1 kbar

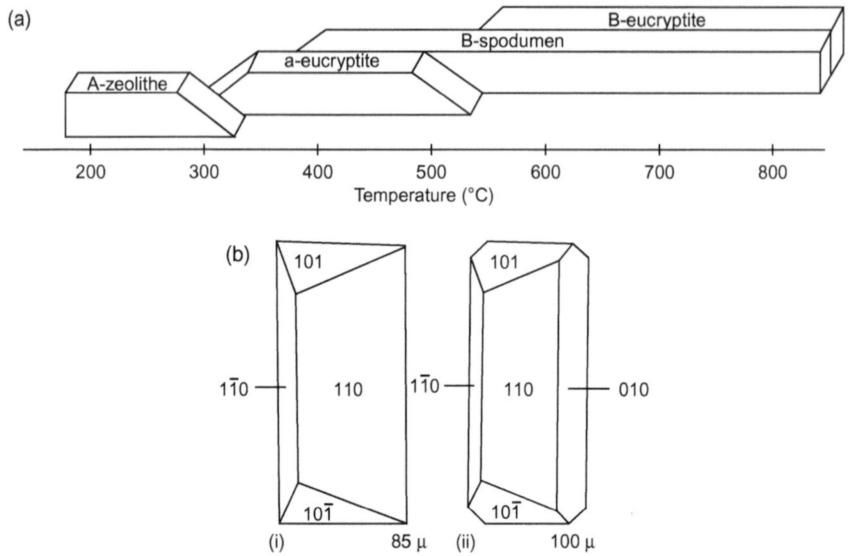

Figure 6.20 (a) Schematic diagrams of zeolite crystals obtained under hydrothermal conditions. Development of mineral phase depending on the temperature. (b) Crystal of A-zeolite at (i) 150°C and (ii) 250°C [117].

pressure and at an experimental duration of 6 weeks maximum, the zeolite phases, thomsonite and edingtonite, were formed in the whole temperature range between 80°C and 230°C, and at 1 kbar pressure and at an experiment duration of 8 weeks maximum, the zeolite phase-natrolite was formed in a temperature range between 80°C and 150°C.

Among all the zeolite groups of materials, only sodalite has been obtained as bulk single crystals, owing to its attractive optical properties as nanocomposites in quantum electronics and nonlinear optics. Sodalites may be represented by the general formula M_8 $[AlSO_4]_6 \cdot X_2$, where M = Na, Li, Ag, . . . , and X = Cl, Br, ClO_4, ... The sodalite framework consists of alternating SiO_4 and AlO_4 corner sharing tetrahedra forming 4 and 6 rings, which make up the basic β-case unit common to several zeolites. A semicondensed structure is formed through the linkage of the β-cases via six rings, with M_4 X clusters located at the center of each cage as shown in Figure 6.24 [124]. The replacement of the framework Al and Si by the larger Ga and Ge affects the structure, the introduction of larger framework atoms may also allow the enclathration of larger anions which cannot be normally incorporated into aluminosilicate sodalites. The most promising method for the production of sodalite single crystals is the hydrothermal method, as any attempt to grow sodalite from the melt simply produces glass of an analogous composition while growing from solution in the melt yields small crystals of <2 mm [123]. Thus, sodalite is normally synthesized from strong basic media under mild hydrothermal conditions; however, by using ethylene glycol as a solvent, Bibby and Dale [125] prepared a purely siliceous

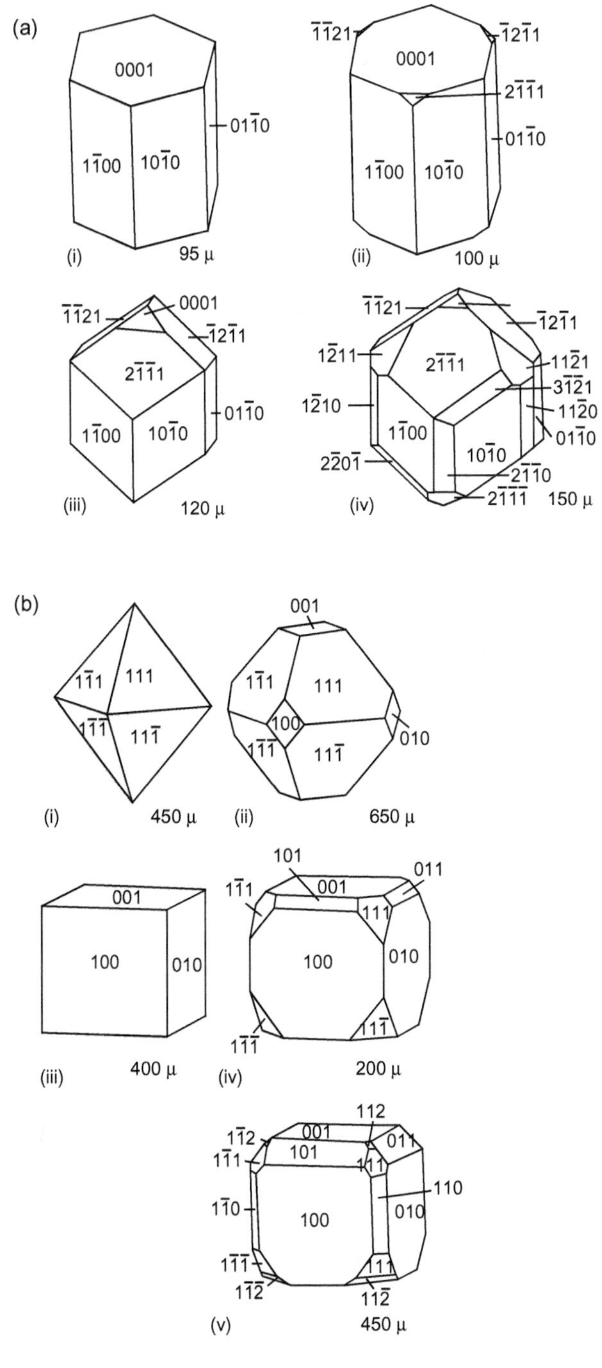

Figure 6.21 (a) Crystal of α-eucryptite at (i) 300°C, (ii) 350°C, (iii) 400°C, and (iv) 450°C. (b) Crystals of β-spodumene at (i) 350°C, (ii) 400°C, (iii) 500°C, (iv) 550°C, and (v) 650°C [117].

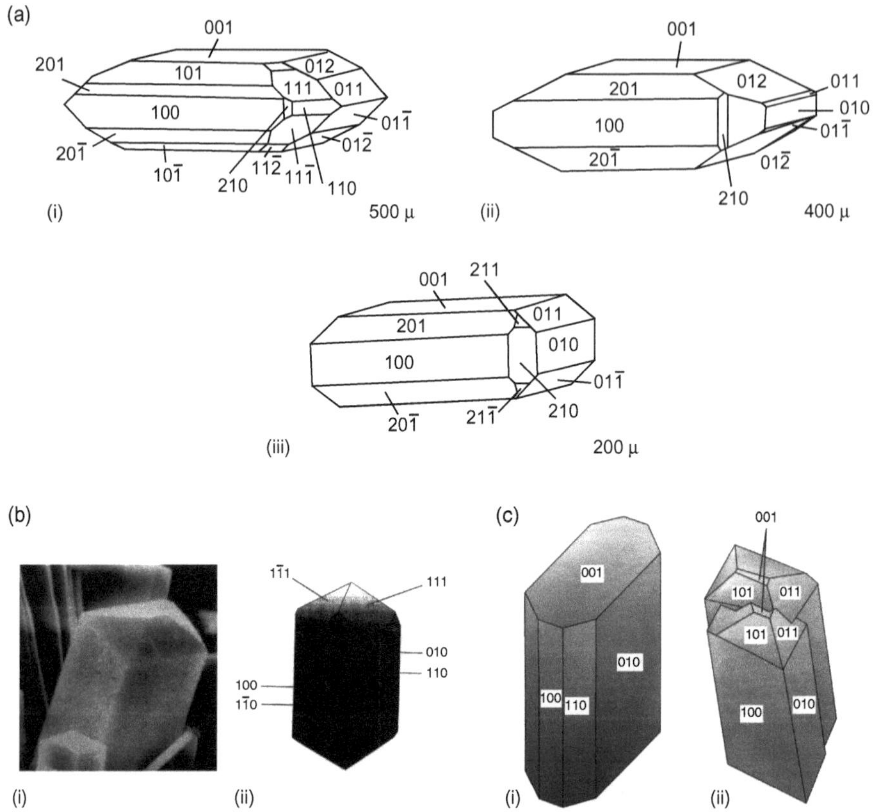

Figure 6.22 (a) Crystal of β-eucryptite at (i) 500°C, (ii) 750°C, and (iii) 850°C. (b) (i) SEM and (ii) schematic diagram of natrolite. (c) Crystal morphology of (i) thomsonite and (ii) edingtonite [119].

sodalite [40]. In fact, the exploration in hydrothermal systems of reactions yielding salt-bearing sodalites began nearly 130 years ago [126,127]. Carlgren and Cleve [127] made sodalites from kaolinite and aqueous caustic soda containing the relevant dissolved salts at temperatures ranging about from 185°C to 220°C. Sodalite can also be made very easily in the absence of any salts; if made so then they contain only zeolitic water and a limited amount of soda. However, the production of large single crystals under hydrothermal conditions is very long and cumbersome. In fact, some experiments have been carried out for 6 months. Demianets et al. [123] have studied the solubility of sodalite in aqueous solutions of NaOH (10–40 wt%) between 200°C and 300°C. Figure 6.25 shows the dependence of solubility of sodalite and the NaOH concentration at constant temperatures. The soluble curve of natural sodalite (hackmanite) lies rather below the curve plotted for the solubility of artificial sodalite. The synthetic mineral dissolves considerably at a rapid rate.

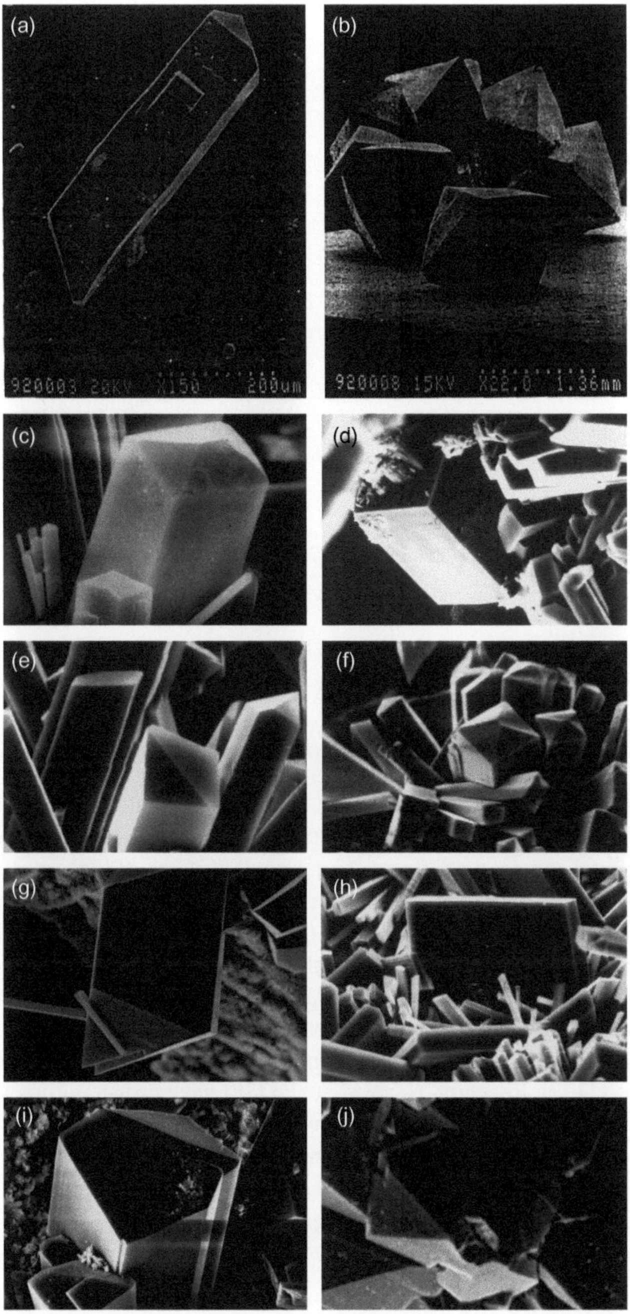

Figure 6.23 SEM of crystals of (a) silicaline-1, (b) dodecasil-3C, (c) natrolite, (d) chabazite, (e) A-zeolite, (f) merlinoite, (g) dachiardite, (h) thomsonite, (i) epistilbite, and (j) gmelinite. *Source*: Photographs courtesy of Prof. H. Ghobarkar.

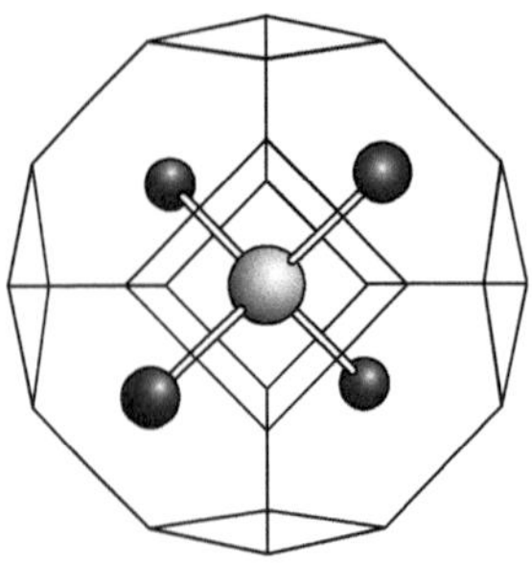

Figure 6.24 Sodalite cage containing M_4X cluster [124].

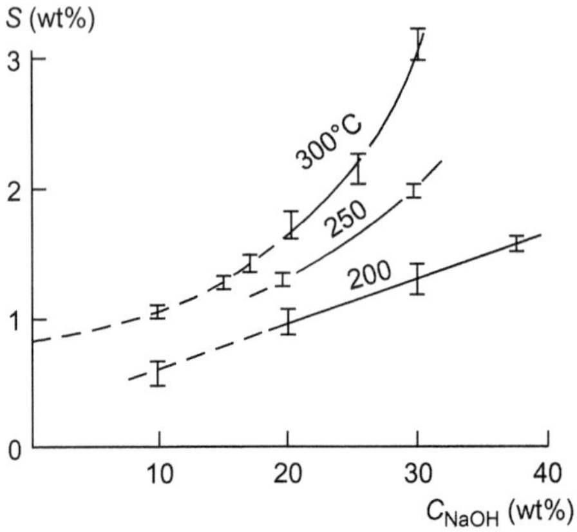

Figure 6.25 Dependence of solubility of sodalite and the NaOH concentration at constant temperature [123].

Mel'nikov et al. [122] have carried out the hydrothermal growth of bulk sodalite crystals using seed crystals and have studied the kinetics of crystallization in detail. The experimental temperature was varied from 150°C to 400°C in steps of 50°C. Figure 6.26 shows the relationship between the growth rates of sodalite crystals in the directions and the temperature gradients: (a) 200°C, (b) 250°C, (c) 300°C, and (d) 350°C. The mass transfer associated with the crystallization of sodalite depends linearly on the temperature gradient for a constant growth temperature. They have found a direct relationship between the growth rates of the sodalite faces and the temperature drop, at constant temperature, and an exponential temperature dependence at constant supersaturation [122]. Recently, Hayashi et al. [128] have grown millimeter-sized aluminosilicate sodalite single crystals under hydrothermal conditions.

In spite of the interesting results obtained by the above authors, the quantum of data available on crystal growth is just meager compared to the data on the synthesis of zeolites. It has to be understood in greater detail.

6.7 Aluminophosphate Zeolites

Aluminophosphate zeolites represent a new class of microporous inorganic solids with the potential to be as useful and as scientifically challenging as the

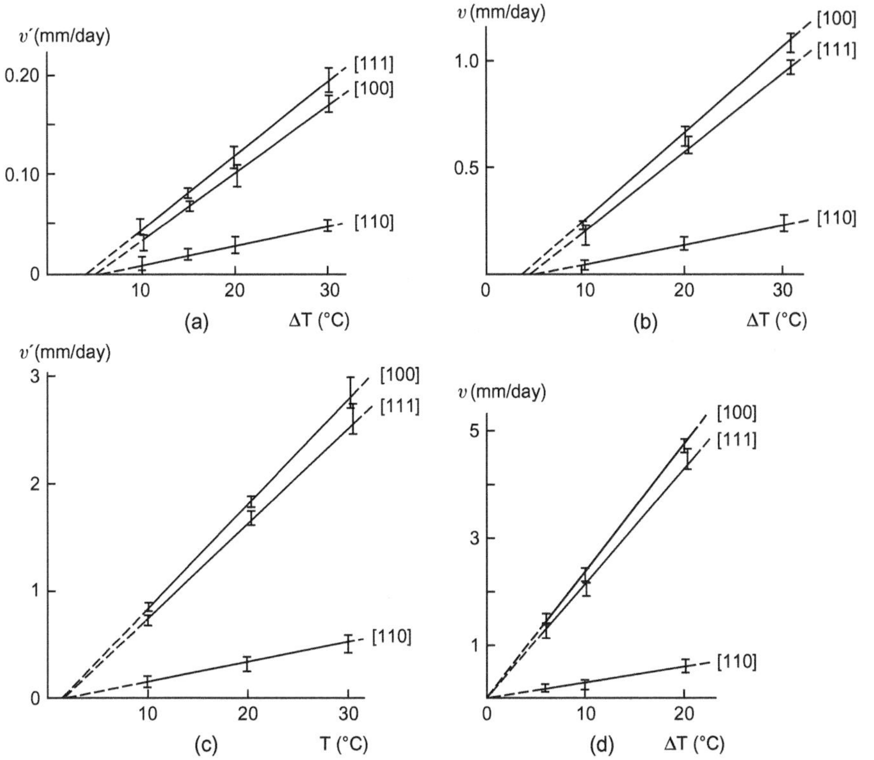

Figure 6.26 Relation between the growth rates of sodalite crystals [122].

aluminosilicate zeolites and form a part of the history of framework of oxide molecular sieves. The basis for the synthesis of such zeolites is the crystal chemical similarities between Si and P, and $AlPO_4$ and SiO_2 in particular. The unique feature of these zeolites is the compositional similarity and a great structural diversity. The basic structural units consist of alternating Al and P with Al exhibiting octahedral coordination and pressure and tetrahedral coordination. The two waters of hydration occupy positions in the coordination sphere of Al.

The synthesis of aluminophosphate zeolites does not differ much from that of the synthesis of aluminosilicate zeolites, and the procedure is as follows: Equimolar amounts of a reactive hydrated alumina, like boehmite, and phosphoric acid are dissolved in water. An aluminophosphate gel is formed to which a templating agent (R), an organic amine or a quaternary ammonium salt is added. This mixture is digested quiescently in the preferred temperature range of 125–200°C. A crystallization product is formed, which is worked up using procedures typical of zeolite synthesis. About 23 structure analogues of three zeolites namely erionite ($AlPO_4$-17), sodalite ($AlPO_4$-20), and analcime ($AlPO_4$-24) have been obtained [129]. Some of the novel structures are $AlPO_4$-5, -11, -14, -18, -31, and -33. Among these, $AlPO_4$-5 was the first $AlPO_4$ molecular sieve synthesized and it was found to be a novel structure belonging to the hexagonal system. Similarly, in the preparation of aluminophosphate zeolites, the precursor gel contains four components, an alumina source, a phosphate source, water, and an organic amine or quaternary ammonium salt. The relative concentrations of the reactants are very important and, similarly, without the presence of organics no aluminophosphate zeolites molecular sieves can be formed. In most of the cases, the ratio of Al:P is 1:0. The commonly used organics are amines and quaternary ammonium cations $i-Pr_2NH$, $n-Pr_2NH$, n-BuEtNH, Et_2NH, $n-Bu_2NH$, $n-Pe_2NH$, Me_2NH, $n-BuNH_2$, and $n-HexNH_2$. A wide variety of templating agents has yielded almost as wide a variety of structure types. The various structure types include a broad range of pore size and volumes as broad as that found among synthetic zeolites. The $AlPO_4$ frameworks have excellent thermal and hydrothermal stabilities. The surface character of the $AlPO_4$ molecular sieves differs from that of the silica molecular sieves even though both framework types are neutral with no extra-framework cations. The molecular sieve silicalite is hydrophobic, and the $AlPO_4$ molecular sieves are moderately hydrophilic. Zeolites are hydrophilic due to the interaction of the dipole of the H_2O molecule with the electrostatic fields of the anionic aluminosilicate framework and the balancing nonframework cations. The hydrophilicity of the $AlPO_4$ materials is apparently due to the difference in electronegativity between Al (1.5) and P (2.1). Neither mechanism is possible with silica molecular sieves. The $AlPO_4$ molecular sieves do exhibit less affinity for H_2O than the hydrophilic zeolites such as type A or type X.

The problems associated with the aluminophosphate zeolites crystallization are traced to the similarity in the composition and diversity in the structure. Therefore, one has to monitor carefully each step in order to prepare a desired zeolite phase and its reproducibility. In the case of aluminophosphate zeolites crystallization, the gel preparation, gel pretreatment, gel aging, pH of the media, experimental duration, experimental temperature of synthesis, molar proportions of the starting material, type of organics, and the time of addition of organics, and alumina source play an important role. The aluminosilicate zeolites are crystallized in a strong basic medium with a pH of the reaction mixture >13, whereas, for the successful crystallization of aluminophosphate zeolites, the pH range of the hydrous gel is found to be in the range of 3.5–6.0 [130]. Any slight change in the above-listed parameters largely influences the resultant product. Figures 6.27 and 6.28 show the possible experimental procedures for treating the $AlPO_4$ gel.

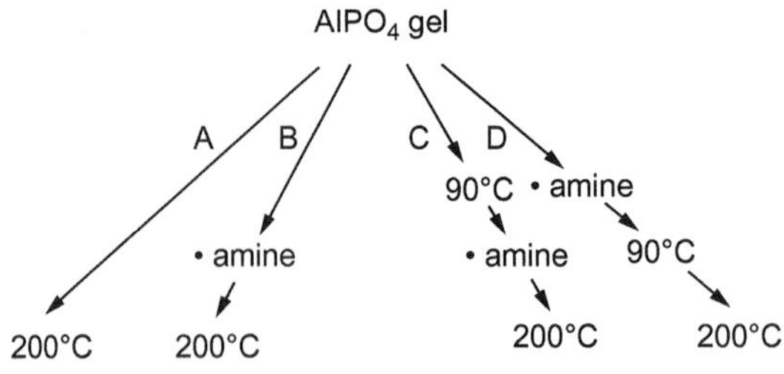

Figure 6.27 Possible experimental procedures for treating the $AlPO_4$ gel [131].

It was well known that the largest ring in silica zeolites consisted of 12 tetrahedral (12 T-atoms), and this concept remained for over 180 years, since the first discovery of natural zeolite with 12-T atom rings, e.g., gmelinite, took place during 1807 [18,19]. Davis et al. [131] have synthesized the first molecular sieve with rings that possess greater than 12 T-atoms. VPI-5 is a family of aluminophosphate-based molecular sieves with the same 3D topology. The extra-large pores of VPI-5 contain unidimensional channels circumscribed by rings that have 18-membered rings and possess free diameters of approximately 12–13 Å [131]. It was only during 1996 that Freyhardt et al. [132] synthesized the first aluminosilicate zeolite with its elliptical 14-tetrahedral-atom pore opening. For the synthesis of VPI-5, pseudoboehmite, 85 wt% H_3PO_4, tetrabutylammonium hydroxide (TBA), and *n*-dipropylamine (DPA) were used. Schmidt et al. [133] have studied the VPI-5 synthesis in detail. Here, we describe only the synthesis of VPI-5 to show the peculiarities involved in the synthesis of aluminophosphate zeolites in general. In this regard, the experimental results of Schmidt et al. [133] are very interesting for a reader. The reaction temperature was varied from 100°C to 200°C using two different gel compositions in order to find the optimum conditions for VPI-5 crystallization. Figure 6.29 shows the

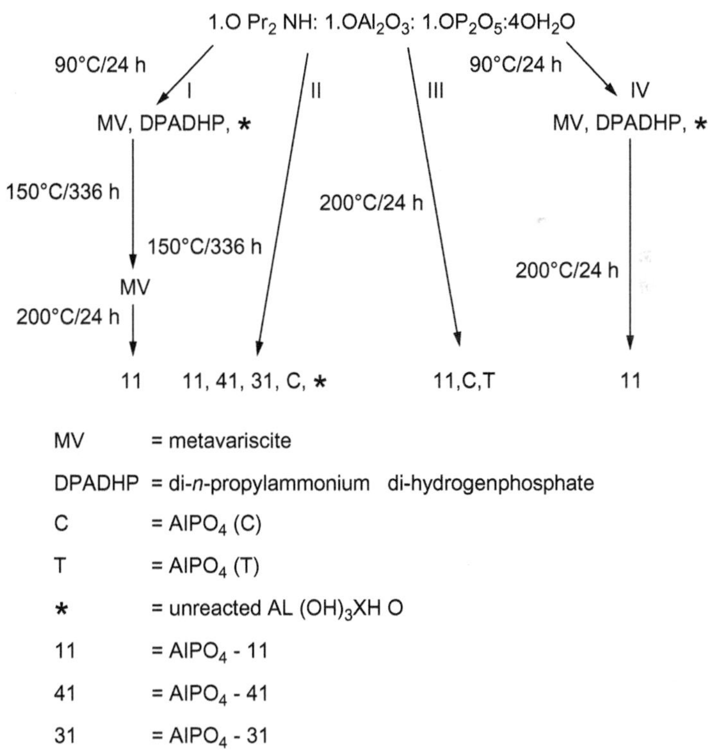

Figure 6.28 Effect of various hydrothermal treatments on the amine containing $AlPO_4$ gel [131].

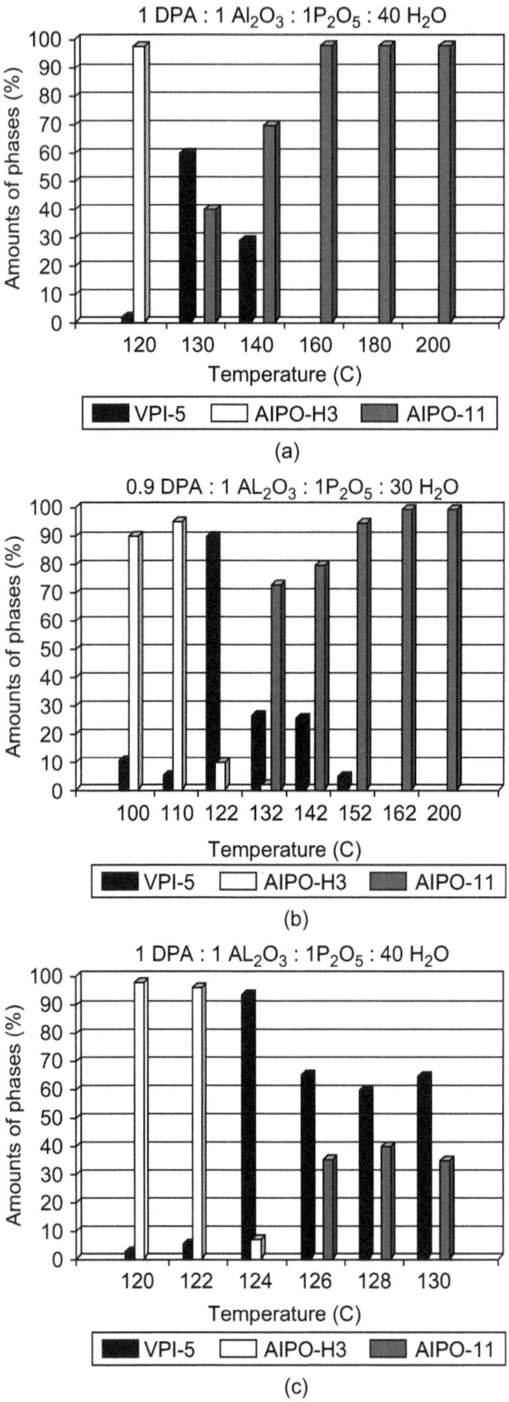

Figure 6.29 Dependence of various $AlPO_4$ phases upon reaction temperature for different molar ratios [133].

dependence of the various $AlPO_4$ phases upon the reaction temperature for different molar ratios.

The change in temperature results in the appearance of various other phases like AlPO-11, AlPO-H3, in addition to VPI-5. Figure 6.29 clearly indicates a very narrow temperature range for the crystallization of VPI-5, which depends essentially upon the molar ratio of DPA = Al_2O_3:P_2O_5:H_2O. It has been shown that the use of bigger autoclaves can reduce the formation of other phases which might be connected with slower heating rate and temperature gradients/thermal currents within the liners.

Similarly, Schmidt et al. [133] have studied the effect of reaction time on the crystallization of $AlPO_4$ phases. Figure 6.30 shows the dependence of the various $AlPO_4$ phases upon reaction time at 124°C for the molar ratio 1 DPA:1 Al_2O_3:1 P_2O_5 = 40 H_2O. Thus, the hydrothermal crystallization of aluminophosphate zeolites from the reaction gel mixture is usually controlled kinetically and the first nucleus that reaches the critical size will determine the final production.

Figure 6.31 shows a schematic representation of the triple helix of water molecules within the 18-ring channel.

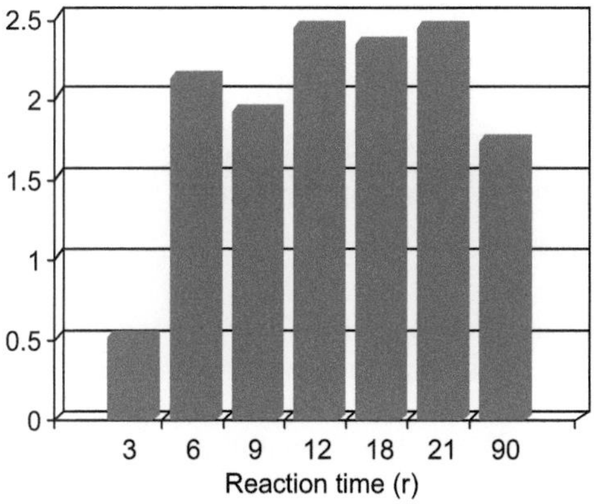

Figure 6.30 Dependence of the various $AlPO_4$ phases upon reaction time at 124°C for the molar ratio—1 DPA:1 Al_2O_3:1 P_2O_5 = 40 H_2O [133].

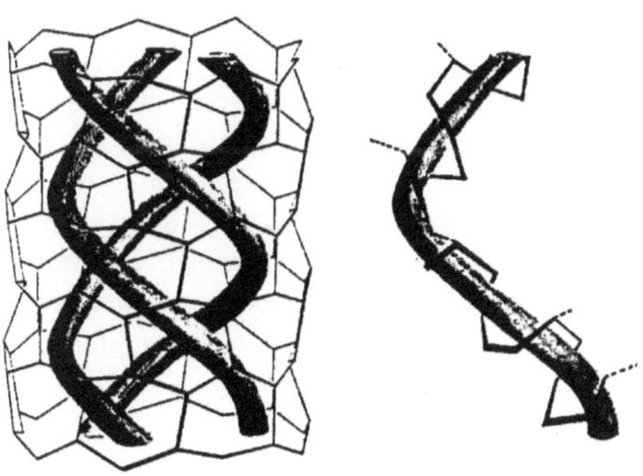

Figure 6.31 Schematic representation of the triple helix of water molecules within the 18-ring channel [133].

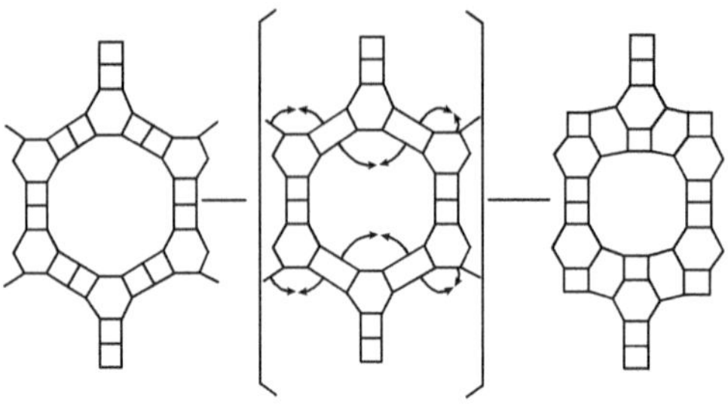

Figure 6.32 Schematic representation of the transformation of VPI-5 into $AlPO_4$-8 [134].

A majority of the aluminophosphate zeolites undergo phase transitions when calcined. For example, when VPI-5 is heated to 130°C under ambient pressure, it leads to a total phase transition of VPI-5 to AlPO-8. Figure 6.32 shows the schematic representation of the transformation of VPI-5 into $AlPO_4$-8 [134]. This transformation takes place after the removal of water under specific washing and drying steps of the as-synthesized materials; the transformation may be complete when occluded water is removed above 100°C.

Thus, the synthesis of aluminophosphate zeolites involves a careful understanding of the influence of several experimental parameters.

6.8 Growth of Zeolite Thin Films and Crystals at Inorganic–Organic Interfaces (Preparation of Zeolite-Based Composites)

There is a growing interest in the development of inorganic films and membranes with controllable porosity owing to their promising applications in asymmetric membranes, catalytic monoliths, and chemical sensors [135–138]. The interest in microporous thin films is primarily motivated by their potential molecular sieving action, large surface areas, and controlled host–sorbate interactions, especially at low vapor pressures. Some authors have explored several possibilities/strategies for thin film formation, including dip-coating of zeolite/silica sol suspensions, and chemical attachment of zeolite crystals via molecular coupling layers [137,138]. Similarly, zeolites also have been embedded into polymers, covered with glazes and extruded into composite monoliths [139,140]. It was observed that in order to enhance control of molecular selectivity in sensors, it would be desirable to tune or switch the pore size of zeolite crystals in the film. Thus, the authors have adopted such a concept by forming zeolite CaA/silica composite thin films and controlling the zeolitic pores through in situ cation exchange and developed several composites. One such composite typically consists of 100 $\mu g/cm^2$ of zeolite crystals and 10 $\mu g/cm^2$ of amorphous silica overlayer. The thickness of the silica layer is estimated to be about 20 nm when assuming a density of 2.2 g/cm^3 and a smooth surface. Systematic studies of the cation exchange in zeolite/silica composite films

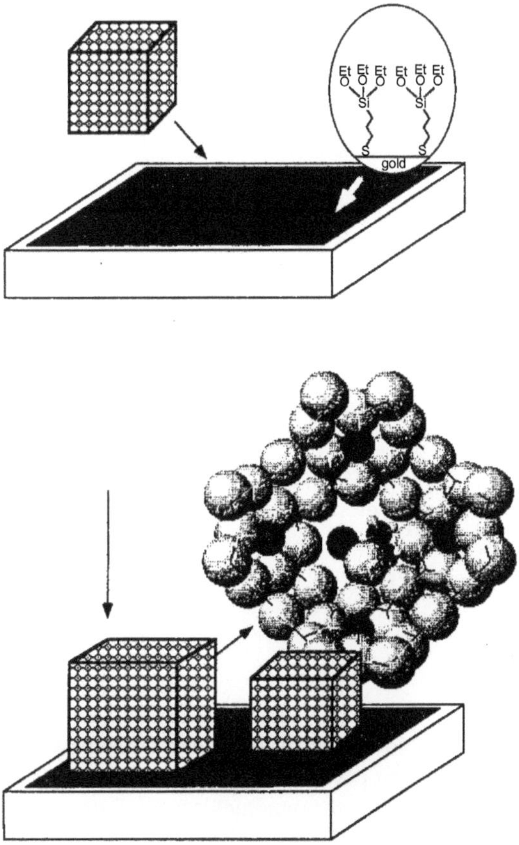

Figure 6.33 Assembly of zeolite CaA crystals on silozane layer on gold surface and stabilization with 20 nm silica layer [138].

and sorption in the composite films have been carried out by Yan and Bein [138]. Figure 6.33a and b shows the assembly of zeolite CaA crystals on a silozane layer on a gold surface and stabilization with 20 nm silica layer. Figure 6.34 shows the mass changes of zeolite silica composite films exchanged with different cations on a piezoelectric substrate. Figure 6.35 shows a schematic diagram of the surface acoustic wave (SAW) device covered with molecular sieve layer. Figure 6.36 shows the proposed reaction scheme for the design of zeolite layers and films on organophosphate films: (a) Aluminophosphate nuclei attach on organophosphonate film and (b) the growth of hexagonal crystals. The SEM pictures have shown that the $AlPO_4$-5 crystals grow with their *c*-axes oriented similar to the gold surface (Figure 6.37). It was also found that changing the water concentration and temperature in the hydrothermal synthesis has a profound effect on the orientation morphology and aspect ratios of the surface-attached crystals. Table 6.18 gives the effect of water concentration on the morphology of surface-attached $AlPO_4$-5 crystals.

Thus, the molecular selectivity of zeolite thin films can be precisely tuned by gating the entry of molecules into the zeolite pores.

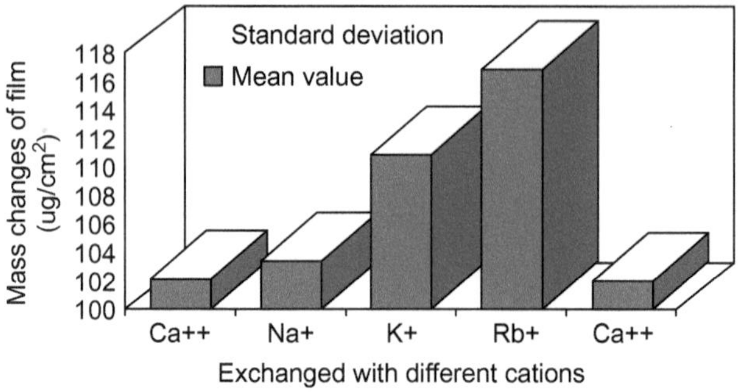

Figure 6.34 Mass changes of zeolite silica composite films exchanged with different cations on a piezoelectric substrate [141].

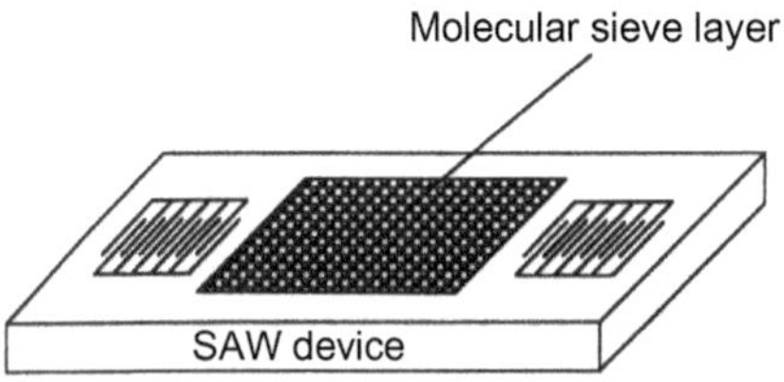

Figure 6.35 Schematic diagram of SAW device covered with molecular sieve layer [141].

Bein [141] has reviewed the zeolite molecular sieve films and membranes. These zeolite-based, nanostructured materials offer the potential to organize matter and to manipulate molecules with high spatial precision at the nanometer level. Pure zeolite membranes (20–250 μm thick) consisting of ZSM-5 crystals have been prepared at 180°C (for 0.3–9 days) on substrates including Teflon, silver, stainless steel, and Vycor glass [142]. Sano et al. [143] have proposed a new preparation method for highly siliceous zeolite films. Similarly, these authors have obtained a pure ZSM-5 zeolite film ~30–100 μm in thickness; it is made up of zeolite crystals, which can be obtained on a Teflon slab. At first, silicalite was hydrothermally synthesized in the presence of cellulose moldings, such as a filter paper. Figure 6.38 shows the SEM photograph of a silicalite–cellulose composite prepared in 48 h. Carbon membranes (in the form of hollow fibers) constitute a novel type of support for zeolite layers. Figure 6.39 shows such hollow-fiber carbon zeolite membranes [144]. Layers with molecular sieving properties were formed by embedding microcrystals of Z, Y, and chabazite, in a glassy silica matrix [145].

Molecular sieve films are the promising candidates for applications in chemical sensors, especially in industrial process control, environmental and indoor air-quality monitoring, effluent and auto-exhaust control, and medical monitoring. Noack et al. [146] have obtained columnar $AlPO_4$-5 crystals aligned and embedded in nickel (by galvanic deposition) and studied in the permeation of heptane and different aromatic hydrocarbons (Figure 6.40a and b). There have been a few works that mimic some features of biomineralization to grow oriented zeolite molecular

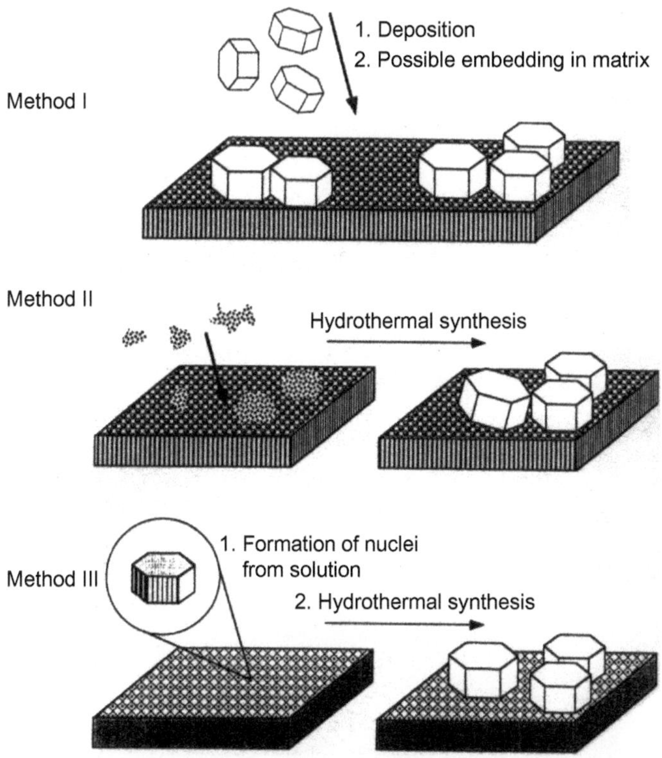

Figure 6.36 Method I—Proposed reaction scheme for design of zeolite layers and films on organophosphate films. Method II—Aluminophosphate nuclei attach on organophosphonate film. Method III—Growth of hexagonal crystals [141].

sieve films. Such self-assembled layers include self-assembled functional layers and they have potential applications in microelectronics and nonlinear optical devices [147].

Thus, there is a great variety of synthesis approaches regarding the design of molecular sieve layers and membranes. These aspects will play a major role in the future zeolite research.

6.9 Applications of Zeolites

Zeolites are conventionally used as adsorbents, in gas separation, in shape-selective catalysis, and in ion exchange [28,40]. In recent years, there is an increased research activity in molecular sieve science and technology toward the development of advanced materials. There are several reports available in the literature dealing with the controlled incorporation of guest compounds into a zeolitic framework to get

Figure 6.37 SEM pictures show that the $AlPO_4$-5 crystals grow with their *c*-axes oriented similar to the gold surface [141].

new optical, electronic, and magnetic materials, such as intrazeolite sulfides, selenides, silver compounds, carbonyls, the intra-crystallization polymerization of acetylene, and pyrrole, the formation of metal clusters, and trapped supramolecular compounds (ship-in-one-bottle) [148,149].

Figure 6.38 SEM photograph of silicalite-cellulose composite prepared in 48 h [143].

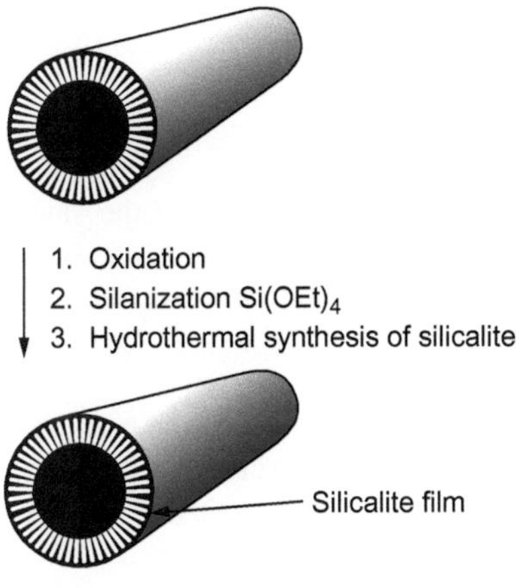

Figure 6.39 Hollow-fiber carbon zeolite membranes [144].

Table 6.18 Effect of Water Concentration on the Morphology of Surface Attached $AlPO_4$-5 Crystals [141]

Water Content (d)	Crystal		
	Length (μm)	Width (μm)	Orientation
300	30–50	40–50	Vertical
400	20–30	3–6	Tilted
600	30–50	2–10	Tilted

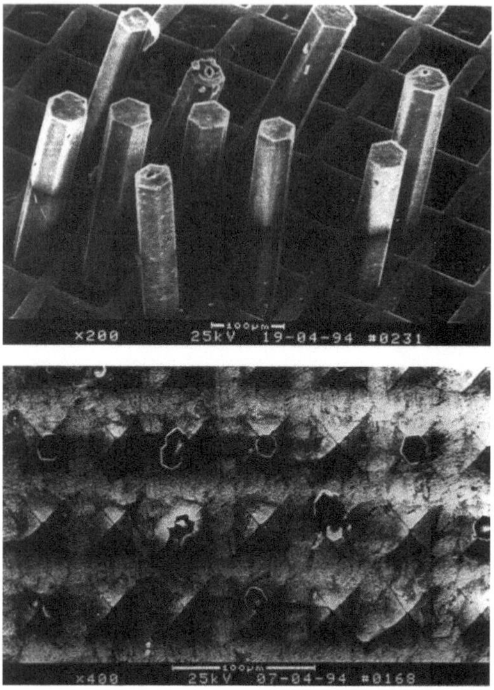

Figure 6.40 Columnar $AlPO_4$-5 crystals aligned (a) and embedded (b) in nickel [146].

The application of zeolites as catalysts is basically dependent upon the accessibility to reactants of intra-crystallization, catalytically active centers, and upon the shape and size of the penetrant molecules. Specific catalytic actions can depend upon zeolite basicity, the generation of acid sites within the crystals, or upon the introduction of particular cations (by ion exchange), or of elements such as S, Te, and Se, or metals like Pt into the crystals.

The catalytic oxidation of H_2S by O_2 exemplifies a process in which the best catalysts are the basic Na- and K-forms of zeolites X and Y [150]. The catalytic activity is directly proportional to the number of AlO_4 tetrahedra and inversely related to the electrostatic potential of the exchange ions introduced (K^+, Na^+, Li^+, and Ca^{2+}). However, it is the hydrocarbon chemistry which has placed zeolites in a unique place in science and technology, especially with H-zeolites providing Brönsted acid sites and reactions proceeding via carbonium ions. The acid centers form the following reversible reaction:

$$H^+ + Al^- \quad {}^-Si \rightleftharpoons Al \quad Si$$

Various shape-selective catalyses have been described in several zeolites. In Ca-A zeolite, chemical dehydration of *n*-butanol (which penetrates the crystals), but not of isobutanol (which does not penetrate), was observed. Whereas Ca-X zeolite does not show this kind of selectivity, but it sorbs both alcohols readily. Similarly, *n*-paraffins and *n*-olefins (which entered the Ca-A) were readily combusted with O_2, while isoparaffins and isoolefins, which did not enter Ca-A, did not react appreciably [151]. Shape-selective catalysis has also been observed in several other zeolites and prominent among them is ZSM-5 and MFI. This sorbs and cracks monomethyl in addition to *n*-paraffins and is used to dewax heavy oils.

Mirajkar et al. [152] have studied the sorption properties of Al-omega and Ga-omega zeolites. Figure 6.41 shows the sorption uptakes ($P/Po = 0.8$, 298 K) of samples of different crystallinity for Al-omega and Ga-omega zeolites synthesized at 398 K. These curves typically show that the sorption uptake increases as the degree of crystallinity increases; water being a polar molecule and comparatively small in size, it is able to become occluded in cavities of samples of low crystallinity. Table 6.19 gives the adsorption (wt%) on Al- and Ga-omega zeolites: $P/Po = 0.8$ and $T = 298$ K.

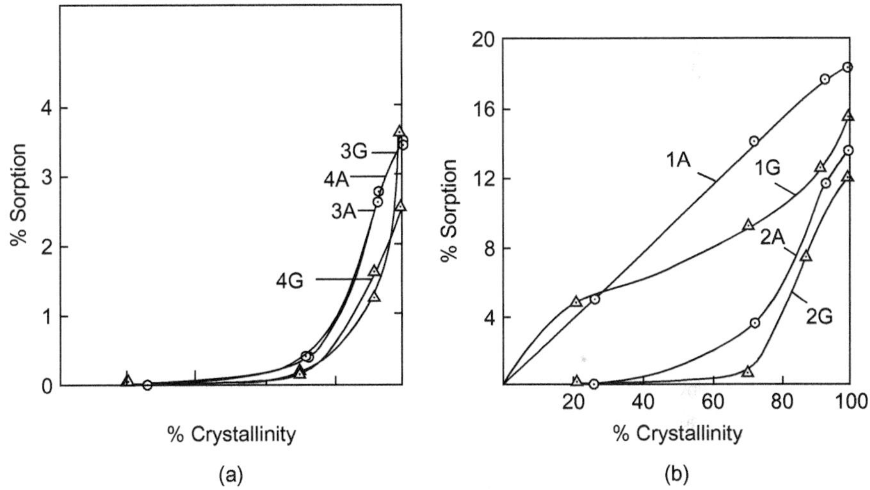

Figure 6.41 (a and b) Sorption uptakes ($P/Po = 0.8$, 298 K) of samples of different crystallinity for Al-omega and Ga-Omega zeolites synthesized at 398 K [152].

Table 6.19 Adsorption (wt%) on Al- and Ga-Omega Zeolites, $P/Po = 0.8$ and $T = 298$ K [152]

Synthesis Temperature (K)	Al-Omega			Ga-Omega		
	383	**398**	**413**	**383**	**398**	**413**
Adsorbate water	20.45	18.10	17.30	17.50	14.89	15.3
n-Butylamine	17.19	13.22	10.10	15.50	11.69	9.6
n-Hexane	6.70	3.51	2.09	4.70	3.57	1.8
Cyclohexane	6.50	3.33	2.25	4.61	2.27	1.8

The sorption uptakes are found to decrease with the increase in experimental temperature and this decrease is consistent with the observed increase in the crystallite size.

As the molecular size of the probe molecules increases, a decrease in sorption uptake with an increase in crystallite size becomes appreciable on account of their different packing in the zeolite voids; Al-omega shows higher sorption uptake than Ga-omega zeolite. These omega zeolites are stable up to a temperature of 1000°C.

The $AlPO_4$ molecular sieves show excellent physical properties. Many are thermally stable and resist loss of structure at 1000°C. Those studied for hydrothermal stability, including AlO_4-5, 11, and 17, show no structure loss when treated with 16% steam at 600°C. The adsorption properties of selected $AlPO_4$ molecular sieves are listed in Table 6.20. The different structure types are arranged in order of decreasing pore size, which is controlled by the ring size. The ring size refers to the number of tetrahedral atoms (Al or P). For the 12-T ring structures, $AlPO_4$-5 and $AlPO_4$-31, adsorption of neopentane has been observed. The 10 or puckered

Table 6.20 Typical Adsorption Pore Volumes for Selected $AlPO_4$-Based Structure Types

		Adsorption Capacity (cc/100 g)[a]						
Adsorbate =		**H_2O**	**O_2**	***n*-Butane**	***n*-Hexane**	**Isobutane**	**Cyclohexane**	**Neopentane**
Kinetic diameter (nm) = 0		0.265	0.346	0.43	0.43	0.50	0.60	0.62
P/Po =		0.9	0.9	0.4	0.3	0.3	0.5	0.6
Temperature (C) =		22	−183	22	22	22	22	22
Structure	Ring size[b]							
5	12	26	16	13				12
36	12	31	21	14				13
46	12	35	30		28			26
11	10	18	13		11		11	2
31	10 or 12	21	13		11		9	11
41	10 or 12	21	12		10		8	11
34	8	33	27	23		2		
35	8	31	17	1				
39	8	23	9	1				
44	8	35	25		10			
47	8	35	27	25				
16	6	28	4	1				
20	6	32	2	1				

[a]Determined by standard McBain–Baker gravimetric techniques after calcination at 500–600°C in air. Samples activated at 350°C, 0.01 torr, prior to measurement.
[b]Number of tetrahedral atoms (Al, P, and Me) in ring controlling the pore size. When structure is not known, it is estimated from adsorption measurements.

12-ring of $AlPO_4$-11 adsorbs cyclohexane but not neopentane. The pore size is similar to that of ZSM-5. Two of the 8-ring structures, $AlPO_4$-17 and $AlPO_4$-18, sorb butane but not isobutane. $AlPO_4$-14 and $AlPO_4$-33 sorb xenon, but not butane, indicating a somewhat puckered 8-ring structures. Similarly, the 16-ring structures of $AlPO_4$-16 and $AlPO_4$-20 absorb only water. The pore volumes are near saturation values—O_2 at 183°C and H_2O at ambient temperature.

In general, the H_2O pore volumes are usually greater than the O_2 pore volumes, reflecting the presence of additional pore volume accessible to H_2O but not to O_2. The model of $AlPO_4$-5 structure shown in Figure 6.42 indicates large pore voids capable of absorbing O_2 or H_2O; the small voids shown in Figure 6.43a and b are outlined by columns of twisted chains of 4- and 6-rings parallel to the *c*-axis, adsorb H_2O only. In $AlPO_4$-5, these small voids apparently sorb about 0.12 cm^3/g. Figure 6.44 shows the O_2 isotherms of $AlPO_4$-5, -11, and -17, compared with those of a typical zeolite, Na-X, and a silica molecular sieve, silicalite. The $AlPO_4$-5 and -17 have saturation pore volumes of ~0.2 cm^3/g, similar to that of silicalite, while $AlPO_4$-11 has a relatively small O_2 pore volume (~0.1 cm^3/g).

Figure 6.45 shows the H_2O adsorption isotherms for $AlPO_4$-5, -11, -17, and -20. For comparison, the hydrophilic Na-X and the hydrophobic silicate are considered. A very unusual shape has been obtained for the $AlPO_4$ adsorption curve for O_2 and paraffins.

In recent years, with the discovery of VPI-5 and other large pore zeolites, several advantages have been found. $AlPO_4$-5 adsorbs O_2 and some hydrocarbons except for triisopropylbenzene. The triisopropylbenzene is too large to penetrate the 12 T-atom ring of $AlPO_4$-5. VPI-5 reveals two phenomena not observed for $AlPO_4$-5: Firstly, triisopropylbenzene is adsorbed; secondly, the adsorption capacity monotonically decreases with increasing adsorbate size. Since VPI-5 contains larger pores of 12–13 Å, all adsorbates other than triisopropylbenzene have the possibility of fitting more than one molecule across the diameter. In other words, packing of adsorbate molecules may be important in these extra-large pores. Table 6.21 gives the adsorption capacity of $AlPO_4$-5 and VPI-5 for various adsorbates [131]. Davis et al. [131] have measured the complete argon adsorption isotherms at liquid argon temperatures. The data obtained from Na-X and VPI-5

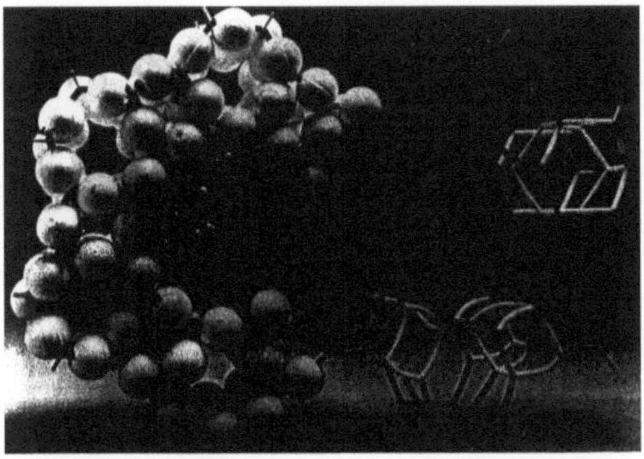

Figure 6.42 The model of $AlPO_4$-5 structure [129].

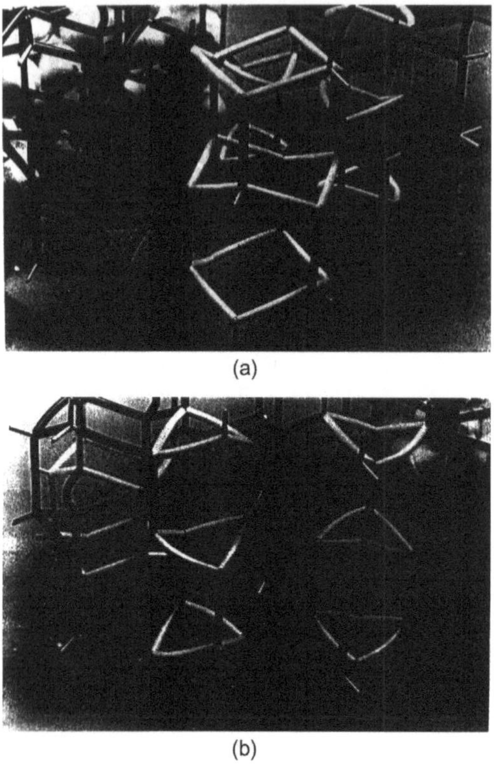

Figure 6.43 (a) and (b) Columns of twisted chains of 4- and 6-rings parallel to the c-axis, adsorb H_2O [129].

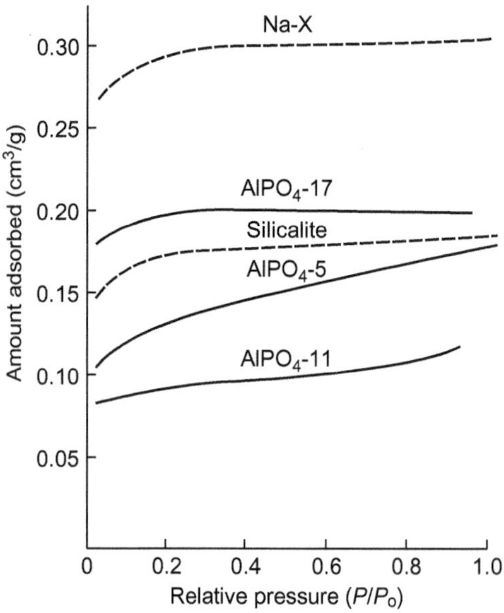

Figure 6.44 Graph of O_2 isotherms of $AlPO_4$-5, -11, and -17 [129].

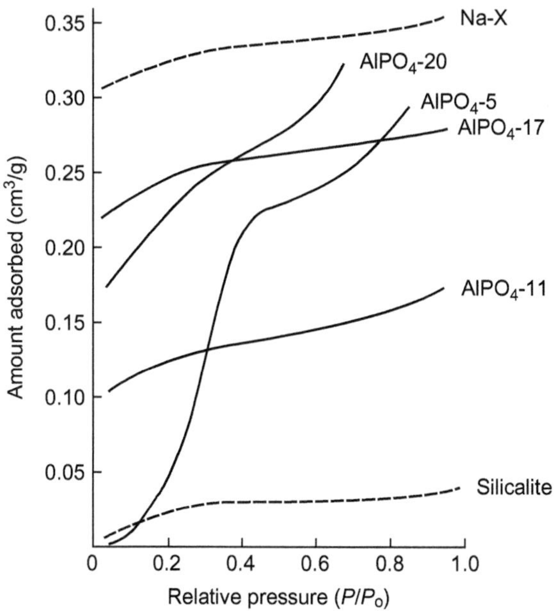

Figure 6.45 The H_2O adsorption isotherms for $AlPO_4$-5, -11, -17, and -20 [129].

Table 6.21 Adsorption Capacity of Molecular Sieves at $P/Po = 0.4$ [131]

Kinetic Diameter (Å)	Adsorbate[a]	Capacity (cm^3/g)	
		$AlPO_4$-5	VPI-5
3.46	O_2	0.146	0.228
4.30	*n*-Hexane	0.139	0.198
6.00	Cyclohexane	0.145	0.156
6.20	Neopentane	0.137	0.148
8.50	Triisopropylbenzene		

[a]Adsorption at room temperature except for O_2 which was performed at either liquid N_2 or O_2 temperatures.

are shown in Figure 6.46. Thus, the zeolites with large pore diameters are becoming essential materials for the adsorption of large molecules, which either absorb or adsorb very slowly into known molecular sieves.

6.10 Oxidative Catalysis on Zeolites

Oxidation processes constitute one of the most common classes of chemical reactions. The endeavor by investigators to create economically more effective and ecologically clean technologies has stimulated the search for new catalysts exhibiting a high selectivity.

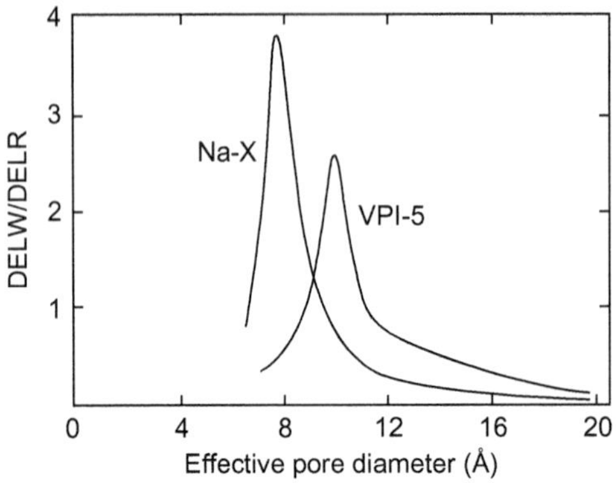

Figure 6.46 Effective pore size distribution [131].

Though zeolites were not regarded as promising catalysts for oxidation reactions, considerable changes have taken place in the last decade. It was during the early 1980s, that titanosilicate zeolites, having a structure of the ZSM-type, were identified and the discovery of their unique properties in oxidation reactions involving H_2O_2 took place. Thus, titanosilicates become the effective catalysts of a wide range of reactions occurring in the liquid phase, such as the oxidation of alcohols, epoxidation of alkenes, hydroxylation of aromatic compounds, and the conversion of cyclohexanone into an oxime [151–157]. These advances in liquid phase oxidation of primary and secondary alcohols to the corresponding carbonyl compounds have been assessed as one of the most important achievements in the field of organic synthesis. The partial oxidation of hydrocarbons is achieved, as a rule, in the presence of transitional metal oxides. Their activity is due to more or less facile electron transfer and their capacity for the reversibility of their oxidation state in interaction with oxygen in the substance. Oxidized zeolites themselves cannot serve as oxidation catalysts, but the possibility of their use as carriers, nevertheless, is of great interest. However, because of the serious environment problems associated with Cr-containing effluent, attention has been focused on the use of catalytic amounts of soluble chromium compounds in conjunction with, for example, tetrabutyl hydroperoxide (TBHP) as the stoichiometric oxidant. The use of heterogeneous catalysts in the liquid phase, on the other hand, offers several advantages compared with their homogenous counterparts, e.g., ease of recovery and recycling and enhanced stability. Kharitonov et al. [157] have studied the reactions involving gasoline oxidation with nitrous oxide, and the best results were obtained in the oxidation of benzene to phenol in the presence of iron-containing ZSM-5 zeolites.

An approach to create solid catalysts with novel activities is to incorporate redox metals by isomorphous substitution into the lattice framework of zeolites and related molecular sieves. Since the discovery of aluminophosphate molecular sieves in 1982, much attention has been focused on the incorporation of various elements into the framework of these molecular sieves [158]. Modified compositions of these silica or aluminophosphate-based materials are expected to expand the scope of their catalytic applications as mesoporous materials and optimized advanced catalysts. These materials have been the most promising class of 3D host frameworks with controlled microporosity. Atomic-scale engineering of catalytic functions of isolated redox sites in the confined environments of zeolitic channels is also of great interest.

Rabo et al. [159,160] showed that Na, Ca, Be, and La forms of Z,Y containing residual protons reacted with NaCl, at 575–675 K, with the evolution of HCl. Clearfield et al. [161] and Clearfield [162] demonstrated the introduction of cations, such as Zn^{2+}, Cu^{2+}, Co^{2+}, Ni^{2+}, Mn^{2+}, Cr^{3+}, Cr^{5+}, Mo^{5+}, Ga^{3+}, and V^{4+}, into zeolites A, X, and Y, by the reaction of their chlorides with the H forms of the zeolites phosphates and molybdates. Figure 6.47 shows the activity of Cr/H-ZSM-5 in CH_4 burnt at 500°C [164,165]. The catalytic activity of the Cr/H-ZSM-5 system increased linearly with the rise of chromium content up to 1.5 wt%, showing considerable site homogeneity. The activity of the catalyst reached a maximum point when the number of isolated Cr^{5+} cations capable of changing valence state upon redox treatment reached maximum. Stabilization of isolated Cr^{5+} ions in a coordinatively unsaturated environment inside H-ZSM-5 may be a reason for such unusually high intrinsic catalytic activity [166]. Similarly, a complete oxidation of CH_4 over Mn/H-ZSM-5, Cu/H-ZSM-5, and V/H-ZSM-5 has also been studied in detail by Kucherov et al. [166]. Ethane oxidation with N_2O over H-ZSM-5 and Cu/H-ZSM-5 has been studied by Kucherov et al. [166,163]. The C_2H_6 oxidation by N_2O, as distinct from O_2, proceeds on both H-ZSM-5 and Cu/H-ZSM-5. At $T > 430$°C, a stable 100% conversion took place on H-ZSM-5, but the decrease in temperature to 410–400°C led to a drastic drop in C_2H_6 and N_2O conversion with the fast cooking of zeolite. The oxidation of ethane with N_2O was treated as a complex heterogeneous–homogeneous process (catalytic decomposition of N_2O and subsequent gas-phase C_2H_6 oxidation with the participation of excited molecules and coupling products) [163,167]. Thus, the modification of pentasils by transition

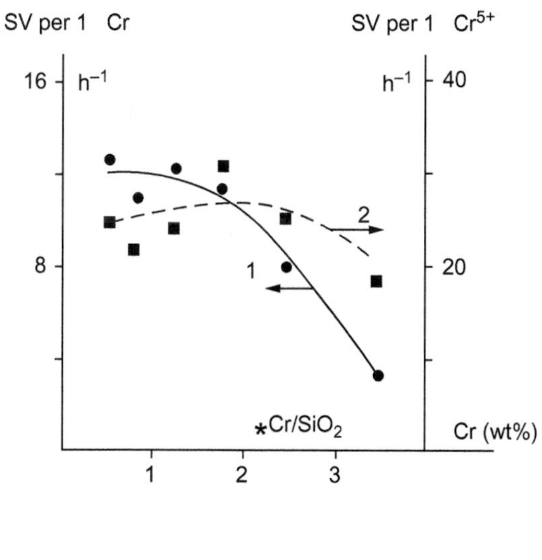

Figure 6.47 Activity of Cr/H-ZSM-5 in CH_4 burnt at 500°C [163].

metal cations permits the preparation of promising catalysts for both complete oxidation of alkanes and NO_x decomposition.

Guth et al. [88] have developed a new route of the synthesis of microporous compounds in the presence of fluorine. The addition of fluoride ions in the reaction medium seems to induce mineralization and allows the crystallization of zeolites at neutral or acidic pH and the partial substitution of silicon by other tri- or tetravalent metals, like B^{3+}, Al^{3+}, Ga^{3+}, Fe^{3+}, Ge^{4+}, and Ti^{4+}. Just like in other phases, fluorine is also incorporated in the open frameworks, and the best example is cloverite, with 3D 20-membered ring channels, in which fluorine is located at the center of a D4R cage [168,169]. A similar structural diversity is associated, in part, with the ability of vanadium-oxygen compounds to form tetrahedral, square-pyramidal, and octahedral coordination polyhedra, and to build larger aggregates by the condensation of polyhedra through shared oxygen atoms [170,171]. Such works produce novel structures/frameworks that encapsulate organic guests to produce large cavities or tunnels. After the organic template is removed by ion exchange or thermal methods, the solid may sorb substrate molecules in a size- and shape-selective fashion [172]. Similarly, there are several mixed tetrahedral–octahedral framework structures which are both novel and different from the traditional aluminosilicate zeolites, where only TO_4 tetrahedra (T = Si, Al, or Ge) are the basic building units. Mixed TO_6 and TO_4 units are also found in the structures of recently synthesized microporous stannosilicates, germanates, and gallophosphates [173,174–176]. These compounds show good ionic conductivity and humidity-sensing property as seen in microporous Na-germanates. Leonowicz et al. [177] have obtained a new boron, MCM-22, zeolite that contains structural features previously unknown among zeolites. The high-resolution electron micrographs and synchrotron X-ray diffraction powder data reveal two independent pore systems, both of which are accessed through rings composed of 10-tetrahedral (T) atoms (such as Si, Al, and B). One of these pore systems is defined by 2-*d*, sinusoidal channels. The other consists of large supercages whose inner free diameter, 7.1 Å, is defined by 12 T-O species (12-rings) and whose inner height is 18.2 Å. These coexisting pore systems may provide opportunities for a wide variety of catalytic applications in the petrochemical and refining industries.

Caro et al. [149] have obtained aligned molecular sieve crystals as nonlinear optical materials. As is well known, the formation of Längmuir–Blodgett films and the crystallization of noncentrosymmetric organic crystals are widely studied as promising techniques to get ordered molecular arrangements for nonlinear optics and optoelectronics [147]. The required bulk alignment of molecular dipoles can also be achieved by their adsorption in the pores of appropriate molecular sieve matrices. Caro et al. [149] have studied the optical second harmonic generation (SHG) by *para*-nitroaniline (PNA) adsorbed in $AlPO_4$-5 crystals. No SHG effects could be observed for the unloaded $AlPO_4$-5 matrix. Figure 6.48 shows the experimental arrangement of the ps-Nd:YAG laser device and the position of the $AlPO_4$-5 crystal in the laser light beam [149]. The relative intensity of the SHG signal as a function of the angle between the polarization plane of the incident laser light and the *c*-axis (length direction [001]) of $AlPO_4$-5 crystal is plotted in Figure 6.49. The vanishing

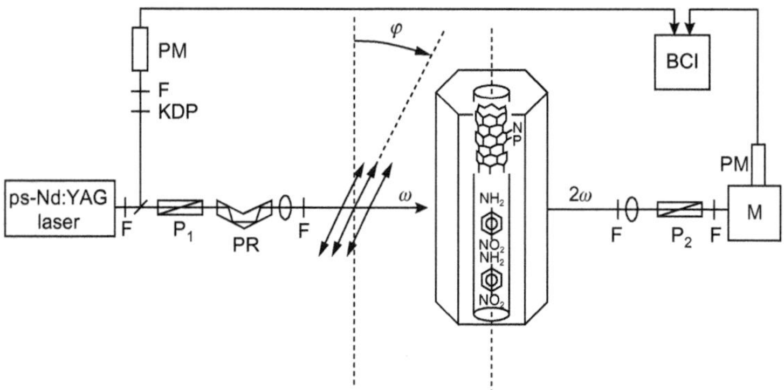

Figure 6.48 Experimental arrangement of the ps-ND:YAG laser device and the position of the $AlPO_4$-5 crystal in the laser light beam [149].

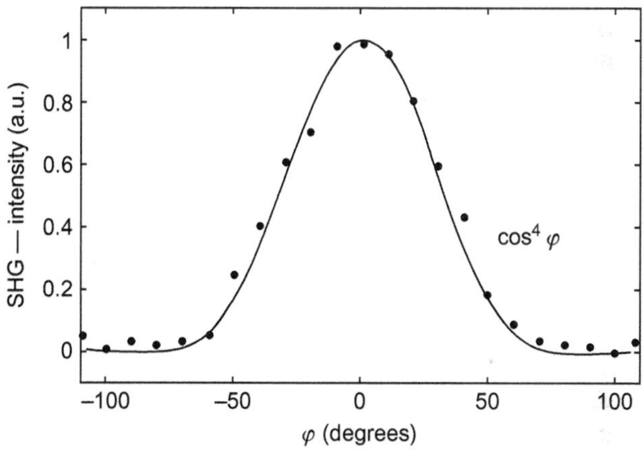

Figure 6.49 Relative intensity of the SHG signal as a function of the angle between the polarization plane of the incident laser light and the *c*-axis (length direction [001]) of $AlPO_4$-5 crystal is plotted [149].

of the SHG intensity for $\phi = 90°$ indicates that the molecules are well aligned by the molecular sieve channels, i.e., their orientational distribution is sharp [147]. Alvaro et al. [178] have prepared a series of four europium complexes incorporated into three different zeolites. Also, the chemiluminescence of Eu^{+3} exchange Y-zeolite upon thermal decomposition of tetramethyl 1,2-dioxetane has been reported.

Smolin et al. [179] have introduced CdS clusters in pores of zeolite-Z. It is shown that these clusters are built up around the S atoms that are situated at the centers of the large cavities of zeolites and coordinated tetrahedrally by Cd atoms. These atoms are bound through sulfur atoms with Cd cations situated near the walls of the cavity in the environment of the aluminosilicate framework oxygen.

Recently, Möller and Bein [180] reported a unique mechanism of creation of pores within pores—how to craft ordered hierarchical zeolites. This has prompted several new avenues for the zeolites' synthesis and applications. A pioneering work has been carried out by Bein [181] on the zeolitic host–guest interactions and building blocks for the self-assembly of complex materials. Zeolite crystals are used as building blocks for the self-assembly of thin films or 3D objects. A combination of weaker and stronger interactions ranging from dispersive forces, hydrogen bonding, and electrostatic interactions to covalent bonding can be used to build functional hierarchical constructs. This has led to the formation of several composites for various novel applications. Some of these aspects are discussed more elaborately in Chapter 10.

Researchers from United Kingdom are working on the application of zeolites in health-related issues. Researchers are producing activated liquid zeolite or activated clinoptilolite in water suspension (ACS), which is popularly being used to safely and effectively increase the urinary excretion of potentially toxic heavy metals in healthy volunteers without negatively impacting the electrolyte profiles, which controls the nutritional levels of the participants [182]. ACS has also undergone 13 years of pharmaceutical research in the United Sates with humans. Zeolite is included on the Food and Drug Administration's generally recognized as safe (GRAS) list and thus is considered to be "completely safe." Zeolite may well be one of the most potent natural preventatives and treatments for cancer ever discovered.

Thus, the study of zeolites is a very popular field. A great diversity, both in terms of compositions and applications, is being reported for zeolites. New structures have been discovered. All these suggest a great future for the zeolites, covering fields of selective and shape catalysis, oxidative catalysis, microelectronics, optoelectronics, health and environmental issues, and so on.

References

[1] A.F. Crönstedt, Om en obekant bärg art, som kallas Zeolites, Akad. Handl. Stockholm 17 (1756) 120.

[2] W.A. Deer, R.A. Howie, J. Zussman, Rock-Forming Minerals, vol. 4, Longmans, New York, London, 1967, p. 351.

[3] E.S. Dana, A Text Book of Mineralogy, John Wiley & Sons, New York, NY, 1955. p. 640.

[4] D. Loew, US Geography and Geological Exploration and Survey, West of 100th Meridian, Part 6, vol. 3, 1875, p. 569.

[4a] Sir J. Murray and A.F. Renyard, Report on deep-sea deposits based on the specimens collected during the voyage of H. M. S. Challenger in the years 1872 to 1876. John Gibson, Pub., Printed for H. M. Stationery off., by Neill and company, 1891.

[5] D.S. Coombs, The nature and alteration of some Triassic sediments, Trans. R. Soc. New Zealand 82 (1954) 65.

[6] H. Eichorn, Ueber die Einwirkung verdünnter Salzlösungen auf Silicate, Ann. Phys. Chem. 105 (1858) 130.

[7] M.H. Hey, Studies on the zeolites part I, general review, Min. Mag. 22 (1930) 422.
[8] C.E. Marshall, The Colloid Chemistry of the Silicate Minerals, Academic Press, New York, NY, 1949.
[9] J.W. McBain, The Sorption of Gases and Vapors by Solids, Rutledge and Sons, London, 1932.
[10] M.E. Davis, R.F. Lobo, Zeolite and molecular sieve synthesis, Chem. Mater. 4 (1992) 756–768.
[11] E.J.P. Feijen, J.A. Martens, P.A. Jacobs, Zeolites and their mechanism of synthesis, in: J. Weitkamp, H.G. Karge, H. Pfeifer, W. Holderich (Eds.), Zeolites and Related Microporous Materials: State of the Art 1994, 84, 1994, Studies in Surface Science and Catalysis.
[12] S.M. Auerbach, K.A. Carrado, P.K. Dutta, Handbook of Zeolites Science and Technology, Marcel Dekker, 2003, 1184 pp.
[13] C.S. Cundy, P.A. Cox, The hydrothermal synthesis of zeolites: precursors, intermediates and reaction mechanism, Microporous Mesoporous Mater. 82 (2005) 1–78.
[14] J. Yu, Synthesis of zeolites. Introduction to Zeolite Science and Practice, third revised ed., in: J. Cejka, H. Van Bekkum, A. Corma, F. Schuth (Eds.), Studies in Surface Science and Catalysis, 168, Elsevier, Amsterdam, 2007, pp. 39–103.
[15] S.T. Wilson, B.M. Lok, C.A. Messina, T.R. Cannon, E.M. Flanigen, Aluminophosphate molecules sieves; a new class of microporous crystalline inorganic solids, J. Am. Chem. Soc. 104 (1982) 1146–1147.
[16] W.M. Meier, D.H. Olson, Atlas of zeolite structure types, Adv. Chem. Ser. 101 (1970) 155.
[17] W.M. Meier, D.H. Olson, Atlas of Zeolite Structure Types, Structure Commission of the IZA, 1978.
[18] W.M. Meier, D.H. Olson, Atlas of Zeolite Structure Types, Structure Commission of the IZA, Butterworth-Heinemann, 1987.
[19] W.M. Meier, D.H. Olson, Atlas of Zeolite Structure Types, Structure Commission of the IZA, Butterworth-Heinemann, 1992.
[20] C. Baerlocher, L.B. McCusker, D.H. Olson, Atlas of Zeolite Framework Types, sixth revised ed., Elsevier, Amsterdam, 2007.
[21] D.W. Breck, Zeolite Molecular Sieves: Structure, Chemistry and Use, John Wiley & Sons, London, 1974.
[22] D.W. Breck, Structure of zeolite, Zeolite Molecular Sieves, Wiley Interscience, New York, NY, 1974, pp. 29–185.
[23] G.O. Brunner, W.M. Meier, Framework density distribution of zeolite type tetrahedral nets, Nature 337 (1989) 146.
[24] J.V. Smith, Felspar Minerals, vol. 2, Springer, Berlin, 1974.
[25] R.M. Barrer, Zeolites and their synthesis, Zeolites 1 (1981) 130–140.
[26] W.M. Meier, Molecular Sieves, Society of Chemical Industry, London, 1968. p. 10.
[27] E.M. Flanigen, Crystal structure and chemistry of natural zeolites, in: A.M. Frederick (Ed.), Reviews in Mineralogy, Vol. 4, Mineralogy and Geology of Natural Zeolites, Mineralogical Society of America, New York, NY, 1986, pp. 19–52.
[28] X. Qinhua, Y. Aizhen, Hydrothermal synthesis and crystallization of zeolites, in: K. Byrappa (Ed.), Hydrothermal Growth of Crystals, 21, 1990, Progress on Crystal Growth and Characterization.
[29] R.M. Barrer, Chemical nomenclature and formulation of compositions of synthetic and natural zeolites, Pure Appl. Chem. 51 (1979) 1091.

[30] R.A. Munson, R.A. Sheppard, A comparative study for basic Characterization of three Clinoptilolite Specimens, Natural zeolites: their Properties, occurrences and uses, Minerals Sci. Eng. 6 (1974) 19.
[31] R.M. Barrer, J.F. Cole, J. Chem. Soc. A (1970) 1516.
[32] L.W. Staples, H.T. Evans Jr., J.R. Lindsey, Cavansite and pentagonite, new dimorphous calcium vanadium silicate minerals from Oregon, Am. Mineral. 58 (1973) 405–411.
[33] H.T. Evans Jr., The crystal structures of cavansite and pentagonite, Am. Mineral. 58 (1973) 412–424.
[34] R. Rinaldi, J.J. Pluth, J.V. Smith, Crystal structure of cavansite dehydrated at 220°C, Acta Crystallogr. B31 (1975) 1598–1602.
[35] K.B. Powar, K. Byrappa, X-ray, thermal, and infrared studies of cavansite from Wagholi Western Maharashtra, India, J. Mineral. Petrol. Sci. 96 (2001) 1–6.
[36] F. d'Yvoire, Etude des phosphates d'aluminium et de fer trivalents, L' - orthophosphate neutre d'aluminium, Further studies on Aluminophosphate Molecular Sieves, Bull. Soc. Chem. (1961) 1762–1776.
[37] H. Robson, Synthesizing zeolites, Chemtech 8 (1978) 176–180.
[38] Wöhler, cited by R. Bunsen, Ann., 65 (1848) 80.
[39] H. Claireville, de St. (1862), cited by G.W. Morey, P. Niggli, The hydrothermal formation of silicates: a review, J. Am. Chem. Soc., 35 (1913) 1086–1130.
[40] R.M. Barrer, Hydrothermal Chemistry of Zeolites, Academic Press, London, 1982.
[41] R.M. Milton, Zeolite synthesis, in: M.L. Occelli, H.E. Robson (Eds.), ACS Symposium Series No. 398, 1989, p. 1.
[42] R.M. Milton, To Union Carbide, US Patent 2882244 (1959).
[43] R.M. Milton, To Union Carbide, US Patent 2996358 (1961).
[44] R.M. Milton, To Union Carbide, US Patent 3010789 (1961).
[45] R.M. Milton, To Union Carbide, US Patent 3030181 (1962).
[46] R.M. Barrer, P. Denney, Hydrothermal chemistry of the silicates, Part IX nitrogenous aluminosilicates, J. Chem. Soc. (1961) 971–1000.
[47] G.T. Kerr, Chemistry of Crystalline Aluminosilicates. II. The Synthesis and Properties of Zeolite ZK-4, Inorg. Chem. 5 (1966) 1537–1539.
[48] J.B. Higgens, Silica zeolites and clathrasils, in: Mineralogy of Silica, Reviews in Mineralogy, Mineralogical Society of America, New York, NY, 1994, pp. 507–543.
[49] R.L. Hay, Geologic occurrence of zeolites in sand, in: L.B. Sand, F.A. Mumpton (Eds.), Natural Zeolites, Pergamon, Oxford, 1978, pp. 135–143.
[50] S.T. Wilson, B.M. Lok, C.A. Messina, T.R. Cannan, E.M. Flanigen, Aluminophosphate molecular sieves, a new class of microporous crystalline inorganic solids, Am. Chem. Soc. (1983) 79–110.
[51] E.M. Flanigen, B.M. Lok, R.L. Patton, S.T. Wilson, in: P.A. Jacobs (Ed.), New Developments in Zeolite Science and Technology, Elsevier, Amsterdam, 1986, p. 103.
[52] M.B. Boisen, G.V. Gibbs, M. O'Keeffe, K.L. Bartelmehs, A generation of framework structures for the tectosilicates using a molecular-based potential energy function and simulated annealing strategies, Microporous Mesoporous Mater. 29 (1999) 219–266.
[53] O.D. Friedrichs, A.W.M. Dress, D.H. Huson, J. Klinowski, A.L. Mackay, Systematic enumeration of crystalline networks, an application of 3D, Nature 400 (1999) 644–647.
[54] B. Zhou, J. Yu, J. Li, Y. Xu, W. Xu, S. Qui, et al., Rational design of 2D layered aluminophosphates with $[Al_3P_4O_{16}]^{3-}$ stoichiometry, Chem. Mater. 11 (1999) 1094–1099.

[55] C.M. Draznieks, J.M., Newsam, A.M. Gorman, C.M. Freeman, G. Férey, De novo prediction of inorganic structures developed through automated assembly of secondary building units (AASBU method), Angew. Chem. Int. Ed. 39 (2000) 2270–2276.
[56] J. Yu, J. Li, R. Xu, Towards rational design and synthesis of aluminophosphates with 2-D layer and 3-D open-framework structures, Microporous Mesoporous Mater. 48 (2001) 47–56.
[57] Y. Li, J. Yu, D. Liu, W. Yan, R. Xu, Y. Xu, Design of zeolite frameworks with defined pore geometry through constrained assembly of atoms, Chem. Mater. 15 (2003) 2780–2785.
[58] J. Li, L. Li, J. Liang, P. Chen, J. Yu, Y. Xu, et al., Template-designed syntheses of open-framework zinc phosphites with extra-large 24-ring channels, Cryst. Growth Des. 8 (2008) 2318–2323.
[59] J. Yu, R. Xu, Rational approaches toward the design and synthesis of zeolitic inorganic open-framework materials, Acc. Chem. Res. 43 (2010) 1195–1204.
[60] Z. Wang, J. Yu, R. Xu, Needs and trends in rational synthesis of zeolitic materials, Chem. Soc. Rev. 41 (2012) 1729–1741.
[61] M. L. Occelli, H.E. Robson (Eds.), Zeolite synthesis, ACS Symposium Series 398, American Chemical Society, Washington, DC, 1989.
[62] K. Varoon, X. Zhang, B. Elyassi, L.F. Francis, A.V. McCormick, A Mkhoyan, M. Tsapatsis, Dispersible exfoliated zeolite nanosheets and their application as a selective membrane, Science 334 (2011) 72–75.
[63] C.E. Gounaris, J. Wei, A. Floudas, R. Ranjan, M. Tsapatsis, Rational design of shape selective separations and catalysis: lattice relaxation and effective aperture size, AIChE 56 (2009) 611–632.
[64] S.P. Zhdanov, Some problems of zeolite crystallization, in: E.M. Flanigen, L.B. Sand (Eds.), Molecular Sieve Zeolite-I, Adv. Chem. Ser., 101, 20, American Chemical Society, Washington, DC, 1971.
[65] S. Ueda, W. Kageyama, M. Koizumi, in: D. Olson, A. Bisio (Eds.), Proceedings of the Sixth International Zeolite Conference, Butterworth, Surrey, 1984, p. 905.
[66] W. Xu, J. Li, W. Li, H. Zhang, B. Liang, Nonaqueous synthesis of ZSM-35 and ZSM-5, Zeolites 9 (1989) 468–473.
[67] F.-Y. Dai, M. Suzuki, H. Takahashi, Y. Saito, New developments in zeolite science and technology, in: Y. Murakami, A. Lijima, J.W. Ward (Eds.), Proceedings of the Seventh International Zeolite Conference, Kodansha, Elsevier, Tokyo, 1986, p. 223.
[68] J. Dwyer, K. Karim, W.J. Smith, N.E. Thompson, R.K. Harris, D.C. Apperley, A comparison of siliceous faujasite zeolites produced by direct synthesis or by secondary synthesis, J. Phys. Chem. 95 (1991) 8826–8831.
[69] F.Y. Dai, M. Suzuki, H. Takahashi, Y. Saitoy, Hydrothermal Synthesis and Growth of Zeolites, Mechanism of zeolite crystallization without using template reagents of organic bases, Stud. Surf. Sci. Catal. 28 (1986) 223.
[70] E.M. Flanigen, Molecular Sieves: Principles of Synthesis and Identification, Syntheises of Zeolites, Adv. Chem. Ser. 121 (1973) 114.
[71] P.K. Dutta, M. Pur, D.C. Shieh, M.M.J. Treacy, J.M. Thomas, J.M. White (Eds.), Formation of zeolites: A molecular description in microstructure and properties of catalysts, Raman spectroscopic study of the synthesis of zeolite Y, MRS Soc. Symp. Proc. 111 (1988) 101.
[72] L.D. Rollman, Templates in zeolite crystallization, Adv. Chem. Ser. (1979) 173–387.

[73] E.M. Flanigen, in: L.V.C. Rees (Ed.), Proceedings of the Fifth International Conference on Zeolites, Heyden, London, 1980, p. 760.
[74] W. Loewenstein, The distribution of aluminum in the tetrahedra of silicates and aluminates, Am. Mineral. 39 (1954) 92–96.
[75] E.M. Flanigen, R.L. Patton, S.T. Wilson, Cracking properties of SAPO-37 and Faujasites, Innovation in Zeolite Materials Science, Stud. Surf Sci. Catal. 37 (1988) 13–27.
[76] T.E. Gier, G.D. Stuckey, Low Temperature Synthesis of Hydrated Zinco(beryllo)-Phosphate and Arsenate Molecular Sieves, Nature 346 (1991) 508.
[77] R. Xu, J. Chen, C. Feng, New families of M(III)X(V)04-type microporous crystals and inclusion compounds, Stud. Surf. Sci. Catal. 60 (1991) 63–72.
[78] A. Katovic, B. Subotic, I. Smit, L.A. Despotovic, M. Curie, in: M.L. Ocelli, H.E. Robson (Eds.), Zeolite Synthesis, ACS Symposium Series, 398, 1989, p. 124.
[79] J.D. Cook, R.W. Thompson, Modeling the effect of gel aging, Zeolites 8 (1988) 322–326.
[80] S.P. Zhdanov, N.N. Samulevich, in: L.V.C. Rees (Ed.), Proceedings of the Fifth International Conference on Zeolites, Heyden, London, 1980, p. 75.
[81] B. Subotic, J. Bronic, in: Y. Murakami, A. Lijima, J.W. Ward (Eds.), Proceedings of the Seventh International Zeolite Conference, Tokyo, 1986, p. 45.
[82] C.N. Wu, K.J. Chao, Synthesis of faujasite zeolites with crown-ether templates, J. Chem. Soc. Farad. Trans. 91 (1995) 167–173.
[83] A.T. Bell, in: M.L. Occelli, H.E. Robson (Eds.), Zeolite Synthesis, ACS Symposium Series, 398, 1989, p. 66.
[84] J.P. Gilson, in: E.G. Derouane, F. Lemos, C. Naccache, F.R. Ribeiro (Eds.), Zeolite Microporous Solids: Synthesis, Structure and Reactivity, Kluwer Academic, Dordrecht, 1992, NATO ASI Series, 352, 19.
[85] K. Byrappa, Hydrothermal growth of crystals, in: D.T.J. Hurle (Ed.), Handbook of Crystal Growth, Elsevier North-Holland, Amsterdam, 1994, p. 465.
[86] R.M. Barre, Trans. Seventh International Ceramic Congress, British Ceramic Society Publisher Clowes, London, 1960.
[87] E.M. Flanigen, R.L. Patton, US Patent 4073864, 1978.
[88] J.L. Guth, H. Kessler, J.M. Higel, J.M. Lamblin, J. Patarin, A. Seive, et al., Synthesis and characterization of defect-free crystals of MFI-type zeolites, Zeolite synthesis, ACS Symp. Ser. 398 (1989) 176.
[89] Xu Qinhua, B. Shullin, D. Jialu, The recrystallization of Zeolite Molecular Sieves-recrystallization Mechanism of 4A to NaY Zeolite, Sci. Sinica (Ser. B) 25 (1982) 349.
[90] D.T. Hayhurt, L.B. Sand, Molecular sieves II, ACS Symp. Ser. 40 (1977) 219.
[91] K.J. Chao, T.G. Tasi, M.S. Chen, I. Wang, Kinetic studies on the formation of zeolites, ZAM-5, J. Chem. Soc. Farad. I 77 (1981) 465–555.
[92] B.D. Witte, J. Patarin, D. Nouen, L. Delmotte, J.L. Guth, T. Cholley, Cyclic polyols: a new class of structure-directing agents? Study of inositol in the synthesis of FAU and LTA-type zeolites, Microporous Mesoporous Mater. 23 (1998) 11–22.
[93] H. Lechert, The pH value and its importance for the crystallization of zeolites, in: H. Robson (Ed.), Microporous Macroporous Material, vol. 22, Karl Petter, Lillerud, 1998, pp. 519–523.
[94] G.A. Ozin, Nano chemistry synthesis in diminishing dimensions, Adv. Mat. 4 (1992) 612–649.
[95] S. Mann, Molecular tectonics in biomaterialization and biometric materials chemistry, Nature 365 (1993) 499–505.

[96] J.L. Guth, P. Caullet, A. Sieve, J. Patarin, F. Delprato, in: Barthomeuf (Ed.), Guidelines For Mastering The Properties of Molecular Sieves, Plenum Press, New York, NY, 1990, p. 69.
[97] E.M. Flanigen, Molecular sieve zeolite technology – The first twenty five years, Pure and Appl.Chem, 52 (1980) 2191–2211.
[98] X. Qiuhua, Z. Huizhu, D. Jialu, W. Rui, Hydrothermal synthesis and growth of Zeolites, Kexue Tongbao 25 (1980) 833.
[99] P. Feng, X. Bu, T.E. Gier, G.D. Stucky, Amine-directed syntheses and crystal structure of phosphate based zeolite analogs, Microporous Mesoporous Mater. 23 (1998) 221–229.
[100] J.C. Jansen, in: H. Van Bekkum, E.M. Flanigen, J.C. Jansen (Eds.), Introduction to Zeolite Science and Practice, Elsevier, Amsterdam, 1991, p. 77.
[101] K. Egeblad, C.H. Christensen, M. Kustova, C.H. Christensen, Templating mesoporous zeolites, Chem. Mater. 20 (2008) 946–960.
[102] J.A. Martens, J. Jammaer, S. Bajpe, A. Aerts, Y. Lorgouilloux, C.E.A. Kirschhock, Simple synthesis recipes of porous materials, Microporous Mesoporous Mater. 140 (2011) 2–8.
[103] K. Na, C. Jo, J. Kim, J. Cho, Y. Seo, R.J. Messinger, et al., Directing zeolite structures into hierarchically nanoporous architectures, Science 333 (2011) 328–332.
[104] B.M. Lok, T.R. Cannan, C.A. Messina, The role of organic molecules in molecular sieve synthesis, Zeolites 3 (1983) 282–291.
[105] A. Moini, K.D. Schmitt, E.W. Valyocsik, R.F. Polomski, The role of diquaternary cations as directing agents in zeolite synthesis, in: J. Weitkamp, H.G. Karge, H. Pfeifer, W. Holderich (Eds.), Zeolites and Related Microporous Materials: State of the Art, 84, 1994, Studies in Surface Science and Catalysis.
[106] M. Yoshimura, W. Suchanek, K. Byrappa, Soft, solution processing—a strategy for advanced materials processing in the 21st century, MRS Bull. (September Issue) (2000).
[107] C.R. Martin, Membrane-based synthesis of nanomaterials, Chem. Mater. 8 (1996) 1739–1746.
[108] P.M. Ajayan, O. Stephan, P. Redlich, C. Colliex, Carbon nanotunes as removable templates for metal oxide nanocomposites and nanostructures, Nature 375 (1995) 564–567.
[109] S. Schact, Q. Huo, I.G. Voigt-Martin, G.D. Stucky, F. Schuth, Oil-Water Interface Templating of Mesophorous Structures, Science 273 (1996) 768–771.
[110] B. Mutaftschiev, Nucleation theory, in: D.T.J. Hurle (Ed.), Handbook of Crystal Growth, 1, Elsevier Science B. V., UK, 1993.
[111] A.S. Myerson, A.F. Izmailov, The structure of supersaturated solutions, in: D.T.J. Hurle (Ed.), Handbook of Crystal Growth, 1, Elsevier Science B. V., UK, 1993.
[112] J. Warzywoda, R.W. Thompson, Zeolite crystallization with autocatalytic nucleation, Zeolites 9 (1989) 341–345.
[113] P.A. Jacobs, in: E.G. Derouane, F. Lemos, C. Naccache, F.R. Ribeiro (Eds.), Zeolite Microporous Solids: Synthesis Structure and Reactivity, Vol. 352, Kluwer Academic, Dordrecht, 1992, NATO ASI Series.
[114] W.H. Chen, H.C. Hu, T.Y. Lee, Transient crystallization and crystal size distribution of zeolite A, Chem. Eng. Sci. 48 (1993) 3683–3691.
[115] R. Roy, Accelerating the kinetics of low-temperature inorganic syntheses, J. Solid State Chem. 111 (1994) 11–17.

[116] J.G. Carmona, R.R. Clemente, J.G. Morales, Comparative preparation of microporous VPI-5 using conventional and microwave heating techniques, Zeolites 18 (1997) 340–346.

[117] H. Ghobarkar, The morphology of some hydrothermally synthesized Li- minerals: A–Z, α-eucryptite, β-spodumene and β-eucryptite, Cryst. Res. Technol. 27 (1992) 181–185.

[118] H. Ghobarkar, O. Schaf, The morphology of hydrothermally synthesized natrolite, Cryst. Res. Technol. 31 (1996) K54–K57.

[119] H. Ghobarkar, O. Schaf, Hydrothermal synthesis and morphology of thomsonite and edingtonite, Cryst. Res. Technol. 32 (1997) 653–657.

[120] H. Ghobarkar, O. Schaf, The morphology of hydrothermally synthesized mesolite, Cryst. Res. Technol. 31 (1996) K67–K69.

[121] A. Kuperman, S. Nadimi, S. Oliver, G.A. Ozin, J.M. Garces, M. Olken, Non-aqueous synthesis of giant crystals of zeolites and molecular sieves, Nature 365 (1993) 239–242.

[122] O.K. Mel'nikov, B.N. Litvin, N.S. Triodina, Crystallization of sodalite on a seed, in: A.N. Lobachev (Ed.), Crystallization Processes Under Hydrothermal Conditions, Consultants Bureau, New York, NY, 1973, pp. 151–172.

[123] L.N. Demianets, E.N. Emelyanova, O.K. Mel'nikov, Solubility of sodalite in various solutions of NaOH under hydrothermal conditions, in: A.N. Lobachev (Ed.), Crystallization Processes Under Hydrothermal Conditions, Consultant Bureau, New York, NY, 1973, pp. 125–149.

[124] G.M. Johnson, M.T. Weller, Synthesis and characterization of gallium and germanium containing sodalites, in: H. Chon, S.K. Ihm, Y.S. Uh (Eds.), Progress in Zeolite and Microporous Materials, 105, 1997, Studies in Surface Science and Catalysis.

[125] D.M. Bibby, P. Dale, Synthesis of silica-sodalite from nonaqueous systems, Nature 317 (1985) 157–158.

[126] J. Lemberg, Ueber Silicatumwandlungen, Z. Deut. Geol. Gesell 33 (1883) 579–611.

[127] O. Carlgren, P.T. Cleve, Uber Einige Ammoniakalische Platinverbindungen, Z. Anorg. Chem. 2 (1892) 65–75.

[128] T. Hayashi, H. Shiga, M. Sadakata, T. Okubo, Hydrothermal growth of millimeter-sized aluminosilicate sodalite single crystals in noble metal capsules, J. Mater. Res. 13 (1998) 891–895.

[129] S.T. Wilson, E.M. Flanigen, Synthesis and characterization of metal aluminophosphate molecular sieves, in: M.L. Occelli, H.E. Robson (Eds.), Zeolite Synthesis, American Chemical Society, Washington, DC, 1989.

[130] X. Ren, S. Komarneni, D.M. Roy, The role of gel chemistry in synthesis of aluminophosphate molecular sieves, Zeolites 11 (1996) 142–148.

[131] M.E. Davis, C. Montes, J.M. Garces, Synthesis of VPI-5, in: Boran Držaj, Stanko Hočevar, Stane Pejovnik (Eds.), Zeolite Synthesis, American Chemical Society, Washington, DC, 1989, pp. 291–304.

[132] C.C. Freyhardt, M. Tsapatsis, R.F. Lobo, K.J. Balkus Jr., M.E. Davis, A high-silica zeolite with a 14-tetrahedra-atom pore opening, Nature 384 (1996) 295–296.

[133] W. Schmidt, F. Schuth, H. Reichert, K. Unger, B. Zibrowius, VPI-5 and related aluminophosphates: preparation and thermal stability, Zeolites 12 (1992) 2–8.

[134] L. Maistriau, Z. Gabelica, E.G. Derouane, E.T.C. Vogt, J. Van Oene, Solid-state NMR study of the transformation of VPI-5/MCM-9 into AlPO4-8/SAPO-8, Zeolites 11 (1991) 583–592.

[135] W. Gopel, T.A. Jones, M. Kleitz, J. Lundstrom, T. Seiyama, Chemical and Biochemical Sensors, Wiley-VCH, New York, NY, 1991.
[136] T. Bein, K. Brown, C.G. Frye, C.J. Brinker, US Patent 5151110, 1992.
[137] S. Feng, T. Bein, Vertical aluminophosphate molecular sieve crystals grown at inorganic–organic interfaces, Science 265 (1994) 1839–1841.
[138] Y. Yan, T. Bein, Zeolite thin films with tuneable molecular sieve function, J. Am. Chem. Soc. 117 (1995) 9990–9994.
[139] H.J.C. Hennepete, D. Bargemen, M.H.V. Mulder, C.A. Smith, Zeolite-filled silicone rubber membranes, part I, membrane preparation and pervaporation results, J. Membr. Sci. 35 (1985) 39–55.
[140] E.R. Geus, J. Schoomian, H. Van Bekkum, in: L. Bedrad, T. Bein, M.E. Davis, J. Garces, V.A. Maroni, G.D. Stucky (Eds.), Synthesis Characterization and Novel Applications of Molecular Sieve Materials, Materials Research Society, Pittsburgh, PA, 1993, p. 231.
[141] T. Bsein, Synthesis and applications of molecular sieve layers and membranes, Chem. Mater. 8 (1996) 1636–1653.
[142] J.G. Tsikoyiannis, W.O. Haag, Synthesis and characterization of a pure zeolite membrane, Zeolites 12 (1992) 126–130.
[143] T. Sano, Y. Kiyozumi, K. Maeda, M. Toba, S. Niwa, F. Mizukami, New preparation method for highly siliceous zeolite films, J. Mater. Chem. 2 (1992) 141–142.
[144] S.P.J. Smith, V.M. Linkov, R.D. Sanderson, L Petrik, C.T. O'Conner, K. Keiser, Preparation of hollow-fibre composite carbon-zeolite membranes, Microporous Mater. 4 (1995) 385–390.
[145] T. Bein, K. Brown, C. Brinker, in: P.A. Jacobs, R.A. Van Santen (Eds.), Zeolites: Facts, Figures, Future, Stud. Surf. Sci. Catal., 49 (1989) 887–896.
[146] M. Noack, P. Koelsch, D. Venzke, P. Toussaint, J. Caro, New one-dimensional membranes-aligned $AlPO_4$-5 molecular sieve crystals in a nickel foil, J. Microporous Mater. 3 (1994) 201–206.
[147] H. Fuchs, H. Ohst, W. Prass, Ultrathin organic films: Molecular Architectures for advanced optical, electronic and bio-related systems, Adv. Mater. 3 (1991) 10.
[148] K. Moller, M.E. Eddy, G.D. Stucky, G. Herron, T. Bein, Stabilization of cadmium seienide molecular clusters in zeolite y: EXAFS and x- ray diffraction studies, J. Am. Chem. Soc. 111 (1989) 2564.
[149] J. Caro, G. Finger, J. Kornatowski, J. Richter-Mendan, W. Lutz, B. Zibrowius, Aligned molecular sieve crystals, Adv. Mater. 4 (1992) 273–276.
[150] M. Ziolek, Z. Dudzik, Catalytic active centers in $H_2O + O_2$ reaction on faujasite, Zeolites 1 (1981) 117–121.
[151] P.B. Weisz, U. Erdol, Molecular sieves as selective catalysts, Kohle 18 (1965) 527.
[152] S.P. Mirajkar, M.J. Eapen, S.S. Tamhankar, B.S. Rao, V.P. Shiralkar, Hydrothermal synthesis of Al and Ga-substituted omega zeolite, J. Inclus. Phenom. Mol. Recog. Chem. 16 (1993) 139–153.
[153] G. Perego, G. Bellussi, C. Corno, et al., New developments in zeolite science and technology, in: Y. Murakami, A. Lijima, J.W. Ward (Eds.), Proceedings of the Seventh International Conference on Zeolites, Tokyo, Elsevier, Amsterdam, 1986, p. 129.
[154] B. Notary, B. Notary (Ed.), Innovation in zeolites materials science, in: J.S. Magee, M.M. Mitchell, (Eds.), Proceedings of the International Symposium, Nieuwpoort, Elsevier, Amsterdam, 1988, p. 413.

[155] U. Romano, A. Esposito, F. Maspero, et al., New developments in selective oxidation, in: Leonard Curtis Luckwill (Ed.), Proceedings of the International Symposium, Rimini, Paper B1, University of Bologna, Bologna, 1989.

[156] P. Roffia, G. Leofanti, A. Casana, et al., New developments in selective oxidation, in: Leonard Curtis Luckwill (Ed.), Proceedings of the International Symposium, Rimini, Paper 42, University of Bologna, Bologna, 1989.

[157] A.S. Kharitonov, V.I. Sobolev, G.I. Panov, Hydrothermal of aromatic compounds with nitrous oxide, new possibilities of oxidative catalysis on zeolites, Uspekhi Khimi 61 (1992) 2062–2077.

[158] S.T. Wilson, E.M. Flanigen, Eur. Pat. Appl. 132, ZSM-5 type zeolites: SZM-5 type zeolites: synthesis and use in gasphase reactions with ammonia, (1985) 708.

[159] J.A. Rabo, M.L. Poutsma, G.W. Skeels, in: J. Hightower (Ed.), Proceedings of the Fifth International Congress Catalogues, North-Holland, New York, NY, 1973, pp. 1353–1361.

[160] J.A. Rabo, Salt occlusion in zeolite crystals, in: J.A. Rabo (Ed.), Zeolite Chemistry and Catalysis, ACS Monograph 171, American Chemical Society, Washington, DC, 1976, pp. 332–349.

[161] A. Clearfield, C.H. Saldarriaga, R.C. Buckley, in: J.B. Uytterhoeven (Ed.), Proceedings of the Third International Conference on Molecular Sieves, University of Leuven Press, Zurich, 1973, pp. 241–245. Paper 130.

[162] A. Clearfield, Hydrothermal synthesis of selected phosphates and molybdates, in: K. Byrappa (Ed.), Hydrothermal Growth of Crystals, Prog. Crystal Growth Charact., 21 (1990) 1–28.

[163] A.V. Kucherov, T.N. Kucherova, A.A. Slinkin, in: Recent Res. Rep. Int. Symp. Zeolites Chemistry Catalysis, Prague, September 8–13, 1991, p. 69.

[164] A.A. Slinkin, A.V. Kucherov, S.S. Goryashenko, K.I. Slovetskaya, in: Julie A Rabo (Ed.), Proceedings of the All-Union Conference on Use of Zeolites in Catalysis, Moscow (in Russian) 1989, p. 191.

[165] A.A. Slinkin, A.V. Kucherov, S.S. Goryaskenko, E.G. Aleshin, K.I. Slovetskaya, in: M.J. Winter (Ed.), Proceedings of the Eighth Soviet-French Seminar Catal., Novosibirsk, 1990, p. 153.

[166] A.V. Kucherov, A.A. Slinkin, Solid-state reaction as a way to transit metal cation introduction into high-silica zeolites, J. Mol. Catal. 90 (1994) 323–354.

[167] A.V. Kucherov, T.N. Kucherova, On the mechanism of nitric oxide decomposition over Cu-ZSM-5, A.A. Slinkin, Kinet. Katal. 33 (1992) 877.

[168] T. Loiseau, G. Ferey, Oxyfluorinated microporous compounds, J. Solid State Chem. Ill (1994) 403–415.

[169] T. Loiseau, R. Retoux, P. Lacorre, G. Ferey, Oxyfluorinated microporous compounds, J. Solid State Chem. 111 (1994) 427–436.

[170] A. Muller, M.T. Pope, Polyoxometalate Chemistry: An old field with new dimensions in several disciplines, Angew. Chem. Intl. Ed. 30 (1991) 34.

[171] V. Soghomonian, Q. Chen, R.C. Haushalter, J. Zubieta, Vanadium phosphate framework solid constructed of octahedra, square pyramids, and tetrahedra with a cavity diameter of 18.4 Å, Angew Chem. Int. Ed. 32 (1993) 610–612.

[172] R.C. Haushalter, L.A. Mundi, Reduced molybdenum phosphates: octahedral tetrahedral framework solids with tunnels, cages and micropores, Chem. Mater. 4 (1992) 31–48.

[173] E.W. Corcoran, J.M. Nawsam Jr., H.E. King Jr., D.E.W. Vaughan, in: M. Occelli, H. E. Robson (Eds.), Zeolites Synthesis, American Chemical Society, Washington, DC, 1989, p. 603.
[174] S. Feng, R. Xu, G. Yang, H. Shun, Chem. J. Chinese Univ. (Engl. Ed.) 4 (1988) 9.
[175] T. Wang, G. Yang, S. Feng, C. Shang, R. Xu, J. Chem. Soc. Chem. Commun. (1989) 2436.
[176] S. Feng, M. Tsai, M. Greenblatt, Preparation, ionic conductivity, and humidity-sensing property of novel, crystalline microporous germanates, $Na_3HGE7O_{16}xH_2O$, x = 0−6.1, Chem. Mater. 4 (1992) 388−393.
[177] M.E. Leonowicz, J.A. Lawton, S.L. Lawton, M.K. Rubin, MCM-22: a molecular sieve with two independent multidimensional channel systems, Science 24 (1994) 1910−1913.
[178] H. Alvaro, V. Fornes, S. Garcia, H. Garcia, J.C. Scaiano, Intrazeolite photochemistry 20, characterization of highly luminescent europium complexes inside zeolites, J. Phys. Chem. B. 102 (1998) 8744−8750.
[179] Y.I. Smolin, Y.F. Shepelev, A.E. Lapshin, E.A. Vasileva, CdS clusters in pores of zeolites X, Crystallogr. Rep. 43 (1998) 387−394.
[180] K. Möller, T. Bein, Pores within pores—how to craft ordered hierarchical zeolites, Science 333 (2011) 297−298.
[181] T. Bein, Zeolitic host−guest interactions and building blocks for the self-assembly of complex materials, MRS Bull. 30 (2005) 713−720.
[182] Available from: http://www.liquidzeolite.org/intro/gabriel-cousens.html.
[183] M. Wark, W. Lutz, M. Bulow, G. Schulz-Ekloff, in: R. von Ballmoos, J.B. Higgins, M.J. Treacy (Eds.), Proceedings of the Third German Workshop on Zeolite Chemistry, Berlin, March 18−19, 1991.
[184] J. Weitkamp, New directions in zeolite catalysis, in: O'hlmann et al. (Eds.), Proceedings of ZEOCAT 90, Catalysis and Adsorption by Zeolites, Studies Surf. Sci. Catal., 65 (1990) 21.
[185] D. Riou, G. Ferey, Oxyfluorinated microporous compounds, J. Solid State Chem. 111 (1994) 422−426.

7 Hydrothermal Synthesis and Growth of Coordinated Complex Crystals (Part I)

7.1 Introduction

Hydrothermal synthesis and growth of complex coordinated crystals are not new especially for silicates. In fact, the hydrothermal technology was clubbed under silicate technology in the earlier days. Silicates have drawn the attention of crystal chemists right from the earliest days owing to their abundance, structural diversity, and widespread applications in various technologies. Thus, the synthesis of complex coordinated complexes began in the nineteenth century. In fact, hydrothermal technique was considered as a most powerful tool to grow these complex coordinated crystals. During the 1960s and 1970s, the importance of the hydrothermal technique for the synthesis of inorganic materials without natural analogs was realized. Since then a large variety of complex coordinated complexes are being obtained with very complex crystal structures. More recently, the hybrid or mixed metal-organic framework (MOF) structures, which are described as crystalline compounds consisting of metal ions or clusters coordinated to often rigid organic molecules to form one-, two-, or three-dimensional structures that can be porous, and can be used for a variety of novel applications like gas separation, gas purification, catalysts, sensors, and hydrogen storage. However, these hybrid MOFs cannot be grown as large bulk crystals, but they are frequently processed as submillimeter size crystalline materials. Although these mixed hybrid framework structures are very popular in modern technology, they cannot be widely synthesized using any processing technique, except hydrothermal and solvothermal techniques. The complex coordinated crystals covers silicates, phosphates, vanadates, arsenates, molybdates, tungstates, fluorides, sulfates, selenides, borates, and so on, with a great variety of cations starting from alkali, alkaline earth, transitional to rare earth metals. The number of all such compounds exceeds several hundreds owing to their academic and technological interest and here it is impossible to discuss each and every compound. Therefore, we have discussed these compounds in two parts for the sake of the reader's convenience. Part I covers silicates, germanates, phosphates, vanadates, and borates (this chapter). The remaining group of complex

Handbook of Hydrothermal Technology.

coordinated compounds is discussed in Part II (Chapter 8). The literature survey shows that among all these compounds, the silicates and germanates have taken a unique place in the hydrothermal research because of their diversified compositions and structures. Here, these compounds will be discussed in two categories: (i) rare earth silicates and phosphates and (ii) transitional metal silicates and phosphates.

In this chapter, the main emphasis is on the crystal growth and crystal chemistry, which in turn provides clues to grow a variety of complex coordinated compounds. It is also well known that hydrothermal, solvothermal and supercritical hydrothermal methods produce very unusual complex structures, which cannot be obtained by other synthesis methods. However, this chapter deals with only the growth of bulk single crystals, and the micro- to nanosize crystals growth will be discussed in Chapter 10.

7.2 Crystal Chemical Background

It is interesting to note that there is a close relationship between the crystal chemistry and crystal growth of these compounds. The primary requirement for any material/crystal/mineral with a device potential is the understanding of its scientific foundation, particularly the consequences of the submicroscopic crystal structure, i. e., the spatial arrangement of atoms, ions, molecules, and the types of chemical bonds between them. It is the chemical bond more than anything else that determines the structure and the properties of a crystal. A crystal chemical knowledge helps in tailoring a given phase or a given structure that is to optimize chemical classification, crystal structure, and defect structure with respect to the desired properties. Therefore, the present authors briefly discuss the structural diversity of silicates and the Si–O chemical bond.

Silicates are the compounds containing Si and O. Their main specific structural feature is the presence of Si-atoms in a fourfold coordination with oxygen, which is favored by the ratio rSi:rO = 0.28. Table 7.1 gives the different types of coordinations of cations as a function of ionic radii ratio. Bond lengths and bond angles in Si-tetrahedra tend to be close to their mean values: d(Si–O) = 1.62 Å; angle

Table 7.1 The Different Types of Coordinations of Cations as a Function of Ionic Radii Ratio [1]

CN	Coordination	Ionic Radii Ratio
2	Dumb-bells	< 0.15
3	Triangle	0.15 – 0.22
4	Tetrahedron	0.22 – 0.41
6	Octahedron	0.41 – 0.73
8	Cube	0.73 – 1.37

O–Si–O = 109°.47′; Si–O–Si = 140°. Si-atoms are found in sixfold coordination in relatively small groups, basically in high-pressure silicates, and in fivefold coordination in some of the organosilicate compounds [1].

The structures of silicates and related compounds contain tetrahedral anions of different configurations (single tetrahedron, pyrogroups, rings, chains, layers, frameworks, and so on). Their comparative analysis considerably extends scientific ideas on the structures of crystalline materials. The attention to silicates is connected to their widespread abundance in the continental crust (Table 7.2), the exceptional variability of tetrahedral anions in their structures, and a widespread application of silicates in modern technology. Silicates are widely being used in ceramic, optical, electronic, piezoelectric, optoelectronic, and other industries [2].

The compounds that are nearer to silicates in the crystal chemical sense are phosphates with over 350 minerals. According to Corbridge [3], the term *phosphate* may be used to include all phosphorus compounds which contain P–O linkages. Similarly, the author [4] calls all the simple and complex oxides obtained from phosphoric acid solutions phosphates. For a free atom of phosphorus, the distribution of electrons in the valence cell is written as follows:

$$\begin{array}{cc} 3s & 3p \\ \hline \downarrow\uparrow & \uparrow\uparrow\downarrow \end{array} \tag{7.1}$$

It means that the phosphorus atom can form the bonds due to 3p electrons and another two owing to 3*s* electrons. Hence, the phosphorus atom participates in the reaction with oxygen to exist in a trivalent state, forming phosphides and in a pentavalent state forming phosphates. As a result of sp^3 hybridization (1s and 3p orbitals of P), four similar energy orbits develop and during the interaction of phosphorus with oxygen, it leads to the formation of a stable tetrahedral group [PO_4]. However, the P–O bonds in such a tetrahedron will not be purely ordinary (σ bonds), but they include a distinct fraction of short bonds (π bonds) as well. It is well known that the P–O tetrahedra are susceptible to condensation reactions due to the formation of bridging bonds –P–O–P–, or P–O (M). The possibility of the development of such bonds make the *p* orbit of the bridging oxygen (which is arranged perpendicular to the plane, POP) interact with the *d* orbits of the P atoms to form π bonds. The bond lengths and bond angles in P tetrahedra are given as

Table 7.2 Bulk Content of Silicates in Continental Crust [2]

Mineral	Percentage	Total
Feldspars	64.0	
Pyroxenes + Amphiboles	9.0	
Quartz	18.0	96.5%
Biotite		
ivine	1.5	

follows: d(P—O) = 1.54 + 0.02 Å, P—O (terminal) = 1.51 Å, angle O—P—O = 109.4 ± 5° and angle P—O—P = 156—167°. P—O (bridging) = 1.64 Å. Although the bond length tends to be close to its mean value, the bond angle, especially angle P—O—P, is obtuse and the rotation flexibility along the P—O—P linkage allows a variety of related orientation of adjacent tetrahedral [5]. From the crystal chemistry of phosphates, it is evident that the variation in the bond lengths and bond angles is quite large and it is difficult to find an analog to [PO_4]-tetrahedra among the inorganic compounds [6]. Hence, a thorough study of the crystal chemistry of silicates and phosphates, in turn, helps to understand the crystal chemistry of analogous compounds like germanates, vanadates, arsenates, sulfates, and borates.

Germanates and phosphates form the nearest analogs of silicates owing to the nearer ionic radii and charge values. All three types of tetrahedra [SiO_4, GeO_4, and PO_4] form condensed radicals, but the ability to undergo polycondensation is high for silica tetrahedra when compared with Ge, P, and other tetrahedra (VO_4, AsO_4, SO_4, BO_4, BeF_4, etc.) having nearer ionic radii. This is probably connected with a significantly lower number of different P:O and Ge:O ratios (10 and 9, respectively) when compared with a similar Si:O, which is 28 as observed in phosphates, germanates, and silicates [7]. This characteristic property of Si-tetrahedra can be explained based on two factors: (i) A suitable value of the valence charge of silicon equivalent to one (this separates silicon from phosphorus), (ii) characteristic radius ratio $rSi^{4+}:rO^{2-}$ for the formation of tetrahedral coordination. In this respect, Si-tetrahedra differ from Ge-tetrahedra for which the radius ratio $rGe^{4+}:rO^{2-}$ (0.38) is nearer to the upper stability limit of the tetrahedral coordination (0.41). The possible valence states and the coordination numbers of basic cations capable of forming oxide complexes are given in Table 7.3. All these elements represented in Table 7.3 form the tetrahedral coordination. Besides, the linking of [TO_4]-tetrahedra (T is the tetrahedral cation) is different for different cations. For example, out of more than 600 structures refined for natural and synthetic sulfates, only in four structures, namely, $K_2S_2O_7$ [8], $nK_2S_2CyV_2O_5$ [9], CdS_2O_7 [10], and $Te(S_2O_7)_2$ [11], have the pyrogroups [S_2O_7] been established recently. Structurally, a very important phosphate mineral, canaphite, having pyrophosphate groups

Table 7.3 Valency States and Coordination Numbers of Basic Oxide Complexes [1]

Group	Element	Valency States					Coordination			
		2	3	4	5	6	3	4	5	6
III	B	−	+	−	−	−	+	+	−	−
IV	Si	−	−	+	−	−	−	+	+	+
	Ge	−	−	+	−	−	−	+	−	+
V	P	−	+	+	+	−	−	+	−	−
	V	+	+	+	+	−	−	+	+	+
	As	−	+	−	+	−	−	+	−	−
VI	S	+	−	+	−	+	−	+	−	−

$[P_2O_7]$ formed by the linking of two tetrahedra at the vertices was reported recently [12]. Before the discovery of this mineral in 1988, there was a unanimous opinion with regard to the absence of condensed phosphate minerals, although more than 320 phosphate minerals are known in nature. However, vanadates, which usually form four-, five,- and sixfold coordinations with oxygen, contain the condensed V-tetrahedra. For example, chervetite, $Pb_2V_2O_7$ having pyrogroup-$[V_2O_7]$ formed by the linking of two $[VO_4]$-teterahedra with one of the corners sharing, has been reported so far [13]. In contrast, such condensed radical groups or pyrogroups, i.e., $[As_2O_7]$ complexes, have not been reported among arsenates.

Main concept of silicate structures. In silicate structures (Si,O)-tetrahedral anions should be adjusted to the cationic polyhedral fragments [14]. In particular, this principle was confirmed by the structures with big cations (K, Na, Ca, RE, etc.), which always contain (Si,O)-complexes on the basis of pyrogroups $[Si_2O_7]$ commensurable with the edges of cationic polyhedra.

Several quantitative correlations actually complete this qualitative approach. For example, there are more than 60 compounds with a general formula $M_a[T_2O_7]_b$, where T = As, Be, Cr, Ge, P, S, Si. They can be subdivided into two groups: (i) Tortveitite-like with angle T—O—T > 140°; (ii) bichromate like with angle T—O—T < 140° Ref [15]. Figure 7.1 illustrates that the frontier between both groups corresponds to the equation $r/T = 1.5r/M - 1.1$, where r is the radii of tetrahedral (T) and nontetrahedral (M) cations [15]. Thus, the increasing number of refined structures allows the establishment of a quantitative relationship between the configuration of tetrahedral complexes and the properties of the nontetrahedral cations.

Among these complex coordinated compounds, silicates have been studied extensively, followed by germanates, phosphates, and others. Although silicates and germanates have a systematic structural classification, their analogous compounds —like phosphates, vanadates, and arsenates— do not have a systematic classification. It is extremely difficult to classify phosphates, unlike silicates, because of high complexity in the phosphates' internal structures [16]. Similarly, the other

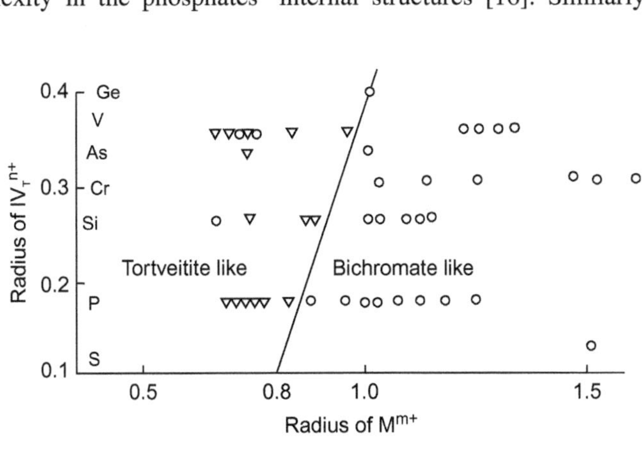

Figure 7.1 Correlation between the stability of tortveitite-like and bichromate-like structures $Ma[T_2O_7]$ and radii of M- and T-cations [1].

analogous compounds, such as borates, vanadates, arsenates, and sulfates, have not been studied in detail and they do not have any systematic structural classification. This is probably connected with their lesser abundance in nature and also lesser degree of structural diversities.

One of the main crystal chemical problems is connected with predictions of structural transformations of elevated temperatures and pressures. The hydrothermal technique is especially very handy in this context.

The diversity of (Si,O)-complexes described in an earlier section has been attributed to silicate stability at relatively lower depths. The lower coordination number of oxygen [2,3] in these structures makes them unstable in the mantle. High-pressure structural studies [17] show the decrease of Si—O distance until the "critical" value—1.59 Å—leads to a change of Si-tetrahedra by Si-octahedra. There is a certain advantage in such transformations, because Si-octahedra with longer distances Si—O can be linked not only by corners, but also by edges and even by faces. Thus, the coordination number of oxygen becomes greater, and at the same time the distribution of Si-atoms becomes more compact [18]. The "critical" value 1.59 Å is not accidental and is confirmed by the correlation between interatomic distances and angles in Si-tetrahedra [19]. $Lg2/d(Si{-}O)avg/ = Lgd(Si{-}Si) - Lg\angle(Si{-}O{-}Si)/2$ with fixed Si—Si distances = 3.06 Å and $\angle(Si{-}O{-}Si)_{max} = 180°$, one gets $d(Si{-}O)_{min} = 1.59$ Å. This value was obtained in the structure of forsterite Mg_2Si0_4 at $P = 14.9$ GPa [20].

The stability of Si-tetrahedra depends on the properties of nontetrahedral M-cations. The presence of M-cations with a high value of electronegativity leads to a decrease in the coordination of oxygen atoms. Consequently, under high pressures, these structures transform into structures with Si-octahedra more easily [21].

McDonald and Cruickshank [22] have shown that in structures with electronegative M-cations, the distances Si—O bridging and Si—O terminal become more equal. According to Brown [23], this effect is followed by a decrease of $(Si{-}O)_{avg}$ distances in Si-tetrahedra. In other words, chemical deformations connected with the presence of M-cations with high electronegativity are similar to the high-pressure structural deformations.

The concept of emphasizing the dominating role of cations in the structure reformation is used for the interpretation of structural transformations under high pressure. The compressibility of cationic polyhedra (β) is inversely proportional to the polyhedral charge density z/d [24], where d is the distance between central cation and O-atoms. The behavior of Si-tetrahedra mainly depends on the compressibility of the M-polyhedra, which can be considered as initial [25]. For example, Mg_2SiO_4 under high pressure successively transforms from α- and β-modifications with decreasing ratios (Mg—O)/(Si—O): 1.291 (α), 1.860 (β), and 1.251 (γ). The significant compressibility of the Mg-octahedra during the transformation from the α- to the γ-form initiates the unexpected expansion of the Si-tetrahedra; d(Si—O) in α-Mg_2SiO_4 = 1.636 Å while in γ-Mg_2SiO_4 = 1.655 Å [26]. It is noteworthy that the structural alterations in these compounds as in some other silicates result in oxygen atom packing ordering [27].

According to Ringwood [28], high-pressure transformations of silicates can be studied on the basis of their germanate analogs. This idea can be understood if one takes into account that the ratio r_{cation}/r_{oxygen} should be increased under high pressure. At atmospheric pressure, rGe^{4+} is greater than rSi^{4+} for 20%. Therefore, the structures of germanates at normal pressure reflect many specific features of high-pressure deformations in silicates. As an example, the authors compare the high-pressure structural changes in quartz [29] with chemical deformations in its structure, where Si-atoms were partially substituted by Ge.

These crystals bear composition $(Si_{0.86}Ge_{0.14})O_2$ which is considered to be unique because, hitherto, the GeO_2 concentration in the synthetic quartz crystals never exceeded 0.13 molar%. The common tendencies in quartz under high pressure and with the substitution of Si for Ge are connected with: (i) A decrease in angle T—O—T (Figure 7.2); (ii) a decrease in intertetrahedral distances O—O; (iii) a similar change in tilt angles between tetrahedra (Figure 7.3a); and (iv) an increase in tetrahedral distortion (Table 7.4) [30].

The structural comparison in silicates and germanates allows explanation of the different crystal chemical phenomena. One of them is connected with the question of why GeO_2 does not crystallize with coesite structure while stishovite—a form of GeO_2—mineral argutite is known even in nature.

The structural changes of coesite, under high pressure are, connected with significant dispersion in Si—O—Si angles. It is noteworthy that the smallest angle sharply decreases. Moreover, under high pressure, the temperature factor of O_1, which participates in bond Si—O_1—Si with an angle equal to 180°, increases up to about 30%, while the temperature factors of other atoms decrease. Substitution of Ge for Si would reinforce the tendencies and would lead to structural distortions, in other words, to the displacement of O_1 and the distortion in angle Si—O_1—Si = 180°.

The tetrahedral complexes are being extensively studied and characterized, owing to their applications in technologies such as laser, superionic, ferroelectric, ion-exchange, piezoelectric, optoelectronics, and a host of others. We initially

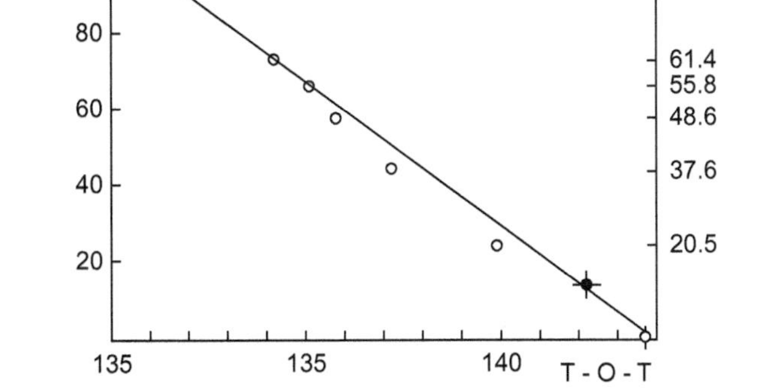

Figure 7.2 Decrease in T—O—T with the substitution of Si for Ge [7].

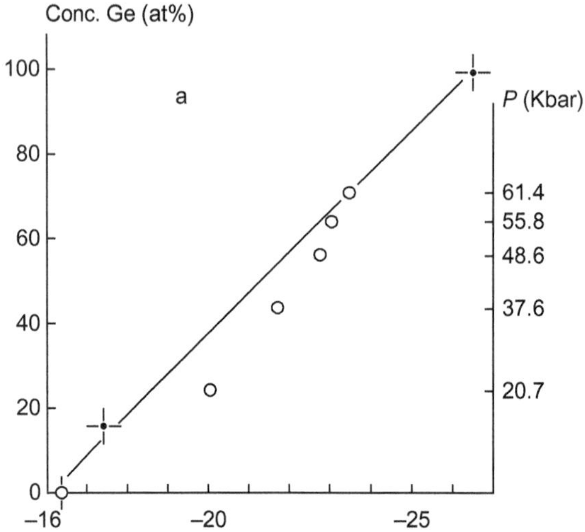

Figure 7.3 Decrease in the tilt angles between tetrahedral [7].

Table 7.4 Comparison of Structural Parameters in Quartz Under High Pressure with the Substitution of Si for Ge [7]

	Pressure		Substitution of Si for Ge	
Structural Parameter	**1 bar**	**61.4 kbar**	**($Si_{0.86} \cdot Ge_{0.14}$)**	**GeO_2**
∠T—O—T	143.73	134.2	142.2	130.1
Intertetrahedral	3.331	2.925	3.304	3.024
Distances O—O Å	3.411	3.064	3.406	3.193
Tilt angle	− 16.37	− 23.47	− 17.41	− 26.55
(O—T—O)-angle's dispersion	0.67	5.51	1.04	10.09

discuss the silicate structure, followed by phosphates, borates, vanadates, sulfates, etc., with reference to their properties. Only some representative compounds have been discussed covering all the tetrahedrally coordinated compounds.

In principle, there are several ways to search new crystalline phases under hydrothermal conditions:

- The investigation of phase diagrams of crystallization for complicated systems and of hydrothermal chemistry of individual elements.
- The direct search for new compounds with the preset physical properties.
- The study of hydrothermal reactions under new pressure−temperature conditions.
- The modification of known compounds with the help of isomorphic displacements or of special doping.

The above aspects have been discussed, in this book, for silicates, germanates, phosphates, borates, arsenates, vanadates, molybdates, tungstates, fluorides, etc.

The study of the phase diagrams of crystallization is the first step in the search of new compounds, the conditions of their synthesis, and the prediction of their physical properties. The next step in the hydrothermal synthesis is the selection of compounds with promising properties and the development of methods providing the synthesis of crystals with sizes appropriate for their practical use in science and technology. The study of phase interactions in multicomponent systems yields a wealth of information for hydrothermal physical and crystal chemistry and simulation of natural processes. In fact, the number of single crystals grown on seeds under hydrothermal conditions is much smaller than that of spontaneously grown single crystals. Today we have the techniques for crystal growth on seeds for the following materials:

Oxides	Quartz SiO_2, GeO_2, $Si_{1-x}Ge_xO_2$, zincite ZnO, cuprite Cu_2O, sapphire $A1_2O_3$, ruby $A1_2O_3$: Cr^{3+}, TeO_2, ABO_4 (A – Sb, B – Sb, Nb), ZnO
Sulfides	ZnS, HgS
Complex oxides	Sillenite $Bi_{12}TO_{20}$ (T = Si, Ge, etc.), garnets $Ln_3Fe_5O_{12}$, ferrites $LnFeO_3$
Silicates	Nepheline (Na, K) $AlSiO_4$, cancrinite $Na_8Al_6Si_6O_{24}$ $(OH,C1)_2 \cdot NH_2O$, sodalite $Na_8Al_6Si_6O_{24}(OH)_2 \cdot NH_2O$; beryl (emerald) $Al_2Be_3Si_6O_{18}$: Cr^3, eulytine Bi_4 $(SiO_4)_3$
Phosphates and family	$A1PO_4$, $GaPO_4$, $KTiOPO_4$, $KTiOAsO_4$, $GaAsO_4$
Carbonates	$CaCO_3$, $CoCO_3$, $MnCO_3$
Germanates	$KAlGeO_4$, $Na_8Al_6Ge_6O_{24}(OH)_2 \cdot nH_2O$ (two modifications), $Bi_4Ge_3O_{12}$
Nitrides	GaN, AlN
Others	$KNbO_3$ (KN), $KTaO_3$ (KT), $KTa_{1-x}Nb_xO_3$ ($x = 0.44$, KTN)

The number of spontaneously nucleated crystals/compounds is very high and it is practically impossible to list the individual phases and their preparation conditions. Therefore, we have made an effort to group them into broader divisions and discuss only some selected phases.

7.3 Rare Earth Silicates

Among silicates obtained under hydrothermal conditions, the rare earth silicates are the most extensively studied ones owing to their academic as well as technological importance. The rare earth elements have a significant influence on the crystallization of various rocks and minerals and in the magmatic processes on the whole. Their distribution in various rocks and minerals clearly demonstrates the physicochemical influence they exert upon the crystallization of the magma. Their distribution and association further depend upon the alkalinity of the melt. Therefore, an extensive work has been carried out on various alkali rare earth metal silicate systems in order to understand the crystallization of a large number of rare earth silicates with and without alkali metals. A majority of these rare earth silicates were found to be new phases, and most of them carry technological significance owing to their unique physical properties. Although a great amount of work has been done on the synthesis of several hundreds of phases of rare earth silicates with and without alkali metals, there are only a few publications dealing with their crystal growth and seeded growth. Nevertheless, a vast literature data accumulated on these compounds yield a great variety of interesting results with reference to the phase formation, pressure–temperature conditions, starting materials, type of rare earth element, alkali metal, the role of solvent, and so on. A large number of structure types have been obtained, which clearly indicates the importance of these rare earth silicates from the academic point of view. Besides, these rare earth silicates exhibit very interesting physical properties, which have been studied by various authors and they have correlated these properties with their molecular structures. Though it is extremely difficult to discuss each and every aspect dealt with in the literature, we have made a sincere attempt to provide an overall picture of these rare earth silicates of hydrothermal origin with a special reference to their phase formation and structure types.

7.4 Phase Formation of Rare Earth Silicates (in Aqueous Solvents)

It is well known that the hydrothermal solutions contain significant amounts of chlorides, carbonates, and other well-soluble components. The distribution and mobility of the rare earth elements greatly depend upon their interaction with Cl, F, SO_4, and CO_3 species. The formation of silicates in nature takes place essentially from fluoride solutions having alkaline character. Among them, the main role is played by the hydroxides of Na and K. In laboratory synthesis of rare earth silicates, essentially aqueous solvents of hydroxides and carbonates of Na and K, and, similarly, chloride, fluorides, and their mixtures are commonly used as solvents. The solubility increases above 60°C in NaOH solutions, especially above the melting point of NaOH (476°C), there exists a complete mixing. Similarly, the increase in the concentration of NaOH with temperature further improves the solubility [31,32]. Such studies have been carried out on carbonate solutions in detail [33,34]. Table 7.5 gives the solubility of Na_2CO_3 in water.

The bonding of the rare earth elements in the form of soluble species could be represented as follows [35]:

$$R^{I}_{y}X_{z} \leftrightarrow R^{+3} \leftrightarrow [R_{en}(AX)m] \quad (7.2)$$

where

R = rare earth ion
X and (AX) = any acidic ions
I and II = electroneutral compounds or complex anions.

Table 7.5 Solubility of Na_2CO_3 in Water [33]

P **(atm)** $\mathbf{Na_2CO_3}$ **in wt%**					
475°C		**500°C**		**540°C**	
900	1.5 − 2	1735	55	1200	<1.0
1300	3.5 − 4	1905	88	1300	1.0
1550	16	**520°C**		1400	1.0
1625	28	1200	1.5		
1637	37	1400	3.5		
1833	44	1500	3.5		
1955	49 − 50				

For example, rare earth elements with carbonate ions would form complex anions of $[R(CO_3)_3]$ type, which can migrate and in the presence of alkali elements would form $Na_3[R(CO_3)_3]_3$. From Lu to La, the stability of such complexes would increase with an increase in the alkali concentration in the medium. Thus, in natural processes, rare earth elements must have migrated as the soluble complexes of carbonates, fluorides, chlorides, and maybe sulfates. Such complexes, in turn, participate in chemical reactions to form more stable complexes with additional anions of CO_3, F, Cl, O_4, and OH, thus, the study of rare earth silicates in aqueous solutions of Na_2CO_3, Na_2SO_4, NaCl, and NaF. In fact, the solubility of Na_2CO_3, Na_2SO_4 and NaF in water decreases with an increase in temperature and increases with an increase in pressure. The solubility of NaCl in water increases with an increase in temperature and pressure. These studies on the behavior of solvents with temperature and pressure have helped in the understanding of the phase formations in various rare earth silicates and analogous systems.

Although the study of various rare earth silicate systems began during the 1970s extensively, the objective was to understand the distribution and mobility of rare earth elements in natural systems. The importance of the addition of alkali elements into the systems was realized in the 1980s, because the alkali and rare earth metals together play an important role in the formation of various phases. Several new phases were discovered and with the availability of a vast amount of structural data on these alkali rare earth silicates, their physical properties became known to the scientific community. Some of these alkali rare earth silicates exhibit very interesting mixed framework structures. Among the cations in the alkali rare earth silicates with mixed framework structures, the alkali ion can serve as a mobile charge carrier. It provides interesting opportunities for the development of potentially new fast ion conductors because of the extensive possibilities for crystal chemical tailoring. The chemical stability for silicates and a large number of related phases are available for investigation [36,37]. High Ionic conductivity has been reported in $Na_5RSi_4O_{12}$ (where R = La−Lu, Sc, Y) ring silicates, almost simultaneously from three different laboratories [38−43]. Hydrothermal synthesis of $Na_5YSi_4O_{12}$,

$Na_5Sc\ Si_4O_{12}$, $Na_5Er\ Si_4O_{12}$, etc. has been reported in the literature. The crystal structure of $Na_5RSi_4O_{12}$ was reported by Maksimov et al. [43] and it is shown in Figure 7.4. Thus, the importance of the rare earth silicates was established during the 1980s. Here, we discuss the behavior of various rare earth silicates and alkali rare earth silicate systems under hydrothermal conditions within a wide range of pressure–temperature conditions. Also, the role of different mineralizers and the starting molar ratios on their crystallization has been studied to obtain a general picture, instead of describing each and every rare earth silicate compound and its synthesis.

In the carbonate solutions and in mixed solutions of $NaOH + Na_2CO_3$, two new phases crystallize and when $RE_2O_3{:}SiO_2 = 1{:}4$, $NaRSi_6O_{14}$ crystallizes, where R = La–Eu. These crystals are quite stable in pure Na_2CO_3 solution and NaOH up to 10%. With an increase in the concentration of the solutions, apatite like phase forms, and with a decrease in the concentration of the solutions, the stability of $NaRSi_6O_{14}$ falls along with the stability field of quartz. In a pure carbonate solution, together with $NaRSi_6O_{14}$, quartz always crystallizes. With an addition of NaOH, the quantity of quartz reduces but does not disappear totally.

When $R_2O_3{:}SiO_2$ approaches 1:1, the system gives another phase $Na_3RSi_6O_{15} \cdot 2H_2O$, which is the characteristic phase for the entire rare earth series, and it persists up to the concentration of NaOH = 30%, and beyond 30% of NaOH, it separates into two types: $Na_3R[Si_2O_7]$ for the second half of the series (yttrium series) and $Na_2R[SiO_4](OH)$ for the first half of the rare earth series (lanthanum series). An overall scheme of crystallization shows that when mixed mineralizers ($NaOH + N_2CO_3$) are used, the crystallization pattern changes and is shown for the Nd representative in Figure 7.5 [44] It is interesting to see that with a decrease in the amount of H_2O in the alkali-carbonate system, the field of $Na_3RSi_6O_{15} \cdot 2H_2O$ also decreases and only one phase NaR_6O_{14} crystallizes. In the $NaOH + Na_2CO_3$ solution and at $T = 550–600°C$ and $P = 1–2$ kbar and only in experiments with NaF and NaCl, it is shown that the surplus silica for the lighter rare earth series in the system, only $NaRSi_6O_{14}$ crystallizes in all cases, of course, the crystal

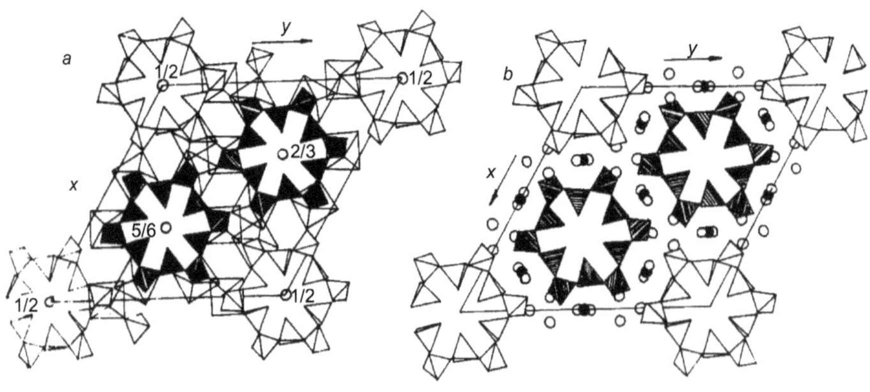

Figure 7.4 Crystal structure of $Na_5RSi_4O_{12}$ [43].

morphology varies from those obtained using carbonate solutions. Figure 7.6 shows the crystallization fields for the system $Na_2O-R_2O_3-SiO_2-H_2O$, $T = 450-650°C$, $P = 400-2500$ atm ((i) surplus R_2O_3; (ii) surplus SiO_2) [35]. In NaF solution, this compound crystallizes as fine crystalline mass. It can be concluded that under hydrothermal experimental conditions, the presence of F, Cl, CO_3, and SO_4 anions in the solvents greatly influences the [SiO_4] anions to form highly complex anions like [Si_6O_{15}] and [Si_6O_{14}]. It is interesting to note that rare earth elements, including the lighter rare earths (La–Nd), form chain silicates with discrete polyhedra. Such discrete rare earth polyhedra were first observed among silicates, and, subsequently, in germanates, phosphates, and so on. [44–46]

Haile et al. [37,47,48] studied the hydrothermal synthesis of new alkali silicates, potassium neodymium phases, and sodium phases, in great detail. The alkali silicates are interesting, both scientifically and technologically, as fast ionic conductors because of their typical open framework structures and the unique possibilities for crystal chemical tailoring. Their studies on the systems $K_2O-Nd_2O_3-SiO_2$ and $Na_2O-Nd_2O_3-SiO_2$ yielded: $K_3NdSi_3O_8(OH)_2$, $K_3NdSi_7O_{17}$, $K_5NdSi_9O_{22}$, $K_8NdSi_{12}O_{32}\cdot OH$, $K_3NdSi_6O_{15}$, $K_3NdSi_6O_{15}$ (P_{bam}), $K_2Nd\cdot Si_4O_{10}\cdot OH$, and $KNd_9(SiO_4)_6O_2$, out of which the first four have no known isomorphs. Similarly, of $NaNdSi_6O_{14}$, $Na_3NdSi_6O_{15}\cdot 2.5H_2O$, $NaNd_9(SiO_4)_6O_2$, $Na_3YSi_6O_{15}$, $Na_3YSi_7O_{17}$, $Na_7YSi_{12}O_{29}$, $Na_3YSi_3O_9$, $Na_5YSi_4O_{12}$, $Na_6YSi_{13}O_{29}(OH)_3$, and $NaYSi_6O_{14}$, six are new. The synthesis has been carried out in aqueous solutions of MOH, M_2CO_3, MF, MHF_2 (where M = Na, K), $K_2B_4O_7$ as solvent, and either a glass of high silica content, $4K_2O$ (Na_2O)$-Nd_2O_3-17SiO_2$, or a mixture of Nd_2O_3, vitreous silica and K_2CO_3(Na_2CO_3) as the precursor material. Pressures, temperatures, and solution molarities utilized ranged from 0.3 to 1.4 kbar, from 350°C to 600°C, and up to 10 M, respectively. The authors found that high-solution molarities, high temperatures, and to some extent, high pressures and long reaction time favored the

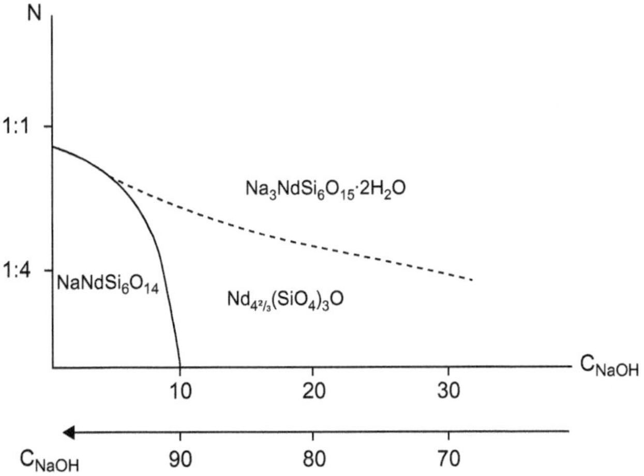

Figure 7.5 Crystallization pattern of Nd silicate [44].

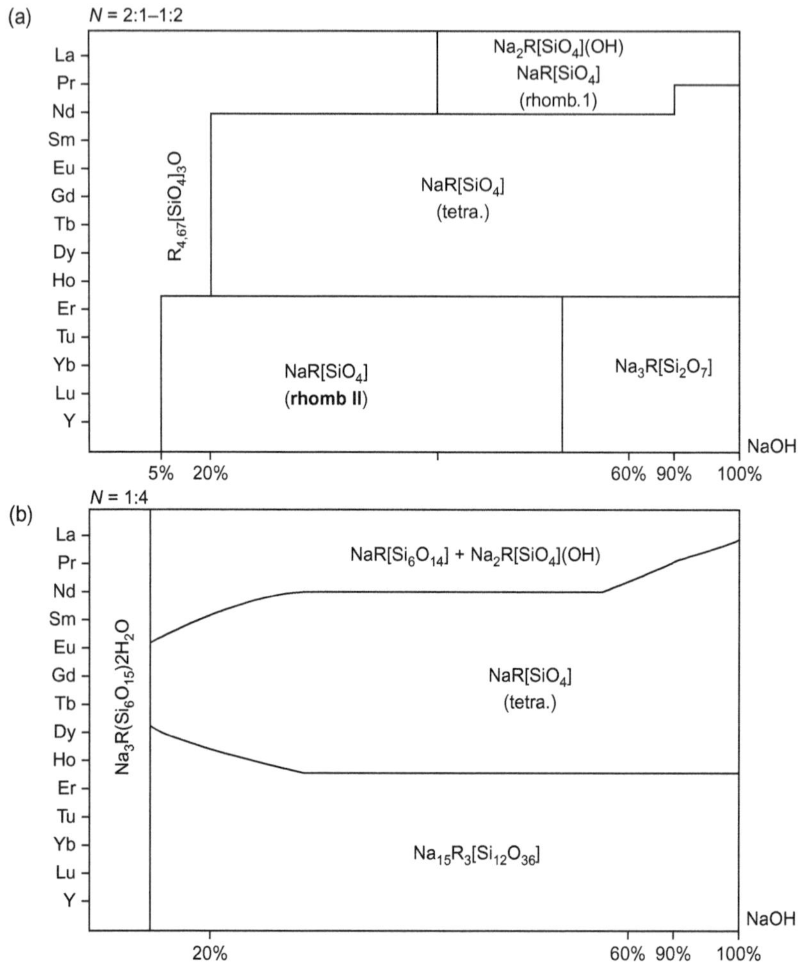

Figure 7.6 Crystallization fields for the system $Na_2O-R_2O_3-SiO_2-H_2O$, $T = 450-650°C$, $P = 400-2500$ atm (i—surplus R_2O_3; ii—surplus SiO_2) [35].

crystallization of phases with low "connectivities" between $[SiO_4]$ tetrahedra in the structure. A decrease in the ratio of interstitial oxygen atoms and hydroxyl groups broadly corresponds to a progressive change from a 3d structure, to one that is layered, to one based on chains or rings and eventually, to one that contains discrete radicals, typically $[SiO_4]^{-4}$ groups. The changes in the structure of the crystallization phase occur via changes in the structure of the silicate species in the solution. The latter, in turn, responds to changes in the experimental conditions in accordance with a generalized reaction:

$$Si_xO_y^{2(2x-y)} + 2zOH^- \leftrightarrow Si_xO_y^{2(2x-y-z)} + H_2O \quad (7.3)$$

The synthesis in the Na_2O—Nd_2O_3—SiO_2—H_2O system is much more sensitive to the basic alkali content than in the K_2O—Nd_2O_3—SiO_2—H_2O system, with the apatite-type phase forming at much lower hydroxide and carbonate concentrations. High SiO_2 solubility in the presence of Na_2O is also responsible for the much greater quantity of gelatinous material that resulted from these experiments.

It is interesting to note that the crystallization of various phases and the stoichiometries of the phases depend mainly on the type and concentration of the alkali metals. The apatite-type phases have been obtained at the highest KOH molarity. The two quantities, Si:O and mol% SiO_2, are necessarily related in a system where the nonsilicon cations are mono- and trivalent. The former has in addition a very direct relationship to the linkage of tetrahedra in the silicate framework. A low Si:O ratio implies that if any or some oxygen atoms are bonded to more than one silicon, the connectivity of the framework is low. The change observed in the nature of the crystallized phase upon an increase in the potassium content, corresponding to a lowering of the connectivity of the structure. The phase encountered at the lowest KOH concentration is an ideal Si_2O_5 layered silicate ($K_3NdSi_6O_{15}$, P_{bam} form) with all SiO_4 tetrahedra sharing apices with three others. The phase encountered at the highest KOH molarity is an orthosilicate ($KNd_9(SiO_4)_6O_2$) structurally composed of isolated SiO_4 tetrahedra. Similarly, $K_3NdSi_7O_{17}$, produced at lower KHF_2 concentrations, has an interconnected, 3d silica framework, whereas $K_2NdSi_4O_{10}OH$, obtained at higher concentrations, is based on infinite (Si_2O_5) tubular units.

The concept of classifying compounds as "chain breakers" (e.g., K_2O) and "chain formers" (e.g., P_2O_5) has been well developed in silicate glass technology, yet rarely used in the context of silicate crystal growth. As the hydrothermal growth involves a solution in addition to the growing silicate phase, it is by no means obvious that such an approach should be applicable in this context. Ikornikova [49] has studied aqueous solutions of SiO_2 and Na_2O, and has observed that at low alkali to silica ratios in the solute, but high overall solute concentrations, silica is dissolved as long chain polymers or colloidal particles (5—20 Å in radius) that have an overall composition ranging between $Si_2O_5^{2-}$ and SiO_3^{2-}. As the alkali to silica ratio in the liquid is increased, such agglomerates break apart, coordination decreases, and the predominant species become more negatively charged pyrosilicate ($Si_2O_7^{6-}$) and orthosilicate (SiO_4^{4-}) groups. In this particular system, we expect reactions such as

$$2Si_6O_{15}^{-6}(aq) + 4OH^- \leftrightarrow Si_{12}O_{32}^{-16}(aq) + 2H_2O \tag{7.4}$$

$$Si_3O_8^{-4}(aq) + 8OH^- \leftrightarrow 3SiO_4^{-4}(aq) + 4H_2O \tag{7.5}$$

to take place. The dissolved species on the left-hand side of Eq. (7.4) has Si:O ratio 0.4 and comprises the basic unit of $K_3NdSi_6O_{15}$, while that on the right-hand side has a ratio of 0.375 and it corresponds to the silicate layer in $K_8Nd_3Si_{12}O_{32}OH$.

Similarly, Eq. (7.5) represents the equilibrium between the aqueous counterparts of $K_8Nd_3Si_{12}O_{23}OH$ and $KNd_9(SiO_4)_6O_2$ (Si:O = 0.25 on the right).

As the OH^- concentration is raised, the right-hand sides of Eqs (7.4) and (7.5) are increasingly favored, Si—O—Si bonding decreases, the fraction of oxygen atoms bonded to only one silicon increases, and the average size of the dissolved silica complexes decreases. During hydrothermal growth, it is expected that these complexes act as precursors to the crystalline precipitate such that, as the aqueous concentration of a particular species is increased, the likelihood of a structure into which it is easily incorporated will precipitate correspondingly increases. It is in this manner "an increase in the alkali content in the liquid phase causes a decrease in the connectivity of the silica framework in the resultant crystalline phase."

The anionic species present in the solvent plays a decisive role in determining the solution molarity required for the synthesis of the various compounds. Much lower concentrations are required to obtain $K_8Nd_3Si_{12}O_{32}OH$ from $K_2B_4O_7$. For example, concentration less than 1 M $K_2B_4O_7$ provides twice as many potassium ions as 1 M KOH. The absence of $KNd_9(SiO_4)_6O_2$ from K_2CO_3 experiments, even at very high concentrations, suggests that, in accordance with reactions shown above, pH is also an important factor. The higher pH values cause not only greater SiO_2 dissolution but also the breakup of dissolved silica complexes into smaller species. The reasons that the fluorine compounds gave rise to entirely different phases remain unclear. The low concentrations at which these fairly pH neutral compounds caused the formation of $K_3NdSi_7O_{17}$ rather than $K_3NdSi_6O_{15}$ (derived when pure water served as the solvent) indicates that some variable other than pH, possibly the high reactivity of F^- with Si^{4+}, is playing a significant role.

Higher temperatures slightly favored the formation of low Si:O phases. This was true in spite of the fact that at high temperatures (at constant molarity), the % fill of liquid was lowered to compensate for the decrease in the density of water, leading to a systematic reduction in the total alkali content. If such a change in experimental conditions was to occur alone, that should result in the synthesis of the high Si:O phases. If we accept that changes in the structure of silica in solution are responsible for changes in the structure of the precipitate, the result implies that high temperatures are particularly effective in shifting the above two reactions to the right, thereby causing dissolved silica complexes to break apart into smaller species.

Higher pressures also favor the low Si:O phases. In this case, however, experiments performed at 1400 bar pressure were given higher percent than those performed at 825 bar (constant molarity), to reflect the increase in solvent density at the higher pressure. It is, therefore, not clear whether the change in crystallized phase resulted from an increase in the pressure or increase in the total alkali content.

Lastly, although only a limited number of experiments have been performed with variation in the crystallization time, a tendency toward the formation of the low Si:O ratio apatite-type phase with increasing time has been observed in the 5 M KOH experiments. The reason may be that the glass, with a relatively high solubility, quickly supersaturates in the solution with respect to the crystalline phases. As the initial silica content in the solution is high, dissolved species are large, and the higher silica content phase—$K_8Nd_3Si_{12}O_{32}OH$—precipitates preferentially over $KNd_9(SiO_4)_6O_2$ in the solution. If more time is given, $K_8Nd_3Si_{12}O_{32}OH$ slowly redissolves, but the total silica content in the solution remains low. Therefore, dissolved species are reduced in size, and the apatite-type phase eventually precipitates.

The conditions (average temperature, estimated pressure, solvent type and molarity, solid-to-liquid ratio, total alkali content, and time) that gave rise to $K_3NdSi_3O_8(OH)_2$ growth using the temperature gradient technique broadly correspond to those that caused $K_8Nd_3Si_{12}O_{32}OH$ and $KNd_9(SiO_4)_6O_2$ co-crystallization isothermally. As $K_3NdSi_3O_8(OH)_2$ has a silicon to (bonding) oxygen ratio of 0.3, a structure based on trisilicate chains [50] and a composition that closely resembles the combined stoichiometry of the nutrient and the mineralizer is formed.

The reasons for the appearance of the different phases under different conditions have been explained in terms of their differing structures, rather than differing stoichiometries, and in terms of silicate solution chemistry. It is generally accepted that during hydrothermal growth, the nature of the dissolved species determines the nature of the crystalline product. We have observed a decrease in the connectivity of the silicate framework in the product phase, as measured by the Si:O ratio (or, alternatively, SiO_2 mol%), with increasing solution molarity (alkali content), temperature and, to some extent, pressure. Therefore, it is concluded that higher solubilities under these conditions coincide with a decrease in the average size of the dissolved complexes and is responsible for the change in the structural nature of the crystalline product. Longer times also appear to favor the crystallization of low Si:O ratio phases. These observations provide guidelines by which new phases with some desired structural feature might be synthesized. Stoichiometric parameter other than the Si:O ratio and the SiO_2 mol% are not easily correlated to the experimental variables. The mole percent alkali in the product phase, for example, does not monotonically increase as the molarity of the solution is increased. Lastly, it is observed that when phases of low silica content form, overall mass balance is maintained via the presence of a high silica content gel. It is most probable that this gel precipitates out of solution as the autoclave is quenched.

Such studies greatly help in understanding the hydrothermal chemistry of silicates and their crystallization mechanism. Hitherto, only zeolites have such a vast amount of data on their hydrothermal chemistry. Another approach to the hydrothermal chemistry of silicates has been treated in the forthcoming sections under degree of silification.

7.5 Crystal Chemical Significance of Phase Formation

The study of the crystallization of various rare earth silicates and their correlation with the crystal chemistry has not been carried out effectively. In laboratory conditions, it is interesting to understand the conditions of synthesis and the structure with the change in the ionic radii of rare earths. Such studies have revealed the mechanism of the formation of some of the fine crystal structures.

Silicates of rare earth elements owing to their ionic radii belong to a large group of silicates with larger cations. During the 1950s, the academican N.V. Belov framed the structural principles for these silicates. Further, the structure formation of silicon–oxygen radicals shall be treated essentially, as a result of cationic motif. Their differentiation as chains, ribbons, cages, and rings are relatively more popular, but a significant variation depending upon their geometric relationship of cationic polyhedra and Si–O radicals can also be observed [51].

Natural rare earth silicates are considered as very complex ones containing a wide range of cations like: Ca, Fe, Be, Al, Ti, Sr, Mn, Ba, and Na. Their crystal chemical elucidation is highly related to the naturally occurring rare earth elements. Thus, we can have two broad groups: alkali rare earth silicates and rare earth silicates without alkalis covering both natural and synthetic rare earth silicates. Figure 7.7 shows the tetrahedral arrangement in silicates with $[Si_6O_{14}]$ radical. The typical compounds are $Na_4[Si_6O_{14}]$ and $NaNd[Si_6O_{14}]$ [49].

Today, over 15 structure types of mixed rare earth silicates without alkalis and 21 structure types of rare earth silicates with alkali cations have been reported. Their crystal chemical characteristics are presented in Tables 7.6–7.9 [35,52].

In these compounds, the following anionic groups have been identified: ortho-, diortho-, mixed ortho-, triortho-, and tetraorthogroups, discrete 4-membered rings, discrete 12-membered rings, spiral chains, layers from 6-, 8-, and 12-membered rings, triple column and Si–O framework. Various types of anionic radicals, found in rare earth silicates are, given in Table 7.10 [52].

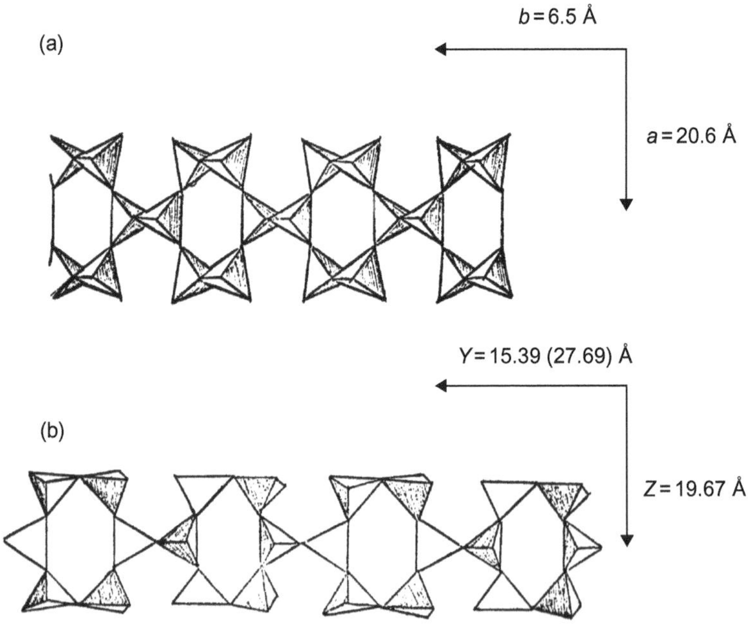

Figure 7.7 Tetrahedral arrangement in silicates with $[Si_6O_{14}]$ radical [7].

Orthotetrahedra (orthosilicates) have a free oxygen not shared with neighboring [Si—O] tetrahedra but are bonded only with the rare earth cations. $R_2[SiO_4]O$ has two structure types: $Eu^{+2}{}_2[SiO_4]$ and $Gd_2[SiO_4]O$; $NaR[SiO_4]$ has three structure types, and similarly in the apatite-type $R_{4.66}[SiO_4]_3O$ and $KNd_9[SiO_4]_6O$.

Diorthogroup (diorthosilicates) contains compounds of $R_2[Si_2O_7]$ (five structure types)—$Y_2[Si_2O_7]$, $Sc_2[Si_2O_7]$, $Na_3R[Si_2O_7]$, $K_3R[Si_2O_7]$, and $Na_6YSc[Si_2O_7]$.

Triorthosilicates are always characteristic of $Y_3[Si_3O_{10}]$ and $K_3Y[Si_3O_8](OH)_2$ (Figure 7.8) [53,54]. Likely triorthogroups in the form is sufficiently rare Si—O radical and twice observed in Gd-silicates: $Na_4Gd_2[Si_3O_{10}]$ [55] and $Na_2Gd_3[Si_3O_{10}]$ [56].

Tetraorthogroups $[Si_4O_{13}]$ as seen in $Na_4Sc_2[Si_4O_{13}]$ is a linear radical which extends to a series of island radicals $SiO_4—Si_2O_7—Si_3O_{10}—Si_4O_{13}$.

Table 7.6 Crystal Chemical Characteristics of Rare Earth Silicates [35]

Crystal Chemical Group	R-element	Symmetry
$Na^{X}R^{IX}[SiO_4]$	La—Nd	Orthorhombic (i)
$Na^{VII}R^{VIII}[SiO_4]$	Pr—Ho	Tetragonal
$Na^{VI}R^{VI}[SiO_4]$	Ho—Lu	Orthorhombic (ii)
$Na^{a}{}_{13}R_3{}^{VI}[Si_{12}O_{36}]$	Ho—Lu,	Y Monoclinic
$Na^{a}R^{VI}[Si_2O_7]$	Er—Lu,	Y orthorhombic
$Na^{a}R^{VI}[Si_6O_{14}]$	La—Eu	Orthorhombic
$Na_2{}^{a}R^{VIII}[SiO_4](OH)$	La, Pr	Orthorhombic
$Na^{a}R^{VII}[Si_6O_{13}](OH)_2$	La—Lu	Orthorhombic

[a]Coordination number of Na in the work is not given.

Table 7.7 Crystal Chemical Characteristics of Silicates [35,52]

Compound	O/Si	SiO_2/ R_2O_3	Si/Na +R	Structural Si—O	Motif R—O	Co. No. R[a]	Co. No. of Na[a]
$R_{4.67}[SiO_4]_3O$	4	0.78	0.64	$[SiO_4]$	3d bond	9.7	–
$NaR[SiO_4]$	4	2	0.5	$[SiO_4]$	$[R_2O_8]_{\infty\infty}$	9	10
$NaR[SiO_4]$	4	2	0.5	$[SiO_4]$	4-Membered ring	8	7
$NaR[SiO_4]$	4	2	0.5	$[SiO_4]$	$[R_2O_8]_{\infty\infty}$	6	6
$Na_2R[SiO_4](OH)$	4	2	0.33	$[SiO_4]$	3d-bond	8	5; 7
$Na_3R[Si_2O_7]$	3.5	4	0.5	$[Si_2O_7]$	$[RO_6]_0$	6	4; 5
$Na_{15}R_3[Si_{12}O_{36}]$	3	8	0.66	$[Si_{12}O_{36}]$	$[RO_6]_0$	6	4; 6
$Na_3R[Si_6O_{15}]$ $2H_2O$	2.5	12	1.5	$[Si_6O_{15}]$		7	–
$NaR[Si_6O_{14}]$	2.33	12	3	$[Si_6O_{14}]$	$[R_3O_{12}]_{\infty}$	8	4; 8

[a]Co. No. = Coordination number.

Table 7.8 Structure Types of Rare Earth Silicates [35,52]

Compound	Representative	Cell Parameter (Å)	Rare Earth, Co. No., Average Bond Distance (Å)			Tetrahedral Radical Form	Rare Earth Polyhedra
$R_2[SiO_4]O$	$Gd_2[SiO_4]O$	$P2_1/c$ $z = 4$	Gd_1	7	2.39	Orthotetrahedra	Framework
R = La–Tb		$a = 9.12$	Gd_2	9	2.49		
		$b = 7.06$					
		$c = 6.73$					
		$\beta = 107.58°$					
$R_2[SiO_4]O$	$Yb_2[SiO_4]O$	$B2/b$ $z = 8$	Yb_1	7	2.33	Orthotetrahedra	Framework
R = Tb–Lu, Y		$a = 14.28$					
		$b = 10.28$	Yb_2	6	2.33		
		$c = 6.653$					
		$\gamma = 122.2°$					
$R_{4.67}[SiO_4]_3O$	$Gd_{4.67}[SiO_4]_3O$	$P6_3/m$ $z = 2$	Gd_1	7	2.41	Orthotetrahedra	Framework
R = La–Lu		$a = 9.431$	Gd_2	9	2.53		
		$b = 9.873$					
	α-$Eu^{2+}{}_2[SiO_4]$	$P2_1/c$ $z = 4$	Eu_1	8	2.62	Orthotetrahedra	Framework
		$a = 5.661$					
		$b = 7.101$	Eu_2	10	2.86		
		$c = 11.518$					
		$\beta = 122.01$					
	β-$Eu^{2+}{}_2[SiO_4]$	$Pnma$ $z = 4$	Eu_1	10	2.86	Orthotetrahedra	Framework
		$a = 7.137$					
		$b = 5.665$	Eu_2	7	2.57		
		$c = 9.767$					
$R_2Si_2O_7$	$Sm_2Si_2O_7$	$P4_1$ $z = 8$	Sm_1	7	2.42	Diortho-	Framework
R = La–Eu		$a = 6.695$	Sm_2	9	2.56		

(Continued)

Table 7.8 (Continued)

Compound	Representative	Cell Parameter (Å)	Rare Earth, Co. No., Average Bond Distance (Å)			Tetrahedral Radical Form	Rare Earth Polyhedra
		$b = 24.40$					
$R_2Si_2O_7$ R = Ho–Lu, Sc	$Yb_2Si_2O_7$ $a = 6.802$ $b = 8.875$ $c = 4.703$ $\beta = 102.12°$	$C2/m$ $z = 2$	Yb	6	2.24	Diortho-	Ribbon
$R_2Si_2O_7$ R = Ho–Er	$Er_2Si_2O_7$	$P2_1/b$ $z = 2$ $a = 4.683$ $b = 5.556$ $c = 10.79$ $\gamma = 96°$	Er	6	2.26	Diortho-	Layer
$R_2Si_2O_7$ R = Eu–Ho	$Gd_2Si_2O_7$	Pna21 $z = 4$ $a = 13.87$ $b = 5.073$ $c = 8.33$	Gd	7	2.40	Diortho-	Framework
$R_2Si_2O_7$ R = Sm–Eu	$Eu_2Si_2O_7$	$P_1 - P_I$ $z = 4$ $a = 8.517$ $b = 12.848$ $c = 5.385$ $\alpha = 91.65°$ $\beta = 92.24°$ $\gamma = 90.44°$				Diortho-	
$R_2Si_2O_7$ R = La–Sm	$Nd_2Si_2O_7$	$P2_I2_I2_I$ $z = 4$ $a = 5.394$	Nd	8	2.53	Diortho-	Framework

(Continued)

Table 7.8 (Continued)

Compound	Representative	Cell Parameter (Å)	Rare Earth, Co. No., Average Bond Distance (Å)			Tetrahedral Radical Form	Rare Earth Polyhedra
		$b = 12.95$ $c = 8.72$					
$R_2[Si_2O_7]$ R = Tm–Lu	$Yb_2[Si_2O_7]$	$P4_12_12$ $z = 4$ $a = 6.574$ $c = 11.98$				Diortho-	Framework
	$(Y,Th)_2[Si_2O_7]$	$P2_1/m$ $z = 2$ $a = 7.50$ $b = 8.06$ $c = 5.02$ $\beta = 112°$	Y	6	2.36	Diortho-	Layer
	$Y_3[Si_3O_{10}](OH)$	$P2_1/n$ $z = 6$ $a = 10.343$ $b = 7.284$ $c = 11.093$ $\gamma = 96.92°$	$Y_{1.2}$ Y_3	7 8	2.32 2.43	Triortho-	Layer
$R_4[Si_3O_{10}]$ $[SiO_4]$ $[SiO_4]$ R = Eu–Er	$Ho_4[Si_3O_{10}]$	P_1 $z = 4$ $a = 6.664$ $b = 6.674$ $c = 12.110$ $\alpha = 94.07°$ $\beta = 89.97°$ $\gamma = 91.66°$	Ho	8	2.44	Triortho- + Orthotetrahedra	Ribbon

Table 7.9 Structure Types of Alkali Rare Earth Silicates [35,52]

Compound	Representative	Cell Parameter (Å)	Rare Earth, Co. No., Average Bond Distance (Å)			Tetrahedral Radical Form	Rare Earth Polyhedra
$NaR[SiO_4]$	$NaY[SiO_4]$	$Pbn2_1$ $z = 4$	Y	6	2.26	Orthotetrahedra	Discrete Polyhedra
R = Er–Lu, Y		$a = 5.09$					
		$b = 10.96$					
		$c = 6.35$					
$NaR[SiO_4]$	$NaSm[SiO_4]$	14/m $z = 8$	Sm	8	2.51	Orthotetrahedra	Framework
R = Nd–Ho		$a = 11.89$					
		$c = 5.45$					
$NaR[SiO_4]$	$NaNd[SiO_4]$	$Pna2_1$ $z = 12$	Nd	8	2.45	Orthotetrahedra	Column
R = Ce–Nd		$a = 20.00$					
		$b = 9.28$					
		$c = 5.45$					
	$KNd_9[SiO_4]_6O_2$	$P6_3$ $z = 1$	Nd	7	2.48	Orthotetrahedra	Framework
		$a = 9.576$	(K,Nd)	9	2.48		
		$c = 7.009$	(K,Nd)	9	2.75		
	$LiY[SiO_4]$	$P2_1/b$ $z = 4$	Y	7	2.01	Orthotetrahedra	Layers from 6-membered ring, seven-sided framework
		$a = 5.368$					
		$b = 10.806$					
		$c = 6.195$					
		$\gamma = 113.40°$					
$Na_2R[SiO_4]OH$		$Pnma$ $z = 4$	La	8	2.59	Orthotetrahedra	
	$Na_2La[SiO_4]OH$	$a = 9.72$					
R = La–Pr		$b = 7.57$					

(*Continued*)

Table 7.9 (Continued)

Compound	Representative	Cell Parameter (Å)	Rare Earth, Co. No., Average Bond Distance (Å)			Tetrahedral Radical Form	Rare Earth Polyhedra
		$c = 6.81$					
I	$4\{NaYSiO_4\}NaF$	$I_4 z = 8$ $a = 11.579$ $c = 5.396$				Orthotetrahedra	Framework
$Na_3R[Si_2O_7]$ R = Er–Lu, Y, Sc	$Na_3Sc[Si_2O_7]$	*Pbnm* $z = 4$ $a = 5.354$ $b = 9.347$ $c = 13.089$	Sc	6	2.10	Diorthohedra	discrete poly
	$Na_6YSc[Si_2O_7]_2$	$P6_3$ $z = 4$ $a = 10.726$ $b = 13.491$	Y	6 6	2.62 2.71	Diorthohedra	Discrete poly
	$K_3Eu[Si_2O_7]$	*P*6c2 $z = 6$ $a = 9.956$ $c = 14.455$	Eu	6	2.31	Diortho-	Discrete polyhedra
	$K_3Y[Si_3O_8(OH)_2]$	*Pmnb* $z = 4$ $a = 13.536$ $b = 13.17$ $c = 5.867$	Y	6	2.23	Triortho-	Discrete polyhedra
	$Na_4Sc_2[Si_4O_{13}]$	*P*na2_1 $z = 4$ $a = 14.442$ $b = 5.288$ c = 14.219	Sc	6	2.11	Tetra-	Polyhedra
	$K_4Sc_2[Si_4O_{12}]$ $(OH)_2$	*P*bam $z = 8$ $a = 12.725$	Sc	6	2.09 −2.16	4-Membered ring	Chain

(*Continued*)

Table 7.9 (Continued)

Compound	Representative	Cell Parameter (Å)	Rare Earth, Co. No., Average Bond Distance (Å)			Tetrahedral Radical Form	Rare Earth Polyhedra
		$b = 12.741$					
		$c = 8.441$					
$Na_5RSi_4O_{12}$ R = Sm–Lu, Y, Sc	$Na_5LuSi_4O_{12}$	R32 $z = 16$ $a = 21.926[4]$ $c = 12.556[2]$	Lu	6	2.20	12-Membered ring	Discrete polyhedra
	$LiScSi_2O_6$	$C2/c$ $z = 4$	Sc	6	2.107	Chains with period of 2 tetrahedral	Layer
		$a = 9.8033$ $b = 8.9581$ $c = 5.3512$ $\beta = 110.281°$					
	$Na_{12}Y_4Si_{12}O_{36}$	$P2_12_12_1$ $z = 16$	Y	6	2.20	Chains with period of 24 tetrahedral	Discrete polyhedra
		$a = 15.033$ $b = 15.142$ $c = 15.213$			−2.36		
$NaRSi_6O_{14}$ R = La–Lu	$NaPrSi_6O_{14}$	Ibam $z = 8$ $a = 7.682$ $b = 15.445$ $c = 19.724$	Pr	6	2.41	Triple layered column	Column
	$NaNdSi_6O_{13}$	Cmm2 $z = 4$	Nd	7	2.44	Layers with 8-, 6-, 5-, and 4-membered rings	Discrete polyhedra
	$(OH)_2 \cdot nH_2O$	$a = 30.87$					

(Continued)

Table 7.9 (Continued)

Compound	Representative	Cell Parameter (Å)	Rare Earth, Co. No., Average Bond Distance (Å)	Tetrahedral Radical Form	Rare Earth Polyhedra
		$b = 7.387$ $c = 7.120$			
$K_3RSi_6O_{15}$	$K_3NdSi_6O_{15}$	Pbam $z = 4$	Nd 6 2.38	Layers with 8-, 6-, 5-, and 4-membered rings	Discrete polyhedra
R = La–Lu		$a = 16.011$ $b = 14.984$ $c = 7.276$			
$K_8R_3[Si_6O_{16}]_2$		$P1$ $z = 1$	Yb 6 2.26	Layers with 12-, 8-, and 6-membered rings	Discrete and bicapped polyhedra
(OH) R = La–Lu	$K_8Yb_3[Si_6O_{16}]_2$ (OH)	$a = 6.808$ $b = 11.434$ $c = 11.449$ $\alpha = 88.52$ $\beta = 90.91$ $\gamma = 100.10$			
	$K_2Ce[Si_6O_{15}]$	12/b $z = 4$ $a = 13.058$ $b = 11.854$ $c = 8.698$ $\gamma = 90.15$	Ce 8 2.23	Framework	Discrete

Co. No. = coordination number

Table 7.10 Anionic Structural Motif in Rare Earth and Alkali Rare Earth Silicates [52]

Radical Form	Representative Structure Type	Rare Earth Polyhedra	Co. No. of Rare Earth Cations
Orthotetrahedra $[SiO_4]$	$Gd_2[SiO_4]O$	Framework	7, 9
	$Yb_2[SiO_4]O$	Framework	6, 7
	$Ge_{4.66}[SiO_4]_3O$	Framework	7, 9
	$Eu_2[SiO_4]$	Framework	8, 10
	$NaNd[SiO_4]$	Column	9
	$NaSm[SiO_4]$	4-Membered ring	6
	$NaY[SiO_4]$	Discrete polyhedra	6
	$4\{NaYSiO_4\}NaF$	4-Membered ring	6
	$Na_2La[SiO_4](OH)$	Framework	8
	$KNd_9[SiO_4]_6O_2$	Framework	7, 9
Diortho-$[Si_2O_7]$	$La_2[Si_2O_7]$	Framework	7, 8, 9
	$Yb_2[Si_2O_7]$	6-Membered ring	6
	$Er_2[Si_2O_7]$	Layer	6
	$Gd_2[Si_2O_7]$	Framework	7
	$Eu_2[Si_2O_7]$	Framework	8
	$Nd_2[Si_2O_7]$	Framework	8
	$Yb_2[Si_2O_7]$	Framework	7
	$(Y,Th)_2\ [Si_2O_7]$	Layer	6
	$Na_3Sc\ [Si_2O_7]$	Discrete polyhedra	6
	$Na_6YSc[Si_2O_7]_2$	Discrete polyhedra	6
	$Na_3Y\ [Si_2O_7]$	Discrete polyhedra	6
	$K_3Eu\ [Si_2O_7]$	Discrete polyhedra	6
Triortho-$[Si_3O_{10}]$	$Y_3[Si_3O_{10}](OH)$	Layer	7, 8
	$K_3Y[Si_3O_8](OH)_2$	Discrete polyhedra	6
Tetraortho-$[Si_4O_{13}]$	$Na_4Sc_2\ [Si_4O_{13}]$	Bicapped polyhedra	6
Mixed ortho- and triortho-$[SiO_4]$ and $[Si_3O_{10}]$	$Ho_4[Si_3O_{10}][SiO_4]$	Ribbon	8
Ring: $[Si_4O_{12}]$			
Four	$K_4Sc_2[Si_4O_{12}](OH)_2$	Column	6
Twelve $[Si_{12}O_{36}]$	$Na_5Lu[Si_4O_{12}]$	Discrete polyhedra	
Spiral chain $[Si_{24}O_{72}]$	$Na_3Y\ [Si_3O_9]$	Discrete polyhedra	6
Layers: from 6,8, 12-rings $[Si_3O_8]$	$K_8Yb_3[Si_6O_{16}]_2(OH)$	Discrete and twinned polyhedra	6
From 4,5,6,8-rings	$NaNd[Si_6O_{13}]$ $(OH)_2;\ nH_2O$	Discrete polyhedra	7
From 4,6,8-rings $[Si_2O_5]$	$Na_2Ce[Si_6O_{14}]$ $(OH)\cdot nH_2O$	Discrete polyhedra	7
	$K_3Nd[Si_6O_{15}]$	Discrete polyhedra	6
Three-storeyed column $[Si_3O_7]$	$NaNd[Si_6O_{14}]$	Column	6
Framework $[Si_2O_5]$	$K_2Ce[Si_6O_{15}]$	Discrete polyhedra	6

Co. No. = coordination number.

Mixed ortho- and triorthogroups $[SiO_4]$ and $[Si_3O_{10}]$ exist simultaneously in $HO_4[Si_3O_{10}]$ $[SiO_4]$. The triorthogroups have similar consideration of diverse tetrahedral anions as in earlier observed $Ca_6[SiO_4]\cdot[Si_3O_{10}]$ [57].

Discrete 4-membered ring $[Si_4O_{12}]$ with 2/m symmetry exists in $K_4Sc(OH)_2\cdot[Si_4O_{12}]$ structure. Likely 4-membered rings with different symmetry were found in several compounds like $Ba_2FeR_2Ti_2O_2\cdot[Si_4O_{12}]_2OH\cdot H_2O$ [54], $Na_3CaPSiO_7$ [58], $Ca_2R_2[Si_4O_{12}]$ [59], $(Ba,K,Na)_8$ $(Ti,Nb)_9$ $(O,OH)_{10}$ $[Si_4O_{12}]_4\cdot nH_2O$ [60], $Ba_2(Fe,Ti,Mg)_2$ $(H_2O)[Si_4O_{12}]$ [61], $KHSiO_3$ [62], $Ba_{12}Mn_6$ $(OH,O)_2$ $[Si_4O_{12}]Cl_9$ $(OH,H_2O)_7$ [63], $(Na,K)_{2-x}$ (Nb,Ti) (O,OH) $[Si_2O_6]\cdot 2H_2O$ [64], Ba_4Ti_4 $(Ti,Nb,Fe)_4$ ClO_6 $[Si_4O_{12}]$ [65].

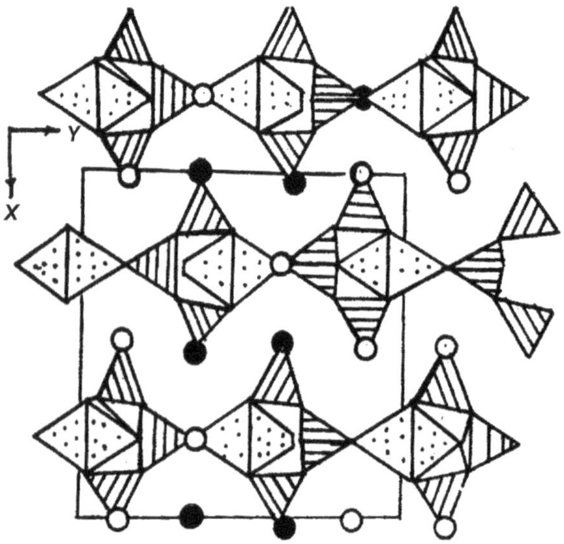

Figure 7.8 Triorthosilicates, $Y_3[Si_3O_{10}]$, and $K_3Y[Si_3O_8](OH)_2$ [53].

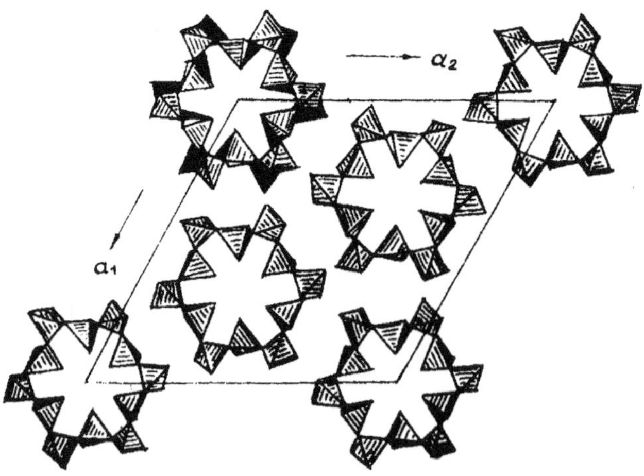

Figure 7.9 Structure of $Na_5Ce[Si_4O_{12}]$ [66].

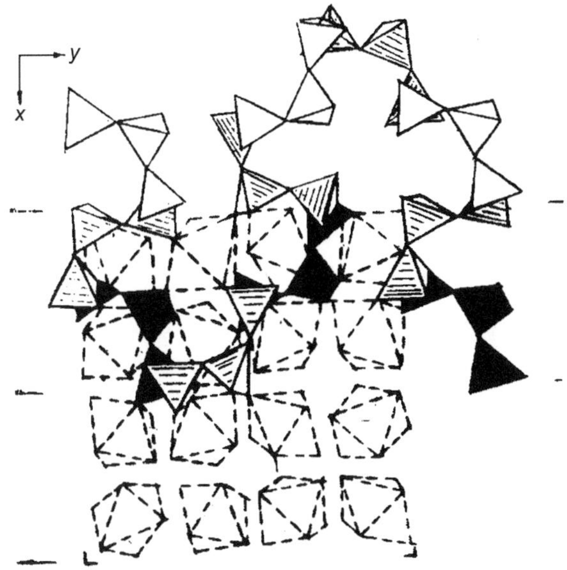

Figure 7.10 Metachain silicate [68].

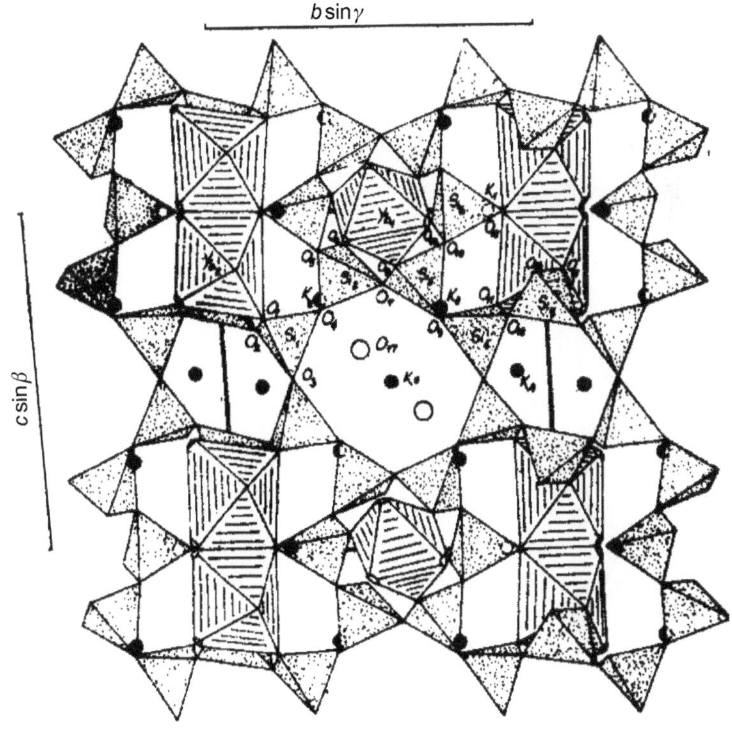

Figure 7.11 Layers from 6-, 8-, and 12-rings of $[Si_3O_8]$ in the structure of $K_8Yb_3[Si_6O_{15}]_2(OH)$ [69].

Discrete 12-membered ring [$Si_{12}O_{36}$]. This was observed in the structure of $Na_5Ce[Si_4O_{12}]$ (Figure 7.9) [66]. The other ring type [$Si_{12}O_{36}$] is found in $Ba_5FeTi(OH)_4$ [Si_6O_{18}] [67]. The rings in either structure differ greatly from one another, e.g., the latter case has higher symmetry and the earlier case has lower symmetry.

Spiral chains [$Si_{24}O_{72}$] are found in the structure of $Na_3Y[Si_3O_9]$. This new type of infinite spiral formation metachain silicate [$Si_{12}O_{36}$] agrees with the formula 12 [SiO_4]-tetrahedra. Such a matachain silicate is shown in (Figure 7.10) [68].

Layers from 6-, 8- and 12-rings [Si_3O_8] have been reported in the structure of $K_8Yb_3[Si_6O_{15}]_2(OH)$ (Figure 7.11) [69].

Layers from 4-, 5-, 6-, and 8-membered rings [Si_2O_5] was found in the structure of $NaNd[Si_6O_{13}](OH)_2 \cdot nH_2O$ (Figure 7.12) [70]. This type of layer forms as a result of the condensation of the ribbon from the reflection of the 8-membered rings. Layered radical from [SiO_4]-tetrahedra with such an arrangement of the rings is unique among silicate structures.

Layers from 4-, 6-, and 8-membered rings [Si_2O_5] was observed in the structure of $K_3Nd[Si_6O_{15}]$ (Figure 7.13) [71]. Such an arrangement has been noticed in several other rare earth and nonrare earth silicates. Usually the layers are extended and structures are highly distorted.

Triple columnar [Si_3O_7] was demonstrated in the structure of $NaR[Si_6O_{14}]$ (Figure 7.14) [72]. The structure is made up of 3d network from 12-membered complexes.

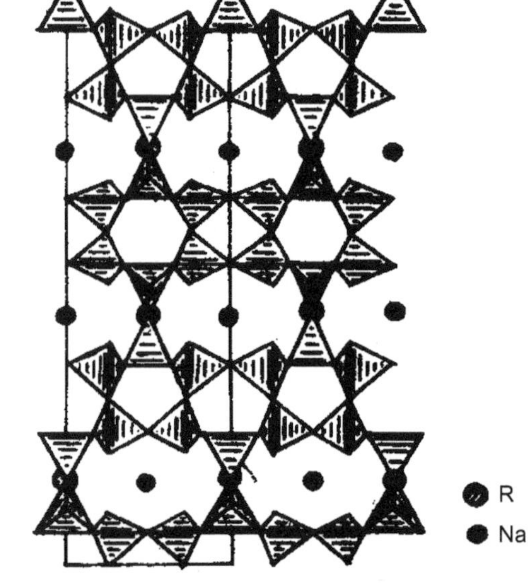

Figure 7.12 Layers from 4-, 5-, 6-, and 8-membered rings, [Si_2O_5], in the structure of $NaNd[Si_6O_{13}](OH)_2 \cdot nH_2O$ [70].

Framework [Si_2O_5] was found in the structure of $K_2Ce[Si_6O_{15}]$ (Figure 7.15) [73].

Thus, a wide variety of Si—O anionic motif forming unique layers, ribbons, chains, rings, and framework is characteristic of only rare earth silicates. Such complex and highly fine structures have been observed in synthetic rare earth silicates.

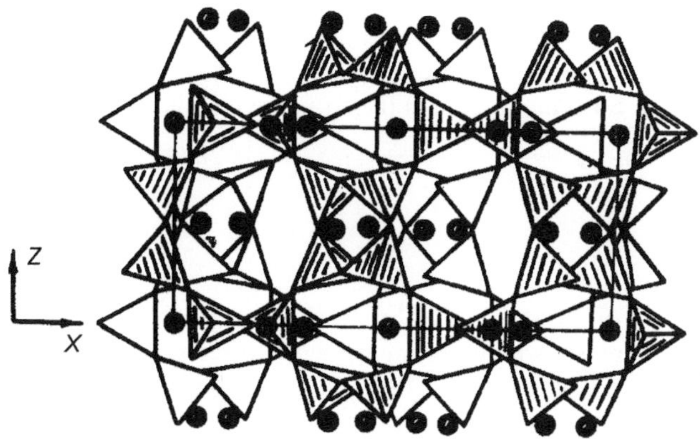

Figure 7.13 Layers from 4-, 6-, and 8-membered rings, [Si_2O_5], in the structure of $K_3Nd[Si_6O_{15}]$ [71].

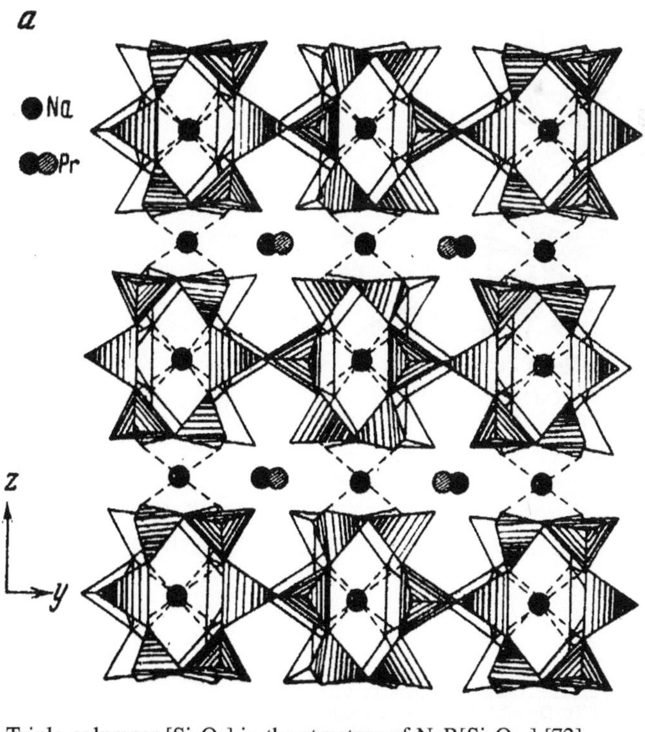

Figure 7.14 Triple columnar [Si_3O_7] in the structure of $NaR[Si_6O_{14}]$ [72].

7.5.1 Phase Formation in Surplus R_2O_3

During the synthesis of rare earth silicates with surplus rare earth elements, the structures of the phases forming and their stability mainly depend upon the characteristics of the rare earth motif, which is closely related to the ionic radii of the rare earth elements. The ionic radii of rare earth elements vary from 1.061 Å (La) to 0.848 Å (Lu). The coordination number also varies from 6 to 8. Similarly, the variation in the ionic radii of the rare earth elements in the structure is compensated by the distortion of the 3d rare earth motif, so that the structure type is retained for the entire rare earth series.

7.5.2 Silicates

In Figure 7.6, the fields of crystallization of various phases in the system $Na_2O-R_2O_3-SiO_2-H_2O$ are shown. In solutions with lower concentrations of NaOH, the representative phase is $R_42/3$ $[SiO_4]_3O$. Moreover, their stability (Figure 7.16) extends toward the lower rare earth elements. The possible reason for this is the search of their structural elucidation.

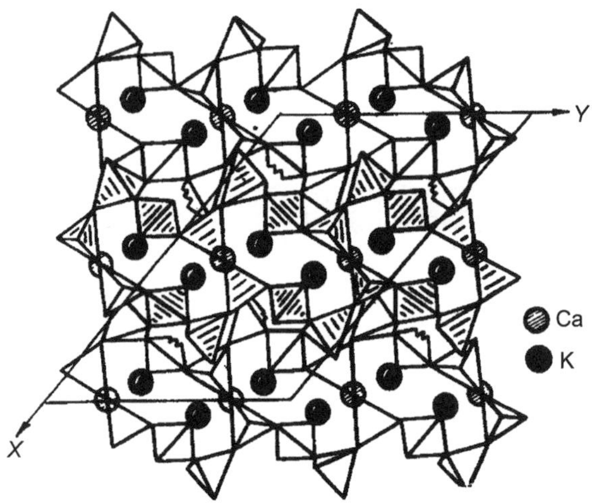

Figure 7.15 Framework $[Si_2O_5]$ in the structure of K_2Ce $[Si_6O_{15}]$ [73].

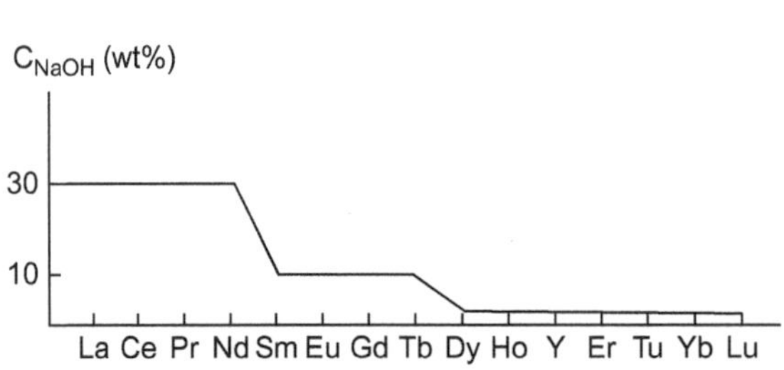

Figure 7.16 Rare earth elements [35].

In Ref. [74], it has been shown that with a decrease in the values of rare earth cations, the stability of the $R_{18}Si_7O_{41}$ decreases especially for smaller elements (Er–Lu). If other elements like Na, Mn, Pb, and Ca go into the structure, the stability of the structures under hydrothermal conditions again rises. However, the stability of rare earth britolite determined mainly from the position of these other elements is in the structures.

Transition from britolite to mixed Na R-silicates takes place during the complete destruction of cationic motif and with a change in the coordination number of rare earth elements.

In highly concentrated aqueous solutions NaOH ($<80\%$) (Table 7.11) mixed silicates like $Na_2O \cdot R_2O_3 \cdot 2SiO_2 = Na.\ R[SiO_4]$ are formed covering a significant part of the rare earth series from Pr to Ho, including Y. The latter one is usually related to the heavier (Eu–Lu) rare earths with the coordination number 7, since the ionic radius of Y is slightly higher (0.99) and it lies between Nd and Sm. This emphasizes sufficiently simple means of synthesis of tetragonal modification of $NaY[SiO_4]$. The field of stability of tetragonal rare earth silicates of type NaR $[SiO_4]$ lies between Ho and Er. This is the traditional boundary for the morphotrophic series of rare earth elements. Its existence is usually based on the geometric representation of the stability of the rare earth octahedra, since the significant part of the R–O bonding is ionic ($\sim$75%). However, under the conditions of high potential elements with lower values of electronegativity, as Na^+, a better model could be obtained giving the influence and an overall covalent bonding in the nature of R–O bond. It can be assumed that in the aqueous solvents, with higher concentration of Na^+, the capability of internal f-orbit for hybridization, which is for Er–Lu, indistinctly becomes closer to zero. Therefore, the Eu–Lu group in NaOH aqueous solvent forms only silicates with octahedral coordination of the rare earth: $NaR[SiO_4]$ and $Na_3R[Si_2O_7]$.

Reduction in the influence of f-orbit at higher potential Na^+ can be seen as the possibility for the distortion of rare earth motifs, particularly leading to the decrease in the coordination number of rare earth elements up to 6 (trigonal prism). In this case, the rare earth polyhedra form discrete groupings of rare earth prism. It is

Table 7.11 Phase Formation in the System $Na_2O–R_2O_3–SiO_2–H_2O$ in the Region of Surplus R_2O_3 Depending upon the NaOH Concentration [35]

	Rare Earth Elements															
CNaOH (%)	**La**	**Ce**	**Pr**	**Nd**	**Sm**	**Eu**	**Gd**	**TB**	**Dy**	**Ho**	**Er**	**Tu**	**Yb**	**Lu**	**Y**	**Sc**
80	A	G	C	C	C	C	C	C	C	C	F	F	F	F	C	G
80 – 55	A	G	A	C	C	C	C	C	C	C	E	E	E	E	H	H
55 – 25	A	A	A	A + C	C	C	C	C	C	C	D	D	D	D	D	G
25 – 5	B	B	B	B	C	C	C	C	C	C	D	D	D	D	D	G
5	B	B	B	B	B	B	B	B	B	B	B	B	B	B	B	H

A, $NaRSiO_4$ (orthorhombic I); C, $NaRSiO_4$ (tetragonal); D, $NaRSiO_4$ (orthorombic II); E, $Na_3RSi_2O_7$; B, britolites $R_{4.67}(SiO_4)_3O$; F, unknown phase; G, experiments not carried out; H, $NaScSi_2O_6$.

emphasized that from the point of view the size of the coordination number of rare earth equals 8, this group is more logical than the coordination number 6.

The stability field of tetragonal modification of $NaR[SiO_4]$ for the lighter lanthanides ends at Pr (Table 7.11), but the La-representative crystallizes in a different structure type "A" (orthorhombic I) and its boundary depends upon the concentration of NaOH. For lighter rare earths, the characteristic feature is the significant contribution of f-orbit that changes the boundary corresponding to the participation of f-orbits in the R–O bonding. With an increase in the NaOH concentration, the coordination number for Pr and Nd decreases from 9 to 8, i.e., the reduction in the participation of f-orbits in the bonding. During the transition of structure type A (orthorhombic I) to type B (tetragonal), the basic nature of the rare earth motif, i.e., columns from rare earth polyhedra, is retained.

In the heavier lanthanide groups (Er–Lu), the change in the NaOH concentration does not speak on the coordination number, but the degree of association of rare earth motif changes from discrete nature in the group $Na_3R[Si_2O_7]$ up to a highly associated nature in $NaR[SiO_4]$ (orthorhombic II). For both of the groups, the characteristic feature is the octahedral surrounding for rare earth elements. Here, the end members of rare earth series are well inscribed as $[YO_6]$ and, therefore, it is easy to obtain $Na_3Y[Si_2O_4]$ (analogous to Sc-diorthosilicate).

7.5.3 Phase Formation in the Rare Earth Silicate Systems in the High Silica Region

It is well known that, in the region of surplus [Si–O], the formation of highly condensed silicates can be expected. Until now, the synthesis under such conditions was carried out in pure water and obviously, the higher the amount of [Si–O] taken, the higher its concentration and more complex silicates were obtained. However, the synthesis did not lead to the differentiation of crystalline individuals and, as a rule, resulted in the formation of coarse crystalline and fine crystalline masses. The action of silicon in the aqueous solvents of salts and alkalies remains unknown. From the works of Dimitrova [35], it was found that in alkaline solutions and solutions with surplus silica in the system, the heavier phases form.

In order to explain the possibility of obtaining silicates in the form of monocrystals in high silica region, a series of experiments have been carried out by Dimitrova [35] and the results are shown in Figure 7.6.

The region in which the crystallization was carried out lies in the stability field of quartz. Therefore, quartz exists in all the experiments. But this occurs only in cases when the NaOH concentration is less than 15%. As the concentration of NaOH increases, quartz disappears and there appears Na silicates and similarly the heavier phases—an amorphous glassy substance. Under these conditions, the formation of silicates takes place with the complexity of the anionic motif. For the lighter rare earths, $NaR[Si_6O_{14}]$ forms and their structural units consist of $[Si_3O_7]_{\infty\infty}$(O/Si = 2.33). With a decrease in the concentration of Si, another structure type having $[Si_6O_{15}]$ radical, e.g., $NaRSi_6O_{15} \cdot 2H_2O$(O/Si = 2.5) crystallizes. With an increase

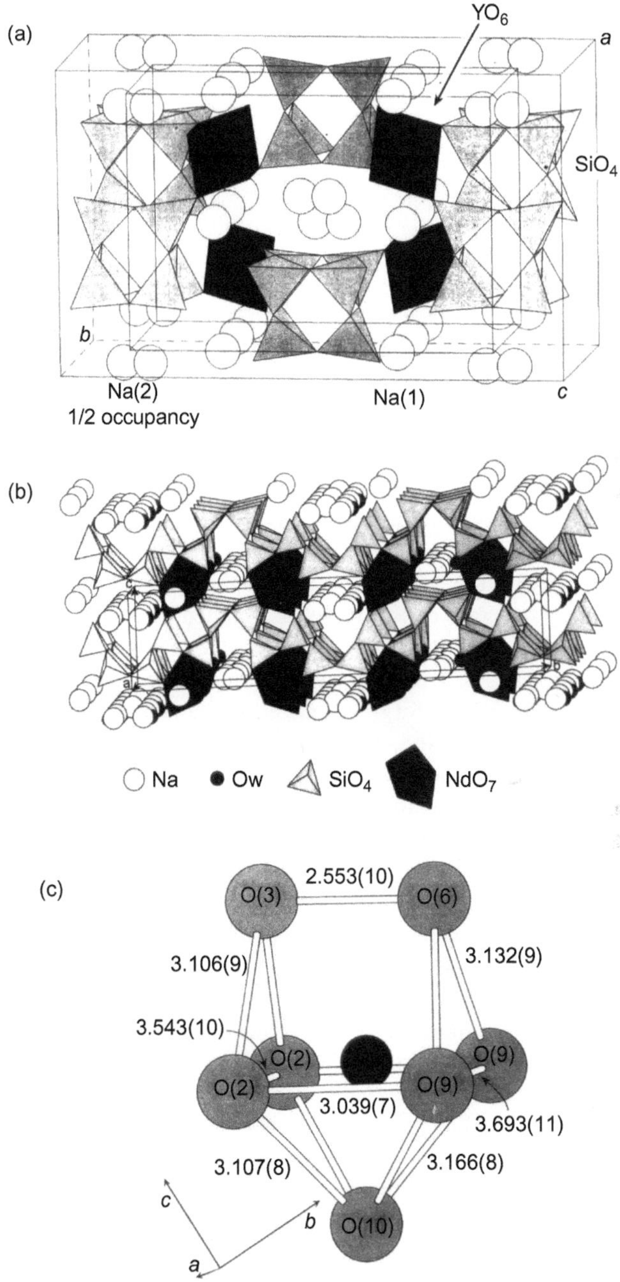

Figure 7.17 Structures of (a) $Na_3YSi_6O_{15}$ as viewed along *c*-axis [75], (b) $Na_3NdSi_6O_{15} \cdot 2H_2O$ viewed along *a*, and (c) isolated NdO_7 polyhedron viewed along "*a*," the figure also gives oxygen–oxygen distances (Å) [70].

Table 7.12 Results of Electron Microprobe Analyses at Three Different Locations Within a Single Crystal of the Present Sodium Neodymium Silicate [76]

Position	Na mol%	Nd mol%	Si mol%	Total wt% oxides
1	10.81	3.49	22.40	94.55
2	10.79	3.50	22.50	94.77
3	10.74	3.46	22.24	94.19
Average	10.78(4)	3.48 (2)	22.4 (1)	94.5 (3)
Expected	10.5	3.5	21.0	94.7

The expected values are those calculated for the composition $Na_3NdSi_6O_{15} \cdot 2H_2O$, deduced from the chemical analysis and the structure determination.

in NaOH concentration, silicates of type $Na_{15}R_3[Si_{12}O_{36}]$ are obtained with Si—O radicals made up of 12-membered rings, $[Si_{12}O_{36}]$(O/Si = 3).

In the region with high silica concentration, R_2O_3 is insufficient for the formation of complexes with 3d cationic motif. When $R_2O_3:SiO_2$ = 1:4 and concentration of NaOH is 5%, $NaR[Si_6O_{14}]$ type was obtained. In this, the cationic motif is represented by rare earth chains. During the increase in NaOH concentration, rare earth chain silicates cease to form with the appearance of compounds with discrete rare earth cations. In $Na_{15}R_5[Si_{12}O_{36}]$, rare earth ion lies as isolated octahedra, whereas in $Na_3R[Si_6O_{15}] \cdot 2H_2O$, the rare earth ion lies as isolated seven-sided polyhedra.

More recently, Haile et al. [75] carried out hydrothermal investigations in the high silica region of the $Na_2O-Y_2O_3-SiO_2$ system and obtained a new phase—trisodium yttrium, Ibmm, with $a = 10.468$, $b = 15.246$, $c = 8.385$ Å; $Z = 4$, and n atoms in the asymmetric unit. The structure is unique in the sense that silica tetrahedra form isolated $[Si_6O_{15}]$ rings, rather than layers as might be expected from the Si:O = 4. No isomorphs to $Na_3YSi_6O_{15}$ have been reported. Figure 7.17a shows the structure of $Na_3YSi_6O_{15}$ as viewed along c-axis. The SiO_4 and YO_6 units are represented as polyhedra.

Similarly, Haile et al. [76] reported another phase rich in sodium, $Na_3NdSi_6O_{15} \cdot 2H_2O$, a layered silicate in the system $Na_2O-Nd_2O_3-SiO_2$ under hydrothermal conditions. Although this compound has very similar lattice constants and diffraction symbols to data reported for $NaNdSi_6O_{13}(OH)_2 \cdot nH_2O$ (special group Cmm2) by Karpov et al [70], the microprobe analysis given by Haile et al. (Table 7.12) shows the composition substantially richer in Na^+. The asymmetric unit of the structure contains 22 atomic sites: 1Nd, 4Na (three of which are only partially occupied). Figure 7.17b shows the structure of $Na_3NdSi_6O_{15} \cdot 2H_2O$ viewed along a. The SiO_4 and NdO_7 units are represented as polyhedra. Figure 7.17c shows an isolated NdO_7 polyhedron viewed along a, and the figure also gives oxygen—oxygen distances (Å).

Tuttle autoclaves have been used in the above works. The solvents used are either pure water or a dilute solution of a sodium base (0.1 M NaOH, Na_2CO_3, or $NaHF_2$), experimental temperature was 500—600°C, and pressure was 8.25×10^5 Pa.

In systems containing Fe^{3+} and weak NaOH solutions, aegirine, $NaFeSi_2O_6$ type silicates were obtained, whereas in the strongly concentrated NaOH solutions,

$Na_{15}Fe_5Si_{12}O_{36}$ forms. The heavier rare earth elements (Ho to Lu) follow Fe^{+3}, whereas the lighter lanthanides form original structures, whose cationic nature has not been observed in natural analogs.

7.6 Degree of Silification

One of the characteristic features of silicates is the molar ratio of rare earth oxides and silicon. Variation in the value of molar ratio varies within a wide range from 0.78 for Na-free silicate (apatite–britolite structures) up to 12 for compounds with highly condensed [Si–O] radicals. The degree of silification (*S*) can be represented as the ratio of Si-atoms to the sum of the cations:

$$S = \frac{K\text{Si}}{(nM^{++} + mMe^{++} + \cdots)} \quad (7.6)$$

The degree of silification decreases with an increase in NaOH concentration. This reflects the interrelation of cationic and anionic motif (crystal chemical factor). It also reflects the phase formation, chemical environment, and conditions of synthesis (genetical factors). In the case of the lower concentration of NaOH, the monocrystals can be obtained in the presence of additional anions like CO_3, Cl, F, and SO_4. Dimitrova [35] has studied the degree of silification for several rare earth silicate systems and found $S \approx 3$ only for one silicate and for a majority of the silicates $S < 1$ with a higher association of cationic motif. Corresponding to the increase in the concentration of Si from 0.27% to 16.7%, the following phase changes have been observed:

$$\begin{array}{llllll} & R_{4\ 2/3}[SiO_4]_3O & \rightarrow & Na_3R[Si_6O_{15}]2H_2O & \rightarrow & NaR[Si_6O_{14}] \\ S = & 0.64 & & 1.5 & & 3.0 \end{array} \quad (7.7)$$

The other example for silification of rare earth silicates in highly concentrated solution of NaOH (30–45%) is the (Ho–Lu) series:

$$\begin{array}{llllll} & NaR[SiO_4] & \rightarrow & NaR[SiO_7] & \rightarrow & Na_{15}R[Si_{12}O_{36}] \\ S = & 0.5 & & 0.5 & & 0.66 \end{array} \quad (7.8)$$

In highly concentrated solutions, the change in the concentration of Si from 48% to 60% does not influence on the degree of silification. For example, at a constant concentration of Si (48%) and increase in NaOH concentration, the following chain sequence is observed:

i. For lighter lanthanides:

$$\begin{array}{llllll} & R_{4\ 2/3}[SiO_4]_3O & \rightarrow & NaR[SiO_4] & \rightarrow & Na_2R[SiO_4](OH) \\ S = & 0.64 & & 0.5 & & 0.5 \end{array} \quad (7.9)$$

ii. For heavier lanthanides:

$$\begin{array}{lccccc} & R_{4\ 2/3}[SiO_4]_3O & \rightarrow & NaR[SiO_4] & \rightarrow & Na_3R[Si_2O_7] \\ S = & 0.64 & & 0.5 & & 0.5 \end{array} \quad (7.10)$$

For more concentrated solutions, Si (>60%) and corresponding NaOH concentration, the lighter rare earths show the following sequence:

$$\begin{array}{lccccc} & Na_3R[Si_6O_{15}] \cdot 2H_2O & \rightarrow & NaR[SiO_4] & \rightarrow & NaR[SiO_4](OH) \\ S = & 3 & & 0.5 & & 0.33 \end{array} \quad (7.11)$$

All these phase transformations show that without depending upon the concentration of Si, the degree of silification of phases decreases with an increase in the NaOH concentration in the system.

It can be assumed that rare earth elements, having high basicity (La–Nd) in the process of phase formation, form more associated rare earth motif with a characteristic 3d bond [35]. Only at higher concentrations of Si, and in the presence of carbonate ions, less associated chains $[R_3O_{12}]_\infty$ as in $NaRSi_6O_{14}$, or discrete polyhedra as in $Na_3RSi_6O_{15} \cdot 2H_2O$, are formed.

7.7 Properties of Rare Earth Silicates

Rare earth silicates exhibit very interesting electrical and optical properties, which make these compounds unique in several respects. For example, the discovery of high ionic conduction in rare earth ring silicates has triggered an intensive search for technically promising superionic conductors [41,77–79]. Because the popular β-alumina had several disadvantages—like its intrinsically high anisotropy, current and stress inhomogeneities—which degrade the mechanical properties. In this respect, the rare earth silicates containing alkali metals offer lower sintering temperatures (1100°C compared to 1600°C for β-alumina), a wide chemical flexibility, and ionic conductivities possibly surpassing, at 300°C, that of β-alumina. Beyeler et al. [79] have grown single crystals of $Na_5YSi_4O_{12}$ using the hydrothermal method and studied its directional ionic conductivity in the frequency range between 1 kHz and 50 MHz. Figure 7.18 shows the Arrhenius plot of $Na_5YSi_4O_{12}$ single crystals. Subsequently, several others have reported the high ionic conductivity in a large variety of rare earth silicates and also proposed the conduction mechanism as there is a close relationship existing between the crystal structure and the properties [48,80–84]. Several authors have reported ionic conductivity in various alkali rare earth silicates [48,82,85].

The ionic conductivity is closely related to the mixed framework structure. The most important compound among rare earth silicates is $Na_5RSi_4O_{12}$, wherein, $[SiO_4]$ tetrahedra are linked to form $[Si_{12}O_{36}]$ rings parallel to the basal plane of the hexagonal cell. Maksimov et al. [43] could locate only 48/90 of the Na^+ atoms

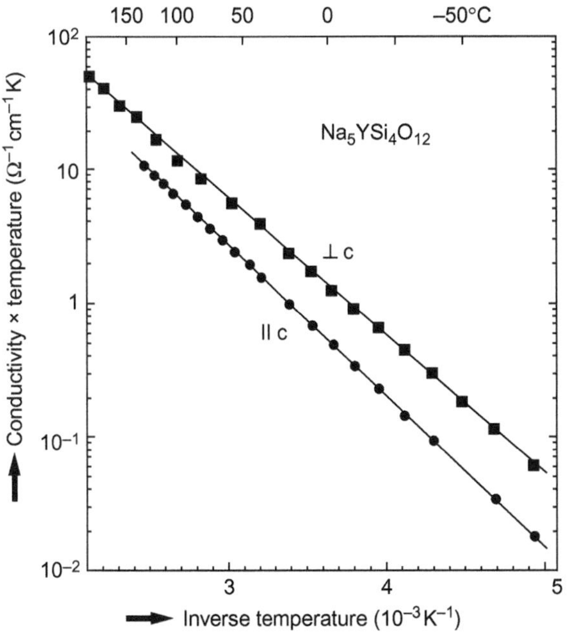

Figure 7.18 Arrhenius plot of $Na_5YSi_4O_{12}$ single crystals [79].

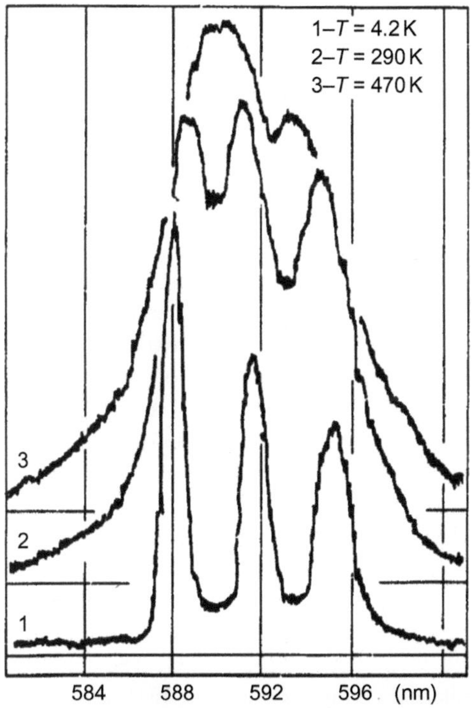

Figure 7.19 Photoluminescence spectra in NaY[SiO_4] crystals [86].

mobile making this compound a good Na^+ ion conductor. The discrete polyhedra of metalayers are mainly responsible for ionic conductivity, as in $K_3R_3(Si_6O_{16})_2OH$, where R = Gd, Ho [69]. Excess rare earth elements lead to the condensation of their polyhedra which gives low conductivity (10^{-4}–10^{-3} ohm^{-1} cm^{-1} at 300°C).

The discrete nature polyhedra in these compounds are responsible for these special properties like photoluminescence. Figure 7.19 shows the photoluminescence spectra in $NaY[SiO_4]$ crystals [86].

7.8 Sodium Zirconium Silicates

Most of the sodium zirconium silicates belong to the NASICON family [87–89]. There are over 20 known sodium zirconium silicates, out of which only six are minerals. The sodium zirconium silicates find extensive applications in modern technology, especially after the discovery of unique properties in these compounds as inorganic adsorbents used in industry and agriculture and for environmental safety. Among the inorganic adsorbents, zeolites were most popular initially. Unfortunately, aluminosilicate-based zeolites have one essential drawback, a low chemical stability in acid and alkaline media, which imposes certain restrictions on their application. Titanium silicates, such as $Na_2Ti_2SiO_7 \cdot 2H_2O$ [90,91], and $Na_2TiSi_2O_7 \cdot 2H_2O$ [92], are among a large family of known titanium containing minerals and synthetic compounds, and they exhibit most promising selective ion-exchange properties suitable for highly alkaline nuclear waste remediation. Clearfield and coworkers [91,93,94] have found that the reason for the unique behavior of crystalline titanium silicates is connected with the correspondence of the geometrical parameters of their adsorption sites (channels, cavities) to the size of the selectively adsorbed Cs^+ or Sr^+ ions. By analogy, the other polyvalent metal silicates—e.g., zirconium silicates—might be as fruitful in yielding the desired exchangers as was the case for titanium silicates. Moreover, zirconium silicates occur widely in nature, and their formation under hydrothermal conditions (in a wide range of temperatures from 300°C to 550°C) has been given considerable attention to solve some of the general geophysical and mineralogical problems. It is reasonable to expect that some of these compounds, due to the presence of ion-exchange functional groups, as well as the existence in their structure of well-defined channels and cavities, could exhibit interesting ion-exchange properties, namely, high selectivity to certain ions and molecules. Unfortunately, the ion-exchange behavior of such zirconium silicates has not been studied in detail yet.

Na_2O–ZrO_2–SiO_2–H_2O system has been studied over a wide range of the initial component ratios, ZrO_2/GeO_2 = 3:1–1:9 (mole); NaOH concentration 0–50 wt %, temperature 300–600°C, pressure 40–300 MPa by several workers [95–98]. The phase composition of zirconium silicates formed in Na-containing media is shown in Figure 7.20 and the crystallization field is shown in Figure 7.21. The analysis of the products of hydrothermal synthesis and X-ray data allows one to

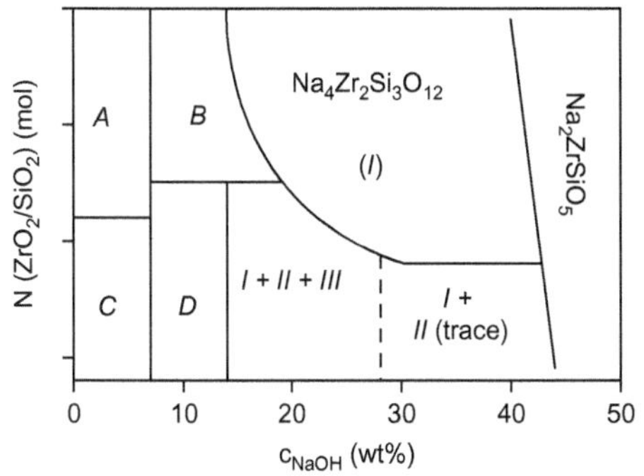

A — $ZrSiO_4$, $Na_2ZrSi_3O_9{\cdot}H_2O$, $Na_4Zr_2Si_5O_{16}{\cdot}H_2O$

B — $Na_2ZrSi_2O_7$, $Na_4Zr_2Si_3O_{12}$;

C — $ZrSiO_4$, $Na_2ZrSi_3O_9{\cdot}H_2O$, $Na_2ZrSi_6O_{15}{\cdot}3H_2O$

D — $Na_8ZrSi_6O_{18}$, $Na_2ZrSi_2O_7$, $Na_2ZrSi_3O_9{\cdot}2H_2O$

II — $Na_8ZrSi_6O_{18}$

III — $Na_3HZrSi_2O_8$

Figure 7.20 Phase composition of zirconium silicates formed in Na-containing media [95].

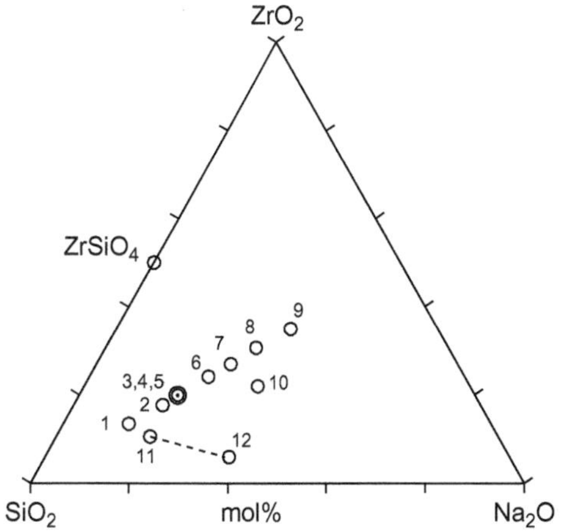

1 — $Na_2ZrSi_6O_{15}{\cdot}3H_2O$

2 — $Na_2ZrSi_4O_{11}$

3 — $Na_2ZrSi_3O_9{\cdot}2H_2O$

4 — $Na_2ZrSi_3O_9{\cdot}3H_2O$

5 — $Na_4Zr_2Si_6O_{18}{\cdot}4H_2O$

6 — $Na_4Zr_2Si_5O_{16}$

7 — $Na_2ZrSi_2O_7$

8 — $Na_4Zr_2Si_3O_{12}$

9 — Na_2ZrSiO_5

10 — $Na_3HZrSi_2O_8$

11 – 12 — $Na_{8-x}H_xZrSi_6O_{18}$

Figure 7.21 Crystallization field in Na_2O—ZrO_2—SiO_2 system [95].

refine the composition of some phases. Thus, the phase earlier described as $Na_{14}Zr_2Si_{10}O_{31}$ based on the data of chemical analysis [97,98] has, in fact, the composition $Na_8ZrSi_6O_{18}$. Its structure is analogous to $Na_8SnSi_6O_{18}$ [99]. $Na_6Zr_2Si_3O_{15}$ (composition based on the data of the chemical analysis) [100] has the same structure as the synthetic germanate $Na_3HZrGe_2O_8$ (Table 7.13).

The general scheme of zirconium silicate crystallization in hydrothermal solution with an increase in alkali concentration and temperature is shown in Figure 7.22.

Analogs of zirconium mineral (zircon, elpidite, vlasovite, catapleite, parakeldyshite) and $Na_4Zr2Si_5O_{16}{\cdot}H_2O$ silicate (still not observed in nature) are formed in the region of solutions with the lowest concentrations. In highly concentrated solutions, the silicates with the lowest silicon content are crystallized: $Na_4Zr_2Si_3O_{12}$ and Na_2ZrSiO_5. This system illustrates the role of the solvent in the formation of

Table 7.13 New Silicates of Zirconium [95]

Compound	Unit Cell Parameters			Space Group
	a (Å)	*b* (Å)	*c* (Å)	
$Na_8ZrSi_6O_{18}$	7.371 $\alpha = 88.34°$	–	–	R3m
$Na_8SnSi_6O_{18}$	7.344 $\alpha = 87.82°$	–	–	R3m
$Na_3HZrSi_2O_8$	9.036 $\alpha = 87.74°$	6.953 $\beta = 89.75$	5.558 $\gamma = 89.16$	12/m

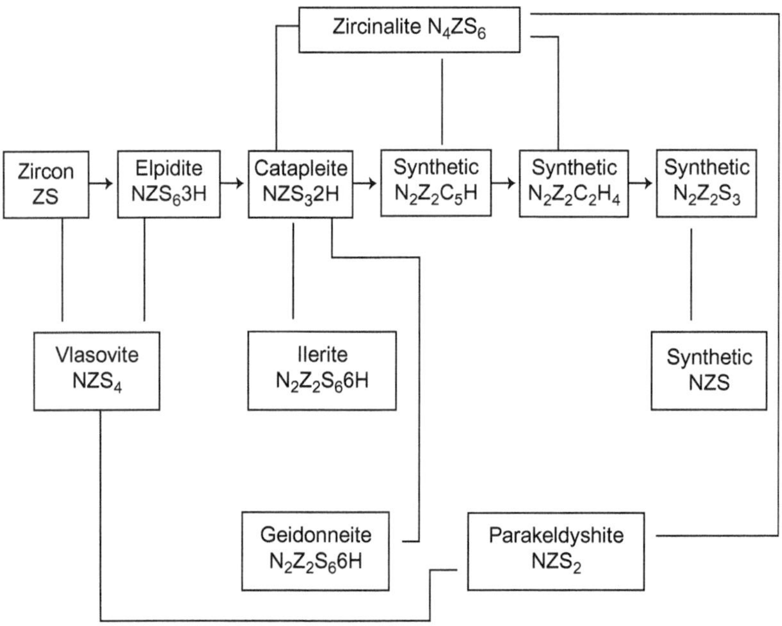

→ denotes the transitions with the temperature rise.
↔ denotes paragenesis, dashed lines denotes natural paragenesis,
N = Na_2O, Z = ZrO_2, S = SiO_2 and H = H_2O.

Figure 7.22 Scheme of zirconium silicate crystallization [95].

crystal phases. With an increase in the NaOH concentration, the process of the formation of alkali-zirconium silicates proceeds with the depolymerization of the Si—O radicals: infinite $[Si_6O_{15}]$ chains in elpidite ⇒ infinite $[Si_4O_{11}]$ chains in vlasovite ⇒ island $[Si_3O_9]$ rings in catapleite ⇒ $[Si_2O_7]$ diorthogroups in parakeldyshite ⇒ $[SiO_4]$ orthogroups in the synthetic $Na_4Zr_2Si_5O_{16}$ and $Na_3HZrSi_2O_8$ phases. Barth et al. [101] studied the structure and ionic conduction in

$Na_4ZrSi_3O_{10}$, belonging to the NASICON family. Figure 7.23 shows part of the system $Na_2O-ZrO_2-SiO_2-P_2O_5$. $Na_4ZrSi_3O_{10}$ has been characterized as a monoclinic compound with lattice parameters which are quite similar to the monoclinic low-temperature modification of NASICON.

In recent years, there has been a general trend to synthesize these zirconium silicates under lower pressure–temperature conditions Yong and Wenqin [102,103] studied this system and obtained $Na_4Zr_2(SiO_4)_3$ at 290–300°C and further reduced temperature conditions (250°C) using $ZrOCl_2$, Na_2SiO_3, and NaOH solution, and the hydrothermal chemical reaction is given below:

$$2ZrOCl_2 + 3Na_2SiO_3 + 2NaOH \rightarrow Na_4Zr_2(SiO_4)_3 + 4NaCl + H_2O \qquad (7.12)$$

Clearfield and coworkers [93,94,104] have obtained a series of zirconium silicates under mild hydrothermal conditions (180–190°C) and studied the influence of the type of silicon-containing reagent, the molar ratio, and concentrations of reagents in the reaction mixture on the yield of resultant products. These authors have studied the formation of several different sodium zirconium silicate phases in the systems: (i) $Zr(OC_3H_7)_4$–silicic acid–NaOH–H_2O; (ii) $Zr(OC_3H_7)_4$-silicon-containing reagent–NaOH(4M)–H_2O under mild hydrothermal conditions (T = 180–190°C, pressure generated from 50% fill). These studies have yielded several novel sodium zirconium silicates. According to these authors, the predominant phase is hydrated sodium zirconium silicates in this system in contrast to mainly anhydrous sodium zirconium silicate (Na_2–$ZrSiO_5$, $Na_2ZrSi_2O_7$, $Na_4Zr_2Si_3O_{12}$, etc.) while carrying out the reactions at much higher temperatures (300–550°C) [95]. This suggests that a relatively low temperature of hydrothermal synthesis

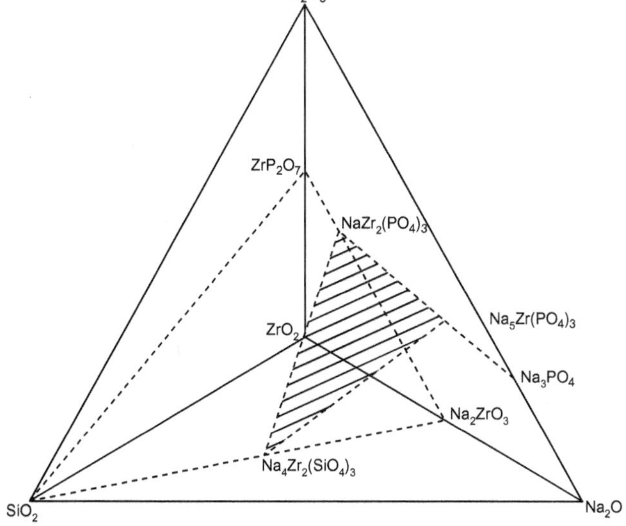

Figure 7.23 Part of the system $Na_2O-ZrO_2-SiO_2-P_2O_5$ [95].

could be one of the factors favorable for the preparation of hydrated zirconium silicates, possessing the ability for ion exchange. Then, the second controlling factor, giving a certain diversity of compounds, is the type of the initial silicon-containing reagent. The comparison shows that in a wide range of Si:Zr ratios (2–5:1) used in the systems $Zr(OC_3H_7)_4-Na_2Si_3O_7-NaOH-H_2O$ and $Zr(OC_3H_7)_4-Na_2SiO_3-NaOH-H_2O$, the main product is $Na_4Zr_2Si_5O_{16}\cdot H_2O$. No formation of $Na_2ZrSi_3O_9\cdot 2H_2O$ or the novel 12.8 Å phase was found. The second feature is the formation of $Na_4ZrSi_3O_{10}$ (as admixture or main phase) even at a Si:Zr ratio of 1:1, which was not observed when silicic acid as the silicon-containing reagent was used.

The hydrothermally grown crystals of the sodium zirconium silicates have a broad size range (from 0.2 to 25 mm) and an irregular form. The lack of a clearly defined morphology correlates with a low crystallinity of the compound.

A majority of the works on the synthesis of alkali-zirconium silicates suggest that all of them are layered compounds. Thermal and chemical stability, different Zr/Si molar ratio, and intercalation ability suggest that these compounds could be promising for pillaring with inorganic and organic species for synthesis of new functionally designed and tailored catalysts, and support for catalyst and adsorbents. High affinity for some transition and toxic heavy metals makes sodium zirconium silicates attractive for high-level purification of certain technological solutions and wastes containing these elements.

7.9 Growth of Selected Silicates

It would be extremely difficult to describe the growth of all the silicates under hydrothermal conditions as there have been a great number of publications in this field since the 1850s. Therefore, we only discuss the synthesis of some selected silicates with an emphasis on the current trends in the growth of silicate crystals.

The first work on silicates on the whole by hydrothermal method was carried out by Wöhler [105]. Followed by this, several other silicate minerals were synthesized during the nineteenth century. Allen et al. [106] carried out a systematic investigation of the hydrothermal synthesis of several pyroxene and amphibole group of minerals. For example, these authors quote an experiment of the synthesis of amphibole at low temperatures by Chroustschoff, who obtained hornblende in which a mixture of silicic acid, alumina, lime, magnesia, ferrous and ferric oxides, and the alkalies, partly in solution partly precipitated were heated in sealed glass vessels at a temperature of 550°C for 3 months. The results of these experiments were used to explain the geological relations. Later, Morey and Niggli [107] published an exhaustive review on the hydrothermal formation of silicates, giving a complete list of silicates obtained till that time, and all the important publications were included. Figure 7.24 shows the equilibrium diagram of the system diopside–anorthite at pressure 1 bar, and it was studied by Bowen (1915) [108]. Soon after

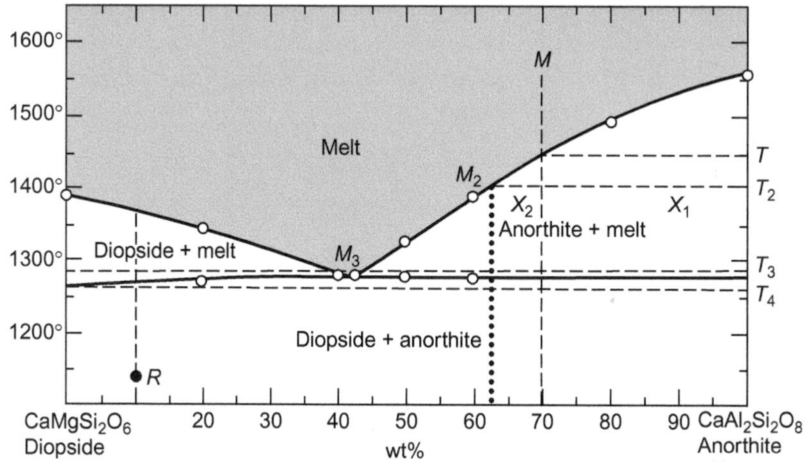

Figure 7.24 Equilibrium diagram of the system diopside–anorthite at pressure 1 bar [108].

this, Morey [109], published a systematic investigation of the ternary system $H_2O-K_2SiO_3-SiO_2$. Here, for the first time, studies were carried out in a systematic fashion with the temperature carefully controlled and measured and the pressure held constant throughout the experiment. Figure 2.6 shows the isothermal polybaric saturation curves and the isobaric polythermal saturation curve at 1 atm pressure of H_2O vapor. This marked the beginning of the phase equilibria studies in the silicate system and, followed by this, several other important publications came out especially from the Geophysical Laboratory in Washington, pertaining to the silicate systems. In fact, Bowen [110] developed his famous petrologic ideas based on these phase equilibrium data. One of the most important phase equilibrium data among the silicate systems is for the system $MgO-SiO_2-H_2O$ given by Bowen and Tuttle [111]. This work, in many respects, laid the foundation for much of the subsequent work on silicate systems. In this work, the authors describe the change of assemblages with temperature and also show the effect of pressure on the temperatures involved. Figure 7.25 shows the pressure–temperature curves of univariant equilibrium in the system $MgO-SiO_2-H_2O$ [112–115]. There are quite a large number of such silicate systems reported in the literature, contributing to the understanding of complex geological processes. Some of the systems like $Al_2O_3-SiO_2-H_2O$ [116–118], $CaO-SiO_2-H_2O$ [119–122], $SrO-SiO_2-H_2O$ [123], $MgO-Al_2O_5-SiO_2-H_2O$ [112–115], and alkali–$Al_2O_3-SiO_2-H_2O$ [124–127] are noteworthy to mention here, as they provide a vast amount of information on the understanding of the natural geological processes leading to the formation of various rocks and minerals.

We give here some of the earliest works on the synthesis of various silicates in general, before going into the systematic synthesis of selected silicates with device potential, gemstones, and of general interest.

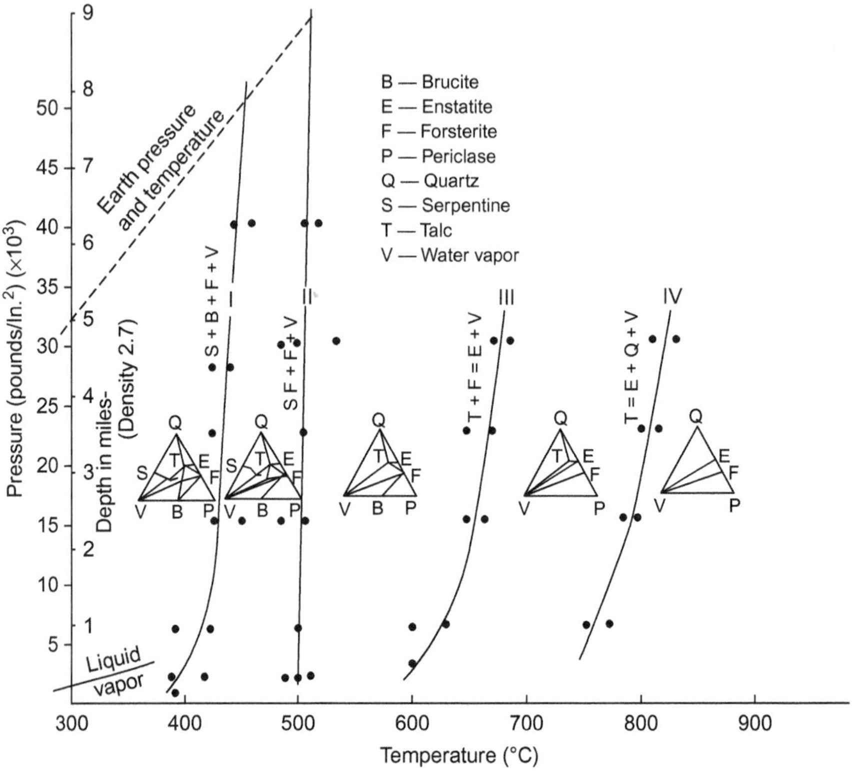

Figure 7.25 Pressure–temperature curves of univariant equilibrium in the system $MgO–SiO_2–H_2O$ [112].

Felspars: The high-temperature varieties of felspars have been successfully prepared by the end of the nineteenth century. However, the complexities in their forms and intergrowth were realized only during the 1950s.

Several hydrous and anhydrous aluminosilicates have been synthesized for the past 150 years using hydrothermal technique. Among these, the dense anhydrous aluminosilicates carried a great significance because of their role in natural metamorphic processes. Though several minerals—like garnets, sillimanite, analusite, kyamte, and epidotes—have been synthesized using hydrothermal method, no reproducible synthesis of these minerals was achieved before the work of Coes [128], whose works are briefly described here for the benefit of the readers in Table 7.14 [120].

The other important minerals synthesized during the nineteenth century by hydrothermal method are staurolite, chloritoid, beryl, vesuvianite, lanosonite, bertrandite, lowsonite, bertrandite, low-temperature forms of the lithium aluminosilicates, spodumene, eucryptite, petalite, micas, clays, tourmaline, uranium minerals, and so on. Therefore, we describe the synthesis of only some technologically important silicates under hydrothermal conditions.

Table 7.14 Early Hydrothermal Synthesis of Various Dense Anhydrous Aluminosilicates [128]

Silicate	Starting Material	Pressure	Temperature (°C)
Jadeite	Sodium sulfate, kaolin + silica	+20,000 kg/cm^4	900
$NaCr(SiO_3)_2$	Sodium dichromate + silica	20,000 kg/cm^2	900
Andalusite	Kaolin + sodium salts	20,000–30,000 kg/cm2	600–900
Kyanite	Alumina + silica	20,000–40,000 kg/cm^2	600–900
Sillimanite	Kaolin + fluorides	20,000–30,000 kg/cm^2	600–900
Topaz	Kaolin + fluorides, but higher fluoride contents or kaoline + NH_4HF_2		
Epidote and zoisite	hydrated calcium and iron nitrates, hydrated alumina and silica	20,000 kg/cm^2	800
Garnet			
Grossularite		15,000 kg/cm^2	600–800
Andradite		20,000 kg/cm^2	600–800
Almandine		20,000 kg/cm^2	600–900
Pyrope		30,000 kg/cm^2	900
$Mg_3Fe_2(SiO_4)_3$		30,000 kg/cm^2	900
$Mn_3Fe_2(SiO_4)$3		20,000 kg/cm^2	800
$Fe_3Fe_2(SiO_4)$3		20,000 kg/cm^2	800
$Mg_3Cr_2(SiO_4)$3		20,000 kg/cm^2	800

7.9.1 Bismuth Silicate

Bismuth silicate, $Bi_{12}SiO_{20}$ (BSO), is of considerable interest for optical signal processing. It is popular as a photorefracting crystal. The *photorefractive effect* can be described as a combination of the photoconductive and the electro-optic effects used in a variety of applications like nonlinear mixing including active spatial filtering phase conjugation by way of degenerating four-wave mixing optical processing devices such as optical oscillators, and as a technique useful for nondestructive materials analysis. Litvin [129] has studied the system $Na_2O{-}Bi_2O_3{-}SiO_2{-}H_2O$ under hydrothermal conditions: $T = 350{-}450°C$, $P = 400{-}500$ atm in NaOH solution up to 40 wt%.

Figure 7.26 shows the fields of crystallization of BSOs in the system $Na_2O{-}Bi_2O_3{-}SiO_2{-}H_2O$. The following phases have been obtained: α-Bi_2O_3 (bismite), Bi_2SiO_{20} (sillenite), and $Bi_4Si_3O_{12}$ (eulytite). Bi_2SiO_{20} crystallizes under hydrothermal conditions in NaOH (1–60 wt%) and KOH (1–40 wt%) solutions and 350 atm pressure [130]. Whereas eulytite crystallizes with NaOH (21 wt%), and it does not depend on the ratio of the starting components ($Bi_2O_3 + SiO_2$). It

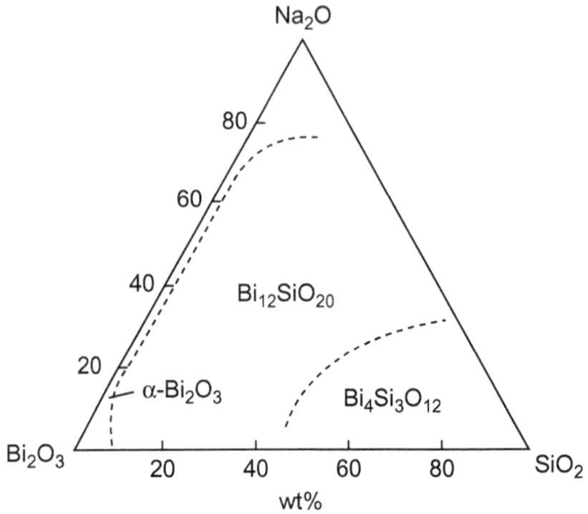

Figure 7.26 Fields of crystallization of BSOs in the system $Na_2O-Bi_2O_3-SiO_2-H_2O$ [130].

recrystallizes well at 380°C. However, the variation in the ratio of the starting components ($Bi_2O_3 + SiO_2$) restricts the field of crystallization of $Bi_{12}SiO_{20}$ on one hand and the appearance of α-Bi_2O_3 on the other hand. The process of the formation of $Bi_{12}SiO_{20}$ is very sensitive to the existence of hydrogen. Surplus hydrogen leads to the establishment of $Bi_{12}SiO_{20} \rightarrow Bi$ metal transformation. The hydrothermal synthesis of BSOs can be carried out in silver, platinum, copper, or titanium liners. In fact, Bi_2O_3 and SiO_2 reactions occur more rapidly than for other silicate systems, so that the entire reaction completes within 2 days.

BSO crystals are usually grown using Czechrolskii and hydrothermal methods. Although commercially available crystals of BSO grown by melt techniques are of sufficiently good quality for most applications, they have limitations in optical uniformity and reproducibility of photorefractive properties. Variations in the process control environment, i.e., temperature, growth rate, ambient atmosphere, rotation rates, and annealing, can affect the segregation of dopants and defects in the melt growth environment. High-temperature solution growth by the hydrothermal process is capable of producing multiple crystals of large size and identical consistency while requiring only a stable temperature control system once the internal environmental conditions have been established [131]. The Cz-grown crystals have higher trap density, whereas the hydrothermally grown crystals have lower trap density. The hydrothermally grown crystals are free of the typical structural defects encountered in commercial material, i.e., veils, bubbles, precipitates, and other strain producing phenomena. BSO crystals show a yellow coloration. The doping studies on both Cz-grown and hydrothermally grown material indicate that Al, Ga, Cr, Mn, and V are effective in altering absorption at different wavelengths. Harris et al. [132] obtained low-defect colorless BSO crystals under hydrothermal conditions using a noble metal liner. Figure 7.27 shows two crystals resulting from one run still mounted on the silver seed holder and the baffle. They have used $1.5 \times 10 \times 10\ mm^3$ seeds taken from a Cz-grown crystal [98]. Centimeter-size pieces of Cz-grown BSO crystals were used as the nutrient. The liner was filled

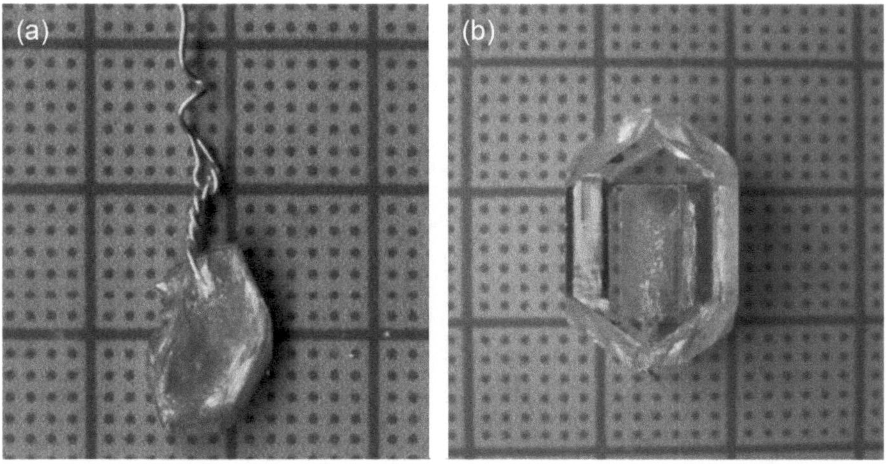

Figure 7.27 (a) Hydrothermally grown BSO (1 wt% SiO_2); (b) Hydrothermally grown BSO (3 wt% SiO_2).
Source: Photos courtesy: Changlong Zhang.

with 4 N NaOH solution and the external void was filled with a sodium-silicate pressure-balancing solution. The following were the experimental conditions:

Dissolution temperature: 390°C
Crystallization temperature: 385°C
ΔT: 5°C
Pressure: 6000–12,000 psi
Growth duration: 30–40 days
Growth rate: 0.1–0.3 mm/day.

Little or no growth occurred at NaOH normalities less than 4 N or nutrient zone temperatures less than 390°C. Excessive, unoriented growth on seeds and spontaneous nucleation on the walls of the liner occurred with normalities above 5°C. Experiments using Czechrolski-grown nutrient produced better crystals. Limited experimental data indicates that the preferred growth orientation is [100]. Most of the crystals were colorless in contrast to the straw color of Czechrolski crystals. Some were pale green, perhaps as a result of minor leaks in the platinum liner [133]. Figure 7.28 shows the absorption spectra of various BSO samples. The solid curves represent the absorption spectra taken at 10 K of a hydrothermally grown sample with those of the Cz-grown undoped and Al-doped samples. The deep-donor absorption shoulder is present in the curve for the undoped-Cz sample. The dashed curves represent the spectra taken on the three samples after they were exposed to intense monochromatic light (0.2 mw/cm^2) for 1 h at 10 K; 3 and 3.3 eV lights were used for the undoped and Al-doped Cz samples, respectively. The additional photochromic absorption in the undoped and Al-doped Cz-BSO samples is caused by the transfer of electrons to as yet unidentified traps. The hydrothermally grown BSO crystals do not produce any photochromic absorption.

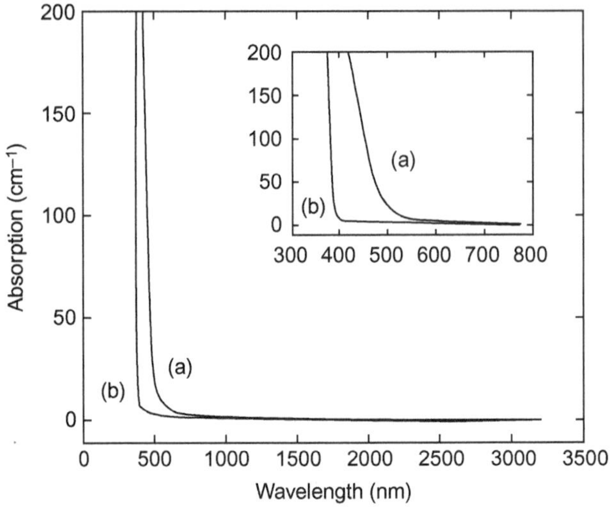

Figure 7.28 Absorption spectra of various BSO samples [134].

Exposing them at 10 K to 3.3–3.45 eV light causes readily visible blue luminescence, but no additional absorption bands are produced. Harris et al. [134] reported the absorption coefficients of BSO crystals doped with V, Cr, Mn, and so on. They conclude that the hydrothermal material shows special promise for optical "tailoring," since the native Czechrolski defect is absent and photoconductive sites and traps may be introduced by doping without interference from this defection. Leigh et al. [135] reported the characterization of Czechrolski and hydrothermally grown BSO crystals. They have observed significantly smaller optically and thermally induced conductivities in the hydrothermal material. A lower defect content also appears to inhibit the conduction processes for the hydrothermal material in terms of missing traps responsible for a hopping conduction process. Slightly higher concentrations of Fe and V were found in the hydrothermal material, but these impurities do not affect the color or contribute to a 500 nm absorption in this material. However, an equivalent concentration of these impurities in the Cz-grown material from a hydrothermal charge does affect both the absorption and the thermally stimulated current Thermally Stimulated Current (TSC) results. Thus, the near-intrinsic BSO can be prepared using hydrothermal growth techniques.

There are several other reports dealing with the study of phase formations in related systems giving compounds isostructural to BSO. For example, Surnina and Litvin [136] have studied the phase formation in the system Na_2O–Me_2O_3–Bi_2O_3–H_2O (Me = Al, Ga, In) under hydrothermal conditions. Abrahams et al. [137] obtained $Bi_{12}GeO_{20}$ (BGO) crystals, germanium sillenite, which exhibit similar properties. The other commonly used cations substituting for Si are Al, Ga, Fe, Ge, and Ti.

7.9.2 Beryl, $Be_3Al_2(SiO_3)_6$

Beryl is one of the most common minerals of hydrothermal origin. It occurs as well-developed crystals of large size. In fact, beryl is the largest crystal produced by nature and is commonly found in pegmatites. It is one of the most important gemstones. The presence of a little Cr^{+3} gives a green color and it is known as *emerald.* The bluish green variety is known as *aquamarine.* The greenish/yellow to iron-yellow and honey-yellow are called *golden beryl.* The rose-colored variety is called *morganite* or *vorobyevite.* The alkali elements like Na^+, Li^+, and Cs^+ are sometimes present replacing the beryllium from 0.25% to 5%.

Artificial growth of beryl crystals began in the nineteenth century. Hautefeuille and Perrey [138] were the first ones to synthesize beryl crystals. Espig [139] studied the crystal structure of beryl. Rogers and Sperisen [140] prepared beryl crystals by the hydrothermal technique. Wyart and Scavnicar [141] synthesized beryl crystals by the hydrothermal technique using $NaHCO_3$ solvents. Nassau [142] has grown large crystals of beryl by the hydrothermal method. Beryl, $Be_3Al_2Si_6O_{18}$, is found with a variety of impurity cations leading to its well-known beautiful colors and its favorable use as a gemstone. Probably the best work on hydrothermal growth of emerald was by Nacken. He made large numbers of synthetic emeralds, using a trace of chromium to produce the color. Hexagonal prisms weighing about 0.2 gm were grown in a few days. Nacken [142] and Hosaka [143] reviewed the hydrothermal growth of emerald. Johann Lechleitner of Innsbruck, Austria, released the first and not completely satisfactory product into the market during 1960 [144–146]. He used faceted beryl gemstones as the seed and grew a thin layer of hydrothermal emerald on the surface, which was subsequently given a light polish (not necessarily on all faces). Such stones were marketed for a short time under the names *Emerita* and *Symerald.* Since the green skin was quite thin, the color of these emeralds was not intense, and the skin could be readily viewed by immersion of the stone in a liquid. Lechleitner emerald can be synthesized under the following conditions:

Growth temperature: 300–400°C
Pressure: 1020 kg/cm^2
Solvent: a weak alkaline solution
Nutrient: silica (SiO_2), alumina (Al_2O_3), beryllium oxide (BeO).

The first completely synthetic hydrothermal emerald was put on the market by the Linde Division, Union Carbide Corp., in 1965. Figure 7.29 shows the emeralds grown hydrothermally by the Linde Co.; the size of the crystal is $50 \times 22 \times 6$ mm^3.

The first patent on beryl revealed that it could be grown in a neutral to alkali medium (pH of 7–12.5) using mineralizers such as alkali or ammonium halides (e.g., $NH_4F + NH_4OH$ or KF) and that Fe, Ni, or Nd could be used as dopants [147]. The second patent dealt with the analogous growth of beryl in an acid medium (pH of 0.2–4.5), using similar halide mineralizers but with an acid reaction (e.g., 8 N NH_4Cl) or with extra acid added [148]. Typical emerald growth conditions included pressures of 10,000–20,000 psi at temperatures of 500–600°C resulting from a 62% fill. Small temperature gradients of 10–25°C were employed. The Al was supplied from gibbsite $Al(OH)_3$, Be from $Be(OH)_2$, Si from crushed crystals of quartz, and Cr from $CrCl_3.6H_2O$. Growth rates as high as 1/3 mm/day could be attained. The pressure vessel was lined with gold.

Yancey [149] could attain a growth rate of 0.8 mm/day using the above experimental conditions. Figure 7.30 shows the schematic diagram for hydrothermal growth of emeralds [142]. In 1971, Vacuum Ventures Inc. bought the installation from Linde and supplied synthetic emeralds using the trade name "Regency emerald." Regency emerald has the same inclusions as Linde emerald, and growth conditions of Regency emerald are considered to be identical to those of Linde emerald [150]. Chlorine was detected in hydrothermally grown Regency and Linde emerald [151,152]. It seems that chlorine must have been derived either from

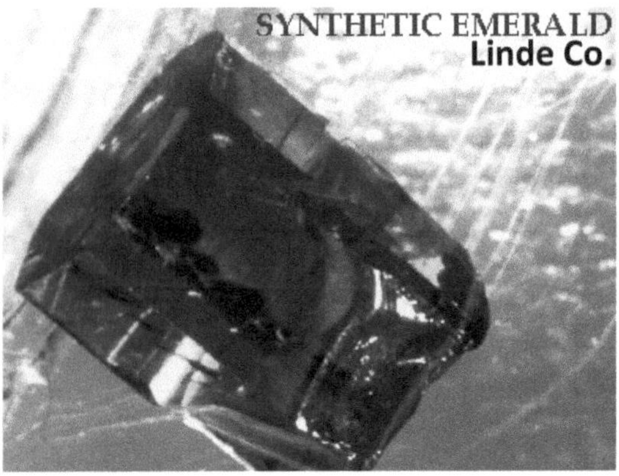

Figure 7.29 Emeralds grown hydrothermally by the Linde Co.; the size of the crystal is $50 \times 22 \times 6\ mm^3$ [143].

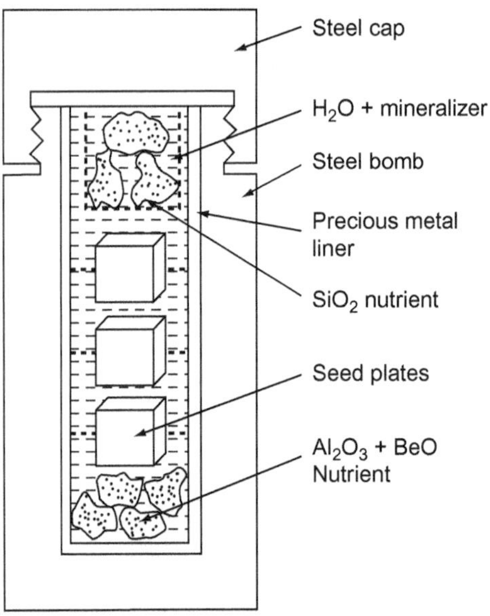

Figure 7.30 Schematic diagram for hydrothermal growth of emeralds. *Source*: After Nassau [142].

chromium chloride ($CrCl_3$) added as a coloring agent or from HCl added for increasing the acid content in the growth solution. Figure 7.31 shows the as-grown crystal of Regency emerald [150]. The crystal size is $14 \times 8 \times 6\ mm^3$. The seed is colorless but is green at the upper growth layer. The green emerald layer has a thickness three to four times that of the seed.

The study of Biron synthetic emerald began in Western Australia during 1977 [153]. In 1985, commercial stones were synthesized and sold through the Perth Lapidary Center under the trade name "Biron." Biron Mineral (Pvt) Ltd. is, at present, known as the Emerald Pool Mining Company (Pvt) Ltd [154]. The growth conditions remain unknown. Chlorine, chromium, and vanadium have been detected in these crystals [153–155]. IR absorption spectrometry has revealed the presence of H_2O. Microscopic examination has revealed the presence of gold crystals [155]. This may indicate that Biron emerald was hydrothermally synthesized in an autoclave lined with gold or gold alloy with chromium and vanadium used as the coloring agents. The crystals have a remarkably large growth sector and are transparent

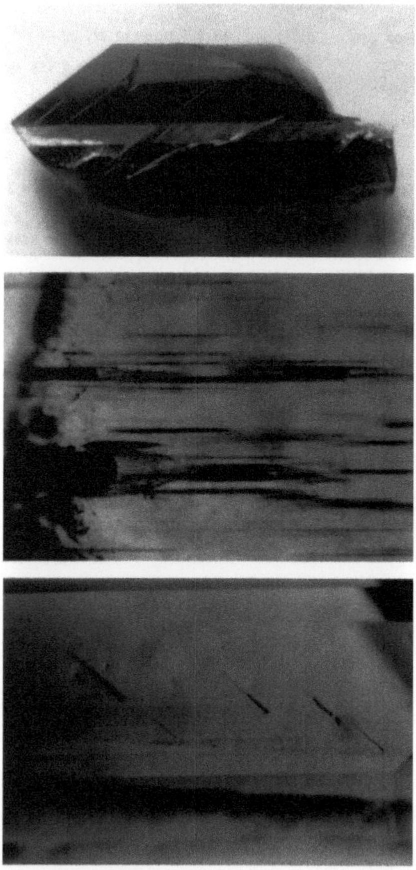

Figure 7.31 As-grown crystal of Regency emerald [150].

[155]. The characteristic inclusions are the particles of gold, grayish white "breadcrumb" inclusions, phenakite crystals, and rare fragments of the seed [156].

Hydrothermal synthesis of emerald in USSR began in 1965 [157]. At present, it has been improved to obtain commercially faceted emeralds [158]. The Russian method of hydrothermal growth of emerald has been reported in detail by Schmetzer [159]. To cite the growth conditions of the Russian synthetic emerald from his paper:

Growth temperature: 590–620°C
Temperature gradient: 45°C, 20–100°C, 70–130°C
Pressure: 816–1530 kg/cm^2
Solvent: acidic solution or flour-bearing solutions of complex components
Nutrient: natural beryl or oxides of Be, Al, and Si
Vessel: stainless steel (200–800 ml)
Seed: plate cut parallel to (5 5 10 6)

Under these growth conditions, a seed plate of 6 cm^2 in size increases its weight by 10 gm in about 20 days, and an emerald layer is 5–7 mm in this period of time. This corresponds to a growth rate of 0.2 – 0.3 mm/day. Cr^{3+}, Fe^{2+}, Fe^{3+}, Ni^{3+}, and Cu^{2+} have been detected through chemical analysis as color-associated transition elements. Fe^{2+} and Cu^{2+} ions were assigned to the tetrahedral sites, and Ni^{3+}, Cr^{3+}, and Fe^{3+} ions to the octahedral sites [159]. Evolution of ions from the inner wall of the autoclave is possible. It is, therefore, unknown what ions were added during the process of growth. The length of the crystal is 6.1 cm, width 1 cm, thickness 0.7 cm. The Russian synthetic emerald crystals grown for commercial aims commented on by Klyakhin are 10–12 cm in length, up to 2–2.5 cm in width and up to 2 cm in thickness of grown layers. Their total thickness varies within 0.8–1.4 cm.

The optical properties and physical features of commercial synthetic emeralds are presented by Sinkankas [160]. Table 7.15 lists the properties of various synthetic emeralds including the Russian hydrothermally grown emerald.

Kodaira et al. [161] obtained hexagonal prism crystals having a maximum size of 0.3 mm in length from 0.1 N NaOH solution at 750°C. Using 0.5 N or 1 N NaOH solution as solvents, beryl crystals were precipitated with phenacite at 500–600°C. Temperatures higher than 650°C resulted in phenacite crystals with a small amount of beryl. Figure 7.32 shows the hexagonal-prism-shaped crystals grown from 0.1 N NaOH solution.

What is most significant in the hydrothermal growth of beryl is that Al and Be hydroxides are highly reactive; the growth process is usually stopped when phenakite crystals start developing at the bottom of the vessel. The growth rate and morphology depend upon temperature, pressure, ΔT, size of the crushed quartz dopants, and so on. High growth rates produce too many inclusions.

As the sources of natural colored gemstones continue to shrink and the population growth of the world continues to increase, the size of the market for gem materials such as synthetic emerald is expected to increase (tempered, however, by occasional reverses in the world economy).

Table 7.15 Properties of Hydrothermally Grown Emeralds [143]

Refractive Index		Birefringence		Specific Gravity	Chelsea Filter	253.7 nm UV	365 nm
Natural (Santa Terezinha)	1.580	1.588	0.008	2.70	Inert to pink	None	None
Lechleitner emerald-coated natural beryl	1.586–1.597	1.580–1.587	0.006–0.010	n.d.	Red	Pale red to bright	Pale red to red
Fully synthetic beryl-emerald composite	1.567–1.573	1.562–1.567	0.005–0.006	2.67–2.69	Bright red	Bright red	Bright red
Linde (0.3–1.2% Cr)	1.567–1.573	1.562–1.567	0.005–0.006	2.67–2.69	Bright red	Bright red	Bright red
Regency	1.568	1.573	0.005	2.68	Bright red	Red	Bright red
Biron	1.577	1.570	0.007	2.72	Red to pink	Inert	Very weak dark green
	1.573	1.569	0.004–0.005	2.68–2.71	Strong pink red	Inert	Inter
Pool	1.574	1.569	0.005	2.75	Red	Inert	Inter
Soviet	1.583–1.586	1.577–1.579	0.006–0.007	2.70–2.71	None to dark orange yellow	None	None to partly dark red
	1.579–1.587	1.574–1.581	0.005–0.006	2.64–2.66	No reaction	Inert	Dull red

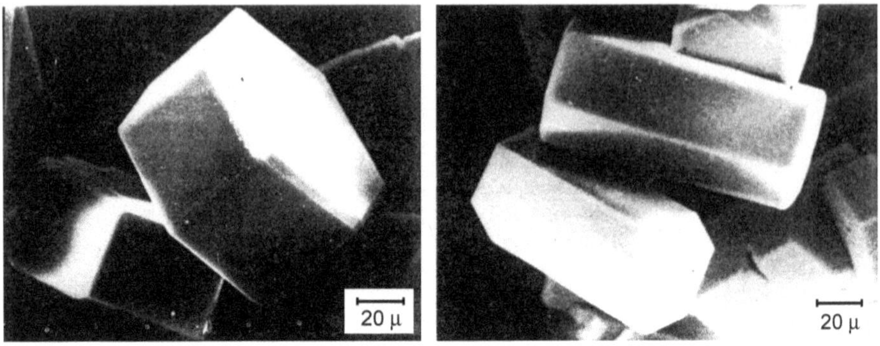

Figure 7.32 Hexagonal prism shaped crystals grown from 0.1 N NaOH solution [161].

There are several other beryllium-bearing silicates—like chkalovite–$Na_2BeSi_2O_6$, iron beryllosilicate–$Fe_3BeSi_3O_9(F,OH)_2$, aluminoberyllosilicate–$Na_3Al \cdot BeSi_2O_8$, barylite-$BaBe_2Si_2O_7$ and so on—have been synthesized as single crystals using the hydrothermal method [162–165]. Sobolev et al. [166] prepared chkalovite and its germainum analog by hydrothermal techniques from the system $BeO–MeO_2–NaOH–H_2O$ (where Me = Si, Ge). Beus and Dikov [167] have studied its field of stability in hydrothermal solutions containing NaOH, Na_2CO_3, NaF, and HF, and prepared from beryllium hydroxide and nitrate with amorphous SiO_2 in hydrothermal NaOH and NaCl solutions. The experimental temperature was 400–600°C, $\Delta T = 40–60°C$, $P = \sim 700$ atm and the experimental duration was 5 days. It is possible that in natural hydrothermal process, where chkalovite is very rarely encountered, undergoes further changes owing to its chemical instability, forming other more stable beryllium compounds. Most of the beryllium-bearing compounds exhibit piezoelectric properties, especially the Na^+- and Bu^{+2}-bearing beryllium silicates. For example, barylite crystals cannot be synthesized by the ordinary solid-state reaction. Maeda et al. [165] obtained this crystal under the following conditions:

P: 500 bar
T: 385°C
Δ*T*: 15°C
Solvent: 3 M KOH
Baffle opening: 15%
Seed: $BaBe_2Si_2O_7$ ceramics.

Likewise, there are several such reports on the hydrothermal synthesis of beryllium-bearing silicates and most of them exhibit interesting electronic properties.

7.9.3 Tourmaline

Tourmaline is a complex silicate of boron and aluminum with either magnesium, iron, or the alkali metals prominent. The general formula can be written as

$XY \cdot Al_6 \cdot [Be_6Si_6O_{27}(OH,F)_4]$, where X = Na, Ca, K; Y = Mg, Fe, Mn, Li, Al, and so on. Tourmaline crystals exhibit pyroelectric properties and the importance of the seeded growth of tourmaline was realized during the 1940s. The basic interest lies in the colorless variety of tourmaline for pyroelectric applications. The earliest attempts to obtain tourmaline single crystals were carried out through proper selection of nutrient and the solvent [168,169]. Robbins et al. [170] have carried out the growth of tourmaline crystals using metastable glass of tourmaline composition and the given pressure–temperature diagrams of stability. Voskresenskaya [171], and Voskresenskaya and Barsukova [172] have carried out the synthesis of colorless tourmaline on the seeds under hydrothermal conditions. The above authors have carried out experiments in autoclaves with a volume of 200 cm^3 at temperatures 550–600°C and pressure 2000 atm, using platinum liners. A mixture of oxides or glass of tourmaline composition was used as the nutrient. Natural tourmaline was used as seed crystals. The growth was carried out in the acidic media from aqueous solutions H_3BO_4–NaCl (NaF). The concentration of boric acid was 5–10 wt%, NaCl ~ 2.5 wt%. The seed orientation was along the (0001) face, polar in character, and crystals usually developed simple forms (1011), (0221), and (0112). However, in recent years, the hydrothermal growth of tourmaline is mainly of academic interest and the focus is on the solid solutions of tourmaline. Besides, there are some reports related to the gemological potential of the synthetic tourmaline.

7.9.4 Nepheline

Nepheline is an important mineral and has the orthosilicate formula $(Na,K)AlSiO_4$, with a framework of aluminosilicate having acentric space group $P6_3$. Nepheline is of interest as piezo- and pyroelectric material. The nepheline structure is related to "tridymite intercalation stones," which allows it to vary the Na/K ratio in a wide interval. In nature, nepheline usually occurs as small glassy crystals or grains. The natural nephelite always contains silica in excess and also small amounts of potassium. Nepheline can be easily prepared artificially by fusing its constituents together in the proper proportions. The earliest experiments on the artificial synthesis began in the nineteenth century. Friedel [173,174] treated muscovite with NaOH or KOH solution and water at 200–500°C and obtained nephelite. The end members of this row are characterized by several polymorphs, and the nepheline structural type is characteristic for Na/K ratios from 1 to 0.5.

Nepheline, $NaAlSiO_4$, is stable in NaOH solutions with concentrations 3–15 wt% at temperatures more than 450°C. Increase in temperature results in the expansion of the stability field to 20–22 wt% at 500°C. At 400°C $< T <$ 450°C, nepheline-hydrate-I and nepheline-hydrate-II form from artificial cancrinate and sodalite. However, under the same conditions, the nepheline phase is in equilibrium with solution if synthetic nepheline is used as the starting material. The boundary nepheline hydrate, cancrinite, seems to be metastable. Nepheline (both synthetic and natural) is replaced by cancrinate at 200–300°C and NaOH concentrations greater than 3 wt% [175,176].

Table 7.16 Chemical Composition of Natural and Artificial Nepheline

Compound	Oxide Contents (wt%)							
	SiO_2	Al_2O_3	Fe_2O_3	FeO	MgO	CaO	Na_2O	K_2O
Nepheline natural (starting material)	43.63	32.37	0.41	0.17	0.76	0.36	15.94	6.0
Nepheline artificial (grown layer)	41.55	34.47	0.74	–	0.10	–	16.77	5.98

Nepheline single crystals can be grown in the field of its thermodynamic stability as in the field of its instability at 300°C $<T<$ 400°C if the growth process is initiated by nepheline seed. In the region of instability, single crystals of nepheline grow on the seed but hydrocancrinate, $Na_8Al_6Si_6O_{24}(OH)_2 \cdot nH_2O$, crystals grow on spontaneous nuclei. At $T>$ 450°C, nepheline and cancrinate grow simultaneously if there are seeds of both nepheline and cancrinate.

Single crystals of (Na,K)$AlSiO_4$ were grown in the system Na_2O–K_2O–Al_2O_3–SiO_2–H_2O using natural nepheline as a nutrient and as a material for seeds. To minimize the mismatch between the seed and the grown layer, the potassium was introduced into the starting NaOH solution. As a result, the composition of the grown layer was close to the composition of seed plates (Table 7.16).

The unit cell parameters are equal to $a=12.64$, $c=5.16$ Å (seed material), $a=12.72$, $c=5.12$ Å (grown layer). At temperatures of 300–350°C in the growth zone, the regeneration of seed plates takes place with the formation of the following simple forms: hexagonal prisms (1010), monohedras (0001) and (0001) and sometimes hexagonal pyramids (1011), (1012). The polar symmetry is observed rarely, and the external habit usually corresponds to the point group 6/m or 6/mmm. In this case, the growth morphology and growth rates of (0001) and (0001) faces are the same, which may be caused also by twinning. A specific cellular microstructure is observed in the transmission light; the cell size is 1.5–3 mm.

The mean values of the growth rates for different crystallographic directions are:

γ(1010): 0.04 mm/day
γ(0001): 0.12 mm/day
γ(1011): 0.4 mm/day
$\gamma(1011) > \gamma(1210) > \gamma(0001) \geq \gamma(0001) > \gamma(1010)$.

The faces (0001) and (10$\bar{1}$0) grow on spiral mechanisms, the faces of pyramid grow on normal mechanisms. Nepheline crystals exhibit interesting unidimensional ionic conductivity. They show anamolously high dielectric constant parallels to the c-axis [177].

7.9.5 Zincosilicates

The earliest work on the hydrothermal synthesis of zincosilicates appeared during 1974 [178]. The study of zincosilicates carries a special significance with reference to their crystal chemistry and phase stability, as they occur in nature under varying geological conditions. Zincosilicates phase formation in the system Na_2O–ZnO–SiO_2–H_2O has been studied in great detail by several workers [178–181]. The following zincosilicates have been identified in this system: $Zn_2[SiO_4]$-willemite, $Na_2Zn[Si_2O_6]$, $NaZn[SiO_4]$, $Na_2Zn[Si_3O_8]$, $Na_2Zn_3(SiO_4)_2$, and $Na_2Zn_2[Si_2O_7]$. Figure 7.33 shows the NC diagram of fields of crystallization of various zincosilicate phases. Figure 7.34 shows the schematic diagram of various zincosilicate phase formations. Litvin [175] proposed the following univariant equilibrium:

$$Zn_2SiO_4 + 2Na_2O + 3SiO_2 \leftrightarrow 2Na_2Zn\ Si_2O_6$$

$$3Zn_2SiO_4 + 2Na_2O + SiO_2 \leftrightarrow 2Na_2Zn_3Si_2O_8$$

$$Zn_2SiO_4 + 2Na_2O + 5SiO_2 \leftrightarrow 2Na_2ZnSi_3O_8$$

$$Zn_2SiO_4 + Na_2O + SiO_2 \leftrightarrow Na_2Zn_2Si_2O_7$$

$$2Na_3Zn_3Si_2O_8 + Na_2O + 2SiO_2 \leftrightarrow 3NaZn_2Si_20_7$$

$$Na_2Zn_2Si_2O_7 + Na_2O + 2SiO_2 \leftrightarrow 2Na_2ZnSi_2O_6$$

$$Na_2ZnSi_2O_6 \leftrightarrow Na_2ZnSiO_4 + SiO_2$$

$$Na_2ZnSi_3O_8 \leftrightarrow Na_2ZnSi_2O_6 + SiO_2$$

$$Na_2Zn_3Si_2O_8 + 2Na_2O + SiO_2 \leftrightarrow 3Na_2ZnSiO_4$$

$$3Na_2ZnSiO_4 \leftrightarrow Na_2Zn_3Si_2O_8 + SiO_2 + 2Na_2O$$

Zn_2SiO_4 (willemite): Willemite crystallizes in pure water, in NaOH solution, and in mixed solvents of $NaOH + Na_2CO_3$. The optimum conditions for spontaneous recrystallization of willemite are NaOH = 4–6 wt%; $T = 375$–$450°C$; $\Delta T = 5$–$8°C$, $P = 400$–600 atm. Single crystals of willemite grow in the form of a hexagonal prism. The fast growing face is the rhombohedral face (1120), the growth rate of which is 0.8–1.0 mm/day. When the concentration of NaOH increases beyond 10 wt%, the solubility of willemite increases and the formation of other phases begins. Also, when the NaOH concentration is 10 wt%, the boundaries of monomineralic synthesis move up to the ratio $ZnO/SiO_2 = 1{:}2.5$. Further increase in the concentration of the solvent increases the field of quartz. Willemite doped with active ions like Mn^{2+} exhibits excellent luminescence properties when exposed to UV radiation, and it is an excellent phosphor.

The field of crystallization of $Na_2ZnSi_3O_8$ falls in the silica-rich region (Figure 7.35). $Na_2ZnSi_3O_8$ crystallizes usually together with willemite, with an increase in the NaOH concentration, especially silica. $Na_2ZnSi_3O_8$ starts melting congruently and recrystallizes in the upper part of the autoclave. The presence of Na_2CO_3 helps crystals and increases the crystal size. The crystals of $Na_2ZnSi_3O_8$

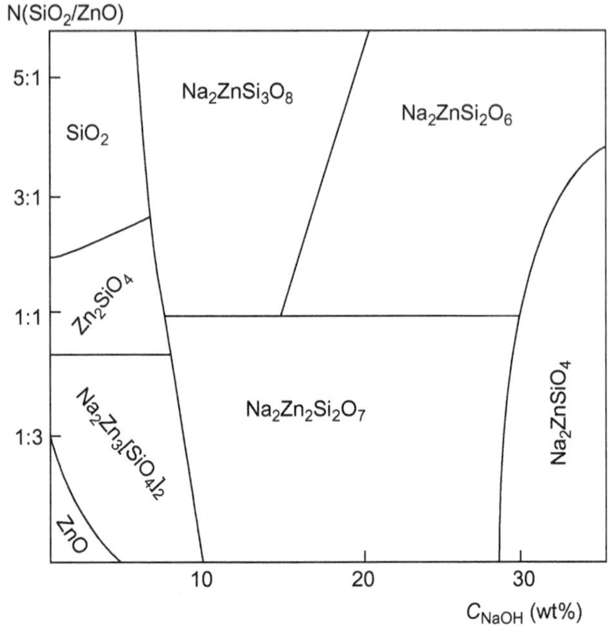

Figure 7.33 NC diagram of fields of crystallization of various zincosilicate phases [175].

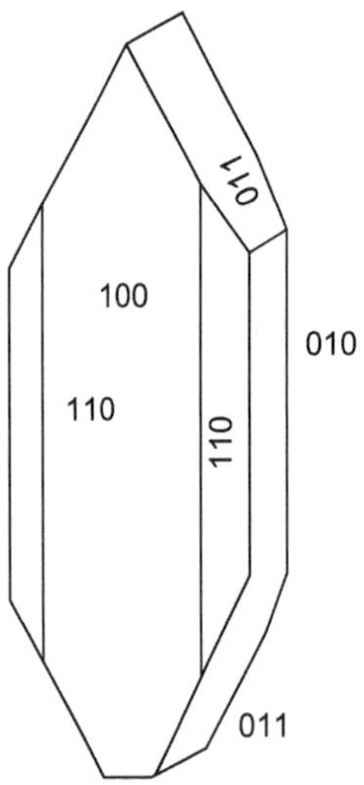

Figure 7.34 Schematic diagram of various zincosilicate phase formation [175].

could be obtained at temperatures 325–400°C, $\Delta T = 8-12°C$, and $P = 400-600$ atm. Dipyramidal crystals with [100] and [110] faces grow under these conditions with a growth rate of 0.08 mm/day. The crystals exhibit piezoelectric effect like quartz. The crystals are colorless, transparent. The crystals doped with Mn^{2+} show good luminescence.

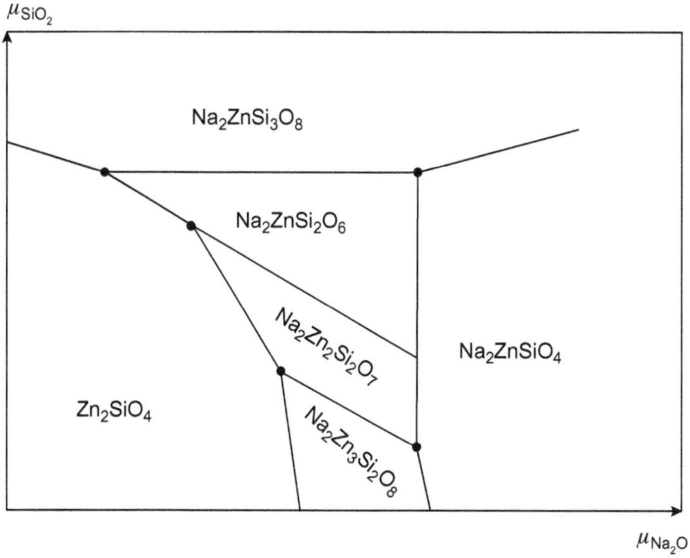

Figure 7.35 Field of crystallization of $Na_2ZnSi_3O_8$ [182].

$Na_2Zn_2Si_2O_7$: This crystallizes in 10–25% NaOH solutions at $ZnO/SiO_2 > 1{:}1$. If the concentration of ZnO increases to 7:1, this phase becomes unstable and transforms to $Na_2Zn_3[SiO_4]_2$. The crystals are thin plates—like at $ZnO/SiO_2 = 2{:}1$ to 1:1. With an increase in the ZnO in the nutrient, the crystals become thicker. The commonly seen faces are (100), (010), (210), (110), (101), and (111). Crystals are colorless, transparent, with glassy surfaces. The crystallization of $Na_2Zn_2Si_2O_7$ takes place in 15–18 wt% NaOH solution at $T = 450°C$, $\Delta T = 4–6°C$, $P = 400–600$ atm. The growth rate is 0.3–0.5 mm/day along the direction [101] [183]. The growth rate of the (001) face increases and becomes equal to the growth rate along other faces. These crystals show luminescence and strong piezoelectric effect. Heating above 1090°C, $Na_2Zn_2Si_2O_7$ starts decomposing to ZnO and melt, and at 1325°C, it completely melts (Figure 7.36) [182].

$Na_2Zn[SiO_4]$: Single crystals of sodium orthozincosilicate have been obtained in highly concentrated NaOH solution (30% NaOH) and it does not depend upon the ratio of the starting components. $Na_2Zn[SiO_4]$ dissolves congruently. The schematic diagram of the crystal morphology is shown in Figure 7.37. The commonly observed forms are (100), (001), (110), and (011). Twinning is common. Crystals are colorless, Mn^{2+}-rich crystals are green, Cu^{2+}-rich crystals are blue. The Mn^{2+} doped $Na_2Zn[SiO_4]$ crystals show luminescence and also significant piezoelectric effect. At $T = 1060 \pm 10°C$, $NaZn[SiO_4]$ undergoes phase transition.

Synthetic sodium zincosilicates are a new group of crystallophosphor materials. Sodium zincosilicates doped with Mn^{2+} exhibit stronger luminescence. Also, $Na_2Zn_2[Si_2O_7]$, $Na_2Zn[Si_3O_8]$, and $Na_2Zn[SiO_4]$ show stronger piezoelectric effect than quartz. However, the basic problem is the growth of large-size single crystals.

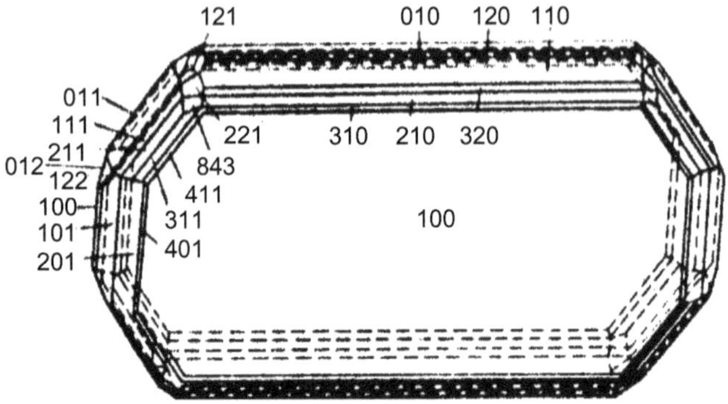

Figure 7.36 Schematic diagram of the crystal morphology [182].

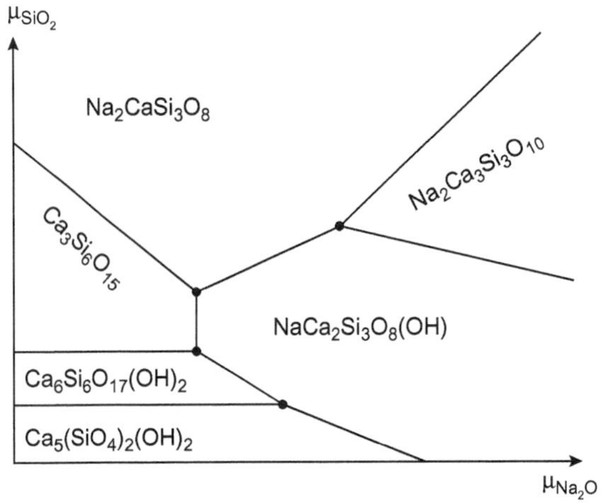

Figure 7.37 NC diagram of crystallization in the system $Na_2O{-}CaO{-}SiO_2{-}H_2O$ [182].

Several other compounds crystallize from the system $Na_2O{-}MO{-}SiO_2{-}H_2O$ (where M = Cd, Ca, Ba, and Mn). Among them, the Cd- and Mn-bearing silicates show interesting physical properties [175,182,184–186]. Similarly, the Ca-bearing silicates carry great geological significance and such systems are being studied extensively [187–190].

Several phases crystallize in the $Na_2O{-}CaO{-}SiO_2{-}H_2O$ under hydrothermal conditions. Figure 7.37 shows the NC diagram of crystallization in the system $Na_2O{-}CaO{-}SiO_2{-}H_2O$. Following are the important phases crystallizing in this system: $Ca[SiO_3]$, $Ca_6[Si_6O_{17}](OH)_2$ (xonolite), $Ca_5[SiO_4]_2 \times (OH)_2$ (khondrodite), $Ca_6[Si_4O_{14}]$ (kilkhoanite), $Ca_8[Si_5O_{18}]$ (orthorolite), $Ca_6[SiO_4]$ $[Si_2O_7](OH)_2$ (dellaite), and so on.

The studies of the crystallization of monocationic calcium silicates reveal that they are formed in at least two ways:

i. Interaction of soluble silica with calcium compound (through neutralization reaction).
ii. Interaction of aqueous solutions of silica with amorphous or crystalline portlandite.

The ratio of SiO_2/CaO (degree of silification) regularly increases 0.5→0.75→1.0. [Si–O] radical leads to the transition from discrete ortho, diortho group to infinite chains and then infinite ribbon $[Si_6O_{17}]_\infty$. In most of the experiments, quartz also crystallizes together with xonotlite.

In the second case, the degree of silification increases insignificantly (0.33→0.4→0.62→0.66) and morphotropic transitions in [Si–O] radical takes place as follows: $[SiO_4]\rightarrow[SiO_4]_2\ [Si_3O_{10}]\rightarrow[SiO_4][Si_3O_{10}]$.

Nd^{3+}-doped $Ca_2Al_2SiO_7$ (gehlenite) is an efficient new laser material [191], since its structure exhibits some disorder and can accept active rare earth ions. The unit cell is tetragonal (melilite type) with special group $P42_{lm}$. Figure 7.38 shows the Na^{3+} emission spectrum corresponding to the $^4F_{3/2}\rightarrow{}^4I_{11/2}$ transition (a) in gehlenite and (b) in YAG.

In the system Na_2O–BaO–SiO_2–H_2O under hydrothermal conditions below 350°C, two phases crystallize: $BaSiO_3 \cdot nH_2O$ [192] and $BaSi_2O_5 \cdot 3H_2O$ [193], in the $BaCl_2 + Na_2H_2SiO_4$ solution, the low temperature of growth and the nature of the solution do not form silicates when $BaO/SiO_2 > 1{:}1$. At temperatures 450–550°C, pressures 400–1200 atm, and BaO/SiO_2 from 7.5:1 to 1:10, the following phases form: $BaSiO_3$, $Ba_2Si_3O_8$, $BaSi_2O_5$ (sanbornite) and $Ba_5Si_4O_{12} \cdot (OH)_2$, $NaBa_3Si_2O_7 \cdot (OH)$ [194]. Figure 7.39 shows NC diagram of crystallization in the system Na_2O–BaO–SiO_2–H_2O, and Figure 7.40 shows the schematic diagram of the crystal morphology of the barium silicates. From the NC diagram, the boundaries of crystallization of $BaSi_2O_5$ and $BaSi_3O_8$ are controlled only by NaOH concentration. Further, the boundaries between monocationic barium silicate and

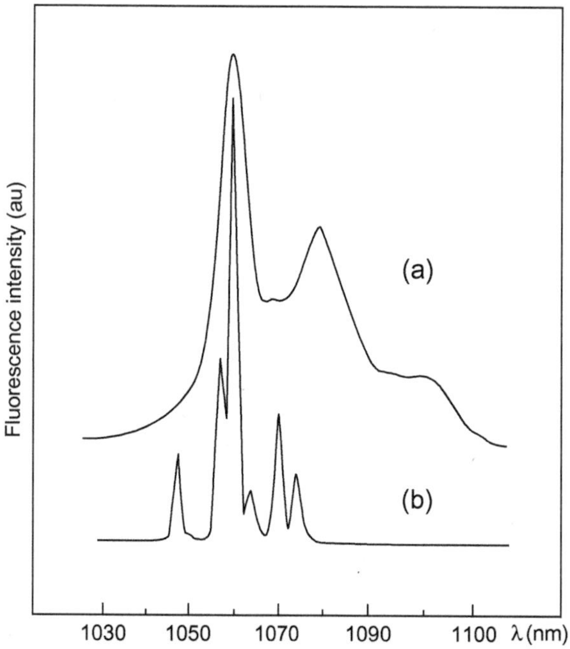

Figure 7.38 Na^{3+} emission spectrum corresponding to the $^4F_{3/2}\ {}^4I_{11/2}$ transition in (a) gehlenite and in (b) YAG [191].

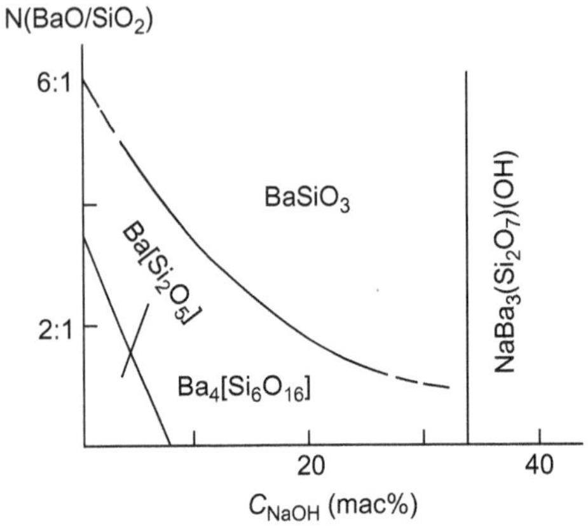

Figure 7.39 NC diagram of crystallization in the system Na_2O–BaO–SiO_2–H_2O [194].

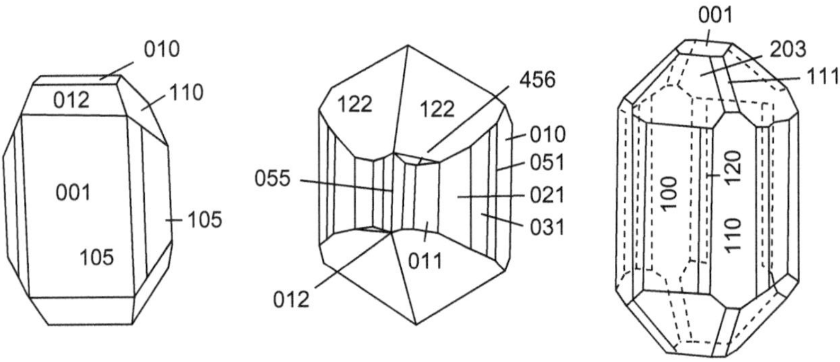

Figure 7.40 Schematic diagram of the crystal morphology of the barium silicates [182].

quartz depends only on the concentration of $Ba(OH)_2(M_{Ba}O)$. All the transitions to $NaBa_3Si_2O_7(OH)$ are controlled mainly by the concentration of NaOH. Mn^{2+}-doped $BaSi_2O_5$ shows luminescence.

The other important system Na_2O–MnO–SiO_2–H_2O gives the following phases under hydrothermal conditions: Mn_7SiO_{12} (braunite), $Mn(OH)_2$ (pyrochroite), Mn_2SiO_4 (tephroite), $Na_2Mn_7Si_8O_{24}$, $Na_2Mn_6Si_7O_{21}$, $Na_2Mn_3Si_4O_{12}$, $Na_2Mn_2Si_2O_7$, Na_2MnSiO_4, $Na_6Mn_3Si_6O_{18}$, $Na_2MnSi_8O_{18}$, $Na_2Mn_7(SiO_3)_8$, $Na_2Mn_8O_{16} \cdot nH_2O$. Several of them occur in nature as minerals. Figure 7.41 shows the NC diagram of crystallization of the system Na_2O–MnO–SiO_2–H_2O [195]. Most of these phases belong to the lower symmetry classes (monoclinic and orthorhombic) and some of them are excellent luminescent materials. Further, these compounds crystallize at $T > 400°C$ and $MnO/SiO_2 > 1$. These compounds can be obtained using alkaline solutions, NaOH (<20%). The major difficulties concerning this system are associated with the change in the valency of Mn^{2-} to Mn^{4+} [196].

A majority of these sodium manganese silicates—$Mn_2[SiO_4]$, $Na_2Mn_2[Si_2O_7]$, $Na_2Mn(SiO_4)$, $Na_2Mn_7(SiO_3)_8$, $Na_6Mn_3(Si_6O_{18})$—show useful physical properties.

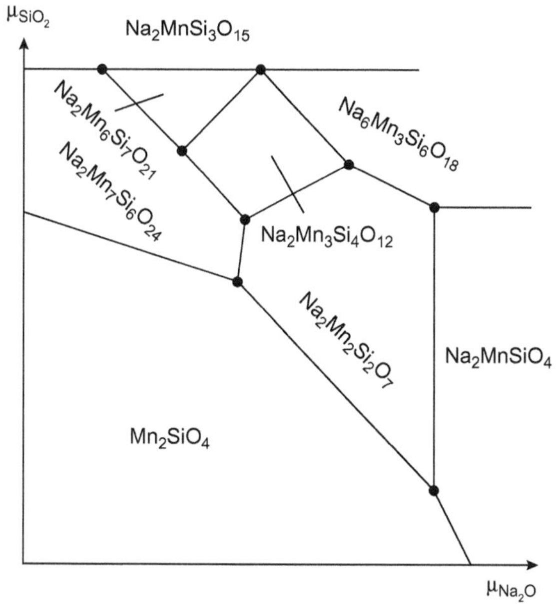

Figure 7.41 NC diagram of crystallization of the system $Na_2O-MnO-SiO_2-H_2O$ [195].

Some of them have been obtained as single crystals using spontaneously grown crystals as seeds [197].

7.10 Hydrothermal Growth of Lithium Silicates

Lithium aluminum silicate systems are being studied extensively because of their ultra-low thermal expansion and their geological importance. Almost all the natural lithium aluminosilicates have been synthesized, together with a number of artificial lithium compounds with no natural analogs. However, the only exception is α-spodumene, which has never been synthesized with an adequate degree of reproducibility, in spite of the high pressure–temperature conditions of synthesis ($T = 600°C$, $P = 4000$ atm) [198,199]. In the $Li_2O-Al_2O_3-SiO_2-H_2O$ system, in temperature range 400–700°C and at pressures of up to 9000 atm, α-eucryptite, β-spodumene, petalite, lithium silicate, and α-spodumene have been synthesized over a wide range of pH [200]. Instead of α-spodumene, others—like eucryptite, β-spodumene, petalite, and a number of artificial lithium compounds—have been obtained.

The results of the foregoing investigations lead to the conclusions that (a) spodumene is formed at pressures of over 4000 atm as indicated, in particular, by the fact that all the phases synthesized have lower densities than that of α-spodumene, and (b) isomorphic impurities reduce the minimum pressure required for the formation of α-spodumene.

Using pure LiCl solutions at 400–700°C and pressures of 750–3000 atm, we may synthesize α-eucryptite $LiAlSiO_4$, β-spodumene β-$LiAlSi_2O_6$, and petalite $LiAlSi_4O_{10}$. The α-eucryptite is mainly formed for high Al_2O_3/SiO_2 ratios in the original charge (~1:1), β-spodumene for smaller ratios (1:2–1:3), and petalite for a high concentration of silica (ratio A/S ~ 1:8). In the neighborhood of the ratio Al_2O_3/SiO_2 ~ 1:5, we observe paragenesis of β-spodumene and quartz.

The concentration of LiCl solutions, and the temperature and pressure within the limits indicated, have little effect on the phase formation in the LiCl–Al_2O_3–SiO_2–H_2O system.

The pH of the solution, together with the Al_2O_3/SiO_2 ratio, constitutes the second most important factor in the crystallization. Reducing the alkalinity of the solution (in LiCl + HCl solutions) has practically no effect on the phase formation, but on increasing the alkalinity (in LiCl + LiOH or LiOH solutions) new compounds appear and the ranges of crystallization of the phases already noted alter. First of all, lithium silicate Li_2SiO_3 appears. For a ratio of Al_2O_3/SiO_2 = 1:1 lithium silicate is synthesized for high values of the alkalinity, but with increasing concentration of SiO_2 in the original mixture the range of its stability moves in the low pH direction (Figure 7.42). In alkaline solutions, the field of formation of petalite also widens, and the latter is synthesized for higher Al_2O_3/SiO_2 ratios (~1:2 for pH > 10) than in LiCl solutions. The range of existence of β-spodumene, on the other hand, contracts slightly on increasing the alkalinity (Figure 7.42), and for pH > 10 it becomes unstable, being replaced by α-eucryptite and lithium silicate.

The densest of all phases synthesized is α-spodumene, which clearly indicates that the stability range lies at much higher pressures. In other words, at pressures of under 4000 atm and in the absence of impurities, no spodumene is formed. The commonly used impurities are Cr^{3+} and Fe^{3+}. The α-spodumene is synthesized in LiCl solutions at pressures over approximately 6000 atm. At lower pressures, it

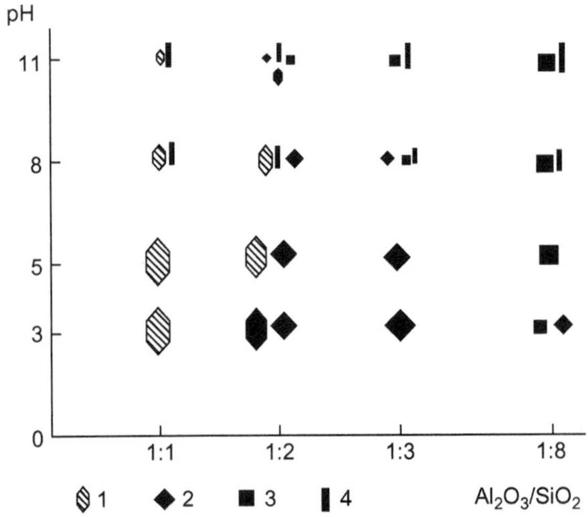

Figure 7.42 Fields of stability of lithium silicates in relation to the pH of the solution and the ratio of the original charge components at T = 600–700° and P = 2000–3000 atm. (1) α-$LiAlSiO_4$; (2) β-$LiAlSi_2O_6$; (3) $LiAlSi_4O_{10}$; (4) Li_2SiO_3 [200].

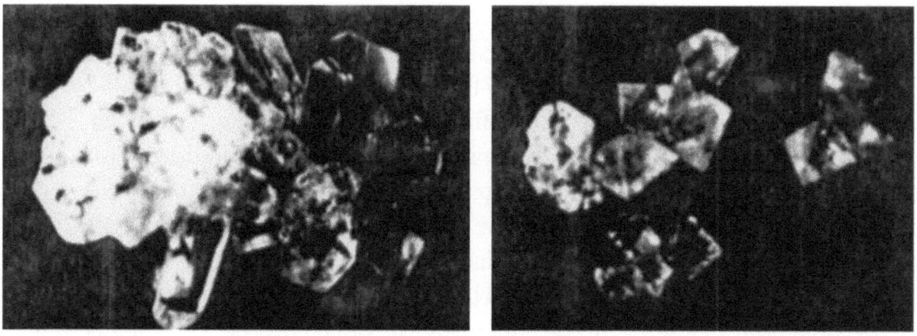

Figure 7.43 Characteristic photographs of α-eucryptite and β-spodumene crystals [200].

decomposes into α-eucryptite and β-spodumene. The partial replacement of Al^{+3} by Cr^{+3} reduces the lower limit of pressures for the formation of spodumene; spodumene containing traces of Cr^{3+} grows on the seed at 3000 atm. A complete replacement of Al_2O_3 by Fe_2O_3 or Cr_2O_3 leads to the synthesis of iron and chromium spodumene in the low-pressure range: approximately 250 and 750 atm, respectively. Figure 7.43 shows the characteristic photographs of α-eucryptite and β-spodumene crystals [200].

Ghobarkar [201] has studied the hydrothermal synthesis of the petalite mineral phases and their morphological changes with temperature in great detail. Here, petalite minerals were synthesized hydrothermally at different temperatures from a glass with the composition Li_2CO_3:Al_2O_3:SiO_2 = 1:1:8. It was found that the development of the mineral phases and the volume and habit of crystals depended on the rise of temperature from 410°C to 490°C, α-petalite crystallized in the monoclinic system with a constant change in the habit. Alpha-petalite splits into quartz and an orthorhombic β_1-petalite from 490°C onward. The tetragonal phase of petalite forms from 560°C to 750°C with changes in the habit.

Thus, the hydrothermal synthesis of lithium silicates is a challenging field in hydrothermal research area. Recently, researchers are more inclined toward the synthesis of the lithium silicate family of materials using hydrothermal method owing to their gemological interest.

7.11 Hydrothermal Growth of Germanates

Germanates form an important group of inorganic compounds having a close proximity to the silicates in many respects. The ionic radii of both Si^{4+} (0.39Å) and Ge^{4+} (0.44 Å) are closer to one another. Researchers are synthesizing the solid solutions of silicates and germinate in order to enhance the specific properties for various applications, e.g., $Si_{1-x}Ge_xO_2$, to enhance the piezoelectric properties. The natural germanates are very few and, hence, their synthetic analogs permit to understand their structures, crystal chemistry, physicochemical conditions, etc. In nature, germanium usually occurs in the form of isomorphous admixtures in some relatively rare minerals (samarskite, tantalite, gadolinite, germantite, and so on). Individual germanium minerals or germanates are rather rare or very few, especially in some sulfides and sulfates [202–204].

Germanates could be obtained by both flux and hydrothermal methods. However, just like silicates, even germanates insists hydrothermal conditions for the formation of a large variety of anionic groups. The flux-grown germanates are very few and are represented by $[GeO_4]$ and $[Ge_2O_7]$ anionic groups, whereas the hydrothermally grown germanates do contain a large variety of anionic groups like $[GeO_4]$, $[Ge_2O_7]$, $[Ge_4O_{13}]$, $[Ge_3O_{12}]$, $[Ge_{10}O_{25}]$, $[Ge_3O_{12}]$, and $[Ge_5O_{16}]$. These germanates with different cations can be obtained at rather lower parameters ($T = 500°C$, $P = 1500–2000$ atm). Under such conditions, germanates can occupy both tetrahedral and octahedral positions in the structure. The fixation of Si in octahedral coordination demands much higher pressure (>10 kbar). This fact reveals an important aspect of the study of germanates crystallization, because the germanates obtained under lower PT conditions may be the analogs of silicates formed in the earth crust at greater depths.

The growth of germanates began as an attempt to synthesize silicates analogs during the middle of this century. Onishi [202] and Wardani [203] studied the geochemistry of germanium in detail. Just like in silicates, even in germanates, we can have two types: (1) Germanates containing rare earth elements and (2) germanates without rare earth elements. The phase formation and crystal chemical elucidation of the germanates have been discussed in general here.

The natural compounds of germanium and rare earth elements are practically absent and only their synthesis allows one to study their structures, crystal chemistry, and physical and chemical characteristics in detail. Under such conditions, germanium can occupy both tetrahedral and octahedral positions in the structure. The fixation of Si in octahedra (coordination number 6) demands much higher pressure (>10 kbar). This fact reveals an important aspect of the study of rare earth germanate crystallization, because the germanates obtained under lower temperatures and pressures may be the analogs of silicate phases formed in the earth crust at large depth (high-pressures phases). The classification of germanates containing rare earth elements differ slightly from that of the silicates containing rare earth elements.

For germanium, like silicon, a tendency to form chains has been noticed. For silicon, both hydrides and halogenides have been obtained, but germanium forms only halogenides with the bond Ge–Ge.

GeO_2 is largely analogous to SiO_2, and it may form glass while cooling from the melt. One of the polymorphic modifications of GeO_2 at 1033°C has a quartz structure. This GeO_2 modification becomes metastable at less than 1053°C. The stable modification of GeO_2 has the characteristic rutile, TiO_2-type structure.

Germanium may exist in solutions as a negatively charged ion and forming sulfides and tiosalts differs from silicon, but at the same time germanates and fluorogermanates are often isostructural with the corresponding silicates. Among rare earth germanates, we have alkali-free germanates and alkali rare earth germanates.

7.11.1 Rare Earth Germanates

There are some known representatives of rare earth germanates: $Ln_{4.67}[GeO_4]_3O$, $Ln_4[GeO_4]O_4$, $Ln_{13}[GeO_4]_6$ $[O, (OH, F)_2]_6$, $Ln_2Ge_2O_7$ (two structure types), Ln_2GeO_5 (three structure types) [44]. The experiments in the system Ln_2O_3–GeO_2–R–H_2O resulted in obtaining new types of germanates (Table 7.17). Two of them correspond to the silicates structures, six germanates belong to new structure types. The crystallization of germanates was carried out at rather high AOH (AF) concentration, but alkali cations took part only in intermediate stages of the synthesis processes (solution, transport) as mineralizers. In the final composition of rare earth germanates, the alkali cations were absent. The common features of all germanates obtained was the prevalence of rare earth component in comparison with germanium, $1 \leq Ln/Ge \leq 4$ (Table 7.17). For most of the new structure types, the presence of morphotropic rows is characteristic (Table 7.18). The compound $Ln_4Ge_3O_9(OH)_6$ has been found in the systems with 10 rare earth elements. The structural type $Ln_3GeO_5(OH)_3$ is characteristic for eight rare earth elements. Structure formation of most rare earth germanates demands some additional anionic groups (OH, F), which usually enter into the coordination polyhedra of rare earth cation. Crystal chemical analysis of new rare earth germanates revealed some interesting features of their structures.

The phase formation in the system $M^+{}_2O$–R_2O_3–GeO_2–H_2O (where M = Li, Na, K, Rb, Cs; R = rare earth elements) has been studied extensively. Like in silicate systems, even in the germanate system, the phase formation is determined by the type of rare earth cation and the ratio of R_2O_3/GeO_2 in the nutrient and the concentration of the solvent. In the system Na_2O–R_2O_3–GeO_2–H_2O, the following phases have been identified: La and Pr form $Na_2RGeO_4(OH)$; Nd and Sm form $NaR_3[GeO_4]_2(OH)_2$, and the rest of the elements (Eu–Lu, Y) form $NaRGeO_4$ crystallizing in the rhombohedral modification (Figures 7.44 and 7.45). For all three phases, the structure–genetic relation is an olivine-like ribbon, which reflects the corresponding rare earth element motif, phase transitions: $Na_2RGeO_4(OH) \rightarrow NaR_3[GeO_4]_2(OH)_2 \rightarrow NaRGeO_4$ from Lu to La (NaOH = 20–40%) is accompanied by the rearrangement of only cationic motif and expressed in the redistribution of Na and R in olivine-like ribbon (Figure 7.46).

Thus, the structural elucidation of the rare earth silicates and germanates in the R_2O_3-rich region of synthesis shows that the structure determining factors are the 3d arrangement of rare earth polyhedra and the stability of which depends upon the NaOH solution. For lighter rare earth elements, the 3d arrangement remains stable even at higher concentration of NaOH. However, the 3d arrangement loosens at the expense of the enrichment of Na-polyhedra, e.g., structures of $NaNdSiO_4$ and $Na_2PrSiO_4(OH)$.

For heavier rare earth elements, in which the amphoteric character is sufficiently strong, the decondensation of the cationic motif occurs from 3d bonding britolite up to 2d network $[R_2O_8]_{\infty\infty}$ and at the end discrete polyhedra–octahedra in $Na_3RSi_2O_7$. The same picture appears even for germanates. However, there is some variation from silicate systems, which are absent in germanates, wherein the rare

Table 7.17 New Structure Types of Alkali-Free Germanates of Rare Earth Elements [44]

Compound	Solvent Concentration (wt%)	Unit Cell Parameters			Space Group	Coordination Number of Ge; Ge–O Radical
		a (Å)	*b* (Å)	*c* (Å)		
$Y_7Ge_2O_{12}(F, OH)_5$	KF 25%	5.551	11.283	12.131 $\gamma = 108.81°$	*A*2/m	4; $[GeO]_4$
$Sm_4Ge_3O_9(OH)_6$	CsOH	15.145	10.909	7.161	*C*mcm	4; 6;$[Ge^{IV}Ge_2{}^{VI}O_9]$
$Y_4GeO_6(OH, F)_4$	KF	5.618 $\alpha = 91.45°$	6.276 $\beta = 109.44°$	6.885 $\gamma = 111.24°$	*P*1	4; $[GeO_4]$
$Gd_3GeO_5(OH)_3$	RbOH 30–50%	5.510	6.011	10.277 $\gamma = 104.89°$	*P*2$_1$/m	5; $[GeO_5]$
$Sm_3GeO_5(OH)_3$	RbOH 30–50%	5.526	6.104	10.328 $\gamma = 104.76°$	*P*2$_1$/m	5; $[GeO_5]$
$Dy_3GeO_5(OH)_3$	CsOH 40%	5.481	5.920	10.215 $\gamma = 105.02°$	*P*2$_1$/m	5; $[GeO_5]$
$Gd_4Ge_3O_{11}(OH)_2$	CsOH 40%	6.713 $\alpha = 62.19°$	9.349 $\beta = 73.67°$	10.128 $\gamma = 69.05°$	P1	4;$[Ge_2O_7] + [GeO_4]$
$Er_{13}Ge_6O_{31}(OH, F)$	CsOH	15.617	–	9.398	R3	4; $[GeO_4]$
Nd_4GeO_8	NaOH	7.475	5.727	17.927	Pmc2$_1$	4; $[GeO_4]$
$Yb_2Ge_2O_7$	KF 25%	9.827	–	–	Fd3m	4; $[GeO_4]$

Table 7.18 New Series of Alkali-Free Rare Earth Germanates [44]

	Ce	Pr	Nd	Pm	Sm	Eu	Gd	Tb	Dy	Ho	Er	Tm	Yb	Lu	Y
$Ln_4GeO_6(OH, F)_4$															++
$Ln_7Ge_2O_{12}(F,OH)_5$															++
$LN_4Ge_3O_9(OH)_6$					++	++	++	++	++	++	++	++	++	++	
$Ln_3GeO_5(OH)_3$					++	++	++	++	++	++	++				
$Ln_4[Ge_2O_7][GeO_4](OH)_2$						++									
Ln_4GeO_8		++										++	++	++	
$Ln_{13}Ge_6O_{31}F$						++					++				
$Ln_2Ge_2O_7$ (cubic)													++	++	++

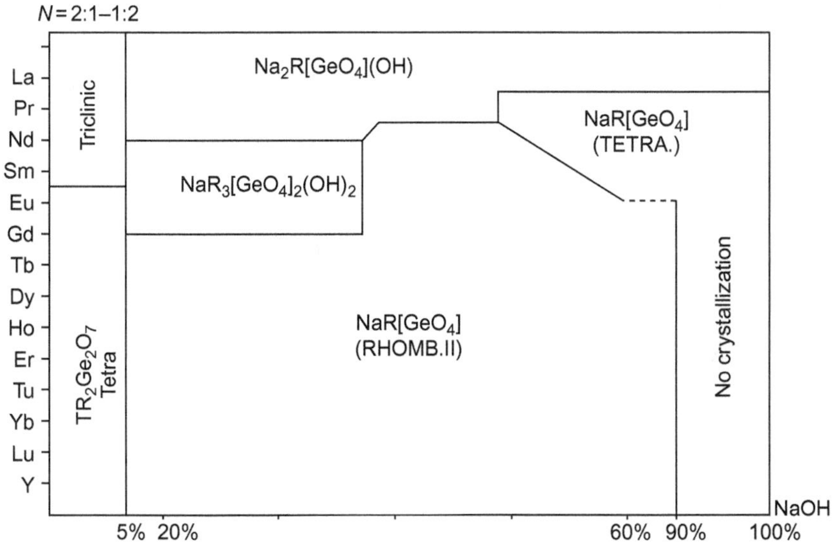

Figure 7.44 NC diagram for sodium rare earth germinate system: $T = 450°C$ [44].

earth elements lie as discrete polyhedra. For example, among germanates, the structure type $Na_3RSi_2O_7$ does not exist. The second important structure elucidation among germanates is connected with more of olivine-type structures. Therefore, rare earth germanates are the closer analogs of silicates.

A detailed work on the system $Na_2O—R_2O_3—GeO_2—H_2O$ indicates that with surplus R_2O_3 in the nutrient ($GeO_2/R_2O_3 = 1:1—1:4$) for groups La—Gd, three types of germanates form: $NaR_3Ge_2O_8(OH)_2$, $NaRGeO_4$, and $Na_2RGeO_4(OH)$. Their synthesis takes place with a temperature gradient. $NaR_3Ge_2O_8(OH)_2$ crystallizes within a narrow interval of GeO_2/R_2O_3 ratio $= 2.5:1—1.75:1$ with NaOH

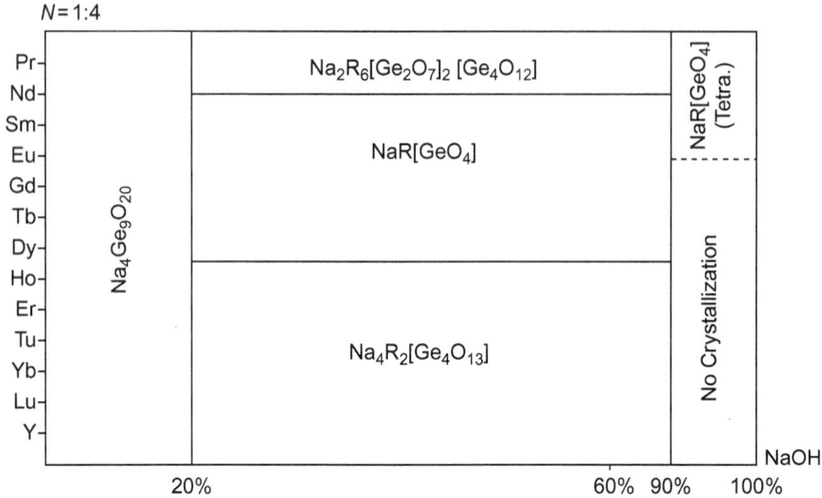

Figure 7.45 NC diagram for sodium rare earth germinate system: $T = 650°C$, $P = 400–2500$ atm [44].

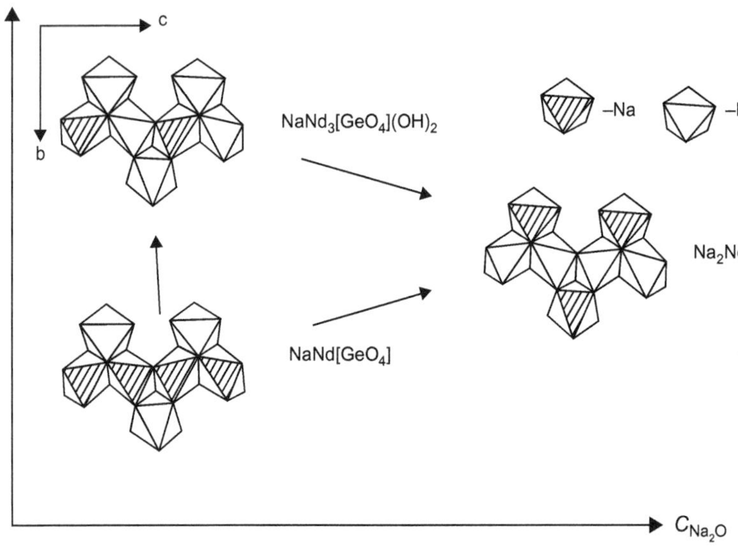

Figure 7.46 Change in rare earth motif during phase transition [44].

concentration 5–42 wt% and 100% output. Crystals of 4–6 mm size, commonly as equidimensional crystals with pinacoidal and rhombohedral prisms belonging to the olivine-type structure, are formed.

$NaRGeO_4$ crystallizes over a broad field and the $GeO_2/R_2O_3 = 2:1–1:1$ in the NaOH solutions of 5–80 wt% with 100% output. Crystals of 5 mm size have been obtained. This compound belongs to the structure-type olivine. At lower pressures and temperatures ($P = 3-20$ atm and $T = 558-650°C$) in the NaOH rich region (>80 wt%) for groups Nd–Eu, phases belong to the tetragonal modification and the crystals are plate like, thick, 1–2 mm in size, and twinning is common.

$Na_2RGeO_4(OH)$ (where R = La–Nd) crystals form at $GeO_2/R_2O_3 = 1:2$ and concentration of NaOH > 6 wt%. Crystals have equidimensional habit. Output of 100% was obtained in solutions when $H_2O < 10\%$. $Na_2R[GeO_4](OH)$ belongs to the olivine-type structures. Thus, for La–Gd group of germanates of olivine-type structures in NaOH solution, the following scheme with an increase in the concentration of NaOH has been proposed:

$$NaR_3[GeO_4]_2(OH)_2 \rightarrow NaR[GeO_4] \rightarrow Na_2R[GeO_4](OH) \quad (7.13)$$

In spite of the rearrangement of Na^+ and rare earth cations, the common olivine-type motif is retained over a wide interval not only on the concentration of NaOH but also on the ratio of GeO_2/R_2O_3. When $GeO_2/R_2O_3 > 3:1$ in 20–90 wt%, NaOH solutions' complex germanates like $Na_2R_6Ge_8O_{26}$ appear for Nd and Pr representatives. However, the output is less than 10% and crystals are small (< 1 mm).

When Dy–Lu are used, a slightly different picture is seen. $NaR_2Ge_4O_{10}(OH)$, $NaRGeO_4$, and $Na_4R_2Ge_4O_{13}$ are the characteristic phases. $NaR_2Ge_4O_{10}(OH)$ phase crystallizes in less than 10 wt% NaOH with $GeO_2/R_2O_3 = 1:2.5-1:3$.

$NaRGeO_4$-orthogermanates crystallize in the R_2O_3 rich region in the system $Na_2O-R_2O_3-GeO_2-H_2O$.

In the GeO_2-rich region, for Dy–Lu groups, $Na_4R_2Ge_4O_{13}$ is the characteristic phase with 100% output in 30–60 wt% NaOH solutions at ratio $GeO_2/R_2O_3 = 2.5:1-3:1$. The crystals are up to 2 mm in size.

Similarly, the rare earth germanates have been synthesized using aqueous solutions of KOH and KF [44]. Unfortunately, output of 100% has not been obtained for several germanates. Quite often two or three phases crystallize simultaneously. All the KR-germanates have been grouped into two structure types: $K_2R_4[Ge_4O_{13}](OH,F)_4$ and $K_4R_2[Ge_4O_{13}](OH,F)_4$. Figure 7.47 shows the phase formation of rare

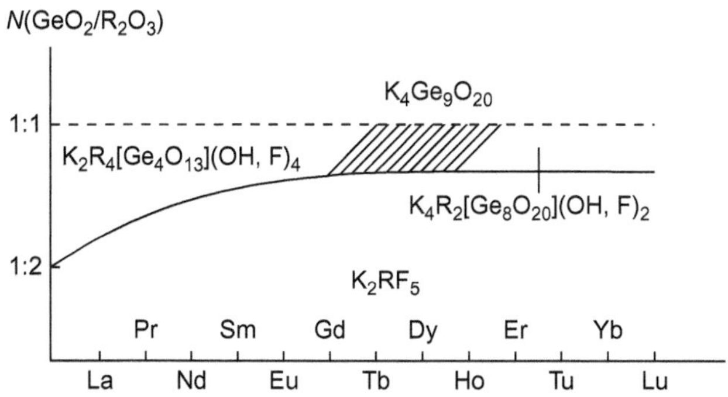

Figure 7.47 Phase formation of rare earth germanates in the aqueous solution of KF [44].

earth germanates in the aqueous solution of KF. $K_2R_4[Ge_4O_{13}](OH,F)_4$ covers Nd–Tb rare earth cations, and the second group covers Er–Lu. When R_2O_3/GeO_2 ratio is less than 1:1, the resultant product always carries $K_4Ge_9O_{20}$ in the form of hexagonal prism and plates. Surplus R_2O_3 gives K_2RF_5 (R = Tb–Lu) and related phases. The boundaries of the appearance of fluorides depend upon the concentration of K and the type of rare earth ion. The expansion of the boundaries of K_2RF_5 and $K_2R[Ge_4O_{13}](OH,F)_4$ slowly moves toward the La and KF solution concentration greater than 45%, and the crystallization of germanates practically ceases. The crystal size decreases sharply toward the heavier rare earth elements, so that the Tb representative cannot be obtained as good single crystals of sufficiently large size. For Er–Lu groups, $K_4R_2[Ge_8O_{20}](OH,F)_2$ type is obtained. The refinement of $K_4Yb_2[Ge_8O_{20}](OH,F)_2$ structure revealed $[Ge_8O_{20}]$ radical. The rare earth motif is represented by the infinite chain from rare earth octahedra shared at the vertices.

The scandium-bearing system has been studied in detail. Despite the fact that it does not cover a wide range of germanates, the commonly observed phases are, for example, $KScGe_2O_6$, $K_2Sc_2Ge_2O_7(OH)_2$, $Sc_2Ge_2O_7$, $Na_4Sc_2Ge_4O_{13}$, $Na_2ScGeO_4(OH)$, and $NaScGe_2O_6$ in the $K_2O-Sc_2O_3-GeO_2-H_2O$ and $Na_2O-Sc_2O_3-GeO_2-H_2O$ systems, respectively [35,44]. Most of these phases crystallize independently depending upon the GeO_2/Sc_2O_3 ratio and the concentration of the alkaline solution. Figure 7.48 shows the NC diagram of the crystallization of various phases in

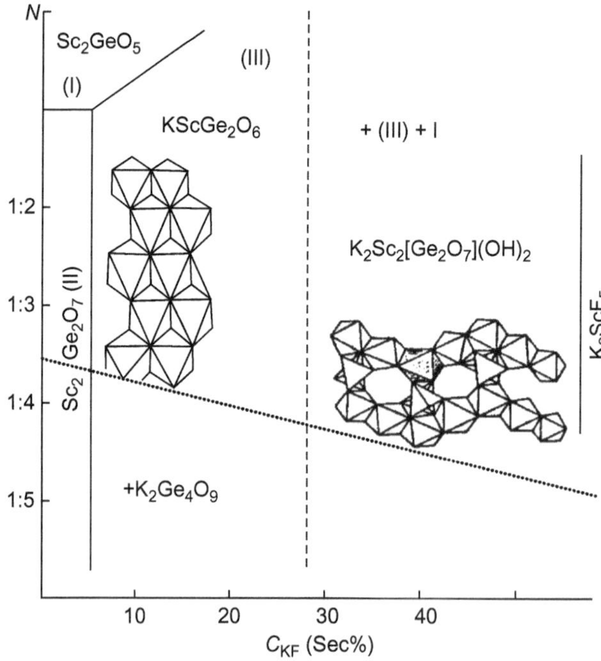

Figure 7.48 NC diagram of crystallization of various phases in the $K_2O-Sc_2O_3-GeO_2-H_2O$ system [35].

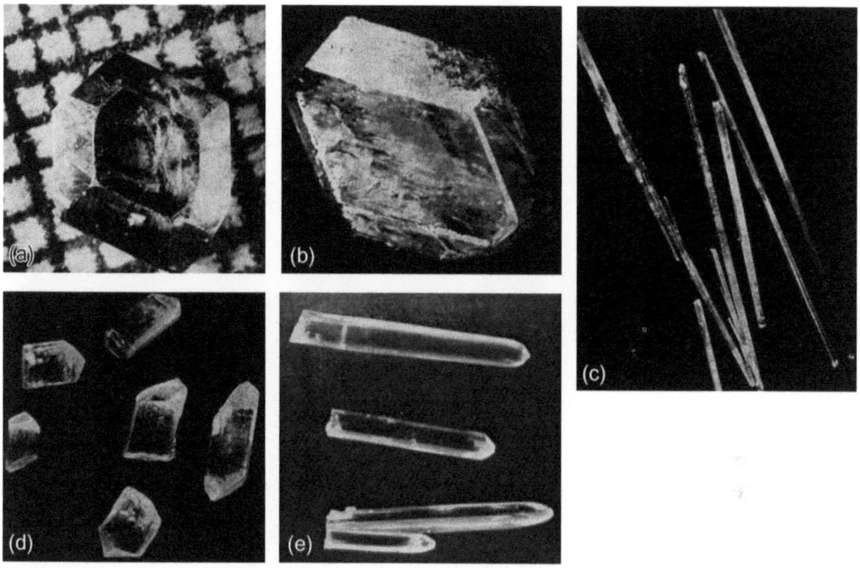

Figure 7.49 The characteristic photographs of rare earth germanates [44].
Source: Courtesy of L.N. Demianets.

the $K_2O-Sc_2O_3-GeO_2-H_2O$ system. Figure 7.49a–f shows the characteristic photographs of rare earth germanates [44]. Tables 7.19 and 7.20 give the structure types of alkali-free rare earth germanates, structure types of alkali and alkaline earth rare earth germanates, respectively.

7.11.2 Zirconium Germanates

The study of zirconium germanate crystallization continues to be popular under hydrothermal conditions. Recently, the number of sodium–zirconium germanates synthesized under hydrothermal conditions has increased enormously. In the $Na_2O-ZrO_2-GeO_2-H_2O$ system, the following germanates can be synthesized (Figure 7.50): $ZrGeO_4$, Na_2ZrGeO_5 (Ge/Zr = 1,Q_1), $Na_2ZrGe_2O_7$ (Q_2), $Na_4Zr_2Ge_3O_{12}$ ($Q_{1.5}$), $Na_3HZrGe_2O_8$ (Q_2Na), and $Na_4Zr_2Ge_5O_{16} \cdot H_2O$ ($Q_{2.5}$).

The subscript of letter Q indicates the value of the Ge/Zr ratio in the solid phase. The crystallization regions of these phases are shown in Figures 7.50 and 7.51, respectively. Five sodium–zirconium germanates were synthesized in the system $Na_2O-ZrO_2-GeO_2-H_2O$. New compounds are marked with asterisks. All the compounds are characterized by the same Zr and Ge coordination and in all the compounds zirconium atoms occupy octahedra, while germanium atoms occupy tetrahedra.

The positions of phase boundaries (crystallization of individual compounds) are determined by the NaOH concentration and the initial ratio of zirconium and

Table 7.19 Structure Types of Alkali-Free Rare Earth Germanates [44]

Compound	Representative	Cell Parameter (Å)	Rate Earth, Co. No., Average Bond Distance (Å)	Tetrahedral Radical Form	Rare Earth Polyhedra
$R_2[GeO_4]O$ R = La–Tb	$Sm_2[GeO_4]O$	$P2_1/c$ $z = 4$ $a = 9.4$ $b = 7.2$ $c = 6.9$ $\beta = 108°$		Orthotetrahedra	Framework
$R_2[GeO_4]O$ R = Dy–Lu, Y	$Er_2[GeO_4]O$	$B2/b$ $z = 8$ $a = 14.86$ $b = 10.56$ $c = 6.83$ $\gamma = 122.10°$		Orthotetrahedra	Framework
	$Sc_2[GeO_4]O$	$B2/b$ $z = 8$ $a = 10.927$ $b = 10.656$ $c = 10.486$ $\gamma = 93.96$	Sc 6 2.09	Orthotetrahedra	Framework
$R_4[GeO_4]_4$ R = La–Gd	$La_4[GeO_4]O_4$	Hexagonal $a = 10.54$ $c = 7.63$		Orthotetrahedra	
$R_4[GeO_4]_4$ R = Tb–Lu, Sc	$Nd_4[GeO_4]O_4$	$Pmc2_1$ $z = 4$ $a = 7.475$ $b = 5.757$ $c = 17.927$	Nd 8 2.50	Orthotetrahedra	Framework

(*Continued*)

Table 7.19 (Continued)

Compound	Representative	Cell Parameter (Å)	Rate Earth, Co. No., Average Bond Distance (Å)			Tetrahedral Radical Form	Rare Earth Polyhedra
$R_{4.67}[GeO_4]_3O$	$Dy_{4.67}[GeO_4]_3O$	$P6_3/m$ $z=2$	Dy	7	2.33	Orthotetrahedra	Framework
		$a=9.200$	Dy	9	2.40		
R = La–Dy		$c=6.808$					
H type $R(OH)_3$ $6R_2.[GeO_4][O,(OH,F)_2]$	$Sm(OH)_3 \cdot 6Sm_2 \cdot [GeO_4][O,(OH,F)_2]$	$P6_3/m$ $z=2$	Sm	7	2.51	Orthotetrahedra	Framework
		$a=15.90$	Sm	9	2.48		
R = Nd–Eu		$c=6.98$					
$R_2Ge_2O_7$	$Er_2Ge_2O_7$	$P4_12_12$ $z=4$	Er	7	2.32	Diorthogroups	Framework
R = Tb–Lu		$a=6.778$					
		$c=12.34$					
	$Sc_2Ge_2O_7$	$C2/m$ $z=2$				Diorthogroups	Layers from 6-membered
		$a=6.56$					
		$b=8.690$					
		$c=4.900$					
		$\beta=102.5°$					
$R_4[Ge_3O_{10}]$ $[GeO_4]$	$La_4[Ge_3O_{10}]$ $[GeO_4]$	$P1$ $z=4$	La	8	2.56	Triorthogroup	Framework
R = La–Pr		$a=12.76$					
		$b=7.070$					
		$c=7.006$					
		$\alpha=90.95°$					
		$\beta=90.35°$					
		$\gamma=94.10°$					
$R_4[Ge_3O_{10}]$ $[GeO_4]$	$Gd_4[Ge_3O_{10}]$ $[GeO_4]$	$P1$ $z=6$	Gd	8	2.48	Triorthogroup	Framework
R = Nd–Gd		$a=18.50$					

(*Continued*)

Table 7.19 (Continued)

Compound	Representative	Cell Parameter (Å)	Rate Earth, Co. No., Average Bond Distance (Å)	Tetrahedral Radical Form	Rare Earth Polyhedra
		$b = 6.800$ $c = 6.858$ $\alpha = 87.88°$ $\beta = 91.60°$ $\gamma = 94.55°$			
$R_2Ge_2O_7$	R = Sc, Yb	Cubic $a = 9.81$		Octahedral framework	Framework
$R_4Ge_3O_9(OH)_6$ R = Sm, Er–Lu	$Sm_4Ge_3O_9(OH)_6$	*C*mcm $z = 4$ $a = 15.195$ $b = 10.909$ $c = 7.161$	Sm 7 2.30 Sm 8 2.39	Chains from octahedral Connected with orthotetrahedral	Framework

Co. No. = Coordination number.

Table 7.20 Structure Types of Alkali and Alkaline Earth, Rare Earth Germanates [44]

Compound	Representative	Cell Parameter (Å)	Rare Earth, Co. No., Average Bond Distance (Å)	Tetrahedral Radical Form	Rare Earth Polyhedra
$NaR[GeO_4]$ R = Sm–Lu, Y	$NaSm[GeO_4]$	$Pbn2_1$ $z = 4$ $a = 5.27$ $b = 11.70$ $c = 6.50$	Sm 6 2.36	Orthotetrahedra	Discrete polyhedra
$NaR[GeO_4]$ R = La–Nd	$NaEu[GeO_4]$	14/*m* $z = 8$ $a = 11.75$ $c = 5.45$		Orthotetrahedra	Framework
$NaR_3[GeO_4]_2(OH)_2$ R = Nd–Gd	$NaSm_3[GeO_4]_2(OH)_2$	*Bb* $z = 4$ $a = 18.21$ $b = 12.15$ $c = 5.27$ $\gamma = 131.88°$	Sm 8 2.46 Sm 7 2.47	Orthotetrahedra	Discrete polyhedra
	$NaHo_4[GeO_4]_2$ $O_2(OH)$	*Pnma* $z = 4$ $a = 6.75$ $b = 21.67$ $c = 6.91$	Ho 7 2.33	Orthotetrahedra	Framework
$Na_2R[GeO_4]$ R = Pr–Gd	$Na_2Pr[GeO_4]$	*Pnma* $z = 4$ $a = 9.52$ $b = 7.48$ $c = 6.98$	Pr 8 2.51	Orthotetrahedra	Framework
	$Na_2Sc[GeO_4](OH)$	$P2_1ab$ $z = 8$ $a = 12.34$ $b = 13.78$ $c = 5.53$	Sc 6 1.99–2.18	Orthotetrahedra	Ribbon

(Continued)

Table 7.20 (Continued)

Compound	Representative	Cell Parameter (Å)	Rare Earth, Co. No., Average Bond Distance (Å)	Tetrahedral Radical Form	Rare Earth Polyhedra
	LiNd [GeO_4]	*Pbcn* $z = 4$ $a = 5.096$ $b = 11.923$ $c = 5.399$	Nd 8 2.486	Orthotetrahedra	Framework
$Me_3R_2[GeO_4]_3$ Me = Sr, Ca R = Ho–Lu, Y	$Ca_3Y_2[GeO_4]_3$	Ia3d $z = 8$ $a = 12.085$		Orthotetrahedra	
$Me_2R_3[GeO_4]_3Cl$ Me = Ca, Ba R = La, Ce	$Ca_2Ce_3[GeO_4]_3$ Cl	$P6_3/m$ $z = 2$ $a = 9.467$ $c = 6.999$		Orthotetrahedra	
$K_2Sc_2[Ge_2O_7](OH)_2$		*Pbcn* $z = 4$ $a = 7.06$ $b = 11.35$ $c = 10.67$	Sc 6 2.11	Diorthogroups	Framework
$Na_4R_2[Ge_4O_{13}]$ R = Yb–Lu, Sc	$Na_4Sc_2[Ge_4O_{13}]$	$Pna2_1$ $z = 4$ $a = 14.30$ $b = 5.46$ $c = 14.65$	Sc 6 2.12	Tetragroups	Twinned polyhedra
$K_2R_4[Ge_4O_{13}]$ (OH, F)$_4$ R = Pr–Gd	$K_2Nd_4[Ge_4O_{13}]$ (OH, F)$_4$	$P2_1$ $z = 4$ $a = 6.821$ $b = 7.438$ $c = 17.639$ $\gamma = 121.33°$	Nd 7 2.40 8 2.53	Tetragroups	Framework
		$P1$ $z = 2$	Pr 7 2.26–2.61		Framework

(Continued)

Table 7.20 (Continued)

Compound	Representative	Cell Parameter (Å)	Rare Earth, Co. No., Average Bond Distance (Å)	Tetrahedral Radical Form	Rare Earth Polyhedra
$Na_2R_6[Ge_2O_7]_2[Ge_4O_{12}]$ R = Pr–Nd	$Na_2Pr_6[Ge_2O_7]_2[Ge_4O_{12}]$	$a = 7.18$ $b = 11.61$ $c = 7.04$ $\alpha = 98.92°$ $\beta = 91.62°$ $\gamma = 102.50°$	8 2.41–2.71	Diorthogroup + 4-membered ring	
	$K_4Yb_2[Ge_8O_{20}](OH, F)_2$	$P2_1/m$ $z = 2$ $a = 12.105$ $b = 11.896$ $c = 8.576$ $\gamma = 112.04°$	Yb 6 2.19	Tubular	Columns ribbons
$MRGe_2O_6$ M = K,Na R = Sc	$NaScGe_2O_6$	Monoclinic $z = 2$ $a = 10.1$ $b = 9.1$ $c = 4.7$ $\beta = 108°$		Metachain with a period of 2-tetrahedra	Ribbons
$Na_5RGe_4O_{12}$ R = Gd–Lu, Y	$Na_5GdGe_4O_{12}$	$R3c$ $z = 16$ $a = 22.781$ $c = 12.994$		12-Membered rings	Discrete polyhedra
$MgLa_2GeO_6$	$a = 3.90$	$Pm3m$		Isolated octahedra	

Co. No. = Coordination number.

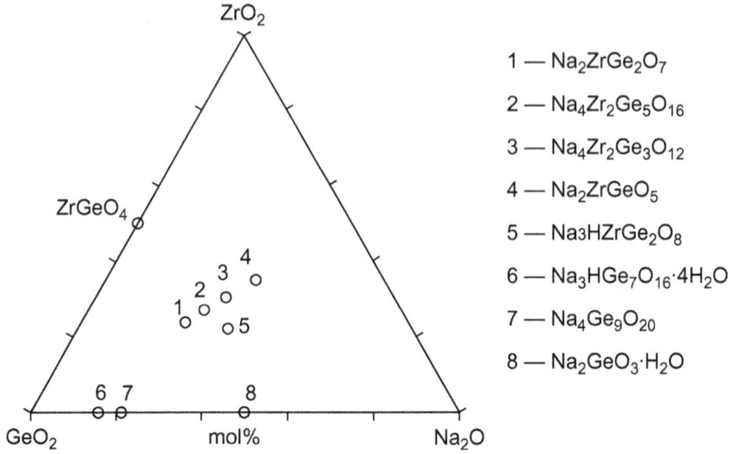

Figure 7.50 Crystallization field of sodium zirconium germanates in the system $Na_2O{-}ZrO_2{-}GeO_2{-}H_2O$ [95].

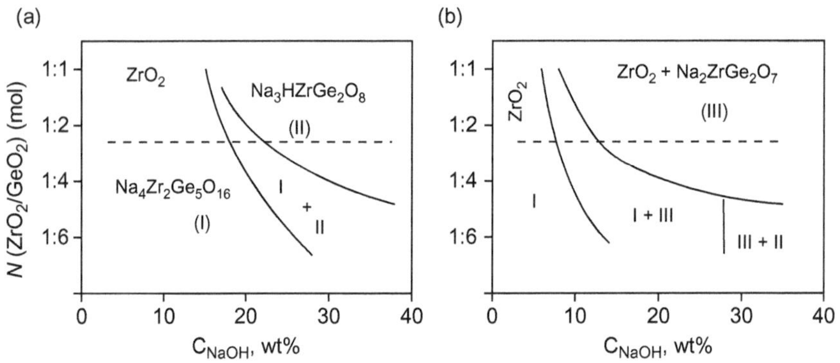

Figure 7.51 Crystallization field of sodium zirconium germanates in the system $Na_2O{-}ZrO_2{-}GeO_2{-}H_2O$ at 300°C (a); and at 500°C (b) [95].

germanium oxides. The sequence of germanate formation in the hydrothermal $Na_2O{-}ZrO_2{-}GeO_2{-}H_2O$ system may be represented by the scheme in Figure 7.52.

The formation of NaZr-germanates is accompanied by a decrease in the coordination numbers of Zr from 8 to 6 in all NaZr-germanates. If we pass from alkali-free to alkali-zirconium germanates, the GeO_2/ZrO_2 ratio first increases in a jump-wise manner and then decreases to the initial minimum value equal to unity:

$$(q)\ 1 \rightarrow 2.5 \rightarrow 2 \rightarrow 1.5 \rightarrow 1 \qquad (7.14)$$

A decrease in the GeO_2/ZrO_2 ratio is reflected in a decrease of the condensation of germanium–oxygen tetrahedra:

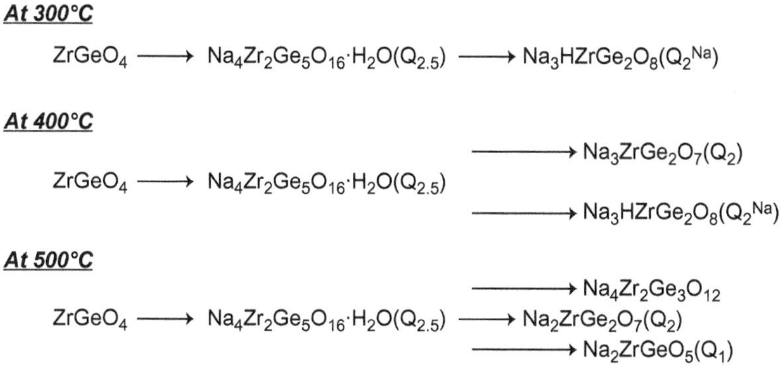

Figure 7.52 Reaction scheme of germanate formation in the hydrothermal $Na_2O-ZrO_2-GeO_2-H_2O$ system [95].

$n = 5$, $[Ge_5O_{16}]$ island chains in $Na_4Zr_2Ge_5O_{16} \cdot H_2O$.
$n = 2$, $[Ge_2O_7]$ diorthogroups in $Na_2ZrGe_2O_7$.
$n = 1$, $[HGeO_4]$ and $[GeO_4]$ type tetrahedra in $Na_3HZrGe_2O_8$.
$n = 1$, $[GeO_4]$ tetrahedra in Na_2ZrGeO_5 and $Na_4Zr_2Ge_3O_{12}$.

The germanate system has no compounds analogous to silicates with the TO_2/ZrO_2 ratio = 3. The maximum GeO_2/ZrO_2 ratio is equal to 2.5. The substitution of Si by Ge results in the change of the structure type of alkali-free compound to $ZrGeO_4$ crystallizing in the scheelite type. The general scheme of the interaction between the zirconium and germanium oxides and NaOH solutions at high temperatures and pressures may be represented in the form:

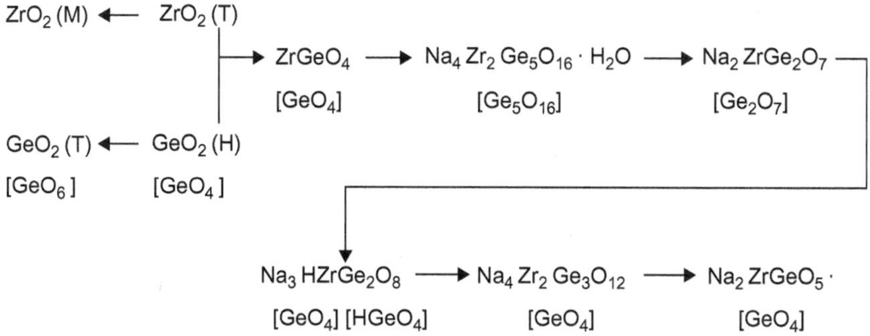

The sequence of the formation of NaZr and Na germanates reflects a general trend in the decondensation of Ge–O radicals with an increase of the alkali content in the system. The formation of each subsequent phase in the chain accompanied by the separation of GeO_2 either participates in the formation of the accessory phase or remains in the solution:

$$Na_4Zr_2Ge_5O_{16} \cdot H_2O \rightarrow 2Na_2ZrGe_2O_7 + GeO_2 + H_2O$$

$$3Na_4Zr_2Ge_5O_{16} \cdot H_2O \rightarrow 4Na_3HZrGe_2O_8 + 2ZrO_2 + 5GeO_2 + H_2O$$

$$3Na_2ZrGe_2O_7 \rightarrow 2Na_3HZrGe_2O_8 + ZrO_2(m_{GeO2} = const)$$

$$2Na_2ZrGe_2O_7 \rightarrow Na_4Zr_2Ge_3O_{12} + GeO_2$$

$$Na_4Zr_2Ge_3O_{12} \rightarrow 2Na_2ZrGeO_5 + GeO_2$$

As the NaOH content increases in the solution we get:

$$2ZrO_2 + 5GeO_2 + 4NaOH \rightarrow Na_4Zr_2Ge_5O_{16} \cdot H_2O + H_2O;$$

$$2Na_4Zr_2Ge_5O_{16} \cdot H_2O + 2NaOH \rightarrow 5Na_2ZrGe_2O_7 + GeO_2 + ZrO_2 + 2H_2O;$$

$$Na_2ZrGe_2O_7 + NaOH \rightarrow Na_3HZrGe_2O_8$$

$$\downarrow\uparrow$$

$$Na_4Zr_2Ge_3O_{12} + GeO_2 + 2NaOH$$

$$(GeO_2 + 2NaOH) \downarrow$$

$$Na_2GeO_3 \cdot H_2O.$$

The above reactions permit one to draw a conclusion that the stable crystallization of different phases in the above sequence which was observed with an increase in the NaOH content will also proceed with a decrease in the GeO_2 content at sufficiently high Na_2O content in the system.

The systematic study of hydrothermal crystallization in the systems $Na_2O{-}MO_2{-}GeO_2{-}H_2O$ (M = Ti, Zr, and Hf) revealed the main regularities of germanate formation and growth. In these systems, compounds having both network and no-network structure can be obtained. For the compound having network, the value $q = Ge/M$ can be equal to 1, 1.5, 2, 2.5; the value $p = Na/M$ being equal to 2. In germanates without network structure $q = 1$ and 1.25, $p = 1$, 1.5, 2, and 3.

In the network NaZr-germanates, the degree of Ge-tetrahedra condensation increases from $n = 1$ till $n = 5$. For the network NaZr-silicates which have no germanate analogs, the new crystallization branch takes place (the formation of the phases with $q = 3$; 3; 6, $n = 3$ and ∞). Such change in n-value is connected with the change of secondary structural units. The analysis of the crystal structure of the synthesized phases leads to the conclusion that the structural unit common to all sodium−zirconium germanates is a centrosymmetric closed dimer grouping of two Zr-octahedra and two Ge-tetrahedra, M_2T_2, which occupy the regular point system in the centrosymmetric groups. During structural rearrangement, such groups link in another way. For silicates, the cyclic dimer M_2T_4 is typical [95,44].

7.11.3 Zincogermanates

The hydrothermal growth of zincogermanates, although not as popular as that of zincosilicates, has been studied in detail. The appreciable solubility of sodium zinc germanates in water solutions of caustic soda at elevated temperatures and pressures allows its preparation in the form of coarse well-formed crystals, the sizes of which amount to some cubic centimeters.

In the majority of the experiments, the synthesis of sodium zinc germanate was carried out in the presence of Mn^{2+} ions in the solution with the aim of preparing crystals activated by Mn^{2+}. It is well known that Mn^{2+} is widely used as an activator for the preparation of various zinc silicate phosphors. Like in zincosilicates, zincogermanates in the presence of Mn^{2+} exhibit photo-, electro-, cathode-, X-ray, and thermoluminescence. Among the zincogermanates, only the Na_2ZnGeO_4 phase is very popular and well studied. Similarly, Na_2CoGeO_4 has been studied in detail owing to its interesting piezoelectric and pyroelectric properties [205–208].

In contrast to the Na_2O–ZnO–GeO_2–H_2O system, the Cd system not only does a change in the alkali concentration in solution leading to an exchange of the crystallization phases, but the phase boundaries are also substantially displaced as the Cd:Ge ratio alters. For surplus GeO_2 in pure water, only the rutile form of the oxide crystallizes, together with a cadmium germanate with $CdO:GeO_2 = 1:2$. As the amount of CdO in the system increases, the amount of free GeO_2 in the crystallization products diminishes. The cadmium germanate $CdGe_2O_5$ remains as a stable solid phase up to a ratio of $CdO:GeO_2 = 1:1$. Further increasing of the cadmium content leads to the formation of an olivine-like phase Cd_2GeO_4, and this remains dominant in the phase diagram of the system [205]. The Na_2O–CdO–GeO_2–H_2O has yielded eight crystalline phases under hydrothermal conditions [205,209,210]. The other germanates of interest are lead germanates, alkali gallium germanates, antimony germanates, neobium germanates, alkali alumino-germanates, barium germanates, bismuth germanates, titano-germanates, and so on [211–217].

Most of these germanates show interesting physical properties. Hence, a systematic understanding of germanate phases obtained from a wide variety of germanate systems will be of great interest in solid-state science.

7.12 Properties of Germanates

Among germanates, rare earth germanates and LISICON [$Li_{14}Zn(GeO_4)_4$] have been studied extensively owing to their unusual physical properties. Most of the germanates, especially rare earth germanates, have very high melting points, high-density and refractive indices. Table 7.21 gives the physicochemical properties of some rare earth germanates [44]. Electrical and magnetic properties have been studied for oxyorthogermanates, R_2GeO_5, and diorthogermanates, $R_2Ge_2O_7$. These studies confirm that R_2GeO_5 and $R_2Ge_2O_7$ belong to a class of dielectrics [218].

Table 7.21 Physicochemical Properties of Some Rare Earth Germantes [44]

Compound	d (g/ cm^3)	Melting temperature T (°C)	Reference Indices
La_2GeO_5	5.97	1800	1.940
Nd_2GeO_5	6.55	1820 ± 25	1.948
Gd_2GeO_5	7.07	1.880	1.941
Y_2GeO_5	4.83	1975 ± 15	1.848
Er_2GeO_5	7.28	2000 ± 50	1.904
La_4GeO_8	6.30	1720[a]	1.973
Gd_4GeO_8	7.77	1650[a]	1.975
Y_4GeO_8	5.10	2000	1.887
Er_4Ge_8	8.02	1600[b]	1.93
$La_{4.67}[GeO_4]_3$	5.89	1770	1.930
$La_2Ge_2O_7$	5.67	1680[a]	1.900
$Nd_2Ge_2O_7$	6.16	1820 ± 25[a]	–
$Gd_2Ge_2O_7$	6.70	1770	1.922
$Y_2Ge_2O_7$	4.69	1800 ± 25	1.840
α-$Er_2Ge_2O_7$	6.99	1875 ± 50	1.890
β-$Er_2Ge_2O_7$	–	–	1.938
$Sc_2Ge_2O_7$	4.46	1850	1.847
$La_2Ge_3O_9$	5.94	1300	1.995

[a]With decomposition.
[b]With decomposition in solid phase.

The Sm- and La-bearing germanates are diamagnetic. The magnetic transition temperatures for oxyortho- and diorthogermanates are as follows [219]:

$$Tb_2Ge_2O_7 = 2.05 \pm 0.05\ \text{K}$$

$$Dy_2Ge_2O_7 = 2.15 \pm 0.05\ \text{K}$$

$$Ho_2Ge_2O_7 = 1.45 \pm 0.05\ \text{K}$$

$$Er_2Ge_2O_7 = 1.15 \pm 0.05\ \text{K}$$

$$Dy_2GeO_5 = 2.50 \pm 0.10\ \text{K}$$

$$Er_2GeO_5 = 1.25 \pm 0.10\ \text{K}$$

Most of the works on the investigations of physical properties of rare earth germanates deal with the luminescence properties. Their emission and absorption spectral characteristics have been studied in great detail. Figure 7.53 shows the luminescence spectra of $NaEuGeO_4$ and $NaTbGeO_4$ [220]. Figure 7.54 shows the excitation spectrum of $NaGd_3[GeO_4]_2(OH)_2$: Tb^{3+} (2.5%) [221].

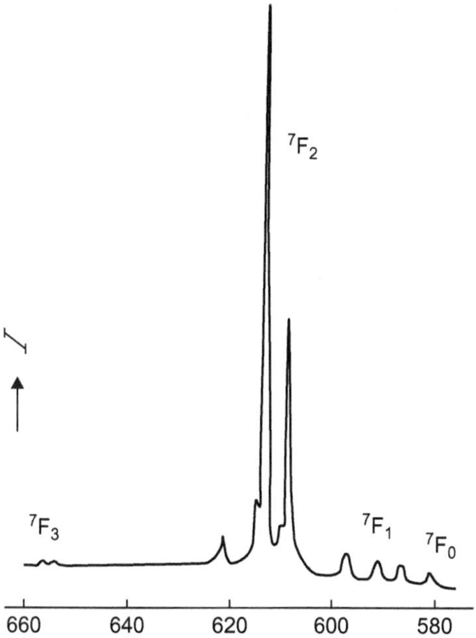

Figure 7.53 Luminescence spectra of $NaEuGeO_4$ and $NaTbGeO_4$ [220].

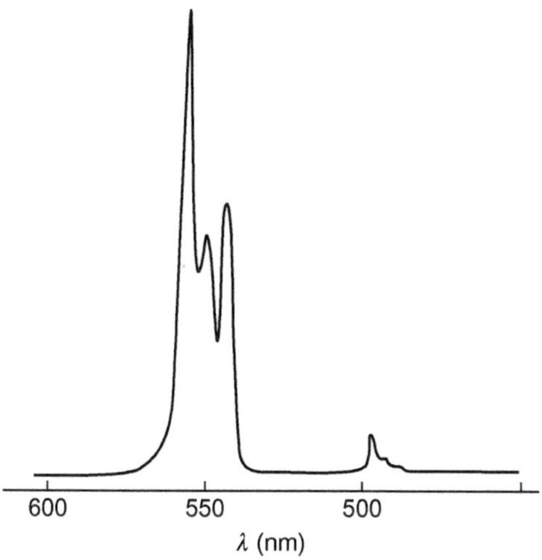

Figure 7.54 Excitation spectrum of $NaGd_3[GeO_4]_2(OH)_2$: Tb^{3+}(2.5%) [220].

Several lithium germanates have been reported to show high ionic conductivity. Among them, $Li_{14}Zn(GeO_4)_4$ is a very well-known Li^+ ion conductor. This one phase of the solid solution phase between Li_4GeO_4–Zn_2GeO_4 is popularly known as LISICON. The electrical conductivity of LISICON is $4 \times 10^{-3}\ \Omega^{-1}\ cm^{-1}$ at 500 K, but it decreases to less than $10^{-6}\ \Omega^{-1}\ cm^{-1}$ at room temperature [205,206]. Similarly, the alkali rare earth germanates exhibit very good ionic conductivity on par with their rare earth silicates counterparts [80,81].

7.13 Hydrothermal Growth of Phosphates

From the time of its discovery, the hydrothermal technique was known for the synthesis of silicates. In the nineteenth century, the hydrothermal technique was included under silicate technology. Next to silicates, it was the oxides and hydroxides of various metals that dominated the hydrothermal technology during the late nineteenth and early twentieth centuries. Subsequently, germanates, sulfides, and carbonates, etc., became popular. However, the growth of phosphates by hydrothermal technique was not attempted for a long time. It was only during the 1950s that hydrothermal technique was tested for the synthesis of phosphates [222], and up to the 1980s there were only scanty reports on the synthesis of phosphates by hydrothermal technique. Today, the situation has changed dramatically. Hydrothermal growth of phosphates is much more popular than the hydrothermal growth of any other compound. A large variety of phosphates starting from simple orthophosphates to the most condensed or ultraphosphates containing a wide range of elements have been obtained. This is probably due to the improved instrumentation techniques and the discovery of aluminophosphate zeolites. Further, the success in the growth of various phosphates has prompted the hydrothermal researchers to synthesize other inorganic compounds hitherto unattempted. Similarly, it is appropriate to state that the success in the growth of phosphates has led to the understanding of hydrothermal crystallization mechanism for various inorganic compounds. It ultimately has set a new trend in the synthesis of inorganic compounds under mild hydrothermal conditions.

There is a growing interest in the study of phosphates in the last two decades owing to their widespread application, e.g., piezoelectric, luminescent, magnetic, superionic, ceramic, solid-state laser, ultra-low thermal expansion, nonlinear optic and moisture sensors as hosts for actinide ions in modern technology. Similarly, phosphate minerals are the important source materials of almost all the rare earth elements. Thus, the phase equilibria studies in phosphate systems help to understand the interaction between various metals and P_2O_5, and it might give possible clues to separate these rare earth elements from phosphate minerals and also to understand their genesis. It is interesting to note that there are over 300 phosphate minerals in nature that have been classified into various groups depending upon the cationic elements present in them [223]. The phosphates containing rare earth elements represent the smallest group having only 15 minerals, out of which 4 are anhydrous: monazite-(Ce, La, Y, Th)PO_4; xenotime-YPO_4; cheralite-(R, Ce, Th)PO_4, and vitosite-$Na_3R(PO_4)$ [224] (a recently reported mineral from Lovozerskii alkaline mass), where R is rare earth element. It is interesting to note that all the naturally occurring phosphates are represented by one anionic group—orthophosphate. Although a wide range of physicochemical conditions exist in nature, we find phosphate minerals without poly-, pyro-, meta-, ultra-, etc. types, unlike the silicate minerals with a very wide variety of condensed silicates like meta-, poly-, soro-, phyllo-, etc. The reasons for the absence of condensed phosphates in nature are not clearly known. In order to understand the peculiarities among natural phosphate minerals, one has to synthesize them in the laboratory by simulating the natural conditions. However, the synthesis of phosphates was started neither to understand the peculiarities in nature nor to use them in technology. The first work on the synthesis of phosphates began in the nineteenth century by Berstelius in 1816 [225]. He synthesized sodium pyrophosphate artificially for the first time. In 1889, Johnson obtained LaP_5O_{14}—the first artificial rare earth phosphate [226]. However, the works on the synthesis of inorganic phosphates or their growth as monocrystals attracted less attention. Even after the beginning of the synthesis of various types of rare earth phosphates from the early 1940s [227,228] up to early 1970s [229,230], the interest was confined only to the studies concerning the X-ray and morphological analyses. The first technological phosphate material was Potassium Dihydrogen Phosphate (KDP) which was first obtained during the 1940s for telecommunication purpose. The interest on this material continues to progress. Except this, virtually phosphates did not carry any technological significance for a very long time. However, a systematic study of rare earth phosphates began in the 1970s with the developments in the field of communication and the problems associated with the miniature sources of light, particularly lasers based on dielectrics. Until recently, these materials contained a low concentration of active ions, which was a major obstacle in the attempts to use them in optoelectronics.

In 1973, laser beams were obtained from NaP_5O_{14} crystal containing 3.86×10^{12} ions of neodymium atoms/cm^3 [231,232]. Such a high concentration of active ions led to the further decrease in size of the laser crystal to tens of micromillimeters. An analogous effect was found in the monocrystals of $MNd(PO_3)_4$ (where M = Li, Na, K) [233,234], $Na_3Nd(PO_4)_2$ [235], $K_3Nd(PO_4)_2$ [236], etc. Prior to the development of these rare earth phosphates, laser beams were obtained in crystals mostly YAG: Nd containing only 2–3% of active ions mostly as dopants, but not as the stoichiometric components. Today it has become a flourishing field in science and technology, particularly in the field of fiber-optic communication. In 1976, the authors [237] reported a three-dimensional high ionic conductivity in NASICON ($Na_{1+x}Zr_2Si_xP_{3-x}O_{12}$; $0 \leq x \leq 3$) type of phosphate which is isostructural with $Na_3Sc_2P_3O_{12}$ [238]. Later on ionic conductivity was reported in some other rare earth phosphates like $RbNdP_4O_{12}$ [239] and $CsNdP_4O_2$ [225,240]. Since NASICON has some problems like nonstoichiometry, nonavailability of single crystals, and highly complex structures, there is a search for new materials with simple structures. The rare earth phosphates containing other elements like sodium, transitional elements, and some alkaline earth elements are found to be the ideal substitutes for NASICON [241]. In all these materials, the ionic conductivity, the optical properties, and so on depend mainly upon the type of cation and the crystalline structure. However, the structures of many new solid electrolytes are yet to be studied because of the nonavailability of these materials in the form of single crystals. It is well known that the rare earth phosphates exhibit island, ribbon, layered, ring, and framework types of structures, which are highly distorted with a very wide variation in P—O bond lengths. The inter-isolation of the rare earth polyhedra in most of the rare earth phosphates and the distortion of the overall structures are mainly responsible for the presence of luminescence and ionic conductivity in these materials. The activation energy and ionic conductivity depend upon the radii of the rare earth cation

[242]. The study of ionic conductivity in condensed phosphates needs a thorough knowledge of crystalline structure, which insists on single crystals. Similarly, rare earth phosphates, particularly orthophosphates have very high melting points (2300–2500°C) and they are known as high-temperature ceramics in addition to their common names like color luminophors and scintillators. More recently, magnetic properties have also been reported in some orthophosphates. All these properties have not been understood precisely. Therefore, the growth of these materials in the form of single crystals and their various physical characteristics need to be understood. There may be over 1000 reports on the synthesis of phosphates bearing rare earth elements, alkali, alkaline earth, and transitional elements. One-third of them might deal with hydrothermal technique.

7.13.1 Structural Chemistry of Rare Earth Phosphates

The formation of a wide variety of phosphates, particularly rare earth phosphates is connected with the reaction susceptibility of phosphoro-oxygen anions to give various degrees of condensation. As a result of such reactions among the phosphoro-oxygen anions, there develops bridging oxygen bonds —P—O—O—P—forming pyrophosphates—$[P_2O_7]$—initially followed by more complex linear anions like $[P_nO_{3n+1}]$, rings like $[P_nO_{3n}]$ ($n < 3$), and so on. It is extremely difficult to classify phosphates unlike silicates, because of high complexity in the phosphates' internal structures. For example, a recently reported structure of $(NH_4)_2Ce(PO_3)_5$ [243] has two infinite crystallographically nonequivalent $(PO_3)^-$ chains, one running parallel to the *a*-axis, the other along *c*-axis, both with a period of five tetrahedra. This compound is the first example of a long chain polyphosphate with crystallographically independent chains and such a peculiarity has not been reported in silicates or germanates so far.

The phosphorus atoms participate in the form of phosphoro-oxygen anions $[H_2PO_4]^{-1}$, $[HPO_4]^{-2}$, and $[PO_4]^{-1}$ in the reactions forming respective phosphates. The $[PO_4]^{-3}$ anions are less reaction susceptible among these three, and that makes the solid-state synthesis difficult. Therefore, ammonium phosphates are used in the solid phase synthesis. The acidic phosphoro-oxygen anions easily take part in the reactions during heating and in fact, high reaction susceptibility of acidic phosphoro-oxygen anions serve as a basis in working out various methods of obtaining rare earth phosphates with a particular reference to the highly condensed ultraphosphates [244]. With a slow dehydration of the acid orthophosphate anion, the removal of the $(OH)^{-1}$ ions and the condensation of phosphoro-oxygen anions occurs. This was a popular method, even in the nineteenth century, but for a long time the mechanism of the process remained unclear. Growth conditions for the required phosphate anions were selected empirically and compelled to participate in the reactions. However, the higher the degree of condensation of phosphate anion, the more difficult it is to ascertain the growth conditions. The process is quite complicated and very often results in the formation of a mixture of various degrees of condensed anions. McGilvery and Scott [245] concluded that water molecules play a distinct role in the dehydration process, particularly its concentration influences, not only on the rate of reaction but also on the stability of various phosphate anions. For the benefit of a reader, we have described the hydrothermal growth of phosphates containing rare earth elements followed by transitional elements.

7.13.2 Hydrothermal Growth of Rare Earth Phosphates

The first work on the hydrothermal synthesis of rare earth phosphates was by Anthony [222], who synthesized monocrystals of monazite, $CePO_4$, from an aqueous solutions of 85% H_3PO_4 at 300°C using Teflon liners. Carron et al. [246] have described the formation of rare earth orthophosphates from HCl solutions at 105–300°C and at pressure of 90 atm. A more systematic work on the hydrothermal synthesis of rare earth orthophosphates was carried out by Carron et al. [247]. However, there was a break in the activity of the hydrothermal synthesis of rare earth phosphates until the 1980s. This was mainly connected with the lack of growth technology, high volatility of phosphorus at higher temperatures, high reaction susceptibility of phosphorus, corrosive nature of phosphorus at higher temperatures, amongst other factors. The experiments on the hydrothermal synthesis of phosphates cannot be carried out at temperatures greater than 500°C, because phosphorus corrodes even the metals like Pt, Ir, Mo, Au, and Ni. Some of these problems could be overcome using Teflon or vitreous carbon glass liners. However, the main drawback in using these liners is sealing. If one can seal the glassy carbon liners by some means, the growth of pure phosphates can be carried out even at higher temperatures by hydrothermal methods. Similarly, the use of Teflon liners is again a problem, because it cannot be used beyond 300°C and 200 atm. Of course, pressure can be slightly increased by means of a counterbalance technique, i.e., by creating the same pressure between the autoclave walls and inside the liner. These problems have been discussed in great detail by Byrappa and Gopalakrishna [248], Byrappa [249], and Rabenau [250]. Byrappa and Gopalakrishna [248] have studied the rare earth phosphate system $Na_2O-R_2O_3-ZrO_2-P_2O_5$.

The first ever report on the study of phase equilibria in the rare earth phosphate system on the whole appeared in 1981 [251], where the phase equilibria in the system $Nd_2O_3-P_2O_5-H_2O$ at 500°C under 100 MPa and the synthesis of NdP_5O_{14} crystals were studied. This is also the first phosphate system studied hydrothermally. All the earlier reports deal with only a part of the phase diagram. The starting materials—Nd_2O_3, P_2O_2, and 85% H_3PO_4 solution—were taken in definite amounts within a sealed golden capsule. Although this method has been included in the hydrothermal method by the Japanese workers, the partial pressure of water under these conditions was always low and often less than 1 atm. Therefore, it can be considered as a special case of the hydrothermal method. This report was followed by another publication [252], which deals with the hydrothermal synthesis of mixed phosphates of neodymium and alkaline metals ($M_2O \cdot Nd_2O3 \cdot 4P_2O_5$). Here, the authors have studied the phase formation in the alkaline rare earth phosphate system ($M_2O-Nd_2O_3-P_2O_5-H_2O$, where M–Li, Na, K, Rb, and Cs) under hydrothermal conditions in the temperature and pressure range 300–700°C and 0.5–600 atm, respectively, using vitreous carbon glass, Teflon, and platinum liners.

The system is a four-component one with temperature and pressure as constants. According to the Gibb's phase rule at the invariant equilibrium, only four phases can exist ($NdPO_4$, NdP_3O_9, NdP_5O_{14}, and $MNdP_4O_{12}$). The variation in the concentration of phosphorus and alkaline metal oxides determines the fields of crystallization of different phases in the composition diagram. According to Le Chatalae's principle, the increase in the potential of M_2O (P_2O_5) provides the reaction taking place with the absorption of this component. The following reactions can explain the various nonvariant points in the ternary phosphate systems:

$$\begin{aligned}
&NdPO_4 + P_2O_5 + G_1 \rightleftharpoons Nd(PO_3)_3 + G_2\\
&2NdPO_4 + M_2O + P_2O_5 + G_3 2 \rightleftharpoons MNdP_4O_{12} + G_4\\
&Nd(PO_3)_3 + P_2O_5 + G_5 \rightleftharpoons NdP_5O_{14} + G_6\\
&2Nd(PO_3)_3 + M_2O + P_2O_5 + G_7 \rightleftharpoons 2MNdP_4O_{12} + G_8\\
&2NdP_5O_{14} + M_2O + G_9 \rightleftharpoons 2MNdP_4O_{12} + P_2O_5 + G_{10}
\end{aligned} \tag{7.15}$$

where G is a liquid phase having transitional composition. The presence of a divariant field of stability of $MNdP_4O_{12}$ (Figure 7.55) indicates the possibility of growing these crystals from solutions with variable concentrations of phosphorus and alkaline metals. The relevant empirical AB-diagrams of fields of crystallization, where $A = M_2O/Nd_2O_3$, $B = P_2O_5/M_2O + Nd_2O_3$, are shown in Figure 7.56. The dependence of PT conditions and the equilibrium state of the system has been studied (Figures 7.57 and 7.58). It is observed that with changes in temperatures, there will be different polymorphic modifications of tetraphosphates. In hydrothermal systems, the pressure measured corresponds to the pressure of aqueous solutions and its influence on the crystallization process is quite important. At a constant temperature, it increases with reference to the molar fraction of the water. In the pressure range of 1–600 atms, the boundaries of phase formation do not change. In this case, the changes in phase formation take place only during changes in the concentration of water in the ampoule. With an increase in the molar fraction of water, the following phase transformation takes place:

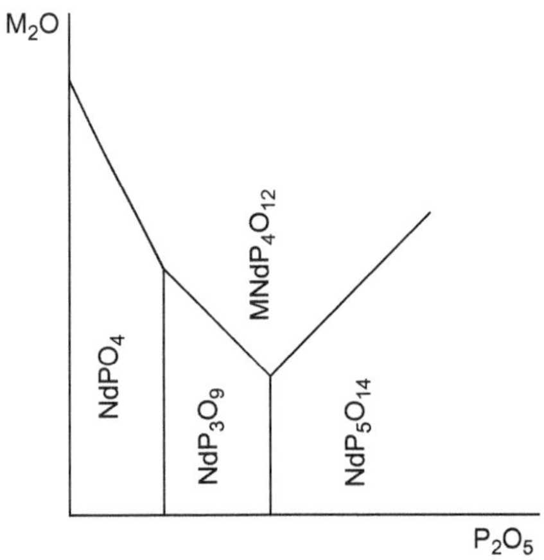

Figure 7.55 Composition diagram for the system Me_2O–Nd_2O_3–P_2O_5–H_2O [248].

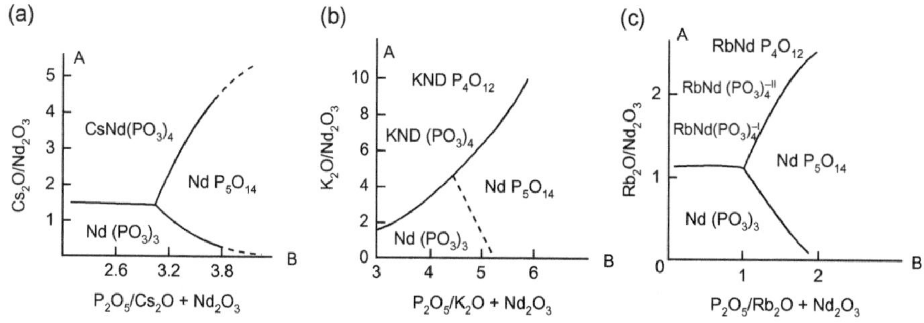

Figure 7.56 AB diagram of crystallization in (a) $Cs_2O{-}Nd_2O_3{-}P_2O_5{-}H_2O$, (b) $K_2O{-}Nd_2O_3{-}P_2O_5{-}H_2O$, and (c) $Rb_2O{-}Nd_2O_3{-}P_2O_5{-}H_2O$ systems [253].

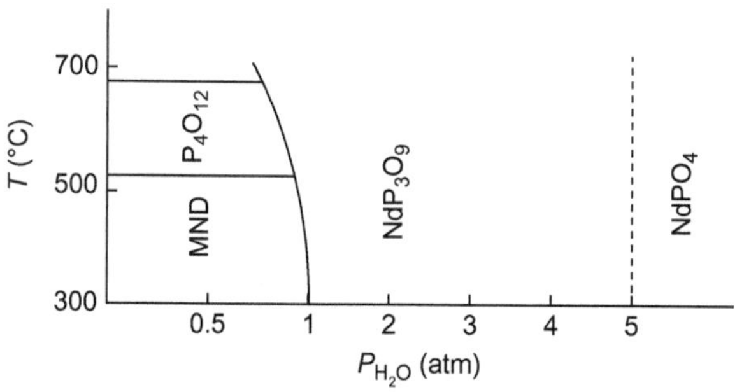

Figure 7.57 Phase transformation with reference to the changes in PT conditions [254].

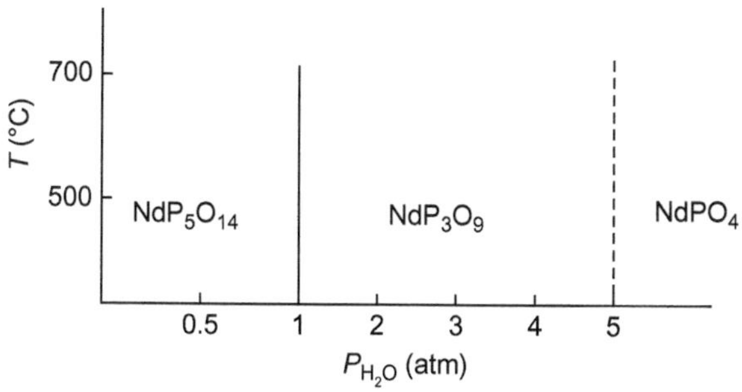

Figure 7.58 Phase transformation with reference to the changes in PT conditions [254].

$$\begin{array}{l} MNdP_4O_{12} \rightarrow NdP_3O_9 \rightarrow NdPO_4 \\ NdP_5O_{14} \rightarrow NdP_3O_9 \rightarrow NdPO_4 \end{array} \tag{7.16}$$

Using these results, Byrappa et al. [253] reported on the hydrothermal growth of TmP_5O_{14} crystals under mild hydrothermal conditions. The crystal size was much higher and crystals were well faceted than the crystals obtained by Yoshimura et al. [251]. The important aspect of this work is the growth of ultraphosphates under mild hydrothermal conditions ($T = 240°C$, $P = < 100$ bar). The autoclaves used were simple Morey type provided with Teflon liners. The crystallization was carried out through spontaneous nucleation.

This is an important conclusion, particularly for geoscientists to explain the possible reasons for the absence of condensed phosphates in nature [254]. The experiments carried out under hydrothermal conditions showed that when $P_{H_2O} > 6$ atm, crystallization of polyphosphates or ultraphosphates ceases and is replaced by crystallization of simple orthophosphates irrespective of the temperature of the system. Similarly, even an increase in the concentration of M_2O in the system leads to decondensation, and ultraphosphates having a highly condensed state or a complex anionic structure $[P_5O_{14}]$ disappear from the products of crystallization and are replaced by polyphosphates with radicals $[P_2O_7]$, $[P_3O_9]$, $[P_4O_{12}]$, and finally $[PO_4]$. These experiments demonstrate that the structures of condensed phosphates are not stable, at least not in the presence of water. It is well known that the magmatic or the postmagmatic processes of mineral formation are always accompanied by a certain amount of water and proceed in the presence of many types of metals. These are the conditions for the formation of orthophosphates and not condensed phosphates [254]. From the above studies, it is clear that only the hydrothermal method of studying alkaline rare earth phosphate systems revealed the formation of various phosphates depending upon the PT conditions.

The growth of rare earth phosphates in the form of single crystals is very interesting owing to the high reaction susceptibility of phosphorus and the easy formation of phosphoro-oxygen radicals in a given system. For example, any anionic group of Nd-phosphate can be readily obtained from phosphoric acid solutions. In the liquid, from which the Nd-phosphates grow, Nd should be dissociated to form individual ions. For this purpose, usually concentrated solutions or alkali melts are used. However, the process of Nd dissociation often carries a reversible character and the phase develops with Nd ions in the condensed pattern during the crystallization—e.g., as in NdP_3O_9 [255]. In order to achieve this, it is necessary that in the process of crystallization, the Nd ions are to be held in a dissociated state, i.e., in the form of a nezoite complex (Figure 7.59). A shell of ligands around Nd prevents the ions from intercondensation and allows them to enter the crystal lattice as isolated ions. These ligands which often occur as tetrahedral oxy-anions $[TO_4]$ is the formation of a nezoite complex of type NdT_8O_{24}, where T = S, P, W, Mo, etc. [256] Thus, to obtain

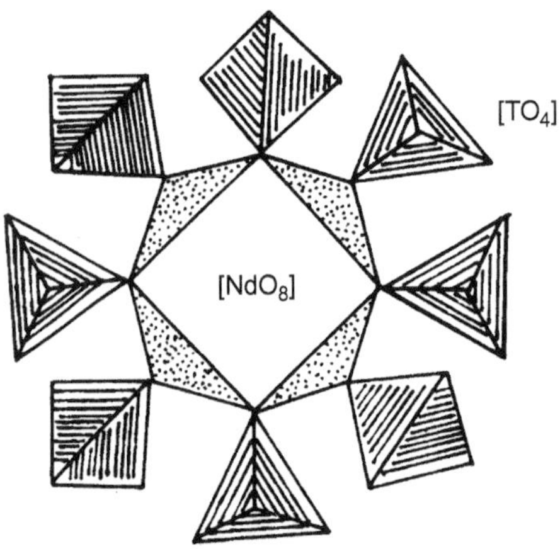

Figure 7.59 Nezoite complex [240].

nezoite crystals, it is first necessary to bring Nd ions into a dissociated state, followed by creating the conditions for the formation of nezoite complexes, and only then does the desired crystallization take place. The most suitable systems for this purpose appear to be Nd containing ternary systems of type $M_2O-Nd_2O_3-T_xO_y$, where M = H, Li, K, Rb, Cs, and Tl or their admixtures. The process of crystallization is much more simple in the multicomponent systems. The quantitative ratio of the component largely determines the stoichiometric nezoitic complexes in these, i. e., the ratio of T_xO_y/Nd_2O_3 should not be less than 8. The main problem of concern is the incorporation of some components in excess than the stoichiometric ratio. As a result, the molar ratio of the nutrient components in the growth of nezoites is not regular and can be expressed approximately as $M_2O(M)/Nd_2O_3/T_xO_y = 2 > 1 < 8$. The additional amount of the components is introduced into the nutrient depending upon the individual cases. Thus, the growth of nezoites requires a definite surplus of T_xO_y components, and they may not be taken from the stoichiometric melt. Therefore, in the growth of nezoites, either high-temperature aqueous solution or flux agents are used. Similarly, in the synthesis of many phosphates in the form of crystals, highly concentrated solution of phosphoric acid is used. It is considered that at over 250°C, orthophosphoric acid converts into pyrophosphoric acid and with a further increase in temperature it transfers to an even more condensed state due to the expulsion of water [257]. Figure 7.60 represents a temperature-composition diagram for the system $H_2O-P_2O_5$, at pressure up to 1 atm [240]. In region (2) of the diagram, crystallization can be carried out at normal pressure, whereas the crystal growth from solution corresponding to the composition of the region (1) requires the use of a hydrothermal apparatus. Here, the authors discuss the crystallization of rare earth phosphates as an example to explain the influence of the physicochemical conditions and the primary structural units. The formation of rare earth phosphates in acidic (pH > 7.0) phosphoro-oxygen media takes place as follows:

$$\begin{aligned} &R_nL + M_xP_yO_z \rightarrow R_nP_yO_z + M_xL \\ &R_nL + M_xP_yO_z \rightarrow M_xR_nP_yO_z \end{aligned} \tag{7.17}$$

where L = anion and it is usually Cl^-, NO_3, or SO_4, M = alkali cation and NH_4.

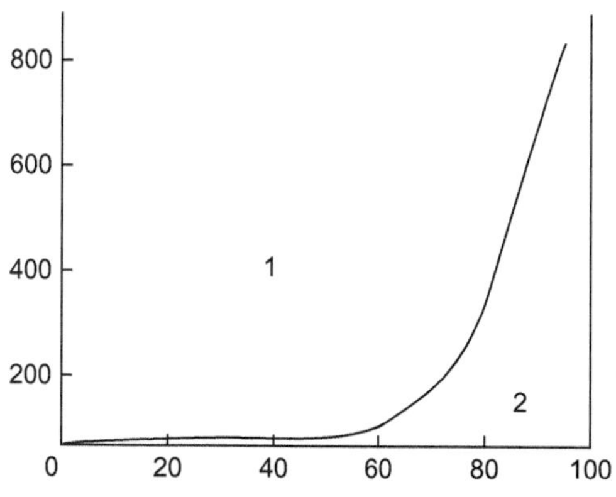

Figure 7.60 Temperature-composition diagram for the system $H_2O-P_2O_5$ at 1 atm pressure [240].

As is evident, the type of rare earth phosphate determines the degree of condensation of phosphoro-oxygen anions, which depends upon the pH of the solution and concentration of P_2O_5. And during the interaction of $NdCl_3$ and YCl_3 with sodium tripolyphosphate at pH > 4.5, formation of $Nd_5(P_3O_{10})H_2O$ was established [258]. The actual process of synthesis from aqueous solution takes place in many stages. The soluble pyrophosphates $M_4P_2O_7$ (where M = NH_4, Na, K) serves as the source of phosphate anions. In the crystallization of $Na_5R(P_4O_{12})_2$, triphosphate anions $(P_3O_{10})^{-5}$ are initially stabilized in the form of solution-$(NH_4)_5P_3O_{10}$ or $Ca_5P_3O_{10}$, in which $ErCl_3$ or $NdCl_3$ dissolves followed by the crystallization of monocationic and mixed triphosphates $Er_5P_3O_{10}$, $MErP_3O_{10} \times H_2O$, where M = Cs and NH_4 [259]. By changing the partial pressure of water and the temperature of the system, the existence of phosphoro-oxygen anions (starting from orthophosphate up to ultraphosphate) can be stabilized. A required $[P_2O_7]$, $[P_4O_{12}]$, or $[P_5O_{14}]$ anion is created first in the solution, and subsequently it was allowed to react with metals to obtain respective metal phosphate complexes. By such a direct method ortho-, pyro-, meta-, and ultraphosphates of rare earth elements have been synthesized.

A series of mixed and simple anhydrous and hydrous rare earth phosphates have been obtained by the Russian and Chinese groups [46,257–260]. Under hydrothermal conditions, the crystal chemistry of these rare earth phosphates have been studied in great detail and correlated with growth conditions. Xu et al. [258–260] and Pang (Personal communication, 1999) have studied $M_2O-CeO_2-P_2O_5-H_2O$ (M = NH_4, Na, K) system under hydrothermal conditions and obtained $M_2Ce(PO_4)_2 \cdot H_2O$, $(Ce = 4^+)$ and $M_3Ce(PO_4)_2$, $M_5Ce(PO_4)_3$.

Byrappa et al. [261,262] synthesized $Na_2(R, Co)Zr(PO_4)_3$ (R = rare earths) crystals under hydrothermal conditions at $T = 200°C$ and pressure 50–60 atm. The experiments were carried out in Morey-type autoclaves provided with Teflon liners. The cross section of the autoclave used in the synthesis is shown in Figure 7.61.

The crystallization was carried out by spontaneous nucleation that was controlled through a systematic rate of heating the alkaline component (Na_2O). It was taken in the form of a solution (NaOH) with a known molarity and this solution acts as a mineralizer. In the growth of double pyrophosphates, the addition of surplus P_2O_5 did not change the resultant product, which can be explained from the influence of H_2O vapor pressure in the system. Surplus P_2O_5 should actually result in the formation of simple or compound orthophosphates.

These rare earth orthophosphates have been obtained using the starting components in the following molar ratio:

$$Na_2O{:}M^{2+}O.\ M_2^{3+}O_3{:}\ M^{+4}O_2{:}P_2O_5 = 4.5{:}1{:}0.8{:}10 \quad (7.18)$$

This has been reviewed by Byrappa [263].

Bykov et al. [264] have obtained $Li_3M_2(PO_4)_3$ (M = Fe, Sc, Cr) using different techniques including hydrothermal technique. However, the resultant product was polycrystalline powder. The crystallization of $Li_3Sc_2(PO_4)_3$ was carried out from the systems: $LiH_2PO_4-Sc_2O_3-H_3PO_4-H_2O$; $Li_3Sc_2(PO_4)_3-NH_3H_2PO_4-H_2O$;

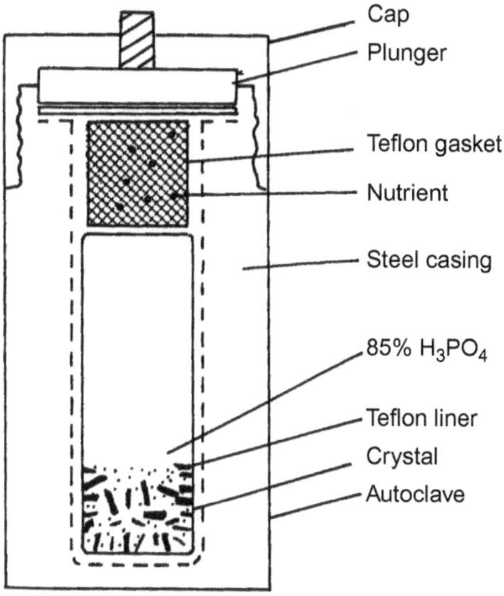

Figure 7.61 Cross section of the autoclave used in the synthesis of superionic phosphates [261].

and $Li_3Sc_2(PO_4)_3$–LiH_2PO_4–H_2O, besides $Li_3Sc_2(PO_4)_3$, the other stable phases like $ScPO_4$, and scandium hydroxides have obtained.

Thus, the hydrothermal technique is becoming a popular technique to synthesize rare earth phosphates. What is most important in the case of hydrothermal synthesis of phosphates in comparison with their analogous compounds—silicates and germanates—is that it is still necessary to clarify why a wide range of structural variations have not been achieved through hydrothermal technique. The other synthesis techniques like aqueous solutions, flux, and phosphoric acid solutions have resulted in very complex structures, whereas the hydrothermal technique has resulted in only a few anhydrous phosphates types like ortho-, diortho-, triorthogroups, pyrophosphates, simple tetra- and ultraphosphates. Therefore, let us discuss the crystal chemistry of rare earth phosphates.

7.13.3 Structure Types of Rare Earth Phosphates

The silicates and germanates have been sufficiently well studied. However, it is only in the recent years that the rare earth phosphates are being studied extensively. As mentioned earlier, the structures of condensed rare earth phosphates in particular do not resemble the structures of silicates and germanates, but the popular representatives such as ortho-, diortho-, ring, layered, chain, ribbon structures, exist in phosphates also, rare earth silicates have 40 structure types, the rare earth germanates have 32 structure types, and rare earth phosphates have over 30 structure types [52]. From the studies on the crystallization of rare earth silicates, germanates, and phosphates, it is clear that a decrease in the R_2O_3/X_xO_y ($X = Si$, Ge, P) ratio leads to the formation of condensed anions. Tables 7.22 and 7.23 give the structure types of alkali-free rare earth phosphates and alkali rare earth phosphates reported so far in the literature.

Table 7.22 Structure Types of Alkali-Free Rare Earth Phosphates [7,52,240]

Compound	Representative	Cell Parameter (Å)	Rare Earth, Co. No., Average Bond Distance Å			Tetrahedral Radical Form	Rare Earth Polyhedra
R $[PO_4]$ R = Gd–Lu, Y, Sc	$Y[PO_4]$	14/and $Z = 4$ $a = 6.885$ $c = 5.982$	Y	8	2.28	Orthotetrahedra	Framework
R $[PO_4]$ R = La–Eu	$Ce[PO_4]$	$P3_121$ $Z = 3$ $a = 7.055$ $c = 6.439$	Ce	8	2.48	Orthotetrahedra	Framework
R $[PO_4]$ R = La–Dy	$Ce[PO_4]$ monazite	$P2_1/n$ $Z = 4$ $a = 6.78$ $b = 7.00$ $c = 6.44$ $\beta = 103.63°$	Ce	8	2.51	Orthotetrahedra	Framework
	$YbH_2[P_3O_{10}]$	A_2 $Z = 2$ $a = 5.617$ $b = 6.666$ $c = 10.011$ $\beta = 97.32°$	Yb	6	2.19	Triortho groups	Discrete polyhedra
R $[PO_3]_3$ R = La–Gd	Nd $[PO_3]_3$	$C222_1$ $Z = 4$ $a = 11.172$ $b = 8.533$ $c = 7.284$	Nd	8	2.43	Chains with a period of 6-tetrahedra	Columns
R $[PO_3]_3$ R = Gd–Lu	Yb $[PO_3]_3$	$P2_1/c$ $Z = 12$ $a = 11.219$ $b = 19.983$ $c = 9.999$	Yb	6	2.20	Chains with a period of 6-tetrahedra	Discrete polyhedra

(*Continued*)

Table 7.22 (Continued)

Compound	Representative	Cell Parameter (Å)	Rare Earth, Co. No., Average Bond Distance Å			Tetrahedral Radical Form	Rare Earth Polyhedra
		$\beta = 97.30°$					
R [P_5O_{14}] R = La–Tb	Nd[P_5O_{14}]	$P2_1/c$ $Z = 4$ $a = 8.771$ $b = 9.012$ $c = 13.057$ $\beta = 89.58°$	Nd	8	2.43	Ribbons from 8-membered rings	Discrete polyhedra
R [P_5O_{14}] R = Dy, Ho, Er, Y	Ho[P_5O_{14}]	$Pcmn$ $Z = 4$ $a = 8.926$ $b = 12.71$ $c = 8.720$	Ho	8	2.32	Ribbons from 8-membered	Discrete polyhedra
R[P_5O_{14}] R = Tb–Lu, Y	Yb [P_5O_{14}]	$C2/c$ $Z = 8$ $a = 12.830$ $b = 12.676$ c = 12.337 $\beta = 91.25°$	Yb	8	2.30	Layer from 20- and 8-membered rings	Discrete polyhedra

Co. No. = coordination number.

Table 7.23 Structure Types of Alkali Rare Earth Phosphates [7,52,240]

Compound	Representative	Cell Parameter (Å)	Rare Earth, Co. No., Average Bond Distance (Å)			Tetrahedral Radical Form	Rare Earth Polyhedra
$Na_3Sc_2[PO_4]_3$		*Bb* Z = 4	Sc	6	2.092	Orthotetrahedra	Discrete polyhedra
	$a = 16.10$						
	$b = 9.109$						
	$c = 8.928$						
	$\gamma = 127.15°$						
$Na_3Sc_2[PO_4]_3$		*R*3c Z = 6	Sc	6	2.080	Orthotetrahedra	Discrete polyhedra
	$a = 8.927$						
	$c = 22.34$						
$Na_{4.5}R_{1.5}[PO_4]_3$	$Na_{4.5}Y_{1.5}[PO_4]_3$	*R*3c Z = 6	Yb, Na	6	2.213	Orthotetrahedra	Discrete polyhedra
R = Tm, Yb, Lu		$a = 9.12$					
		$c = 21.81$					
$Na_3R[PO_4]_2$	$Na_3Nd[PO_4]_2$	$Pbc2_1$ Z = 24	Nd	7	2.41	Orthotetrahedra	Discrete polyhedra
R = La–Er, Y		$a = 15.874$	Nd	6	2.43		
		$b = 13.952$	Nd	8	2.48		
		$c = 18.470$	Nd				
$K_3Nd[PO_4]_2$		$P2_1/m$ Z = 2	Nd	7	2.44	Orthotetrahedra	Discrete polyhedra
	$a = 9.532$						
	$b = 5.631$						
	$c = 7.444$						
	$\beta = 90.95°$						
$Na_3Ce[PO_4]_2$		Orthorhombic	Ce	–	–	–	––
	$a = 14.07$						
	$b = 16.04$						
	$c = 18.61$						

(*Continued*)

Table 7.23 (Continued)

Compound	Representative	Cell Parameter (Å)	Rare Earth, Co. No., Average Bond Distance (Å)			Tetrahedral Radical Form	Rare Earth Polyhedra
	$v = 4200$ Å^3						
$MR[PO_4]_2.H_2O$ $K_2Ce[PO_4]_2.H_2O$ $M = NH_4$, K, Na R = Yb, Ce	$a = 12.793$ Å $b = 14.015$ Å $c = 15.945$ Å $v = 2854$ Å^3	Orthorhombic	–	–	–	–	–
$M_3R_2[PO_4]_3$ M = Na,K R = Ce–Gd	$K_3Ce_2[PO_4]_3$	$P3ml$, $P_{31}m$ $a = 9.30$ $c = 14.41$	Z = 3	–	–	–	–
$KR[PO_3]_4$ R = Eu, Ho	$KHo[PO_3]_4$	$C2/c$ $Z = 4$ $a = 7.798$ $b = 12.310$ $c = 10.551$ $\beta = 112.63°$	Ho	8	2.34	4-Membered rings	Discrete polyhedra
β-$CsR[PO_3]_4$ R = Pr, Nd	β-$CsNd[PO_3]_4$	143d $Z = 12$ $a = 15.233$	Nd	8	2.46	4-Membered rings	Discrete polyhedra
$LiNd[PO_3]_4$	$a = 16.408$ $b = 7.035$ $c = 9.726$ $\beta = 126.38°$	$C2/c$ $Z = 4$	Nd	8	2.46	Chains with a period of 4-tetrahedra	Discrete polyhedra
α-$KR[PO_3]_4$ R = Nd, Sm, Dy, Ho, Er	α-$KNd[PO_3]_4$	P_{21} $Z = 2$ $a = 8.007$ $b = 8.433$	Nd	8	2.44	Chains with a period of 4-tetrahedra	Discrete polyhedra

(Continued)

Table 7.23 (Continued)

Compound	Representative	Cell Parameter (Å)	Rare Earth, Co. No., Average Bond Distance (Å)	Tetrahedral Radical Form	Rare Earth Polyhedra
		$c = 7.280$			
		$\beta = 91.97°$			
$NaNd[PO_3]_4$		P_{21}/n $Z = 4$	Nd 8 2.45	Chains with a period of 4-tetrahedra	Discrete polyhedra
	$a = 9.907$				
	$b = 13.10$				
	$c = 7.201$				
	$\beta = 90.51°$				
$A\text{-}CsR[PO_3]_4$	$CsNd[PO_3]_4$	$P2_1$ $Z = 2$	Nd 8 2.45	Chains with a period of 8-tetrahedra	Discrete polyhedra
R = Tb, Pr, Nd		$a = 7.145$			
		$b = 8.809$			
		$c = 11.007$			
		$\beta = 105.86°$			
$MR[PO_3]_4$	$T1Nd[PO_3]_4$	$P2_1/Z = 4$	Nd 8 2.441	Chains with a period of 8-tetrahedra	Discrete polyhedra
M = K,Tl,Rb,Cs		$a = 10.440$			
R = Nd, Ho, Er		$b = 8.050$			
		$c = 11.007$			
		$\beta = 105.86°$			
$KYb[PO_3]_4$		$P2_1/n$ $Z = 4$	Yb 7 2.27	Chains with a period of 8-tetrahedra	Discrete polyhedra
	$a = 7.776$				
	$b = 8.853$				
	$c = 14.831$				
	$\beta = 96.36°$				

Co. No. = coordination number.

7.14 Hydrothermal Growth of Mixed Valent Metal Phosphates

The interest in the synthesis of phosphates, other than rare earth phosphates, is connected with the publication of NASICON, whose ionic conductivity is equivalent to that of Na β-alumina. It has a mixed framework structure and the general formula is $Na_{1+x}Zr_2Si_xP3_{-x}O_{12}$; $0<x<3$ [265]. It is a solid solution between $NaZr_2P_3O_{12}$ and $Na_4Zr_2Si_3O_{12}$. NASICON posed a challenge to materials scientists in understanding its structure and conduction mechanism due to the lack of single crystals, nonstoichiometry in the composition, zirconium deficiency, and so on. Many variations have been investigated by appropriate substitutions in the NASICON system [261]. The basic structure of most of these derivatives remains that of $Na_3Sc_2P_3O_{12}$. NASICON has two structure types, namely, NASICON and anti-NASICON [266,267].

The structure of NASICON is highly complicated because of its solid solution nature. There are nearly a hundred atoms per unit cell, which makes it rather difficult to describe. However, simplification can be introduced by assuming that this type of structure is derived from that of glaserite $K_3Na(SO_4)_2$—a rhomobohedral super-structure [268] (Figure 7.62).

Hong [266] studied the structure of NASICON, $Na_{1+x}Zr_2Si_xP_{3-x}O_{12}$, solid solution series and reported a monoclinic deformation. Space group is C_2/c in the region of $x = 1.8-2.2$. The first crystal structure of NASICON-type material was performed in 1968 by Hagman and Kierkegaard [269], who studied the structure of $NaM_2(PO_4)_3$ (where M = Ge, Ti, and Zr) and found them to be isomorphous. $Na_4Zr_2(SiO_4)_3$ structure has been further studied by Sizova et al. [270] and Tranqui et al. [271] and reported multiple Na^+ sites leading to the nonstoichiometry in NASICON compound. Delbecq et al. [272], Susman et al. [273], Collin et al. [274], Boilot et al. [275], Clearfield et al. [276], and Baur et al. [277] have reported five different lattice sites for Na^+, out of which only the Na^+ is mobile.

Stoichiometry, structure, and conductivity of NASICON are being studied by several authors consistently from time to time. Since the structure of NASICON is

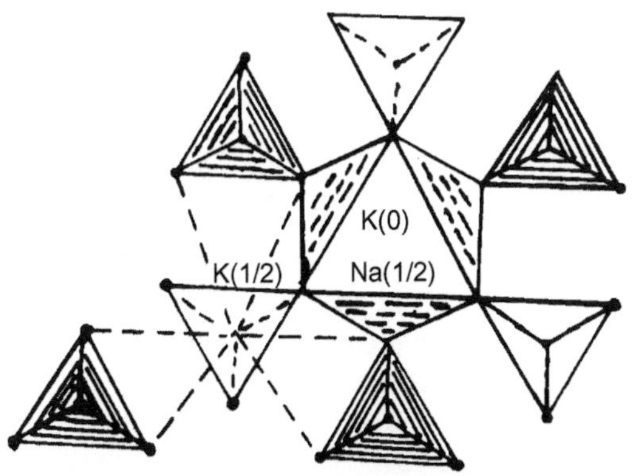

Figure 7.62 Structure of glaserite $K_3Na(SO_4)_2$ [268].

highly complicated, scientists continue to strive to develop new superionic conductors with simple structures. Phosphates are found to be the most suitable ones. The structure of these phosphates consists of a three-dimensional framework. A series of new NASICON analogs like $Na_3M^{2+}(PO_4)_3$, (M = Sr, Mg, Fe, Mn) [278]. $Na_2(R, M^{3+})M^{4+}(PO_4)_3$ [263]. $A_3M^{3+}(PO_4)_3$, (A = Li, Na, Ag, K; M = Cr, Fe) [279], $ABC(PO_4)_3$ (A = K, Rb, Cs; B = Ca, Sr; C = Ln and Bi) [280] (M = Sc, Cr, Fe) [281] were synthesized with an intention to develop compounds having stoichiometric composition and relatively simple structures. However, all these compounds contained only the orthogroup of radicals, irrespective of the wide variety. Byrappa and coworkers [282–284] have reported, for the first time, pyrophosphates showing high ionic conductivity grown by a hydrothermal method.

It has not been attempted, using the hydrothermal method, to grow phosphates widely, except for rare earth phosphates and aluminum orthophosphates during the 1950s [246,251,252,280–289]. Only a few reports on the hydrothermal growth of superionics are seen in literature. Clearfield et al. [276] and Genet and Barj [290] adopted this technique to process NASICON compounds. According to them, an insufficient amount of Si^{+4} enters the composition, but it is possible to develop stoichiometric NASICON. The problem arises when Si^{+4} is added to the system. By adopting the hydrothermal method, almost all the problems quoted earlier have been overcome by Byrappa and Gopalakrishna [248].

In the case of hydrothermal growth, the selection of the nutrient materials, including solvents, is the most important aspect. Such studies are not readily available in the literature as the hydrothermal crystal growers touch upon the solvent effect, solvent–solute interaction, complex formation, etc. only in the recent years. Keeping this in mind, the authors have made an attempt for solvent selection with reference to the superionic pyrophosphates containing different divalent metals. A full understanding of the complex system formed in solutions is inconceivable without a study of the effects of the different solvents on dissolved ionic species. The properties of a solvent are determined jointly by its general chemical nature and physical properties. These properties are involved in such a complex interaction that it is difficult to establish unambiguously how they contribute individually to the general behavior of the solvents. Complex formation reactions occurring in solutions are known to be, in effect, substitution reactions in which the legend replaces solvent molecules bound in the solvate sphere of the metal ion. Thus, it is understandable that the rates, kinetics, and mechanism of such reactions are dependent on the solvents. The magnitude of the interaction between the coordinated solvent molecules and the cation (the strength of the coordinate bond) influences the rate of exchange of the coordinated solvent with some other legend.

In the most complex-equilibrium investigations in solvent mixtures, the choice of the solvent, was determined by the solubilities of the components of the system. The aim of the investigations, therefore, was not the detection of the solvent effect, but the determination of the compositions and stabilities of the complexes formed in solution. The linkage between the solvent and solute is essentially a coordination type and hence varied acid–base complexes have been observed in the solution. Byrappa and Umesh Dutt [291] and Byrappa and Nirmala [292] have studied in great detail the hydrothermal crystallization mechanism of $HNaMP_2O_7$ (where M = Mn, Co, Ni, Zn, Cd, Pb) crystals. Table 7.24 gives the experimental conditions of the growth of these novel phosphates with very interesting structures. In all the experiments, 85% H_3PO_4 and NaOH solutions have been used as the solvents and only in a very few experiments, other solvents like HNO_3, HCl, H_2SO_4, and HCOOH have been used.

The hard bases H_2O, OH^-, PO_4^{-3}, NO_3, and SO_4^{2-} act upon the dissolution of the solvated complexes determining the solubility of the final resultant stable complex [293]. The following reaction series can be proposed for the formation of $HNaCoP_2O_7$:

$$3NaOH + H_3PO_4 \rightarrow Na_3PO_4 + 3H_2O$$

$$3Co(NO_3)_2 + 2H_3PO_4 \rightarrow Co_3(PO_4)_2 + 3H_2O + 6NO_2\uparrow + 3[O]\uparrow$$

$$Zr(NO_3)_4 + 4NaOH \rightarrow 4NaNO_3 + 2H_2O + ZrO_2\downarrow$$

$$2H_3PO_4 \rightarrow H_4P_2O_7 + H_2O$$

$$Na_3PO_4 + H_4P_2O_7 \rightarrow NaH_3P_2O_7 + HNO_3$$

$$Na_3PO_4 + H_4P_2O_7 \rightarrow NaH_3P_2O_7 + Na_2HPO_4$$

$$Co_3(PO_4)_2 + H_4P_2O_7 + 4H_2O \rightarrow H_2CoP_2O_7 + 2H_3PO_4 + 2Co(OH)_2$$

$$Co(NO_3)_2 + H_4P_2O_7 \rightarrow H_2CoP_2O_7 + 2HNO_3$$

$$Co(NO_3)_2 + H_4P_2O_7 \rightarrow H_2CoP_2O_7 + 2HNO_3$$

$$NaOH + H_2CoP_2O_7 \rightarrow HNaCoP_2O_7 + H_2O$$

$$Co(OH)_2 + NaH_3P_2O_7 \rightarrow HNaCoP_2O_7 + 2H_2O$$

The formation of a stable complex, i.e., $HNaCoP_2O_7$ at 250°C and 100 bar pressure, must have undergone several intermediate stages of solvation, forming stable and unstable solvated complexes like $NaNO_3$, having very low values of lattice energy constant, 755 kJ/mole, and Co, which has a desolvation energy value around 15 eV, etc. that have further interacted with each other because of their weak ionic bonding of very low charge to give a stable coordinated compound with a very high of lattice energy constant and stronger ionic bonding. But at temperatures around 250°C, the polymerization of the orthogroup of phosphates to form the pyrogroup with higher lattice energy may be expected. Also, the reaction between the alkali pyrophosphate complexes with metal complexes results in the formation of stable complexes of pyrophosphate, i.e., $HNaCoP_2O_7$.

Table 7.24 Experimental Conditions for Low-Temperature Hydrothermal Synthesis

Compound	Nutrient Components				T (°C)	P (bar)	Duration (days)	Size (mm)	Prominent Faces	Morphology
$Na_2H_3Al(P_2O_7)_2$	H_3PO_4 (ml)	NaOH (ml)	$A1_2O_3$ (gm)	$Zr(NO_3)_4$ (gm)						Rhombohedral, tabular, rods, shaped, prismatic
	6	5(5M)	1	0.5	250	85	8	0.15–2	010	
$NaHCoP_2O_7$	H_3PO_4 (ml)	NaOH (ml)	$Co(NO_3)_4$ (gm)	$Zr(NO_3)_4$ (gm)						Perfect rhombohedral, prismatic red shaped
	5	5(5M)	1	0.3	250	80	8	0.5–2	010	
$NaHNiP_2O_7$	H_3PO_4 (ml)	NaOH (ml)	$NiCl_2$ (gm)	$Zr(NO_3)_4$ (gm)						
	5	5(5M)	1.0	0.5	250	85	8	0.5	110	Prismatic, rod shaped
	5	5(5M)	1.25	0.5	250	85	8	0.5		
	5	5(5M)	1.5	0.5	250	90	8	0.5–1.5		
$NaHMnP_2O_7$	H_3PO_4 (ml)	NaOH (ml)	$MnCl_2$ (gm)	$CaCo_3$ (gm)						
	5	5(5M)	1.5	0.75	250	85	8	1	100	Rod shaped, prismatic
	5	5(5M)	1.5	1.00	250	90	8	1–3		
	5	5(5M)	1.5	1.25	250	90	8	2–3		
	5	5(5M)	1.5	1.5	250	90	8	2		
$NaHZnP_2O_7$	H_3PO_4 (ml)	NaOH (ml)	ZnO (gm)	$Zr(NO_3)_4$ (gm)						
	5	5(5M)	0.5	0.5	250	80	8	1–2	101	Prismatic, needles
	5	5(5M)	0.75	0.5	250	80	8	1–2		
	5	5(5M)	1.00	0.5	250	85	8	2–5		
	5	5(5M)	1.50	0.5	250	90	8	1–3		
$NaHCdP_2O_7$	H_3PO_4 (ml)	NaOH (ml)	CdO (gm)	$Zr(NO_3)_4$ (gm)						
	5	5(5M)	1	0.5	250	70	8	2–5	111	Subrounded tabular, dome shaped

Source: From the works of K. Byrappa.

The hydrothermal crystallization of these superionic phosphates has been studied through another approach, which is quite effective from the point of hydrothermal solution chemistry.

The crystallization process involves many chemical interactions leading to the formation of a stable complex. To understand the growth of title compounds, the authors have described the following possible stages of interactions [291]:

1. Acid–base interactions
2. Formation of metal–aqua complexes
3. Interaction between acid–base and metal–aqua complexes.

Acid–base interactions. Components such as H_2O, OH^-, and PO_4^{3-} are considered as the hard and form pyrophosphate complexes at around 250°C:

$$2H_3PO_4 \rightarrow H_4P_2O_7 + 2H_2O \quad (7.19)$$

$$NaOH + H_4P_2O_7 \rightarrow NaH_3P_2O_7 + H_2O \quad (7.20)$$

As the pH increases, the stability of $NaH_3P_2O_7$ decreases (pH > 3), and this leads to the formation of $Na_2H_2P_2O_7$:

$$NaOH + NaH_3P_2O_7 \rightarrow Na_2H_2P_2O_7 + H_2O \quad (7.21)$$

The occurrence of Eq. (7.21) also increases the pH of the system, but the compound $Na_2H_2P_2O_7$ is stable in the pH range 3–9. The formation of $Na_3HP_2O_7$ is favored if pH > 9.

$$NaOH + Na_2H_2P_2O_7 \rightarrow Na_3HP_2O_7 + H_2O \quad (7.22)$$

Formation of metal–aqua complexes. The zinc metal–aqua complexes in the hydrothermal system have been studied in detail. The tendency of zinc toward a tetrahedral coordination is evidently responsible for the existence of $[Zn(H_2O)_x(OH)_{4-x}]^{x-2}$ type aqua complexes in aqueous solutions, where the value of x depends on the pH of the medium [294]. In strong alkaline solutions, $Zn(OH)_4^{2-}$ occurs (and here $x = 0$). By reducing the solution pH to neutral, a dizincate $[Zn_2O(H_2O)_x(OH)_{6-x}]^{x-4}$ may be formed as a result of the association of zincate ion.

The zinc oxide forms aqua complexes when dissolved in alkaline solutions [295]. The distribution of zinc aqua ions in alkaline solution is shown in Figure 7.63.

Similar aqua complexes can be expected for the other divalent cations, like Mn^{2+}, Ni^{2+}, Fe^{2+}, Cu^{2+}, Pb^{2+}, and Cd^{2+}, under hydrothermal conditions.

Although the metal cations form octahedral complexes, tetrahedral complexes are the more stable ones, and a few octahedral complexes retain their constitution for a long time depending upon their half-life of aqua-ligands [296].

Interactions between acid–base and metal–aqua complexes. Pyrophosphoric acid forms salts due to the sequestering action of the pyrophosphate radical, the

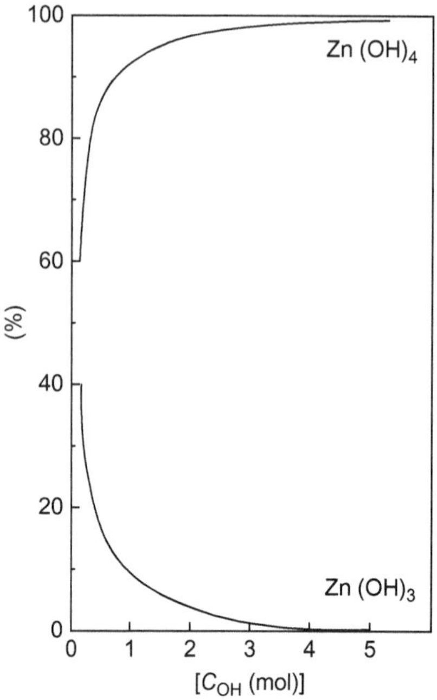

Figure 7.63 Distribution of zinc aqua ions in alkaline solution [295].

sequestering ability decreases with higher concentrations of salts, or with increasing pH. Interactions between metal–aqua complexes with the sodium salts of pyrophosphoric acid can be written as follows:

$$NaH_3P_2O_7 + M(OH)_2 \rightarrow HNaMP_2O_7 + 2H_2O \quad (pH = 2\text{–}3) \tag{7.23}$$

$$Na_2H_2P_2O_7 + M(OH)_2 \rightarrow HNaMP_2O_7 + NaOH + H_2O \quad (pH = 3\text{–}9) \tag{7.24}$$

$$Na_3HP_2O_7 + M(OH)_2 \rightarrow HNaMP_2O_7 + 2NaOH \quad (pH > 9) \tag{7.25}$$

where M = Mn, Fe, Co, Ni, Cu, Zn, Pb.

The aqua complexes can form more stable complexes because of the higher desolvation energy of metal cations when compared to sodium (3.97 eV). Hence, the formation of $HNaMP_2O_7$ is expected in the system instead of stable sodium salts of pyrophsophoric acid, namely $Na_2H_2P_2O_7$ and $Na_4P_2O_7$.

The hydrothermal growth of $HNaMP_2O_7$ (where M = Mn, Co, Ni, Zn, Cd, Pb) crystals clearly shows that there is a gradual change in the crystallinity and morphology from Mn end to the Pb end. Table 7.25 gives the cations in superionic pyrophosphates. As evident from Table 7.25, higher the entropy values, the lower will be the degree of crystallinity and the crystal morphology. For example,

Table 7.25 Cations in Superionic Pyrophosphates [297]

Element	At. No.	At. Wt.	At. Radii[a] (A°)	Entropy at 298 °K (e.u.)
Al	13	65.38	1.82	6.769
Mn	25	54.93	1.79	7.590
Co	27	58.93	1.67	6.800
Ni	28	58.70	1.62	7.137
Zn	30	65.38	1.53	9.950
Cd	48	112.41	1.71	12.300
Pb	82	207.20	1.81	15.490

[a]Table of periodic properties of the elements, Sargent–Welch Scientific Company.

$HNaMP_2O_7$ (M = Mn, Co, Al, Ni) show good crystallinity and well-developed crystal morphology, whereas the crystals $HNaMP_2O_7$ (M = Zn, Cd, Pb) with higher entropy values have a lower degree of crystallinity. These crystals show poor morphology and are often polycrystalline [297].

The authors have obtained good crystals up to 6 mm size. Twinning was absent in general, except in crystals doped with additives. In the case of $(Na_{2/3}Zr_{1/3})_2P_2O_7$, Bi and Ce were added as additives even in higher concentration, but they did not enter the composition. The cations play a major role in the crystallization of pyrophosphates. Also, divalent metals enter the composition more easily than trivalent metals, particularly in condensed phosphates. A large number of experiments have been carried out using Al, Cr, Fe, Bi, Co, and rare earths even in surplus amount. Only Al enters the composition easily and others need higher temperature of synthesis. The volatiles and type of cations play a major role in the crystallization of condensed phosphates. The morphology of these crystals depends upon the concentration of P_2O_5 and H_2O in the system [263,298]. For example, in $NaHCo(P_2O_7)$ crystals, the prism faces dominate over the pyramidal faces with an increase in P_2O_5 and H_2O concentration. The morphology of these superionic pyrophosphates is very interesting and it is shown together with their schematic diagrams in Figure 7.64a–e. The crystallization was carried out by spontaneous nucleation, which was controlled by a systematic rate of heating. Hence, the crystals were well developed and of excellent quality.

The X-ray diffraction studies (both powder and single crystal) show that these crystals belong to the lower symmetry and show very interesting morphology. A general morphology of these pyrophosphates is given in Table 7.25.

In order to grow these crystals, it is very important to know their solubility, internal atomic structure, macro- and micromorphology. The solubility of these pyrophosphates was found to be positive. Similarly, the crystals obtained by the hydrothermal technique show very interesting morphology. Hence, it becomes rather easy to grow them as bulk single crystals of several centimeters in length, despite the phosphoric acid media in which the growth occurs.

The most common morphological habits observed in pyrophosphates are perfect rhombohedral, prismatic, rod shaped, amongst others. The crystal faces are more or

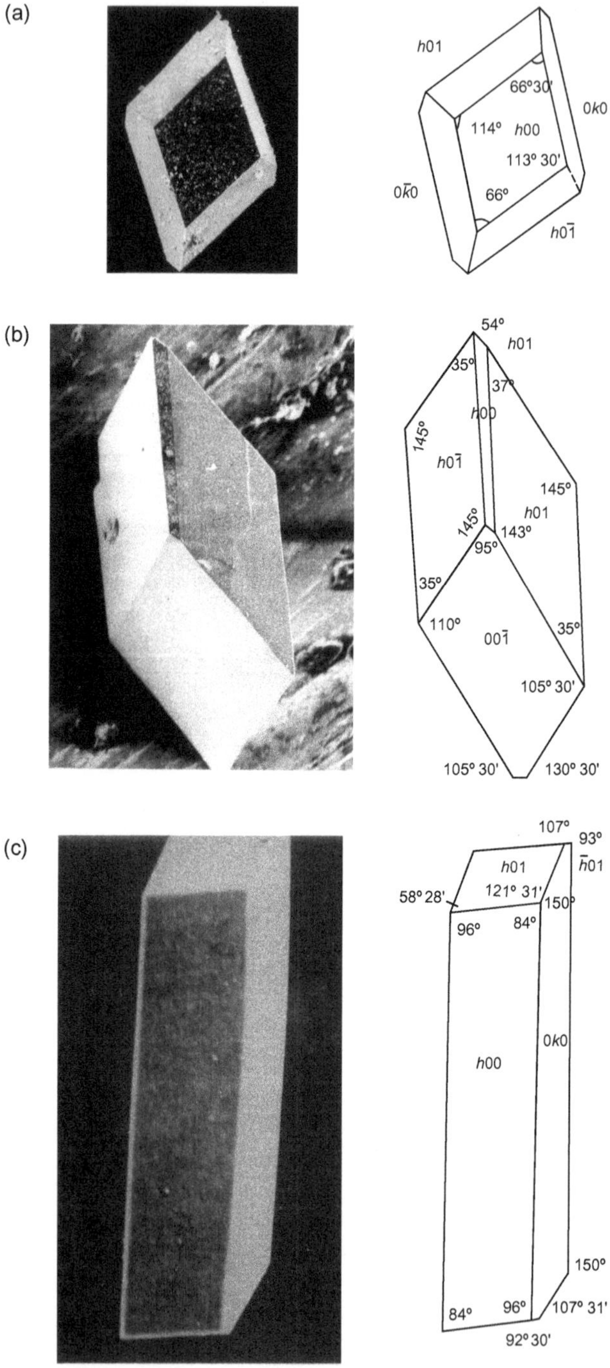

Figure 7.64 Morphology of superionic pyrophosphates (magnification $\times 20$) [298].

(d)

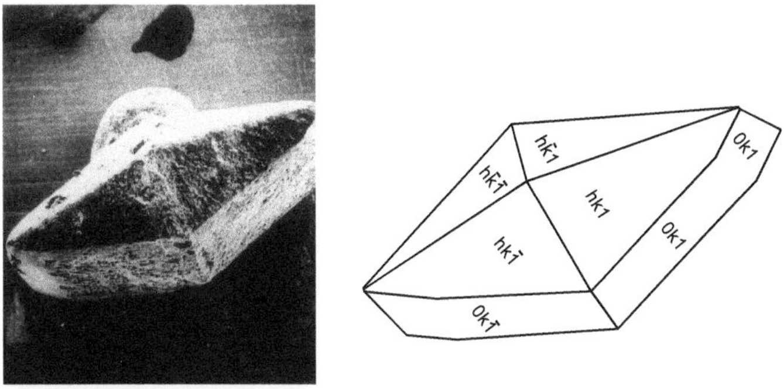

(e)

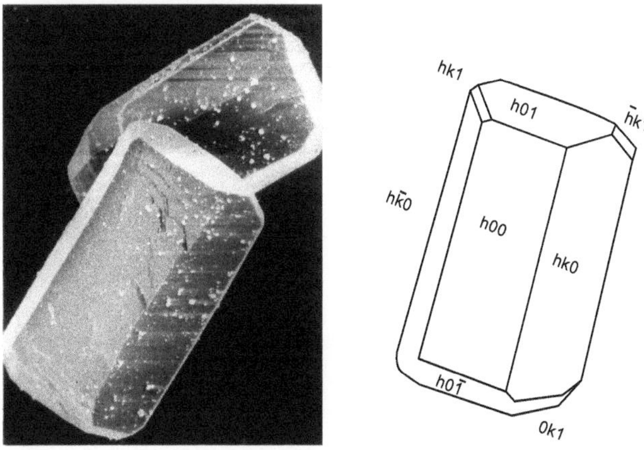

Figure 7.64 Continued

less smooth and vitreous in luster, with a high degree of transparency. The prism faces dominated over all other faces in all the crystals followed by pinacoidal and dome faces. There is a clear-cut distinction in the morphology of these pyrophosphates. In most of the experiments, the entry of the selective dopants was controlled in order to observe the morphological variations. The morphology of these pyrophosphates varies from one to another, depending on the cations present.

To gain an understanding of the morphological variation with respect to the growth conditions, it is essential to study particularly the role of the degree of supersaturation, the concentration of water and Na_2O in the system. $NaHCo(P_2O_7)$ and $NaHZn(P_2O_7)$ crystals were selected for such a systematic study, because these compounds show a very well-developed morphology and are sensitive to such variations. In fact, both these compounds show similar morphological variations. Figure 7.64a–e shows how the morphology varies with the concentration of P_2O_5. This is further represented in the schematic diagram of Figure 7.65. The interfacial angles for the representative pyrophosphates have been measured using a Leitz phase-contrast microscope and these angles are shown in the schematic diagrams

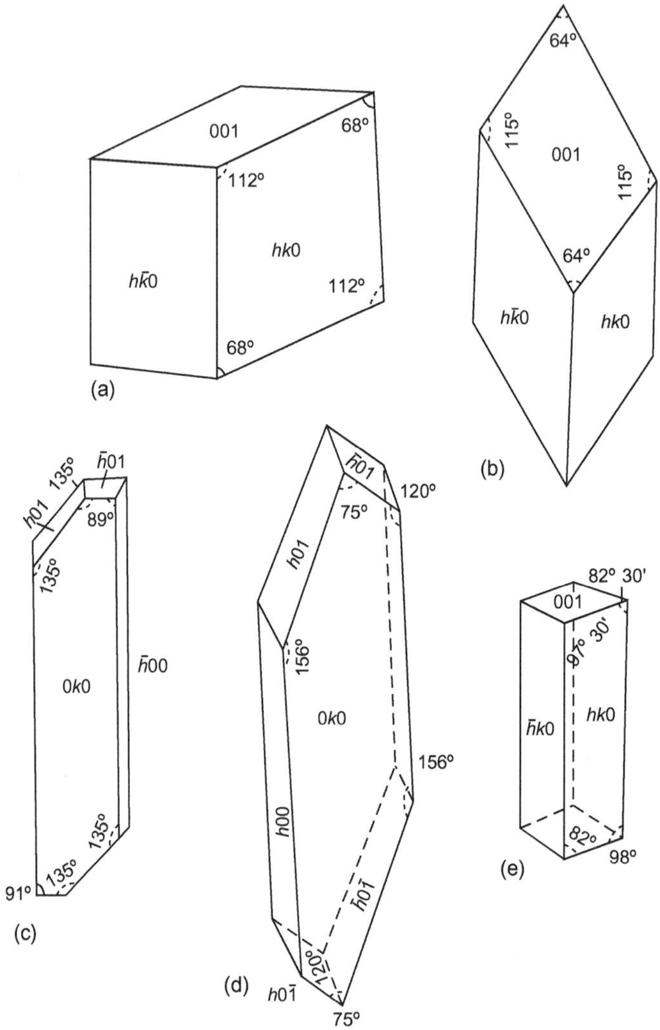

Figure 7.65 Schematic diagrams of the variation in morphology of superionic pyrophosphates with P_2O_5 content in the system.

of Figure 7.65. As the concentration of P_2O_5 and H_2O increases, the pyramidal faces are replaced by the prismatic and the pinacoidal faces [298]. Similarly, an attempt has been made to study the variation of the growth rate of $NaHCo(P_2O_7)$ crystals with concentration of P_2O_5, Na_2O, and H_2O (Figure 7.66a–c). These figures show critical points, beyond which the growth rate falls sharply. This helps to fix the optimum conditions for the growth of $NaHCo(P_2O_7)$ [298].

Similar features were observed in the case of $NaHZn(P_2O_7)$ crystals. Representative photographs (Figure 7.67) of $NaHZn(P_2O_7)$ show how morphology

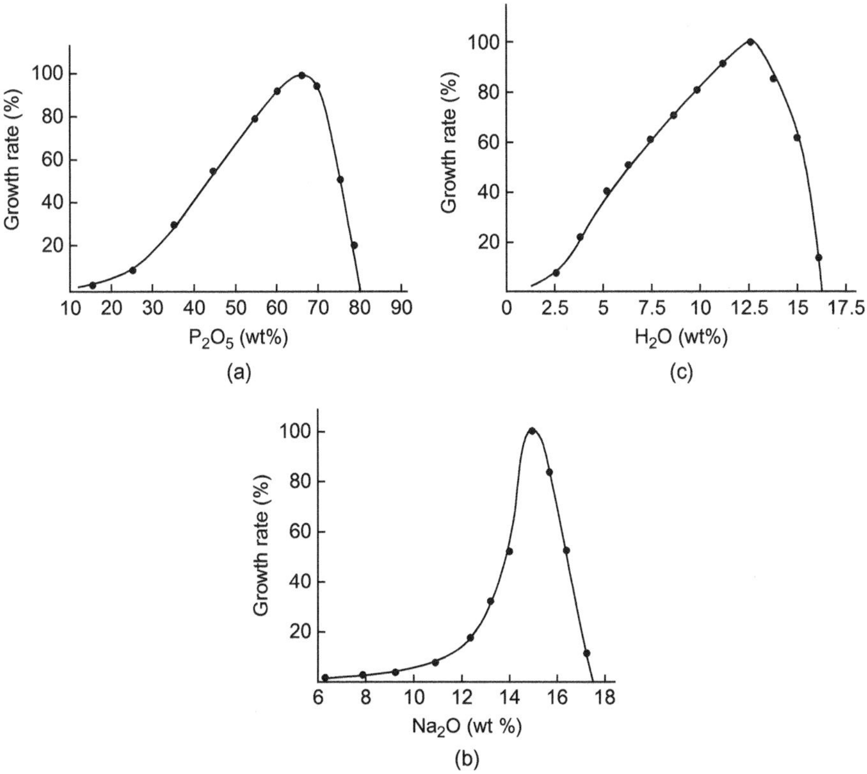

Figure 7.66 Growth rate of superionic pyrophosphate, $NaHCo(P_2O_7)$ versus (a) P_2O_5 concentration, (b) Na_2O concentration, and (c) H_2O concentration [298].

varies with the above-mentioned growth parameters. These experiments indicate that the degree of supersaturation, concentration of cations, and dopants determine the morphology of these pyrophosphates. Such studies help in the successful growth of superionic phosphates in the form of single crystals with a desired morphology. They help us to understand the growth technology of these phosphates in detail.

A number of factors such as the degree of supersaturation, temperature, pressure, accidental/deliberate entry of impurities and their concentration, type of the solvent, and pH of the mineralizer affect the habit of crystals obtained under hydrothermal conditions. Habit modifications are also observed when significant changes in the growth temperature occur, since an increase in temperature hastens the growth kinetics and in turn growth rate.

The crystal faces of most of the superionic pyrophosphates are more or less smooth and vitreous in luster with a high degree of transparency. The morphology of these pyrophosphates varies from one another depending upon the cations present.

It is interesting to observe the morphological variations in the pyrophosphates with reference to the variation in the cation. The cations used are essentially the

Figure 7.67 Photographs of NaHZn (P_2O_7) crystals (magnification × 30) [297].

divalent and trivalent metals such as Al, Mn, Co, Ni, Zn, Cd, and Pb. Table 7.25 gives cation properties. As evident from Table 7.25 the Al is the smallest ion and it shows excellent morphology (Figure 7.64a). The crystals are very well developed and highly transparent. This is followed by Mn, which gives good crystals of an excellent crystal habit. However, due to the susceptibility of Mn for the changes in the valency with sudden changes in the experimental growth parameters, there is a tendency for the formation of polymorphic modifications of Mn bearing pyrophosphates. But both the polymorphic modifications of Mn superionic pyrophosphates show excellent morphology with well-developed habits, and smooth and vitreous surfaces (Figure 7.64a).

The cobalt bearing superionic pyrophosphate shows probably the best morphology (Figure 7.64b). The crystals are very well developed with vitreous and smooth surfaces having a high degree of transparency. The size of the cobalt bearing pyrophosphate is higher than the other pyrophosphates of this class.

It is believed that all these superionic pyrophosphates belong to the same structure type, i.e., the triclinic, P1 space group. As the ionic size of the cations inserted into the structure increases, there develops a general structural distortion, which in turn affects the crystal morphology. Table 7.26 gives the variation in bond lengths of these pyrophosphates. As evident from Table 7.26, the difference in maximum and minimum bond lengths of Na–O polyhedra increases gradually toward the Zn end member. Thus, toward the Pb end member, the crystals become poorly

Table 7.26 Bond Lengths for Superionic Pyrophosphates Crystals [297]

Compound	Na–O (max)	Na–O (min)	Na–O difference (max–min)	M^{+2}–O (max)	M^{+2}–O (min)	M^{+2}–O difference (max–min)	P_1O (max)	P_1O (min)	P_1O difference (max–min)	P_2–O (max)	P_2–O (min)	P_2–O difference (max–min)
$Na_2AlH_3(P_2O_7)_2$	2.929	2.397	0.532	1.940	1.855	0.085	1.599	1.495	0.104	1.587	1.500	0.087
$NaMnHP_2O_7$	2.79	2.32	0.47	2.20	2.00	0.2	1.64	1.52	0.12	1.58	1.49	0.09
$NaCoHP_2O_7$	3.097	2.445	0.652	2.196	2.035	0.161	1.587	1.502	0.85	1.602	1.498	0.104
$NaNiHP_2O_7$	3.385	2.315	1.07	2.157	2.016	0.141	1.593	1.505	0.085	1.607	1.497	0.11
$NaZnHP_2O_7$	3.446	2.308	1.138	2.283	2.020	0.263	1.591	1.504	0.087	1.595	1.488	0.107

developed. The poor morphology of Cd and Pb bearing pyrophosphates is due to the changes in the bond lengths and bond angles leading to the slightly higher degree of structural disorder as indicated by the preliminary X-ray single crystal diffraction studies. Also, it is evident from Table 7.25 that the values of entropy for Zn, Cd, and Pb gradually increase and these values are quite high compared to those of Mn, Al, Co, or Ni. Entropy is directly related to the structural disorder. Thus, the thermodynamic properties of the cations directly influence the morphology of these new novel pyrophosphates.

There are several publications, but the majority of these publications essentially deal with the general synthesis of a large variety of phosphates under hydrothermal conditions and hence, in most of the cases, the resultant product contained either crystalline powder or tiny crystals. However, one can grow these as single crystals if systematic investigations are carried out as in the case of $NaHMP_2O_7$ (M = Ni, Co, Mn, Zn, Cd, Pb) crystals.

7.15 Properties of Rare Earth and Mixed Valent Metal Phosphates

The mixed framework rare earth phosphates are known as the laser-host materials, and mixed valent metal phosphates show very interesting physicochemical properties, especially the superionic properties. Laser beams have been obtained from condensed rare earth phosphates containing active elements like Nd, Pr, Er, and Eu. In this respect, the ultraphosphates RP_5O_{14} and tetraphosphates MRP_4O_{12} (where M = Li, Na, K, Rb, Cs; R = Nd, Pr, Eu, Er) are the most popular ones. Figures 7.68 and 7.69 show the luminescence spectra of NdP_5O_{14} and α-$CsNd(PO_3)_4$ crystals of

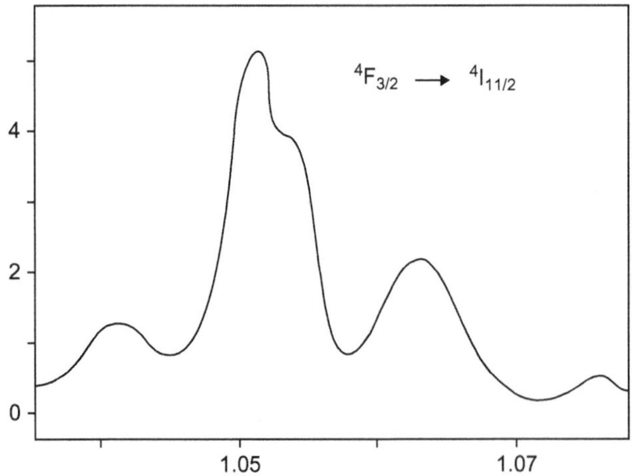

Figure 7.68 Luminescence spectra of NdP_5O_{14} crystals [299].

hydrothermal origin [299,300]. Similarly, ionic conductivity has been reported in some of the crystals of MRP_4O_{12} [301]. However, the ionic conductivity values are not high and not comparable to mixed valent metal phosphates. Since the majority of these compounds either belong to the NASICON family or are analogous to NASICON structures, they obviously exhibit high ionic conductivity. $Na_3M^{2+}(PO_4)_3$ (M = Sr, Mg, Fe, Mn); $A_3M^{3+}(PO_4)_3$ (A = Li, Na, Ag, K; M = Cr, Fe); $ABC(PO_4)_3$ (A = K, Rb, Cs; B = Ca, Sr; C = Ln and Bi) $NaHMP_2O_7$ (M = Mn, Co, Ni, Zn, Pb, Cd); $(Na_{.66}\ Zr_{.33})_2\ (P_2O_7)_2$, and so on, exhibit high ionic conductivity. The unique feature of these NASICON analogs is the thermal stability. Unlike the NASICON compound, whose ionic conductivity is hampered by the presence of multiple phase transitions from greater than 60°C, these compounds show only one or two phase transitions, usually greater than 400°C [302]. Figure 7.70 shows the Differential Thermal Analysis (DTA) curves for the representative superionic pyrophosphates. Similarly $MSn_2(PO_4)_3$ (M = Na, K, NH_4) grown hydrothermally is stable up to 1200°C without phase transitions [303–305]. Also, some of the rare earth phosphates show an ultra-low thermal expansion coefficient, which has a potential application in several fields [306]. Byrappa et al. [307]) and Sanjeev [308] have studied the ionic conductivity in superionic pyrophosphates within a wide range of

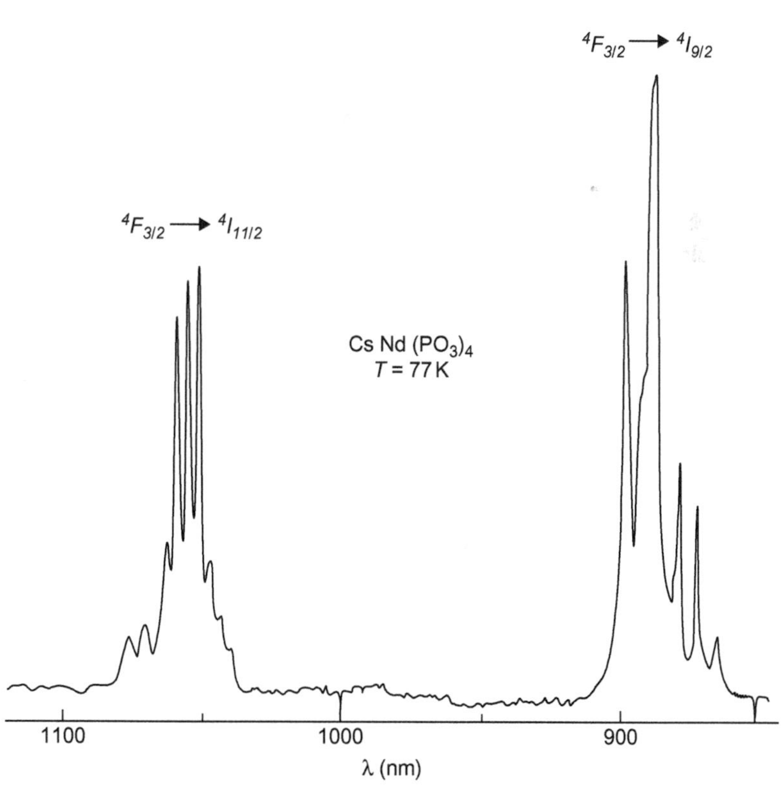

Figure 7.69 Luminescence spectra of α-CsNd $(PO_3)_4$ crystals [300].

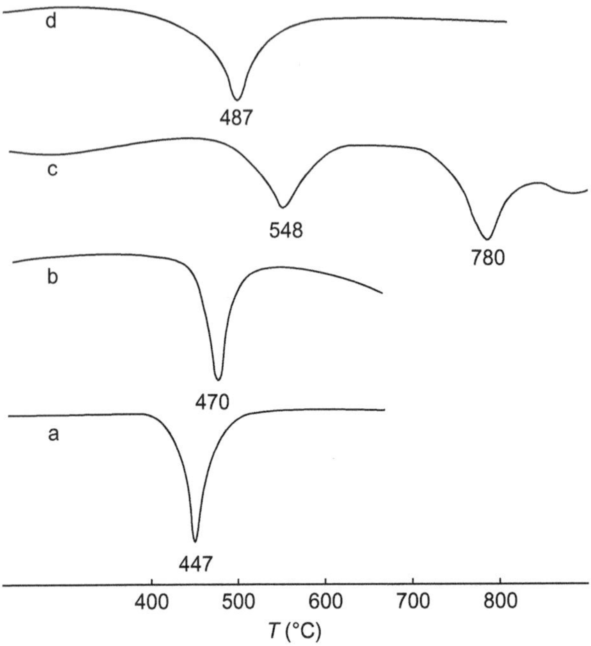

Figure 7.70 DTA curves for the representative superionic pyrophosphates: (a) $Na_2CaMn_2(P_2O_7)_2$, (b) $Na_2CoZr(P_2O_7)_2$, (c) $Na_2NiZr(P_2O_7)_2$, and (d) $Na_2CaMn_2(P_2O_7)_2$ [302].

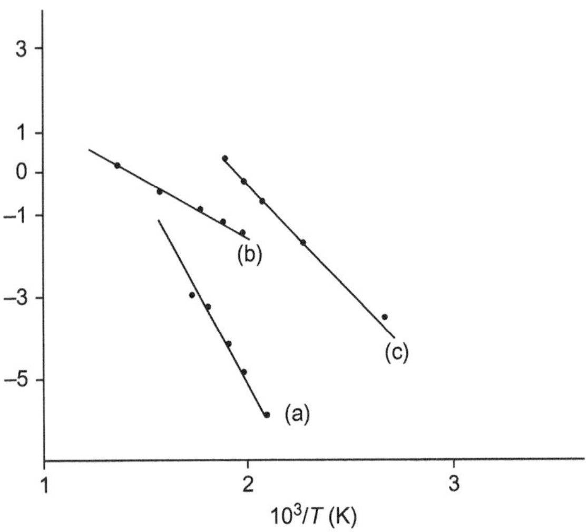

Figure 7.71 Arrhenius plot for the representative superionic pyrophosphates: (a) $HNaCoP_2O_7$; (b) $HNaNiP_2O_7$; and (c) $HNaZnP_2O_7$ [307].

temperature and internal frequency. Figures 7.71 shows the Arrhenius plots for the representative superionic pyrophosphates.

The $AVOPO_4$ (A = Li, Na, K) group of compounds are very interesting and isostructural with the well-known nonlinear optical material $KTiOPO_4$. These compounds have high thermal stability and magnetic susceptibility. The experimental results show that the Curie–Weiss behavior: $\lambda_M = C/CT - \theta)$, where

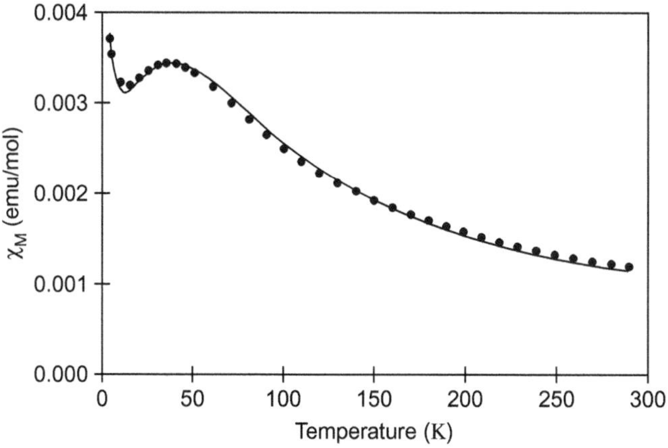

Figure 7.72 Thermal stability and magnetic susceptibility of $Mn_2VO(PO_4)_2 \cdot H_2O$ [309].
Source: Courtesy of the Academic Press, Orlando, FL.

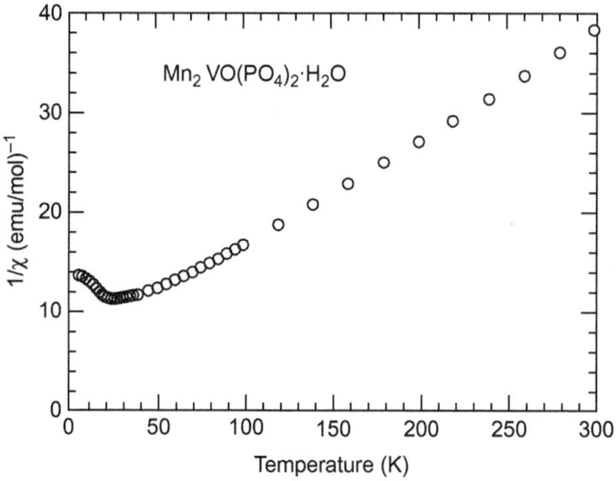

Figure 7.73 Plot of the reciprocal magnetic susceptibility versus temperature for $Mn_2VO(PO_4)_2 \cdot H_2O$ [311].
Source: Courtesy of the Academic Press, Orlando, FL.

$C = 0.4168$ cm^3 K/mol is the curie constant and $\theta = -67.5$ is the Weiss temperature with increasing temperature, λ_M first decreases rapidly but reaches a minimum at $T \sim 15$ K, increases up to approximately 40 K, then smoothly decreases again with temperature (see Figure 7.72) [309]. Similarly very interesting magnetic susceptibility has been reported in several other hydrothermally prepared metal vanadyl phosphates like $K_2(VO)_3(HPO_4)_4$, $Mn_2VO(PO_4)_2$,H_2O, a-$NH_4(VO_2)(HPO_4)$, and $CsMo_2O_3(PO_4)_2$ [310–313]. Figure 7.73 shows a plot of the reciprocal magnetic susceptibility versus temperature for $Mn_2VO(PO_4)_2 \cdot H_2O$. This compound displays Curie–Weiss behavior down to $T_N = 20$ K, at which point an antiferromagnetic transition takes place. The high-temperature portion (150–300 K) was fit to the Curie–Weiss expression, yielding a Curie constant of $C = 9.0$ emu/K/mol and a Weiss constant of $\theta = -46$ K. The effective magnetic moment $\mu_{eff=}8.5$ BM. The origin of the antiferromagnetic interaction most likely arises between Mn(2) and V atoms, which are 3.150 Å apart in their edge-sharing polyhedral arrangement.

7.16 Hydrothermal Synthesis of Vanadates

The synthesis of vanadates is not as popular as silicates, phosphates, germanates, or sulfates, because of their limited use in the modern technology unlike its analogous materials silicates or phosphates. Further, the synthesis of vanadates is a relatively recent field. The first systematic synthesis of vanadates was during the 1950s, when ABO_4 (A = Al, Ga, B, Bi, Fe, etc.; B = P, V, As) isostructural to $AlPO_4$ was obtained [314]. However, the solid-state synthesis was the most popular one followed by the flux method. The main objective of the synthesis of AVO_4 (A = Al, Cr, B, Fe, Ga, Mn, etc.) was academic, and it was quite recently their technological potential was explored. For example, $FeVO_4$ shows very interesting magnetic properties. This compound shows successive phase $FeVO_4$-I→$FeVO_4$-II→$FeVO_4$-III→$FeVO_4$-IV type structures with increasing pressures [315]. During the 1970s, a series of simple and complex rare earth vanadates were synthesized with a motivation to develop new luminophors, because a majority of these vanadates, particularly mixed rare earth phosphates, contain isolated rare earth ions [316]. Whereas simple rare earth orthovanadates show very high thermal stability owing to their very high melting points and also exhibit unique optical and magnetic properties [317–319]. All these simple rare earth orthovanadate single crystals have been obtained by flux and melt techniques. Byrappa et al. [320] attempted the synthesis of vanadates for the first time by hydrothermal technique, and they have obtained $NH_4Zr_2V_3O_{12}$ crystals, which show interesting proton conducting properties. However, during the 1990s, it was shown that the hydrothermal technique helps in synthesizing a new class of compounds with unique structures, especially from the phosphate, vanadate, and arsenate groups. This made the hydrothermal technique popular for the synthesis of novel vanadates, which exhibited interesting physicochemical properties owing to their unique structural peculiarities. Similarly, the growth of simple rare earth orthovanadates doped with active elements was realized by hydrothermal technique quite recently [321].

Here, we describe the growth of simple rare earth orthovanadates and mixed metal vanadates of recent importance by hydrothermal method.

7.16.1 Growth of $R = MVO_4$ ($R = Nd, Eu$; $M = Y, Gd$)

Single crystals of yttrium orthovanadate, YVO_4, with zircon structure, $ZrSiO_4$, having tetragonal space group 14/amd [322] find extensive applications owing to their most valuable unusual magnetic, optical, and luminescent properties. Hence, they are used as highly efficient laser diode-pumped microlasers [323], an efficient phosphor [324] as very attractive polarizer material [325], and low-threshold laser host [326]. The laser emission cross section of Nd:YVO_4 at 1064 nm is 2.7 times greater than that of Nd:YAG [327]. As a laser-host material, Nd:YVO_4 offers attractive properties such as highly polarized emission, large absorption cross section, high thermal fracture limit, and good mechanical strength. Together, these characteristics are favorable for a broad range of diode-pumped laser applications including visible lasers. Single crystals of yttrium orthovanadates have been obtained by various techniques like Czochralski [328,329], zone melting [318,330], Verneuil method [329], dissolution in melts [331], from flux [331,332], and top seeded solution growth [333]. However, there are no reports dealing with the hydrothermal synthesis of these crystals. This is probably connected to the high-temperature and pressure conditions involved in their growth and also to the strong belief that their growth under mild hydrothermal conditions will introduce water into the structure or to the interstices. Further, these crystals are known to have high melting points and very low solubility. Therefore, the high-temperature solutions and melt techniques were preferred over the others. In fact, several problems have been encountered in the growth of yttrium orthovanadates by the above said conventional methods. These problems might be connected to the instability of the pentavalent vanadium at high temperatures and the loss of oxygen through surface encrustation by the reaction of the melt with the crucible material. The unstable melt behavior of YVO_4 leads to the appearance of zoning in crystals due to the variation in the dopant concentration. Recently, it has been observed that YVO_4 crystals are very difficult to grow with a stoichiometric or near stiochiometric composition. Further, the most critical aspect is the appearance of metastable phases which precipitate within YVO_4 crystals [334]. The detection of correct Y—V—O stoichiometry in YVO_4 crystals is difficult [335]. Moreover, as the temperatures involved in the flux growth is about 1300°C and in melt growth is above 1800°C, there is a high probability of vanadium attacking the container. In order to overcome most of these difficulties in melt and high-temperature solution techniques, Byrappa et al. [321] have preferred hydrothermal technique as a solution to this problem. Since the experiments in hydrothermal technique are carried out in a closed system, the loss of oxygen could be readily prevented. In addition, the experimental conditions adopted by the authors involve the use of Teflon liners for the autoclaves and the interaction between the container and vanadium is avoided.

The experiments on the hydrothermal growth of R:MO_4 (R = Nd, Eu; M = Y, Gd) were carried out using the Morey type of autoclaves provided with Teflon liners. The use of Teflon liners has helped in overcoming the entry of inclusions from the autoclave material. The schematic diagram of hydrothermal autoclave used in the present work is shown in Figure 7.74. The starting materials such as Y_2O_3, V_2O_5, and R_2O_3 (where R = Nd, Eu,) were weighed in appropriate proportions in Teflon liners. A suitable mineralizer solution of known concentration was added into the Teflon liners and the entire mixture was stirred well until a homogeneous and relatively viscous solution was obtained. The temperature was kept constant at 240°C and the autogeneous pressure generated depended on the fill, usually

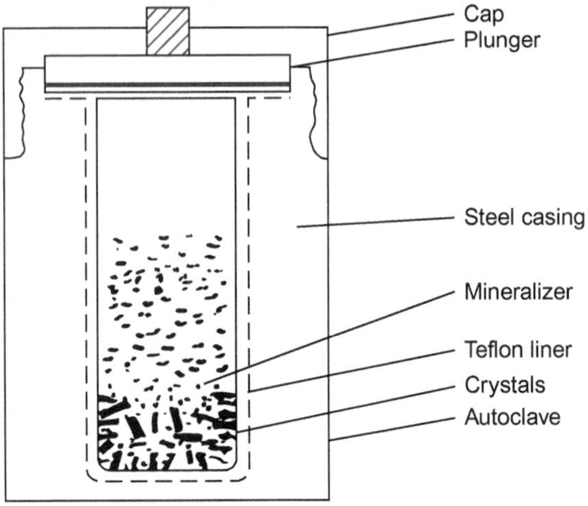

Figure 7.74 Schematic diagram of a hydrothermal autoclave [321].

50–60%. The pH of the growth medium was measured before and after each experimental run. In the present work, the crystallization was carried out in all the experiments through spontaneous nucleation. Hence, an effort was made to reduce the number of nucleation centers by increasing the temperature of the crystallization reactor slowly at the rate of 10°C/h up to 100°C and beyond at the rate of 5°C/h. The nutrient materials were held at 240°C for a period of 3 days without any fluctuations in the temperature. From the fourth day onward, the temperature fluctuation to the tune of ±10°C to ±20°C were introduced periodically to reduce the number of nucleation centers. By this means, small crystallites, which had developed earlier, dissolve leaving only bigger crystallites to grow. The authors had previously employed an oscillating temperature technique to obtain bigger crystals of yttrium orthovanadates from high-temperature solutions [331]. More details can be obtained from this work. In these experiments, it was observed that as the size of the crystal increases, the crystal quality is reduced. The formation of stable complex R:YVO_4 at 240°C and around 100 bar pressure must have undergone several intermediate stages of solvation forming stable and unstable solvated complexes from room temperature to 240°C. The representative chemical reactions are given below:

$$V_2O_5 + 10HCl \rightarrow 2\underline{VCl_4} + 5H_2O + Cl_2\uparrow$$

$$Y_2O_3 + 3HCl \rightarrow \mathit{Y(OH)_3} + YCl_3$$

$$Y_2O_3 + 6HNO_3 \rightarrow 2\underline{Y(NO_3)_3} + 3H_2O$$

$$V_2O_5 + 6HNO_3 \rightarrow 2H_3VO_4 + 3N_2O_5\uparrow$$

$$2VCl_4 + 4HNO_3 \rightarrow 2\underline{VOCl_3} + 2HCl + H_2\uparrow + 2N_2O_5\uparrow$$

$$2VOCl_3 + 8Y(OH)_3 \rightarrow \underline{Y_8V_2O_{17}} + 6HCl + 9H_2O$$

$$2VOCl_3 + 2Y(NO_3)_3 \rightarrow 2YVO_3 + 3N_2O_5 + 3Cl_2\uparrow$$

$$YCl_3 + H_3VO_4 \rightarrow YVO_4 + 3HCl$$

$$VOCl_3 + Y(OH)_3 \rightarrow YVO_4 + 3HCl$$

$$2H_3VO_4 + 2Y(NO_3)_3 \rightarrow 2YVO_4 + 3H_2O + 3N_2O_5\uparrow$$

$$H_3VO_4 + Y(NO_3)_3 \rightarrow YVO_4 + 3HNO_3$$

$$H_3VO_4 + Y(OH)_3 \rightarrow YVO_4 + 3H_2O$$

Underlined formulae represent the solid precipitation. The viscosity of nitric acid and sulfuric acid greater than that of HCl to HNO_3 would form a more efficient solvent for the growth of R:YVO_4. The dissolution of rare earth vanadates in HCl and HNO_3 is a process of complex formation between R^{3+}, V^{5+}, and the ligands present in the solution OH^-, H_2O, Cl^-, NO^{3-}. With a rise in temperature, the association of Cl^- with H^+ increases the yield of HCl and destabilizes the R^{3+} complex. Under these circumstances, nitrates replace chloride ligands. Finally, the nucleation takes place by the polymerization of tetrahedral R^{3+} complexes. The yttrium orthovanadate, YVO_4, crystals were doped with active ions like Nd or Eu and obtained crystals of Nd:YVO_4 and Eu:YVO_4. The rare earth concentrations are in the range 5–10%. After the experimental run, the autoclaves were quenched initially using an air jet followed by water, and the products were washed thoroughly in double distilled water in an ultrasonic cleaner to remove all excess acid media. The crystals obtained were well-faceted single crystals of 2–4 mm in size.

Solubility. Successful growth of any crystal insists on the availability of solubility data. In fact, the earlier workers did not attempt the hydrothermal growth of rare earth vanadate crystals due to the lack of the solubility data on these compounds. Mineralizers like HCl, HNO_3, HCOOH, H_2SO_4, NaOH, NaF, and mixed acid mineralizers like $HCl + HNO_3$, $HCl + H_2SO_4$, $HCl + HCOOH$ have been tried. The solubility measurements were carried out by the weight loss method. Here a crystal is kept in equilibrium with an appropriate solution at the desired pressure and temperature conditions for a period of time greater than that is necessary to establish equilibrium. All the solubility measurements were carried out in Morey-type autoclaves. Figure 7.75 shows the solubility of R:YVO_4 (in wt%) as a function of temperature and at a pressure of 100 bar in 1.5 M HCl + 3.0 M HNO_3. R:YVO_4 shows a negative temperature coefficient of solubility under hydrothermal conditions. The enthalpy of dissolution calculated from an Arrhenius plot log S versus $10^3/T$ was found to be 4.250 kJ/mole (Figure 7.76). The experimental temperature to study the solubility was up to 350°C. It was found that above 300°C,

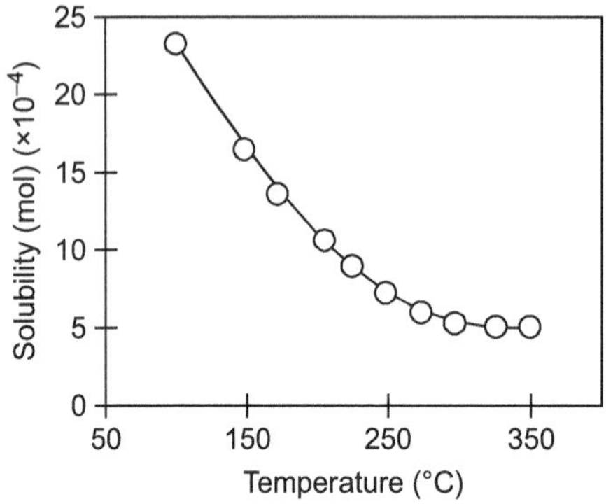

Figure 7.75 Solubility curve of Nd:YVO_4 [321].

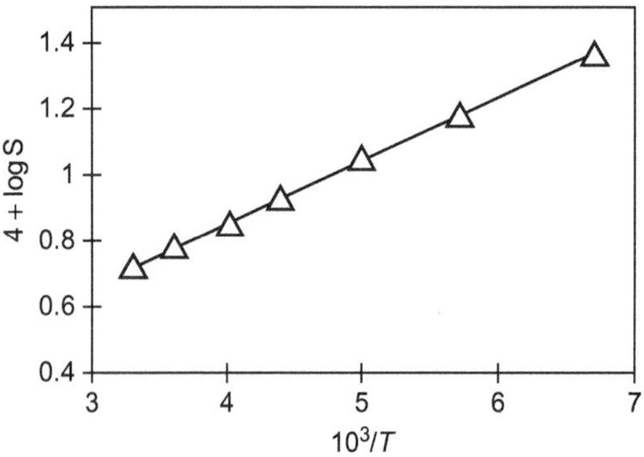

Figure 7.76 Arrhenius plot log *S* versus $10^3/T$.

the solubility does not alter much, which suggests that the crystal growth beyond this temperature is not practical. Hence, from the solubility data, it is found that the temperature range between 180° and 300°C is the most ideal, and 240°C particularly is the best growth temperature. The solvents like HCl, HNO_3, H_2SO_4, and HCOOH when used alone were not good solvents since fine crystalline products having no morphology were obtained. The use of alkaline solvent did not result in the crystallization of yttrium orthovanadates but instead yttrium hydroxides, Y $(OH)_3$, or hydrous yttrium vanadates, $YVO_4 \cdot nH_2O$, were formed. Since the use of single acid solvents did not substantially enhance the solubility, mixed acid mineralizers were tried (HNO_3 + HCl, HCl + H_2SO_4, HNO_3 + HCOOH, and so on). It was found that 1.5 M HCl + 3.0 M HNO_3 taken in 2:1 ratio is the best solvent for the growth of R:YVO_4 crystals.

Rare earth orthovanadate, Nd:YVO_4, crystals were obtained under hydrothermal conditions. The earlier workers used mostly the melt technique to obtain rare earth vanadates. The study of the morphology of crystals is very important since the crystal habit is governed by the kinetics rather than the equilibrium considerations with a number of factors like degree of supersaturation, solvent type, and the pH of the media [336]. Thus, the authors have studied the influence of these parameters on the growth morphology of Nd:YVO_4 and Eu:YVO_4. The most commonly observed

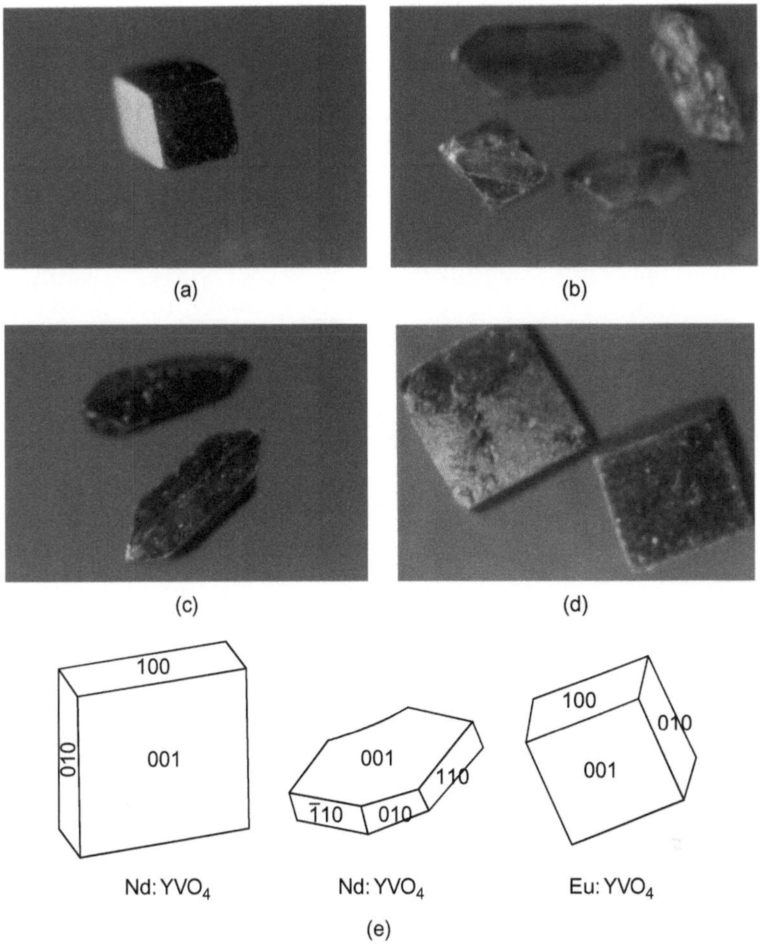

Figure 7.77 (a–d) Photographs of doped YVO_4 crystals and (e) schematic diagrams [321].

morphology is rod shaped or prismatic habit. Since the requirement of a crystal for solid-state laser host is a thin tabular platelet of $5 \times 5 \times 2\ mm^3$ dimension, the crystals obtained either by flux or melt technique exhibiting highly facetted morphological forms have to be cut and polished to the desired shape before the device fabrication. Hence, an emphasis has been made on the influence of pH of the growth media to get the desired morphology. Figure 7.77a shows photographs of doped YVO_4 crystals, and the corresponding schematic diagrams are shown in Figure 7.77b. Table 7.27 gives the morphology of the rare earth vanadates. The morphology could be changed from elongated broad hexagonal platelets to more or less cubic shape by a systematic variation in the pH of the growth media, and the thickness was controlled to approximately 2 mm. Thus, the largest crystals obtained

Table 7.27 Morphology of Yttrium Orthovanadates

Composition	Growth Temperature (°C)	pH Range	Common Faces	Growth Rate
Nd:YVO_4	240	0.05 to 0.10	(001) (010) (110)	γ(110) > γ(010)
Nd:YVO4	240	−0.03 to 0.15	(001) (100) (010)	γ(100) > γ(010)

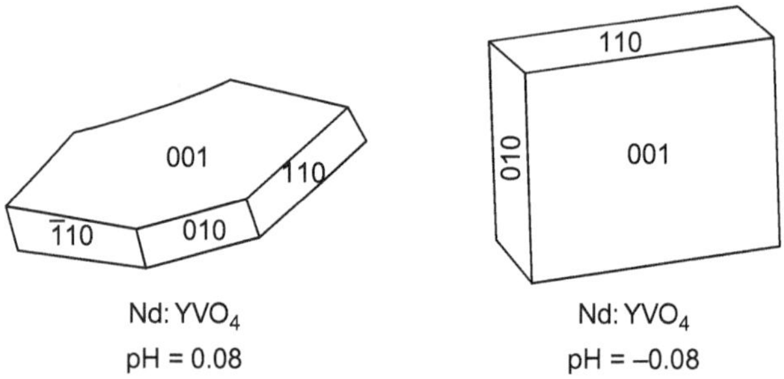

Figure 7.78 Morphology of rare earth vanadates [321].

through spontaneous nucleation provided by temperature oscillations were $4.2 \times 4.2 \times 2.0\ mm^3$ in about 8 days of experimental duration. Simple rare earth vanadate crystals were obtained when the pH was kept low (<4.0), and $Y(OH)_3$ and $YVO_4 \cdot nH_2O$ crystallized when pH > 4.0. Well-developed platelets were obtained in the pH range −0.03 to −0.15 (see morphology shown in Figure 7.78). In situ growth of the desired morphology under moderate hydrothermal conditions has shown in general that yttrium orthovanadate crystals have a strong tendency to flattening. The authors have shown that some rare earth orthovanadates obtained by the flux method, bearing Pr, Eu, and Tm, are almost isometric. A study of the growth rate of the rare earth vanadate crystals was attempted and Figure 7.79 shows the dependence of growth rate on the pH of the growth media. Higher growth rate was obtained in the pH range −0.08 to 0.08, in which the crystals grow thicker and the optical quality suffers. It has been reported earlier that the rare earth vanadates belonging to the zircon type show a tendency for banding/zoning, particularly on the (001) face [337]. The yttrium orthovanadate crystals generally show dark brown to light brown colors with a change in the dopant level. The darker colors appear when the crystals grow thicker and molar ratio is not maintained properly because the growth depends on the control of oxygen activity and surplus V_2O_5. A small quantity of hydrogen peroxide was added in order to create oxygen rich atmosphere and that in turn improved the quality of the crystals. They became honey-yellow in color, uniform throughout, and the transparency also increased. The hydrothermally grown crystals do not show zoning or banding, unlike the rare earth orthovanadates obtained by other techniques.

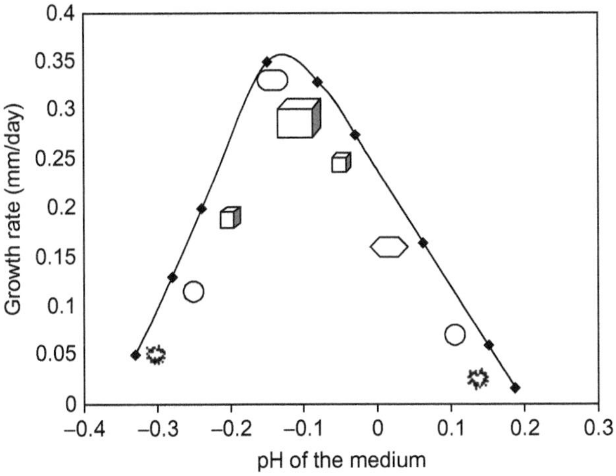

Figure 7.79 Dependence of growth rate on the pH of the growth media [321].

7.16.2 Growth of Mixed Valent Vanadates

The first hydrothermally grown mixed valent vanadate is $NH_4Zr_2V_3O_{12}$—a proton conductor [320]. In comparison with flux and melt techniques, the hydrothermal growth of vanadates has been carried out at fairly lower temperature−pressure conditions ($T < 300°C$ and $P < 100$ bar).

Although the method is not very popular for vanadates, it has helped in synthesizing several new vanadates with unique structures. Oka and coworkers [315] have prepared a series of vanadates under mild hydrothermal conditions.

A series of experiments were conducted by hydrothermal method to obtain $NH_4Zr_2V_3O_{12}$ crystals, the initial experiments with a varying molar ratio of the nutrient components yielded only a crystalline powder without any experimental morphology, the authors were able to finally synthesize small needle-like crystals of $NH_4Zr_2V_3O_{12}$ under the following conditions: $T = 250°C$, $P = 200$ atm; NH_4VO_3 5 gm, and HCOOH (2.0 M) 6 ml [320]. The experiments were carried out for 8 days using Teflon liners in Morey-type autoclaves of capacity 50 ml. Since crystallization occurred due to spontaneous nucleation, the furnace temperature was slowly increased to control the nucleation rate (~5°C/h.). The main drawback of this method is the small size of the crystals (Figure 7.80).

Here it is appropriate to discuss the growth of some selected vanadate crystals like $FeVO_4$ and MV_3O_7 (M = Cd, Ca, Sr).

Iron vanadate, $FeVO_4$, exhibits four polymorphs ($FeVO_4$-I, $FeVO_4$-II, $FeVO_4$-III, $FeVO_4$-IV) belonging to triclinic, orthorhombic, orthorhombic, and monoclinic, respectively [338,339]. In the preparation of $FeVO_4$-II, Yao et al. [340] used the aqueous solutions of $VOCl_2$ and $FeCl_3$ with V/Fe molar ratios of 0.5−1.0, where the concentration of $FeCl_3$ was fixed to 0.1 mol/l. The solutions were sealed in Pyrex ampoules and treated hydrothermally at 280°C for 40 h. Similarly, Oka and coworkers prepared several layered structures of hydrated vanadium oxides, and bronzes,

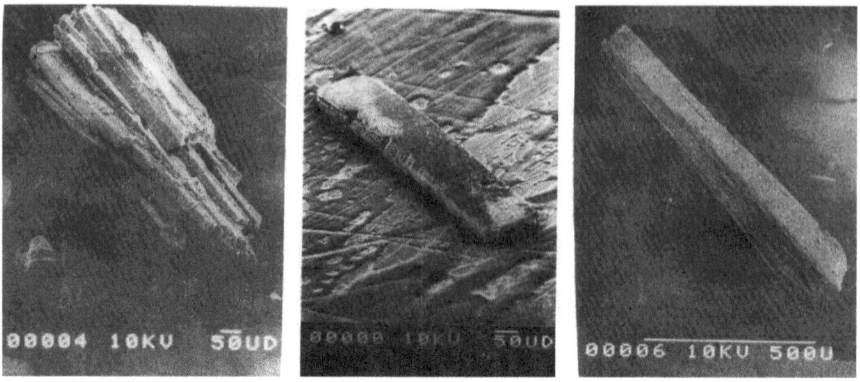

Figure 7.80 Photographs of $NH_4Zr_2V_3O_{12}$ crystals obtained under hydrothermal conditions [320].

as these oxides in hydrated states have a tendency to adopt layered structures accommodating foreign cations in the interlayer regions. Layered phases containing alkali-metal ions, the alkali-metal intercalates $A_{0.3}V_2O_5 \cdot nH_2O$ (A = alkali ion) have been successfully synthesized under hydrothermal conditions. These vanadates are usually prepared hydrothermally at 250°C from aqueous $VOSO_4$ solutions under autogenous pressure [340–342].

7.17 Hydrothermal Synthesis of Borates

Borates have been studied for the past 85 years. Over 85 phases have been established with 58 known compounds belonging to over nine different structure types. The study of borates in general was purely of academic interest and did not carry any great significance until the late 1970s when mixed alkali borates were first used in miniature laser applications [343]. Subsequently, lithium boracite was reported to be exhibiting superionic properties [344]. However, these applications did not generate any special interest among materials scientists, but it is the β-barium borate, which opened a new chapter for borates [345]. The recent discovery of piezoelectric $Li_2B_4O_7$ and nonlinear optical LiB_3O_5 has attracted a great deal of attention from materials scientists, and today it is the alkali borates that are being studied extensively worldwide [346,347]. The growth of alkali borates is being carried out mostly by melt and flux techniques. There are some reports about the solid-state reactions, but the work on borates, in general by hydrothermal technique, was almost untouched until the past decade. Only a few alkali borates have been obtained by the hydrothermal technique. The $Li_2O-B_2O_3$ system was first studied in 1958 by Sastry and Hummel [348] by the flux method.

In general, the alkali borates exhibit essentially framework structures made up of [BO_4] tetrahedra and [BO_3] triangles. Among alkali borates, only one compound shows a chain arrangement, some four compounds show islands, and two compounds show a layered arrangement. The ring arrangement in general is absent. On the other hand, the phosphates, silicates, and sulfates show extensive chain and ring arrangements and the crystals essentially belong to the lower symmetry. The alkali borates, because of their basic structural units made up of [BO_3] triangles and [BO_4] tetrahedra, crystallize both in higher and in lower symmetry systems and show wide morphological variations. Similarly, the growth experiments clearly show that any slight change in any of the growth parameters would drastically change the resultant products. Hence, the study of morphology of alkali borates is very interesting and a wide range of morphological variations has been observed which exhibit some definite relationship with reference to the type of the structure, compound, and growth conditions.

Borates were almost ignored by hydrothermal crystal growers. It was only during the last decade that we become acquainted with an in-depth knowledge about the behavior of boron in the hydrothermal solutions and its interaction with various metal cations. Although not much has been done on the alkali borate systems, several other borate systems like alkali transitional metal borate systems have been studied in detail. Today we have data available on several borate systems like: $Na_2O-CdO-B_2O_3-H_2O$, $Li_2O-CdO-B_2O_3-H_2O$ [349], $Na_2O-ZnO-B_2O_3$, $Li_2O-CdO-B_2O_3-H_2O \cdot K_2O-ZnO-B_2O_3-H_2O$, $ZnO-B_2O_3-H_2O$ [350], $MgO-CdO-B_2O_3-H_2O$ [351], $Li_2O-ZnO-B_2O_3-H_2O$ [352], under hydrothermal conditions. More than 25 different borate components have been obtained by hydrothermal technique and many of them show interesting luminescence, nonlinear, superionic, properties. For example, the $Li_2O-ZnO-B_2O_3-H_2O$ system has been studied under hydrothermal conditions within the temperature range 250–450°C (see Figure 7.81a–c) [352]. A series of lithium borates like $LiBO_2$, $Li_2B_4O_7$, $Li_3B_5O_8(OH)_2$ and mixed borates like $Li_4Zn_3B_4O_{11}$, $Li_8Zn_4B_6O_{17}$ have been obtained in this system. Further, the authors [352] have added MnO into the nutrient in a very small quantity and the experiments have been carried out using copper ampoules. It is well known that all the zinc compounds activated with Mn_2^+ show strong luminescence in green or orange regions [353]. Hence, the authors have added a small quantity (0.5 mol %) of Mn^{2+} into the system $M_2O-ZnO-B_2O_3-H_2O$ (M = Li, Na, K) [352].

Further, it is known that Mn and Cu are the activators of $Li_2B_4O_7$ for thermally stimulated luminescence dosimeters [354,355]. Here it is appropriate to quote the earlier work of Garret et al. [346] who while growing $Li_2B_4O_7$ observed that only small crystals of $Li_2B_4O_7$ could be obtained because of the highly corrosive nature of $Li_2O-B_2O_3$ that melts the platinum crucibles. Therefore, the use of copper ampoules for the borate system is doubtful and copper might contribute to the crystallization of various other phases just like the addition of MnO in the system $Li_2O-ZnO-B_2O_3-H_2O$ as reported earlier [352]. The authors have studied this $Li_2O-B_2O_3-H_2O$ system under lower PT conditions ($T = 200-300$°C, $P = 50-200$ bar) and discovered several new phases like $Li_4H_2B_2O_6$ and $LiH_2B_5O_9$.

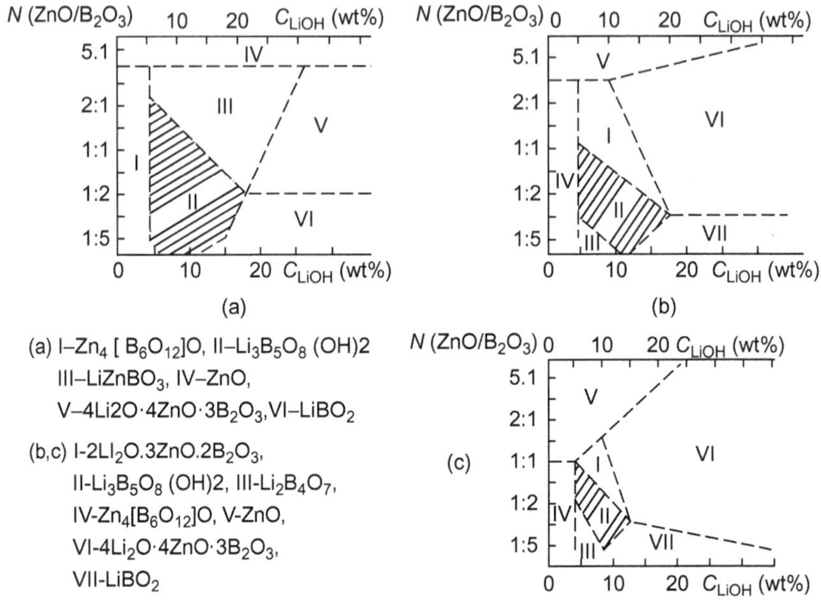

Figure 7.81 (a–c) Phase diagrams of $Li_2O-Ln_2O_3-B_2O_3-H_2O$ [352].

The crystal structures of these compounds show that they are the potential technological materials. It is observed that in the system $M_2O-CdO-B_2O_3-H_2O$ (M = Li, Na), the alkali concentration determines the sequence of crystallizing phases. Figure 7.82 shows the phases crystallizing in the system $Li_2O-B_2O_3-H_2O$ under hydrothermal condition [356]. At low alkali concentrations, alkali-free borates are formed. In the cadmium systems it is CdB_2O_4, cubic borate with a framework structure, and at 300°C it is aqueous borate with the composition $6CdO \cdot 3B_2O_3 \cdot 4H_2O$ that are formed. In the right upper part of the diagram, another cubic borate is formed—$Cd_4(BO_3)_2(OH)_2$. In the lithium system, the region between pure cadmium and pure lithium borates is the region where Li- and Cd-borates are crystallized. These borate structures provided a wealth of information for the further development of hydrothermal and crystal chemistry of borates. This data illustrates the trend of boron atoms to occupy tetrahedral positions for higher concentration of boron and "triangle" positions if the boron concentration is low. In the former case the B–O fragments prevail, whereas large cations play less important role (CdB_4O_7), in the latter case the B–O fragments are the links connecting Cd(Ca, K, Li) fragments. Such triangles in the $Cd_4(BO_3)_2(OH)_2$ structure impart additional rigidity to the strong Cd–O framework. Some intermediate variants are also observed—both boron coordinations are possible in a compound, e.g., in $K_2CaB_4O_5(OH)_4 \cdot 8H_2O$ and $Li_3B_5O_8(OH)_2$-II [349–351]. The crystals of lithium rare earth borates $Li_3Nd_2(BO_3)_3$ and $LiNd[BO_3(OH)]$ were synthesized in the systems $Li_2O-Ln_2O_3-B_2O_3-H_2O$ at LiOH concentrations 20–30 wt% [350].

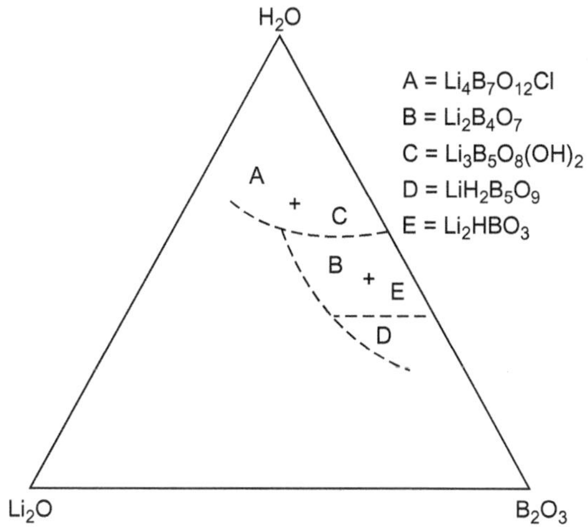

Figure 7.82 Phases crystallizing in the system Li_2O—B_2O_3—H_2O under hydrothermal conditions [356].

7.17.1 Hydrothermal Growth of Selected Borates

It is practically impossible to describe the hydrothermal growth of each and every type of borate crystals. Therefore, the authors have described the hydrothermal growth of only some important borate materials and their overall growth elucidation.

$Li_2B_4O_7$. Lithium tetraborate ($Li_2B_4O_7$) is an important piezoelectric and nonferroelectric crystal which has attracted great attention in research and industry for its application in SAW and bulk acoustic wave (BAW) devices. It has a lower temperature coefficient of delay than lithium niobate and tantalate and a higher electromechanical coupling constant than α-quartz. Since berlinite has several major problems in its growth as large single crystals, lithium tetraborate is becoming an attractive prospect as a piezoelectric crystal [357]. The raw materials for the growth of $Li_2B_4O_7$ in large quantities are abundant, and the ready availability of isotopically pure ^{6}Li, ^{10}B, and ^{11}B as the nutrient materials make $Li_2B_4O_7$ growth both practical and economical.

Before Garret et al. [346], only small crystals or grains of $Li_2B_4O_7$ were obtained. Although an attempt was made to synthesize $Li_2B_4O_7$ by the hydrothermal method, the size of the crystal was too small (Figure 7.83). It always had some divalent metal oxide like MnO or CuO [350]. Perhaps this is mainly due to the corrosive nature of the Li_2O—B_2O_3 melt toward the platinum crucible. Hence, Garret et al. [346] used the nickle crucibles. However, the method adopted by these authors is quite cumbersome because of the complexity of the technique that has a graphite susceptor and inert atmosphere. The only method advocated up to now by various authors to grow $Li_2B_4O_7$ crystals is the melt technique, particularly the Czochralski method [346,358,359]. More recently, a group of Chinese researchers have successfully obtained large crystals (2 in. diameter and 4 in. length) of high quality $Li_2B_4O_7$ for the first time by the Bridgman technique [360]. Almost at the

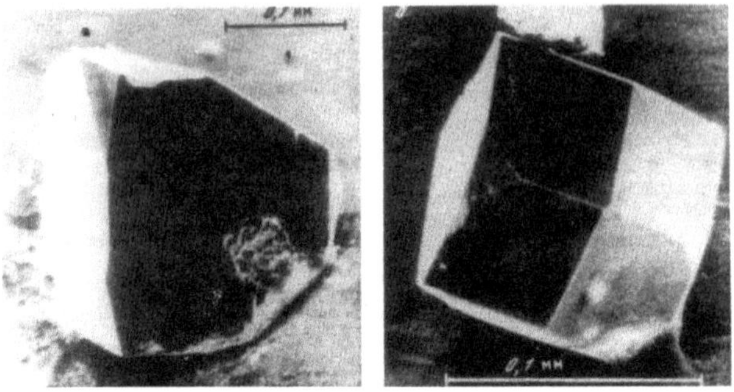

Figure 7.83 Characteristic photographs of $Li_2B_4O_7$ crystals [350].

same time a Japanese group has reported the preparation of $Li_2B_4O_7$ thin films by the sol–gel method [343]. The $Li_2B_4O_7$ crystals obtained by the Czochralski method always show coring which is the main macrodefect thereby hindering wide industrial applications of $Li_2B_4O_7$ in the manufacture of SAW and BAW devices, i. e., a continuous or intermittent opaque region parallel to the growth direction at the central core of the crystal often occurs and is related to the growth rate, deviation of the melt composition from stoichiometry (e.g., becomes Li rich) and contents of H_2O or O_2 in the melt of $Li_2B_4O_7$ [358,359]. In the Bridgman growth of $Li_2B_4O_7$, the authors have observed a similar opaque core in the $Li_2B_4O_7$ crystal and considered this defect to be a cluster of holes and a cellular structure or dendritic [360]. A few white inclusions of impurities and evidence of constitutional supercooling in the holes or on the walls of cells have been observed. Similarly, the cracking by thermal stress frequently occurs in Cz growth because of the anisotropy of the expansion coefficient of $Li_2B_4O_7$. No cracks due to thermal stress occur in the Bridgman growth of $Li_2B_4O_7$ for various orientations, since the temperature gradient in the $Li_2B_4O_7$ crystal is smaller. However, if the residual melt on the top of an as-grown crystal in ⟨100⟩ or ⟨110⟩ directions changes into polycrystalline phase by fast cooling, cracks in the $Li_2B_4O_7$ occur often in the top region. Similarly, the sol–gel technique has some disadvantages although the method has several advantages like: It allows high purity, low processing temperature, precise composition control, and the formation of bulk monoliths, fibers and coating films. The important conclusions drawn by the authors based on their work on sol–gel method is that a larger amount of water is necessary for the crystallization of $Li_2B_4O_7$ thin films, but also to suppress the formation of undesirable crystalline phases other than $Li_2B_4O_7$ [361]. Obviously, hydrothermal technique helps not only in the crystallization of $Li_2B_4O_7$ crystals at fairly lower temperatures, but also in solving many of the problems encountered in other methods as described earlier. Contrary to the result of Garret et al. [346], no difficulties were experienced using platinum crucibles to contain the melt by the authors [359]. Hence, in order to avoid these controversies, Byrappa and Shekar [362] used Telfon liners such that the incorporation of impurities could be totally avoided.

The experiments on the hydrothermal synthesis of $Li_2B_4O_7$ were carried out using Morey-type autoclaves provided with Teflon liners of capacity 25 ml. The use of Teflon liners has helped in overcoming the corrosive nature of the melt ($Li_2O-B_2O_3$). The nutrient materials such as $LiBO_2$ or LiOH and Boric acid were taken in Teflon liners. The molar ratio of the starting materials was selected through theoretical calculations. The % fill was varied from 50% to--60%. Several mineralizers like HCl, HCOOH, H_2SO_4, and HNO_3 were tried in the synthesis of $Li_2B_4O_7$ crystals. The experimental conditions are given in Table 7.28. It was found that only HCOOH acts as a suitable mineralizer, whereas the other solvents like HCl and HNO_3 did not yield good crystals. Further, the molarity of HCOOH was varied from 1 to 3.5 M, and it was found that good results were obtained in 1.5 M HCOOH solution.

The formation of $Li_2B_4O_7$ takes place according to the following reactions:

$$2\ LiBO_2 + 2\ HCOOH \rightarrow 2\ LiHCOO + B_2O_3 + H_2O$$

$$LiHCOO + H_3BO_3 \rightarrow LiB(OH)_4 + CO$$

$$2\ LiB(OH)_4 + 2\ H_3BO_3 \rightarrow Li_2B_4O_7 + 7\ H_2O$$

The formation of the hydroxy borate ion on an ionization of boric acid is spontaneous. It is possible that the intermediate solvated complexes like boric acid (it is a weak electrolyte and the complexes formed are rather stable), LiHCOO, and $LiBO_2$ interact with each other because of their ionic bonding or very low charge in order to give a stable coordinated compound having a high lattice energy and relatively stronger ionic bonding, $Li_2B_4O_7$.

The crystallization was carried out in all the experiments through spontaneous nucleation. As such, the crystals obtained under the above said conditions are colorless, highly transparent, and vary in size from 5 to 6 mm. The crystals are transparent and the faces are also well developed (Figure 7.84a and b). In general, the morphology of the hydrothermally grown $Li_2B_4O_7$ is very interesting and it varies widely depending upon the experimental conditions, particularly the starting materials and the % fill. Even the size of the crystals varies depending upon the above parameters from 1 up to 12 mm. As the pressure in the system was decreased slowly (up to 40 atm), the size of the crystals increased. In some experiments, the authors have obtained well-developed and perfect tetragonal crystals (Figure 7.85) of small size and in some other experiments wedge shaped, massive, and irregular crystals [363]. When the size of the crystals is large, the crystals become less transparent. $Li_2B_4O_7$ crystals can be obtained within the following molar ratio: $Li_2O:B_2O_3 = 1:7-9$. The chemical analysis of the $Li_2B_4O_7$ obtained shows that the ratio $Li_2O:B_2O_3$ is higher than the theoretical ratio (69.2 molar% B_2O_3 experimental ratio as against 66.6 M% of B_2O_3 theoretical) as in the case of Ref. 360. Figure 7.86 shows the schematic diagram of morphological variations in $Li_2B_4O_7$ crystals with the % fill in the autoclaves [360].

In order to reduce the number of nucleation centers, the temperature of the crystallization reactor was increased very slowly at the rate of 10°C/h up to 100°C and beyond which at the rate of 5°C/h. The nutrient materials were held at 250°C

Table 7.28 Experimental Conditions for the Growth of $Li_2B_4O_7$ Crystals [362]

No.	Nutrient Composition	Temperatue (°C)	Pressure (bar)	% Fill	Duration (days)	Size (mm)	Remarks
1.	$LiBO_2$—1.668 B_2O_3—2.332 HCOOH (3 M)—8 ml	250	120	70	8	4–14	Bulk crystals of tetragonal and regular shapes
2.	$LiBO_2$—1.668 B_2O_3—2.332 HCOOH (2.5 M)—8 ml	250	115	65	9	1–8	Polycrystalline and wedge-shaped crystals
3.	$LiBO_2$—1.668 B_2O_3—2.332 HCOOH (2 M)—8 ml	250	100	60	8	3–10	Wedge-shaped crystals
4.	$LiBO_2$—1.668 B_2O_3—2.332 HCOOH (1.5 M) —6.5 ml	250	100	55	7	1–8	Wedge-shaped crystals
5.	$LiBO_2$—1.668 B_2O_3—2.332 HCOOH (1.5 M)—6 ml	250	90	50	8	0.5–2	Well-developed tetragonal crystals
6.	$LiBO_2$—1.668 B_2O_3—2.332 HCOOH (3 M)—8 ml	250	120	70	8	4–14	Bulk crystals of tetragonal and regular shapes

for a period of 3 days without any fluctuations in the temperature. From the fourth day onward, the temperature fluctuations to the tune of $\pm 20°C$ were introduced periodically in order to reduce the nucleation centers. By this means, the earlier developed small crystallites dissolve and only the bigger crystallites remain. This provides the growth of bigger crystals ($12 \times 12 \times 10$ mm^3). The big crystals do

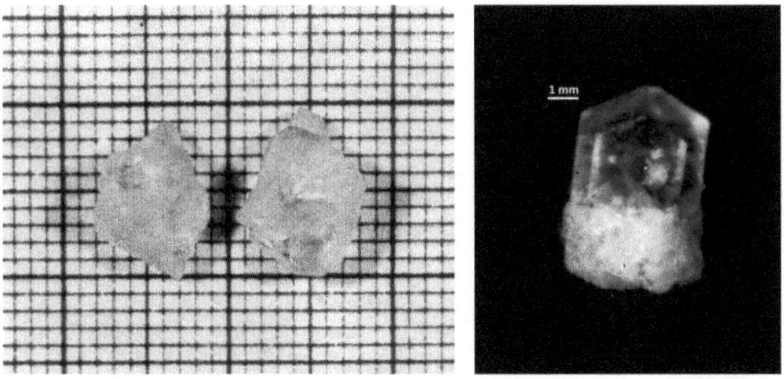

Figure 7.84 Photographs of $Li_2B_4O_7$ crystals [363].

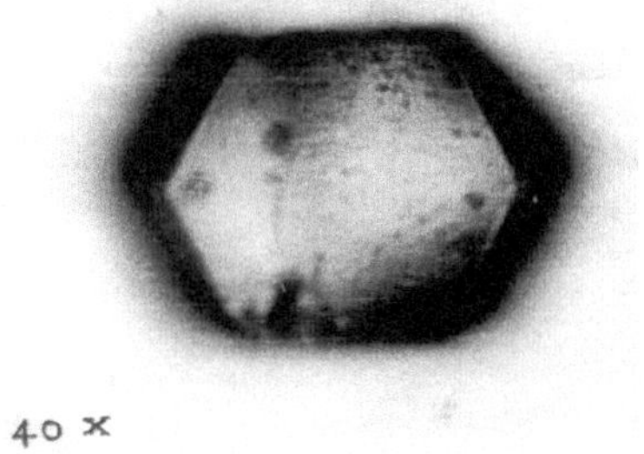

Figure 7.85 Photograph of tetragonal crystal of $Li_2B_4O_7$ [363].

show cracking due to the thermal stress during the sudden quenching of the autoclaves after the experiments. Striations and thermal cracking are commonly noticed in $Li_2B_4O_7$ crystals obtained by Czochralski and Bridgman techniques. However, the greatest advantage in the hydrothermal growth is that the thermal stress either during the growth or during the quenching can be easily overcome. Further, the hydrothermal method helps in growing the pure and large single crystals. This work has clearly demonstrated the advantages of growing $Li_2B_4O_7$ crystals devoid of all the macrodefects. Even the separately crystallized $Li_2B_4O_7$ crystals with the experimental duration of 8–10 days have yielded crystals of $12 \times 12 \times 10\ mm^3$ size which shows fairly higher growth rates. Therefore, a careful study of the growth parameters and an introduction of seed will make this technique a more attractive one for $Li_2B_4O_7$. It is well known that both $Li_2B_4O_7$ and berlinite are the most suitable piezoelectric materials having high SAW coupling coefficients for high-frequency and -temperature compensation. Here, it is appropriate to mention that the growth of berlinite has been proved it to be extremely difficult to obtain large single crystals of high quality, owing to the difficulties associated with the highly corrosive nature of P_2O_5, its high reaction susceptibility, and the negative temperature coefficient of solubility. Therefore, the successful growth of high-quality $Li_2B_4O_7$ by the hydrothermal technique would solve the existing problems with respect to the other techniques adopted so far. Also, the use of Teflon liners

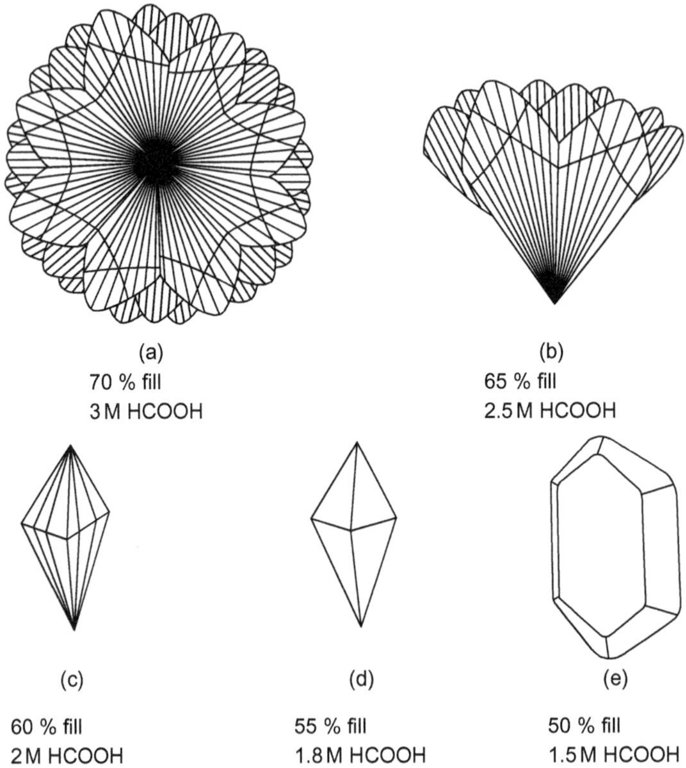

Figure 7.86 Schematic diagram of morphological variations [363].

and the availability of isotopically pure starting materials help in getting highly pure crystals. Further, the temperature and pressure condition involved in the growth of $Li_2B_4O_7$ is relatively low and this insists upon very simple autoclave designs and higher practicability. Byrappa et al. [363] have studied the solubility of Lithium Triborate, LiB_3O_5 (LBO) under hydrothermal conditions with reference to the molarity, % fill, pressure, etc. in different solvents. The solubility is found to be positive. The solubility measurements were carried out by the weight loss method in different solvent media and these studies were performed with utmost care to obtain good results by taking the average values. The possible errors, if any, in our measurements are due to the fluctuations in temperature (±1°C), weight and volume determination. Figure 7.87a–c shows the solubility curves for LBO crystals with varying molarity of LiOH, HCOOH, pressure, and temperature. As evident from Figure 7.87, the ideal temperature for the growth of LBO crystals is within the temperature range of 200–260°C (Figure 7.87a). Similarly, the ideal molarity range of the mineralizer in the growth of LBO crystals is 1.5–1.8 M (Figure 7.87b).

$Li_3B_5O_8(OH)_2$. The synthesis of $Li_3B_5O_8(OH)_2$ crystals has been reported earlier by the authors while investigating the system $Li_2O–ZnO–B_2O_3–H_2O$ under

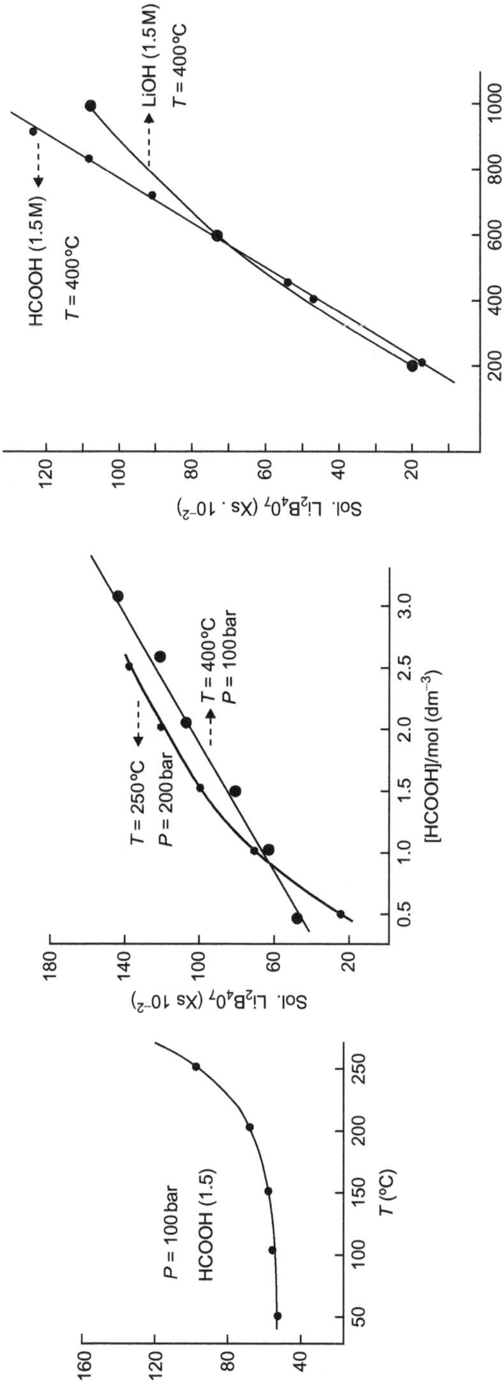

Figure 7.87 Solubility curves for LBO crystals [363].

hydrothermal conditions within the temperature range 250–450°C [350]. They obtained the crystals through the following chemical reactions:

$$3LiOH + 5H_3BO_3 \rightarrow Li_3B_5O_8(OH)_2 + 8H_2O \qquad (7.26)$$

This has two polymorphic modifications: (i) $Li_3B_5O_8(OH)_2$ and (ii) $Li_3B_5O_8(OH)$. They have obtained a series of lithium borates: $LiBO_2$, $Li_2B_4O_7$, $Li_3B_5O_8(OH)_2$, and mixed Zn borates: $LiZnBO_3$, $2Li_2O \cdot 3ZnO \cdot 2B_2O_3$, $4Li_2O \cdot 4ZnO \cdot 3B_2O_3$. Among these, $Li_3B_5O_8(OH)_2$ crystallizes within a narrow region (as shown in Figure 7.82) when MnO was added into the nutrient in a very small quantity. Contrary to these earlier works, Byrappa et al. [364] obtained $Li_3B_5O_8(OH)_2$ crystals in the system Li_2O–B_2O_3–H_2O under lower PT conditions ($T = 240°C$, $P = 100$ atm) in the absence of any other components like MnO or CuO.

The experiments were carried out using a Morey-type autoclave provided with Teflon liners. The starting materials such as LiOH and H_3BO_3 were taken in Teflon liners. The % fill was varied from 50% to 60%. Several mineralizers were tried and the HCOOH was found to be a better mineralizer with its molarity varying from 1.5 to 5 M. In most of the earlier works, LiOH has been taken as a mineralizer. A change in the concentration of LiOH in the system changes the resultant lithium compound. Byrappa et al. [364] observed that LiOH exists in the system, the presence of additional mineralizer namely HCOOH enhances the solubility of this compound and the solvation processes. Figure 7.88 shows the solubility of $Li_3B_5O_8(OH)_2$ in HCOOH [365]. The formation of $Li_3B_5O_8(OH)_2$ takes place according to the following reactions:

$$LiOH + H_3BO_3 \rightarrow LiBO_2 + 2H_2O \qquad (7.27)$$

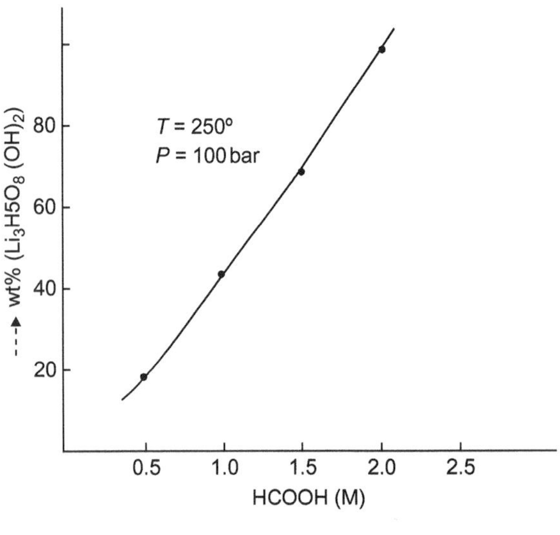

Figure 7.88 Solubility of $Li_3B_5O_8(OH)_2$ in HCOOH [365].

$$3LiBO_2 + 2H_3BO_3 \rightarrow Li_3B_5O_8(OH)_2 + 2H_2O \quad (7.28)$$

Water is considered as the hard base and strongly acts upon the dissolution of the solvanted particles of $LiBO_2$ and determines the solubility of the final resultant acid–base complex, i.e., $Li_3B_5O_8(OH)_2$. It must have been formed through several intermediate stages of solvation and formed several stable and unstable ions through an active interaction between the solute and solvent depending upon the dissolution energy of various cations and donors. Usually when HCOOH is taken as a mineralizer, LiOH forms LiHCOO which later forms $LiBO_2$ and interacts with H_2BO_3. Since the system studied is a quarternary system, it becomes extremely difficult to assess the exact stage and nature of crystallization. The reactions given are more or less hypothetical without much speculation. It is possible that the intermediate solvated complexes like LiHCOO and $LiBO_2$ interact with each other because of their weak ionic bonding or very low charge in order to give a stable coordinated compound with a high lattice energy and a relatively strong ionic bonding, $Li_3B_5O_8(OH)_2$.

The experimental conditions in the hydrothermal synthesis of $Li_3B_5O_8(OH)_2$ are given in Table 7.29. The size varies up to 15 mm and the crystals are good in quality (Figure 7.89). The present study also shows that $Li_3B_5O_8(OH)_2$ crystallizes within a narrow region in the system $Li_2O-B_2O_3-H_2O$ (Figure 7.82). The crystals were occasionally twinned and are transparent. Since, $Li_3B_5O_8(OH)_2$ crystals of size 15 mm could be obtained in an experimental run of 8–10 days, a careful study of growth parameters and an introduction of the seed will enable one to grow bigger crystals.

$Li_4B_7O_{12}Cl$. In the recent years, boracites have attracted attention of materials scientists owing to their variety of applications in modern technology, particularly after the successful application of transitional metal rare earth borates—$NdAl_3(BO_3)_4$, $Nd(Al, Cr)_3(BO_3)_4$, $Nd(Ga, Cr)_3(BO_3)_4$, etc.—as miniature laser materials [343]. Almost at the same time, high ionic conductivity was observed in lithium boracite ($Li_4B_7O_{12}Cl$) [366]. This compound was successfully synthesized in 1972 [367]. Later the structure of the compound was reported in 1973 [368]. This compound has been reported to be obtained earlier by two techniques: (1) through sublimation and (2) hydrothermally. Although the hydrothermal technique has been successfully adopted to the synthesis of this compound, the PT conditions were quite high and crystal size was too small and less then 1 mm. Byrappa et al. [369] selected the suitable composition region in the phase diagram where $Li_4B_7O_{12}Cl$ could be crystallized in the presence of Cl under hydrothermal condition. They took chlorine in the starting mixture as HCl with a varying molarity. Figure 7.82 shows the crystallization boundary for this compound, $Li_4B_7O_{12}Cl$. The crystallization field is rather narrow and any slight change in the composition will give rise to other varieties of lithium borates like $Li_2B_4O_7$, LiB_3O_5, and $Li_3B_5O_8(OH)_2$. Therefore, one has to be extremely cautious in selecting the nutrients. The experiments were taken in a Teflon liner, which was placed inside the autoclave. The capacity of the autoclave was 30–50 ml. These autoclaves are normally provided with a safety blowout disk. The percent filling of the liners and the experimental temperatures were used in

Table 7.29 Experimental Conditions for the Growth of $Li_3B_5O_8(OH)_2$ Crystals [364]

No.	Nutrient Composition	Temperature (°C)	Pressure (bar)	% Fill	Duration (days)	Size (mm)	Remarks
1.	LiOH—0.800 H_3BO_3—3.200 HCOOH (1.5 M)—8 ml	250	120	70	7	3–4	Well-developed tetragonal crystals
2.	LiOH—0.800 H_3BO_3—3.200 HCOOH (1.8 M)—8 ml	250	120	70	10	8.0–10	Wedge-shaped crystals
3.	LiOH—0.800 H_3BO_3—3.200 HCOOH (1.5 M)—7 ml	250	100	60	15	8.0–10	Wedge-shaped polycrystalline crystals
4.	LiOH—0.800 H_3BO_3—3.200 HCOOH (1.5 M)—8 ml	250	120	70	8	8.0–10	Wedge-shaped crystals
5.	LiOH—0.800 $H_3BO_3$3.200 HCOOH (2 M)—7 ml	250	100	60	8	8.0–10	Well-developed tetragonal crystals
6.	LiOH—0.800 H_3BO_3—3.200 HCOOH (1.5 M)—7 ml	250	100	60	7	8.0–10	Well-developed tetragonal crystals

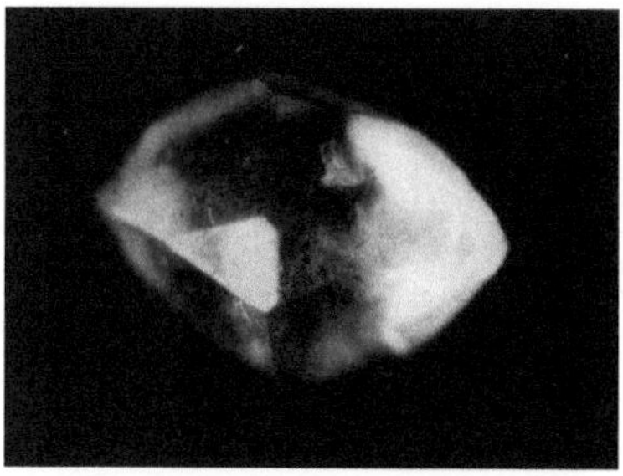

Figure 7.89 Photograph of $Li_3B_5O_8(OH)_2$ crystal (magnification $\times 25$) [364].

calculating the pressure. The experiments were carried out by spontaneous crystallization. The starting materials such as lithium carbonate and boric acid were taken in the following molar ratio:

$$Li_2O{:}B_2O_3 = (0.4 - 0.7){:}(2.5 - 4.3) \tag{7.29}$$

The mineralizer solution HCl was taken with a definite molarity (1.5–2). Since the crystallization took place through spontaneous nucleation, the temperature of the furnace was slowly increased to control the rate of nucleation. In the majority of the experiments, the percent filling varied from 50 to 65. The experimental conditions under which $Li_4B_7O_{12}Cl$ crystals obtained are given in Table 7.30 [369].

These experiments yielded crystals up to 2 mm in size. In the majority of the experiments, $Li_4B_7O_{12}Cl$ crystals obtained were twinned and the size varied up to 3 mm. The twinning was also controlled and the transparency of the crystals was also enhanced by means of increasing the growth temperature, quenching of the autoclaves, and accurate monitoring of the experimental conditions. The rate of heating and the rate of cooling of the autoclaves and the establishment of equilibrium conditions in the system control the defect formation in these crystals.

$LiH_2B_5O_9$. The synthesis of $LiH_2B_5O_9$ crystals has been carried out for the first time by Byrappa et al. [370]. $LiH_2B_5O_9$ crystallizes within a narrow region in the $Li_2O-B_2O_3-H_2O$ system.

This new compound belonging to the monoclinic system space group ($P2_1/a$) has been obtained under hydrothermal conditions at 250°C and 100 bar pressure. The starting materials such as $LiBO_2$, B_2O_3, and solvent 1.5 M HCOOH were taken in Teflon liners. The $Li_2O{:}B_2O_3{:}H_2O$ molar ratio was 1–1.5:.5.5–6:3–3.5. The experiments were carried out using Morey type of autoclaves. The % fill was varied from 50 to 60. Several mineralizers were tried and the HCOOH was found to be a better mineralizer with its molarity varying from 1.5 to 2 M.

The formation of $LiH_2B_5O_9$ must have taken place according to the following reaction:

$$2LiBO_2 + 2HCOOH \rightarrow 2HCOOLi + B_2O_3 + H_2O \tag{7.30}$$

$$HCOOLi + 5H_3BO_3 \rightarrow LiH_2B_5O_9 + CO + 7H_2O \tag{7.31}$$

Table 7.30 Experimental Conditions for the Growth of $Li_4B_7O_{12}Cl$ Crystals [369]

No.	Nutrient Composition	Temperature (°C)	Pressure (bar)	% Fill	Duration (days)	Size (mm)	Remarks
1.	Li_2CO_3—1.000 B_2O_3—3.000 HCl (5 M)—8 ml	250	90	60	8	1–4	Transparent cubic crystals
2.	Li_2CO_3—0.9 B_2O_3—3.000 HCl (5 M)—8 ml	250	90	60	7	1–2	Transparent cubic crystals
3.	Li_2CO_3—0.8 B_2O_3—3.000 HCl (5 M)—8 ml	250	90	58	7	1–2	Transparent cubic crystals
4.	Li_2CO_3—1.000 B_2O_3—3.000 HCl (5 M)—8 ml	250	90	60	8	0.5–1	Small transparent cubic crystals
5.	Li_2CO_3—1.2 B_2O_3—3.000 HCl (5 M)—8 ml	250	90	60	8	0.5–1	Small transparent cucic crystals
6.	Li_2CO_3—1.000 B_2O_3—3.000	250	90	60	8	0.1	Very small crystals

The reaction of $LiBO_2$ and HCOOH resulted in the formation of HCOOLi. When HCOOLi reacts with H_3BO_3 with 1:3 ratio, $Li_2B_4O_7$ was formed but with an increase in the amount of H_3BO_3 (HCOOLi: H_3BO_3 = 1:5) the new compound $LiH_2B_5O_9$ has resulted.

The experimental conditions for the growth of $LiH_2B_5O_9$ are given in Table 7.31. During the crystallization process, hydrogen from HCOOH has entered the structure indicating hydrolysis process. The crystals are colorless with vitreous luster and pseudoorthorhombic symmetry [356,365].

It is interesting to note that the growth of borates by hydrothermal method is not yet popular, and even today the flux method followed by meet method are the most popular ones. A large group of compounds with general formula $RM_3(BO_3)_4$ (where R = Y, La–Lu, Bi, In, but M = Al, Sc, Fe, Ga, Cr) have been obtained mainly by flux method. However, the hydrothermal technique is slowly catching up because of the advantages as listed earlier in the growth of $Li_2B_4O_7$.

Applications of borates. Borates exhibit very interesting properties like piezoelectric, pyroelectric and ferroelectrics, acousto-electric, luminescent, and

Table 7.31 Experimental Conditions for the Growth of $LiH_2B_5O_9$ Crystals [356]

No.	Nutrient Composition	Temperature (°C)	Pressure (bar)	% Fill	Duration (days)	Size (mm)	Remarks
1.	$LiBO_2$—2.0 B_2O_3—4.0 HCOOH (1.5 M)—6 ml	250	100	60	7	1–4	Transparent crystals
2.	$LiBO_2$—2.0 B_2O_3—4.0 HCOOH (1.5 M)—5 ml	250	90	55	7	1–3	Transparent crystals
3.	$LiBO_2$—2.0 B_2O_3—4.0 HCOOH (1.5 M)—7 ml	250	110	65	7	5	Massive crystals
4.	$LiBO_2$—2.0 B_2O_3—4.0 HCOOH (2 M)—6 ml	250	100	60	7	0.5–1	Small transparent crystals
5.	$LiBO_2$—2.0 B_2O_3—4.0 HCOOH (3 M)—6 ml	250	100	60	8	0.5	Small transparent crystals
6.	$LiBO_2$—2.2 B_2O_3—4.0 HCOOH (1.5 M)—6 ml	250	100	65	7	1–3	Transparent crystals

superionic. Also, they possess good thermal and chemical stability. Owing to the possibility of wide isomorphous substitutions, one can consider these borates as polyfunctional materials with device potential. However, borates of hydrothermal origin are being used essentially as piezoelectric, ferroelectric, and superionic borates.

$Li_2B_4O_7$ is known as a Li superionic conductor with its ionic conductivity of the order of $10^{-4}\ W^{-1}\ cm^{-1}$ at about 500 K and has a large piezoelectric constant. A combination of piezoelectric and superionic properties should give rise to electro-acoustic anomalies in $Li_2B_4O_7$ crystals. The complex impedance spectroscopic (CIS) studies reveal the physical nature of the acousto-ionic interaction and also on the characteristics of ion transport in solid electrolytes on the whole. $Li_2B_4O_7$ may be used as a dielectric resonator at about 10 MHz. The noise observed in the CIS data for $Li_2B_4O_7$ is not a thermal noise, but it appears to be 1/f quantum noise [371].

$Li_3B_5O_8(OH)_2$ shows poor ionic conductivity and behaves as a dielectric material below 498 K, with the electronic equivalent circuit representation indicating a pure capacitor type behavior. Between 299 and 523 K, the material exhibits piezoelectric behavior and at about 523 K, it shows superionic conducting behavior. The complex impedance becomes almost zero at 10 MHz [372].

The ionic conductivity has been studied by several workers. The ionic conductivity generally varies from 10^{-2} to 10^{-3} $(\text{ohm cm})^{-1}$ at about 300°C. The activation energy value is around 0.4 eV (380).

Alkali borates, α-$LiBO_2$, crystals are doped with Mn^{2+} ions, and they show weak yellow luminescence (UV excitation). New alkali-free borates of divalent metals doped with Mn^{2+} ions show interesting luminescence properties. For example, Mn-activated cadmium borates show yellow-green luminescence and $Zn_4B_6O_{12}$ borate yields green luminescence [373,374]. similarly, Mn^{2+}-doped $LiZnBO_3$, $2Li_2O \cdot ZnO \cdot 2B_2O$, and $Na_2Zn_2MnB_{4.67}O_{11}$ show crimson luminescence, and $4Li_2O \cdot 4ZnO \cdot 3B_2O_3$ shows yellow luminescence [374].

The complex coordinated complexes are best obtained under hydrothermal conditions. Such complex structures have not been able to be obtained by other crystal growth methods. Especially with the reduction in the PT conditions of synthesis, the method is becoming more attractive for these complexes. New structures being obtained are throwing light upon new and interesting physicochemical properties of these compounds.

Thus among the complex coordinated compounds, today it is the mixed framework structures from the rare earth phosphates, silicates, germanates, vanadates, and borates, which have been studied extensively. However, in recent years, except a few selected crystals of these mixed framework structures, the focus has been shifted toward the MOF structures popularly known as hybrid framework structures owing to their exceptional application potential, ease of their synthesis, chemically tunable structures, and exceptionally high specific surface area. In spite of their overwhelming popularity, their growth is limited only to the nanosize and not the bulk single crystals. Readers can find more valuable information on these MOFs in Refs. [375–379]

References

[1] K. Byrappa, D.Yu. Pushcharovsky, Crystal chemistry and its significance on the growth of technological materials: Part 1; silicates, phosphates and their analogs, Prog. Cryst. Growth Charact. 24 (1992) 269–359.

[2] K.H. Wedeponl, Geochemistry, Holt, Reinhart, and Winston, New York, NY, 1971. p. 231.

[3] D.C.E. Corbridge, The structural chemistry of phosphorous. 4 preparative methods and growth of rare earth phosphates, Bull. Soc. Fr. Crystallogr. 94 (1971) 271.

[4] K. Byrappa, Preparative methods and growth of rare earth phosphates, Prog. Cryst. Growth Charact. 3 (1986) 163–198.

[5] D.C.E. Corbridge, Topics in Phosphorus Chemistry, Plenum Press, New York, NY, 1968.

[6] K.K. Palkina, Izvestiya. Akad. Nauk SSSR, Inorg. Mater. 14 (1978) 789 (in Russian).
[7] D.Yu. Pushcharovsky, Structural Mineralogy of Silicates and Their Analogs, Nedra, Moscow, 1986160 (in Russian).
[8] H. Lyton, M.R. Truter, J. Chem. Soc. (1960) 5112.
[9] N.N. Bushuev, V.V. Illarionov, S.B. Kapilevich, J. Inorg. Chem. 26 (1981) 2908 (in Russian).
[10] M.A. Simonov, S.V. Shkovrov, S.I. Troyanov, Kristallographia 33 (1988) 502.
[11] F.W. Einstein, A.C. Willis, Acts. Cryst. B37 (1981) 218.
[12] R.C. Rouse, D.R. Peacor, R.L. Freed, Am. Mineral 73 (1988) 168.
[13] A. Kawahara, Bull. Soc. Mineral Cryst. 90 (1967) 279.
[14] N.V. Belov, Crystal Chemistry of Large Cation Silicates, Consultants Bureau, New York, NY, 1959.
[15] G.M. Clark, R. Morley, Chem. Soc. Revs. 5 (1976) 269–274.
[16] K. Byrappa, I.I. Plyusnina, G.I. Dorokhova, Growth, structure and IR-spectra of $CsRP_4O_{12}$Crystals, J. Mater. Sci., U.K. 17 (1982) 1847–1853.
[17] R.M. Hazen, L.W. Finger, Comparative Crystal Chemistry Temperature and Pressure, Composition and the Variation of Crystal Structure, John Wiley & Sons, London, 1982.
[18] A. Clearfield, M.A. Subramanian, P.R. Rudolf, A. Moin, Solid State Ionics 18/19 (1986) 13.
[19] R.J. Hill, G.V. Giffs, Acta Cryst. B53 (1979) 25.
[20] Y. Kudoh, Y. Takenchi, Z. Kristalogr. 171 (1985) 291.
[21] F. Liebau, Structural Chemistry of Silicates, Springer-Verlag, Berlin, Heidelberg, 1985. p. 347
[22] W.S. McDonald, D.W.J. Cruickshank, Acta Cryst. 22 (1967) 37.
[23] I.D. Brown, Chem. Soc. Revs. 7 (1978) 359.
[24] R.M. Hazen, Ch. Prewitt, Am. Mineral 62 (1977) 309.
[25] V.S. Urusov, D.Yu. Pushcharovsky, Crystal Chemical Principles of High-Pressure Mineral, Wiley-VCH, Berlin, 1984, p. 23 (in Russian).
[26] S. Sasaki, Ch. Prewitt, Y. Sato, E. Ito, J. Geophys. Res. 87 (1982) 7829.
[27] H. Horiuchi, H. Sawamoto, Am. Mineral. 66 (1981) 568.
[28] A.E. Ringwood, Germanates as the model of silicates, stable at high pressures, Problems of Petrology and Genetic Mineralogy, Nauka, Moscow, 1969 (in Russian), p. 41.
[29] L. Levien, Ch. Prewitt, D.J. Weidner, Am. Mineral 65 (1980) 920.
[30] D.Yu. Pushcharovsky, S.V. Vyatkin, N.A. Yamnova, S.L. Sorokina, Kristallographia 35 (1990) 1172–1176.
[31] L.A. Samailovich, Dependence of Pressure, Temperature, and Density of Water-Salt Solutions, BHUUCUMC (VNIISIMS), Moscow, 1969 (in Russian), p. 47.
[32] I.V. Zakirov, Investigations of the thermodynamic parameters of hydrothermal systems, in: Oganov A.R. (Ed.), Problems for Geology and Metallogenesis of Volcanic Materials., Abstracts volume of the Geology Congress, Vladivostok, 1968, pp. 155–156.
[33] M.I. Ravich, F.E. Borovaya, Solubility of sodium carbonate in water at higher pressure and temperature interval 475–540°C, Doklady Akad. Nauk SSSR 180 (1968) 1372–1375.
[34] M.I. Ravich, F.E. Borovaya, E.G. Smirnova, Solubility and vapor pressure in potassium carbonate–water systems under high temperature and pressures, J. Inorg. Chem. 13 (1968) 1422–1427.

[35] O.V. Dimitrova, Ph.D. Thesis, Moscow State University, Moscow, 1977.
[36] S.M. Haile, B.J. Wuensch, R.A. Laudise, Hydrothermal synthesis of new alkali silicates I. Potassium neodymium phases, J. Cryst. Growth 131 (1993) 352–372.
[37] S.M. Haile, B.J. Wuensch, R.A. Laudise, Hydrothermal synthesis of new alkali silicates II. Sodium neodymium and sodium yttrium phases, J. Cryst. Growth 131 (1993) 373–386.
[38] R.D. Shannon, B.E. Taylor, T.E. Gier, H.Y. Chen, T. Berzine, Am. Chem. Soc. Inorg. Chem. 17 (1978) 958–963.
[39] R.D. Shannon, H.Y. Chen, T. Berzine, Mater. Res. Bull. 12 (1977) 959.
[40] H.Y.P. Hong, J.A. Kafalas, M. Bayard, Mater. Res. Bull. 13 (1978) 757–760.
[41] H.U. Beyeler, T. Hibma, Solid State Commun. 27 (1978) 641–645.
[42] N.L. Bowen, J.F. Schairer, H.W.V. Willems, Am. J. Sci. 20 (1930) 405.
[43] B.A. Maksimov, Yu.A. Kharitonov, N.V. Belov, Dok. Acad. Nauk USSR 213 (1973) 1072.
[44] L.N. Demianets, A.N. Lobachev, G.A. Emelchenko, Germanates of Rare Earth Elements, Nauka, Moscow, 1980.
[45] B.N. Litvin, K. Byrappa, L.G. Bebikh, Growth and properties of monocrystals for miniature lasers, Prog. Cryst. Growth Charact. 3 Oxford, UK (1981) 257–271.
[46] E.A. Pabedimskaya, D.Yu. Pushcharovsky, O.G. Karpov, Structure Types of Rare Earth Silicates, Germanates and Phosphates, Moscow State University Press, Moscow, 1984.
[47] S.M. Haile, B.J. Wuensch, T. Siegrist, R.A. Laudise, Hydrothermal synthesis of new alkali silicates I. Potassium neodymium phases, J. Cryst. Growth 13 (1993) 352–372.
[48] S.M. Haile, B.J. Wuensch, T. Siegrist, R.A. Laudise, Conductivity and crystallography of new alkali rare earth silicates synthesized as possible fast-ion conductors, Solid State Ionics 53–56 (1992) 1292–1301.
[49] N.Yu. Ikornikova, in: A.N. Lobachev (Ed.), Crystallization Processes Under Hydrothermal Conditions, Consultants Bureau, New York, NY, 1971, pp. 93–111.
[50] S.M. Haile, Ph.D. Structure of $Cs_3(HSO_4)_2(H_2PO_4)$ - a new compound with a superprotonic phase transition, Thesis, Massachusetts Institute of Technology, USA, 1992.
[51] N.V. Belov, Crystal Chemistry of Silicates with Large Cations, Nauka, Moscow, 1961.
[52] E.A. Pobedimskaya, D.Yu. Pushcharovsky, O.G. Karpov, Structure Types of Rare Earth Silicates, Germanates and Phosphates, Moscow State University Press, Moscow, 1984.
[53] B.A. Maksimov, V.V. Ilyukhin, N.V. Belov, Crystalline structure of $K_3Y[Si_3O_8(OH)_2]$, Doklady Akad. Nank USSR 181 (1968) 400–403.
[54] E. Dowty, Crystal structure of joaquinite, Am. Mineral 60 (1975) 872–878.
[55] M.A. Simonov, Yu.K. Egorov–Tismenko, N.V. Belov, New Si–O radical $[Si_3O_{10}]$ in $Na_4Gd_2[Si_3O_{10}]$, Doklady Akad. Nauk USSR 179 (1968) 860–863.
[56] M.A. Simonov, Yu.K. Egorov–Tismenko, N.V. Belov, Crystal structure of Na, Gd-triorthosilicate $Na_2Gd_3[Si_3O_{10}]$, Doklady Akad. Nauk USSR 181 (1968) 38–42.
[57] H.F.W. Taylor, The crystal structure of kilchoanite $Ca_6(SiO_4)$ (Si_3O_{10}) with some comments on related phases, Miner. Mag. 38 (1971) 67–71.
[58] V.M. Krutik, D.Yu. Pushcharovsky, A.P. Khomyakov, E.A. Pobedimskaya, N.V. Belov, Anionic radical mixed type $[Si_4O_{12}]$ and P-orthotetrahedra) in monodinic phosphate, Kristallografia 25 (1980) 178–181.
[59] G.F. Volodina, I.M. Rumanova, N.V. Belov, Crystal structure of $Ca_2R_2[Si_4O_{12}]$ $CO_3 \cdot H_2O$, Doklady Akad. Nauk USSR 149 (1963) 23–26.
[60] N.I. Golovastikov, Crystal structure of alkali titanosilicate labuntzovite, Kristallografia 18 (1973) 771–774.

[61] F. Mazzi, G. Rossi, The crystal structure of taramellite $Ba_2(Fe,Ti,Mg)_2 \cdot H_2/O_2(Si_4O_{12})/$, Z. Kristallogr. 121 (1965).
[62] W. Hilmer, Die Kristallstructure des sauren Kaliummetasilikates $K_4(HSiO_3)_4$, Acta Crystallogr. 17 (1964) 1063–1065.
[63] A.R. Kampf, A.A. Khan, W.H. Baur, Barium cloride silicate with an open framework: verplanckite, Acta Crystallogr. 29 (1973) 379–382.
[64] G. Perrault, C. Boucher, J. Vicat, et al., Structure cristalline du nenadkevichite $(Na, K)_{2-x}$ (Nd,Ti) (O,OH) $Si_2O_6 \cdot 2H_2O$, Acta Crystallogr. 29 (1973) 7.
[65] Yu.A. Nekrasov, V.I. Ponomarev, B.I. Simonov, D.M. Kheiker, Establishment atomic structures biotite and isomorphous ratios in this mineral, Kristallagrafia 14 (1969).
[66] B.A. Maksimov, B.V. Merinov, V.S. Borovkov, A.K. Ivanov–Shits, Yu.A. Kharitonov, N.V. Belov, Specialities of construction of rare earth silicates family, $Na_5RSi_4O_{12}$—a new class of solid electrolytes, Kristallografia 24 (1979).
[67] Yu.A. Malinovski, E.A. Pobedimskaya, N.V. Belov, Crystal structure of traskite, Doklady Akad. Nauk USSR 229 (1976).
[68] B.A. Maksimov, V.R. Kalinin, B.V. Merinov, V.V. Ilyukhin, N.V. Belov, Crystal structure rare earth NaY—metasilicate $\{Na_3YSi_3O_9\} \times 4Na_{12}Y_4\ [Si_{12}O_{36}]_{\infty}$, Doklady Akad. Nauk USSR (1980).
[69] D.Yu. Pushcharovsky, A.M. Dago, E.A. Pobedimskaya, N.V. Belov, Crystal structure $K_8Yb[Si_6O_{16}](OH)$ polymorphism radicals $[Si_3O_8]$, Doklady Akad. Nauk USSR 258 (1979).
[70] O.G. Karpov, D.Yu. Pushcharovsky, E.A. Pobedimskaya, I.F. Burshtein, N.V. Belov, Crystal structure of rare earth silicate $NaNdSi_6O_{13}(OH)_2 \cdot nH_2O$, Sov. Phys. Dokl. 22 (1977) 464–466.
[71] D.Yu. Pushcharovsky, O.G. Karpov, E.A. Pobedimskaya, N.V. Belov, Crystal structure of $K_3NdSi_6O_{15}$, Doklady Akad. Nauk 234 (1977).
[72] O.G. Karpov, D.Yu. Pushchrovsky, E.A. Pobedimskaya, N.V. Belov, Crystal structure of rare earth silicates $NaPrSi_6O_{14}$ and $NaNdSi_6O_{14}$. New infinite silica radical 3d column $[Si_3O_7]_{\infty\,\infty}$, Doklady Akad. Nauk USSR 226 (1976).
[73] O.G. Karpov, E.A. Pobedimskaya, N.V. Belov, Crystal structure of K · Ce-silicate with 3d anion framework $K_2CeSi_6O_{15}$, Kristallografia 22 (1977).
[74] J. Felsche, The crystal chemistry of the rare-earth silicates, Struct. Bond. 13 (1973).
[75] S.M. Haile, J. Maier, B.J. Wuensch, R.A. Laudise, Structure of $Na_3YSi_6O_{15}$—a unique silicate based on discrete Si_6O_{15} units, and a possible fast ion conductor, Acta Crysta. B51 (1995) 673–680.
[76] S.M. Haile, B.J. Wuensch, R.A. Laudise, J. Maier, Structure of $Na_3NdSi_6O_{15} \cdot 2H_2O$—a layered silicate with paths for possible fast ion conduction, Acta Crysta. B53 (1997) 7–17.
[77] R.D. Shannon, H.Y. Chen, T. Berzins, Mater. Res. Bull. 12 (1977) 969.
[78] R.D. Shannon, B.E. Taylor, T.E. Gier, H.-Y. Chen, T. Berzins, Inorg. Chem. 17 (1978) 958–962.
[79] H.U. Beyeler, R.D. Shannon, H.Y. Chen, Ionic conductivity of single-crystal $Na_5YSi_4O_{12}$, Appl. Phys. Lett. 37 (1980) 934–935.
[80] L.O. Atomyan, E.A. Ukshe, in: Crystal structure of the solid electrolyte Na_3- springer. Physical Chemistry Modern Problems, Moscow, Chemistry, 1983, p. 90 (in Russian).
[81] O.S. Filipenko, L.O. Atomyan, V.I. Panomerov, B.N. Litvin, O.V. Dimitrova, L.S. Leonova, et al., Heterrogeneous 10-sided crystal framework—a desired structure factor for ionic conductivity in complex oxides, Phyl. Chem. 283 (1985) 430–433.

[82] S.M. Haile, J. Maier, B.J. Wuensch, R.A. Laudise, in: B. Scrosati, et al. (Eds.), Fast Ion Transport in Solids, Kluwer Academic Publishers, 1993, pp. 315–326.
[83] S.M. Haile, B.J. Wuensch, R.A. Laudise, J. Maier, Structure of $Na_3NdSi_6O_{15} \cdot 2H_2O$ —a layered silicate with paths for possible fast-ion conduction, Acta Crysta. B53 (1997) 7–17.
[84] S.M. Haile, J. Maier, B.J. Wuensch, R.A. Laudise, Structure of $Na_3YSi_6O_{15}$—a unique silicate based on discrete Si_6O_{15} units, and a possible fast-ion conductor, Acta Crysta. B51 (1995) 673–680.
[85] L.S. Leonova, E.A. Ukshe, Crystal structure and ionic conductivity of solid electrolytes $M_5RESi_4O_{12}$, where M = Na, Ag; RE = Sm→Lu, Solid State Ionics 14 (1984) 137–142.
[86] B.A. Maksimov, B.N. Litvin, V.V. Ilyukhin, N.V. Belov, Kristallografia 14 (1968) 490–493.
[87] J.P. Boilot, G. Collin, R. Comes, J. Solid State Chem. 50 (1983) 91–94.
[88] D. Tranqui, J.J. Capponi, J.C. Joubert, R.D. Shannon, J. Solid State Chem. 39 (1981) 219–223.
[89] R.A. Huggins, A. Rabenau, Mat. Res. Bull. 31 (1978) 1315–3120.
[90] R.G. Anthony, C.V. Philip, R.G. Dosch, Waste Manag. 13 (1993) 503–508.
[91] A. Clearfield, A.I. Bortun, L.N. Bortun, Solvent Extr. Ion Exch. 14 (1996) 341.
[92] A.I. Bortun, L.N. Bortun, R.A. Cahill, A. Clearfield, Solvent Extr. Ion Exch. 15 (1997) 285–290.
[93] D.M. Poojary, R.A. Cahill, A. Clearfield, Chem. Mater. 6 (1994) 2364–2369.
[94] D.M. Poojary, A.I. Bortun, L.N. Bortun, A. Clearfield, Inorg. Chem. 35 (1996) 6131–6135.
[95] L.N. Demianets, in: K. Byrappa (Ed.), Hydrothermal Growth of Crystals, Prog. Cryst. Growth Charact. 21(291–355) (1991).
[96] G.D. Iljushin, L.N. Demianets, V.V. Iljukhin, N.V. Belov, Doklady Akad. Nauk USSR 271 (1983) 1133–1135.
[97] K.V. Alyamovskaya, V.G. Chukhlantsev, Izvestiya Akad. Nauk SSSR, Neorgan. Mater. 6 (1970) 8 (in Russian).
[98] K.V. Alyamovskaya, V.G. Chukhlantsev, Izvestiya VUZ'ov, Khimiya i Khim. Tekhnologiya 13 (1970) 7 (in Russian).
[99] A.N. Safronov, N.N. Nevskii, V.V. Iljukhin, N.V. Belov, Doklady. Akad. Nauk SSSR 225 (1980) 1114–1118 (in Russian).
[100] F. Genet, M. Barj, Solid State Ionics 9–10 (1983) 891–895.
[101] S. Barth, A. Felitz, C. Jager, Structure and ionic conduction in solids: VI. Evidence for the non-existence of $Na_4ZrSi_3O_{10}$ as a definite crystalline phase, Solid State Ionics 26 (1988) 189–195.
[102] Yue Yong, Pang Wenqin, J. Chem. Soc. Chem. Commun. (1990) 1166–1169.
[103] Yue Yong, Pang Wenqin, Synthesis and crystallization of $Na_4Zr_2(SiO_4)_3$ under lower temperature and pressure hydrothermal conditions, J. Mater. Sci. Lett. 10 (1991) 1009–1010.
[104] A.I. Bortun, L.N. Bortun, A. Clearfield, Hydrothermal synthesis of sodium zirconium silicates and characterization of their properties, Chem. Mater. 9 (1997) 1854–1864.
[105] K. Wöhler (1948), cited by Bunsen, Apophyllite Heated with H_2O at 180°–190°C Under a Pressure of 10–12 atm, Ann. 65, 1948, pp. 80–96.
[106] E.T. Allen, F.E. Wright, J.K. Clement, Part I formation and properties of the four crystal forms, Am. J. Sci. IVth Series XXII (1906) 385–438.
[107] G.W. Morey, P. Niggli, The hydrothermal formation of silicates, a review, J. Am. Chem. Soc. 35 (1913) 1086–1131.

[108] N.L. Bowen, The crystallization of haplobasaltic, haplodioritic, and related magmas, Am. J. Sci. 40 (1915) 161–185.
[109] G.W. Morey, The ternary system H_2O–K_2SiO_3–SiO_2, J. Am. Chem. Soc. 39 (1917) 1173–1229.
[110] N.L. Bowen, The behavior of inclusions in igneous magmas, J. Geol. 30 (1922) 513–520.
[111] N.L. Bowen, O.F. Tuttle, The system MgO–SiO_2–H_2O, Bull. Geol. Soc. Am. 60 (1949) 439–460.
[112] H.S. Yoder, The MgO–Al_2O_3–SiO_2–H_2O system and the related metamorphic facies, Am. Sci. Bowen Volume (1952) 569–627.
[113] D.M. Roy, R. Roy, Studies in the system MgO–Al_2O_3–SiO_2–H_2O, Bull. Geol. Soc. Am. 63 (1952) 1293–1294 (Abstract).
[114] D.M. Roy, R. Roy, Synthesis and stability of minerals in the system MgO–Al_2O_3–SiO_2–H_2O, Am. Miner. 40 (1955) 147–178.
[115] B.W. Nelson, R. Roy, Polymorphic 7 Å and 14 Å Phases in the MgO–Al_2O_3–SiO_2–H_2O System and Their Identification, Proceedings of the Second National Clay Conference, Columbia, Mo. (See Publication of the National Research Council, No. 327), 1953.
[116] W. Noll, Mineralbildung im system Al_2O_3–SiO_2–H_2O, Neues Jahrbuch Mineralogie 70 (1935) 65–115.
[117] W. Noll, Zur Kenntnis der siallitischen Mineralumbildung. Fortschritte der Mineralogie, Kristallographite und Petrographie 21 (1937) 85–88.
[118] W. Noll, Neuere Untersuchungen von reaktionen in Wesserhaltigen Silicat-systemen, Die Chemie 57 (1944) 90–94.
[119] H.F.W. Taylor, Hydrated calcium silicates. Part I. Compound formation at ordinary temperatures, J. Chem. Soc. 726 (1950) 3682–3690.
[120] H.F.W. Taylor, Hydrothermal reactions in the system CaO–SiO_2–H_2O, in: S. Stürwald (Ed.), Proceedings of the International Symposium on Reactivity of Solids, Gothenburg. Germany, Auspices of Ingeniörvetenskapsakademien and Chalmers Tekniska Högskola, Part 2, 1952, pp. 677–682.
[121] H.F.W. Taylor, Hydrated calcium silicates. Part V. The water content of calcium silicate hydrate—I, J. Chem. Soc. 33 (1953) 163–171.
[122] H.F.W. Taylor, G.E. Bessey, Review of hydrothermal reactions in the system lime––silica–water, Magazine Concrete Res. 4 (1950) 15–26.
[123] E.T. Carlson, L.S. Wells, Hydrothermal preparation of some strontium silicates, J. Res. Natl. Bur. Stds. 51 (1953) 2.
[124] R.M. Barrer, E.A.D. White, The hydrothermal chemistry of silicates. Part II. Synthetic crystalline sodium aluminosilicates, J. Chem. Soc. 286 (1952) 1561–1571.
[125] L.B. Sand, R. Roy, E.F. Osborn, Stability relations of some minerals in the system Na_2O–Al_2O_3–SiO_2–H_2O, Bull. Geol. Soc. Amer. 64 1469–1470 (Abstract).
[126] I. Friedman, Some aspects of the system H_2O–Na_2O–SiO_2–Al_2O_3, J. Geol. 59 (1951) 19–31.
[127] H.S. Yoder, H.P. Eugster, Phlogopite synthesis and stability range, Geochim. et Cosmochim. Acta. 6 (1954) 157–185.
[128] L. Coes, A new dense crystalline silica, Science 118 (1953) 131–133.
[129] B.N. Litvin, Investigations of the Hydrothermal Crystallization in Some Silicate Systems, Ph.D. Thesis, Institute of Crystallography, Acad. Nauk USSR, 1964.
[130] V.S. Surnina, B.N. Litvin, Crystallization in the systems Na_2O–Bi_2O_3–SiO_2–H_2O and Na_2O–Bi_2O_3–GeO_2–H_2O, Izvestiya Akad. Nauk USSR, Inorg. Mater. 6 (1970) 1695–1697.

[131] A.F. Armington, Recent advances in the growth of high quality quartz, in: K. Byrappa (Ed.), Hydrothermal of Crystals, Prog. Cryst. Growth Charact. 21 (1991) 97–111.
[132] X. He, H. Zhou, W. Zhou, Z. Hu, C. Zhang, H. Huo, et al., Solubility and optical quality of $Bi_{12}SiO_{20}$ crystals, J. Crystal Growth 351 (2012) 182–187.
[133] J. Larkin, M.T. Harris, J.E. Cormier, A.F. Armington, Hydrothermal growth of bismuth silicate (BSO), J. Cryst. Growth 128 (1993) 871–875.
[134] M.T. Harris, J.J. Larkin, J. Cormier, A.F. Armington, Optical studies of czochralski and hydrothermal bismuth silicate, J. Cryst. Growth 137 (1994) 128–131.
[135] W.B. Leigh, J.J. Larkin, M.T. Harris, R.N. Brown, Characterization of czochralski- and hydrothermal grown $Bi_{12}SiO_{12}$, J. Appl. Phys. 76 (1994) 660–666.
[136] V.S. Surnina, B.N. Litvin, Investigations of phase formations in the system $Na_2O–Me_2O_3–Bi_2O_3–H_2O$ (where Me = Al, Ga, In) under hydrothermal conditions, Kristallografia 15 (1970) 604–607.
[137] S.C Abrahams, P.B. Jamieson, J.L. Bernstein, Crystal structure of piezoelectric bismuth germanium oxide $Bi_{12}GeO_{20}$, J. Chem. Phys. 747 (1967) 4034–4041.
[138] P. Hautefeuille, A. Perrey, Crystal Research. Acad. Sci. 106 (1888) 1800–1810Paris
[139] H.–Z. Espig, Kristallografia 92 (1935) 387–392.
[140] A.F. Rogers, F. Sperisen, J., Am. Miner. 27 (1942) 762–765.
[141] J. Wyart, S. Scavnicar, Synthèse hydrothermale du beryl, Bull. Soc. Franc. Minèr. Crist. 80 (1957) 395–396.
[142] R. Nacken, cited by K. Nassau. Synthetic emerald: The confusing history and the current technologies, J. Cryst. Growth 35 (1976) 211–217.
[143] M. Hosaka, Hydrothermal growth of gem stones and their characterization, in: K. Byrappa (Ed.), Hydrothermal Growth of Crystals, Prog. Cryst. Growth Charact. 21 (1990) 71–96.
[144] B.W. Anderson, Gem Testing, eighth ed., Butterworths, London, 1971, pp. 97–106, 223–229
[145] R.T. Liddicoat Jr., Handbook of Gem Identification, 10th ed., Gemnological Institute of America, Los Angeles, CA, 1975, pp. 106–108, 138–144, 308–313
[146] R. Webster, Gems, third ed., Newnes-Butterworths, London, 1975, pp. 83–100, 359–366
[147] E.M. Flanigen, US Patent 3567642, March 2, 1971.
[148] E.M. Flanigen, N.R. Mumbach, US Patent 3567643, March 2, 1971.
[149] P.J. Yancey, US Patent 3723337, 1973.
[150] G. Brown, J. Snow, Aust. Gemmol. 15 (2) (1983) 57.
[151] H.A. Hanmi, J. Gemmol. 18 (2) (1982) 138.
[152] C. Stockton, Gems Gemol. 20 (3) (1984) 141.
[153] P.J. Darragh, M.J. Willing, Aust. Gemmol. 14 (12) (1982) 344.
[154] K. Scarratt, J. Gemmol. 21 (5) (1989) 294.
[155] R.E. Kane, R.T. Liddicoat Jr., Gems Gemol. 21 (3) (1985) 156.
[156] G. Brown, J. Snow, Aust. Gemmol. 16 (12) (1988) 437.
[157] E.N. Emeĺyanova, S.V. Grum–Grzhimailo, O.N. Boksha, T.M. Varina, Sov. Phys. Cryst. 10 (1965) 10.
[158] H. Takubo, J. Gemmol. Soc. Jpn. 6 (4) (1979) 3.
[159] K. Schmetzer, J.Gemmol. 21 (3) (1988) 145.
[160] J. Sinkankas, Emerald and Other Beryl, Chilton Book Company, PA, 1981.
[161] K. Kodaira, Y. Iwase, A. Tsunashima, T. Matsushita, High-pressure hydrothermal synthesis of beryl crystals, J. Cryst. Growth 60 (1982) 172–174.
[162] Ya. Bauér, D. Rykl, Synthetic Chkalovite, Sov. Phys. Crystallogr. 17 (1972) 615–616.

[163] V.V. Bakakin, L.P. Solov'eva, Crystal structure of $Fe_3BeSi_3O_9$ $(F,OH)_2$, an example of a wollastonite-like silicon–oxygen chain based on Fe, Sov. Phys. Crystallogv. 15 (1971) 999–1005.

[164] Yu.A. Kharitonov, V.M. Golyshev, R.K. Rastsvetaeva, N.V. Belov, Crystal structure of synthetic sodium aluminoberyllosilicate $Na_3AlBeSi_2O_8$, Sov. Phys. Crystallogr. 19 (1975) 667–668.

[165] M. Maeda, T. Uehara, H. Sato, T. Ikeda, Hydrothermal syntheses of $BaBe_2Si_2O_7$, Jap, J. Appl. Phys. 30 (1991) 2240–2242.

[166] B.P. Sobolev, L.N. Dern'yanets, Yu.P. Dikov, V.V. Ilyukhin, N.V. Belov, Geokhimiya 6 (1966) 634–639.

[167] A.A. Beus, Yu.P. Dikov, Geochemistry of Beryllium in Processes of Endogenic Mineral Formation, Nedra, Moscow, 1967.

[168] C. Frondel, C.S. Hurlbut, R.C. Collete, Synthesis of tourmaline, Am. Miner. 32 (1947) 680–681.

[169] M.C. Michel-Levy, Synthese de la tourmaline et de la jeremeiwite, Crystal Acad. Sci. 228 (1949) 1814–1816.

[170] K.P. Robbins, Kh.S. Yoder, J.P. Shairrer, Synthesis of tourmaline, Problems in Theoretical and Experimental Petrology, Mir Publishers, Leningrad, 1963. p. 523

[171] I.E. Voskresenskaya, Properties of synthetic tourmaline, Mineralogy, Lvov University Press, Lvov, 1966, pp. 164–171.

[172] I.E. Voskresenskaya, M.L. Barsukova, Synthesis and properties of some iron and no iron bearing tourmaline, Hydrothermal Synthesis of Crystals, Nauka, Moscow, 1967, pp. 175–192

[173] C. Friedel, L. J. S. Boleite and Related Minerals, G. Bull. Soc. Min. 13 (1890) 129–139.

[174] C. Friedel, G. Bull, Mineralogie de la France et des ses colonies, Soc. Min. 19 (1896) 5–14.

[175] B.N. Litvin, Hydrothermal Chemistry of Silicates of Germanats, D.Sc. Thesis, Institute General and Inorganic Chemistry, Akad. Nauk USSR, 1978, p. 33.

[176] B.N. Litvin, O.K. Melnikov, Crystallization in the system $Na_2O–Al_2O_3–SiO_2–H_2O$, Kristallografia 14 (1969) 101–105.

[177] G. Roth, H. Bhm, Ionic conductivity of sodium–nepheline single crystals, Solid State Ionics 18/19 (1986) 553–556.

[178] E. Ingerson, O.F. Tuttle, Artificial willemite needles, Am. J. Sci. 245 (1947) 313–319.

[179] D.M. Roy, F.A. Mupton, Stability of minerals in the system $ZnO–SiO_2–H_2O$, Econ. Geol. 51 (1956) 432–443.

[180] A.E. Ringwood, A. Major, High-pressure transformations on the zinc germanates and silicates, Nature 215 (1967) 1367–1368.

[181] B.N. Litvin, O.S. Tarasenkova, N.V. Belov, Synthesis and crystal chemical elucidation of compounds in the system $Na_2O–ZnO–SiO_2–H_2O$, at 400–500°C, Physico-Chemical Petrology Series, vol. 4, Nauka, Moscow, 1974, pp. 89–101.

[182] B.N. Litvin, V.I. Popolitov, Hydrothermal Synthesis of Inorganic Compounds, Nauka, Moscow, 1984, pp. 66–81.

[183] L.M. Belyaev, B.N. Litvin, I.M. Dianova, O.K. Melnikov, Hydrothermal synthesis crystallophosphors of $Na_xZn_ySi_pO_q$ type, Kristallografia 11 (1966) 285–287.

[184] M.A. Simonov, O.K. Melnikov, B.N. Litvin, Synthesis of NaCl silicates, Hydrothermal Synthesis of Crystals, Nauka, Moscow, 1968, pp. 155–159.

[185] M.A. Simomov, Yu.K. Egorov–Tismenko, N.V. Belov, Confirmation of crystal structure of Na, Cd–triorthosilicate–$Na_4Cd_2Si_3O_{10}$, Doklady Akad. Nauk USSR 238 (1978) 348–351.

[186] B.N. Litvin, I.M. Dianova, L.A. Kagon, Synthesis and properties of monocrystals of composition $Na_2O \cdot 2MnO \cdot 2SiO_2$, Kristallografia 9 (1964) 571–574.
[187] F.W. Taylor, Hydrosilicates of calcium, Proceedings of the Fifth International Congress on the Chemistry of Cement, Stroizdat, Moscow, 1973, pp. 114–136.
[188] V.V. Ilyukhin, V.A. Kuznestov, A.N. Lobachev, V.S. Bakshutov, Hydrosilicates of Calcium, Nauka, Moscow, 1979. p. 184.
[189] B.N. Litvin, O.V. Kudryasteva, Structural–genetic elucidation of phase formation in the system $CaO-SiO_2-H_2O$, Proceedings of the Sixth International Congress on Chemistry of Cement, Stroizdat, Moscow, 1976, pp. 211–213.
[190] H.F.W. Taylor, The crystal structure of kilchoanite, $Ca_6(SiO_4)$ (Si_3O_{10}) with some comments on related phases, Miner. Mag. 38 (1971) 26–31.
[191] A.M. Lejus, A. Kahn-Harari, J.M. Benitez, B. Viana, Crystal growth, characterization and structure refinement of neodymium doped gehlenite, a new laser material, $[Ca_2Al_2SiO_7]$, Mat. Res. Bull. 29 (1994) 725–734.
[192] H. Funk, Uber barium silicathydrate und ihre hydrothermal-produkte, Ztschr. Anorg. Und allg. Chem. 296 (1958) 46–62.
[193] P. Eskola, Synthetic barium silicate, Am. J. Sci. 4 (1922) 331–332.
[194] O.S. Philipenko, B.N. Litvin, E.A. Pobedimskaya, N.V. Belov, Hydrothermal synthesis of some barium silicates and their X-ray data, Kristallografia 15 (1970) 863–865.
[195] I. Ismail-Zade, B.N. Litvin, Hydrothermal synthesis of monocrystals of Na–Mn–silicates and germanates, Azerbaizan Chem. J. 6 (1969) 135–146.
[196] B.N. Litvin, USSR Patent 1305288, February 19, 1969, Growth of Bivalent Mn-Silicates and Germanates Crystals, and Ismail-Zade, I.
[197] B.N. Litvin, V.S. Fonin, D.Yu. Pushcharovskii, E.A. Pobedimskaya, Crystallization of silicates, Crystal Growth, Nauka, Moscow, 1977, pp. 96–100.
[198] R.A. Hatch, Am. Mineral. 28 (1943) 471–478.
[199] T. Isaacs, R. Roy, Geochim. Cosmochim. Acta. 15 (1958) 213–219.
[200] V.A. Kuznetsov, A.A. Shternberg, T.K. Ivanova, Hydrothermal crystallization of lithium silicates, synthesis of spodumene, in: A.N. Lobachev (Ed.), Crystallization Processes Under Hydrothermal Conditions, Consultants Bureau, New York, NY, 1973, pp. 173–184.
[201] H. Ghobarkar, The hydrothermal synthesis of petalite mineral phases and their morphological changing by increasing of temperature, Cryst. Res. Technol. 27 (1992) 451–456.
[202] H. Onishi, Notes on the geochemistry of germanium, Bull. Chem. Soc. Jpn. 29 (1956) 686–694.
[203] S.A. Wardani, On the geochemistry of germanium, Geochim et Cosmochim. Acta. 13 (1957) 5–19.
[204] A.P. Vinogradov, Geochemistry 1 (1956) 6–52.
[205] N.G. Duderov, L.N. Dem'yanets, E.L. Belokoneva, A.N. Lobachev, N.V. Belov, New crystalline phases in the $Na_2O-CdO-GeO_2-H_2O$ system, Sov. Phys. Crystallogr. 17 (1973) 916–920.
[206] N.G. Duderov, L.N. Demianets, A.N. Lobachev, Yu.V. Pisarevsky, I.M. Silvestrova, Na_2COGeO_4 single crystal growth and study of their piezoelectric and elastic properties, J. Cryst. Growth 44 (1978) 483–491.
[207] L.N. Demianets, N.G. Duderov, A.N. Lobachev, F Lefaucheux, M.C. Robert, A. Authier, X-ray topographic study of growth defects in hydrothermal Na_2CoCeO_4 single crystals in relation to their growth conditions, J. Cryst. Growth 44 (1978) 570–580.

[208] I.P. Kuzmina, A.N. Lobachev, V.M. Vinokurov, N.M. Nizamutdinov, L.A. Volkova, Zn–Mn isomorphism in Na_2ZnGeO_4 crystals, Sov. Phys. Crystallogr. 18 (1973) 110–111.
[209] E.L. Belokoneva, Yu.A. Ivanov, M.A. Simonov, N.V. Belov, Crystal structure of cadmium orthogermanate Cd_2GeO_4, Sov. Phys. Crystallogr. 17 (1972) 177–179.
[210] A.N. Kornev, L.N. Dem'yanets, B.A. Maksimov, V.V. Ilyukhin, N.V. Belov, Hydrothermal synthesis in the $KCl-CdO-GeO_2-H_2O$ system and the crystal structure of $Cd_2[Ge_2O_6]$, Sov. Phys-Crystallogr. 17 (1972) 244–246.
[211] S.A. Ivanov, A.A. Bush, S.V. Fedotov, V.E. Zavodnik, Single crystals of new phase $Na_{0.67}Pb_2Ge_{2.67}O_{7.33}(OH)_{0.67}$ in $PbO-GeO_2-NaOH-H_2O$ system: conditions of preparation, atomic structure, and properties, J. Struct. Chem. 32 (1991) 13–17.
[212] T.G. Uvarova, L.N. Dem'yanets, A.N. Lobachev, Hydrothermal synthesis of sodium aluminogermanate single crystals, Sov. Phys. Crystallogr. 21 (1976) 497–499.
[213] Yu.A. Malinovskii, E.A. Pobedimskaya, N.V. Belov, Synthesis and crystal structure of the germanate–carbonate apatite $Ba_5[(Ge,C)(O,OH)_4]_3$ (OH), Sov. Phys. Crystallogr. 20 (1975) 394–396.
[214] Yu.A. Malinovskii, E.A. Pobedimskaya, V.A. Kuznetsov, N.V. Belov, Phase formation in the system $BaO-GeO_2-K_2O-H_2O$ under hydrothermal conditions, Sov. Phys. Crystallogr. 20 (1976) 725–728.
[215] V.S. Kurazhkovskaya, Ph.D., Hydrothermal Growth of Selected Borates, Thesis, Moscow State University, Moscow, 1971.
[216] M.N. Tseitlin, G.F. Plakhov, A.N. Lobachev, V.I. Popolitov, M.A. Simonov, N.V. Belov, Study of crystallization in the hydrothermal system $GeO_2-Sb_2O_3-KF-H_2O$, Sov. Phys. Crystallogr. 18 (1974) 525–527.
[217] V.Ya. Verkovskii, E.A. Kuzmin, V.V. Ilyukhin, Belov, Crystal structural of Na-titanogermanate, $Na_2(TiO)$ $[GeO_4]$, Doklady Akad. Nauk USSR, Kristallografia 190 (1970) 91–93.
[218] V.V. Popov, M.A. Petrova, Synthesis and electric properties of germanates of rare earth elements, Izvestiya Akad. Nauk USSR, Inorg. Mater. 9 (1973) 61–63.
[219] J. Fournier, R. Kohlmuller, Surles systemes $GeO_2-Ln_2O_3-C$, R. Acad. Sci. (C) 266 (1968) 136–362.
[220] G.A. Emelchenko, B.N. Grechushnikov, V.V. Ilyukhin, Spectroscopic investigations of rare earth germanates, Spectroscopy of Crystals, Nauka, Moscow, 1976, pp. 265–269
[221] H.Y.-P. Hong, Crystal structure and ionic conductivity of $Li_{14}Zn(GeO_4)_4$ and other new Li^+ superionic conductors, Mat. Res. Bull. 13 (1978) 117–124.
[222] J.W. Anthony, Am. Min. 42 (1957) 904–910.
[223] I. Kostov, Mineralogy, Mir Publishers, Moscow, 1971.
[224] J.C. Ronsbo, A.P. Khomyakov, E.I. Semenov, A.A. Voronkov, V.K. Garanin, Neues Jahrb. Mineral. Abhandl. 1 (1979) 42–45.
[225] J.J. Berstelius, Ann. Physik 53 (1816) 393.
[226] K.R. Johnson, Ber. Deut. Chem. Ges. 22 (1889) 976.
[227] R.C.L. Mooney, J. Chem. Phys. 16 (1948) 1003.
[228] R.C.L. Mooney, Acta Crystallogr. 3 (1950) 337.
[229] S.C. Jaulmes, R. Acad. Sci. C 268 (1969) 935 Paris.
[230] M. Beucher, Europium in the ultraphosphate lattice: Polarized spectra and structure of EuP_5O_{14}, Les Elements des Terres Rares, Coll No. 180. Vol. 1 (1970) pp. 331–336, Paris C.N.R.S, France
[231] H.G. Danielmeyer, G. Huber, W.W. Kruhler, J.F. Jesser, Appl. Phys. 2 (1973) 335–339.

[232] H.P. Weber, T.C. Damen, H.G. Danielmeyer, B.C. Tofield, Appl. Phys. Lett. 22 (1973) 534–536.
[233] K. Otsuka, T. Yamada, J Nakano, T. Kimura, M. Seruwateri, J. Appl. Phys. 46 (1975) 4600.
[234] K. Otsuka, S. Miyazawa, T. Yamada, H. Iwasaki, J. Appl. Phys. 48 (1977) 2099–2103.
[235] W.W. Krühler, R.D. Plattner, W. Fabian, P. Mockel, J.G. Grabmaier, Opt. Commun. 20 (1977) 354–360.
[236] S.R. Chin, H.Y.-P. Hong, Opt. Commun. 15 (1975) 345–349.
[237] J.B. Goodenough, H.Y.-P. Hong, J.A. Kafalas, Mater. Res. Bull. 11 (1976) 203–210.
[238] H.Y.P. Hong, Crystal structures and crystal chemistry in the system $Na_1 + {}_xZr_2 Si_xP_{3-x}O_{12}$, Mater. Res. Bull. 11 (1976) 173–180.
[239] J.B. Goodenough, Fast ionic conductors in: Proceedings of the International Course on Materials Science, (E. Kaldis, ed.) Erice, Italy, April 14–24, 1980.
[240] K. Byrappa, Ph.D., Fluorescence in $CsNdP_4O_{12}$. Thesis, Moscow State University, Moscow, 1982.
[241] K. Byrappa, G.S. Gopalakrishna, M.N. Viswanathiah, V. Venkatachalapathy, B. Puttaraj, Hydrothermal growth and properties of Na_2 (La, CO) Zr $(PO_4)_3$ crystals, J. Mater. Sci. Lett. 4 (1985) 565–570.
[242] L.O. Atomyan, O.S. Philepenko, B.I. Ponomarev, L.S. Leonov, N.S. Tkacheva, E.A. Ukshe, et al., Dokl. Acad. Nauk USSR 261 (1981) 874–879.
[243] M. Rzaigui, K. Ariguib, M.T. Averbuch-Pouchot, A. Durif, J. Solid State Chem. 50 (1983) 240–246.
[244] N.N. Chudinova, D.Sc., Chemistry of condensed phosphates of Trivalent Metals Hydrothermal-Growth-of-Selected-Borates, Thesis, Institute of General and Inorganic Chemistry, Academy of Science, USSR, Moscow, 1980.
[245] J.D. McGilvery, A.E. Scott, Can. J. Chem. Soc. 79 (1957) 3647.
[246] M. Carron, M.E. Mrose, K.J. Murata, Relation of ionic radius to structures of rare earth phosphates, arsenates and vanadates, Am. Min. 43 (1958) 985–989.
[247] M.K. Carron, C.R. Naeser, H.J. Rose Jr., F.A. Hildebrand, Fractional precipitation of rare earths with phosphoric acid, U.S.Geol, Survey Bull. No. 1036-N (1958) 253–273.
[248] K. Byrappa, A. Gopalakrishna, Critical survey on the study of alkaline rare earth phosphate and NASICON systems with a special reference to the hydrothermal method, Prog. Cryst. Growth Character. 11 (1985) 89–107.
[249] K. Byrappa, Preparative methods and growth of rare earth phosphates, Prog. Cryst. Growth Character. 13 (1986) 163–196.
[250] A. Rabenau, Lithium nitride and related materials, in: Proceedings of the International Course on Physics of Materials, (R. Srinivasan, ed.) I.I.T. Madras, India, September 3–22, 1984.
[251] M. Yoshimura, K. Fujii, S. Somiya, Phase equilibria in the system Nd_2O_3–P_2O_5–H_2O at 500°C, under 100 MPa and synthesis of NdP_5O_{14} crystals, Mat. Res. Bull. 16 (1981) 327–333.
[252] K. Byrappa, B.N. Litvin, Hydrothermal synthesis of mixed phosphates of neodymium and alkaline metals ($Me_2O \cdot Nd_2O_3 \cdot 4P_2O_5$), Mater. Sci. 18 (1983) 703–708.
[253] K. Byrappa, S. Srikantaswamy, S. Gali, Hydrothermal synthesis and structure of TmP_5O_{14}, J. Mater. Sci. Letts. 9 (1990) 235–236.
[254] K. Byrappa, The possible reasons for the absence of condensed phosphates in nature, Phys. Chem. Miner. 10 (1983) 94–95.

[255] Yu.E. Gorbunov, V.V. Ilyuhin, V.G. Kuznestov, et al., Doklady Akad. Nauk USSR 222 (1976) 1329 (in Russian).
[256] A.V. Chiganov, N.V. Belov, Geochemistry 12 (1968) 1456.
[257] K.K. Palkina, Crystal chemistry of rare earth phosphates, Inorg. Mater. 18 (1982) 1199–1221.
[258] Y. Xu, S. Feng, W. Pang, Hydrothermal synthesis and characterization of$Na_3Ce(PO_4)_2$, Mat. Lett. 28 (1996) 313–315.
[259] Y. Xu, S. Feng, W. Pang, Hydrothermal synthesis and characterization of $K_2Ce(PO_4)_2 \cdot H_2O$, Mat. Lett. 28 (1996) 499–502.
[260] B.N. Litvin, Chemical processes in the growth of single crystals, Inorg. Mater. 20 (1984) 898–904.
[261] K. Byrappa, G.S. Gopalakrishna, V. Venkatachalapathy, B. Puttaraju, Crystallization and characterization of $Na_2(La, Me) Zr (PO_4)_3$, J. Mater. Sci. 20 (1985) 1419–1424.
[262] K. Byrappa, G.S. Gopalakrishna, A.B. Kulkarni, V. Venkatachalapathy, Synthesis and characterization of Na_2 (R,Co)$Zr(PO_4)_3$ crystals, J. Less Common Metals 110 (1985) 441–444.
[263] K. Byrappa, Growth and characterization of some new superionic phosphates, Trans. Mater. Res. Soc. Jpn. (1990) 433–456.
[264] A.B. Bykov, A.P. Chirkin, L.N. Demyanets, S.N. Doronin, E.A. Genkina, A.K. Ivanov-Shits, et al., Superionic conductors $Li_3M_2(PO_4)_3$ (M = Fe, Sc, Cr): synthesis, structure and electrophysical properties, Solid State Ionics 38 (1990) 31–52.
[265] J.B. Goodenongh, H.Y.P. Hong, J.A. Kafalas, Fast Na^+ ion transport in skeleton structures, Mater. Res. Bull. 11 (1976) 203–220.
[266] H.Y.-P. Hong, Crystal structures and crystal chemistry in the system $Na_{1+x}Zr_2Si_xP_{3-x}O_{12}$, Mater. Res. Bull. 11 (1976) 173–176.
[267] V.A. Efremov, V.B. Kalinin, Determination of crystal structure of $Na_3Sc_2(PO_4)_3$, Kristallografia 23 (1978) 703–708.
[268] M. Vlasse, C. Parent, R. Salmon, Le.G. Flem, P. Hagenmuller, The structures of the $Na_3Ln(XO_4)_2P$ phases (Ln = rare earth X = P, V, As), J. Solid State Chem. 35 (1980) 318–324.
[269] L.O. Hagman, P. Kierkegaard, The crystal structure of $NaMe_2^{IV}(PO_4)_3$, Acta Chem. Scandinav. 22 (1968) 1822–1832.
[270] R.G. Sizova, A.A. Voronkov, N.V. Shumyatskaya, V.V. Illyakhin, N.N. Belov, DokladyAkad. USSR Physics Series 17 (1973) 618.
[271] D. Tranqui, J.J. Copponi, J.C. Joubert, R.D. Shannon, Crystal structure and ionic conductivity in $Na_4Zr_2Si_3O_{12}$, J. Solid State Chem. 39 (1981) 219.
[272] C.J. Delbecq, S.A. Marschall, S. Susman, Evidence of a structural phase change in the fast ionic conductor Na3$Sc_2P_3O_{12}$, Solid State Ionics 1 (1980) 145–149.
[273] S. Susman, C.J. Delbecq, T.O. Bran, Structure and conductivity of the NASICON analog $Na_3Sc_2(PO_4)_3$, Proc. Electro Chemical Soc. (1982) 1172–1173 (Spring Meet May 9–14, Mautreal Extd. Abst.
[274] G. Collin, J.P. Boilot, M. dt la Rochere, R. Comes, in: F.J. Salzano (Ed.), Proc. Cont. on High temperature Solid Oxide Electrolytes, Broolthaven National Lab., New York, NY, 1983.
[275] J.P. Boilot, G. Collin, Ph. Colomban, Nasicon and related compounds: a review, in: T.W. Wheat, A. Ahmad., A. Kauahox (Eds.), Progress in Solid Electrolytes, Energy, Mines and Resources, Ottawa, ON, 1983, pp. 91–122.
[276] A. Clearfield, M.A. Subramanian, P.R. Rudolf, A. Moini, Stoichiometry, structure and conductivity of NASICON, Solid State Ionics 18/19 (1986) 13–20.

[277] W.H. Baur, J.R. Bygar, D.H. Whitemore, Faber, Solid State Ionics 18/19 (1986) 93–99.
[278] A.B. Feltz, S. Barth, Preparation and conductivity behavior of $Na_3M^{II}Zr(PO_4)_3$, (M^{II}: Mn, Mg, Zn), Solid State Ionics 9/10 (1983) 867.
[279] F.D. Y'voire, M. Pintard-Screpel, E. Bretey, Les phasphates $Na_3M_2^{+3}(PO_4)_3$ (Cr, Fe = M) et $Na_3Cr_3(PO_4)_4$: donnees cristallographiques et conductivite lonique, CR Acad Sci. Pris 290 (1980) C185–188.
[280] L.P. Keller, G.J. Mc Carthy, R.G. Garvey, Mater. Res. Bull. 20 (1985) 459.
[281] M.de la Rochera, A. Kahu, d' Yvoire, E. Bretey, Crystal structure and cation transport properties of the ortho-diphosphates $Na_7(MP_2O_7)PO_4$ (M = Al, Cr, Fe), Mat. Res. Bull. 20 (1985) 27–34.
[282] S. Gali, K. Byrappa, G.S. Gopolakrishna, Structure of $Na_2Zr(P_2O_7)$, (M = Ni, Co), Acta Crist. C45 (1989) 1667–1669.
[283] K. Byrappa, G.S. Gopalakrishna, S. Gali, Synthesis and characterization of new superionic pyrophosphates, Indian J. Phys. 63A (1989) 321–325.
[284] S. Gali, K. Byrappa, Structure of $(Na_{2/3}Zr_{1/3})_2P_2O_7$, Acta Cryst. C46 (1990) 2011.
[285] V.M. John, S. Kordes, Chem. Earth 70 (1953) 75.
[286] J.M. Stanley, Hydrothermal synthesis of large aluminum phosphate crystals, Ind. Eng. Chem. 468 (1954) 1684.
[287] H.M. Kurbanov, V.Yu. Kara-ushnov, B.C. Halikov, Orthophosphates of Rare Earth Elements, Donish Publishers, Dushanbe, 1981 (in Russian).
[288] L. Yonghua, H. Ninghai, Z-Qinglian, S. Endong, W. Mingyi, L. Shuzhen, W. Shixue, Kexue Tongbao 27 (1982) 1126 (in Chinese).
[289] F.D. Y'voire, M. Pintard-Screpel, E. Brety, M. De La Rochere, Solid State Ionics 9/10 (1983) 851.
[290] F. Genet, M. Barj, Hydrothermal, synthesis and recrystallization of compounds belonging to the NASICON family: synthesis and crystallization of $Na_4Zr_2Si_3O_{12}$, Solid State Ionics 9/10 (1983) 891.
[291] K. Byrappa, U. Dutt, Crystal growth processes of formation of $HNaMP_2O_7$ crystals, J. Mater. Sci. 29 (1994) 6468–6472.
[292] K. Byrappa, B. Nirmala, Study of crystallization process in some rare earth vanadate, tungstate and phosphate systems under hydrothermal conditions, Indian J. Phys. 73 (A) (1999) 1–9.
[293] K. Burger, Solvation, Ionic and Complex Formation Reaction in Non-Aqueous Solvents, Akademiai Kiaddo, Budapest, 1983. p. 42
[294] N.V. Belov, Miner. Sb. 15 (1961).
[295] T.P. Dirke, J. Electrochem. Soc. 101 (1954) 328.
[296] C.K. Jorgenson, Werner Centennial Advances in Chemistry Series, American Ceramic Society, New York, NY, 1967. p. 161.
[297] K. Byrappa, B. Sanjeeva Ravi Raj, Crystal growth, morphology and properties of $NaHMP_2O_7$ (M = Ni, Co, Mn, Cd, Zn, Pb), Ind. J. Phys. 72 (A) (1998) 1–10.
[298] K. Byrappa, G.S. Gopalakrishna, B.V. Umesh Dutt, Morphology of new superionic pyrophosphates, J. Mater. Sci. 27 (1992) 4439–4446.
[299] H.G. Danielmeyer, H. Huber, W.W. Kruhler, J.F. Jesser, Appl. Phys. 2 (1973) 335.
[300] G.I. Dorokhova, Ph.D., Hydrothermal synthesis and growth of coordinated complex crystals, Thesis, Moscow State University, Moscow, 1985.
[301] J.N. Vashishta, J.N. Mundy, G.K. Shenoy (Eds.), Proceedings of the International Conference on Fast Ion Transport in Solids, Elsevier Science Publishers, Lake Geneva, 1979.

[302] K. Byrappa, Fast ionic conductors—new perspectives, Indian J. Phys. 66A (1992) 233–261.
[303] Y. Yue, W. Pany, Hydrothermal synthesis and characterization of $NaSn_2(PO_4)_3$, J. MSL 11 (1992) 148–149.
[304] Y. Xu, S. Feng, W. Pany, Hydrothermal synthesis and characterization of $KSn_2(PO_4)_3$, J. Chem. Soc. Chem. Commun. (1993) 1428–1429.
[305] Y. Xu, S. Feng, W. Pang, Hydrothermal synthesis characterization of $NH_4Sn_2(PO_4)_3$, J. Solid State Chem. 119 (1995) 197–198.
[306] K. Byrappa, G.S. Gopalakrishna, D.S. Mahadevappa, J. Shashidharaprasad, in: M. Yussouff (Ed.), Physics of Materials, World Scientific Publishers, Singapore, 1987, p. 222.
[307] B.R. Sanjeev, Ph.D., Probabilistic checking of proofs and the hardness of approximation problems, Thesis, University of Mysore, India, 1996.
[308] K. Byrappa, B.V. Umesh Dutt, A. Clearfield, M. Damodara Poojary, Crystal growth, morphology, structure and properties of $HNaMP_2O_7$, where M = Co, and Ni crystals, J. Mater. Res. 9 (1994) 1519–1525.
[309] K.H. Lii, C.H. Li, C.Y. Cheng, S.L. Wang, Hydrothermal synthesis structure and magnetic properties of a new polymorph of lithium vanadyl (iv) orthophosphate: $LiVOPO_4$, J. Solid State Chem. 95 (1991) 352–359.
[310] K.H. Lii, H.J. Tsai, Hydrothermal synthesis and characterization of vanadyl (iv) hydrogenphosphate, $K_2(VO)_3$ (HPO_4), J. Solid State Chem. 91 (1991) 331–338.
[311] A. Mar, F. Leroux, D. Guyomard, A. Verbaere, Y. Piffard, Hydrothermal synthesis and structure of $Mn_2VO(PO_4)_2 \cdot H_2O$, J. Solid State Chem. 115 (1995) 76–82.
[312] P. Amoros, A. Bail Le, Synthesis and crystal structure of a-$NH_4(VO_2)(HPO_4)$, J. Solid State Chem. 97 (1992) 283–291.
[313] T. Hoareau, A. Leclaire, M.M. Borel., A. Grandin, B. Raveau, A mixed valent molybdenum monophosphate with a layer structure: $CsMO_2O_3(PO_4)_2$, J. Solid State Chem. 116 (1995) 87–91.
[314] S. Srikantaswamy, Ph.D., Phases and crystallization in the system aluminum orthophosphate, Thesis, University of Mysore, Mysore, India, 1989.
[315] Y. Oka, T. Yao, N. Yamamoto, Y. Ueda, S Kawasaki, M. Azuma, et al., Hydrothermal synthesis crystal structure and magnetic properties of $FeVO_4$-II, J. Solid State Chem. 123 (1996) 54–59.
[316] M. Vlasse, C. Parent, R. Salmon, Le Ge. Flem, The structures of $Na_3Ln(XO_4)_2$ phases (Ln = rare earth, X = P, V,As), Rare-Earths in Modern Sciences and Technology, 2, Plenum Press, New York, NY, 1980195–201
[317] J.J. Rubin, L.G. Utiert, Growth of large yttrium vanadate single crystals for optical maser studies, J. Appl. Phys. 37 (1966) 2920–2924.
[318] K. Muto, K. Awasu, Growth of yttrium vanadate crystals by a modified floating zone technique, J. Appl. Phys. 8 (1969) 1360–1364 Jpn.
[319] B.K. Tanner, S.H. Smith, A study of the perfection of flux-grown rare earth vanadates by X-ray topography, J. Cryst. Growth 28 (1975) 77–84.
[320] K. Byrappa, A.B. Kulkarni, N.B. Desai, H.P. Jagadeesh, $NH_4Zr_2V_3O_{12}$ proton conductor, Bull. Mater. Sci. 9 (1987) 323–330.
[321] K. Byrappa, B. Nirmala, M. Yoshimura, Crystal growth of Nd:RVO_4 (where R = Y, Gd) under mild hydrothermal conditions, Materials Forum 315–317 (1999) 506–513.
[322] B.C. Chakoumakos, M.M. Abraham, L.A. Boatner, J. Solid State Chem. 109 (1994) 194.

[323] K. Shimamura, S. Uda, V.V. Kochauriklin, T. Taniuchi, T. Fukuda, J. Appl. Phys. 35 (1996) 320jpn, (in Japanese)
[324] M. Prasad, A.K. Pandit, T.H. Ansari, R.A. Singh, B.M. Wanklyn, Electrical transport properties of $EuVO_4$ single crystals, Phys. Lett. A 138 (1989) 61–64.
[325] E.A. Maunders, E.G. Deshaser, J. Opt. Soc. 61 (1971) 68.
[326] B.G. Chai, G. Loutts, X.X. Chang, P. Hong, M. Bass, I.A. Sherbakov, et al., Advanced Solid-State Lasers, Optical Society of America, Washington, DC, 1994Technical Digest, p. 230
[327] M. Ross, IEEE J. Quantum Electron. QE-11 (1975) 938.
[328] J.J. Rulin, L.G. Utert, Growth of large yttrium vanadate single crystals for optical maser studies, J. Appl. Phys. 37 (1966) 2920.
[329] Kh.S. Bagdasarov, V.S. Karylor, V.I. Popov, New yttrium and gadolinium vanadates, High Temperature Chemistry of Silicates and Oxides, Nauka, Leningrad, 1972 (in Russian)
[330] Yu.P. Udalov, Z.S. Appen, Izvestiya Akad. Nauk USSR, Neorganicheskie Materially 18 (1982) 1143.
[331] Kh.S. Bagdasarov, E.I. Getman, N.N. Mikhailichenko, I.V. Mokhosoeva, M.V. Mokhosoev, V.I. Popov, Gadolinium and yttrium orthovanadates, Izvestiya Akad. Nauk USSR Inorg. Mater. 5 (1969) 1581.
[332] Hintzmann, G. Müller-Vögt, Xes characterization of rare earth vanadates, J. Cryst. Growth 5 (1969) 274.
[333] Y. Terado, K. Shimamura, V.V. Kochurikhin, L.V. Barashov, M.A. Ivanov, T. Fukuda, Hydrothermal Growth of Selected Borates, J. Cryst. Growth 167 (1996) 369.
[334] S. Erdei, B.M. Jin, F.W. Ainger, Influence of defects in the output power of c- and a-axis growth, J. Cryst. Growth 174 (1997) 328.
[335] S. Erdei, M. Milmkiewicz, F.W. Ainger, B. Kaszei, J. Vandlik, A. Surverges, 7 Hydrothermal Synthesis and Growth of Coordinated Complex, Mater. Letts. 24 (1995) 301.
[336] J.W. Gibbs, Collected Works, Longman, Green, London, 1925.
[337] M. Sato, T. Utsunomiya, Crystal growth of YVO_4 and distribution of rare earth ions in the growth crystals, Bull. Tokyo Inst. Tech. 98 (1970) 15.
[338] B. Robertson, E. Kostiner, J. Solid State Chem. 4 (1972) 29–35.
[339] A.P. Young, C.M. Schwarz, Acta Cryst. 15 (1962) 1305.
[340] T. Yao, Y. Oka, N. Yamamoto, Layered structures of hydrated vanadium oxides, part-1, alkali-metal intercalates $A_{0.3}V_2O_5 \cdot nH_2O$ (A = Na, K, Rb, Cs and NH_4), J. Mater. Chem. 2 (1992) 331–336.
[341] T. Yao, Y. Oka, N. Yamamoto, Layered structures of hydrated vanadium oxides, part-2, vanadyl intercalates $(VO)_x\ V_2O_5 \cdot nH_2O$, J. Mater. Chem. 2 (1992) 337–340.
[342] Y. Oka, T. Yao, N. Yamamoto, O. Tamada, Hydrothermal synthesis of hydrous vanadium oxide bronzes $M_xV_3O_8(VO)_y \cdot nH_2O$ (M = K, Rb, Ba), Mater. Res. Bull. 32 (1997) 59–66.
[343] F. Lutz, M. Leiss, Recent Nd-Miniature Lasers, Extended Abstract C57, ECCG-2, Lancaster, UK, September 10–15, 1979.
[344] J.M. Reau, A. Levasseur, G. Maniez, B. Cales, C. Fouassier, P. Hagenmuller, Sur de nouveaux conducteurs inoiques du lithium de structure boracite, Mater. Res. Bull. 11 (1976) 1087.
[345] C.T. Chen, B.C. Wu, A.D. Jiang, G.M. A new ultraviolet SHG crystal β-BaB_2O_4, Yov, Sci. Sin. B28 (1985) 238.
[346] J.D. Garrett, M. Natarajan-Iyer, J.E. Greedan, The crystal growth of $LiBO_2$ and $Li_2B_4O_7$, J. Cryst. Growth 41 (1977) 225.

[347] V.H. Konig, R. Hoppe, Z. Anorg, Local structure of LiB3O5 single crystal from 7Li nuclear magnetic resonance, Allg. Chem. 439 (1978) 71.
[348] B.S.R. Sastry, F.A. Hummel, Studies in Lithium Oxide Systems: I, $Li_2O{\cdot}B_2O_2$–B_2O_3, J. Am. Ceram. Soc. 41 (1958) 7.
[349] E.V. Sokolova, Ph.D. Asymmetric catalysis, Hydrogenation on metals, Thesis, Moscow State University, Moscow (in Russian), 1980.
[350] N.V. Belov, A.I. Ivaschenko, O.S. Bondareva, A.N. Lobachev, Yu.A. Malinovskii, O. K. Melnikov, et al., in: A.N. Lobachev (Ed.), Hydrothermal Synthesis and Growth of Single Crystals, Nauka, Moscow, 1982. p. 158.
[351] O.V. Yakubovich, Ph.D. Experimental Investigation of Transport Robots Control, Thesis, Moscow State University, Moscow, 1978.
[352] O.S. Bonderava, Y.K. Egorov-Tismenko, M.A. Simonov, M.V. Belov, Sov. Phys. Dokl. 23 (1978) 806.
[353] S. Terol, M.J. Otero de le Gaudaka, Naturforsch, Z. 16A (1961) 920.
[354] L.A. Dewerd, Mater. Res. Bull. 11 (1976) 1413.
[355] M. Takanaga, Nucl. Instrum. Methods 175 (1980) 77.
[356] K. Byrappa, K.V.K. Shekar, Phases and crystallization in the system $Li_2O-B_2O_3-H_2O$ under hydrothermal conditions, J. Mater. Res. U.S.A. 8 (1993) 864–870.
[357] K. Byrappa, S. Srikantaswamy, Recent progress in the growth of aluminum orthophosphate, Prog. Cryst. Growth Charact. 21 (1990) 119–254.
[358] M. Adachi, Jpn., J. Appl. Phys. 24 (1985) 72.
[359] D.S. Robertson, I.M. Young, J. Mater. Sci. 17 (1982) 1729.
[360] J.S. Fan, G.S. Shen, W. Wand, J.L. Li, X. –H Le, Bridgman growth of $Li_2B_4O_7$ crystals, J. Cryst. Growth 99 (1990) 811–817.
[361] H. Yamashita, T. Yoko, S. Sakka, J. Mater. Sci. Lett. 9 (1990) 796–798.
[362] K. Byrappa, K.V.K. Shekar, Hydrothermal synthesis and characterization of piezoelectric $Li_2B_4O_7$ crystals, J. Mater. Chem. Royal Soc. Chem. 2 (1992) 13–18.
[363] K. Byrappa, K.V.K. Shekar, R.R. Clemente, Hydrothermal synthesis and morphology of $Li_2B_4O_7$ crystals, Mater. Res. Bull. 7 (1993) 864–870.
[364] K. Byrappa, K.V.K. Shekar, S. Gali, Synthesis and characterization of $Li_3B_5O_8(OH)_2$ crystals, Cryst. Res. Technol. 27 (1992) 767–772.
[365] K.V.K. Shekar, Ph.D. Synthesis and characterization of lithium borates, Thesis, University of Mysore, India, 1993.
[366] J.M. Reau, A. Levasseur, G. Maniez, B. Cales, C. Fouassier, P. Hagenmuller, Hydrothermal Synthesis and Crystal Growth of Fluorides, Sulfides, Mat. Res. Bull. 11 (1976) 1087.
[367] W. Jettschko, T.A. Bithetr, Naturaf. Z. 27B (1972) 1423–1426.
[368] Mm.A. Lavasseur, D.J. Lloyd, C. Fouassier, P. Hagenuller, Determination structurale de la boracite $Li_4B_7O_{12}Cl$, J. Solid State Chem. 8 (1973) 318–323.
[369] K. Byrappa, K.V.K. Skekar, A.B. Kulkarni, S. Gali, Hydrothermal synthesis and characterization of $Li_4Bi_7O_{12}Cl$ crystals—fast ionic conductors, Ind. J. Phys. 66A (1992) 263–271.
[370] A. Cardenas, J. Solans, K. Byrappa, K.V.K. Shekar, Structure of lithium catena-poly [3,4-dihydroxopentaborate-1:5-m-oxo], Acta Cryst. C49 (1993) 645–647.
[371] K. Byrappa, V. Rajeev, V.J. Hanumesh, A.R. Kulkarni, A.B. Kulkarni, Crystal growth and electrical properties of $Li_2B_4O_7$, J. Mater. Res. 11 (1996) 2616–2621.
[372] K. Byrappa, V. Rajeev, A.R. Kulkarni, A.B. Kulkarni, $Li_3B_5O_8(OH)_2$: crystal growth ionic conductivity studies, J. Mater. Sci. 32 (1997) 1599–1602.

[373] E.V. Sokolova, Hydrothermal Synthesis and Crystal Structures of Cd– and Li–Cd Borates, Ph.D. Thesis, Moscow State University, Moscow (in Russian), 1980.

[374] O.S. Bondareva, Hydrothermal Synthesis and Crystal Structures of Li, Zn and Li, Zn-Borates, Ph.D. Thesis, Moscow State University, Moscow (in Russian), 1979.

[375] O.M. Yaghi, M. O'Keeffee, N. Ockwig, H.K. Chae, M. Eddaoudi, J. Kim, Reticular synthesis and the design of new materials, Nature 423 (2003) 705–714.

[376] C.N.R. Rao, A. Mjiller, A.K. Cheetham, A. Muller, The Chemistry of Nanomaterials: Synthesis, Properties and Applications, John Wiley & Sons, Oxford, UK, 2004. p. 761.

[377] G. Ferey, Hybrid porous solids: past, present, future, Chem. Soc. Rev. 37 (2008) 191–214.

[378] A.K. Cheetham, C.N.R. Rao, There's room in the middle, Science 318 (2007) 58–59.

[379] A.U. Czaja, N. Trukhan, U. Muller, Industrial applications of metal-organic frameworks, Chem. Soc. Rev. 38 (2009) 1284–1293.

[380] H. Ohtsuka, A. Yamaji, Preparation and characteristic of $Li_{14}Zn(GeO_4)_4$ thin films for solid state batteries, IEEE (1983)CH 1855–1856:353–359.

[381] J. Müller, J.C. Joubert, Synthese sous haute pression d' oxygene d'une forme dense or d'une varie te allotropique de structure $CrVO_4$ J, Solid State Chem. 14 (1975) 8–13.

8 Hydrothermal Synthesis and Crystal Growth of Fluorides, Sulfides, Tungstates, Molybdates, and Related Compounds (Coordinated Complex Crystals, Part II)

8.1 Introduction

The hydrothermal synthesis of complex coordinated compounds covers a vast group of compounds. In Chapter 7, which deals with Part I of these complexes, we covered silicates, germanates, phosphates, vanadates, and borates. Here we deal with fluorides, sulfides, tungstates, titanates, molybdates, tantalates, neobates, selenides, and related compounds.

The synthesis of these compounds did not carry much geological significance in the past; hence, their synthesis began in the middle of the twentieth century. However, the interest was purely academic. Today, these compounds have a great technological significance. Although the synthesis of these compounds is being studied extensively worldwide, interest is more confined to the nanosize crystals rather than bulk single crystals growth. Some of these coordinated complexes with reductions in the size and the establishment of core-shell structures carry special characteristics useful as sensors, quantum dots, imagers, semiconductors over a wide band gap region, and so on. Since the publication of the first edition of this book, publications are seldom found on the bulk single crystal growth of these complex coordinated compounds, as interest is mainly confined to the nanosize crystals, which are discussed separately in Chapter 10.

8.2 Fluorides

Fluorides form a large group of inorganic compounds; from the simplest to the most complicated types. In addition, we have fluorophosphates, fluorocarbonates, fluoroarsenates, oxyfluorides, and so on. These compounds contain a great variety of cations of the elements from groups 1, 2, 3, and 4 of the Periodic Table. For simple fluorides like the AF type (where A = Li, Na, K, Rb, Cs), the optimum

Handbook of Hydrothermal Technology.

growth method is from flux, or melt, or gas transport, whereas for more complicated fluorine-bearing compounds, hydrothermal synthesis is the most convenient method. The hydrothermal method helps in developing new crystal structures which are difficult to obtain by other conventional methods. Among fluorides, those containing rare earth elements, particularly the active elements like Nd^{3+}, and Eu^{3+} are the most popular because of their excellent luminescent properties. The fluorides bearing Fe, Mn, and other transitional metals show good magnetic properties; however, it is the rare earth elements bearing fluorides that carry greater importance as laser host materials. Kaminskii et al. [1] have reviewed in detail new CW-laser (Continuous Wave-laser) crystals with semiconductor laser pumping on the anisotropic fluorides with Nd^{3+} ions obtained by the hydrothermal method. Similarly, Demianets [2] reviewed the hydrothermal growth of alkali, rare earth fluorides and alkali zirconium fluorides and discussed their luminescent and ionic conductivity properties. Bridenbangh et al. [3] discussed the hydrothermal growth of alkali fluoride crystals and studied the phase equilibrium of the $KF-GdF_2-H_2O$ system in detail under hydrothermal conditions.

8.2.1 Hydrothermal Synthesis of Rare Earth Fluorides

Rare earth fluorides have attracted materials scientists for more than 40 years, owing to their unusual crystal structures and physical properties, but even today the problems concerning the polymorphism of double fluorides and the nature of morphotropic transitions in these compounds have not yet been understood. The physical properties of compounds formed in the $AF-LnF_3$ systems (*A* is the single-charged cation) have not been studied, which may be explained by the absence of sufficiently simple methods of synthesis of double fluorides of alkali and rare earth elements single crystals. The use of hydrothermal solutions provides the study of the behavior of the whole series of rare earth elements in the presence of F-ions under the same conditions. Under hydrothermal conditions, it is easy to synthesize fluorides of compositions $ALnF_4$, A_3LnF_6, A_2LnF_5 (A = Li, Na, K, Rb, Cs), amongst others. These compounds form the morphotropic series (K_2LnF_5) embracing most of the rare earth elements except La, Ce, Pr, and Sc. Rare earth fluorides are being studied extensively owing to their unusual spectroscopic characteristics, but these crystals are grown as small nanosize crystals rather than the bulk single crystals.

The hydrothermal phase equilibria of the $Ln_2O_3-KF-H_2O$ system has been studied in detail [2,3]. The stability fields of several compounds, such as $KLnF_4$, K_2LnF_5, K_3LnF_6, K_2LnF_5, and KLn_2F_7, have been fixed. Figure 8.1 shows the composition diagram of the system $KF-GdF_3-H_2O$ at 450°C and 10 kpsi. The region A in Figure 8.1 is close to the limit of $KGdF_4$ solubility, relative to KF concentration, as estimated from the solubility measurements. The single-phase field for KGd_2F_7 should not be considered to extend to the H_2O-GdF_3 join. The only information on the right side of this field was obtained at very high H_2O contents (98 mol%); this was deemed insufficient to allow projection of a phase boundary; however, the single-phase stability of KGd_2F_7 likely does not persist much farther to the right than the individual single-phase results shown. An additional two-phase

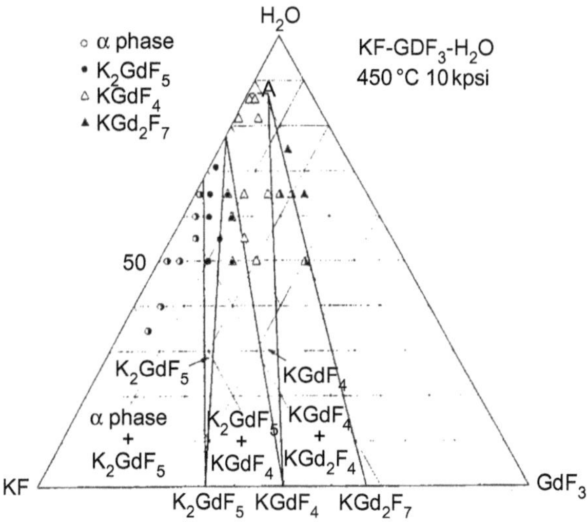

Figure 8.1 Composition diagram of the system KF–GdF_3–H_2O at 450°C and 10 kpsi [3].

field is shown along the left side of the diagram. A general point of particular interest from Figure 8.1 is that in this hydrothermal system, the phases K_2GdF_5, $KGdF_4$, and KGd_2F_7 are formed and are stable only where KF is in excess in the stoichiometric solid, thus, they are incongruently saturating. KF must be between 17.5 and 27 total mol% for $KGdF_4$ stability where water is at 65 total mol%. A 17.5 total mol% corresponds to the stoichiometry of $KGdF_4$, indicating that the phase is probably marginally congruently soluble. However, the other boundary of the $KGdF_4$ phase field, positioned at 27 mol% total KF, indicates that 17 mol% KF is required to stabilize 8 mol% $KGdF_4$. Thus, it can be concluded that a large excess of KF is required to stabilize the $KGdF_4$ phase. Analogous reasoning can be applied to choose conditions for the growth of other phases in Figure 8.1.

By considering the nature of the rare earth oxides' interaction with KF solutions, one can single out four rare earth subgroups. The first one (La–Pr), which showed no formation of K_2LnF_5, begins at $q = M_{KF}/MH_2O = 0.56$. In the third subgroup (Sm, Gd), the values of the q-ratio necessary for the formation of such compounds were 0.36 and 0.26, respectively; and in the fourth subgroup (Tb–Lu), the crystallization of the K_2LnF_5 phases begins for q-values smoothly, varying from 0.16 to 0.10. Pentafluorides are crystallized as transparent, faceted, prismatic, crystals up to 15 mm in length, of colors typical for the corresponding rare earth elements.

The conditions for the investigation of the phase relations in the fluoride systems are KF/Ln_2O_3 (molar) ratio up to 100, the $M(H_2O)/M(Ln_2O_3)$ ratio 200, temperature 450–500°C, and pressure 100–150 MPa. Under these conditions, fluoride crystals were grown for spectroscopic studies. It has been shown that the K_2GdF_5: Nd^{3+} compounds may be considered as potential media for the excitation of a stimulated emission in the range 1.06–1.35 μm, whereas the terbium-activated yttrium and gadolinium compounds may serve as luminophores with saturated green luminescence.

The solubility of $KGdF_4$ is a function of T in 12 mol% KF (defined as mol KF/(mol of KF + H_2O)) at a pressure of 10 kpsi. Figure 8.2 shows the solubility of $KGdF_4$ as a function of temperature and concentration of KF. Assuming a linear relationship between solubility and mol% KF, a slope of 0.02 mol solute dissolved/mol KF suggests the following generalized multistaged reaction:

$$KGdF_4 + nF^- \rightleftharpoons KGdF_{4+n} - n \tag{8.1}$$

Since the reaction does not proceed easily to the right, 50 mol of KF are required to dissolve 1 mol of $KGdF_4$ [1]. Past experience in growing crystals hydrothermally indicates that the magnitudes of solubility and temperature coefficients of solubility are appropriate for growth with a KF mineralizer. The log of solubility versus the reciprocal of absolute temperature is approximately linear. From the slope of a log(s) versus l/T plot, the heat of solution (or the heat of reaction of Eq. (8.1)) is approximately 12 kcal/mol [3]. In order to obtain reasonably sized crystals, controlled transport in a temperature gradient of nutrient that is thermodynamically stable is essential. Mineralizer concentration is chosen to be within the $KGdF_4$ phase field, high enough to provide reasonable solubility. The optimum conditions are as follows: KF solvent: 16 mol%, 80% initial fill, dissolution temperature: 450°C, growth temperature: 425°C. The previously synthesized seeds of size $4 \times 0.5 \times 0.5\ mm^3$ were placed on a point gauze basket mounted at the top of the growth zone.

The other alkali rare earth fluorides of interest are $LiKYF_5$, CsY_2F_7, KYF_4, and K_2YF_5 doped with Nd^{3+} ions up to 10 at%. These crystals were obtained at $T = 450-700$°C and $P = 100-150$ MPa, over a period of 15–20 days, and the crystal size was 5 mm [1]. Some of these single-centered laser fluorides such as $LiKYF_5:Nd^{3+}$ have better physical characteristics in comparison with oxygen-containing substances. Most of these compounds crystallize in the lower symmetry. Figure 8.3 shows the fragments of the $LiKYF_5$ crystal structure [4].

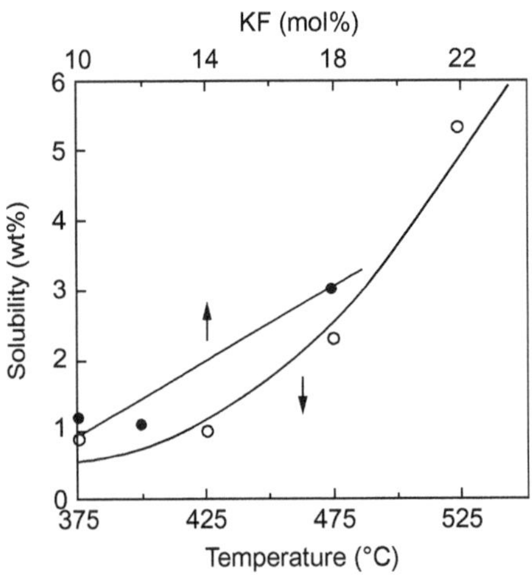

Figure 8.2 Solubility of $KGdF_4$ as a function of temperature and concentration of KF [3].

8.2.2 Spectroscopic Properties of Rare Earth Fluorides

There are over 250 Nd^{3+}-doped insulating laser crystals and anisotropic single-centered fluorides with ordered structures which are unique owing to their excellent spectroscopic properties [5,6]. The most important ones are tetragonal $LiLnF_4$ crystals with a space group $I4_{1/a}$, as well as monoclinic BaY_2F_8 (space group *C2/m*) crystals. Similarly multicentered disordered cubic $5NaF-9LnF_3$: Nd^{3+} crystals with space group *Fm3m*, may be considered as potential media for the excitation of a stimulated emission in the range 1.06−1.35 μm [7]. Being transparent in a wide spectral region, these fluorides reveal a negative d*n*/d*T* coefficient, possess a high stability with reference to short wavelength radiation of pumping Xe-flashlamps, and have satisfactory thermophysical characteristics. This has led to the fact that these materials are rather widely used in quantum electronics. However, low distribution coefficients of Nd^{3+} ions prevent obtaining single crystals of a majority of these compounds with characteristics that are optimal for solid-state lasers of the new generation, i.e., lasers using semiconductor laser-diode excitation. That is why the search for new anisotropic single-centered generating fluorides is an extremely urgent problem. These new crystals are able to generate stimulated emission at 300 K in free-running pulse operating mode on inter-stark transitions of two $^4F_{3/2} \rightarrow {}^4I_{11/2}$ and $^4F_{3/2} \rightarrow {}^4I_{13/2}$ laser channels of Nd^{3+} ions with Xe-flashlamp excitation as well as in CW mode ($^4F_{3/2} \rightarrow {}^4I_{11/2}$) under laser-diode pumping [1,8−10]. The radiation from these transitions promotes the formation of mainly one line, which in turn provides the conditions favorable for the excitation of the stimulated emission at frequencies of the main intense lines. The spectroscopic data obtained have been used in the synthesis of these rare earth fluorides by the hydrothermal technique [4]. Figure 8.4 shows the luminescence spectra ($^4F_{3/2} \rightarrow {}^4I_{13/2}$) at 77 K (a) and 300 K (b) and the scheme of crystal field splitting of $^4F_{3/2}$ and $^4I_{13/2}$ transitions (c) of Nd^{3+} ions in

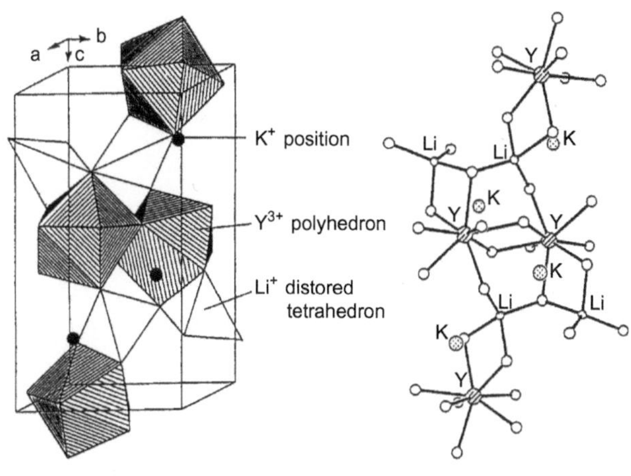

Bond	Distance (10^{-1} nm)
Y polyhedron	
Y–F	2.202(3)
	2.215(2)
	2.244(3)
	2.304(3) × 3
	2.313(3)
	2.323(3)
Li tetrahedron	
Li–F	1.842(10)
	1.870(9)
	1.948(9)
	1.986(10)
K polyhedron	
K–F	2.647 to 3.112 (9 contacts)

Figure 8.3 Fragments of the $LiKYF_5$ crystal structure [4].

K_2GdF_5. These crystals may be a prospective material for solid-state lasers in the range of 3–4 μm [2]. Figure 8.5 shows the fluorescent emission spectrum and fluorescent lifetime for the $Pr(^1G_4 \rightarrow {}^3H_5)$ transition in $KGd_{0.98}Yb_{0.01}Pr_{0.01}F_4$ when the $^1G_4$ level is pumped directly at 1.02 μm. In $KGdF_4$, this transition is centered near 1.3 μm and is broad enough for wavelength multiplexing, the lifetime is 150 μs. Similar measurements have been made on $KGd_{0.98}Yb_{0.01}Pr_{0.01}F_4$ and $KGd_{0.94}Yb_{0.05}Pr_{0.01}F_4$ using three different pumping schemes: (i) pump Yb $(^2F_{7/2} \rightarrow {}^2F_{5/2})$ at 0.94 μm and observed fluorescent emission at 0.97 μm from Yb $(^2F_{5/2} \rightarrow {}^2F_{7/2})$; (ii) pump $Yb(^2F_{7/2} \rightarrow {}^2F_{5/2})$ at 0.970 μm and observe emission from

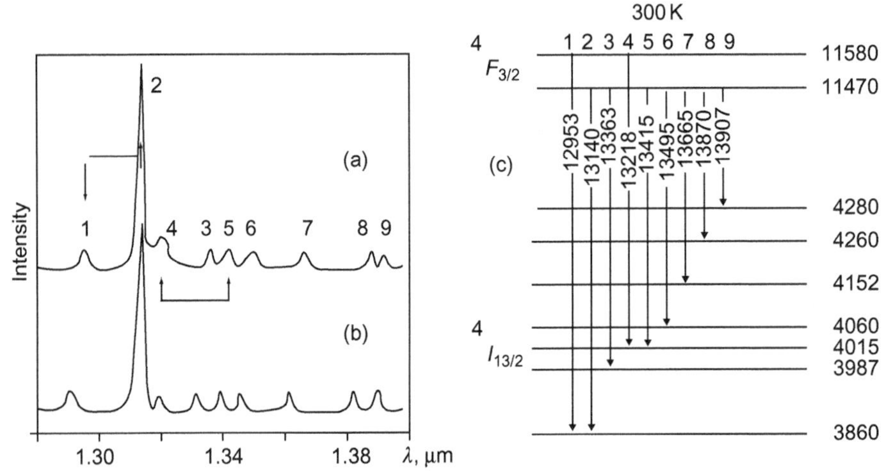

Figure 8.4 Luminescence spectra $(^4F_{3/2} \rightarrow {}^4I_{13/2})$ at (a) 77 K and (b) 300 K and (c) the scheme of crystal field splitting of $^4F_{3/2}$ and $^4I_{13/2}$ transitions of Nd^{3+} ions in K_2GdF_5 [4].

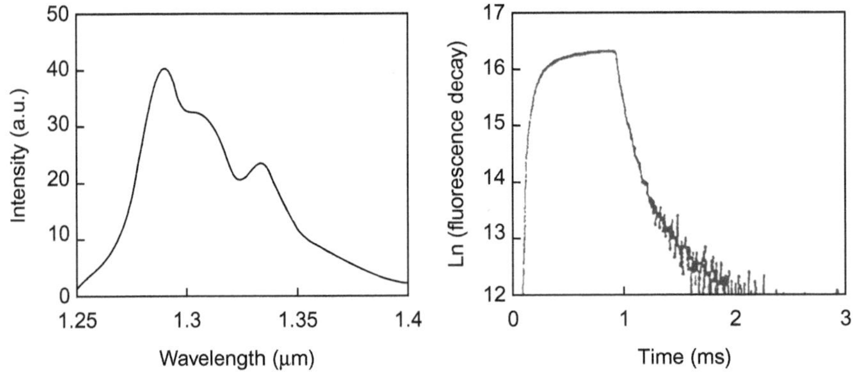

Figure 8.5 Fluorescent emission spectrum and fluorescent lifetime for the $Pr(^1G_4 \rightarrow {}^3H_5)$ transition in $KGd_{0.98}Yb_{0.01}Pr_{0.01}F_4$ [5].

$Pr(^1G_4 \rightarrow ^3H_5)$ at 1.3 μm; and (iii) pump $Pr(^3H_4 \rightarrow ^1G_4)$ at 1.02 μm and observed emission from $Pr(^1G_4 \rightarrow ^3H_5)$ at 1.3 μm. The results are given in Table 8.1.

From the data in Table 8.1, the following observations can be made. First, the lifetime of the Yb fluorescence is sufficiently long to permit energy transfer from the Yb $^2F_{5/2}$ to the Pr $^1G_4$ levels. At both doping levels, the lifetime of the Yb $^2F_{5/2}$ state is longer than the lifetime of the Pr $^1G_4$ level. Second, both the intensity and the lifetime of the $Pr(^1G_4 \rightarrow ^3H_5)$ transition increase when Yb is pumped and compared to the Pr pumped.

Thus, the rare earth fluorides have found unique technological applications. Their synthesis is highly challenging.

8.3 Hydrothermal Synthesis of Transition Metal Fluorides

The hydrothermal method permits us to synthesize single crystals of some transitional metals and other ternary fluorides. The majority of these compounds show superionic, magnetic, and other important physical properties. Besides, these compounds show very interesting structural characteristics which contribute to their superior physical properties. Thus some researchers focus mainly on their structural variations than their bulk single crystal growth. Aldous [11] has synthesized about 57 novel crystal structures belonging to this family using mild solvothermal route. Ti, Zr, Hf, V, Nb, Ta, and Mo have been used to make fluoride or oxyfluoride materials at temperatures ranging from 60°C to 200°C. Both inorganic and organic moieties ("templates") have been used to direct the formation of the (oxy)fluoride materials, usually as an anionic species. These structures have a tendency to form isolated polyhedral units under the mild solvothermal conditions. Several examples of isolated $[MF_6]^{n-}$ and $[MF_7]^{n-}$ monomers are reported and also one example of a transconnected $[TiOF_4]_2$ infinite chain structure.

Among alkali zirconium fluorides, compounds of special interest are those belonging to the A_3ZrF_7 family (A = Na, R, Rb, Cs, NH_4^+) with the characteristic ordered–disordered (OD) structures. Except for Na_3ZrF_7 (its structure is a

Table 8.1 Fluorescent Intensity and Lifetime of $KGdF_4$ Crystals [4]

Crystal Fluorescent Rare Earth	**Yb**[a]		**Pr**[b]		**Pr**[c]	
	Intensity (a.u.)	**Lifetime (μs)**	**Intensity (a.u.)**	**Lifetime (μs)**	**Intensity (a.u.)**	**Lifetime (μs)**
$KGd_{0.98}Yb_{0.01}Pr_{0.01}F_4$	240	1120	40	150	90	600
$KGd_{0.94}Yb_{0.05}Pr_{0.01}F_4$	250	400	90	115	280	500

[a]Test conditions: pump $Yb(^2F_{5/2})$ at 0.94 μm; emission from $(^2F_{5/2} \rightarrow ^2F_{7/2})$ Yb at 0.97 μm.
[b]Test conditions: pump $Pr(^1G_4)$ at 1.02 μm; emission from $(^1G_4 \rightarrow ^3H_5)$ Pr at 1.3 μm.
[c]Test conditions: pump $Yb(^2F_{5/2})$ at 0.97 μm; emission from $(^1G_4 \rightarrow ^3H_5)$ Pr at 1.3 μm.

derivative of a fluorite-type structure), the above fluorides are related to the K_3UF_7 type (usually characteristic of heavy metals). Heptafluorides are usually obtained using the methods of chemical interaction at room temperature by sintering in inert a fluorinating atmosphere and crystallization from KF flux.

Illyushin et al. [12] have obtained several millimeters large K_3ZrF_7 single crystals under hydrothermal conditions at temperatures of 450–500°C and pressures of 100–150 MPa. The initial materials were aqueous zirconium dioxide and potassium fluoride solutions with the concentration 10–60 wt%. Potassium–zirconium heptafluoride is a stable fluorine-containing crystalline phase within the range of temperatures and pressures used in hydrothermal experiments.

Conductivity (σ) of K_3ZrF_7 single and polycrystals was measured in the temperature range 300–700 K. The values of σ were calculated from the impedance (admittance) frequency plots of an electrochemical $Ag/K_3ZrF_7/Ag$ cell. The conductivity of single crystals is somewhat higher than that of polycrystals, yet no noticeable anisotropy in conductivity was observed [12]. At 300°C, conductivity is 3.9×10^{-2} and $1.17 \times 1{-}2\ ohm^{-1}\ cm^{-1}$ for single and polycrystals, respectively (Figure 8.6).

A synthetic Fe^{3+} variety of elpasoite, a rare natural mineral (K_2NaAlF_6), was synthesized in the course of the investigation of the phase formation in the $AF–Fe_2O_3–(NH_4)_2\ HPO_4–H_2O$ system with A = Na, K, Cs (T = 400°C, fill = 0.7, C_{KF} = 30–50 wt%) [13]. The structural analysis yielded the composition $K_2NaH_{1.5}Fe_{0.6}F_6$ for the cubic crystals of this system (space group $Fm3m$). Under hydrothermal conditions, complex fluorides of the system $A^{1+}\ F–A^{2+}\ F_2–A^{3+}F_3$ (A^{1+} = Li, Na, K; A^{2+} = Ca, Sr, Ba, Mg, Zn; A^{3+} = B, Al, Ga) were also obtained [14].

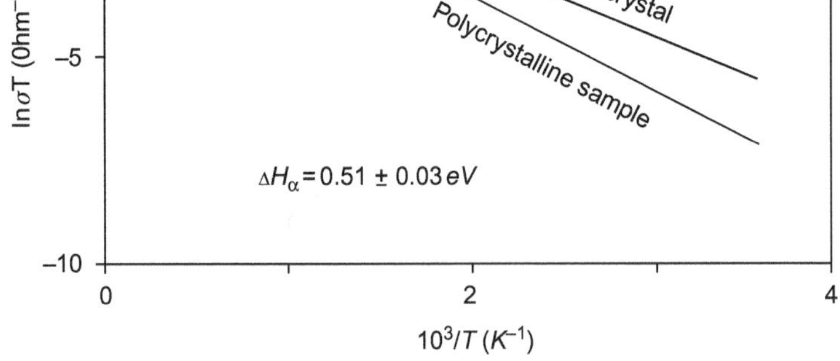

Figure 8.6 Conductivity (σ) for single and polycrystals [12].

According to their structures, these compounds are close to well-known minerals elpasoite, K_2NaAlF_6, cryolite, Na_3AlF_6, and veberite, Na_2MgAlF_7. In these compounds, A^{2+} and A^{3+} ions are displaced by cations of iron and rare earth subgroups, respectively. Such compounds were synthesized from aqueous solutions with high HF concentrations; the stoichiometric mixtures of oxides and hydroxides served as starting materials. Such crystalline products were used as nutrients for single crystal growth though the temperature gradient technique at $T = 120-280°C$, ($\Delta T = 20-40°C$) [14]. Other fluorides, like $Cs_2FeF_5H_2O$, Na_3FeF_6, and $Ba_3Ta_2O_2F_{12}$, were also synthesized under hydrothermal conditions [15–17]. The hydrothermal technique provides the possibility of growing hydrous transition metal fluorides in HF solutions under PT conditions [18]. This includes several fluoride systems like $AF-AlF_3$, $AF-FeF_3$, $AF-M^{2+}F_2$, AF_2-FeF_3, where A is an alkaline, ammonium, or Hg_2^{2+} ion, M^{2+}, a divalent transition metal (Mn, Fe, Co, Zn, Ni, Cu) [19,20]. Some of the ternary and quaternary systems like $H_2O-AF-MF_2$ or MF_3 (A = Li, Na, Cs, NH_4, M = Mn, Cu, V, Cr, Fe), $H_2OvBaF_2-(ZnF_2$ or $VF_3)$, $H_2O-AF-MF_2-MF_3$ (A = Li, Na, K, Rb, Cs, NH_4; M = Mn, Fe, Ni, Al, V, Cr, Fe), and H_2O-BaF_2 (MnF_2 or NiF_2) (CrF_3 or FeF_3) have been studied [21]. The major advantages of the hydrothermal technique for fluorides are the ability to obtain low-temperature varieties, rhombohedral α-FeF_3, phases with slow kinetics of formation, hexagonal $CaAlF_4$, and compounds with incongruent melting, like NH_4AlF_4.

The phases which appear in the system H_2O-FeF_3 are classified according to the growing temperature and the hydrothermal medium, H_2O or 49% HF. The iron coordination polyhedra are indicated. It is clear that high temperature and higher concentration of hydrofluoric acid favor the dehydration of the fluorides in the order Fe^{III}, Fe^{II}. The FeF_2 is grown only at extreme conditions (450°C, 49% HF).

The Mn^{II} compounds, $MnFeF_5(H_2O)_2$ and $MnFe_2F_8(H_2O)_2$, analogous to $Fe^{II}Fe^{III}F_5(H_2O)_2$ and $Fe^{II}Fe_2{}^{III}F_8(H2O)_2$, are synthesized as byproducts in the presence of sodium fluoride [22].

$NaMnFeF_6$, synthesized under hydrothermal conditions, cannot be obtained by solid-state reaction at any temperature: weberite, Na_2MnFeF_7, always forms [23]. $NaMnFeF_6$, trigonal ($a = 9.042$ Å, $c = 4.995$ Å; [5]) is isotopic of $NaMnCrF_6$ [3,24] and undergoes endothermic decomposition at $T = 546°C$ [5].

$$2NaMnFeF_6 \rightarrow Na_2MnFeF_7 + MnF_2 + FeF_3 \quad (8.2)$$

N-$RbMnFeF_6$ crystallizes with the new structural form of NH_4MnFeF_6 described below. Above 608°C, N-$RbMnFeF_6$ adopts the modified pyrochlore structure; the irreversible phase transition N-$RbMnFeF_6 \rightarrow$ P-$RbMnFeF_6$ is endothermic (N and P are the two structure forms in $RbMnFeF_6$) [5,25].

The growth of ammonium phases becomes particularly accessible by the hydrothermal method. Crystals of $NH_4M^{III}F_4$ (M = Fe, V, Cr) compounds grow easily. In spite of frequent twinning, their parameters and space groups are derived. For NH_4FeF_4 ($a = 7.559$ Å, $c = 12,754$ Å), the space group is *Pnma* [4,8,25].

Single crystals of $NH_4Fe^{II}Fe^{III}F_6$ are produced when the reduction of Fe^{3+} to Fe^{2+} occurs. Thus, the hydrothermal synthesis helps in a large variety of growth solutions, and the variation of the temperature leads to numerous new fluorinated phases of structural and magnetic interest. For stable compounds, the hydrothermal method represents an alternative way of crystal growth; its specificity is enhanced when low-temperature (HTB-FeF_3, N-$RbMnFeF_6$) or thermally unstable phases (hydrates, ammonium compounds) are desired as single crystals. Much of the work in recent years on the transition metal fluorides is mostly focused toward the nano-size crystals owing to the difficulties associated with the fluoride systems under hydrothermal conditions to grow bulk single crystals.

8.4 Hydrothermal Synthesis of Fluorocarbonates and Fluorophosphates

A new class of fluorides is being studied extensively in recent years, owing to their potential physical properties like nonlinear, magnetic, and electrical conductivity [26–28]. These materials have a high thermal and chemical stability. The hydrothermal synthesis at high-temperature and -pressure conditions yields these compounds. Among these, fluorides and fluorocarbonates are the most popular. These compounds contain rare earth 3d transition and alkaline earth metals in their composition. The following are the most popular compounds:

$$M_3Ln_2(CO_3)_4F \quad (M = Na, K;\ Ln = rare\ earth\ elements) \tag{8.3}$$

$$M^{1+}M^{2+}(CO_3)_4F \quad (M = Na, K;\ M^{2+} = Cu, Fe, Zn) \tag{8.4}$$

$$M^{2+}Ln(CO_3)_2F \quad (M^{2+} = Ba, Mg;\ Ln = rare\ earth) \tag{8.5}$$

$$M_3^{2+}Ln_2(CO_3)_5F_5 \quad (M = Ba, Mg;\ Ln = rare\ earth) \tag{8.6}$$

$$M^{2+}M^{2+}(CO_3)F_2 \quad (M^{2+} = Ba, Mg;\ M^{3+} = Mn, Cu) \tag{8.7}$$

$$M_2^{2+}PO_4F \quad (M^{2+} = Mn, Co) \tag{8.8}$$

$$M^{+}MPO_4F \quad (M^{+} = Na, K, Cs, NH_4;\ M = Fe) \tag{8.9}$$

The crystal growth of rare earth, alkaline earth, and 3d transition metal fluorocarbonates is difficult, and numerous earlier attempts involving alkaline fluorides or carbonates with cobalt, iron, or zinc, rare earth carbonates or fluorides failed. However, the association of alkali/alkaline earth carbonate with rare earth or 3d transition metal fluoride was successful [29–32]. Hydrothermal synthesis at high temperature and high pressure is well suited for the crystal growth of rare earth or yttrium fluorocarbonates. Several interesting compounds with unique structures

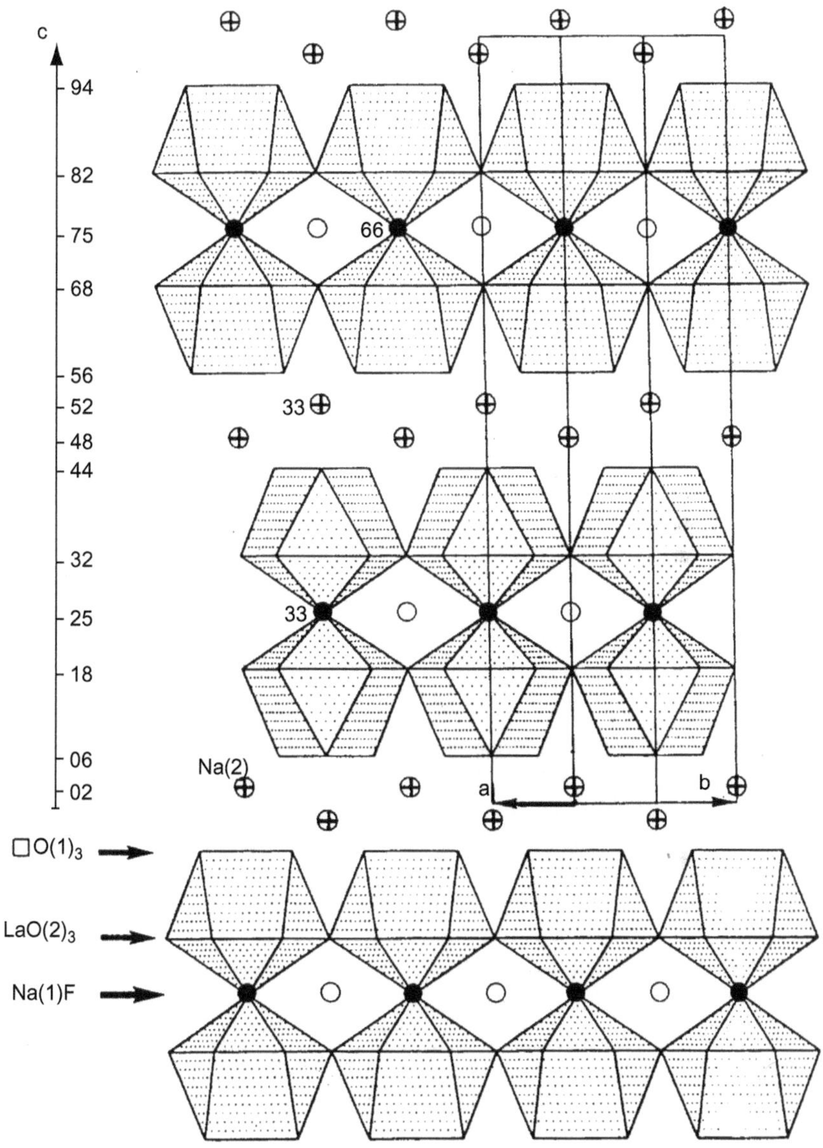

Figure 8.7 Structural projection of $Na_3La_2(CO_3)_4F$ [22].
Source: Courtesy of the Academic Press, Orlando, FL.

have been obtained by the hydrothermal method [31–35]. These compounds have been synthesized by hydrothermal growth at $T = 740°C$ and $P = 200$ MPa in 20 h.

The starting materials are usually the carbonates of alkali or alkaline earth, rare earth, or 3d transition metal fluorides, and alkali fluoride. Figure 8.7 shows the structural projection of $Na_3La_2(CO_3)_4F$, along Ref. 22. The structure shows double layers along "*c*" of the lanthanum coordination polyhedra, which adopts the 10-fold (30 + 6O + 1F) coordination, similar to that of $Ba_3La_2(CO_3)_5F_2$, $BaSm(CO_3)_2F$, and $Ba_2Ce(CO_3)_2F$, for La or rare earth ions. The structure exhibits a 2d character with infinite $(LaO_6)_2F$; double layers parallel to the (*a*,*b*) plane are separated by Na^+ ions.

The mineral-type structures obtained are $PrCO_3F$ (bastnaesite type), $Ba_2Ce(CO_3)_3F$ (zhonghuacerite), $BaSm(CO_3)_2F$ (huanghoite type), $Ba_2La_2(CO_3)_5F_2$ (cebaite type), and several new structures like $Na_3La_2(CO_3)_4F$ exhibiting interesting physical properties.

The 3d transitional metal fluorocarbonates could be synthesized at a slightly lower temperature than the rare earth fluorocarbonates. The typical experimental conditions are $T = 700°C$ and $P = 200$ MPa. Several new compounds like $BaMn(CO_3)F_2$, $BaCu(CO_3)F_2$, and $KCu(CO_3)F$ have been obtained. Under hydrothermal conditions, the association of alkaline carbonates with 3d transition metal fluorides leads, in most cases, to anionic exchange and to the formation of 3d transition metal carbonates and alkaline fluorides. However, when the starting materials are K_2CO_3 and CuF_2 in 1:1 ratio, green crystals of $KCu(CO_3)F$ which can reach 4 mm^3 are obtained. Thus, the fluorocarbonates are being studied extensively, especially by the French group (Leblanc and coworkers [18])[26–32].

The hydrothermal synthesis of fluorophosphates is slowly attracting the attention of materials scientists because the magnetic effects have been investigated in fluorides which present triangular cationic planes or K atom, lattices and where the coordination MF6 polyhedra are mainly connected by vertices, and superexchange interactions occur through 180° type M–F–M bridges, which frequently lead to antiferromagnetism. Several phosphate minerals adopt the general formula M_2PO_4X, where M^{2+} = 3d cation, Mg^{2+}, Ca^{2+} and $X^- = F^-, OH^-$; an arsenate or vanadate series also exist. The occurrence of the triploidite substructure is said to be related to the ordered F–OH substitution on the anionic sublattice. The coordination of the M^{2+} cations is octahedral; in triploidite, the evolution of the M–F distances implies that half of the cations M^{2+} adopt a fivefold coordination, consequently, the magnetic superexchange pathway in triploidite results from that of triplite by breaking one of the three $M^{2+}–F^-–M^{2+}$ magnetic bridges and strengthening the others [33].

The majority of these fluorophosphates, particularly the oxyfluorinated prepared under hydrothermal conditions, have been quite popular in recent years as microporous compounds belonging to the systems $AF–M_2O_3–P_2O_5$–amine ($A = H^+$, M = Al, Ga, 3d transition metal). The experiments are usually carried out at 217°C under autogenous pressure, for 36–40 h. These compounds obtained under hydrothermal conditions show interesting crystal structures which are again difficult to obtain by other conventional methods.

8.5 Oxyfluorinated Compounds

Oxyfluorinated compounds are being studied extensively in recent years owing to their interesting physical properties. Several phases in the systems

BaF_2–TiO_2–FH_{aq} and BaF_2–M_2O_5–HF_{aq} (M = Ta or Nb) under hydrothermal conditions have been identified [36,37]. Among them, $Ba_3Nb_2O_2F_{12} \cdot 2H_2O$, $Ba_4Nb_2O_3F_{12}$, and $Ba_3Ta_2O_2F_{12}$ have interesting structures. $Ba_3Ta_2O_2F_{12}$ is a new oxyfluoride with sevenfold coordinated Ta^{5+}. In the monocapped trigonal prism $TaOF_6$, the distances Ta–X (X = O_2 or F^-) are close to those found in an octahedral environment, which led to X–X distances smaller than the calculated values. The isolated $TaOF_6$ groups are connected together by two types of Ba^{2+} polyhedra. The $Ba_3Nb_2O_2F_{12} \cdot 2H_2O$ is a new acentric hydrated niobium oxyfluoride. Its structure is built from isolated $NbOF_6$ pentagonal bipyramids. The $Ba_4Nb_2O_3F_{12}$ is a 3d oxyfluoride, which exhibits two kinds of coordination polyhedra for Nb atoms: The octahedral form with (NbO_2F_4), (NbO_3F_3) formulations, and the pentagonal bipyramid ($NbOF_6$). Four (NbX_6) octahedra are cris-oxygen connected to form separated $[Nb_4O_7F_{14}]^{8-}$ condensed blocks, while ($NbOF_6$) pentagonal bipyramids are isolated but linked together by Ba^{2+} cations.

In all these cases, single crystals have been prepared by the hydrothermal technique and not by other conventional methods.

The other materials of technological importance among fluorides are α-PbF_2 single crystals. Lead fluoride stands out as an important optical material because of its chemical and mechanical stability, wide transparency range (from UV to medium IR), and rich substitutional chemistry (permitting the variation of its composition and, consequently, its main operational parameters), and also because its single crystals can be readily obtained from both melt and solution. Lead fluoride is poorly soluble and can be easily prepared by the reaction of HF acid with lead hydroxide or carbonate or by precipitation from aqueous solutions of these reagents with NH_4F. The α–β transformation, orthorhombic-to-cubic in PbF_2, takes place at 460°C. Beta-PbF_2 is chemically more stable and can be readily obtained as large single crystals from the melt [38]. However, the α-PbF_2 cannot be prepared by melt or solid-state reacts or flux methods. Nikolskaya and Demianets [39] have carried out the growth of α-PbF_2 at T = 280°C using Teflon liners and at T = 450°C using copper ampoules. Extra-pure-grade PbF_2, 6–10% NH_4F and 10–20% KF solutions have been used as the starting materials. The temperature gradient has been maintained at 35–42°C. At a lower growth temperature of 280°C, the quantity of α-PbF_2 is higher than the crystals obtained at 450°C. The crystals are highly transparent, well faceted, elongated, and up to 8 mm in length. In some cases, the crystal length was up to 10 mm, but in this case the crystals were fragile. Figure 8.8 shows the micrographs of α-PbF_2 crystals obtained under hydrothermal conditions by Nikolskaya and Demianets [39].

8.6 Physical Properties of Transition Metal Fluorides and Fluorocarbonates/Fluorophosphates/Oxyfluorides

These compounds exhibit essentially magnetic properties, ionic conducting, and nonlinear optical properties. The oxyfluorides exhibit microporous structures, which

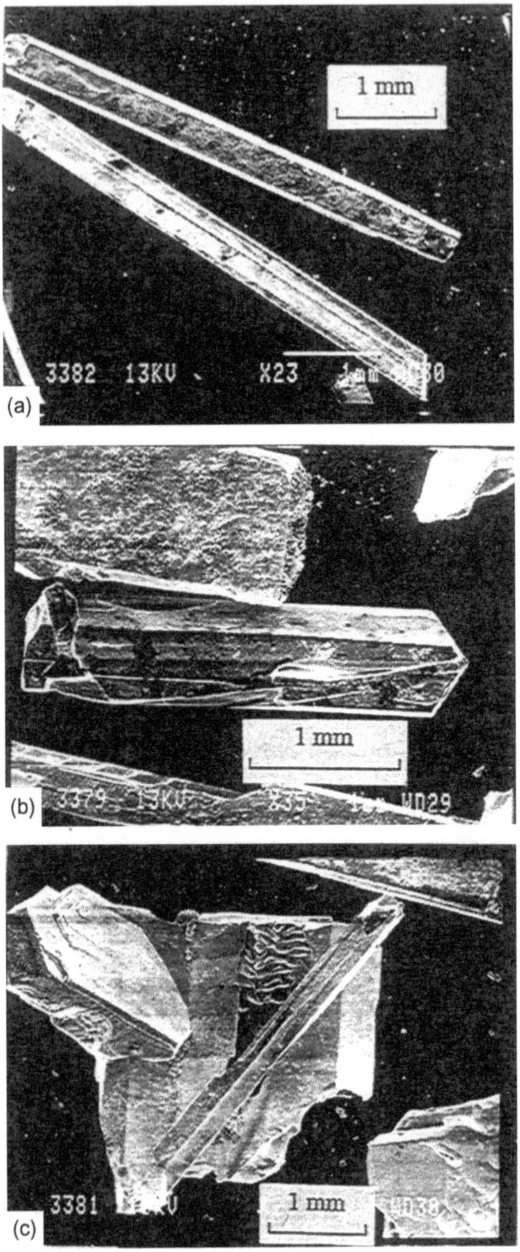

Figure 8.8 Micrographs of α-PbF_2 crystals obtained under hydrothermal conditions [39].

are highly useful as selective gas absorbers, catalysts, and so on. Hence, the hydrothermal and solvothermal syntheses of nanostructures of these compounds carry a greater significance in recent years than the growth of bulk single crystals. Here we discuss only a few representative compounds and their physical properties.

Among the fluorocarbonates, $KCu(CO_3)F$ and $BaCu(CO_3)F_2$ exhibit excellent magnetic properties.

The variation of the reciprocal susceptibility of $KCu(CO_3)F$, in the temperature range 4.2–300 K, is given in Figure 8.9. The magnetic susceptibility is weak and nearly independent of the temperature between 300 and 100 K, it then increases continuously from 100 to 4.2 K.

Strong antiferromagnetic interactions must be present between nearest Cu^{2+} neighbors in order to explain the magnetic behavior at high temperature; however, this hypothesis does not appear really acceptable for two reasons:

1. Superexchange interactions Cu–F–Cu possible in the infinite files along *a* are not generally very strong ($J/k \approx 100$ K).
2. Super-superexchange interactions Cu–O–O–Cu (through the carbonate ions) are probably even weaker.

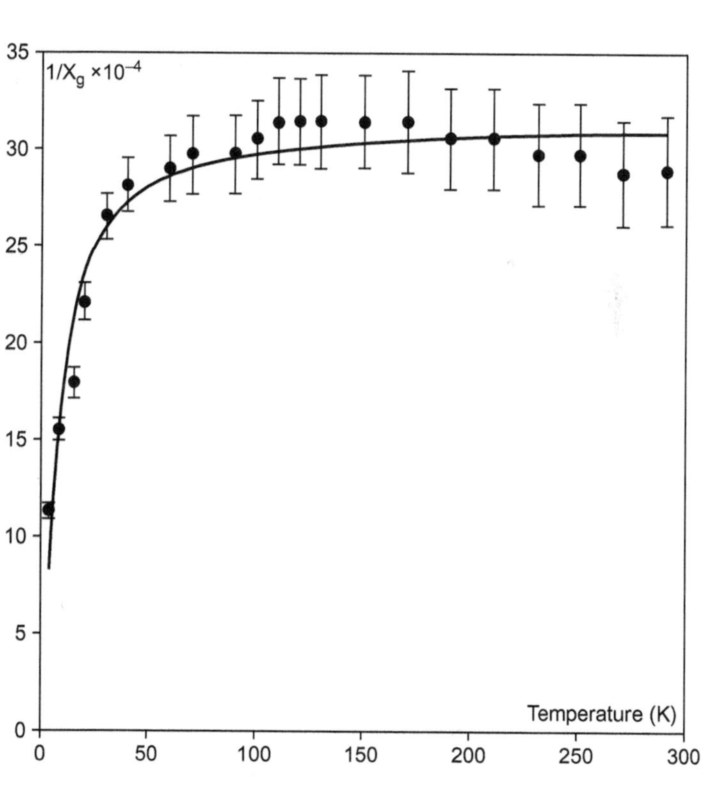

Figure 8.9 Reciprocal susceptibility of $KCu(CO_3)F$ [32].
Source: Courtesy of the Gauthier-Villars/ESME, Paris.

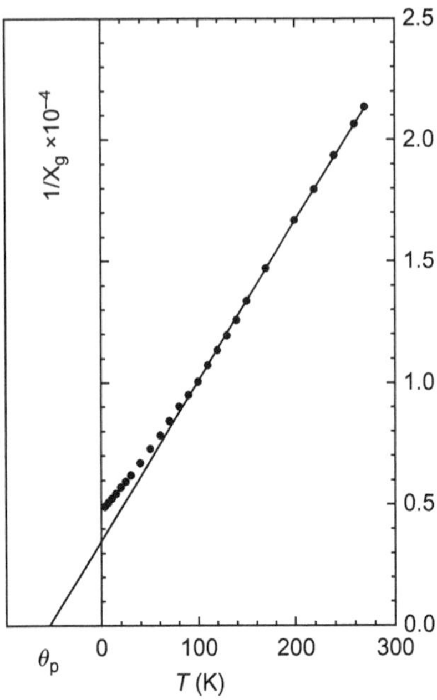

Figure 8.10 Variation of the inverse magnetic susceptibility with temperature [32].
Source: Courtesy of the Gauthier-Villars/ESME, Paris.

Ferromagnetic interactions must also take place at low temperature ($T < 60$ K). The variation of the inverse magnetic susceptibility with the temperature is given in Figure 8.10 for $BaMn(CO_3)F_2$. Above $T \approx 90$ K, the linear Curie–Weiss law is well obeyed, leading to a molar Curie constant $C_M = 4.39$ (as compared to the theoretical value $C_M = 4.375$ for Mn^{2+}) and a Curie paramagnetic temperature $\theta_p = -52$ K. This value of θ_p, together with the deviation of $1/x$ from the Curie–Weiss law at $T < 90$ K indicates that antiferromagnetic interactions dominate; however, no clear evidence of magnetic order can be found at a low temperature [32].

Magnetic susceptibility of Mn_2PO_4F and CO_2PO_4F has been studied by Leblanc et al. [33]. The variation of the inverse magnetic susceptibility Xg^{-1} with temperature (Figure 8.11) and the absence of field dependence of the susceptibility are indicative of antiferromagnetic behavior with $T_N = 15$ and $T_N = 14$ K for Mn_2PO_4F and CO_2PO_4F, respectively. Above approximately 100 K, the Curie–Weiss law is well obeyed and yields $\theta_p = -50$ K, $C_M = 8.2$ and $\theta_p = -53$ K, $C_M = 6.4$. These molar Curie constants, which give $\mu_{eff} = 5.70$ and 5.1 μ_B, are in good agreement with the calculated value $C_M(th) = 8.75$ (spin only value) for Mn^{2+} and with the value usually found for CO^{2+}.

8.7 Hydrothermal Synthesis of Tungstates

Tungstates form an important class of materials with a wide range of divalent and trivalent metals in their composition. Among the trivalent metals, the rare earth and

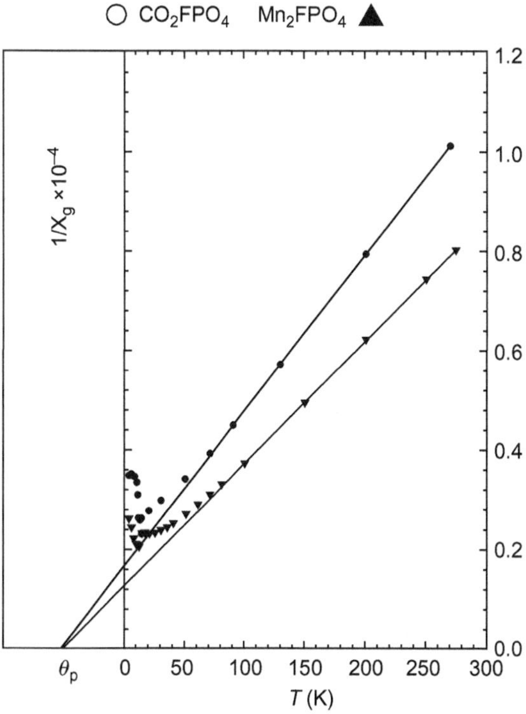

Figure 8.11 Variation of the inverse magnetic susceptibility Xg^{-1} with temperature [33].

transitional metals are the common ones. Tungstates are either simple or mixed types but exhibit high thermal and chemical stability in general. The important physical properties are excellent luminescent and electrical conductivity.

Tungstates can be obtained by different methods, like solid-state reactions, flux, and hydrothermal; however, it is the flux method that is very popular in the growth of a large variety of tungstates. With the recent advances in the hydrothermal chemistry of various inorganic systems and the influence of organic precursors or solvents in the growth of various inorganic compounds, the hydrothermal synthesis of tungstates is becoming popular. Several simple and complex tungstates, like $Me^{2+}WO_4$ (Me = Ca, Cd, Mn, Fe, Ca, Sr, Ba, Cd, Mn, Zn), M_2^{2+} $M(WO_4)_2$, MLn $(WO_4)_2$ (M = Li, Na, K, Rb; Ln = rare earths), have been reported. In general, the synthesis of Sr and Ba tungstates are carried out using either an (NaOH) solution or chloride solution, whereas it is only possible to obtain growth of Ca, Cd, Mn and Fe, in the chloride solutions with 100% output [40–42]. Wolframites of divalent metals belong to two structural groups, one crystallizing in the scheelite type covering $CaWO_4$, $SrWO_4$, and $BaWO_4$; the second one crystallizing in the wolframite structure type covering $CdWO_4$, $MnWO_4$, and $FeWO_4$. The first group usually forms bipyramidal crystals of 3–5 mm size. The second group of crystals are usually prismatic in form and their synthesis is carried out from a mixture of $M^{2+}O$ and WO_3 (M^{2+} = Cd, Mn, Fe) in 1:1 ratio at temperature 375–550°C and pressure up to 2000 atm. As mineralizers, aqueous solutions of chlorides and more complex solutions like the mixture of chlorides and fluorides, sodium wolframite and sodium chloride and sodium nitrate are used [41,42]. Among the mixed wolframites, $MM^{2+}(WO_4)_2$ (M = Li, Na, K; M^{2+} = Cd, Mn, Fe) are the common ones prepared by the hydrothermal method. The wolframites of trivalent metals form two large groups: (i) simple trivalent metal wolframites and (ii) mixed trivalent metal wolframites. Besides, there are several hydrous wolframites, $Me^{3+}WO_4(OH)$ [43], oxychloride wolframite $Me^{3+}WO_4Cl$, and oxywolframites $n Me_2O_3-m WO_3$. The ratio of *n*:*m* varies in this case between 1:1 and 1:3 [44].

Most of these simple or mixed tungstates containing divalent and trivalent metals exhibit magnetic properties. However, it is the mixed rare earth tungstates which occupy a unique place in the tungstate family owing to their excellent optical, electrical, and magnetic properties. Among these high-temperature polymorphs of mixed rare earth tungstates, $Mr(WO_4)_2$ (M = Li, Na, K, Rb, Cs; R = rare earth elements) crystals have a common crystal structure which is similar to that of tetragonal scheelite, ($CaWO_4$) under atmospheric pressure; however, the low-temperature polymorphs of these mixed rare earth tungstates belong to the monoclinic system [45]. Klevstov and coworkers [46–50] have worked out in detail the synthesis of simple and mixed rare earth tungstates and one of the important systems is $Pr_2O_3-WO_3-LiCl-H_2O$, with a concentration of LiCl-20 wt%. Figure 8.12 shows the TN-diagram of crystallization ($N = Pr_2O_3/WO_3$). With a decrease in $T < 450°C$ and increase >550°C, the field of crystallization of monocationic tungstates, $PrWO_4 \cdot OH$ and $Pr_2W_2O_9$, occupy relatively lower values of $N \cdot RWO_4(OH)$: hydroxyl wolframates containing almost the entire series of rare earth elements (Pr–Lu) have been obtained by Klevstov and coworkers [46]. The lighter lanthanides crystallize in the aqueous solutions LiCl, as LiR $(WO_4)_2$, and a similar picture is obtained for KCl, RbCl giving rise to $KR(WO_4)_2$ or $RbR(WO_4)_2$.

These mixed alkali rare earth tungstates are most important among tungstates in general. Although these tungstates were reported during the late 1950s and 1960s, not much has been studied with reference to their crystal growth and phase relations in the respective systems, and there exists some lacunae in the characterization of alkali rare earth tungstates. The probable reason for this may be the non-availability of good quality single crystals due to the high temperature of synthesis involved. There are several reports on the synthesis and characterization of alkali rare earth tungstates, but most of them deal with the flux growth and solid-state reactions [45,50–52]. The crystals are fine grained and poor in quality without a proper morphological development. Similarly, there are quite a few reports on the hydrothermal growth of alkali rare earth tungstates, but the PT conditions of synthesis were quite high ($T = 700-900°C$, $P = 2-3$ kbar) [40,41]. Byrappa and Jain [53] have reported the hydrothermal growth of $NaLa(WO_4)_2$ crystals under fairly low PT conditions ($T = 240°C$, $P = 80-100$ bar), using a suitable mineralizer solution. Carobbi and Tancredi [54] have studied the system $Ce_2(WO_4)_3-Na_2WO_4-H_2O$ at 25°C and this work has formed some basis for the work of Byrappa and Jain [53] to grow $NaLn(WO4)_2$ (where Ln = La, Ca, Nd) crystals at moderately lower PT conditions.

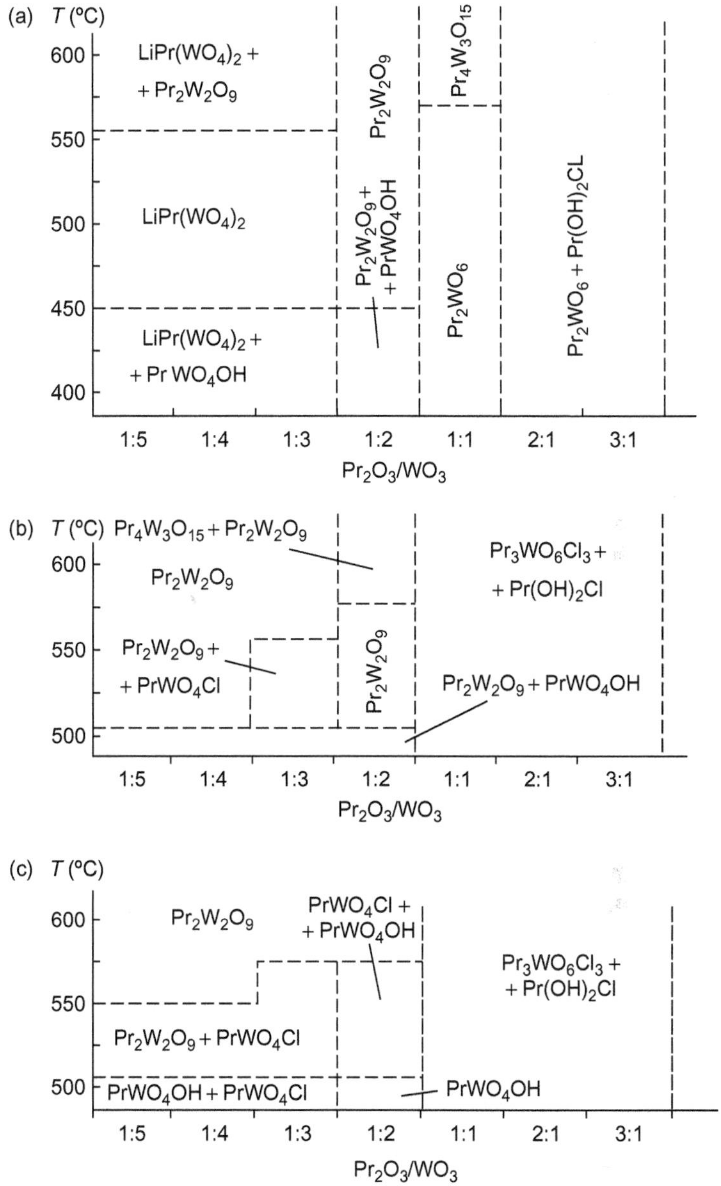

Figure 8.12 TN-diagram of crystallization ($N = Pr_2O_3/WO_3$) [46].

On the whole, the growth of alkali rare earth tungstates by the hydrothermal method is quite complicated due to the high temperature of synthesis involved and the lack of solubility data. Moreover, the crystals obtained by earlier workers were fine grained or small rounded or subrounded.

A systematic study on the search of a suitable mineralizer has been carried out on the empirical basis. Further, some selected crystal growth experiments have been carried out at lower PT conditions using suitable mineralizers. The experiments were carried out using Morey-type autoclaves and Teflon liners of 20, 30, 40, and 50 ml capacity. The starting materials such as WO_3 and $La(NO_3)_3$ were taken in desired molar proportions: $Na_2O{:}La_2O_3{:}WO_3 = 12-18{:}1{:}18{-}23$ and placed inside the Teflon liner. The alkaline component of the starting materials was taken in the form of a NaOH solution with a definite molarity and this solution acts as a mineralizer. The entire mixture was stirred well until a homogeneous solution was obtained. In addition, several other mineralizers, such as HCl, HCOOH, H_2SO_4, HNO_3, and mixed mineralizers like HCl + HCOOH, HCl + H_2SO_4, and HCOOH + HNO_3, with varying concentrations have been tried. However, the majority of these experiments yielded mainly crystalline powders and also mixed phases. Therefore, NaOH has been proven to be the most suitable mineralizer for the growth of alkali rare earth tungstates, whereas the earlier workers had concluded that the chloride solutions were the best for the growth of these tungstates.

The crystallization was carried out in all experiments through spontaneous nucleation by Byrappa and Jain [53]. In order to control the nucleation centers, the temperature of the crystallization reactor was increased very slowly at a rate of 10°C/h up to 100°C and beyond this at 5°C/h. The nutrient materials were held at 240°C for a period of 3 days without any fluctuations in the temperature. From the fourth day onward, temperature fluctuations of ± 10°C were introduced periodically in order to reduce the nucleation centers. By this means, the small crystallites that had developed previously dissolved and only the larger crystallites remained. This resulted in the growth of larger crystals ($7 \times 5 \times 5$ mm^3). The pH of the medium was varied from 7 to 11, and good quality single crystals were obtained in the pH range 8.4–8.8. The crystals obtained by this method were of good quality, showing high transparency, vitreous luster, and well-developed crystal morphology, with the size varying from 1.0 to 7.0 mm. Table 8.2 gives the experimental conditions of the growth of $NaLa(WO_4)_2$ crystals. The crystals obtained by the earlier workers at higher PT conditions were smaller and more equidimensional, experimental temperature, $T = 240$°C, duration = 10 days, pH of the nutrient measured before the experimental run.

The probable reason for the small crystals may be the higher ionic mobility, low viscosity, and higher concentration of solute ions. The crystals obtained by this means were well developed, bigger in size, and of good quality. The experimental conditions are ideal because more controlled diffusion of ions exists at a relatively low temperature due to high viscosity, high density, and high dielectric constant, and less thermal strain, which leads to well-developed crystals. The crystal morphology varies significantly with a change in the growth parameters.

The formation of $NaLa(WO_4)_2$ takes place under hydrothermal conditions according to the following reactions:

$$2NaOH(aq) + WO_3(s) \rightarrow Na_2WO_4(aq) + H_2O(aq) \tag{8.10}$$

$$La(NO_3)_3 + 3NaOH \rightarrow La(OH)_3(s) + 3NaNO_3(aq) \tag{8.11}$$

$$2La(OH)_3 + 2WO_3(s) \rightarrow La_2(WO_4)_3 + 3H_2O \tag{8.12}$$

$$La_2(WO_4)_3 + Na_2WO_4(aq) \rightarrow 2NaLa(WO_4)_2 \tag{8.13}$$

$$4NaOH(aq) + 2WO_3(s) + La(NO_3)_3(s) \rightarrow NaLa(WO_4)_2 + 3NaNO_3 + 2H_2O \tag{8.14}$$

Table 8.2 Experimental Conditions for the Growth of $NaLa(WO_4)_2$ Crystals [53]

No.	Nutrient	pH	*P* (bar)	Size (mm)	Remarks
1.	WO_3:6.0 g La $(NO_3)_3$:0.75 g 4.5 M NaO^{II}:7 ml	7.0	120		White crystalline powder
2.	WO_3:6.0 g La $(NO_3)_3$:0.75 g 4.7 M NaOH:7 ml	7.5	120	2–3	Small, broad, tabular, transparent, rhombohedral, colorless crystals
3.	WO_3:6.0 g La $(NO_3)_3$:0.75 g 5.0 M NaOH:7 ml	8.0	120	3–4	Rhombohedral, colorless crystals
4.	WO_3:6.0 g La $(NO_3)_3$:0.75 g 5.4 M NaOH:7 ml	8.4	120	4–5	Well-developed, tabular, prismatic, highly transparent, colorless crystals
5.	WO_3:6.0 g La $(NO_3)_3$:0.75 g 5.4 M NaOH:7 ml	8.4	60		White crystalline powder
6.	WO_3:6.0 g La $(NO_3)_3$:0.75 g 5.4 M NaOH:7 ml	8.4	80	2–3	Thin needle-like crystals
7.	WO_3:6.0 g La $(NO_3)_3$:0.75 g 5.4 M NaOH:7 ml	8.4	100	3–4	Rhombohedral, transparent, colorless crystals
8.	WO_3:6.0 g La $(NO_3)_3$:0.75 g 5.9 M NaOH:7 ml	8.8	120	5–6	Tabular, prismatic, transparent
9.	WO_3:6.0 g La $(NO_3)_3$:0.75 g 6.2 M NaOH:7 ml	9.3	120	6–7	Rod, prismatic subtransparent, colorless crystals
10.	WO_3:6.0 g La $(NO_3)_3$:0.75 g 6.2 M NaOH:7 ml	9.8	120	4–5	Thin, long, translucent, needle-like crystals
11.	WO_3:6.0g La $(NO_3)_3$:0.75 g 6.9 M NaOH:7 ml	10.2	120	2–3	Small, thin, translucent, needle-like crystals

The trioxides of tungsten (VI) dissolve in aqueous alkali metal hydroxide to form tungstates and water. These normal tungstates contain the discrete tetrahedral ions WO_4^{2-}. It is not certain that the WO_4^{2-} ions are also tetrahedral in an aqueous solution. Tungstate ions polymerize in an acid solution to give meta- and paratungstate ions. The degree of polymerization in the solution increases as the pH is lowered, and a number of tungstates $M_2O \cdot nWO_3M \cdot H_2O$, differing in the value of n, crystallize from the solution at different pHs. It has been observed that the diffusion coefficient of tungstate ions in aqueous solutions decreases considerably with the increasing acidity of the solution, which is probably due to an increase in the polymerization of ions. In all the experiments, pH was measured at the beginning and at the end of the experimental runs and it was found that the system was moderately alkaline, which avoids the possibility of polymerization. Lanthanide ions, in spite of high charges (+3), have low charge densities because of their large size. Therefore, they cannot cause much polarization and, consequently, they do not have much tendency to form complexes. The tendency to form complexes and their stability increase with increasing atomic number. Since ionic size decreases from La^{3+} to Lu^{3+}, the basicity of hydroxides decreases in the same order. Thus, $La(OH)_3$ is the strongest and $Lu(OH)_3$ is the weakest base. $La(OH)_3$ precipitates during the reaction as follows:

$$La(OH)_3 + n - 3 \rightleftharpoons La(OH)_n^{(n-3)} \qquad (8.15)$$

and later dissolves in an NaOH solution through any one of the earlier quoted reactions. Like La complexes, we can also expect NaLa complexes such as $Na_xLa(OH)_{3-x}$ in the solution. Being a good base, $La(OH)_3$ reacts with WO_3 to form lanthanum tungstates, which in turn dissolves in normal sodium tungstate to form $NaLa(WO_4)_2$ crystals under controlled conditions. In all the previous works on the synthesis of alkali rare earth tungstates, mineralizers such as NH_4Cl and alkali halides were used at very high PT conditions, but in the work of Byrappa and Jain [53], they have essentially used water and NaOH as mineralizer, with an emphasis on controlling the release of hydration shell water in the system through moderately lower PT conditions of synthesis. This permits a considerable reduction in the free energy of the system, leading to the complexation process. Therefore, the use of NaOH in the system and the formation of $La(OH)_3$ in the course of the reaction are quite useful in the complexation process, as it is well known that hydration shell water experiences less competition from bulk solvent water at moderate temperature. They have also carried out a series of experiments using heavier lanthanides under the same experimental conditions; however, it was found that the solvents used are suitable for the first group of lanthanides (La, Ce, Nd) but not for the other lanthanides. The reason for this may be the decrease in the basicity with an increase in the atomic number. Hence, the synthesis of $NaLn(WO_4)_2$ (where Ln = Pr–Lu) insists on the use of other solvents.

The most common morphological habits observed in $NaLa(WO_4)_2$ crystals are prismatic, rod shaped, rhombohedral, needle like, and so on. The prism faces dominated over all other faces in the crystals followed by pinacoidal and dome faces. $NaLa(WO_4)_2$ crystals belong to the tetragonal system with scheelite structure type; hence, their morphology is supposed to be simple, at least in the case of single crystals.

However, the crystals exhibit simple twinning and pseudotriclinic symmetry. Figure 8.13 represents the schematic diagrams of $NaLa(WO_4)_2$ crystals. Figures 8.14 and 8.15 show the variation in the growth rate and morphology of $NaLa(WO_4)_2$ crystals with the pH of the medium and internal pressure, respectively. Good quality crystals were obtained in the pH range 8.4–8.8. Figure 8.16 shows the representative photographs of the alkali rare earth tungstates obtained by Byrappa and Jain [53]. The size of the crystals increases with an increase in the pH of the medium, but the quality of the crystals becomes poor and beyond pH 11.5, $NaLa(WO_4)_2$ does not crystallize. The growth rate of $NaLa(WO_4)_2$ crystals was found to increase with an increase in pressure, and good quality crystals were obtained at about 120 bar. The twinning is

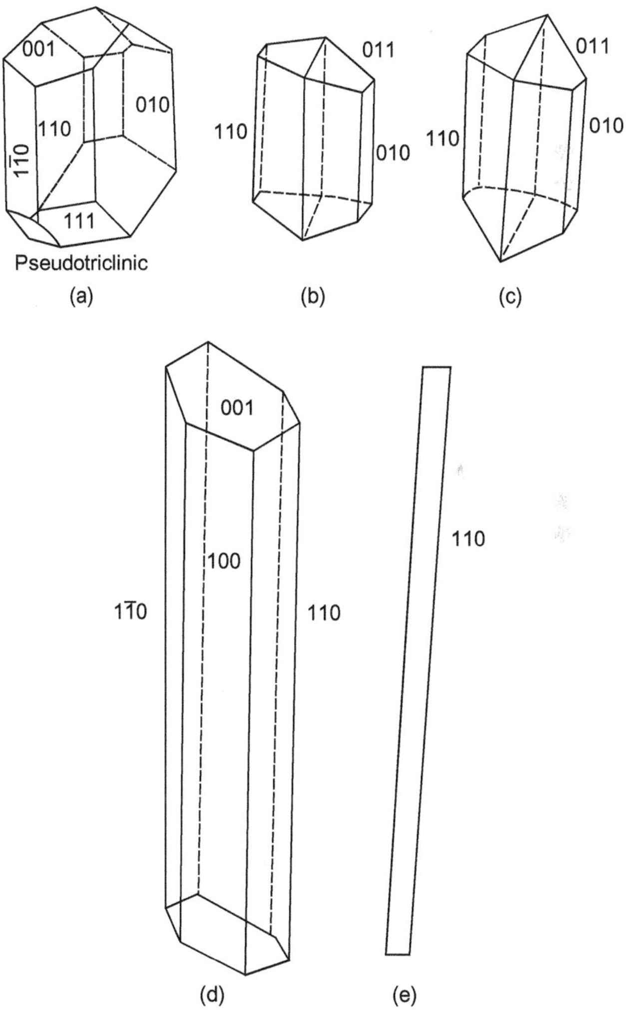

Figure 8.13 Schematic diagrams of $NaLa(WO_4)_2$ crystals [53].

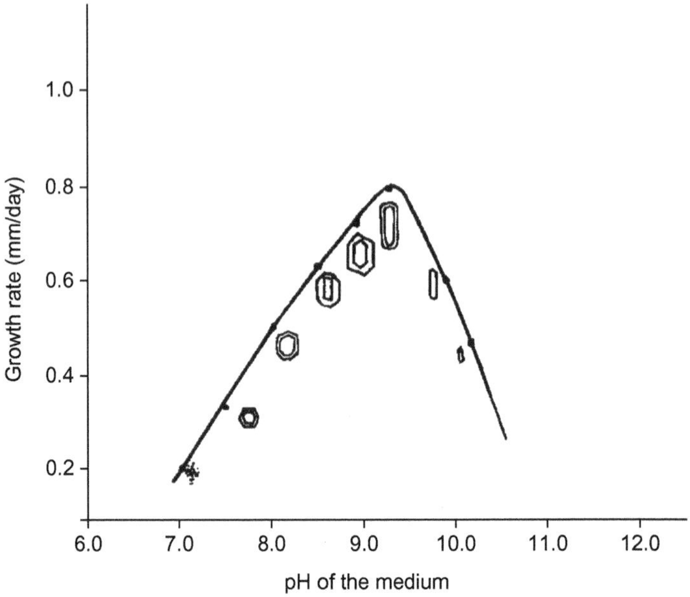

Figure 8.14 Variation in the growth rate and morphology of $NaLa(WO_4)_2$ crystals with the pH [53].

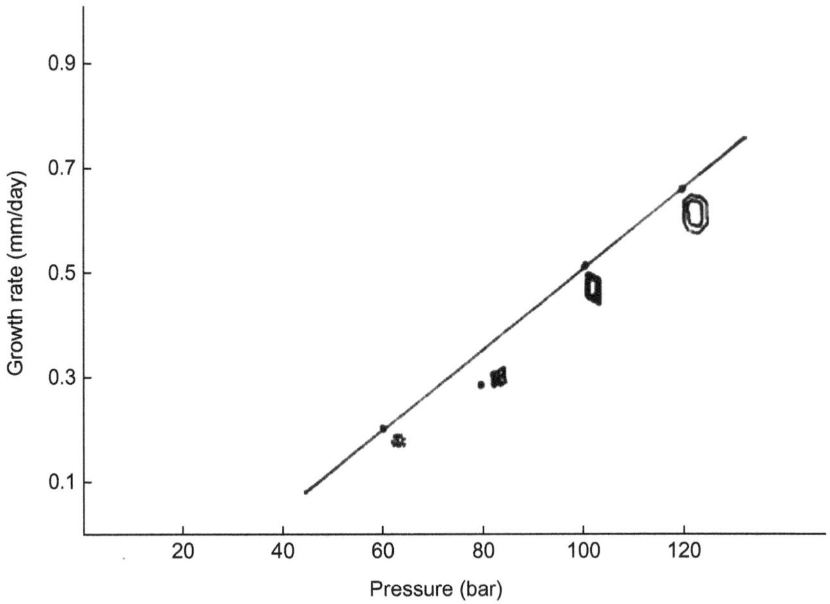

Figure 8.15 Variation in the growth rate and morphology of $NaLa(WO_4)_2$ crystals with internal pressure [53].

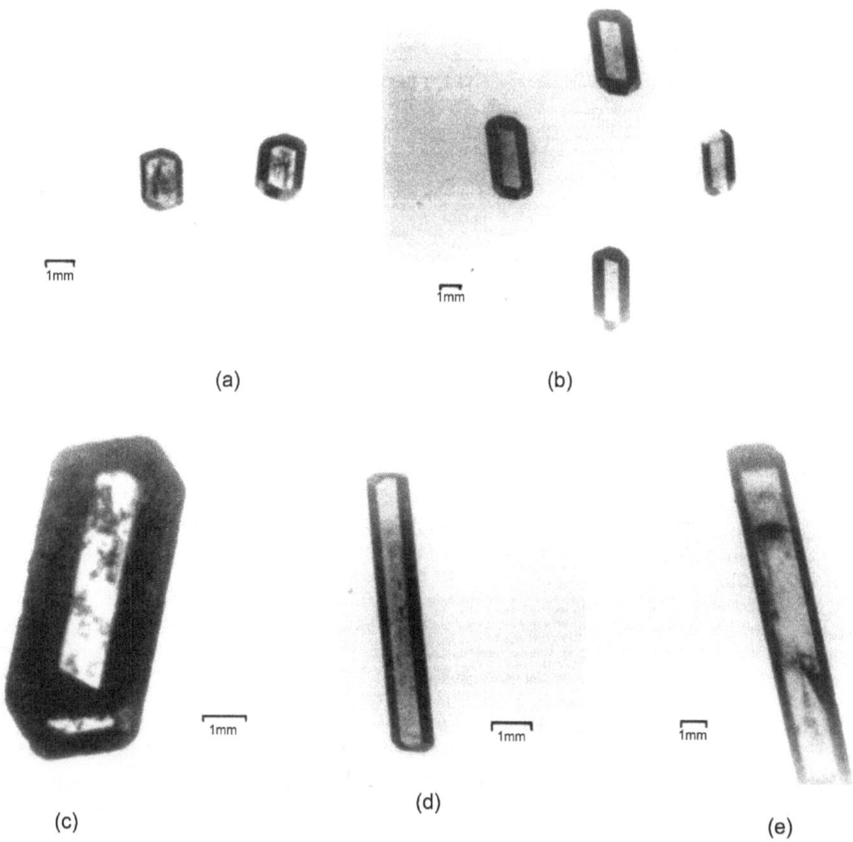

Figure 8.16 Photographs of the alkali rare earth tungstates [53].

more in experiments with pH values varying from 8.0 to 9.0. The pH of the mineralizer and the % fill have a considerable influence on the morphology and growth rate of $NaLa(WO_4)_2$ crystals. With an increase in the pH of the system, the morphology varies from well-developed bulk crystals of pseudolower symmetry to long needles of even 7 mm long in an experimental run, often of days. Similarly, with an increase in the pressure of the system, the morphology varies from needles (at lower pressure conditions) to bulk crystals of pseudolower symmetry (at moderate pressure conditions); thus, the hydrothermal growth of tungstates, particularly alkali rare earth tungstates, carries a great significance in hydrothermal research.

8.8 Hydrothermal Synthesis of Molybdates

There are not many molybdenum bearing minerals in nature, and the most important mineral is wolfenite, $PbMoO_4$, belonging to the scheelite group. Hence, molybdates are closely related to tungstates in their formation, crystal chemistry, and also physical properties. Several molybdates have been synthesized artificially by various techniques, like melt, solid state, and flux. Their synthesis is closely related to the synthesis of tungstates and hence, several isostructural compounds between tungstates have been obtained. These molybdates show interesting optical, electrical, and magnetic properties. Only a small number of molybdates have been obtained under hydrothermal conditions and that too during 1960s and 1980s, especially from the erstwhile Soviet Union. However, during 1990s, an extensive study on the hydrothermal/solvothermal chemistry and the structural chemistry of a large variety of molybdates was initiated. Klevstov and coworkers [46–50][44,46] have worked out the growth technology for molybdates along with tungstates. Crystals of $CdMoO_4$ have been obtained in 5–15 wt% NH_4Cl, the nutrient materials are CdO and MoO_3 in stoichiometric ratio [55]. The crystallization was carried out at 350–500°C, $\Delta T = 10–25°C$, in titanium-lined autoclaves.

The hydrothermal synthesis of molybdates is usually carried out at a high temperature, greater than 500°C. The most well-studied system is R_2O_3–MoO_3–K2MoO4–H_2O [56]. Figure 8.17 shows the TN-diagram of crystallization in the system R_2O_3–K_2MoO_4–H_2O. Four major phases crystallize in this system: $RMoO_4(OH)$, R_2MoO_6, $KR(MoO_4)_2$, and $K_5R(MoO_4)_4$.

Hydroxyl molybdates, $RMoO_4(OH)$. This has been obtained for La–Eu at temperatures of 450–550°C and pressures of 1200–1400 atm [56]. Rare earth oxides and molybdenum in the mol ratios 1:2, 1:3, 1:4, and 1:5 were taken as the nutrient. Good quality crystals have been obtained in aqueous solutions of LiCl and KCl (20 wt%) with an addition of NH_4Cl (5 wt%). With a rise in the temperature, the crystal output decreases up to 50% at 525°C, and the tabular and needle-like crystals of polyoxides of molybdenum increase.

Oxymolybdates, (R_2MoO_6). These are obtained in the ratio $R_2O_3/MoO_3 = 1:1$, and oxymolybdates with rare earth elements from Pr to Er have been obtained under hydrothermal conditions [57]. The synthesis is carried out at 500–600°C in aqueous solutions of K_2MoO_4 with concentrations of less than 35 wt% and the starting R_2O_3/MoO_3 ratios = 1:1 to 1:2. The oxymolybdates under these conditions crystallize in the low-temperature monoclinic modification isostructural to Nd_2WO_6 [57]. Kodama and Izumi [58] have obtained Bi_2WO_6 crystals in pure water at $T = 300–500°C$ and $P = 500–800$ atm. The nutrients Bi_2O_3 and MoO_3 in the ratio 1:1 have been used. Similarly, these authors have obtained Bi_2MoO_6 in Li, Na, K, NH_4 ($T = 200–400°C$, $P = 175–400$ atm) chlorides, in KF solution ($T = 100–300°C$, $P = 50–300$ atm) and NaF solution ($T = 200–400°C$, $P = 100–420$ atm). Several compounds exist in the Bi_2O_3–MoO_3 system because they are efficient catalysts (for instance, for oxidation of unsaturated hydrocarbons). Bi_2MoO_6 is one of the compounds in the system and is found in natural deposits as the mineral koechlinite, which is one of the polymorphs of Bi_2MoO_6. This compound has three polymorphic forms. The γ-form transforms quickly and irreversibly into the high-temperature form when it is heated above 600°C. Therefore, the hydrothermal method is very useful for preparing the γ-form. Under hydrothermal conditions, the presence of water vapor enabled reactions between Bi_2O_3 and MoO_3 to occur at much lower temperatures than those required

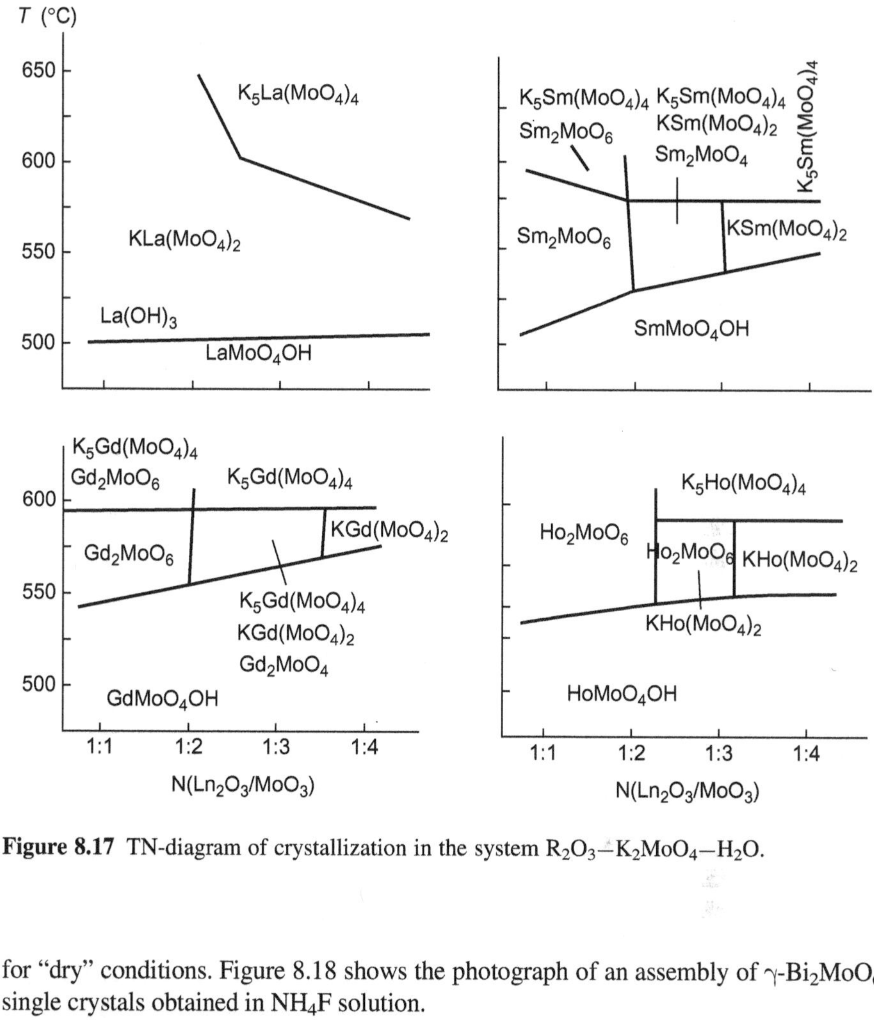

Figure 8.17 TN-diagram of crystallization in the system R_2O_3–K_2MoO_4–H_2O.

for "dry" conditions. Figure 8.18 shows the photograph of an assembly of γ-Bi_2MoO_6 single crystals obtained in NH_4F solution.

Several oxymolybdates of 1:3 ratio have been obtained and the prominent ones are $Fe_2(MoO_4)_3$, $In_2(MoO_4)_3$, and $Fe_2(MoO_4)_3$ obtained in 5–10 wt% LiCl solutions at $T = 400$–$500°C$ and $P = 400$–1000 atm.

Among the mixed rare earth molybdates, the prominent ones are $M^+R(MoO_4)_2$ ($M^+ = Li, Na, K$); R = rare earth elements [59–61]. These double molybdates are known for every series of rare earth elements (Ln = La–Lu, Y). They are prepared by crystallization from solution, in the melt, solid-phase synthesis, and by the hydrothermal method. The entire series of compounds are isostructural to the tetragonal scheelite, $CaWO_4$. The experimental conditions are $T = 525$–$600°C$, $P = 1000$–1500 atm. Aqueous solutions of 5–50 wt% of K_2MoO_4 were used as solvents and KOH was added to obtain the desired pH of the growth media. The other molybdates are α-$KBi(MoO_4)_2$, $KY(MoO_4)_2$, $K_5R(MoO_4)_4$, $K_2Pb(MoO_4)_2$, $K_2(K_{0.5}R_{0.5})(MoO_4)_2$, $LiLa(MoO_4)_2$, and so on [62]. The experimental conditions,

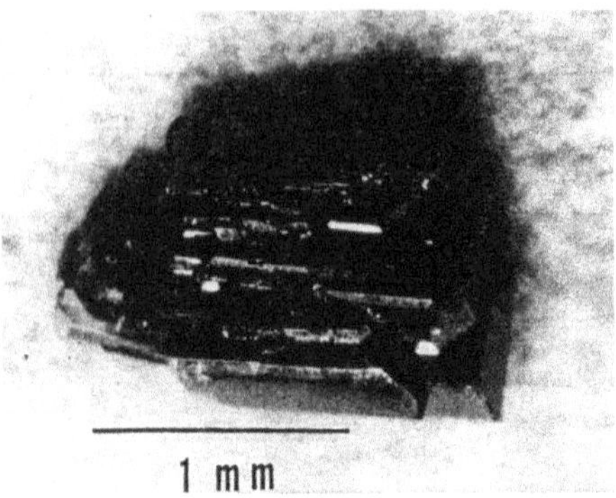

Figure 8.18 Photograph of an assembly of γ-Bi_2MoO_6 single crystals obtained from NH_4F solution [58].

like pressure, temperature, and solvent, are the same except for the type and ratio of the starting components which determine the crystallization of other phases.

In recent years, there have been several new molybdates synthesized under hydrothermal conditions exhibiting good ionic conductivity and some zeolitic structures. Among them, the copper molybdates are very interesting superionic conductors. The $Cu^{1+}_6Mo^{6+}_5O_{18}$ and $Cu^{1+}_4Mo^{6+}_5O_{17}$ have been synthesized by the hydrothermal technique using stoichiometric quantities of cuprous oxide and molybdenum trioxide taken in a gold capsule which is filled with a known quantity of distilled water. The capsule was sealed and held at 3 kb pressure and 500°C temperature for 12 h, followed by a slow cooling (10°C/h) to room temperature. Under the microscope, very thin crystals appeared and were reddish-brown in transmitted light [63,64]. Figure 8.19 shows the resistivity data for $Cu_4Mo_5O_{17}$ crystals. The resistivity measurements were made on single crystals using indium contacts soldered on the crystals in a nitrogen atmosphere to avoid oxidation.

Mecarron III and Calabrese [136] reported many reduced molybdenum phosphates with octahedral–tetrahedral framework. Following this, there has also been much interest in molybdates and molybdenum oxides due to their catalytic activity and their capability of forming crystal structures which allow rich intercalation chemistry using secondary and tertiary amines [65–67]. These compounds yield microporous structures. The hydrothermal method helps to synthesize molybdates in an acidic medium by using various templates as structure-directing units. The structures of $(Mo_xO_y)^{n-}$ can be "tailored" to a certain degree by altering the size, conformation, and properties of templates and the acidity of the reaction system. This results in a series of chain and layer-type polymeric molybdenum oxides with intercalating cations. Thus, these new molybdates are the most perspective materials which can only be obtained by the hydrothermal technique.

8.9 Hydrothermal Synthesis of Titanates

Titanium is one of the most abundant elements in the earth's crust and takes the ninth place in the order of abundance. It can be considered in two parts. The first

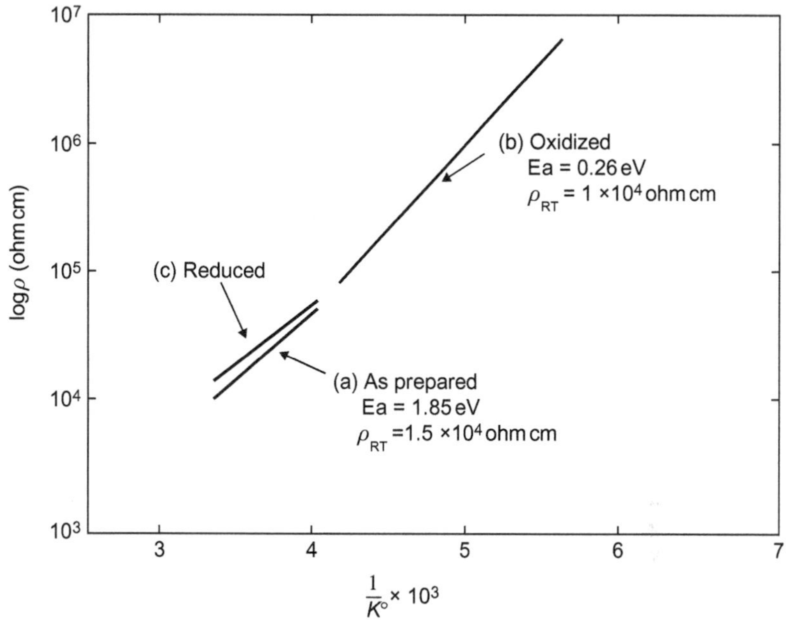

Figure 8.19 Resistivity data for $Cu_4Mo_5O_{17}$ [64].
Source: Courtesy of the Academic Press, Orlando, FL.

part is its concentration in only a few mineral structures, like ilmenite, sphene, rutile, and a few others. In the second part, it is an isomorphous substituting cation in a majority of silicates and other complexes in spite of the heterovalency, large difference in the ionic radii, and so on. The presence of Ti^{4+} and Zr^{4+} in the silicates usually leads to the formation of mixed framework structures.

A lot of chemical elements can form the octahedral coordination in crystal structures. Among these compounds, it is possible to enlist a rather numerous group of oxides of Ti, Ta, Nb, Zr, and of transitional elements (Cu, Mn, Fe, Cr, Al, amongst others). Among these, the titanates are the most important materials, not only because of their structural diversity, but also their increased technological applications, like electrooptic, laser, superionic, and high-temperature superconductors.. In contrast to silicates and their analogs, the titanates have octahedral coordination and have layer or framework configurations, but not isolated rings, islands, chains, and so on.

8.9.1 Crystal Chemistry of Titanates

In titanates, formally in each $[TiO_6]$ unit, each oxygen atom gets about $4/6 = 0.67$ unit from the central cation. Therefore, it is quite certain that during the crystallization of titanates (Ti^{4+}), the oxygen anions should be bonded with additional cations in order to attain the desired local charge balance. However, under surplus concentration of Ti^{4+} in the starting composition, the function of other cations is usually met by Ti^{4+} itself. In the natural systems, titanates usually have Na^+, K^+, Ca^{2+}, Sr^{2+}, Ba^{2+}, and other metals in their composition and the charge balance can be well established in such compounds. In the earth's crust, silicon is more abundant with a higher charge (+4) and lower coordination number (4). Hence, it forms a more common partnership with Ti^{4+} in the earth's crust to form titanosilicates. The other cations which play a major role in the charge compensation in titanium minerals are: Zr^{4+}, Fe^{3+}, Fe^{2+}, Mn^{2+}, Mg^{2+}, and also the proton (H^+). Table 8.3 gives the common association of $[TiO_6]^{8-}$ in the structure of oxide compounds. The said formula unit of column 1 of Table 8.3 can be easily obtained directly from the association symbol. For example, the radical formula corresponding to the

Table 8.3 Common Association of $[TiO_6]^{8-}$ in the Structure of Oxide Compounds

Symbol and Radical Formula	Association Character	Structure Type
6(1) $(TiO_6)^{8-}$	Discrete octahedra	$K_2TiSi_3O_9$, $NaTi_2(PO_4)_3$, $K_2TiSi_6O_{15}$
5(1)−1 (2) $(Ti_2O_{11})^{14-}$	Octahedra sharing common vertices	$NaTi_2Al_5O_{12}$, $Ba_4Ti_2PtO_{10}$, $TiTe_3O_8$
4(1)−2(2)−$(TiO_5)^{6-}$	Para octahedra sharing common edges, infinite chain of octahedra sharing transvertices	Leucosphenite, neptunite tinakite
	a. Linear	Pseudobrookite sphene
	b. Zigzag	
3(1)-3(2) $(Ti_2O_9)^{10-}$	Para octahedra sharing common faces	Hexagonal modification of $BaTiO_3$
2(1)-4(2) $(TiO_4)^{4-}$	Infinite pillars of edge-shared octahedral	Rutile
	a. Linear	Brookite
	b. Zigzag layered network of vertices shared octahedral	
1(1)-5 (2) $(Ti_2O_7)^{6-}$	Double layered vertices shared shared octahedra	$Bi_2(NbTi)O_9$, $Bi_3(TaTi)O_9$ $Sr_3Ti_2O_7$
6(2) $(TiO_4)^{4-}$	Framework vertices shared octahedra edge-shared octahedra, vertices and faces linked octahedra	Ilmenite, perovskite, pyrochlore $Ca_3Ti_2O_6$—$ThTi_2O_6$ $CaTi_2O_4$ $BaTiO_3$ hexagonal modification
3(2)-3(3) $(Ti_2O_5)^{2-}$	–	–
6(3) $(TiO_2)^0$ Ba $(Ti_7Mg)O_{16}$	Close packed layered octahedra framework with vertices edges shared octahedra	$TiCl_2$, TiI_2, TiS_2 Rutile, Brookite, anatase
6(4) $(Ti_2O_3)^0$	Framework with vertices, edges and faces shared octahedra	Ti_2O_3 corundum

association 5(1)−1(2) takes the form TiO(5/1 + 1/2) = $TiO_{5.5}$ or Ti_2O_{11}. It is observed that the sequence of all the Ti−O radicals given in Table 8.3 is made up of homological series with a general formula $[Ti_2O_{12-n}]^{-2(8-n)}$, where $n = 0, 1, 2, 3,\ldots, 8$. The most simple titano-oxygen motif is $[TiO_6]^{8-}$ radical with all the six free vertices. Instead of discussing each radical unit given in Table 8.3, only a few selected and most widespread radical groups are discussed here. The most popular among these is the Ti−O framework consisting of the basic structure related to perovskite-type $CaTiO_3$ [68]. Similarly, it is interesting to consider perovskite-type framework in ilmenite, pyrochlore, and spinel types; however, the most important association in the crystal chemistry of titanates is the TiO_2 in three polymorphic modifications, of which the rutile type is the most stable one, with linear columns of octahedra sharing the edge. The most complex association is that where each oxygen atom is linked with four Ti^{4+} octahedra and it is considered to be a highly unique structure since it indicates an anomalously high supersaturation of the corresponding oxygen ligands with Ti^{4+}. Such complex structures are formed in $K(NbTi)O_5$, $Na_2Ti_3O_7$, $A_2Ti_6O_{13}$, where A = Na, K, Rb, and similarly $Na_xTi_4O_8$ [69−72]. Here, it is appropriate to discuss the groups consisting of titanium coordination with five vertices, i.e., tetragonal pyramids and trigonal pyramids. In this case also, it is possible to use the principle of character association. With the help of simple and complex symbols, for Ti−O polyhedra in the discrete form, only tetragonal pyramids have been reported and such structures can be expected in titanates, silicates, germanates, and rare earth titanates, $RTiO_5$, where R = Y, La [73,74]. All the above-mentioned different motifs of Ti−O octahedra—either isolated or discrete pairs—are linked with each other to form chains, ribbons, layers, etc., clearly indicating the ability of Ti^{4+} to decrease its coordination number up to five and, thus, giving new motifs corresponding to Pauling's polyhedra. Obviously, intercondensation of Ti−O polyhedra makes a part of their edges and vertices remain free. This is attributed to the higher probability of the severely disturbed charge balance, and it becomes important when the Ti−O distances in coordination polyhedra fall from the normal values. Reduction in the coordination number of Ti^{4+} up to 5 can be considered due to the limited manifestation of the tendency of Ti^{4+} ions under the given conditions by giving out one of its ligands.

In summary, the above described crystal chemical elucidation of titanium, particularly Ti octahedra, undergo condensation, their varying associations, abundance of the edge contacts between these polyhedra, an anomalous case of their faces linked, reduction in the coordination number of Ti, the possibility of reaching a complete charge compensation, etc., and justify a wide structural diversity, both in natural and synthetic compounds.

In comparison with the other anion forming elements, Ti^{4+} stands close to Si^{4+} in several respects. Probably the main reason for such closeness is the repulsive forces between the higher valent ions like Ti^{4+} and, in general, Ti^{4+} can form complexes easily with the group IV elements of the periodic table. Since Ti^{4+} has a higher coordination number, it is not as easy to understand the absence of analogous phase, $TiSiO_4$, in the system TiO_2−SiO_2 [75]. Out of 32 known titano-silicate compounds in nature, 18 consist of one or the other groups of titanium octahedra. Under these circumstances, it leads to an unexpected variation in their association, like isolated, infinite motifs of chains and layers. Here it is necessary to emphasize the abundance of Ti octahedra sharing their edges in silicates, making it closer to its neighbor diagonally in the periodic table, i.e., Al^{3+}, which can form tetrahedral and octahedral coordinations with both Al octahedra and Ti octahedra sharing their edges (epidote, kyanite, sillimanite, and andalusite). Such a formal substitution of Al^{3+} in tetrahedra and Ti^{4+} octahedra does not change the charge balance in the silicate structures. Considering the closeness of Al–Si and Ti–Si frameworks, it is necessary to show their principal difference of bonding with different valency and coordination of silicon substitutes.

It is well known that all the variations in the framework silicates are connected with the heterovalent substitution: $Si^{4+} \rightarrow Al^{3+}$. If the framework is made up of pure (SiO_2), then in the isovalent substitution (but not isomorphic) of $Si^{4+} \rightarrow Ti^{4+}$, the Ti^{4+} gets the additional oxygen atoms to form an octahedral coordination.

It is interesting to note that Zr–Hf forms an excellent solid solution because of its unique chemical and crystal chemical closeness, which brings in the neighboring elements, including the lanthanide series. The other elements which can form the isomorphic substitution with Ti^{4+} are Nb^{5+}, Ta^{5+}, Zr^{4+}, Fe^{3+}, and Al^{3+}. All these substitutions are obtained experimentally; however, in nature such a large substitution has not been found, e.g., Al^{3+} rarely substitutes for Ti^{4+} in minerals or vice versa. Here, it is necessary to remember that both Al and Ti show chemically amphoteric behavior which changes their coordination numbers significantly, so also the ionic radii ($r\text{Ti}^{4+} = 0.68$ Å, $r\text{Al}^{3+} = 0.51$ Å) plays an important role. Therefore, a part of the Al^{3+}, octahedrally coordinated, can form the isomorphous substitution with octahedrally coordinated Ti^{4+}.

One of the most common schemes observed in the heterovalent substitution of $Ti^{4+} \rightarrow Nb^{5+}$ shows $2Ti^{4+} \rightarrow Nb^{5+} + Fe^{3+}$. There are many such examples for $Ti^{4+} \rightarrow Nb^{5+}$ in the synthetic compounds like $TiNbO_7$, $Ba_6Ti_2Nb_8O_{30}$, and, similarly, $YNbTiO_6$, a symmetric analog of the metamict mineral [76–78]. Only in two mineral structures—epistolite, $Na_2(Nb, Ti)_2Si_2O_9 \cdot nH_2O$, and vuonnemite, $Na_4TiNb_2Si_4O_{17} \cdot 2Na_3PO_4$—has such an isomorphism between Ti^{4+} and Nb^{5+} been reported.

In spite of the isovalent behavior of these Ti^{4+} and Zr^{4+} atoms, their crystal chemical relation is not as close as that of $Ti^{4+} \rightarrow Nb^{4+}$. The main reason is higher ionic radii of Zr^{4+} (0.79 Å) and also the different nature of bonding of Zr–O and Ti–O, particularly the latter, which is appreciably different. Moreover, there is a greater variation in the Ti–O bond distances in the Ti–O octahedra, and also the Ti^{4+} octahedra has a tendency to undergo condensation, and it is well known that the coordination number of Ti^{4+} in quite a few structures goes down from 6 to 5 and even to 4. In contrast, the Zr–O bond distances do not vary much in the Zr octahedra, which show no tendency for undergoing condensation. Besides, in a series of oxides (both simple and complex), the Zr^{4+} cations increase their coordination number from 6 to 7 and up to 8.

In recent years, a large number of experiments have been carried out on the isomorphism of various titanium bearing compounds, which have technological applications. $BaTiO_3$, $SrTiO_3$, alkali titanates, $LiNbO_3$, ABO_4 (A = Sb, Bi; B = Nb,

Ta, Sb), etc., are the important examples. Alkali titanates with layered structures can form suitable examples for the nonstoichiometry on the guest cations. $KTiNbO_5$ reported by Wadsley [69] is one such compound, where each sheet $(TiNbO_5)^{n+}$ consists of double zigzag strings of octahedra sharing corners, which suggests the possibility of nonstoichiometry on the K ion sites. Rebbah et al. [79] have studied the nonstoichiometric $A_{1-x}(Ti_{1-x}M_{1+x})O_5$ (where A = K, Rb, Te, Cs; M = Nb, Ta).

8.9.2 Hydrothermal Synthesis of Selected Titanates

Titanates have great technological applications because of their wide range of structural variations and the possible substitution of Ti^{4+} by various other cations. In the last two decades, the research activities on the synthesis of a variety of titanates have increased exponentially owing to their technological significance especially in the nanosize regime, and their growth as bulk single crystals does not carry much significance in recent years. Here, instead of going into the growth aspects of all the titanates, we discuss only the alkali titanates and rare earth titanates which have some important applications. TiO_2 and alkaline earth titanates are discussed separately in Chapters 9 and 10.

Alkali titanates exhibit layered structures, and the host cations can easily accept the guest cations giving rise to a wide range of nonstoichiometric compounds [80]. All these compounds are very important owing to their ionic conductivity and exchange properties. Usually, the growth of alkali titanates is carried out by solid-state reactions or exchange reactions, flux growth, and melt growth. The structural/stereochemical behavior of the titanium does not favor epitaxial methods. Similarly, the melt technique has not been adapted to the growth of alkali titanates even though this is one of the best techniques for the growth of various titanoniobates. The most popular method for all the titanates, in general, to obtain single crystals is the flux method, because at high temperature, these flux agents become highly active and the titanium forms various complexes depending upon the availability of other cations and anions and undergoes condensation.

The alkali titanates could be obtained by reacting carbonates/oxalates of alkali metals with titanium dioxide (anatase) at higher temperatures (usually above 900°C in an inert atmosphere). The introduction of fluoride and sometimes boron containing molten salts leads to the condensation of the titanium octahedra. Also, the crystallization temperature and starting molar ratio of the components play a major role in the preparation of titanates with various degrees of condensation. Although a large variety of alkali titanates have been obtained in the series $A_2O \cdot nTiO_2$ $(1 < n < 6)$, where A = Li, Na, K, Rb, Cs, Ti, Ag, by solid-state reactions, the member $(n = 5)$ has not been obtained [81–83]. Keeping this in mind, Byrappa et al. [84] made an effort to obtain this member $(n = 5)$ of the series by the hydrothermal technique. An attempt to obtain a member $(n = 5)$ by the hydrothermal technique was successful. The probable reason might be the stabilities of $Na_2Ti_3O_7$, $Na_2Ti_6O_{13}$, and $K_2Ti_4O_7$ which are higher and they may not allow the formation of other stable phases, although such members might be present during the course of crystallization reactions. Figure 8.20 shows the TN-diagram of crystallization of alkali titanates. The experiments were carried out using Tuttle cold-cone sealed autoclaves provided with platinum liners. The starting materials, such as TiO_2 and NaOH, were taken in desired molar proportions and placed inside the platinum liner. The suitable mineralizer solution with a definite molarity was added into the platinum liner which was sealed later. The platinum liner was later placed inside an autoclave which was kept at $T = 500°C$ and $P = 1$ kbar for a period of 3 days. Long needle-like crystals of $Na_2Ti_3O_7$ were obtained. The crystals were about 5–8 mm long, colorless, and transparent. Such good quality crystals of $Na_2Ti_3O_7$ have not been obtained by the solid-state reactions method; the resultant product was mostly a crystalline powder.

The ideal experimental conditions in the growth of $Na_2Ti_3O_7$ crystals are given below:

NaOH	5.0 gm
TiO_2	2.5 gm
HCOOH	6 ml (3 M)
Temperature	600°C
Pressure	1 kb
Duration	4 days

In experiments with high molarity of HCOOH, the crystal quality was poor, and beyond 10 M HCOOH, the $Na_2Ti_3O_7$ did not crystallize. In the present work, the other mineralizers like HCl, HNO_3, H_2SO_4, and H_2O were tried, but these experiments did not yield good results and instead produced either mixed or other phases. The crystallization of $Na_2Ti_3O_7$ probably takes place through the following reactions:

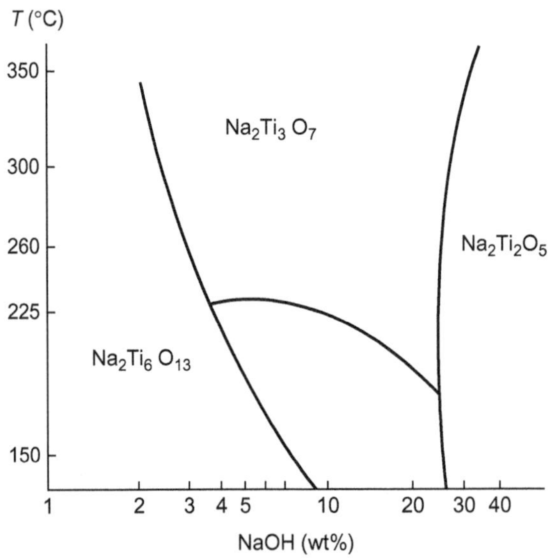

Figure 8.20 TN-diagram of crystallization of alkali titanatses [75].

$$NaOH + HCOOH \rightarrow NaHCOOH + H_2O \tag{8.16}$$

$$5NaHCOO + TiO_2 \rightarrow NaTi(OH)_5 + 2Na_2O + 5CO \tag{8.17}$$

$$Na_2O + HCOOH \rightarrow NaHCOOH + NaOH \tag{8.18}$$

$$5NaOH + TiO_2 \rightarrow NaTi(OH)_5 + 2Na_2O \tag{8.19}$$

$$2NaTi(OH)_5 + TiO_2 \rightarrow Na_2Ti_2O_7 + 5H_2O \tag{8.20}$$

The solubility of $Na_2Ti_3O_7$ was determined under hydrothermal conditions in the HCOOH media with a varying temperature and at a constant pressure of 1 kb. The solubility curve is shown in Figure 8.21. It is evident from Figure 8.21 that the hydrothermal synthesis of $Na_2Ti_3O_7$ can be carried out within the temperature range of 350–600°C.

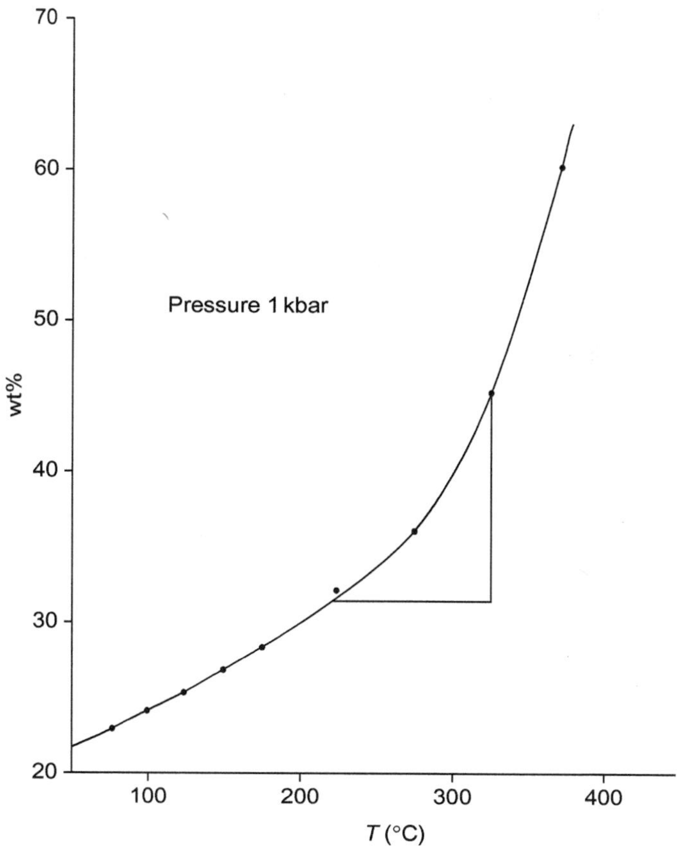

Figure 8.21 The solubility curve of $Na_2Ti_3O_7$ [84].

Byrappa [85] has reviewed the hydrothermal synthesis of various titanates and the reader can find more information in his work.

Besides alkali titanates, trivalent metal titanates, like bismuth titanate, have been studied in great detail owing to their high dielectric permeability. The $Bi_2O_3-TiO_2$ system has been studied widely, and several phases like $Bi_2Ti_2O_7$, $Bi_2Ti_3O_9$, $Bi_4Ti_3O_{12}$, Bi_8TiO_{14}, $Bi_{12}TiO_{20}$, and $Bi_2Ti_4O_{11}$ are obtained under the varying Bi_2O_3/TiO_2 ratios from 25:1 to 8:1. The solvents consisted of aqueous KF solutions while the initial charge was a mixture of the oxides Bi_2O_3 and TiO_2. The experiments were conducted in the range 450–600°C with a percentage filling factor of 0.5–0.8.

Among these titanates, the Ti sillenite, $Bi_{12}TiO_{20}$, was almost independent of the composition of the initial charge. When the TiO_2 content in the charge was more than 5%, $Bi_4Ti_3O_{12}$ (77.00% Bi_2O_3 and 20.80% TiO_2) began to crystallize together with the Ti-selenite. As the TiO_2 content of the charge was increased, the amount of $Bi_4Ti_3O_{12}$ enhanced and this compound became the first predominant and also the only phase over the range 17–25% TiO_2. Due to the temperature gradient in the autoclave, the bismuth titanates crystallize in the upper cooler region. The minimum recrystallization temperatures (500°C) were characteristic of the Ti-selenite. The other bismuth titanates required higher temperatures, but sufficiently rapid transfer of all the compounds synthesized occurred at 550–600°C. Figure 8.22(a)–(c) shows the schematic diagrams of the bismuth titanate crystals [86]. Similarly, Barsukova

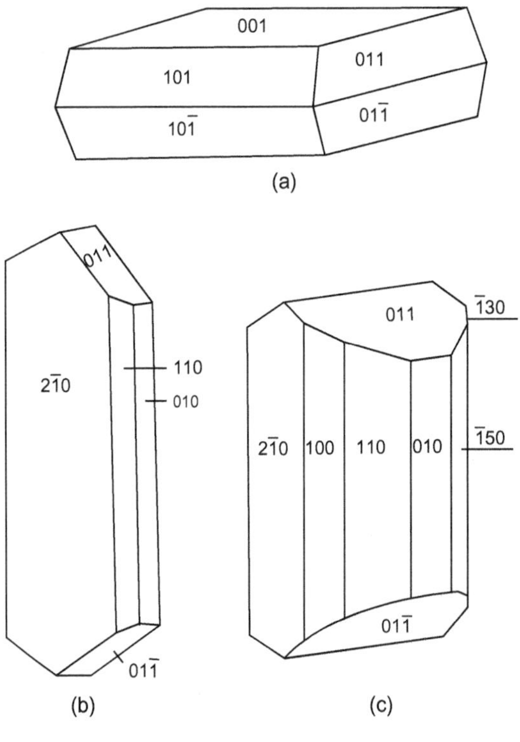

Figure 8.22 Schematic diagram of the bismuth titanate crystals: (a) crystal of $Bi_4Ti_3O_{12}$, (b) plate-like crystal of $Bi_2Ti_4O_{11}$, and (c) primatic crystal of $Bi_2Ti_4O_{11}$ [86].

et al. [87] have studied the system $Bi_2O_3-TiO_2-KF-H_2O$ under hydrothermal conditions and a part of the phase diagram is shown in Figure 8.23.

Muto and Taki [88] have synthesized potassium lithium hexatitanate $K_2Li_2Ti_6O_{14}$ crystals by the hydrothermal technique using TiO_2 and KOH + LiOH. The compound could be obtained in the following molar ratios: TiO_2:LiOH:KOH = 1.5:1.0:7.5, and concentration of alkali 10 N, filling of solutions 35–50%, $T \geq 400°C$, time of run ≥ 70 h. The interesting feature is that potassium lithium hexatitanate is not produced through the direct reaction of lithium ion with potassium hexatitanate already formed.

The growth of rare earth titanates has not been reported in the literature so far by the hydrothermal method, but there are some scanty reports by the flux method [89].

Titanates, in general, show interesting physical properties like electrical, nonlinear, dielectric, ferroelectric properties. Figure 8.24 shows the Arrhenius plot log σT

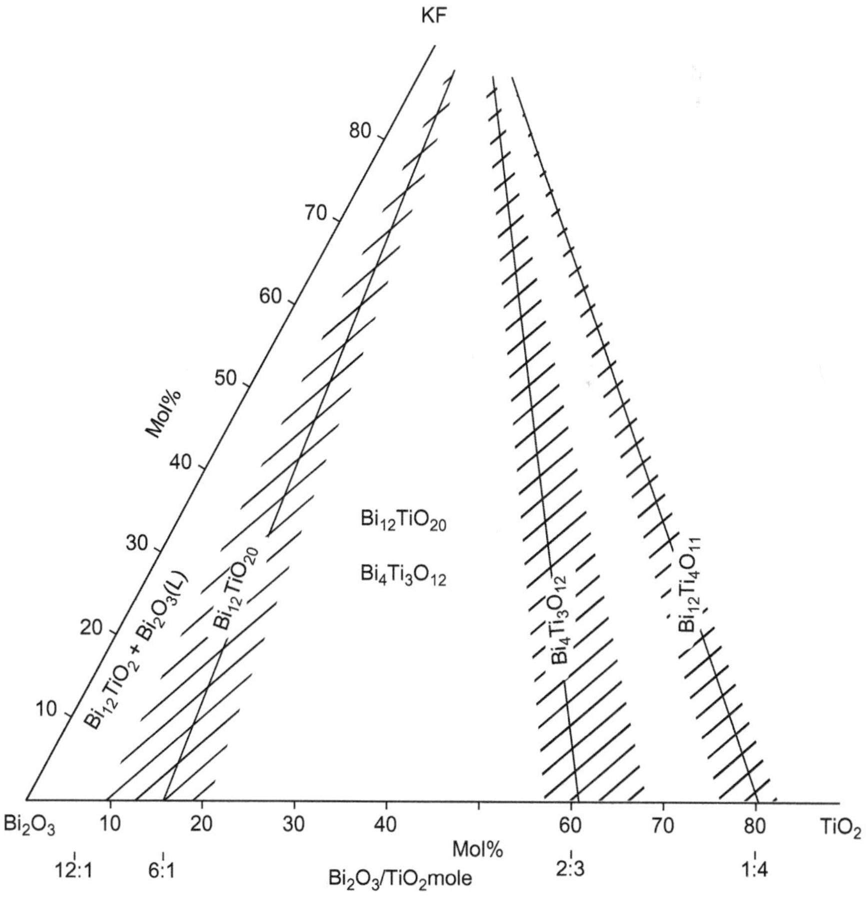

Figure 8.23 Part of the phase diagram of the $Bi_2O_3-TiO_2-KF-H_2O$ system under hydrothermal conditions [87].

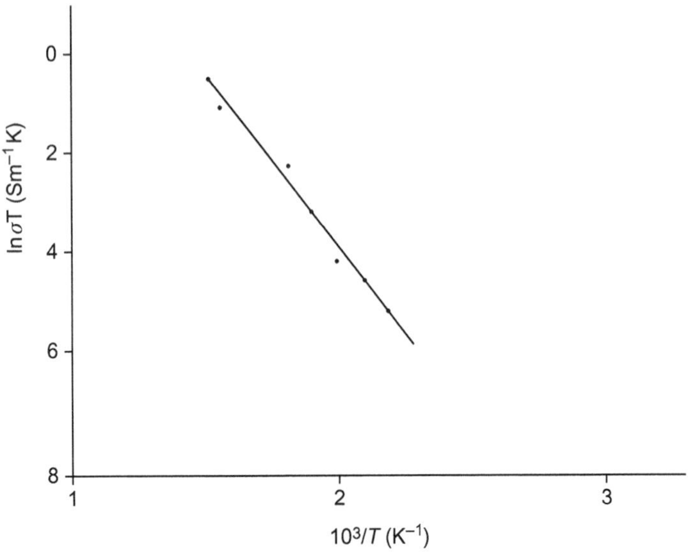

Figure 8.24 Arrhenius plot log σT versus $10^3/T$ for $Na_2Ti_3O_7$ crystals [90].

versus $10^3/T$ for $Na_2Ti_3O_7$ crystals [90]. The activation energy calculated from the Arrhenius plot is 1.3 eV. The complex impedance measurements show that $Na_2Ti_3O_7$ represents resonance at about 10 MHz. With a possible modification of the material, it could be used as a dielectric resonator at 10 Hz.

Nonlinear optical properties of Ti-selenite and $Bi_4Ti_3O_{12}$ crystals have been studied in detail and these results have been correlated with other compounds like Ge- and Si-selenites to investigate the effects associated with photoconductivity.

8.10 Hydrothermal Growth of Lithium Metagallate Crystals

Lithium metagallate, $LiGaO_2$, is of special interest because of its high electromechanical coupling constant and acoustic Q. The technology of the growth of lithium metagallate has been worked out in detail by Kolb et al. [91]. The $LiGaO_2$ can be obtained by the Czochralskii pulling and flux methods. However, the flux method does not result in the growth of large-size single crystals, and also the preferential loss of Li_2O at the melting point makes it difficult to control stoichiometry in melt-grown crystals. The initial hydrothermal experiments prompted Marshal and Laudise [92] to study the hydrothermal solubility of $LiGaO_2$ crystals in NaOH solutions. The experiments have been carried out in platinum-lined Morey-type autoclaves using 3.5 M NaOH at $T = 360–425°C$, $\Delta T = 5–40°C$, percentage fill, 0.7–0.8. Czochralskii-grown crystals have been used as seed. The trial experiments at $T = 400°C$ and $\Delta T = 25°C$, % fill 0.7 show that under these conditions, the seed melts. The $LiGaO_2$ nutrient was hydrothermally synthesized by reacting 5 M LiOH and 2.5 M Ga_2O_3 in a short platinum

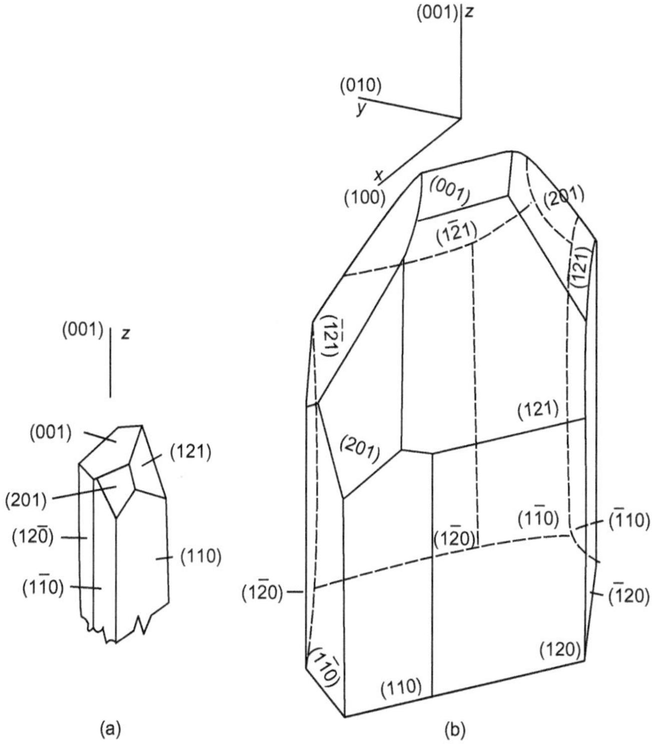

Figure 8.25 Habit of typical hydrothermal $LiGaO_2$ crystals (a) grown on a spurious seed and (b) grown on (010) seed [92].

autoclave for 10 days at the dissolving temperature of 390°C, 360°C growth temperature, and 76% fill. In addition, the nutrient was made by melting Li_2CO_3 and Ga_2O_3 as a part of Czochralskii-grown crystals or from hydrothermal crystals with a diamond-impregnated wire saw. Growth rates have been measured by determining the increase in seed thickness upon the completion of the experimental run. Figure 8.25 shows the habit of typical hydrothermal $LiGaO_2$ crystals (a) grown on a spurious seed, (b) grown on (010) seed.

The perfection of hydrothermally grown $LiGaO_2$ was excellent as evidenced by the absence of electrical twinning. The material grown in (001) is virtually free of (OH).

8.11 Hydrothermal Synthesis of Sulfides

The hydrothermal technique has been popularly used for the synthesis of sulfides for the past several decades. The main advantage of this technique is the production of fairly large-size single crystals of sulfides. From the literature survey, it is evident that sulfides are next only to silicates and oxides. Similarly, in nature also, sulfides are one of the most abundant minerals whose origin is closely related to hydrothermal—high pressure and temperature. The experimental studies on the growth of sulfides have shown that their crystallization can occur within a large variety of solutions and also strong acid. Thus, the study on the synthesis of sulfides began in the nineteenth century; however, a systematic study began after the discovery of their technological potential, especially as photosemiconductors and considering their piezoelectric properties. However, in recent years, the interest in the growth of bulk single crystals has declined significantly, but in contrary, the interest in the synthesis of nanosize sulfide crystals has increased exponentially owing to their unique properties in the nanosize regime. This is discussed in detail in Chapter 11.

Sulfides can be divided as simple or complex, containing a wide range of elements (uni-, di-, tri-, and tetravalent metals).

The easiest way of obtaining sulfides under hydrothermal conditions is through the following reaction:

$$\mathrm{Me} + \mathrm{XS}^{-2} + 2\mathrm{H_2O}(\mathrm{MeSx} + 2\mathrm{OH}^- + \mathrm{H_2}) \tag{8.21}$$

However, the presence of hydrogen hinders the sulfides formation reaction and hence, the following reaction would be more practical:

$$\mathrm{MeSO_4} + \mathrm{Na_2Sr} \rightarrow \mathrm{MeS} + \mathrm{Na_2SO_4} \tag{8.22}$$

By this exchange reaction, one can obtain a variety of sulfides with 100% output [93–96]. Latimer [97] has observed that the monophase synthesis usually occurs through hydrolysis and oxidation of the nutrient with the formation of sulfates.

$$\mathrm{S}^{-2} + 4\mathrm{H_2O} \leftrightarrow \mathrm{SO}_4^{-4} + 4\mathrm{H_2}\uparrow \tag{8.23}$$

i.e., with a raise in partial pressure of hydrogen, the sulfate ion becomes unstable leading to the process of hydrolysis:

$$\mathrm{MeS} + 5\mathrm{H_2O} \leftrightarrow \mathrm{MeO} + \mathrm{SO}_4^{-4} + 5\mathrm{H_2O} \tag{8.24}$$

$$\mathrm{Me_2S_3} + 7\mathrm{H_2O} \leftrightarrow \mathrm{Me_2O_3} + \mathrm{SO}_4^{-4} + 7\mathrm{H_2} + 2\mathrm{S} \tag{8.25}$$

The hydrolysis reaction usually takes place at $\mathrm{pH} > 7$ [97].

8.11.1 Hydrothermal Synthesis of Sulfides of Univalent Metals

Simple sulfides of univalent metals, like silver and copper, with 100% output, could be obtained in chloride solutions (1–7 wt%) and the lower partial pressure of hydrogen at $T = 400–500°\mathrm{C}$ and $P = 1000–1500$ atm. The growth is usually carried out in titanium-lined autoclaves with a temperature gradient (40–60°C). For the monophase synthesis of the above compounds, the surplus amount of the nutrient is taken. The Cu_2S is abundant in nature as chalcogenides. The other univalent metal sulfides are not common in the literature.

8.11.2 Hydrothermal Synthesis of Divalent Metal Sulfides

Divalent metal sulfides form the largest group of sulfides obtained by the hydrothermal technique. A large number of sulfides like ZnS, CdS, HgS, PbS, MnS, FeS, Fe_3S_4, NiS, CoS, and CuS have been obtained within a wide range of pH [93,96,98]. Kuznetsov [98] has studied the solubility and stability fields of these sulfides under hydrothermal conditions in detail. Figure 8.26 shows the solubility of HgS, ZnS, CdS in KOH and NaOH solutions [98–100].

In all cases the dependencies are nonlinear. This means that the composition of complexes is changed with an increase in the solvent concentration. The analysis

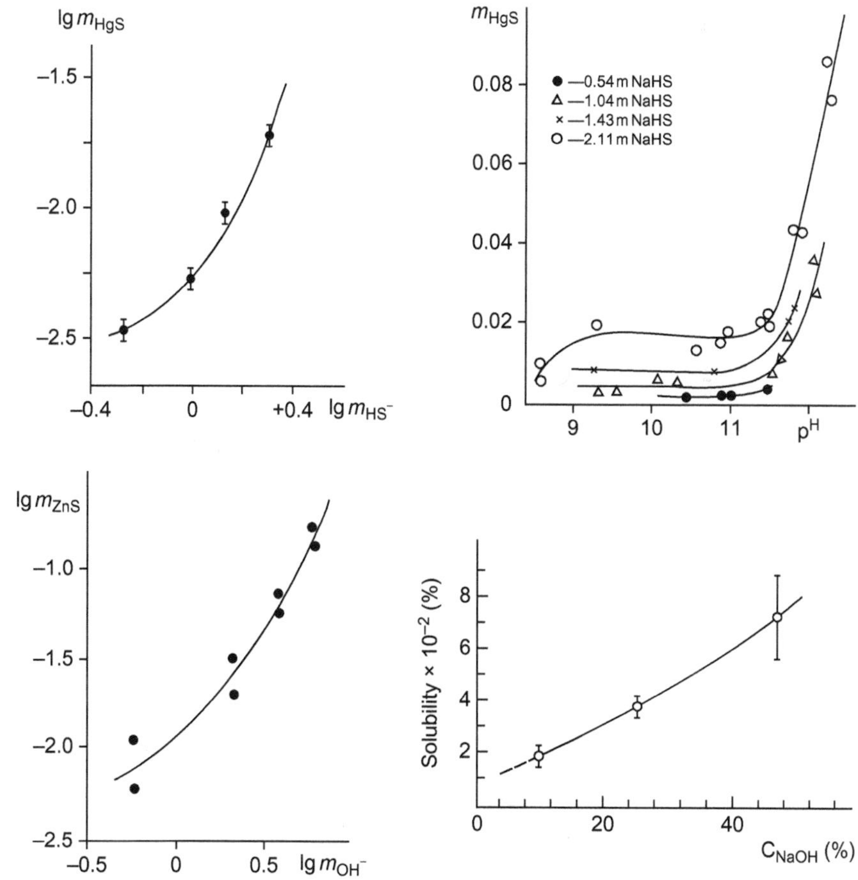

Figures 8.26 Solubility of HgS, ZnS, CdS in KOH and NaOH solutions [98].

of above dependencies allows the inference of the reactions responsible for the dissolution of HgS, ZnS, and ZnSe, at different solvent concentrations:

$$HgS + HS^- \rightarrow HgS(HS)^- \quad C_{NaHS} < 1\ M \tag{8.26}$$

$$HgS + 2HS^- \rightarrow HgS(HS)_2^{2-} \quad C_{NaHS} > 1\ M \tag{8.27}$$

$$ZnS + OH^- \rightarrow Zn(S, OH)^- \quad C_{KOH} \leq 1\ M \tag{8.28}$$

$$ZnS + 3(OH)^- + H_2O \rightarrow Zn(OH)_4^{2-} + HS^- \quad C_{KOH} > 1\ M \tag{8.29}$$

$$ZnSe + H_2O \rightarrow Zn(H, Se)OH - \quad C_{KOH} < 0.1\ M \tag{8.30}$$

$$ZnSe + H_2O + OH^- \rightarrow Zn(H, Se)(OH)_2^- \quad 0.1\ m < C_{KOH} < 0.5\ M \tag{8.31}$$

$$ZnSe + H_2O + 2(OH) \rightarrow Zn(H, Se)(OH)_3^{2-} \quad C_{KOH} > 0.5\ M \tag{8.32}$$

It can be concluded that the dissolution of HgS (in HgS—NaHS—H_2O system) is described by reactions, just as ZnS (in ZnS—KOH—H_2O system) is dissolved by a reaction at a low solvent concentration and a reaction at a higher one. As a result, the transfer of HgS and ZnS (in low concentrated solutions) takes place in the form of heterocomplexes, and ZnS is transferred in higher concentration solutions by the separated ions ($Zn(OH)_2^{-4}$ and HS^-). These differences are displayed in the kinetics of crystal growth. Figure 8.27a and b gives examples of the growth rate of HgS and ZnS and its dependence on the solvent concentration [101,102]. From Figure 8.27b, one can see that $V_{(III)}$ of ZnS hardly depends on the solvent concentration and remains practically unchanged at $C_{KOH} > 20$ wt%. The solubility of ZnS under such conditions comes up to approximately 1.5 wt%. The authors associate the unusual

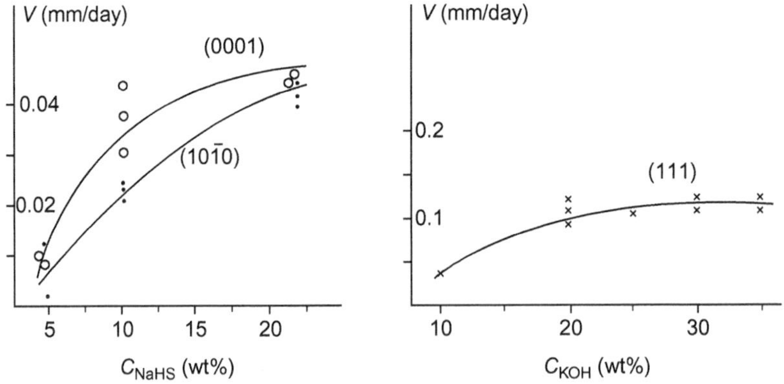

Figure 8.27 Growth rate of HgSd and ZnS and its dependence on the solvent concentration [101,102].

situation of ZnS growth with the above specific complex formation in the system ZnS–KOH–H_2O.

The interaction of the sulfides with the aqueous solutions can be accompanied by their hydrolysis and redox reactions. In an effort to estimate the stability of the above sulfides under hydrothermal conditions, Kuznetsov [98] has constructed the theoretical diagrams for Zn–S–H_2O, Cd–S–H_2O, Pb–S–H_2O, Zn–Se–H_2O, Hg–S–H_2O systems with the Eh–pH coordinates at high temperatures. In Figure 8.28, the upper lines limiting the stability fields of CdS correspond to the equilibria of these compounds with the solutions containing the oxidized sulfur; the lowest lines correspond to their equilibria with the solutions enriched by the reduced forms of the sulfur.

CdS and ZnS are stable in a wide range of Eh–pH values (from acid to the most alkaline solutions). The fields of the metallic Cd and Zn are located outside the stability field of water; therefore, the "decomposition" reaction of CdS, ZnSe, ZnS cannot take place during the usual hydrothermal experiments. Another situation characterizes the Pb–S–H_2O system. Unlike the above sulfides, lead sulfide is unstable at high temperatures in alkaline solutions and the fields of metallic lead appear at both low and high redox potentials.

Among the considered compounds, mercury sulfide has the most narrow field of stability at elevated temperatures which lies inside the water stability field. Moreover, the field of HgS is surrounded by metallic mercury. It means that the

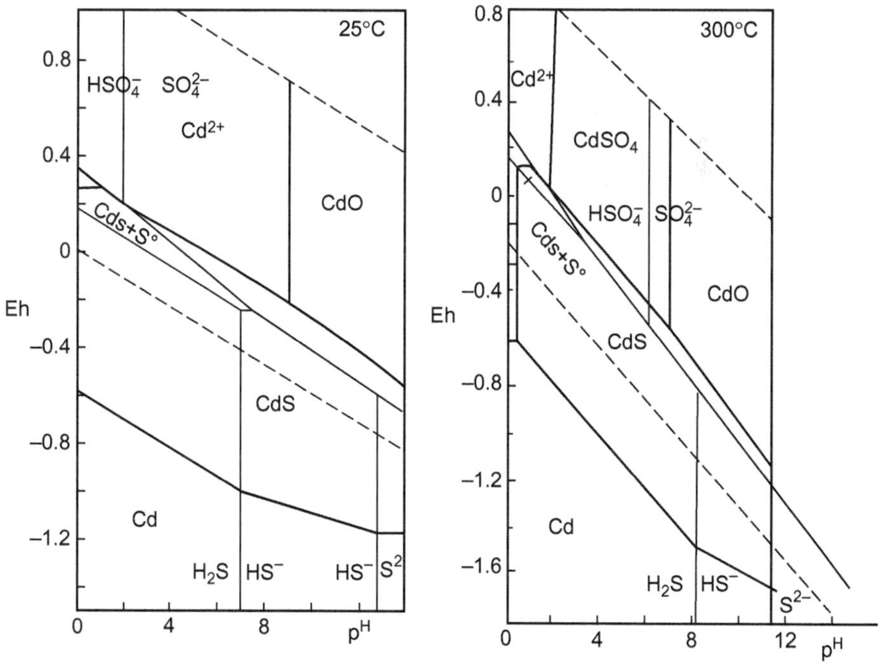

Figure 8.28 Stability fields of CdS [98].

insignificant changing of redox potential in the solution can result in the "decomposition" of HgS.

As noted above, the differences in the stability of CdS and ZnS, on the one hand, and PbS and HgS, on the other hand, are very important for choosing the growth conditions of these substances. It is evident that the stability of CdS and ZnS in a wide range of Eh and pH magnitudes gives the possibility to use various media for its crystallization. In the case of lead sulfide, the growth of crystals in alkaline solutions is only possible in certain interval of Eh values. The narrow field of stability of mercury sulfide makes great demands on growth conditions.

The growth of CdS crystals was realized in acid (H_3PO_4), chlorine (NH_4Cl) and alkaline (KOH, NaOH) solutions. In acid and chloride solutions ($H_3PO_4 = 10-30$ wt %, $NH_4Cl = 1-25$ wt%), very intensive transfer of CdS takes place at temperatures above 400°C and spontaneous crystals of about 3–5 mm in size were grown [100]. In alkaline solution, the intensive transfer of the sulfide was observed at higher temperatures (480–500°C), but at lower temperatures partial hydrolysis takes place. At 500°C and KOH, NaOH concentrations of 30–50 wt%, spontaneous CdS crystals reached $10-20 \times 0.5-1.0$ mm^2 in size.

The intensity of hydrolysis of the sulfides is distinctly inversely proportional to their crystallization. Thus, in KOH solutions, crystallization of CdS takes place stably only at 500°C, and hydrolysis is largely suppressed above 500°C. At 530–535°C, a most intense transfer of the substance from the lower to the upper zone of the autoclave was observed, and circular CdS crystals up to 8–10 mm long are readily formed.

In NaOH solutions, owing to the weaker hydrolysis of CdS and PbS, their crystallization takes place more intensely at lower temperatures than in KOH solutions. Transfer of considerable amounts of substance takes place, even at 400–450°C. However, as in the case of KOH solutions, maximum transfer of the substance was observed at 500–535°C, when sulfide hydrolysis in NaOH was practically absent.

The properties of crystals obtained in alkaline and acid solutions strongly differ. The hydrothermal method is the only technique to obtain the large-size single-crystal cinnabar, HgS. Toudic et al. [103] obtained large-size HgS crystals using Na_2S solutions at 265°C, considerably below the temperature of the $\alpha \rightarrow \beta$-HgS transition (344°C). These authors have found that, at temperatures above 200°C, cinnabar is unstable in solutions with the high pH values characteristic of Na_2S, hence, it is desirable to crystallize HgS from weak alkaline solutions with lower pH values in which the cinnabar is more stable.

The experiments have been carried out in autoclaves provided with Teflon liners. The pressure in these experiments is maintained at between 100 and 200 atm and temperature between 230 and 265°C. The pH is usually maintained at 9–13, $\Delta T = 28-40$°C. The crystals are usually prismatic and often needle like. The growth rate is nearly 0.1 mm/day in the pinnacoidal face and less than 0.02 mm/day on the prism face. Cambi and Elli [96] have studied in detail the morphological aspects of HgS with reference to the growth conditions.

The sphalerite ZnS crystals grown by different methods differ from one another in concentration and, in this respect, the hydrothermal sphalerites are more perfect. The growth of ZnS has been carried out with varying parameters, like temperature,

Figure 8.29 Spahalerite crystals grown under hydrothermal conditions [98].

the solvent composition, and pH. The experiments are carried out in autoclaves provided with platinum linings in KOH solutions and the growth temperature varies from 355°C to 400°C; however, the sphalerite crystals of lowest degree of perfection are grown under hydrothermal condition at high temperature (~400°C, Figure 8.29). Bryatov and Kuzmina [104] and Ikornikova [105] have studied the hydrothermal growth of ZnS crystals in acidic chloride solutions. When 25 wt% ammonium chloride aqueous solution was used in the experiments, intensive recrystallization of sphalerite takes place at 400°C and $\Delta T = 2$°C. Mininzon et al. [106] have studied in detail the solubility of sphalerite in phosphoric acid solutions and the morphology of the crystals. It is interesting to note that the solubility of sphalerite is highest in phosphoric acid solutions.

Greigite, Fe_3S_4, has been synthesized under hydrothermal conditions using Na_2S solution [107]. Mohr's salt solution is mixed with Na_2S solution at room temperature, to form black particles of iron sulfide precipitates, and upon heating together with the mother solution yields greigite (Fe_3S_4, cubic). The optimum heating temperature for preparing greigite depends on the pH of the solution; it is about 180°C for a solution with pH 9 and lowers to 80°C with a decrease in pH. Above these temperatures, greigite particles disappear by decomposition and alternatively, pyrrhotite ($Fe_1{-}XS_1$ hexagonal) or pyrite (FeS_2, cubic) forms, depending on the pH of the solution.

8.11.3 Hydrothermal Synthesis of Complex Sulfides

The complex sulfides include compounds containing multiple cations. The difficulties in obtaining these complex sulfides are connected to the existence of a large number of stable phases in these systems; hence, the earliest hydrothermal experiments usually do not cover these complex sulfides. However, in the last few decades, much work has been done on the synthesis of these natural analogs of complex sulfides [108]. For obtaining these complex sulfides with 100% output, synthesis through ion exchange reactions is the most useful technique. The more complex the cationic composition, the more difficult it is to fix the stoichiometry of the nutrient material. Under such circumstances, the trial-and-error experiments are more useful.

8.11.4 *Hydrothermal Synthesis of Chalcohalides*

The chalcohalides include very complex inorganic compounds in which the sulfide and halogenides simultaneously go as anions in their composition. The interest in these compounds is connected to the photosemiconductor and piezoelectric properties. Hydrothermal synthesis of chalcohalides of type ABC, where A = Sb, Bi; B = S, Se, Te; C = Br, I, have been carried out by several authors [109–113]. The principal experimental difficulty arising while studying the conditions of synthesis of the ternary chalcohalides A^{v} B^{vi} C^{vii} under hydrothermal conditions lies in creating conditions that would eliminate the possibility of oxidation or, in other words, it is essential that the original components would have the valence state A^{3+} B^{2-} C^{1-} in the solution. However, a specific value of the redox potential is merely necessary, not a sufficient condition for the synthesis of A^{v} B^{vi} C^{vii}. The probability of the formation of any of these compounds will be determined by the concentration of the original components as well as the pH of the medium and the presence of a halogen.

$$\left.\begin{array}{l} Sb + 4HI \rightarrow [SbI_4^-] + 2H_2\uparrow \\ 2[SbI_4^-] + 4[H_2S] \rightarrow [Sb_2I_2(H_2S)^4_{-4}] + 6I \\ [Sb_2I_2(H_2S)^4_{-4}] \rightarrow 2SbSI + 2H_2S + 2H_2\uparrow \end{array}\right\} pH < 6 \qquad (8.33-8.35)$$

$$\left.\begin{array}{l} Sb + 4HI \rightarrow [SbI_4^-] + 2H_2\uparrow \\ [SbI_4^-] + 2[HS^-] \rightarrow [SbS_2^-] + 2I + 2HI \\ Sb + 2[HS^-] \rightarrow [SbS_2^-] + H_2\uparrow \\ 2[SbS_2^-] + 4[HS^-] \rightarrow 2Sb_2S_3 + 2H_2S\uparrow \end{array}\right\} pH = 7-10 \qquad (8.36-8.39)$$

$$\left.\begin{array}{l} Sb + 2[HS^-] \rightarrow [SbS_2^-] + H_2\uparrow \\ 2[SbS_2^-] + 4[HS^-] \rightarrow 2Sb_2S_3 + 2H_2S\uparrow \\ 2[SbS_2] + [S^{2-}] \rightarrow Sb_2S_3 + 2S \end{array}\right\} pH > 10 \qquad (8.40-8.42)$$

Hence, SbSI will chiefly be formed in solutions with pH < 7. The range of formation of SbSI in solutions with low pH values may be associated with the stability of a hypothetical complex of the $[Sb_2I_2(H_2S)_4^{4-}]$ type, which may be formed as a result of an interaction between the two genuinely existing complexes. The $[SbS_2^-]$ is well known to be a stable ion interacting with other sulfide complexes; it necessarily leads to the formation of Sb_2S_3. Popolitov and Litvin [111] have synthesized $Pb_5S_2I_6$ single crystals in the system Pb–S–I–R–H_2O (R = solvent) under hydrothermal conditions. The experiments were carried out in Teflon-lined autoclaves at 200–300°C with a temperature gradient of 0.6–1.8°C/cm. Figure 8.30 shows a photograph of $Pb_5S_2I_6$ single crystals [113]. Note that $Pb_5S_2I_6$ belongs to the monoclinic system and the phase transition occurs at 290 ± 5°C. Rabenau and Rau [113] have reported the growth of $Pb_7S_2Br_{10}$ single crystals. The typical experimental conditions are T = 300–450°C, liquid S pressure (2.02–20.2) 107 Pa, ΔT = 0.4–1.6°/cm. Elemental Pb and S, liquid Br_2, $PbBr_2$, and PbS, in various molar ratios were used as starting materials.

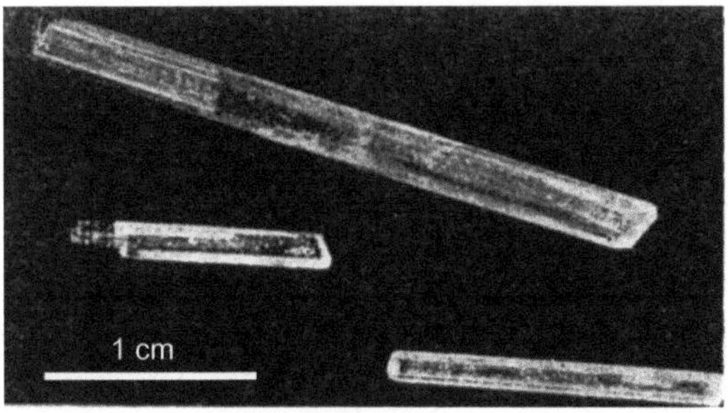

Figure 8.30 Photograph of $Pb_5S_2I_6$ single crystals [111].

The method of producing arsenic and bismuth sulfohalides differs very little from that producing the corresponding antimony sulfohalides. Above 200°C, the arsenic varieties are characterized by the "igneous" melting of the charge as a result of which a reddish-brown viscous liquid is formed. In experiments at about 300°C, the liquid turns into a vitreous mass on cooling and this is amorphous with reference to X-rays. The important chalcohalides obtained by the hydrothermal technique are SbSI, SIAs, and BiSI.

8.12 Hydrothermal Synthesis of Selenides, Tellurides, Niobates, and Tantalates

Selenides and tellurides are popularly grown by two methods: melt and hydrothermal. Although the melt method has been used as the main one for preparing single crystals of selenides and tellurides, some of these belong to the group of minerals which have been deposited in nature under hydrothermal conditions. Among the selenides and tellurides, HgSe, ZnSe, CdSe, HgTe, ZnTe, CdTe, $FeTe_2$, $FeSe_2$, $NiSe_2$, $NiTe_2$, $CoSe_2$, CoTe, Sb_2Se, Bi_2Se_3, Bi_2Te_3, and Sb_2Te_3 are the most popular ones [98,113–115]. Their stability range, within a wide interval of pH and Eh values, allows one to use various solutions for crystallization and to investigate the influence of physicochemical conditions on the properties of crystals. In fact, several solutions have been proposed, and the most commonly used ones are the alkaline solutions (5 M NaOH, for ZnSe, 1 M NaOH or KOH for ZnTe, and 20 M NaOH or KOH for CdSe and CdTe).

Polycrystallization HgSe and HgTe or elements in stoichiometric proportions are used as the starting materials. The temperature of growth is 350–380°C, $P = 1000–2000$ atm [116]. Several experiments have been carried out in acid media (HCl and HBr) in vitreous silica ampoules. Rau and Rabenau [116] have used CO_2 pressure balance gas by pumping CO_2 from a cylinder with a high-pressure intensifier into the autoclave at room temperature. In seeded growth, the temperature gradient was usually 25°C. Figure 8.31 shows the morphology of ZnSe in NaOH and LiOH solutions. The chemical reactions responsible for the crystallization of HgSe and HgTe, for example, can be written as follows:

$$HgSe + 2HCl \rightarrow H_2Se + HgC_{12} \tag{8.43}$$

$$HgTe + 2HCl \rightarrow H_2Te + HgCl_2 \tag{8.44}$$

Table 8.4 gives the optimum conditions for the crystallization of HgSe and HgTe.

In the hydrothermal synthesis of $FeTe_2$, $FeSe_2$, $NiSe_2$, $NiTe_2$, $CoSe_2$, and $CoTe_2$ crystals, Na_2Se and Na_2Te are used as the nutrient materials. The synthesis of diselenide and ditelluride of Ni and Co is carried out in quartz ampoules, whereas the synthesis of diselenide and ditelluride of iron is carried out in titanium ampoules or liners.

The physical properties of these hydrothermal selenides and tellurides are much superior as they are grown in the controlled atmosphere.

The solvothermal and hydrothermal growth of tellurides and selenides as nanosize crystals carries a special interest in recent years owing to their application potential in various technological fields.

The hydrothermal growth of niobates and tantalates began during the 1960s, at the time when the hydrothermal synthesis of inorganic compounds without natural

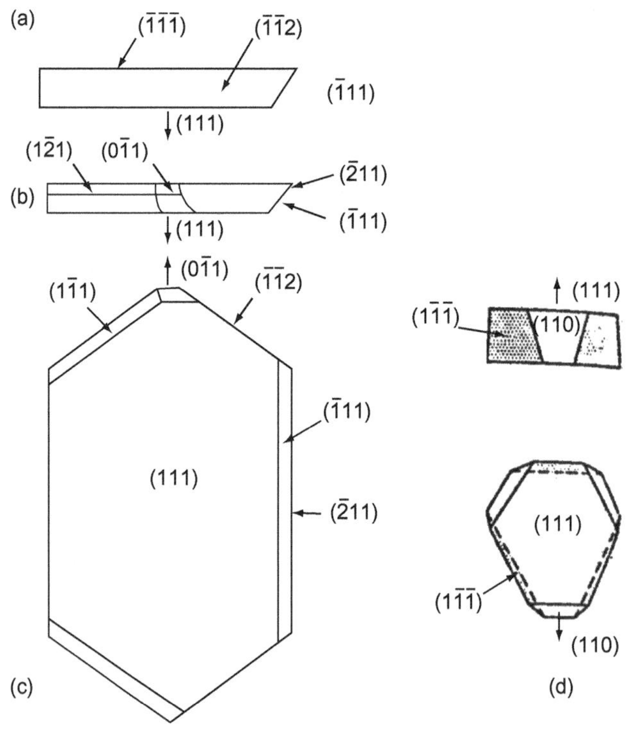

Figure 8.31 Morphology of ZnSe in NaOH and LIOH solutions [111].

analogs was extremely popular. The interest in niobates and tantalates is connected to their excellent ferroelectric and piezoelectric properties. In fact, the hydrothermal method is the only perspective method of obtaining the monocrystals of niobates and tantalates. Compared to the majority of the inorganic compounds, these materials have very low solubility. In aqueous solutions, both the elements form the polymeric oxyanions of type $[H_2Nb_6O_{19}]^{6-}$, $[HNb_6O_{19}]^{7-}$, $[Nb_6O_{19}]^{8-}$, $[Ta_6O_{19}]^{6-}$, etc., however, most of these polymeric oxyanions are not stable and, with the variation in pH and with an increase in the growth temperature, they form hydrated pentoxides. These complexes are readily soluble in strong hydrofluoric acid or strong alkaline solutions [117–119]. In weak acidic solutions, the formation of anions like $[NbOF_6]^{3-}$ and $[TaOF_6]^{3-}$ takes place, so that the synthesis of niobates and tantalates is possible, always within limited and narrow boundaries; thus, the synthesis of these compounds is always a challenging one. Marshall and Laudise [120] have studied the systems $K_2O–Ta_2O_5–H_2O$ and $K_2O–Nb_2O_5–H_2O$ at 400°C and approximately 560 atm. Mixed crystals of K(Ta, Nb)O_3 have been grown on a seed crystal. The rates of growth of the face reached 0.25 mm/day in 15 M KOH at 650°C, $\Delta T = 40$°C, and ~1100 atm [99]. K(Ta, Nb)O_3 crystals with dimensions of up to 1 cm were grown. Rare earth orthotantalates, $LnTaO_4$ (Ln = La–Lu) have been obtained at $T = 1500$°C, $P = 8 \times 10^9$ Pa. These are the high-temperature and -pressure compounds and exhibit intense polymorphism [120–122].

The discovery of piezo- and ferroelectric properties in the natural mineral stibiotantalite generated a great interest in obtaining a series of synthetic monocrystals of the group ABO_4, where A = Sb, Bi; B = Nb, Ta, Sb^{5+}. Their structures are made up of weakly formed oxygen octahedra and, in the middle of these octahedra, the cations A and B are located. These crystals can be obtained essentially by the hydrothermal technique. Although some attempts have been made to obtain these crystals by other techniques like flux and melt, good quality crystals are obtained only by the hydrothermal technique.

The hydrothermal growth technology for ABO_4 crystals has been worked out in great detail by Popolitov. Usually, alakaline solutions like MOH, MF, MHF_2, or H_2O_2 (where M = Li, Na, K, Rb, Cs) are used as mineralizers in the hydrothermal growth of these crystals. The growth of these compounds is carried out in the systems $A_2O_3–B_2O_5–R–H_2O$ and $ABO_4–R–H_2O$ (where R = LiF, NaF, KF,

Table 8.4 Optimum Experimental Conditions of Crystallization of HgSe and HgTe [111]

T (°C)	Δ*T* (°C)	Fill (%)	Experimental Duration (days)	HCl Concentration (M)	Transport of Nutrient (%)
300	100	82.92	7	12	64
316	78	82.92	10	12	82
318	82	83.34	12	12	100
300	50	85.3	6	12	22.5
350	50	80.43	6	12	29

RbF, CsF). The crystallization depends upon the growth temperature, temperature gradient, type and concentration of alkali solutions, ratio of the nutrient components, and a series of other parameters.

According to the authors, Nb_2O_5 or Ta_2O_5 exists in the alkaline solutions as hydroxyl complexes like $[Nb_6(OH)_{38}]_{8-}$ or $[Ta_6(OH)_{38}]^{8-}$ [123].

The bonding between the individual atoms of metals in such a structure is obtained through OH-bridging which is necessary for the formation of the polymerized solutions. The entire Nb or Ta is bonded in the above form to develop $SbNbO_4$ or $SbTaO_4$.

The composition of these single crystals shows that under distinct concentrations of aqueous solutions KF, RbF, CsF, there occurs the breaking of valent bonds Nb $(Ta)F_7$. These complexes, in turn, combine with the alkali cations to form Me_2Nb $(Ta)F_7$. The thermal hydration is different for different salts. MeF and the aqueous solutions of KF, RbF, and CsF, consisting of larger cations, are less hydrated. Under higher concentration of aqueous solutions of (K, Rb, Cs)F, the activity of the free ions of F is so high that it forms valent bond with Nb(Ta)—O, resulting in the formation of $Me_2Nb(Ta)F_7$. These studies show that the ability to dissolve the Nb_2O_5 or Ta_2O_5 and Sb_2O_3 increases as follows: LiF→NaF→CsF→RbF→KF; therefore, the crystallization of ABO_4 crystals depends upon the solvent—solute interaction under hydrothermal conditions.

When CO_3 or C_2O_4 is introduced, in the form of potassium carbonate and potassium oxalate, into the mineralizer solutions (aqueous solutions), it was found that the formation of $Nb(Ta)F_7$ becomes very simple and effective because the anions CO_3 and C_2O_4, with their larger size, strongly polarize the Nb(Ta) in the solution. It can be explained through the following reactions:

$$Nb_2O_5 + 15KF + 7H_2O \rightarrow 2(NbF_7)^{2-} + 15K^+ + 12OH^- + 2H^+ + F^- \quad (8.45)$$

$$2[NbF_7]^{2-} + 5[CO_3]^{2-} + H_2O \rightarrow Nb_2(CO_3)_5 + OH^- + H^+ + 14F^- \quad (8.46)$$

$$Nb_2(CO_3)_5 \rightarrow Nb_2O_5(\text{solution}) + 5CO_2 \quad (8.47)$$

When KHF_2 mineralizer (5 wt%) was used at $T = 530°C$ and $\Delta T = 40°C$, good single crystals of ABO_4 could be obtained. An increase in the concentration of KHF_2 in the system and change in the molar ratio of Sb_2O_3/Nb_2O_5 lead to the crystallization of intermediatory phases like $K_2SbNb_5O_{15}$. The hydrogen peroxide fulfills the following function: It increases the oxidation potential in the given system, stabilizes the soluble form of Nb^{5+}, and leads to the transition of less soluble Nb_2O_5 in the solution, consequently, the formation of peroxide complexes takes place [124].

The growth of $BiNbO_4$ has been carried out in the system Bi_2O_3—Nb_2O_5—H_2O using solvents like KHF_2 and H_2O_2. Figure 8.32a shows the growth temperature dependence of the crystal output of $BiNbO_4$ and Figure 8.32b shows the NC-diagram of fields of crystallization. The hydrothermal crystallization of the ABO_4 crystals can be carried out in an acidic media, for example, in the system

A_2O_3–B_2O_5–H_2SO_4–H_2SO_4–H_2O_2–H_2O within a wide range of concentrations of aqueous solutions and sulfuric acid [$C_{H_2SO_4} = 10 - 15$ wt%, $C_{H_2O} = 5 - 6$ wt%] and the volume ratio $V_{H_2SO_4}/V_{H_2O} = 2.5{:}1$ [125].

There are several publications on the hydrothermal growth of Bi and Sb niobates, and tantalates. These compounds strongly decompose at higher temperatures and, hence, the flux or melt techniques are not suitable to grow these compounds; and obviously, the hydrothermal method was found to be the most suitable one [126–128]. The synthesis is carried out in two types of solvents: (i) In aqueous solutions of alkali metals, LiOH, NaOH, and KOH and (ii) in aqueous solutions of alkali fluorides, Li–Cs. The autoclaves used for the synthesis of these bismuth and antimony niobates and tantalates are provided with copper, silver, or titanium linings. The growth takes place at $T = 450$–650°C and $\Delta T = 15$–30°C. The nutrient materials are the mixtures of Sb_2O_3 and Nb_2O_5 (Ta_2O_5). The analysis of the products of hydrothermal reaction shows that the alkali solutions bring in the reaction $2MeOH + Sb_2O_3 + 2Nb_2O_5 \rightarrow 2SbNbO_4\downarrow + 2Me\,NbO_3\downarrow + H_2O$.

Synthesis in strong fluoride solutions KF, RbF, and CsF increases the growth rate; however, further increase in the concentration of F leads to the formation of fluoroniobates and fluorotantalates of alkali metals Me_2NbF_7 and Me_2TaF_7. Obviously, such concentrated solutions do not give 100% niobates and tantalates of Bi and Sb. It has been found that in aqueous solution of KHF_2, even the formation of complex niobate$K_2SbNb_2O_{15}$ takes place.

Thus, the hydrothermal method is a most valuable tool for the synthesis of a variety of simple and complex niobates and tantalates. However, a complete understanding of the growth technology for these compounds insists on systematic investigations of various aspects of the growth parameters, like temperature, pressure, pH, type of nutrient, and solvent.

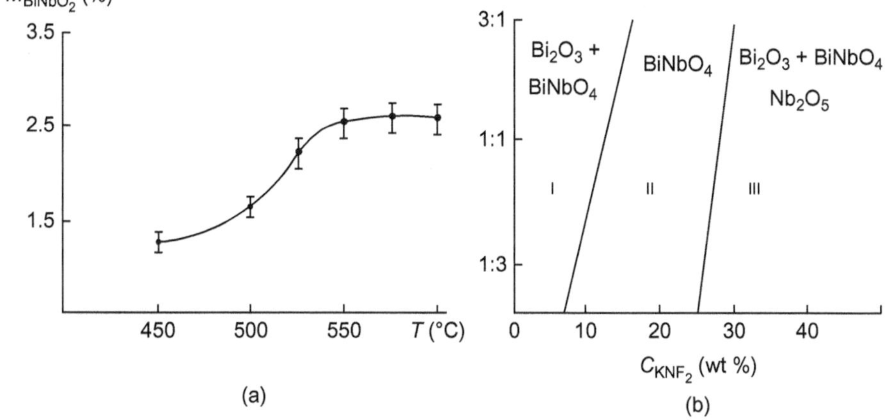

Figure 8.32 (a) Dependence of the growth temperature and crystal output of $BiNbO_4$ and (b) the NC-diagram of fields of crystallization of $BiNbO_4$ [125,126].

8.13 Hydrothermal Synthesis of Arsenates

Arsenates are closer to phosphates and vanadates in their mineralogy and crystal chemistry. There are several arsenate minerals occurring in nature, like mimetite, (PbCl) Pb_4 $(AsO_4)_3$, tilasite (MgF) $CaAsO_4$; adelite [(MgOH)$CaAsO_4$], sarkinite [(MnOH) $MnAsO_4$], olivenite [Cu_2(OH) AsO_4], adamite [Zn_2(OH)AsO_4], clinoclasite ($6CuO \cdot As_2O_3 \cdot 3H_2O$), soselite $(Ca,Co,Mg)_3$, and $As_2O_8 \cdot 2H_2O$. However, the number of arsenate minerals is much less than the number of vanadates and phosphates. The same is the case with reference to the synthetic analogs, i.e., the number of synthetic arsenates is much less than the number of synthetic phosphates and vanadates. Hence, in comparison with the rich structural chemistry of transition metal phosphates, reports on the arsenates are rare. Although a few arsenates have stoichiometries similar to those of the phosphate analogs, their structures are considerably different. Most of these arsenates adopt new structural types and have stoichiometries not found in the phosphates. The hydrothermal technique helps immensely to obtain new structures among arsenates. Some of the arsenates synthesized belong to the NASICON family $Na \cdot Zr_2(AsO_4)_3$; $A^{2+}B^{2+}As_2O_7$ (A = Ca, Sr, Ba; B = Cu), both exhibiting superionic properties, $NaNb_2AsO_8$ with nonlinear optical properties, $ABXO_4$ type [A = Mn, Zn, Cu^{2+}, Co^{2+}, Mg^{2+}, etc.; B = (OH); X = As], several complex arsenates showing unique structures related to zeolites. In spite of a wide structural diversity, the hydrothermal synthesis of arsenates is not very popular owing to the best known reasons, that arsenic is highly corrosive and highly reaction susceptible. Also, it contaminates the system unless the experiments are carried out at $T < 300°C$ using Teflon liners [129–133].

The majority of experiments in recent years have been carried out in stainless steel autoclaves provided with Teflon lining. However, the high-temperature experiments, like in the case of $NaNb_2AsO_8$, are carried out in sealed gold liners, at $T = 750°C$ and $P = 32{,}000$ psi.

For example, the copper diarsenates ($ACuAs_2O_7$, A = Ca, Sr, Ba) synthesis has been carried out in Teflon-lined autoclaves at 230°C for 4 days, followed by a slow cooling to room temperature at 50°C/h. The pure phase is obtained by heating a mixture of $Me^{2+}(OH)_2$ (Me^{2+} = Ca, Sr, Ba), CuO, KCl, H_3AsO_4, pyridine, and water, in the following molar ratios:

Ca^{2+}:Cu:As = 1:2:7.5
Sr^{2+}:Cu:As = 1:4:15
Ba^{2+}:Cu:As = 1:2:7.5

The crystal chemical study of these $A^{2+}B^{2+}As_2O_7$ compounds shows that the As_2O_7 group is very adaptive to the bonding requirements of other groups in the structure by adjusting the As—O—As bond angle and the torsional angle between the two AsO_4 tetrahedra. The size of the alkali earth metal cation seems not to play an important role in deciding the structural types of the arsenates as it does on the corresponding phosphates.

The complex arsenates, like (Mg, Ni)$_2$(OH)(AsO_4), could be prepared using hydrothermal conditions and starting from the $(Mg_{3-x}Ni_x)$ $(AsO_4)_2 \cdot 8H_2O$ (X = 0,

0.5) precursors. The starting compounds were at 180°C in a Teflon-lined steel autoclave 90% filled [129].

Chou and Kanatzidis [133] have synthesized the novel type of transition and main group metal polychalcogenides, and some of them are inaccessible by traditional solution methods. These authors have used organic counterions $R4E^+$ (E = P, R = Ph; E = N, R = alkyl) to synthesize these polychalcoarsenates.

With the recent discovery of several nonlinear optical properties among arsenates, like in $KTiOAsO_4$, there is a growing interest for new NLO materials from the arsenate family [134]. Among these, $NaNb_2AsO_8$ is an important material showing a great promise as second harmonic generator material [135].

Thus, the growth of complex coordinated compounds like sulfides, selenides, tellurides, and arsenates under hydrothermal conditions is very important.

References

[1] A.A. Kaminskii, G.R. Verdun, N.M. Khaidukov, New CW-laser crystals with semiconductor laser pumping in anisotropic fluorides containing Nd^{3+} ions, obtained by hydrothermal method, Doklady Akad. Nauk. Physika 328 (1993) 181–195.

[2] L.N. Demianets, Hydrothermal synthesis of new compounds, in: K. Byrappa (Ed.), Hydrothermal Growth of Crystals, Prog. Cryst. Growth Charact. 21 (1990) 299–355.

[3] P.M. Bridenbaugh, J.O. Eckert, G. Nykolak, Thomas, W. Wilson, L.N. Demianets, R.E. Riman, R.A. Laudise, Hydrothermal growth and properties of $KGdF_4$, J. Cryst. Growth 144 (1994) 243–252.

[4] A.A. Kaminskii, V.S. Mironov, S.N. Bagaev, N.M. Khaidukov, M.F. Joubert, B. Jacquier, et al., Spectroscopy and laser action of anisotropic single-centered $LiKYF_5$:Nd^{3+} grown by the hydrothermal method, Phys. Stat. Solidi. 145 (1994) 177–195.

[5] A.A. Kaminskii, Laser Crystals, Their Physics and Properties, Springer-Verlag, Berlin/Heidelberg/New York/London/Paris/Tokyo, 1981, 1990.

[6] M.J. Weber (Ed.), Handbook of Laser Sciences and Technology, Supply, 1, CRC Press, Boca Raton/Ann Arbor/Boston, 1991.

[7] A.A. Kaminskii, S.E. Sarkisov, K.A. Kurbanov, L.N. Demianets, N.K. Khaidukov, Hydrothermal Synthesis and Crystal Growth of Fluorides, Sulfides, Tungstates, Molybdates, and related compounds, Inorg. Mater. 23 (1987) 2049–2054 (in Russian).

[8] A.A. Kaminskii, N.M. Khaidukov, Phys. Stat. Solidi. 129 (1992) K65.

[9] A.A. Kaminskii, N.M. Khaidukov, New low-threshold LiKYF5:$Nd3^+$ laser crystal, Kvantovaya Electronika 19 (1992) 213 (in Russian).

[10] A.A. Kaminskii, N.M. Khaidukov, W. Koechner, H.R. Verdun, New Laser-Diode-Pumped CW Crystalline Lasers Based on $Nd3^+$-Ion Doped Hydrothermal Fluorides, Phys. Stat. Solidi. 132 (1992) K105.

[11] D.W. Aldous, Solvothermal Chemistry of Early Transition Metal Fluorides, Ph.D. Thesis, University of St. Andrews, UK, 2008.

[12] L.N. Iljushin, L.N. Demianets, N.I. Sorokin, in: Collected Abstracts International Symposium on the System with Fast Ion Transport, Czechoslovakia, Prague, 1985, p. 162.

[13] O.V. Yakubovich, M.A. Simonov, O.K. Melnikov, U.S. Urusov, Crystal Chemistry of Titanates, Doklady Akad. Nauk USSR 287 (1986) 126 (in Russian).

[14] R.L. Dunin-Barkovskii, S.S. Kolyago, E.V. Pestryakov, A.M. Razheu, in: 7th USSR Conference on Crystal Growth, Moscow, Collected Abstracts (in Russian), 2 (166) (1988).
[15] N.V. Belov, N.I. Golovastikov, A.N. Ivaschenko, B.Ya. Kotjuzanskii, O.K. Melnikov, V.I. Filippov, Hydrothermal Synthesis Crystal Structure and Magnetic Properties of $Cs_2FeF5 \cdot H_2O$, Kristallografiya 27 (1982) 511 (in Russian).
[16] E.N. Matvienko, O.V. Yakubovich, M.A. Simonov, A.N. Ivaschenko, O.K. Melnikov, N.V. Belov, Containing Nd3+ Ions, Obtained by Hydrothermal Method, Doklady Akad. Nauk USSR 257 (1981) 105 (in Russian).
[17] M.P. Crosnier-Lopez, Y. Laligand, J.L. Fourquet, A new barium tantalum oxyfluoride $Ba_3Ta_2F_{12}$, structural approach, Eur. J. Solid State Inorg. Chem. 30 (1993) 155–163.
[18] G. Ferey, M. Leblanc, R. de Pape, M. Passaret, M.P. Bothorel-Razazi, Cristallisation par voie hydrothermale des Fluorures FeF_3; FeF_3H_2O; $FeF_3 \cdot 3H_2O$ and NH_4FeF_4, J. Cryst. Growth 29 (1975) 209–211.
[19] J.M. Ravez, J. Viollet, R. dePape, P. Hogenmuller, Bull. Soc. Chim. Fr. 4 (1967) 1325.
[20] M. Leblanc, G. Ferey, R. dePape, Hydrothermal synthesis of transition metal fluorides in binary, ternary, and quaternary systems, Mat. Res. Bull. 19 (1984) 1581–1590.
[21] N.V. Belov, N.I. Golovastikov, A.N. Ivashchenko, B.Ya. Kotyuzhanskii, O.K. Mel'nikov, V.I. Filippov, Hydrothermal synthesis crystal structure and magnetic properties of $Cs_2FeF_5 \cdot H_2O$, Sov. Phys. Crystallogr. 27 (1982) 309–311.
[22] M. Leblanc, G. Ferey, Y. Calage, R. de Pape, Idle spin behavior of the shifted hexagonal tungsten bronze type compounds $FeIIFeII_2F8(H_2O_2)$ and $MnFe_2F8(H_2O)_2$ J. Solid State Chem. 53 (1984) 360.
[23] W. Verschaeren, D. Babel, The Crystal Structure of the Weberite Na2MgInF7, J. Solid State Chem. 24 (1978) 405.
[24] G. Courbion, C. Jacoboni, R. dePape, Application to transition metal fluoride glasses, Acta Cryst. B 33 (1977) 1405–1408.
[25] C. Jacoboni, Crystal Growth and Structures of Rare Earth Fluorocarbonates, These d'Etat Paris VI (1976).
[26] N. Mercier, M. Leblanc, Crystal growth and structures of rare earth fluorocarbonates: II. Structures of zhonghuacerite $Ba_2Ce(CO_3)_3F$; correlations between huanghoite, cebaite, and zhonghuacerite type structures, Eur. J. Solid State Inorg. Chem. 30 (1993) 207–216.
[27] N. Mercier, M. Leblanc, A scandium fluorocarbonate, $Ba_3Sc(CO_3)F_7$, Acta Cryst. 50 (1994) 1862–1864.
[28] N. Mercier, M. Leblanc, A new rare earth fluorocarbonate, $Na_2Eu(CO_3)F_3$, Acta Cryst. 50 (1994) 1864–1865.
[29] N. Mercier, F. Taulelle, M. Leblanc, Growth, structure NMR characterization of a new fluorocarbonate $Na_3La_2(CO_3)_4F$, Eur. J. Solid State Inorg. Chem. 30 (1993) 609–617.
[30] N. Mercier, M. Leblanc, Synthesis characterization and crystal structure of a new copper fluorocarbonate $KCu(CO_3)F$, Eur. J. Solid State Inorg. Chem. 31 (1994) 423–430.
[31] N. Mercier, M. Leblanc, Crystal growth and structures of rare earth fluorocarbonates: I. Structures of $BaSm(CO_3)_2F$ and $Ba_3La_2(CO_3)_5F_2$: revision of the corresponding huanghoite and cebaite type structures, Eur. J. Solid State Inorg. Chem. 30 (1993) 195–205.
[32] N. Mercier, M. Leblanc, Existence of 3d transition metal fluorocarbonates: synthesis characterization of $BaM(CO_3)F_2$ (M = Mn, Cu) and crystal structure of $BaCu(CO_3)F_2$, Eur. J. Solid State Inorg. Chem. 30 (1993) 217–225.

[33] M. Lablanc, I. Fevre-Collin, G. Ferey, The magnetic structures of Mn_2PO_4F and Co_2PO_4F at 1.2 K, J. Magn. Magn. Mater. 167 (1997) 71–79.
[34] Th. Loiseau, Y. Calage, P. Lacorre, G. Ferey, NH_4FePO_4F: structural study and magnetic properties, J. Solid State Chem. 111 (1994) 390–396.
[35] N. Mercier, M. Leblanc, Eur. J. Solid State Inorg. Chem. 28 (1991) 727–736.
[36] M.P. Crosnier, J.L. Fourquet, Eur. J. Solid State Inorg. Chem. 29 (1992) 199.
[37] M. Crosnier-Lopez, J.L. Fourquet, $Ba_3Nb_2O_2F_{12.2}H_2O$: synthesis and crystal structure, J. Solid State Chem. 105 (1993) 92–99.
[38] C. Barta, O. Gilar, E. Krcova, Lead halide crystal growth, Chem. Abstr. 109 (1988) 220138K.
[39] O.K. Nikolskaya, L.N. Demianets, Hydrothermal synthesis of α-PbF_2 single crystals, Izvestia Akad. Nauk USSR Inorg. Mater. 30 (1994) 1097–1102.
[40] R. Caruba, A. Baumer, Methode d'Obtention de Nin, Raux de Dungstene (Husbnarit, Ferberite et Scheelite), in: 8 Reen Annu. Sci. Terre, Marseille, France 1960, p. 1980.
[41] A.G. Malyavka, L.I. Lavrinenko, T.B. Seistunova, I.R. Kolyatkevich, Investigations of the influence of composition ore forming solutions on the physical properties of ferferite, in: Use of Mineralogical Method of Investigations, Soviet Alma-Ata, Erstwhile USSR, 2 (1981) 63–54.
[42] L.F. Yastrebova, A.F. Borina, M.I. Ravich, Solubility of molybdate and wolframite in aqueous solutions of chlorides of potassium and sodium at high temperatures, J. Inorg. Chem. 8 (1963) 208–217.
[43] L.Yu. Kharchenko, V.I. Protasova, R.L. Gainatutbinova, Synthesis and X-ray characteristics of basic rare earth wolframite crystals of yttrium group, J. Inorg. Chem. 25 (1980) 851–852.
[44] L.Yu. Kharchenko, P.V. Klevstov, L.P. Solov'eva, Crystallization wolframite chloride, $LnWO_4Cl$ (Ln = Ce, Pr, Nd), Doklady Akad. Nauk USSR 176 (1967) 179.
[45] S. Oishi, I. Tate, Some effects of R (R = Li, Na and K) ions on crystal growth of RNd $(WO_4)_2$ crystals from ternary systems Nd_2O_3–WO_3–R_2CO_3, Chem. Lett. (1980) 1449–1452.
[46] P.V. Klevstov, L.Yu. Kharchenko, Hydrothermal synthesis of crystals and crystallographic elucidation of some double wolframites of rare earth elements, Cryst. Growth, 7, N.N. Sheftal (Ed.), Nauka, Moscow, 1967.
[47] L.Yu. Kharchenko, P.V. Klevstov, Hydrothermal synthesis in some chloride systems containing oxides of Pr and W, Kristallografia 14 (1969) 1117–1119.
[48] L.Yu. Kharchenko, T.M. Polyanskaya, P.V. Klevstov, Synthesis and crystallographic studies of oxywolframites $Pr_4W_3O_{15}$ and $Nd_4W_3O_{15}$, Izvestia Akad. Nauk USSR Inorg. Mater. 6 (1970) 1720–1722.
[49] L.Yu. Kharchenko, P.V. Klevstov, Hydrothermal synthesis of wolframites Sm, Eu, and Gd, Izvestia Akad. Nauk USSR Inorg. Mat. 6 (1970) 594–596.
[50] P.V. Klevtsov, L.P. Kozeeva, P.V. Klevtsova, Synthesis and polymorphism of crystals of rubidium–lanthanum tungstate $RbLa(WO_4)_2$, Sov. Phys. Crystallogr. 21 (1976) 54–58.
[51] S. Oishi, I. Tate, , Crystal Growth of Na(La, Nd) $(WO_4)_2$ solid solution from high-temperature solutions, Chem. Lett. (1980) 1227–1228.
[52] A.G. Postupaeva, L.L. Petrov, D.D. Nikhtyanova, L.Yu. Kharchenko, Morphology of epitaxial layers of double alkali rare earth tungstates, Inorg. Mater. 28 (1992) 1191–1196.
[53] K. Byrappa, A. Jain, Hydrothermal growth and characterization of $NaLa(WO_4)_2$ crystals, J. Mater. Res. 11 (1996) 2869–2875.
[54] G. Carobbi, G. Tancredi, Gass. Chim. Ital. 58 (1928) 45.

[55] A.V. Chichagov, L.N. Demianets, V.V. Ilyukhin, N.V. Belov, Synthesis and structure of Cd-molybdate, $CdMoO_4$ 11 (1966) 686–689.

[56] V.I. Protasova, L.Yu. Kharchenko, P.V. Klevstov, Hydrothermal synthesis and crystallographic properties hydroxy molybdates rare earth elements $LuMoO_4 \cdot OH$, Izvestia Akad. Nauk USSR Inorg. Mater. 9 (1973) 421–423.

[57] T.M. Polyanskaya, S.V. Borisov, N.V. Belov, New modification of structure type scheelite—crystalline structure Nd_2WO_6, Doklady Akad. Nauk USSR 193 (1970) 83–86.

[58] H. Kodama, F. Izumi, Crystallization of γBi_2MoO_6 under hydrothermal conditions, J. Cryst. Growth 50 (1980) 515–520.

[59] L.Yu. Kharchenko, V.I. Protasova, P.V. Klevstsov, Investigation of the hydrothermal conditions of synthesis of rare earth molybdates in aqueous solutions of potassium molybdate, J. Inorg. Chem. 22 (1977) 986–990.

[60] P.V. Klevstova, V.I. Protasova, L.Yu. Kharchenko, R.F. Klevstova, Hydrothermal synthesis of a new crystalline modification of lithium lanthanum molybdate, α-LiLa $(MoO_4)_2$, Sov. Phys. Crystallogr. 18 (1974) 523–524.

[61] R.F. Klevstova, The crystal structure of lithium lanthanum molybdate α-$LiLa(MoO_4)_2$, Sov. Phys. Crystallogr. Group 20 (1976) 456–458.

[62] P.V. Klevstov, L.P. Koz'veva, V.I. Protasova, Synthesis and X-ray studies of double molybdate, $K_5Ln(MoO_4)_4$ Ln = La–Tb, crystals, Kristallografia 20 (1975) 57–62.

[63] E.M. McCarron III, J.C. Calabrese, Hydrothermal synthesis and structure of $Cu^{1+}{}_6Mo^{6+}{}_{0.5}O_{18}$, J. Solid State Chem. 62 (1986) 64–74.

[64] E.M. McCarron III, J.C. Calabrese, Synthesis and structure of $Cu^{1+}{}_{0.4}Mo^{6+}{}_{0.5}O_{17}$, J. Solid State Chem. 65 (1986) 215–224.

[65] N. Guillou, G. Ferey, Hydrothermal synthesis and crystal structure of anhydrous ethylene diamine trimolybdates $(C_2H_{10}N_2)$ $[Mo_3O_{10}]$, J. Solid State Chem. 132 (1997) 224–227.

[66] Y. Xu, Li-H. An, L.L. Koh, Investigations into the engineering of inorganic/organic solids: hydrothermal synthesis and structure characterization of one-dimensional molybdenum oxide polymers, Chem. Mater. 8 (1996) 814–818.

[67] J. Guo, P. Zavalij, M.S. Whittingham, Hydrothermal synthesis of a new molybdate with a layered structure (NMe_4) $Mo_{4-\delta}O_{12}$, Chem. Mater. 6 (1994) 357–359.

[68] Noray-Szabo, The Role of Fluid Mechanics in Modern Warfare, Naturwissenschaften 31 (1943) 202.

[69] A.D. Wadsley, The crystal structure of the high temperature form of niobium pentoxide, Acta Cryst. 17 (1964) 623.

[70] S. Andersson, A.D. Wadsley, Crystallographic Shear and Diffusion Paths in Certain Higher Oxides , The crystal Structure of K2Ti2O5 Acta Chem. Scand. 15 (1961) 663.

[71] S. Andersson, A.D. Wadsley, The crystal structure of lithium diborate, $Li_2O._2B_2O_3$,The structures of Na2Ti6O13 and Rb2Ti6O13 and the alkali metal titanates Acta Cryst. 15 (1962) 194–201.

[72] S. Andersson, A.D. Wadsley, The crystal structure of lithium diborate, Li2O.2B2O3, NaxTi4O8 an alkali metal titanium dioxide bronze, Acta Cryst. 15 (1962) 201–206.

[73] W.G. Mumme, A.D. Wadsley, A new class of compound Mx+Ax3+Ti2-xO4, The structure of orthorhombic Y2TiO5, an example of mixed seven- and five fold coordination, Acta Cryst. 24 (1968) 1327–1333.

[74] M. Guillen, E.F. Bertant, Über die Beziehungen zwischen Gauss und Bessel, Acad. C. R., Sci. Paris Ser. 262 (1966) 962.

[75] N.A. Toropov, V.P. Barzakovskii, V.V. Lapin, N.N. Kursteva, Composition Diagrams of Silicate Systems, Nauka, Moscow, 1965.
[76] A.D. Wadsley, Mixed Oxides of Titanium and Niobium. II, Acta Cryst. 14 (1961) 660.
[77] P.B. Jamieson, S.C. Abrahams, Ferroelectric Tungsten Bronze-Type Crystal Structures, Acta Cryst. B24 (1968) 984.
[78] V.B. Aleksandrov, Experimental Investigations in the Field of Mineralogy and Geochemistry of Rare Earth Elements, Nauka, Moscow, 19673−11
[79] H. Rebbah, G. DesGardin, B. Raveau, Nonstoichiometric oxides with a layer structure: The compounds A1-x(Ti1- × M1+ ×)O5 J. Solid State Chem. 31 (1980) 321.
[80] S. Andersson, A.D. Wadsley, The crystal structure of Na2Ti3O7, Acta Cryst. 14 (1961) 1245−1249.
[81] S. Anderson, A.D. Wadleg, Crystal Chemistry of Titanates, The crystal structure of Na2Ti3O7 Acta. Cryst. 14 (1961) 1245−1249.
[82] M. Dion, Y. Piffard,, M. Toumoux, The tetratitanates M2Ti4O9 (M = Li, Na, K, Rb, Cs, Tl, Ag), J. Inorg. Nucl. Chem. 40 (1978) 917−918.
[83] G.S. Gopalakrishna, Synthesis and Characterization of NASICON analogues, The Hydrothermal growth of Lithium, Ph.D. Thesis, University of Mysore, Mysore, India, 1996, pp. 76−77.
[84] K. Byrappa, B. Sanjeeva Ravi Raj, V. Rajeev, A.B. Kulkarni, R.R. Clemente, Hydrothermal growth and characterization of $Na_2Ti_3O_7$, Indian J. Phys. 71A (1997) 131−142.
[85] K. Byrappa, Hydrothermal processing of advanced materials, Kirk−Othmer Encyclopedia of Chemical Technology, John Wiley, UK, 2005.
[86] M.L. Barsukova, V.A. Kuznetsov, A.N. Lobachev, T.N. Tanakina, Crystallization of bismuth titanates, Sov. Phys. Crystallogr. 17 (1973) 739−742.
[87] M.L. Barsukova, V.A. Kuznetsov, A.N. Lobachev, Yu.V. Shaldin, Hydrothermal crystallization and some properties of bismuth titanates, J. Cryst. Growth 13/14 (1972) 530−534.
[88] F. Muto, S. Taki, Hydrothermal synthesis of $K_2Li_2Ti_6O_{14}$ and its crystal structure, Bull. Chem. Soc. (1976) 77−82.
[89] M. Sato, T. Utsunomiya, Crystal growth and some properties of $R_2Ti_2O_7$, Bull. Tokyo Inst. Technol. 98 (1970) 25−35.
[90] B. Sanjeev Ravi Raj, The Hydrothermal growth of Lithium, Crystal growth, structure and properties of new superionic conductors, Ph.D. Thesis, University of Mysore, 1996, p. 146.
[91] E.D. Kolb, A.J. Caporaso, R.A. Laudise, The hydrothermal growth of lithium metagallate, J. Cryst. Growth 8 (1971) 354−358.
[92] D.J. Marshall, R.A. Laudise, The hydrothermal solubility of lithium metagallate in NaOH solutions, J. Cryst. Growth 1 (1967) 88−92.
[93] L. Cambi, M. Elli, Processi Idrotermali: Sintesi di Solfuri e Solfosali da Ossidi di Metalli e di Metalloidi, per Azione del Solfo e dell' Acqua, Nota 1 Chim. e Industr. (Ital.) 46 (1964) 1275−1279.
[94] L. Cambi, M. Elli, Processi Idrotermali Nota 9 Sintesi di Solfuri e Solfosali di Metalli Trivalent Chim. e Industr. (Ital.) 48 (1966) 944−951.
[95] L. Cambi, M. Elli, Processi Idrotermali Sintesi di Solfosali Derivati da Solfuro d'Argentd e Solfuri d'Anti-Monio, Nota 3 Chim. e Industr. (Ital.) 47 (1965) 282−290.
[96] V.A. Kuznetsov, E.P. Efremova, Hydrothermal synthesis of sulphides of divalent metals, Hydrothermal Synthesis and Growth of Monocrystals, Nauka, Moscow, 1982, pp. 78−119

[97] V.M. Latimer, Oxidation State of Elements and Their Potentials in Aqueous Solutions, Inostrannya Literature, Moscow, 1954. p. 400
[98] V.A. Kuznetsov, The physical chemistry of the hydrothermal crystal growth and the crystallization of some A_2B_6 crystals, in: K. Byrappa (Ed.), Hydrothermal Growth of Rare Crystals, 21, 1990Prog. Cryst. Growth Charact.
[99] R.A. Laudise, E.D. Kolb, I.P. Deneutoille, Am. Miner. 50 (3/4) (1965) 981.
[100] A.R. Kotel'nikov, V.A. Kuznetsov, A.L. Labachev, Stability and crystallization of Cds and Pbs in alkaline high-temp. solutions, Izvestiya Akad. Nauk USSR Inorg. Mater. 12 (1974) 827–829.
[101] V.A. Kuznetsov, D.V. Kostomarov, Krystallography 28 (1) (1983) 151.
[102] E.P. Efremova, V.A. Kuznetsov, Growth and Characterization of KDP and Its Analogs, Solubility and crystallization of cinnabar under hydrothermal conditions, in: Abstract of the First International Symposium on Hydrothermal Reactions, Ed. S. Somiya, Tokyo, 1982, p. 317.
[103] Y. Toudic, A. Regreny, M. Passaut, R. Aumont, J.F. Bayon, Nauverau type d'Appareillage Applique la Croissance Hydrothermale de Monocristaux de Cinabare de Grande Dimensions, J. Cryst. Growth 13/14 (1972) 519–523.
[104] L.V. Bryatov, I.P. Kuzmina, Crystallization of Pb and Zn sulphides from aqueous solutions of chloride salts, Cryst. Growth, 3, Nauka, Moscow, 1961, pp. 416–420
[105] N.Yu. Ikornikova, Hydrothermal Synthesis of Crystals in Chloride System, Nauka, Moscow, 1975. p. 227
[106] Yu.M. Mininzon, L.N. Demianets, A.N. Lobachev, On the solubility of synthetic sphalerite in aqueous H_3PO_4 solution, Kristallografia 19 (1974) 435–437.
[107] S. Horiuch, H. Wada, T. Moori, Morphology and imperfection of hydrothermally synthesized greigite (Fe_3S_4), J. Cryst. Growth 24/25 (1974) 624–626.
[108] I.P. Voronina, B.N. Litvin, Hydrothermal Synthesis of Inorganic Compounds, Bibliography, VINITI, Moscow, 1971. p. 188
[109] B.N. Litvin, V.I. Popolitov, Hydrothermal method of obtaining $A^{v}B^{vi}C^{vii}$ compounds, Izvestia Akad. Nauk USSR Inorg. Mater. 6 (1970) 575–576.
[110] V.I. Popolitov, Crystallization of BiSI, Kristallografia 15 (1970) 1271–1272.
[111] V.I. Popolitov, B.N. Litvin, Synthesis of monocrystals of ternary chalcohalogenides (AvBviCvii), Investigation of Crystallization Processes Under Hydrothermal Conditions, Nauka, Moscow, 1970, pp. 55–68
[112] V.I. Popolitov, A.N. Lobachev, Investigation of processes crystallization in the system Sb–Se–I–R–H_2O (R = Solution) under hydrothermal conditions, Izvestia Akad. Nauk USSR Inorg. Mater. 8 (1972) 1229–1232.
[113] A. Rabenau, H. Rau, Lead sulfide halides and Pb_4 $SeBr_4$, Z. Inorg. Allg. Chem. 369 (1969) 295–305.
[114] V.A. Kuznetsov, A.N. Labochev, Hydrothermal method for the growth of crystals, Sov. Phys. Crystallogr. 17 (1973) 775–804.
[115] B.N. Litvin, V.I. Popolitov, Hydrothermal Synthesis of Inorganic compounds, Nauka, Moscow, 1984.
[116] H. Rau, A. Rabenau, Hydrothermal synthesis von CuS and CuSe in 48%:ger HBr, Mater. Res. Bull. 2 (1967) 609–614.
[117] E.D. Kolb, R.A. Laudise, Hydrothermal crystallization of zinc selenide, J. Cryst. Growth 7 (1970) 199–202.
[118] G. Neumann, On the hydrolysis of niobates in 3 M KCl medium, Acta Chem. Scand. 18 (1964) 278–280.

[119] A.A. Shternberg, Ya.E. Lapsker, V.A. Kuznetsov, Hydrothermal synthesis of niobates and tantalates of potassium, Kristallografia 12 (1967) 957–958.

[120] D.J. Marshall, R.A. Laudise, The hydrothermal phase diagrams $K_2O–Nb_2O_5$ and $K_2O–Ta_2O_5$ and the growth of single crystals of K(Ta, Nb)O_3, J. Phys. Chem. Solids Supply 1 (1967) 557–561.

[121] H.P. Rooksby, E.A.D. White, The structures of 1:1 compounds of rare earth oxides with Niobia and Tantala, Acta Cryst. 16 (1963) 888–890.

[122] V.S. Stubican, High temperature transitions in rare earth niobates and tantalates, J. Am. Ceram. Soc. 47 (1964) 55–58.

[123] Yu. Titov, A.M. Sych, A.N. Sokolov, A.A. Kapshuk, V.P. Yashchuk, Polymorphism of $NdTaO_4$, Inorg. Mater. 33 (1997) 60–63.

[124] V.I. Popolitov, V.F. Peskin, A.N. Lobachev, Investigations of the conditions of crystallization of orthoniobate and orthotantalate of antimony, Crystal Growth from High Temperature Aqueous Solutions, Nauka, Moscow, 1977, pp. 217–227

[125] V.I. Popolitov, Growth of monocrystals of orthoniobate of antimony and their physical properties, Inorg. Mater. 26 (1990) 367–371.

[126] V.I. Popolitov, Growth of $BiNbO_4$ monocrystals and some properties, Inorg. Mater. 26 (1990) 336–360.

[127] V.I. Popolitov, A.N. Lobachev, V.F. Peskin, Antiferroelectrics, ferroelectrics, and pyroelectrics of a stibiotantalite structure,ferroelectrics. 40 (1982) 9–16.

[128] V.I. Popolitov, Some physico-chemical principles of crystallization under hydrothermal conditions, Hydrothermal Synthesis and Growth of Mono crystals, Nauka, Moscow, 1982. p. 145

[129] T.C. Chen, S.L. Wang, Synthesis and structural study of copper diarsenates, J. Solid State Chem. 121 (1996) 350–355 [11]

[130] M.J. Rojo, J.L. Mesa, J.L. Pizarro, L. Lezama, M.I. Arriortua, T. Rojo, Structural and spectroscopic study of the $(Mg, Ni)_2 \cdot (OH) \cdot (AsO_4)$ arsenates, J. Solid State Chem. 132 (1997) 107–112.

[131] H. Wang, W. Pang, Hydrothermal synthesis and characterization of $NaZr_2(AsO_4)_3$, J. Mater. Chem. 3 (1993) 989.

[132] H. Effenberger, R. Miletich, $Na_2[Bi_2Cu_3O_4 \cdot (AsO_4)_2] \cdot H_2O$ and $K_2[Bi_2Cu_3O_4(AsO_4)_2] \cdot 2H_2O$ two related crystal structures with topologically identical layers, Zeit. Kristallogr. 210 (1995) 421–426.

[133] J.H. Chou, M.G. Kanatzidis, Hydrothermal synthesis and characterization of (Me_4N) $[HgAsSe_3]$, (Et_4N) $[HgAsSe_3]$ selenoargsenates, J. Solid State Chem. 123 (1996) 115–122.

[134] R.F. Belt, J.B. Ings, Hydrothermal growth of potassium titanyl arsenate (KTA) in large autoclaves, J. Cryst. Growth 128 (1993) 956–962.

[135] W.T.A. Harrison, C.S. Liang, T.M. Nenoff, G.D. Stucky, Synthesis and crystal structure of $NaNb_2AsO_8$, J. Solid State Chem. 113 (1994) 367–372.

9 Hydrothermal Synthesis of Native Elements and Simple Oxides

9.1 Introduction

The hydrothermal technique is very useful for growing crystals of native elements and simple oxides. The native element crystals popularly grown under hydrothermal conditions are gold, silver, platinum, copper, arsenic, cobalt, nickel, tellurium, and bismuth-related compounds. Similarly, among the hydrothermally produced oxides, zincite, rutile, zirconia, and hafnia are very popular. In fact, the hydrothermal method is the only suitable method for the growth of these compounds for various reasons like: high melting point, very low solubility, acid resistance, and so on. In this chapter, we describe the hydrothermal synthesis of selected native element crystals.

9.2 Hydrothermal Synthesis of Native Elements

The hydrothermal technique is especially suitable for obtaining pure and high-quality crystals of several elements. The interest in the synthesis of these elements has rejuvenated, particularly after the discovery of noble metals in seafloor hydrothermal mineralization [1,2]. Gold, silver, iridium, palladium, platinum, etc., have been reported in hydrothermal mineralization from two mid-oceanic ridges: Juan de Fuca and mid-Atlantic at 26°N [2]. In fact, this has also led to the development of a new branch of science called geomicrobiology, and the biologists have begun to synthesize nanoparticles of several metals, particularly gold, silver, copper, and so on combining the hydrothermal solution medium and the bacterial activity based on the nature-inspired processes. The solubilities of noble metals in hydrothermal fluids are highly relevant to the redistribution and possible concentration of these metals by the action of water-rich fluids on magmatic occurrences.

The hydrothermal synthesis of noble metals takes place through two types of reactions:

$$2\mathrm{Me}(s) + 2(\mathrm{OH})^-(1) \rightarrow 2\mathrm{MeO}_x(1) + \mathrm{H}_2\uparrow \quad (9.1)$$

$$2\mathrm{Me(s)} + 2(\mathrm{HI})(1) \rightarrow 2\mathrm{MeL}(1) + \mathrm{H}_2\uparrow \quad (9.2)$$

where L is the acid anion (ligand).

Handbook of Hydrothermal Technology.

The first reaction is characteristic of alkali solutions, and the second reaction is characteristic of acids. In both the cases, hydrogen helps in establishing equilibrium. As a result of this, the desired metal crystallizes. If the temperature gradient is maintained in the autoclave, in one case, the crystallization of native metal crystals takes place in the hotter zone, and in the other case, in the cooler zone depending upon the hydrogen concentration (Exception handling (Eh)), of the system. Thus, the hydrothermal crystallization of metals is possible in both acid and alkali solutions. It is possible to obtain the entire series of metals as single crystals when an oxidizing agent like halogenide is used [3]. The crystallization takes place in quartz ampoules with external control of the CO_2 gas at temperatures 440–600°C and temperature gradients 30–120°C in aqueous solutions of HCl, HI, HBr and the crystals obtained are Au, Ag, Pt, Co, Ni, Te, and As. In this regard, it is appropriate to emphasize the work of Rabenau and Rau [4] on the hydrothermal synthesis of gold crystals. In aqueous solutions of HI at 500°C, the well-soluble $[AuI_4]^-$ complexes are formed, their stability depends upon the temperature, and it falls with destabilization of $[AuI_4]$ complexes and the formation of gold crystals:

$$Au + 4HI \rightarrow [AuI_4]^- + 2H_2\uparrow \tag{9.3}$$

The first attempt of gold crystallization from aqueous solutions was demonstrated by Bujor [5] under hydrothermal conditions at 390°C and 300 atm and cubic crystals of gold were obtained.

Honma et al. [6] have studied, in detail, the dissolution and precipitation of gold and silver at 150°C in various strong acid solutions and aqueous NaHS over periods of 60–120 days. The alloy $Au_{50}Ag_{50}$ was used as the starting material. For the first time, millimeter-sized free-growing enhedral crystals of pure gold were obtained from 6 N HCl solution after 120 days, and no precipitation of either gold or silver was obtained from the sulfuric acid and nitric acid solutions. Small crystals of both electrum ($Au_{71}Ag_{29}$) and argentite were precipitated onto the starting alloy sample after 120 days in 5.2 M aqueous NaHS [6].

Gold solubility has been studied extensively by several authors in HCl aqueous solution and others [7–12]. Seward and coworkers [12–14] have done excellent work on the complexing and solubility of gold and other noble metals under hydrothermal conditions, especially in the aqueous sulfide solutions at temperatures between 150°C and 500°C and pressure of 500–1500 bar over a wide range of pH. They have derived stability constants for the complexes AuHS and $Au(HS)_2^-$. It can be used to quantitatively model the transport of gold in natural hydrothermal fluids to understand the mechanisms and conditions of gold deposition. It has been shown that, in sulfide solutions at near neutral pH, the complex $Au(HS)_2^-$ is the dominant gold species. The solubility of gold reaches a maximum where pH = pK of H_2S. However, with increasing temperatures, the pK_1 of H_2S shifts toward more alkaline values [15] and, therefore, the maximum solubility due to this species also shifts to more alkaline pH. In acidic pH high-temperature solutions, the neutral species AuHS°, therefore, will be the dominant gold transporting species and, although the gold solubility may be 1–2 orders of magnitude lower than the solubility at pH > pK_1 (H_2S), this complex has to be taken into account when such systems are modeled. The main reason for the decreasing solubility of gold in high-temperature sulfide solutions is the change in pH and/or total reduced sulfur concentration. The effect of temperature on the solubility of gold in sulfide solutions is very strongly related to the total sulfur concentration and the hydrothermal fugacity.

Crystallization of copper, silver, selenium, and tellurium in alkaline solutions takes place in aqueous solutions of hydroxides of alkali metals at partial pressure of hydrogen, 0.5–40 atm, water pressure > 600 atm, and temperature 350–550°C. For obtaining fairly large crystals, the experiments are carried out with temperature gradients and in autoclaves with or without different liners. Glass carbon ampoules are most effective in this respect. In experiments at $T < 300°C$, Teflon liners are used. As nutrient, respective metal oxide powders or metal pieces are used.

Popolitov [16] and Popolitov and Litvin [17] have studied the systems Sb–Te–I(Br)–H_2O and obtained the crystals of Se and Te. They proposed a simple method of synthesis of Se and Te crystals in aqueous solutions of hydrochloric acid in autoclaves provided with quartz ampoules.

The octahedral or more complex form of single crystals of copper have been obtained in the cooler zone of the autoclave containing aqueous solutions of NaOH KOH. In highly concentrated aqueous solutions of hydroxides of alkali metals and at $P_{H_2O} > 50$ atm, often the platelike crystals form with (110) faces exhibiting highly vitreous luster, and the crystal size varies from 1 to 3 mm [18].

Silver crystallization conditions are very close to those of the crystallization of copper. Crystals are usually platelike and show twinning. The crystal surfaces exhibit metallic luster with a rise in temperature and silver dendrites form, resembling its appearance in nature. The size varies from 0.1 to 0.5 mm [19].

Selenium crystals have been obtained under hydrothermal conditions by Kolb and Laudise [20]. The crystals are needlelike, prismatic, with metallic luster, colored light gray. The size of the individual needles or crystals is up to 10–12 mm with a thickness of 0.2–0.3 mm.

Tellurium crystals have been obtained in aqueous NH_4Cl solutions (8–15 wt%) at temperatures of 450–550°C, the partial pressure of hydrogen was ~80 atm, and P_{H_2O} 600 atm with a rise in the concentration of the solvent (up to 20 wt%). The skeleton-like forms appear in Ref. [21].

In recent years, the synthesis and growth of bulk crystals of native elements under hydrothermal or solvothermal conditions does not carry a great significance, and the present trend is the synthesis of nanoparticles of several elements for specific applications.

Diamond is the most precious of the gemstone. The word "Diamond" originally comes from the Greek word "Adams" which means "Unconquerable." Diamonds were formed at greater depths of 150–300 km below the surface of the earth's crust under enormous pressure and very high temperature, millions and millions of years ago. Diamond synthesis has become a very important activity. Several research organizations and corporations have taken up diamond synthesis as a challenge in solid-state science. Man started using diamond several centuries BC. At that time, diamond was thought of as a hard glass. When cutting technology was discovered in the seventeenth century, man began to use diamond as an aristocratic jewel. Today, an estimated one billion US dollars of diamonds are bought annually via the Internet.

Industrial application of diamonds started in the beginning of the twentieth century and the first application of diamond was in a grinding wheel. The wheel consisted of diamond powder with metal (or polymer) binder, which can be called a refined material.

In 1955, diamond was produced artificially and the success was closely related to the discovery of high-pressure technology by General Electric. However, several thousand crystals were required to produce one carat. Thus, these small crystals were found to be more practical for grinding purpose than the crushed bort that had been used for years before. But in 1970, General Electric announced the production of jewelry grade synthetic diamond in sizes up to slightly over one carat in the rough. This technology was later applied to the production of sintered diamond, which can be categorized as a refined artificial material. This technology was also applied to the production of cubic boron nitride, which is a completely artificial material. The greatest discovery related to diamond was its synthesis from the gas phase for the first time by Deriaguin et al. [22] in 1976 and Matsumoto et al. [23] in 1982. Although many methods have been reported for diamond synthesis from the gas phase, all methods are basically modified Chemical Vapor Deposition (CVD) methods. Diamond made by these methods, therefore, is referred to as CVD diamond. Some three decades ago, no one could have conceived that diamond would become an electronic material. People considered a diamond mainly as a jewel, but the discovery of diamond synthesis from the gas phase changed that image drastically.

The authors believe that the unique properties of diamond contributed to the enhancement of research on the material, resulting in a wide range of research with hope for new applications.

There are three major processes of diamond synthesis: (i) high-pressure–high-temperature (HPHT), (ii) CVD (which are at least 50 and 40 years old, respectively), and (iii) low-pressure solid-state source (LPSSS). Bundy et al. [24] at General Electric Co. announced their success in synthesizing diamond under HPHT conditions [25]. A few tons of industrial diamond abrasive grains are made this way each year in various factories around the world. De Vries [26] reviewed CVD diamond synthesis in detail. Roy and coworkers [27,28] have worked out the growth technology for diamond using LPSSS method in detail. Figure 9.1 shows the schematic diagram comparing three popular methods of diamond synthesis [28].

It is interesting to note that the nucleation and growth kinetics had won out over thermodynamics. The CVD process represents the field outside the PT stability region for diamond. Figure 9.2 shows the phase diagram for carbon crystalline phases. The equilibrium data for material trapped in inclusions in diamonds indicates final temperatures of 600–800°C and pressures up to 10 Mpa [29–31].

Several researchers have argued a case for the possibility of growing diamonds under hydrothermal conditions [31,32]. Szymanski et al. [29] have carried out diamond synthesis using relatively low-pressure liquid phase epitaxy on seeds, in an

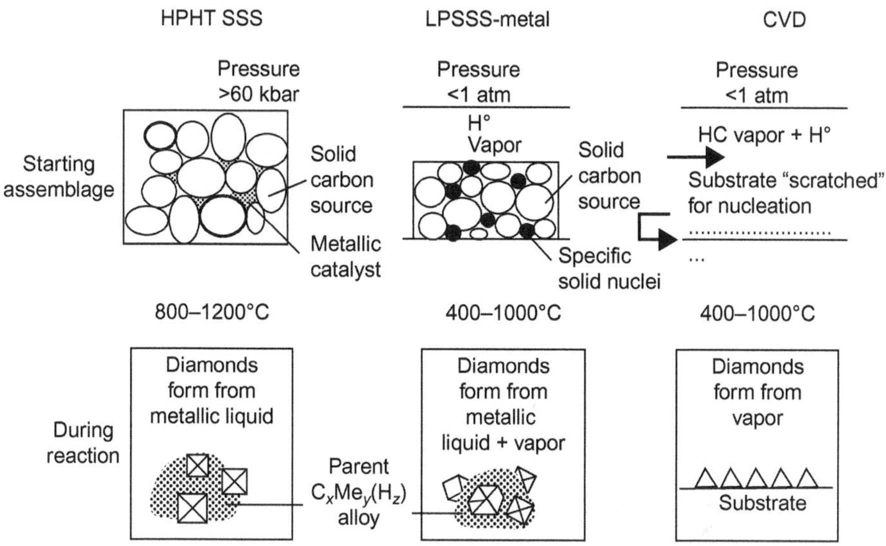

Figure 9.1 Schematic diagram comparing three popular methods of diamond synthesis processes [28].

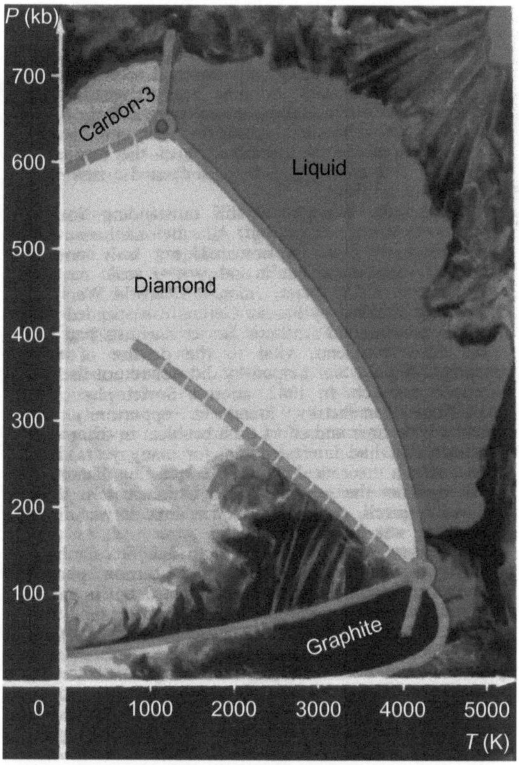

Figure 9.2 Phase diagram for the carbon crystalline phases.

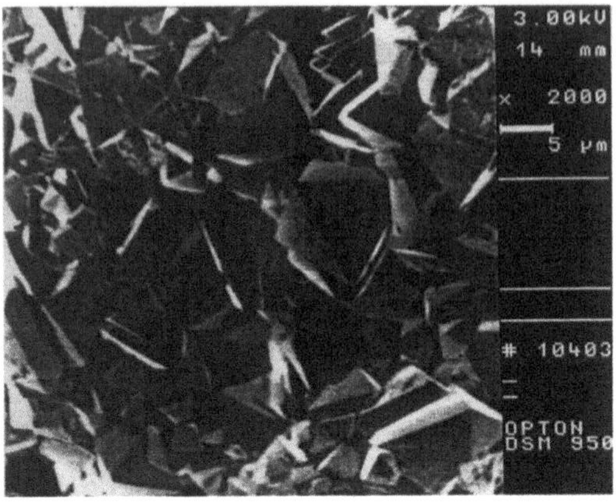

Figure 9.3 Polycrystalline diamond [29].

autoclave filled with a specially prepared water solution. Natural diamonds of 0.22 and 0.49 g were used as seeds. The solution and seeds were subjected to 170 MPa and 400°C for 21 days. A thin irregular (15–14 μm) colorless film of polycrystalline diamond on the natural (111) faces of the seed crystals was obtained (Figure 9.3). This work shows that hydrothermal synthesis may be an effective way for homo- and heteroepitaxy of diamond as a continuous layer on seed material surfaces of diversified shape and possibly for the growth of large diamonds on seeds.

Recently, Yamaoka and Akaishi [33] reported the growth of diamond under HPHT water conditions. The basic principle used by these authors is that carbon reacts with high-temperature water (vapor) producing so called "water gas." It is natural to expect that graphite reacts with water and precipitates as diamond under HPHT conditions in the thermodynamically stable region of diamond. The authors have used three types of capsules to carry out the diamond synthesis under HPHT conditions: (i) Ta cylindrical containers, (ii) Mo double capsules, and (iii) conventional point-sealed capsules. The selection of these capsules depends upon the duration of the experiments and the pressure–temperature conditions. Figure 9.4 shows the diamond growth under HPHT conditions. This work shows that graphite reacts with HPHT water and precipitates as diamond in the regions of 1300–2200°C and 5.5–7.7 GPa. The reaction proceeded more rapidly with increasing pressure and temperature.

Suito and Onodera [34] have synthesized diamond using phenolic resins between 1.6 and 4 GPa and cobalt at 10 GPa with calcium carbonate as catalyst-solvents. In this case, the pressure and the temperature conditions of diamond synthesis depended upon the prefiring temperatures. Well-defined single crystals of 0.3–0.7 mm in size were obtained at 4 GPa and 1500°C (Figure 9.5).

However, the diamond synthesis in the presence of metal powder, molten high-molar alkali solution, and water at 800°C and 1.4 kbar pressure range is the most popular technique. The growth rate is quite low. The crystals obtained are usually 5–10 μm in size. Several types of starting materials have been tried for the

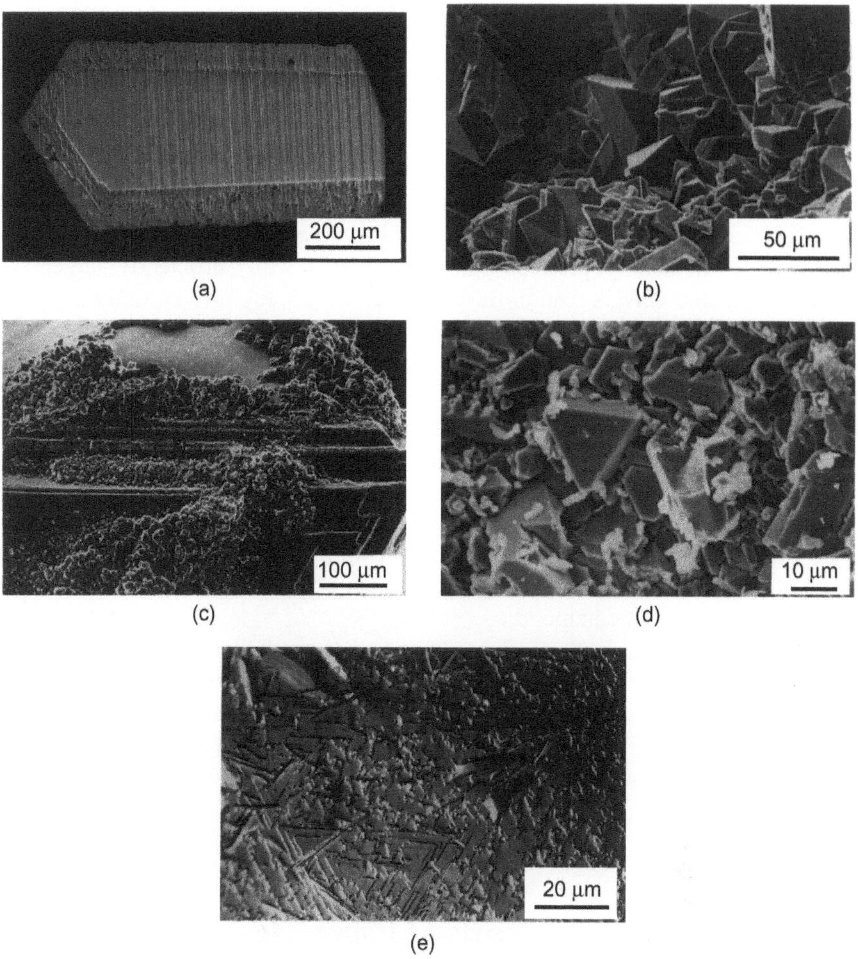

Figure 9.4 Diamond growth under HPHT conditions [33].

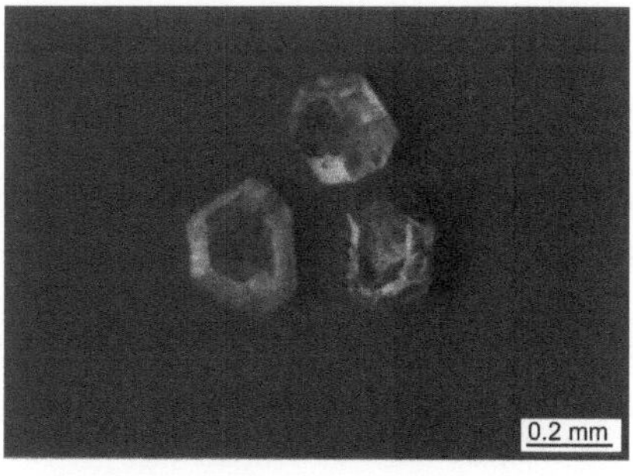

Figure 9.5 Diamond single crystals [34].

synthesis of diamond. The most important aspects are the behavior of carbon under HPHT conditions, the presence of a matching substrate, then preferential oxidation of sp-bonded C by water. Thermodynamically diamond is stable at temperatures <800°C. Graphitization of the diamond surface should be prevented. The most important thing is to bring carbon into the high-energy state under hydrothermal conditions and arrest it.

The growth of diamond crystals is still a developing area of research, and an extensive research is being carried out on the synthesis of diamonds for a highly cost-effective method to produce diamond crystals with sufficiently larger size without noticeable defects. Dunin-Barkovsky et al. [35,36] have carried out hydrothermal growth of diamond at seedings at atmospheric pressure and temperatures below 100°C and later carried out low-temperature growth of diamond seedings in acid solutions of nitric, sulfuric and acetic acids with graphite as the source of carbon. These authors also proposed an oxidative dissolution of graphite under hydrothermal conditions with the formation of various intermediate carbon compounds, and subsequent transfer of carbon into the growth zone and the growth of diamond on the seedings. There are several other reports on the diamond synthesis in the wet systems, but the success is limited towards the growth of bigger diamond crystals of several millimeters in size, and much of the work is on the nanoforms of carbon including nanodiamond.

On the whole, the growth of diamond crystals is a challenging task because of the lack of knowledge of the pH of the growth media, halogen problem, very slow growth rates, lack of solubility data, the solvent chemistry, and so on. Therefore, the current situation in diamond synthesis can be compared to that of quartz growth some one hundred years ago. By 2020 the future of diamond research will become clear.

9.3 Hydrothermal Synthesis of Hydroxides

The hydrothermal method is popularly used in the synthesis of single crystals of hydroxides of several divalent and trivalent metals. Usually, the field of stability is expanded by the presence of surplus water during the transition reaction oxide $\rightleftharpoons$ hydroxide and temperature of the system. In this regard, the alkali metal hydroxide solutions are more effective. Table 9.1 gives the list of the metal hydroxides obtained by hydrothermal methods and their experimental conditions [37]. The experimental temperatures and pressures vary from 100 to 600°C and 6 to 4000 atm.

Hydrothermal synthesis of $Mn(OH)_2$, $Cd(OH)_2$, $Co(OH)_2$, and $Ni(OH)_2$ crystals depends upon the partial pressure of hydrogen ($P_{H_2O} > 10$ atm). Some of the hydroxide crystals like $Mg(OH)_2$, $Mn(OH)_2$, and $Co(OH)_2$ form fairly large-size crystals of $5-10 \times 5-10$ mm size with a thickness of 0.1 mm. The other hydroxide crystals are small (0.1 mm). The double-distilled water is usually used in the synthesis of hydroxides. The temperature gradient method gives single crystals of

Table 9.1 Hydrothermal Experimental Conditions of Synthesis of Hydroxides [37]

Hydroxide	Solvent	*T* (°C)	*P* (atm)
$Be(OH)_2$	H_2O	100–175	4000
$Mg(OH)_2$	3% NaOH	450	600
$Ca(OH)_2$	3% NaOH	550	650
$Cd(OH)_2$	35% NaOH	450	1000
$Mn(OH)_2$	5% NaOH	450	1000
$Co(OH)_2$	1% NaOH	280	3500
$Ni(OH)_2$	1% NaOH	280	3500
$TR(OH_3)$	15% NaOH	350	600
TR = La–Gd			
TR = Dy, Er, Yb	15% NaOH	450	600
$Cr(OH)_3$	3% NaOH	300	1500
$In(OH)_3$	40% NaOH	400	400
$Al(OH)_3$	H_2O	<150	39
$In(OH)_3$	H_2O	185–275	50–75
$In(OD)_3$	D_2O	180–210	18–50
InOOH	H_2O	325–410	120–800
InOOD	D_2O	360–380	320–400
α-ScOOH	H_2O	162–350	6–170
β-ScOOH	H_2O	350	160
$Y(OH)_3$	H_2O	300	85
CrOOH	H_2O	450	2700
GdOOD	D_2O	600	1400
SmOOH	H_2O	600	1400
$Sr_3Cr_2(OH)_{12}$	H_2O	150–200	

hydroxides. The oxyhydroxides (MeOOH) often crystallize at relatively higher temperatures. Among these, the polymorphism is very common, and it depends upon the temperature, for example, in rare earth hydroxides [38].

Similar to the alkali solutions in the synthesis of metal hydroxides, acid solutions, particularly chloride solutions, are also used. Thus, in this case, an increase in the concentration of the solution forms hydroxyl chlorides, often with 100% output. Their synthesis takes place as a result of the substitution of one hydroxyl group with chloride as follows:

$$Me(OH)_3 + HCl \rightarrow Me(OH)_2Cl + H_2O \tag{9.4}$$

In this way, Demianets and Emelyanova [39] have obtained hydroxyl chlorides of neodymium, $Nd(OH)_2Cl$, under hydrothermal conditions using aqueous solutions of NH_4Cl (concentration 7 wt%) at temperatures 450–480°C. At 500°C, rare earth hydroxyl chlorides, $R(OH)_2Cl$, where R = La, Ce, Pr, Nd, Sm, and Gd, have been obtained by Kletsov et al. [40]. Similarly, Carter and Levinson [41] have obtained hydroxyl chlorides containing La, Nd, Sm, Gd, and Y, and the more complex hydroxyl chloride of ytterbium, $Yb_3O(OH)_5Cl_2$, using acidic chloride solutions at 1400 atm pressure and 550°C. Under hydrothermal conditions, the chloride group can be replaced by carbonate ions with the formation of hydroxyl carbonate, R $(OH)CO_2$ [42]. Similarly, Christensen and Hazeil [43] have obtained mixed hydroxides of type $MeSn(OH)_6$, where Me = Ca, Mn, Fe, Co, and Zn, through the interaction of aqueous solutions of $NaSnO_3$ with chlorides or nitrates of the respective metals. Wolski et al. [44] have obtained $Mn_xFe_{2-2x}(OH)_{6-4x}$ which finds numerous applications in technology. For certain x values, the initially amorphous mixtures, $Mn_xFe_{2-2x}(OH)_{6-4x}$, are converted into ferrimagnetic species when stored as water suspensions well below 100°C, and the higher the temperature, the shorter the time of transformation. Aqueous solutions of titrated 0.25 M Fe (III) nitrates, mixed together in the desired proportions to give the hydroxide mixtures $Mn_xFe_{2-2x}(OH)_{6-4x}$ in 0.1 × intervals, were treated by constant stirring with 2 M NaOH till pH 9.5 was reached to ensure full Mn^{2+} precipitation and at the same time to avoid oxidation to Mn_3O_4 and β-MgOOH at higher pH. The washed and filtered hydroxides were placed in Teflon vessels and were stored in the form of water suspensions in autoclaves at 100°C for 24 h.

Thus, the hydrothermal technique provides unique opportunities for a preparative chemist to obtain a large variety of hydroxides, hydroxyl carbonates, hydroxyl chlorides, etc., of various metals. Also hydrothermal and solvothermal methods help in the structural stabilization of a variety of hydroxides, which cannot be stabilized through other preparative chemistry routes.

9.4 Hydrothermal Synthesis of Selected Oxides

Oxides form one of the largest groups of inorganic compounds both in nature and in the laboratory. The hydrothermal technique is one of the most popular techniques for growing these high-temperature oxides, complex oxides, and low-temperature modifications of the oxides. The work on the hydrothermal synthesis of oxides began during the nineteenth century with quartz in 1845 [45]. Followed by this, large-scale research activity began during the 1890s on corundum and other related high-temperature oxides [46,47]. Today, we find a large number of oxides obtained by the hydrothermal method: TiO_2, ZrO_2, HfO_2, Cu_2O, BeO, Bi_2O_3, In_2O_3, Al_2O_3, ZnO, and Fe_2O_3, to mention a few, and a great variety of mixed oxides. Some of these carry great significance as synthetic gemstones, and some as technological materials both in nanosize and bulk size. Here, we describe the hydrothermal growth of a few selected oxides of both gemstone and technological grades.

9.4.1 Cu_2O (Cuprite)

The growth of cuprite monocrystals under hydrothermal conditions can be carried out using LiOH, NaOH, and KOH in aqueous solutions of alkali ammonium halogenides. Kuzmina and Kaidukov (1977) have worked out a detailed technology of hydrothermal synthesis of cuprite, and the solubility of cuprite can be written as follows [48]:

$$Cu_2O + (2a - 2)OH^- \rightarrow 2CuOa(2a - 2)^- + (a - 1)H_2O \quad (9.5)$$

The solubility of hemioxide of cuprite under HPHT conditions is less stable with respect to supersaturation, and with a slight variation in the temperature gradient, spontaneous crystallization of cuprite takes place such that the growth of cuprite crystals become difficult.

Cuprite crystal growth is carried in autoclaves with copper or silver linings. The growth temperature is 250–450°C, $\Delta T = 25-75$°C, and % fill 70–80%. For chloride solutions, Teflon liners have been used and the temperature was maintained at $T = 300$°C and $\Delta T = 20-45$°C, and the pH varied from 5.0 to 5.5. The crystals obtained are 4–5 mm in size with ruby-red color having octahedral or often dodecahedral cover. In recent years, much focus is on the synthesis of nanosize crystals of cuprite for a variety of applications.

9.4.2 BeO (Bromelite)

Extremely high toxicity of beryllium makes the crystal growth of beryllium-bearing compounds very difficult. Therefore, the hydrothermal method becomes most useful in this regard. The first attempt to grow bromelite was made by Hartmann [49]. Seeded growth of bromelite was carried out using a 5 mm BeO crystal oriented along $[\bar{1}0\bar{1}0]$ and $[\bar{1}0\bar{1}1]$. The average growth rate was ~0.07 mm/day. The crystallization was carried out at 500–530°C using NaOH solutions.

The autoclaves are usually provided with copper or silver lining, though there are several other lining materials. The experimental conditions are as follows:

Mineralizer	–	25 wt% NaOH
Nutrient	–	Beryllium oxide
% Fill	–	80%
Crystallization temperature	–	450°C
ΔT	–	20–50°C

Under the above conditions, colorless hexagonal crystals of size 3–4 mm with the basic pinacoidal, pyramidal, and prismatic forms are formed. The work on bromelite is mostly of academic interest today.

9.4.3 Zinc Oxide

Zinc oxide is well known as *n*-type wide band gap semiconductor ($\Delta E = 3.37$ eV at 300 K) with a large exciton energy of 60 meV and thermal energy of 27 meV, and is a subject of research owing to its unique mechanical, electrical, and optical properties with a combination of high stability, very high melting point with valuable device potential for piezoelectric transducers, gas sensors, optical waveguide, transparent conductive films, varistors, solar cell windows, bulk acoustic wave devices, and so on. Due to bright UV luminescence, ZnO is a perspective material for the manufacture of UV-light-emitting diodes, UV lasers operating at room temperature, and display devices. Moreover, ZnO quantum dots with very low toxicity, high photostability, biofriendly, and biodegradable have been demonstrated [50–52]. Zincite crystals may be grown by different methods: From gas phase (it produces needles of ZnO), from high-temperature solution using PbF_2 (it produces platelike crystals), and hydrothermally. It is interesting to note that the properties of crystals grown by the first two methods are technologically unsatisfactory owing to the imperfections in the crystal, and to unsuitable habits. Therefore, the most promising method of producing isometric zinc oxide crystals of good quality is the hydrothermal method [53]. Hence, the process of crystallization represents an important technique for many applications. In this respect, understanding, controlling, and optimizing crystal morphology is of fundamental importance in those fields where an incorrectly defined crystal morphology may have an impact upon a number of important technological applications. Therefore, the hydrothermal technique is found to be the most suitable one for the growth of ZnO crystals. Seeded crystal growth of zincite has been carried out using NaOH and KOH solutions and, similarly, aqueous solutions of chlorides [54–58]. The most useful solvent is 2–10 M KOH, in which the growth takes place within a wide temperature interval 200–500°C and $P = 150–600$ atm. During 1970's, Lobachev and his group have studied the solubility of zincite in detail [53,54]. The solubility is less dependent upon the temperature when the concentration of the solution is <3 M, but for higher concentrations, it is distinct. Figure 9.6 shows the solubility of zincite in KOH solution. With a raise in the concentration of KOH, the solubility of zincite in KOH solution increases nonlinearly, but with a raise in the concentration of KOH at lower temperature (<200°C), it becomes linear.

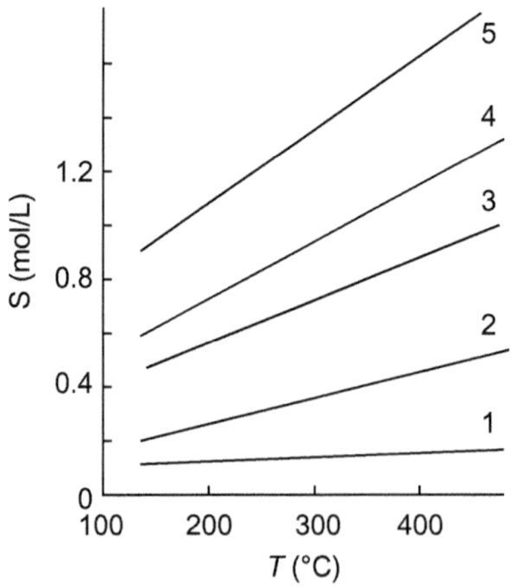

Figure 9.6 Solubility of zincite in KOH solution [54].

In the growth of zincite, the use of alkalis sometimes leads to the formation of hydrogen arising as a result of the oxidation of the autoclave material. The amount of hydrogen formed depends greatly on the temperature, the concentration of the alkali, and the duration of the experiment. In fact, the dissolution of the crystals, the transport processes, and crystallization in turn depend greatly on the redox potential of the medium. Sakagami [59] has obtained zincite crystals of high purity by the hydrothermal method under a partial pressure of oxygen, using platinum-lined autoclaves and ultrapure reagents. The growth conditions are as follows:

Growth temperature	–	370–400°C
Temperature difference	–	10–15°C
Total pressure	–	700–1000 kg/cm
Partial pressure	–	10–30 kg/cm
Solvent	–	KOH 3.0 M + LiOH 1.5 M
Oxidizer	–	H_2O_2 0.1–0.3 M
Nutrient	–	ZnO sinter
Lining tube	–	Point (0.3 mm thickness)
Growth run	–	15–20 days/run

Figure 9.7 shows high-purity crystals of zincite obtained at a growth rate of 0.2 mm/day in a direction normal to a prismatic face ($10\bar{1}0$). The degree of coloration depends upon the growth conditions, e.g., solutions containing a H_2O_2 oxidizer and without an oxidizer. The author has used the coulometric technique to detect a small deviation in the stoichiometric constituent toward $Zn_{1+x}O$. The concentration of excess Zn in the crystals grown under excess partial pressures of oxygen decreased to the minimum of 0.8 ppm, and their resistivity increased to about 10^8 Ωm. Thus, obtaining the composition of the solution in which the zincite crystallizes and changes the habit and properties of the developing crystals. In KOH solutions, ZnO crystals or prismatic habits, drawn out sharply along the *c*-axis, are formed in caustic soda solutions. The crystals are more isometric in LiOH solutions. The growth rate of the monohedral face is very slow, while that of the pyramid face is commensurable with or slightly exceeds that of the prism, and the crystals assume a tabular form. The presence of F^+ ions in the solution increases the growth rate of the monohedral (0001) face. If the solution contains iron, characteristic zincite twins grow along the face of the monohedron. In aqueous

Figure 9.7 High-purity crystals of zincite obtained at a growth rate of 0.2 mm/day [59].

ammonium chloride solutions (1–7 wt%), recrystallization of zincite takes place at 450°C. The crystal habit varies with the concentration of the solution [60,61]. In highly concentrated solutions, hexagonal platelets are formed. Hence, in order to obtain good crystals of zincite, NaOH and KOH solutions are used. The resistance of zincite crystals grown in alkali solutions containing lithium ions is small: 10^5–10^8 Ωcm. Annealing in air or oxygen increases the resistance of the samples to 10^5–10^8 Ωcm. This is due to the diffusion of the Li ions, and to the motion of the hyper-stoichiometric proportion of zinc to the surface of the crystal faces, with its subsequent oxidation to zinc oxide.

Zincite crystals obtained under hydrothermal conditions are normally activated by Mn^{2+} and Ni^{2+}, which emit red and green radiation. However, the resistance of the crystals remains unchanged. Similarly, the Mn^{2+} and Ni^{2+} ions also have no appreciable effect on the kinetics of the crystallization of zincite.

Wang et al. [62] have carried out a systematic investigation of zincite crystals and their morphology through a crystal chemical approach to obtain a desirable morphology. Also, they have explained the effects of the additive OH^- on the crystal morphology of ZnO crystallites. ZnO microcrystallites are very important, having wide application in ceramics, coatings, and electronic devices. Thus, hydrothermal zincite crystals have great device potential.

Ehrentraut et al. [50] have successfully grown large-size single crystals of ZnO using the hydrothermal method. These authors used the crystal growth mechanism of ZnO proposed by earlier workers [63–65]. Almost 100 specimens of size 2 in. have been grown on mainly (0001)- and (1010)-oriented seeds. All crystals developed excellent facets and the average crystal thickness was about 1 cm in the ⟨0001⟩ direction (Figure 9.8). A ZnO crystal of 3 in. diameter and a wafer processed from it are shown in Figure 9.9a and b, respectively. These are the largest ZnO single crystals grown by the hydrothermal method.

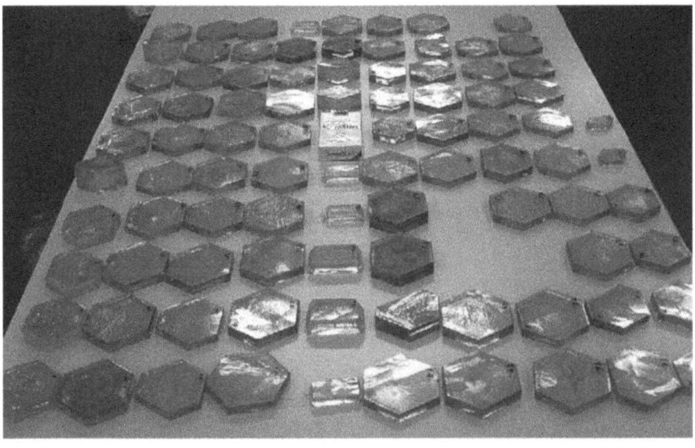

Figure 9.8 ZnO crystals of size 2 in. produced during one growth run.
Source: Photograph courtesy of Dirk Ehrentraut [65a], Tohoku Univ. Japan.

9.4.4 Hydrothermal Growth of Corundum

The hydrothermal synthesis of corundum began during the 1890s. Corundum is stable under hydrothermal conditions at $T > 400°C$ [66,67] and is soluble in alkali and carbonate solutions [67a]. Corundum shows a lower temperature coefficient of solubility in alkaline solutions and that is the reason for the unsuccessful growth of corundum in these solutions [68]. Although, hydrothermally grown corundum is commercially available at present, as gemstones it has not been reported. The hydrothermal growth of corundum is achieved using a metastable-phase technique because of its very low solubility. Gibbsite ($Al(OH)_3$) is used as a nutrient, α-Al_2O_3 as a seed, and an alkaline solution as a solvent. A more rapid growth rate is obtained in KOH or K_2CO_3 solution than in NaOH or Na_2CO_3 solution. The solubility is lower in KOH or K_2CO_3, but the interaction between the impurity absorbed beforehand onto the crystal surface and K^+ ion existing in the solution acts as a factor to augment the growth rate. The lower the pH level in the solution, the thicker the growth layer along the *c*-axis. The growth of corundum in extremely low pH 6 N HCl has been reported. In this case, the nutrient is dissolved at the lower temperature side and is deposited on the seed crystal at the higher temperature side, because the enthalpy in the α-Al_2O_3–HCl reaction is negative. Isothermal studies of the system Al_2O_3–H_2O have been carried out by many workers and have shown that corundum could be formed from diaspore, boehmite, or gibbsite in situ at temperatures >450°C. The pressure varied from 15,000 to 30,000 psi. Figure 9.10 shows the phase relations in Al_2O_3 systems [68]. Kashkurov et al. [69] have studied the growth of large corundum crystals at pressures up to 2000 atm and at temperatures up to 550°C in alkali solutions of various concentrations. Crystals weighing up to 1 kg were prepared by the hydrothermal technique, and the imperfect state of these crystals was apparently associated with internal stresses and the mosaic structure of seed crystals which were prepared by

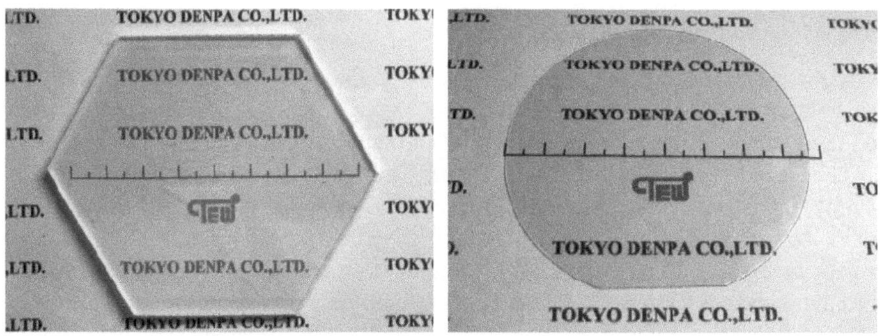

(a) (b)

Figure 9.9 View down *c*-axis of (a) 3 in. ZnO crystal and (b) 3 in. (0001) ZnO wafer (Cytidine 5-monophosphate). Scale bar is 80 mm.
Source: Photograph courtesy of Dirk Ehrentraut [65a].

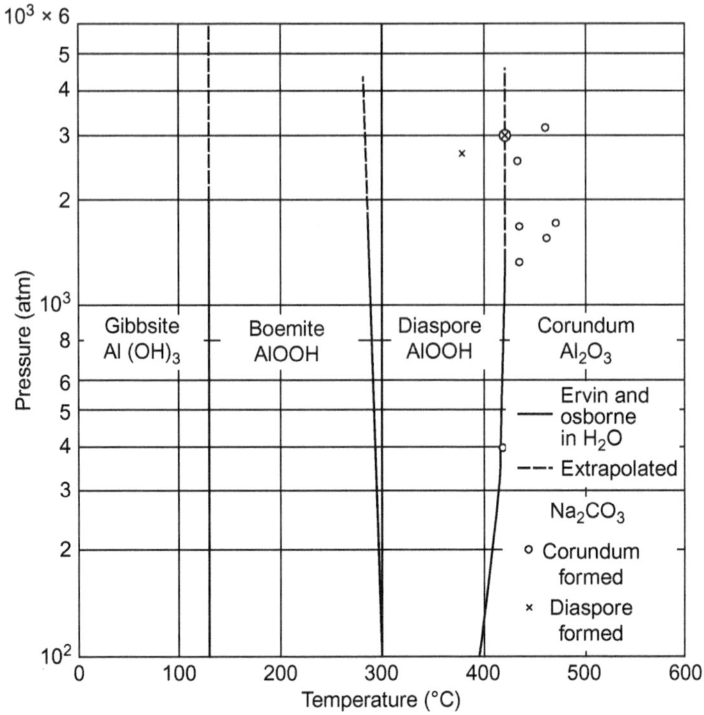

Figure 9.10 Phase relations in Al_2O_3 systems [68].

Verneuil technique. Rumyanetsev et al. [70] have shown that among alkali and carbonate solutions, the alkali solutions dissolve corundum and form aluminates.

$$Al_2O_3 + 2NaOH + 3H_2O \rightarrow 2Na[Al(OH)_2] + 4OH^- \tag{9.6}$$

In the absence of carbonate ions, CO_3^-, the aluminate ions are always stable and the reaction becomes irreversible and, at the same time, with the use of carbonate solutions, the reaction becomes reversible. In fact, this is a characteristic feature of lower concentrations of sodium bicarbonate. At higher concentrations of solutions, usually, the heavier phases appear. In aqueous solutions of K_2CO_3 and Pb_2CO_3, the heavy phase does not appear; therefore, the experimenters emphasize the corundum growth in such solutions. Figure 9.11 shows the dependence of the solubility of corundum on temperature; Figure 9.12 shows the dependence of the growth rate of different faces on the temperature [70]. With a raise in temperature, the growth rate nonlinearly increases within the temperature interval 400–500°C. Good-quality corundum crystals were grown on the seeds oriented along the face (122) and were obtained from the melt. The growth was carried out in a 4 M K_2CO_3 solution at 480°C, pressure of 1330 atm, 85% fill, $\Delta T = 30$°C, and colorless and transparent

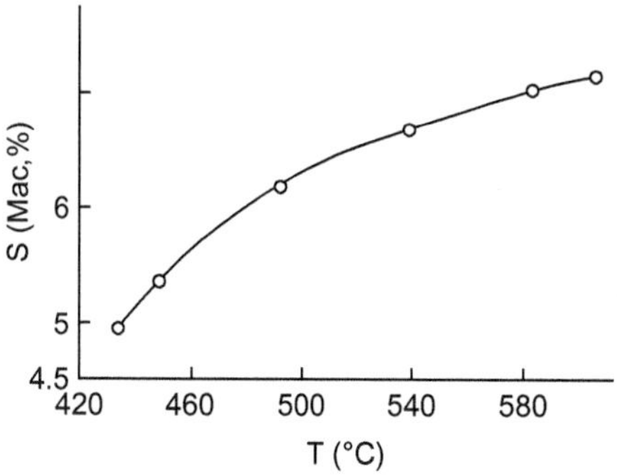

Figure 9.11 Dependence of solubility of corundum [70].

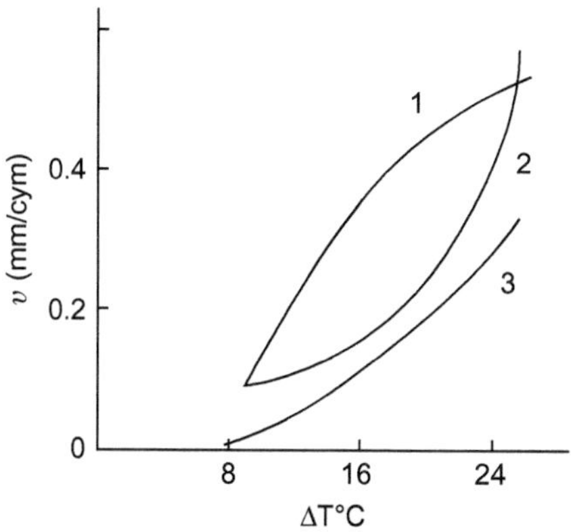

Figure 9.12 Dependence of the growth rate of different faces and the temperature [70].

crystals with (211), (311), and (111) faces were obtained. Thomas [71] has studied the influence of KOH/CO_2 ratio in alkaline-bicarbonate solutions on the growth rate of corundum crystals under hydrothermal conditions. It has been shown that the concentration of CO_2 in the system has a main influence on the growth rate of corundum crystals; as for the KOH content in the bicarbonate solutions, it influences growth velocity with only a small concentration of the alkaline component. A mechanism for aluminum transfer with the growth of corundum crystals in the carbonate and bicarbonate solutions has been discussed. The corundum of hydrothermal origin is used both as gemstone and technological material.

The hydrothermal technique is useful for obtaining large crystals of sapphire and ruby. The corundum modifications of Al_2O_3, which are relatively strain free, can be prepared at moderate temperature conditions and without the excessive gradients characteristic of flame fusion and melt techniques. These are the gemstone varieties of corundum and named according to their colors: sapphire blue and ruby red. Probably, sapphire is the second material after quartz to be grown in any size by hydrothermal methods. Similarly, large ruby crystals can be grown using carbonate solutions. The solubility of sapphire and ruby is the same as that of corundum and it increases with temperature. Oxides of chromium and iron, when taken in the nutrient along with Na_2CO_3 solution, more strongly influence the crystallization than when taken with K_2CO_3 solution. However, when the concentration of these components in the nutrient is higher than 1.6% in 10% Na_2CO_3 solution, crystals practically do not grow. Kuznetsov and Shternberg [72] have investigated the crystallization of ruby under hydrothermal conditions by three methods: (1) Recrystallization of aluminum oxide in solutions containing chromium; (2) Joint recrystallization of Al_2O_3 and Cr_2O_3, which are placed in the autoclave as separate components; and (3) finally, recrystallization of ruby crystals having a given chromic oxide concentration. The last method makes possible the uniform and controlled introduction of the impurity into the solution and the growing crystal throughout the experiment. Figure 9.13 shows the growth rate of the $(10\bar{1}1)$ face versus the Cr_2O_3 concentration in the initial charge; 10% Na_2CO_3 solution at 550°C, autoclave 60% filled. These authors conclude that in carbonate and bicarbonate solutions, chromic oxide and aluminum oxide have substantially different solubilities in carbonate solutions and different rates of dissolution. Also, the effect of solution redox potential on the transport of chromic oxide under hydrothermal conditions has been studied by these authors. Laudise and Ballman [73] and Monchamp et al. [74] have carried out very large-scale growth of sapphire and ruby with considerable success. Figure 9.14 shows hydrothermally grown ruby crystal [75]. Green crystals containing iron have been

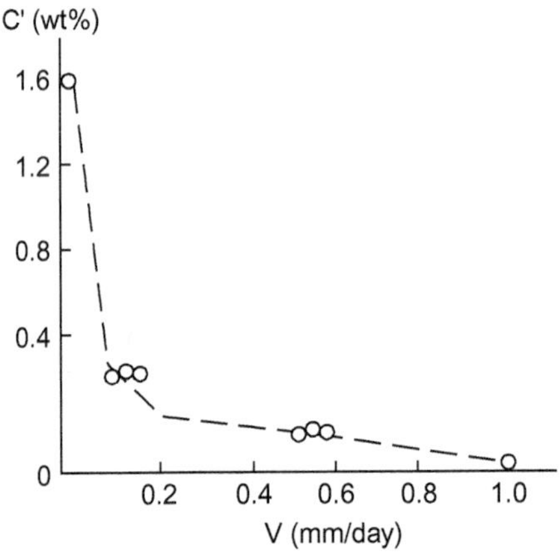

Figure 9.13 Growth rate of $(10\bar{1}1)$ face versus Cr_2O_3 concentration in the initial charge [72].

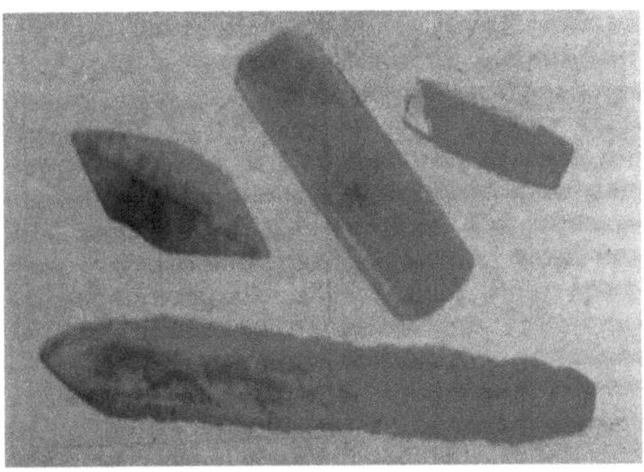

Figure 9.14 Hydrothermally grown ruby crystals.
Source: Courtesy of Balitskii [75].

grown by carrying out the crystallization in the welded liners directly, and colorless crystals which were essentially pure have been prepared by the use of a silver tube [74]. The overall features are similar to the growth of corundum crystals. Recently, there have been several reports on the growth of crystals of ruby, sapphire and, α-alumina for various applications. Transition metal (chromium and titanium) doped α-alumina crystals have attracted attention for many years not only as gemstones but also due to their application as high power laser host media [76]. Bliss et al. [77] have grown Ti^{3+}:Al_2O_3 (titanium-doped sapphire) single crystals using solvothermal method for laser applications. The growth experiments were carried out with $T = 430-550°C$ and $P = 15-25$ kpsi. Several workers have carried out the growth of α-alumina under milder pressure–temperature conditions using organic solvents like 1.4-butendiol and other glycol-based solvents [78–81].

9.4.5 Hydrothermal Growth of Oxides of Ti, Zr, and Hf

The oxides of titanium, zirconium, and hafnium belong to a group of chemically stable compounds with high melting points. These properties make them very important technological materials, but at the same time restrict the possibility of growing them in the form of single crystals. Both ZrO_2 and HfO_2 are popular as solid oxide fuel cells, high K gate dielectrics, thermal barrier coatings, nuclear waste storage materials, and so on. Among these three compounds, TiO_2 melts easily (1850°C) and can be grown by direct melting (Verneuil method), but analogous methods have not yet found successful practical application for ZrO_2 and HfO_2, since the melting points are rather high, 2700°C and 2780°C, respectively. Hence, Li_2MoO_4, PbF_2, and $Na_2B_2O_4$ fluxes are used to grow these crystals, as their high melting points make them difficult to grow directly from the melt. Although the crystallization of rutile by the Verneuil method is frequently employed, the method often yields crystals of a nonstoichiometric nature. Kuznetsov [82] has reviewed the growth of oxides of the titanium subgroup metals in detail. Anikin et al. [83] have obtained TiO_2 at 550°C and 1000 atm. Subsequently, Harville and Roy [84] have obtained TiO_2 crystals in H_2SO_4 solutions at 700°C and 1000–4000 atm. However, the growth rate was very low. The commonly used solvents are solutions of alkalis, lithium, potassium, sodium, and ammonium carbonates and bicarbonates, sodium sulfide, lithium, potassium and ammonium chlorides, sulfates (Na_2SO_4, K_2SO_4), borax, boric acid, and fluorides (KF, NaF, NH_4F). The concentrations of the solutions were 5–40%.

TiO_2

Kuznetsov and coworkers [82,85,86] have worked out the growth technology of TiO_2 (rutile phase) under hydrothermal conditions using 7–10% KF or NaF and 5–10% NH_4F solutions at temperatures over 500–550°C and pressures of 500–800 atm. The amount of material transported increases with the increasing temperature of the experiment. Crystals are prismatic with dark-brown or black color. When a small amount of $KClO_3$ is added to the nutrient, crystals become light green in color. Rutile synthesized by the hydrothermal method has properties closer to the natural rutile. In pure water and aqueous solutions of NaOH or Na_2CO_3 at temperatures below 350–400°C, the anatase phase of TiO_2 crystallizes and, at temperatures further down, the brookite phase crystallizes [86].

A rutile crystal grown by the Verneuil method was used as a seed and experiments were carried out at a solubility zone temperature 550–600°C, with respective autoclave filling coefficients of 0.6–0.5 and ($T = 20–35$°C) [86]. Under these conditions the growth rate of the (100) and (110) faces was 0.15–0.21 mm/day in the 5-day experiments. Typical conditions for crystallization were 570°C autoclave filling coefficient = 0.6, $\Delta T = 25$°C. The given directions are not directions of rapid growth. Judging by the morphology of the spontaneous rutile crystals (they often have a typical long-prismatic form), the growth rate along (001) should be higher. In recent years, the growth of nanosize TiO_2 both as anatase and rutile is very popular for a variety of applications, and its growth is more popular in this size range than the growth of TiO_2 as bulk single crystals. The growth of nanosize TiO_2 crystals will be dealt in detail in Chapter 10.

ZrO_2 (Baddeleyite)

Both ZrO_2 and HfO_2 adopt the monoclinic $P2_1/c$ baddeleyite structure, which is a defect fluorite structure type [87]. The small ionic radii of tetravalent zirconium and hafnium are not capable of supporting the eight coordinate environment associated with the fluorite (CaF_2) structure. Instead, a coordination sphere of seven oxygen ligands surrounds the metal center. Structurally very similar, both zirconia and hafnia undergo phase transitions to a tetragonal structure when heated above 1200°C and 1750°C, respectively [88]. Finally, above 2600°C, they both adopt the cubic fluorite structure and remain stable until melting at 2710°C for zirconia and 2900°C for hafnia. However, the high-temperature cubic phases of zirconia and hafnia can be stabilized at room temperature by incorporating additives such as rare earth and alkaline earth metals into the crystal lattice [89]. Kuznetsov [82] was the first to synthesize ZrO_2 crystals under hydrothermal conditions using fluoride solutions NaF, or KF and NH_4F with concentration 7–10% at temperatures 520–690°C. In KF solutions, ZrO_2 is transported only at temperatures over 600°C, and small crystals of 0.5 mm size are formed in the upper zone of the autoclave. Below 600°C, all that happens is that the original material is made coarser, and lamellar crystals up to 0.3 mm in size are formed in the lower zone of the autoclave. In NH_4F solution, there occurs an intensive recrystallization of ZrO_2 at relatively low temperatures. However, there are some practical difficulties in the growth of ZrO_2 due to the retrograde solubility in NH_4F solutions. In order to simplify the problem, a furnace with a horizontal autoclave, a "hot" zone, was created at the obturator. By this means, ZrO_2 crystals in the form of thin plates $2.5 \times 2.5 \times 0.5$ mm with (001), (010), and (110) faces were obtained. Increase in the temperature gradient raises the growth rate, and the crystals become thicker and form twinning. Growth rate is maximum in the *c*-direction and minimum in the *a*-direction and slower in the *b*-direction. The growth rate of the (100) face is roughly 10 times lower than that of the (001). The basic properties of this synthetic ZrO_2 are closer to the mineral baddeleyite. Figure 9.15 shows the photograph of ZrO_2 crystals grown under hydrothermal conditions. Today, several variations in the hydrothermal, solvothermal, and supercritical fluid growth of ZrO_2 crystals include thin films and the nanosize crystals, for a variety of application have been proposed.

HfO_2

According to Kuznetsov [86], the growth of HfO_2 does not differ much from that of ZrO_2. The crystallization of HfO_2 takes place at a higher temperature in KF and NaF solutions and at still higher temperatures in NH_4F solution (700°C). Again, the major disadvantage is the higher temperature of crystallization and retrograde solubility. Hence, in this case also, a horizontally disposed autoclave was used by Kuznetsov and coworkers [82,86]. The autoclave had a transverse baffle in which the orifice area was ~15% of the internal cross-sectional area. Crystals up to $2 \times 2 \times 0.5$ mm

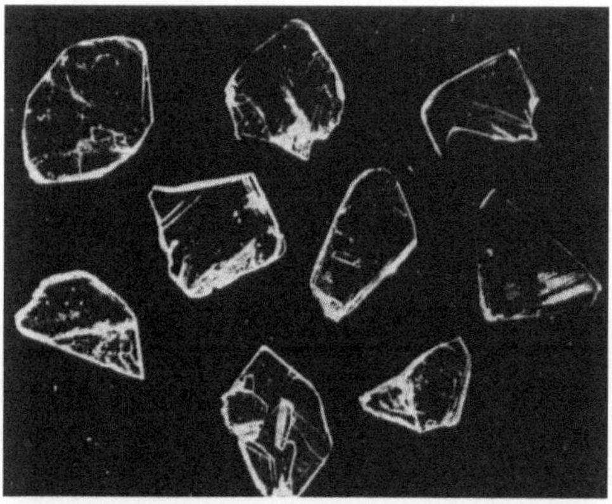

Figure 9.15 ZrO_2 crystals grown under hydrothermal condition.
Source: Courtesy of Kuznetsov [82].

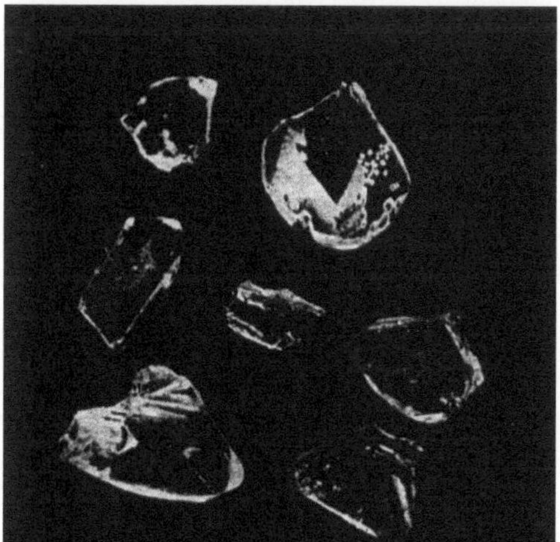

Figure 9.16 HfO_2 crystals.
Source: Courtesy of Kuznetsov [86].

of HfO_2 (Figure 9.16) were obtained. Recently, Mann and Kolis [90] have obtained single crystals of HfO_2 stabilized with yttrium, neodymium, holmium, and erbium up to the size range of 0.25 mm from hydrothermal solutions by spontaneous nucleation in 20 M KOH at 750°C. Single crystals of HfO_2 form only with high concentrations of the stabilizer dopants. As per the literature survey, this is the lowest temperature hydrothermal synthesis of hafnia. Figure 9.17a and b shows the yttrium-stabilized cubic hafnia crystals and neodymium-stabilized cubic hafnia crystals [90]. However, in recent years, the solvothermal synthesis of small nanoparticles of zirconia and hafnia is very popular under much lower temperatures.

TeO_2

Tellurium dioxide has two modifications: tetragonal (paratellurite) and orthorhombic (tellurite) [91,92]. Among these two, the paratellurite is of great interest for solid-state physicists, because of its valuable piezoelectric and acousto-optical properties. In particular, paratellurite crystals may be used in acousto-optical modulators of light deflectors and delay lines. At present, there are several methods of growing these crystals. The most popular is the Czochralski method but, in this method, the growth temperature is high and there are strong temperature gradients in the growth region [93]. This, together with the high vapor pressure of tellurium dioxide, often results in the production of defected and stressed crystals. In addition, the method is rather cumbersome and only about 20% of crystals produced have acceptable quality.

The hydrothermal method of crystallization of paratellurite started almost at the same time as the Czochralski method [91,92]. However, the effects of various physical and chemical parameters on the growth of paratellurite crystals were not properly studied. This technology is not advantageous because of its poor reproducibility and the poor quality of crystals. The possibility of obtaining spontaneous paratellurite monocrystals from solutions of hydrohalogenic acids was demonstrated in Ref. [71]. It was found that the mass crystallization decreases in the following sequence: HCl > HBr > HI > HF, while all other parameters are the same.

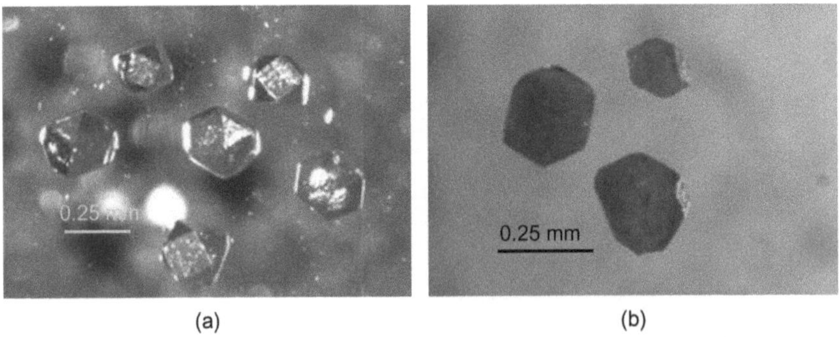

Figure 9.17 (a) Y^{3+}-stabilized and (b) Nd^{3+}-stabilized cubic HfO_2 crystals [90].

The experiments done in platinum-lined autoclaves showed that platinum dissolves in hydrohalogenide complexes; hence, the process of dissolution of TeO_2 in the acid deteriorates and goes out of control. Thus, effective usage of hydrothermal technology of paratellurite could not be developed owing to the difficulties in observing and controlling the parameters which determine the crystallization of α-TeO_2 in platinum-lined steel autoclaves.

Popolitov et al. [95] have studied the solubility of paratellurite in detail in hydrochloric acid, which is an effective solvent for the recrystallization of TeO_2 [94]. Figure 9.18 shows the dependence of solubility of tellurium dioxide and HCl concentrations measured by titration with sodium hydroxide. The starting material used was monocrystals of paratellurite obtained through spontaneous crystallization from water solutions of HCl.

The hydrothermal growth of monocrystals on seeds is known to include the following stages: Dissolution of solid substance, transport of dissolved materials to the growth region, and the growth of seed crystals. To reveal the mechanism of transport, it is of great importance to use the data on solubility of paratelluride in HCl solutions and to consider the possible mechanism of transport from the dissolution region to the growth region in the process of α-TeO_2 growth on the seed.

The synthesis of paratellurite with 100% output occurs in aqueous solutions of nitric acid with concentrations up to 25% at temperatures of 260–340°C and pressures of 50–370 atm [91]. Element tellurium serves as the nutrient, so that the synthesis reaction occurs as follows:

$$3Te + 4HNO_3 \rightarrow 3TeO_2(solid) + 2H_2O + 4NO\uparrow \quad (9.7)$$

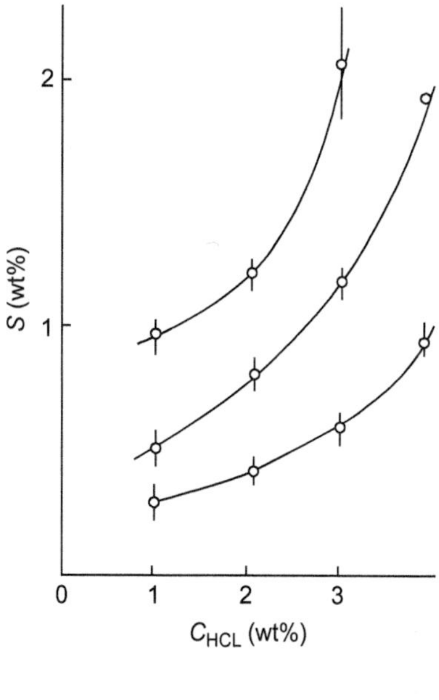

Figure 9.18 Dependence of solubility of tellurium dioxide and HCl concentration [94].

With further rise in the concentration of aqueous nitric acid solution ($>25\%$), there occurs the oxidation of the tellurium with the formation of crystals of telluric acid:

$$Te + 6HNO_3 \rightarrow H_2TeO_4(\text{solid}) + 6NO_2 + 2H_2O \quad (9.8)$$

The morphology of paratellurite crystals is very interesting. The crystals grown in hydrochloric acid solutions exhibit more complex forms than the crystals obtained in nitric acid solutions. The paratellurite monocrystals grow in the form of a tetragonal prism (110) and tetragonal dipyramids (101) and (102). The growth forms of tellurium oxychloride monocrystals are pinacoids (001) and (010), rhombic prisms (110) and (130), and rhombic dipyramids (111), (112), (113), and (114). Figure 9.19 shows the habits of (a) tellurium dioxide and (b) tellurium oxychloride crystals.

Popolitov and coworkers [95] have studied, in detail, the kinetics of crystallization of paratellurite under hydrothermal conditions with reference to the mass transport, temperature gradient, solution concentration, growth rate, and so on [94]. On the basis of the kinetic study of growth conditions for α-TeO_2 crystals, it was found that the growth rate in the homogeneous solution is limited by the diffusion mass transport of the dissolved substance toward the growing faces, while in the binary phase, the surface, processes are limited. A method for the optimal growth of the uniform α-TeO_2 monocrystals was developed wherein the crystals are grown on seeds and in 3.65 wt% water solution of HCl at 150–200°C.

Acoustic characteristics of α-TeO_2 monocrystals obtained by the hydrothermal method are similar to those of Czochralski grown crystals. A reader can find more information in Ref. [96].

Iron Oxides

Iron oxides and oxyhydroxides attract researchers owing to their functional properties like magnetic and catalytic properties, pigments, coatings, and gas sensors.

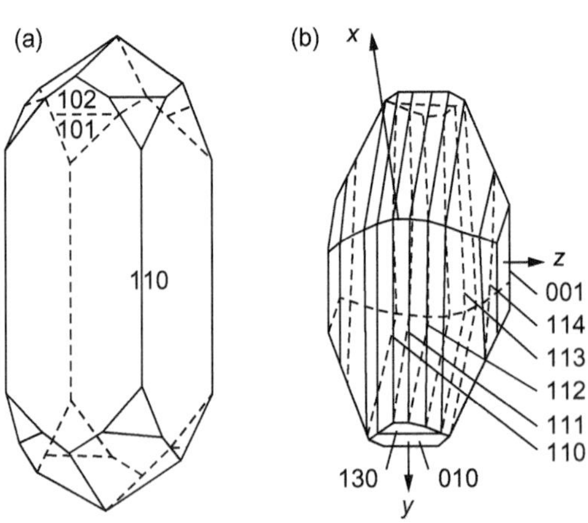

Figure 9.19 Habits of (a) tellurium dioxides and (b) tellurium oxychloride crystals [94].

Among iron oxides, Fe_2O_3 has three modifications: α-Fe_2O_3 (hematite type), β-Fe_2O_3, and γ-Fe_2O_3 (maghemite type) and some authors even predict the monoclinic phase ε-Fe_2O_3. There are various methods of synthesizing iron oxides, and the simplest way is forced hydrolysis of Fe^{3+} salt solutions. But the good-quality crystals are obtained by the hydrothermal method or to some extent the flux method. Several authors [97–100] have carried out the preparation of fine crystals of magnetite. In the pure Fe_2O_3–H_2O system, α-Fe_2O_3 (hematite) is stable up to about 550°C. Above this temperature, up to 710°C, mixtures of α-Fe_2O_3 and Fe_3O_4 are stable. Runs above 710°C yield magnetite (Fe_3O_4) as the only stable phase. Viswanathiah et al. [97] have conducted experiments in the temperature interval 100–600°C. Traces of magnetite are found to be stable even at temperatures as low as 100°C. From 100°C up to 200°C, iron formate ($C_2H_2FeO_4 \cdot 2H_2O$) and hematite (α-Fe_2O_3) are the major phases along with traces of magnetite. HCOOH is used as the mineralizer, and it contributes to the formation of siderite phase also. Further, the liberation of carbon monoxide into the system creates a reducing environment, which promotes the stabilization of Fe_3O_4 rather than of Fe_2O_3. Thus, perfect octahedral crystals of magnetite are formed at 250°C. Figure 9.20 shows hydrothermally grown magnetite crystals, as well as aggregates.

Although there are several methods of preparing crystalline iron oxide powders, hydrothermal and sol–gel techniques are supposed to be the most popular ones. Hirano and Somiya [99] have obtained magnetite crystals under hydrothermal conditions in the presence of hydrogen. The hydrogen for the control of the reduced atmosphere was supplied by the reaction of charged metallic iron with water [100]. The experiments were carried out at 550°C and 1000 kg/cm^2. As the concentration of hydrogen was increased, the (110) faces of magnetite were found to predominate over the (111) faces. The authors have used 10 M NaOH as the solvent. No hematite formation was reported. Dendritic magnetite crystals were grown in 5 M

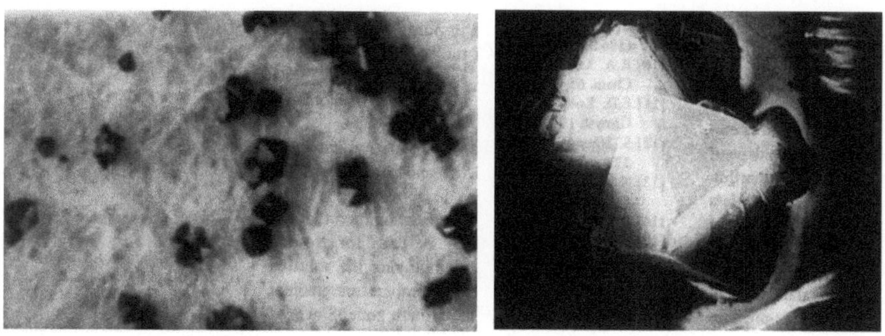

(a) (b)

Figure 9.20 Hydrothermally grown magnetite crystals: (a) Magnetic crystals along with other products of the system; (b) magnified image of magnetite crystals.
Source: Photograph courtesy of Dr. J.A.K. Tareen, Univ. of Mysore, India.

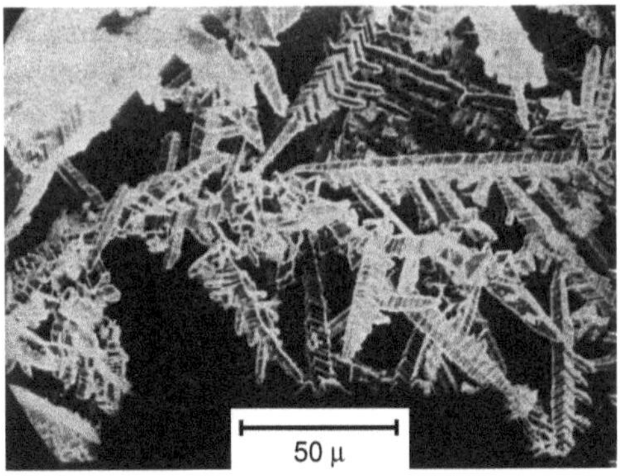

Figure 9.21 Dendrite magnetite crystals grown in 5 M NH_4Cl solution 5.3 M mol hydrogen [100].

NH_4Cl solution at 500°C under 1000 kg/cm^2. Figure 9.21 shows the dendritic magnetite crystals grown in 5 M NH_4Cl solution in 5.3 M mol hydrogen.

Chen and Xu [101] have prepared Fe_3O_4 particles of nanometer size hydrothermally from iron (II) 2-methoxyethoxides ($Fe(OMOE)_2$), using 2-methoxyethanol (MOE)−water mixed solvent as the medium. As in the TiO_2 preparation, here also the reaction conditions such as solvent, temperature, and time usually have important effects on the resultant products. These authors have shown that the particle size of Fe_3O_4 powders increase with prolonged reaction time.

In recent years, hydrothermal synthesis of magnetite in the presence or absence of organic chelating agents, oxidation of ferrous solutions using KNO_3, and solvothermal methods have become popular [102,103]. The ability to direct the morphology of magnetite particles may be a key feature in the development of new applications, particularly if other physical, chemical, and electronic properties are retained. The formation of iron oxide phases is often influenced by the addition of secondary additives to aqueous systems containing either soluble or insoluble precursors of the iron oxides. There have been quite a few reports dealing with the influence of transition metals on the dissolution−precipitation reactions in which iron oxides can participate under hydrothermal conditions. McGarvey and Owen [100] have obtained magnetite under hydrothermal conditions at 150°C and 175°C in the presence of hydrazine and copper (II) oxide. In the absence of CuO, octahedral crystals of magnetite corresponding to the literature data are found. In the presence of CuO, magnetite crystals with distorted morphology were prepared [102] in which the crystal faces were absent, leaving a framework structure of corners and edges. (See Figure 9.22.) Substitution of NiO, ZnO, or Ag_2O produced somewhat different results. Thus copper, as a result of having three accessible oxidation states (Cu(O), Cu^{+1}, and Cu^{+2}) and rapid electron transfer between the species, is able to participate in a redox catalytic mechanism involving hydrazine to facilitate the reduction of hematite to magnetite as well as directing its morphology. Uchida et al. [103] have prepared micaceous iron oxide by the oxidation of iron with pressurized oxygen in concentrated sodium hydroxide solution at elevated temperature. Thus, the preparation of iron oxide under hydrothermal conditions is an attractive field.

Figure 9.22 Magnetite crystals with distorted morphology (a) in the presence of CuO; (b) in the absence of CuO [102].

Demianets et al. [104] have obtained high-quality single crystals of hematite under hydrothermal conditions ($T = 420–450°C$; $P = 1200–1600$ atm; $T = 25–50°C$; mineralizer = alkali hydroxide or carbonate solutions) with self-sealed closure-type autoclaves provided with silver liners. The authors proposed two growth mechanisms responsible for the growth morphologies in hematite single crystals: layer-by-layer growth and island growth.

Oxides of rare earth metals receive special attention as solid-state laser host materials because of their excellent thermal properties. However, these rare earth oxides are difficult to grow as single crystals. Perhaps for the first time, Joseph Kolis and coworkers [105] have succeeded in the hydrothermal growth of bulk single crystals of Lu_2O_3 doped with Er^{3+} and Yb^{3+}. These bulk single crystals were grown at temperature 600–650°C and pressure 2 Kbar in aqueous solutions of KOH as mineralizer. Figure 9.23 shows hydrothermally grown Lu_2O_3 and $Er:Lu_2O_3$ single crystals [105].

There are several other reports in the literature on hydrothermal synthesis of simple oxides like Mn_3O_4, PbO, VO_2, GeO_2, SnO_2, V_2O_3, R_2O_3, In_2O_3, Bi_2O_3, CdO, HgO, and so on.

9.5 Hydrothermal Synthesis of Mixed Oxides

Mixed oxides form one of the most important modern materials. The group includes aluminates, garnets, spinel, metal oxides with complex structures, and so

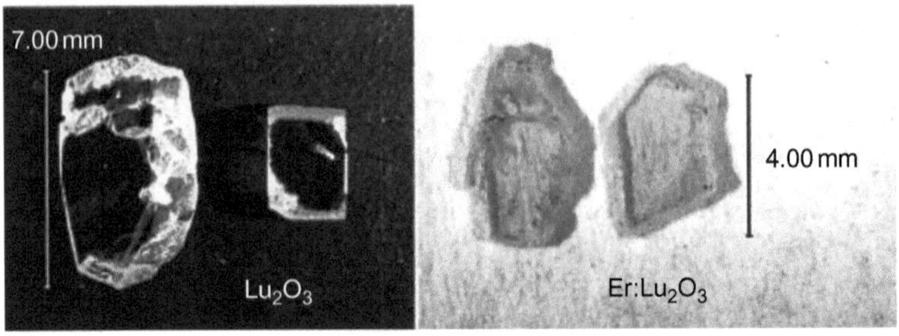

Figure 9.23 Hydrothermally grown (a) Lu_2O_3 and (b) $Er:Lu_2O_3$ single crystals [105].

on. Especially, from 2005 the number of these complex oxides is increasing. Here, we discuss the hydrothermal synthesis of only some important complex oxides.

9.5.1 Hydrothermal Synthesis of Aluminates

Aluminates show special properties like ceramics, electronic, catalytic, and even when doped with active rare earth metal show luminescent properties. Several growth methods like conventional ceramic processing routes, sol–gel, co-precipitation, calcinations, spray pyrolysis, flame synthesis, and alumina impregnating method are used. But the hydrothermal method provides good-quality crystals with minimum possible defects. The majority of the aluminates are formed in pure water at elevated temperatures (>500°C) and pressures (>1000 atm). The hydrothermal reaction involving divalent metals takes place through $MeO + Al_2O_3 \rightarrow MeAl_2O_4$.

Since pure water is used in the synthesis of these oxides, their solubility is very low, and to obtain 100% output, it is necessary to maintain the ratio of MeO/Al_2O_3 in the nutrient at 1:1—i.e., to the near stoichiometry. Aluminates crystallize as small crystals of <0.1 mm. The aqueous solutions of NaOH are preferable in the synthesis; however, crystallization of aluminates of transitional metals (Fe, Co, and Ni) takes place with the formation of corresponding oxides. Litvin and Popolitov [37] have carried out the synthesis of aluminates of Ca, Be, Zn, Fe, Co, and Ni under hydrothermal conditions.

For alkaline earth metals ($MeO/Al_2O_3 > 2{:}1$), the characteristic feature is the formation of hydrogarnets of type $Me_3Al_2O_6 \cdot 6H_2O$. Crowley [106] has carried out the synthesis of $Me_3Al_2O_6 \cdot 6H_2O$ in the system $CaO–Al_2O_3–H_2O$, at temperatures less than 250°C and partial pressure of water 500–800 atm [107]. In pure water, 100% output is achieved at MeO/Al_2O_3 ratio in the nutrient equal to 3:1. The stoichiometry of the nutrient may be strongly inclined toward the surplus MeO while synthesizing in the aqueous solutions of NaOH. Moreover, in the synthesis of strontium hydrogarnet, it is shown that the monophase synthesis sharply increases in the highly concentrated solutions as shown in Figure 9.24 for the system $SrO–Al_2O_3–NaOH–H_2O$ at 500°C [108]. The crystals are either cubic or rhombic dodecahedral in habit. Synthesis takes place at $CaO/Al_2O_3 > 5{:}1$ within the

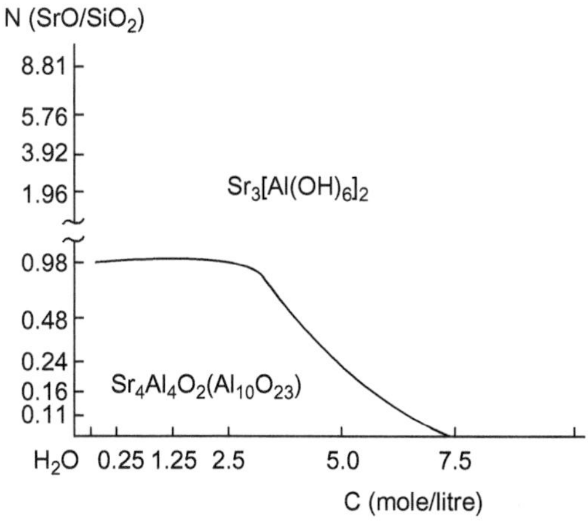

Figure 9.24 Phase diagram for the system $SrO-Al_2O_3-NaOH-H_2O$ at 500°C [108].

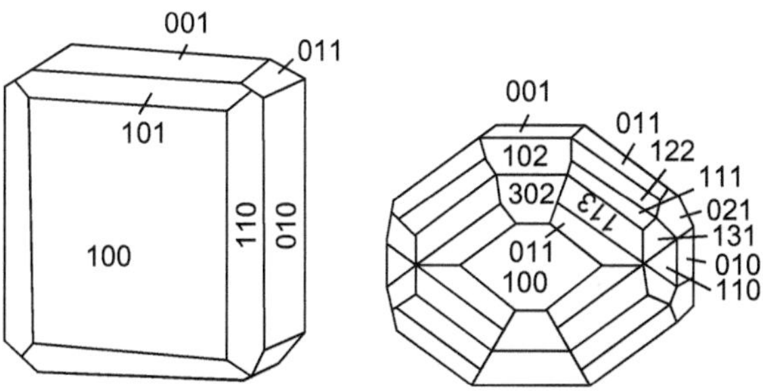

Figure 9.25 Habit of the calcium aluminate crystals [107].

temperature interval 300–450°C [108]. Figure 9.25 shows the habit of the calcium aluminate crystals. Similarly, the strontium aluminates are also formed.

In weakly concentrated solutions, new phases like $Sr_4Al_{14}O_{25}$, platelike crystals of aluminates belonging to the rhombohedral symmetry with space group *Pmma* are obtained.

In the system $Na_2O(K_2O)-Bi_2O_3-Al_2O_3-H_2O$ bismuth tetra-aluminate, $Bi_2Al_4O_9$ crystals have been synthesized under hydrothermal conditions [109]. The crystallization occurs at a higher concentration of Al_2O_3 in the nutrient in the order of 45–90 wt%. The pure phase occurs only when the molar ratio of $Bi_2O_3:Al_2O_3:H_2O = 0.8:0.05:1$, and with a rise in concentration of NaOH, the monophase synthesis reduces. This is connected with the structure becoming unstable with the rise in the concentration of NaOH. These crystals belong to the rhombic system with special group *Pbam*, and the crystals have (110) and (001) forms. However, the size of the crystals is <1.0 mm. Ivakin et al. [110] have synthesized zinc aluminate (gahnite) single crystals and regulated the crystal size using the hydrothermal technique. The experimental temperature was in the range 220–400°C and they doped the gahnite with Eu^{3+} to obtain luminescence in the crystals.

9.5.2 Hydrothermal Synthesis of Antimonites and Antimonates

Oxides of trivalent antimony belong to the group of antimonates. Among hydrothermally synthesized rare earth penta-antimonites, $R_3Sb_5O_{12}$ and $In_3Sb_5O_{12}$ are the important ones [111–113]. Following are the penta-antimonites obtained by hydrothermal technique:

$$In_3Sb_5O_{12},\ La_3Sb_5O_{12},\ Pr_3Sb_5O_{12},\ Tb_3Sb_5O_{12},\ \text{and}\ Yb_3Sb_5O_{12}$$

The synthesis of these compounds is usually carried out by the hydrothermal technique within the temperature interval 450–550°C ($\Delta T = 0.5-1.0$°C and $P = 110-1100$ atm).

Variation in the ratio of $R_2O_3(In_2O_3)/Sb_2O_3$ in the nutrient limits the crystallization of the product with the appearance of Sb_2O_3 crystals. Higher concentration of KF or CsF results in the formation of K_3RF_6, and both of these phases often reduce the output of penta-antimonites field.

Penta-antimonite, $In_2Sb_5O_{12}$, forms in the aqueous solutions of KHF_2, KF (45–55 wt%), and CsF (20 wt%). The crystals are 1–2 mm in size. The basic forms are hexagonal $(11\bar{2}0)$ and trigonal $(01\bar{1}0)$ prisms, pyramids $(01\bar{1}2)$, $(10\bar{1}1)$, $(02\bar{2}1)$, and $(\bar{1}3\bar{2}\bar{1})$. This compound has three polymorphic modifications and two are reversible transformations: trigonal α-form (< 140°C), trigonal β-form (140–450°C), and cubic γ-form (> 450°C) [112]. Figure 9.26 shows the $In_3Sb_5O_{12}$ crystals obtained under hydrothermal conditions [113]. These crystals show interesting piezoelectric and ferroelectric properties.

Similarly, $Sc_3Sb_5O_{12}$ crystallizes in the aqueous solutions of KF with a concentration of up to 60 wt%. The synthesis of scandium penta-antimonite takes place simultaneously with the formation of K_3ScF_6 and attains 70% output in 30% KF solution. In aqueous solutions of CsF with concentrations of 12–60 wt% and

Figure 9.26 $In_3Sb_5O_{12}$ crystals obtained under hydrothermal conditions [113].

Sc_2O_3/Sb_2O_3 ratio >1:1, $Sc_3Sb_5O_{12}$ crystallizes with 100% output. The morphology of these crystals is similar to that of indium analogue [113].

Rare earth antimonites with a general formula $R_3Sb_5O_{12}$ (R = La, Pr, Ti, and Yb) are the important compounds in this group [114]. It is interesting to note that the lanthanum penta-antimonite crystallization field does not exist and its crystallization in aqueous solutions of KF with concentrations of 5–50 wt% is always accompanied by the related phases. Stefanovich et al. [115] have studied the phase transformations in these rare earth antimonites in detail. In the presence of an oxidizing agent, trivalent antimony oxidizes up to the pentavalent state forming antimonates and producing H $[Sb(OH)_6]_6$ acid. They are poorly soluble in water. There are several antimonate minerals; among the synthetic varieties, only a few compounds have been obtained: $Na[SbO_3]$, $Na_3Sb_3[Sb_2O_{11}]$, and $Sb[SbO_4]$ [116,117]. Figure 9.27 shows (a) the field of crystallization of antimonites in the system $Na_2O-Sb_2O_3-Sb_2O_5-H_2O$ and (b) the field of synthesis of antimonites. Figure 9.28 shows the field of crystallization of

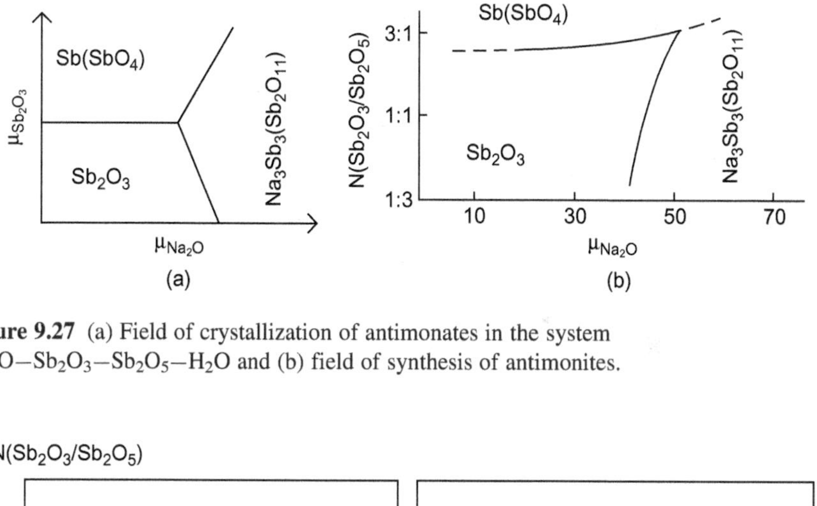

Figure 9.27 (a) Field of crystallization of antimonates in the system $Na_2O-Sb_2O_3-Sb_2O_5-H_2O$ and (b) field of synthesis of antimonites.

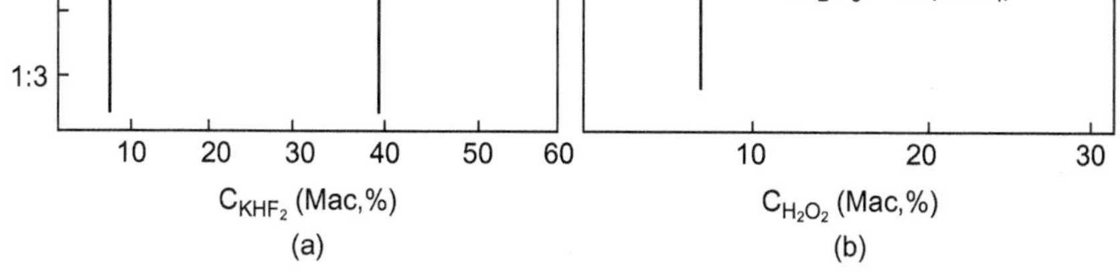

Figure 9.28 Field of crystallization of antimonates at (a) lower and (b) higher oxygen potentials [116].

antimony antimonates at (a) lower and (b) higher oxygen potential [116]. Synthesis is carried out in aqueous solutions in 70% NaOH, 60% KF, and 13% KHF_2, at temperature 550°C, with % fill 60–70%, and at a temperature gradient of 20–35°C. In order to stabilize the pentavalent antimony, 10–15% H_2O_2 is added. In Sb_2O_5, the rich region (2.5:1) forms $NaSbO_3$; with the change in the oxygen potential in the system, other phases crystallizing in the system are $Sb^{3+}Sb^{5+}O_4$ and $Na_3Sb_3{}^{3+}Sb_2{}^{5+}O_{11}$. In a highly concentrated aqueous solution of NaOH (>42%), $Na_3Sb_3{}^{3+}Sb_2{}^{5+}O_{11}$ crystals are formed into prisms up to 5 mm size. In this system, the field of crystallization of $Sb^{3+}Sb^{5+}O_4$ with 100% output does not exist. Antimonates can be crystallized in the aqueous KF solutions with concentration 60% and 98% KHF_2 at temperatures 450–550°C and 60–70% fill [116]. In aqueous KHF_2 solutions, only two phases are formed: Sb_2O_3 and $Sb^{3+}Sb^{5+}O_4$. The antimony antimonates exhibit antiferroelectric properties. The solvothermal synthesis of complex antimonites is very popular in the nanosize range owing to their unique physical properties.

9.5.3 *Hydrothermal Synthesis of Garnets*

Garnet, $Y_3Al_5O_{12}$, has extensively been used as a host lattice for luminescent ions and as substrates for the epitaxial deposition of magnetic garnets for magnetic bubbles. For these applications, large defect-free crystals are requested. Though garnets are ordinarily grown from the flux, the hydrothermal technique is particularly suitable for the preparation of these mixed oxide crystals of high quality and large size. Such growth is easier to control on a seed and it takes place at a lower temperature than the flux growth [68,118,119].

Yttrium aluminum garnet, $Y_3Al_5O_{12}$, crystallization has been studied in a variety of solvents including $(CO_3)^-$, $(OH)^-$, and halides. Among these, higher

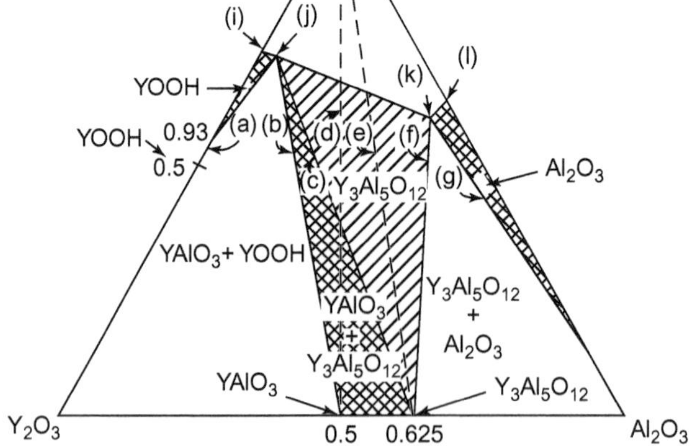

Figure 9.29 Phase diagram of Al_2O_3–Y_2O_3–H_2O between 350°C and 450°C [121].

temperatures of synthesis are required for growth. Laudise and Kolb [120] and Kolb and Laudise [121] have studied the phase equilibria of $Y_3Al_5O_{12}$ and $Y_3Fe_5O_{12}$ in the systems Y_2O_3–Al_2O_3–H_2O and Y_2O_3–Fe_2O_3–H_2O, respectively. Brochier et al. [122] and Toudic and Passaret [123] have studied the heteroepitaxial growth of thin $Y_3Fe_{5-x}Ga_xO_{12}$ and $Gd_yY_{3-x}Fe_{5-x}Ga_xO_{12}$ films under hydrothermal conditions using GdGaG as seed plates. Similarly, there are several other reports on the hydrothermal epitaxial growth of garnet films [124–127]. Figure 9.29 shows the part of the phase diagram Al_2O_3–Y_2O_3–H_2O between 350°C and 450°C at a pressure of 25,000 psi, where the K_2CO_3 concentration is constant at 6 M. $Gd_3Ga_5O_{12}$ appears to be incongruently saturating in contrast to $Y_3Al_5O_{12}$. Both growth rate and quality were poorer in $(CO_3)^{=}$than in $(OH)^-$ at comparable concentrations. Indeed, in 6.0 M K_2CO_3, under conditions where growth in $(OH)^-$ was appreciable, seeds grew a negligible amount or dissolved slightly. This is in contrast to Al_2O_3 (corundum) where $(CO_3^{=})$, because of the higher coefficient of solubility, gives much larger rates than $(OH)^-$. Thus, $(CO_3^{=})$ is preferable to $(OH)^-$. The optimum growth conditions for GdGa are as follows [121]:

Excess Ga_2O_3	:	2–3 M
Solvent	:	420°C
Growth temperature	:	80°C
% Fill	:	70–72%
Seed orientation	:	(100) (111)
Growth rate	:	As high as 0.6 mm/day flawed; improved quality at rates near 0.25 mm/day

These authors have obtained a wide range of garnets using flux-grown nutrient:

$Gd_{2.34}Tb_{0.66}Fe_5O_{12}$
$Eu_{1.9}Ga_{1.1}Al_{0.5}Fe_{4.5}O_{12}$
$Er_{2.0}Eu_{1.0}Fe_{4.3}Ga_{0.7}O_{12}$
$Er_{1.0}Eu_{2.0}Fe_{4.3}Ga_{0.7}O_{12}$
$Ca_{1.8}Bi_{1.2}V_{0.9}Fe_{4.1}O_{12}$
$Ca_{2.0}Bi_{0.95}V_{0.95}Fe_{4.05}O_{12}$

In most of the experiments on the hydrothermal growth of garnets, hydrothermal tipping is used to avoid the seed attack during the warm-up stage.

For the epitaxial film growth of $Er_{1.0}Eu_{2.0}Fe_{4.3}Ga_{0.7}O_{12}$ on (110) seeds of $Gd_3Ga_5O_{12}$, the following conditions have been employed by Laudise and Kolb [122]:

Growth temperature	:	330–445°C
ΔT	:	40–80°
% Fill	:	60–65%
Solvent	:	10–20 M KOH
Ga_2O_3	:	0.75 M

Nijs et al. [128] have obtained $Mn_3Al_2Si_3O_{12}$ under hydrothermal conditions at temperature 550°C and pressure 2 kbar, and in the experimental duration of 14 days. The starting material was $Al_2O_3 \cdot 3SiO_2$ gel, into which ground metallic manganese was added to obtain a fine powder with the composition $Al_2O_3 \cdot 3SiO_2 \cdot 3Mn$. This starting material, plus excess water, was taken in a gold capsule which was placed in a Turtle vessel. After 14 days, the capsule was opened to get small crystals of manganese garnet. This material shows red luminescence from the Mn^{2+} ions, which is quenched at room temperature due to energy migration to quenching centers. The growth of nanosize crystals of garnets doped with active elements find valuable applications in modern technology.

9.5.4 Hydrothermal Synthesis of Ferrites

Ferrites are important technological materials with great success in the development of magnetic oxide powders with a small particle size from several microns to nanometer range [129]. Among them, barium hexaferrite and manganese ferrite are the most important ones. Both could be prepared through solid-state sintering, chemical precipitation, melting, and hydrothermal methods [44,130–133]. Of these methods, the hydrothermal method is probably the most attractive one, because the resulting ferrite has a perfect crystal structure with a definite composition and very fine particle size (usually smaller than 1 μm) and is almost a monodispersed crystal. Therefore, the product obtained from hydrothermal synthesis can be used directly for ceramic processes without a calcination step.

Kumazawa et al. [134] have synthesized barium ferrite fine particles from an aqueous suspension containing goethite and barium hydroxide by the hydrothermal method and systematically examined the relation between the particle size and the reaction conditions commercially available, and also synthetic goethite particles are used in the hydrothermal preparation of ferrites. The experimental temperature ranges from 200 to 300°C. Both with and without stirring techniques have been employed under hydrothermal conditions. Alpha-FeOOH is dehydrated to form α-Fe_2O_3. The dehydration rate may increase with the increasing outer surface area of the particle. The barium ferrite particle may be formed by the incorporation of barium ion under such conditions that α-FeOH is dehydrated to form α-Fe_2O_3. Similarly, a barium hexaferrite particle is formed by dissolution followed by the crystallization. Usually, the starting materials in the form of either hydroxides or nitrates of iron and barium, and mineralizer NaOH and water, are taken in the hydrothermal reactor, and the slurry mixture is heated to a desired temperature. After the experimental run, the suspension is withdrawn from the reactor, neutralized by a diluted HCl solution and then filtered and washed thoroughly with water. The reaction can be written as follows:

$$12FeOOH + Be(OH)_2 \rightarrow BaO \cdot 6Fe_2O_3 + 7H_2O \tag{9.11}$$

Wang et al. [135] have studied reaction mechanism of producing barium hexaferrites from α-Fe_2O_3 and $Ba(OH)_2$ under hydrothermal conditions. Wang et al.

[136] have studied the kinetics of the crystallization of barium hexaferrite in detail. The reaction rate of producing barium hexaferrite was highly dependent on the concentrations of $Fe(OH)_4^-$ (aq.), $Ba(OH)_2$, and NaOH in the aqueous solution. These authors have found that the reaction should be carried out at a temperature greater than 280°C to obtain the main desired product, $BaO \cdot 6Fe_2O_3$. A kinetic model, based on the homogenous phase reaction has been built to describe the dissolution of γ-FeOOH (s) and the precipitation of the dissolved γ-FeOOH (s) {or $Fe(OH)_4^-$ (aq.)} and $Ba(OH)_2$ in sequence to produce the desired product, $BaO \cdot 6Fe_2O_3$. The reaction rate at 280°C is expressed as a function of $Fe(OH)_4^-$ (aq.) and Ba^{2+} concentration, i.e.,

$$\mathrm{Rp} = \frac{0.19[\mathrm{Fe(OH)}_4^-\mathrm{aq}]^2[\mathrm{Ba}^{2+}]}{1 + 8.6[\mathrm{Fe(OH)}_4^-(\mathrm{aq})]} \tag{9.12}$$

Figure 9.30 shows the effect of an alkali concentration on the extent of conversion of the $BaO \cdot 6Fe_2O_3$: The reaction rate increases with the increase of the concentration of NaOH solution. Higher concentration of NaOH also results in a higher yield of the production. The maximum yield of $BaO \cdot 6Fe_2O_3$ achieved is 81% for $[OH]^- = 2$ N. In case of the nitrate source for the nutrient, the particle size decreased with an increase in alkali molar ratio defined as $[OH^-]/[NO_3^-]$. Particles of about 0.1 μm in size were produced at molar ratios above six. As the concentration of NaOH increases, the mean size of the particles decreases. These experimental evidences may support the concept that during hydrothermal treatment at a constant temperature (e.g., 300°C), an increase in the crystallinity of precipitated particles mainly proceeds instead of particle growth driven by Ostwald ripening.

The synthesis of manganese ferrites is carried out in the same way as that of barium hexaferrites. Wolski et al. [44,133] have studied the hydrothermal synthesis and treatment of manganese ferrites in detail. The main problem associated with the manganese ferrites is the tendency of manganese to form an enormous number of individuals with various valencies, different O/Mn ratios, and with the vacancies

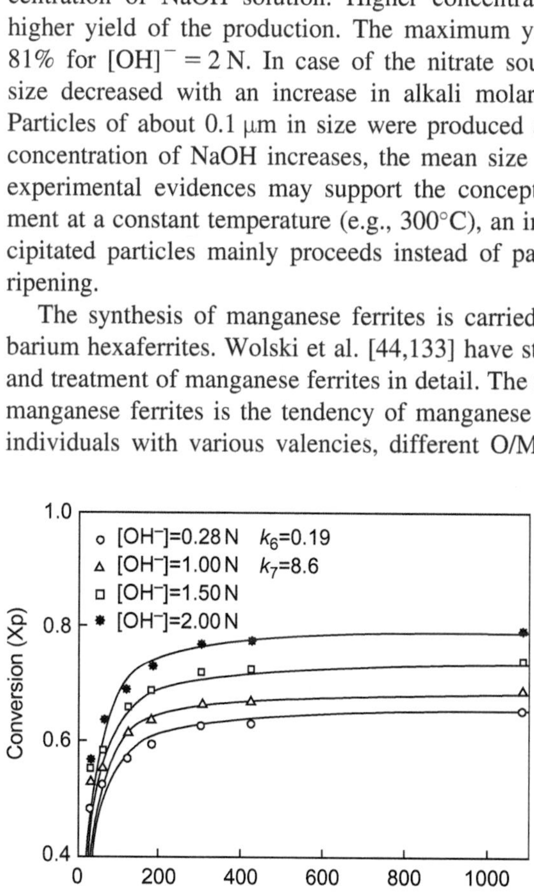

Figure 9.30 Effect of alkali concentration on the conversion of the $BaO \cdot 6Fe_2O_3$ [136].

spread over different positions of the sublattices. Wolski et al. [44] have studied the hydrothermal treatment of such phases to eliminate the nonmagnetic traces from the product and also suggested the probable stages of conversion of $Mn_{0.5}Fe_{1.0}(OH)_4$, for example, up to the product $MnFe_2O_4$, which is a ferrimagnetic spinel.

9.5.5 Hydrothermal Synthesis of Complex Oxides

The mixed/complex oxides represent only a small but important and diversified part of the entire family of inorganic compounds. Mixed oxides are widely used as superionics, fillers, or pigments in pulp and paper, paint, and ceramic electronic industries, and so on. Most of these mixed or complex oxides have highly complex structures, and are highly heat-resistant and chemically inert. In recent years, there has been a growing demand for these complex oxides, especially from the manganese family and other related transitional metal families owing to their unique layered structures [137,138]. These advanced mixed oxides are those which present specific characteristics in their composition, such as pigments for electronics, or in their morphology, such as nanometer-size particles. The hydrothermal method is potentially superior for low-cost production of advanced mixed oxides because complex oxide powders are formed directly. The temperature of synthesis lies between the boiling point of water and its critical temperature (374°C), whereas the pressure can go as high as 15 MPa. The major advantages are the use of inexpensive raw materials such as oxides, hydroxides, chlorides and nitrates, control of growth rate, particle size, stoichiometry, particle shape, elimination of impurities, and so on.

Among these mixed oxides, manganese-bearing mixed oxides are very important as the majority of them show layered structures. Feng et al. [139] have synthesized birnessite-type lithium manganese oxide, by reacting a $Mn(NO_3)_2$ solution with a mixed solution of H_2O_2 and LiOH at room temperature. This manganese oxide has a layered structure with a single sheet of crystal water and lithium ions between 2D edge-shared MnO_6 octahedral sheets. The interlayer spaces can have Na^+, K^+, and Cs^+. Such manganese oxides can be used as metal ion adsorbents and cathodes for lithium rechargeable batteries [140,141]. Figure 9.31 shows the structure of birnessite-type lithium manganese oxides. The lithium ions may locate between the MnO_6 octahedral sheets, and they can be topotactically exchanged with metal ions.

Manganese oxides like $LiMn_2O_4$ are of particular interest because they readily intercalate lithium into their structures and are therefore potentially useful as the cathode of lithium batteries. For example, Sony's lithium ion cell uses the very expensive $LiCoO_2$ cathode. Hence, extensive research is currently underway to find

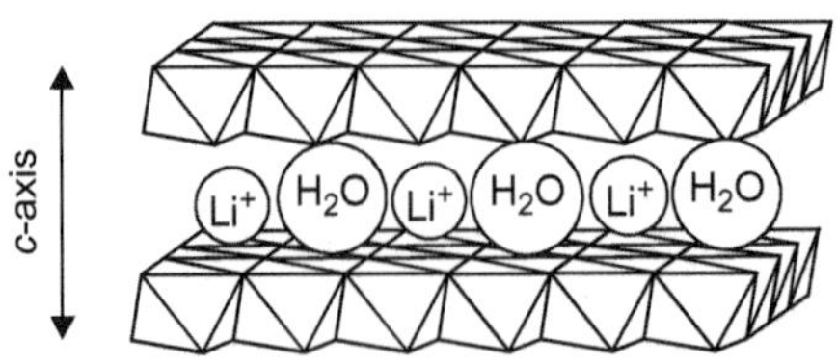

Figure 9.31 Structure of birnessite-type lithium manganese oxides [139].

promising candidates for cathode materials in lithium secondary batteries to replace the expensive cobalt used in present commercial cells. A manganese oxide that behaved like the layered $LiCoO_2$ would be a prime candidate for this application because of its high free energy of reaction with Li and relatively low cost [142]. The experiments involved reacting potassium permanganate with hydrochloric acid and then hydrothermally treating the reaction mixture. The overall reaction takes place as follows:

$$KMnO_4 + (1 - x + 2y)/2H_2O \rightarrow K_2MnO_2 \cdot yH_2O + (1 - x)KOH + (3 + x)/4O_2 \quad (9.13)$$

Figure 9.32 shows the schematic diagram of layered manganates showing (a) anhydrous material and (b) one water layer and two water layers.

Demazeau et al. [143] have obtained La_2MnIrO_6 belonging to the perovskite family using a simple approach of ferromagnetic interactions through a 180°C super exchange. The structural and magnetic studies have confirmed the

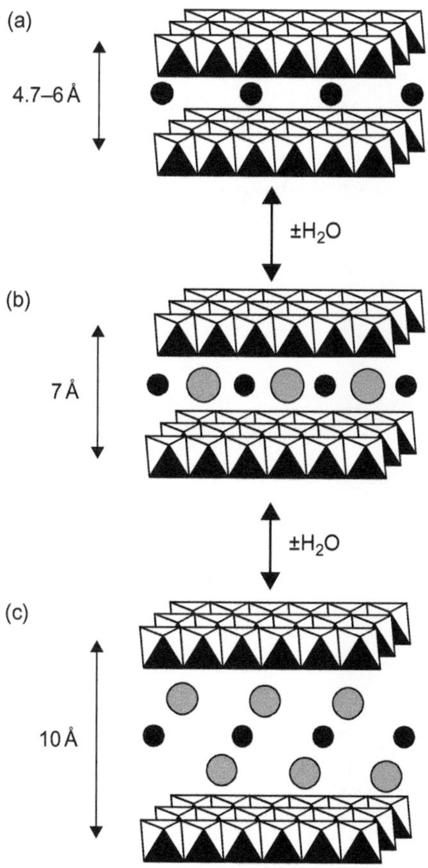

Figure 9.32 Schematic diagram of layered manganates showing (a) anhydrous material, (b) one water layer, and (c) two water layers [142].

ferromagnetic behavior of such an oxide. Turrillas et al. [144] have synthesized $Ca_2Pt_3O_8$ under hydrothermal conditions ($T = 528°C$, $P = 170$ MPa). The platinum ternary oxides have potential uses owing to their good ternary catalytic and electrochemical properties, and also semiconductor, insulator, and metallic-conductor properties. Schwartz and Prewitt [145] have studied the structure and properties of a large number of binary and ternary oxides of platinum. Usually KOH and water are used as mineralizers. These platinum compounds/oxides are insoluble in water, nitric acid, hydrochloric acid, or hot aqua regia but soluble in hot hydrobromic acid. The most prominent feature of $Ca_2Pt_3O_8$ is that it has three polymorphic modifications (tetragonal, orthorhombic, and hexagonal).

In recent years, there is a growing interest in complex oxides of yttria, ceria, and zirconia stabilized with a wide range of cations showing excellent catalytic properties, which also act as structural and electronic promoters of heterogeneous catalytic reactions and as oxide ion conducting solid electrolytes in electrochemical cells [146–148]. A series of solid solutions based on the perovskite-type structure showing the above-mentioned physical and chemical properties are being hydrothermally prepared. The prominent members among them are as follows:

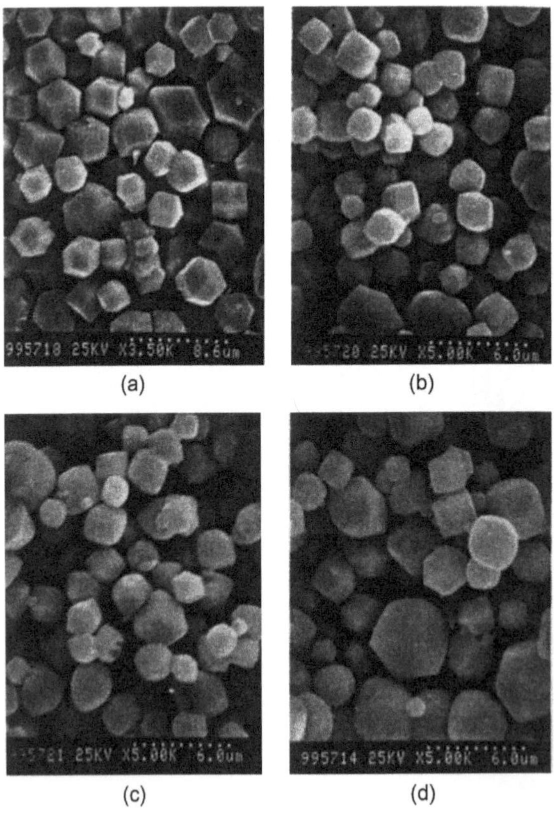

Figure 9.33 SEM photographs of $BaZr_{1-x}M_xO_{3-x}$ (M = Al, Ga, In, $x = 0.20$): (a) $BaZrO_3$, (b) $BaZr_{0.9}Al_{0.1}O_{3-\alpha}$, (c) $BaZr_{0.85}Ga_{0.15}O_{3-\alpha}$, and (d) $BaZr_{0.80}In_{0.20}O_{3-\alpha}$.[149].

$BaZr_{1-x}M_xO_{3-x}(M = Al,\ Ga,\ In;\ \ x = 0.20)$

$SrZrO_{3-\alpha}(M = Al,\ Ga,\ In;\ \ x = 0.20)$

$\mathbf{Ce_{1-x}Pr_xO_{2\cdot\delta}}$

$Ce_{1-x}Bi_xO_{2-\delta}$

and so on [149–152].

The preparation of these solid solutions is carried out usually at temperatures of 150–250°C under autogeneous pressure. Morey-type autoclaves, or other commercially available autoclaves like Parr or Berghoff autoclaves provided with Teflon liners, are used for the synthesis of these oxides. The pH is maintained usually at higher levels (>10). The commonly used solvents are KOH or NH_4OH. Figure 9.33 shows the SEM photographs of $BaZr_{1-x}M_xO_{3-\alpha}(M = Al,\ Ga,\ In;\ \ x = 0.20)$.

There are several other varieties of mixed oxides synthesized under hydrothermal conditions. Hydrothermal technique is especially suitable for obtaining such unique structures, which are impossible to obtain by other conventional methods.

References

[1] J.H. Crocket, Noble metals in sea-floor hydrothermal mineralization from the Juan de Fuca and mid-Atlantic ridges: a fractional on of gold from platinum metals in hydrothermal fluids, Can. Miner. 28 (1990) 639–648.

[2] M.D. Hannington, J.M. Peter, S.D. Scott, Gold in sea-floor polymetallic sulfide deposits, Econ. Geol. 81 (1986) 1867–1883.

[3] H. Rau, A. Rabenau, Hydrothermal growth of some elements, J. Cryst. Growth 3/4 (1968) 417–421.

[4] A. Rabenau, H. Rau, Hydrothermal synthesis of gold, Naturwissenschaften 55 (1968) 336–339.

[5] D.I. Bujor, Synthetische Versuche uber die Bildung der Golder Zlagerstatten zu brad in rumanien, Ark. Kemi Miner. Och. Geol., 26A, 1948.

[6] H. Honma, N. Shikazono, M. Nakata, Hydrothermal synthesis of gold, electrum and argentite, Can. Miner. 29 (1991) 217–221.

[7] S.P. Ogryzlo, Hydrothermally experiments with gold, Econ. Geol. 30 (1935) 400–424.

[8] R.W. Henley, Soluble of gold in hydrothermal chloride solutions, Chem. Geol. 11 (1973) 73–87.

[9] D.S. Glyuk, A.A. Khlebnikova, Gold solubility in water, HCl, HF and sodium and potassium chloride, fluoride, carbonate and bicarbonate solutions at a pressure of 1000 kg/cm^2, Dokl. Acad. Sci. USSR Earth Sci. Sect. 254 (1980) 190–194.

[10] N. Shikazono, Selenium content of acanthite and the chemical environments of Japanese vein-type deposits, Econ. Geol. 73 (1978) 524–533.

[11] P. Pan, S.A. Wood, The solubility of Pt and Pd sulfides and Au metal in bisulfide solutions II. Results at 200–350°C and at saturated vapor pressure, Mineral. Deposita 29 (1994) 373–390.

[12] L.G. Benning, T.M. Seward, Hydro-sulphide complexing of Au(I) in hydrothermal solutions from 150–400°C and 500–1500 bar, Geochim. Cosmochim. Acta 60 (1996) 1849–1871.

[13] P.J. Renders, T.M. Seward, The stability of hydrosulphido- and sulphido complexes of Au (I) and Ag (I) at 25°C, Geochim. Cosmochim. Acta 53 (1989) 244–253.

[14] L.G. Benning, T.M. Seward, Hydrosulphide complexes of gold (I) At high pressures and temperatures: equilibrium and kinetics problems, Mineral. Mag. 58A (1994) 75–76.

[15] O.M. Suleimenov, T.M. Seward, Spectrophotometric determination of the first ion: sat. constant of hydrogen sulphide at high temperatures, in: Y.K. Kharaka, O.V. Chudaev (Eds), Water–Rock Interaction – 8, Balkema Press, The Netherlands, 1995, pp. 113–115.

[16] V.I. Popolitov, Hydrothermal synthesis of semiconductor compounds group $A^{V}{}_{2}B^{III}{}_{3}$ and $A^{V}B^{VI}C^{VII}$, Ph.D. Thesis, Institute of Crystallography, Akademy Nauka, USSR, 1969.

[17] V.I. Popolitov, B.N. Litvin, Synthesis of monocrystals of ternary chalcohalogenides ($A^{V}B^{VI}C^{VII}$), Investigations of Processes of Crystallization Under Hydrothermal Conditions, Nauka, Moscow, 1970, pp. 55–68.

[18] A. Rabenau, H. Rau, Kristalgróeu en chemische synthese under hydrothermale conditities, Philips Techn. Tijdschr. 30 (1969) 94–101.

[19] A. Christoph, P. Gorlich, U. Ludke, Synthese Au, Krist. Techn. 1 (1966) 563–568.

[20] E.D. Kolb, R.A. Laudise, Solubility of Se in Na_2S, J. Cryst. Growth 8 (1971) 191–196.

[21] V.I. Popolitov, A.Y. Shapiro, G.F. Plakhov, Chemical interaction in systems Te–R–H_2O (R = HF, HCl, HBr, CH_3COOH, CHOOH, $C_2H_2O_3$) under high pressures temperatures, J. Phys. Chem. 55 (1982) 1229–1233.

[22] B.V. Derjaguin, B.V. Spitsyn, L.L. Bouliov, A.A. Klochkov, A.E. Gorodetsk, A.E. Smolyanov, Synthesis of diamond crystals on non-diamond substrates, Sov. Phys. 21 (1976) 676–679.

[23] S. Matsumoto, Y. Sato, M. Kamo, N. Setaka, Vapor deposition of diamond particles from methane, J. Appl. Phys. 42 (1982) L183–L186.

[24] F.P. Bundy, H.T. Hall, H.M. Strong, R.H. Wentorf, Man-made diamonds, Nature 176 (1955) 51–54.

[25] W.A. Bassett, Diamond anvil cell, 50th birthday, High Pres. Res. 29 (2009) 163–186.

[26] R.C. De Vries, Synthesis of diamond under metastable conditions, Ann. Rev. Mater. Set. 17 (1987) 161–187.

[27] D. Ravichandran, R. Roy, Growth of diamond on diamond substrates in presence of an alkali and metal under hydrothermal conditions, Mater. Res. Bull. 31 (1996) 1075–1082.

[28] R. Roy, K.A. Cherian, J.P. Cheng, A. Badzian, C. Langlade, H. Dewan, W. Drawl, Precipitation of diamond from $Me_xC_yH_z$ solutions at 1 atm, Mater. Res. Innov. 1 (1997) 117–129.

[29] A. Szymanski, E. Abgarowicz, A. Bakon, A. Niedbalska, R. Slalcinski, J. Sentek, Diamond formed at low pressures and temperatures through liquid phase hydrothermal synthesis, Diamond and Relat. Mater. 4 (1995) 234–235.

[30] A.A. Chaidarov, D.S. Gafitullina, K.P. Argunov, Nuclear–Physical Methods of Diamond Quality Control, FAN, Tashkent, 1986, pp. 115–133 (Russian).

[31] W.L. Tauson, M.G. Abramowicz, Physicochemical Transformations of Real Crystals in Mineral Formations, Nauka, Novosibirsk, 1988. p. 110 (Russian).

[32] R.C. De Vries, R. Roy, S. Somiya, S. Yamada, A review of liquid phase systems pertinent to diamond synthesis, Trans. Mater. Res. Soc. Japan 19B (1994) 641–647.
[33] S. Yamaoka, M. Akaishi, Growth of diamond under HP/HT water condition, in: Proceedings of the sixth NIRIM International Symposium in Advanced Materials (ISAM 99), Tsukuba, Japan, February 28–March 3, 1999, pp. 1–2.
[34] K. Suito, A. Onodera, Synthesis of diamond using phenolic resin, in: Proceedings of the sixth NIRIM International Symposium on Advanced Materials (ISAM 99), Tsukuba, Japan, February 28–March 3, 1999, pp. 7–8.
[35] R.L. Dunin-Barkovsky, Author's Certificate 1522789, USSR, 1988.
[36] R.L. Dunin-Barkovsky, A.R. Dunin-Barkovsky, O.V. Drozdova, B.I. Kidyarov, S.S. Kolyago, E.M. Kozhbakhteyev, et al., Low-temperature growth of diamond seedings in acid solutions, Chem. Sustainable Dev. 8 (2000) 147–153.
[37] B.N. Litvin, V.I. Popolitov, Hydrothermal Synthesis of Inorganic Compounds, Nauka, Moscow, 1984, pp. 37, 43.
[38] E.D. Kolb, A.Z. Caporaso, R.A. Laudise, Hydrothermal crystallization of some II–VI compounds, J. Cryst. Growth 3/4 (1968) 422–425.
[39] L.N. Demianets, E.N. Emelyanova, Crystallization of hydroxylchloride of neodymium $Nd(OH)_2O$ under hydrothermal conditions, Kristallografia 14 (1969) 753–754.
[40] P.V. Kletsov, T.G. Lisvnina, L.Y. Kharchenko, Hydrothermal synthesis and investigations of hydroxylchlorides, $Ln_3Cl_2(OH)_5$ and $Ln(OH)_2Cl$, rare earth elements of yttrium group, J. Spec. Chem. 14 (1973) 87–91.
[41] F.L. Carter, S. Levinson, The hydrothermal preparation single crystal lattice parameters, and decomposition data for some lanthanide dihydroxy chlorides and the related hydroxy chloride, $Yb_3O(OH)_5Cl_2$, Inorg. Chem. 8 (1969) 2788–2791.
[42] J.A.K. Tarren, T.R.N. Kutty, Hydrothermal phase equilibria in $Ln_2O_3–H_2O–CO_2$ systems, J. Cryst. Growth 50 (1980) 527–532.
[43] A.N. Christensen, R.G. Hazeil, Hydrothermal synthesis of stanats: structure of $MnSn(OH)_6$, Acta Chem. Scand. 23 (1969) 1219–1224.
[44] W. Wolski, E. Wolska, J. Kaczmarek, P. Piszora, Ferrimagnetic spinels in hydrothermal and thermal treatment of $Mn_xFe_{2-2X}(OH)_{6-4x}$, J. Therm. Anal. 48 (1997) 247–258.
[45] Schafthaul, Münchener gelehrte Anzeigen, p. 575 (1845), cited by L. Bourgeois, Reproduction Artificielle des Minerals, Paris, 1884, p. 80.
[46] G. Friedel, Hydrothermal Synthesis of Corundum, Bull. Soc. Min. 14 (1891) 7–10.
[47] S.J. Thugutt, Hydrothermal Synthesis of Native Elements and Simple Oxides, Z. Anorg. Chem. 2 (1891) 64–107.
[48] I.P. Kuzmina, N.M. Khaidukov, Crystallization of copper oxide under hydrothermal conditions, Crystal Growth from High Temperature Aqueous Solutions, Nauka, Moscow, 1977, pp. 178–189.
[49] H. Hartmann, Optical properties of Cu_2O, Phys. Status Solidi 2 (1962) 585–597.
[50] D. Ehrentraut, H. Sato, Y. Kagamitani, H. Sato, A. Yoshikawa, T. Fukuda, Solvothermal growth of ZnO, Prog. Cryst. Growth Charact. Mater. 52 (2006) 280–335.
[51] T.L. Yang, D.H. Zhang, J. Ma, H.L. Ma, Y. Chen, Transparent conducting ZnO:Al films deposited on organic substrate deposited by R.F. magnetron sputtering, Thin Solid Films 326 (1998) 60–65.
[52] K. Namratha, K. Byrappa, Novel solution routes of synthesis of metal oxide and hybrid metal oxide nanocrystals, Prog. Cryst. Growth Charact. Mater. 58 (2012) 14–42.
[53] I.P. Kuz'mina, A.N. Lobachev, N.S. Triodina, Synthesis of zincite by the hydrothermal method, in: A.N. Lobachev (Ed.), Crystallization Process Under Hydrothermal Conditions, Consultants Bureau, New York, NY, 1973, pp. 27–41.

[53a] Lobachev et al., Factors Affecting Involved Repeat Stimulation in Recomination and Deletion, In: T. Hiep Tran.: in Saccharomyces Cerevisiae, Gen. Soc. Am., (1982) 1–18.

[54] A.N. Lobachev, I.P. Kuz'mina, Y.V. Shaldin, Zinc oxide: growth and some physical properties, Crystal Growth From High Temperature Aqueous Solutions, Nauka, Moscow, 1977, pp. 158–177.

[55] J.W. Nielsen, E.E. Dearborn, The growth of large single crystals of zinc oxide, J. Phys. Chem. 64 (1960) 1762–1763.

[56] R.A. Laudise, A.A. Ballman, Hydrothermal synthesis of zinc oxide and zinc sulphide, J. Phys. Chem. 64 (1960) 688–691.

[57] M.M. Lyukina, V.E. Khaji, Kinetics of crystallization of zincite under hydrothermal conditions, Crystal Growth, vol. 4, Nauka, Moscow, 1964, pp. 151–156.

[58] D.E. Croxall, R.C. Ward, C.A. Wallace, R.C. Kell, Hydrothermal growth and investigation of highly doped zinc oxide crystals of high purity and perfect, J. Cryst. Growth 22 (1974) 117–124.

[59] N. Sakagami, Hydrothermal growth and characterization of ZnO single crystals of high quality, J. Cryst. Growth 99 (1990) 905–909.

[60] I.P. Kuz'mina, Investigation of the Crystallization of Germanates and Zincogermanates of Na and K Under Hydrothermal Conditions, Ph.D. Thesis, Institute of Crystallography, Academy Nauka, USSR, 1968.

[61] D. Rykl, J. Bauer, Hydrothermal synthesis von zinkit, Krist. Techn. 3 (1968) 375–384.

[62] B.G. Wang, E.W. Shi, W.Z. Zhong, Understanding and controlling the morphology of ZnO crystallites under hydrothermal conditions, Cryst. Res. Technol. 32 (1997) 659–667.

[63] R.A. Laudise, E.D. Kolb, A.J. Caporaso, Hydrothermal growth of large sound crystals of zinc oxide, J. Am. Ceram. Soc. 47 (1964) 9–12.

[64] I.L. Khodakovsky, A.E. Elkin, Experimental determination of zincite solubility in water and NaOH aqueous solutions at temperatures, 100, 150, and 200°C, Geochemistry 10 (1975) 1490–1497.

[65] L.N. Demianets, D.V. Kostomarov, Mechanism of zinc oxide single crystal growth under hydrothermal conditions, Ann. Chim. Sci. Mat. 26 (2001) 193–198.

[65a] D. Ehrentraut, Technology of Gallium Nitride Crystal Growth, D. Ehrentraut, E. Meissner, M. Bockowski (Eds), Springer-Verlag, New York, LLC, (2010).

[66] R.L. Barnes, R.A. Laudise, R.M. Shields, The solubility of corundum in basic hydrothermal solvents, J. Phys. Chem. 67 (1963) 835–840.

[67] G. Yamaguchi, H. Yanagida, S. Sojma, Hydrothermal Growth of Polyscale Crystals, Bull. Soc. Chem. Japan 35 (1962) 1789–1791.

[67a] R.A. Laudise and R.L. Parker, The Growth of Single Crystals, Rost monokristallov, Mir. Moscow, 540 (1974) 2.

[68] R.A. Laudise, Growth of Single Crystals, Prentice-Hall, New Jersey, 1970, pp. 289, 291–292.

[69] K.F. Kashkurov, P.I. Nikitichev, V.V. Osipov, L.D. Sizova, A.V. Simonov, Growth of large corundum crystals by the hydrothermal method, Sov. Phys. Crystal. 12 (1968) 837–839.

[70] V.N. Rymyanstev, I.G. Ganeev, I.C. Rez, SiO_2 in alkali and carbonate solutions, Crystal Growth, vol. 9, Nauka, Moscow, 1972, pp. 51–54.

[71] V.G. Thomas, On a mechanism of substance transfer under the hydrothermal growth of corundum crystals in bicarbonate solutions, Geol. Geophys. 37 (1996) 96–99.

[72] V.A. Kuznetsov, A.A. Shternberg, Crystallization of ruby under hydrothermal conditions, Sov. Phys. Crystal. 12 (1967) 280–285.

[73] R.A. Laudise, A.A. Ballman, Hydrothermal synthesis of sapphire, J. Am. Chem. Soc. 80 (1958) 2655–2657.
[74] R.R. Monchamp, R.C. Puttbach, J.W. Nielson, Hydrothermal growth of iapphite, J. Cryst. Growth 2 (1968) 178–187.
[75] V.S. Balitskii, E.E. Lisinstina, Synthetic Analog and Imitation of Natural Gemstones, Nedra, Moscow, 1981.
[76] Wiley-VCH Encyclopedia of Laser Physics and Technology, 2011. <http://www.rp-photonics.com/titanium_sapphire_lasers.html>.
[77] D.F. Bliss, B. Wang, M.J. Callahan, Hydrothermal synthesis and growth of Ti:sapphire (Ti^{3+}:Al_2O_3) laser crystals, J. Cryst. Growth 311 (2009) 443–447.
[78] J.H. Adair, S.B. Cho, N.S. Bell, A.J. Perrotta, Recent developments in morphological control of alpha-Al_2O_3 particles synthesized in 1,4-butanediol solution, J. Dispersion Sci. Tech 22 (2001) 143–165.
[79] J. Yang, S. Mei, J.M.F. Ferreira, Hydrothermal synthesis of submicrometer alpha-alumina from seeded tetraethylammonium hydroxide-peptized aluminum hydroxide, J. Amer. Ceram. Soc. 86 (2003) 2055–2058.
[80] W.L. Suchanek, Development of Hydrothermal Production of New Materials, Crystal Illuminations, A Publication of Sawyer Research Products, Inc., May 2003.
[81] G. Demazeau, Hydrothermal and solvothermal processes: the main physico-chemical factors involved and new trends, Res. Chem. Intermed. 37 (2011) 107–123.
[82] V.A. Kuznetsov, Oxides of titanium subgroup metals, Crystallization Processes Under Hydrothermal Conditions, Consultants Bureau, New York, NY, 1973, pp. 43–55.
[83] I.N. Anikin, I.A. Naumova, G.V. Rumyantseva, Hydrothermal Synthesis of Native Elements and Simple Oxides, Kristallografia 10 (1965) 231–235.
[84] M.L. Harvill, R. Roy, H.S. Peiser (ed.) Crystal Growth, J. Phys. Chem. Solids (Suppl. 1) (1967) 563–568.
[85] V.A. Kuznetsov, V.V. Panteleev, Hydrothermal synthesis of rutile, Kristallografia 10 (1965) 445–448.
[86] V.A. Kuznetsov, Crystallization of titanium, zirconium and hafnium oxides and some titanate and zirconate compounds under hydrothermal conditions, J. Cryst. Growth 3/4 (1968) 405–410.
[87] J.D. McCullough, K.N. Trueblood, The crystal structure of baddeleyite (monoclinic ZrO_2), Acta Crystallogr. 12 (1959) 507–511.
[88] A.M. Alper, Refractory Materials, 2, Academic Press, London, 1970.
[89] K. Nassau, Cubic zirconia, the latest diamond imitation and skull melting, Lapidary J. 31 (1977) 900–904922–926.
[90] M. Mann, J. Kolis, Hydrothermal crystal growth of yttrium and rare earth stabilized hafnia, J. Cryst. Growth 312 (2010) 461–465.
[91] R.D. Schuilling, B.W. Vink, Stability relation of some titanium minerals (sphene, perovskite, rutile, anatase), Geochim. Cosmochim. Acta 31 (1967) 2399–2411.
[92] V.I. Popolitov, A.N. Lobachev, Crystallization of paratellurite under hydrothermal conditions, Izv. Akad. Nauk USSR Inorg. Mater. 18 (1972) 960–961.
[93] E.D. Kolb, R.A. Laudise, The hydrothermal crystallization of paratellurite, α-TeO_2, Mater. Res. Bull. 8 (1973) 1123–1130.
[94] J. Zubertz, Hydrothermal Synthesis of Native Elements and Simple Oxides, Krist. Techn. 4 (1969) 221.
[95] V.I. Popolitov, Kh.M. Kurbanov, O.N. Zverkova, M.N. Zeitlin, Hydrothermal Synthesis of Complex Oxides, Dokl. Akad-Nauk Tazhakisthan USSR 18 (1975) 21–24.

[96] V.I. Popolitov, Hydrothermal growth under visual examination, in: K. Byrappa (Ed.), Hydrothermal Growth of Crystals, 21, 1990 Progress in Crystal Growth Characterization.
[97] M.N. Viswanathiah, J.A.K. Tareen, K.V. Krishnamurthy, Low temperature hydrothermal synthesis of magnetite, J. Cryst. Growth 49 (1980) 189–192.
[98] G. Visalakshi, G. Venkateswaran, S.K. Kulkreshtha, P.N. Moorthy, Compositional characteristics of magnetite synthesized from aqueous solutions at temperatures up to 523 K, Mater. Res. Bull. 28 (1993) 829–836.
[99] S. Hirano, S. Somiya, Hydrothermal crystal growth of magnetite in the presence of hydrogen, J. Cryst. Growth 35 (1976) 273–278.
[100] G.B. McGarvey, D.G. Owen, Copper (II) oxide as a morphology directing agent in the hydrothermal crystallization of magnetite, J. Mater. Sci. 31 (1996) 49–53.
[101] D. Chen, R. Xu, Hydrothermal synthesis and characterization of nanocrystallization Fe_3O_4 powders, Mater. Res. Bull. 33 (1998) 1015–1021.
[102] A. Okuwaki, Synthesis of several inorganic materials under hydrothermal conditions, in: M. Yoshimura (Ed.), Proceedings of the International Workshop on Soft and Solution Processing for Advanced Inorganic Materials, Tokyo Institute of Technology, Tokyo, Japan, 1996, pp. 64–73.
[103] S. Uchida, T. Sato, A. Okuwaki, Preparation of micaceous iron oxide by the oxidation of iron with pressurized oxygen in concentrated sodium hydroxide solution at elevated temperatures, J. Mater. Sci. 27 (1992) 1332–1336.
[104] L.N. Demianets, S.V. Pouchko, R.V. Gaynutdinov, Fe_2O_3 single crystals: hydrothermal growth, crystal chemistry and growth morphology, J. Cryst. Growth 259 (2003) 165–178.
[105] C. McMillen, D. Thompson, T. Tritt, J. Kolis, Hydrothermal single-crystal growth of Lu_2O_3 and lanthanide-doped Lu_2O_3, Cryst. Growth Des. 11 (2011) 4386–4391.
[106] M.S. Crowley, Effect of starting materials on phase relations in the system, $CaO-Al_2O_3-H_2O$, J. Am. Ceram. Soc. 47 (1964) 144–148.
[107] V.I. Ponomerev, B.N. Litvin, N.V. Belov, Hydrothermal synthesis of hydroaluminates of calcium and their X-ray data, Izv. Akad. Nauk USSR Inorg. Mater. 6 (1970) 1657–1659.
[108] T.N. Nadedjina, V.A. Kuznestov, E.A. Pobedimskaya, N.V. Belov, Crystallization of silicates, aluminates and germanates of strontium under hydrothermal conditions, Crystal Growth for High-Temperature Aqueous Solutions, Nauka, Moscow, 1977, pp. 120–135.
[109] T.N. Nadejdina, Synthesis and Crystallization Chemical Studies of Sr-Silicates and Their Analogs, Ph.D. Thesis, Moscow State University, Moscow, 1975.
[110] Y.D. Ivakin, M.N. Danchevskaya, G.P. Muravieva, Regulation of gahnite crystal size during hydrothermal synthesis, J. Phys.: Conf. Ser. 121 (2008) 082007.
[111] Kh.M. Kurbanov, Hydrothermal synthesis and crystal data of $LuSbO_3$, $PrSbO_3$, $DySbO_3$ monocrystals, Kristallografia 23 (1978) 1266–1287.
[112] R.C. Buchurin. Hydrothermal Synthesis Antimonites of Trivalent Elements and Their Physicochemical Properties, Ph.D. Thesis, Institute of Crystallography, Akademi Nauka, USSR, 1981.
[113] R.Ch. Bichurin, V.I. Popolitov, S.Yu. Stefanovich, Growth and physico-chemical properties of $In_3Sb_5O_{12}$ and $Sc_3Sb_5O_{12}$ monocrystals, J. Theor. Phys. 7 (1981) 1292–1295.
[114] V.I. Popolitov, R.Ch. Bichurin, Hydrothermal crystallization in the system $Sb_2O_3-In_2O_3-R-H_2O$ (R-solvent), Izv. Akad. Nauk USSR Inorg. Mater. 20 (1984) 1535–1539.

[115] S.Yu. Stefanovich, R.Ch. Bichurin, V.I. Popolitov, Yu.N. Venevstev, Identification of phase transits in rare earth antimonite crystals, $Ln_3Sb_5O_{12}$, J. Solid State Phys. 24 (1982) 616–618.
[116] V.I. Popolitov, Some physicochemical principles crystallization under hydrothermal conditions, Hydrothermal Synthesis and Growth of Monocrystals, Nauka, Moscow, 1982, pp. 119–149
[117] V.I. Popolitov, A.N. Lobachev, M.N. Steitlin, Crystallization of orthoantimonites of antimony under hydrothermal conditions, Crystal Growth from High Temperature Solutions, Nauka, Moscow, 1977, pp. 198–216.
[118] R.C. Puttbach, R.R. Monchamp, J.W. Nielsen, Hydrothermal growth of Y3Al5O12, in: H.S. Peiser (Ed.), Crystal Growth, Pergamon Press, Oxford, UK 1967, pp. 569–571.
[119] C.P. Khattak, F.F. Wang, in: K.A. Gschneidner Jr., L. Eyring (Eds), Handbook on the Physics and Chemistry of Rare Earths, vol. 3, North-Holland, Amsterdam, 1979.
[120] R.A. Laudise, E.D. Kolb, Growth of Single-Crystal Garnets by a Modified Pulling Technique, J. Am. Ceram. Soc. 45 (1962) 51–54.
[121] E.D. Kolb, R.A. Laudise, Phase equilibria of $Y_3Al_5O_{12}$, hydrothermal growth of $Gd_3Ga_5O_{12}$ and hydrothermal epitaxy of magnetic garnets, J. Cryst. Growth 29 (1975) 29–39.
[122] A. Brochier, P. Coeure, B. Ferrand, J.C. Gay, J.C. Joubert, J. Mareschal, et al., Heteroepitaxie de Couches Minces de Grenat de Feryttrium par Methode de Flux et Synthese Hydrothermale, J. Cryst. Growth 13/14 (1972) 571–575.
[123] Y. Toudic, M. Passaret, Croissance Parvoie Hydrothermale de Films Ferrimagnetiques, Epitaxies sur des Substrats de GdGaG, J. Cryst. Growth 24/25 (1974) 621–623.
[124] B. Ferrand, Heteroepitaxie par Syntheses Hydrothermale de Films Minces Monocristallins de Grenats Ferrimagnetiques, High Temp. High Press. 6 (1974) 619–628.
[125] B. Ferrand, J. Daval, J.C. Joubert, Heteroepitaxial growth of single crystal films of YIG on GdGaG substrates by hydrothermal synthesis, J. Cryst. Growth 17 (1972) 312–314.
[126] B. Ferrand, J. Geynet, D. Challeton, J. Daval, J.C. Joubert, Growth of epitaxial substituted garnet films by hydrothermal synthesis, Mater. Res. Bull. 9 (1974) 495–506.
[127] M.J.G. Van Hout, J.C. Verplanke, J.M. Robertson, Hydrothermal synthesis of single crystal thin films of magnetic garnets and their analysis, Mater. Res. Bull. 10 (1975) 125–132.
[128] O. Nijs, J.M.W. Verweij, G. Blasse, A divalent manganese garnet with red luminescence, Mater. Chem. Phys. 30 (1992) 199–203.
[129] J. Smit, H.P.J. Wijn, Ferrites, John Wiley & Sons, New York, NY, 1959. p. 194.
[130] G.C. Bye, C.R. Hooward, The synthesis of barium hexaferrite from iron oxide and barium carbonate, 1. General features of the reaction, J. Appl. Chem. Biotechnol. 21 (1971) 319–323.
[131] H. Hibst, Hexagonal ferrites from melts and aqueous solutions magnetic recording materials, Angew. Chem. Int. Ed. Engl. 21 (1982) 270–275.
[132] M.A. Farhat Hadj, J.C. Jabnert, Hydrothermal synthesis and characterization of the hexagonal ferrite Fe–Y, J. Magn. Magn. Maters. 62 (1986) 353–360.
[133] W. Wolski, E. Wolska, J. Kaczmarek, P. Piszora, Formation of manganese ferrite by modified hydrothermal method, Phys. Status Solidi 152 (1995) K19–K22.
[134] H. Kumazawa, H.M. Cho, E. Sada, Hydrothermal synthesis of barium ferrite fine particles from goethite, J. Mater. Sci. 28 (1993) 5247–5250.

[135] M.L. Wang, Z.W. Shih, C.H. Lin, Preaction mechanism of producing barium hexaferrites from α-Fe_2O_3 and $Ba(OH)_2$ by hydrothermal method, J. Cryst. Growth 114 (1991) 435–445.
[136] M.L. Wang, Z.W. Shin, C.H. Lin, Kinetic study of producing barium hexaferrite from goethite and barium hydroxide by hydrothermal method, Chem. Eng. Comm. 126 (1993) 79–95.
[137] C.N.R. Rao, Novel materials, materials design and synthetic strategies: recent advances and new directions, J. Mater. Chem. 9 (1998) 1–14.
[138] P. Cousin, R.A. Ross, Preparation of mixed oxides: a review, Mater. Sci. Engg. A130 (1990) 119–125.
[139] Q. Feng, K. Yanagisawa, N. Yamasaki, Synthesis of birnessite-type lithium manganese oxide, J. Ceram. Soc. Japan 104 (1996) 897–899.
[140] M. Tsuji, S. Komarneni, Y. Tamaura, M. Abe, Cation exchange properties of a layered manganic acid, Mater. Res. Bull. 27 (1992) 741–751.
[141] P. Strobel, C. Mouget, Electrochemical lithium insertion into layered manganates, Mater. Res. Bull. 28 (1993) 93–100.
[142] R. Chen, P. Zavalij, S. Whittingham, Hydrothermal synthesis and characterization of $K_xMnO_2 \cdot YH_2O$, Chem. Mater. 8 (1996) 1275–1280.
[143] G. Demazeau, B. Siberchicot, S. Matar, C. Gayet, A. Largeteau, New ferromagnetic oxides derived from the perovskite structure prepared under high pressures, High Pres. Res. 12 (1994) 337–341.
[144] X. Turrillas, C. Laviron, H. Vincent, J. Pannetier, J.C. Joubert, Synthesis and characterization of a new calcium platinum oxide: $Ca_2Pt_3O_8$, J. Solid State Chem. 67 (1987) 297–307.
[145] K.B. Schwartz, C.T. Prewitt, Structural and electronic properties of binary and ternary platinum oxides, J. Phys. Chem. Solids 45 (1984) 1–12.
[146] S.P.S. Badwal (Ed.), Science and Technology of Zirconia, Technomic, Lancaster, 1993.
[147] H. Inaba, H. Tagawa, Ceria – based solid electrolytes, Solid State Ionics 83 (1996) 1–8.
[148] P. Shuk, H.D. Wiemhöfer, U. Guth, W. Göpel, M. Greenblatt, Oxide ion conducting solid electrolytes based on Bi_2O_3, Solid State Ionics 89 (1996) 179.
[149] W. Zheng, C. Liu, Y. Yue, W. Pang, Hydrothermal synthesis and characterization of $BaZr_{1-x}M_xO_{3-x}$, (M = Al, Ga, In, x 0.20) series oxides, Mater. Lett. 30 (1997) 93–97.
[150] W. Zheng, W. Pang, G. Meng, Hydrothermal synthesis of $SrZrO_{3-x}$ (M = Al, Ga, In, x 0.20) series oxides, Solid State Ionics 108 (1998) 37–41.
[151] P. Shuk, M. Greenblatt, Hydrothermal synthesis and properties of mixed conductors based on $Ce_{1-x}Pr_xO_{2-x}$, solid solutions, Solid State Ionics 116 (1999) 217–223.
[152] S. Dikmen, P. Shuk, M. Greenblatt, Hydrothermal synthesis and properties of $Ce_{1-x}Bi_xO_{2-x}$ solid solutions, Solid State Ionics 112 (1998) 299–307.

10 Hydrothermal Technology for Nanotechnology—A Technology for Processing of Advanced Materials

10.1 Introduction

The term *materials processing* is used in a very broad sense to cover all sets of technologies and processes for a wide range of industrial sectors. Obviously, it refers to the preparation of materials with a desired application potential. Among various technologies available today in advanced materials processing, the hydrothermal technique occupies a unique place owing to its advantages over conventional technologies. It covers processes like: hydrothermal synthesis, hydrothermal crystal growth leading to the preparation of fine to ultrafine crystals, bulk single crystals, hydrothermal transformation, hydrothermal sintering, hydrothermal decomposition, hydrothermal stabilization of structures, hydrothermal dehydration, hydrothermal extraction, hydrothermal treatment, hydrothermal phase equilibria, hydrothermal–electrochemical reactions, hydrothermal recycling, hydrothermal microwave-supported reactions, hydrothermal–mechanochemical, hydrothermal–sonochemical, hydrothermal–electrochemical processes, hydrothermal fabrication, hot pressing, hydrothermal metal reduction, hydrothermal leaching, hydrothermal corrosion, and so on. The hydrothermal processing of advanced materials has lots of advantages and can be used to give high product purity and homogeneity, crystal symmetry, metastable compounds with unique properties, narrow particle size distributions, a lower sintering temperature, a wide range of chemical compositions, single-step processes, dense sintered powders, submicron to nanoparticles with a narrow size distribution using simple equipment, lower energy requirements, fast reaction times, lowest residence time, as well as for the growth of crystals with polymorphic modifications, the growth of crystals with low to ultra-low solubility, and a host of other applications [1].

In the twenty-first century, hydrothermal technology, on the whole, will not be just limited to the crystal growth, or leaching of metals, but it is going to take a very broad shape covering several interdisciplinary branches of science. Therefore, it has to be viewed from a different perspective. Further, the growing interest in enhancing the hydrothermal reaction kinetics using microwave, ultrasonic, mechanical, and electrochemical reactions will be distinct [2]. Also, the duration of

Handbook of Hydrothermal Technology.

experiments is being reduced at least by three to four orders of magnitude, which will in turn, make the technique more economic. With an ever-increasing demand for composite nanostructures, the hydrothermal technique offers a unique method for coating of various compounds on metals, polymers, and ceramics as well as for the fabrication of powders or bulk ceramic bodies. It has now emerged as a front-line technology for the processing of advanced materials for nanotechnology. On the whole, hydrothermal technology in the twenty-first century has altogether offered a new perspective which is illustrated in Figure 10.1. It links all the important technologies like geotechnology, biotechnology, nanotechnology, and advanced materials technology. Thus, it is clear that the hydrothermal processing of advanced materials is a highly interdisciplinary subject, and the technique is popularly used by physicists, chemists, ceramists, hydrometallurgists, materials scientists, engineers, biologists, geologists, technologists, and so on.

10.2 Current Trends in Hydrothermal Technology

There is a great difference between the hydrothermal research carried out during the previous century and the early twenty-first century. During mid-twentieth century, hydrothermal technology—which was in its peak—was mainly focusing on the high-temperature and high-pressure regime of materials processing because of the lack of knowledge on the hydrothermal chemistry of the medium, solubility of several compounds, and also on the selection of an appropriate solvent. The First International Hydrothermal Reactions Symposium (1982) held at the Tokyo Institute of Technology, Tokyo, Japan, brought together specialists from interdisciplinary branches of science. Since then, the knowledge on the physical chemistry, PVT relationship in the hydrothermal systems greatly improved, which helped in

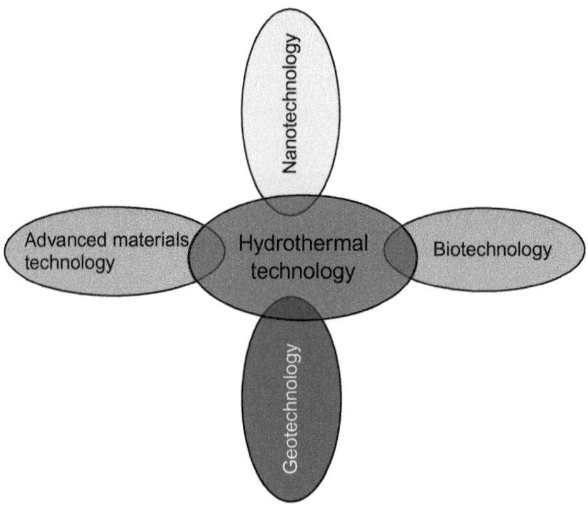

Figure 10.1 Hydrothermal technology in the twenty-first century.

drastically reducing the temperature and the pressure conditions of processing. Similarly, the solvothermal and supercritical processing which used a variety of other solvents like organic, organometallic complexes in materials processing, thereby taking this technology toward Green Chemistry. Table 10.1 gives the trends in hydrothermal processing of materials, and these trends take hydrothermal technology toward green technology for sustained human development. Hydrothermal technology consumes less energy with no or little solid waste, or liquid or gas waste, or no further treatment to recover materials, without involving any hazardous materials to process, and has a high selectivity to process materials in a closed system [1]. The important subjects of technology in the twenty-first century are predicted to be the balance of environmental and resource and or energy problems. This has led to the development of a new concept related to the processing of advanced materials in the twenty-first century, namely, industrial ecology or the science of sustainability [3]. Hydrothermal chemistry has to be understood precisely in order to process the materials under soft and environmentally benign conditions. The behavior of the solvents under hydrothermal conditions dealing with aspects like structure at critical, supercritical, and subcritical conditions, dielectric constant, pH variation, viscosity, coefficient of expansion, and density are to be understood with respect to pressure and temperature. Today, much of the hydrothermal research is done based on the intelligent modeling of the hydrothermal reactions prior to the actual experiments. This greatly helps in predicting the experimental conditions to obtain a desired phase with a controlled shape and size [4,5].

10.3 New Concepts in Hydrothermal Technology

More recently, the addition of external energy like microwave energy, sonar, mechanochemical, electrical, and magnetic into hydrothermal technology has

Table 10.1 Current Trends in Hydrothermal Technology

Compound	Earlier Work	Byrappa's work
$Li_2B_4O_7$	$T = 500–700°C$ $P = 500–1500$ bar	$T = 240°C$ $P = <100$ bar
$Li_3B_5O_8(OH)_2$	$T = 450°C$ $P = 1000$ bar	$T = 240°C$ $P = 80$ bar
$NaR(WO_4)_2$ R = La,Ce,Nd	$T = 700–900°C$ $P = 2000–3000$ bar	$T = 200°C$ $P = <100$ bar
$R{:}MVO_4$ R = Nd,Eu,Tm;M = Y,Gd	Melting point >1800°C	$T = 100°C$ $P = <30$ bar
$LaPO_4$	Synthesized at >1200°C	$T = <120°C$ $P = <40$ bar
Diamond	$T > 1000°C$ $P = 10$ Kbar	$T < 800°C$ $P = <3$ Kbar

opened up a new chapter in materials processing. It is being popularly referred to as multienergy processing of materials. Hitherto, in the hydrothermal technique, the researchers were dealing only with the temperature, pressure, and chemical potential as the three main variables in materials processing, and much of the thermodynamic issues have been understood more or less precisely. But with the additional energy variables in the system, the thermodynamic relationship reads completely differently and becomes more complicated. This forms the future of materials processing, which can be termed as novel methods of advanced materials processing [6,7]. From 2010, researchers are using the concept of instant hydrothermal reactions to obtain the desired nanoparticles in a shortest possible time and some even imagine a system like a vending machine to produce the desired nanoparticles with definite physical properties [8]. This leads us to the concept visualized by an eminent American Energy Department consultant as *chemistry at the speed of light* [9]. There are so many advantages in such a multienergy concept of advanced materials processing. The credit for this multienergy processing goes to the researchers at the Tokyo Institute of Technology, who first attempted hydrothermal reactions with electrochemical and mechanical energies, which firmly established a new trend in materials processing in 1970s and 1980s [10]. Followed by this, the Materials Research Laboratory at the Penn State well explored the possibilities of microwave and sonar in the hydrothermal reactions in 1990s and went on to become the world leaders in this fascinating area of science [11,12].

The discovery of hydrothermal activity in the deep sea during 1970s [13] has led to a new thinking in marine biology and geochemistry, which set a new trend in advanced materials processing. Now it is strongly believed that the roots of life on earth can be found in hydrothermal ecosystems. Thus, the organic synthesis under hydrothermal conditions was established, which has now become an important area of research. The organic–inorganic hybrid materials are forming the core of the nanotechnology, which insist on the precise control over the size and morphology of the nanoparticles that influence directly the physical properties because of size quantization [14]. As we know, earth is a blue planet of the universe where water is an essential component. Circulation of water and other components such as entropy (energy) are driven by water vapor and heat (either external or internal). Water has a very important role in the formation of material or transformation of materials in nature, and hydrothermal circulation has always been assisted by bacteria, photochemical, and other related activities. Table 10.2 gives some representative bioassisted materials [15]. Such an understanding helps in the processing of advanced inorganic materials with the assistance of biomolecules, e.g., proteins, organic ligands, DNA, and amino acids. A great variety of biomaterials can be fabricated under ambient conditions by employing the nature-inspired conditions. Yoshimura [16–18] is popularizing a new concept called *Soft Solution Processing*, from 2000 one decade, as the process which is inspired basically by the natural processes. It covers all set of materials processes, which can be operated under ambient or near ambient conditions or just above the ambient conditions to prepare a wide range of materials including even complex and high melting materials. The basis of the soft solution processing concept of Yoshimura is based on the nature-inspired processes

Table 10.2 Bioprocessed Materials

Metallurgical materials	None (except for Au peromicrobium)			
Organic materials	Plenty and versatile			
Inorganic materials evolutionally	Limited and selected			
1. Silica (Opal)	Diatoms, Radiolaria	Skeleton	Silica blade of pampas grass	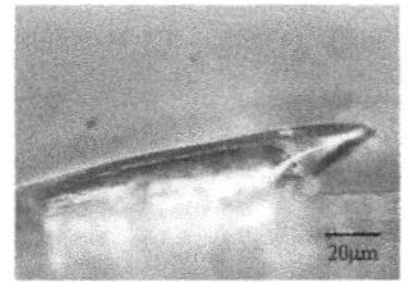
	Plants	Leaves, cell wall		
2. Iron oxides	Bacteria, tuna, salmon	Sensors		
	Chitons, limpets	Teeth		
	Beaver, rat, fish	Teeth surface		
3. Iron sulfides	Gastropod	Dermal sclerite		
4. Ca carbonates	Mollusks, gastropod	Shells		
	Corals	Cell wall	Gastropod with iron sulfide armor	
	Cocolithophoridae	Cell wall scales		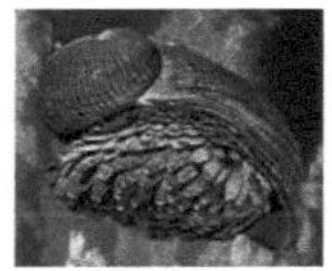
	Aves	Egg shells		
5. Ca phosphates	Fish	Scales		
	Vertebrates	Bones		
	Mammals	Teeth		
	Chitons	Teeth		
6. Ca sulfate	Jellyfish	Gravity receptor		
Ba sulfates	Loxodes, Xenophyophora	Gravity receptor		
Sr sulfates	Acantharea	Skeleton		

Other minerals (<100) are formed outside the bodies by bacterial process.
Source: After S. Mann [15] and modified by M. Yoshimura.

and the energy required in such processing. Bioprocessing can produce very useful materials using minimal energy but for only limited species, size, shape, and location. Although the artificial processing can produce almost all materials, it consumes high energy. Figure 10.2 shows the PT map of various materials-processing techniques [17]. According to Figure 10.2, the hydrothermal processing of advanced materials can be considered as environmentally benign. Besides, for processing nanomaterials, the hydrothermal technique offers special advantages because of the highly controlled diffusivity in a strong solvent media in a closed system.

The features involved in the in situ fabrication of materials by hydrothermal processing cover the direct or single-step formation of shaped/sized/deposited/oriented particles or ceramics with a minimized consumption of energy, resulting in any desired shape and size. Since the system acts as a closed flow system, thus makes it easy for charging, separation, cycling, and recycling, we can achieve high deposition rate, and the entire process is highly versatile. The processes falling into the category of soft solution processing can be seen in Figure 10.3. It shows advanced materials processing involving a single-step versus multistep processes [18]. The in situ fabrication of the materials is the present trend in materials processing without involving postsynthesis treatment. This is the greatest advantage of hydrothermal processing, which facilitates the in situ size, morphology control, and also surface modification.

The important step in the processing of fine particles of advanced materials is the use of surfactants and chelates to control the nucleation of a desired phase, such that the phase homogeneity, size, shape, and dispersibility could be achieved

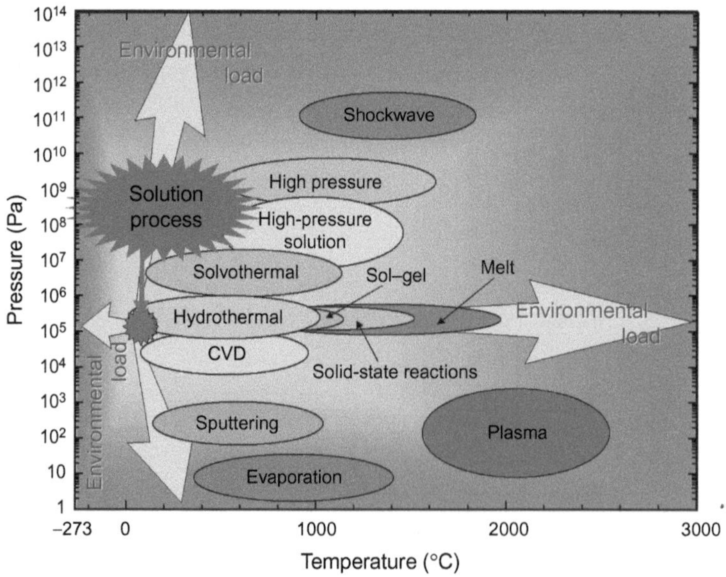

Figure 10.2 Pressure–temperature map of materials processing techniques [17].

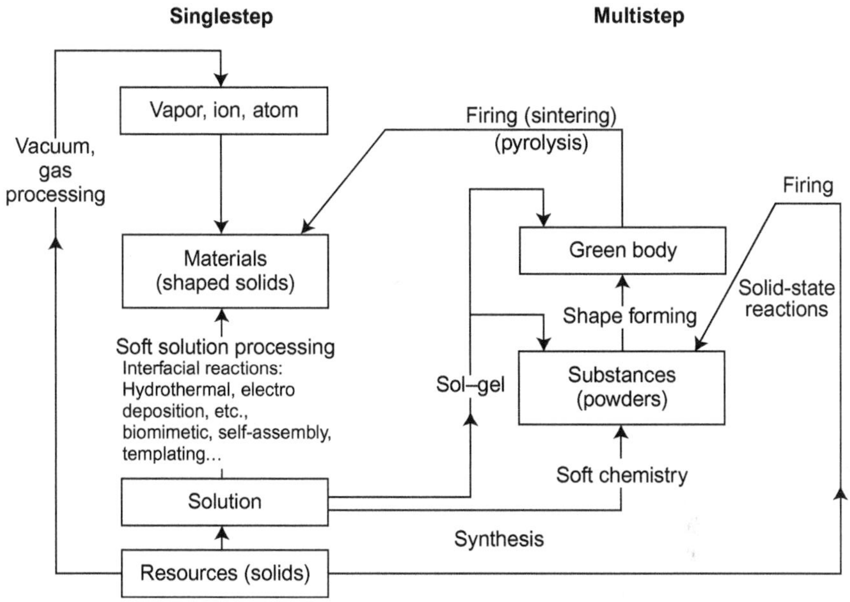

Figure 10.3 Schematic diagram of advanced materials processing showing single-step versus multistep process [18].

during the crystallization of the fine particles. This marked the beginning of the study of precursor preparation for different systems, the surface interactions with the capping agents or surfactants, and polymerized complexes. The surfaces of the particles could be altered to hydrophobic or hydrophilic depending upon the applications [19,20]. Today, this approach is playing a key role in nanotechnology to prepare highly dispersed, oriented, and self-assembled particles. Such an approach leads to the processing of highly controlled and also self-assembled particles of even complex and multicomponent materials. Using such an approach, a wide range of advanced materials like the PZT family of ceramics, ferrites, phosphates, sulfides, oxides, and HAps and composites have been prepared as fine particles for technological applications with preferred morphology such as whiskers, rods, needles, plates, and spheres, depending upon the applications. This precursor-based chemical approach to hydrothermal synthesis has made tremendous progress in recent years and also drastically reduced the temperature and the pressure conditions of processing of materials. The addition of surface modifiers also helps to inhibit the crystal growth that facilitates the smaller particle size with a narrow particle-size distribution. Figure 10.4 shows the Transmission Electron Microscopy (TEM) images of nanoparticles synthesized under supercritical hydrothermal conditions [19]. As is clearly seen from Figure 10.4, those nanoparticles without modifier are aggregated and they do not disperse in the aqueous solvents. Also the particle size is larger compared to the modified particles. When the pH of the

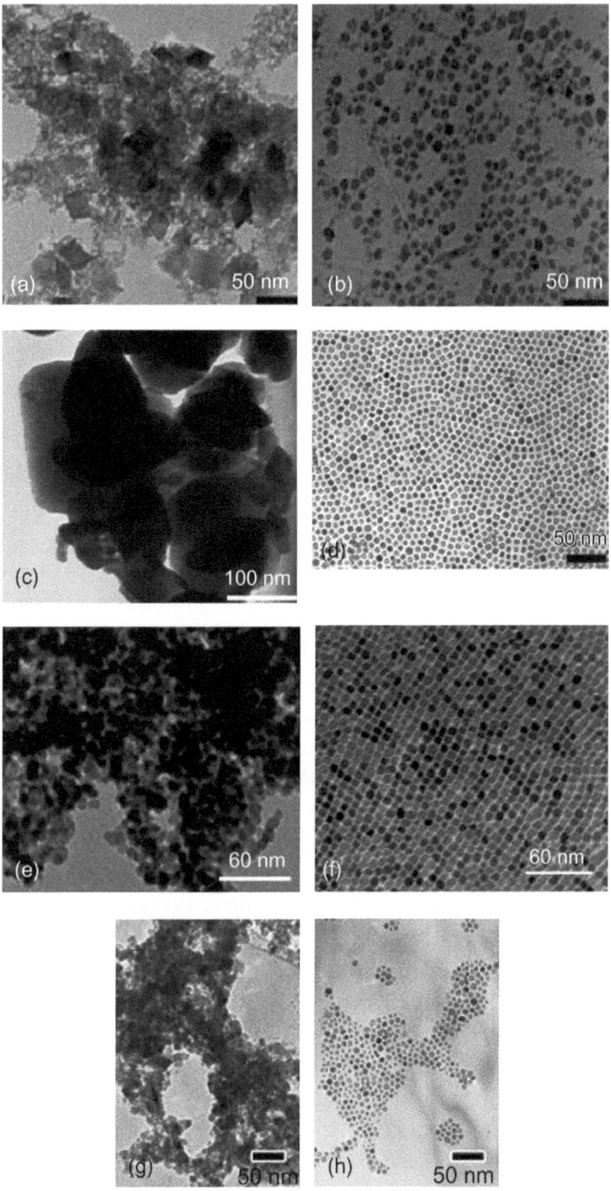

Figure 10.4 TEM images of nanoparticles synthesized under supercritical hydrothermal conditions: (a) Fe_2O_3 (without modifier); (b) Fe_2O_3 (with modifier); (c) Co_3O_4 (without modifier); (d) Co_3O_4 (with modifier); (e) CeO_2 (without modifier); (f) CeO_2 (with modifier); (g) TiO_2 (without modifier); and (h) TiO_2 (with modifier) [19]. (Photos from the works of Adschiri and Byrappa).

reaction medium is highly acidic or highly basic, very small particles along with the large particles are formed leading to a broader size distribution. It seems that because of redissolving of nanoparticles at very high and low pH, Ostwald repining occurs. Therefore, in surface modification, pH of the medium, isoelectric point (iep), and another important parameter, namely, the dissociation constant (pK_a) of the modifiers are very important. At pH below pK_a, the modifier does not dissociate. Moreover, below iep, the surface of metal oxide nanoparticles is surrounded by positive charges (major) and hydroxylic groups (minor). Under these conditions, there is no chemical reaction occurring between the modifier and the metal oxide nanoparticle surface, but it is only through a strong hydrogen bonding the modifier can attach to the nanoparticles surface. On the contrary, at higher pH than pKa, dissociation of modifier takes place and results in a chemical reaction between dissociated part of modifier and OH_2^+ from particles' surface. Thus, by dehydration reaction, the modifier attaches to the surface of the particles. By considering the and chemical reactions, mass balances, and charge balance in the actual system, pH and the modifier can be fixed for most of the systems.

10.4 Hydrothermal Processing of Fine Particles

Processing of fine particles under hydrothermal conditions has been known ever since hydrothermal technology was born. During the late-nineteenth and mid-twentieth centuries, lots of such experiments were carried out on the synthesis of fine particles of zeolites, clays, some silicates, and hydroxides [21,22]. When Barrer reported the hydrothermal synthesis of fine particles of zeolites during 1940s, it opened a new branch of science, namely, molecular sieve technology. During the late 1960s and 1970s, attempts were made to synthesize fine ceramic particles, especially metal oxides using hydrothermal method. It was a most popular field of research under hydrothermal technology [23–26]. A great variety of ceramic materials were synthesized, and the significance of the hydrothermal technique was realized in the processing of highly crystalline fine ceramic particles [27]. This also showed the advantages of hydrothermal technique over other conventional techniques like firing, heat treatment, molding, and hot pressing. The hydrothermal research during 1990s marks the beginning of the work on the processing of fine ultrafine particles with a controlled size and morphology. Today, it has evolved as one of the most efficient methods of soft chemistry in processing the advanced materials like fine nanomaterials with a controlled size and shape. Also, evident from Table 1.7, the hydrothermal technique is ideal for the processing of very fine powders having high purity, controlled stoichiometry, high quality, narrow particle size distribution, controlled morphology, uniformity, less defects, dense particles, high crystallinity, excellent reproducibility, controlling of microstructure, high reactivity/sinterability, and so on. Figure 10.5 shows the major differences in the products obtained by ball milling or sintering or firing, and hydrothermal methods [1]. Currently, the annual market value of electronic ceramics is over a billion dollar, and the market for nanoparticles processing in 2002 was 120 billion dollars and is

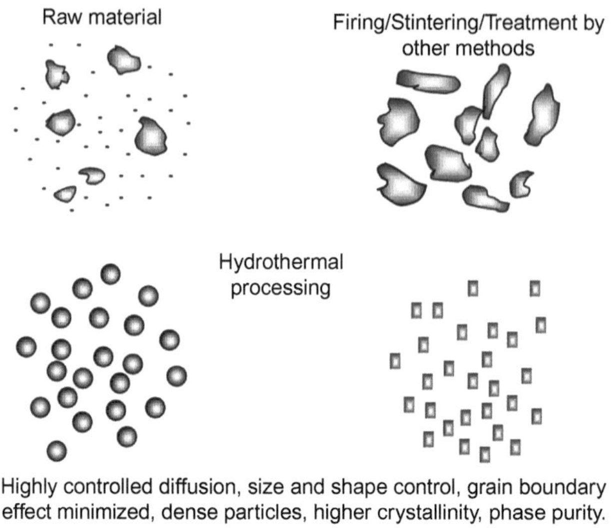

Figure 10.5 The major differences in the products obtained by ball milling or sintering or firing and hydrothermal methods [1].

now raising at the rate of 15% annually and reach 370 billion dollars by 2010, and as per National Science Foundation (NSF) prediction it would jump into a trillion dollar industry by 2015.

Of all the ceramics, the PZT family of ceramics has been studied extensively using the hydrothermal technique. From early 1980s, several thousands of reports have appeared on the preparation of these ceramics. Thermodynamic calculation and kinetics of these systems have been studied extensively [4,5,28]. Several new variants/approaches in the processing of these electronic ceramics have been reported to enhance the kinetics, to shorten the processing time, to control the size and shape, to maintain the homogeneity of the phases, and to achieve reproducibility. A great variety of precursors and also the solvents have been attempted in the processing of these ceramics. Similarly, fine film formation of these ceramics on an appropriate substrate has been accomplished by several workers [29,30].

10.5 Hydrothermal Technology for Nanotechnology

The nanomaterials are known for their unique mechanical, chemical, physical, thermal, electrical, optical, magnetic, and also specific surface area properties, which in turn define them as nanostructures, nanoelectronics, nanophotonics, nanobiotechnology, and nanoanalytics. In the last one decade, a large variety of nanomaterials and devices with new capabilities have been generated employing nanoparticles based on metals, metal oxides, ceramics (both oxide and nonoxide), silicates, organics, and polymers. One of the most important properties of materials in nanosize regime is the changing physical properties. Nanoparticles possess unique optical and electronic properties not observed for corresponding bulk samples owing to a substantial increase in the fraction of surface atoms and the increasing role of the surface effects. The optical properties, and also other characteristics of materials (structure of electronic energy levels and transitions, electron affinity, conductivity, phase transition temperature, magnetic properties, melting points, etc.) become dependent on the nanoparticle size and shape. Such size-dependent properties have been exploited for biological tagging, for example, as fluorescent biological labels. Hence, it is extremely important to control the size and also shape of the nanoparticles/nanocrystals in order to obtain a desired physical property. Therefore, the field of functionalization is almost endless and represents an enormous space to explore. This in combination with molecular imaging can provide a new branch of science, namely, nanobiotechnology imaging research. A major challenge in the nanomaterials science is the accurate control of the size and shape, which in turn is directly linked with the nanomaterials processing method. Nanoparticles can be obtained from a great variety of processes involving the conversion of solid to solid or liquid to solid or gas to solid. The stringent requirements for the biological applications like therapeutic, bioimaging, hyperthermia, targeted drug delivery system (DDS), biosensors, magnetic resonance imaging (MRI), and microelectronics, insist on the control of the size and shape of nanomaterials. Hence, the solution techniques like hydrothermal are becoming the most valuable nanomaterials fabrication tool in the recent years, and they have an edge over all other processing methods because of the high quality of products. By choosing the appropriate capping agents, the surface properties of the nanoparticles can be significantly altered from hydrophilic to hydrophobic and vice versa. Also a perfect dispersion of the nanoparticles in the given solvent can be achieved for making the self-assembly structures. The knowledge on the nucleation, crystallization, self-assembly, and the growth mechanism of the nanocrystals in hydrothermal solution media are rather complicated and are still not well understood [31].

Gold nanoparticles have been around since Roman times. As per the literature data, Michael Faraday [32] was the first to seriously experiment with gold nanoparticles starting in 1850s. The development of new sophisticated tools like Scanning tunneling microscope (STM), Transmission electron microscope (TEM), Atomic force microscope (AFM), to observe, measure, and manipulate processes at the nanoscale level gave a breakthrough to the nanotechnology. Majority of the early hydrothermal experiments carried out during 1840s to early 1900s mainly dealt with the nanocrystalline products, which were discarded as failures due to the lack of sophisticated tools to examine the fine to nanoproducts except in some chemical techniques [21,33–35]. Until the works of Giorgio Spezia [36] in 1900, the hydrothermal technology did not gain much importance in the growth of bulk crystals, as the products in the majority of the cases were very fine grained and submicroscopic in size without any X-ray data. But today, it has evolved as one of the most efficient methods of soft chemistry in processing the advanced materials like nanomaterials with a controlled size, shape, and physical characteristics.

A great variety of nanomaterials have been obtained using the hydrothermal method. Amongst them the native elements, metal oxides, hydroxides, silicates, carbonates, phosphates, sulfides, tellurides, nitrides, and selenides both as particles and nanostructures like nanotubes, nanowires, and nanorods are the most common ones. The method has been popularly used for the synthesis of a variety of nanoforms of carbon like sp^2, sp^3, and intermediate types.

Several researchers [37–40] have pioneered the low temperature soft hydrothermal processing technique for preparing highly oriented, dispersed, self-assembled nanoparticles with a great control over size and morphology, for a wide variety of compounds ranging from native metals, sulfides, selenides, metal oxides, hydroxides, ferrites, PZTs, carbonates, silicates, tantalates, titanates, vanadates, and carbon polymorphs. Their approach is unique, and soft processing methods have been employed through an appropriate solvent and precursor selection. Also some of them have used the external energy like microwave, sonar, electrochemical, mechanochemical, milling, and magnetic with hydrothermal to prepare some of the above-said variety of nanomaterials. Similarly, Adschiri and Arai [42,44] have pioneered the hydrothermal technique under supercritical conditions for processing a wide range of metal oxides, hydroxides, and so on. They have extensively studied the theoretical and experimental aspects of this technique and proposed a systematic mechanism for the formation of various metal oxide nanoparticles under supercritical hydrothermal conditions. In recent years, Adschiri and coworkers have extended this technique for a wide range of other materials like organic–inorganic hybrid materials with a perfect control over the particle formation, size, morphology, and self-assembly [42–44]. Since, the subject of nanomaterials processing under hydrothermal conditions is a vast one in the present day context, here the authors discuss only a selected group of nanomaterials processing to show the present trend and the future direction in hydrothermal processing of materials [31].

The technique is especially handy for the preparation of nanoparticles of native elements, metal oxides, ceramics (PZT, alumina, zirconia, yttria, ceria, and several bioceramics), epitaxial growth of crystalline thin films, composites, fine, ultrafine, and nanoparticles with a desired shape, size, and dispersability. There are several thousands of publications on the processing of advanced materials, using hydrothermal technology, and it is impossible to discuss in detail every aspect of it. Accordingly, the processing of a few important selected advanced materials like native elements, oxides, carbon-based materials, selenides, tellurides, and sulfides, and so on including their nanostructures will be selected. Also the processing of composites, coatings, thin films, fine particles, ultra-fine particles, nanoparticles,ceramics, whiskers, and reinforcement under hydrothermal, solvothermal and supercritical hydrothermal conditions will be discussed.

10.6 Hydrothermal Processing of Selected Advanced Materials

The term *advanced materials* refers to a chemical substance whether organic or inorganic or biological or mixed in composition possessing desired physical or chemical or biological properties. In this chapter, the processing of selected advanced materials using hydrothermal, solvothermal, and supercritical processes is dealt with.

Luis Brus et al. [45] first explained that hydrothermally prepared nanoparticles of cadmium sulfide, CdS, in an aqueous suspension, had a blue shift in the visible absorption and emission spectra compared with bulk CdS. Particles whose radius is less than the exciton Bohr radius exhibit discrete energy levels similar to single atoms. Unlike the band energies observed in bulk materials, every unique crystal diameter on the nanoscale corresponds to a discrete energy. Materials that exhibit this characteristic are called "artificial atoms" or QDs. Recent reviews [46–48] elucidate the degree to which solvothermal synthetic techniques are now an essential technique for controlling the size of the II–VI and III–V semiconductor materials. Synthesis of the QDs typically requires a cation source material that is soluble in the chosen solvent and a surfactant that caps or stabilizes the QD, arresting its growth.

10.6.1 Processing of Native Metals

In recent years, noble metal particles (like Au, Ag, and Pt), magnetic metals (Co, Ni, and Fe), metal alloys (like FePt and CoPt), and multilayers (like Cu/Co and Co/Pt) have attracted the attention of researchers owing to their new interesting fundamental properties and potential applications as advanced materials with electronic, magnetic, optical, thermal, and catalytic properties [49–52]. Although metallic particles can also exhibit the QD behavior, their exciton Bohr radius is much smaller than the semiconductors, resulting in significant synthetic challenges. However, metallic nanoparticle synthesis is of current interest for applications in nanocircuits and devices. The size, shape, and type of material desired depend on the application. For example, the desire for higher density magnetic recording devices initiated the development of a new nanosized ferromagnetic material based on the 3D self-assembly of $Fe_{58}Pt_{42}$ (4 nm) into a superlattice colloidal crystal [53].

The intrinsic properties of noble metal nanoparticles strongly depend upon their morphology and structure. The synthesis and study of these metals have implications for the fundamental study of the crystal growth process and shape control. Majority of the nanostructures of these metals alloys and multilayers form under far-from-equilibrium conditions [54]. Among these metals, alloys, and multilayers, shape anisotropy exhibits interesting properties. Both hydrothermal and hydrothermal supercritical water techniques have been extensively used in the preparation of these nanoparticles.

Zhu et al. (2003) have reported the synthesis of silver dendrite nanostructures using anisotropic nickel nanotubes [50] via mild hydrothermal reactions. The nickel nanotubes acted as a reducing agent. The crystal morphologies which changed from dendrite to compact crystals were investigated during the evolution of the reaction system from nonequilibrium to quasi-equilibrium conditions. Here, the strong shape anisotropy of the Ni nanotube has influenced the formation of Ag dendritic nanostructures. When a poly - vinyl pyrrolidone (PVP) surfactant was used, the nanostructures were replaced by bulk or compact particles. Figures 10.6 and 10.7 show the characteristic photographs of Ag nanocrystals and Ag compact crystals [50].

Several magnetic nanoparticles have been reported in the literature. Xie et al. [52] and Liu et al. [55] have reported the hydrothermal synthesis of cobalt nanorods and nanobelts with and without surfactants. When a microemulsion was used, cobalt nanorods with hcp structures have been obtained at 90°C, with an average particle size of 10 nm diameter and 260 nm length [55]. Similarly, Co-nanobelts via a surfactant-assisted hydrothermal reduction process at 160°C for 20 h have been reported by Xie et al. [52]. Liu et al. [56] have reported a complex surfactant-assisted hydrothermal route to ferromagnetic nickel nanobelts at about 110°C in 24 h. These Ni-nanobelts show remarkably enhanced ferromagnetic properties. Here the key factors in the preparation of these Ni-nanobelts are the pre-formation of the Ni complex $Ni(C_4H_2O_6)^{2-}$, the presence of surfactant sodium dodecylebenzene sulfonate (SDBS–Spectral Database for Organic Compounds), and the selective use of the reducing agent NaH_2PO_2. Such an approach can be extended to the hydrothermal preparation of nanobelts of several other transitional metals and their alloys.

Niu et al. [57] have prepared Ni–Cu alloy nanocrystallites at low temperatures under hydrothermal conditions. These nanoscale metallic alloys such as CuNi,

Figure 10.6 TEM images of Ag dendrites.
Source: Photographs Courtesy Prof. Y.T. Qian.

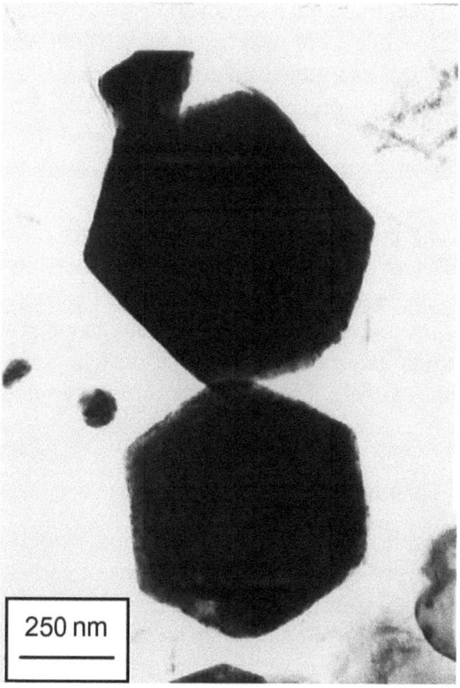

Figure 10.7 TEM image of Ag compact with the addition of PVP in the reaction system.
Source: Photograph Courtesy Prof. Y.T. Qian.

AgPd, and AuPt can be applied in small-scale electronic devices. The authors have used a polymer–surfactant to obtain these Ni–Cu alloy nanoparticles at about 80°C. The average diameter of the particles is about 12 nm. The most vital factor in the preparation of these nanoparticles is the simultaneous reduction of nickel and copper metals, which enables the ready interdiffusion of the different atoms.

In recent years, supercritical conditions have provided reactions for synthesizing nanoparticles of Ag, Au, Pd, In, Pt, Si, Ge, Cu, etc., and are becoming very popular as a consequence of fast kinetics and rapid particle production with the shortest residence time. There are several reports on the preparation of nanoparticles under supercritical water conditions. The reader can refer to Refs [58–61].

Metallic nanoparticles and QDs are finding applications in biosensors. These nanoparticles require hydrophilic surface moieties in order to be compatible with biomolecules. Hydrothermally prepared nanoparticles are particularly suited to biotech applications because the nanoparticles are hydrophilic due to surface hydroxyl groups. However, these hydroxyls often influence the properties of interest in the nanoparticle (e.g., reduce the quantum yield of QDs or oxidize the surface of metals). Other solvothermal routes, however, can be used to prepare nanoparticles which, upon the addition of surfactants, are made hydrophilic. Gold nanoparticles are of particular interest because of their inert nature. Monosized gold nanoparticles have been synthesized under solvothermal conditions by several researchers [58,59]. Here, the hydrogen tetrachloroaurate tetrahydrate was reduced with sodium borohydride, and mercaptosuccinic acid was used as a stabilizer. Figure 10.8 shows the TEM micrograph of self-assembled gold nanoparticles.

Similarly, the coating of nanocrystalline films of Cu, Ni, Ag, Au, Pt, Pd, Rh, etc. on silicon wafers for microelectronics and data storage has been reported [62]. Such an approach has been extended to several other materials like the coating of nanocrystalline carbon on Si wafers.

Thus, the hydrothermal–solvothermal and hydrothermal-supercritical water offer unique advantages over the preparation of these metal nanoparticles over other conventional methods. More recently, the researchers are aiming at imitating the deep-sea hydrothermal conditions in the laboratory to synthesize various metal nanoparticles in the presence of hydrothermal solution and biological molecules, which has opened up new branches of science, namely, hydrogeomicrobiology and geomicrobiology.

10.6.2 Hydrothermal Processing of Carbon Nanoforms

Elemental carbon is known to exist in two well-known polymorphic forms, namely, graphite (sp^2 hybridized) and diamond (sp^3 hybridized). These two forms of carbon exist in nature under highly contrasting conditions and possess very different physical and chemical properties. Besides graphite and diamond, there are reports on new solid forms of carbon [63,64] which have unusual shape and size and are mainly sp^2 hybridized. They are graphene, CNTs (bucky tubes), fullerenes (bucky balls), carbon onions (bucky onions), filamentous carbon, nanocells, and

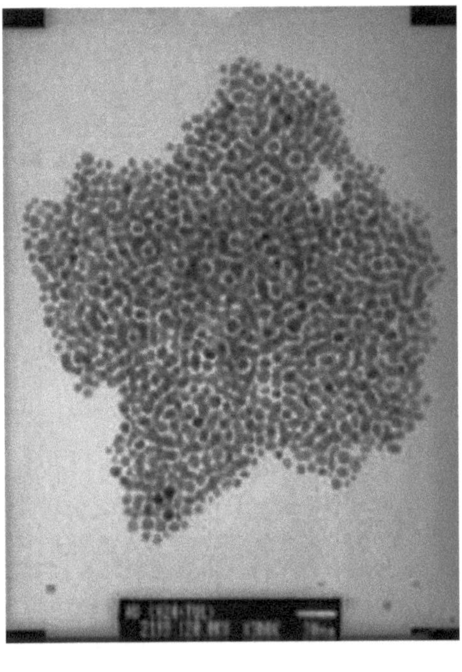

Figure 10.8 TEM micrograph of self-assembled gold nanoparticles.
Source: Photograph courtesy of B.L. Gersten.

nanobeads. These new solid forms of carbon have attracted intense interest owing to their several outstanding properties such as: high specific surface area, good mechanical stability, chemical inertness, and porous nature having large pore volume. Amongst these, graphene, as the fundamental 2D carbon structure with exceptionally high crystal and electronic quality, has emerged as a rapidly rising star in the field of materials science. Its sudden discovery in 2004 led to an explosion of interest in the study of graphene with respect to its unique physical, chemical, and mechanical properties, opening up a new research area for materials science and condensed-matter physics, and aiming for wide-ranging and diversified technological applications [65].

Nanometer-sized diamond has been found in meteorites, protoplanetary nebulae and interstellar dusts, as well as in residues of detonation and in diamond films. It is known that primitive chondritic meteorites contain up to approximately 1500 ppm of nanometer-sized diamonds, containing isotopically anomalous noble gases, nitrogen, hydrogen, and other elements. These isotopic anomalies indicate that meteoritic Nano-diamond (ND) probably formed outside our solar system prior to the Sun's formation (they are thus presolar grains) [65]. Nanoscale diamond particles were first produced by detonation in the USSR in the 1960s [66], but they remained essentially unknown to the rest of the world until the end of the 1980s [67]. Presser et al., have reviewed that beginning in the late 1990s, a number of important breakthroughs led to wider interest in these particles, which are now known as nanodiamonds [68]. First, colloidal suspensions of individual diamond particles with diameters of 4–5 nm ("single-digit" nanodiamonds) became available. Second, researchers started to use fluorescent nanodiamonds as a nontoxic alternative to semiconductor QDs for biomedical imaging. Third, nanoscale magnetic sensors based on nanodiamonds were developed. Fourth, the chemical reactivity of the surface of nanodiamonds allowed a variety of wet and gas chemistry techniques to be employed to tailor the properties of nanodiamonds for use in composites and also for other applications, such as attaching drugs and biomolecules. Fifth, new environmentally benign purification techniques were developed, and these allowed high-purity nanodiamond powders with controlled surface chemistry to be produced in large volumes at a low cost.

Finally, ND was found to be less toxic than other carbon nanoparticles and, as a result, is currently being considered for applications in biomedical imaging, drug delivery, and other areas of medicine. It appears that some interstellar emission bands approximately in the 21 μm spectral regions could originate from NDs [68].

There is an array of carbon nanoforms synthesis available today for researchers and industrialists with relatively fair amount of success [69]. Most of them need high-energy processes like high-temperature arc discharge/laser ablation and the chemical routes like plasma-assisted CVD, chemical vapor transport, and supercritical hydrothermal fluids. The other methods that are equally popular are detonation technique, high-energy ball milling of HPHT diamond microcrystals, chlorination of carbides, and ion irradiation of graphite. Some researchers are even envisaging the moderate conditions of various forms of carbon synthesis with a relatively high yield compared to the other high-energy methods using supercritical fluids [70].

Although the carbon system has been investigated for the past several decades by researchers worldwide, most of the solid forms of carbon-like diamond such as carbon (amorphous carbon), carbon nanocells, fullerenes, and graphene still do not find a place in the carbon phase diagram, which gives only the stable fields of phases like graphite, diamond, and liquid phase. Graphite and diamond occur in contrasting geochemical environments. The exact processes that control their formation in nature are still debatable. But it is very well known that the P–T–fO_2 and C–O–H fluid systems have played a significant role in the formation of these two pure forms.

Here, we shall consider only the wet chemical routes, particularly hydrothermal, solvothermal, and supercritical fluid techniques of synthesis of nanoforms of carbon. Ever since De Vries [64] first suggested the possibilities of hydrothermal synthesis of diamonds based on some of the geological evidences like syngenetic C–H–O fluid and silicate mineral inclusions in natural diamonds, several researchers attempted the synthesis of nanosize and bigger size diamond crystals and other carbon-based nanomaterials using the hydrothermal method [70]. Different carbon sources were attempted with varying success rate. The hydrothermal and solvothermal techniques are highly promising for reactions involving volatiles as they attain the supercritical fluid state, and supercritical fluids are known for their greater ability to dissolve nonvolatile solids. Silicon carbide powder has been used for the synthesis of carbon polymorphs, and Gogotsi et al. [72] have reported decomposition of silicon carbide in supercritical water and discussed the formation of various carbon polymorphs. In 1998, Li et al. (1998) [73] succeeded in synthesizing diamond particles through a metallic reduction–pyrolysis–catalysis route based on the Wurtz-like reaction of carbon tetrachloride and sodium at 700°C. This inspired the synthesis of CNTs using a carbon precursor like hexachlorobenzene, with a planar hexagon configuration [74]. The scheme below illustrates the catalytic assembly benzene-thermal route to multiwalled CNTs with an average diameter of 40 nm at a moderate temperature. Co/Ni catalyst was used at a temperature of 350°C. This is perhaps a breakthrough in the reduction of experimental temperature of synthesis of CNTs, though yield of CNTs was relatively low (~10%).

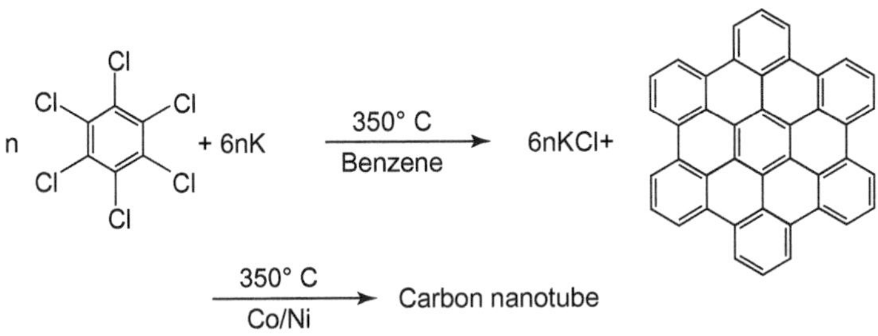

The catalytic assembly benzene-thermal route to multiwalled CNTs with an average diameter of 40 nm at a moderate temperature. Co/Ni catalyst was used at a temperature of 350°C [74].

Later, several researchers explored the possibility of solvothermal synthesis of carbon nanoforms. Basavalingu et al. (2001) have synthesized carbon polymorphs under solvothermal conditions through decomposition of silicon carbide in the presence of

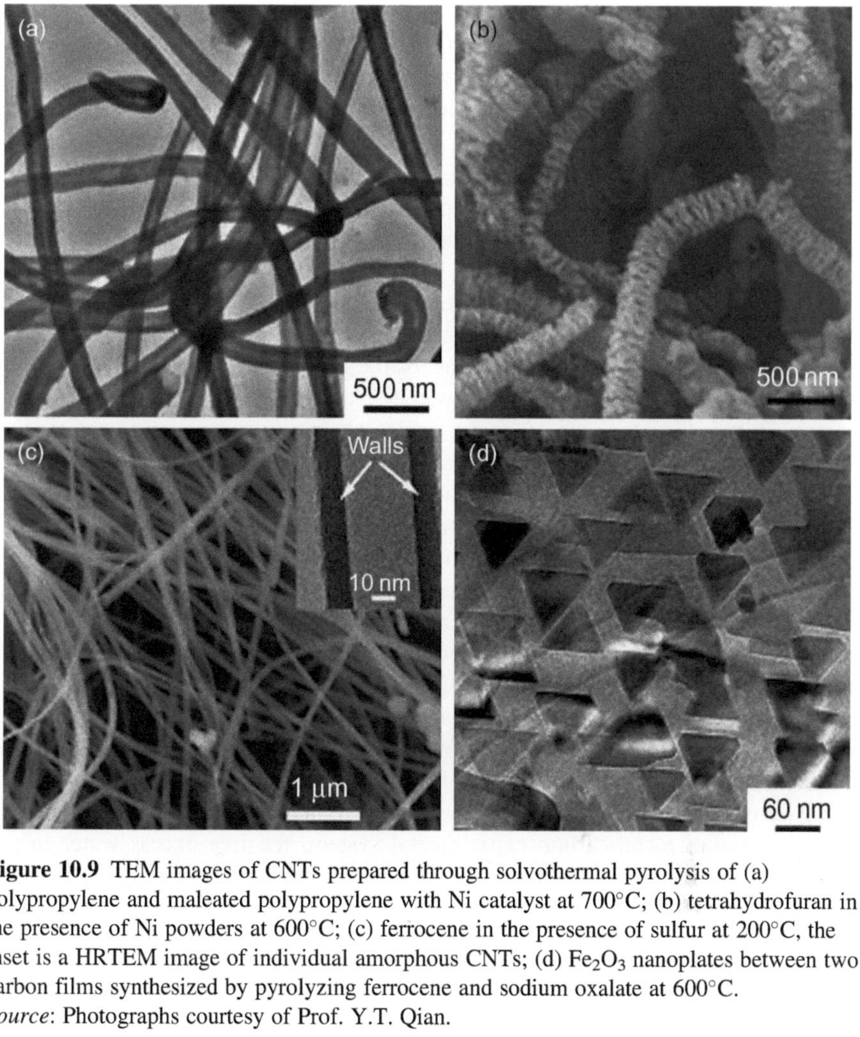

Figure 10.9 TEM images of CNTs prepared through solvothermal pyrolysis of (a) polypropylene and maleated polypropylene with Ni catalyst at 700°C; (b) tetrahydrofuran in the presence of Ni powders at 600°C; (c) ferrocene in the presence of sulfur at 200°C, the inset is a HRTEM image of individual amorphous CNTs; (d) Fe_2O_3 nanoplates between two carbon films synthesized by pyrolyzing ferrocene and sodium oxalate at 600°C.
Source: Photographs courtesy of Prof. Y.T. Qian.

organic compounds [63]. The organic compounds decompose into various C–O–H fluids (it is also popular as solvothermal pyrolysis, since the organic compounds decompose at high temperatures); the main components are CO, OH, CO_2, and C_1H_x radicals. It is very well known that these fluids play a significant role in creating a highly reducing environment in the system and also assist in the dissociation of silicon carbide and precipitation of the carbon phase. Figure 10.9 shows TEM images of CNTs prepared through solvothermal pyrolysis. Using such solvothermal pyrolysis, CNTs with outer diameters of ~50 nm were prepared from ethanol at 550°C with NiO and Co_2O_3 as catalysts [75], from the pyrolysis of polypropylene and maleated polypropylene in the presence of Ni catalyst at 700°C [76]. Carbon rods and fibers using microsized catalyst particles were produced through the pyrolysis of tetrahydrofuran at 500°C in the presence of Fe and Ni powders that led to carbon nanofibers with an average diameter of 100 nm [77], and the pyrolysis of tetrahydrofuran at 600°C in the presence of Ni powders that resulted in carbon fibrils that were formed with stacking graphite sheets of 10–40 nm in thickness [78]. Amorphous CNTs with outer diameters of 50 nm and lengths up to 100 μm were prepared through the pyrolysis of ferrocene in the presence of sulfur at 200°C. When sodium oxalate was used instead of sulfur, the pyrolysis of ferrocene at 600°C generated a novel sandwich structure made of carbon film/Fe_3O_4 triangular nanoplates/carbon film.

In conjunction with solvothermal treatment, hydrothermal carbonization is yet another way to extract carbon from carbonaceous matter, and the carbohydrates are the best precursors to synthesize nanostructured carbon materials or carbon-based hybrid nanomaterials via a low-temperature hydrothermal carbonization process [79,80].

Basavalingu, Gogotsi, Roy, and others have used metal carbides in the synthesis of carbon nanoforms [63,70–72,81]. The starting charge was β-SiC powder having specific area less than 8 m^2/g^{-1} was used along with organic compounds. The organic compounds used in the present investigation are of reagent grade formic acid, oxalic acid, malonic acid, maleic acid, glycolic acid, and citric acid [58]. The starting materials comprising of β-SiC and organic compound without water were sealed in annealed gold capsules (50–60 mm length and 4.5 mm i.d. having a wall thickness of 0.1 mm), thus restricting the amount of water in the system to the water released through the dissociation of organic compounds. We found that the excess water in the system would decrease the yield of carbon precipitation and the thermodynamic calculation of Jacobson et al. [82] indicated that the formation of free carbon is expected in the low water to carbide ratio. Further, in the high-pressure metal-carbon experimental system, the free excess water in the system inhibits the formation of diamond, and the formation of graphite is more favorable. It was found that the silicon carbide decomposes to either quartz or cristobalite along with free carbon particles in the presence of both water and organic compounds, but the yield of carbon particles has improved when there was no excess water in the system. The carbon particles formed were discrete or linked spherical shaped particles having pores, the pores were elongated irregular in shape with pore diameter of 20–30 nm. Thus, demonstrating that the C–O–H supercritical fluids produced through decomposition of organic compounds will have great influence in decomposing the silicon carbide and precipitating the free elemental carbon. The increased concentration of atomic hydrogen and the C_1H_x radicals, by dissociating the organic compound in a closed system, is an ideal environment for the stabilization of the carbon phase especially for the sp^3-hybridized carbon under the subnatural conditions. The authors noticed not only the improvement in the yield of carbon particles but also the change in the shape of the carbon particles precipitated, spherical, ovoid, and scaly material having metallic luster (Figure 10.10). After careful examination of these carbon particles under high-resolution SEM, we found that some of the spherical particles are porous and hollow, and the broken pieces of these carbon particles exhibit the growth of very

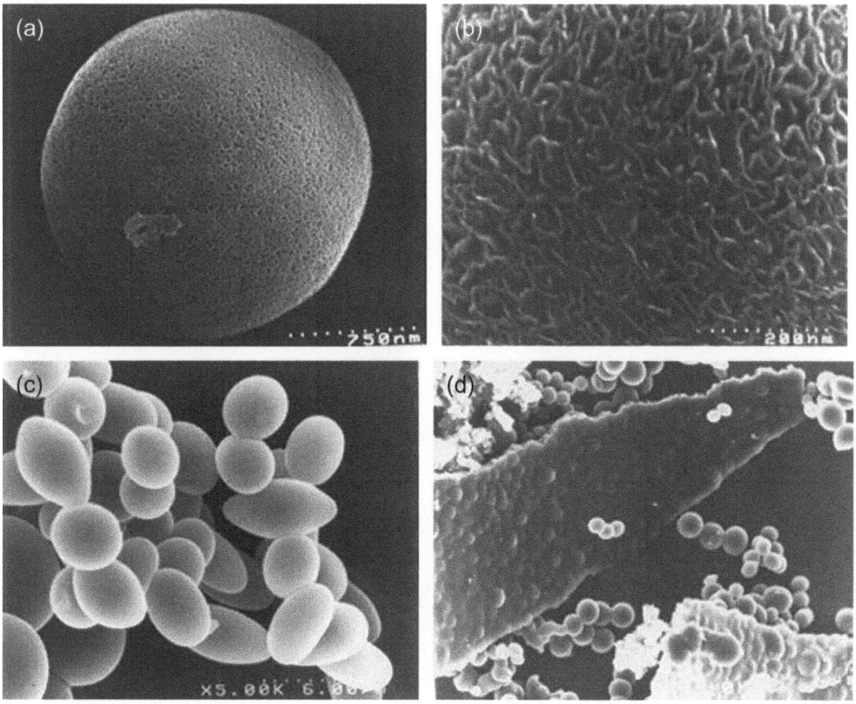

Figure 10.10 SEM images showing (a) spherical particles, (b) enlarged image of pores, (c) ovoid-shaped carbon, and (d) spherules with scaly material.
Source: Photographs courtesy of B. Basavalingu.

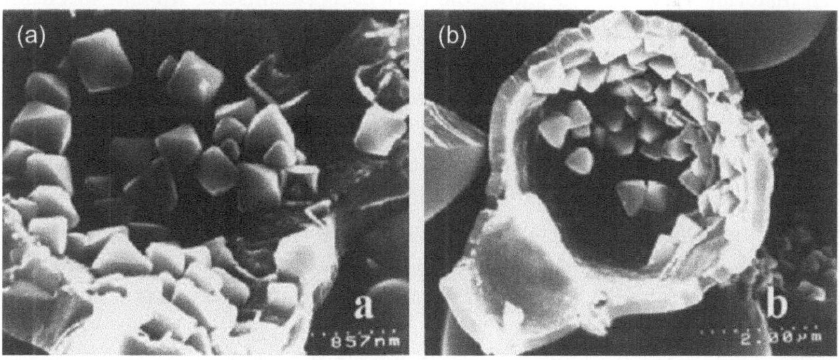

Figure 10.11 (a and b) SEM images of carbon particles showing well-developed octahedral facets adhered to the inner walls of the broken spherical particles.
Source: Photographs courtesy of B. Basavalingu.

minute crystallites which are adhered to the inner walls' spheres, and these crystals show well-developed octahedral facets (Figure 10.11a and b). The growth of these crystallites resembles the zeolite crystals grown in vugs and cavities of volcanic

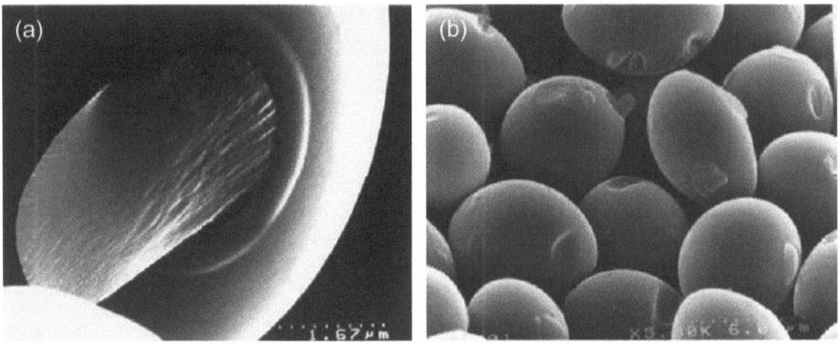

Figure 10.12 (a and b) SEM images showing fluid-like material coming out of solid spherical and ovoid particles.
Source: Photographs courtesy of B. Basavalingu.

flows. We have also noticed different stages in the development of hollow and porous carbon phase formation, i.e., fluid like material is coming out of the solid spherical or ovoid particles through a vent and resulting in hollow ovoid or spheres (Figure 10.12a and b).

Raman spectroscopic study of the run products indicates that the bulk of the carbon particles exhibit the broad spectrum with prominent peaks at 1350 and 1580 cm^{-1}, which are referred to D- and G-band of C–C stretching vibrations. The G-band is from the carbon atoms forming hexagonal lattice, and the D-band is associated with vibrations of carbon atoms in dangling bond in plane terminations which is characteristic of sp^2 hybridization corresponding to graphitic or disordered carbon material. Figure 10.13 shows the typical representative Raman spectrum for spherical and ovoid-shaped aggregates of carbon particles along with the Specpure graphite sample. The Raman spectra obtained for a few selected scaly materials having metallic luster and for the nanosized crystallites adhered to the inner walls of the porous spherical particles is shown in Figure 10.14 along with the spectrum of commercially available diamond powder (Hyprez—Engis, USA). This spectrum has a sharp peak at 1332 cm^{-1} and a shallow broad peak at 1590 cm^{-1}.

When chromium carbide was used as the source of carbon, under similar experimental conditions, it yielded filamentous-type carbon along with some spherical shaped carbon. Figure 10.15 shows SEM images of filamentous carbon and spherical particles.

Presser et al. [68] have studied in detail the thermodynamics of carbide reactions with halogens in the formation of carbide-derived carbons—from porous networks to nanotubes and graphene. Several carbides like SiC, Ti_3AlC_2, TiC, Ti_2AlC, and ZrC have been considered in such studies. Wang et al. [83] have studied the thermodynamics of diamond nucleation on the nanoscale. They conclude that the size of the diamond-critical nuclei should be limited within several nanometers in the hydrothermal synthesis and there should be a reduction of carbide (HSRC) supercritical fluid systems [84]. According to the thermodynamic model, they first

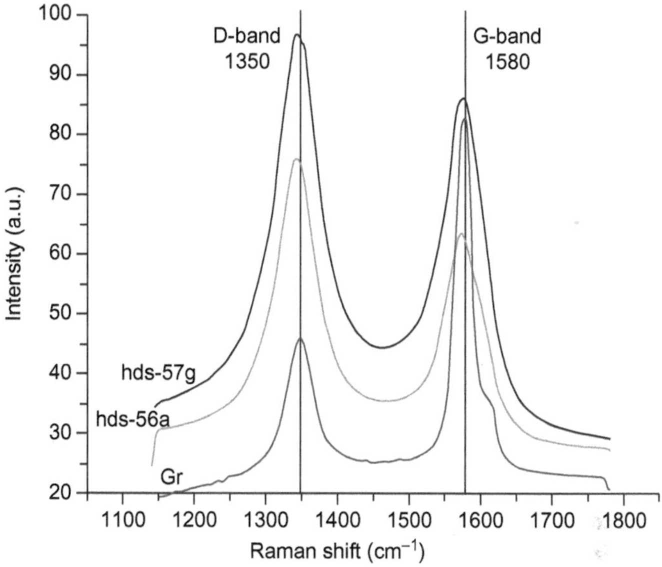

Figure 10.13 Micro-Raman spectra of spherical (hds-56a) and ovoid-shaped (hds-57g) aggregates of carbon particles compared with that of spec pure graphite (Gr).

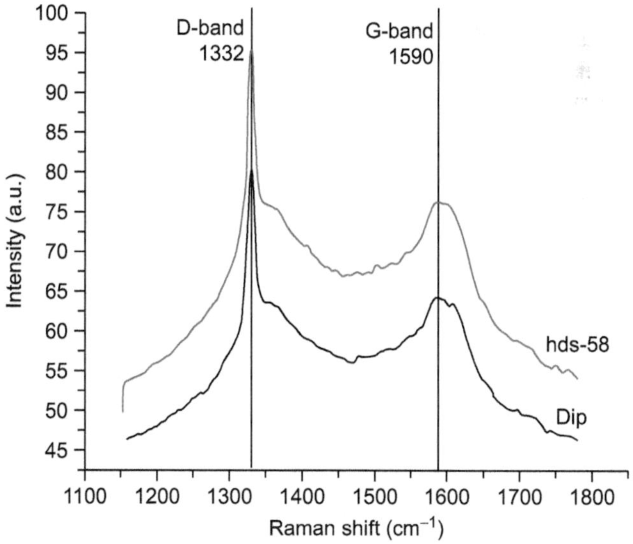

Figure 10.14 Micro-Raman spectra of porous spherical particles (hds-58) compared with that of the commercially available diamond powder (Dip).

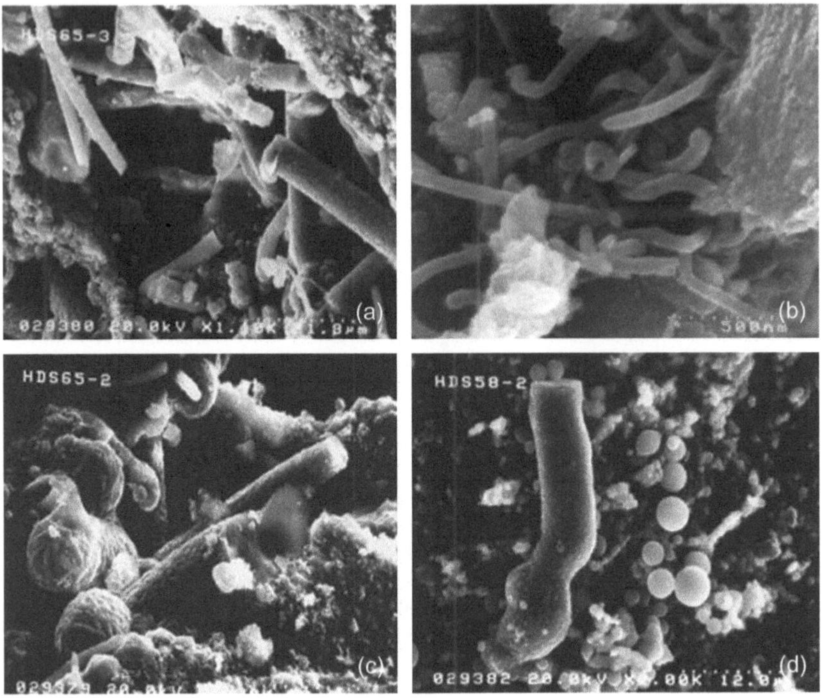

Figure 10.15 SEM images of filamentous carbon (a–c); and spherical particles (d).
Source: Photographs courtesy of B. Basavalingu.

calculated the size and the energy of the critical nucleation of diamonds, respectively, in which all data are from the securable literatures about the diamond synthesis in the HSRC supercritical fluid systems. More importantly, our theoretical results are in excellent agreement with experiment data and other calculations from the first principles.

In recent years, fluorescent nanomaterials are often called QDs, which have wide application potential in biomedical imaging and sensing [85–87]. The semiconductor based QDs often pose cytotoxic and other environmental hazards, and in this respect the fluorescent carbon nanoparticles (FCNs) are the best possible replacement for the semiconductor-based QDs like InAs, CdS, and CdSe. Sun et al. [88] first synthesized these FCNs and they are considered as a brand-new class of fluorescent materials. Several researchers are employing novel one-step approach to synthesize FCNs. Similarly, one-pot hydrothermal synthesis of graphene QDs is equally popular. In both cases, a variety of surfactants are popularly used to alter the surface characteristics of these QDs and their photoelectric conversion the under the near-infrared region [87]. Small graphene oxide sheets and polyethylene glycol are used as the starting materials. This provides a highly cost-effective material for solar energy conversion.

10.6.3 Hydrothermal Processing of Advanced Ceramics

From 1990s onwards, the science of ceramics has undergone a revolution almost as dramatic as the more familiar ones in electronics. Novel approaches in preparing and processing ceramic solids have been developed, ingenious ways of circumventing the age-old problem of brittleness have been introduced, and new markets have opened up in such areas as: electronics, sensors, photonics, orthopedics, catalysis mixed ionic and electronic conducting ceramics, advanced nitride ceramics, advanced cements, mineralizers, heat engines, functional ceramics related to energy conservation environmental issues and so on [88,89].

Recently advanced ceramics represent developments well beyond the imagination of even the few far-sighted scientists of 30 years ago who first perceived the remarkable potential of ceramic solids and established "ductile" engineering ceramics as a suitable objective for material researchers to pursue. Figure 10.16 shows the interactions of ceramics science with other technical fields [88]. Since 1980, much attention has been paid to hydrothermal processing of fine zirconia, ceria, and titania powders. Several methods were used to prepare fine zirconia hafnia, titania, ceria, and PZT particles (powders) under hydrothermal conditions.

Applications of hydrothermal reactions in ceramics include the following aspects: phase equilibria, ultrafine single crystals, ultrafine amorphous, single crystal growth, hydrothermal reaction sintering, hydrothermal sintering, hydrothermal crystallization, dissolving, corrosion, etching, composites (inorganic + organic, inorganic + inorganic), testing, thin films, radioactive waste management, hydrothermal oxidation, hydrothermal decomposition, hydrothermal anodic oxidation,

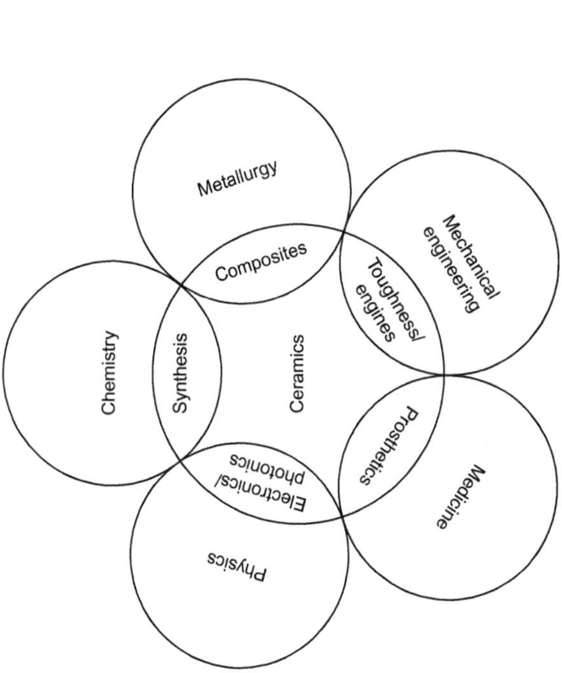

Figure 10.16 Interactions of ceramics science with other technical fields [88].
Source: Courtesy of National Academy Press, Washington, DC.

and reactive electrode submerged arc process. A comparison of different ceramic processing methods have already been discussed in Table 1.6.

The hydrothermal preparation of very fine powders is an excellent approach to ideal powders. The ideal powder should have the following parameters:

i. Fine powder less than 1 μm
ii. Soft or no agglomeration
iii. Narrow particle-size distribution
iv. Morphology sphere or equiaxed
v. Chemical composition controllable
vi. Microstructure controllable
vii. Uniformity
viii. Free flowing
ix. Less defects dense particle
x. Less stress
xi. Reactivity sinterability
xii. Crystallinity
xiii. Reproducibility
xiv. Process control.

The shape of ceramic products obtained under hydrothermal conditions is highly varied and some common hydrothermal products are listed below:

i. Fine powder (single crystals or amorphous)
ii. Fiber
iii. Hydrate cement
iv. Large single crystal
v. Sintered body
vi. Film.

The major advantages of hydrothermal processing of ceramics are as follows:

i. High quality
ii. High purity
iii. High rate of reaction
iv. Dispersion
v. Better shape control
vi. Pollution free
vii. Energy saving
viii. Low temperature operation
ix. Use of large volume of the equipment
x. New products
xi. Better nucleation control and so on.

However, there are some disadvantages like equipment—autoclaves which are of complicated design, and the fact that it is expensive; cumbersome operations like assembling and dissembling; the fact that it is impossible to observe the actual process; solubility aspects; and problems related to surface chemistry. Table 10.4 lists some of the salient features of the ceramic synthesis by commonly used techniques.

Table 10.4 Ceramics Synthesis

Process	Advantages	Disadvantages	Composites	Particle Size
Sol gel (hydrolysis of metal alkoxides), precipitation, coprecipitation	High product purity and homogeneity, crystal symmetry; metastable compounds with unique properties; narrow particle size distribution; lower sintering temperatures	May require calcination, sometimes milling; mixed alkoxides can cause inhomogeneity, nonstoichiometry; sometimes expensive raw materials	Oxides; LAS, mullite, spinel, cordierite	Down to 5 nm; 0.1–5 μm
Evaporative decomposition solids	Wide range of chemical compositions; single-step process; no separate calcination or milling required	Hollow aggregates formed; precursor must decompose at low temperatures; excess carbon impurities may require calcinations	Oxides, nonoxides, composites, fibers	1–20 μm
Hydrothermal processing	High surface area; 99% dense sintered powders; submicron particles with narrow size distribution; simple equipment; continuous; no milling or calcination; short reaction times; lower energy requirements	Requires moderate temperatures/ pressures; requires additives (seed crystals), surface treatments	Y_2O_3–PSZ, Eu_2O_3-doped HfO_2 lead-zirconate titanate, other oxides	Down to 8 nm

(Continued)

Table 10.4 (Continued)

Process	Advantages	Disadvantages	Composites	Particle Size
Rapid expansion of supercritical fluids	Wide range of compositions including amorphous powders and nonequilibrium materials; fast reaction times (10^{-5} s); lower temperatures than plasma process	Requires high pressure and high temperature; corrosive solvents; some agglomeration; machined nozzles to improve morphology	Oxides; (SiO_2, GeO_2, SiO_2–GeO_2) fiber precursors (polycarbosilane); labile, explosive materials	0.01–1.3 μm (ceramics) <0.1 μm fiber, 1 μm particle (polymers)
Reductive dehalogenation of elemental halides	Low temperature, any combination of elemental halides that can be reduced by alkali metals	Crystallization requires higher temperatures; high-speed, high-shear stirring needed; starting materials can be expensive; reactions are explosive or combustible	Amorphous precursors of TiB_2, SiC, B_4C, and other borides and carbides, SiC/TiC, SiC/TiN	

Source: From the works of M. Yoshimura and S. Somiya.

10.6.4 Hydrothermal Preparation of Simple Oxide Ceramics

Zirconia ceramics are remarkable materials due to their excellent mechanical, electrical, thermal, and optical properties [89,90]. Ceria- and zirconia-based oxide nanoparticles are being studied extensively for a variety of applications like catalysis, UV adsorbers, gas sensors, abrasive, electrochemistry, optics, and also biological applications. Structural, optical, electronic, and catalytic properties of ceria-based oxide nanocrystals are strongly dependent on their crucial geometrical parameters like size and shape. Therefore, the preparation of high quality of ceria nanocrystals of desired morphology is of great fundamental and technological interest. Physical and chemical properties associated with these characteristics are closely related to structures and phase changes. Therefore, numerous studies have been undertaken on phase diagrams and structural changes in the zirconia systems [91–93]. There are many contradictions and discrepancies, especially around 1000°C. One of the most important steps in solving this problem is the study of *metastable phase diagram* [94–96]. Yashima et al. [92] have studied the problem of diffusionless cubic–tetragonal (c–t') phase transition, where t' emphasizes a metastable tetragonal phase. Thus, it can be distinguished from the stable t, which is formed diffusionally. Figure 10.17 shows a metastable–stable phase diagram of the ZrO_2–CeO_2 system that is depicted in a composition temperature field. On cooling from 1650°C, the high-temperature cubic phase (B) transforms without cationic diffusions into a metastable tetragonal phase below the tetragonal cubic transformation. Such a metastable phase boundary is drawn by dashed and dotted lines [97]. There are several groups in the world working on the hydrothermal processing of ceria-based oxide nanoparticles [98–100].

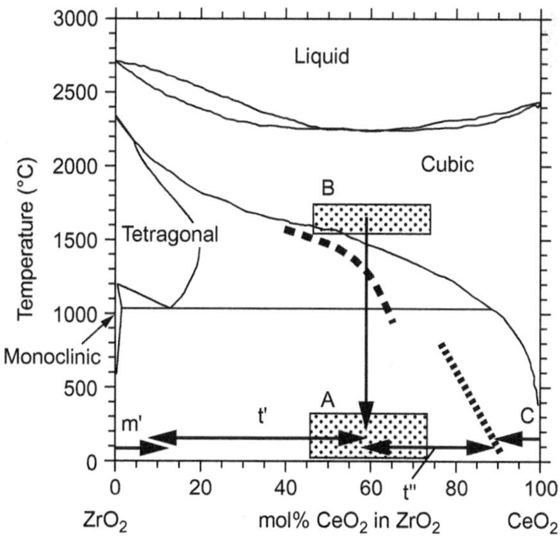

Figure 10.17 Metastable–stable phase diagram of the ZrO_2–CeO_2 system [97].

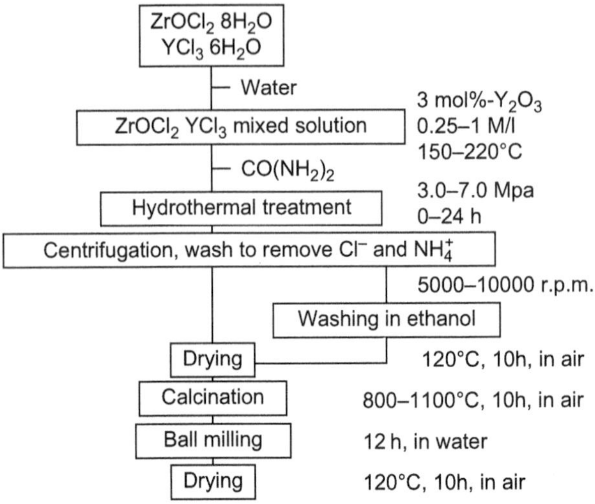

Figure 10.18 Hydrothermal homogeneous precipitation method [99].

Tani et al. [97a] have studied in detail the effect of mineralizers on the crystallization of solid solutions in the system ZrO_2-CeO_2 under hydrothermal conditions [98]. Samples prepared by the coprecipitation method were treated under 100 MPa at 600°C for 2–72 h using distilled water or a solution of alkaline metal (Li, Na, K) fluorides, chlorides, bromides, carbonates, nitrates, sulfates, or hydroxides as mineralizers.

Somiya et al. [121] have given the following flowchart for the hydrothermal homogeneous precipitation method Figure 10.18 [99]. The autoclaves used were (i) Turtle type, (ii) 500 cl, (iii) 1000 ml, (iv) 5000 ml, and (v) 20,000 ml. The lining metals were Pt, Zr, and Ti. By this means, the authors could obtain very fine powders of ZrO_2 and $Y_2O_3-ZrO_2$ of size 20 nm, with a narrow size distribution, very high purity, excellent sinterability, less defects, and controllable shapes. Hence, hydrothermal processing provides an excellent mean fraction by obtaining controlled shape and size from the ceramic particles.

In the hydrothermal oxidation of metals, pulverization and oxidation proceed simultaneously during the reaction with high-temperature and high-purity water. These phenomena have been applied in the preparation of fine oxide powders (20–30 nm) of Fe_3O_4, Cr_2O_3, ZrO_2, HfO_2, Al_2O_3, and others at relatively lower temperatures (400–600°C) [100–102]. Figure 10.19a–d shows the fine crystals of zirconia and ceria. Figure 10.20a–c shows the representative photographs of hydrothermally processed ceramic powders [99].

In order to obtain surface modification and a desired size and shape, several modifiers are being used by researchers. A variety of organic acids and amines are being used for this purpose. Figure 10.21a shows the TEM image of CeO_2 nanoparticles obtained through such an approach by Ahniyaz et al. (2005) [103]. This figure shows the self-assembly of fine nanoparticles of several nanometers size. The valency state of the Ce ions alter with the size of the nanoparticles, and a control over such an alteration is quite important in nanomaterials processing. Similarly, Ahniyaz et al. (2005) have obtained fine nanoparticles of 5 ± 1 nm size particles of

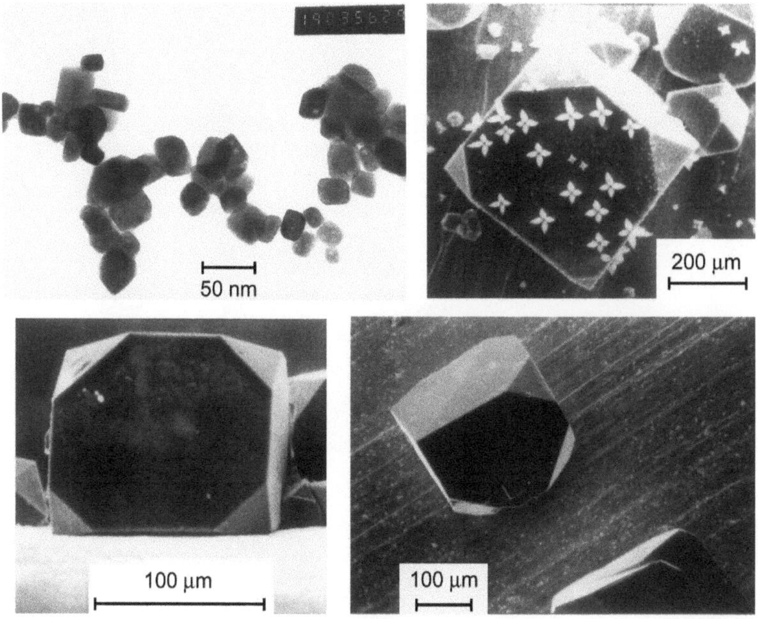

Figure 10.19 Fine crystals of zirconia and ceria.
Source: Photographs by M. Yoshimura.

ceria–zirconia solid solutions ($Zr_{0.5}Ce_{0.5}O_2$). A clear solution of 0.2 M $Ce(NO_3)_3$ and $ZrO(NO_3)_2$ was treated hydrothermally at 120°C for 6 h in the higher pH region (>9.0). Figure 10.21b shows the TEM photographs of ceria–zirconia solid solution nanoparticles [103]. The most significant aspect of these reports is the much lower temperature of ceria nanoparticles processing compared to the earlier published literature data and also without any additional external energy source or sophisticated equipment. A simple soft chemical processing route has been employed here. When ball milling was employed to the hydrothermal technique in the preparation of ceria–zirconia nanoparticles, the particles size was further reduced.

It is interesting to note that the structural, optical, electronic, and catalytic properties of ceria nanocrystals are strongly dependent on their crucial geometrical parameters like size and shape. Therefore, the preparation of high-quality ceria nanocrystals with desired morphology is of great fundamental and technological interest. Adschiri and Arai have done extensive work on the supercritical hydrothermal synthesis of ceria particles [41,104,105]. Figure 10.22a and b shows typical ceria particles obtained under subcritical and supercritical conditions. As it could be seen from Figure 10.22a and b, the ceria particles crystallized under subcritical conditions are larger than the ceria particles obtained under supercritical conditions [106]. It was also confirmed by these authors that the rise in experimental temperature and experimental duration influence on the particle size. At subcritical conditions, particle size varied with the reaction time even above conversions of 95%. The contribution of the residual 5% of ion to the particle growth is negligibly small. The growth of the

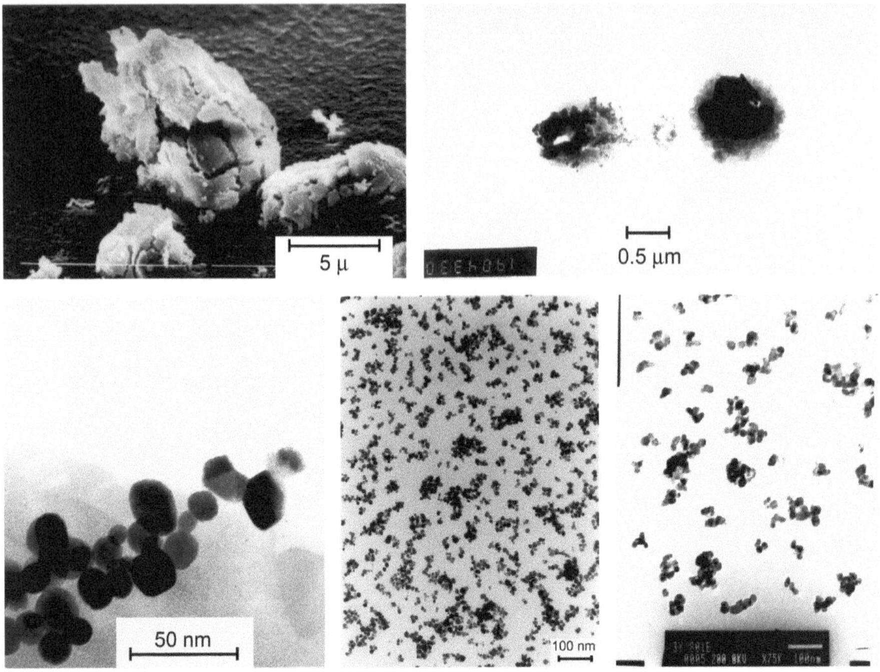

Figure 10.20 Representative photographs of hydrothermally processed ZrO_2 ceramic powders.
Source: Photographs courtesy of S. Somiya.

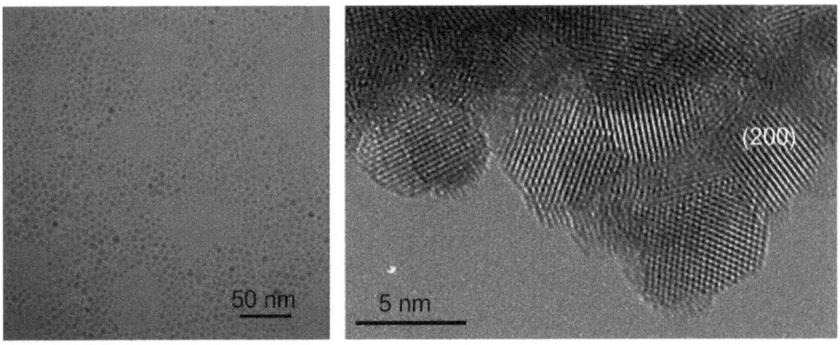

Figure 10.21 TEM photographs of ceria–zirconia solid solution nanoparticles.
Source: Photographs courtesy of M. Yoshimura.

crystal is probably due to Ostwald ripening, namely dissolution of smaller particles and recrystallization into larger crystals.

Sun et al. [102b] have reported the solvothermal preparation of CeO_2 nanorods 40–50 nm in diameter and 0.3–2.2 μm in length by adding ethylene diamine [107]. The morphology was controlled by adjusting solvent composition, surfactant, cerium

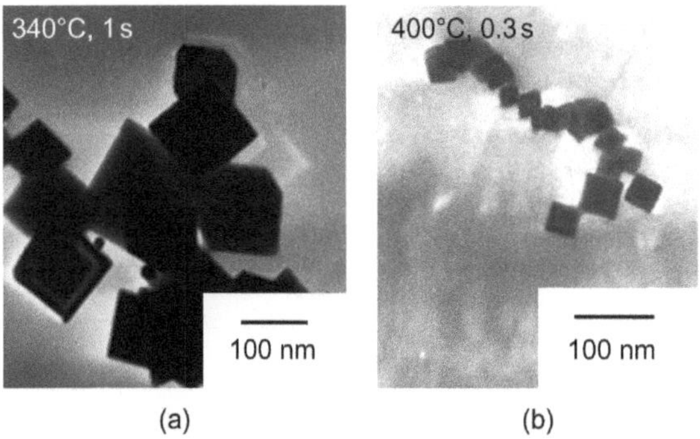

Figure 10.22 TEM photographs of CeO_2 crystals obtained under (a) subcritical conditions and (b) supercritical conditions.
Source: Photographs courtesy: Prof. T. Adschiri, Tohoku University, Japan.

source, reaction temperature, and duration. The UV–Vis absorption and photoluminescence spectra of CeO_2 nanorods show unusual redshift and enhanced light emission respectively, compared with that of bulk CeO_2. This might be due to the abundant defects in CeO_2 nanorods and the shape-dependent effect.

Recently, a simple and green chemistry approach based on the organic ligand assistance to achieve the shape control and self-assembly of the ceria particles under supercritical conditions has been proposed [43,44]. A simple strategy for the synthesis of metal oxide nanocrystals in the organic ligand-assisted SCF technique is shown in Figure 10.23.

Figure 10.24 shows the TEM photographs of the ceria nanoparticles obtained through such an approach. The morphology and the particle size of the ceria vary with the mole ratios of decanoic amine to decanoic acid, which suggests that the self-assembly is influenced by nanocrystals shape, especially, the strong interactions from the exposed crystal planes can control the superlattice pattern, when the average size of nanocrystals and surface ligand molecules are similar. Even the carboxylic acid can be a good surfactant for ceria particle growth under supercritical hydrothermal conditions.

Figure 10.25 shows the pronounced effect of organic ligand molecules on the morphology of the nanocrystals formed by the SCF technique. The transformation of the shape of the ceria nanocrystals from truncated octahedral to cubic was mostly caused by the suppression of crystal growth on the (001) surface as the organic ligand molecules were likely to interact preferentially with the (001) surface (Figure 10.25).

A wide range of metal oxide nanoparticles have been obtained under supercritical hydrothermal conditions for applications not only in ceramics, coatings, catalysts, sensors, semiconductors, magnetic data storage, solar energy devices, ferrofluids but also in medical fields such as hyperthermia, bioimaging, cell labeling, special drug delivery system (DDS), and so on. However, their application potential is dictated by their surface nature, particle size, and also shape. Although the synthesis of metal oxides is not new, their applications were limited. With the discovery of size quantization effect in these materials during 1980s, there is a seminal progress in the synthesis of these metal oxides with desired properties for

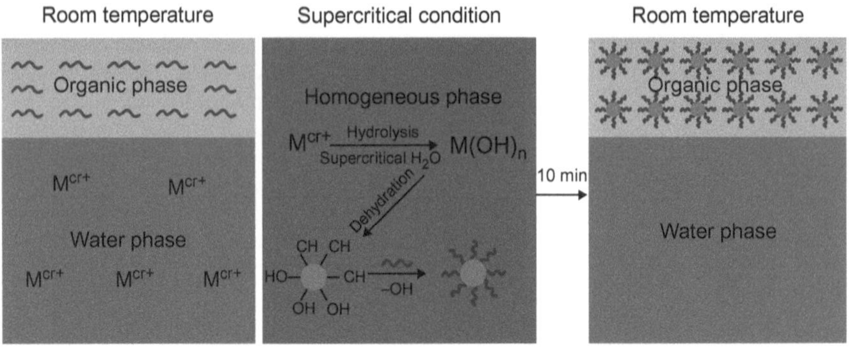

Figure 10.23 The strategy for the synthesis of metal oxide nanocrystals in the organic ligand-assisted SCF technique [44].

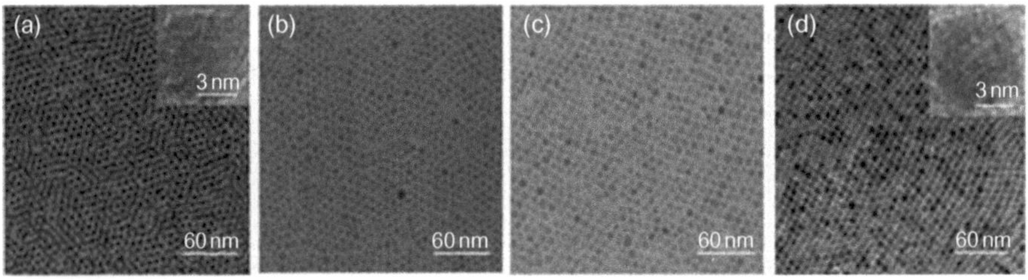

Figure 10.24 TEM photographs of ceria nanocrystals obtained through organic ligand-assisted supercritical hydrothermal synthesis. The mole ratios of decanoic amine to decanoic acid were (a) 0, (b) 1:5, (c) 1:1, and (d) 2:1.
Source: Photographs courtesy: Prof. T. Adschiri, Tohoku University, Japan.

several new applications. Researchers like Adschiri, Arai, Reverchon, Byrappa, Lester, and Poliakoff have contributed extensively on the synthesis of metal oxides and proposed a systematic mechanism [8,41,42,58,105,108,109]. Metal oxides like: Al_2O_3, Ga_2O_3, In_2O_3, SiO_2, GeO_2, ZnO, V_2O_5, VO_2, V_2O_3, TiO_2, CeO_2, ZrO_2, CoO, α-Fe_2O_3, γ-Fe_2O_3, NiO, Co_3O_4, Mn_3O_4, γ-MnO_2, Cu_2O, $CoFe_2O_4$, $ZnFe_2O_4$, $ZnAl_2O_4$, Fe_2CoO_4, $BaZrO_3$, $BaTiO_3$, $BaFe_{12}O_{19}$, $LiMn_2O_4$, $LiCoO_2$, and La_2O_3 have been prepared by the above method. Usually, the particles obtained in the subcritical water conditions are larger than those in the scH_2O, because there is a particle growth with an increase in the residence time, whereas under supercritical conditions such a phenomenon has not been observed. The hydrothermal reaction rate in scH_2O is higher, and the solubility of the metal oxides is much lower than that in subcritical water. This leads to the generation of higher degree of supersaturation. The nucleation rate is expected, by the function of degree of supersaturation, and the surface energy according to the nucleation theory. Thus, extremely high nucleation rate can be expected at supercritical conditions, which leads to the formation of nanosize particles. Figure 10.26 shows the mechanism for the fine

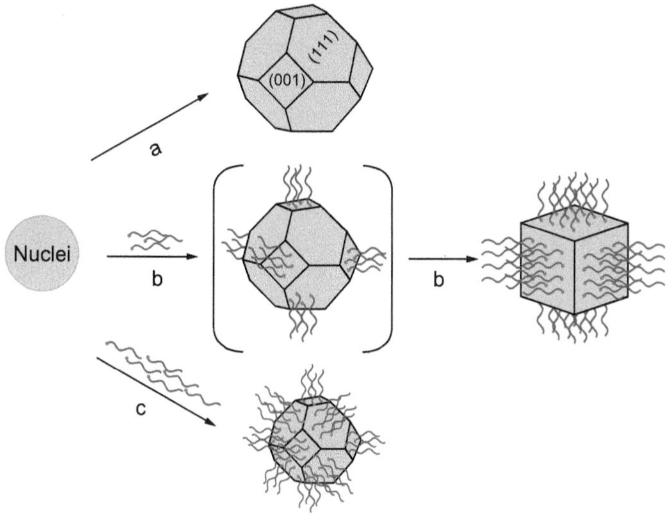

Figure 10.25 Shape control in ceria nanoparticle fabrication: (a) truncated octahedron in the case when no organic ligand molecules are used; (b) at a low decanoic acid to ceria precursor ratio, the preferential interaction of the ligand molecules with the ceria {001} planes slows the growth of {001} faces relative to (111) faces, which leads to the formation of nanocubes; and (c) at a high deconoic acid to ceria precursor ratio, organic ligand molecules block growth on both (001) and (111) faces, which leads to the formation of truncated octahedral and smaller crystals [44].

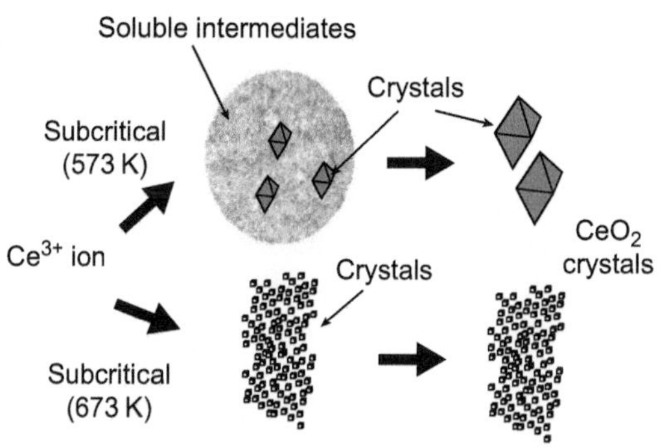

Figure 10.26 Mechanism of CeO_2 nanoparticles formation under supercritical hydrothermal conditions.

CeO_2 particles formation in scH_2O. More or less a complete list of the materials obtained under supercritical hydrothermal conditions is available in the works of Reverchon and Adami [58] and Byrappa and Adschiri [20]. Basically, the hydrothermal synthesis method is available for the metal oxides by conventional hydrothermal synthesis method. The point to be noted here is that under supercritical hydrothermal conditions, nanometer-sized metal oxides could be synthesized, and crystallinity of the nanoparticles would be much higher when compared to the metal oxides obtained under conventional hydrothermal conditions, wherein bulk single crystals are formed. This sometimes leads to the specific characteristics of the products.

Jiao et al. [110] have reported the hydrothermal preparation of ZrO_2 nanocrystallites using organic additives. Phase-pure tetragonal and monoclinic zirconia nanocrystallites of various particle sizes and morphologies were prepared in the presence of polyhydric alcohols such as glycerols and di- and tri-ethanolamine, which gave a tetragonal phase, while alkyl halides favored the formation of monoclinic ZrO_2. The as-prepared tetragonal zirconia particles were spherical or elliptical in shape and ~8–30 nm in size, whereas the monoclinic zirconia particles were spindle-like and ~20–40 nm in size.

10.6.5 Hydrothermal Processing of TiO_2 and ZnO Nanoparticles

Metal oxides such as TiO_2 and ZnO find extensive applications in modern technology exhibiting a number of attractive characteristics such as chemical stability, nontoxicity, low cost, and the highest oxidation rate. Both have high melting temperatures. The unique physicochemical properties of these metal oxides also offer an exciting spectrum of applications having some additional advantages of being biocompatible and environmentally friendly. They have several advantages due to their low cost, ease of handling, and high resistance to photoinduced decomposition [111]. TiO_2 shows maximum light scattering with virtually no absorption. It is nontoxic and chemically inert. It also has a special photocatalytic sterilization function, which can be used for antibacterial applications [112]. It is a widely used photoprotective component of various cosmetic products. Hence, the penetration of titania nanoparticles through the epidermis of human foreskin to understand its possible biological effects in vivo and in vitro is an important study [113]. The titania composites are most popular as implants and hence their biological response of tissues with macrophagic activity to titania has been investigated extensively [114]. Major breakthrough occurred during 1970s when Fujishima and Honda [111a] reported electrochemical photolysis of water at semiconductor electrode (TiO_2) [115]. TiO_2 has several other specific applications and the common ones are as photocatalyst, dye-sensitized solar cells, white pigment, ceramic glazes, sun screen, and UV absorbers, also as electronic data storage, and so on. Zinc oxide is well known as n-type wide band gap semiconductor ($\Delta E = 3.37$ eV at 300 K) with a large exciton energy of 60 meV and thermal energy of 27 meV, and is a subject of research owing to its unique mechanical, electrical, and optical properties with a combination of high stability, very high melting point with valuable device potential for piezoelectric transducers, gas sensors, optical waveguide, transparent conductive films, varistors, solar cell windows, bulk acoustic wave devices, and so on. Due to bright UV luminescence, ZnO is a perspective material for the manufacture of UV-light-emitting diodes, UV lasers operating at room temperature, and display devices. Moreover, ZnO QDs with very low toxicity, high photostability, biofriendly, and biodegradable have been demonstrated [116–118].

During 1960s, the growth of TiO_2 and ZnO as bulk single crystals was very popular, and much of the earlier literature survey shows such attempts, and hydrothermal method was popularly used to synthesize both these metal oxides under high-temperature and high-pressure conditions [120–121]. However, researchers experienced several difficulties in the large-sized crystals of these metal oxides, and their popularity was highly limited.

With the discovery of quantization effect during the late 1970s, the size reduction in metal oxide semiconductors became a major objective to achieve higher efficiency in their applications. Difficulties in the growth of metal oxide nanocrystals were associated with the control of size, morphology, coagulation, reproducibility of results, and dispersability. During 1990s, several such attempts were made on the use of organic molecules as ligands or capping agents or surfactants. Also ex situ surface modification was tried with limited success. Thus, an alternate route was envisaged to obtain high-quality nanocrystals of TiO_2 and ZnO with controlled size, morphology, dispersability, and without any coagulation using in situ surface modification through novel solution processing consisting of hydrothermal, solvothermal, and supercritical hydrothermal methods. The size could be reduced significantly to a few nanometers without any coagulation, and the organic coating was so much uniform and thin that did not alter the inherent properties of metal oxide core. The organics that are insoluble and exist as separate phases under ordinary conditions became homogeneous phases under hydrothermal/solvothermal conditions. The small size and high surface-to-volume ratio of the individual nanoparticles impart distinct size-tunable physical and electronic properties that have prompted some to refer to them as "artificial atoms." A highly controlled self-assembly of these hybrid nanocrystals, when dispersed in organic solvents into 2D and/or 3D ordered structures or superlattice structures, remains a relatively unexplored area. In recent years, the significant reduction in the metal oxide semiconductors particle size has resulted in newer applications such as QDs for biological imaging. A proper understanding of the cytotoxicity and genotoxicity of these metal oxides would be highly useful for their applications in biological systems [122]. The researchers are exploring the possibilities of their applications in food packing and storage. In finding appropriate applications for these metal oxides, the control of size, morphology, and dispersibility to some extent are most important. Several approaches are being adopted to achieve higher efficiency in this respect like the synthesis methods, doping with an appropriate metal, surface modification, and organic additives. Although several methods of synthesis are employed for both TiO_2 and ZnO, such as sol–gel, microemulsion, hydrothermal, microwave, and homogeneous precipitation, the novel solution processing consisting of hydrothermal, solvothermal, and supercritical hydrothermal methods are more effective owing to their controlled diffusion, higher activity of the solvent, and the homogenization of the surfactant with the media under higher temperature and pressure conditions [19,20,31]. The hydrothermal synthesis has been clearly identified as an important technology for materials synthesis [1,20,31]. Hydrothermal technique is a promising alternative synthesis method because of the low process temperature and very easy to control the crystal size.

The hydrothermal process has several advantages over other growth processes such as: The use of simple equipment, ease of operation, catalyst-free growth, low cost, and environmentally friendly process. The low reaction temperatures make this method an attractive one for microelectronics and plastic electronics [6]. Also the method is especially useful for the synthesis of nanotubes, nanowires, nanoribbons, and other nanostructures of these two metal oxides, which find some special applications in the nanostructure forms. In spite of the fact that doping has been one of the oldest approaches crystal growers adopted in the earlier times to modify the crystal morphology, it is still popular, even in the processing of nanoparticles. The researchers are also employing coupled doping to alter the semiconductor properties of these metal oxides, and it has become an active area of research. Several parameters are being worked out theoretically in this respect [123,124]. Further, the use of in situ surface modification has yielded enormous success in the processing of these metal oxide nanoparticles under hydrothermal conditions [125]. Some researchers are discussing the advantages of organic additives over the surface modifiers not only to control the size and morphology but also to enhance the reaction kinetics.

Contrary to the growth of TiO_2 and ZnO metal oxide bulk single crystals, the synthesis of TiO_2 and ZnO nanoparticles is usually carried out under mild hydrothermal and solvothermal conditions (usually $T = <120-180°C$, but in some cases up to 250°C and P = autogeneous) and at around 400°C in case of supercritical hydrothermal method. Generally, the experiments are carried out in general purpose autoclaves using Teflon liners or batch reactors using platinum or gold liners or capsules. Figure 10.27 shows the flow chart of the experimental methodology adopted in our laboratory. A wide range of solvents and surfactants has been used for both TiO_2 and ZnO nanoparticle fabrication. Also the molar concentration of the raw material TiO_2 and ZnO was varied to investigate the quality of the resultant products. The experimental duration was varied from 24 h to just 4 h depending upon the precursors used, especially the organic additives and surfactants. The dopants were introduced in the solution form in the desired mole concentration, and in some cases they were introduced as metals directly into the hydrothermal and solvothermal system. After the experimental run, the products were washed and freeze-dried.

Solvents

Acidic (HNO_3, HCl, and HCOOH), alkaline (KOH and NaOH), and organic solvents (methanol, diethylene glycol, *n*-butyl alcohol, and diphenylether) were used as solvents, and the pH of the system was measured before and after the experimental run to check the pH variation. The organic solvents were very active under hydrothermal conditions, and the experimental temperature was significantly low when organic solvents were used.

Also the organic additives played a prominent role in the crystallization mechanism of these two metal oxides, and in controlling their particle size and agglomeration. Hexylamine has been used as additive in small quantity. The role of organic solvents has not been understood precisely, although there are some excellent publications

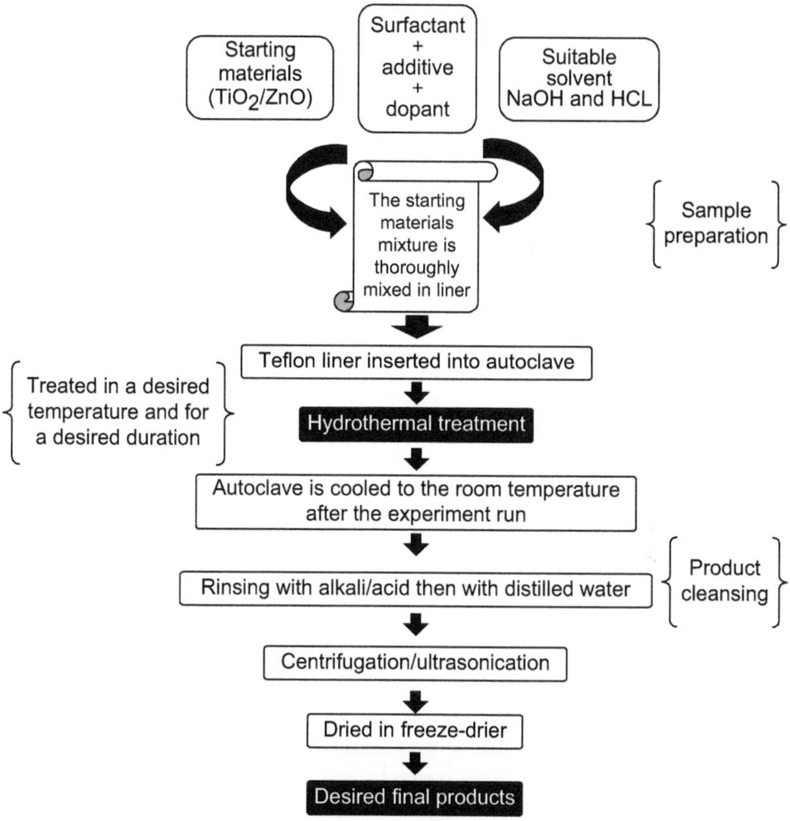

Figure 10.27 Flow chart of the experimental methodology adopted in the synthesis of TiO_2 and ZnO nanoparticles.

available in the literature. Several surface modifiers like *n*-butylamine, caprylic acid, oleic acid, gluconic acid, benzylaldehyde, olcylamine, sodium dodecyl sulfate, citric acid, triton-X, and *n*-hexanol were used. A major disadvantage in using some surfactants, which mask the properties of the inorganic core material and also when used in surplus, can produce in some cases thick and nonuniform outer coatings, leading to the loss of desired properties of the inorganic core material.

Doping

An important part of the crystallization process whether it is the growth of bulk materials or nanomaterials, dopants play an important role in controlling the crystallization processes to some extent, particularly in controlling the morphology, as they hinder growth along some directions in the crystal. There have been several explanations put forward to explain doping difficulties in wide band gap semiconductors [123,124]. First, there can be compensation by native point defects or dopant atoms that locate on interstitial sites. The defect compensates for the substitutional impurity level through the formation of a deep level trap. In some cases, strong lattice relaxations can drive the dopant energy level deeper within the gap. In other systems, one may simply have a low solubility for the chosen dopant limiting the accessible extrinsic carrier density. Each dopant has a different character in this respect. A wide range of dopants like silver, indium, tin, tungsten, palladium, chromium, manganese, iron, antimony, neodymium, and in some cases coupled doping, have been introduced to achieve strong lattice relaxations, which in turn result in a large-scale atomic level defects that enhances the application potential of these compounds. The SCF method is the most popular preparative method for titania nanoparticles. TiO_2 (rutile or anatase) nanoparticles have been prepared frequently starting from stabilized $TiCl_4$ solutions. Some researchers have used the combination of hydrolysis and polycondensation of titanium tetra-isopropoxide, $Ti(OR)_4$, used it in the presence of tetramethyl ammonium hydroxide, with a preliminary heat treatment under reflux. Usually, the titania prepared in supercritical ethanol exhibits a higher degree of crystallinity and contains less hydroxide. The smaller particles have proved to be the best candidates for the photocatalytic and biological applications owing to the larger surface area. However, the conventional hydrothermal methods yield larger particles, and hence the SCF technology is the most viable one for producing nanoparticles with a shortest residence time [120]. Also the properties of titania particles could be easily enhanced through a systematic approach to the synthesis methods. Mousavand et al. [43,125] have done in situ surface modification during supercritical hydrothermal synthesis of titania particles. A $Ti(SO_4)_2$ solution was heated to 200°C or 400°C in the presence of hexaldehyde, which resulted in a perfect dispersion of the synthesized nanoparticles in isooctane, implying more efficient immobilization of hexaldehyde on the titania nanoparticles for the in situ surface modification during the SCF synthesis than for postsurface modification. The binding of the organic molecules used for surface modification might not be physical adsorption, but rather covalent binding. The titania particles without hexaldehyde were aggregated with a broad size distribution, whereas those synthesized with hexaldehyde were more dispersed and with a size range up to 10 nm. The immobilization of the organic molecules immediately after the nucleation of the particles probably suppressed the growth nuclei, and in addition, the formation of the organic layer also suppressed the aggregation of nanoparticles in the solution. Using in situ surface modification, the particles' surfaces can be changed to hydrophilic or hydrophobic, and also the surface energy can be altered, which are the key factors in the biomedical application of these nanoparticles. Figure 10.28 shows the TEM images of the TiO_2 particles synthesized at 400°C in the presence of hexaldehyde. The titania particles without hexaldehyde in the reaction were aggregated and the size distribution was broad ranging from several nanometers to 70 nm, while with the hexaldehyde, the particles were relatively dispersed on the grid, and the size was in the range of several nanometers to 10 nm. It should be noted that the addition of the organic molecules resulted in the reduction of particle size and the dispersion of the particles. This clearly implies that the immobilization of the organic molecules must have occurred

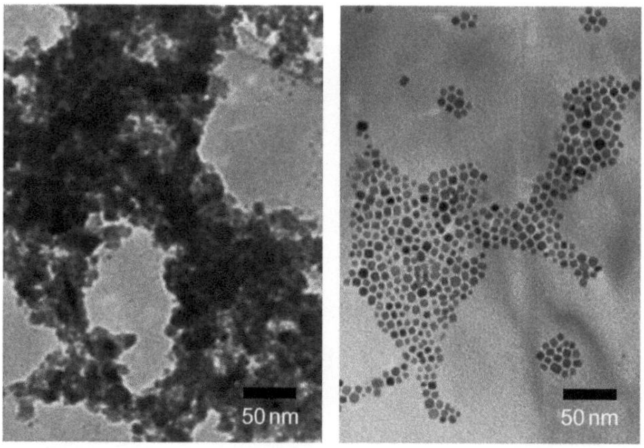

Figure 10.28. TEM images of TiO_2 nanoparticles synthesized at 400°C: (a) without and (b) with surface modifier.
Source: Photographs courtesy of T. Adschiri.

immediately after the nucleation of the particles, which suppressed the growth nuclei, and in addition, the formation of the organic layer also suppressed the aggregation of nanoparticles in the solution. Similarly, titania particles have been obtained using the reverse micelles with $scCO_2$ as solvent catcher to eliminate the organic phase [126].

SEM and high-resolution TEM (HRTEM) images of some representative samples of selectively doped TiO_2 and ZnO nanomaterials are shown in Figure 10.29. When surplus dopants are used as in the case of silver, they form a separate phase, and particles are attached to the host TiO_2 and ZnO particles. In most cases, depending upon the experimental parameters, dopant metals, and surfactants, the particles are unagglomerated with a definite morphology. Without surfactants bigger particles and agglomerated particles were obtained although the other process parameters were carefully monitored.

10.6.6 Other Metal Oxides

Among this class, the magnetic nanoparticles form an important class of metal oxides exhibiting unusual behaviors as a result of size effects. Several new uses of these nanoparticles have been proposed in high-density information storage [127] and in a number of biomedical applications like MRI, hyperthermia, targeted drug delivery, and cellular imaging. Also several clinical applications like detection of liver metastases, metastatic lymph nodes, and inflammatory and/or degenerative diseases have been carried out. The iron oxide nanoparticles synthesized and derivatized with human transferring are used on human dermal fibroblasts in vitro and found that they appear to localize to the cell membrane without instigating receptor-mediated endocytosis, and also induce up-regulation in the cells for many genes, particularly in the area of cytoskeleton and cell signaling [128]. Such special

Figure 10.29 Electron microscopy images of selectively doped TiO_2 and ZnO nanomaterials: (a) manganese doped TiO_2 with caprylic acid as surfactant; (b) neodymium-doped TiO_2 with *n*-butylamine as surfactant; (c) tungsten-doped TiO_2 with caprylic acid as surfactant; (d) ZnO with caprylic acid as surfactant; (e) indium-doped ZnO without surfactant; and (f) indium-doped ZnO with *n*-butylamine as surfactant.
Source: From the works of K. Byrappa.

properties insist on a precise control of properties such as particle size, shape, and variables that affect these properties are the focus of attention among researchers. It is also very important that the particles be superparamagnetic for DDS. In targeted drug delivery, a drug is bound to the surface of a coated nanoparticle and an external magnetic field is applied to attract the particle to a specific site in the body. At the site, the chemical unbinds from the nanoparticle due to specific chemical interactions. The external magnetic field is then turned off and the nanoparticle freely circulates throughout the body until it is naturally eliminated. Therefore, it is important to know the biocompatibility of the particles and possible coating materials [129,130]. Currently, paramagnetic metal ion complexes, such as Gd-EDTA, are used as contrast agents. However, the superparamagnetic nanoparticles offer higher molar relaxivities, which would enhance image contrast [131]. Magnetic nanoparticles in hyperthermia involve heating certain tissues or organs to between 41°C and 46°C for cancer therapy [132]. Hence, the synthesis of magnetic nanoparticles is an important subject of study in nanomaterials science and technology. Several magnetic nanoparticles based on ferrites and perovskites (such as manganates) are in use for such applications, among them iron oxide (α-Fe_2O_3, γ-Fe_2O_3), cobalt oxide and cobalt iron oxide are the most important ones. The present authors discuss these materials in more detail in this review.

Amyn Teja and coworkers have done an extensive work on the synthesis of a variety of magnetic nanoparticles using continuous hydrothermal system

[5,133,134]. Both iron oxide and cobalt oxide nanoparticles were obtained via the following reactions:

$$2Fe(NO_3)_3 + 3H_2O = Fe_2O_3 + 6HNO_3$$

$$6Co(NO_3)_2 + 12NaOH + O_2 = 2Co_3O_4 + 12NaNO_3 + 6H_2O$$

The effect of operating parameters such as experimental temperature, feed concentration, residence time, and salt:sodium hydroxide ratio on particle size, size distribution, and morphology have been examined for these materials. Also thermodynamic modeling has been carried out on these systems to understand the relationship between particle morphology and species distribution [134]. However, most of these works yielded the nanoparticles which were not well dispersed because of the coagulation factor with the size reduction. In order to overcome that problem, several researchers have tried to use the surfactants route to synthesize nanoparticles of very high quality with smaller particle size and highly dispersed under SCF conditions [135,136]. Use of surfactants does not alter the crystal structure, but greatly alters the surface morphology and surface charge of nanoparticles, by controlling the nucleation and crystal growth. Several modifiers like decanoic acid, oleic acid, and hexaldehyde have been used. Figure 10.30 shows the cobalt oxide nanoparticles synthesized by Adschiri and coworkers with and without surface modifiers. The size of the particles can be very effectively controlled by appropriately maintaining the molar ratio of the modifier to starting material and the type of surfactant. Figure.10.31 shows the size control for Co_3O_4 nanoparticle fabrication under SCF conditions.

There are several reports dealing with the deposition of magnetic nanoparticles on activated carbon, HAp, porous media, and several other templates [133]. However, the fabrication of magnetite nanoparticles in conjugation with several biodegradable polymers and coatings using the SCF technology is still a developing field.

Perrotta [136], Laudise and Ballman (1958) [137], and Al'myasheva et al. [138] have reviewed the hydrothermal synthesis of corundum nanoparticles under hydrothermal conditions. A high specific surface area corundum has been synthesized

Figure10.30 TEM images of Co_3O_4 nanoparticles prepared using SCF technology: (a) without surface modifier and (b) with surface modifier C_9COOH at 300°C and 20 MPa *Source*: Photographs courtesy Prof. T. Adschiri, Tohoku University, Jaan.

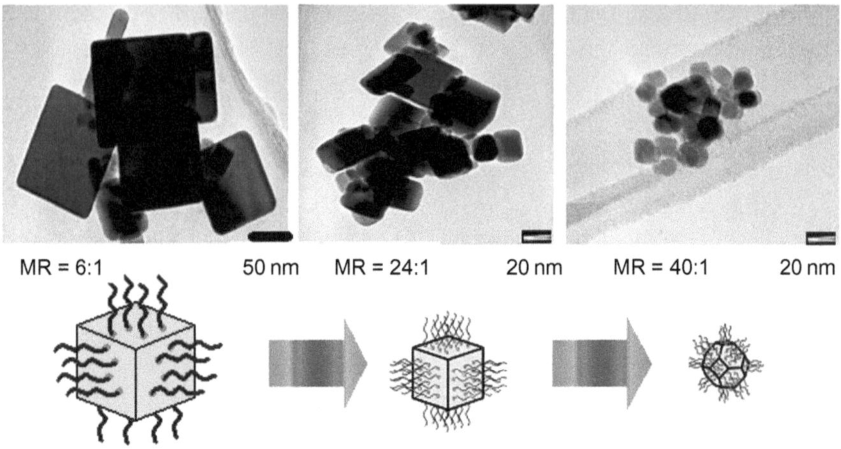

Figure 10.31 Mechanism for size control in Co_3O_4 nanoparticle fabrication under SCF conditions ($T = 300°C$, $P = 20$ MPa, surfactant is hexaldehyde) with molar ratio (MR) of starting material to the modifier.
Source: Courtesy of T. Adschiri, Tohoku University, Japan.

through the conversion of diaspore to corundum under hydrothermal conditions. This nanosized alumina has great application potential. The authors were able to develop a new transitional alumina reaction sequence that gave rise to an alpha intermediate structure, α'-Al_2O_3 with a very high surface area. Also they have investigated the thermodynamic basis and equilibrium relationships for the nanocrystalline phases.

Wang et al. [139] have reported the synthesis of Dy_2O_3 nanorods under hydrothermal conditions at 180°C in about 24 h. Dy_2O_3 was dissolved in concentrated HNO_3, and the pH was adjusted to 7–8 using 10% KOH solution. Then the precipitate was transferred to an autoclave for hydrothermal treatment. The thermal decomposition of $Dy(OH)_3$ gave rise to Dy_2O_3 nanorods.

Sorescu et al. [137a] have synthesized nanocrystalline rhombohedral In_2O_3 under hydrothermal conditions at about 200°C in 4 h [140]. This In_2O_3 has a corundum structure and is a high-pressure phase crystallizing with a rhombohedral structure. The hydrothermally treated product was postannealed at 500°C.

Several workers have prepared the α-Fe_2O_3 (hematite) phase as nanoparticles under hydrothermal conditions (using both aqueous and nonaqueous solvents) with or without surfactants [133–137, 141]. These hematite particles find extensive applications such as catalysts, pigments, recording medium, and sensors. Hydrothermal method shows advantages over conventional methods like sol–gel and hydrolysis of iron salts. Surfactants like sodium dodecylsulfonate, SDBS, cetyltrimethyl ammonium bromide (CTAB), and hexadecylpyridinium chloride have been used. $Fe(NO_3)_3 \cdot 9H_2O$ or FeC_2O_4 was used as the source of iron. NaOH or *N*, *N*-dimethylformamide was used as a solvent. The experimental temperature ranges from 180°C to 250°C in most of the cases. The typical size of the products varies from 20 nm to 200 nm depending upon the starting materials and the experimental temperature. Iron oxides of spinel and magnetic structures are very important for their unique magnetic properties, which can be varied systematically through dopants like Co, Ni, Zn, and Mn.

Wu et al. [142] have prepared nanowire arrays of Co-doped magnetite under hydrothermal conditions at 200°C using ferrous chloride, cobalt chloride, and sodium hydroxide. These nanowires are believed to possess a single magnetic domain which can be regarded as small wire-like magnets.

Wan et al. [136] have proposed a soft-template-assisted hydrothermal route to prepare single crystal Fe_3O_4 nanorods with an average diameter of 25 nm and length of 200 nm at 120°C in 20 h. The formation of these Fe_3O_4 nanorods has been ascribed to ethylenediamine, which plays a crucial role not only as a base source but also as a soft template to form single crystal Fe_3O_4 nanorods. Figure 10.32 shows the Fe_3O_4 nanorods obtained through a soft-template-assisted hydrothermal route. Teja and Koh [134] have extensively reviewed the synthesis of all the magnetic phases of magnetite under hydrothermal conditions. Hematite nanoparticles <10 nm in size were produced from ferric nitrate and ferric ammonium sulfate solutions using this method [143]. Magnetite nanoparticles were also produced using ferrous sulfate, with the addition of urea [144]. It was noted that the particle size and morphology were strongly dependent on operating temperature and processing time. The particle size increased as the operating temperature increased and different forms of iron oxide were produced from the same precursor by changing the temperature. For instance, mixtures of 6-line ferrihydrite (a weakly crystalline form of iron oxide) and hematite were formed at 250−350°C, while pure hematite was obtained at temperatures >350°C.

Figure 10.33 shows very interesting results on the iron oxide nanoparticles modification with different organic ligand molecules. By adding different amounts of the same functional groups as modifier reagents, affects on the particle size, arrangement and morphology of the particles obtained under supercritical hydrothermal conditions (400°C and 30 MPa) can be altered. By adding decanoic and oleic acids to the starting precursors, cubic and spherical magnetite particles were obtained with a mean size of about 25 nm and 5 nm, respectively. Without the modifier it yielded hematite particles. However, under subcritical conditions, the addition of oleic acid to the precursor resulted in the formation of nanowires up to 85 nm length.

Kominami et al. [145] have prepared Ta_2O_5 nanoparticles through solvothermal routes and have studied their photocatalytic properties. They used tantalum pentabutoxide in toluene at 200−300°C in the presence of water. Ta_2O_5 powder of 20−100 nm size showing high surface area of >200 m^2/g was obtained.

Adschiri and coworkers [146−150] have worked out in detail a continuous synthesis (flow technique) of fine metal oxide particles using supercritical water as the reacting medium. They have shown that fine metal oxide particles are formed when a variety of metal nitrates are contacted with supercritical water in a flow system. They postulated that the fine particles were produced because supercritical

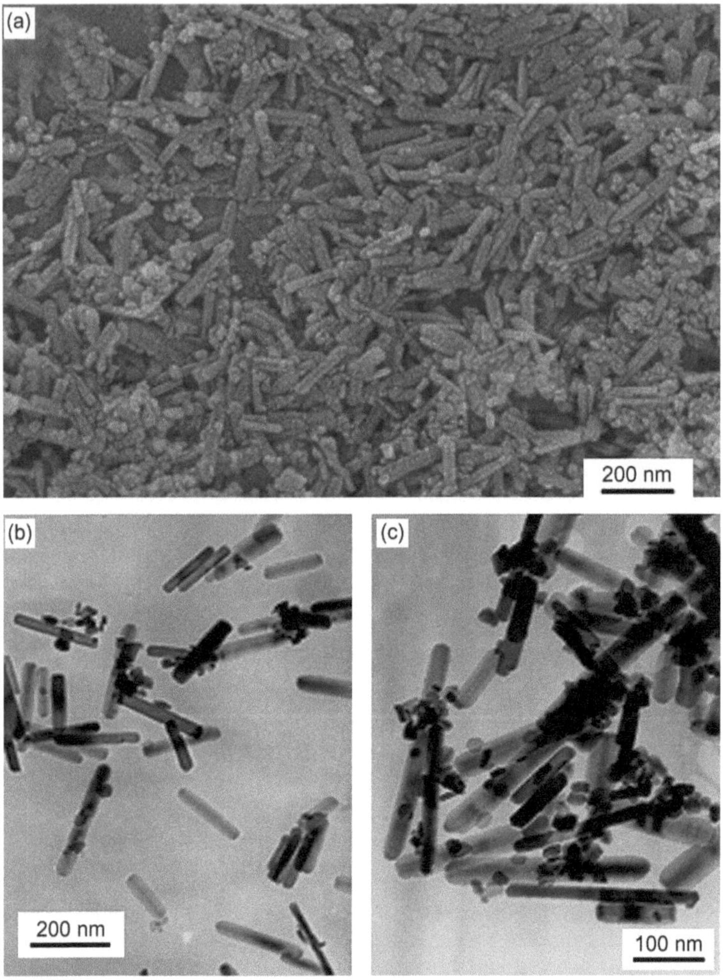

Figure 10.32. FESEM and TEM photographs of Fe_3O_4 nanorods.
Source: Photographs courtesy of Prof. Y.T. Qian.

water causes the metal hydroxides to rapidly dehydrate before significant growth takes place. The two overall reactions that lead from metal salts to metal oxides are hydrolysis and dehydration:

$$M(NO_3)_2 + xH_2O \rightarrow M(OH)_x + xHNO_3$$

$$M(OH)_x \rightarrow MO_{x/2} + 1/2xH_2O$$

Recently, Lester et al. [151] reported a new design that employs a nozzle mixer to exploit differences in the densities of supercritical water and precursor solutions

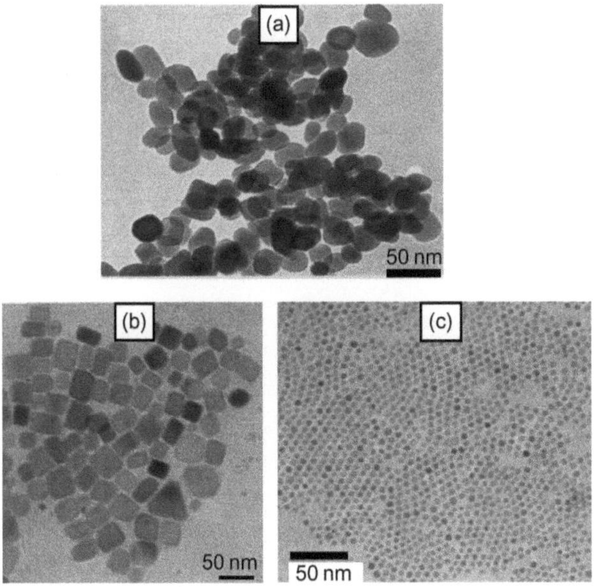

Figure 10.33. TEM images of iron oxide nanoparticles synthesized under supercritical hydrothermal conditions (400°C and 30 MPa) without and with organic ligand molecules: (a) without modifier (hematite), (b) modified with decanoic acid (magnetite), and (c) modified with oleic acid (magnetite).
Source: From the works of T. Adschiri, Tohoku, univ, Sendai, Japan.

to improve mixing inside the reactor. A diagram of their reactor is shown in Figure 10.34. Supercritical water is introduced into the reactor from the top and the precursor metal salt stream from the bottom. Mixing of the two streams is almost instantaneous and makes use of buoyancy-induced eddies to produce "ideal" mixing conditions. This leads to very low residence times and limits subsequent particle growth. Lester et al. synthesized many metal oxides in the size range of 6–64 nm in their system, thereby demonstrating the effective separation of nucleation and growth steps in continuous hydrothermal processing. The continuous hydrothermal technique offers many opportunities for controlling particle size and morphology by keeping residence times low and mixing processes efficient. It is also easy to scale up. However, the engineering of particle surfaces cannot be accomplished in situ and requires additional postprocessing steps.

A major challenge for all methods is the design of magnetic nanoparticles with effective surface coatings that provide optimum performance in vitro and in vivo biological applications. Additional challenges include scale-up, toxicity, and safety of large-scale particle production processes.

Recently, considerable attention has been focused on low-dimensional nanomaterials owing to their unique physical properties and potential applications in sensors, magnetic, electric transportation, optics, and even as building blocks for nanoscale devices. In this regard, the hydrothermal method has become more popular for the synthesis of low-dimensional nanoarchitectures of several metal oxides like titania, zinc oxide, and iron oxide in the form of nanotubes, nanorods, nanowires, nanoribbons, and nanofilms. Yao and Yu [152] have reviewed the hydrothermal synthesis low-dimensional nanoarchitectures. However, the rationale for the hydrothermal synthesis of such nanomaterials with specific nanostructures is still in

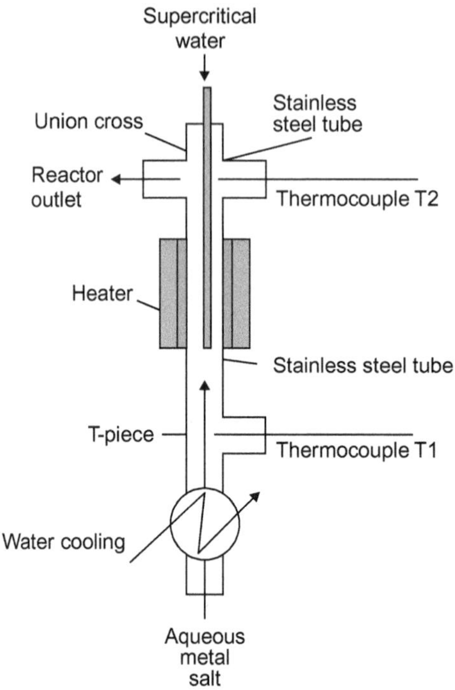

Figure 10.34 Schematic of the nozzle mixer.
Source: Courtesy of E. Lester.

its infancy and far beyond its maturity even though this hydrothermal method has been studied for many years. Yu and coworkers have reported the hydrothermal synthesis of such nanostructures of ZnO, CuO, VO_2, MnO_2, MoO_3, h-WO_3, Fe_2O_3, SnO_2, and oxides and hydroxides of rare earths [153].

10.6.7 Hydrothermal Processing of Mixed Oxides

Several mixed oxide nanoparticles like $CoFe_2O_4$, $ZnFe_2O_4$, $ZnAl_2O_4$, Fe_2CoO_4, $BaFe_{12}O_{19}$, $LiMn_2O_4$, and $LiCoO_2$ have been synthesized using hydrothermal and supercritical hydrothermal methods. Similarly, $MgFe_2O_4$ nanoparticles in the size range 20–50 nm have been prepared under supercritical hydrothermal conditions ($T = 600°C$; $P = 30$ MPa; Mg/Fe mole ratio = 1.5; experimental duration = 10 min) showing greater capability for magnetic hyperthermia. The coercivity force and saturation magnetization at room temperature of these nanoparticles are 61.3 Oe and 15.3 emu/g, respectively. Figure 10.35 shows nanoparticles of $LiMn_2O_4$ with a particle size ranging from 10 to 20 nm synthesized from LiOH, $Mn(NO_3)_2$, and H_2O_2 at 400°C and 30 MPa. These particles do not show the decay of its capacity even after the charge–discharge cycles, which has been considered as a major breakthrough point of these solid electrolyte materials [154,155]. Kanamura et al. [156] have discussed this mechanism in detail and concluded that these particles are single crystals of $LiMn_2O_4$ and are totally different from those obtained by

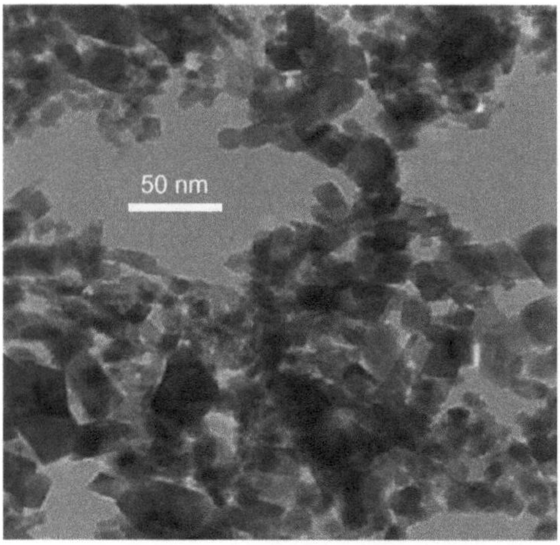

Figure 10.35 TEM image of $LiMn_2O_4$ nanoparticles synthesized at 400°C and 30 MPa.
Source: Photograph courtesy of T. Adschiri.

other methods. In the synthesis of $LiMn_2O_4$ nanoparticles, oxidizing reaction atmosphere has to be controlled by regulating oxygen gas partial pressure in the system. In this case, Mn^{2+} of $Mn(NO_3)_2$ should be oxidized into Mn^{3+}. In order to achieve this, H_2O_2 was fed into the system. H_2O_2 decomposes at supercritical conditions into oxygen gas, which forms a homogeneous phase with supercritical water to provide an excellent oxidizing atmosphere [154–156]. Ni nanoparticles have been formed from nickel acetate and formic acid at 400°C and 30 MPa. In this case, HCOOH was introduced as a reducing agent with nickel acetate solution [157,158]. In supercritical water, HCOOH is decomposed into hydrogen and carbon dioxide. An important point is that these gases and supercritical water form a homogeneous phase, and this mixture of gas (H_2 and CO_2) shows higher reducing ability than H_2 gas, as was reported in the literature [159,160].

Although supercritical hydrothermal technology gives highly crystalline nanoparticles with a homogeneous composition, still there is a problem of larger size, coagulation, and poor dispersibility of nanoparticles in aqueous solutions like water as seen from Figure 10.36.

Poor dispersibility of such nanoparticles is due to the fact that there is an aggregation between nanoparticles which leads to worse dispersion and, after a while, precipitated particles will appear at the bottom of the bottles as shown in Figure 10.36. Hence, there is a trend to obtain small nanoparticles with a perfect control over the size and morphology and high dispersibility with a modern approach in materials processing like the use of organic ligands, capping agents, surfactants, and chelating agents, which generate a new class of nanomaterials, namely organic–inorganic hybrid nanomaterials. This strategy is based on the miscibility of the organic ligand molecules with scH_2O due to the lower dielectric constant of the water; and the nanocrystal shape control by selective reaction of organic ligand molecules to the specific inorganic crystal surface. For this purpose, even some amino acids, peptides, proteins, or DNA are used to modify the particle surfaces and such particles can be specifically combined with other proteins or

Figure 10.36 Dispersibility of nanoparticles: $CoAl_2O_4$ (left); ZnO (center); Fe_2O_3 (right) in water.

DNA, which leads to the applications to new biotechnology or IT devices using directed assembly of semiconductor nanoparticles.

10.6.8 Hydrothermal Preparation of Perovskite Type of Mixed Oxide Ceramics

Alkaline earth titanates have always appeared in two types of stable phases: the pyrochlore and the perovskite structural phases. Between these, the high-purity, fine perovskite titanates with exact stoichiometry are widely used as ferroelectric materials, owing to their excellent electric, pyroelectric, semiconducting, and electro-optic properties. Hydrothermal processing is a very convenient technique for the preparation of various multicomponent oxide materials that have maximum utility in present-day electronics applications [161–163]. These multicomponent oxides from electronic ceramic or catalytic applications can be produced by hydrothermal synthesis at moderate temperatures and pressures (>100°C and 0.1 MPa). The success of hydrothermal synthesis depends on the selection of precursors that are both reactive and cost-effective as well as appropriate process condition variables, which include temperature, pH, and reagent concentration [4]. From 2000 onwards, hundreds of publications have appeared in literature devoted to various aspects for preparation of these perovskite-type oxides owing to their potential applications in modern technology. However, it is impossible to discuss every aspect of these perovskite types of oxides in this handbook which is devoted to all general aspects of hydrothermal technology.

The general formula for perovskite can be written as ABO_3, where the A cation is relatively large and of low valence (such as Ba^{2+}, Sr^{2+}, Ca^{2+}, Pb^{2+}, La^{3+}, Sm^{3+}, Nd^{3+}, Bi^{3+}, and K^{+}), and the B cation is relatively small (such as Ti^{4+}, Zr^{4+}, Sn^{4+}, W^{6+}, Nb^{5+}, Ta^{5+}, Fe^{3+}, Mn^{3+}, Mg^{2+}, Zn^{2+}, and Ni^{2+}), and for the solid solution $A(B_xC_{1-x})O_3$, where $0 < x < 1$, A = Ca, Sr, Ba, Pb, and Bi; B = Ti; and C = Zr.

Kutty and Balachandran [164] have synthesized crystallization of perovskite ($x = 0.5$) at 573 K starting with crystallization of PbO and mixed Zr–Ti gels, obtained by hydrolyzing $TiOCl_2$ and $ZrOCl_2$ and NH_3 (aq.). Beal [165] obtained perovskite ($x = 0.5$) at 573 K from mixtures of zirconia–titania gels, crystalline PbO, and various mineralizers. The author found that the purity and morphology of the product depended on the chemical identity of the mineralizers.

Dawson and Swartz [166] have synthesized several solid solutions ($0 < x < 1$) from the PZT family using aqueous gels obtained by hydrolyzing $TiCl_4$ and $ZrOCl_2$ in basic solutions. The slurry obtained, along with lead oxide and precursors of other elements (dopants), was hydrothermally treated at 573 K. It was found that the obtained solids had the same metal ion stoichiometry as the feeding material.

Riman and coworkers [167–169] have done excellent work on the preparation of these PZT types of ceramics and have studied in detail the thermodynamics and kinetics of these systems. The thermodynamic model proposed by them has been well accepted for this perovskite system. Also, they have developed a new approach—intelligent engineering—in order to transform hydrothermal synthesis from an empirically based technology to one that revolves around engineering principles. They approached this problem from a multidisciplinary perspective of chemistry, chemical engineering, and physical chemistry, which all embrace principles of thermodynamics and kinetics [170,171]. Thermodynamic principles enable one to determine how to design a reaction to yield phase-pure materials. Without this knowledge, it is impossible to distinguish a process that is being controlled by thermodynamics versus kinetics. These authors have studied all the possible reactions that may occur in the hydrothermal medium, more of a typical PZT system, for example, in the Ba–Ti and Pb–Ti systems. The following are the relevant equilibria in the Ba–Ti and Pb–Ti hydrothermal systems [167]:

i. $H_2O = H^+ + OH^-$
ii. $H_2O_{(g)} = H_2O$
iii. $TiO_{2(s)} = OH^- - HTiO_3^-$
iv. $Ba(OH)_{2(s)} = Ba^{2+} + 2OH^-$
v. $BaOH^+ = Ba^{2+} + OH^-$
vi. $BaTiO_{3(s)} + H_2O = Ba^{2+} + 2OH^- + TiO_{2(s)}$
vii. $Ba(OH)_2 \cdot 8H_2O_{(s)} = Ba^{2+} + 2OH^- + 8H_2O$
viii. $BaO_{(s)} + 2H^+ = Ba^{2+} + H_2O$
ix. $Ba_2TiO_{4(s)} + 2H_2O = 2Ba^{2+} + 4OH^- + TiO_{2(s)}$

x. $CO_{2(g)} = CO_{2(aq.)}$
xi. $CO_{2(aq.)} + H_2O = H^+ + HCO_3^-$
xii. $HCO_3^- = H^+ + CO_3^{2-}$
xiii. $BaCO_{3(s)} = Ba^{2+} + CO_3^{2-}$
xiv. $BaCO_{3(aq.)} = Ba^{2+} + CO_3^{2-}$
xv. $BaHCO_3^+ = Ba^{2+} + HCO_3^-$
xvi. $Ti_4^+ + H_2O = TiOH^{3+} + H^+$
xvii. $TiOH^{3+} + H_2O = Ti(OH)_2^{2+} + H^+$
xviii. $Ti(OH)_2^{2+} + H_2O = Ti(OH)^{3+} + H^+$
xix. $Ti(OH)^{3+} + H_2O = Ti(OH)_{4(aq.)} + H^+$
xx. $Ti(OH)_{4(aq.)} = TiO_{2(s)} + 2H_2O$
xxi. $PbO^+_{(s)}H_2O = Pb^{2+} + 2OH^-$
xxii. $PbO^+_{(aq.)}H_2O = PbOH^+ + OH^-$
xxiii. $PbOH^+ = Pb^{2+} + OH^-$
xxiv. $Pb(OH)_{2(s)} = Pb^{2+} + 2OH^-$
xxv. $Pb^{2+} + 2OH^- = H^+ + HPbO_2^-$
xxvi. $Pb_2OH^{3+} + H^+ = 2Pb^{2+} + H_2O$
xxvii. $Pb_3(OH)_4{}^{2+} = 3Pb^{2+} + 4OH^-$
xxviii. $Pb_4(OH)_4{}^{4+} = 4Pb^{2+} + 4OH^-$
xxix. $Pb_6(OH)_8^{4+} = 6Pb^{2+} + 8OH^-$
xxx. $PbTiO_{3(s)} + H_2O = Pb^{2+} + 2OH^- + TiO_{2(s)}$

About 30 independent reactions consisting of 31 species are shown above. The thermodynamic modeling also helps in estimating the morphology of the products to some extent, and the work needs some more validation for a large number of variables. Figure 10.37 shows the representative images of PZT and other compounds with a control over their morphology as obtained by Riman and coworkers.

Yin and Alivisatos [172] have made similar attempt for kinetic size control of nanomaterials in a colloidal system with organic–inorganic interface. The critical size depends on the monomer concentration, with low monomer concentration favoring a larger critical size.

Aksay et al. [173] have prepared nanometer-sized $BaTiO_3$ particles under hydrothermal conditions by dispersing TiO_2 powders in a concentrated aqueous solution of $Ba(OH)_2$.The TiO_2 particles dissolve in the aqueous $Ba(OH)_2$ solution and lead to the nucleation of nanometer-sized cubic phase $BaTiO_3$ particles. In concentrated solutions, the $BaTiO_3$ particles grow through multiple clustering. These authors have used a similar approach to obtain $BaTiO_3$ films from organometallic precursors at 80°C and below. It was found that the grain size of the film depends on the nucleation rate of the $BaTiO_3$ particles.

For the preparation of fine particles of this perovskite-type mixed oxide, knowledge of stability diagrams under hydrothermal conditions is very important. The stability diagram shows the regions of reagent concentrations and pH at which various species predominate in the system (Figure 10.38). Thus, the stability diagrams indicate the optimum synthesis conditions for which desirable products are thermodynamically stable. However, the stability diagrams were considered for a limited number of hydrothermal systems on the assumption that the aqueous solutions were ideal [174,175]. They are especially inaccurate when concentrated electrolyte

Figure 10.37 Characteristic images of (a and b) PZT particles, (c) potassium titanate, and (d) $LiMn_2O_4$ obtained under mild hydrothermal conditions in 24 h duration.
Source: Photographs courtesy of Prof. R.E. Riman.

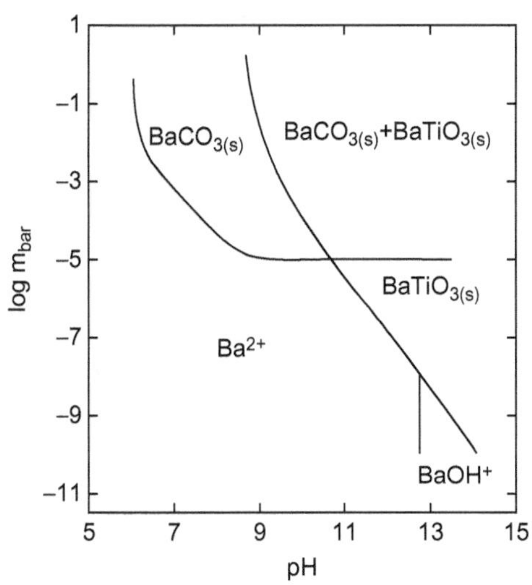

Figure 10.38 Stability diagram for barium titanate system [170].

solutions are utilized or when a multitude of competing reactions occur in a solution, thus making the equilibrium concentrations of various species strongly dependent on activity coefficients. Similarly, the speciation and the yield diagrams help

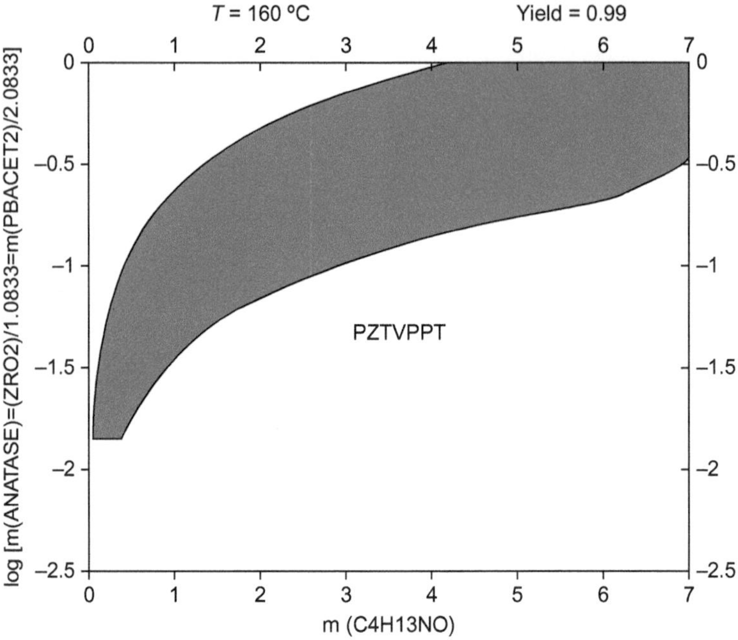

Figure 10.39 Calculated yield of $PbTiO_3$ in the Pb–Ti–H_2O complexing agent system for various Pb/Ti ratios [170].

greatly in the hydrothermal synthesis of phase-pure ceramics. Figure 10.39 shows the calculated yield of $PbTiO_3$ in the Pb–Ti–H_2O complexing agent system for various Pb/Ti ratios. A majority of the PZT systems incorporate intolerable amounts of alkaline metals, which are introduced in the form of mineralizers. In recent years, organic mineralizers are being used by a large number of workers. For example, Riman et al. [170] have found that tetramethylammonium hydroxide [$N(CH_3)_4OH$] is a favorable substitute for alkaline metal hydroxide mineralizers in producing phase-pure PZT. Phase-pure $MeTiO_3$ (Me = Ca, Sr, and Ba) can be obtained at input molalities of Ba, Sr, and Ca greater than 7×10^{-5}, 10^{-6}, and 5×10^{-5}, respectively. Otherwise, the relative location of the 99.995% yield regions for the three titanates will be similar to the pattern noted for stability diagrams [163]. In concentrated solutions, the consumption of OH^- ions is caused by the following predominant reactions:

$$Me^{2+} + TiO_2 + 2OH^- = MeTiO_3 + H_2O \qquad (10.1)$$

Thus, 2 M of OH^- is consumed for the synthesis of 1 M of $MeTiO_3$, and only a relatively small amount of OH^- is necessary to ensure correct pH for respective alkaline earth. Unlike the synthesis with nitrates, the use of metal hydroxide precursors at high concentrations does not require the addition of a mineralizer because the necessary concentration of OH^- groups is readily provided by the hydroxide precursor. To the contrary, at dilute concentrations identical amounts of mineralizers are needed, irrespective of whether a nitrate or a hydroxide is used as precursor. However, the required mineralizer concentration differs substantially for the three metals. This may be caused by the strong, specific effects of the chemical identity of cations on activity coefficients due to their high concentration to form alkaline earth titanates.

Eckert et al. [169] have proposed an in situ reaction mechanism and dissolution–precipitation reaction mechanism (Figure 10.40a and b). As evident from these figures, the in situ transformation model assumes that TiO_2 reacts initially with dissolved barium. This produces a continuous layer of $BaTiO_3$ through which additional barium must diffuse in order to continue reaction until the TiO_2 supply is exhausted. The product layer may be either a dense or a porous layer or of monocrystalline or polycrystalline nature. The dissolution–precipitation model involves multiple steps. For an anhydrous TiO_2 precursor, Ti–O bonds must be broken via hydrolytic attack, to form hydroxy-titanium complexes ($Ti(OH)_x^{4-x}$) capable of dissolution and reaction with barium ions or complexes (Ba^{2+} or $BaOH^+$) in solution to precipitate $BaTiO_3$. In contrast, use of a hydrous TiO_2 reactant bypasses some, or most, of the hydroxylation steps. $BaTiO_3$ nuclei may either originate at the TiO_2 substrate (heterogeneous nucleation) or form directly in the bulk solution (homogeneous nucleation). When hetereogeneous nucleation occurs, the dissolving TiO_2 particle can be encapsulated, thereby limiting the supply of soluble TiO_2 species available to react with the barium species. As with the in situ transformation model, this diffusional barrier serves to slow if not to halt the hydrothermal reaction. Such mechanisms of formation of $BaTiO_3$ can also be applied to other perovskite-type titanates. Figure 10.41 shows the perovskite-type alkaline earth titanates prepared under hydrothermal conditions [163,172]. The pH of the medium, precursor, and the ratio of Me/Ti play an important role in determining the morphology of these particles.

The preparation of alkaline earth titanates has been carried out using nonaqueous solutions, which come under solvothermal synthesis. This has some advantage, as the solvothermal synthesis might allow the product to be free from foreign atoms because the organic solution, having a low relative permittivity, is free of ionic species. Chen and Xu [176] have synthesized $PbTiO_3$ powder under solvothermal conditions. The starting materials involved are amorphous xerogels consisting of a mixture of equivalent molar amounts of PbO and TiO_2 (1:1 ratio), prepared by using lead acetylacetate and tetrabutyl titanate. The precursor xerogel was poured into the solvent to form a suspension solution and 30.0 cm^3 of the suspension was fed into a 40 ml capacity stainless steel autoclave with a Teflon liner. The required amount of methanol was poured into the autoclave, and the autoclave was held at 240°C for 10–60 h. The crystallinity increased with the increasing reaction time.

Although the crystallization of $PbTiO_3$ in an alcohol solution required higher temperatures for longer times, the nanometer-sized particles, in comparison with the micrometer-sized ones derived from an aqueous solution, exhibited a lower

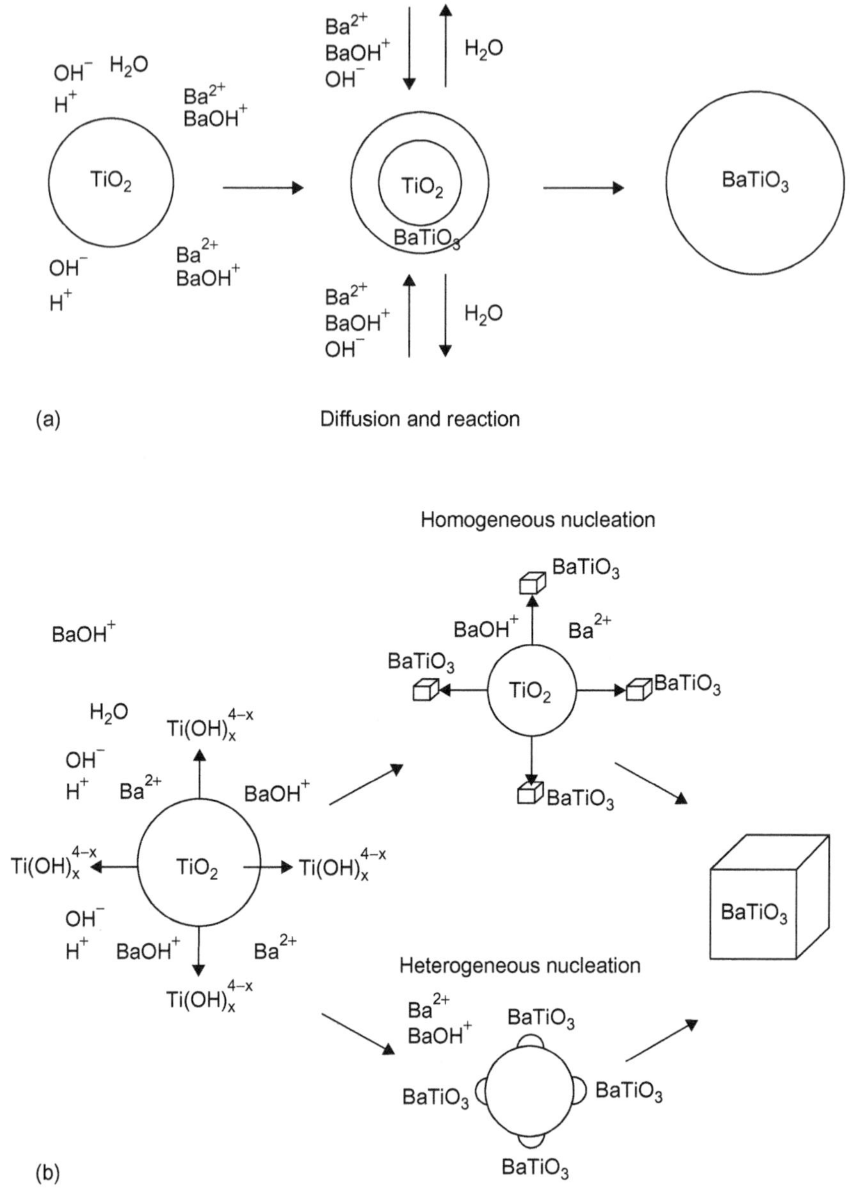

Figure 10.40 (a) In situ reaction mechanism [169]and (b) dissolution–precipitation reaction mechanism [169].

agglomeration due to the organic materials having a lower relative permittivity and a simple mode of size distribution.

Like the hydrothermal process, the formation of $PbTiO_3$ proceeds in alcohol solution by a dissolution–reprecipitation mechanism, in which the dissolution of

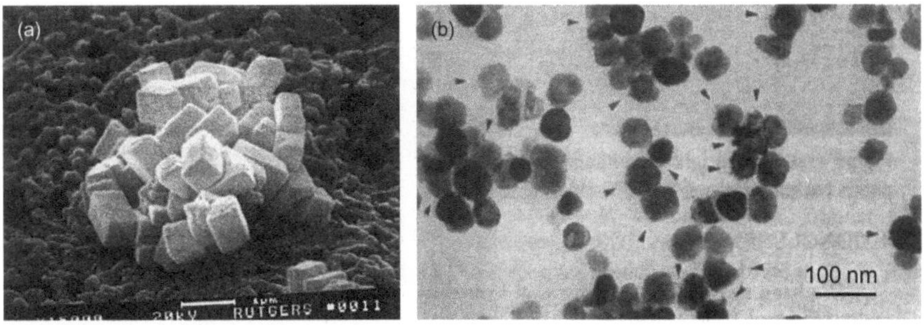

Figure 10.41 Perovskite-type alkaline earth titanates. (a) See Ref. [163] and (b) see Ref. [172].

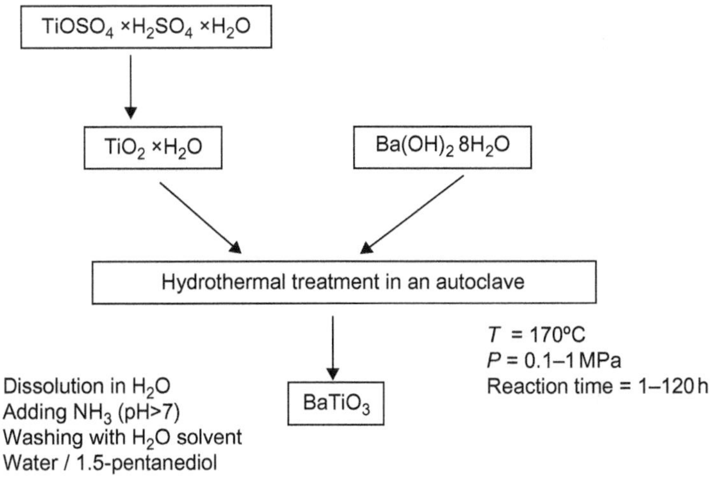

Figure 10.42 Schematic preparation route for lyothermal synthesis of $BaTiO_3$ [177].

precursor is the rate controlling stage. In comparison with those in methanol, the reactions in MOE for the syntheses of $PbTiO_3$ powders required higher temperatures or longer time because of MOE having a lower vaporizing pressure and relative permittivity.

Kaiser et al. [177] have synthesized $BaTiO_3$ powders at a temperature of 175°C and pressures between 0.1 and 1 MPa, starting from $TiO_2 \cdot xH_2O$ and Ba $(OH)_2 \cdot 8H_2O$ in solvents of pure water, 1,5-pentanediol and in different mixtures of both using the lyothermal method. Figure 10.42 shows the schematic preparation route for lyothermal synthesis of $BaTiO_3$.

The authors could control the size and habit of $BaTiO_3$ particles through the viscosity of the solvent and the concentration of the starting materials. The lyothermal method offers a promising means of producing various nanocrystalline materials of alkaline earth titanates. Table 10.5 lists fine powders synthesized under hydrothermal conditions.

In recent years, glycothermal and solvothermal syntheses are popular to prepare fine powders of nitrides, alumina, iron oxide, hexaferrites, and so on [178–181]. These methods are very useful in obtaining shaped and sized particles under relatively lower pressure–temperature conditions.

10.6.9 Synthesis of II–VI Semiconductor Nanoparticles

There is a growing interest in the synthesis of II–VI semiconductor nanoparticles with a precise control over their shape and size owing to their unique properties, which make them useful in solar cells, light-emitting diode, nonlinear optical materials, optoelectronic, and electronic devices, and also for medical and biorelated applications. Further, these compounds exhibit varying structures such as zincblende, wurtzite, and halite. Prior to the usage of the term QDs, these compounds were known as colloidal particles, or semiconductor cluster molecules. These are also popularly known as chalcogenides, which include sulfides, selenides, and tellurides. The synthesis and properties of these II–VI semiconductors have been extensively studied especially after their promising applications in biomedical science. There are several reviews published on these materials [182–185]. Several methods of synthesis have been attempted on these materials to prepare them as fine particles since the discovery of quantum size effect in 1980s. CdS and ZnS are the earliest chalcogenides prepared, both uncapped and polymer-capped nanoparticles through wet chemical methods. The success of the earlier methods depended essentially on the ability to stop the crystal growth process immediately after nucleation begins by controlling the equilibrium between solid CdS (ZnS) and solvated metal ions in the solution [186]. Later, SCF technology was also employed to obtain bulk quantities of these chalcogenides. However, several problems were faced by the earlier workers as the products were aggregated and not uniformly distributed. Nonaqueous solvents were employed to prepare fine particles of these chalcogenides under subcritical to supercritical conditions [187–189]. Heating of $CdCl_2$ and S, Se, or Te for 24 h in an autoclave filled with ethylenediamine (80% of the total volume) resulted in CdS, CdSe, or CdTe nanorods, respectively [190]. These compounds are especially interesting for their semiconducting and thermoelectric properties. There are several hundreds of reports on these sulfides, such as CdS, CdSe, PbS, ZnS, CuS, NiS, NiS_2, NiS_7, Bi_2S_3, $AgIn_5S_8$, MoS, FeS_2, InS, and Ag_2S, prepared with or without capping agents/surfactants/additives to alter their morphologies and sizes as desired. Most of them are known for their high-performance photovoltaic solar cells based on nontoxic and earth abundant materials.

Among II–VI group semiconductor nanomaterials, AX (A = Cd, Pb, Zn; X = S, Se, Te), CdS is an important one. Li et al. [191] have used thioacetamide as the sulfide source, as it easily releases sulfide ions, which are beneficial to lower the reaction rate and shorten the reaction period. The experiments are usually carried out in the temperature range 150–200°C using nonaqueous solvents, whose critical temperatures are generally within this range.

Table 10.5 List of Fine Powders Synthesized Under Hydrothermal Conditions

A1O−OH	**ZrO_2**	**Al_2O_3**	**ZrO_2-HfO_2**	**$BaFe_{12}O_{19}$**	**Sb_2S_3**	**Sb**	**$ZnSiO_4$**
FeO−OH	HfO_2	Cr_2O_3	SiO_2-ZrO_2	$CaSiO_3$	CdS	As	$ZrSiO_4$
SiO_2-H_2O	Fe_3O_4	CrO_2	UO_2-ThO_2	$BaTiO_3$	FeS	Bi	$CaCO_3$
$Cr_2O_3-H_2O$	Fe_2O_3	SiO_2	$Al_2O_3-ZrO_2$	$BaZrO_3$	MnS	Au	$kTiO_4$
$MnO \cdot H_2O$	MnO_2	Nb_2O_5	$Al_2O_3-TiO_2$	$LaCrO_3$	ZnS	Pb	
$Mn_3O_4 \cdot H_2O$	UO_2	TiO_2	$Al_2O_3-HfO_2$	(La,Sr) CrO_3	Hg		
	FeO	ZnO	$NiO-Fe_3O_4$	KMF_3	Ni		
	Cu_2O	CoO_3	$Y_2O_3-ZrO_2$	M = Co,Mn,Fe	Co		
	NiO	MgO	$CaO-ZrO_2$	$AlPO_4$	Cu		
	CdO	V_2O_3	$MgO-ZrO_2$	$NdP5O_{14}$	Pt		
	CeO_2	Ta_2O_5	$Nb_2O_5-TiO_2$	$NdPO_4$	Ag		
		PbO_2	$Ta_2O_5-TiO_2$	SrF_2-LaF_3	Sm		
			$V_2O_5-ZrO_2$	YAG	Ga		
				$kNbO_3$	$Na_8[AlSi_4O_4]_6$ $(NO_2)_2$		
Ba-Ferrites	$NiFe_2O_4$	Niobates	$PbZr_xTi_{1-x}O_3$ $(0.46 < x < 0.75)$				
Sr-Ferrite	$ZmFe_2O_4$	Tantalates	Co_{10} $(PO_4)_6(OH)_2$				
	$NixZm_{1-x}Fe_2O_4$	Vanadates	Ba Titanate				
	$CoFe_2O_4$	Titanate	Pb Titanate				
	$Mn_{0.5}Zn_{0.5}Fe_2O_4$		$Rb_2(MoO_3)_3SeO_3$				
			$Tl_2(MoO_3)_3SeO_3$				
			Na Beidellite				

Source: Courtesy of Prof. S. Somiya.

Figure 10.43 TEM and SEM images of CdS products obtained at 180°C for 5 h in mixed solvents with different volume ratios: (a) of en, TEM image with SEM image as inset; (b) 15%, TEM image with SEM image as inset; (c) 65%, TEM image; (d) 100%, SEM image with the upper right inset showing a magnified picture of the hexagonal ends of the long rods, and the lower left inset showing the HRTEM image of a nanorod. The scale bars in the TEM and SEM images all represent 100 nm. The scale bar in the HRTEM image is 5 nm. *Source*: Photographs courtesy of Prof. Yan Li.

Yao et al. [192] have obtained nanowires of CdS using ethylenediamine as the reaction medium. The experimental temperature was 140°C. The shape-controlled synthesis of CdS nanocrystals in mixed solvents has been carried out by some researchers [193]. They could obtain CdS nanotetrahedron, pencil-shaped nanorods, tetrapod, prickly spheres, high aspect ratio hexagonal nanoprisms (Figure 10.43) through adjusting the ratio of two solvents, namely ethylene diamine and ethylene glycol under SCF conditions, and the experimental time and temperature. They did not use the surfactants or other templates in the preparation of CdS nanoparticles. The experiments were carried out at 180°C for 5 h. The morphology of the resultant CdS nanoparticles could be determined by shifting the reactions between thermodynamically and kinetically controlled parameters [194].

Zhou et al. [195] have reported the synthesis of ZnS nanoplates in the temperature range 160–200°C in 24 h. They have used sulfur powder and $ZnCl_2$ in the presence of ethylenediamine. The use of zinc acetate and sodium diethyldithiocarbamite with an experimental temperature of 150–200°C for 12–72 h produces ZnS nanocrystallites with a different morphology.

The growth of ~70 nm monodispersed PbS nanocubes to dendrites has been achieved using $Pb(NO_3)_2$ and dithizone as reagents and ethylene diamine as solvent at about 140°C for 5 h [196]. PbS and PbSe nanocrystals have been prepared under near supercritical conditions [197]. According to the authors [198], the reaction of S and Se with a Pb salt in a ethylenediamine solvent is reported to give the respective phase with particle sizes in the range of 20–100 nm.

The synthesis of α-MnS nanocrystallites with rock salt structure has been reported in Ref. [199]. The crystals are well faceted, pyramid-like single crystals of α-MnS. The experimental temperature was varied from 100°C to 250°C to get the optimum condition for the formation of α-MnS crystals with a defined morphology. Manganese acetate and thiourea were used as precursors with benzene as the solvent. CdS_xSe_{1-x} ($0 < x < 1$) nanorods with a diameter of 10–20 nm and length up to 100–150 nm were synthesized at 140°C in about 10 h [200] (Figure 10.44).

In recent years, it has been proved that a shell of one composition (e.g., ZnS) can be synthesized over a core of another nanocrystal (e.g., CdS). The core can also be used as a seed to grow larger particles by adjusting the concentration after the initial growth. Such core-shell QD fabrication from these sulfide-based semiconductor materials is becoming very popular. Several such nanoparticles core-shell structures like CdSe/CdS, CdS/PbS, CdS/HgS, Cd/HgS/CdS, CdSe/CdS/ZnS, CdTe/CdSe, CdSe/ZnTe, and CdSe/ZnS have been fabricated for applications in optoelectronic devices and as efficient fluorescent labels or photoelectrochemical tests for various biomedical applications. The development of this avenue of research is due to not only the quantum size effects but also the fact that the average sizes of nanoparticles used are comparable with the size of biomacromolecules.

Wang et al. [201] have synthesized zinc tellurides with zinc blende structure, under mild hydrothermal conditions within the temperature range 100–160°C using NaOH as the solvent. Ji et al. [202] have synthesized Bi_2Te_3, $CoSb_3$, PbTe, Bi_2S_3, etc. using solvothermal and hydrothermal methods. PbSe and Bi_2S_3 show unique morphological features. The thermoelectric properties of these compounds have been studied in detail.

The processing of nanostructures of these sulfides, tellurides, and selenides is very popular in recent years owing to their unique physical characteristics. Tai et al. [203] have synthesized thermoelectric indium telluride nanostring-cluster hierarchical structures under solvothermal conditions using mixed solvents of ethylenediamine and ethylene glycol at 200°C in 24 h. The indium telluride hierarchal structures have been explained through a diffusion-limited reaction mechanism. Denac et al. [204] have reported the solvothermal synthesis of nanocrystalline CdSe and CdTe. Williams [205] has carried out a systematic investigation of the hydrothermal synthesis and characterization of CdSe nanocrystals.

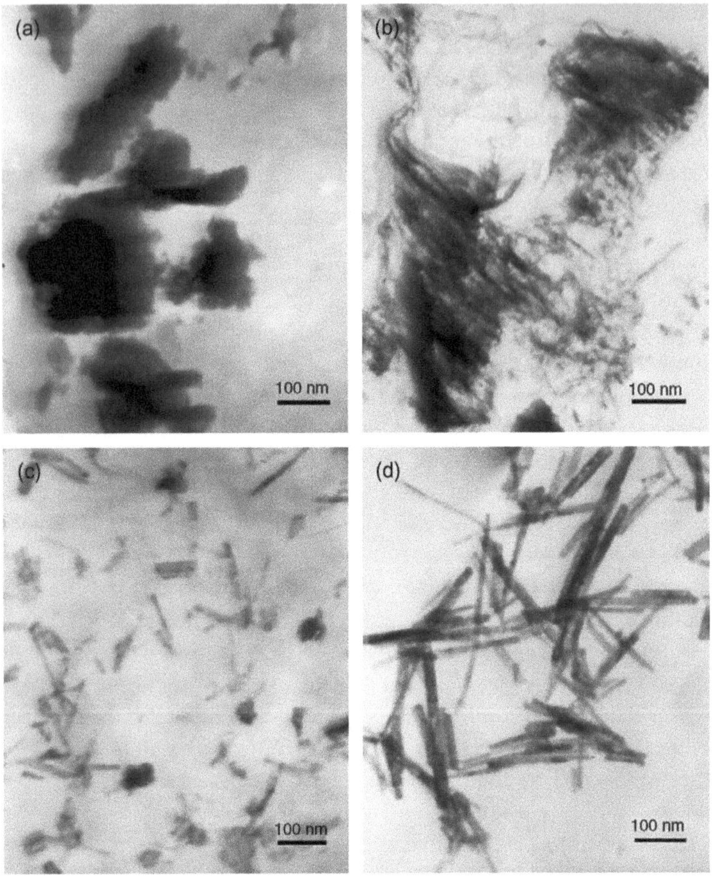

Figure 10.44 TEM images of CdS_xSe_{1-x} products with different reaction time: (a) 1 h, (b) 2 h, (c) 3 h, and (d) 4 h.
Source: Photographs courtesy of Prof. Yong Liu.

10.6.10 Synthesis of Phosphor Nanoparticles

This is one of the most attractive topics in nanotechnology owing to the device fabrication and bioimaging (medical use) capabilities of phosphor nanoparticles. There are several phosphor nanoparticles prepared under near-critical to supercritical hydrothermal conditions. Among them, the most common ones are rare earth vanadates ($RVO_4:Nd^{3+}$, $RVO_4:Eu^{3+}$), rare earth phosphates ($LaPO_4:Nd^{3+}$, $LaPO_4:Eu^{3+}$, $LaPO_4:Er^{3+}$), rare earth–doped garnets ($YAG:Eu^{3+}$, $YAG:Nd^{3+}$, $YAG:Ce^{3+}$), rare earth–doped perovskite-based compounds ($YAlO_3:Eu^{3+}$, $YAlO_3:Er^{3+}$), and rare earth oxides ($Lu_2O_3:Eu^{3+}$, $Y_2O_3:Eu^{3+}$, $Y_2O_3:Er^{3+}$, $R_xLu_{2-x}O_3:Eu^{3+}$, where R = Y, Gd) [206–209]. Depending upon the active rare earth ion dopant, the color of the light-emitting diode varies. Here we discuss the synthesis of only some selected phosphor nanoparticles, because, the supercritical hydrothermal synthesis of these phosphor nanomaterials is growing fast in recent years. Much of the literature data on these materials deal with the other conventional methods like sol–gel, solid-state reactions, melt, hydrolysis, and other wet chemical methods. As per the literature survey data, the flow reactor is being used only by the authors' group to synthesize nanoparticles of these materials.

10.6.11 Rare Earth Vanadates

$YVO_4:R^{3+}$ and $GdVO_4:R^{3+}$ (where R = Nd, Er, Eu) are known as efficient laser host materials, excellent polarizer, and as a phosphor in its powder form [210]. These compounds show high melting and no phase transitions. The synthesis of the rare earth vanadates is popular by the conventional methods. The materials prepared by these conventional methods encountered several problems like oxygen vacancies resulting in major structural defects, presence of mixed phases in the products, and loss of vanadium during crystallization leading to the loss of stoichiometry of the product. In order to overcome these problems, the supercritical or near-critical hydrothermal technique was proposed by Byrappa et al. [211]. There is only one recent conference proceeding so far on the use of SCF technology (continuous flow reactor) to synthesize rare earth vanadates on the whole in the literature. Both acidic and basic solvents are used in the synthesis of yttrium and vanadium compounds, which result in macro- and microcrystals. The addition of EDTA, or ammonium metavanadate, into the system stunts the growth of larger particles and produces nanoparticles of rare earth vanadates. The synthesis is normally carried out in the temperature range 220–400°C. Figure 10.45 shows what the rare earth vanadates produced under subcritical conditions. The use of SCF technology with a residence time of 2.08 s yielded highly dispersed nanoparticles of $GdVO_4:Eu^{3+}$ with 10–15 nm size. NH_4VO_3, KOH, $Gd(NO_3)_3$, and $Eu(NO_3)_3$ were used as the starting materials with gluconic acid as a modifier to inhibit the particle aggregation. Figure 10.46 shows the nanoparticles and nanostructures prepared using SCF technology. These results have shown a great potential for the synthesis of these materials using SCF technology.

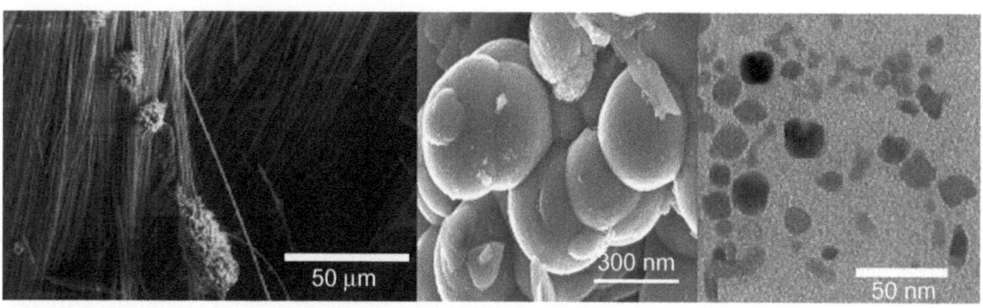

Figure 10.45 $YVO_4:Nd^{3+}$ nanostructures and nanoparticles.
Source: Photographs courtesy of K. Byrappa.

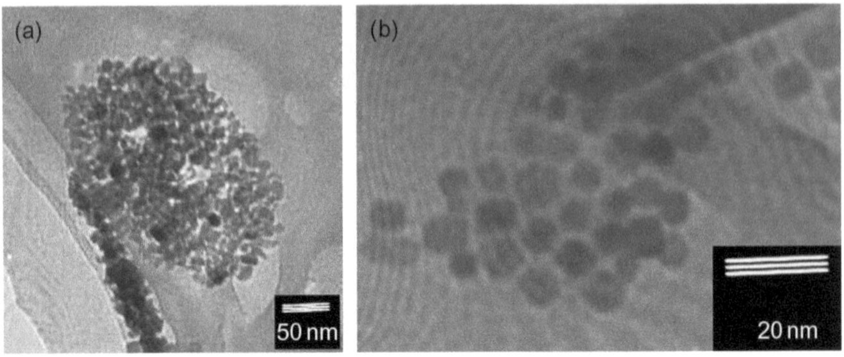

Figure 10.46. $GdVO_4:Eu^{3+}$ nanoparticles: (a) without modifier and (b) with modifier.
Source: Photographs courtesy of T. Adschiri.

Figure 10.47 SEM photographs of (a) $LaPO_4$ and (b) $LaPO_4:Nd^{3+}$.
Source: Photographs courtesy of K. Byrappa.

10.6.12 Rare Earth Phosphates

The synthesis of rare earth phosphates is very old and the most commonly used techniques were solid-state chemical reactions, aqueous solution methods, flux, and vapor deposition. However, all the works deal with the growth of bulk or fine crystals and not the nanocrystals. It was only in recent years the synthesis of nanoparticles of these rare earth phosphates began with their new potential applications in bioimaging [207]. Figure 10.47 shows the nanoparticles of $LaPO_4$ doped with Nd. The experiments are usually carried out in the temperature range above 240°C using batch reactors. Both acidic and basic solvents are employed in their synthesis. But the use of nonaqueous solvents for their synthesis is better because of the presence of structural water in the case of products prepared using aqueous solvents owing to the affinity of $[PO_4]^{3-}$ radicals to the OH^- molecules. Also the use of surface modifiers would greatly assist in synthesizing unagglomerated monodispersed fine nanoparticles. Such work is being carried out in the laboratories of the present authors. There are no reports in the literature on the synthesis of anhydrous rare earth phosphate nanoparticles using the continuous flow reactor under SCF conditions. This is an unexplored area at the moment for researchers.

10.6.13 Rare Earth Garnets

The preparation of rare earth garnets using SCF technology is popular. The alkaline solvents are used in most of the works in the temperature range of 420–450°C. Over the past few years, the active rare earth–doped garnet fine particles have been prepared by several workers using SCF technology [212,213]. The nanophosphor garnet is better than fine-powder phosphors with a larger particle size because nanophosphors could reduce internal scattering when they were coated onto a bare light-emitting diode surface. Using the continuous flow type of reactor with SCF technology, only fine particles of rare earth garnets have been obtained. Fine powders of YAG were prepared using glycothermal method with alkoxides (aluminum isopropoxide) as the starting material. The whole problem in all these experiments is the larger particle size and agglomeration. The results of the synthesis of nanoparticles of YAG:Ce in the size range 60 nm using SCF technology with batch reactors is perhaps the best in the literature [212]. The use of surface modifier in the supercritical hydrothermal conditions would be the best solution to synthesize high-quality small nanoparticles of YAG.

More recently, the nanoparticles of $YAlO_3$ (YAP) doped with active rare earth ions like Nd, Er, and Eu belonging to the perovskite family with cubic, hexagonal, and orthorhombic polymorphs are becoming popular materials for bioimaging purposes. The orthorhombic YAP has several advantages over the conventional YAG owing to the ideal distribution coefficients for active rare earth elements. The present authors have successfully synthesized this orthorhombic garnet for the first time using SCF technology. However, the particle size is still a problem. The use of surface modifiers and capping agents would provide some solutions to these problems.

Rare earth oxides and hydroxides doped with active rare earth ions like Er and Eu (Lu_2O_3:Eu^{3+}; Y_2O_3:Eu^{3+}; Y_2O_3:Er^{3+}; $R_xLu_{2-x}O_3$:Eu^{3+}; where R = Y,Gd; Y$(OH)_3$:Eu^{3+}) are some of the potential phosphors in recent years with a capability in biological applications [214–218]. However, the use of SCF technology has been seldom reported. Presently also, using conventional hydrothermal and solvothermal syntheses routes, several bioimaging phosphors, upconverting nanophosphors for biological labels like $LaPO_4$:Nd, NiP, RF_3 (where R = rare earths), and Gd_2O_3:Eu have been reported in the literature.

Bioimaging using phosphor attracts keen interest among researchers. To recognize how a drug delivers inside the body is essential to design a DDS. Since the organic probes used in bioimaging cannot survive too long in the body, the replacement was found in some semiconductor materials showing strong photoluminescence behavior, whose wavelength could be controlled with its particle size ("quantum size effect") [219]. With a same wavelength of excitation, wide range of colors can be obtained by changing the particle size of the nanoparticles of CdSe,

CdTe, CdS, etc., the so called QDs [220,221]. Recently, nontoxic QDs for biomedical applications have been prepared by liquid phase synthesis [222,223]. We have also prepared nontoxic metal oxide nanoparticles with a photoluminescence property using SCF technology. $GdVO_4$:Eu nanoparticles synthesized showed sphere-like morphology with particle size ~10 nm (Figure 10.48). These $GdVO_4$:Eu nanoparticles showed strong red light emission and their photoluminescence property is very stable in comparison with QD (Figure 10.49).

On the other hand, some groups have reported that metal oxides show unique photoluminescence properties. The perovskite, $BaSnO_3$, exhibits strong near-infrared luminescence [224]. Eu^{2+}-activated strontium aluminate phosphor possesses an afterglow property [225]. YVO_4 crystals doped with ytterbium and holmium show upconversion photoluminescence, i.e., conversion of infrared radiation into visible emission [226]. Hence, these nanoparticles, prepared by supercritical

Figure 10.48. Luminescence and stability of synthesized $GdVO_4$:Eu nanoparticles.

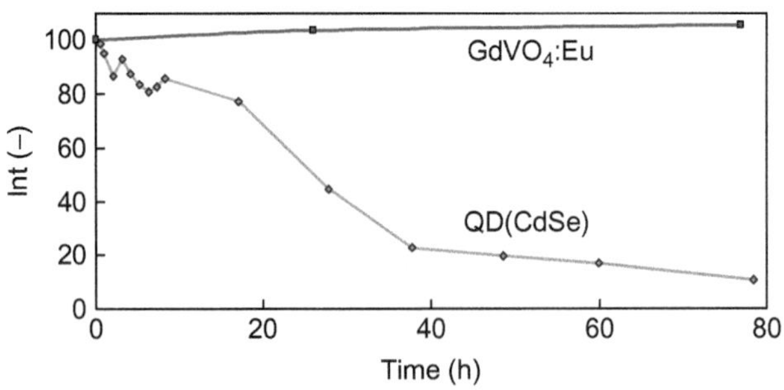

Figure 10.49 Comparison of photoluminescence property of $GdVO_4$:Eu nanoparticles with QD (CdSe).

hydrothermal synthesis, can be expected to perform as advanced bioimaging phosphors.

The magnetic nanoparticles are utilized in biotechnology such as in magnetic separation [227] and MRI [228]. Fine Fe_3O_4 nanoparticles with high crystallinity and narrow size distribution were prepared by supercritical hydrothermal synthesis at 200°C and 20 MPa with hexaldehyde as the surfactant (Figure 10.50). The mean size of the particles obtained is about 3.4 nm. Superparamagnetic iron oxide particles have been used in cellular imaging with imaging of in vivo macrophage activity.

$Gd(OH)_3$ and $GdVO_4$:Eu nanoparticles can be also used for bioimaging with MRI. Also, these nanoparticles are able to be used as X-ray imaging agent, because Gd presents high X-ray absorption property. For this purpose, it is highly essential to synthesize these inorganic nanoparticles with controlled particle size. Another important point is to modify its surface with organic ligands to disperse the particles perfectly in water (blood) or to attach some biofunctions (like antibodies of cancers and drugs) on the surface of nanoparticles. Supercritical hydrothermal synthesis is a promising method to synthesize various metal oxides nanoparticles and their in situ surface modification.

10.7 Hydrothermal Processing of Organic–Inorganic Hybrid Nanoparticles

The organic–inorganic hybrid nanoparticles are considered to be the most promising new class of materials that show the trade-off functions between polymers/organics and inorganics (light and high mechanical strength, high thermal conductivity, and electrical resistance, transparent and flexible electroconducting films,

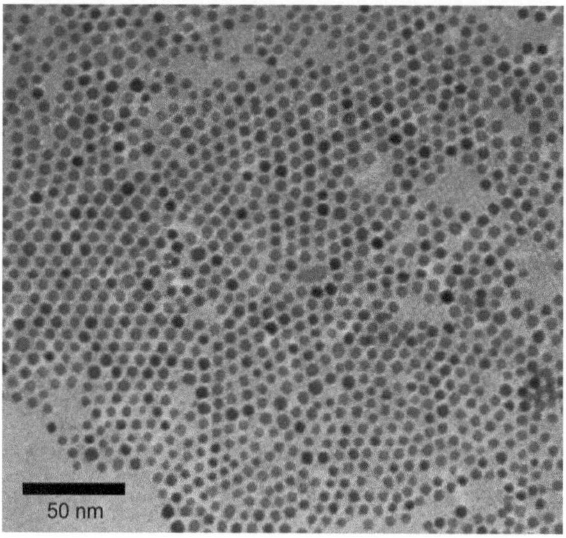

Figure 10.50 TEM image of Fe_3O_4 nanoparticles synthesized using a surface modifier. *Source*: Photograph courtesy of T. Adschiri.

and so on). There is a huge opportunity for discovering hybrid analogues of classical oxide systems that exhibit a wide range of physical properties. Future targets should include metallic hybrids (analogues of conducting polymers), lasers, and even superconductors.

10.7.1 Organic–Inorganic Hybrid Nanoparticles

Although the study of hybrid nanoparticles is considered to be a recent development, the literature survey shows that the approach to process organic–inorganic materials began during the 1970s. Several chemical routes were employed to prepare such hybrid nanoparticles and disperse them in the organic solvents. Silane-coupling agents have been in common use for decades providing enhanced adhesion between a variety of inorganic and organic agents. The general formula of these organosilane coupling agents is $R_nSiX_{(4-n)}$, having dual functionality. The majority of silane-coupling agents contain a hydrolyzable group (X), typically methoxy or ethoxy which readily reacts with a proton to give methanol or ethanol as by-products of the coupling. Metal oxide has hydroxyl groups which provide the necessary proton for the coupling reaction. The "R" group is a nonhydrolyzable organic group designed to provide a hydrophobicity nature for the surface of metal oxide nanoparticles. In this case, as shown in Figure 10.51, by dehydration reaction between the silane-coupling agent and OH groups of metal oxide, stable bonds of M–O–Si–C will be formed, resulting in the surface modification and changing the surface property of the metal oxides from hydrophilic to hydrophobic [229–232]. The greatest disadvantage of this route is the presence of Si shell in the structure of hybrid organic–inorganic metal oxide particles introducing significant changes in the original properties of these metal oxides and limiting their applications. For example, silane is very sensitive to UV radiation, and this type of hybrid metal oxides, like ZnO, cannot be used for such applications as sun block. Further, the silica shell around the metal oxide significantly increases with the size of the modified particles. Thus, by considering such an ex situ surface modification with functional groups and a silane-coupling agent, the problem of surface modification

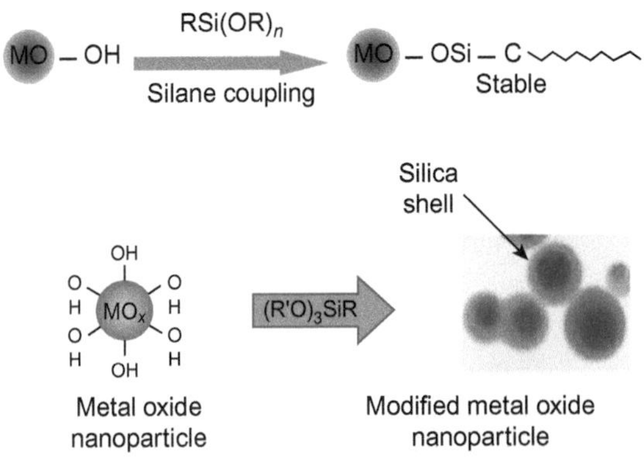

Figure 10.51 Schematic representation of the modified metal oxide particles with silane coupling agent.

of individual particles may not be possible, also there is a poor dispersibility of the modified particles and there is no control over the size and shape of the particles. Hence, researchers proposed an in situ surface modification of the nanoparticles to overcome all the above mentioned problems. Perhaps, the first steps in this direction leading to a rational nanoparticle assembly strategy were taken by Mirkin and coworkers [233,234] and Alivisatos and coworkers [235], who demonstrated that DNA-modified colloidal gold nanoparticles could be assembled into superstructures by hybridization of the complementary base sequences in the surface-bound DNA molecules. The motive behind such studies was to obtain a perfect dispersion of inorganic nanoparticles in solvents and plastics and to achieve a change in surface properties to the nanoparticles to hydrophobic, then, it is necessary to control the surface of the nanoparticles by organic modifiers and make a new generation of advanced materials, namely hybrid organic–inorganic nanoparticles. Thus, a combination of inorganic materials at the nanosize with the organic molecules could solve several problems encountered in the application of nanoparticles, and it could led to the emergence of the in situ surface modification of nanoparticles with a great variety of organic surface modifiers, which would bring in a perfect dispersion of the nanoparticles in solvents or in polymers. So far, tremendous efforts have been made to fabricate nanoparticle dispersed polymer, but it has been considered a difficult task to disperse the nanoparticles in organic solvents or in polymers, especially for the particles synthesized under hydrothermal conditions. This is because the metal oxide particle surface is hydrophilic, and for the case of nanoparticles, it shows extremely high surface energy, which leads to the formation of aggregates.

10.7.2 Supercritical Hydrothermal Organic–Inorganic Hybrid Nanoparticles

In order to resolve the problem of particle aggregation, to achieve the perfect control over the size and morphology of the particles, and to obtain a desired surface property of the nanoparticles, a new processing strategy has been proposed by Adschiri and coworkers [216,236–240] utilizing the supercritical hydrothermal technology. Figure 10.52 shows a schematic representation of highly effective strategy for the synthesis of metal oxide nanocrystals in the organic ligand-assisted supercritical hydrothermal technology [236]. The method yields perfect hybrid organic–inorganic nanocrystals with very high dispersibility and a precise control over the size and the shape of the nanoparticles. The organic components are introduced into the system during the hydrothermal synthesis, and in situ surface modification is obtained with an ultrathin layer of organics surrounding the inorganics unlike the case of silane coupling on the metal oxides. The organic ligands and supercritical water form a homogeneous phase, and it is known that under these conditions water molecules themselves work as acid or base catalyst for various organic reactions. Depending upon the applications of nanoparticles, one can select suitable functional groups to introduce hydrophobicity or hydrophilicity property to

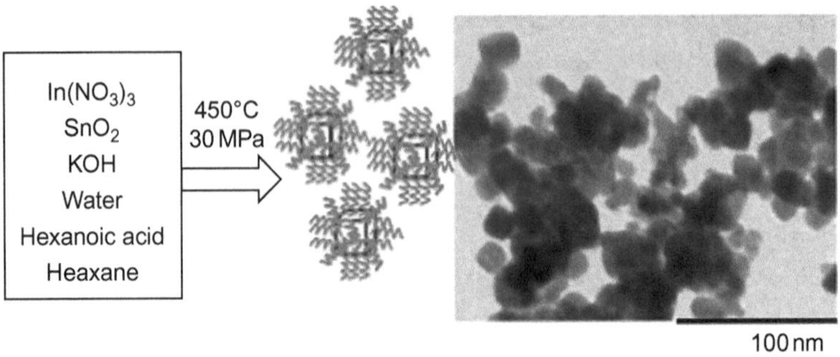

Figure 10.52 Schematic representation of strategy for indium tin oxide nanoparticles synthesis.
Source: Photograph courtesy of T. Adschiri.

the surface of the modified nanoparticles. This has been discussed in detail earlier in Section 10.3.

10.7.3 Mechanism of Formation of Organic–Inorganic Hybrid Nanoparticles

A knowledge on the mechanism of the formation of organic–inorganic hybrid nanoparticles is very important and it deals with the interaction of the organic ligand molecules with the inorganic metal oxide surfaces. Adschiri and coworkers have worked out, in detail, the theory and mechanism of the hybrid organic–inorganic nanoparticle formation under supercritical hydrothermal conditions.

10.7.4 Self-Assembly of Organic–Inorganic Hybrid Nanoparticles

Self-assembly is a generic term used to describe a process leading to the ordered arrangement of molecules and small components such as small particles like nanoparticles occurring spontaneously under the influence of certain forces such as chemical reactions, capillary forces, and electrostatic attraction. Self-assembly and more generally self-organization of particles in a solvent is considered to be a powerful process for building patterns up to nanoscopic level through multiple interactions among the components of the system under consideration. Ordered assemblies of nanometer-sized particles represent an interesting class of nanomaterials that provide exceptional potential for a wide variety of applications. These structures would be useful for various applications like displays, sensors, data storage, and photonic band gap materials owing to their exceptional physical, chemical, and electronic properties [241–244]. Common approaches to noncovalent assembly strategies employ van der Waals–packing interactions, hydrogen bonding, ion pairing, and host–guest inclusion chemistry [245–248]. These self-assembly methods provide direct access to extended structures from appropriately designed nanoparticle building blocks. The assembly of nanoparticles from solutions into close-packed monolayers and superlattice structures on solid surfaces has met with a fair degree of success [249–251], as oppossed to the controlled assembly of nanoparticles in the organic solvent which has not been understood precisely. In order to meet the major challenges in nanotechnology in realizing many of the desired technological goals, self-assembly provides a possible convenient route, but controlling size, size distribution, shape, and surface chemistries of the nanohybrid particles is critical to achieving desired structures. Initial syntheses have yielded nearly spherical shapes due to the thermodynamic driving force of minimizing surface area, and self-assembly has been limited to the close-packing of spheres [252,253]. There are several reviews published on this aspect [254,255]. However, it is to be noted that the self-assembly of hybrid organic–inorganic nanoparticles synthesized through supercritical hydrothermal routes are seldom found in the literature.

The mechanism of making self-assembly structures involves two steps: The first step is the positioning of perfectly dispersed particles (in solvent) on the substrate by controlling interactions between substrate and particles' surfaces (hydrophilicity and hydrophobicity). In the second step, capillary forces between the particles and the surface laterally displace the particles during drying. Steric repulsion between particles because of a capping agent on the surface prevents the aggregation of nanoparticles. After complete evaporation in the third stage, an irreversible reorganization of the particle–substrate interface occurs, which prevents the displacement of the nanoparticles [256]. Figure 10.53 shows the HRTEM image of the self-assembly ceria of nanoparticles obtained under supercritical hydrothermal conditions in the presence of decanoic acid as the modifier. It is clearly seen from these images that the organic ligand molecules are bonded to the surface of ceria nanocubes. With the increasing concentration of organic ligands, particle size not only decreases but the morphology also changes from the cubic to the truncated octahedron. The organic layer around the ceria particle is quite thin and it is shown with a yellow loop. Figure 10.54 shows the self-assembly of Co_3O_4 and Fe_3O_4 nanoparticles obtained under supercritical hydrothermal conditions in the presence of organic ligand molecules C_9COOH and $C_{17}COOH$, respectively. The decanoic acid has a carbon chain length of 1.4 nm, and it can be seen that the distance between

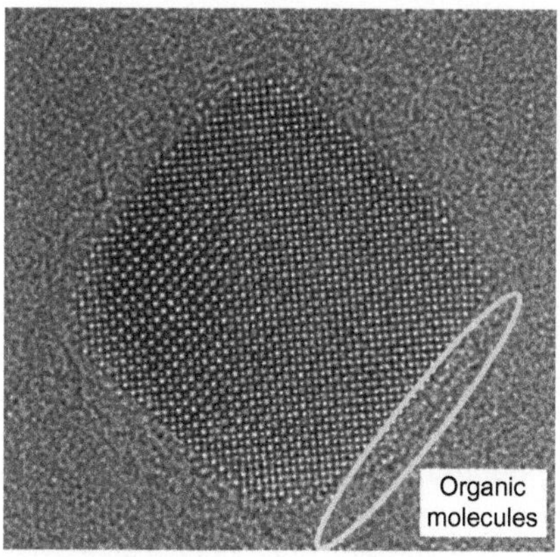

Figure 10.53 HRTEM image of 7 nm size decanoic acid modified ceria nanocubes synthesized at 400°C and 30 MPa.
Source: Photograph courtesy of T. Adschiri.

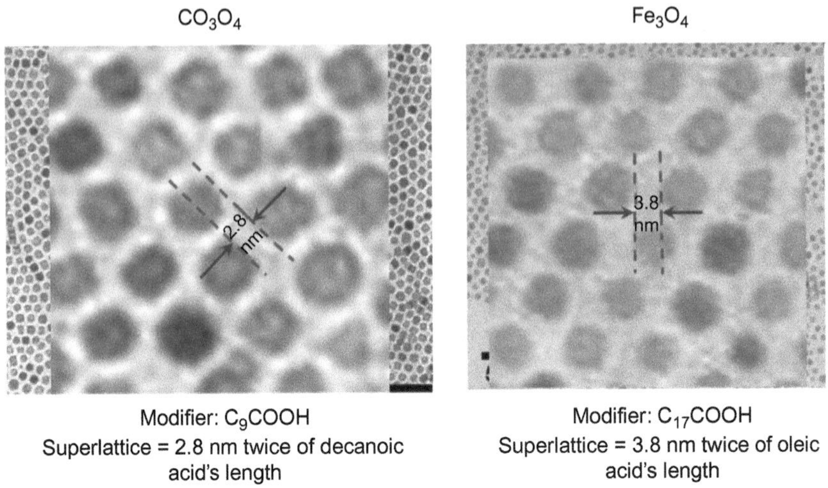

Figure 10.54 Superlattices of modified Co_3O_4 and Fe_3O_4 synthesized by supercritical hydrothermal method. The gap between the nanoparticles is twice the length of the organic modifier carbon chain.
Source: Photograph Courtesy of T. Adschiri.

self-assembled nanoparticles is about 2.8 nm. Similarly, oleic acid has a carbon chain length of 1.9 nm, but the distance between the oleic acid–modified magnetite particles is about 3.8 nm. Thus, by choosing different types of modifiers with different carbon chain lengths, the superlattice of the self-assembled nanoparticles can be monitored precisely.

A large number of such superlattice structures have been obtained for modified titania, zinc oxide, boehmite, cobalt spinels, and so on [237–240]. This area of self-assembly of the hybrid organic–inorganic nanoparticles is growing quickly and it has several advantages in obtaining the desired superlattice structures with different packing arrangement for specific applications.

10.8 Hydrothermal Processing of Bioceramics

Bioceramics represent a broad spectrum of ceramic materials designed for chemical compatibility and optimal mechanical strength with the physiological environment. These materials are used for the repair and reconstruction of diseased or damaged parts of the musculoskeletal system. Bioceramics may be bioinert (alumina, zirconia), resorbable (tricalcium phosphate), bioactive (HAp), bioactive glasses, and glass ceramics, or porous for tissue ingrowth (HAp-coated metals, alumina). Applications include: replacements for hips, knees, teeth, tendons, and ligaments, and repair of periodontal disease, maxillofacial reconstruction, augment and stabilization of the jaw bone, spinal fusion, and bone fillers after tumor surgery [257,258]. Table 10.6

Table 10.6 Types of Bioceramic Tissue Attachment and Bioceramic Classification [257]

Type of Attachment	Type of Bioceramic
Dense, nonporous, almost inert ceramics attach by bone growth by cementing the device into the tissue, or by press-fitting into a defect (morphological fixation)	Al_2O_3 ZrO_2
For porous implants, bone ingrowth occurs, which mechanically attaches the bone to the material (biological fixation)	Porous HAp and HAp-coated porous metals
Surface-reactive ceramics, glasses, and glass–ceramics attach directly by chemical bonding with the bone (bioactive fixation)	Bioactive glasses, bioactive glass–ceramics, dense HAp
Resorbable ceramics and glasses in bulk or powder form designed to be slowly replaced by bone	Calcium sulfate (plaster of Paris), tricalcium phosphate, calcium phosphate salts, bioactive glasses

gives the types of bioceramics and also the tissue attachment and bioceramic classification [259]. Bioceramics are also widely used in dentistry as restorative materials, gold porcelain crowns, glass-filled ionomer cements, dentures, and so on.

High-density, high-purity (>99.5%) Al_2O_3 (α-alumina) was the first bioceramic, widely used for clinical purposes during the 1960s. It is used in total hip prostheses and dental implants because of its combination of excellent corrosion resistance, good biocompatibility, low friction, high wear resistance, and high strength [260,261]. Other clinical applications of Al_2O_3 include knee prostheses, bone screws, alveolar ridge (jaw bone), and maxillofacial reconstructs, ossicular (middle ear) bone substitutes, keratoprostheses (corneal replacements), segmental bone replacements, and blade and screw and post-type dental implants [260,261].

Zirconia (ZrO_2), in tetragonal form, stabilized by either magnesium or yttrium, has also been developed as a medical-grade bioceramic for use in total joint prostheses. The interest in ZrO_2 is derived from its high fracture toughness and tensile strength. These improved properties make it possible to manufacture femoral heads for total hip prostheses that are smaller than the present generation of Al_2O_3 heads. ZrO_2 implants are now used clinically; however, only implant survivability data over a 10-year period will establish clinical advantages [261,262].

These materials are prepared by hydrothermal techniques such as hydrothermal sintering, HHP, and under hot isostatic pressure.

The most significant area of growth for bioceramics, however, involves a more complex material—HAp $\{Ca_{10}(PO_4)_6(OH)_2\}$, which is the main mineral constituent of teeth and bones, representing 69% by weight. HAp-based bioceramics have been in use in medicine and dentistry for 20 years [261–264]. It has the physicochemical advantages of stability, inertness, and biocompatibility. However, its relatively low strength and toughness produced little interest among researchers searching for bulk structural materials. HAp ceramics do not exhibit any cytotoxic effects, and

HAp can directly bond to the bone. Unfortunately, due to low reliability, especially in wet environments, the HAp ceramics cannot be used for heavy load-bearing applications, like artificial teeth or bones. Nevertheless, there has been a lot of research aiming to fabricate more mechanically reliable bioactive ceramics including, of course, the HAp materials. Suchanek and Yoshimura [265] have reviewed in detail the past, present, and future of the HAp-based biomaterials from the point of view of the preparation of hard tissue replacement implants. The chemical components of the mineral constituents of teeth and bones are very important in the synthesis of HAp-based biomaterials. The inorganic phases present in the hard tissues contain mostly Ca^{2+} and P, considerable amounts of Na^+, Mg^{2+}, K^+, also CO_3^{2-}, F^-, Cl^-, and H_2O. All these species, if applied in appropriate quantities, should be well tolerated in the implant by the surrounding tissues.

Presently, the HAp-bioceramics are at the pinnacle stage of their development. Powder processing, formation, and densification have been understood quite well, allowing the control of chemical composition and microstructures of both dense and porous HAp ceramics. Any new developments concerning powder preparation/shaping/densification may affect only the price of the products but are not expected to affect their medical applications, which are restricted, due to the nature of HAp [265].

Several techniques have been used for the preparation of HAp powders [265–268]. Two main methods of preparation of the HAp powders are wet methods and solid-state reactions. In the case of HAp fabrication, the wet methods can be divided into three groups: precipitation, hydrothermal technique, and hydrolysis of other calcium phosphates. Depending upon the technique, materials with various morphology, stoichiometry, and level of crystallinity can be obtained. Solid-state reactions usually give a stoichiometric and well-crystallized product, but they require relatively high temperatures and long heat-treatment times. Moreover, the sinterability of such powders is usually low. In the case of precipitation, nanometer-sized crystals can be prepared, and they have the shapes of blades, needles, rods, or equiaxed particles. Their crystallinity and Ca/P ratio depend mainly upon the preparation conditions which are, in many cases, lower than for well-crystallized stoichiometric HAp. The hydrothermal technique usually gives HAp materials, with a high degree of crystallinity and with a Ca/P ratio close to the stoichiometric value, a better outcome. Their crystal size is in the range of nanometers to millimeters. Hydrolysis of tricalcium phosphate, monetite, brushite, or octacalcium phosphate requires low temperatures (usually below 100°C) and results in HAp needles or blades the size of microns.

The authors have extensively studied the hydrothermal synthesis of HAp by adapting the intelligent engineering approach based on thermodynamic principles [28,269–271]. Experimental conditions were planned based on calculated phase boundaries in the system $CaO-P_2O_5-NH_4NO_3-H_2O$ at 25–200°C. HAp powders were then hydrothermally synthesized in stirred autoclaves at 50–200°C and by the mechanochemical–hydrothermal method in a multiring media mill at room temperature. The synthesized powders were characterized using X-ray diffraction, infrared spectroscopy, thermogravimetry, chemical analysis, and electron microscopy. Hydrothermally synthesized HAp particle morphologies and sizes were controlled through thermodynamic and nonthermodynamic processing variables, e. g., synthesis temperature, additives, and stirring speed. Hydrothermal synthesis yielded well-crystallized needle-like HAp powders (size range 20–300 nm) with minimal levels of aggregation. The thermodynamic model appears to be applicable for both stoichiometric and nonstoichiometric compositions. Experimental conditions for hydrothermal synthesis of HAp were based upon calculated phase boundaries in the system $CaO-P_2O_5-NH_4NO_3-H_2O$ between 25°C and 200°C. Phase diagrams were calculated at each experimental temperature using commercial thermochemical process simulation software. Briefly, standard state chemical potentials at the temperature of interest are calculated either from temperature-dependent equilibrium constant functions for each species. The standard state heat capacities were both used in conjunction with solute and solvent activity coefficients. The equations used to calculate the latter quantities were documented by Lencka and Riman [171]. The standard state quantities (ΔG_f^o, ΔH_f^o, S_f^o) for the solute species were generally taken from the standard references and data bank [167–171].

Changes in solute-free energies were calculated as functions of temperature and pressure using the modified HKF model.

Field emission scanning electron microscopy (FESEM) photographs of selected batches of HAp crystals synthesized at 200°C in 1 wt% KCl (aq.) and 50 vol.% 2-propanol (aq.) are shown in Figure 10.55. HAp crystals synthesized in 50 vol.% 2-propanol (aq.) had low aspect ratios ranging between 2 and 3 and diameters between 20 and 40 nm Figure 10.55a. Conversely, uniform nanosized needles (dimensions of about 20 × 100–160 nm, aspect ratio of 5–8) (Figure 10.55b) were formed when 1 wt% KCl additive was used. HAp crystals prepared under similar conditions but without additives were ~20 × 50–100 nm in size, yielding aspect ratios between 3 and 5. Here, the authors explain the possible mechanism of the morphology control for HAp [28,272].

Formation of amorphous calcium phosphates Acepromazine (ACP) prior to hydrothermal reaction may explain why either equiaxed nanoparticles or anisotropic needles, over the range of experimental conditions, are formed. Since ACP has

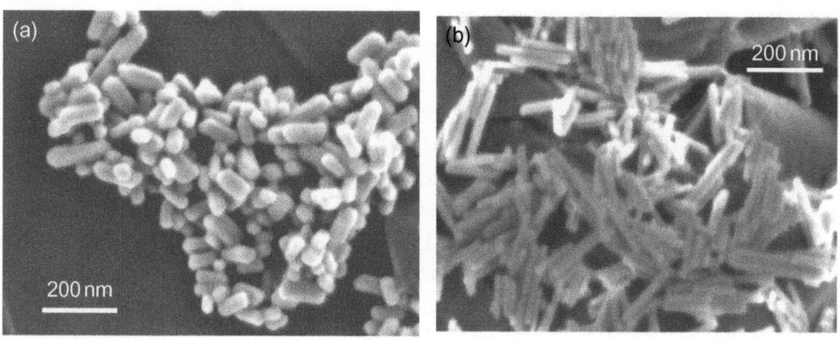

Figure 10.55 HAp crystals prepared hydrothermally at 200°C for 24 h using moderate stirring. Room temperature pH of precursor slurries was 10. (a) Powders crystallized in 50 vol% 2-propanol in H_2O (aq.) and (b) powders crystallized with 1 wt% KCl (aq.).

a sufficiently low solubility such that the initial precipitate does not completely dissolve under the experimental conditions, it is likely that the ACP particles acted as templates for HAp crystallization via interface reaction rate control. The formation of anisotropic particles may be due to reaction conditions that promote a greater degree of dissolution–precipitation that competes with interface reaction rate control. Greater flexibility in tailoring HAp crystal size and morphology by the hydrothermal technique could be achieved through precipitation from homogeneous solutions containing both Ca and P. Use of chelating agents for Ca, such as lactic acid or EDTA, prevents the formation of ACP upon mixing sources of Ca and P at room temperature. However, published work with these chelating agents shows that anisotropic fibers are also formed [273,274]. Thus, future work will need to identify additives that inhibit anisotropic particle growth.

There are also alternative methods of HAp powders preparation, like sol–gel, flux method, electrocrystallization, spray-pyrolysis, freeze-drying, biomimetic, microwave irradiation, mechanochemical method, or emulsion processing [275–281].

Many HAp powders can be sintered up to theoretical density, without pressure, at moderate temperatures (1000–1200°C). Processing at higher temperatures may lead to exaggerated grain growth and decomposition of HAp and, subsequently, to strength degradation HP, hot isostatic pressing (HIP), or HIP—postsintering makes it possible to decrease the temperature of the densification process, decrease the grain size, and achieve higher densities. This leads to finer microstructures, higher thermal stability of HAp, and subsequently, better mechanical properties of the prepared HAp ceramics [265]. HAp ceramics, in a porous form, have been widely applied as bone substitutes. Porous HAp exhibits strong bonding to bone. The classical way to fabricate porous HAp ceramics (pore size of 100–600 μm) is through hydrothermal sintering of the HAp powder with appropriate pore-creating additives, like naphthalene, paraffin, and hydrogen peroxide, which evolve gases at elevated temperatures. Natural porous materials, like coral skeletons made of $CaCO_3$, can be converted into HAp under hydrothermal conditions (250°C, 24–48 h) [283]. Microstructure, undamaged porous HAp structures can also be obtained by HHP [286,288]. This technique allows solidification of HAp powder at 100–300°C, 30 MPa, for 2 h.

Calcium phosphate bone cements find extensive applications as important biomaterials. These are mixtures of various calcium phosphate powders, such as $CaHPO_4 \cdot 2H_2O$, $Ca_4(PO_4)_2O$, $CaHPO_4$, $Ca_8H_2(PO_4)_6 \cdot 5H_2O$, $Ca(H_2PO_4)_2H_2O$, or tri calcium phosphate (TCP), and water, or another liquid (e.g., H_3PO_4 or Na_2HPO_4). The mixture transforms into HAp during setting, forming a porous body, even at 37°C [286–288]. The setting time of calcium phosphate cements can be reduced to a few minutes. Similarly, the decay of cements, when in contact with blood, can be prevented. Several types of cements like HAp clays, consisting of HAp granules in a saline solution of calcium alginate, bioactive glass bone cements, HAp-, TCp-, or bioactive glass-reinforced polymeric bone cements, have also been developed.

The advantages of the calcium phosphate bone cements are high biocompatibility, bioactivity, and osteoconductivity. The only serious disadvantage is their relatively poor mechanical strength. Easy shaping of bone cements enables using them to fill bone defects much better than HAp solid blocks, which are difficult to shape, or HAp powders/granules, which are difficult to keep in place. Calcium phosphate bone cements may, in the future, replace Polymethyl methacrylate (PMMA) cements as bone/implant fixation, if their mechanical properties can be improved. Moreover, they can be used as fillings of tooth root canals or as Drug delivery system (DDS) [265,289]. The hydrothermal method has several advantages in treating these calcium phosphate bone cements and slurries in a simple and effective way.

Several authors have contributed greatly on the synthesis of HAp. But what is most challenging to the materials scientist is the control of aspect ratio in HAp. In this respect, the hydrothermal technology has many advantages over other techniques. In recent years, there has been a growing tendency to prepare carbonate/magnesium/strontium coupled with magnesium and carbonate substitutions in calcium phosphate, which imitates the inorganic portion of the natural bone composition.

The future of HAp-based bioceramics lies in HAp-based composites HAp-ceramic, HAp-metal, HAp-polymer, HAp-collagen, and so on. However, there remain several unanswered questions related to HAp bioceramics [265]: Is it possible to make HAp-based ceramics applicable to heavy-loaded implants? Hydrothermal processing plays a key role in the preparation of bioceramics. A combination of methods, like sol–gel and hydrothermal processing also help greatly in the development or processing of future bioceramics.

10.8.1 Hydrothermal Preparation of Thin Films

This is one of the fast developing areas in hydrothermal research. The epitaxial growth of thin films under hydrothermal conditions began during the 1970s. The heteroepitaxial growth of single crystal films of YIG on GGG substrates by hydrothermal synthesis has been reported by many workers [290–292]. In present day context, the method has been modified and is popularly known as the hydrothermal–electrochemical technique, which is a very convenient method of preparing a wide range of films on substrates, coating of materials, preparation of multilayered compounds, functionally gradient materials, and so on. The method is especially versatile and convenient because of the lower temperature and pressure conditions involved compared to the conventional hydrothermal technique involving very HPHT conditions. Among the compounds obtained through hydrothermal–electrochemical technique, perovskite-type alkaline earth titanates dominate, followed by tungstates, molybdates, and a series of solid solutions of alkaline earth titanates [295,296].

Synthesis of large $PbTiO_3$ single crystals and polycrystallized ceramics is very difficult because it cracks spontaneously when passing through its phase transformation temperature ($T_c = 490°C$) during cooling. Possible applications in electronic and optical devices have brought much attention to the preparation of pure $PbTiO_3$ film. Most of the films have been fabricated by either RF sputtering or CVD. However, these methods have several disadvantages, such as the stoichiometric change of the film composition and residue of source material, such as $PbCl_2$, in the film. Furthermore, to form ferroelectric $PbTiO_3$ films during or after processing requires high-temperature treatment, which causes a reaction of the film with the substrate and sometimes leads to cracking and/or peeling of the film. MOCVD and sol–gel methods have also been applied to the preparation of $PbTiO_3$ film.

During processing, high temperatures above 500°C are also needed to remove organics. On the other hand, hydrothermal synthesis has frequently been used in preparing ceramic powders for a variety of applications. It is superior to the other powder preparation methods since high-temperature calcination is not necessary for the synthesis of crystallized ceramic powder. Verification can be supported by the experimental results of hydrothermal synthesis of $PbTiO_3$ powders.

During the 1990s, several authors developed hydrothermal or hydrothermal–electrochemical methods to prepare various double oxides such as $BaTiO_3$- and $SrTiO_3$-crystallized films directly on Ti-metal substrates by taking advantage of the hydrothermal reactions in high-temperature water (200°C) [302–310]. This method, in general, enables us to synthesize perovskite-type compound ABO_3, on a B-site metal in an aqueous alkaline solution containing A-site elements. It is increasingly being used for the preparations of $BaTiO_3$ film by other groups as well.

In the hydrothermal–electrochemical method, $BaTiO_3$ films are produced on a titanium electrode, which is anodized in the presence of an electrolyte containing $Ba(OH)_2$ under pressurized conditions. Recent results indicate that polycrystalline $BaTiO_3$ films can be prepared at temperatures as low as 55°C, using electrochemical activation in a barium acetate electrolyte in an oxygen atmosphere [311]. This method has been used as a highly versatile one to prepare perovskite films of several materials such as: $BaTiO_3$ and $SrTiO_3$ [303–305], $CaTiO_3$ [307], $BaFeO_3$ [305], $LiNbO_3$ [305], $BaNbO_3$ [307], $Ba_2Nb_5O_3$ [307], $PbTiO_3$ [296,312], $BaWO_4$ [298], TiO_2 [313], ZnO [314], and TiO_3 [315].

A closely related technique for film formation is hydrothermal deposition in which perovskite films are produced under similar conditions but without electrochemical activation. By this technique, $BaTiO_3$ [303,316], $SrTiO_3$ [303], (Ba, Sr) TiO_3, and $BaZrO_3$ [316] films have been obtained.

The hydrothermal–electrochemical film formation technique is not yet well established, and there is not much information on the electrochemical mechanism involved in the formation of these films. Moreover, there is not a clear understanding of the relationship between the film formation mechanisms for the hydrothermal method and those for the hydrothermal–electrochemical method. Most of the earlier works on $BaTiO_3$ hydrothermal–electrochemical formation have been performed in two-electrode cells, titanium anode and platinum cathode. In such a system, it is not possible to obtain accurate, specific, information on the electrochemical phenomena occurring at the titanium anode as the measured voltage drop on the cell is the result of processes at the anode, cathode, and in the electrolyte.

Kajiyoshi et al. [29] have described the autoclave designed for hydrothermal–electrochemical treatment/preparation of $ATiO_3$ (A = Ba, Sr) thin films. It has a three-electrode technique [309]. The electrolytic cell assembled in the autoclave is shown schematically in Figure 10.56. It accommodates an Ag/AgCl external reference electrode (Toshin Kogyo, Tokyo, Japan) [317], which enables one to measure potentials of electrodes on a thermodynamically meaningful scale,

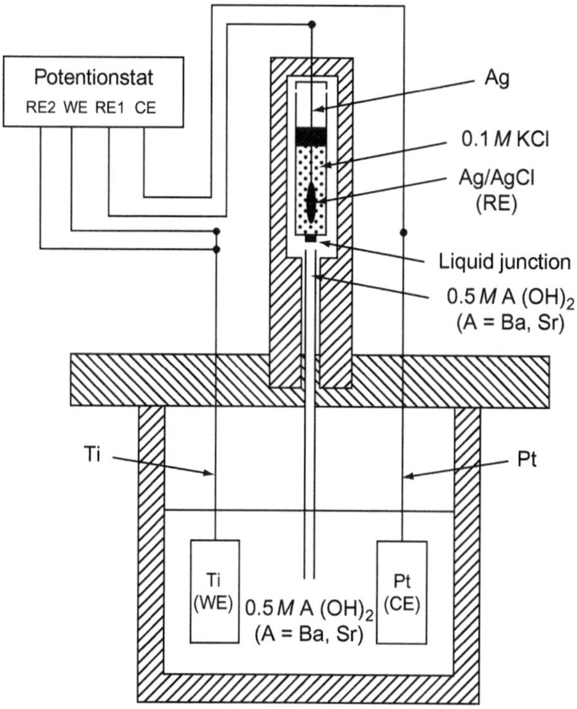

Figure 10.56 Electrolytic cell assembly in the autoclave [309].

regardless of the electrolysis conditions [318]. The preprocessed Ti substrate and the platinum plate are suspended as the working electrode (anode) and the counter electrode (cathode), respectively, by 0.5 mm diameter wires of the same metal as the respective plates, keeping an interval of 30 mm between them in the electrolytic cell containing 500 ml of the solution. The experiments were carried out at 150°C, an approximate heating rate being 1.5°C/min. The Ti working electrode was polarized potentiostatically from 50°C in the heating region to the end of the 150°C isothermal region, keeping an anodic potential of +12.0 V versus Ag/AgCl. The electrolysis current varies characteristically in the heating process and then remains almost constant in the isothermal process. After the experiment, the Ti substrate was washed in distilled water with an ultrasonic cleaner and dried at 120°C. Kajiyoshi et al. [29] have also studied, in detail, the transport mechanism of film-constituting elements. They propose two models: the first based on the principle of tracer technique: (i) substitutional transport and (ii) interstitial transport. The second is the short-circuit path model. The latter one is more appropriate here to describe the mass transport in $ATiO_3$ (A = Ba, Sr). Figure 10.57 illustrates schematically the mass transport and film growth mechanisms of the short-circuit path model. In this model, all the observed results can be explained comprehensively, together with an atom-mixing mechanism accompanied by the dissolution–recrystallization process. Figure 10.58 shows representative SEM micrographs of a fractured cross section of

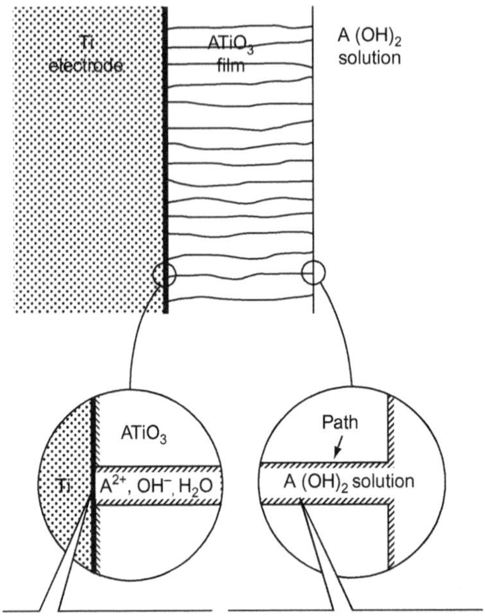

Figure 10.57 Schematic representation of mass transport and the film growth mechanism [309].

growth films. All the films have a similar cross sectional microstructure, regardless of their treatment processes. The advantage thickness of the thin layer is ~0.1–0.2 μm.

Such studies have been carried out by several other workers on different compounds. For example, Cho and Yoshimura [295] have studied, in detail, the hydrothermal–electrochemical synthesis of highly crystallized barium tungstate films on tungsten metal substrates at fairly low temperatures. The hydrothermal method did not produce a film because only discrete $BaWO_4$ particles appeared on the tungsten substrate. The electrolysis experiments were carried out using tungsten substrates as an anode (i.e., working electrode) and a platinum substrate as a cathode (i.e., counter electrode). A reference electrode (KCl-saturated Ag/AgCl) was put in a glass tube with a capillary tip close to the working electrode. $Ba(OH)_2$ solutions were made from redistilled water. Figure 10.59 shows the schematic diagram of the autoclave assembly used in the hydrothermal–electrochemical method. The authors propose a model that interprets the mechanism for the formation and growth of the amorphous tungsten oxide film. The schematic diagram in Figure 10.60 illustrates the mass transport and the amorphous film formation during anodic oxidation. Similarly, Figure 10.61 shows the SEM photograph of sample surface prepared by the hydrothermal–electrochemical treatment.

In general, the morphology, grain size, and quality of the film can be controlled through several parameters, like temperature, pH of the solution at the interface, electrode current density, impurities (dopants), and duration of the experiment. Figure 10.62a–d shows the $BaWO_4$ film morphology in 0.01 M $Ba(OH)_2$ formed during conduction after (a) 1 min, (b) 15 min, (c) 20 min, and (d) 25 min. The crystallization was characterized by 3D nucleation and growth. Thus, the

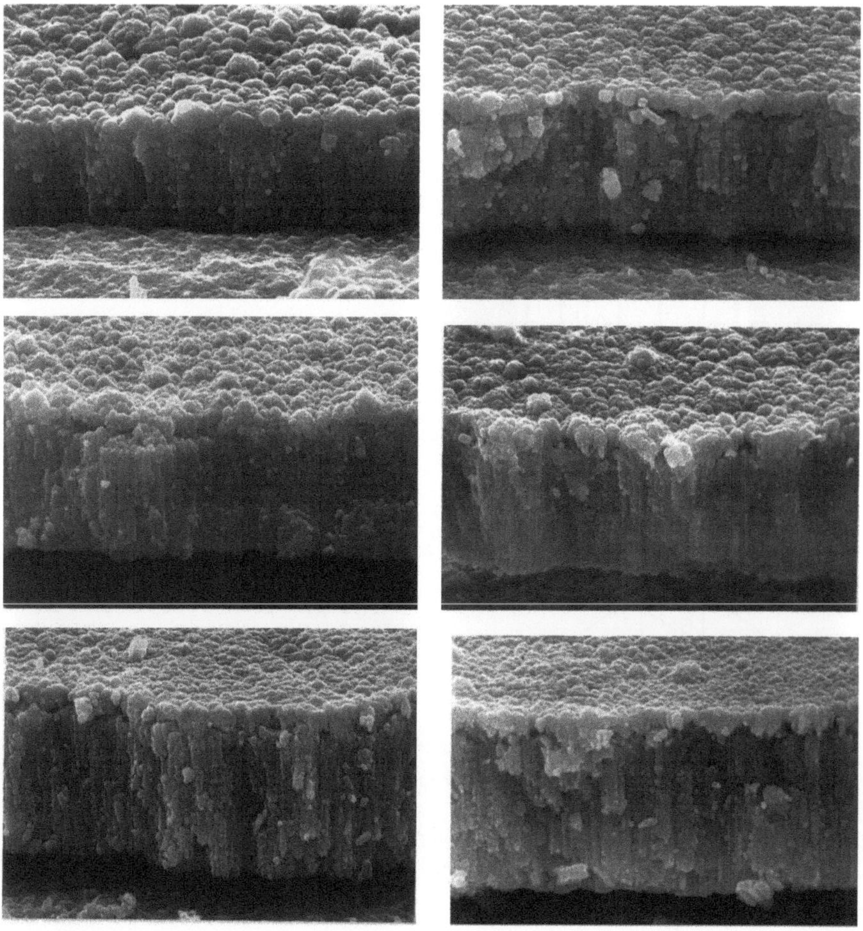

Figure 10.58 Representative SEM micrographs of fractured cross section of growth films. *Source*: Photographs by M. Yoshimura.

hydrothermal–electrochemical processing/treatment of materials is going to play a pivotal role in futuristic hydrothermal research.

10.8.2 Hydrothermal Processing of Composites

The nature invented composites—wood (cellulose and lignin) and bone (the polymer collagen and the mineral HAp)—are specific examples that a composite can, and often does, have much more desirable properties than the individual pure or virgin materials from which it was made. Man-made composites have also been successful, as in the case of the rubber tire, which in its most common modern form is a composite of vulcanized rubber (the synthetic or natural polymer), carbon

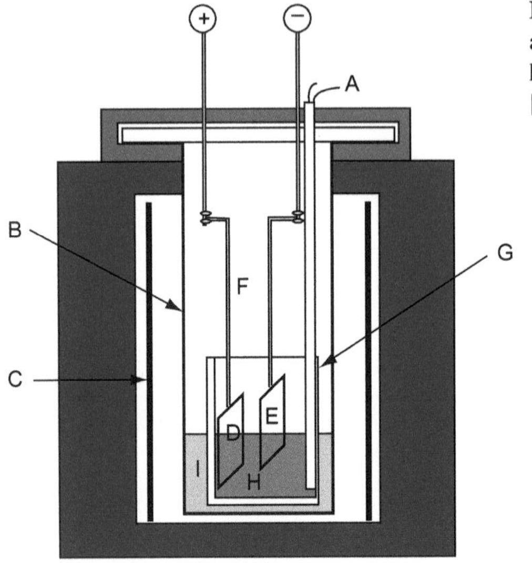

Figure 10.59 Schematic diagram of the autoclave assembly used in the hydrothermal–electrochemical method [298].

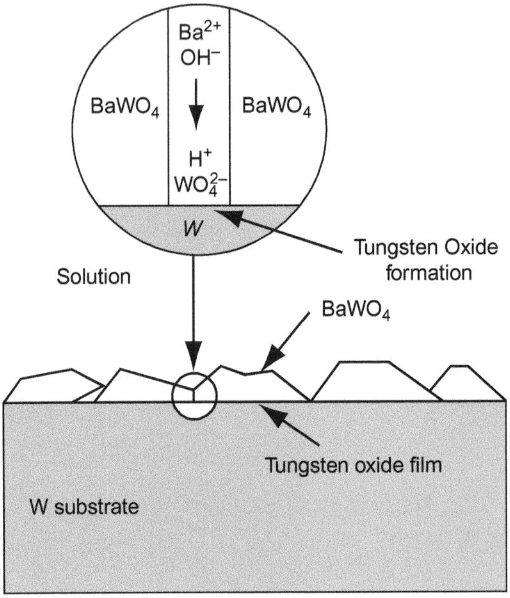

Figure 10.60 Mass transport and the amorphous film formation during anodic oxidation [298].

filler, and steel or polymeric fibers. One reason for interest in other man-made composites is indicated in Figure 10.63 [319]. The high strength-to-weight ratio of composites is more favorable than their ratio of strength to size. The general concept of composite materials has great appeal. It offers the prospect of microstructurally combining various classes of materials, by design, to realize superior sets of properties not obtainable from any of the constituent materials serving alone.

Figure 10.61 SEM photograph of the sample surface prepared by the hydrothermal–electrochemical treatment.
Source: Photographs by M. Yoshimura.

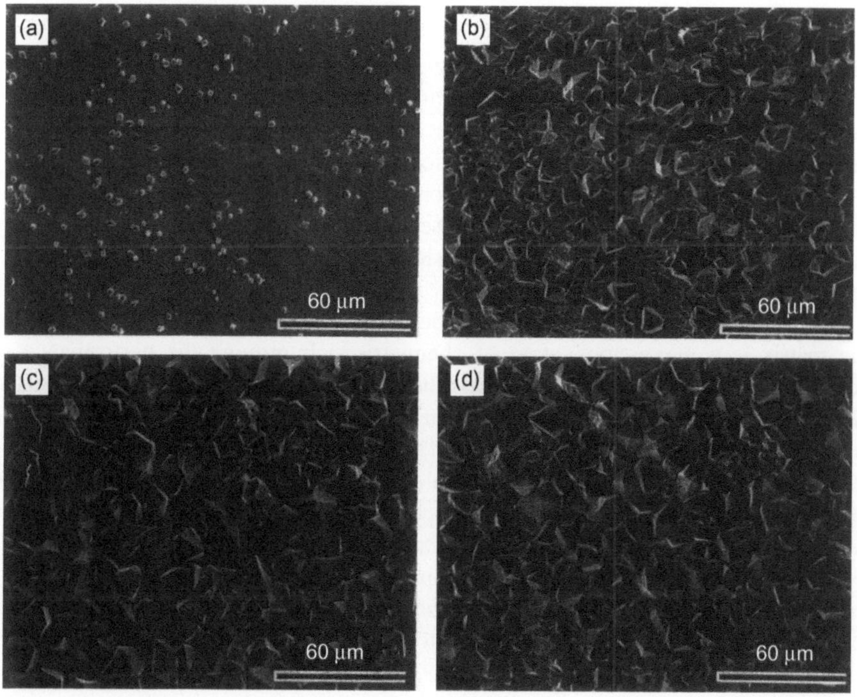

Figure 10.62 $BaWO_4$ film morphology in 0.01 M $Ba(OH)_2$ formed during conduction after (a) 1 min, (b) 15 min, (c) 20 min, and (d) 25 min.
Source: Photographs by M. Yoshimura.

Actually, this is an old idea (witness straw in bricks, horsehair in plaster, and reinforcing bars in concrete), but modern engineering requirements have become much more demanding, particularly, with regard to specific strength and stiffness for aerospace, military, and similar critical applications. By 1947, the filament-wound glass composite rocket motor case had been successfully flown, and the associated industrial contractors supported the Navy's decision to use fiberglass motor cases for the Polaris missiles. Since then, composites have served in successive generations

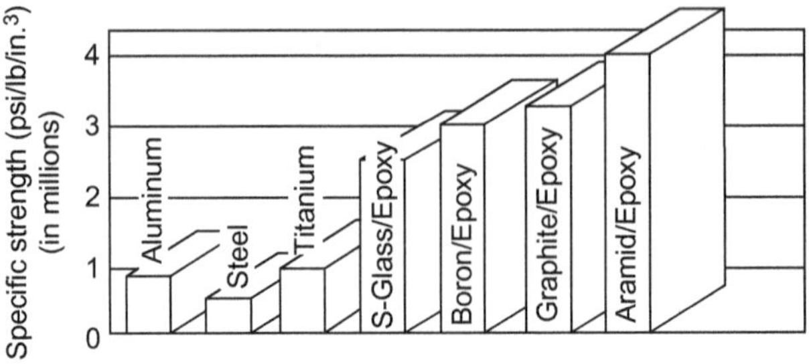

Figure 10.63 Man-made composites [319].
Source: Courtesy of the National Academy Press, Washington, DC.

of rockets, reducing weight and providing strength and durability [319]. Composites are highly useful in the fabrication of high-performance microelectronic devices consisting of multilayer substrates—multiple layers of ceramic (alumina), metal, and thin-film organic insulators. The fabrication of these substrates is made difficult by a number of problems, prominent among which is that of ensuring adhesion between the different components. These components commonly exhibit widely divergent coefficients of thermal expansion and fundamentally incompatible surface chemistries.

Tailoring the properties of the interface between the reinforcing component and the matrix is a major application of chemistry in improving the performance of composites. Failure in multilayer fiber-reinforced composite structures often occurs either at the fiber–matrix interface or at the matrix–matrix interface. The surface treatments, now used to modify the surface properties of reinforcing fibers in composites, are largely empirical. In this regard, the hydrothermal method of processing the materials to obtain composites and multilayers of ceramics and coating of substrates on other materials is very significant. The literature data available on composites are so vast that it is impossible to discuss each and every material in this chapter. Therefore, we have chosen the most important material, namely HAp-based composites, owing to their potential applications in modern technology.

10.8.3 Hydrothermal Processing of HAp Coatings and HAp-Based Composites

Industrial application of HAp requires its use in various forms, e.g., porous or dense sintered body. Catalytic and chromatographic applications utilize powder HAp. Recently, film HAp has become important as a coating for metal or ceramic bodies, and tiny electronic devices such as sensors.

In order to overcome the inferiority of HAp in mechanical properties, the coating of HAp films has been attempted on bioinert materials of strength and/or toughness, such as polycrystalline alumina (Al_2O_3), zirconia (ZrO_2), titanium metal (Ti), and titanium carbide (TiC). Some methods are reported to be available to produce thin films of HAp. HAp coatings on Ti plates have been reported to yield by electrophoretic deposition.

One of the most important clinical applications of HAp is as a coating on metal implants, such as hip joint prostheses. This concept combines the mechanical advantages of metal alloys with an excellent biocompatibility and bioactivity of HAp. Uncoated metal implants do not integrate with bone because bioinert materials are encapsulated by dense fibrous tissues which prevent proper distribution of stresses and thereby cause loosening of the implant. In the case of HAp-coated metal, bone tissue integrates itself completely with the implant, even during early functional loading [265].

HAp coatings provide stable fixation of the implant to bone and minimize adverse reaction by provision of a biocompatible phase. Moreover, the HAp coatings decrease the release of metal ions from the implant to the body and shield the metal surface from environmental attack. In the case of porous metal implants, the HAp coating enhances bone ingrowth into pores. HAp coatings have been applied to metals like Ti alloys, or Ca–Cr–Mo alloy, to carbon implants, to sintered ceramics like ZrO_2 and Al_2O_3, and even to polymers (*Pmma*) [265]. There are various methods to fabricate HAp coatings. The common ones are: HIP, spray painting, oxyfuel combustion spraying magnetron sputtering, flame spraying, ion-beam deposition, chemical deposition under hydrothermal conditions, electrochemical deposition, metal-organic CVD, and sol–gel. The coating under hydrothermal conditions has been carried out effectively by many workers and all these references are listed by Suchanek and Yoshimura [265].

The thickness of the HAp coatings is usually in the range of 40–200 μm. With increasing thickness of coating, concentration of metal ions released to the body decreases; the coatings must be thick enough to resist resorbability of HAp, which can be as much as 15–30 μm per year. Fixation to the bone can be improved if the HAp coating has an appropriate porosity, which promotes bone intergrowth. It has been found that the interface between HAp and the metal, ceramics, or polymers often contain several undesired phases, like phosphides and amorphous phosphates, which decrease chemical stability and enhance degradation of the coatings [321]. Fujishiro et al. [321] have carried out a detailed study of the coating of HAp on metal plates using thermal dissociation of calcium-EDTA chelate in phosphate solutions under hydrothermal conditions. Figure 10.64a and b illustrates schematically the mechanism of HAp deposition on iron and titanium plates. The mechanism of deposition, according to the authors, is different in both the cases. Fujishiro et al. [293] have studied the thermodynamics of the homogeneous precipitation of HAp in $Ca(EDTA)^{2-}$– NaH_2PO_4 solution. Figure 10.65 shows SEM photographs of HAp films formed on the surface of iron plates in various concentrations of Ca $(EDTA)^{2-}$–NaH_2PO_4 solutions at pH = 5 and 150°C for 4 h [321]. The direct coating of HAp and double coating of HAp on titanium plates have been investigated with the varying pH of the NaH_2PO_4 solutions. Depending upon the pH and temperature of deposition, different morphologies of the HAp and monetite particles were obtained.

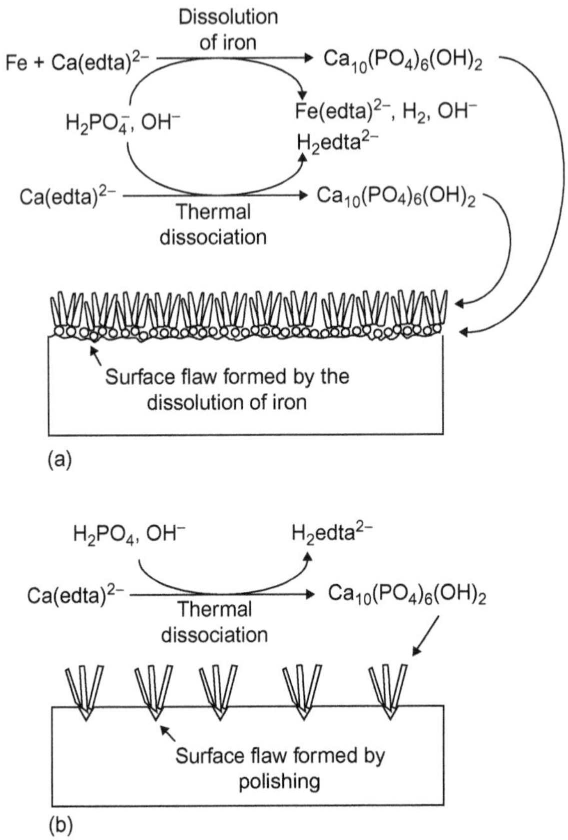

Figure 10.64 Mechanism of HAp deposition on iron and titanium plates [321].

Figure 10.66a–c shows SEM photographs of the surfaces of titanium plates coated with HAp (experimental conditions 0.05 M $Ca(EDTA)^{2-}$ – 0.05 M NaH_2PO_4 solution at 160°C for 2 h) at pH (a) 4, (b) 5, and (c) 6.

Figure 10.67 represents SEMs of the surfaces and cross sections of titanium plates after the second coating in 0.05 M $Ca(EDTA)^{2-}$ – 0.05 M NaH_2PO_4 solution at 160°C for 4 h at pH (a) 6 and (b) 9. The first coating was carried out with a solution of the same composition at initial pH 5 and 160°C for 2 h [321].

Thick films of HAp are also attempted for the use of gas sensors. There are several varieties of HAp-based composites like HAp/bioactive glass composites, HAp/polymer composites, and HAp/HAp (whiskers). Among the commonly used organics for HAp/polymer composites are phosphorylated cotton fibers, polymeric substrate, polymethyl methacrylate, and poly [bis (sodium carboxylatophenoxy) phosphase] [321].

A comprehensive study undertaken at the Tokyo Institute of Technology, Japan, under the leadership of Prof. M. Yoshimura, has moved HAp whisker-reinforced HAp (HAw/HA) composites closer to real-world use. Researchers have long eyed HA implants as replacements for natural bone. The two materials share the same

Figure 10.65 SEM photographs of the HAp films formed on the surface of the iron plates [293].
Source: Courtesy of the Academic Press, Orlando, FL.

chemistry, but the mesostructure of bone gives it two to six times more fracture toughness than HA. This limits current HAp usage to small, unloaded implants and coatings.

HA whisker-reinforced composites might solve the problem. Their production requires hot pressing and/or HIP at 1000–1100°C. Even these processes fail to achieve fully dense samples.

Unfortunately, they also give rise to a whole new set of problems. Many sintering additives form tricalcium phosphate, which increases HA biodegradability and reduces implantation lifespan. Others form CaO, a reactive material whose volume

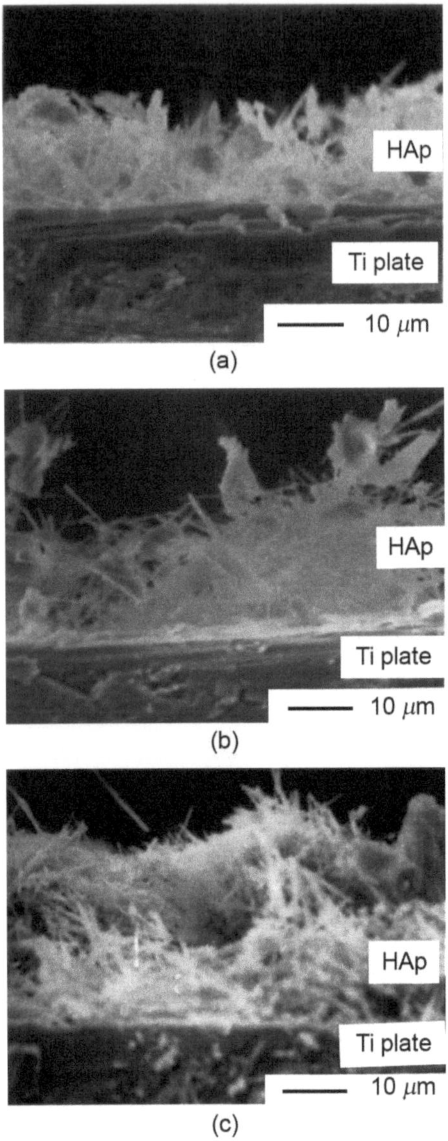

Figure 10.66 SEM photographs of the surfaces of titanium plates coated with HAp at (a) pH 4, (b) pH 5, and (c) pH 6 [288].

changes once inside the body and causes the implant to crumble and fail. In addition, any sintering aid must promote weak binding between the HA whiskers and monolith in order to improve toughness by allowing fibers to pull free under stress.

Researchers selected a very broad range of potential sintering aids that shared the same constituents as hard tissues and bioactive glasses. These were narrowed down by analyzing their phase diagrams and physicochemical properties, then tested in 5 wt%

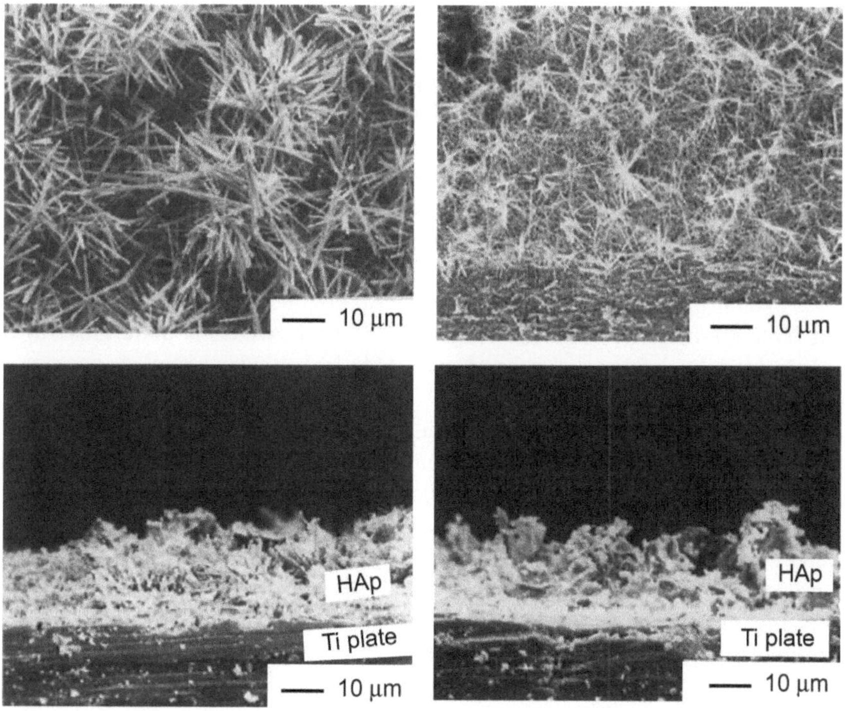

Figure 10.67 Scanning electron micrographs of the surfaces and cross sections of titanium plates [130].

additions to HAw/HA composites. They found that all sodium phosphates improved sinterability without forming tricalcium phosphate or CaO. Beta-$NaCaPO_4$ proved the only sodium phosphate to provide the weak fiber–matrix interface needed for composite toughness [273,274,320].

There are several kinds of HAp/bioactive glass composites. The first one is also called bioactive glass ceramics. In these composites, HAp and/or wollastonite or other crystalline phases crystallize from the glassy matrix during an appropriate heat treatment [322–326]. The bioactive glass ceramics exhibit strength of 100–200 MPa, K_{Ic} of 1.0–2.6 MPa, $m^{1/2}$, fracture energy of 6–26 J/m^2, and Weibull modulus of nine. Coefficient of subcritical crack growth (η) is reported to be in the range of 18–33. Bioactive glass ceramics maintain high strength for a longer time than HAp, both under in vitro and in vivo conditions [265]. HAp/bioactive glass composites can also be prepared by simple sintering of appropriate HAp/bioactive glass powder mixtures. In another approach, small quantities of bioactive glass are added to HAp ceramics in order to improve densification and/or mechanical properties.

In spite of high bioactivity, high biocompatibility, and superior (but still insufficient) mechanical properties of HAp ceramics, the HAp/bioactive glass composites did not find wide applications as bone substitutes.

In contrast with HAp/bioactive glass composites, the HAp/polymer composites are very interesting. Bonfield and coworkers [172,327–329]. developed HAp/polyethylene composites. The mechanical properties of the polyethylene-HA composition are similar to those of bone. Implant tests of the polyethylene-0.4 volume fraction HA composites demonstrated development of bone bonding between the natural hard tissue and the synthetic implantation. These composites have the advantages of ease of shaping of the implant to meet the patient's needs, the time of surgery, formation of a bioactive bond to hold the implant in place, and mechanical properties that closely match those of the host tissues. HAp/polyethylene composites exhibit brittle/ductile transition at a HAp volume content of 40–45% [330]. Their Young's modulus is in the range of 1–8 MPa, which is quite close to the Young's modulus of bone. Unfortunately, HAp/polyethylene composites are not biodegradable.

There are several publications in recent years on the fabrication of HAp/collagen composites, which are similar to bone from the point of view of chemical composition, but do not have such a complex microstructure. The composites can be prepared by mixing HAp with a collagen solution with subsequent hardening by different methods. Suchanek and Yoshimura [265] have reviewed all the mechanical properties of various HAp/polymer composites. The hydrothermal method of processing these composites is very useful to achieve the optimum mechanical properties.

10.8.4 Hydrothermal Processing of Whisker Crystals

A whisker is a long filamentary crystal, often containing a single screw dislocation. A core of the dislocation usually follows the whisker axis. Whiskers are thus single crystals with a very low dislocation density, and are of special interest for studies of the influence of dislocation on properties. Whiskers have unusually high tensile properties because of their low dislocation density [331].

The growth of whisker crystallites is not new. In ancient times, the use of fire allowed the development of a ceramic industry. In the heat of pottery kilns (pottery is an older technology, dating from about 15,000 BC), the clays transformed so that needle-like crystallites of aluminum silicate mullite could form as a dense crack-free mass in the shapes of vases, lamps, amphoras, and tiles [332]. However, scientific thinking on whiskers began during the nineteenth century in order to understand their morphology, dislocation density, and other related defects. Vapor growth is one of the favorite techniques for the formation of whiskers, followed by growth from solutions. Whisker crystals are commonly observed in metals and intermetallic alloys. However, interest in whiskers remained purely academic. It was only during the 1990s that interest in whisker growth was rejuvenated, owing to their applications. Therefore, a wide range of whisker crystals has been prepared, even among nonmetallic, ceramic, and other inorganic compounds. For many years, inorganic fibers and whiskers have been used mostly as reinforcements in composites and in thermal insulation. Various fibrous materials such as glass, carbon, HAp, SiC, Si_3N_4, Al_2O_3, and ZrO_2 have been prepared for such purposes. Among natural fibers, asbestos has been widely used, due to its high tensile strength, chemical stability, and low cost. Unfortunately, its application has a great health risk as it causes many serious diseases, including lung cancer. The toxicity of widely used silicon carbide whiskers is stated to be only slightly lower. The carcinogenic effect of the fibrous materials was restricted to long and thin fibers (diameter $<1\ \mu m$, length $>10\ \mu m$). Many commercially available whiskers and fibers may be safe from this point of view; however, their chemical compositions must also be taken into account. Alumina, zirconia, titania, silicon carbide, and silicon nitride are known as bioinert materials [259]. If the material is toxic, the surrounding tissue dies. If the material is nontoxic and biologically inactive (bioinert), a fibrous tissue of variable thickness is formed. If the material is nontoxic and biologically active, an interfacial bond forms. In this respect, HAp is known as the most biocompatible material. It has been used in medicine for many years in the form of small unloaded implants, powders, and coatings.

Fibrous HAp can be used as insulating agents, packing media for column chromatography, etc. It is a very promising material for preparation of composites. In recent years, because of their excellent biocompatibility, HAp whiskers are the most studied whiskers. HAp contains nontoxic species, such as Ca, P, OH^-, usually CO_3^{2-}, and, even if some of the crystals have a dangerous shape, they should dissolve in a human body without causing any health problems. It has been reported that different kinds of proteins are absorbed on different planes of HAp crystals. Therefore, if HAp crystals, whose particular planes grow selectively, are obtained, they can separate various kinds of proteins. Nagata et al. [333] has obtained plate-like HAp crystals synthesized hydrothermally with methanol.

In recent years, several reports concerning the fabrication of HAp fibers and whiskers appeared in literature [257,266,268,334–337]. All preparation techniques for HAp whiskers can be divided into two main groups: (1) the homogeneous precipitation method, using urea, and (2) the decomposition of chelating agents. The first method utilizes a continuous increase of pH in a solution containing calcium and phosphate ions at a high temperature. In the second case, chelating agents like EDTA, lactic acid, or citric acid are used. During the heat treatment, which is usually carried out under hydrothermal conditions, Ca-complexes with chelating agents decompose, followed by the precipitation of HAp whiskers.

In the preparation of HAp whiskers by the first method, the resultant product is always contaminated with large quantities of carbonate ions and they have not been verified, as yet, to be single crystals (i.e., whiskers) [257]. Yoshimura and coworkers prepared, under hydrothermal conditions, HAp materials, such as fine needle-like crystals and whiskers [265,266,268]. They prepared HAp whiskers of diameters $0.1-5\ \mu m$ and length $30-100\ \mu m$, using the decomposition of chelating agents. Christiansen and Riman [257] also adopted a similar procedure to prepare HAp whiskers [335–338]. The most significant feature of these whiskers is the absence of the CO_3^{2-} species in larger quantities, but their Ca/P molar ratio was in the range of 1.59–1.63, deviating from the stoichiometric value of 1.67.

Suchanek et al. [335] have obtained several batches of HAp crystals under hydrothermal conditions. Starting chemicals like H_3PO_4, $Ca(OH)_2$, and lactic acid were taken in a Teflon beaker, inserted into an autoclave and hydrothermally treated at 200°C for 5 h under a pressure of 2 MPa. The molar ratios of lactic acid/Ca and Ca/P were 2.0, 4.0, 6.0, and 1.43, 1.80, 3.00, respectively. Figure 10.68 shows all the steps involved in the experimental procedure concerning the preparation of HAp whiskers [334]. Yoshimura has proposed the mechanism of formation of HAp whiskers under hydrothermal conditions (Figure 10.69) [335].

The morphology of HAp crystals was controlled by both the lactic acid/Ca and Ca/P molar ratios in the starting solution. The SEM studies revealed that the crystals were not aggregated and had a shape of hexagonal rods or whiskers, elongated along the *c*-axis (Figure 10.70a–d) [335]. As shown in Figure 10.70a–b, the diameter increases with increasing Ca/P molar ratio in the starting solution and is generally larger for high lactic acid/Ca molar ratios. The aspect ratio of HAp crystals is in the range of 5–20. It decreases with increasing Ca/P ratio and is lower in the case of high lactic acid/Ca ratios Figure 10.70c and d. When the lactic acid/Ca and Ca/P starting ratios are low, crystals have the shapes of whiskers, while in other cases large elongated grains are formed.

Distributions of the length and aspect ratio of selected whiskers are shown in Figure 10.71a and b. In this case, the aspect ratio for the majority of the crystals is in the range of 10–25. The length of the whiskers shows a big scattering, suggesting that some of them might be broken during and/or after the hydrothermal treatments.

The Ca/P molar ratios of HAp crystals are in the range of 1.59–1.62. Deviation from the stoichiometric value of 1.67 decreases slightly with the increasing lactic acid/Ca ratio. It is difficult to explain why whiskers are nonstoichiometric. Several theories have been proposed to explain the phenomenon of nonstoichiometry in HAp. They include adsorbed phosphate species on the surface, the presence of additional phases as amorphous calcium phosphate or octacalcium phosphate, and various combinations of lattice defects. This is in agreement with the model involving Ca and OH vacancies and the substitution of some phosphate groups by HPO_4^{2-}.

To check the presence of various species of phosphates, carbonates, hydroxides, etc., the IR-spectra are very useful. Absence of any distinct bands in the range of $1400-1500\ cm^{-1}$ indicates that the HAp whiskers do not contain large quantities of carbonate ions [336]. Figure 10.72 shows the IR-spectra of the whiskers with improved stoichiometry (Ca/P molar ratios 1.64 and 1.67).

HAp whiskers have the following lattice parameters: $a = 9.4317\ Å \pm 0.0007\ Å$ and $c = 6.8822 \pm 0.0012\ Å$. These values correspond to HAp prepared in aqueous systems. Thus, after a critical evaluation of the experimental conditions, Yoshimura and coworkers have proposed that in the system $Ca(OH)_2-H_3PO_4$–citric acid, HAp whisker (30 μm length, 0.1–1 μm width) could be synthesized preferentially under the conditions of:

i. lower concentration of Ca^{2+} (0.2 M),
ii. lower Ca/P molar ratio (1.67 for $[Ca^{2+}] = 0.2$ M),

iii. restricted citric acid/Ca molar ratio (1.2–1.31),
iv. higher temperature (180°C) and longer duration (2.75 h at 200°C),
v. final pH of 3.4–3.8.

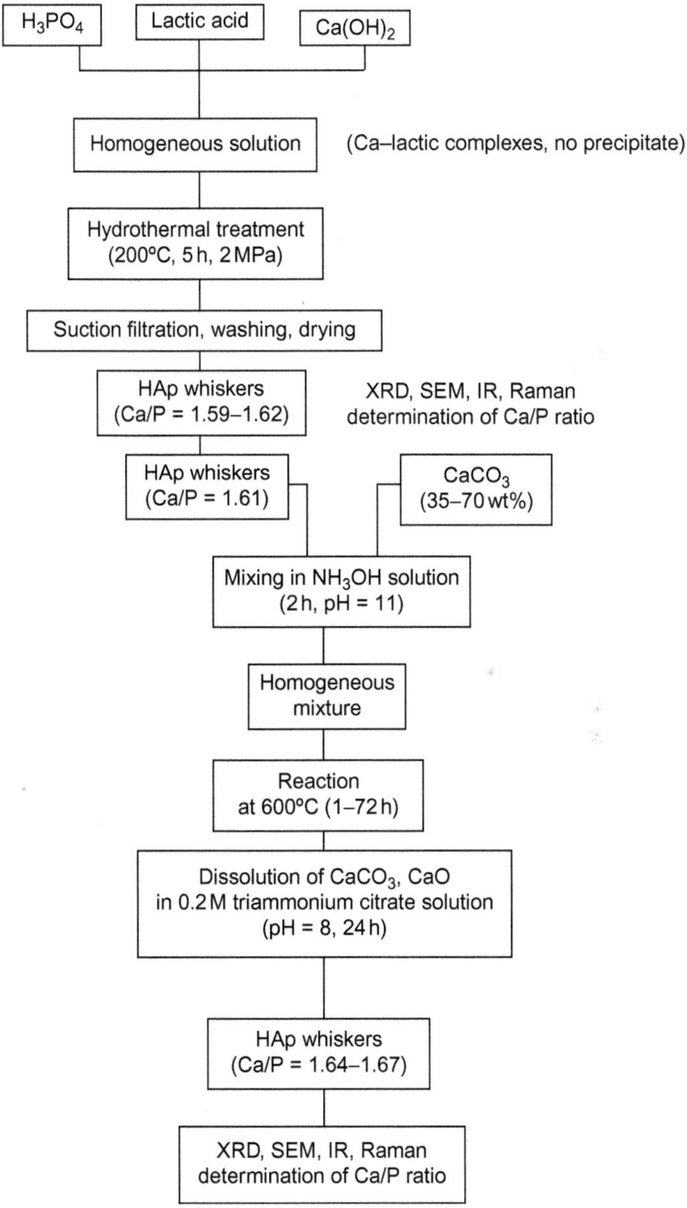

Figure 10.68 Steps involved in the experimental procedure for the preparation of HAp [335].

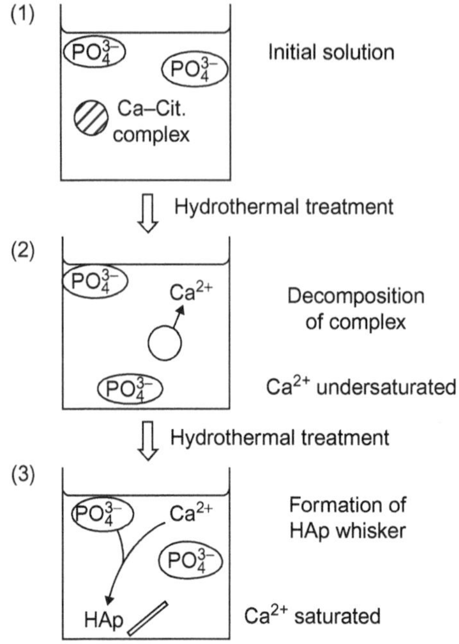

Figure 10.69 HAp whiskers under hydrothermal conditions [335]. *Source*: By M. Yoshimura.

Suchanek and Yoshimura [336] have studied the thermal stability and densification of HAp whiskers in great detail.

The HAp whiskers have already been used as a biocompatible reinforcement in the HAp/HAp (whisker) composites, Figure 10.73a. Consequently, the fracture toughness of pure HAp ceramics has been improved even to the value of 2.0 MPa—this is the highest value in 25 years. In addition, a variety of microstructurally controlled HAp-based composites with improved reliability and high biocompatibility can be prepared using β-rhenenite (β-$NaCaPO_4$), a newly discovered weak interphase material for HAp ceramics (Figure 10.74a–c). The fibrous HAp can also be used to fabricate porous HAp ceramics or porous HAp/β-TCP composites as shown in Figure 10.73b. Moreover, the HAp fibrous skeleton should be an appropriate reinforcement for HAp/polymer biodegradable bone substitutes. Thus, HAp whiskers are very important in materials science and technology.

The preparation of TiO_2-based composites with SiO_2, activated carbon, Al_2O_3, zeolites, etc. have been reported in the literature. Several authors have worked extensively on the preparation of TiO_2–HAp composites. Recently, the authors [337] have prepared anatase nanocrystals deposited on HAp under hydrothermal conditions. The HAp was prepared initially using $CaCO_3$, H_3PO_4, and HNO_3. Ammonia solution was used to adjust the pH. Later, to synthesize TiO_2-deposited HAp, 0.2 g of the above-synthesized HAp or commercial HAp particles were added into a pH-adjusted neutral solution of 4 ml titanium amine complex and distilled water. The precursor was treated in an autoclave at 180°C or 120°C for various reaction times. The anatase crystals were nucleated by heterogeneous nucleation and grew on the HAp surface.

Figure 10.70 SEM photographs of selected HAp crystals.
Source: Photographs courtesy of W. Suchanek.

A higher experimental temperature of 180°C was effective for producing highly crystallized anatase with rod-like crystals of 100–150 nm in length on HAp crystals.

In recent years, several new composites for enhancing photocatalytic properties have been reported [337,338]. Wu et al. [337] have reported the synthesis of $HNbWO_6$/Mo nanocomposites for photocatalytic applications. This nanocomposite was obtained in several steps involving the preparation of $HNbWO_6$, $LiNbWO_6$, then $[Fe_3(CH_3COO)_7(OH)(H_2O)_2]NO_3$ was added with Deggusa P-25 grade unsupported TiO_2. M^{2+}(M = Mn, Ni, and Cu) or M^{3+} (M = Cr and Fe) ions were incorporated into the interlayer of $HNbWO_6$ by hydrothermal reactions of $HNbWO_6$ with 1 M $M(NO_3)_2$ or $M(NO_3)_3$ aqueous solution in an autoclave at 120°C for 12 h. After being filtered and washed with water, the precipitate was heated at 250°C for 3 h so as to decompose any water remaining in the interlayer of $HNbWO_6$. The samples obtained thus were designated as $HNbWO_6/Cr_2O_3$, $HNbWO_6$/MnO, $HNbWO_6$/NiO, and $HNbWO_6$/CuO.

Byrappa et al. [338,339] have prepared an activated carbon:titania (A.C:TiO_2) nanocomposite photocatalyst under hydrothermal conditions. Activated carbon has long been recognized as one of the most versatile adsorbent materials used for the effective removal of low concentrations of organic and inorganic species from solution and industrial wastewater.

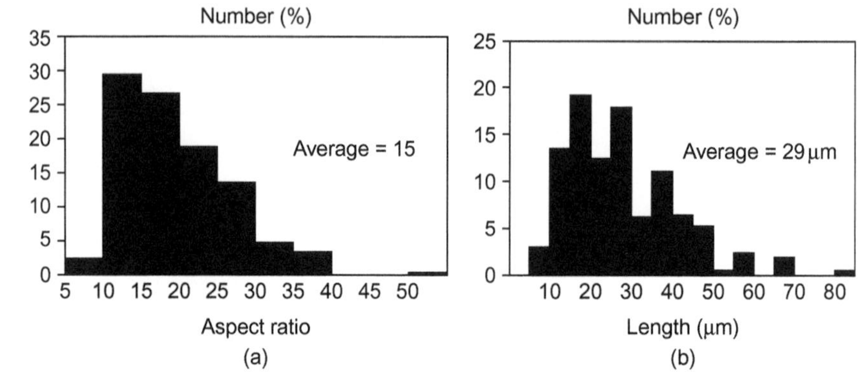

Figure 10.71 Distributions of (a) the length and (b) the aspect ratio of selected whiskers. *Source*: Courtesy of W. Suchanek.

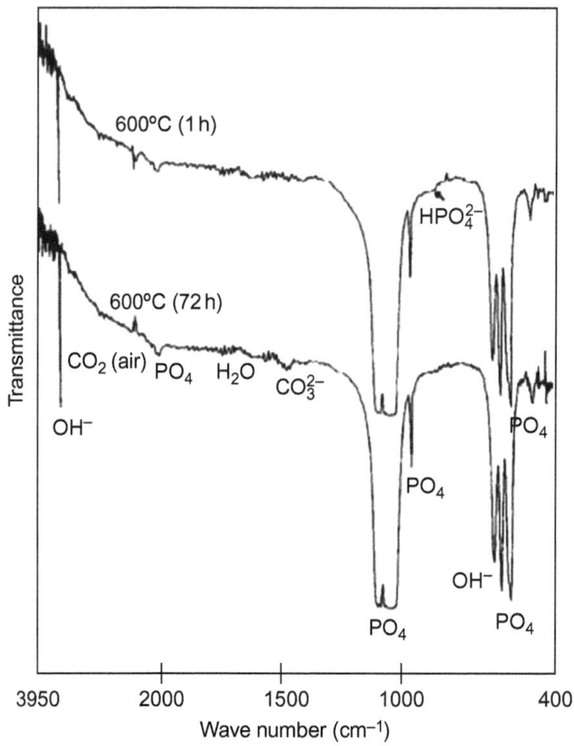

Figure 10.72 IR-spectra of the whiskers which improved stoichiometry (Ca/P molar ratios 1.64 and 1.67) [335].

There is much scope for the impregnation of a suitable semiconductor onto the activated carbon surface layers to prepare a carbon/semiconductor photocatalyst. The adsorption capacities and the feasible removal rates could be substantially boosted by the impregnation of the activated carbon with suitable semiconductors.

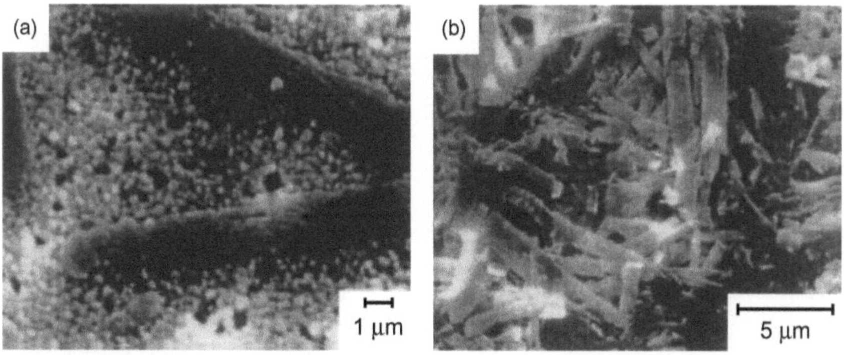

Figure 10.73 (a) Biocompatible reinforcement in the HAp/HAp (whisker) composites and (b) HAp whiskers with Ca/P—1.66.
Source: Photographs courtesy of W. Suchanek.

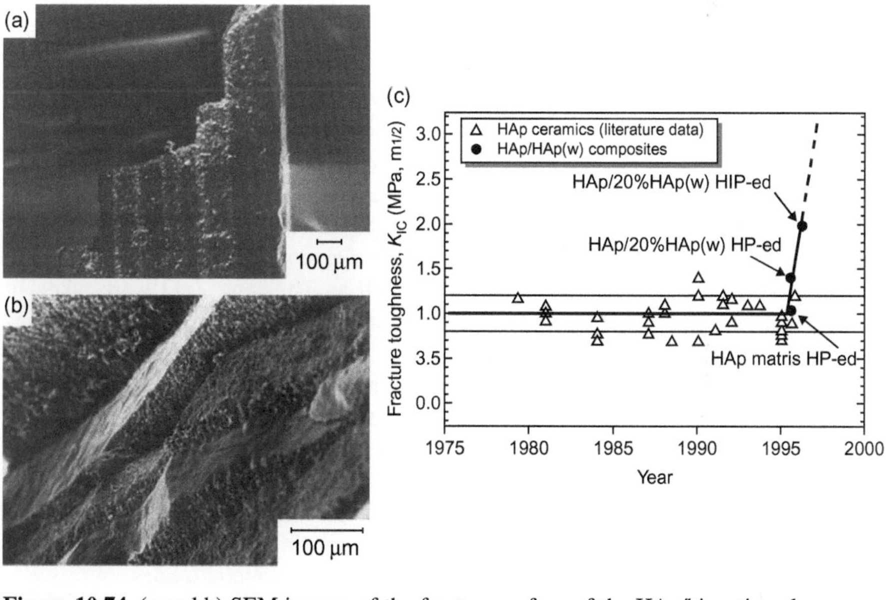

Figure 10.74 (a and b) SEM images of the fracture surface of the HAp/bioactive glass composite. Notice the crack deflection on the β-$NaCaPO_4$ interphase layers. (c) Fracture toughness of HAp ceramics.
Source: Parts (a and b) photographs courtesy of W.L. Suchanek and Part (c) courtesy of M. Yoshimura and W.L. Suchanek.

Here the authors have used activated carbon as an inert porous carrier material for distributing TiO_2 to be accessible to reactants for photocatalytic degradation. Commercially available activated carbon was used along with coconut shell-based activated carbon. Activated carbon was crushed into small particles and separated

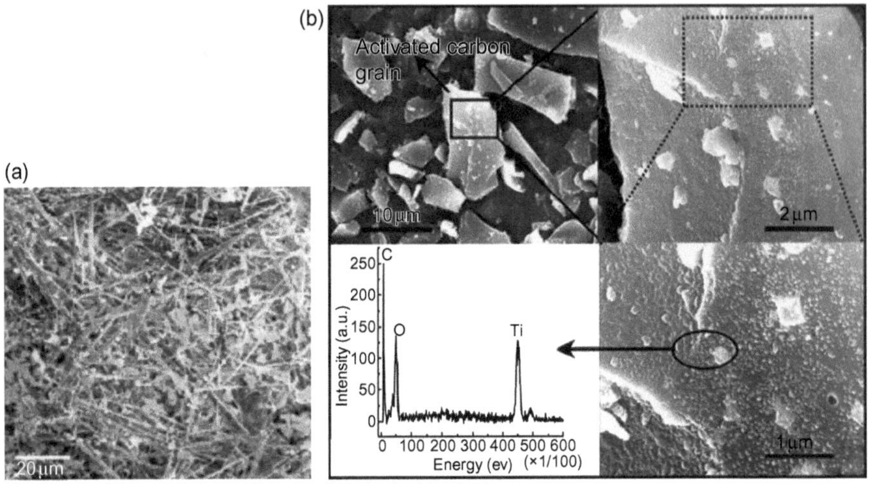

Figure 10.75 (a) ZnO:CNT and (b) TiO_2:activated carbon [19,339].

by 50–80 mesh. First activated carbon was washed with double distilled water until the black color of washings disappeared, followed by soaking in 5% HCl solution with constant shaking for 24 h. Further, it was washed with distilled water till the pH of washings became neutral. Finally, the product was dried at about 80°C. Then a required amount of activated carbon (2 g) was taken in a Teflon liner containing a desired amount (10 ml) of different molar concentrations of HNO_3 and NaOH as solvents. The active metal oxides such as TiO_2 and ZnO were taken in the form of respective oxides or gels. This mixture was stirred well using a magnetic stirrer for 2 h. Later, the Teflon liner was placed in an autoclave, which was kept inside a furnace provided with a temperature programmer controller. The temperature of the furnace was raised slowly up to a predetermined temperature (150–200°C) for a period of 8–24 h. A.C:ZnO shows better catalytic properties than A.C:TiO_2. Figure 10.75 shows the ZnO:CNT and TiO_2:A.C composites [19,339]. However, the greatest disadvantage of these composites is the high cost of activated carbon, but definitely, activated carbon:metal oxide–based composites have been proven to be better photocatalysts than the pure metal oxides.

Recently, the authors [338] have reported the hydrothermal preparation of Nd_2O_3-coated titania composite particles for photocatalytic applications. They used 1 M NaOH at 250°C and pressure ~80 bar with an experimental duration of 5–72 h. A different wt% of Nd_2O_3 was used for the coating in order to reveal the role of Nd_2O_3. Highly monodispersed nanocomposite particles were obtained. Figure 10.76 shows the schematic representation of the formation of Nd_2O_3-coated titania designer composite particulates. Similarly, there are several reports on the hydrothermal coating of functional materials on some ceramic supports. For example, ZnO and TiO_2 nanoparticulates are coated on calcium aluminum silicate beads as supports under hydrothermal conditions (Figure 10.77) with 1 M

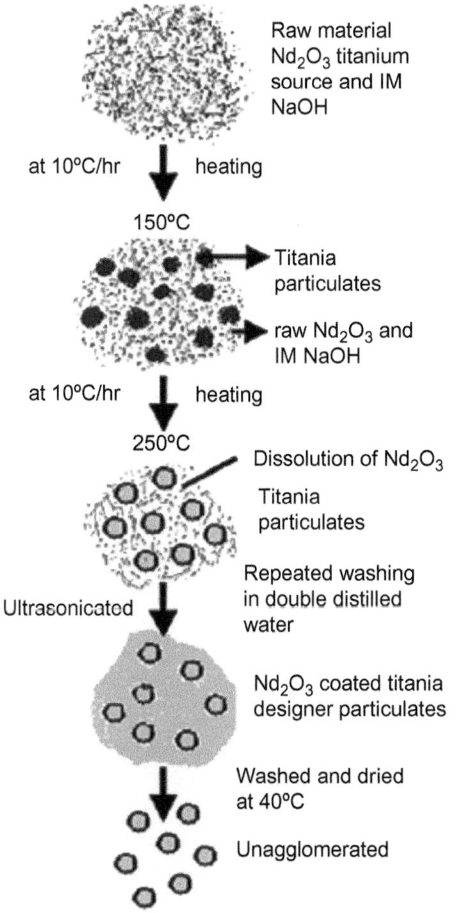

Figure 10.76 The schematic representation of the formation of Nd_2O_3-coated titania designer composite particulates [338].

Figure 10.77 ZnO and TiO_2 coatings on CASB.
Source: Photographs by K. Byrappa.

HCl or 1 M NaOH as solvent at temperature (200–220°C) with autogeneous pressure in the autoclave to obtain strong coating of nanocrystals of ZnO and TiO_2 having high crystallinity. This has some special advantages in that the coating is done through chemical bonding, and the beads can be recovered easily after the degradation and used again without losing the photocatalysts and also its efficiency [340,341].

In recent years, functionalized surfaces based on CNTs and polymers, DNA-assisted dispersion and separation of CNTs, and some special HAp based composites for biomedical and optoelectronic applications have been prepared [342–347]. Hydrothermal technique has been effectively used to synthesize such nanocomposites. Similarly, the encapsulation of active molecules such as ZnO, TiO_2, and zeolites into CNTs has also been achieved.

10.8.5 Related Methods of Hydrothermal Processing of Materials

Modern methods of the hydrothermal processing of materials cover several other processing techniques related to the hydrothermal method. Although, these new methods are not used widely in the routine hydrothermal processing, they have special applications in the processing of some selected technological materials, like hydrothermal transformation alteration, recycling, densification, solidification, strengthening, and sintering. The most commonly used processing techniques are HHP, HIP, hydrothermal sintering, microwave hydrothermal, hydrothermal leaching, hydrothermal decomposition of toxic organic materials, amongst others. These processes have helped in the processing/treatment of materials, increasing their mechanical properties, enhancing the yield, reaction kinetics, and so on.

HHP and HIP

HHP is one of the good processing routes for preparing a ceramic body at relatively low temperature (below 300°C). Yamasaki and coworkers [348] have used this technique popularly to process a wide range of materials. The compression of samples under hydrothermal conditions accelerates densification of inorganic materials. This technique is applied to solidify many kinds of materials such as glass, silicate, titania, calcium aluminate-phosphate cement, calcium carbonate, and HAp sewage sludge [344–353], and the method is expected to provide energy-effective processing to fabricate new engineering materials. Calcium alumina reacts with phosphate to form strong bonding. Using HHP techniques, the strength of solidified bodies with mixtures of alumina cement or calcium aluminates, sodium phosphates, and silica fume could be enhanced. Yanagisawa et al. [354] have solidified three kinds of silica powders, crystalline silica, glass, and gel using the HHP method. With the addition of an alkaline solution, mixtures of low-quartz and other oxides were densified by this method. Reaction products of the mixtures under alkaline hydrothermal conditions are bonded to quartz grains. The addition of pure water-densified silica glass powders containing network-modifying oxides. In this case, a hydrated reaction layer was produced by the reaction of the glass particles. The silica gel (5 g) was placed in the autoclave and compressed at 20 MPa. The autoclave was heated to 300°C at the rate of 2°C/min, while the pressure was kept constant at

20 MPa during HHP. The temperature was kept constant at 300°C for 1 h and then the autoclave was cooled down to room temperature. The solidified compact was removed from the autoclave and dried at 110°C.

Yanagisawa et al. [355] have studied the formation of anatase porous ceramics by HHP of amorphous titania spheres prepared by hydrolysis of titanium tetraethoxide. After fine anatase crystals were formed in the original amorphous spheres by HHP, the spherical particles were deformed and fine anatase crystals were flowed into the interstices among the original spheres by compression from outside the autoclave, to form a compact with homogeneous distribution of fine pores. The fine anatase crystals in the compacts were bonded together by dissolution and deposition to form a compact with high mechanical strength. The presence of water accelerated the crystallization of the starting amorphous titania to anatase, and anatase compacts were produced, even at 100°C [355].

Yamasaki et al. [348] have worked out a method to solidify sewage sludge incinerated ash by HHP and investigated superior conditions to solidify ash for producing recycled products. The solidified sample had a high compressive strength of 98 MPa by HHP at $T = 300°C$, $P = 49$ MPa, NaOH dosage approximately 6%. The product could be used as building materials, such as tiles and bricks [356]. These authors have also worked out a method of immobilizing and a high-volume reducing technique for low-level radioactive waste from nuclear power plants or reprocessing plants, through the hydrothermal solidification method [357]. It is confirmed through evaluation tests and a full-scale mock-up test that this technique can be adapted to an actual system for spent solvent treatment in commercial reprocessing plants.

HIP allows rapid densification with mineral grain growth and is one of the most effective ceramic densification processes. This method is being popularly employed to process HAp-based bioceramics [358–360]. Implant materials require not only biocompatibility, but also mechanical strength and porosity to promote the connection with tissues. Therefore, microstructure designing, i.e., grain size, pore size, and porosity, are necessarily tailored in their application to bioceramics. HAp single crystals of about 255 nm × 90 nm in size synthesized hydrothermally at 200°C under 2 MPa for 10 h were normally sintered in air for 3 h. The ceramics obtained were hot-isostatically pressed at temperatures of 900–1100°C under 200 MPa of Ar for 1 h, without any capsules. This postsintering brought about densification up to ~100% for the samples. The fully dense ceramics, with a grain size of about 0.54 μm, showed transparency. Furthermore, dense/porous layered HAp ceramics could be prepared by the same technique, from the fine crystals and coarse powders with relatively low sinterability. Uematsu et al. [350a] have synthesized HAp powder and have formed it into a compact in an aqueous medium using a filter-cake method [360]. The compact was hot isostatically pressed at 700–1000°C and 100 MPa for 2 h. Fully dense, transparent materials were obtained above 800°C. Both forming and densification methods were found to be important in obtaining transparent materials. Figure 10.78 shows a transparent HAp block obtained through HHP [358].

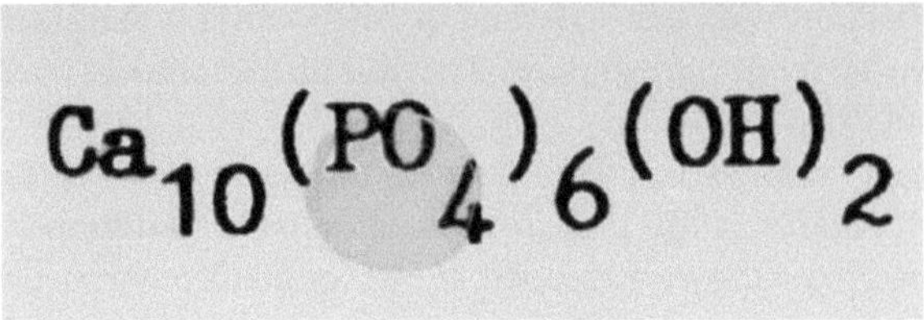

Figure 10.78 Transparent HAp block obtained through HHP.
Source: Photograph by M. Yoshimura.

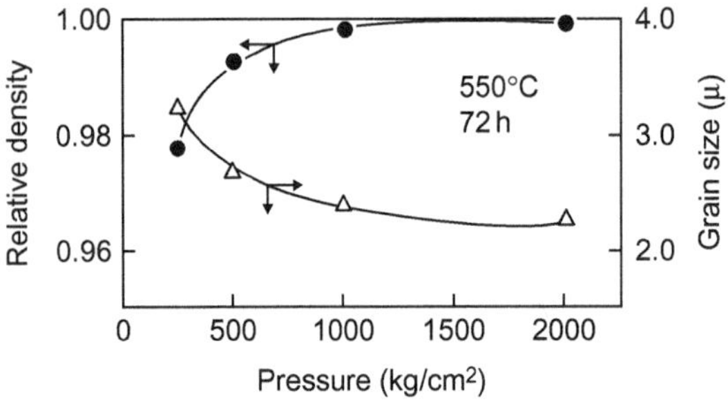

Figure 10.79 Relation between pressure and grain size or relative density hydrothermal reaction sintering of magnetite [362].

Hydrothermal Reaction Sintering of Processing Materials

Hydrothermal reaction sintering is one of the methods used to produce high-density oxide ceramics. A large variety of oxide ceramics have been processed by this method. Oxide powders were obtained by reactions between solutions or nonaqueous liquid and metals. Hydrothermal reaction sintering is a kind of reaction with hot isostatic processing. The hydrothermal reaction sintering of oxides is as follows:

a. low-temperature sintering—able to make sintered body even if materials have high vapor pressure, decomposition, and/or transit,
b. able to make a very fine grain-sized body,
c. able to make high-purity body,
d. able to make high-density body,
e. able to make uniform microstructures,
f. able to save energy due to low-temperature sintering,
g. able to control valency.

Factors, such as purity, grain size and shape of starting materials, reaction temperature, pressure, duration time, ratio of metal and water, and kinds of salt in solution, have made hydrothermal reaction sintering unclear. Several ceramic powders have been processed using this technique. The important ceramic materials obtained by this method are: magnetite, wüstite, monoclinic ZrO_2, chromia, monoclinic HfO_2, alumina, and so on [361–366]. Figure 10.79 shows the relation between pressure and grain size or relative density hydrothermal reaction sintering of magnetite [362]. Figure 10.80 illustrates the SEM fracture surface of high-density hydrothermal reaction sintered Cr_2O_3 (100 MPa, 1000°C for 3 h, 0.01 M HNO_3 solution) with a relative density of 99.1% [363]. Table 10.7 gives selected results of hydrothermal reaction sintering of monoclinic ZrO_2 for

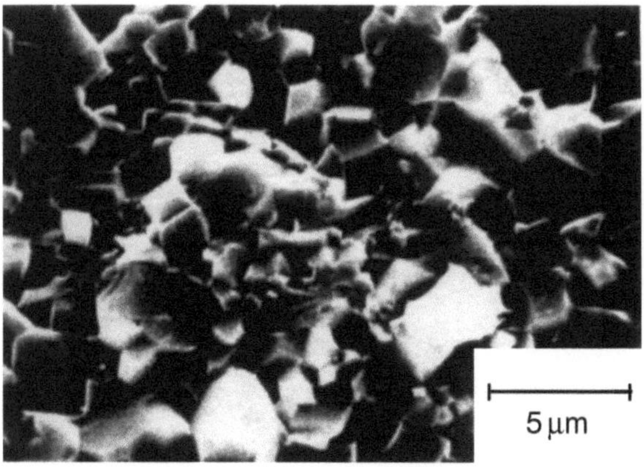

Figure 10.80 SEM fracture surface of high-density hydrothermal reaction sintered Cr_2O_3 [363].

Table 10.7 Selected Results of Hydrothermal Reaction Sintering of Monoclinic ZrO_2 for 3 h [362]

Sample No.	H_2O/ Zr	Degree of Fill (%)	Temperature (°C)	Pressure (MPa)	Bulk Density (g/cm^3)	Relative Density (%)	Average Grain Size (μm)
3	1.773	19.3	1000	98	5.85	99.0[a]	3
5	2.020	15.5	1000	98	5.51	93.2	1
29	2.063	14.8	1000	98	5.17	87.5	0.5
30	2.095	15.0	1000	98	5.38	89.2	0.5
10	2.415	18.4	1000	98	Not sintered		
25	2.038	15.1	1000	196	5.51	93.2	0.3
28	1.985	15.1	1000	294	5.96	100.9	0.1
8	2.012	22.7	1000	490	5.83	98.7	3
7	2.010	20.4	1000	686	5.77	97.7	3
23	2.038	41.2	1000	686	5.70	96.4	
32	2.016	14.8	1000	98	5.68	96.1	2
40	2.142	14.3	1100	98	Not sintered		
41	2.089	14.2	1200	98	Not sintered		
39	2.091	14.1	1000	490	Not sintered	0.5	

[a]With gray skin; gray fine body, rare case; capsule leaked slightly.

3 h [362]. Yin et al. [367] have studied the low-temperature sintering and mechanical properties of ceria and yttria co-doped zirconia crystallized in supercritical methanol.

The preparation and processing of perovskite-type titanates is becoming an attractive field owing to their unique electronic properties. However, their electrical properties are closely linked to their microstructural features, such as porosity and grain size [368]. These authors have prepared $BaTiO_3$ powders by the hydrothermal synthesis technique and have fired via a fast sintering process

Figure 10.81 SEM fracture surface of high-density hydrothermal reaction sintered Cr_2O_3 [368].

or so-called zone sintering. During fast sintering, a powder compact undergoes a rapid high-temperature thermal treatment by being brought to a hot zone within a very short time. In this way, the densification process (e.g., lattice diffusion) can be enhanced while limiting the grain growth process (e.g., surface diffusion). With proper control of time and temperature, it can result in a high-density ceramic with a fine grain size. Figure 10.81 shows the typical SEM micrographs of fast-sintered $BaTiO_3$ samples for 5 min at: (a) 1250°C and (b) 1300°C, of the fast-sintered samples for 15 min at different temperatures. Table 10.8 and Figure 10.82 give density, grain size, and dielectric properties of the sintered $BaTiO_3$ samples.

Similarly, sintering effects on HAp bioceramics have been studied in detail by several workers [369,370]. Thus, hydrothermal sintering reactions find interesting applications in ceramic processing.

Multienergy Hydrothermal Processing of Advanced Materials

Earlier in Chapter 3, Section 3.5.2, Novel Autoclaves, we have treated the equipment used in the multienergy hydrothermal processing of advanced materials. The multienergy covers the combination of hydrothermal, solvothermal reactions in combination with microwave, ultrasonic, mechanochemical, photochemical, electrochemical, or the presence of biomolecules, and this takes hydrothermal technology to a new direction and opens an enormous potential which is yet to be explored by the humankind. The experimental duration is being reduced by at least 4–6 orders of magnitude which in turn makes the technique more economic. Also it evolves a new concept like chemistry at the speed of light. This has also led to the other concepts called Instant Hydrothermal System and Automatic Hydrothermal Vending Machine to synthesize particles instantly.

The use of microwaves in solid-state materials processing is at least 25 years old and is presently being employed very widely to process a large variety of inorganic materials [371–373]. For the past few years, a number of groups have demonstrated that the kinetics of organic and inorganic chemical synthesis may be significantly accelerated using 2.45 GHz microwaves. The acceleration of chemical reactions by microwave dielectric heating has been demonstrated for organic

Table 10.8 Density, Grain Size, and Dielectric Properties of the Sintered $BaTiO_3$ Samples [368]

Type of Sintering	Sintering *T* (°C)	*k*′ (30°C)	tan δ (30°C)	Grain Size[a] (μm)	Density[b] (g/cm)	T_o (°C)	*C* (×10⁵ K)
Conventional	1250	2814	0.038	~4	5.61 (93)	108	1.8
	1300	2663	0.039	~20	5.63 (93)	109	1.9
	1350	2558	0.037	~28	5.59 (94)	112	2.0
Fast—5 min	1250	3676	0.020	~1	5.53 (92)	84	2.1
	1300	3274	0.020	~2	5.65 (94)	97	1.7
	1350	2892	0.009	~4	5.71 (95)	110	1.8
Fast—15 min	1250	2947	0.009	~6	5.76 (96)	111	1.9
	1300	2721	0.006	~12	5.89 (98)	111	2.0

[a]Average grain size determined by taking the mean diagonal values of about 30 grains.
[b]Values in parentheses are the percentage relative densities compared to the theoretical density.

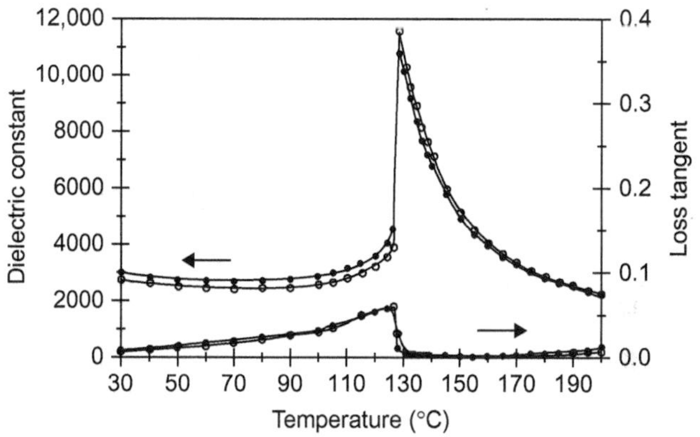

Figure 10.82 Dielectric properties (*k*′ and tan δ) of $BaTiO_3$ [368].

synthesis [374,375], organometallic synthesis [376,377], and preparation of intercalation compounds [378] where organic molecules have been intercalated into inorganic phases. Hydrated inorganic zeolites and other molecular sieves have been synthesized rapidly, using microwaves [379,380].

Komarneni and coworkers, however, were the first to show the rapid synthesis of anhydrous ceramic oxides [38,381,382], hydroxylated phases [383,384], metal powders [383,385], and metal-intercalated clays [386,387]. Also, the rapid synthesis of some inorganic phases, such as ZrO_2 and Fe_2O_3, have been reported [388–390]. In addition to all the above-quoted inorganic phases, microwave-hydrothermal processing is useful in synthesizing novel phases [383,391]. Komarneni et al. [392] prepared

Figure 10.83 SEM photographs of $BiFeO_3$ powders [392].
Source: Courtesy of S. Komameni.

$BiFeO_3$ and $CsAl_2PO_6$ using microwave-hydrothermal conditions and found an enhancement of kinetics. $BiFeO_3$ has a perovskite structure and can be prepared at 194°C under microwave-hydrothermal conditions only. The other phase $CsAl_2PO_6$ is useful in nuclear waste disposal and is accomplished at 138°C under microwave-hydrothermal conditions. Figure 10.83 shows SEM photographs of $BiFeO_3$ powders: (a) highly crystallized agglomerated $BiFeO_3$ powder prepared by microwave-hydrothermal process at ~194°C with a duration of 2–3 h. Figure 10.84 shows SEM photographs of $CsAl_2PO_6$ [392]. These results clearly indicate that the use of microwave field catalyzes crystallization of inorganic phases under hydrothermal conditions.

Recently, D'Arrigo et al. [393] and Kim et al. [394] have synthesized nanophase ferrites such as $ZnFe_2O_4$, $NiFeO_4$, $MnFe_2O_4$, and $CoFe_2O_4$ under microwave-hydrothermal conditions. Multicomponent oxides like spinel phase Co-, Co–Zn and Ni–Zn ferrite nanoparticles have been prepared using microwave-hydrothermal method. The average particle size obtained by such a process is about 10 nm. In the Co-ferrite system, single-phase ferrites with a spinel structure began to form at a relatively low temperature (100°C) in a short holding time (30 min). Figure 10.85 shows TEM images of $Co_{1-x}Zn_xFe_2O_4$ and $Ni_{1-x}Zn_xFe_2O_4$ nanoparticles obtained using the microwave-hydrothermal method [395]. Nanophase ferrites with high surface areas, in the range of 72–247 m^2/g, have been synthesized in a matter of a few minutes at temperatures as low as 164°C. The rapid synthesis of nanophase ferrites via an acceleration of reaction rates under microwave-

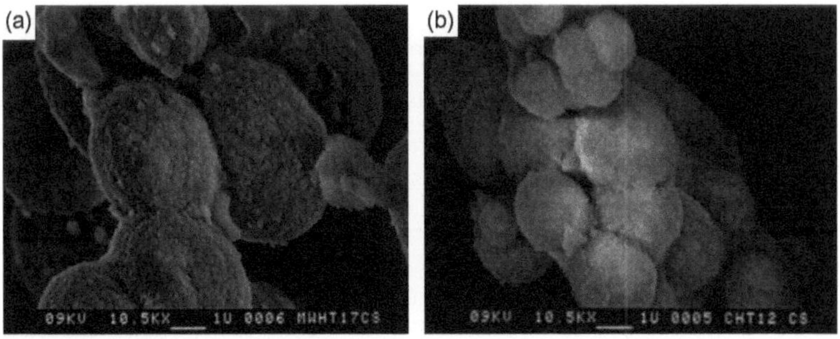

Figure 10.84 SEM photographs of $CsAl_2PO_6$ [392].
Source: Courtesy of S. Komameni.

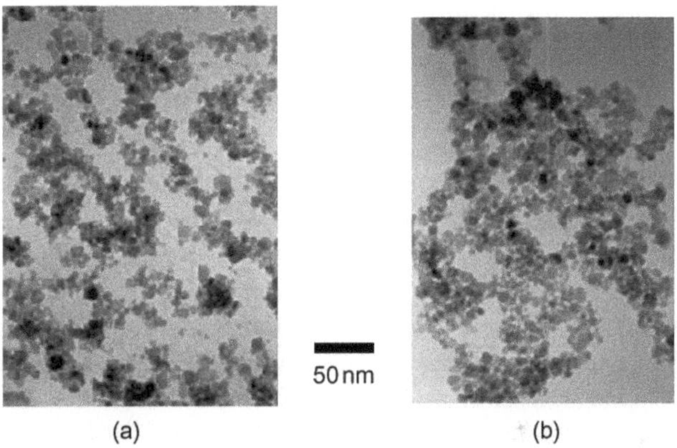

Figure 10.85 TEM images: (a) $Co_{1-x}Zn_xFe_2O_4$ and (b) $Ni_{1-x}Zn_xFe_2O_4$ nanoparticles obtained using microwave-hydrothermal method [395].

hydrothermal conditions is expected to lead to savings in energy and the cost of production of these materials.

A majority of the experiments on microwave-hydrothermal processing are done using a commercially available microwave digestion system, MDS-2000, which is designed by the CEM Corporation, USA. This system produces a microwave frequency of 2.45 GHz and the maximum pressure is ~200 psi.

Komarneni et al. [391] have carried out the synthesis of $BaTiO_3$, $SrTiO_3$, $Ba_{0.5}Sr_{0.5}TiO_3$, $BaZrO_3$, $SrZrO_3$, $PbZrO_3$, and $Pb(Zr_{0.5}2Ti_{0.48})O_3$, using both conventional and microwave-hydrothermal techniques. Conventional-hydrothermal processing, using $TiO_2 \cdot xH_2O$ gel and $Sr(OH)_2$, usually yields platy, needles, or irregular or subrounded agglomerated particles. However, microwave-hydrothermal processing yields agglomerate-free, narrow particle size (0.1–0.2 μm) distribution with spherical morphology, which is expected to have better sintering properties. Tables 10.9 and 10.10 give X-ray diffraction (XRD) analysis of $BaTiO_3$, $SrTiO_3$, $Ba_{0\,5}Sr_{0\,5}TiO_3$,

Table 10.9 XRD Analysis of $BaTiO_3$, $SrTiO_3$, and $Ba_{0.5}Sr_{0.5}TiO_3$ Produced by the Microwave-Hydrothermal Technique [391]

Concentration (M)				T (°C)	Duration (h)	Reaction Products in Order of Abundance as Determined by XRD
$Ba(NO_3)_2$	$Sr(NO_3)_2$	$TiCl_4$	KOH			
$BaTiO_3$						
0.35	0	0.33	10	109	2	$BaTiO_3$
0.35	0	0.33	10	194	0.5	$BaTiO_3$, trace $BaCO_3$
0.35	0	0.33	10	194	2	$BaTiO_3$
$SrTiO_3$						
0	0.1334	0.13	10	164	2	$SrTiO_3$
0	0.1334	0.13	10	194	2	$SrTiO_3$
$Ba_{0.5}Sr_{0.5}TiO_3$						
0	0.1334	0.26	10	194	2	$Ba_{0.5}Sr_{0.5}TiO_3$

$BaZrO_3$, $SrZrO_3$, and $PbTiO_3$ powders produced by microwave-hydrothermal and conventional-hydrothermal techniques. Komarneni et al. [38] had carried out earlier, similar studies on ceramic powders synthesis (titania, zirconia, iron oxide, chromia, etc.), using both conventional- and microwave-hydrothermal conditions and had obtained similar results. Thus, the crystal size, morphology, and level of agglomeration of the different ceramic oxides can be controlled by parameters such as concentration of the chemical species, pH, time, and temperature. Submicron powders of TiO_2, ZrO_2, Fe_2O_3, KNO_3, and $BaTiO_3$ have been obtained by this means.

Cheng et al. [396] have studied in detail the microwave processing of tungsten carbide-cobalt composites and ferroic titanates. This helps the sintering of tool bits and related parts with better properties in about one-tenth the cycle time required by conventional means, and most significantly, it can help in retaining very fine microstructures when starting with very fine powders.

Komarneni et al. [397] have studied the microwave-hydrothermal processing of metal clusters supported in and/or on montmorillonite. The authors could obtain novel Ag^+, Pt^+- and Pd^+-metal intercalated clays, which may be useful in catalysis. These are prepared by using the combination of microwave-hydrothermal and polyol reduction processes. Fang et al. [398,399] have carried out extensive studies on HAp ceramics using microwave-hydrothermal conditions. This not only enhances the reaction kinetics but also gives better densification and mechanical properties [273].

The application of microwave-hydrothermal techniques to process modern materials is becoming highly popular and is a rapidly expanding area of research and development. A reader can get more useful information in the reviews of Komarneni [383,384] and Ehsani et al. [273a].

The mechanochemical–hydrothermal synthesis utilizes aqueous solution as a reaction medium. Mechanochemical activation of slurries can generate local zones of high

Table 10.10 XRD Analysis of Titanates Produced by Microwave-Hydrothermal and Conventional-Hydrothermal Techniques [391]

Compound	Concentration (M)			*T* (°C)	Duration (h)	Reaction Products in Order of Abundance as Determined by XRD
$PbTiO_3$	$Pb(NO_3)_2$	$TiCl_4KOH$				
	Microwave-hydrothermal					
	0.4079	0.4	6	138	0.5	PbO, $PbTiO_3$
	0.4079	0.4	6	138	1.0	PbO, $PbTiO_3$
	0.4079	0.4	6	164	0.5	PbO, $PbTiO_3$
	0.4079	0.4	6	164	1.0	PbO, $PbTiO_3$
	0.3	0.25	2	194	0.5	$PbTiO_3$, PbO
	0.3	0.25	3	194	0.5	$PbTiO_3$, trace PbO
	0.3	0.25	4	194	0.5	$PbTiO_3$, trace PbO
	Conventional-hydrothermal					
	0.4079	0.4	4	195	24	$PbTiO_3$
$BaZrO_3$	$Ba(NO_3)_2$	$Zr(NO_3)_2$	KOH			
	Microwave-hydrothermal					
	0.3113	0.3144	10	194	2	$BaZrO_3$
	0.3113	0.3144	10	130	2	$BaZrO_3$
	Conventional-hydrothermal					
	0.3113	0.3144	10	194	2	$BaZrO_3$
$SrZrO_3$	$SrZrO_3$	$Zr(NO_3)_2$	KOH			
	Microwave-hydrothermal					
	0.3113	0.3113	10	138	0.5	$SrZrO_3$
	0.3113	0.3113	10	138	1	$SrZrO_3$
	0.3113	0.3113	10	164	0.5	$SrZrO_3$
	0.3113	0.3113	10	164	1	$SrZrO_3$
	0.3113	0.3113	10	194	2	$SrZrO_3$
	Conventional-hydrothermal					
	0.3113	0.3113	10	138	24	$SrZrO_3$
	0.3113	0.3113	10	138	24	$SrZrO_3$

temperatures (up to 450–700°C) and high pressures due to friction effects and adiabatic heating of gas bubbles (if present in the slurry), while the bulk system is close to room temperature [400]. Consequently, the thermodynamics of the local reaction environment favor reactions which may otherwise be kinetically inhibited at the bulk system temperature and pressure. Low-cost raw materials can be used for most hydrothermal and mechanochemical–hydrothermal processes which, when coupled with the use of conventional autoclaves and mills, can lead toward the development of low-cost powder synthesis processes. Mechanochemical–hydrothermal synthesis was performed at room temperature utilizing system compositions so that the model calculations indicated would yield phase-pure HAp powders. These conditions were effective for synthesis of phase-pure undoped HAp, carbonate-substituted HAp (CO_3HAp), or coupled sodium carbonate-substituted HAp ($NaCO_3HAp$). Figure 10.86 shows as-prepared carbonated HAp powders with aggregates of nanosized HAp crystals. It is well known that magnesium and carbonate are the two main ionic substitutions in biological apatites, and their preparations carry a great significance for biomedical applications. Such magnesium- and carbonate-substituted HAps have been prepared using via heterogeneous chemical reactions under ambient temperature with the help of mechanochemical–hydrothermal route [401].

During mid-1990s, the influence of biomolecules in nanomaterials synthesis was reported by several researchers [402,403]. In this special edition, Komarneni et al. [387] are reporting the synthesis of 1D nanorods/nanowires and assemblies of inorganic materials with the assistance of biomolecules (such as sugars and their derivatives; amino acids and their polymers, peptide, and protein) under conventional- or microwave-hydrothermal conditions [404]. Similarly, a great variety of magnetic nanoparticles, phosphors, and QDs have been prepared using the hydrothermal processing method.

Hydrothermal Treatment/Recycling/Alteration

This is probably one of the most important areas of research in the field of hydrothermal technology, wherein, the supercritical water properties are exploited for

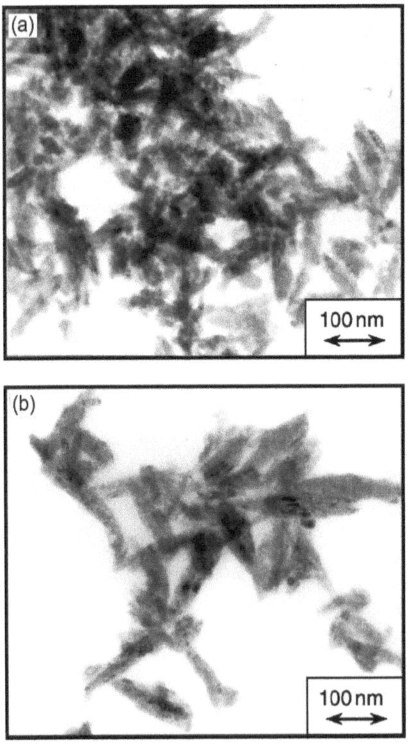

Figure 10.86 Carbonated HAp powders with aggregates of nanosized HAp crystals [401].

effective detoxification and disposal of problematic industrial, nuclear, military, and municipal wastes. In supercritical water, at 450–700°C, many organic compounds are rapidly (0.1–100 s) and efficiently (99.9–99.99 + %) oxidized (supercritical water oxidation or SCWO), with their carbon, hydrogen, and nitrogen, almost completely converted to CO_2, H_2O (mineralization), and N_2. These attributes make supercritical water an attractive medium for chemical reactions and physical separations, i.e., for hydrothermal processing. Environmental applications include the rapid and efficient destruction of hazardous organic substances, e.g., aqueous wastes, and decontamination and/or separation of inorganic pollutants [405–408]. Thus, SCWO is an emerging technology for the treatment of aqueous waste streams, so that they can be recycled as process streams and for the ultimate destruction of organic wastes. Recycling of waste plastics, such as polyethylene, polystyrene, polypropylene and polyethylene terephthalate, radioactive waste, and concrete wastes, have received special attention [409]. Similarly, the decomposition of chlorocarbons, chlorofluorocarbons, polymers, polymer additives, nitroaromatics, and so on, under HPHT conditions, have been well understood [409,410]. The conventional method of treating most of these waste materials is oxidation pyrolysis in incinerators, but this method is not effective as to the dechlorination of chlorofluorocarbons, for example, they need high-temperature plasma destruction, requiring large and expensive apparatus, and the reactors are easily corroded by HCl or Cl_2 gas (the decomposed productions). Similarly, reductive decomposition is not a cost-effective method, requiring expensive reductive agents.

Savage and coworkers have worked out the reaction models for SCWO processes in detail, based on molecular dynamics studies of supercritical water, in order to understand better the potential roles of water in influencing elementary chemical reaction rates [411]. The study of hydrogen bonding in supercritical water and its dependence on temperature and density has been carried out by many workers to understand fundamental issues connected with structure dynamics and thermodynamics of pure water [411–414].

Hydrothermal decomposition of organics or recycling of waste materials is usually carried out in small autoclaves, or Turtle cold-cone seal autoclaves or batch reactors/flow reactors, depending on the experimental conditions and purpose.

Adschiri et al. [415] have studied the conversion of lignin, polystyrene, and polyethylene under supercritical conditions of water and found that polystyrene could be completely decomposed into ethylbenzene, toluene, benzene, styrene, and xylene in 5 min. However, the conversion yield of polyethylene was fairly low, even at a longer reaction time of 2 h at 35 MPa, 400°C but, by the addition of oxygen (about 0.013 mol), conversion increased to 60% at the same temperature. The advantages of such conversion are that lesser char and more aldehyde, ketone, and acid production in supercritical water are observed, as compared to neat pyrolysis reaction.

The above works clearly indicate that a new trend is being set in hydrothermal technology. This technology is going to be the one used in materials processing in the twenty-first century, and it is not only human friendly and environmentally safe but also economical cost wise. Already, several groups have cropped up all over the world to tackle the existing problems and search for new avenues in materials synthesis and processing under a new concept called green materials, which deals with industrial ecology, environmentally friendly methods, recycling, and so on.

In this regard, hydrothermal technique occupies a unique place because of its pollution-free nature as the reactions take place within a closed system. Also, the reactions take place at relatively lower pressure/temperature conditions, which make the technique more popular and easier to operate.

10.9 Hydrothermal Technology for the Twenty-First Century

Belching smokestacks were viewed as a sign of progress in the nineteenth century, worldwide, and even now, in some developing countries. Industrialization, indeed, should be viewed as a sign of progress. It leads to higher living standards, greater longevity, and a better quality of life.

Of late, we have begun to understand the implications of pollution and its prevention, of population growth, and increased industrialization in the third world and in the vibrant Pacific Rim economies. By the middle of the twenty-first century, the population will increase by at least by 50% and then probably stabilize as most countries will have relatively high standards of living. Industrial production will probably increase at least five fold as the newly industrialized nations will have formed a mainstream [416].

Modern human society has been sustained both by a remarkable development of advanced materials and by a huge consumption of energy and resources. As we cannot withdraw ourselves from using them now and even in the near future, the wastes of materials, chemicals, energy, and heat would increase markedly to cause environmental problems on the earth (Figure 10.87) [417]. The global environment has great effects on human lives. Far-reaching changes in social structure are being forced upon us already by environmental issues such as global warming, desertization, depletion of the ozone layer, and acid rain, which stem from two main problems: global scale expansion of the economic activities of industrially advanced nations and population expansion in developing countries. Global environmental problems will lead to great changes in the social structure of the twenty-first century. These changes will not only affect human life and industrial activity but also force significant reforms in fundamental concepts of manufacturing goods [418]. As the earth is almost a closed system with limited biosphere, lithosphere, hydrosphere, and atmosphere (Figure 10.88), all materials should be cycled, and moreover, heat and energy (entropy) loss should be minimized [417,419]. Considering these specifications, we must search for materials (1) that are less hazardous to human life, and preferably compatible with human beings and other living species, and (2) which have environmentally friendly processing to fabricate, to manipulate, to treat, to reuse, to recycle, and to dispose of those materials. The term *materials cycle* is generally used to designate raw materials synthesis, their fabrication, functional products, and their disposal. It is well known that all materials are extracted from the earth and then converted into functional products through various means of materials processing, usually

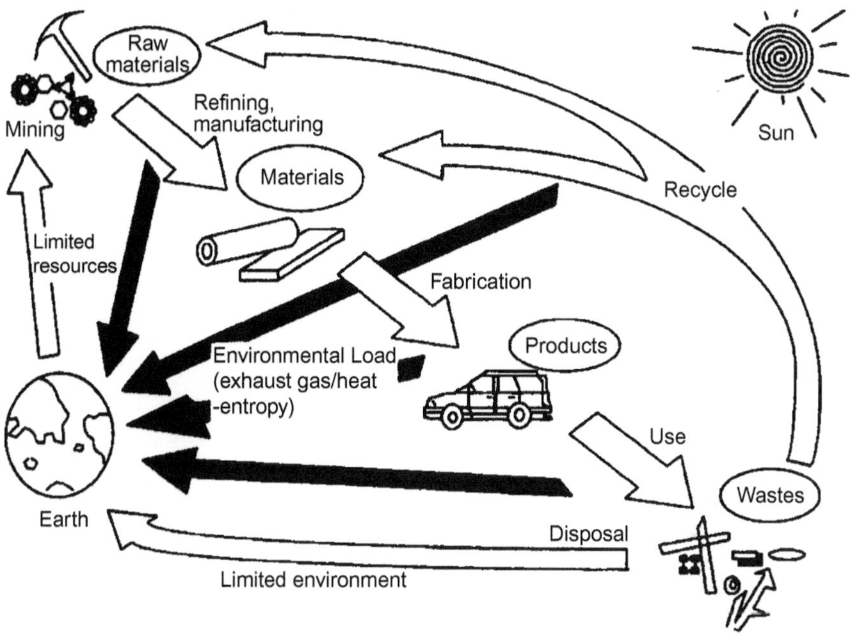

Figure 10.87 Life cycle of artificial materials with environment/resources on the earth [417].

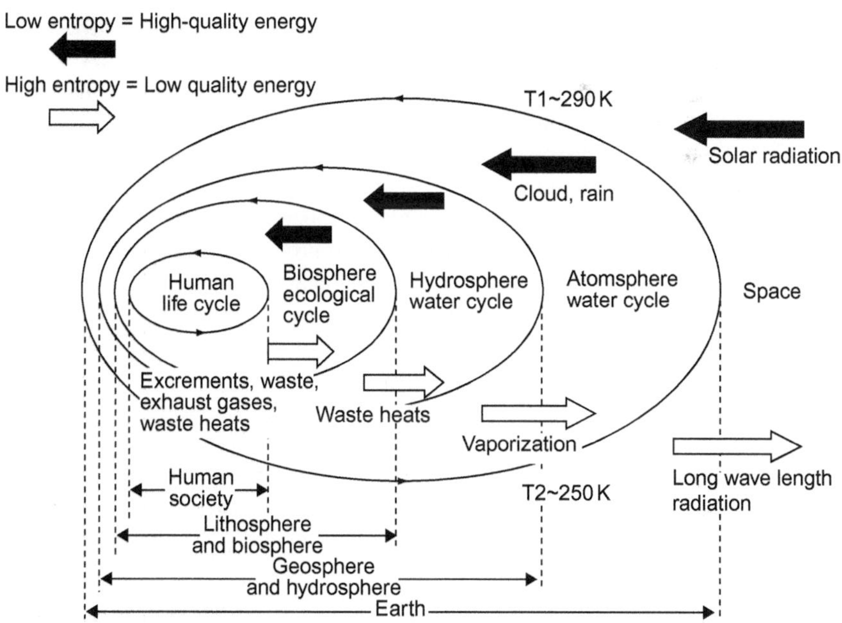

Figure 10.88 Entropy transfer in energy cycle on the earth [417,419].

involving very high temperature/energy and cost, which in turn contribute to global warming. These materials may be disposed off on the earth or recycled [420].

A number of substances with particular compositions, crystal structures, and specified properties have been investigated in detail by various workers all over the world. However, only a few materials can be considered as useful materials, because some substances which have desired composition structures and properties cannot be used as they are extremely difficult to give desired shapes or forms (this important point has often been overlooked). Particularly, it is difficult to give desired shapes, forms, and size, to inorganic materials owing to their high brittleness. Organic materials, such as polymers and plastics, or metallic materials can be generally deformed when local stresses (above their yield stresses) are applied to them, but inorganic materials, particularly ceramics, are likely to break due to brittle fracture. Because ceramics have generally been fabricated by a rather special "ceramic processing" which consists of two steps: (1) Synthesis of powders and (2) shape forming by firing/sintering of the powders or melting (in the case of glasses), both the steps usually require high temperatures and consume a lot of energy.

10.9.1 Thermodynamic Principles of Advanced Materials Processing

Processing of advanced materials generally consists of two steps: (1) Synthesis of substances (ceramic, metallic, organic) which have a particular chemical composition, structure, and properties and (2) materials fabrication (i.e., shape forming by firing/sintering, melting, molding, or casting). Organic materials, like polymers and plastics, or metallic materials can generally be deformed when local stresses (above their yield stresses) are applied to them [421,422], but ceramics are likely to break due to brittle fracture [423]. The two steps in "classical" processing usually require high temperatures, thus consuming a lot of energy, particularly in the case of ceramics [424]. More recent processing routes, using a gaseous phase, like CVD and MOCVD [425,426], or vacuum systems such as sputtering and MBE [425,427], require much more energy than standard high-temperature processing. Generally, all these techniques have resulted in environmental problems because the consumed energies are emitted as exhaust gas(es) or exhaust heat (entropy) except for the part involved in the production. Especially, the vacuum systems seem to be worse because they need continuous pumping to maintain the vacuum and their exhaust gas(es) cannot be cycled due to their diluted huge volumes.

The total energy consumption among all the mentioned processing routes should be the lowest in aqueous solution systems, because an excess of energy is necessary to create melts, vapor, gas, or plasma, than to form aqueous solutions at the same temperature (Figure 1.9) [417]. This idea can be demonstrated using an example of $BaTiO_3$, which is one of the most important materials for the electronic industry. Driving force (ΔG) for representative synthesis reactions of the $BaTiO_3$ Eqs (10.2)–(10.6) are 38 kcal/mol, 3685 kcal/mol, 17 kcal/mol, and -14 kcal/mol, respectively, at room temperature [167,428]. Figure 10.40 shows the energy diagram for the formation of $BaTiO_3$ from various precursors.

$$BaO_{(crystal)} + TiO_{2\ (crystal)} \leftrightarrow BaTiO_{3\ (crystal)} \quad (10.2)$$

$$Ba_{(vapor)} + Ti_{(vapor)} + 3/2O_{2\ (gas)} \leftrightarrow BaTiO_{3\ (crystal)} \quad (10.3)$$

$$Ba^{2+}_{(gas)} + Ti^{4+}_{(gas)} + 3O^{2-}_{(gas)} \leftrightarrow BaTiO_{3\ (crystal)} \quad (10.4)$$

$$TiO_{2\ (crystal)} + Ba^{2+}_{(aq.)} + 2OH^{-}_{(aq.)} \leftrightarrow BaTiO_{3\ (crystal)} \quad (10.5)$$

$$Ba^{2+}_{(aq.)} + Ti(OH)_{4\ (aq.)} - H_2O_{(aq.)} - 2H^{+}_{(aq.)} \leftrightarrow BaTiO_{3\ (crystal)} \quad (10.6)$$

Because the raw materials of Ba and Ti must be solid oxides or carbonate ores, processing using gas/vapor requires a huge energy expenditure of 727–3685 kcal/mol to make solid $BaTiO_3$ and this energy must be discarded into the environment. On the other hand, as the lattice energy of BaO and TiO_2 is almost equal, the hydration (solvation) energy of Ba^{2+} and Ti^{4+} ions, solution processing consumes very little energy if the synthesis activation energy (ΔG^*) can be overcome. Generally, ΔG^* is inversely proportional to $(\Delta G)^{\eta}$, where ΔG and the activation energy (ΔG^*) are sufficiently provided for reaction to yield crystallization compounds with the desired shape/size via several steps, such as diffusion, adsorption, reaction, nucleation, and growth [429]. On the other hand, species in aqueous solutions are hydrated (or chelated by some complexing agents). Thus, they have only a small driving of force (ΔG) for the reaction and rather high activation energies are necessary for reaction to occur by defeating the hydration (chelation) energies of ions. Electro- or electroless-plating for metals is achieved by reducing metal ion (s) electrochemically or chemically. However, in the case of ceramics, anions must be oxidized at the same time as the reduction of cations. Because, some particular activation processes, such as electro-, photo-, sono-, complexo-, organo-, and mechanoactivation are required to accelerate the kinetics of synthesis of crystallized single/multicomponent ceramic materials from the solution (Figure 10.89).

Schematic diagram of a *temperature–pressure map* for various kinds of materials processing is shown in Figure 1.10. Solution processing is located in the pressure temperature range characteristic for conditions of life on earth. All other processing routes are connected with increasing temperature and/or increasing (or decreasing) pressure; therefore, they are environmentally stressed.

Although environmental problems have been argued from various points of view, ecologically, biologically, technologically, economically, and even politically, the most scientific arrangements, thus universally acceptable, are the thermodynamic ones.

Thus, important subjects of technology in the twenty-first century are predicted to be the balance of environmental and resource and/or energy problems. This has led to the development of a new concept, related to the processing of advanced materials in the twenty-first century, namely, industrial ecology—science of sustainability [430]. The first textbook on the subject, written by two AT&T authorities, provides the following definition:

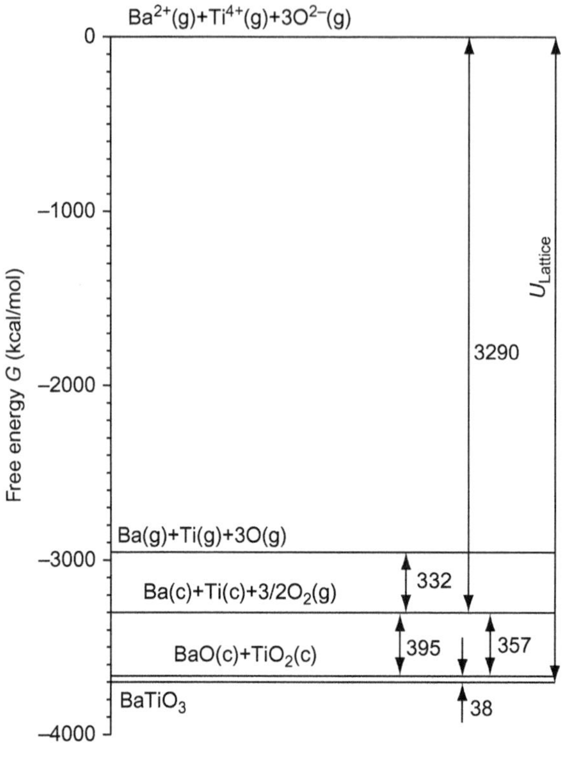

Figure 10.89 Energy diagram for the formation of $BaTiO_3$ from various precursors. *Source*: Courtesy of M. Yoshimura.

> Industrial ecology *is the means by which humanity can deliberately and rationally approach and maintain a desirable carrying capacity, when given continued economic, cultural and technological evolution. The concept requires that an industrial system be viewed not in isolation from its surrounding systems, but in concert with them. It is a system's view in which one seeks to optimize the total materials cycle, from virgin material to finished material, to component, to product, to obsolete product, and to ultimate disposal. Factors to be optimized include resources, energy, and capital [431].*

Many believe that implementing industrial ecology will be a principle challenge for business and society in the twenty-first century. The race is to become the most innovative, most visionary, and most effective company understanding industrial ecology, and implementing designs for the environment.

With all of the above discussed aspects in mind, Yoshimura proposed a new concept of materials processing, namely *soft solution processing*, which meets all the demands of materials processing in the twenty-first century. This soft solution processing for high performance inorganic materials is fast catching on with materials scientists worldwide, because it deals with low energy processing using solutions which minimize environmental impact (sometimes called *chimie douce*), it will be the key to environmental improvement. Temperature for preparation will affect energy consumption and a low temperature leads to "soft processing." A significant example is biological processes that produce magnetite oxides at room temperature.

The solution process also has merit in that some unique inorganic materials can be synthesized only by this process. There are many kinds of processes related to soft solution processing, and the important ones are soft solution electrochemical process, liquid phase deposition, sol–gel processing, hydrothermal technique, coprecipitation, emulsions, the polymerizable complex method, biomimetic, solvothermal, self-assembly, wet chemical processing, templating, and so on. Development of simple patterning methods with nanometer resolution, acceleration of kinetics of synthesis, framing (deducing) the theoretical approaches to the growth of materials from solutions, and the development of in situ observation techniques are some examples of emerging research subjects. One of the most important technological issues of the soft solution processing is its integration with functional device technology, because the possibility of fabrication of a variety of materials and microstructures from solutions has been already been demonstrated. In these processes, solutions are always used and therefore, many factors, such as pH, concentration, ligand type, temperature, and electrode potential, are controlled in order to design advanced materials. Additionally, the technique using the solution flow may be used to synthesize various single phase and multilayered thin films and this is an important step in the application of soft solution processing in the technology of the future integrated functional devices. From this point of view, fabrication of ceramic thin films in the solution flow in the recycled system [432], below 200°C, is very important. In this connection, a flow cell for hydrothermal–electrochemical synthesis and its applicability to fabricate single phase thin films, as well as multilayered structures in the system $BaTiO_3$–$SrTiO_3$, has been demonstrated by Yoshimura and coworkers [31,37,295,296]. The technique using the solution flow under the hydrothermal–electrochemical conditions is an important step in the integration of the solution processing with the functional devices technology and may find applications for various single and/or multilayered thin films.

During the twenty-first century, hydrothermal technology, on the whole, will not just be limited to the crystal growth, or leaching of metals, but it is going to take on a very broad shape, covering several interdisciplinary branches of science. Therefore, it has to be viewed from a different perspective, as it offers several new advantages, like homogeneous precipitation using metal chelates, decomposition of hazardous and/or refractory chemical substances, monomerization of high polymers, like polyethylene or tetraphthalate, and other environmental engineering and chemical engineering issues dealing with recycling of rubbers and plastics instead of burning. Further, the growing interest to enhance the hydrothermal reaction kinetics using microwave, ultrasonic, mechanical, and electrochemical reactions will be distinct [433]. Also, duration of the experiments is being reduced at least by two orders of magnitude which, in turn, makes the technique more economic. With an ever-increasing demand for composite nanostructures, the hydrothermal technique offers a unique method for the coating of various compounds on metals, polymers, and ceramics, as well as the fabrication of powders or bulk ceramic bodies.

In recent years, activated carbons showing high surface area are specially impregnated with active metals, like Cd, Cu, W, Cr, and Mo. Such impregnated activated carbons have additional characteristics, like antibacterial effects, absorb harmful gases in industrial and military applications [434]. The hydrothermal technique is very effective for the impregnation of active metals into activated carbon. Activated carbon impregnated with tungsten metal has been successfully employed for the decomposition of toxic organic compounds, like phenols and nitroarenes [435,436].

Feng et al. [437–439] have employed the hydrothermal soft chemical process for the synthesis of tunnel manganese oxides from a layer of manganese oxide. This process comprises of two steps: (1) Preparation of a framework precursor with layered structure and insertion of template ions or molecules (structure directing agents) into its interlayer space by a soft chemical reaction and (2) transformation of the template inserted precursor into a tunnel structure by hydrothermal treatment.

Hydrothermal–solvothermal processing helps greatly in the structure stabilization of several new compounds [440,441]. Natural phyllosilicates are characterized by a low thermal stability due to the presence of $(OH)^-$ groups in the lattice. Through hydrothermal treatment, the replacement of $(OH)^-$ groups by oxygen atoms takes place, and this, in turn, improves the thermal stability of phyllosilicates. Similarly, the replacement of metals can be carried out under hydrothermal conditions to obtain new structures.

Recovery, recycling, decomposition, and treatment processes, under hydrothermal conditions, are going to play a major role in the twenty-first century [442–446]. Figure 10.90 shows the author's imagination—a new generation, complex industry constructed underground which will exhaust no toxic or hazardous wastes [415]. It is an ideal closed system, a combination of waste treatment, energy recovery, and formation of resource. This deals with the recovery process of human waste to energy resource using hydrothermal technology. The authors have decomposed the night soil without any insecticides and deodorizers using the hydrothermal

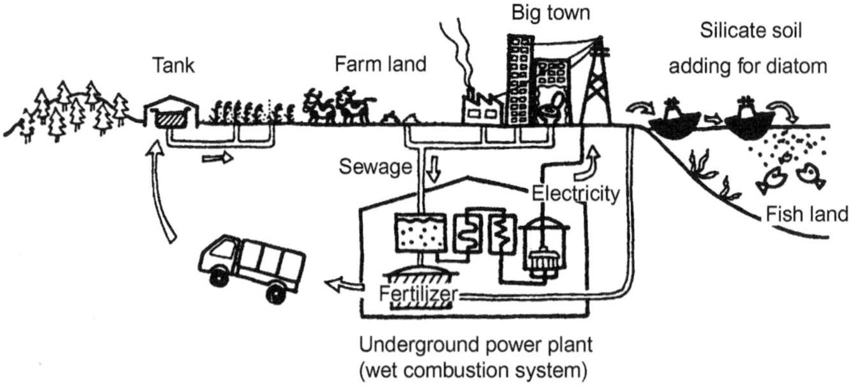

Figure 10.90 Author's imagination—a new generation complex industry, constructed underground and exhausting no toxic or hazardous wastes.
Source: Courtesy of N. Yoshimura.

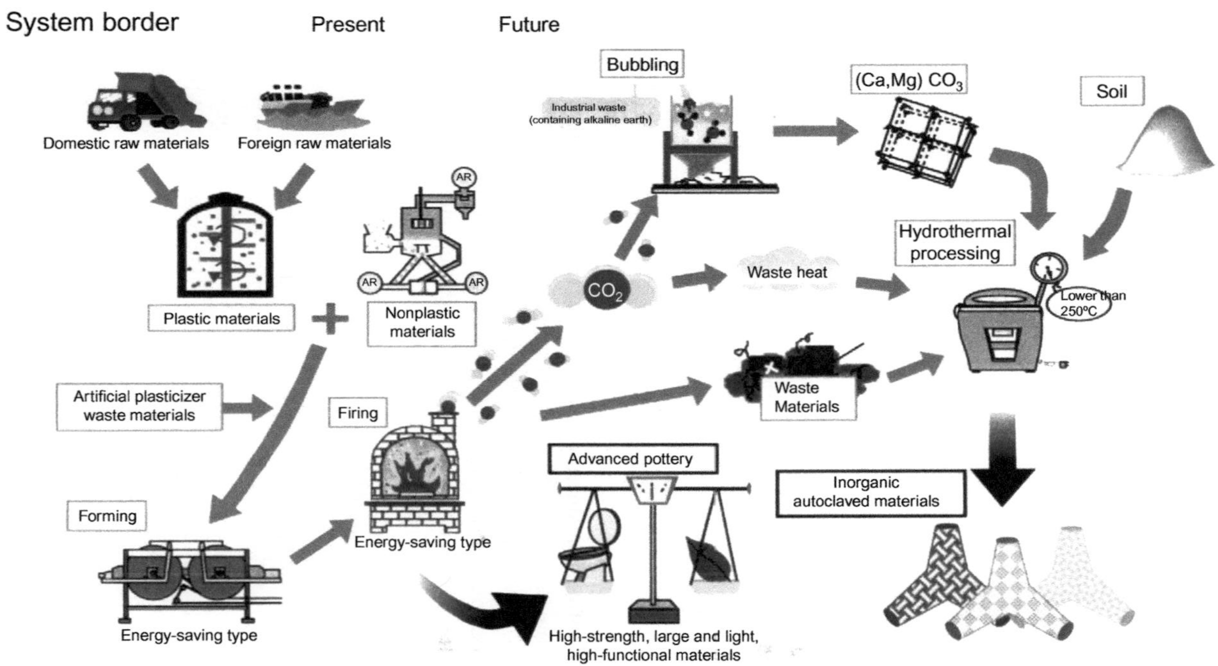

Figure 10.91 Closed manufacturing system.
Source: Courtesy of E.H. Ishida.

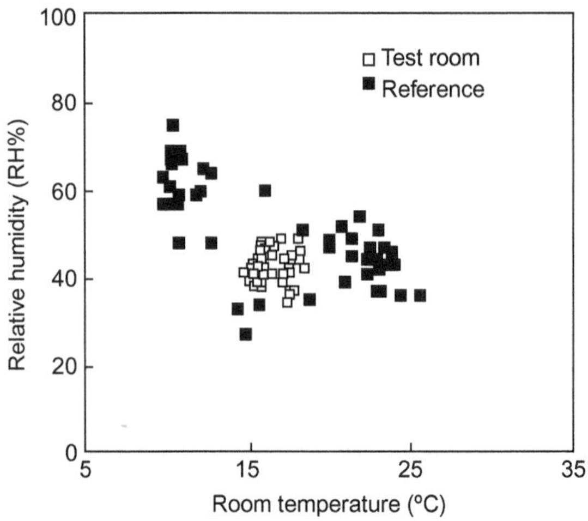

Figure 10.92 Influence of earth ceramics on the room temperature and relative humidity [418].

technology. The decomposition of the night soil was defined by a ratio of Chemical oxygen demand (COD) Mn before and after the reaction. The reaction time was just 10 min. The decomposition ratio increased with increasing temperature, and oxygen pressure over 2 MPa accelerated the decomposition. Night soil was over 90% decomposed when the reaction temperature of the hydrothermal treatment was at 300°C for 10 min. Under these conditions, hydrothermal decomposition of night soil occurred with an offensive smelling gas. Figure 10.90 shows the conversion of organic wastes to bioresource and energy. Furthermore, CO_2 gas is not discharged to atmosphere in this system and it is possible to change CH_3OH for the Cl chemical industry. In the future, a closed system such as this type must be attained in all industrial fields for the survival of modern civilization and for earth ecology. To quote an example of such a system, at IAx Corp., Japan, a new closed system for the collection of waste, separation, and recycling has been introduced, and achieved a reduction in waste output by 85% without increasing the intake (Figure 10.91). In order to achieve the objective of a waste zero system, the corporation is working out new technology based on the *closed manufacturing system*. Soil, solidified hydrothermally, is a new material born from the development of a closed production system [418]. The strength of the hydrothermally processed soil is equal to or greater than that of concrete building materials (flexural strength = 4–6 MPa). Its heat capacity is greater than that of wood flooring (1090 kJ/m^3, K), tatami mat (430 kJ/m^3, K), and carpet (330 kJ/m^3, K). It also exhibits high humidity absorption and desorption ability, similar to wood because of its pore sizes, which are extremely small (10–20 nm) when compared to concrete blocks or conventional chinaware. Figure 10.92 shows the influence of earth ceramics on the room temperature and relative humidity (temperature and relative humidity were measured for 1 month in winter) [418]. The extremely small pores are the result of the pores in the raw material (soil) itself and the pores that are formed during hydrothermal solidification. This hydrothermally treated soil product, with a trade name *Earth Ceramics*, is being

Figure 10.93 Earth ceramics used for floors.
Source: Photographs courtesy of E.H. Ishida.

widely used as flooring material in Japan because of the improved properties compared to the conventional materials (Figure 10.93). The amount of energy used for air conditioning after the use of soil ceramics was 25% less than that without earth ceramics. Figure 10.94 shows the flexural strength per manufacturing energy (MPa m^3/GJ) for the hydrothermally solidified waste soil (earth ceramics) and other common ceramic products. Similarly, earth ceramics have less allergic problems than vinylon cloth and carpets. Thus, the hydrothermal processing helps greatly in the production of ecofriendly materials, namely ecomaterials or green materials.

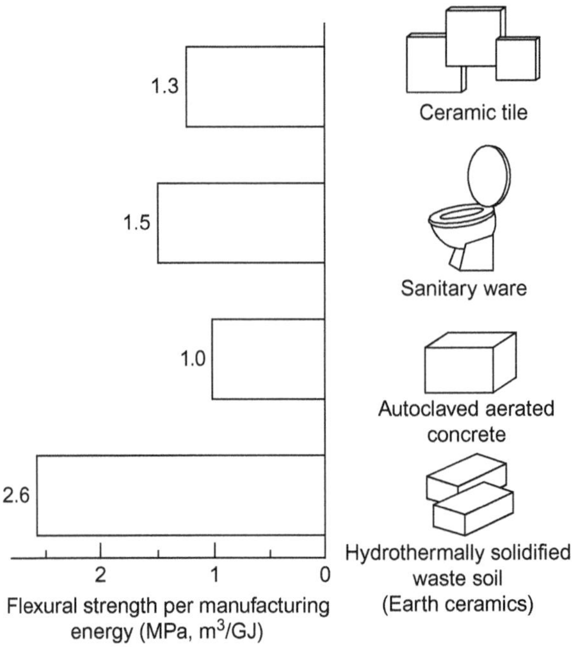

Figure 10.94 Flexural strength per manufacturing energy (MPa, m^3/GJ) for the hydrothermally solidified waste soil (earth ceramics) and other common ceramic products [418].

10.10 Future Trends in Hydrothermal Research

Hydrothermal technology, whether it is hydrothermal or solvothermal or supercritical, has a great perspective owing to its multifaceted advantages in processing of a wide variety of advanced materials starting from bulk single crystals to fine and ultrafine crystals and finally the nanocrystals or nanoparticles. The hydrothermal technology has been revived significantly in the last few years for the growth of most strategic crystals like ZnO, $GaPO_4$, langasite, and GaN, as bulk single crystals, followed by the synthesis of diamond, silicon carbide, silicon nitride, ruby, emerald, zoisite, etc. The processing of fine particles and nanoparticles is growing fast as a most attractive tool owing to several advantages in comparison with the conventional methodologies. More recently the concept of multienergy hydrothermal technology has been introduced and it is going to play a vital role in future materials processing because of its speed, cost, and convenience and environmentally benign conditions. Here, it is very important to note the role played by the chemistry, especially the precursor's preparation for materials processing whether it is hydrothermal or other conventional methods. For industrial scale production of materials, the supercritical hydrothermal, particularly the supercritical CO_2 and H_2O will have a greater role to play for a wide range of materials, as they are part of the green chemistry. Also the wasteless processing is the future of materials processing with an eye on the environmentally benign conditions without leading to the global warming. In this context, the soft solution processing, which embraces all the set of processes that operate without involving extreme pressure or heat, or vacuum, etc., has a great potential as a materials processing tool. In soft solution processing what is most important is the precursor solutions. Several researchers predict that the combination of polymerizing sol–gel and hydrothermal has a great potential to process materials under environmentally benign conditions. For example, a methodology proposed by Kakihana and Yoshimura [419] like a water-soluble titanium complex for the selective synthesis of brookite, rutile, and anatase under hydrothermal conditions will lay a firm foundation for environmentally friendly processing in future. Such an approach has yielded fruitful results to the processing of even high-temperature electroceramics like $BaTiO_3$ and $SrTiO_3$. Similarly, the gel chemistry of metal alkoxides would provide new avenues for precursor preparation for the hydrothermal processing of advanced materials. Hence, an interdisciplinary approach will be the most effective solution for the future materials processing strategies, which are environmentally benign and highly cost-effective. Hydrothermal technology will be one such most powerful tool in combination with the sol–gel and multienergy even for multicomponent systems [447].

References

[1] K. Byrappa, Hydrothermal processing of advanced materials, Kirk–Othmer Encyclopedia of Chemical Technology, John Wiley & Sons, London, 2005.

[2] R. Roy, Accelerating the kinetics of low-temperature inorganic syntheses, J. Solid State Chem. 111 (1994) 11–17.

[2a] S-H. Feng, Z. Shi, J-S. Chen, Proceedings of the first International Symposium on Hydrothermal Reactions, March 22–28, Shigeyuki Sōmiya (Ed.), Association for Science Documents Information, 1982, Japan.

[3] L.W. Jelinski, T.E. Graedal, R.A. Laudise, D.W. McCall, C.K.N. Patel, Industrial ecology: concepts and approaches, Proc. Natl. Acad. Sci. USA 89 (1992) 793–800.

[4] M.M. Lencka, A. Andreko, R.E. Riman, Hydrothermal precipitation of lead zirconate titanate solid solutions: thermodynamic modeling and experimental synthesis, J. Am. Ceram. Soc. 78 (1995) 2609–2615.

[5] Y. Hao, A.S. Teja, Continuous hydrothermal crystallization of Fe_2O_3 and Co_3O_4 nanoparticles, J Mater. Res. 18 (2003) 415–422.

[6] K. Byrappa, M. Yoshimura, Photocatalytic Properties, A novel method of advanced materials processing, J. Mater. Sci., 41, (2006).

[7] R. Roy, Solvothermal as One Example of Multienergy Processing—History and Current Status, Plenary Talk, ISHR & ICSTR 2006, Sendai, Japan, August 5–9, 2006.

[8] E. Lester, P. Blood, J. Li, M. Poliakoff, Reactor Geometry and Supercritical Water Reactions, Invited Talk, ISHR & ICSTR 2006, Sendai, Japan, August 5–9, 2006.

[9] R. Varma, Chemistry at the speed of light, in: D. Agarwal (Ed.), Immediate Energy Savings via Microwave Usage in Major Materials Technologies, Spectrum 1 Richmond, Australia, (2006) 2.

[10] S. Somiya (Ed.), Proceedings of the First International Symposium on Hydrothermal Reactions, 1982, Gakujutsu Bunken FUKYU-KAI Publications, Tokyo, Japan, (1983) 1–965.

[11] R. Roy, Diamond Jubilee of Hydrothermal/Solvothermal Research at Penn State, Lifetime Achievement Award Lecture, ISHA-2008, Nottingham, UK, 2008.
[12] R. Roy, Fifty year perspective on hydrothermal research, in: Proceedings of the Workshop on Solvothermal and Hydrothermal Reactions, Sun Mess Kagawa, Japan, January 22–24, 1996.
[13] J.B. Corliss, Metallogenesis at oceanic spreading centres, J. Geol. Soc. Lond. 136 (1979) 621–626.
[14] A.I. Ekimov, A.L. Efros, A.A. Onushchenko, Quantum size: effect in semiconductor microcrystals, Solid State Commun. 56 (1985) 921–924.
[15] S. Mann (Ed.), Biomimetic Materials Chemistry, Wiley-VCH, Germany, 1996.
[16] M. Yoshimura, Importance of soft solution processing for advanced inorganic materials, J. Mater. Res. 13 (1998) 796–802.
[17] M. Yoshimura, W.L. Suchanek, K. Byrappa, Soft, solution processing—a strategy for materials processing in twenty-first century, MRS Bull. 25 (2000) 17–25.
[18] M. Yoshimura, Soft solution processing: concept and realization of direct fabrication of shaped ceramics (nano-crystals, whiskers, films, and/or patterns) in solution without post-firing, J. Mater. Sci. 41 (2006) 1299–1306.
[19] K. Namratha, K. Byrappa, Novel solution routes of synthesis of metal oxide and hybrid metal oxide nanocrystals, Prog. Cryst. Growth Charact. Mater. 58 (2012) 14–42.
[20] K. Byrappa, T. Adschiri, Hydrothermal technology for nanotechnology, Prog. Cryst. Growth Charact. Mater. 53 (2007) 117–166.
[21] K. Byrappa, Hydrothermal growth of polyscale crystals, in: G. Dhanaraj, K. Byrappa, V. Prasad, M. Dudley (Eds), Springer Handbook of Crystal Growth, Springer-Verlag, Germany, 2010, pp. 599–644.
[22] G.W. Morey, Hydrothermal synthesis, J. Am. Ceram. Soc. 36 (1953) 279–285.
[23] K. Byrappa (Ed.), Hydrothermal growth of crystals, Prog. Cryst. Growth Charact. Mater. 21 (1990). pp. 1–370.
[24] T. Mitsuda, Hydrothermal reaction and industry of calcium silicates, Ceram. Japan 15 (1980) 184–189.
[25] S. Somiya, Historical developments of hydrothermal works in Japan, especially in ceramic science, J. Mater. Sci. 41 (2006) 1307–1318.
[26] A.N. Lobachev (Ed.), Crystallization Processes Under Hydrothermal Conditions, Consultants Bureau, New York, NY, 1973. p. 225.
[27] S. Somiya (Ed.), Hydrothermal Preparation of Fine Powders, Advanced Ceramics III, Elsevier, UK, 1990.
[28] R.E. Riman, W.L. Suchanek, K. Byrappa, C.W. Chen, P. Shuk, C.S. Oakes, Solution synthesis of hydroxyapatite designer particulates, Solid State Ionics 151 (2002) 393–402.
[29] K. Kajiyoshi, K. Tomono, Y. Hamaji, T. Kasanami, M. Yoshimura, Short-circuit diffusion of Ba, Sr, and O during $ATiO_3$ (A = Ba,Sr) thin-film grown by the hydrothermal electrochemical method, J. Am. Ceram. Soc. 78 (1995) 1521–1527.
[30] W.S. Cho, M. Yashima, M. Kakihana, A. Kudo, T. Sakata, M. Yoshimura, Active electrochemical dissolution of molybdenum and application for room-temperature synthesis of crystallized luminescent calcium molybdate film, J. Am. Ceram. Soc. 80 (1997) 765–769.
[31] M. Yoshimura, K. Byrappa, Hydrothermal processing of materials: past, present and future, J. Mater. Sci. 43 (2008) 2085–2103.
[32] M. Faraday, Experimental relations of gold (and other metals) to light, Philos. Trans. R. Soc. London 147 (1857) 145–181.

[33] K. Von Chroustshoff, History of Hydrothermal Technology, Ann. Chem. 3 (1873) 281–286.
[34] G.W. Morey, P. Niggli, The hydrothermal formation of silicates, a review, J. Am. Ceram. Soc. 35 (1913) 1086–1130.
[35] J.B. Hannay, On the artificial formation of the diamond, Proc. R. Soc. London 30 (1880) 178–189.
[36] G. Spezia, La Pressione e' chimicamente inattive nella solubilite e riecostituzione del quarzo, Atti. Accad. Sci. Torino 40 (1904 – 1905) 254–262.
[37] M. Yoshimura, Soft solution processing: concept and realization of direct fabrication of shaped ceramics (nano- crystals, whiskers, films, and/or patterns) in solutions without post firing, J. Mater. Sci. 41 (2006) 1299–1306.
[38] S. Komarneni, R. Roy, Q.H. Li, Microwave-hydrothermal synthesis of ceramic powders, Mater Res. Bull. 27 (1992) 1393–1405.
[39] S.H. Yu, Y.T. Qian, Soft synthesis inorganic nanorods, nonowires, and nanotubes, in: M. Adachi, D.J. Lockwood (Eds), Nanostructure Science and Technology, Self-Organized Nanoscale Materials, John Wiley & Sons, USA, 2006, pp. 101–158.
[40] G. Demazeau, Solvothermal synthesis, in: Proceedings of the First International Conference on Solvothermal Reactions, N. Yamasaki, and K. Yanasagawa (Eds) Takamatsu, Japan, December 5–7, 1994.
[41] T. Adschiri, K. Arai, Hydrothermal synthesis of metal oxide nanoparticles under supercritical conditions, in: Y.-P. Sun (Ed.), Supercritical Fluid Technology in Materials Science and Engineering, Marcel Dekker, New York, NY, 2002, pp. 311–325.
[42] K. Byrappa, S. Ohara, T. Adschiri, Nanoparticles synthesis using supercritical fluid technology—towards biomedical applications, Adv. Drug Deliv. Rev. 60 (2007) 299–327.
[43] T. Mousavand, Synthesis of organic–inorganic hybrid nanoparticles by in situ surface modification under supercritical hydrothermal conditions, Ph.D. Thesis, Tohoku University, Sendai, Japan, 2007.
[44] J. Zhang, S. Ohara, M. Umetsu, T. Naka, Y. Hatakeyama, T. Adschiri, Novel approach to colloidal ceria nanocrystals: tailor-made crystal shape in supercritical water, Adv. Mater. 19 (2007) 203–206.
[45] L.E. Brus, Electron-electron and electron-hole interactions in small semiconductor crystallites: The size dependence of the lowest excited electronic state, J. Chem. Phys. 80 (1984), pp. 4403.
[46] M. Bruchez, M. Moronne, P. Gin, P. Weiss, A.P. Alivisatos, Semiconductor nanocrystals as fluorescent biological labels, Science 281 (1998) 2013–2016.
[47] M. Green, Solution routes to III–V semiconductor quantum dots, Curr. Opin. Solid State Mater. Chem. 6 (2002) 355–363.
[48] M. Rajamathia, Seshadri, Oxide and chacogenide nanoparticles from hydrothermal/solvothermal reaction, R. Curr. Opin. Solid State Mater. Chem 6 (2002) 337–345.
[49] S. Forster, M. Anjtonietti, Amphiphilic block copolymers in structure-controlled nanomaterials hybrids, Adv. Mater. 10 (1998) 195–217.
[50] Y. Zhu, H. Zheng, Y. Li, L. Gao, Z. Yang, Y.T. Qian, Synthesis of Ag dendritic nanostructures by using anisotropic nickel nanotubes, Mater. Res. Bull. 38 (2003) 1829–1834.
[51] V.F. Puntes, K.M. Krishnan, A.P. Alivisatos, Colloidal nanocrystal shape and size control: the case of cobalt, Science 291 (2001) 2115–2117.
[52] Q. Xie, Z. Dai, W. Huang, J. Lianjg, C. Jiang, Y.T. Qian, Synthesis of ferromagnetic single-crystalline cobalt nanobelts via a surfactant-assisted hydrothermal reduction process, Nanotechnology 16 (2005) 2958–2962.

[53] H. Zeng, J. Li, J.P. Liu, Z.L. Wang, S. Sun, Exchange-coupled nanocomposite magnets by nanoparticle self-assembly, Nature 420 (2002), pp. 395–398.
[54] Z. Tian, J. Liu, J.A. Voigt, H. Xu, M.J. McDermott, Dendritic growth of cubically ordered nanoporous materials through self-assembly, Nano. Lett. 3 (2003) 89–92.
[55] W. Liu, W. Zhong, X. Wu, N. Tang, Y. Du, Hydrothermal microemulsion synthesis of cobalt nanorods and self assembly into square shaped nanostructures, J. Cryst. Growth 284 (2005) 446–451.
[56] Z. Liu, S. Li, Y. Yang, S. Peng, Z. Hu, Y.T. Qian, Rectangular single crystal mullite microtubes, Adv. Mater. 15 (2003) 1946–1949.
[57] H.L. Niu, Q.W. Chen, Y.S. Lin, Y.S. Jia, H.F. Zhu, M. Ning, Hydrothermal formation of magnetic Ni–Cu alloy nanocrystallites at low temperatures, Nanotechnology 15 (2004) 1054–1058.
[58] E. Reverchon, R. Adami, Nanomaterials and supercritical fluids, J. Supercrit. Fluid 37 (2006) 1–22.
[59] A. Kameo, T. Yoshimura, K. Esumi, Preparation of noble metal nanoparticles in supercritical carbon dioxide, Colloids Surf. A Physicochem. Eng. Aspects 215 (2005) 181–189.
[60] P.S. Shah, S. Husain, K.P. Johnston, B.A. Korgel, Nanocrystal arrested precipitation in supercritical carbon dioxide, J. Phys. Chem. B-105 (2001) 9433–9440.
[61] M.C. McLeod, W.F. Gale, C.B. Roberts, Metallic nanoparticle production utilizing a supercritical carbon dioxide flow process, Langmuir 20 (2004) 7078–7082.
[62] J.M. Blackburn, D.P. Long, A. Cabanas, J.J. Watkins, Deposition of conformal copper and nickel films from supercritical carbon dioxide, Science 294 (2001) 141–145.
[63] B. Basavalingu, J.M.C. Moreno, K. Byrappa, Y.G. Gogotsi, M. Yoshimura, Decomposition of silicon carbide in the presence of organic compounds under hydrothermal conditions, Carbon 39 (2001) 1763–1767.
[64] R.C. De Vries, Hydrothermal carbon: a review from carbon in "Herkimer Diamonds" to that in real diamonds, in: S. Somiya (Ed.), Advanced Ceramics, vol. 3, Elsevier, The Netherlands, 1990.
[65] O. Guillois, G. Ledoux, C. Reynaud, Diamond infrared emission bands in circumstellar media, Astrophys. J. 521 (1999) L1333–L1336.
[66] V.V. Danilenko, On the history of the discovery of nanodiamond synthesis, Phys. Solid State 46 (2004) 595–599.
[67] N.R. Greiner, D.S. Phillips, J.D. Johnson, E. Volk, Diamonds in detonation soot, Nature 333 (1988) 440–442.
[68] V. Presser, M. Heon, Y. Gogotsi, Carbide-derived carbons from porous network to nanotubes and graphene, Adv. Funct. Mater 21 (2011) 210–833.
[69] Z. Liu, X. Zhou, Y.T. Qian, Synthetic methodologies for carbon nanomaterials, Adv. Mater. 22 (2010) 1963–1966.
[70] Y.G. Gogotsi, K.G. Nickel, P. Kofstad, Hydrothermal synthesis of diamond from diamond-seeded β-SiC powder, J. Mater. Chem. 5 (1995) 2313–2314.
[71] R. Roy, D. Ravindranathan, A. Badzian, Evidence for hydrothermal growth of diamond in the C–H–O and C–H–O–halogen system, J. Mater. Res. 11 (5) (1996) 1164–1168.
[71a] Basavalingu, et al., Structure and electronic properties of carbon onions, J. Chem. Phys. 114 (2001) 1–6.
[72] Y.G. Gogotsi, J.D. Jeon, M.J. McNallan, Carbon coatings on silicon carbide by reaction with chlorine-containing gases, J. Mater. Chem. 7 (9) (1997) 1841–1848.

[73] Y. Li, Y.T. Qian, H. Liao, Y. Ding, L. Yang, C. Xu, et al., A reduction–pyrolysis–catalysis synthesis of diamond, Science 281 (1998) 246.
[74] Y. Jiang, Y. Wu, S. Zhang, C. Xu, W. Yu, Y. Xie, et al., A catalytic assembly solvothermal route to multiwall carbon nanotubes at a moderate temperature, J. Am. Chem. Soc. 122 (2000) 12383–12384.
[75] W. Zhang, D. Ma, J. Liu, L. Kong, W. Yu, Y. Qian, Solvothermal synthesis of carbon nanotubes by metal oxide and ethanol at mild temperature, Carbon 42 (2004) 2143–2341.
[76] J. Zhang, J. Li, J. Cao, Y. Qian, Synthesis and characterization of larger diameter carbon nanotubes from catalytic pyrolysis of polypropylene, Mater. Lett. 62 (2008) 1839–1842.
[77] Y.M. Kim, H.W. Lee, S.H. Lee, S.S. Kim, S.H. Park, J.K. Jeon, et al., Pyrolysis properties and kinetics of mandarin peel, Korean J. Chem. Eng. 28 (2011) 2012–2016.
[78] H. Li, Y. Zhu, Z. Mao, J. Gu, J. Zhang, Y. Qian, Synthesis and characterization of carbon fibrils formed by stacking graphite sheets of nanometer thickness, Carbon 47 (2009) 328–330.
[79] B. Hu, K. Wang, L.H. Wu, S.H. Yu, M. Antonietti, M.M. Titirici, Engineering carbon materials from the hydrothermal carbonization process of biomass, Adv. Mater. 22 (2010) 813–828.
[80] B. Hu, S.H. Yu, K. Wang, L. Liu, X.W. Xu, Functional carbonaceous materials from hydrothermal carbonization of biomass: an effective chemical process, Dalton Trans. 40 (2008) 5414–5423.
[81] B. Basavalingu, P. Madhusudan, A.S. Dayananda, K. Lal, K. Byrappa, M. Yoshimura, Formation of filamentous carbon through dissociation of chromium carbide under hydrothermal conditions, J. Mater. Sci. 43 (2008) 2153–2157.
[82] N.S. Jacobson, Y.G. Gogotsi, M. Yoshimura, Thermodynamic and experimental study of carbon formation on carbides under hydrothermal conditions, J. Mater. Chem. 5 (1995) 595–601.
[83] C.X. Wang, Y.H. Yang, N.S. Xu, G.W. Yang, Thermodynamics of diamond nucleation on the nanoscale, J. Am. Chem. Soc. 126 (2004) 11303–11306.
[84] T. Kraft, K.G. Nickel, Carbon formed by hydrothermal treatment of α-SiC crystals, J. Mater. Chem. 10 (2006) 671–680.
[85] B. Zhang, C.Y. Liu, Y. Liu, A novel one-step approach to synthesize fluorescent carbon nanoparticles, Eur. J. Inorg. Chem. (2010) 4411–4414.
[86] T. Kraft, K.G. Nickel, Carbon formed by hydrothermal treatment of α-SiC crystals, J. Mater. Chem. 10 (2000) 671–680.
[87] J. Shen, Y. Zhu, X. Yang, J. Zong, J. Zhang, C. Li, One-pot hydrothermal synthesis of graphene quantum dots surface-passivated by polyethylene glycol and their photoelectric conversion under near-infrared light, New J. Chem. 36 (2012) 97–101.
[88] Y.P. Sun, B. Zhou, Y. Lin, W. Wang, K.A.S. Fernando, P. Pathak, et al., Quantum sized carbon dots for bright and colorful luminescence, J. Am. Chem. Soc. 128 (2006) 7756–7757.
[89] D.W. Stickler, W.G. Carlson, Electrical conductivity in the ZrO_2-rich region of several M_2O_3–ZrO_2 systems, J. Am. Ceram. Soc. 48 (1965) 286–289.
[90] D.J. Green, R.H.J. Hannink, M.V. Swain, Transformation Toughening of Ceramics, CRS Press, Boca Raton, FL, 1989.

[91] M. Yashima, N. Ishizawa, H. Fujimori, M. Kakiyana, M. Yoshimura, In situ observation of the diffusionless tetragonal cubic phase transition and metastable phase diagram in the $ZrO_2-ScO_{1.5}$ system, Eur. J. Solid State Inorg. Chem. 32 (1995) 761–770.

[92] M. Yashima, M. Kakihana, M. Yoshimura, Metastable–stable phase diagrams in the zirconia-containing systems utilized in solid-oxide fuel cell application, Solid State Ionics 86–88 (1996) 1131–1149.

[93] M.R. Thornber, D.J.M. Bevan, E. Summerville, Mixed oxides of the type MO_2 (fluorite)-M_2O_3 very phase studies in the system $ZrO_2-M_2O_3$ (M = Sc, Yb, Er, Dy), J. Solid State Chem. 1 (1970) 545–553.

[94] M. Yoshimura, Phase stability of zirconia, Ceram. Bull. 67 (1988) 1950–1955.

[95] M. Yashima, S. Sasaki, M. Kakihana, Y. Yamaguchi, H. Arashi, M. Yoshimura, Oxygen-induced structural change of the tetragonal phase around the tetragonal-cubic phase boundary in $ZrO_2-YO_{1.5}$ solid solutions, Acta Crystallogr. B50 (1994) 663–673.

[96] M. Yashima, N. Ishizawa, M. Yoshimura, High-temperature X-ray study of the cubic-tetragonal diffusionless phase transition in the $ZrO_2-ErO_{1.5}$ system: II, temperature dependences of oxygen ion displacement and lattice parameter of compositionally homogeneous 12 mol% $ErO_{1.5}-ZrO_2$, J. Am. Ceram. Soc., 76, 1993.

[97] M. Yashima, K. Ohtake, M. Kakihana, H. Arashi, M. Yoshimura, Determination of tetragonal-cubic phase boundary of $Zr_{1-x}R_xO_{2-x}$ (R = Nd, Sm, Y, Er and Yb) by Raman scattering, J. Phys. Chem. Solids 57 (1996) 17–24.

[97a] I. Tani, et al., Structure-Retention relationship of sterols and triterpene alcohols in gas chromatography on a glass capillary column, J. Chromatogr 234 (1982) 65–76.

[98] E. Tani, M. Yoshimura, S. Somiya, Effect of mineralizers on the crystallization of solid solutions in the system ZrO_2-CeO_2 under hydrothermal conditions, Yogyo Kyokai Shi 90 (1982) 195–201 (Japan).

[98a] Yoshimura, et al., GSK-3beta regulates phosphorylation of CRMP-2 and neuronal polarity, 1, Cell 120 (2005) 137–149.

[99] Z. Wu, M. Yoshimura, Electrochemical studies on the mechanism of the fabrication of ceramic films by hydrothermal–electrochemical technique, Bull. Korean Chem. Soc. 20 (1999) 869–874.

[100] N. Sogo, A. Kato (Eds), Ceramics: Toward the Twenty-First Century, Ceramic Society of Japan, Tokyo, 1991.

[100a] K. Kajiyoshi, M. Yoshimura, Y. Hamaji, K. Tomono, T. Kasanami, Growth of (Ba, Sr) TiO_3 thin films by the hydrothermal electrochemical method and effect of oxygen evolution on their microstructure, J. Mater. Res. 11 (1996) 169–175.

[101] S. Somiya (Ed.), Hydrothermal Reactions for Materials Science and Engineering: An Overview of Research in Japan, Elsevier, London, 1989.

[102] G. Wang, Z. Wang, Y. Zhang, G. Fei, L. Zhang, Controlled synthesis and characterization of large scale uniform $Dy(OH)_3$ and Dy_2O_3 single crystal nanorods by hydrothermal method, Nanotechnology 15 (2004) 1307–1311.

[102a] S. Hirano, S. Somiya, Hydrothermal crystal growth of magnetite in the presence of hydrogen, J. Cryst. Growth 35 (1976) 273–278.

[102b] Sun, et al., Gamma-S crystallin gene (CRYGS) mutation causes dominant progressive cortical cataract in humans, J. Med. Genet 42 (2005) (2005) 706–710.

[103] A. Ahniyaz, T. Watanabe, M. Yoshimura, Tetragonal nanocrystals from the $Zr_{0.5}Ce_{0.5}O_2$ solid solution by hydrothermal method, J. Phys. Chem. B109 (2005) 6136–6139.

[104] T. Adschiri, K. Kanazawa, K. Arai, Rapid and continuous hydrothermal crystallization of bohemite particles in subcritical and supercritical water, J. Am. Ceram. Soc. 75 (1992) 2615–2620.

[105] T. Adschiri, K. Kanazawa, K. Arai, Rapid and continuous hydrothermal crystallization of metal oxide particles in supercritical water, J. Am. Ceram. Soc. 75 (1992) 1019–1025.

[106] S. Takami, S. Ohara, T. Adschiri, Y. Wakayama, T. Chikyow, Continuous synthesis of organic–inorganic hybridized cubic nanoassemblies of octahedral cerium oxide nanocrystals and hexanedioic acid, Dalton Trans. 40 (2008) 5442–5446.

[107] C. Sun, H. Li., H. Zhang, Z. Wang, L. Chen, Controlled synthesis of Ceo2 nanorods by a solvothermal method, Nanotechnology 16 (2005) 1454–1463.

[108] T. Chudoba, E. Lester, W. Lojkowski, M. Poliakoff, J. Li, E. Grzanka, et al., Synthesis of nano-sized yttrium–aluminum garnet in a continuous-flow reactor in supercritical fluids, Z. Naturforsch. B 63 (2008) 756–764.

[109] H. Hayashi, Y. Hakuta, Hydrothermal synthesis of metal oxide nanoparticles in supercritical water, Materials 3 (2010) 3794–3817, doi:10.3390/ma3073794.

[110] X. Jiao, D. Chen, L. Xiao, Effects of organic additives on the hydrothermal zirconia nanocrystallites, J. Cryst. Growth 258 (2003) 158–162.

[111] R. Wang, K. Hashimoti, A. Fujishima, Light induced amphiphilic surfaces, Nature 388 (1997) 431–432.

[111a] A. Fujishima, K. Honda, Electrochemical photolysis of water at a semiconductor electrode, letters to nature, Nature 238 (1972) 37–38.

[112] T. Saito, T. Iwase, J. Horie, T. Morioka, Mode of photocatalytic bactericidal action of powdered semiconductor TiO_2 on mutans streptococci, J. Photochem. Photobiol., B Biol. 14 (1992) 369–379.

[113] Z. Kertesz, Z. Szikszai, E. Gontier, P. Moretto, J.E. Surleve-Bazeille, B Kiss, et al., Nuclear microprobe study of TiO_2—penetration in the epidermis of human skin xenografts, Nucl. Instrum. Methods Phys. Res B231 (2005) 280–285.

[114] S.H. Lee, H.W. Kim, E.J. Lee, L.H. Li, H.E. Kim, Hydroxyapatite-TiO_2 hybrid coating on Ti implants, J. Biomater. Appl. 20 (2006) 195–208.

[115] D.W. Stickler, W.G. Carlson, Electrical conductivity in the ZrO_2-rich region of several M_2O_3–ZrO_2 systems, J. Am. Ceram. Soc. 48 (1965) 286–289.

[116] C. Guiot, Utilization de cristaux liquides minerourx comme "template" pour, L'elaboration de solide hybride. Mesostructure, Ph.D. Thesis, Universite de Montpellier, France, 2009.

[117] J. Peral, X. Domenech, D.f. Ollis, Photocatalytic functional coating of TiO_2 thin film deposited by cyclic plasmal chemical vapor deposition at atmospheric pressure, Chem. Tech. Biotech 70 (1997) 117–122.

[118] A.L. Linsebigler, G. Yates, Photocatalysis on TiO_2 surfaces: Principles, mechanisms, and selected results, Chem. Rev. 95 (1995) 735–758.

[119] M.R. Hoffman, S. Martin, W. Choi, D.W. Bahnmann, Environmental applications of semiconductor photocatalysis, Chem. Rev. 95 (1995) 69–96.

[120] H. Hayashi, K. Torii, Hydrothermal synthesis of titania photocatalyst under subcritical and supercritical water conditions, J. Mater. Chem. 12 (2002) 3671–3676.

[121] S. Somiya, T. Kumaki, K. Hishinuma, Z. Nakai, T. Akiba, Y. Suwa, Hydrothermal precipitation of ZrO_2 powder, in: K. Byrappa (Ed.), Hydrothermal Growth of Crystals, in: Prog. Cryst. Growth Charact. 21 (1990) 195–198.

[122] H. Yin, P.S. Casey, M.J. McCall, M. Fenech, Effects of surface chemistry on cytotoxicity, genotoxicity, and the generation of reactive oxygen species induced by ZnO nanoparticles, Langmuir 26 (2010) 15399–15408.

[123] Y.W. Heo, D.P. Norton, L.C. Tein, Y. Kwon, B.S. Kang, F. Ren, et al., ZnO nanowire growth and devices, Mater. Scie. Engg R47 (2004) 1–47.

[124] C. Woll, The chemistry and physics of zinc oxide surfaces, Prog. Surf. Sci, 82 (2007) 55–120.

[125] K. Namratha, M.B Nayan, K Byrappa, Hydrothermal synthesis and photocatalytic properties of modified and unmodified zinc oxide nanoparticle, Mater. Res. Innov 15 (2011) 36–42.

[125a] T. Mousavand, J. Zhang, S. Ohara, M. Umetsu, T. Naka, T. Adschiri, Organic ligand-assisted supercritical hydrothermal synthesis of titanium oxide nanocrystals, J. Nanopart. Res. 69186 (2007).

[126] D. Liu, J. Zhang, B. Han, J. Chen, Z. Li, D. Shen, et al., Recovery of TiO_2 nanoparticles synthesized in reverse micelles by antisolvent CO_2, Colloids Surf. A227 (2003) 465–470.

[127] D.E. Speliotis, Magnetic recording beyond the first 100 years, J. Magn. Magn. Mater. 193 (1999) 29–35.

[128] C.C. Berry, S. Charles, S. Wells, M.J. Dalby, A.S.G. Curtis, The influence of transferring stabilized magnetic nanoparticles on human dermal fibroblasts in culture, Int. J. Pharm. 269 (2004) 211–225.

[129] A.S. Lubbe, C. Bergemann, J. Brock, D.G. McClure, Physiological aspects in magnetic drug-targeting, J. Magn. Magn. Mater. 194 (1999) 149–155.

[130] Z.G.M. Lacava, R.B. Azevado, E.V. Martins, L.M. Lacava, M.L.L. Freitas, V.A.P. Garcia, et al., Biological effects of magnetic fluids: toxicity studies, J. Magn. Magn. Mater. 201 (1–3) (1999) 431–434.

[131] C.W. Jung, J.M. Rogers, E.V. Groman, Lymphatic mapping and sentinel node location with magnetite nanoparticles, J. Magn. Magn. Mater. 194 (1–3) (1999) 210–216.

[132] A. Jordan, R. Scholz, P. Wust, H. Schirra, T. Schiestel, H. Schmidt, et al., Endocytosis of dextran and silan-coated magnetic nanoparticles and the effect of intracellular hyperthermia on human mammary carcinoma cells in vitro, J. Magn. Magn. Mater. 194 (1–3) (1999) 185–196.

[133] C. Xu, A.S. Teja, Supercritical water synthesis and deposition of ion oxide (α-Fe_2O_3) nanoparticles in activated carbon, J. Supercrit. Fluids 39 (2006) 135–141.

[134] A.S. Teja, P.Y. Koh, Synthesis, properties, and applications of magnetic iron oxide nanoparticles, Prog. Cryst. Grow. Charact. L. Mater. 55 (2009) 22–45.

[135] G.S. Li, R.L. Smith Jr., H. Inomata, K. Arai, Preparation and magnetization of hematite nanocrystals with amorphous iron oxide layers by hydrothermal conditions, Mater. Res. Bull. 37 (2002) 949–955.

[135a] Y.-C. Zhang, et al., Calculation of radiative fluxes from the surface to top of atmosphere based on ISCCP and other global data sets: Refinements of the radiative transfer model and the input data, J. Geophys. Res 109 (2004) D19105.

[136] A.J. Perrota, Nanosized corundum synthesis, Mater. Res. Innovat. 2 (1998) 33–38.

[137] R.A. Laudise, A.A. Ballman, Hydrothermal synthesis of sapphire, 80 (1958) 2655–2657.

[137a] Sorescu, et al., A Mossbauer study of manganese doped magnetite, Matt, Lett 57 (2003) 1867–1869.

[138] O.V. Al'myasheva, E.N. Korytkova, A.V. Maslov, V.V. Gusarov, Preparation of nanocrystalline alumina under hydrothermal conditions, Inorg. Mater. 41 (2005) 460–467.

[139] Z. Jing, S. Wu, Synthesis and characterization of monodisperse hematite nanoparticles modified by surfactants via hydrothermal approach, Mater. Lett. 58 (2004) 3637–3640.
[139a] G. Wang, Z. Wang, Y. Zhang, G. Fei, L Zhang, Controlled synthesis and characterization of large - scale, uniform Dy(OH)3 and Dy2O3 single - crystal nanorods by hydrothermal method, Nanotechnology 15 (2004) 1307–1311.
[140] M. Sorescu, L. Diamandescu, D. Tarabasanu-Mihaila, V.S. Teodorescu, Nanocrystalline rhombohedral In2O3 synthesized by hydrothermal and postannealing pathways, J. Mater. Sci. 39 (2) (2004) 675–677.
[140a] L. Suber, D. Fiorani, P. Imperatory, S. Foglia, A. Montone, R. Zysler, Effects of thermal treatments on structural and magnetic properties of acicular α-Fe_2O_3, Nanostruct. Mater. 11 (1999) 797–803.
[141] S. Takami, T. Sato, T. Mousavand, S. Ohara, M. Umetsu, T. Adschiri, Hydrothermal synthesis of surface-modified iron oxide nanoparticles, Mater. Letts. 61 (2007) 4769–4772.
[142] M. Wu, Y. Xiong, Y. Jia, J. Ye, K. Zhang, Q. Chen, Co-doped magnetite nanowire arrays prepared hydrothermally, Appl. Phys. A81 (2005) 1355–1358.
[142a] J. Wan, X. Chen, Z. Wang, X. Yang, Y.T. Qian, A soft-template-assisted hydrothermal approach to single-crystal Fe_3O_4 nanorods, J. Cryst. Growth 276 (2005) 571–576.
[143] D.W. Matson, J.C. Linehan, J.G. Darab, M.F. Buehler, Nanophase iron-based liquefaction catalysis: synthesis, characterization, and model compound reactivity, Energy Fuels 8 (1994) 10–18.
[144] E. Lester, P. Blood, J. Denger, D. Giddings, B. Azzopard, M. Pohakoff, Synthesis of nanoparticulate yttrium aluminum garnet in supercritical water–ethanol mixtures, J. Supercrit. Fluids 37 (2006) 209–214.
[145] H. Kominami, M. Miyakawa, S. Murakami, T. Yasuda, M. Kohno, S. Onoue, et al., Solvothermal synthesis of tantalum(V) oxide nanoparticles and their photocatalytic activities in aqueous suspension systems, Phys. Chem. Chem. Phys. 3 (2001) 2697–2703.
[146] K. Matsui, T. Noguchi, N.M. Islam, Y. Hakuta, H. Hayashi, Rapid synthesis of $BaTiO_3$ nanoparticles in supercritical water by continuous hydrothermal flow reaction system, J. Cryst. Growth. 310 (2008) 2584–2589.
[147] H. Hayashi, A. Ueda, A. Suino, K. Hiro, Y. Hakuta, Hydrothermal synthesis of yttria stabilized ZrO_2 nanoparticles in subcritical and supercritical water using a flow reaction system, J. Solid State Sci. 182 (2009) 2985–2990.
[148] Y. Hakuta, T. Adschiri, T. Suzuki, T. Chida, K. Seino, K. Arai, Flow method for rapidly producing barium hexaferrite particles in supercritical water, J. Am. Ceram. Soc. 81 (1998) 2461–2464.
[149] T. Adschiri, Y. Hakuta, K. Sue, K. Arai, Hydrothermal synthesis of metal oxide nanoparticles at supercritical conditions, J. Nanopart. Res. 3 (2001) 227–235.
[150] Y. Hakuta, S. Onai, H. Terayama, T. Adschiri, K. Aria, Production of ultra-fine ceria particles by hydrothermal synthesis under supercritical conditions, J. Mater. Sci. Lett. 17 (1998) 1211–1213.
[151] E. Lester, G. Aksomaityte, Jun Li, Sara Gomez, Lose Gonzalez - Gonzatez, Martyn Poliakoff, Controlled continuous hydrothermal synthesis of cobalt oxide (Co_3O_4) nanoparticles, Prog, Cryst. Mater 58 (2012) 3–13.
[152] W.T. Yao, S.H. Yu, Recent advances in hydrothermal syntheses of low dimensional nanoarchitectures, Int. J. Nanotech. 4 (2007) 21–31.

[153] S.H. Yu, Hydrothermal/Solvothermal process: powerful routes for controlled synthesis of functional materials, ISHA Newsletter 5 (2) (2010) 9–31.
[154] T. Adschiri, Y. Hakuta, K. Kanamura, K. Arai, Continuous production of $LiCoO_2$ fine crystals for lithium batteries by hydrothermal synthesis at supernatural conditions, High Pressure Res. 20 (2001) 373–380.
[155] K. Kanamura, A. Goto, R.Y. Ho, T. Umegaki, K. Toyoshima, K. Okada, et al., Preparation and electrochemical characterization of $LiCoO_2$ particles prepared by super critical water synthesis, Electrochem. Solid State Lett. 3 (2000) 256.
[156] K. Kanamura, T. Umegaki, K. Toyoshima, K. Okada, Y. Hakuta, T. Adschiri, et al., Electroceram. Jpn. III Key Eng. Mater. 181 (2001) 147.
[157] M. Cleroq Le, T. Adschiri, K. Arai, Hydrothermal processing of nickel containing biomining or bioremediation, Biomass Energy 21 (2001) 73–80.
[158] K. Sue, N. Kakinuma, T. Adschiri, K. Arai, Continuous production of nickel fine particles by hydrogen reduction in supercritical hydrothermal conditions, Ind. Eng. Chem. Res. 43 (2004) 2073–2078.
[159] T. Adschiri, R. Shibata, T. Sato, M. Watanabe, K. Aria, Catalytic hydrodesulphurization of dibenzothiophene through partial oxidation and a water–gas shift reaction in supercritical water, Ind. Eng. Chem. Res. 37 (1998) 2634–2638.
[160] K. Sue, A. Suzuki, Y. Hakuta, H. Hayashi, K. Arai, Y. Takebayashi, et al., Hydrothermal-reduction synthesis of Ni nanoparticles by super-rapid heating using a micromixer, Chem. Letts. 38 (2009) 1018–1020.
[161] R.E. Riman, Chemical precipitation of ceramic powders, in: R.A. Williams (Ed.), Colloid Eng., Buttersworths, New York, NY, 1992, pp. 140–167.
[162] M.M. Lencka, R.E. Riman, Hydrothermal synthesis of perovskite materials: thermal dynamic modeling and experimental verification, Ferroelectrics 151 (1994) 159–164.
[163] M.M. Lencka, R.E. Riman, Thermodynamics of the hydrothermal synthesis of calcium titanate with reference to other alkaline earth titanates, Chem. Mater. 7 (1995) 18–25.
[164] T.R.N. Kutty, R. Balachandran, Direct precipitation of lead zirconate titanate by the hydrothermal method, Mater. Res. Bull. 19 (1984) 1479–1488.
[165] K.C. Beal, Precipitation of lead zirconate titanate solid solutions under hydrothermal conditions, in: G.L. Messing, K.S. Mazdiyasni, J.W. McCauley, R.A. Haber (Eds), Advances in Ceramics: Ceramic Powder Science, American Ceramic Society, Westerville, OH, 21 (1987) 33–41.
[166] W.J. Dawson, S.L. Swartz, Process for Producing Submicron Ceramic Powders of Perovskite Compounds with Controlled Stoichiometry and Particle Size, US Patent No. 5112433, May 12, 1992.
[167] M.M. Lencka, R.E. Riman, Thermodynamic modeling of hydrothermal synthesis of ceramic powders, Chem. Mater. 5 (1993) 61–70.
[168] M.M. Lencka, R.E. Riman, Synthesis of lead titanate: thermodynamic modeling and experimental verification, J. Am. Ceram. Soc. 76 (1993) 2649–2659.
[169] J.O. Eckert Jr., C.C. Hung-Houston, B.L. Gersten, M.M. Lencka, R.E. Riman, Kinetics and mechanisms of hydrothermal synthesis of barium titanate, J. Am. Ceram. Soc. 79 (1996) 2929–2939.
[170] R.E. Riman, M.M. Lencka, L.E. Me Candlish, B.L. Gersten, A. Andrenko, S.B. Cho, Intelligent engineering of hydrothermal reactions, in: Proceedings of the Second International Conference Solvothermal Reactions, (N. Yamasaki, and K. Yamagisawa Eds) Takamatsu, Japan, December 18–20, 1996, pp. 148–151.
[171] M.M. Lencka, R.E. Riman, Estimation of thermochemical properties for ceramic oxides: a focus on $PbZrO_3$, Thermochim. Acta 256 (1995) 193–203.

[172] Y. Yin, P. Alivisatos, Colloidal nanocrystal synthesis and the organic–inorganic interface, Nature 437 (2005) 664–670.
[173] I.A. Aksay, C.M. Chun, T. Lee, Mechanism of $BaTiO_3$ formation by hydrothermal reactions, in: Proceedings of the Second International Conference on Solvothermal Reactions, Takamatsu, Japan, December 18–20, 1996, pp. 76–79.
[174] J.H. Adair, R.P. Denkiewicz, F.J. Arriagada, K. Osseo-Asare, in: G.L. Messing, E.R. Fuller Jr., H. Hausner (Eds), Ceramic Transactions, Ceramic Powder Science II A, vol. 1, American Ceramic Society, Westerville, OH, 1998, p. 135.
[175] K. Osseo-Asare, F.J. Arriagada, J.H. Adair, in: G.L. Messing, E.R. Fuller Jr., H. Hausner (Eds), Ceramic Transactions, Ceramic Powder Science II A, vol. 1, American Ceramic Society, Westerville, OH, 1988.
[176] D. Chen, R. Xu, Solvothermal synthesis and characterization of $PbTiO_3$ powders, J. Mater. Chem. 8 (1998) 965–968.
[177] A. Kaiser, A. Berger, D. Sporn, H. Bertagnolli, Lyothermal synthesis of nanocrystalline $BaTiO_3$ and TiO_2—powders, Ceram. Process. Sci. Technol. (1994) 51–55.
[178] T. Dubois, G. Demazeau, Preparation of Fe_3O_4 fine particles through a solvothermal process, Mater. Letts. 19 (1994) 38–47.
[179] G. Demazeau, A. Wang, S. Matar, J.D. Cillard, Solvothermal synthesis of nitrides as fine particles, High Press. Res. 12 (1994) 343–346.
[180] M. Inoue, T. Nishikawa, T. Nakamura, T. Inui, Glycothermal reaction of rare earth acetate and iron acetylacetonate: formation of hexagonal $REFeO_3$, J. Am. Ceram. Soc. 80 (1997) 2157–2160.
[181] M. Inoue, H. Otsu, H. Komonami, T. Nakamura, T. Inu, Thermal stability of phosphorus-modified alumina prepared by the glycothermal method, J. Mat. Sci. Letts. 13 (1994) 787–789.
[182] C.N.R. Rao, A.S. Govindaraj, R.C. Vivekchand, Inorganic nanomaterials: current status and future prospects, Annu. Rep. Prog. Chem. Sect. A102 (2006) 20–45.
[183] A.C.C. Esteves, T. Trindade, Synthetic studies on II/VI semiconductor quantum dots, Curr. Opin. Solid State Mater. Sci. 6 (4) (2002) 347–353.
[184] O. Masala, R. Seshadri, Synthesis routes for large volumes of nanoparticles, Annu. Rev. Mater. Res. 34 (2004) 41–81.
[185] J.R. Ota, S.K. Srivastava, A new hydrothermal route for synthesis of molybdenum disulphide nanorods and related nanostructures, J. Nanosci. Nanotech. 6 (2006) 168–174.
[186] R. Rossetti, R. Hull, J.M. Gibson, L.E. Brus, Excited electronic states and optical spectra of ZnS and CdS crystallites in the ≈ 15 to 50 Å size range: evolution from molecular to bulk semiconducting properties, J. Chem. Phys. 82 (1) (1985) 552–559.
[187] Y. Li, Y. Ding, Y. Qian, Y. Zhang, L. Yang, A solvothermal elemental reaction to produce nanocrystalline ZnSe, Inorg. Chem. 37 (1998) 2844–2845.
[188] S.H. Yu, J. Yang, Z.H. Han, Y. Zhou, R.Y. Yang, Y.T. Qian, et al., Controllable synthesis of nanocrystalline CdS with different morphologies and particle sizes by a novel solvothermal process, J. Mater. Chem. 9 (1999) 1283–1287.
[189] Q. Peng, Y. Dong, Z. Deng, X.Y. Sun, Li, low-temperature elemental direct-reaction route to II–VI semiconductor nanocrystalline ZnSe and CdSe, Inorg. Chem. 40 (2001) 3840–3841.
[190] Z.X. Deng, L. Li, Y. Li, Novel inorganic–organic-layered structures: crystallographic understanding of both phase and morphology formations of one-dimensional CdE (E = S,Se,Te) nanorods in ethylenediamine, Inorg. Chem. 42 (2003) 2331–2341.

[191] Y. Li, F. Huang, Q. Zhang, Z. Gu, Solvothermal synthesis of nanocrystalline cadmium sulfide, J. Mater. Sci. 35 (2000) 5933–5937.
[192] J. Yao, G. Zhao, D. Wang, G. Han, Solvothermal synthesis and characterization of CdS nanowires/PVA composite films, Mater. Letts. 59 (28) (2005) 3652–3655.
[193] H. Chu, X. Li, G. Chen, W. Zhou, Y. Zhang, Z. Lin, et al., Shape-control-led synthesis of CdS nanocrystals in mixed solvents, Crystal Growth Design. 5 (2005) 1801–1806.
[194] Y.C. Li, X.H. Li, C.H. Yang, Y.F. Li, Controlled synthesis of CdS nanorods and hexagonal nanocrystals, J. Mater. Chem. 13 (10) (2003) 2641–2648.
[195] G.T. Zhou, X. Wang, J.C. Yu, A low-temperature and mild solvothermal route to the synthesis of wurtzite-type ZnS with single-crystalline nanoplate-like morphology, Cryst. Growth Design. 5 (5) (2005) 1761–1765.
[196] W. Zhang, Q. Yang, L. Xu, W. Yu, Y. Qian, Growth of PbS crystals from nanocubes to eight-horn-shaped dendrites through a complex synthetic route, Mater. Letts. 59 (27) (2005) 3383–3388.
[197] U.K. Gautam, R. Seshadri, Preparation of PbS and PbSe nanocrystals from a new solvothermal route, Mater. Res. Bull. 39 (2004) 669–676.
[198] Y. Li, Z. Wang, Y. Ding, Room temperature synthesis of metal chalcogenides in ethylenediamine, Inorg. Chem. 38 (1999) 4737–4740.
[199] S. Biswas, S. Kar, S. Chaudhuri, Solvothermal synthesis of α-MnS single crystals, J. Cryst. Growth 284 (2005) 129–135.
[200] Y. Liu, Y. Xu, J.P. Li, B. Zhang, D. Wu, Y.H. Sun, Synthesis of CdS_xSe_{1-x} nanorods via a solvothermal route, Mater. Res. Bull. 41 (2006) 99–105.
[201] B.B. Wang, M.K. Zhu, H. Wang, G.B. Dong, Study on growth and photoluminescence of zinc telluride crystals synthesized by hydrothermal method, Opt. Mater. 34 (2011) 42–47.
[202] X. Ji, T.M. Tritt, X. Zhao, J.W. Kolis, Solution chemical synthesis of nanostructured thermoelectric materials, J. South. Carolina Acad. Sci. 6 (2008) 1–9.
[203] G. Tai, C. Miao, Y. Wang, Y. Bai, H. Zhang, W. Guo, Solvothermal synthesis and thermoelectric properties of indium telluride nanostring-cluster hierarchical structures, Nanoscale Res. Lett. 6 (2011) 329.
[204] B. Denac, M. Kristl, M. Drofenik, Thermal behavior of nanocrystalline cadmium selenides and tellurides, Chacogenide Lett. 8 (2011) 427–434.
[205] J.V. Williams, Hydrothermal Synthesis and Characterization of Cadmium Selenide Nanocrystals, Ph.D. Thesis, University of Michigan, USA, 2008.
[206] H. Wu, H. Xu, Q. Su, T. Chen, M. Wu, Size- and shape-tailored hydrothermal synthesis of YVO_4 crystals in ultra-wide pH range conditions, J. Mater. Chem. 13 (2003) 1223–1228.
[207] H. Meyssamy, K. Riwotzki, A. Kornowski, S. Naused, M. Haase, Wet-chemical synthesis of doped colloidal nanomaterials: particles and fibers of $LaPO_4$: Eu, $LaPO_4$:Ce and $LaPO_4$:Ce,Tb, Adv. Mater. 11 (1999) 840–844.
[208] I.F. Elder, M.J.P. Payne, YAP vs. YAG as a diode pumped host for thulium, Opt. Commun. 148 (1998) 265–269.
[209] D. Jia, Y. Wang, X. Guo, K. Li, Y.K. Zou, W. Jia, Synthesis and characterization of YAG:$Ce3^+$ LED nanophosphors, J. Electrochem. Soc. 154 (2007) J1–J4.
[210] M. Yu, J. Lin, Z. Wang, J. Fu, S. Wang, H.J. Zhang, et al., Fabrication, patterning, and optical properties of nanocrystalline YVO_4:A (A = Eu^{3+}, Dy^{3+}, Sm^{3+}, Er^{3+}) phosphor films via sol–gel soft lithography, Chem. Mater. 14 (2002) 2224–2231.

[211] K. Byrappa, K.M.L. Rai, B. Nirmala, M. Yoshimura, Crystal growth of Nd:RVO_4 (where R = Y, Gd) under mild hydrothermal conditions, Mater. Sci. Forum 315–317 (1999) 506–513.
[212] X. Li, H. Liu, J.Y. Wang, H.M. Cui, F. Han, YAG:Ce nano-sized phosphor particles prepared by a solvothermal method, Mater. Res. Bull. 39 (2004) 1923–1930.
[213] M. Inoue, H. Otsu, H. Kominami, I. Inui, Synthesis of yttrium garnet by the glycothermal method, J. Am. Ceram. Soc. 74 (6) (1991) 1452–1454.
[214] G.S. Yi, G.M. Chow, Synthesis of hexagonal-phase $NaYF_4$: Yb, Er and $NaYF_4$: Yb, Tm nanocrystals with efficient up-conversion fluorescence, Adv. Funct. Mater. 16 (2006) 2324–2329.
[215] K. Byrappa, M.K. Devaraju, J.R. Paramesh, B. Basavalingu, K. Soga, Hydrothermal synthesis and characterization of $LaPO_4$ for bio-imaging phosphors, J. Mater Sci. 43 (2008) 2229–2233.
[216] Sun Zhang Chao, Zhang Lingdong, Yawen, Yan Chunhua, Rare earth upconversion nanophosphors: synthesis, functionalization and application as biolabels and energy transfer donors, J. Rare Earths 28 (6) (2010) 807.
[217] G. Liu, S. Zhang, X. Dong, J. Wang, Solvothermal synthesis of Gd_2O_3:Eu^{3+} luminescent nanowires, J. Nanomater. (2010) 5.
[218] C. Bertail, S Maron, V. Buissette, T. Le Mercier, T. Gacoin, J.P. Boilot, Structural and photoluminescent properties of $Zn_2 \cdot SiO_4$:Mn^{2+} nanoparticles prepared by a protected annealing process, Chem. Mater. 23 (2011) 2961–2967.
[219] B.O. Dabbousi, J.R. Viejo, F.V. Mikulec, J.R. Heine, H. Mattoussi, R. Ober, et al., (CdSe) ZnS core-shell quantum dots: synthesis and optical and structural characterization of a size series of highly luminescent materials, J. Phys. Chem. B. 101 (1997) 9463–9475.
[220] I.L. Medintz, H.T. Uyeda, E.R. Goldman, H. Mattoussi, Quantum dot bioconjugates for imaging, labelling and sensing, Nat. Mater. 4 (2005) 435–446.
[221] A. Hoshino, K. Hanaki, K. Suzuki, K. Yamamoto, Applications of T-lymphoma labeled with fluorescent quantum dots to cell tracing markers in mouse body, BBRC 314 (2004) 45–53.
[222] J.H. Warner, A. Hoshino, K. Yamamoto, R.D. Tilley, Water-soluble photoluminescent silicon quantum dots, Angew. Chem. Int. Ed. 44 (2005) 4550–4554.
[223] R.D. Tilley, J.H. Warner, K. Yamamoto, I. Matsui, H. Fujimori, Micro-emulsion synthesis of monodisperse surface stabilized silicon nanocrystals, Chem. Comm. (2005) 1833–1835.
[224] H. Mizoguchi, P.M. Woodward, C.H. Park, D.A. Keszler, Strong near-infrared luminescence in $BaSnO_3$, J. Am. Chem. Soc. 126 (2004) 9796–9800.
[225] C. Chang, Z. Yuan, D. Mao, Eu^{2+} activated long persistent strontium aluminate nano scaled phosphor prepared by precipitation method, J. Alloys Comp. 415 (2006) 220–224.
[226] R. Lisiecki, G.D. Dzik, W.R. Romanowski, T. Lukasiewicz, Conversion of infrared radiation into visible emission in YVO_4 crystals doped with ytterbium and holmium, J. Appl. Phys. 96 (2004) 6323–6330.
[227] H. Furukawa, R. Shimojyo, N. Ohnishi, H. Fukuda, A. Kondo, Affinity selection of target cells from cell surface displayed libraries: a novel procedure using thermo-responsive magnetic nanoparticles, Appl. Microbiol. Biotechnol. 62 (2003) 478–483.
[228] W.Y. Jun, M.Y. Huh, S.J. Choi, H.J. Lee, H.-T. Song, S. Kim, et al., Nanoscale size effect of magnetic nanocrystals and their utilization for cancer diagnosis via magnetic resonance imaging, J. Am. Chem. Soc. 127 (2005) 5732–5733.

[229] H. Fan, Z. Chen, C.J. Brinker, J. Clawson, T. Alam, Synthesis of organo-silane functionalized nanocrystal micelles and their self-assembly, J. Am. Chem. Soc. 127 (2005) 13746–13747.

[230] J.D. Miller, H. Ishida, Quantitative monomolecular coverage of inorganic particulates by methacryl-functional silanes, Surf. Sci. 148 (1984) 601–622.

[231] J.D. Miller, H. Ishida, Quantitative analysis of covalent bonding between substituted silanes and inorganic surfaces, J. Chem. Phys. 83 (1987) 1593–1600.

[232] Y. Kera, M. Kamada, Y. Hanada, H. Kominami, Chemical modification of metal oxide carrier's surfaces with a silane coupling agent and an application of the method to catalyst preparation, Compos. Interfaces 8 (2001) 109–119.

[233] C.A. Mirkin, R.L. Letsinger, R.C. Mucic, J.J. Storhoff, A DNA-based method for rationally assembling nanoparticles into macroscopic materials, Nature 323 (1996) 607–609.

[234] J.J. Storhoff, C.A. Mirkin, Programmed materials synthesis with DNA, Chem. Rev. 99 (1999) 1849–1862.

[235] A.P. Alivisatos, K.P. Johnsson, X. Peng, T.E. Wilson, C.J. Loweth, M.P. Bruchez Jr., et al., Organization of nanocrystal molecules using DNA, Nature 382 (1996) 609–613.

[236] J. Lu, K. Minami, S. Takami, M. Shibata, Y. Kaneko, T. Adschiri, Supercritical hydrothermal synthesis and in situ organic modification of indium tin oxide nanoparticles using continuous flow reaction system, Appl. Mater. Interfaces 4 (2012) 351–354.

[237] T. Mousavand, S. Takami, M. Umetsu, S. Ohara, T. Adschiri, Supercritical hydrothermal synthesis of organic–inorganic hybrid nanoparticles, J. Mater. Sci. 41 (2006) 1445–1448.

[238] T. Togashi, M. Umetsu, T. Naka, S. Ohara, Y. Hatakayama, T. Adchiri, One-pot hydrothermal synthesis of an assembly of magnetite nanoneedles on a scaffold of cyclic-diphenyalanine nanorods, J. Nanopart. Res. 13 (2011) 3991–3999.

[239] K. Kaneko, K. Inoke, B. Freitag, A.B. Hungria, P.A. Midley, T.W. Hansen, et al., Structural and morphological characterization of cerium dioxide nanocrystals prepared by hydrothermal synthesis, Nano Lett. 7 (2007) 421–425.

[240] T. Mousavand, S. Ohara, M. Umetsu, J. Zhang, S. Takami, T. Naka, et al., Hydrothermal synthesis and in situ surface modification of boehmite nanoparticles in supercritical water, J. Supercrit. Fluids 40 (2007) 397–401.

[241] M.P. Pileni, Nanocrystal self assemblies: fabrication and collective properties, J. Phys. Chem. B105 (2001) 3358–3372.

[242] J. Dutta, H. Hofmann, Self-organization of colloidal nanoparticles, Encycl. Nanosci. Nanotechnol. 9 (2003) 617–640.

[243] A.C. Templeton, W.P. Wuelfing, R.W. Murray, Monolayer-protected cluster molecules, Acc. Chem. Res. 33 (2000) 27–36.

[244] A.R. Tao, S. Habas, P. Yang, Shape control of colloidal metal nano-crystals, Small 4 (2008) 310–325.

[245] J.E. Martin, J.P. Wilcoxon, J. Odinek, P. Provencio, Control of the interparticle spacing in gold nanoparticle superlattices, J. Phys. Chem. B104 (2000) 9475–9486.

[246] A.K. Boal, V.M. Rotello, Intra- and inter-monolayer hydrogen bonding in amide functionalized alkanethiol SAMs on gold nanoparticles, Langmuir 16 (2000) 9527–9532.

[247] J. Kolny, A. Kornowski, H. Weller, Self-organization of cadmium sulfide and gold nanoparticles by electrostatic interaction, Nano Lett. 2 (2002) 361–364.

[248] M.C. Daniel, D. Astruc, Gold nanoparticles: assembly, supramolecular chemistry, quantum-size-related properties, and applications toward biology, catalysis, and nanotechnology, Chem. Rev. 104 (2004) 293–346.
[249] M. Brust, M. Walkder, D. Bethell, D.J. Schiffrin, R.J. Whyman, Synthesis of thiol derivatized gold nanoparticles in a 2-phase liquid–liquid system, Chem. Soc., Chem. Commum. (1994) 801–802.
[250] J.H. Fendler, F. Meldrum, Colloid chemistry approach to nanostructured materials, Adv. Mater. 7 (1995) 607–631.
[251] S. Yamamuro, D.G. Farrell, S.A. Majetich, Direct imaging of self-assembled magnetic nanoparticle arrays: phase stability and magnetic effects on morphology, Phys. Rev. B65 (2002) 224431.
[252] C.B. Murray, D.J. Norris, M.G. Bawendi, Synthesis and characterization of nearly monodisperse CdE (E = sulfur, selenium, tellurium) semiconductor nanocrystallites, J. Am. Chem. Soc 115 (1993) 8706–8715.
[253] C.B. Murray, C.R. Kagan, M.G. Bawendi, Self-organization of CdSe nanocrystallites into three-dimensional quantum dot superlattices, Science 270 (1995) 1335–1338.
[254] C.B. Murray, Watching nanocrystals grow, Science 324 (2009) 1276–1277.
[255] J. Park, J. Joo, S.G. Kwon, Y. Jang, T. Hyeon, Synthesis of monodisperse spherical nanocrystals, Angew. Chem. Int. Ed. 46 (2007) 4630.
[256] P.A. Kralchevsky, N.D. Denkoy, Capillary forces and structuring in layers of colloid particles, Curr. Opin. Colloid Interface Sci. 6 (2001) 383–401.
[257] N. Christiansen, R.E. Riman, Bioceramics: a future through microstructural and chemical design, in: Proceedings of the Fifth Scandinavian Symposium Materials Science, New Materials and Processes, (N. Yamasaki, and K. Yamagisawa Eds) May 22–23, 1989, pp. 209–220.
[258] L.L. Hench, Bioceramics: from concept to clinic, J. Am. Ceram. Soc. 74 (1991) 1487–1510.
[259] L.L. Hench, Bioceramics, J. Am. Ceram. Soc. 81 (1998) 1705–1728.
[260] F. Hubert, J.C Bokros, L.L. Hench, J. Wilson, G. Heimke, Ceramics in clinical applications: past, present and future, in: P. Vincenzini (Ed.), High Tech Ceramics, Elsevier, Amsterdam, 1987, pp. 189–213.
[261] L.L. Hench, J. Wilson, An Introduction to Bioceramics, World Scientific, London, UK, 1993.
[262] J.F. Shackelford, Bioceramics—current status and future trends, Mater. Sci. Forum. 293 (1999) 99–106.
[263] de K. Groot, Bioceramics of Calcium Phosphate, CRC Press, Boca Raton, FL, 1983.
[264] T. Yamamuro, L.L. Hench, J. Wilson (Eds), Handbook of Bioactive Ceramics, Vol. II., Calcium Phosphate and Hydroxylapatite Ceramics, CRC Press, Boca Raton, FL, 1990.
[265] W. Suchanek, M. Yoshimura, Processing and properties of hydroxyapatite-based biomaterials for use as hard tissue replacement implant, J. Mater. Res. 13 (1998) 1–24.
[266] M. Yoshimura, H. Suda, Hydrothermal processing of HAp: past, present and future, in: P.W. Brown, B. Constantz (Eds), Hydroxyapatite and Related Compounds, CRC Press, Cleveland, OH and Boca Raton, FL, 1994, p. 45.
[267] J.C. Elliott, Structure and Chemistry of the Apatites and Other Calcium Orthophosphates, Elsevier, Amsterdam, 1994.
[268] K. Ioku, M. Yoshimura, S. Somiya, Microstructure-designed HAp ceramics from fine single crystals synthesized hydrothermally, in: H. Oonishi, H. Aoki, K. Sawai (Eds), Bioceramics, Proceedings of the First International Bioceramics, Symposium, Ishiyaku Euro-America Inc., Tokyo–St. Louis, 1989, pp. 62–67.

[269] M. Lencka, R.E. Riman, Thermodynamics of the hydrothermal synthesis of calcium titanate with reference to other alkaline earth titanates, Chem. Mater. 7 (1995) 18–25.

[270] E.L. Shock, H.C. Helgeson, D.A. Sverjensky, Calculation of the thermodynamic and transport properties of aqueous species at high pressures and temperatures: standard partial molal properties of inorganic neutral species, Geochim. Cosmochim. Acta 53 (1989) 2157–2183.

[271] D.A. Sverjensky, E.L. Shock, H.C. Helgeson, Prediction of the thermodynamic properties of aqueous metal complexes to 1000°C and 5 kb, Geochim. Cosmochim. Acta 61 (1997) 1359–1412.

[272] W. Suchanek, H. Suda, M. Yashina, M. Kakihana, M. Yoshimura, Biocompatible whiskers with controlled morphology and stoichiometry, J. Mater. Res. 10 (1995) 512–529.

[273] K. Yamashita, T. Kanazawa, Hydroxyapatite, in: T. Kanazawa (Ed.), Inorganic Phosphate Materials, Kodansha and Elsevier, 1989, pp. 23–24. Materials Science Monograph, 52, p. 15.

[273a] G.N. Ehsani, A.J. Ruys, C.C. Sorrell, Microwave sintering of hydroxyapatite, in: C.C. Sorrell, A.J. Ruys (Eds.), International Ceramic Monographs, Proceedings of the International Ceramics Conference 1, 1994, pp. 714–721.

[274] M.R. Mucalo, Y. Yokogawa, M. Toriyama, T. Suzuki, Y. Kawamoto, F. Nagata, et al., Growth of calcium phosphate on phosphorylated cotton fibers, Apatite 2 (1997) 71–73.

[274a] T. Yoshioka, A. Hasame, S. Watanabe, S.M. Shin, A. Okuwaki, Determination of oxalic acid and benzene carboxylic acids from PVC materials by oxygen—oxidation in NaOH solutions at elevated temperatures, in: Proceedings of the Second International Conference on Solvothermal Reactions, Takamatsu, Japan, December 18–20, 1996, pp. 46–49.

[275] R.L. SmithJr., P. Atmaji, Y. Hakuta, T. Adschiri, K. Arai, Hydrothermal crystallization of metals from simulated high level liquid waste, in: Proceedings of the Second International Conference on Solvothermal Reactions, Takamatsu, Japan, December 18–20, 1996, pp. 42–45, 50–53.

[275a] H. Takahashi, M. Yashima, M. Kakihana, M. Yoshimura, Synthesis of stoichiometric (Ca/P = 1.67) HAp by a gel route from the aqueous solution of citric and phosphonoacetic acids, Eur. J. Solid State Inorg. Chem. 32 (1995) 829–835.

[276] T. Hashida, H. Takahashi, Development of solvothermal methods for recycling concrete wastes, in: Proceedings of the Second International Conference on Solvothermal Reactions, Takamatsu, Japan, December 18–20, 1996, pp. 152–155.

[276a] H. Aoki, Science and Medical Applications of HAp, JAAS, Tokyo, 1991.

[277] N. Yamasaki, Y. Yamasaki, Hydrothermal decomposition of chlorinated organic compounds, in: Proceedings of the Second International Conference on Solvothermal Reactions, Takamatsu, Japan, December 18–20, 1996, pp. 139–142.

[277a] K. Inoue, H. Ohgush, T. Yoshikawa, M. Okumura, T. Sempuku, S. Tamai, et al., The effect of aging on bone formation in porous hydroxyapatite: biochemical and histological analysis, J. Bone Miner. Res. 12 (1997) 989–994.

[278] K. Itatani, O. Takahashi, A. Kishioka, M. Kinoshita, Properties of hydroxyapatite prepared by spray-pyrolysis technique, J. Gypsum Lime 213 (1988) 77 (Japan)

[279] M. Aizawa, K. Itatani, F.S. Howell, A. Kishioka, Formation of porous calcium phosphate films on partially stabilized zirconia substrates by the spray–pyrolysis technique, J. Mater. Sci. 30 (1995) 4936–4945.

[280] M. Aizawa, K. Itatani, F.S. Howell, A. Kishioka, Some properties of carbonate containing hydroxyapatite powder prepared by spray pyrolysis technique using urea as a foaming agent, J. Ceram. Soc. Jpn. 103 (1995) 1214–1219.

[281] C. Mossaad, M. Starr, S. Patil, R.E. Riman, Thermodynamic modeling of hydroxyapatite crystallization with biomimetic precursor design considerations, Chem. Mater. 22 (2010) 26–46.

[282] D.M. Roy, S.K. Linnehan, Hydroxyapatite formed from coral skeletal carbonate by hydrothermal exchange, Nature 247 (1974) 220–221.

[283] K. Ioku, T. Kai, M. Nishioka, in: M.A.H. Aoki, N. Nagai, T. Tsuji (Eds), Apatite, Proceedings of the First International Symposium Apatite, vol. 1, Takayama Press System Center, Inc., 1992.

[284] K. Ioku, Y. Eguchi, S. Goto, in: Transactions of the Twelfth Symposium on Apatite, Nagoya, Japan, December 12–13, 1996, p. 29.

[285] W.E. Brown, L.C. Chow, in: P.W. Brown (Ed.), Cements Research Progress—1987, American Ceramic Society, Westerville, OH, 1988, p. 351.

[286] P.W. Brown, R.I. Martin, K.S. Tenhuisen, in: Bioceramics: Materials and Applications, R.P. Rusin, G.S. Fishman (Eds), Ceram. Trans. 48 (1995) 37–46.

[287] W.E. Brown, L.C. Chow, Combinations of Sparingly Soluble Calcium Phosphate in Slurries and Pastes as Mineralizers and Cements, US Patent 4612053, 1986.

[288] M. Otsuka, Y. Nakahigashi, Y. Matsuda, J.L. Fox, W.I. Higuchi, In vitro drug release behaviors from self-setting calcium phosphate cement loaded on bovine bone, Apatite 2 (1997) 141–144.

[289] M. Otsuka, Y. Matsuda, Y. Suwa, J.L. Fox, W.I. Higuchi, A novel skeletal drug delivery system using self-setting calcium phosphate cement 4: effect of the mixing solution volume on the drug release rate of heterogeneous aspirin loaded cement, J. Pharm. Sci. 83 (1994) 259–263.

[290] B. Ferrand, J. Daval, J.C. Joubert, Heteroepitaxial growth of single crystal films of YIG on GGG substrates by hydrothermal synthesis, J. Cryst. Growth 17 (1972) 312–314.

[291] B. Ferrand, Heteroepitaxie par syntheses hydrothermale de films minces monocristallins de gernats ferrimagnetiques, High Temperature–High Pressure 6 (1974) 619–628.

[292] Y. Toudic, M. Passareti, Croissance par voie hydrothermale de films ferrimagnetiques; epitaxies sur des substrats de GGG, J. Cryst. Growth 24/25 (1974) 621–623.

[293] Y. Fujishiro, A. Fujimoto, T. Sato, A. Okuwaki, Coating of HAp on titanium plates using thermal dissociation of calcium-EDTA chelate complex in phosphate solutions under hydrothermal conditions, J. Colloid Interface Sci. 173 (1995) 119–127.

[294] T. Iton, Q. Zhang, M. Abe, Y. Tamaura, Preparation of $CoxFe_{3-x}O_4$ films in aqueous solution at 120–200°C by "Hydrothermal Ferrite Plating", Jpn. J. Appl. Phys. 31 (1992) 1236–1238.

[295] W.S. Cho, M. Yoshimura, Hydrothermal, hydrothermal–electrochemical and electrochemical synthesis of highly crystallized barium tungstate films, Jpn. J. Appl. Phys. 36 (1997) 1216–1222.

[296] W.S. Cho, M. Yoshimura, Preparation of highly crystallized $BaMoO_4$ film using a solution reaction assisted by electrochemical dissolution of molybdenum, Solid State Ionics 100 (1997) 143–147.

[297] J.J. Xu, A.S. Shaikh, R.W. Vest, High K $BaTiO_3$ films from metal organic precursors, IEEE Trans. Ultrason. Ferroelectr. Freq. Control 36 (1989) 307–312.

[298] P.C. Van Buskirk, R. Gardiner, P.S. Kirlin, S. Nutt, Reduced pressure MOCVD of highly crystalline $BaTiO_3$ thin films, J. Mater. Res. 7 (1992) 542–546.

[299] M.N. Kamalasanan, S. Chandra, P.C Joshi, A. Mansingh, Structural and optical properties of sol−gel processed $BaTiO_3$ ferroelectric films, Appl. Phys. Letts. 59 (1991) 3547−3556.

[300] W.S. Cho, M. Yoshimura, Hydrothermal synthesis of Pb TiO_3 films, J. Mater. Res. 12 (1997) 833−839.

[301] V.S. Dharmadhikari, W.W. Grannemann, AES study on the chemical composition of ferroelectric $BaTiO_3$ thin films rf-sputter deposited on silicon, J. Vac. Sci. Technol. A1 (1983) 483−489.

[302] M. Hayashi, N. Ishizawa, S.E. Yoo, M. Yoshimura, Preparation of barium titanate thin film on titanium deposited glass substrate by the hydrothermal electrochemical method, J. Ceram. Soc. Jpn. 98 (1990) 930−934.

[303] N. Ishizawa, H. Banno, M. Hayashi, S.E. Yoo, M. Yoshimura, Preparation of $BaTiO_3$ and $SrTiO_3$ polycrystallization thin films on flexible polymer film substrate by the hydrothermal method, Jpn. J. Appl. Phys. 29 (1990) 2467−2471.

[304] S.E. Yoo, M. Hayashi, N. Ishizawa, M. Yoshimura, Preparation of strontium titanate thin film on titanium metal substrate by the hydrothermal electrochemical method, J. Am. Ceram. Soc. 73 (1990) 2561−2567.

[305] S.E. Yoo, N. Ishizawa, M. Hayashi, M. Yoshimura, Review on preparation of films of multifunctional complex oxide by the hydrothermal−electrochemical method, Rep. Res. Lab. Eng. Mater. 16 (1991) 39−44Tokyo Inst. Technol.

[306] R. Basca, P. Ravindranathan, J.P. Dougherty, Electrochemical, hydrothermal and electrochemical hydrothermal synthesis of barium titanate thin films on titanium substrates, J. Mater. Res. 7 (1992) 423−428.

[307] M. Yoshimura, Hydrothermal electrochemical technique: a new fabrication method of thin films of perovskite-type oxides, in: Proceedings of the Ninety-Fifth Annual Meeting of the American Ceramic Society, Cincinnati, OH, April 20, 1993.

[308] M.E. Pilleux, V.M. Fuenzalida, Hydrothermal $BaTiO_3$ films on silicon: morphological and chemical characterization, J. Appl. Phys. 74 (1993) 4664−4670.

[309] A.V. Alvarez, V.M. Fuenzalida, Evidence of transition metal diffusion during hydrothermal ceramic film growth: $Ba(Ti,Zr)O_3$ on layered Ti−Zr alloy, J. Mater. Res. 14 (1999) 4136−4139.

[310] I.S. Escobar, C.I. Silva, T. Vargas, V.M. Fuenzalida, Nucleation and inhibition control during the hydrothermal−electrochemical growth of barium titanate films, J. Am. Ceram. Soc. 83 (2000) 2673−2676.

[311] P. Bendale, S. Venigalla, J.R. Ambrose, E.D. Verink, J.H. Adair, Preparation of barium titanate films at 55°C by an electrochemical method, J. Am. Ceram. Soc. 76 (1993) 2619−2627.

[312] Y. Ohba, M. Miyauchi, E. Sakai, M. Daimon, Hydrothermal synthesis of lead zirconate titanate thin films fabricated by a continuous-supply autoclave, Jpn. J. Appl. Phys. 34 (1995) 5216−5219.

[313] Q. Chen, Y. Qian, Z. Chen, W. Wu, Z. Chen, G. Zhou, et al., Hydrothermal epitaxy of highly oriented TiO_2 thin films on silicon, Appl. Phys. Lett. 66 (1995) 1608−1610.

[314] Q. Chen, Y. Qian, Z. Chen, G. Zhou, Y. Zhang, Hydrothermal preparation of highly oriented polycrystallization ZnO thin films, Mater. Letts. 22 (1995) 93−95.

[315] T. Hoffman, M. Gultzow, C.M.S. Torres, T. Doll, V.M. Fuenzalida, Microstructures in $BaTiO_3$ thin films by hydrothermal growth and lift-off technique, Mater. Sci. Semicond. Process. 2 (1999) 335−340.

[316] M.E. Pilleux, C.R. Grahmann, V.M. Fuenzalida, R. Avila, Hydrothermal ABO_3 ceramic thin films, Appl. Surf. Sci. 65/66 (1993) 283–290.
[317] K. Tachibana, A new pressure-balanced outer reference electrode with oxidized zircaloy tube container for corrosion studies in high-temperature–high-pressure aqueous solution and its application to obtaining polarization curves, Boshoku Gijutsu 34 (1985) 125–131.
[318] D.D. Macdonald, A.C Scott, P. Wentrcek, External reference electrodes for use in high-temperature aqueous systems, J. Electrochem. Soc. 126 (1979) 908–911.
[319] P.A. Psaras, H.D. Langford (Eds), Advancing Materials Research, National Academy Press, Washington, DC, 1987. p. 274.
[320] Y. Yokogawa, F. Nagata, M. Toriyama, Y. Kawamoto, T. Suzuki, K. Nishizawa, et al., HAp formation on polymeric substrate in the presence of polyacrylic acid, Apatite 2 (1997) 67–70.
[321] Y. Fujishiro, T. Sato, A. Okuwaki, Coating of hydroxyapatite on metal plates using thermal dissociation of calcium-EDTA chelate in phosphate solutions under hydrothermal conditions, J. Mater. Sci. Mater. Med. 6 (1995) 172–176.
[322] S.F. Hulbert, J.H. Dove, Fatigue properties of a polymethylmethacrylate–HAp composite bone cement, Apatite 2 (1997) 57–60.
[323] S. Ban, T. Hanaichi, S. Maruno, Microstructure of electrochemically deposited apatite on the HA–G–Ti composite, Apatite 2L (1997) 11–14.
[324] K.S. Tenhuisen, P.W. Brown, C.S. Reed, H.R. Allcock, Low temperature synthesis of a self-assembling composite: HAp-poly [bis (sodium carboxylatophenoxy) phosphazene], J. Mater. Sci. Mater. Med. 7 (1996) 673–682.
[325] A.B. Loyall, HAp composites move closer to reality, Bio. Med. Technol. Alert, 3 (40) (1997) 4–5.
[326] W. Höland, W. Vogel, Machineable and phosphate glass – ceramics, in: L.L. Hench, J. Wilson (Eds), An Introduction to Bioceramics, World Scientific, London, 1993Adv. Ser. Ceram., 1, 125.
[327] W. Höland, V. Rheinberger, M. Frank, S. Wegner, in: T. Kokubo, T. Nakamura, F. Miyaji (Eds), Bioceramics, in: Proceedings of the Ninth International Symposium Ceramics in Medicine, 1996, Otsu, Japan, Pergamon Press, Oxford, 1996, p. 445.
[328] W. Bonfield, Hydroxyapatite-reinforced polyethylene as an analogous materials for bone replacement, in: P. Ducheyne, J. Lemons (Eds), Bioceramics: Materials Characteristics Versus in vivo Behavior, Annals of New York Academy of Science, New York, NY, 1988, pp. 173–177.
[329] W. Bonfield, C. Doyle, K.E. Tanner, In vivo evaluation of hydroxyapatite-reinforced polyethylene composites, in: P. Christel, A. Meunier, A.J.C. Lee (Eds), Biological and Biological and Biomechanical Performance of Materials, Elsevier, New York, NY, 1986, p. 153.
[330] J. Huang, L. Di Silvo, M. Wang, K.E. Tanner, W. Bonfield, In vitro mechanical and biological assessment of hydroxyapatite-reinforced polyethylene composite, J. Mater. Sci. Mater. Med. 8 (1997) 775–779.
[331] R.A. Laudise, Growth of Single Crystals, Prentice-Hall, New Jersey, NJ, 1970.
[332] H.J. Scheel, Historical introduction, in: D.T.J. Hurle (Ed.), Handbook of Crystal Growth, vol. 4, Elsevier, Amsterdam, 1993.
[333] F. Nagata, Y. Yokogawa, M. Toriyama, Y. Kawamoto, T. Suzuki, K. Nishizawa, et al., Hydrothermal synthesis of plate-like hydroxyapatite crystals, in: H. Aoki (Ed.), Advanced Materials, '93, IIA: Biomaterials, Organic and Intelligent Materials, Elsevier, 1994Trans. Mat. Res. Soc. Jpn., vol. 15A

[334] W. Suchanek, H. Suda, M. Yashima, M. Kakihana, M. Yoshimura, Biocompatible whiskers with controlled morphology and stoichiometry, J. Mater. Res. 10 (1995) 521–529.

[335] W. Suchanek, M. Yoshimura, Preparation of fibrous, porous hydroxyapatite ceramics from hydroxyapatite whiskers, J. Am. Ceram. Soc. 81 (1998) 765–767.

[336] P. Sujaridworakum, F. Koh, T. Fujiwara, D. Pongkao, A. Ahniyaz, M. Yoshimura, Preparation of anatase nanocrystals deposited on hydroxyapatite by hydrothermal treatment, Mater. Sci. Eng C25 (2005) 87–92.

[336a] W. Suchanek, M. Yoshimura, M. Kakihana, M. Yoshimura, β-Rhenanite (β-$NaCaPO_4$) as weak interphase for hydroxyapatite ceramics, J. Eur. Ceram. Soc. 18 (1998) 1923–1929.

[337] J. Wu, S. Yin, Y. Lin, J. Lin, M. Huang, T. Sato, Hydrothermal synthesis of HNbWO6/MO series nanocomposites and their photocatalytic properties, J. Mater. Sci. 36 (2001) 3055–3059.

[338] K. Byrappa, M.H. Sunitha, A.K. Subramani, S. Ananda, K.M.L. Rai, B. Basavalingu, et al., Hydrothermal preparation of neodymium oxide coated titania composite designer particulates and its application in the photocatalytic degradation of procian red dye, J. Mater. Sci. 41 (2006) 1369–1375.

[339] K. Byrappa, R. Dinesh, M. Yoshimura, Photocatalytic degradation of nitroarenes using activated carbon/TiO_2 photocatalyst, Trans. Jpn. Mater. Res. Soc. 29 (2004) 2407–2411.

[340] H.P. Shivaraju, C.P. Sajan, T. Rungnapa, M.S. Vijay Kumar, C. Ranganathaiah, K. Byrappa, Photocatalytic treatment of organic pollutants in textile effluent by using hydrothermally prepared photocatalytic composite, Mater. Res. Innov. 14 (2010) 80–86.

[341] H.P. Shivaraju, K. Byrappa, M.B. Shayan, T. Rungnapa, S. Pakamard, Vijaya Kumar, et al., Hydrothermal coating of ZnO onto calcium alumino-silicate beads and their application in the photodegradation of amaranth dye, Mater. Res. Innov. 14 (2010) 73–79.

[342] T. Murota, Economy of Entropy and Energy, Toyo Keizai Shinposha Ltd., Tokyo, 1983 (in Japanese).

[343] M. Yoshimura, What, why, and how soft and solution processing for ceramics?, in: Proceedings of the World Conference on The Role of Advanced Materials in Sustainable Development, IUPAC Seoul, Korea, September 1–6, 1996, pp. 389–397.

[344] W. Suchanek, M. Yashima, M. Kakihana, M. Yoshimura, Hydroxyapatite/hydroxyapatite whisker composites without sintering additives: mechanical properties and microstructural evolution, J. Am. Ceram. Soc. 80 (1997) 2805–2813.

[345] K. Hosoi, T. Hashida, H. Takahashi, N. Yamasaki, T. Korenaga, New processing technique for hydroxyapatite ceramics by the hydrothermal hot-pressing method, J. Am. Ceram. Soc. 79 (1996) 2771–2774.

[346] K. Hosoi, T. Hashida, H. Takahashi, N. Yamasaki, T. Korenaga, Solidification behavior of calcium carbonate via aragonite–calcite wet transformation with hydrothermal hot pressing, J. Mater. Sci. 16 (1997) 382–385.

[347] K. Hosoi, T. Hashida, H. Takahashi, N. Yamasaki, T. Korenaga, Low temperature solidification of calcium carbonate through vaterite–calcite wet transformation, J. Mater. Sci. Letts. 15 (1996) 812–814.

[348] N. Yamasaki, T. Kai, M. Nishioka, K. Yanagisawa, K. Ioku, Porous hydroxyapatite ceramics prepared by hydrothermal hot pressing, J. Mater. Sci. Letts. 9 (1990) 1150–1151.

[349] K. Ioku, K. Yanagisawa, T. Kai, K. Yamamoto, N. Yamasakii, Apatite ceramics, prepared by hydrothermal hot pressing, in: Proceedings of the First International Conference on Solvothermal Reactions, (N. Yamasaki, and K. Yanagisawa Eds) Takamatsu, Japan, December 5–7, 1994, pp. 118–121.
[350] K. Hosoi, T. Korenaga, H. Takahashi, T. Weiping, H. Lei, K. Yanagisawa, et al., Solidification of carbonate at low temperature by HHP technique, in: Proceedings of the First International Conference on Solvothermal Reactions, (N. Yamasaki, and K. Yanagisawa Eds) Takamatsu, Japan, December 5–7, 1994, pp. 94–97.
[350a] Uematsu, et al., Pentafluoroethyl Chloride, NIST Standard Reference Database 69, Chem. WebBook (1989) 1–4.
[351] K. Hosoi, T. Korenaga, T. Hashida, H. Takahashi, N. Yamasaki, Strengthening and toughening in HHP—pressed $CaCO_3$ composite, in: Proceedings of the Second International Conference on Solvothermal Reactions, (N. Yamasaki, and K. Yanagisawa Eds) Takamatsu, Japan, December 18–20, 1996, pp. 85–88.
[352] K. Sato, T. Hashida, H. Takahasi, N. Yamasaki, Solidification method for calcium aluminate-phosphate cement by HHP and fiber reinforcement, in: Proceedings of the Second International Conference on Solvothermal Reactions, (N. Yamasaki, and K. Yanagisawa Eds) Takamatsu, Japan, December 18–20, 1996, pp. 89–92.
[353] K. Sato, T. Hashida, H. Takahashi, N. Yamasaki, Development of high strength calcium aluminate-phosphate cement by hydrothermal hot pressing, J. Mater. Sci. Letts. 16 (1997) 1464–1468.
[354] K. Yanagisawa, M. Nishioka, K. Ioku, N. Yamasaki, Densification of silica gels by hydrothermal hot pressing, J. Mater. Sci. Letts. 12 (1993) 1073–1075.
[355] K. Yanagisawa, K. Ioku, N. Yamasaki, Formation of anatase porous ceramics by HHP of amorphous titania spheres, J. Am. Ceram. Soc. 80 (1997) 1303–1306.
[356] S. Ogawa, S. Minamic, N. Yamasaki, Solidification of sewage sludge incinerated ash by hydrothermal hot pressing, in: Proceedings of the First International Conference on Solvothermal Reactions, (N. Yamasaki, and K. Yanagisawa Eds) Takamatsu, Japan, December 5–7, 1994, pp. 126–129.
[357] K. Mori, M. Katakura, N. Yamasaki, Development of hydrothermal solidification for low level radioactive wastes, in: Proceedings of the First International Conference on Solvothermal Reactions, Takamatsu, Japan, December 5–7, 1994, pp. 150–153.
[358] K. Ioku, M. Yoshimura, Microstructure designing of hydroxyapatite ceramics by hip post-sintering, in: M. Koizumi (Ed.), Hot Isostatic Pressing—Theory and Applications, Proceedings of the Third International Conference, Osaka, Japan, June 10–14, 1991, pp. 123–128.
[359] K. Hirota, T. Hasegawa, H. Monma, Densification of hydroxyapatite by hot isostatic pressing, Yogyo Kyokai Shi 90 (1982) 62–64.
[360] K. Uematsu, M. Takagi, T. Honda, N. Uchida, K. Saito, Transparent hydroxyapatite prepared by hot isostatic processing of filter cake, J. Am. Ceram. Soc. 72 (1989) 1476–1478.
[361] S. Somiya (Ed.), Advanced Ceramics I., Elsevier, Amsterdam, 1990.
[362] S. Somiya, Hydrothermal reaction sintering of high-density sintered oxides, in: Proceedings of the First International Symposium on Hydrothermal Reactions, (Ed. S. Somiya) Gakujutsu Bunken Fukyu- Kai, Tokyo, Japan, March 22–26, 1982, pp. 887–903.

[363] S. Somiya, Hydrothermal reaction sintering of oxides, in: G.S. Upadhyaya (Ed.), Sintered Metal-Ceramic Composites, Elsevier, Amsterdam, 1984, pp. 97–105.
[364] S. Somiya, Hydrothermal reactions in inorganic systems, in: Advanced Materials, 1993, VI/Frontiers in Materials Science and Engineering, S. Somiya et al., (Eds), Trans. Mat. Res. Soc. Jpn., 19B (1994) 1105–1134.
[365] H. Toraya, M. Yoshimura, S. Somiya, Densification and grain growth in hydrothermal reaction sintering of Hafnia (HfO_2), Yogyo Kyokai Shi 91 (1983) 235–240.
[366] G. Pfoff, A. Feltz, Solid state reactivity and mechanisms in oxide systems. VI. Sintering behavior of hematite prepared by the hydrothermal method, Solid State Ionics 38 (1990) 25–29.
[367] S. Yin, Y. Fujishiro, S. Uchida, T. Sato, Low temperature sintering and mechanical properties of ceria and yttria co-doped zirconia crystallized in supercritical methanol, in: Proceedings of the Second International Conference on Solvothermal Reactions, (N. Yamasaki, and K. Yanagisawa Eds) Takamatsu, Japan, December 18–20, 1996, pp. 27–30.
[368] W. Zhu, C.C Wang, S.A. Akbar, R. Asiale, Fast-sintering of hydrothermally synthesized $BaTiO_3$ powders and their dielectric properties, J. Mater. Sci. 32 (1997) 4303–4307.
[369] A.J. Ruys, M. Wei, C.C. Sorrell, M.R. Dickson, A. Brandwood, B.K. Milthorpe, Sintering effects on the strength of hydroxyapatite, Biomaterials 16 (1995) 409–415.
[370] W. Suchanek, M. Yashima, M. Kakihana, M. Yoshimura, Hydroxyapatite ceramics with selected sintering additives, Bioceramics 18 (1997) 923–933.
[371] M.K. Krage, Microwave sintering of ferrites, Am. Ceram. Soc. Bull. 60 (1981) 1232–1234.
[372] R. Roy, S. Komarneni, L.J. Yang, Controlled microwave heating and melting of gels, J. Am. Ceram. Soc. 68 (1985) 392–395.
[373] S. Komarneni, R. Roy, Anomalous microwave preparation of mullite powders, in: M. H. Brooks, I.J. Chabinsky, W.H. Sutton (Eds), Proceedings of the Materials Research Society Symposium Microwave Processing of Materials, vol. 124, Materials Research Society, Pittsburgh, PA, 1988.
[374] R. Gedye, F. Smith, K. Westaway, H. Ali, L. Baldisera, L. Laberge, et al., The use of microwave ovens for rapid organic synthesis, Tetrahedron Lett. 27 (1986) 279–282.
[375] R.J. Gigure, T.L. Bray, S.M. Duncan, G. Majetich, Application of commercial microwave ovens to organic synthesis, Tetrahedron Lett. 27 (1986) 4945–4948.
[376] D.R. Baghurst, D.M.P. Mingos, M.J. Watson, Application of microwave dielectric loss heating effects for the rapid and convenient synthesis of organometallic compounds, J. Organomet. Chem. 368 (1989) C43–C45.
[377] D.L. Greene, D.M.P. Mingos, Application of microwave dielectric loss heating effects for the rapid and convenient synthesis of ruthenium (11) polypyridine complexes, Transit. Met. Chem. 16 (1991) 71–72.
[378] K. Chatakondu, M.L.H. Green, D.M.P. Mingos, S.M. Reynolds, Application of microwave dielectric loss heating effects for the rapid and convenient synthesis of intercalation compounds, J. Chem. Soc. Chem. Commun. (1989) 1515–1517.
[379] V.C. Virtuli, P. Chu, F.G. Dwyer, Crystallization Method Using Microwave Radiation, US Patent Application 4778666, 1988.
[380] A. Arafat, J.C Jensen, A.R. Ebaid, H. Van Bekkum, Microwave preparation of zeolite Y and ZSM-5, Zeolites 13 (1993) 162–165.

[381] S.F. Liu, I.R. Abothu, S. Komarneni, Barium titanate ceramics prepared from conventional and microwave hydrothermal powders, Mat. Letts. 38 (1999) 344–350.
[382] S. Komarneni, V.C. Menon, Q.H. Li, Synthesis of ceramic powders by novel microwave-hydrothermal processing, in: J.J. Kingsley, C.H. Schilling, J.H. Adair (Eds), Ceramic Transactions, vol. 2, Science, Technology and Commercialization of Powder Synthesis and Shape Forming Processes, American Ceramic Society, Westerville, OH, 1995, pp. 37–46.
[383] S. Komarneni, Novel microwave-hydrothermal processing for synthesis of ceramic and metal powders, in: J. Singh, J.M. Copley (Eds), Novel Techniques in Synthesis and Processing of Advanced Materials, The Minerals, Metals and Materials Society, Warrendale, PA, 1995, pp. 103–117.
[384] S. Komarneni, Environmentally benign microwave-hydrothermal processing for synthesis of ceramic powders, in: H. Yanagida, M. Yoshimura (Eds), Proceedings of the International Symposium Environmental Issues of Ceramics, Ceramic Society of Japan, Tokyo, 1995, pp. 199–206.
[385] S. Komarneni, R. Pidugu, Q.H. Li, R. Roy, Microwave-hydrothermal processing of metal powders, J. Mater. Res. 10 (1995) 1687–1692.
[386] S. Komarneni, Microporous metal intercalated clay nanocomposites, in: K. Ishizaki, L. Sheppard, S. Okada, T. Hamasaki, B. Huybrechts (Eds), Porous Materials, American Ceramic Society, Westerville, OH, 1993Ceram. Trans. 31 155–168
[387] S. Komarneni, N. Kozai, R. Roy, Novel function for anionic clays: selective transition metal cation uptake by diadochy, J. Mater. Chem. 8 (1998) 1329–1331.
[388] Y.T. Moon, D.K. Kim, C.H. Kim, Preparation of monodisperse ZrO_2 by the microwave heating of zirconyl chloride solutions, J. Am. Ceram. Soc. 78 (1995) 1103–1106.
[389] D. Daichuan, H. Pinjie, D. Shushan, Preparation of uniform α-Fe_2O_3 particles by microwave-induced hydrolysis of ferric salts, Mater. Res. Bull. 30 (1995) 531–535.
[390] D. Daichuan, H. Pinjie, D. Shushan, Preparation of uniform α-FeO (OH) colloidal particles by hydrolysis of ferric salts under microwave irradiation, Mater. Res. Bull. 30 (1995) 537–541.
[391] S. Komarneni, Q.H. Li, K.M. Stefanson, R. Roy, Microwave-hydrothermal processing for synthesis of electroceramic powders, J. Mater. Res. 8 (1993) 3176–3183.
[392] S. Komarneni, V.C. Menon, Q.H. Li, R. Roy, F. Ainger, Microwave-hydrothermal processing of $BiFeO_3$ and $CsAl_2PO_6$, J. Am. Ceram. Soc. 79 (1996) 1409–1412.
[393] M.C. D'Arrigo, C. Leonelli, G.C. Pellacani, Microwave-hydrothermal synthesis of nanophase ferrites, J. Am. Ceram. Soc. 81 (1998) 3041–3043.
[394] C.K. Kim, J.H. Lee, S. Katoh, R. Murakami, M. Yoshimura, Synthesis of Co-, Co–Zn and Ni–Zn ferrite powders by the microwave-hydrothermal method, Mater. Res. Bull. 36 (2001) 2241.
[395] M. Aizawa, K. Itatani, F.S. Howell, A. Kishioka, Hydrothermal Technology for the 21st Century, J. Ceram. Soc. 104 (1995) 126–128 (Japan).
[396] J.P. Cheng, D.K. Agarwal, S. Komarneni, M. Mathis, R. Roy, Microwave processing of WC–Co composites and ferroic titanates, Mat. Res. Innovat. 1 (1997) 44–52.
[397] S. Komarneni, M.Z. Hussein, C. Liu, E. Breval, P.B. Malla, Microwave-hydrothermal processing of metal clusters supported in and/or on montmorillonite, Eur. J. Solid State Inorg. Chem. 32 (1995) 837–849.
[398] Y. Fang, D.K. Agarwal, D.M. Roy, R. Roy, Fabrication of porous hydroxyapatite ceramics by microwave processing, J. Mater. Res. 7 (1992) 490–494.
[399] Yi Fang, D.K. Agarwal, D.M. Roy, R. Roy, Microwave sintering of HAp ceramics, J. Mater. Res. 9 (1994) 180–187.

[400] N.V. Kosova, A.Kh. Khabibullin, V.V. Boldyrev, Hydrothermal reactions under mechanochemical treating, Solid State Ionics 53 (1997) 101–103.
[401] W.L. Suchanek, P. Shuk, K. Byrappa, R.E. Riman, K.S. Ten Huisen, V.F. Janas, Mechanochemical–hydrothermal synthesis of carbonated apatite powder at room temperature, Biomaterials 23 (2002) 699–710.
[402] C.A. Mirkin, R.L. Letsinger, R.C. Mucic, J.J. Storhoff, A DNA-based method for rationally assembling nanoparticles into macroscopic materials, Nature 382 (1996) 607–609.
[403] D. Baranov, L. Mannav, A.G. Kanaras, Chemically induced self-assembly of spherical and anisotropic inorganic nanocrystals, J. Mater. Chem. 21 (2011) 16694–16703.
[404] S. Komarneni, Synthesis of nanorods and nanowires using biomolecules under conventional and microwave hydrothermal conditions, J. Mater. Sci. 43 (2008) 2377–2386.
[405] R.W. Shaw, T.R. Brill, A.A. Clifford, C.A. Eckert, E.U. Franck, Supercritical water: a medium for chemistry, Chem. Eng. News 69 (1991) 26–30.
[406] M. Modell, Supercritical water oxidation, in: H.M. Freeman (Ed.), Standard Handbook of Hazardous Waste Treatment and Disposal, McGraw-Hill, New York, NY, 1989, pp. 8.153–8.168.
[407] A. Shanableh, E.F. Gloyna, Supercritical water oxidation waste waters and sludges, Water Sci. Technol. 23 (1991) 389–393.
[408] W.A. Peters, Supercritical water technology: process systems research for new applications, in: Proceedings of the Second International Conference Solvothermal Reactions, (N. Yamasaki, and K. Yamagisawa Eds.) Takamatsu, Japan, December 18–20, 1996, pp. 35–41.
[409] N. Yamasaki, K. Yanagisawa (Eds.), Proceedings of the Second International Conference on Solvothermal Reactions, Takamatsu, Japan, Dec. 18–20, (1996), pp. 1–280.
[410] N. Yamasaki, K. Yanagisawa, M. Fujita, Hydrothermal decomposition of chlorofluorocarbons: CFC 11 and CFC 113, in: Proceedings of the First International Conference on Solvothermal Reactions, Takamatsu, Japan, December 5–7, 1994, pp. 134–141.
[411] P.E. Savage, Reaction models for supercritical water oxidation processes, in: Proceedings of the Second International Conference on Solvothermal Reactions, Takamatsu, Japan, December 18–20, 1996, pp. 2–4.
[412] M. Nakahara, T. Yamaguchi, H. Ohtaki, The structure of water and aqueous electrolyte solutions under extreme conditions, Recent Res. Develop. Phys. Chem. 1 (1997) 17–49.
[413] H. Ohataki, T. Radnai, T. Yamaguchi, Structure of water under subcritical and supercritical conditions studied by solution X-ray diffraction, Chem. Soc. Rev. (1997) 41–51.
[414] M. Nakahara, N. Matubayasi, C. Wakai, NMR study of hydrogen bonds and water structure in super- and subcritical conditions, in: Proceedings of the Second International Conference on Solvothermal Reactions, Takamatsu, (N. Yamasaki, and K. Yanagisawa Eds) Japan, December 18–20, 1996, pp. 9–12.
[415] T. Adschiri, R. Malaluan, K. Machida, O. Sato, K. Arai, Waste polymer conversion to chemicals in supercritical water, in: Proceedings of the First International Conference on Solvothermal Reactions, Takamatsu, Japan, December 5–7, 1994, pp. 138–141.
[416] R.A. Laudise, Green materials: industrial ecology, in: S. Somiya, R.P.H. Chang, M. Doyama, R. Roy (Eds), Materials Science and Engineering Serving Society, Elsevier, The Netherlands, 1998, pp. 23–27.

[417] M. Yoshimura (Ed.), Proceedings of the Fourth International Workshop on Design and Soft Solution—Processing for Advanced Inorganic Materials, Yokohama, Japan, February 2000.
[418] E.H. Ishida, Environmentally Friendly Manufacturing, Processing and Materials, 1998.
[419] M. Kakihana, M. Yoshimura, Synthesis and characteristics of complex multicomponent oxides prepared by polymer complex method, Bull. Chem. Soc. Jpn. 72 (1999) 1427–1443.
[420] K. Tomita, V. Petrykin, M. Kobayashi, M. Shiro, M. Yoshimura, M. Kakihana, A water-soluble titanium complex for the selective synthesis of nanocrystalline brookite, rutile, and anatase by a hydrothermal method, Angew. Chem. Int. Ed. 45 (2006) 2378–2381.
[421] M.D. Baijal, Plastic Polymer Science and Technology, John Wiley & Sons, New York, NY, 1982.
[422] R.A. Higgins, Engineering Metallurgy, Edward Arnold, London, 1993.
[423] W.D. Kingery, H.K. Bowen, D.R. Uhlmann, Introduction to Ceramics, John Wiley & Sons, New York, NY, 1976.
[424] R.A. Ring, Fundamentals of Ceramic Powder Processing and Synthesis, Academic Press, San Diego, CA, 1996.
[425] M. Ohring, The Materials Science of Thin Films, Academic Press, San Diego, CA, 1992.
[426] C.E. Morosanu, Thin Films by Chemical Vapor Deposition, Elsevier, Amsterdam, 1990.
[427] E.H.C Parker, The Technology and Physics of Molecular Beam Epitaxy, Plenum Press, New York, NY, 1985.
[428] JANAF Thermo Chemical Tables, National Bureau of Standards, American Chemical Society, American Institute of Physics, 1968–1982.
[429] B.R. Pamplin (Ed.), Crystal Growth, Pergamon Press, Oxford, UK, 1980.
[430] T.E.H. Graedel, B.R. Allen, Industrial Ecology, Prentice-Hall, Englewood Cliffs, NJ, 1995.
[431] T.E.H. Graedal, B.R. Allen, Industrial Ecology and Sustainable Engineering, Prentice-Hall, Englewood Cliffs, NJ, 2009.
[432] T. Itoh, S. Hori, M. Abe, Y.J. Tamaura, Light-enhanced ferrite plating of $Fe_{3-x}M_xO_4$ (M = Ni, Zn, Co and Mn) films in an aqueous solution, Appl. Phys. 69 (1991) 511–514.
[433] S. Komarneni, Nanophase materials by hydrothermal, microwave-hydrothermal and microwave-solvothermal methods, Curr. Sci. 85 (2003) 1730–1734.
[434] H. Marsh, E.A. Heintz, F. Rodriguez-Reinoso, Introduction to Carbon Technologies, Publications in Spain, 1997. p. 67
[435] V. Kasivisweswari, Catalytic Decomposition of Phenols Using Impregnated Activated Carbon, M.Sc., Dissertation, University of Mysore, India, 1999.
[436] A. Muralidhar, Catalytic Decomposition of Nitroarenes using Impregnated Activated Carbon, M.Sc., Dissertation, University of Mysore, India, 1999.
[437] Q. Feng, K. Yanagisawa, N. Yamasaki, Synthesis of manganese oxides with tunnel structure by hydrothermal soft chemical process, in: Proceedings of the Second International Conference on Solvothermal Reactions, Takamatsu, Japan, December 18–20, 1996, pp. 235–238.
[438] Q. Feng, C. Honbu, K. Yanagisawa, N. Yamasaki, Synthesis of lithiophorite by hydrothermal soft chemical process, in: S. Somiya, R.P.H. Chang, M. Doyama, R. Roy (Eds), Materials Science and Engineering Serving Society, Elsevier, The Netherlands, 1998, pp. 121–124.

[439] Q. Feng, E.H. Sun, K. Yanagisawa, N. Yamasaki, Synthesis of birnessite-type sodium manganese oxide and the behavior under hydrothermal conditions, in: S. Somiya, R. P.H. Chang, M. Doyama, R. Roy (Eds), Materials Science and Engineering Serving Society, Elsevier, The Netherlands, 1998, pp. 125–128.
[440] G. Demazeau, The solvothermal synthesis: a way to a new chemistry, in: Proceedings of the International Workshop on Soft and Solution Processing for Advanced Inorganic Materials, Tokyo Institute of Technology, Japan, February 22–27, 1996, pp. 4–12.
[441] G. Demazeau, Solvothermal reaction: an original route for the synthesis of novel materials, J. Mater. Sci. 43 (2008) 2024–2029.
[442] D.M. Roy, Concrete, the gigaphase ceramic: its key role in environmentally friendly technology, in: S. Somiya, R.P.H. Chang, M. Doyama, R. Roy (Eds), Materials Science and Engineering Serving Society, Elsevier, The Netherlands, 1988, pp. 81–88.
[443] N. Yamasaki, Possibility of materials recycling using hydrothermal process, in: S. Somiya, R.P.H. Chang, M. Doyama, R. Roy (Eds), Materials Science and Engineering Serving Society, Elsevier, The Netherlands, 1988, pp. 107–112.
[444] T. Yoshioka, T. Sato, A. Okuwaki, Hydrolysis of water PET by sulfuric acid at 150°C for a chemical recycling, J. Appl. Polymer Sci. 52 (1994) 1353–1355.
[445] N. Yamasaki, Recovery process of human waste to energy resource using hydrothermal technology, in: Proceedings of the World Conference on the Role of Advanced Materials in Sustainable Development, Seoul, Korea, September 1–6, 1996, pp. 199–205.
[446] K. Kugimiya, Ceramics for the next millennium, in: S. Somiya, R.P.H. Chang, M. Doyama, R. Roy (Eds), Materials Science and Engineering Serving Society, Elsevier, The Netherlands, 1988, pp. 28–33.

Index

Note: Page numbers followed by "*f*" and "*t*" refer to figures and tables, respectively.

Kolibri 360 GmbH
Ferdinand-Jühlke-Straße 7
99095 Erfurt, Deutschland
produktsicherheit@kolibri360.de